Genetics

FROM GENES TO GENOMES

SECOND EDITION

McGraw
Hill

Boston   Burr Ridge, IL   Dubuque, IA   Madison, WI   New York   San Francisco   St. Louis
Bangkok   Bogotá   Caracas   Kuala Lumpur   Lisbon   London   Madrid   Mexico City
Milan   Montreal   New [   ]re   Sydney   Taipei   Toronto

The McGraw·Hill Companies

GENETICS: FROM GENES TO GENOMES, SECOND EDITION

International   4 5 6 7 8 9 0 DOW/DOW 0 9 8 7 6 5
Domestic   4 5 6 7 8 9 0 DOW/DOW 0 9 8 7 6 5

ISBN 0–07–291930 – 2
ISBN 0–07–111328 – 2 (ISE)

Publisher: *Martin J. Lange*
Senior sponsoring editor: *Patrick E. Reidy*
Senior developmental editor: *Jean Sims Fornango*
Marketing manager: *Tami Petsche*
Lead project manager: *Peggy J. Selle*
Senior production supervisor: *Laura Fuller*
Media project manager: *Jody K. Banowetz*
Associate media technology producer: *Janna Martin*
Design manager: *Stuart Paterson*
Cover/interior designer: *Christopher Reese*
Cover image: *Christos Magganas*
Lead photo research coordinator: *Carrie K. Burger*
Photo research: *Karen Pulgiano*
Supplement producer: *Brenda A. Ernzen*
Compositor: *Carlisle Communications, Ltd.*
Typeface: *10/12 Times Roman*
Printer: *R. R. Donnelley Willard, OH*

Library of Congress Cataloging-in-Publication Data

Genetics : from genes to genomes / Leland Hartwell . . . [et al.]. — 2nd ed.
       p. ; cm.
   Includes index.
   ISBN 0–07–246248–5 (hard copy : alk. paper)
   1. Genetics.
   [DNLM: 1. Genetics. QH 430 G3277 2004] I. Hartwell, Leland.

QH430 .G458   2004
576.5—dc21

                                    2002012767
                                    CIP

www.mhhe.com

Dr. Leland Hartwell received his Ph.D from Massachusetts Institute of Technology. Dr. Hartwell held assistant and associate professorships at the University of California before joining the faculty of the University of Washington, where he continues as a full professor. In 1996, Dr. Hartwell joined the Fred Hutchinson Cancer Research Center as a full member and senior advisor for scientific affairs and was named president and director of the center in July 1997.

Combining mutants and time-lapse photomicroscopy, Dr. Hartwell identified 32 genes in yeast that regulate the cell cycle with specific defects in spindle pole body duplication and segregation, DNA replication, mitosis, cytokinesis, and budding. He discovered a control point in the cell cycle, Start, where yeast cells exit the cell cycle to mate, arrest after nutritional starvation, and integrate growth with division. He used genetics to define many of the steps in the signal transduction pathway that feed into Start, including the cell surface receptor for mating pheromone. The gene controlling Start, *CDC28,* was cloned in his lab and was the first CDK identified. He investigated the fidelity of chromosome transmission in the cell cycle, discovering that limitation or overexpression of many essential cell-cycle components lead to errors in chromosome transmission. Studies on how cells integrate the repair of DNA damage and cell division led to the discovery of cell-cycle checkpoints and the identification of six genes that control the DNA damage checkpoint.

Dr. Hartwell has received numerous awards and honors in the course of his career. Among them are the Brandeis University Rosenteil Award in 1993, the Sloan-Kettering Cancer Center Katherine Berkan Judd Award, and the Genetics Society of America Medal in 1994. In 1995 he was awarded the MGH Warren Triennial Prize and in 1996 the Columbia University Horwitz Award and the Passano Award. In 1998 Dr. Hartwell received the Albert Lasker Award for medical research. In 2001 he was awarded the Nobel Prize for Physiology or Medicine in recognition of his discovery of key regulators of the cell cycle.

Dr. Lee Hood received an M.D. from the Johns Hopkins Medical School and a Ph.D. in biochemistry from the California Institute of Technology. His research interests include immunology, development, and the development of biological instrumentation (for example, the protein sequenator and the automated fluorescent DNA sequencer). His research played a key role in unraveling the mysteries of antibody diversity.

Dr. Hood has taught molecular evolution, immunology, molecular biology, and biochemistry. He was one of the advocates for the Human Genome Project and directed one of the federal genome centers that sequenced the human genome. Dr. Hood is currently the president and director (and cofounder) of the cross-disciplinary Institute for Systems Biology in Seattle, Washington.

Dr. Hood has received a variety of awards including the Albert Lasker Award for Medical Research (1987), Dickson Price Award (1987), Cefas Award for Biochemistry (1989), and Distinguished Service Award from the National Association of Teachers (1998). He is the 2002 recipient of the Kyoto Prize in Advanced Biotechnology. Dr. Hood received this award in recognition of his pioneering work in developing the protein and DNA synthesizers and sequencers that provide the technical foundation of modern biology. He is deeply involved in K–12 science education. His hobbies include running, mountain climbing, and reading.

Dr. Michael Goldberg is a professor at Cornell University, where he teaches introductory genetics. He was an undergraduate at Yale University and received his Ph.D. in biochemistry from Stanford University. Dr. Goldberg performed postdoctoral research at the Biozentrum of the University of Basel (Switzerland) and at Harvard University, and he received an NIH Fogarty Senior International Fellowship for study at Imperial College (England) and at the University of Rome (Italy). His current research uses the tools of *Drosophila* genetics to investigate the mechanisms that ensure proper chromosome segregation during mitosis and meiosis.

Dr. Ann Reynolds is an educator and author. She began teaching genetics and biology in 1990. An affiliate faculty member of the Genetics Department at the University of Washington, her research has included studies of gene regulation in *E. coli,* chromosome structure and DNA replication in yeast, and chloroplast gene expression in marine algae. She is a graduate of Mount Holyoke College and received her Ph.D. from Tufts University. Dr. Reynolds was a postdoctoral research fellow with the Harvard University Department of Molecular Biology. She was also an author and producer of the laserdisc and CD-ROM *Genetics: Fundamentals to Frontiers.*

Dr. Lee M. Silver received his Ph.D in biophysics from Harvard University. He was a research fellow in genetics at the Sloan-Kettering Institute for Cancer Research and a senior scientist at Cold Spring Harbor Laboratory before joining the faculty of Princeton University. Currently, Dr. Silver is a professor of molecular biology and public affairs at Princeton University, with appointments in Princeton's Department of Molecular Biology, the Woodrow Wilson School of Public and International Affairs, the Program in Science, Technology, and Environmental Policy, and the Office of Population Research. During his career in genetic research, Dr. Silver made intensive use of the mouse as a model organism for the study of mammalian reproduction, development, evolution, and behavior. His current research lies at the interface of biotechnology and society and focuses on the political, legal, religious, and ethical issues raised by advances in human genetics and stem cell research. These issues are discussed in Dr. Silver's book *Remaking Eden: How Genetic Engineering and Cloning Will Transform the American Family,* which has been published in 15 languages. Dr. Silver is also the author of *Mouse Genetics: Concepts and Applications.*

In 1993, Dr. Silver was elected a Fellow of the American Association for the Advancement of Science (AAAS). In 1995, he received a MERIT Award from the National Institutes of Health. Dr. Silver has been elected to the governing boards of the Genetics Society of America and the International Mammalian Genome Society. He was also a member of the New Jersey Bioethics Commission Task Force formed to recommend reproductive policy for the New Jersey state legislature.

Ruth C. Veres is a science writer and editor with 30 years of experience in textbook publishing. She received her B.A. from Swarthmore College, obtained M.A. degrees from Columbia University in New York and Tufts University, and taught writing and languages at the University of California at Berkeley.

In addition to developing and editing more than 30 texts in the fields of political science, economics, psychology, nutrition, chemistry, and biology, Veres has coauthored a book on the immune system and an introductory biology text. She is currently working on a book with Dr. Lee Hood that looks at biology as an information science.

Contributors

Genetics research tends to proceed down highly specialized paths. A number of experts in specific areas generously provided information in their areas of expertise. We thank them for their contributions to this edition of our text.

Anthony Bleeker, *University of Wisconsin*
Michael Culbertson, *University of Wisconsin*
Ian Duncan, *Washington University, St. Louis*
Nancy Hollingsworth, *SUNY Stony Brook*

Patrick H. Mason, *University of Wisconsin*
Debra Nero *Cornell University*
William Wood, *University of Colorado*

BRIEF CONTENTS

CONTENTS

PART II

What Genes Are and What They Do

P A R T   I I I
Genomes

P A R T I V
How Genes Travel

**P A R T V
How Genes Are Regulated**

A Note from the Authors

Completion of a rough draft of the human genome in February of 2001 changed the way we think about genetics and biology. The ability to sequence a genome quickly and efficiently has generated enormous amounts of genetic data and opened up avenues of inquiry that were previously closed to scientific investigation. The explosion of information and the potential for deciphering it with recently developed technologies and computational tools, in conjunction with hypothesis-driven experiments, make it possible to carry out large-scale molecular analyses of the complex systems that compose plants, animals, and fungi. Today's students are poised on the brink of an astonishing scientific frontier whose main features we hope to convey in this new edition of *Genetics: From Genes to Genomes*.

It was during the twentieth century that genetics emerged as a central discipline in biology. In 1900 Gregor Mendel's laws of heredity were rediscovered; in the 1950s, James Watson and Francis Crick revealed the structure of DNA, the molecule of heredity, as a double helix that encodes digital information; and in the 1990s, the Human Genome Project provided a nearly complete sequence of the human genome. For much of the century, the study of genetics focused on the identification of individual genes and their function. In the last decade of the century, however, another idea gained currency—the concept that no gene acts alone; instead it is through complex molecular interactions within and among vast networks of genes, proteins, and other molecules that organisms ultimately live and die. By the end of the twentieth century tools for analyzing genes, genomes, and their expression within these complex biological systems had progressed beyond the most optimistic expectations.

Our Focus—An Integrated Approach

Genetics: From Genes to Genomes reflects this new perspective. The book presents a new approach to an undergraduate course in genetics. It represents the way we, the authors, currently view the molecular basis of life. We integrate:

- **Formal genetics:** the rules by which genes are transmitted.

- **Molecular genetics:** the structure of DNA and how it directs the structure of proteins.

- **Genomics and information science:** the new technologies that allow a comprehensive analysis of the entire gene set and its expression in an organism.

- **Human genetics:** how genes contribute to health and disease.

- **The unity of life-forms:** the synthesis of information from many different organisms into coherent models that explain many biological systems.

- **Molecular evolution:** the molecular mechanisms by which biological systems and whole organisms have evolved and diverged.

The strength of this integrated approach is that students who complete the book will have a strong command of genetics as it is practiced today by academic and corporate researchers. These scientists are rapidly changing our understanding of living organisms, including ourselves; increasing our ability to prevent, treat, and diagnose disease and to engineer new life-forms for food and medical uses; and, ultimately, creating the ability to replace or correct detrimental genes.

The Genetic Way of Thinking

To encourage a genetic way of thinking, we begin the book with a presentation of Mendelian principles and the chromosomal basis of inheritance. From the outset, however, the integration of Mendelian genetics with fundamental molecular mechanisms is central to our approach. Chapter 1 presents the foundation of this integration. In Chapter 2, we tie Mendel's studies of pea-shape inheritance to the action of an enzyme that determines whether a pea is round or wrinkled. In the same chapter, we point to the relatedness of patterns of heredity in all organisms by using Mendelian principles to look at heredity in humans. Starting in Chapter 6, we focus on the physical characteristics of DNA, the implications and uses of mutations, and how the double helix structure of DNA encodes, copies, and transmits biological information. Beginning in Chapter 9 we look at modern genetic techniques, including such biotechnology tools as gene cloning, hybridization, PCR, and microarrays, exploring how researchers use them to reveal the modular construction and genetic relatedness of genomes. We then show how the complete genome sequences of humans and model organisms provide insights into the architecture and evolution of genomes; how modular genomic construction has contributed to the relatively rapid evolution of life and helped generate the enormous diversity of life-forms we see around us. References at the end of the book contain detailed discussions of model organisms, which clarify that their use in the study of human biology is possible only because of the genetic relatedness of all organisms. Throughout our book, we present the scientific reasoning of some of the ingenious researchers who have carried out genetic analysis, from Mendel, to Watson and Crick, to the collaborators on the Human Genome Project.

Student-Friendly Features

We have taken great pains to help the student make the leap to this deeper level of understanding. Numerous features of this book were developed with that goal in mind.

- **One Voice** The role of our science writer, Ruth Veres, is to create one voice for our author team. With 30 years experience in life science textbook publishing, Ms. Veres is uniquely suited to this task. By working closely with everyone on the team, she has created the friendly, engaging reading style that helps students master the concepts throughout this book. This team approach provides the student with the focus and continuity required to make the book successful in the classroom.

- **Visualizing Genetics** The highly specialized art program developed for this book integrates photos and line art in a manner that provides the most engaging visual presentation of genetics available. Our Feature Figure illustrations break down complex processes into step-by-step illustrations that lead to greater student understanding. All illustrations are rendered with a consistent color theme—for example, all presentations of phosphate groups are the same color, as are all presentations of mRNA.

- **Problem Solving** Developing strong problem-solving skills is vital for every genetics student. The authors have carefully created problem sets at the end of each chapter that allow students to improve upon their problem-solving ability.

  - **Social and Ethical Issues** questions require critical thinking analysis of the scientific issues that impact our society.

  - **Solved Problems** provide insight into the step-by-step process of problem solving.

  - **Review Problems** offer a variety of levels of questions that develop excellent problem-solving skills.

- **Accessability** Our intention is to bring cutting-edge content to the student level. A number of more complex illustrations have been revised and segmented to help the student follow the process. Legends have been streamlined to highlight only the most important ideas, and throughout the book, topics have been revised to focus on the most critical information.

New to the Second Edition

- **"Genetics: The Study of Biological Information:"** A completely new Chapter 1 presents the book's approach to the study of genetics and genomics. We describe the new tools and concepts (such as the use of DNA chips or microarrays) used to examine complex biological systems or develop individual health profiles. The chapter also raises

several social concerns in the context of future challenges, for example, aging and the treatment of older people, and the use or abuse of genetic information.

- **Part III "Genomes"** (Chapters 9, 10, 11), focuses on the dramatic explosion of information that is the result of the completion of the Human Genome Project. It also presents the latest ways to carry out genomic research on a wide variety of organisms. These chapters are completely restructured and revised.

  "Deconstructing the Genome: DNA at High Resolution" (Chapter 9) covers the topics *fragmenting complex genomes for analysis, cloning fragments of DNA, PCR, DNA sequence analysis,* with a comprehensive example, and *understanding the genes for hemoglobin.*

  "Reconstructing the Genome Through Genetic and Molecular Analysis" (Chapter 10) covers the topics *analyses of genomes; high-density linkage maps; the integration of linkage, physical, and sequence maps; major insights from the human and model organism genome sequences; high-throughput genomic and proteomic platforms permit global analyses of gene products; systems biology;* and applications in *predictive/preventative medicine.*

  "The Direct Detection of Genotype Distinguishes Individual Genomes" (Chapter 11) includes topics such as *DNA variation, detecting DNA genotypes of different polymorphisms, positional cloning, genetic dissection of complex traits,* and *haplotype association studies for high-resolution mapping in humans.*

- **"Using Genetics to Study Development"**— **An entirely new chapter.** Chapter 19 covers the general principles of development with examples from the most commonly used model organisms, as well as humans. The chapter unfolds in an orderly, concept-based manner and again reflects how the information gained by the Human Genome Project has changed the way geneticists think.

- **Updates** throughout make this the most current and modern book available. Every chapter reflects the updated information generated by the breakthroughs of the past few years.

- **Expanded and refined problem sets.** The problem sets at the end of each chapter have been reviewed, revised, and expanded to provide an even stronger opportunity for students to develop their problem-solving skills.

A Word About the Portraits of Model Organisms

Five **Genetic Portraits** are included in the reference section at the back of the book, each one profiling a different model organism whose study has greatly contributed to genetic research. Th five selected were the ones chosen as the focus of the Human Genome Project. They are:

Saccharomyces cerevisiae: Genetic Portrait of Yeast
Arabidopsis thaliana: Genetic Portrait of a Model Plant
Caenorhabditis elegans: Genetic Portrait of a Simple Multicellular Animal
Drosophila melanogaster: Genetic Portrait of the Fruit Fly
Mus musculus: Genetic Portrait of the House Mouse

We anticipate that instructors will choose to cover one or two portrait references during the semester. Students may then use the specifics of the selected model organism to build an understanding of the principles and applications discussed in the book. The unique genetic manipulations and properties of each of the models make them important for addressing different biological questions using genetic analysis. In the portraits, we explain how biologists learned that the evolutionary relatedness of all organisms permits the extrapolation from a model to the analysis of other living forms. The portraits should thus help students understand how insights from one model organism can suggest general principles applicable to other organisms, including humans.

Media and Supplements

For the Instructor

- **Digital Content Manager CD-ROM.** Instructor resource CD-ROM containing easy-to-use JPEG files for all line art, tables, and most photos from the book; as well as animations, live art files, lecture outlines, and PowerPoint presentations for each chapter. See page xxvi for further details!

- **Instructor Testing and Resource CD-ROM.** 2000 questions available on a cross-platform CD-ROM, this test bank uses Brownstone Diploma testing software to quickly create customized exams. The user-friendly program allows instructors to search for questions by topic, format, or difficulty level; edit existing questions or add new ones; and scramble questions and answer keys for multiple versions of the same test.

- **Transparencies.** 150 four-color illustrations from the book will be available to adopters as well as 50 key illustrations with labels removed (for use in testing).

For the Student

- **Solutions Manual/Study Guide.** Extensively revised by Dr. Debra Nero of Cornell University, presents the solutions to the end-of-chapter problems and questions along with the step-by-step logic of each solution.

- **Genetics: From Genes to Genomes CD-ROM.** Developed with the content of this book, covers the most challenging concepts in the introductory genetics course. The CD presents animations of basic genetic processes and interactive exercises and simulations involving fundamental principles. Additional quizzing options allow students to

self-test and identify those areas needing additional study. Glossary definitions can be reached via hot links. A correlation guide linking book topics to the related CD material is included on the CD.

Web-Based Material

Online Learning Center

The *Genetics: From Genes to Genomes* Online Learning Center (OLC) offers access to a wide array of premium content to fortify the learning and teaching experience. This web site may be reached at **www.mhhe.com/hartwell2** and provides additional materials for both students and instructors.

Student Center
The Student Center of the OLC features quizzes, study tools, web links library, and problem-solving exercises to help the student master genetics.

Instructor Center
Instructor's Manual and Integration Guide. Prepared by John Kemner of the University of Washington, this manual provides a guide to integrating all the available resources for *Genetics: From Genes to Genomes* into your course presentations.
PageOut. McGraw-Hill's exclusive tool for creating your own web site for your genetics course. It requires no knowledge of coding and is hosted by McGraw-Hill.
Course Management Systems. OLC content compatible with online course management systems like WebCT and Blackboard makes putting together your course web site easy. Contact your local McGraw-Hill sales representative for details.

Acknowledgments

The creation of a project of this scope is never solely the work of the authors. We are grateful to our colleagues around the world who took the time to review this manuscript and make suggestions for its improvement. Their willingness to share their experiences and expertise was a tremendous help to us.

Second-Edition Reviewers:

Lawrence R. Aaronson, *Utica College*
Ruth Ballard, *California State University, Sacramento*
Mary Bedell, *University of Georgia*
Michelle Bell, *Xavier University*
Michael Benedik, *University of Houston*
Susan Bergeson, *University of Texas*
David Carroll, *Florida Institute of Technology*
Helen Chamberlin, *Ohio State University*
Ruth Chesnut, *Eastern Illinois University*
Bruce Cochrane, *University of South Florida*
Claire Cronmiller, *University of Virginia*
Mike Dalbey, *University of California, Santa Cruz*
David Durica, *University of Oklahoma*
David Duvernell, *Southern Illinois University*
Sarah Elgin, *Washington University*
Johnny El-Rady, *University of South Florida*

Victor Fet, *Marshall University*
Janice Fisher, *University of Texas, Austin*
David Foltz, *Louisiana State University*
Jim Ford, *Stanford University*
David Fromson, *California State University, Fullerton*
Anne Galbraith, *University of Wisconsin, La Crosse*
Peter Gergen, *SUNY Stony Brook*
Elliot Goldstein, *Arizona State University*
James Haber, *Brandeis University*
Ralph Hillman, *Temple University*
Nancy Hollingsworth, *SUNY Stony Brook*
Jackie Peltier Horn, *Houston Baptist University*
Shelley Jansky, *University of Wisconsin, Stevens Point*
Eric N. Jellen, *Brigham Young University*
Cheryl Jorcyk, *Boise State University*
Kathleen Karrer, *Marquette University*
Sean Kimbro, *Clark Atlanta University*
Rebecca Kohn, *Ursinus College*
Stanley Maloy, *University of Illinois*
Steve McCommas, *Southern Illinois University, Edwardsville*
Elliot M. Meyerowitz, *California Institute of Technology*
Leilani Miller, *Santa Clara University*
Roderick M. Morgan, *Grand Valley State University*
Nick Norton, *Southeastern Louisiana University*
Valerie Oke, *University of Pittsburgh*
Greg Podgorski, *Utah State University*
Jim Prince, *California State University, Fresno*
Inder Saxena, *University of Texas*
Daniel Schoen, *McGill University*
Mark Seeger, *Ohio State University*
Malcolm Shields, *University of West Florida*
David Smith, *University of Delaware*
Harold Vaessin, *Ohio State University*
Dan Wells, *University of Houston*
Andrew Wood, *Southern Illinois University*

First-Edition Reviewers:

Ken Belanger, *University of North Carolina, Chapel Hill*
John Belote, *Syracuse University*
Anna Berkovitz, *Purdue University*
John Botsford, *New Mexico State University*
Michael Breindl, *San Diego State University*
Bruce Chase, *University of Nebraska, Omaha*
Lee Chatfield, *University of Central Lancashire*
Alan Christensen, *University of Nebraska*
Bruce Cochrane, *University of South Florida*
James Curran, *Wake Forest University*
Rowland Davis, *University of California, Irvine*
Paul Demchick, *Barton College*
Stephen D'Surney, *University of Mississippi*
Rick Duhrkopf, *Baylor University*
Susan Dutcher, *University of Colorado*
DuWayne Englert, *Southern Illinois University*
Bentley Fane, *University of Arkansas*
Victoria Finnerty, *Emory University*
David Foltz, *Louisiana State University*

David Futch, *San Diego State University*
Ann Gerber, *University of North Dakota*
Richard Gethmann, *University of Maryland, Baltimore County*
Mike Goldman, *San Francisco State University*
Elliott Goldstein, *Arizona State University*
Nels Granholm, *South Dakota State University*
Charles Green, *Rowan College of New Jersey*
Poonam Gulati, *University of Houston*
Stephen Hedman, *University of Minnesota*
Ralph Hillman, *Temple University*
Christine Holler-Dinsmore, *Fort Peck Community College*
Martin Hollingsworth, *Tallahassee Community College*
Nancy Hollingsworth, *State University of New York, Stony Brook*
Andrew Hoyt, *Johns Hopkins University*
Lynne Hunter, *University of Pittsburgh*
Robert Ivarie, *University of Georgia*
R. C. Jackson, *Texas Technological University*
Duane Johnson, *Colorado State University*
Chris Kaiser, *Massachusetts Institute of Technology*
Kenneth J. Kemphues, *Cornell University*
Susan Kracher, *Purdue University*
Alan Koetz, *Illinois State University*
Andrew Lambertsson, *University of Oslo*
Don Lee, *University of Nebraska*
John Locke, *University of Alberta*
Larry Loeb, *University of Washington*
Robertson McClung, *Dartmouth College*
Peter Meacock, *University of Leicester*
John Merriam, *University of California, Los Angeles*
Beth Montelone, *Kansas State University*
Patricia Moore, *Transylvania University*
Gail Patt, *Boston University*
Michael Perlin, *University of Louisville*
Richard Richardson, *University of Texas, Austin*
Mary Rykowski, *University of Arizona*
Mark Sanders, *University of California, Davis*
Randall Scholl, *Ohio State University*
David Sheppard, *University of Delaware*
Anthea Stavroulakis, *Kingsborough Community College*
John Sternick, *Mansfield University*
David Sullivan, *Syracuse University*
William Thwaites, *San Diego State University*
Akfi Uzman, *University of Houston*
Peter Webster, *University of Massachusetts*
Dean Whited, *North Dakota State University*
John Williamson, *Davidson College*
John Zamora, *Middle Tennessee State University*
Stephan Zweifel, *Carleton College*

We would also like to thank the highly skilled publishing professionals who firmly guided the development of the second edition of *Genetics: From Genes to Genomes:* Jim Smith for his continuing support and understanding of the project; Jean Fornango for her almost daily communications, outstanding organizational skills, and spirited good humor; and Peggy Selle and her production team for their careful attention to details. We appreciate their input.

Integrating Genetic Concepts

Genetics: From Genes to Genomes takes an integrated approach in its presentation of genetics, thereby giving students a strong command of genetics as it is practiced today by academic and corporate researchers. Principles are related throughout the text in examples, essays, case histories, and Connections sections to make sure students fully understand the relationships between topics.

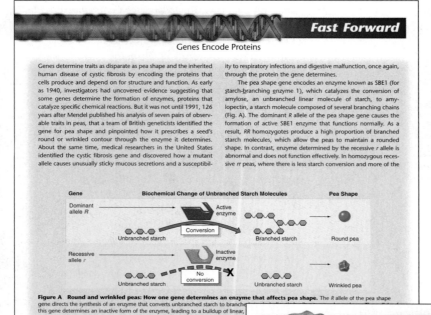

Fast Forward

Genes Encode Proteins

Genes determine traits as disparate as pea shape and the inherited human disease of cystic fibrosis by encoding the proteins that cells produce and depend on for structure and function. As early as 1940, investigators had uncovered evidence suggesting that some genes determine the formation of enzymes, proteins that catalyze specific chemical reactions. But it was not until 1991, 126 years after Mendel published his analysis of seven pairs of observable traits in peas, that a team of British geneticists identified the gene for pea shape and pinpointed how it prescribes a seed's round or wrinkled contour through the enzyme it determines. About the same time, medical researchers in the United States identified the cystic fibrosis gene and discovered how a mutant allele causes unusually sticky mucous secretions and a susceptibility to respiratory infections and digestive malfunction, once again, through the protein the gene determines.

The pea shape gene encodes an enzyme known as SBE1 (for starch-branching enzyme 1), which catalyzes the conversion of amylose, an unbranched linear molecule of starch, to amylopectin, a starch molecule composed of several branching chains (Fig. A). The dominant *R* allele of the pea shape gene causes the formation of active SBE1 enzyme that functions normally. As a result, *RR* homozygotes produce a high proportion of branched starch molecules, which allow the peas to maintain a rounded shape. In contrast, enzyme determined by the recessive *r* allele is abnormal and does not function effectively. In homozygous recessive *rr* peas, where there is less starch conversion and more of the

Figure A Round and wrinkled peas: How one gene determines an enzyme that affects pea shape. The *R* allele of the pea shape gene directs the synthesis of an enzyme that converts unbranched starch to branched this gene determines an inactive form of the enzyme, leading to a buildup of linear,

Fast Forward Essays
This feature is one of the methods used to integrate the Mendelian principles presented early in the book with the molecular principles that will follow.

Genetics and Society

Amplified Trinucleotide Repeats May Have Medical Consequences

Expansions of the base triplet CGG cause a heritable disorder known as *fragile X syndrome*. Adults affected by this syndrome manifest several physical anomalies, including an unusually large head, long face, large ears, and in men, large testicles. They also exhibit moderate to severe mental retardation. Fragile X syndrome has been found in men and women of all races and ethnic backgrounds. The fragile X mutation is, in fact, a leading genetic cause of mental retardation worldwide, second only to the trisomy 21 that results in Down syndrome.

Specially prepared karyotypes of cells from people with fragile X symptoms reveal a slightly constricted, so-called fragile site near the tip of the long arm of the X chromosome (Fig. A). In some pictures of these karyotypes, the X has actually broken at this site, releasing a small piece containing the end of the chromosome (not shown). The long tracts of CGG trinucleotides, which make up the fragile X mutation, apparently produce a localized constricted region that breaks easily. Geneticists named the fragile X disorder for this specific pinpoint of fragility more than 20 years before they identified the mutation that gives rise to it.

The gene in which the fragile X mutation occurs is called *FMR-1* (for fragile-X-associated mental retardation). Near one end of the gene, different people carry a different number of repeats of the sequence CGG, and geneticists now have the molecular tools to quantify these differences. Normal alleles contain 5–54 of these triplet repeats, while the *FMR-1* gene in people with fragile X syndrome contains 200–4000 repeats of the exact same triplet (Fig. B.1). The rest of the gene's base sequence is the same in both normal and abnormal alleles.

The triplet repeat mutation that underlies fragile X syndrome has a surprising transmission feature. Alleles with a full-blown mutation are foreshadowed by *premutation alleles* that carry an intermediate number of repeats—more than 50 but fewer than 200 (Fig. B.1). Premutation alleles do not themselves generate fragile

Figure A A karyotype reveals a fragile X chromosome. The fragile X site is seen on the bottom of both chromatids of the X chromosome at the *right*.

X symptoms in most carriers, but they show significant instability and thus forecast the risk of genetic disease in a carrier's progeny. The greater the number of repeats in a premutation allele, the higher the risk of disease in that person's children. For example, if a woman carries a premutation allele with 60 CGG repeats, 17% of her offspring run the risk of exhibiting fragile X syndrome. If she carries a premutation allele with 90 repeats, close to 50% of her offspring will show symptoms. In short, the change that produces a premutation allele increases the likelihood that the *FMR-1* gene will incur more mutations; and the larger the original number of

Genetics and Society Essays
Dramatic essays explore the social and ethical issues created by the multiple applications of modern genetic research.

Comprehensive Examples

Comprehensive Examples are extensive case histories or research synopses that, through text and art, summarize the main points in the preceding section or chapter and show how they relate to each other.

Figure 7.30 How the world looks to a person with tritanopia. Compare with Fig. 4.22 on p. 103.

How Gene Mutations Affect Light-Receiving Proteins and Vision: A Comprehensive Example

Researchers first described anomalies of color perception in humans close to 200 years ago. Since that time, they have discovered a large number of mutations that modify human vision. By examining the phenotype associated with each mutation and then looking directly at the DNA alterations inherited with the mutation, they have learned a great deal about the genes influencing human visual perception and the function of the proteins they encode.

Several attributes of human subjects facilitate the experi-... of vision. First, people can recognize and de-... in the way they see, from trivial differences in ...ed looks like, to not seeing any difference be-...reen, to not seeing any color at all. Second, the ...d science of psychophysics provides sensitive, ...s for accurately defining and comparing pheno-...nostic test, for example, is based on the fact that ...each color as a mixture of three different wave-...—red, green, and blue—and ca...blue light of different intensities ...fourth wavelength such as yell-...does not combine to form the fo...at way to the eye. A person wit...ll select a well-defined propor...atch a particular yellow, but a ...en will permit any proportion o...e same match. Finally, since inh... in the visual system rarely affect fecundity or l... ern human societies, mutations generating m... alleles that change visual perception remain ... over time.

(a) Photoreceptor-containing cells

(b) Photoreceptor proteins

(c) Red/green pigment genes
X chromosomes from normal individuals:

(d) Evolution of visual pigment genes
Primordial gene

Figure 7.28 The cellular and molecular basis of vision.
(a) Rod and cone cells in the retina carry membrane-bound photoreceptors. **(b)** The photoreceptor in rod cells is rhodopsin. The blue-, green-, and red-receiving proteins in cone cells are related to rhodopsin. **(c)** One red photoreceptor gene and one to three green photoreceptor genes are clustered on the X chromosome. **(d)** The genes for rhodopsin and the three color receptors probably evolved from a primordial photoreceptor gene through three gene duplication events followed by divergence of the duplicated copies.

Connections

Each chapter closes with a Connections section that serves as a bridge between the topics in the just-completed chapter and those in the upcoming chapter or chapters.

CONNECTIONS

The Watson-Crick model for the structure of DNA, the single most important biological discovery of the twentieth century, clarified how the genetic material fulfills its primary function of carrying information: Each long, linear or circular molecule carries one of a vast number of potential arrangements of the four nucleotide building blocks (A, T, G, and C). The model also suggested how base complementarity could provide a mechanism for both accurate replication and changes in sequence combinations that arise from recombination events.

Unlike its ability to carry information, DNA's capacities for replication and recombination are not solely properties of the DNA molecule itself. Rather they depend on the cell's complex enzymatic machinery. But even though they rely on the complicated orchestration of many different proteins acting on the DNA, replication and recombination both occur with extremely high fidelity—normally not a single base pair is gained or lost. Rarely, however, errors do occur, providing the genetic basis of evolution. While most errors are detrimental to the organism, a very small percentage produce dramatic changes in phenotype without killing the individual. For example, although most parts of the X and Y chromosomes are not similar enough to recombine, occasionally an "illegitimate" recombination does occur. Depending on the site of crossing-over, such illegitimate recombination may give rise to an XY individual who is female or an XX individual who is male (Fig. 6.23). The explanation is as follows. In the first six weeks of development, a human embryo has the potential to become either male or female, but in the critical seventh week, information from a small segment of DNA— the sex-determining region of the Y chromosome containing the *SRY* gene—determines the embryo's sex. An illegitimate re-

Figure 6.23 Illegitimate recombination may produce an XY female or an XX male. The *SRY* gene, normally located on the Y chromosome, dictates male development; in the absence of *SRY*, an embryo develops as a female. Rare, illegitimate recombination between the X and Y chromosome can create a Y chromosome without *SRY* and an X chromosome with *SRY*. Fertilization with gametes containing these unusual chromosomes will produce XY females or XX males.

with an X chromosome containing *SRY* produces XX individuals that develop as males.

Visualizing Genetics

Full-color illustrations and photographs bring the printed word to life. These visual reinforcements support and further clarify the topics discussed throughout the text.

Feature Figures

Special multipage spreads integrate line art, photos, and text to summarize in detail important genetic concepts.

In animal cells

Centriole

Microtubules
Centrosome
Centromere
Chromosome
Sister chromatids
Nuclear envelope

(a) Prophase: (1) Chromosomes condense and become visible; (2) centrosomes move apart toward opposite poles and generate new microtubules; (3) nucleoli begin to disappear.

Astral microtubules
Kinetochore
Kinetochore microtubules
Polar microtubules

(b) Prometaphase: (1) Nuclear envelope breaks down; (2) microtubules from the centrosomes invade the nucleus; (3) sister chromatids attach to microtubules from opposite centrosomes.

Metaphase plate

(c) Metaphase: Chromosomes align on the metaphase plate with sister chromatids facing opposite poles.

Separating sister chromatids

(d) Anaphase: (1) Centromeres divide; (2) the now separated sister chromatids move to opposite poles.

Re-forming nuclear envelope

Chromatin — Nucleoli reappear

(e) Telophase: (1) Nuclear membranes and nucleoli re-form; (2) spindle fibers disappear; (3) chromosomes uncoil and become a tangle of chromatin.

(f) Cytokinesis: The cytoplasm divides, splitting the elongated parent cell into two daughter cells with identical nuclei.

Figure 4.8 Mitosis maintains the chromosome number of the parent cell nucleus in the two daughter nuclei. In the photomicrographs of newt lung cells, chromosomes are stained *blue* and microtubules appear either *green* or *yellow.* The artist's sketches illustrate the stages of mitosis in the nematode *Ascaris,* where $2n = 4$.

(a)

(b)

(c)

(d)

(e)

Figure 18.25 Mutations in three types of glial cells cause brain cancer. (a) Astrocyte brain cells from the cortex of a mammalian brain. **(b)** Oligodendrocyte cell (*orange structure in the middle*) attaching to an axon nerve cell at the start of myelination. **(c)** Ependymal cell. The astrocyte cells in (a) mutate to form astrocytomas (low-grade) shown in **(d).** The oligodendroglioma shown in **(e)** is the result of mutated oligodendrocyte cells (b). The tumors resulting from mutated ependymal cells are extremely rare and are not shown here.

Visualizing Genetics

(a)

1. Drop cells onto a glass slide.

2. Gently denature DNA by treating briefly with DNase.

3. Add hybridization probes labeled with fluorescent dye and wash away unhybridized probe.

Fluorescent probes

Fluorescent dye

Fluorescence microscope

Eyepiece

Barrier filter 2 (further blockage of stray UV rays)

UV source

Mirror to UV light; transparent to visible light

Barrier filter 1 (blocks dangerous short UV rays, allows needed long UV rays to pass through)

Objective lens

Object

4. Expose to ultraviolet (UV) light. Take picture of fluorescent chromosomes.

(b)

Figure 10.6 The FISH protocol. (a) The technique. (1) [___] slide. The force of the droplet hitting the slide causes the cel[___] and gently denature the DNA within them such that the ove[___] up at numerous points. (3) Label a DNA probe with a fluore[___] hybridization to occur, and wash away unhybridized probe. [___] on the chromosomes. The UV light causes the bound probe[___] through the eyepiece and photograph it. **(b)** A fluorescence [___] shows the microtubular structure.

Experiment and Technique Figures
Illustrations of performed experiments and genetic analysis techniques highlight how scientific concepts and processes are developed.

Experiment and Technique Figures
Illustrations of performed experiments and genetic analysis techniques highlight how scientific concepts and processes are developed.

Comparative Figures
Comparison illustrations lay out the basic differences of often confusing principles.

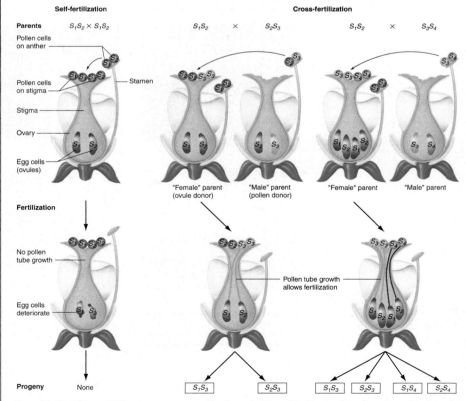

Figure 3.8 Plant incompatibility systems promote outbreeding and allele proliferation. A pollen grain carrying an allele of the self-incompatibility gene that is identical to either of the two alleles carried by a potential female parent is unable to grow a pollen tube; as a result, fertilization cannot take place. Because all the pollen grains produced by any one plant have one of the two alleles carried by the female reproductive parts of the same plant, self-fertilization is impossible.

Solving Genetics Problems

The best way for students to assess and increase their understanding of genetics is to practice through problems. Found at the end of each chapter, problem sets assist students in evaluating their grasp of key concepts and allows them to apply what they have learned to real-life issues.

Social and Ethical Issues

These challenging problems stir discussion and debate. The issues are presented within the context of real-life case studies and require the student to consider not only scientific issues but legal and ethical issues as well.

Review Problems

Revised and expanded, these problems offer a variety of levels of questions that develop strong problem-solving skills. The answers to selected problems can be found in an appendix in the back of the text.

SOCIAL AND ETHICAL ISSUES

1. The bacterium *Bacillus thuringiensis* (Bt) produces a toxin capable of killing many insects that attack crop plants. For more than 20 years, Bt has served as a biological pesticide that is sprayed on crops. The constant spraying, however, has led to the development of resistant bacterial strains in the field. A cloned version of Bt toxin consists of the toxin gene fused to a regulatory region that allows expression of the toxin only when crops carrying the gene are under stress—as in an insect infestation. Should farmers be forced to plant the transgenic crops rather than spray their fields so that this biological pesticide will remain effective? Are there ways to enforce this policy in the United States or worldwide?

2. Many scientists who are employed as professors at universities have become involved in biotechnology ventures outside the university. Companies often set up their laboratories near universities to take advantage of the intellectual resources there. Industries that develop near universities employ nonscientists in the community as well as scientists and are generally looked upon as an asset to the community. The president of a major state

university is encouraging faculty members to become active in the development of associated industries, viewing their mission as academics to include community related activities. But legislators and the public, whose tax dollars fund the university, believe that the professors should concentrate on teaching. Should faculty be encouraged to participate in outside ventures because it is a service to the community and potentially to humankind through development of new biotechnology, or should they be held more strictly to a solely academic, teaching mission?

3. In studying virulence in bacteria, a group of researchers stumbled on ways of modifying a particular bacterial species in a way that increased the virulence. While their fi...
vi...
w...
th...
bi...
to...
ha...

PROBLEMS

16-1 For each of the terms in the left column, choose the best matching phrase in the right column.

a. induction	1. glucose prevents expression of catabolic operons
b. repressor	2. protein undergoes a reversible conformational change
c. operator	3. often fused to regulatory regions of genes whose expression is being monitored
d. allostery	4. stimulation of protein synthesis by a specific molecule
e. operon	5. site to which repressor binds
f. catabolite repression	6. gene regulation involving premature termination of transcription
g. reporter gene	7. group of genes transcribed into one mRNA
h. attenuation	8. negative regulator

16-2 All mutations that abolish function of the Rho (termination) protein in *E. coli* are conditional mutations. What does this tell you about the *rho* gene?

16-3 Bacteriophage λ, after infecting a cell, can integrate into the chromosome of the cell if the repressor protein, cI, binds to and shuts down phage transcription immediately. (A strain containing a bacteriophage

is a lysogen was crossed with a lysogenic F⁻ recipient cell and no phages were produced. However, when the Hfr lysogen donor strain transferred its DNA to a nonlysogenic F⁻ recipient cell, the recipient cell burst, releasing a new generation of phages. Why did infection of a nonlysogenic cell result in phage growth and release but infection of a lysogenic recipient did not?

16-4 The promoter of an operon is the site to which RNA polymerase binds to begin transcription. Some base changes in the promoter result in a mutant site to which RNA polymerase cannot bind. Would you expect mutations in the promoter that prevent binding of RNA polymerase to act in *trans* on another copy of the operon on a plasmid in the cell or only in *cis* on the copy immediately adjacent to the mutated site?

16-5 For each of the *E. coli* merodiploid strains containing the *lac* operon alleles listed, indicate whether the strain is inducible, constitutive, or unable to express β-galactosidase and permease.
a. $I^+ O^+ Z^+ / I^+ O^c Z^+ Y^-$
b. $I^+ O^+ Z^+ Y^+ / I^- O^c Z^+ Y^-$
c. $I^s O^+ Z^- Y^+ / I^+ O^c Z^+ Y^-$
d. $I^- P^- O^+ Z^+ Y^- / I^+ P^+ O^c Z^- Y^+$
e. $I^s O^+ Z^+ / I^- O^+ Z^+ Y^-$

16-6 Mutants were isolated in which the constitutive phenotype of a missense *lacI* mutation was suppressed.

SOLVED PROBLEMS

I. In the galactose operon in *E. coli*, a repressor, encoded by the *galR* gene, binds to an operator site, *O*, to regulate expression of three structural genes, *galE*, *galT*, and *galK*. Expression is induced by the presence of galactose in the media. For each of the merodiploids listed, would the cell show constitutive, inducible, or no expression of each of the structural genes? (Assume that *galR⁻* is a loss of function mutation.)

a. *galR⁻ galO⁺ galE⁺ galT⁺ galK⁺*

b. *galR⁺ galO^c galE⁺ galT⁺ galK⁺*

c. *galR⁻ galO⁺ galE⁺ galT⁺ galK⁻ /galR⁺ galO⁺ galE⁻ galT⁺ galK⁺*

d. *galR⁻ galO^c galE⁺ galT⁺ galK⁻ /galR⁺ galO⁺ galE⁻ galT⁺ galK⁺*

Answer

This problem requires an understanding of how regulatory sites and proteins that bind to regulatory sites behave. To predict expression in merodiploid strains, look at each copy of the operon individually, and then assess what effect alleles present in the other copy of the operon could have on the expression. After doing that for each copy of the operon, combine the results.

a. The *galR* gene encodes a repressor, so the lack of a *GalR* gene product would lead to constitutive expression of the *galE, T,* and *K.*

b. The *galO^c* mutation is an operator site mutation. By analogy with the *lac* operon, the designation *galO^c* indicates that repressor cannot bind and there is constitutive expression of *galE, T,* and *K.*

c. The first copy of the operon listed has a *galK⁻* mutation. Alone, this would lead to constitutive synthesis *galE* and *galT.* (*galK* is mutant, so there will not be constitutive expression of this gene.) The other copy is wild type for the *galR* gene, so it produces a repressor that can act in *trans* on both copies of the operon, overriding the effect of the *galR⁻* mutation. Overall, there will be *inducible expression of the three gal genes.*

d. The first copy of the operon contains an *galO^c* mutation, leading to constitutive synthesis of *galE* and *galT.* The other copy has a wild-type operator so it is inducible, but neither operator has effects on the other copy of the operon. The net result is constitutive *galE* and *galT* and inducible *galK* expression.

II. The *araI* site is required for induction of *araBAD.* I⁻ mutants do not express *araBAD.* In an I⁻ mutant, a second mutation arose that resulted in constitutive arabinose synthesis. A Southern blot using a probe from the regulatory region and early part of the *araB* gene showed a very different set of restriction fragments than were seen in the starting strain. Based on the altered

Solved Problems

Solved problems offer step-by-step guidance needed to understand the problem-solving process.

A Powerful Lecture Resource

This multimedia collection of visual resources allows instructors to utilize artwork from the text in multiple formats to create customized classroom presentations, visually-based tests and quizzes, dynamic course website content, or attractive printed support materials. The digital assets on this cross-platform CD-ROM are grouped within the following easy-to-use folders.

Art and Photo Library

Full-color digital files of all the tables and illustrations and many of the photos in *Genetics: From Genes to Genomes* can be readily incorporated into lecture presentations, exams, or custom-made classroom materials.

G_1: Cell grows in size, prepares for DNA replication

Centrosome
Cell membrane
DNA
Nuclear membrane

S: DNA replication

Daughter cells

G_2: Cell prepares for division

Spindle pole
Condensed chromosome
Mitotic spindle

M: Mitosis
1. Breakdown of nuclear membrane
2. Condensation of chromosomes
3. Attachment of chromosomes to mitotic spindles

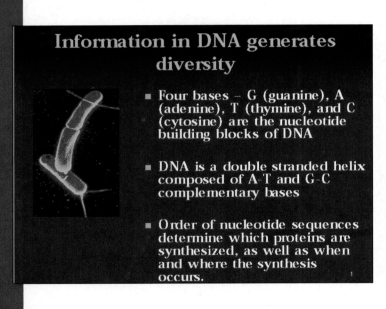

Information in DNA generates diversity

- Four bases – G (guanine), A (adenine), T (thymine), and C (cytosine) are the nucleotide building blocks of DNA

- DNA is a double stranded helix composed of A-T and G-C complementary bases

- Order of nucleotide sequences determine which proteins are synthesized, as well as when and where the synthesis occurs.

PowerPoint Lecture Outlines

Ready-made presentations that combine art and lecture notes cover each of the 21 chapters of the text. These lectures can be used as they are or can be tailored to reflect your preferred lecture topics and sequences.

Animations

Instructive, full-color animations are available to harness the visual impact of processes in motion. Import these animations into classroom presentations or online course materials.

TBP

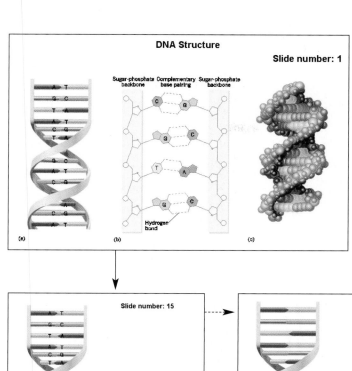

DNA Structure

Slide number: 1

Sugar-phosphate backbone Complementary base pairing Sugar-phosphate backbone

Hydrogen bond

(a) (b) (c)

Active Art

Active Art allows you to customize key figures from the text to complement your lecture! Figures have been broken into smaller digestable parts so each piece can be used in whatever order or format chosen. Using Microsoft PowerPoint's ungroupable art feature, instructors can also move, resize, or change the color of any or all objects in a piece of art.

Slide number: 15

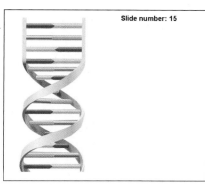

Slide number: 15

Remove Labels

Labels and leader lines can be easily repositioned, edited, or removed. Your customized images can then be used for course assignments or additional quizzing for your students.

Sugar-phosphate backbone Complementary base pairing Sugar-phosphate backbone Slide number: 26

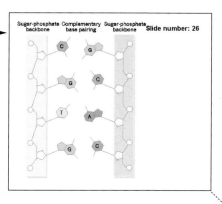

Sugar-phosphate backbone Complementary base pairing Sugar-phosphate backbone Slide number: 26

Change Colors

Colors can be removed and/or changed from any Active Art slide or object. Black and white images can be created for use in lecture supplements or exams.

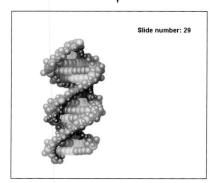

Slide number: 29

Resize Objects

The entire image or parts of the image can be made larger or smaller depending on what you choose to emphasize.

Sugar-phosphate backbone Complementary base pairing Sugar-phosphate backbone Slide number: 25

1

Genetics: The Study of Biological Information

Genetics, the science of heredity, is at its core the study of biological information. All living organisms—from single-celled bacteria and protozoa to multicellular plants and animals—must store, replicate, transmit to the next generation, and use vast quantities of information to develop, grow, reproduce, and survive in their environments (Fig. 1.1). Geneticists examine how organisms pass biological information on to their progeny and how they use it during their lifetime.

This book introduces the field of genetics as it exists in the first decade of the twenty-first century. Seven overarching themes recur throughout our presentation:

- The biological information fundamental to life is encoded in the DNA molecule.

- Biological function emerges primarily from protein molecules.

- Complex biological systems emerge from the functioning of regulatory networks that specify the behavior of genes.

- All living forms are closely related.

- The modular construction of genomes has allowed the relatively rapid evolution of biological complexity.

- Genetic techniques permit the dissection of biological complexity.

- Our focus is on human genetics.

In the remainder of this chapter, we introduce these themes. It will help to keep them in mind as you delve into the details of genetics.

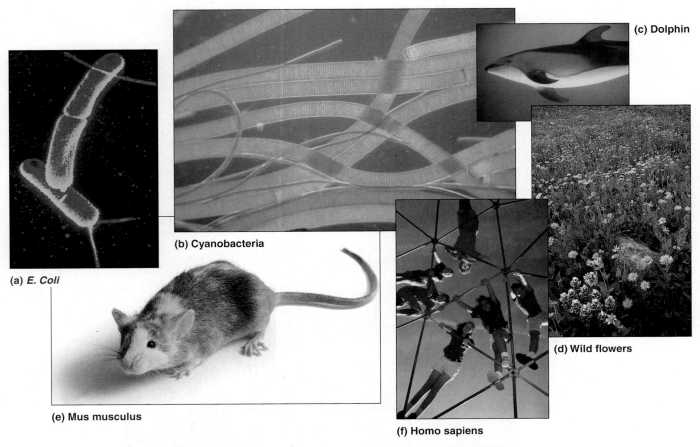

Figure 1.1   The biological information in DNA generates an enormous diversity of living organisms.

(a) *E. Coli*

(b) Cyanobacteria

(c) Dolphin

(d) Wild flowers

(e) Mus musculus

(f) Homo sapiens

The Biological Information Fundamental to Life Is Encoded in the DNA Molecule

The process of evolution has taken close to 4 billion years to generate the amazingly efficient mechanisms for storing, replicating, expressing, and diversifying biological information seen in organisms now inhabiting the earth. The linear DNA molecule stores biological information in units known as nucleotides. Within each DNA molecule, the sequence of the four letters of the DNA alphabet—G, C, A, and T—specify which proteins an organism will make as well as when and where protein synthesis will occur. The letters refer to the bases that are components of the nucleotide building blocks of DNA. The DNA molecule itself is a double strand of nucleotides carrying complementary A–T or G–C base pairs (Fig. 1.2). The molecular complementarity of double-stranded DNA is its most important property and the key to understanding how DNA replicates, transmits, mutates, and stores genetic information.

Although the DNA molecule is three-dimensional, most of the information within the molecule is one-dimensional and

digital. It is one-dimensional because it is encoded as a specific sequence of letters along the length of the molecule. It is digital because each unit of information—one of the four letters of the DNA alphabet—is discrete. Because genetic information is digital, it can be stored as readily in a computer memory as in a DNA molecule. Indeed, the combined power of DNA sequencers (Fig. 1.3), computers, and DNA synthesizers makes it possible to determine the sequence of, store, replicate, and transmit genetic information electronically from one place to another anywhere on this or any other planet. Such electronic wizardry works something like this: A DNA sequencer reads the base sequence of a DNA molecule. The sequence information is stored in a computer. The computer transmits the information via satellite from New York to a receiver in Hong Kong or from Paris to a receiver on Mars. There, the information is fed into a DNA synthesizer, which makes an exact replica of a portion of the originally sequenced DNA molecule.

The DNA regions that encode proteins are called *genes.* Just as the limited number of letters in a written alphabet places no restrictions on the stories one can tell, so too the limited number of letters in the genetic code alphabet places no restric-

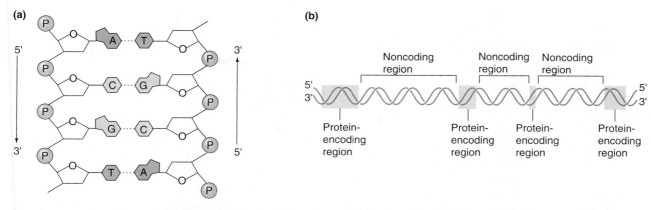

Figure 1.2 Complementary base pairs are a key feature of the DNA molecule. (a) A single strand of DNA is composed of nucleotide subunits each consisting of a deoxyribose sugar (depicted here as a *white pentagon*), a phosphate (depicted as a *yellow circle*), and one of four nitrogenous bases—adenine, thymine, cytosine, or guanine (designated as *lavender* or *green* As, Ts, Cs, or Gs). The chemical structure of the bases enables A to associate tightly with T and C to associate tightly with G. As a result, A and T form one kind of complementary base pair, while C and G form another kind of complementary base pair. The association through base pairing of two complementary DNA strands produces a DNA double helix. The *arrows* labeled *5′* to *3′* show that the two strands of the double helix have opposite orientations relative to chemically distinct 5′ and 3′ ends. **(b)** Along the DNA double helix, protein-encoding regions are separated by noncoding regions. In bacteria, close to 90% of the DNA encodes proteins; in humans, the proportion of protein-coding DNA is less than 5%.

Figure 1.3 An automated DNA sequencer. This instrument can sequence about 350,000 base pairs a day.

Figure 1.4 One of 24 different types of human chromosomes. Each chromosome contains thousands of genes.

tions on the kinds of proteins and thus the kinds of organisms genetic information can define. As mentioned, these organisms range from single-celled bacteria to complex multicellular animals such as fruit flies, mice, and humans. The basic genetic language is virtually the same for all. The differences are in the content and amount of information, and in when and where that information is *expressed,* that is, converted to RNA and protein.

Within the cells of an organism, DNA molecules carrying the genes are assembled into *chromosomes:* organelles that package and manage the storage, duplication, expression, and evolution of DNA. The entire collection of chromosomes in each cell of an organism is its *genome.* Human cells, for example, contain 24 distinct kinds of chromosomes carrying approx-

imately 3×10^9 base pairs and roughly 40,000–60,000 genes (Fig. 1.4). The amount of information that can be encoded in this size genome is equivalent to 6 million pages of text containing 250 words per page, with each letter corresponding to one *base pair,* or pair of nucleotides. It may seem incredible that it takes only 3 billion base pairs of genetic information to develop a human being, from its basic body plan to the initiation of consciousness. To appreciate the long journey from a finite amount of genetic information (easily storable on a computer disk, now that the Human Genome Project has sequenced it) to producing a human being, it is necessary to follow the flow of information from DNA to proteins, the molecules that determine how complex systems of cells, tissues, and organisms function.

Biological Function Emerges Primarily from Protein Molecules

Although there is no single characteristic that distinguishes living organisms from inanimate matter, you would have little trouble deciding which entities in a group of 20 objects are alive. Over time, these living organisms, governed by the laws of physics and chemistry as well as a genetic program, would be able to reproduce themselves; bacteria would do so through a form of fission; flies or fish would lay eggs, and so forth. Most of the organisms would have an elaborate and complicated structure. Consider the fly. It lays eggs, which hatch into larvae, which metamorphose at the appropriate time into adult flies. Not only are these adults complex at the resolution of what the unaided human eye can see, but the closer you look, using powerful microscopes, the more intricate their structure becomes. Yet another characteristic of life is the ability to move. Animals swim, fly, walk, or run, while plants grow toward or away from light. Whenever you see something move, you suspect that it is alive. Still another characteristic of living organisms is the capacity to adapt selectively to the environment, whether it be a robin choosing materials to build a nest or a vine weaving its way up a fence. Finally, a key characteristic of living organisms is the ability to use sources of energy and matter to grow, that is, the ability to convert foreign material into their own body parts. The chemical and physical reactions that carry out these conversions are known as *metabolism*. A captured bird, for example, if unfed and thus unable to support its metabolism, will die within a few days.

Most properties of living organisms ultimately arise from the class of molecules known as proteins. *Proteins* are large polymers composed of hundreds to thousands of amino-acid subunits strung together in long chains; each chain folds into a specific three-dimensional conformation dictated by the sequence of its amino acids (Fig. 1.5). There are 20 different amino acids. The information in the DNA of genes dictates, via a genetic code, the order of amino acids in a protein molecule. You can think of proteins as constructed from a set of 20 different kinds of snap beads distinguished by color and shape; if you were to arrange the beads in any order, make strings of a thousand beads each, and then fold or twist the chains into shapes dictated by the order of their beads, you would be able to make a nearly infinite number of different three-dimensional shapes. The astonishing diversity of three-dimensional protein structure generates the extraordinary diversity of protein function that is the basis of each organism's complex and adaptive behavior. The structure and shape of the hemoglobin protein, for example, allow it to transport oxygen in the bloodstream and release it to the tissues. The proteins myosin and actin can slide together to allow muscle contraction. Chymotrypsin and elastase are enzymes that help break down other proteins. Most of the properties associated with life emerge from the constellation of protein molecules that an organism synthesizes according to instructions contained in its DNA.

(a)

Chymotrypsin — ANTPORLQQASLPLLSNTNCKK--YWGTKIKDAMICAGAS—GVS

Elastase ———— GQLAQTLQQAYLPTVDYAICSSSSYWGSTVKNSMVCAGGDGVRS

SCMGDSGGPLVCKKNGAWTLVGIVSWGSS—TCSTS—TPGVYARVTALVNWVQQTLAAN

GCQGDSGGPLHCLVNGQYAVHGVTSFVSRLGCNVTRKPTVFTRVSAYISWINNVIASN

A = Ala = alanine	G = Gly = glycine	M = Met = methionine	S = Ser = serine
C = Cys = cysteine	H = His = histidine	N = Asn = asparagine	T = Thr = threonine
D = Asp = aspartic acid	I = Ile = isoleucine	P = Pro = proline	V = Val = valine
E = Glu = glutamic acid	K = Lys = lysine	Q = Gln = glutamine	W = Trp = tryptophan
F = Phe = phenylalanine	L = Leu = leucine	R = Arg = arginine	Y = Tyr = tyrosine

(b)

Hemoglobin β chain   Lactate dehydrogenase   NAD-binding domain

Figure 1.5 Proteins are polymers of amino acids that fold in three dimensions. The specific sequence of amino acids in a chain determines the precise three-dimensional shape of the protein. **(a)** A comparison of equivalent segments in the chains of two digestive proteins, chymotrypsin and elastase. The *red lines* connect sites in the two sequences that carry identical amino acids; the two chains differ at all the other sites shown. Thus, even though these two proteins are evolutionarily related to each other, they differ at enough amino acids that their structures (and functions) are not identical. **(b)** Schematic drawings of the hemoglobin β chain (*green*) and the lactate dehydrogenase NAD-binding domain (*purple*) show the different three-dimensional shapes determined by different amino-acid sequences. The distinct three-dimensional shapes enable the two molecules to perform distinct tasks. The β chain is part of the complex hemoglobin molecule, which binds and delivers oxygen to body tissues. The NAD-binding domain is part of the lactate dehydrogenase molecule, an enzyme that catalyzes energy conversions in microorganisms such as yeast and in the muscle cells of strenuously exercising animals.

Complex Systems Arise from DNA-Protein and Protein-Protein Interactions

A third level of biological information encompasses dynamic interactions among DNA, protein, and other types of molecules as well as interactions among cells and tissues. These complex interactive networks represent **biological systems** that function

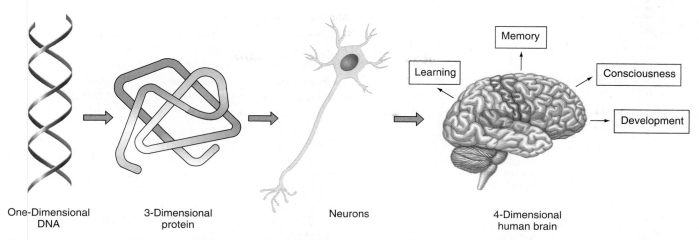

One-Dimensional
DNA

3-Dimensional
protein

Neurons

4-Dimensional
human brain

Figure 1.6   Diagram of the conversion of biological information from a one- to a three- and finally a four-dimensional state.

both within individual cells and among groups of cells within an organism. Here we use *biological system* to mean any complex network of interacting molecules or groups of cells that function in a coordinated manner through dynamic signaling. There are several layers of biological systems. The human pancreas, for example, is an isolated biological system that operates within the larger biological system of the human body and mind. A whole community of animals, such as a colony of ants that functions in a highly coordinated manner, is also a biological system.

The information that defines any biological system is four-dimensional because it is constantly changing over the three dimensions of space and the one dimension of time. One of the most complex examples of this level of biological information (other than an entire human being) is the human brain with its 10^{11} (100,000,000,000) neurons connected through perhaps 10^{18} (1,000,000,000,000,000,000,) synapses. From this enormous biological network, based ultimately on the information in DNA and protein, arises systems and emergent properties such as memory, consciousness, and the ability to learn (Fig. 1.6).

All Living Things Are Closely Related

The evolution of biological information is a fascinating story spanning the 4 billion years of earth's history. Many biologists think that RNA was the first information-processing molecule to appear. RNA molecules are very similar to DNA and are also composed of four subunits: the bases G, C, A, and U. Like DNA, RNA has the capacity to store, replicate, mutate, and express information; like proteins, RNA can fold in three dimensions to produce molecules capable of catalyzing the chemistry of life. RNA molecules, however, are intrinsically unstable. Thus, it is probable that the more stable DNA took

over the linear information storage and replication functions of RNA, while proteins, with their far greater capacity for diversity, preempted the functions derived from RNA's three-dimensional folding. The information contained in the sequence of DNA nucleotides then came to specify the sequence of amino acids in the proteins. With this division of labor, RNA became an intermediary in converting the information in DNA into the sequence of amino acids in protein. The DNA letters G, C, A, and T are informationally equivalent to the RNA letters G, C, A, and U (Fig. 1.7a). The separation that placed information storage in DNA and biological function in proteins was so successful that all organisms alive today descend from the first organisms that happened upon this molecular specialization.

The evidence for the common origin of all living forms is present in their DNA sequences. All living organisms use essentially the same arbitrary genetic code in which various triplet groupings of the 4 letters of the DNA and RNA alphabets encode the 20 letters of the amino-acid alphabet. Via the code, the order of bases in any organism's DNA specifies the amino-acid sequence of its proteins (Fig. 1.7b).

The relatedness of all living organisms is also evident from comparisons of genes with similar functions in very different organisms. For example, there is striking similarity between the genes for many proteins in bacteria, yeast, plants, worms, flies, mice, and humans (Fig. 1.8). Moreover, it is often possible to place a gene from one organism into the genome of a very different organism and see it function normally in the new environment. Human genes that help regulate cell division, for example, can replace related genes in yeast and enable the yeast cells to function normally.

One of the most striking examples of relatedness at this level of biological information was uncovered in studies of eye development. Both insects and vertebrates (including humans) have eyes; but they are of very different types (Fig. 1.9).

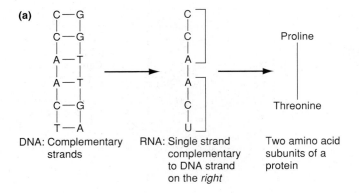

(a)

DNA: Complementary strands → RNA: Single strand complementary to DNA strand on the *right* → Two amino acid subunits of a protein

Proline

Threonine

(b)

Second letter

	U	C	A	G	
U	UUU UUC } Phe / UUA UUG } Leu	UCU UCC UCA UCG } Ser	UAU UAC } Tyr / UAA Stop / UAG Stop	UGU UGC } Cys / UGA Stop / UGG Trp	U C A G
C	CUU CUC CUA CUG } Leu	CCU CCC CCA CCG } Pro	CAU CAC } His / CAA CAG } Gln	CGU CGC CGA CGG } Arg	U C A G
A	AUU AUC } Ile / AUA / AUG Met	ACU ACC ACA ACG } Thr	AAU AAC } Asn / AAA AAG } Lys	AGU AGC } Ser / AGA AGG } Arg	U C A G
G	GUU GUC GUA GUG } Val	GCU GCC GCA GCG } Ala	GAU GAC } Asp / GAA GAG } Glu	GGU GGC GGA GGG } Gly	U C A G

First letter / Third letter

Figure 1.7 RNA is an intermediary in the conversion of DNA information into protein via the genetic code. (a) The linear bases of DNA are copied through molecular complementarity into the linear bases of RNA. The bases of RNA are read three at a time, that is, as triplets, to encode the amino-acid subunits of proteins. **(b)** The genetic code dictionary specifies the relationship between RNA triplets and the amino-acid subunits of proteins. (Note that this table uses three-letter abbreviations of the 20 amino acids.)

S. cerevisiae	---PGSAKKGATLFKTRCQQCHTIEEGGPNKV
A. thaliana	----GDAKKGANLFKTRCAQCHTLKAGEGNKI
C. elegans	---AGDYEKGKKVYKQRCLQCHVVDS-TATKT
D. melanogaster	---AGDVEKGKKLFVQRCAQCHTVEAGGKHKV
M. musculus	---MGDVEKGKKIFVQKCAQCHTVEKGGKHKT
H. sapiens	---MGDVEKGKKIFIMKCSQCHTVEKGGKHKT
	* * ** .. . * * * * * *

S. cerevisiae	GPNLHGIFGRHSGQVKGYSYTDANINKNVKW
A. thaliana	GPELHGLFGRKTGSVAGYSYTDANKQKGIEW
C. elegans	GPTLHGVIGRTSGTVSGFDYSAANKNKGVVW
D. melanogaster	GPNLHGLIGRKTGQAAGFAYTDANKAKGITW
M. musculus	GPNLHGLFGRKTGQAAGFSYTDANKNKGITW
H. sapiens	GPNLHGLFGRKTGQAPGYSYTAANKNKGIIW
	** ***. **. ** . * . ** * . . *

S. cerevisiae	DEDSMSEYLTNPKKYIPGTKMAFAGLKKEKDR
A. thaliana	KDDTLFEYLENPKKYIPGTKMAFGGLKKPKDR
C. elegans	TKETLFEYLLNPKKYIPGTKMVFAGLKKADER
D. melanogaster	NEDTLFEYLENPKKYIPGTKMIFAGLKKPNER
M. musculus	GEDTLMEYLENPKKYIPGTKMIFAGIKKKGER
H. sapiens	GEDTLMEYLENPKKYIPGTKMIFVGIKKKEER
	... *** *********** * . ** . *

S. cerevisiae	NDLITYMTKAAK---
A. thaliana	NDLITFLEEETK---
C. elegans	ADLIKYIEVESA---
D. melanogaster	GDLIAYLKSATK---
M. musculus	ADLIAYLKKATN---
H. sapiens	ADLIAYLKKATN---
	*** .. .

* Indicates identical and . indicates similar

Figure 1.8 Comparisons of gene products in different species provide evidence for the relatedness of living organisms. This chart shows the amino-acid sequence for equivalent portions of the cytochrome C protein in six species: *Saccharomyces cerevisiae* (yeast), *Arabidopsis thaliana* (a weedlike flowering plant), *Caenorhabditis elegans* (a nematode), *Drosophila melanogaster* (the fruit fly), *Mus musculus* (the house mouse), and *Homo sapiens* (humans). Consult Fig. 1.5 for the key to amino-acid names. The cytochromes are a family of heme-containing proteins that function as electron donors and acceptors during cellular respiration and photosynthesis. Cytochrome C is the most abundant and most stable of the cytochromes; its form and function have been conserved throughout evolution. As the chart shows, there are many sequence similarities among the six types of organisms. Not included is a bacterial cytochrome C, whose sequence would show significant but less pronounced similarity with the cytochrome Cs in the chart.

Biologists had long assumed that the evolution of eyes occurred independently in the lineages leading to present-day insects and present-day vertebrates. Indeed, in many evolution textbooks, eyes are used as an example of *convergent evolution,* that is, of evolution in which structurally unrelated but func-

tionally analogous organs emerge in different species as a result of natural selection. Studies of a gene called the *Pax6* gene have turned this view upside down.

Mutations in the *Pax6* gene lead to a failure of eye development in both people (with a condition known as anirida) and

Figure 1.9 The eyes of insects and humans have a common ancestor. (a) A fly eye and **(b)** human eye.

mice, and molecular studies have suggested that *Pax6* might play a central role in the initiation of eye development in all vertebrates. Remarkably, when the human *Pax6* gene is expressed in cells along the surface of the fruit fly body, it induces numerous little eyes to develop there. This result demonstrates that there was a single origin of the eye in an ancestor common to flies and people and that, after 600 million years of divergent evolution, both vertebrates and insects still share the same main control switch for initiating eye development. Studies of many other genes have shown that the entire developmental program of flies and people uses many of the same genes. These genes have duplicated and evolved divergent functions, but they still retain their ancestral markings.

The utility of the finding of relatedness and unity at all levels of biological information cannot be overstated. It means that in many cases, the experimental manipulation of organisms known as *model organisms* can shed light on complex networks in humans. If genes similar to human genes function in simple model organisms such as fruit flies (*Drosophila melanogaster*), roundworms (*Caenorhabditis elegans*), yeast (*Saccharomyces cerevisiae*), and the bacterium *Escherichia coli*, one can determine gene function and regulation in these experimentally manipulable organisms and bring these insights to an understanding of the human organism. The same is true of the shared informational pathways such as DNA replication and protein synthesis. The mouse (*Mus musculus*), a more complex model organism than those just mentioned, shares many of the sophisticated informational networks found in humans.

In short, the close relatedness of all living organisms at the molecular level has great significance for an understanding of biology. It makes it possible to combine bits and pieces learned from different organisms into a global understanding of molecular and cellular biology that is valid for all organisms. And even though controlled experimentation with humans is usually impossible, the relatedness of all organisms allows us to learn about human biology from mice, flies, worms, peas, yeast, and other organisms that are accessible to experimentation.

The Modular Construction of Genomes Has Allowed the Relatively Rapid Evolution of Complexity

We have seen that roughly 40,000–60,000 genes direct human growth and development. How did such complexity arise? Recent technical advances have enabled researchers to complete structural analyses of the entire genome of more than 100 organisms, including many microbes. The information obtained reveals that families of genes have arisen by duplication of a primordial gene; after duplication, mutations and rearrangements may cause the two copies to diverge from each other (Fig. 1.10a). In both mice and humans, for example, five different hemoglobin genes produce five different hemoglobin molecules at successive stages of development, with each protein functioning in a slightly different way to fulfill different needs for oxygen transport. The set of five hemoglobin genes arose from a single primordial gene by several duplications followed by slight divergences in structure.

Duplication followed by divergence underlies the formation of new genes with new functions. In addition, the genomes of multicellular organisms have evolved an organization that facilitates the construction of new genes from the modular components of existing genes. Many genes in multicellular organisms are assemblages of protein-encoding parts, known as *exons,* separated from each other by noncoding parts, known as *introns.* The exons often correspond to a protein's independent functional domains. When genes with this kind of modular construction undergo duplication, one of the resulting copies can ensure production of the original protein, while a shuffling of exons in the other copy could generate new genes that produce new proteins. Thus, duplications followed by rearrangements within the genome may permit old genes to continue to function and new combinations of exons to generate new genes (Fig. 1.10b).

The tremendous advantage of the duplication and divergence of existing pieces of genetic information is evident in the history of life's evolution (Table 1.1). *Prokaryotic* cells such as bacteria, which do not have a membrane-bounded nucleus, evolved about 3.7 billion years ago; *eukaryotic* cells such as yeast, which have a membrane-bounded nucleus, emerged around 2 billion years ago; and multicellular eukaryotic organisms appeared 600–700 million years ago. Then, at about 570 million years ago, within the relatively short evolutionary time of roughly 20–50 million years known as the Cambrian explosion, the multicellular lifeforms diverged into a bewildering array of organisms, including primitive vertebrates. A fascinating question is: Since it took eukaryotic cells almost 2 billion years to evolve from prokaryotic cells and multicellular organisms three-quarters of a billion years to evolve from single-celled eukaryotes, how could the multicellular forms achieve such enormous diversity in only 20–50 million years? The answer lies, in part, in the hierarchic organization of the information encoded in chromosomes. Exons are arranged into genes;

Figure 1.10 How genes arise by duplication and divergence.
(a) Duplications of ancestral *gene A* followed by mutations and DNA rearrangements generate a family of related genes. (The *blue* and *red* bands indicate different regions of the genes.) **(b)** A duplication, followed by chromosomal changes that rearrange exons, generates new genes from an exact copy of ancestral *gene B.*

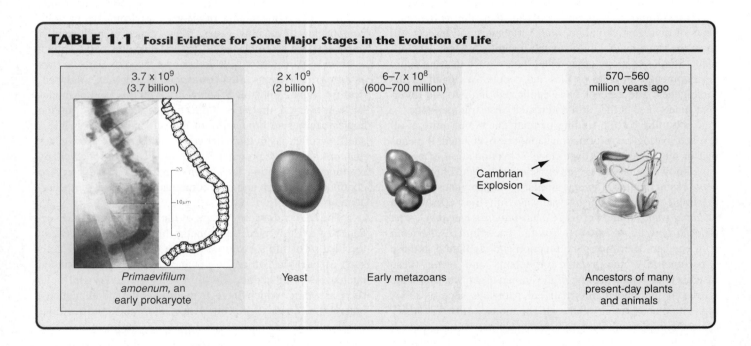

TABLE 1.1 **Fossil Evidence for Some Major Stages in the Evolution of Life**

3.7×10^9 (3.7 billion)	2×10^9 (2 billion)	$6-7 \times 10^8$ (600–700 million)		570–560 million years ago
Primaevifilum amoenum, an early prokaryote	Yeast	Early metazoans	Cambrian Explosion	Ancestors of many present-day plants and animals

genes duplicate and diverge to generate multigene families; and multigene families sometimes rapidly expand to supergene families containing hundreds of related genes. In both mouse and human adults, for example, the immune system is encoded by a supergene family composed of hundreds of closely related but slightly divergent genes. With the emergence of each successively larger informational unit, evolution gains the ability to duplicate increasingly complex informational cassettes through single genetic events.

Even more important for the evolution of complexity is the rapid change of regulatory networks that specify how genes behave (that is, when, where, and to what degree they are expressed) during development. For example, the two-winged fly evolved from a four-winged ancestor not because of changes in structural genes, but rather because of a rewiring of the regulatory network, which converted one pair of wings into two balancing organs known as haltere (Fig. 1.11).

Figure 1.11 Two-winged and four-winged flies. Geneticists converted a contemporary two-winged fly to a four-winged insect resembling the fly's evolutionary antecedent. They accomplished this by mutating a key element in the fly's regulatory network. Note the club-shaped halteres behind the wings of the fly on the left.

Organism	E. coli	S. cerevisiae	C. elegans	D. meanogaster	Mus musculus
Genome size: (in megabases)	4.5 Mb	16 Mb	100 Mb	130 Mb	3000 Mb
Number of genes	4500	6200	19,200	13,900	40,000–60,000

Figure 1.12 Five model organisms whose genomes were sequenced as part of the Human Genome Project. The chart indicates genome size in millions of base pairs, or megabases (Mb). It also shows the projected number of genes for each organism.

Genetic Techniques Permit the Dissection of Complexity

The complexity of living systems has developed over 4 billion years from the continuous amplification and refinement of genetic information. The simplest bacterial cells contain about 1000 genes that interact in complex networks. Yeast cells, the simplest eukaryotic cells, contain about 6000 genes. Nematodes (roundworms) and fruit flies, the simplest multicellular organisms, contain roughly 14,000–19,000 genes; humans may have as many as 60,000 genes. With genetic techniques, researchers can dissect this complexity piece by piece, although the task is daunting. The logic used in genetic dissection is quite simple: inactivate a gene and observe the consequences. For example, loss of a gene for visual pigment produces fruit flies with white eyes instead of eyes of the normal red color. Over the years, geneticists have performed this type of dissection on several species, including the bacterium *Escherichia coli,* the yeast *Saccharomyces cerevisiae,* the nematode *Caenorhabditis elegans,* the fruit fly *Drosophila melanogaster,* the plant *Arabidopsis thaliana,* and the mouse *Mus musculus.* From their study of these so-called *model organisms,* they are amassing a detailed picture of the beauty and complexity of living systems. Moreover, the Human Genome Project, in addition to completing the sequencing of the entire human genome, has sequenced the genomes of *E. coli,* yeast, the nematode, the fruit fly, and the mouse (Fig. 1.12). Each of these organisms has provided valuable insights into biology in general and human biology in particular.

However, even though the power of genetic techniques is astonishing, the complexity of biological systems is difficult to comprehend. We have seen that the human nervous system, for example, is a network of 10^{11} neurons with perhaps 10^{18} connections. Knowing everything there is to know about an individual neuron provides little insight into learning, memory, emotional stability, and personality—traits that emerge from interactions among intricate networks of neurons and the environment. Thus, after defining the single units (whether genes or neurons), it is necessary to look at how multiple units interact to generate the characteristics of the complex system. The shift from the analysis of single units to the analysis of complex systems is a major challenge in modern biology. Genetics provides unique tools for meeting this challenge, making it possible to sort out the multiplicity of genes dictating the diversity of proteins that generate complex systems.

The new global tools of genomics—such as high-throughput DNA sequencers, genotypers, and large-scale DNA arrays (also called DNA chips)—have the capacity to analyze thousands of genes rapidly and accurately. These global tools are not specific to a particular system or organism; rather, they can be used to study the genes of all living things.

The DNA chip is a powerful example of a global genomic tool. Produced by several techniques, individual chips are subdivided into arrays of microscopic blocks that each contain a unique string of DNA units (Fig. 1.13). When a chip is exposed to a complex mixture of fluorescently labeled nucleic acid—such as DNA or RNA from any cell type or sample—the unique string in each microscopic block can bind to and detect a specific complementary sequence. A computer-driven microscope can then analyze the bound sequences of the hundreds of thousands of blocks on the chip, and special software can enter this information into a database.

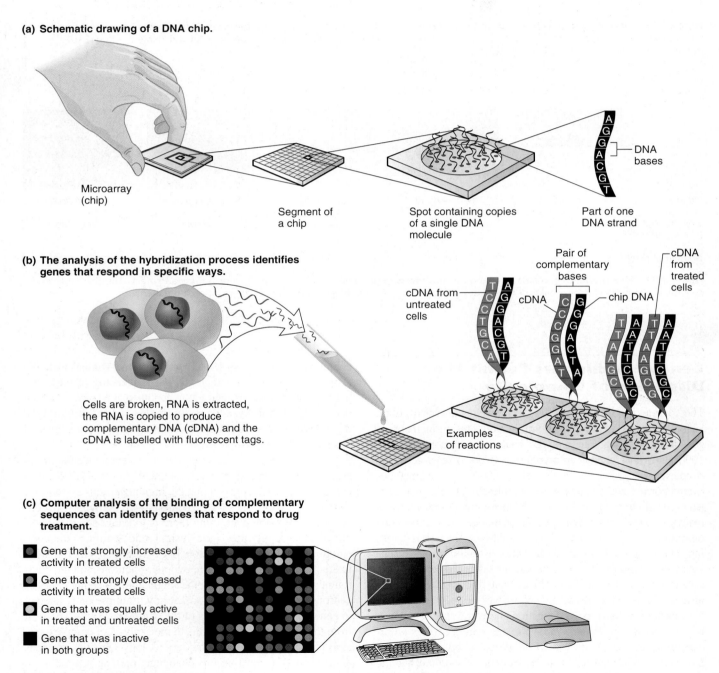

(a) Schematic drawing of a DNA chip.

Microarray (chip)

Segment of a chip

Spot containing copies of a single DNA molecule

Part of one DNA strand

DNA bases

(b) The analysis of the hybridization process identifies genes that respond in specific ways.

Cells are broken, RNA is extracted, the RNA is copied to produce complementary DNA (cDNA) and the cDNA is labelled with fluorescent tags.

cDNA from untreated cells

Pair of complementary bases

cDNA

chip DNA

cDNA from treated cells

Examples of reactions

(c) Computer analysis of the binding of complementary sequences can identify genes that respond to drug treatment.

Gene that strongly increased activity in treated cells

Gene that strongly decreased activity in treated cells

Gene that was equally active in treated and untreated cells

Gene that was inactive in both groups

Figure 1.13   A DNA chip. (a) Schematic drawing of a DNA chip. **(b)** cDNA is the left strand and chip DNA is the right strand.
(c) Computerized analysis of a DNA chip hybridized (paired by complementarity) to cDNA that has been labeled with a fluorescent tag.

The potential of DNA chips is enormous for both research and clinical purposes. Already chips with over 400,000 different detectors can provide simultaneous information on the presence or absence of 400,000 discrete DNA or RNA sequences in a complex sample. And they can do it within hours. Here is one example. Now that the sequence of all human genes is known, unique stretches of DNA representing each of the 40,000–60,000 human genes can be placed on a chip and used to determine the complete set of expressed genes, as well as their level of expression, in any human cell type at any stage of development or differentiation. Such computer-driven comparisons could be used to contrast gene activation in normal and cancerous cells, making it possible to pinpoint the genes involved in establishing the cancerous phenotype. Medical researchers have already used DNA chip technology to assess the genetic differences between breast cancers that respond well to treatment and those that do not (that is, that recur despite treatment). Knowledge of these differences could lead to improved therapies.

Our Focus Is on Human Genetics

In the mid-1990s, a majority of scientists who responded to a survey conducted by *Science* magazine rated genetics as the most important field of science for the next decade. One reason is that the powerful tools of genetics open up the possibility of understanding biology, including human biology, from the molecular level up to the level of the whole organism. In combination with an appreciation of the relatedness of all living organisms, the potential of genetic analysis heralds an era that promises to help reveal more about who we are.

The Human Genome Project, by changing the way we view biology and genetics, has led to a significant paradigm change: the systems approach to biology and medicine. The systems approach enables one to study the relationships of all the elements in a biological system as it undergoes genetic perturbation or biological activation (see Chapter 10). This is a fundamental change from the study of complex systems one gene or protein at a time.

How Human Genetics Is Leading Us Toward Predictive and Preventive Medicine

Over the next 25 years, geneticists will identify hundreds of genes with variations that predispose people to many types of disease: cardiovascular, cancerous, immunological, mental, metabolic. As we know, some mutations will always cause disease; others will only predispose to disease. For example, a change in a specific single nucleotide (that is, a change in one DNA unit) in the *β-globin* gene will nearly always cause sickle-cell anemia, a painful, life-threatening condition that leads to severe anemia. By contrast, a mutation in the *breast cancer 1 (BRCA1)* gene has only a 70% chance of causing breast cancer in a 60-year-old woman carrying one copy of the mutation; this conditional state arises either because the *BRCA1* gene interacts with environmental factors that affect the probability of activating the cancerous condition or because various forms of other genes modify expression of the *BRCA1* gene. Defining

and analyzing the multiple factors contributing to genetic predispositions may be an important element in understanding and designing therapies for some diseases. Physicians may be able to use DNA diagnostics—a collection of techniques for characterizing genes—to analyze an individual's DNA for genes that predispose to some diseases. With this genetic profile, they may be able to write out a probabilistic health history for some medical conditions. This will move us into the era of predictive medicine.

As scientists come to understand the complex systems in which disease genes operate, they will be able to design therapeutic drugs to block and/or reverse the effects of mutant genes. If taken before the onset of disease, such drugs could prevent occurrence or minimize symptoms of the gene-based disease. This will usher in the era of preventive medicine in which the focus of medical practice is on keeping individuals healthy rather than on trying to heal the sick.

In this new era of predictive and preventive medicine, many individuals could live into their 80s and 90s, physically fit and mentally alert (Fig. 1.14). At this point, society will have to rethink how it treats the elderly, as they will represent an enormous reservoir of able, talented people capable of making significant contributions to society. Society will also have to redesign the education of physicians, other health professionals, and the public to include a perspective on predictive and preventive medicine.

The New Scope of Human Genetics and the New Potential of Predictive and Preventive Medicine Intensify the Need to Confront Many Social Issues

Genetics began as a separate biological discipline dedicated to determining the rules governing the frequency of appearance of alternative traits in siblings and other related individuals. At the beginning of the twenty-first century, it has become the central focus and tool in the study of complex biological systems created by the interactions of molecular entities. Although biological information is similar to other types of information from a strictly technical point of view, it is as different as can be in its meaning and impact on individual human beings and human

Figure 1.14 Options at age 95?

society as a whole. The difference lies in the personal nature of the unique genetic profile carried by each person from birth. Within this basic level of biological information are complex life codes that provide greater or lower susceptibility or resistance to many diseases, as well as greater or lesser potential for the expression of many physiologic, physical, and neurological attributes that distinguish people from each other. Until now, almost all this information has remained hidden away. But if research continues at its present pace, in less than a decade it will become possible to read a large part of a person's genetic profile; and with this information will come the power to make some limited predictions about future possibilities and risks.

As we will see in many of the Genetics and Society boxes throughout this book, society can use genetic information not only to help people but also to restrict their lives (for example, by denying insurance or employment). We believe that just as our society respects an individual's right to privacy in other realms, it should also respect the privacy of an individual's genetic profile and work against all types of discrimination.

Another issue raised by the potential for detailed genetic profiles is the interpretation or misinterpretation of that information. Without accurate interpretation, the information becomes useless at best and harmful at worst. Proper interpretation of genetic information requires some understanding of statistical concepts such as risk and probability. To help people understand these concepts, widespread education in this area will be essential. Since the media play an enormous role in the lives of most people, public education could begin with media reports on new genetic findings that are well reasoned and accurate. It will also be essential to bring kindergarten through high school education up-to-date so that children can learn the concepts and implications of modern human biology as a science of information.

Yet another pressing issue concerns the regulation and control of the new technology. With the sequencing of the entire human genome, government funds appropriated for the sequencing project can be redirected toward analyzing genetic variation among humans as well as various aspects of genome structure and organization. The question of whether the government should establish guidelines for the use of genetic and genomic information, reflecting society's social and ethical values, remains in open debate.

To many people, the most frightening potential of the new genetics is the development of technology that can alter or add to the genes present within the germ line of human embryos. This technology (referred to as "transgenic technology" in scientific discourse and "genetic engineering" in public discussions) has become routine in hundreds of laboratories working with various other animals.

Some people caution that developing the power to alter our own genomes is a step we should not take, arguing that if genetic information and technology are misused (as they certainly have been in the past), the consequences could be horrific. Attempts to use genetic information for social purposes was prevalent in the early twentieth century, leading to enforced sterilization of individuals thought to be inferior, to laws that prohibited interracial marriage, and to immigration laws prohibiting immigration of certain ethnic groups. The scientific basis of these actions has been thoroughly discredited. Others agree that we must not repeat the mistakes of the past, but warn that if the new technologies could help children and adults lead healthier, happier lives, we need to think very carefully about whether the reasons for objecting to their use outright are valid. Most agree that the biological revolution we are living through will have a greater impact on human society than any technological revolution of the past, and that education and public debate are the key to preparing for the consequences of this revolution.

The focus on human genetics in this book looks forward into the new era of biology and genetic analysis. As we gain increasingly sophisticated knowledge about the human genetic makeup, it will not only become possible to cure human diseases that now resist therapy; it will also become possible to have an impact on our own evolution. We have seen that these new possibilities raise serious moral and ethical issues that will demand wisdom and humility. It is in the hope of educating young people for the moral and ethical challenges awaiting the next generation that we write this book.

CONNECTIONS

Genetics, the study of biological information, is also the study of the DNA and RNA molecules that store, replicate, transmit, and evolve information for the construction of proteins. With their extraordinary diversity of structure and function, proteins generate the complex and adaptive behaviors of all living organisms. At the molecular level, all living things are closely related. As a result, observations of model organisms as different as yeast and mice can provide insights into the general biological principles as well as human biology.

Remarkably, more than 75 years before the discovery of DNA, Gregor Mendel, an Augustinian monk working in what is now Brno in the Czech Republic, delineated the basic laws of gene transmission with no knowledge of the molecular basis of heredity. He accomplished this by following simple traits, such as flower or seed color, that come in two discrete forms, such as white and purple or yellow and green, through several generations. He used the pea plant (*Pisum sativum*) as his experimental organism and set up carefully controlled matings between plants that differed in one or a few traits. We now know that his findings apply to all sexually reproducing organisms. Chapter 2 describes Mendel's studies and insights, which became the foundation of the field of genetics.

2

Mendel's Breakthrough: Patterns, Particles, and Principles of Heredity

Although Mendel's laws can predict the probability that an individual will have a particular genetic makeup, the chance meeting of particular male and female gametes determines an individual's actual genetic fate.

A quick glance at an extended family portrait is likely to reveal children who resemble one parent or the other or who look like a combination of the two, with perhaps wavy hair from the father, a broad nose from the mother, and a skin color in between the two parents' (Fig. 2.1). Some children, however, look unlike any of the assembled relatives and more like a throwback to a great, great grandparent. What causes the similarities and differences of appearance and the skipping of generations?

The answers lie in our **genes,** the basic units of biological information, and in **heredity,** the way genes transmit biochemical, physical, and behavioral traits from parents to offspring. Each of us starts out as a single fertilized egg cell that develops, by division and differentiation, into a mature adult made up of 10^{14} (a hundred trillion) specialized cells, including muscle cells capable of contraction, brain cells structured for rapid communication, red blood cells tailored for transporting oxygen, and hair cells that carry pigment for black, brown, blond, or flaming red hair. By initial estimates from the Human Genome Project, only 40,000–60,000 genes control this amazing developmental process. Passed from parents to offspring through egg and sperm, these genes underlie

the formation of every heritable trait. Such traits are as diverse as the shape of your hairline, the tendency to bald as you age, the timbre of your voice, the way you clasp your hands, even your susceptibility to heart disease and certain cancers. And they all run in families in predictable patterns that impose some possibilities and exclude others.

Genetics, the science of heredity, pursues a precise explanation of the biological structures and mechanisms that determine what is inherited and how. Geneticists seek to identify genes, to learn how they determine particular traits, and to understand how genes work together to create a person, a plant, or a protozoan. In some instances, the relationship between gene and trait is remarkably simple. A change in a single gene, for example, results in sickle-cell anemia by causing construction of a defective hemoglobin molecule, the oxygen-carrying protein in red blood cells; when oxygen is in short supply, red blood cells carrying the abnormal hemoglobin become sickle-shaped and clog small vessels. In other instances, the correlations between genes and traits are bewilderingly complex. An example is the genetic basis of facial features, in which many genes determine a

Figure 2.1   A family portrait. The extended family shown here includes members of four generations.

Figure 2.2   Gregor Mendel. Photographed around 1862 holding one of his experimental plants. His work formed the basis for our understanding and continued exploration of the science of genetics.

large number of molecules that interact in a variety of ways to generate the combination we recognize as a friend's face.

Gregor Mendel (1822–1884; Fig. 2.2), a stocky, bespectacled, highly respected Augustinian monk and expert plant breeder, trained in both physics and plant physiology, discovered the basic principles of genetics in the mid–nineteenth century. He published his findings in 1866, just seven years after Darwin's *On the Origin of Species* appeared in print. Mendel lived and worked in Brunn, Austria (now Brno in the Czech Republic), a nineteenth-century center of learning in the sciences and humanities, situated in the rich agricultural valley of the province of Moravia. Here he examined the inheritance of such clear-cut alternative traits in pea plants as purple versus white flowers or yellow versus green seeds. In so doing, he discovered why some of these traits disappeared in one generation and then reappeared in another. By rigorously analyzing the patterns of transmission through generations, he inferred genetic laws that allowed him to make verifiable predictions about which traits would appear, disappear, and then reappear in which generations. Simple and straightforward, Mendel's laws are based on the hypothesis that observable traits such as seed color are determined by independent units of inheritance not visible to the naked eye. We now call these units *genes.* The concept of the gene continues to change as research deepens and refines our understanding of genetic phenomena. Today, a gene is recognized as a region of DNA that encodes a specific protein or a particular type of RNA. In the beginning, however, it was an abstraction—an imagined particle with no physical features, whose function was to control a visible trait—that Mendel proposed as the explanation for the results of his plant breeding experiments.

Because Mendel's laws are basic to an understanding of genetics, we begin our study of genetics with a detailed look at what these laws are and how they were discovered. In subsequent chapters, we discuss logical extensions to Mendel's laws and describe how his successors grounded the abstract concept of hereditary units (genes) in an actual biological molecule (DNA). With knowledge of that molecule, geneticists devised techniques to isolate and identify the units of heredity. Today, they integrate the analytical tools of Mendel with modern molecular techniques

as they continue to probe the nature of the hereditary material and examine exactly how it is passed from parent to offspring, how it acts in organisms to produce the visible and invisible traits that define individuals, and how it evolves over time.

Four general themes emerge from our detailed discussion of Mendel's work. The first is that variation, as expressed in alternative forms of a trait (a high-pitched voice or a low one, a green pea or a yellow one), is widespread in nature. This genetic diversity provides the raw material for the continuously evolving variety of life we see around us. Second, observable variation is essential for following genes. If all the traits of all offspring resembled their parents', Mendel would have had no basis for discerning and analyzing patterns of transmission. Third, variation is not distributed solely by chance; rather, it is inherited according to genetic laws that explain why like begets both like and unlike. Dogs beget other dogs, pea plants beget pea plants, and people, people. But there are hundreds of breeds of dogs and even within a breed—Labrador retrievers, for instance—two black dogs could have a litter of black, brown, and tan puppies (Fig. 2.3). Mendel's

Figure 2.3   Like begets like and unlike. A Labrador retriever with her litter of pups.

insights help explain why this is so. Fourth, the laws Mendel discovered about heredity apply equally well to all sexually reproducing organisms, from protozoans to peas to people.

Our presentation of Mendelian genetics examines:

■ The background: The historical puzzle of inheritance and how Mendel's innovative experimental approach helped resolve it.

■ The work itself: Genetic analysis according to Mendel, including a discussion of Mendel's seminal experiments and related analytical tools of genetics.

■ Repercussions: A comprehensive example of Mendelian inheritance in humans.

■ ■ ■

Background: The Historical Puzzle of Inheritance

There are several steps to understanding genetic phenomena: the careful observation over time of groups of organisms, such as human families, herds of cattle, or fields of corn or tomatoes; the rigorous analysis of systematically recorded information gleaned from these observations; and the development of a theoretical framework, a way of thinking about the phenomena, that can explain their origin and relation. In the mid–nineteenth century, Gregor Mendel became the first person to combine the three approaches and reveal the true basis of heredity. For many thousands of years before that, observation was the main road traveled, and elementary selective breeding of domesticated plants and animals, with no guarantee of what a particular mating would produce, was the only genetic practice.

Artificial Selection Was the First Applied Genetic Practice

A rudimentary use of genetics was the driving force behind a key transition in human civilization, allowing hunters and gatherers to settle in villages and survive as shepherds and farmers. Even before recorded history, people practiced applied genetics as they domesticated plants and animals for their own uses. From a large litter of semitamed wolves, for example, they sent the savage and the misbehaving to the stew pot while sparing the alert sentries and friendly companions for longer life and eventual mating. As a result of this **artificial selection**—purposeful control over mating by choice of parents for the next generation—the domestic dog *(Canis domesticus)* slowly arose from ancestral wolves *(Canis lupus)*. The oldest bones identified indisputably as dog (and not wolf) are a skull excavated from a 20,000-year-old Alaskan settlement. Many millennia of evolution guided by artificial selection have produced massive Great Danes and minuscule Chihuahuas as well as hundreds of other modern breeds of dog. The amazing range of size, shape, and behavior bears witness to the enormous amount of genetic variation in ancient canines and the degree of differentiation that artificial selection can produce. By 10,000 years ago, people had used this same kind of genetic manipulation to develop economically valuable herds of reindeer, sheep, goats, pigs, and cattle that produced life-sustaining meat, hides, and wools. By the time civilization flourished in ancient Greece, such selective animal breeding had acknowledged social value. According to Homer

and other Greek poets, the wealth and power of ancient Troy arose in part from the culture's horse-breeding skills.

Similarly, farmers carried out artificial selection of plants, storing seed from the healthiest individuals with the tastiest and most succulent leaves, roots, or stems for the next planting, and eventually producing strains that grew better, produced more, and were easier to cultivate and harvest. In this way, scrawny grasses and other weedlike plants gradually, with human guidance, turned into rice, wheat, barley, lentils, and dates in Asia; corn, squash, tomatoes, potatoes, and peppers in North and South America; yams, peanuts, and gourds in Africa. Later, plant breeders recognized male and female organs in plants and carried out artificial pollination. An Assyrian frieze carved in the ninth century B.C., pictured in Fig. 2.4, is the oldest known visual record of this kind of genetic experiment. It depicts priests brushing the flowers of female date palms with selected

Figure 2.4 The earliest known record of applied genetics. In this 2800-year-old Assyrian relief from the Northwest Palace of Assurnasirpal II (883–859 B.C.), priests wearing bird masks artificially pollinate flowers of female date palms.

male pollen. By this method of artificial selection, early practical geneticists produced several hundred varieties of dates, each differing in specific observable qualities, such as the fruit's size, color, or taste. A 1929 botanical survey of three oases in Egypt turned up 400 varieties of date-bearing palms, twentieth-century evidence of the degree of natural and artificially generated variation among the date palms.

The Puzzle of Passing on Desirable Traits

In 1822, the year of Mendel's birth, what people in Moravia understood about the basic principles of heredity was not much different from what the people of ancient Egypt and Greece had understood, even though some advances in artificial selection had occurred. By the nineteenth century, precise techniques for controlling matings in plants had given rise to the systematic creation of strains in which offspring often carried a prized parental trait. Using such strains, breeders could produce plants with desired characteristics for food and fiber, but they could not always predict why a valued trait would sometimes disappear and then reappear in only some offspring. The same problem plagued animal breeders. For example, selective breeding practices had resulted in valuable flocks of merino sheep producing large quantities of soft, fine wool, but at the 1837 annual meeting of the Moravian Sheep Breeders Society, one breeder's dilemma epitomized the state of the art. He possessed an outstanding ram that would be priceless "if its advantages are inherited by its offspring," but "if they are not inherited, then it is worth no more than the cost of wool, meat, and skin." Which would it be? According to the meeting's recorded minutes, current breeding practices offered no assurances, no definite answers. In the concluding remarks at this same sheep-breeders meeting, the Abbot Cyril Napp pointed to a possible way out. He proposed that breeders could improve their ability to predict what traits would appear in the offspring by finding the answers to three basic questions: What is inherited? How is it inherited? What is the role of chance in heredity?

This is where matters stood in 1843 when 21-year-old Johann Mendel entered the monastery in Brno, presided over by the same Abbot Napp. The young Johann's name was changed to Gregor at this time, and thereafter that was the name he used. Although Gregor Mendel was a monk trained in theology, he was not a rank amateur in science, discovering new and exciting things in a vacuum. Moravia was a center of learning and scientific activity, providing access to the latest techniques and literature; Mendel, for instance, was able to acquire a copy of Darwin's *On the Origin of Species* shortly after it was translated into German in 1863. As a youth, Mendel had been a student of such outstanding caliber that even when his father was disabled in an accident, Mendel, the only male child of a relatively poor peasant couple, was allowed to enter the monastery to pursue his studies, while a brother-in-law took over management of the family farm. Abbot Napp, recognizing Mendel's intellectual abilities, sent him to the University of Vienna—all expenses paid—where he prescribed his own course of study based on his particular interests. His choices were an unusual mix: physics

(a)

(b)

Figure 2.5 Mendel's garden and microscope. (a) Gregor Mendel's garden was part of his monastery's property in Brno. **(b)** Mendel used this microscope to examine plant reproductive organs and to pursue his interests in natural history.

(with the renowned Christian Doppler, discoverer of the "Doppler effect"), mathematics, chemistry, botany, paleontology, and plant physiology. The cross-pollination of ideas from several disciplines would play a significant role in Mendel's discoveries. One year after he returned to Brno, he began his series of seminal genetic experiments. Figure 2.5 shows where Mendel worked and the microscope he used.

A New Experimental Approach

Before Mendel, many misconceptions clouded people's thinking about heredity. Two of the prevailing myths were particularly misleading. The first was that one parent contributes most to an offspring's inherited features; Nicolaas Hartsoeker, one of the earliest microscopists, contended in 1694 that it was the male, by way of a fully formed homunculus inside the sperm (Fig. 2.6). Another deceptive notion was the concept of *blended inheritance,* the idea that parental traits become mixed and for-

Figure 2.6   The homunculus: A myth. Well into the nineteenth century, many prominent microscopists believed they saw a fully formed, miniature fetus crouched within the head of a sperm.

ever changed in the offspring, as when blue and yellow pigment merge to green on a painter's palette. The theory of blending may have grown out of a natural tendency for parents to see a combination of their own traits in their offspring. While blending could account for children who look like a combination of their parents, it could not explain obvious differences between biological brothers and sisters nor the persistence of variation within extended families.

The experiments Mendel devised would lay these myths to rest by providing precise, verifiable answers to the three questions Abbot Napp had raised almost 15 years earlier: What is inherited? How is it inherited? What is the role of chance in heredity? A key component of Mendel's breakthrough was the way he set up his experiments.

What did Mendel do differently from those who preceded him? First, he chose the garden pea *(Pisum sativum)* as his experimental organism (Fig. 2.7a and b). Peas grew well in Brno and, with male and female organs in the same flower, they were normally self-fertilizing. In **self-fertilization** (or *selfing*), both egg and pollen come from the same plant. The particular anatomy of pea flowers, however, makes it easy to prevent self-fertilization and instead to **cross-fertilize** (synonyms: to *cross* or to *outbreed*) by brushing pollen from one plant onto a female organ of another plant, as illustrated in Fig. 2.7c. Peas offered yet another advantage. For each successive generation, Mendel could obtain large numbers of individuals within a relatively short growing season. By comparison, if he had worked with sheep, each mating would have generated only a few offspring and the time between generations would have been several years.

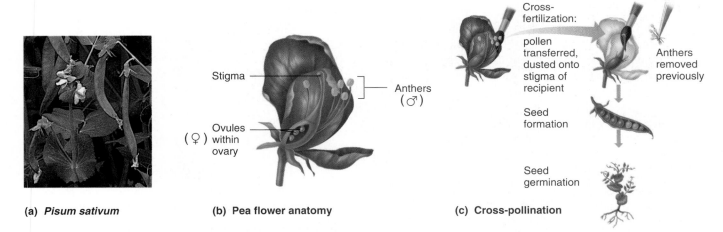

(a) *Pisum sativum*   **(b)** Pea flower anatomy   **(c)** Cross-pollination

Figure 2.7   Mendel's experimental organism: The garden pea. (a) Pea plants with white flowers. **(b)** Pollen is produced in the anthers. Mature pollen lands on the stigma, which is connected to the ovary (which becomes the pea pod). After landing, the pollen grows a tube that extends through the stigma to one of the ovules (immature seeds), allowing fertilization to take place. **(c)** To prevent self-fertilization, breeders remove the anthers from the female parents (here, the white flower) before the plant produces mature pollen. They then transfer pollen with a paintbrush from the anthers of the male parent (here, the purple flower) to the stigma of the female parent. Each fertilized ovule becomes an individual pea (mature seed) that can grow into a new pea plant. All of the peas produced from one flower are encased in the same pea pod, but these peas form from different pollen grains and ovules.

Second, Mendel examined the inheritance of clear-cut alternative forms of particular traits—purple versus white flowers, yellow versus green peas. Using such "either-or" traits, he could distinguish and trace unambiguously the transmission of one or the other observed characteristic, because there were no intermediate forms. (The opposite of these so-called *discrete traits* are *continuous traits* such as height and skin color in humans. Continuous traits show many intermediate forms.)

Third, Mendel collected and perpetuated lines of peas that bred true. Matings within such **pure-breeding lines** produce offspring carrying specific parental traits that remain constant from generation to generation. Here is how Mendel himself defined a pure-breeding line: "Permit me to state that, as an empirical worker, I must define constancy of type as the retention of a character during the period of observation." Mendel observed his pure-breeding lines for up to eight generations. Plants with white flowers always produced offspring with white flowers; plants with purple flowers produced only offspring with purple flowers. Mendel called constant but mutually exclusive, alternative traits, such as purple versus white flowers or yellow versus green seeds, "antagonistic pairs" and settled on seven such pairs for his study (Fig. 2.8). In his experiments, he not only perpetuated pure-breeding stocks for each member of a pair, but he also cross-fertilized pairs of plants to produce **hybrids,** offspring of genetically dissimilar parents, for each pair of antagonistic traits. Figure 2.8 shows the appearance of the hybrids he studied.

Fourth, Mendel, the expert plant breeder, carefully controlled his matings, going to great lengths to ensure that the progeny he observed really resulted from the specific fertilizations he intended. Thus he painstakingly prevented the intrusion of any foreign pollen and assured self- or cross-pollination as the experiment demanded. Not only did this allow him to carry out controlled breedings of selected traits, he could also make **reciprocal crosses.** In such crosses, he reversed the traits of the male and female parents, thus controlling whether a particular trait was transmitted via the egg cell within the ovule or via a sperm cell within the pollen. For example, he could use pollen from a purple flower to fertilize the eggs of a white flower and also use pollen from a white flower to fertilize the eggs of a purple flower. Because the progeny of these reciprocal crosses were similar, he disproved the idea that one parent contributes more to the next generation, demonstrating instead that they contribute equally to inheritance. "It is immaterial to the form of the hybrid," he wrote, "which of the parental types was the seed or pollen plant."

Fifth, Mendel worked with large numbers of plants, counted all offspring, subjected his findings to statistical analysis, and then compared his results with predictions based on his models. He was the first person to study inheritance in this manner, and no doubt his background in physics and mathematics contributed to this quantitative approach. Just as cutting and polishing a dull gray stone can reveal astonishing patterns that reflect its chemical composition, Mendel's careful quantitative analysis revealed patterns of transmission that reflected basic laws of heredity.

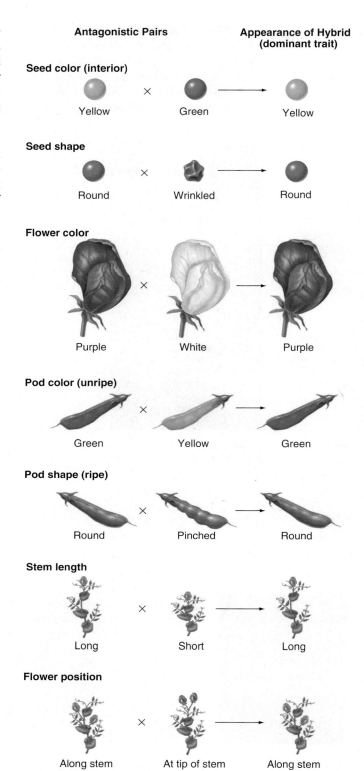

Figure 2.8 The mating of parents with antagonistic traits produces hybrids. Note that each of the hybrids for the seven antagonistic traits studied by Mendel resemble only one of the parents. The parental trait that shows up in the hybrid is known as the "dominant" trait.

Finally, Mendel was a brilliant practical experimentalist. When comparing tall and short plants, for example, he made sure that the short ones were out of the shade of the tall ones so their growth would not be stunted. Eventually he focused on certain traits of the pea seeds themselves, such as their color or shape, rather than on traits of the plants arising from the seeds. In this way, he could cull and observe many more individuals from the limited space of the monastery garden, and he could evaluate the results of a cross in a single growing season.

In short, Mendel purposely set up a simplified "black-and-white" experimental system and then figured out how it worked. He did not look at the vast number of variables that determine the development of a prize ram nor at the origin of differences between species. Rather, he looked at discrete traits that came in two mutually exclusive forms and asked questions that could be answered by observation and computation. Mendel was the only botanist in the nineteenth century to conceive of and carry out these kinds of experiments.

Genetic Analysis According to Mendel

In early 1865 at the age of 43, Gregor Mendel presented a paper entitled "Experiments on Plant Hybrids" before the Natural Science Society of Brno. Despite its modest heading, it was a scientific paper of uncommon clarity and simplicity that summarized a decade of original observations and experiments. In it Mendel describes in detail the transmission of visible characteristics in pea plants, defines unseen but logically deduced units (genes) that determine why these visible traits appear, and analyzes the function of these discrete determinants in simple mathematical terms to reveal precise patterns and heretofore unsuspected principles of heredity. He also tries to integrate his ideas with the newly emerging cell theory of life and the hotly debated theory of evolution.

Published the following year, the paper would become the cornerstone of modern genetics. Its stated purpose was to see whether there is a "generally applicable law governing the formation and development of hybrids." Let us examine its insights.

Monohybrid Crosses Reveal Units of Inheritance and the Law of Segregation

Once he had isolated pure-breeding lines for several sets of characteristics, Mendel carried out a series of matings between individuals that differed in only one trait, such as seed color or stem length. In each cross, one parent carries one form of the trait, and the other parent carries an alternative form of the same trait. Figure 2.9 illustrates one such mating. Early in the spring of 1854, for example, Mendel planted pure-breeding green peas and pure-breeding yellow peas and allowed them to grow into the **parental (P) generation.** Later that spring when the plants had flowered, he dusted the female stigma of "green-pea" plant flowers with pollen from "yellow-pea" plants. He also performed the reciprocal cross, dusting "yellow-pea" plant

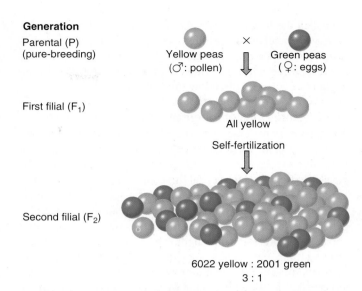

Generation

Parental (P) (pure-breeding)

Yellow peas (♂: pollen) × Green peas (♀: eggs)

First filial (F₁)

All yellow

Self-fertilization

Second filial (F₂)

6022 yellow : 2001 green
3 : 1

Figure 2.9 Analyzing a monohybrid cross. Cross-pollination of pure-breeding parental plants produces F_1 hybrids, all of which resemble one of the parents. Self-pollination of F_1 plants gives rise to an F_2 generation with a 3:1 ratio of individuals resembling the two original parental types. For simplicity, we do not show the plants that produce the peas or that grow from the planted peas.

stigmas with "green-pea" pollen. In the fall, when he collected and separately analyzed the progeny peas of these reciprocal crosses, he found that in both cases, the peas were all yellow. These yellow peas, progeny of the P generation, were the beginning of what we now call the **first filial (F_1)** generation. To learn whether the green trait had disappeared entirely or remained intact but hidden in these F_1 yellow peas, Mendel planted them and allowed the F_1 plants that grew from them to self-fertilize. Such experiments involving hybrids for a single trait are often called **monohybrid crosses.** He then harvested and counted the peas of the resulting **second filial (F_2)** generation, progeny of the F_1 generation. Among the progeny of one series of F_1 self-fertilizations, there were 6022 yellow and 2001 green F_2 peas, an almost perfect ratio of 3 yellow : 1 green. F_1 plants derived from the reciprocal of the original cross produced a similar ratio of yellow to green F_2 progeny.

Reappearance of the Recessive Trait Disproves Blending

These results were irrefutable evidence that blending had not occurred. If it had, the information necessary to make green peas would have been irretrievably lost in the F_1 hybrids. Instead, the information remained intact and was able to direct the formation of some 2000 green peas actually harvested from the second filial generation. These green peas were indistinguishable from their green grandparents. Mendel concluded that there must be two types of yellow peas: those that breed true like the yellow peas of the P generation, and those that can yield some green offspring like the yellow F_1 hybrids. This second type somehow contains latent information for green peas.

He called the trait that appeared in all the F₁ hybrids—in this case, yellow seeds—**dominant** (see Fig. 2.8) and the "antagonistic" green-pea trait that remained hidden in the F₁ hybrids but reappeared in the F₂ generation **recessive.** But how did he explain the 3:1 ratio of yellow to green F₂ peas?

The Discrete Units of Inheritance Are Alleles of Genes

To account for his observations, Mendel proposed that for each trait, every plant carries two copies of a unit of inheritance, receiving one from its maternal parent and the other from the paternal parent. Today, we call these units of inheritance *genes*. Each unit determines the appearance of a specific characteristic. The pea plants in Mendel's collection had two copies of a gene for seed color, two copies of another for seed shape, two copies of a third for stem length, and so forth. Mendel further proposed that each gene comes in alternative forms, and it is these alternative forms that determined the contrasting characteristics he was studying. Today we call the alternative forms of a single gene **alleles.** The gene for pea color, for example, has yellow and green alleles; the gene for pea shape has round and wrinkled alleles. (The Fast Forward box "Genes Encode Proteins," pp. 22–23, describes the biochemical and molecular mechanisms by which different alleles determine different traits.) In Mendel's monohybrid crosses, one allele of each gene was dominant, the other recessive. In the P generation, one parent carried two dominant alleles for the trait under consideration, the other parent two recessive alleles. The hybrids of the F₁ generation carried one dominant and one recessive allele for the trait. Individuals having two different alleles for a single trait are **monohybrids.**

The Law of Segregation Explains How Genes Are Transmitted

If a plant has two copies of every gene, how does it pass only one copy of each to its progeny? And how then do the offspring end up with two copies of these same genes, one from each parent? Mendel drew on his background in plant physiology and answered these questions in terms of the two biological mechanisms behind reproduction: gamete formation and the random union of gametes at fertilization. **Gametes** are the specialized cells—eggs within the ovules of the female parent and sperm cells within the pollen grains—that carry genes between generations. He imagined that during the formation of pollen and eggs, the two copies of each gene in the parent separate (or *segregate*) so that each gamete receives only one allele for each trait (Fig. 2.10a). Thus, each egg and each pollen grain receives only one allele for pea color (either yellow or green). At fertilization, pollen with one or the other allele unites at random with an egg carrying one or the other allele, restoring the two copies of the gene for each trait in the fertilized egg, or **zygote** (Fig. 2.10b). If the pollen carries yellow and the egg green, the result will be a hybrid yellow pea like the F₁ monohybrids that resulted when pure-breeding parents of opposite types mated. If the yellow pollen unites with a yellow egg, the result will be a yellow pea that grows into a pure-breeding plant like those of the P generation that produced only yellow peas. And finally, if

(a) The two alleles for each trait separate during gamete formation.

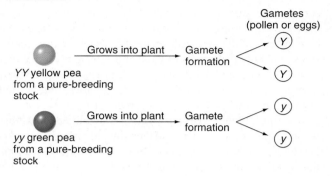

(b) Two gametes, one from each parent, unite at random at fertilization.

Y = yellow-determining allele of pea color gene
y = green-determining allele of pea color gene

Figure 2.10 The law of segregation. (a) The two identical alleles of pure-breeding plants separate (segregate) during gamete formation. As a result, each pollen grain or egg carries only one of each pair of parental alleles. **(b)** Cross-pollination and fertilization between pure-breeding parents with antagonistic traits result in F₁ hybrid zygotes with two different alleles, one from each type of parent, for each trait. For the seed color gene, a *Yy* hybrid zygote will develop into a yellow pea.

pollen carrying the allele for green peas fertilizes a green-carrying egg, the progeny will be a pure-breeding green pea.

Mendel's **law of segregation** encapsulates this general principle of heredity: *The two alleles for each trait separate (segregate) during gamete formation, and then unite at random, one from each parent, at fertilization.* Throughout this book, the term **segregation** refers to such *equal segregation* in which one allele, and only one allele, of each gene goes to each gamete.

Figure 2.11 shows a simple way of visualizing the results of the segregation and random union of alleles during gamete formation and fertilization. Mendel invented a system of symbols that allowed him to analyze all his crosses in the same way. He designated dominant alleles with a capital *A, B,* or *C* and recessive ones with a lowercase *a, b,* or *c.* Modern geneticists have adopted this convention for naming genes in peas and many other organisms, but they often add some reference to the trait in question—a *Y* for yellow or an *R* for round. Throughout this book, we present gene symbols in italics. In our diagram, we denote the dominant yellow allele by a capital *Y* and the

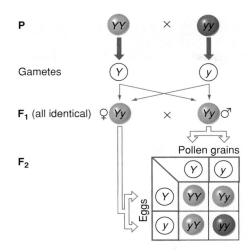

Figure 2.11 The Punnett square: Visual summary of a cross. This Punnett square illustrates the combinations that can arise when an F$_1$ hybrid undergoes gamete formation and self-fertilization. The F$_2$ generation should have a 3:1 ratio of yellow to green peas.

recessive green allele by a lower case *y*. The pure-breeding plants of the parental generation are either *YY* (yellow peas) or *yy* (green peas). The *YY* parent can produce only *Y* gametes, the *yy* parent only *y* gametes. You can see from the diagram why every cross between *YY* and *yy* produces exactly the same result—a *Yy* hybrid—no matter which parent (male or female) donates which particular allele.

Next, to visualize what happens when the *Yy* hybrids self-fertilize, we set up a Punnett square (named after British mathematician Reginald Punnett, who introduced it in 1906; Fig. 2.11). The square provides a simple and convenient method for tracking the kinds of gametes produced as well as all the possible combinations that might occur at fertilization. As the Punnett square shows, each hybrid produces two kinds of gametes, *Y* and *y*, in a ratio of 1:1. Thus half the pollen and half the eggs carry *Y*, the other half *y*. At fertilization, 1/4 of the progeny will be *YY*, 1/4 *Yy*, 1/4 *yY*, and 1/4 *yy*. Since the gametic source of an allele (egg or pollen) for the traits Mendel studied had no influence on the allele's effect, *Yy* and *yY* are equivalent. This means that 1/2 of the progeny are yellow *Yy* hybrids, 1/4 *YY* true-breeding yellows, and 1/4 true-breeding *yy* greens. The diagram illustrates how the segregation of alleles during gamete formation and the random union of egg and pollen at fertilization can produce the 3:1 ratio of yellow to green that Mendel observed in the F$_2$ generation.

Mendel's Results Reflect Basic Rules of Probability

Though you may not have realized it, the Punnett square illustrates two simple rules of probability that are central to the analysis of genetic crosses. These rules predict the likelihood that an event will occur. The **product rule** states that the probability of two or more *independent events* occurring together is the *product* of the probabilities that each event will occur by itself.

With independent events:

Probability of event 1 *and* event 2 =

Probability of event 1 $\times$ probability of event 2

Consecutive coin tosses are obviously independent events; a heads in one toss neither increases nor decreases the probability of a heads in the next toss. If you toss two coins at the same time, the results are also independent events. A heads for one coin neither increases nor decreases the probability of a heads for the other coin. Thus, the probability of a given combination is the product of their independent probabilities. For example, the probability that both coins will turn up heads is

$$1/2 \times 1/2 = 1/4$$

Similarly, the formation of egg and pollen are independent events; in a hybrid plant, the probability is 1/2 that a given gamete will carry *Y* and 1/2 that it will carry *y*. Since fertilization happens at random, the probability that a particular combination of maternal and paternal alleles will occur simultaneously in the same zygote is the product of the independent probabilities of these alleles being packaged in egg and sperm. Thus, to find the chance of a *Y* egg (formed as the result of one event) uniting with a *Y* sperm (the result of an independent event), you simply multiply 1/2 $\times$ 1/2 to get 1/4. This is the same fraction of *YY* progeny seen in the Punnett square of Fig. 2.11, which demonstrates that the Punnett square is simply another way of depicting the product rule.

While we can describe the moment of random fertilization as the simultaneous occurrence of two independent events, we can also say that two different fertilization events are mutually exclusive. For instance, if *Y* combines with *Y*, it cannot also combine with *y* in the same zygote. A second rule of probability, the **sum rule,** states that the probability of either of two such *mutually exclusive events* occurring is the *sum* of their individual probabilities.

With mutually exclusive events:

Probability of event 1 *or* event 2 =

Probability of event 1 + probability of event 2

To find the likelihood that an offspring of a *Yy* hybrid self-fertilization will be a hybrid like the parents, you add 1/4 (the probability of maternal *Y* uniting with paternal *y*) and 1/4 (the probability of the mutually exclusive event where paternal *Y* unites with maternal *y*) to get 1/2, again the same result as in the Punnett square. In a slightly different use of the rule of the sum, you could predict the ratio of yellow to green F$_2$ progeny. The fraction of F$_2$ peas that will be yellow is the sum of 1/4 (the event producing *YY*) plus 1/4 (the mutually exclusive event generating *Yy*) plus 1/4 (the third mutually exclusive event producing *yY*) to get 3/4. The remaining 1/4 of the F$_2$ progeny will be green. So the yellow-to-green ratio is 3/4 to 1/4, or more simply, 3:1.

Genes Encode Proteins

Genes determine traits as disparate as pea shape and the inherited human disease of cystic fibrosis by encoding the proteins that cells produce and depend on for structure and function. As early as 1940, investigators had uncovered evidence suggesting that some genes determine the formation of enzymes, proteins that catalyze specific chemical reactions. But it was not until 1991, 126 years after Mendel published his analysis of seven pairs of observable traits in peas, that a team of British geneticists identified the gene for pea shape and pinpointed how it prescribes a seed's round or wrinkled contour through the enzyme it determines. About the same time, medical researchers in the United States identified the cystic fibrosis gene and discovered how a mutant allele causes unusually sticky mucous secretions and a susceptibil-

ity to respiratory infections and digestive malfunction, once again, through the protein the gene determines.

The pea shape gene encodes an enzyme known as SBE1 (for starch-branching enzyme 1), which catalyzes the conversion of amylose, an unbranched linear molecule of starch, to amylopectin, a starch molecule composed of several branching chains (Fig. A). The dominant *R* allele of the pea shape gene causes the formation of active SBE1 enzyme that functions normally. As a result, *RR* homozygotes produce a high proportion of branched starch molecules, which allow the peas to maintain a rounded shape. In contrast, enzyme determined by the recessive *r* allele is abnormal and does not function effectively. In homozygous recessive *rr* peas, where there is less starch conversion and more of the

Figure A   Round and wrinkled peas: How one gene determines an enzyme that affects pea shape. The *R* allele of the pea shape gene directs the synthesis of an enzyme that converts unbranched starch to branched starch, indirectly leading to round pea shape. The *r* allele of this gene determines an inactive form of the enzyme, leading to a buildup of linear, unbranched starch that ultimately causes seed wrinkling.

Further Crosses Confirm Ratios Predicted by the Law of Segregation

Although Mendel's law of segregation explains the data from his pea crosses, he performed additional experiments to rule out other possibilities. In the rigorous check of his hypotheses illustrated in Fig. 2.12, he allowed self-fertilization of all the plants in the F_2 generation, counted the types of F_3 progeny, and then allowed self-fertilization of the F_3 generation, counted the F_4 progeny, and so on. What kind of plants would you anticipate from a planting of green peas of the F_2 generation, if Mendel's laws do, in fact, apply? Mendel found that, as predicted, the plants that developed from F_2 green peas all produced only F_3 green peas, and when the resulting F_3 plants self-fertilized, the next generation also produced green peas. This is what we (and Mendel) would expect of pure-breeding lines carrying two

copies of the recessive allele. The yellow peas were a different story. When Mendel allowed 518 F_2 plants that developed from yellow peas to self-fertilize, he observed that 166, roughly 1/3 of the total, were pure-breeding yellow through several generations, but the other 352 (2/3 of the total) gave rise to yellow and green F_3 peas in a ratio of 3:1. All succeeding generations where the plants from hybrid yellow peas were allowed to self-fertilize also produced a 3:1 ratio of yellow to green progeny. Of the yellow progeny, 1/3 were true-breeding yellow, while the remaining 2/3 were yellow hybrids.

It took Mendel years to conduct such rigorous experiments on seven pairs of pea traits, but in the end, he was able to conclude that the segregation of dominant and recessive alleles during gamete formation and their random union at fertilization could indeed explain the 3:1 ratios he observed whenever he linear, unbranched starch, sucrose builds up. The excess sucrose modifies osmotic pressure, causing water to enter the young seeds. As the seeds mature, they lose water, shrink, and wrinkle. The single dominant allele in *Rr* heterozygotes apparently produces enough of the normal enzyme to prevent wrinkling. Although the biochemistry may seem complex, the genetic process is straightforward. A specific gene determines a specific enzyme whose activity affects pea shape. Knowledge of how SBE1 works to determine pea shape provides a physical reality for Mendel's concept of hereditary units.

The human disease of cystic fibrosis (CF) was first described in 1938 (see p. 31), but doctors and scientists did not understand the biochemical mechanism that produced the serious respiratory and digestive malfunctions associated with the disease. As a result, treatments could do little more than relieve some of the symptoms, and most CF sufferers died before the age of 30. In 1989, molecular geneticists identified and cloned the cystic fibrosis gene (see Chapters 9 and 10 for specific techniques and procedures). Once they had located the gene, they could find the protein it encodes and clarify how a defect in just one protein can cause a cascade of biochemically determined symptoms. They could also work toward a cure.

The cystic fibrosis gene determines a protein that forges a channel through the cell membrane (Fig. B). Known as the cystic fibrosis transmembrane conductance regulator (CFTR), the protein regulates the flow of chloride ions into and out of the cell. The normal allele of this gene produces a CFTR protein that correctly regulates the back-and-forth exchange of ions, which in turn determines the cell's osmotic pressure and the flow of water through the cell membrane. In people with cystic fibrosis, however, the cells carry two recessive alleles that produce only an abnormal form of the CFTR protein. The abnormal protein creates nonfunctional chloride channels that remain shut. Since chloride ions do not flow in and out of the cells as required, the control of osmotic pressure breaks down, the cells retain water, and a thick, dehydrated mucus builds up outside the cells. In cells lining the airways and the ducts of secretory organs such as the pancreas, this single biochemical

Carbohydrate side chains

Lipid bilayer of cell membrane

CFTR protein

Figure B   The cystic fibrosis gene encodes a cell membrane protein. A model of the normal CFTR protein that regulates the passage of chloride ions through the cell membrane. A small change in the gene that codes for CFTR results in an altered protein that prevents proper flow of chloride ions, leading to the varied symptoms of cystic fibrosis.

defect in an ion channel produces clogging and blockages that result in respiratory and digestive malfunction.

Cloning of the cystic fibrosis gene brought not only a protein-based explanation of disease symptoms but also the promise of a cure. In the early 1990s, medical researchers successfully used the cloned cystic fibrosis gene to relieve disease symptoms in mice. They did this by placing the normal allele of the gene into respiratory tissue, which could then produce a functional CFTR protein. Such encouraging results in these small mammals suggested that in the not-too-distant future gene therapy might bestow relatively normal health on people suffering from this once life-threatening genetic disorder. Unfortunately, human trials of CFTR gene therapy have not yet achieved clear success.

Although Mendel defined genes as independent units of heredity that reside in cells, he did not have the biochemical information even to speculate on precisely what they are and how they work. Chapters 6–8 describe the chemical composition of genes and clarify how they carry information for making proteins.

---

allowed hybrids to self-fertilize. His results, however, raised yet another question, one of some importance to future plant and animal breeders. Plants showing a dominant trait, such as yellow peas, can be either pure-breeding (*YY*) or hybrid (*Yy*). How can you distinguish one from the other? For self-fertilizing plants, the answer is simply to observe the appearance of the next generation. But how would you distinguish pure-breeding from hybrid individuals showing the dominant trait in species that do not self-fertilize?

Testcrosses Establish Genotype

Before describing Mendel's answer, we need to define a few more terms. An observable characteristic, such as yellow or green pea seeds, is a **phenotype,** while the actual pair of alleles present in an individual is its **genotype.** A *YY* or a *yy* genotype

is called **homozygous,** because the two copies of the gene that determine the particular trait in question are the same. In contrast, a genotype with two different alleles for a trait is **heterozygous;** in other words, it is a hybrid for that trait (Fig. 2.13). An individual with a homozygous genotype is a **homozygote;** one with a heterozygous genotype is a **heterozygote.** If you know the genotype and the dominance relation of the alleles, you can accurately predict the phenotype. The reverse is not true, however, because some phenotypes can derive from more than one genotype. For example, the phenotype of yellow peas can result from either the *YY* or the *Yy* genotype.

With these distinctions in mind, we can look at the method Mendel devised for deciphering the unknown genotype, we'll call it *Y–*, responsible for a dominant phenotype;

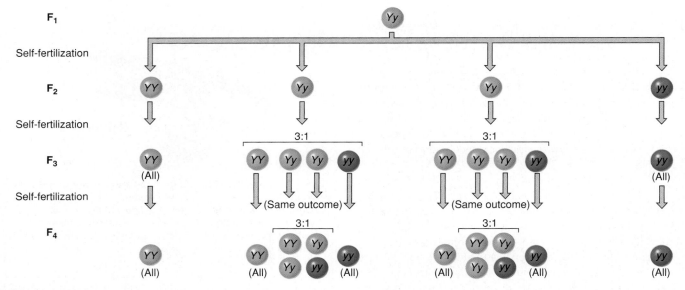

Figure 2.12 Yellow F₂ peas are of two types: Pure breeding and hybrid. The distribution of a pair of contrasting alleles (*Y* and *y*) after four generations of self-fertilization. The homozygous individuals of each generation breed true, whereas the hybrids do not.

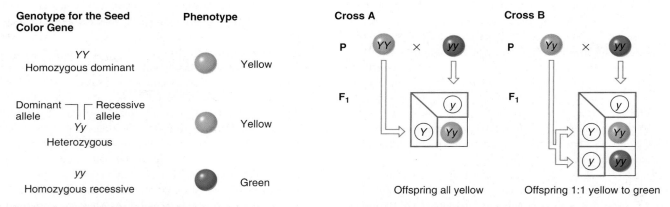

Figure 2.13 Genotype versus phenotype in homozygotes and heterozygotes. The relationship between genotype and phenotype with a pair of contrasting alleles where one allele (*Y*) shows complete dominance over the other (*y*).

Figure 2.14 How a testcross reveals genotype. An individual of unknown genotype, but dominant phenotype, is crossed with a homozygous recessive. If the unknown genotype is homozygous, all progeny will exhibit the dominant phenotype, as shown in *cross A*. If the unknown genotype is heterozygous, half the progeny will exhibit the dominant trait, half the recessive trait, as shown in *cross B*.

the dash represents the unknown second allele, either *Y* or *y*. Called the **testcross,** it is a mating in which an individual showing the dominant phenotype, for instance, a *Y*– plant grown from a yellow pea, is crossed with an individual expressing the recessive phenotype, in this case a *yy* plant grown from a green pea. As the Punnett squares in Fig. 2.14 illustrate, if the dominant phenotype in question derives from a homozygous *YY* genotype, all the offspring of the testcross will show the dominant phenotype and thus be yellow peas. But if the dominant parent of unknown genotype is a heterozygous hybrid (*Yy*), 1/2 of the progeny will be yellow peas, the other half green. In this

way, the testcross establishes the genotype behind a dominant phenotype, resolving any uncertainty.

As we mentioned earlier, Mendel deliberately simplified the problem of heredity, focusing on traits that come in only two forms. He was able to replicate his basic monohybrid findings with corn, beans, and four o'clocks (plants with tubular, white or bright red flowers). As it turns out, his concept of the gene and his law of segregation can be generalized to all sexually reproducing organisms.

Dihybrid Crosses Reveal the Law of Independent Assortment

Having determined from monohybrid crosses that genes are inherited according to the law of segregation, Mendel turned his attention to the simultaneous inheritance of two or more apparently unrelated traits in peas. He asked how two pairs of alleles would segregate in a **dihybrid** individual, that is, in a plant that is heterozygous for two genes at the same time. To construct such a dihybrid, he mated true-breeding plants grown from yellow round peas (*YY RR*) with true-breeding plants grown from green wrinkled peas (*yy rr*). From this cross he obtained a dihybrid F₁ generation (*Yy Rr*) showing only the two dominant phenotypes, yellow and round (Fig. 2.15). He then allowed these F₁ dihybrids to self-fertilize to produce the F₂ generation. Mendel could not predict the outcome of this mating. Would all the F₂ progeny show one or the other of the parental combinations of traits in a 3:1 ratio of yellow round to green wrinkled? Or would some new combinations of phenotypes occur that were not seen in the parental lines, such as yellow wrinkled or green round peas? New phenotypic combinations like these are called **recombinant types.** When Mendel counted the F₂ generation of one experiment, he found 315 yellow round peas, 101 yellow wrinkled, 108 green round, and 32 green wrinkled. There were, in fact, yellow wrinkled and green round recombinant phenotypes, providing evidence that some shuffling of alleles had taken place.

From the observed ratios, Mendel inferred the biological mechanism of that shuffling—the **independent assortment** of gene pairs during gamete formation. Because the genes for pea color and for pea shape assort independently, the allele for pea shape in a *Y* carrying gamete could with equal likelihood be either *R* or *r*. Thus, the presence of a particular allele of one gene, say, the dominant *Y* for pea color, provides no information whatsoever about the allele of the second gene. Each dihybrid of the F₁ generation can therefore make four kinds of gametes: *Y R, Y r, y R,* and *y r.* In a large number of gametes, the four kinds will appear in an almost perfect ratio of 1:1:1:1, or put another way, roughly 1/4 of the eggs and 1/4 of the pollen will contain each of the four possible combinations of alleles. That "the different kinds of germinal cells [eggs or pollen] of a hybrid are produced on the average in equal numbers" was yet another one of Mendel's incisive insights.

At fertilization then, in a mating of dihybrids, 4 different kinds of eggs can combine with any one of 4 different kinds of pollen, producing a total of 16 possible zygotes. Once again, a Punnett square is a convenient way to visualize the process. If you look at the square in Fig. 2.15, you will see that some of the 16 potential allelic combinations are identical. In fact, there are only nine different genotypes—*YY RR, YY Rr, Yy RR, Yy Rr, yy RR, yy Rr, YY rr, Yy rr,* and *yy rr*—because the source of the alleles (egg or pollen) does not make any difference. If you look at the combinations of traits determined by the nine genotypes, you will see only four phenotypes—yellow round, yellow wrinkled, green round, and green wrinkled—observed in a ratio of 9:3:3:1. If,

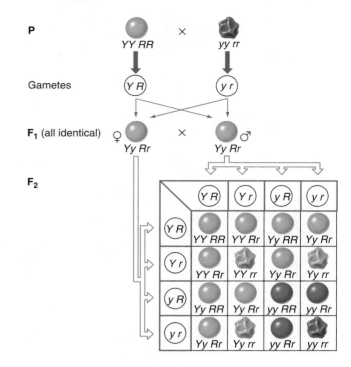

Type	Genotype	Phenotype		Number	Phenotypic ratio
Parental	*Y– R–*		yellow round	315	9/16
Recombinant	*yy R–*		green round	108	3/16
Recombinant	*Y– rr*		yellow wrinkled	101	3/16
Parental	*yy rr*		green wrinkled	32	1/16

Ratio of yellow (dominant) to green (recessive) = 12:4 or 3:1
Ratio of round (dominant) to wrinkled (recessive) = 12:4 or 3:1

Figure 2.15 A dihybrid cross produces parental types and recombinant types. In this dihybrid cross, pure-breeding parents (P) produce a genetically uniform generation of F₁ dihybrids. Self-pollination or cross-pollination of the F₁ plants yields the characteristic F₂ phenotypic ratio of 9:3:3:1.

however, you look at just pea color or just pea shape, you can see that each trait is inherited in the 3:1 ratio predicted by Mendel's law of segregation. In the Punnett square, there are 12 yellows for every 4 greens and 12 rounds for every 4 wrinkleds. In other words, the ratio of each dominant trait (yellow or round) to its antagonistic recessive trait (green or wrinkled) is 12:4, or 3:1. This means that the inheritance of the gene for pea color is unaffected by the inheritance of the gene for pea shape and vice versa.

The preceding analysis became the basis of Mendel's second general genetic principle, the **law of independent assortment:** *During gamete formation, different pairs of alleles*

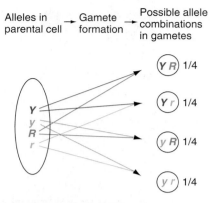

Figure 2.16 The law of independent assortment. In a dihybrid or multihybrid cross, each pair of alleles assorts independently during gamete formation. In the gametes, *Y* is equally likely to be found with *R* or *r* (that is, *Y R* = *Y r*); the same is true for *y* (that is, *y R* = *y r*). As a result, all four possible types of gametes (*Y R*, *Y r*, *y R*, and *y r*) are produced in equal frequency.

Figure 2.17 Following crosses with branched-line diagrams. A branched-line diagram, which uses a series of columns to track every gene in a cross, provides an organized overview of all possible outcomes. This branched-line diagram of a dihybrid cross generates the same phenotypic ratios as the Punnett square in Fig. 2.15, showing that the two methods are equivalent ways of looking at the same process.

segregate independently of each other (Fig. 2.16). The independence of their segregation and the subsequent random union of gametes at fertilization determine the phenotypes observed. Using the product rule for assessing the probability of independent events, you can see mathematically how the 9:3:3:1 phenotypic ratio observed in a dihybrid cross derives from two separate 3:1 phenotypic ratios. If the two sets of alleles assort independently, the yellow-to-green ratio in the F_2 generation will be 3/4 : 1/4, and likewise, the round-to-wrinkled ratio will be 3/4 : 1/4. To find the probability that the two independent events of yellow and round will occur simultaneously in the same plant, you multiply as follows:

Probability of yellow round = 3/4 × 3/4 = 9/16

Probability of yellow wrinkled = 3/4 × 1/4 = 3/16

Probability of green round = 1/4 × 3/4 = 3/16

Probability of green wrinkled = 1/4 × 1/4 = 1/16

Thus, in a population of F_2 plants, there will be a 9:3:3:1 phenotypic ratio of yellow round to yellow wrinkled to green round to green wrinkled.

A convenient way to keep track of the probabilities of each potential outcome in a genetic cross is to construct a **branched-line diagram** (Fig. 2.17), which shows all the possibilities for each gene in a sequence of columns. In Fig. 2.17, the first column shows the two possible pea color phenotypes; and the second column demonstrates that each pea color can occur with either of two pea shapes. Again, the 9:3:3:1 ratio of phenotypes is apparent.

An understanding of dihybrid crosses has many applications. Suppose, for example, that you work for a wholesale nursery that has only F_1 dihybrid plants, and your assignment is to grow pure-breeding plants guaranteed to produce yellow

round peas. How would you proceed? One answer would be to allow the plants to self-fertilize and produce F_2 peas. You would then plant the F_2 peas with the desired yellow round phenotype. Only one out of nine of the resultant F_2 plants—those grown from peas with a *YY RR* genotype—will be appropriate for your uses. To find these plants, you could subject each yellow round F_2 plant to a testcross for genotype with a green wrinkled (*yy rr*) plant, as illustrated in Fig. 2.18. If the testcross yields all yellow round offspring (testcross A), you can sell your test plant or its progeny from self-fertilization, because you know it is homozygous for both pea color and pea shape. If your testcross yields 1/2 yellow round and 1/2 yellow wrinkled (testcross B), or 1/2 yellow round and 1/2 green round (testcross C), you know that the F_2 plant in question is genetically homozygous for one trait and heterozygous for the other and must therefore be discarded. Finally if the testcross yields 1/4 yellow round, 1/4 yellow wrinkled, 1/4 green round, and 1/4 green wrinkled (testcross D), you know that the plant is a heterozygote for both the pea color and the pea shape genes.

Using Mendel's Laws: Probabilities and Predictions

Mendel performed several sets of dihybrid crosses and also carried out **multihybrid crosses:** matings between the F_1 progeny of true-breeding parents that differed in three or more traits. In all of these experiments, he observed numbers and ratios very close to what he expected on the basis of his two general biological principles: the alleles of a gene segregate during the formation of egg or pollen, and the alleles of different genes assort independently of each other. Mendel's laws of inheritance, in conjunction with the mathematical rules of probability, provide geneticists with powerful tools for predicting and interpreting the results of genetic crosses. But as with all tools, they have their limitations. We examine here both the power and the limitations of Mendelian analysis.

First the power: Using simple Mendelian analysis, it is possible to make accurate predictions about the offspring of

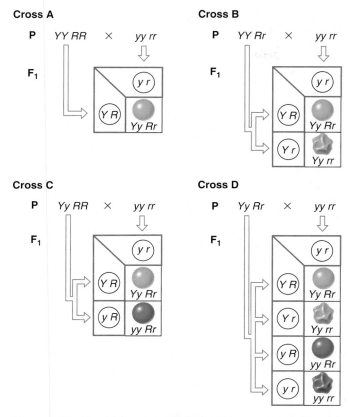

Figure 2.18 Testcrosses on dihybrids. Testcrosses involving two pairs of independently assorting alleles yield different, predictable results depending on the tested individual's genotype for the two genes in question.

extremely complex crosses. Suppose you want to predict the occurrence of one specific genotype in a cross involving several independently assorting genes. For example, if hybrids that are heterozygous for four traits are allowed to self-fertilize—*Aa Bb Cc Dd* × *Aa Bb Cc Dd*—what proportion of their progeny will have the genotype *AA bb Cc Dd*? You could set up a Punnett square. Since for each trait there are two different alleles, the number of different eggs or sperm is found by raising 2 to the power of the number of differing traits (2^n, where *n* is the number of traits). By this calculation, each hybrid parent in this cross would make 16 different kinds of gametes: 4 different traits implies $2^4 = 16$ kinds of gametes. The Punnett square depicting such a cross would thus contain 256 boxes (16 × 16). This may be fine if you live in a monastery with a bit of time on your hands, but not if you're taking a 1-hour exam. It would be much simpler to analyze the problem by breaking down the multihybrid cross into four independently assorting monohybrid crosses. Remember that the genotypic ratios of each monohybrid cross are 1 homozygote for the dominant allele, to 2 heterozygotes, to 1 homozygote for the recessive allele = $\underline{1}/4 : \underline{2}/4 : \underline{1}/4$. Thus you can find the probability of *AA bb Cc Dd* by multiplying the probability

of each independent event: *AA* (1/4 of the progeny produced by *Aa* × *Aa*); *bb* (1/4); *Cc* (2/4); *Dd* (2/4):

$$1/4 \times 1/4 \times 2/4 \times 2/4 = 4/256 = 1/64$$

The Punnett square approach would provide the same answer, but it would require much more time.

If instead of a specific genotype, you want to predict the probability of a certain phenotype, you can again use the product rule as long as you know the phenotypic ratios produced by each pair of alleles in the cross. For example, if in the multihybrid cross of *Aa Bb Cc Dd* × *Aa Bb Cc Dd*, you want to know how many offspring will show the dominant A trait (genotype *AA* or *Aa* = 1/4 + 2/4, or 3/4), the recessive b trait (genotype *bb* = 1/4), the dominant C trait (genotype *CC* or *Cc* = 3/4), and the dominant D trait (genotype *DD* or *Dd* = 3/4), you simply multiply

$$3/4 \times 1/4 \times 3/4 \times 3/4 = 27/256$$

In this way, the rules of probability make it possible to predict the outcome of very complex crosses.

You can see from these examples that particular problems in genetics are amenable to particular modes of analysis. As a rule of thumb, Punnett squares are excellent for visualizing simple crosses involving a few genes, but they become unwieldy in the dissection of more complicated matings. Direct calculations of probabilities, such as those in the two preceding problems, are useful when you want to know the chances of one or a few outcomes of complex crosses. If, however, you want to know all the outcomes of a multihybrid cross, a branched-line diagram is the best way to go as it will keep track of the possibilities in an organized fashion.

Now the limitations of Mendelian analysis: Like Mendel, if you were to breed pea plants, or corn or any other organism, you would most likely observe some deviation from the ratios you expected in each generation. What can account for such variation? One element is chance, as witnessed in the common coin toss experiment. With each throw, the probability of the coin coming up heads is equal to the likelihood it will come up tails. But if you toss a coin 10 times, you may get 30% (3) heads and 70% (7) tails or vice versa. If you toss it 100 times, you are more likely to get a result closer to the expected 50% heads and 50% tails. The larger the number of trials, the lower the probability that chance significantly skews the data. This is one reason Mendel worked with large numbers of pea plants. Mendel's laws, in fact, have great predictive power for populations of organisms, but they do not tell us what will happen in any one individual. With a garden full of self-fertilizing monohybrid pea plants, for example, you can expect that 3/4 of the F_2 progeny will show the dominant phenotype and 1/4 the recessive, but you cannot predict the phenotype of any particular F_2 plant. This is similar to physicists' predictions about an element's radioactivity; they can predict the half-life of a radioactive element and expect to see half of the atoms decay within the specified time, but they cannot say which specific atoms will decay and which will not. In Chapter 5, we discuss mathematical methods for assessing whether the chance variation observed in a sample of individuals within a population is compatible with a genetic hypothesis.

Why Mendel's Work Was Unappreciated Before 1900

Mendel's insights into the workings of heredity were a breakthrough of monumental proportions. By counting and analyzing data from thousands of pea plant crosses, he inferred the existence of genes—independent units that determine the observable patterns of inheritance for particular traits. His work explained the reappearance of "hidden" traits, disproved the idea of blended inheritance, and showed that mother and father make an equal genetic contribution to the next generation. The model of heredity that he formulated was so specific that he could test predictions based on it by observation and experiment. In the process of developing his model, Mendel deduced the two basic principles of gene transmission: the segregation of each gene's alleles during gamete formation followed by a random union at fertilization, and the independent assortment of the alleles for two or more different genes. Finally, he tied the transmission of traits to the function of particular cells—the gametes—and also recognized that knowledge of the general laws of heredity were a prerequisite for understanding "the evolutionary history of organic forms [organisms]."

With the exception of Abbot Napp, none of Mendel's contemporaries appreciated the importance of his research, and Napp died less than two years after the publication of "Experiments on Plant Hybrids." Mendel, it seems, was far ahead of his time. Sadly, despite written requests from Mendel that others try to replicate his studies, no one repeated his experiments. Several citations of his paper between 1866 and 1900 referred to his expertise as a plant breeder but made no mention of his laws. Moreover, at the time Mendel presented his work, no one had yet seen the structures within cells, the *chromosomes,* that actually segregate and assort independently. That would happen only in the next few decades (as described in Chapter 4).

In an ironic twist of history, Mendel revealed a heritable source of variation—the occurrence, at fertilization, of new combinations of independently assorting alleles, which may give rise to recombinant phenotypes—but Charles Darwin (1809–1882), who was unfamiliar with Mendel's work, was plagued in his later years by critics who found his explanations for the persistence of variation in organisms insufficient. Darwin considered such variation a cornerstone of his theory of evolution, maintaining that natural selection would favor particular variants in a given population in a given environment. If the selected combinations of variant traits were passed on to subsequent generations, this transmission of variation would propel evolution. He could not, however, say how that transmission might occur. Had he been aware of Mendel's laws, he might not have been backed into such an uncomfortable corner. Nearly a century after Mendel published his findings, historians found an uncut copy of Mendel's paper in Darwin's study; Darwin had received but never read it.

For 34 years, Mendel's laws lay dormant—untested, unconfirmed, and unapplied. Then in 1900, 16 years after Mendel's death, Carl Correns, Hugo de Vries, and Erich von Tschermak independently rediscovered and acknowledged his work (Fig. 2.19). The scientific community had finally caught up with Mendel. Within the decade, investigators had coined many of the modern terms we have been using: phenotype, genotype, homozygote, heterozygote, gene, and genetics, the label given to the twentieth-century science of heredity. Mendel's paper provided the new discipline's foundation. His principles and analytic techniques endure today, guiding geneticists and evolutionary biologists in their studies of genetic variation. It took less than a century for genetics to become a central discipline of biology, with its concepts and tools influencing fields as disparate as immunology and ecology, medicine and agriculture.

(a) Gregor Mendel **(b) Carl Correns** **(c) Hugo de Vries** **(d) Eric von Tschermak**

Figure 2.19 The science of genetics begins with the rediscovery of Mendel. Working independently near the beginning of the twentieth century, Correns, de Vries, and von Tschermak each came to the same conclusions as those Mendel summarized in his laws.

Mendelian Inheritance in Humans: A Comprehensive Example

Although many human traits clearly run in families, most do not show a simple Mendelian pattern of inheritance. Suppose, for example, that you have brown eyes, but both your parents' eyes are blue. Since blue is normally considered recessive to brown, does this mean that you are adopted or that your father isn't really your father? Not necessarily, because eye color is influenced by more than one gene.

Like eye color, most common and obvious human phenotypes arise from the interaction of many genes. In contrast, there are few confirmed single-gene traits in people and they often involve an abnormality that is disabling or life-threatening. Examples are the progressive mental retardation and other neurological damage of Huntington disease and the clogged lungs and potential respiratory failure of cystic fibrosis. A defective allele of a single gene gives rise to Huntington disease; defective alleles of a different gene are responsible for cystic fibrosis. There were roughly 9000 such single-gene traits known in humans in 1997, and the number continues to grow as new

studies confirm the genetic basis of more traits. Table 2.1 lists some of the most common single-gene traits in humans.

Determining a genetic defect's pattern of transmission is not always an easy task since people make slippery genetic subjects. They do not base their choice of mates on purely genetic considerations. Their generation time is long, and the families they produce are relatively small, which makes statistical analysis difficult. There are no pure-breeding lines and no controlled matings. And there is rarely a true F_2 generation (like the one in which Mendel observed the 3:1 ratios from which he derived his rules) because brothers and sisters almost never mate. Geneticists circumvent these difficulties by working with a large number of families, or with several generations of a very large family. This allows them to study the large numbers of genetically related individuals needed to establish the inheritance patterns of specific traits. A family history, known as a **pedigree,** is an orderly diagram of a family's relevant genetic features, extending back to at least both sets of grandparents and preferably through as many more generations as possible. From systematic pedigree analysis in the light of Mendel's laws, geneticists can tell if a trait is determined by alternative

TABLE 2.1   Some of the Most Common Single-Gene Traits in Humans

Disease	Effect	Incidence of Disease
Caused by Recessive Allele		
Thalassemia (chromosome 16 or 11)	Reduced amounts of hemoglobin; anemia, bone and spleen enlargement	1/10 in parts of Italy
Sickle-cell anemia (chromosome 11)	Abnormal hemoglobin; sickle-shaped red cells, anemia, blocked circulation; increased resistance to malaria	1/625 African-Americans
Cystic fibrosis (chromosome 7)	Defective cell membrane protein; excessive mucus production; digestive and respiratory failure	1/2000 Caucasians
Tay-Sachs disease (chromosome 15)	Missing enzyme; buildup of fatty deposit in brain; buildup destroys mental development	1/3000 Eastern European Jews
Phenylketonuria (PKU) (chromosome 12)	Missing enzyme; mental deficiency	1/10,000 Caucasians
Albinism (chromosome 11)	Missing enzyme; unpigmented skin, hair, and eyes	1/10,000 in Northern Ireland
Caused by Dominant Allele		
Hypercholesterolemia (chromosome 19)	Missing protein that removes cholesterol from the blood; heart attack by age 50	1/122 French Canadians
Huntington disease (chromosome 4)	Progressive mental and neurological damage; neurologic disorders by ages 40–70	1/25,000 Caucasians

alleles of a single gene and whether a single-gene trait is dominant or recessive. We now look at two pedigrees, one for the dominant trait of Huntington disease, the other for the recessive condition of cystic fibrosis. Because Mendel's principles are so simple and straightforward, a little logic can go a long way in explaining how traits are inherited in humans. Clear vertical or horizontal patterns in a trait's appearance in the pedigree lead to hypotheses of great predictive power.

A Vertical Pattern of Inheritance Indicates a Rare Dominant Trait

Huntington disease is named for George Huntington, the New York physician who first described its course. This illness usually shows up in middle age and slowly destroys its victims both mentally and physically. Symptoms include intellectual deterioration, severe depression, and jerky, irregular movements, all caused by the progressive death of nerve cells. If one parent develops the symptoms, his or her children have a 50% probability of suffering from the disease, provided they live to adulthood. Because symptoms are not present at birth and manifest themselves only later in life, Huntington disease is known as a **late-onset** genetic condition. The pedigree in Fig. 2.20 illustrates the transmission of Huntington disease through five generations of a family living in the village of San Luis on Lake Maracaibo in Venezuela. This village was founded by a small number of immigrants, including one with the Huntington trait. Isolated by their geography, the villagers are like one large extended family. For generations, most descendants have remained concentrated in one place, and the gene for Huntington disease has remained unusually prevalent among them.

Before analyzing the transmission pattern of Huntington disease, you need to know how to interpret a family pedigree (Fig. 2.20). Squares (□) represent males, circles (○) are females, diamonds (◊) indicate that the sex is unspecified; family members affected by the trait in question are indicated

by a filled-in symbol (for example, ■). A single horizontal line connecting a male and a female (□─○) represents a mating, a double connecting line (□═○) designates a **consanguineous mating,** that is, a mating between relatives, and a horizontal line above a series of symbols (○□○) indicates the children of the same parents (a *sibship*) arranged and numbered from left to right in order of their birth. Roman numerals to the left or right of the diagram indicate the generations.

How would you proceed in assigning genotypes to the individuals in the pedigree depicted in Fig. 2.20? First, you would need to find out if the disease-producing allele is dominant or recessive. Several clues suggest that Huntington disease is transmitted by a dominant allele of a single gene. Everyone who develops the disease has at least one parent who shows the trait, and in several generations, approximately half of the offspring are affected. The pattern of affected individuals is thus vertical: If you trace back through the ancestors of any affected individual, you would see at least one affected person in each generation, giving a continuous line of family members with the disease. When a disease is rare in the population as a whole, a vertical pattern is strong evidence that a dominant allele causes the trait. (A recessive trait that is extremely common might also show up in every generation.) In tracking a dominant allele through a pedigree, you can view every mating between an affected and an unaffected partner as analogous to a testcross. If some of the offspring do not have Huntington's, you know the parent showing the trait is a heterozygote. You can check your genotype assignments against the answers in the caption to Fig. 2.20.

As yet no effective treatment exists for Huntington disease, and because of its late onset, there was until the 1980s no way for children of a Huntington's parent to know before middle age—usually until well after their own childbearing years—whether they carried the Huntington disease allele (*HD*). Children of Huntington's parents have a 50% probability of

Figure 2.20   Huntington disease: A rare dominant trait. A pedigree of a Venezuelan family is shown at *left;* symbols used in the pedigree are defined at *right.* The vertical pattern of inheritance is typical of Huntington disease and other conditions caused by a rare dominant allele. All individuals represented by filled-in symbols are heterozygotes (with the possible exception of individual I-1, who could have been homozygous for the *HD* disease allele); all individuals represented by open symbols are homozygotes for the normal *HD⁺* allele of the *HD* gene. Among the 14 children of the consanguineous mating between affected individuals, DNA testing shows that some are *HD HD*, some are *HD HD⁺*, and some are *HD⁺ HD⁺*. Human geneticists employ the diamond designation to mask information about sex or genetic status so as to preserve patient confidentiality, or to indicate a lack of knowledge.

inheriting *HD* and a 25% probability of passing the defective allele on to one of their children. In the mid-1980s, with new knowledge of the gene, molecular geneticists developed a test that determines whether an individual carries the *HD* allele (see the Fast Forward box "The Direct Analysis of Human Genotype" on p. 32). Because of the lack of effective treatment for the disease, some young adults whose parents died of Huntington's prefer not to be tested so that they will not prematurely learn their own fate. However, many at-risk individuals employ the test for the *HD* allele to guide their decisions about having children. If someone whose parent had Huntington disease does not have *HD*, he or she has no chance of developing the disease or of transmitting it to offspring. If the test shows the presence of *HD*, the at-risk person and his or her partner might chose to conceive a child, obtain a prenatal diagnosis of the fetus, and then, depending on their beliefs, elect an abortion if the fetus is affected. The Genetics and Society box "Developing Guidelines for Genetic Screening" on pp. 34–35 discusses significant social and ethical issues raised by information obtained from family pedigrees and molecular tests.

A Horizontal Pattern of Inheritance Indicates a Rare Recessive Trait

Unlike Huntington disease, most confirmed single-gene traits in humans are recessive. This is because, with the exception of late-onset traits, deleterious dominant traits are unlikely to be transmitted to the next generation. For example, if people affected with Huntington disease died by the age of 10, the trait would disappear from the population. In contrast, individuals can carry one allele for a recessive trait without knowing they have it. Figure 2.21 shows three pedigrees for cystic fibrosis (CF), the most commonly inherited recessive disease among Caucasian children in the United States. A double dose of the recessive *CF* allele causes a fatal disorder in which the lungs, pancreas, and other organs become clogged with a thick, viscous mucus that can interfere with breathing and digestion. One in every 2000 white Americans is born with cystic fibrosis, and only 10% of them survive into their 30s.

There are two salient features of the CF pedigrees. First, the family pattern of people showing the trait is often horizontal: The parents, grandparents, and great-grandparents of children born with CF do not themselves manifest the disease, while several brothers and sisters in a single generation may. A horizontal pedigree pattern is a strong indication that the trait is recessive. The unaffected parents are heterozygous **carriers:** they bear a dominant normal allele that masks the effects of the recessive abnormal one. An estimated 12 million Americans are carriers of the recessive *CF* allele.

The second salient feature of the CF pedigrees is that many of the couples who produce afflicted children are blood relatives, that is, their mating is consanguineous (as indicated by the double line). In Fig. 2.21a, the consanguineous mating in generation V is between third cousins. Of course, children with cystic fibrosis can also have unrelated carrier parents, but because relatives share genes, their offspring have a much

greater than average chance of receiving two copies of a rare allele. Whether or not they are related, carrier parents are both heterozygotes. Thus among their offspring, the proportion of unaffected to affected children will be 3:1. To look at it another way, the chances are that one out of four children of two heterozygous carriers will be homozygous CF sufferers. You can gauge your understanding of this inheritance pattern by assigning a genotype to each person in Fig. 2.21 and then checking your answers against the caption. Note that for several individuals, such as the generation I individuals in part a of the figure, it is impossible to assign a full genotype. We know that one of these people must be the carrier who supplied the original *CF* allele, but we do not know if it was the male or the female. As with an ambiguous dominant phenotype in peas, the unknown second allele is indicated by a dash.

In Fig. 2.21a, a mating between unrelated carriers, VI-1 and VI-2, produced a child with cystic fibrosis. How likely is such a marriage between unrelated carriers for a recessive genetic condition? The answer depends on the gene in question and the particular population into which a person is born. As Table 2.1 on p. 29 shows, the incidence of genetic diseases (and thus the frequency of their carriers) varies markedly among populations. Such variation reflects the distinct genetic histories of different groups. The area of genetics that analyzes differences among groups of individuals is called *population genetics,* a subject we cover in detail in Chapter 20. Notice that

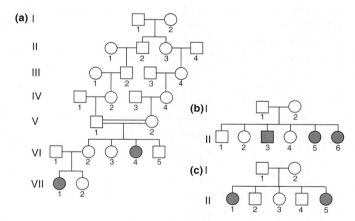

Figure 2.21 Cystic fibrosis: A recessive condition. In **(a)**, the two affected individuals (VI-4 and VII-1) are *CF CF*; that is, homozygotes for the recessive disease allele. Their unaffected parents must be carriers, so V-1, V-2, VI-1, and VI-2 must all be *CF CF⁺* (where *CF⁺* is the dominant normal allele). It is probable that II-2, II-3, III-2, III-4, IV-2, and IV-4 are also carriers of the same *CF* allele inherited from a common ancestor in the first generation. We cannot determine which of these founders (I-1 or I-2) was a carrier, so we must designate their genotypes as *CF⁺–*. Because the *CF* allele is relatively rare, it is likely that II-1, II-4, III-1, III-3, IV-1, and IV-3 are *CF⁺CF⁺* homozygotes. The genotype of the remaining unaffected people (VI-3, VI-5, and VII-2) is uncertain (*CF⁺–*). (**b** and **c**) Two families in which carrier parents produce multiple children with cystic fibrosis demonstrate horizontal patterns of inheritance. Without further information, the unaffected children in each pedigree must be regarded as having a *CF⁺–* genotype.

Fast Forward

The Direct Analysis of Human Genotype

Part of Mendel's genius was the ability to infer a *hidden* genotype from the phenotype expressed by not just one individual but by that individual's relatives as well. Human pedigrees follow this tradition, revealing genotypes on the basis of phenotypic information observed over several generations of a family. But despite the information they reveal, pedigrees do not always provide enough data for prediction. Phenotypically normal prospective parents, for example, may not know, even from an extensive family history, whether they both carry an allele for a deleterious recessive trait; they are thus unable to assess their chances of producing an afflicted child. Today, however, geneticists can go beyond deduction and inference and analyze genotype directly. With techniques developed in the mid-1970s, it is possible to tell exactly what a particular gene "looks like," pick out specific gene pieces, and assess comparable pieces for their differences or similarities. Such comparisons may reveal the presence or absence of disease-causing alleles, such as the dominant allele underlying Huntington disease or the recessive alleles giving rise to cystic fibrosis and sickle-cell anemia. For the prospective parents just alluded to, if just one of them does *not* carry a disease-causing recessive allele, the couple does not have to worry about conceiving an afflicted child.

The power of the new technology arises from its resolution and sensitivity. The resolution, or ability to detect differences between two similar substances, is as high as it can get. A combination of procedures allows detection of differences at the level of single nucleotides, the elementary building blocks of genes (see Chapters 6, 9, and 10). The sensitivity of the new technology is also as great as it can be. Today researchers can detect and analyze the one copy of each gene present in a single sperm cell.

In one application of the new technology, identification of the beta-globin (β-globin) genotype provides the basis for diagnosing sickle-cell anemia, a recessive genetic disease that afflicts roughly 1 in 600 African-Americans. Figure A shows the potential results of a test for the normal and sickle-cell alleles of the ß-globin gene. The protein determined by this gene is one component of the oxygen-carrying hemoglobin molecule. In the generation of disease, the normal β-globin allele (*Hbß^A*, abbreviated *A*) is dominant, and the abnormal allele (*Hbß^S*, or simply *S*) is recessive. Because the normal *A* allele leads to the production of fully functional hemoglobin, people possessing at least one copy of this allele are healthy under most conditions. By contrast, because the abnormal *S* allele leads to the formation of defective hemoglobin molecules, people with two copies of the *S* allele suffer from sickle-cell anemia. With no normal hemoglobin, these *SS* homozygotes have a decrease in oxygen supply, tire easily, and often develop

Individual genotypes

Probes for alternative β-globin alleles

Figure A   A direct look at the ß-globin genotype. Two probes, one specific for the normal *A* allele, the other specific for the abnormal sickle-cell *S* allele, are necessary to determine the genotype of a sample of DNA amplified by the polymerase chain reaction. The presence of a particular allele is indicated by reaction of the corresponding probe with the DNA sample to produce a darkened signal visible on X-ray film. In this diagram of a β-globin genotype analysis, the *green circles* show that a probe reacted with a specific allele, the *white circles* indicate no reaction. Note that only the carrier with both *A* and *S* alleles has two green circles.

heart failure from stress on the circulatory system. The test whose possible results are depicted in Fig. A was based on the polymerase chain reaction (PCR), which can replicate a single gene, or parts thereof, many times over (see Chapter 9 for details). Geneticists can find out what alleles are present in the replicated material through probes that show whether or not a particular allele is present. For example, with two allele-specific probes for the β-globin gene—one for *A,* the other for *S*—they can distinguish a normal *AA* homozygote from a healthy heterozygous *AS* carrier, and both of these from a homozygous *SS* individual afflicted with sickle-cell anemia.

Like the mutant alleles for the β-globin gene, the mutant alleles that determine Huntington disease (HD) and cystic fibrosis (CF) are distinguishable in powerful molecular tests. With HD, both heterozygotes and homozygotes for the mutant allele will eventually show symptoms of the disease. With CF, heterozygotes will be carriers that do not show symptoms, while homozygotes for the mutant allele will have the disease.

The ability to analyze genotype directly has profound social implications. This is particularly true because geneticists can use PCR and other modern techniques on fetal cells obtained from a pregnant woman, and thereby diagnose the genotype of the fetus even before it is born. The Genetics and Society box on pp. 34–35 looks at some of the issues related to potential uses of the new technology.

in Fig. 2.21a, several unrelated, unaffected people, such as II-1 and II-4, married into the family under consideration. Although it is highly probable that these individuals are homozygotes for the normal allele of the gene (CF^+CF^+), there is a small chance (whose magnitude depends on the population) that any one of them could be a carrier of the disease.

Genetic researchers identified the cystic fibrosis gene in 1989, but they are still in the process of developing a gene therapy that would cure the disease's debilitating symptoms (review the Fast Forward box "Genes Encode Proteins" on pp. 22–23).

Genetics and Society

Developing Guidelines for Genetic Screening

In the early 1970s, the United States launched a national screening program for carriers of the sickle-cell trait based on a simple test of hemoglobin mobility; normal and "sickling" hemoglobins move at different rates in a gel. People who participated in the screening program could use the test results to make informed reproductive decisions. A healthy man, for example, who learned he was a carrier, would not have to worry about having affected children if his mate were a noncarrier. If, however, they were both carriers, they could choose either not to conceive or to conceive in spite of the 25% risk of bearing an afflicted child. In the 1980s, the possibility of direct prenatal diagnosis of the fetal genotype, as described in the Fast Forward box "The Direct Analysis of Human Genotype" on p. 32, provided additional options. Depending on their beliefs, a couple could decide to continue a pregnancy only if the fetus were not a homozygote for the *S* allele, or knowing that their child would have sickle-cell anemia, they could learn how to deal with the symptoms of the condition.

The original sickle-cell screening program, based on detection of the abnormal hemoglobin protein, was unfortunately not an unqualified success, largely because of insufficient educational follow-through. Many who learned they were carriers mistakenly thought they had the disease. Moreover, because employers and insurance companies obtained access to the information without receiving sufficient instruction as to its meaning, some *AS* heterozygotes were denied jobs or health insurance for no acceptable reason. Problems of public relations and education thus made a reliable screening test into a source of dissent and alienation.

Today, with the ability to look directly at the genotype of individuals born or unborn, it is becoming feasible to screen families at risk not only for sickle-cell anemia but for a growing number of other genetic disorders as well. The need to establish guidelines for genetic screening thus becomes more and more pressing. Several related questions reveal the complexity of the issue.

1. *Why carry out genetic screening at all?* There are two basic reasons. The first is to obtain information that will benefit individuals. For example, if you learn at an early age that you have a genetic predisposition to heart disease, you can change your lifestyle if necessary to include more exercise and a low-fat diet, thereby improving your chances of staying healthy. Or, you can use the results from genetic screening to make informed reproductive decisions that reduce the probability of having children affected by a genetic disease. In Brooklyn, New York, for example, among a community of Hasidic Jews of Eastern European descent, there used to be a high incidence of a fatal neurodegenerative syndrome known as Tay-Sachs disease. In this traditional, Old World community, marriages are arranged by rabbis or matchmakers who, by encouraging testing for the abnormal allele, helped eradicate the disease. With confidential access to test results, a rabbi could counsel against marriages between two carriers. The second reason for genetic screening, which often conflicts with the first, is to benefit groups within society. Insurance companies and employers, for example, would like to be able to find out who is at risk for various genetic conditions.

2. *When is a test accurate and comprehensive enough to be used as the basis for screening?* The accuracy of standard genetic tests for cystic fibrosis is more than 90%. Because it is not 100%, a few people who test negative may actually be carriers. In contrast, the tests for Huntington disease and the sickle-cell trait pick up close to 100% of those who carry the abnormal allele. In addition to the problem of false negatives, all genetic tests occasionally produce a false positive. What all this means is that some people might decide not to have children on the basis of inaccurate information. While the accuracy of the tests continues to

CONNECTIONS

Mendel answered the three basic questions about heredity as follows: To "What is inherited?" he replied "alleles of genes." To "How is it inherited?" he responded "according to the principles of segregation and independent assortment." And to "What is the role of chance in heredity?" he said "for each individual, inheritance is determined by chance, but within a population, this chance operates in a context of strictly defined probabilities." Knowing all this, we can understand why a trait may disappear in the F_1 hybrid generation and reappear in the F_2 generation or why with individuals that are pure-breeding, like begets like, while with hybrids, like can beget both like and unlike.

Within a decade of the 1900 rediscovery of Mendel's work, numerous breeding studies had shown that Mendel's laws hold true not only for seven pairs of antagonistic characteristics in peas, but for an enormous diversity of traits in a wide variety of improve, the question remains: Do the benefits outweigh both the costs and the risks of inaccuracies?

3.   *Once an accurate test becomes available at reasonable cost, should screening be required or optional?* This is partly a societal decision because the public treasury bears a large part of the cost of caring for the sufferers of genetic diseases. But it is also a personal decision. For most inherited diseases there is as yet no cure. Since the psychological burden of anticipating a fatal late-onset disease for which there is no treatment can be devastating, some people might decide not to be tested. Other reasons people may not want to be tested include religious beliefs and concerns about confidentiality. On the other hand, timely information about the presence of an abnormal gene that causes a condition for which there is a therapy can save lives. An example is the genetic test for hemochromatosis, an inherited disorder affecting 1 of every 200 people in the United States. The disease deposits iron in the heart, liver, and other organs of the body, and by the fifth or sixth decade of life, those organs break down. The simple, ancient practice of bloodletting, if begun early in life, prevents the complications of hemochromatosis. Timely information may also affect childbearing decisions and thereby reduce the incidence of a disease in the population.

4.   *If a screening program is established, who should be tested?* The answer depends on what the test is trying to accomplish as well as on the fact that genetic screening is expensive. Ultimately, the cost of a procedure must be weighed against the usefulness of the data it provides. For the 1 in 200 people inheriting the hemochromatosis mutation, widespread testing could reduce health-care costs of affected individuals later in life. By contrast, to reduce the risk of a child being born with a rare inherited disease, a program might try to target groups with the highest incidence. In the United States, only one-tenth as many African-Americans as Caucasians are affected by cystic fibrosis, and Asians almost never have the disease. Should all racial groups be tested or only Caucasians? Because of the expense, this kind of analysis has not yet been applied to large populations; instead, such direct testing has been reserved for couples or individuals whose family history puts them at risk for a severely debilitating disease.

5.   *Should private employers and insurance companies be allowed to test their clients and employees?* Some employers advocate genetic screening to reduce the incidence of occupational disease, arguing that they can use data from genetic tests to make sure employees are not assigned to environments that might cause them harm. People with sickle-cell disease, for example, may be at increased risk for a life-threatening episode of severe sickling if exposed to carbon monoxide or trace amounts of cyanide. Critics of this position say that screening violates workers' rights, including the right to privacy, and increases racial and ethnic discrimination in the workplace. Many critics also oppose informing insurance companies of the results of genetic screening, as these companies may deny coverage to people with inherited medical problems or just the possibility of developing such problems. According to many medical ethicists, discrimination of this sort will grow unless countries pass laws, similar to one enacted in France, that ensure genetic information is confidential, to be given out only at the discretion of the tested individual. In 2000, President Clinton issued an executive order banning genetic discrimination by federal agencies in the hiring or promotion of any employee. By 2002, several states, including Connecticut and North Carolina, had passed laws prohibiting genetic discrimination in employment practices.

6.   *Finally, how should people be educated about the meaning of test results?* In one small-community screening program, people identified as carriers of the recessive, life-threatening blood disorder known as β-thalassemia were ostracized and as a result, ended up marrying one another. This only made medical matters worse as it greatly increased the chances that their children would be born with two copies of the defective allele and thus the disease. By contrast, in Ferrara, Italy, where there used to be 30 new cases of β-thalassemia every year, extensive screening was so successfully combined with intensive education that the 1980s passed with no more than a few new cases of the disease.

Given all of these considerations, what kind of guidelines would you like to see established to ensure that genetic screening reaches the right people at the right time and that information gained from such screening is used for the right purposes?

---

sexually reproducing plant and animal species, including four o'clock flowers, beans, corn, wheat, fruit flies, chickens, mice, horses, and humans. Some of these same breeding studies, however, raised a challenge to the new genetics. For certain traits in certain species, the studies uncovered unanticipated phenotypic ratios, or the results included F_1 and F_2 progeny with novel phenotypes that resembled neither pure-breeding parent's. Mendel himself observed such "puzzling phenomena" when he repli-

cated his pea experiments with beans. A mating between bean plants with white flowers and those with crimson flowers produced a whole range of colors in the offspring, from purple to pale violet to white.

These phenomena could not be explained by Mendel's hypothesis that for each gene, two alternative alleles, one completely dominant, the other recessive, determine a single trait. Mendel, in fact, correctly explained the unexpected outcomes he observed by proposing that two genes (instead of one) determine flower color in beans. In humans, we now know that most common traits, including skin color, eye color, and height, are determined by interactions between two or more genes. We also know that within a given population, there may be more than two alleles for some of those genes. In Chapter 3 we show how the genetic analysis of such complex traits, that is, traits produced by complex interactions between genes and between genes and the environment, extended rather than contradicted Mendel's laws of inheritance. Each trait is still determined by the alleles of genes that behave during gamete formation and fertilization exactly as Mendel envisioned. What is new is the relationship between genotype and phenotype. It is not always as straightforward as in the analysis of pea color or of human genetic diseases such as cystic fibrosis.

E S S E N T I A L   C O N C E P T S

1. Discrete units called *genes* control the appearance of inherited traits.

2. Genes come in alternative forms called *alleles* that are responsible for the expression of different forms of a trait.

3. Body cells of sexually reproducing organisms carry two copies of each gene. When the two copies of a gene are the same allele, the individual is *homozygous* for that gene. When the two copies of a gene are different alleles, the individual is *heterozygous* for that gene.

4. The *genotype* is a description of the allelic combination of the two copies of a gene present in an individual. The *phenotype* is the observable form of the trait that the individual expresses.

5. A cross between two parental lines (P) that are pure-breeding for alternative alleles of a gene will produce a *first filial (F_1) generation* of hybrids that are heterozygous. The phenotype expressed by these hybrids is determined by the *dominant* allele of the pair,

and this phenotype is the same as that expressed by individuals homozygous for the dominant allele. The phenotype associated with the *recessive* allele will reappear only in the F_2 generation in individuals homozygous for this allele. In crosses between F_1 heterozygotes, the dominant and recessive phenotypes will appear in the F_2 generation in a ratio of 3:1.

6. The two copies of each gene segregate during the formation of gametes. As a result, each egg and each sperm or pollen grain contains only one copy, and thus, only one allele, of each gene. Male and female gametes unite at random at fertilization. Mendel described this process as the *law of segregation*.

7. The segregation of alleles of any one gene is independent of the segregation of the alleles of other genes. Mendel described this process as the *law of independent assortment*. According to this law, crosses between *Aa Bb* F_1 dihybrids will generate F_2 progeny with a phenotypic ratio of $9(A- B-) : 3(A- bb) : 3(aa B-) : 1(aa bb)$.

S O C I A L   A N D   E T H I C A L   I S S U E S

1. Judy and Tom were unable to have children because Judy had fibroids (benign tumors) in her uterus, making implantation of an embryo impossible. Judy was, however, able to ovulate healthy eggs. Judy and Tom decided they wanted to have a child that was genetically related to both of them. They contacted a surrogate mother, Sara, to carry an embryo conceived from their gametes by *in vitro* fertilization. A month before the child was born, Judy and Tom were killed in a plane crash. Sara decided she wanted to raise the child herself. She thought this wouldn't pose any legal problem since the genetic parents were no longer alive. But Judy's sister Ellen filed a petition with the court to gain custody of the

child. Should the courts award the child to the surrogate mother or to the family of the genetic parents?

2. Louise, a single mother with four children, has a job in an aeronautics factory where she is exposed to the element beryllium. Some individuals are susceptible to the damaging effects of this element, developing a respiratory illness called chronic beryllium disease. Workers in the factory were recently tested for susceptibility and Louise tested positive. She needs the job and wants to continue working at the factory until she finds another job. The company wants her to quit immediately. Should the individual or the company have

the right to decide whether a susceptible person continues to work in a potentially damaging environment?

3. Angela and Andre had difficulties conceiving and decided to have a child using *in vitro* fertilization methods. They have since gotten divorced and Angela, now a single mother of one, would like to have a second child by implantation of an embryo that was frozen when she and Andre had the first procedure. When Andre heard of Angela's plan through a mutual friend, he was upset and opposed the idea, stating that the embryos are half his. He feels that Angela needs his permission to use the embryos, and he will not grant that permission. Who has rights to these embryos? Should the embryos be destroyed? Should there be a limit to how long embryos are maintained?

S O L V E D   P R O B L E M S

Solving Genetics Problems

The best way to evaluate and increase your understanding of the material in the chapter is to apply your knowledge in solving genetics problems. Genetics word problems are like puzzles. Take them in slowly—don't be overwhelmed by the whole problem. Identify useful facts given in the problem, and use the facts to deduce additional information. Use genetic principles and logic to work toward the solutions. The more problems you do, the easier they become. In doing problems you will not only solidify your understanding of genetic concepts, but you will also develop basic analytical skills that are applicable in many disciplines.

Solving genetics problems requires more than simply plugging numbers into formulas. Each problem is unique and requires thoughtful evaluation of the information given and the question being asked. The following are general guidelines you can follow in approaching these word problems:

a. Read through the problem once to get some sense of the concepts involved.

b. Go back through the problem, noting all the information supplied to you. For example, genotypes or phenotypes of offspring or parents may be given to you or implied in the problem. Represent the known information in a symbolic format—assign symbols for alleles; use these symbols to indicate genotypes; make a diagram of the crosses including genotypes and phenotypes given or implied. Be sure that you do not assign different letters of the alphabet to two alleles of the same gene, as this can cause confusion.

c. Now, reassess the question and work toward the solution using the information given. Make sure you answer the question being asked!

d. When you finish the problem, check to see that the answer makes sense. You can often check solutions by working backwards; that is, see if you can reconstruct the data from your answer.

e. After you have completed a question and checked your answer, spend a minute to think about which major concepts were involved in the solution. This is a critical step for improving your understanding of genetics.

For each chapter, the logic involved in solving two or three types of problems is described in detail.

I. In cats, white patches are caused by the dominant allele *P*, while *pp* individuals are solid-colored. Short hair is caused by a dominant allele *S*, while *ss* cats have long hair. A long-haired cat with patches whose mother was solid-colored and short-haired mates with a short-haired, solid-colored cat whose mother was long-haired and solid-colored. What kinds of kittens can arise from this mating, and in what proportions?

Answer

The solution to this problem requires an understanding of dominance/recessiveness, gamete formation, and the independent assortment of alleles of two genes in a cross.

First make a representation of the known information:

Mothers:	solid, short-haired	solid, long-haired
Cross:	cat 1	cat 2
	patches, long-haired × solid, short-haired	

What genotypes can you assign? Any cat showing a recessive phenotype must be homozygous for the recessive allele. Therefore the long-haired cats are *ss;* solid cats are *pp*. Cat 1 is long-haired, so it must be homozygous for the recessive allele (*ss*). This cat has the dominant phenotype of patches and could be either *PP* or *Pp*, but since the mother was *pp*, and could only contribute a *p* allele in her gametes, the cat must be *Pp*. Cat 1's full genotype is *Pp ss*. Similarly, cat 2 is solid-colored, so it must be homozygous for the recessive allele (*pp*). Because this cat is short-haired, it could have either the *SS* or *Ss* genotype. Its mother was long-haired (*ss*) and could only contribute an *s* allele in her gamete, so cat 2 must be heterozygous *Ss*. The full genotype is *pp Ss*.

The cross is therefore between a *Pp ss* (cat 1) and a *pp Ss* (cat 2). To determine the types of kittens, first establish the types of gametes that can be produced by each cat and then set up a Punnett square to determine the genotypes of the offspring. Cat 1 (*Pp ss*) produces *Ps* and *ps* gametes in equal proportions. Cat 2 (*pp Ss*) produces *pS* and *ps* gametes in equal proportions. *Four types of kittens can result from this mating with equal probability:* Pp Ss *(patches, short-haired),* Pp ss *(patches, long-haired),* pp Ss *(solid, short-haired), and* pp ss *(solid, long-haired).*

Cat 1

	P s	p s
p S	Pp Ss	pp Ss
p s	Pp ss	pp ss

Cat 2 (label at left)

You could also work through this problem using the product rule of probability instead of a Punnett square. The principles are the same: gametes produced in equal amounts by either parent are combined at random.

Cat 1 gamete		Cat 2 gamete		Progeny	
1/2 *P s*	×	1/2 *p S*	→	1/4 *Pp Ss*	patches, short-haired
1/2 *P s*	×	1/2 *p s*	→	1/4 *Pp ss*	patches, long-haired
1/2 *p s*	×	1/2 *p S*	→	1/4 *pp Ss*	solid-colored, short-haired
1/2 *p s*	×	1/2 *p s*	→	1/4 *pp ss*	solid-colored, long-haired

II. In tomatoes, red fruit is dominant to yellow fruit and purple stems are dominant to green stems. The progeny from one mating consisted of 305 red fruit, purple stem plants; 328 red fruit, green stem plants; 110 yellow fruit, purple stem plants; and 97 yellow fruit, green stem plants. What was the genotype of the parents in this cross?

Answer

This problem requires an understanding of independent assortment in a dihybrid cross as well as the ratios predicted from monohybrid crosses.
      Designate the alleles:

R = red, r = yellow

P = purple stems, p = green stems

In genetics problems, the ratios of offspring can indicate the genotype of parents. You will usually need to total the number of progeny and approximate the ratio of offspring in each of the different classes. For this problem, in which the inheritance of two traits is given, consider each trait independently. For red fruit there are 305 + 328 = 633 red-fruited plants out of a total of 840 plants. This value (633/840) is close to 3/4. About 1/4 of the plants

have yellow fruit (110 + 97 = 207/840). From Mendel's work, you know that a 3:1 phenotypic ratio results from crosses between plants that are hybrid (heterozygous) for one gene. Therefore the genotype for fruit color of each parent must have been *Rr*.

      For stem color, 305 + 110 or 415/840 plants had purple stems. About half had purple stems and the other half (328 + 97) had green stems. A 1:1 phenotypic ratio occurs when a heterozygote is mated to a homozygous recessive (as in a testcross). The parents' genotypes must have been *Pp* and *pp* for stem color.
      The complete genotype of the parent plants in this cross was Rr Pp × Rr pp.

III. Tay-Sachs is a recessive lethal disease in which there is neurological deterioration early in life. This disease is rare in the population overall but is found at relatively high frequency in Ashkenazi Jews from Central Europe. A woman whose maternal uncle had the disease is trying to determine the probability that she and her husband could have an affected child. Her father does not come from a high-risk population. Her husband's sister died of the disease at an early age.
  a. Draw the pedigree of the individuals described. Include the genotypes where possible.
  b. Determine the probability that the couple's first child will be affected.

Answer

This problem requires an understanding of dominance/recessiveness and probability. Designate the alleles:

T = normal allele; t = Tay-Sachs allele

The genotypes of the two affected individuals, the woman's uncle (II-1) and the husband's sister (III-3) are *tt*. Because the uncle was affected, his parents must have been heterozygous. There was a 1/4 chance that these parents had a homozygous recessive (affected) child, a 2/4 chance that they had a heterozygous child (carrier), and a 1/4 chance they had a homozygous dominant (unaffected) child. However, you have been told that the woman's mother (II-2) is unaffected, so the mother could only have had a heterozygous or a homozygous dominant genotype. Consider the probability that these two genotypes will occur. If you were looking at a Punnett square, there would be only three combinations of alleles that are possible for the normal mother. Two of these are heterozygous combinations and one is homozygous dominant. There is a 2/3 chance (2 out of the 3 possible cases) that the mother was a carrier. The father was not from a high-risk population so we can assume that he is homozygous dominant. There is a 2/3 chance that the wife's mother was heterozygous and if so, a 1/2 chance that the wife inherited a recessive allele from her mother. Because both conditions are necessary for inheritance of a recessive allele, the individual probabilities are multiplied and the probability that the wife (III-1) is heterozygous is $2/3 \times 1/2$.

The husband (III-2) has a sister who died from the disease, therefore his parents must have been heterozygous. The probability that he is a carrier is 2/3 (using the same rationale as for II-2). The probability that the man and woman are both carriers is $2/3 \times 1/2 \times 2/3$. Since there is a 1/4 probability that a particular child of two carriers will be affected, *the overall probability that the first child of this couple (III-1 and III-2) will be affected is $2/3 \times 1/2 \times 2/3 \times 1/4 = 4/72$, or 1/18.*

PROBLEMS

2-1 For each of the terms in the left column, choose the best matching phrase in the right column.

a. phenotype	1. having two identical alleles of a given gene
b. alleles	2. the allele expressed in the phenotype of the heterozygote
c. independent assortment	3. alternate forms of a gene
d. gametes	4. observable characteristic
e. gene	5. a cross between individuals both heterozygous for two genes
f. segregation	6. alleles of one gene separate into gametes randomly with respect to alleles of other genes
g. heterozygote	7. reproductive cells containing only one copy of each gene
h. dominant	8. the allele that does not contribute to the phenotype of the heterozygote
i. F_1	9. the cross of an individual of ambiguous genotype with a homozygous recessive individual
j. testcross	10. an individual with two different alleles of a gene
k. genotype	11. the heritable entity that determines a characteristic
l. recessive	12. the alleles an individual has
m. dihybrid cross	13. the separation of the two alleles of a gene into different gametes
n. homozygote	14. offspring of the P generation

2-2 Describe the characteristics of the garden pea that made it a good organism for Mendel's analysis of the basic principles of inheritance. Evaluate how easy or difficult it would be to make a similar study of inheritance in humans by considering the same attributes you described for the pea.

2-3 In a particular population of mice, certain individuals display a phenotype called "short tail," which is inherited as a dominant trait. Some individuals display a recessive trait called "dilute," which affects coat color. Which of these traits would be easier to eliminate from the population by selective breeding? Why?

2-4 How many genetically different eggs could be formed by women with the following genotypes?
a. *Aa bb CC DD*
b. *AA Bb Cc dd*
c. *Aa Bb cc Dd*
d. *Aa Bb Cc Dd*

2-5 A mouse sperm of genotype *a B C D E* fertilizes an egg of genotype *a b c D e*. What are all the possibilities for the genotypes of (a) the zygote and (b) a sperm or egg of the baby mouse that develops from this fertilization?

2-6 In humans, a dimple in the chin is a dominant characteristic.
a. A man who does not have a chin dimple has children with a woman with a chin dimple whose mother lacked the dimple. What proportion of their children would be expected to have a chin dimple?
b. A man with a chin dimple and a woman who lacks the dimple produce a child who lacks a dimple. What is the man's genotype?
c. A man with a chin dimple and a nondimpled woman produce eight children, all having the chin dimple. Can you be certain of the man's genotype? Why or why not? What genotype is more likely, and why?

2-7 Two short-haired cats mate and produce six short-haired and two long-haired kittens. What does this information suggest about how hair length is inherited?

2-8 Piebald spotting is a condition found in humans in which there are patches of skin that lack pigmentation. The condition results from the inability of pigment-producing cells to migrate properly during development. Two adults with piebald spotting have one child who has this trait and a second child with normal skin pigmentation.

a. Is the piebald spotting trait dominant or recessive? What information led you to this answer?

b. What are the genotypes of the parents?

2-9 Among native Americans, two types of earwax (cerumen) are seen, dry and sticky. A geneticist studied the inheritance of this trait by observing the types of offspring produced by different kinds of matings. He observed the following numbers:

Parents	Number of mating pairs	Offspring	
		Sticky	Dry
Sticky × sticky	10	32	6
Sticky × dry	8	21	9
Dry × dry	12	0	42

a. How is earwax type inherited?

b. Why are there no 3:1 or 1:1 ratios in the data shown in the chart?

2-10 If you roll a die (singular of dice), what is the probability you will roll: (a) a 6? (b) an even number? (c) a number divisible by 3? (d) If you roll a pair of dice, what is the probability that you will roll two 6s? (e) an even number on one and an odd number on the other? (f) matching numbers? (g) two numbers both over 4?

2-11 Galactosemia is a recessive human disease that is treatable by restricting lactose and glucose in the diet. Susan Smithers and her husband are both heterozygous for the galactosemia gene.

a. Susan is pregnant with twins. If she has fraternal (nonidentical) twins, what is the probability both of the twins will be girls who have galactosemia?

b. If the twins are identical, what is the probability that both will be girls and have galactosemia?

For parts c–g, assume that none of the children is a twin.

c. If Susan and her husband have four children, what is the probability that none of the four will have galactosemia?

d. If the couple has four children, what is the probability that at least one child will have galactosemia?

e. If the couple has four children, what is the probability that the first two will have galactosemia and the second two will not?

f. If the couple has three children, what is the probability that two of the children will have galactosemia and one will not, regardless of order?

g. If the couple has four children with galactosemia, what is the probability that their next child will have galactosemia?

2-12 What is the probability of producing a child that will phenotypically resemble either one of the two parents in the following four crosses? How many phenotypically different kinds of progeny could potentially result from each of the four crosses?

a. *Aa Bb Cc Dd* × *aa bb cc dd*

b. *aa bb cc dd* × *AA BB CC DD*

c. *Aa Bb Cc Dd* × *Aa Bb Cc Dd*

d. *aa bb cc dd* × *aa bb cc dd*

2-13 As a *Drosophila* research geneticist, you keep stocks of flies of specific genotypes. You have a fly that has normal wings (dominant phenotype). Flies with short wings are homozygous for a recessive allele of the wing-length gene. You need to know if this fly with normal wings is pure-breeding or heterozygous for the wing-length trait. What cross would you do to determine the genotype and what results would you expect for each possible genotype?

2-14 A mutant cucumber plant has flowers that fail to open when mature. Crosses can be done with this plant by manually opening and pollinating the flowers with pollen from another plant. When closed × open crosses were done, all the F_1 progeny were open. The F_2 plants were 145 open and 59 closed. A cross of closed × F_1 gave 81 open and 77 closed. How is the closed trait inherited?

2-15 Imagine you have just purchased a black stallion of unknown genotype. You mate him to a red mare, and she delivers twin foals, one red and one black. Can you tell from these results how color is inherited, assuming that alternative alleles of a single gene are involved? What crosses could you do to work this out?

2-16 A pea plant from a pure-breeding strain that is tall, has green pods, and has purple flowers that are terminal is crossed to a plant from a pure-breeding strain that is dwarf, has yellow pods, and has white flowers that are axial. The F_1 plants are all tall and have purple axial flowers as well as green pods.

a. What phenotypes do you expect to see in the F_2?

b. What phenotypes and ratios would you predict in the progeny from crossing an F_1 plant to the dwarf parent?

2-17 The following chart shows the results of different matings between jimsonweed plants that had either purple or white flowers and spiny or smooth pods. Determine the dominant allele for the two traits and indicate the genotypes of the parents for each of the crosses.

Parents	Offspring phenotypes			
	purple spiny	white spiny	purple smooth	white smooth
a. purple spiny × purple spiny	94	32	28	11
b. purple spiny × purple smooth	40	0	38	0
c. purple spiny × white spiny	34	30	0	0
d. purple spiny × white spiny	89	92	31	27
e. purple smooth × purple smooth	0	0	36	11
f. white spiny × white spiny	0	45	0	16

2-18 A cross between two pea plants, both of which grew from yellow round seeds, gave the following numbers of seeds: 156 yellow round and 54 yellow wrinkled. What are the genotypes of the parent plants? (Yellow and round are dominant traits.)

2-19 A third-grader decided to breed guinea pigs for her school science project. She went to a pet store and bought a male with smooth black fur and a female with rough white fur. She wanted to study the inheritance of those features and was sorry to see that the first litter of eight contained only rough black animals. To her disappointment, the second litter from those same parents contained seven rough black animals. Soon the first litter had begun to produce F_2 offspring, and they showed a variety of coat types. Before long the child had 125 F_2 guinea pigs. Eight of them had smooth white coats, 25 had smooth black coats, 23 were rough and white, and 69 were rough and black.
   a. How are the coat color and texture characteristics inherited?
   b. What phenotypes and proportions of offspring should the girl expect if she mates one of the smooth white F_2 females to an F_1 male?

2-20 A pea plant heterozygous for plant height, pod shape, and flower color was selfed. The progeny consisted of 272 tall, inflated pods, purple flowers; 92 tall, inflated, white flowers; 88 tall, flat pods, purple; 93 dwarf, inflated, purple; 35 tall, flat, white; 31 dwarf, inflated, white; 29 dwarf, flat, purple; 11 dwarf, flat, white. Which alleles are dominant in this cross?

2-21 The achoo syndrome (sneezing in response to bright light) and trembling chin (triggered by anxiety) are both dominant traits in humans.
   a. What is the probability that the first child of parents who are heterozygous for both the achoo gene and trembling chin will have achoo syndrome but lack the trembling chin?
   b. What is the probability that the first child will not have achoo syndrome or trembling chin?

2-22 The self-fertilization of an F_1 pea plant produced from a parent plant homozygous for yellow and wrinkled seeds and a parent homozygous for green and round seeds resulted in a pod containing seven F_2 peas. (Yellow and round are dominant.) What is the probability that all seven peas in the pod are yellow and round?

2-23 For each of the following human pedigrees, indicate whether the inheritance pattern is recessive or dominant. What feature(s) of the pedigree did you use to determine the inheritance? Give the genotypes of affected individuals and of individuals who carry the disease allele.

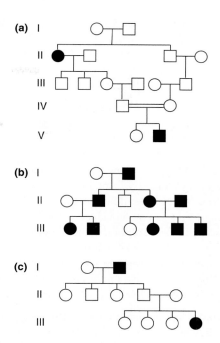

2-24 Albinism is a condition in which pigmentation is lacking. In humans, the result is white hair, nonpigmented skin, and pink eyes. The trait in humans is caused by a recessive allele. Two normal parents have an albino child. What are the parents' genotypes? What is the probability that the next child will be albino?

2-25 A young couple went to see a genetic counselor because each had a sibling affected with cystic fibrosis. (Cystic fibrosis is a recessive disease and neither member of the couple nor any of their four parents is affected.)
   a. What is the probability that the female of this couple is a carrier?
   b. What are the chances that their child will be affected with cystic fibrosis?
   c. What is the probability that their child will be a carrier of the cystic fibrosis mutation?

2-26 Huntington disease is a rare fatal, degenerative neurological disease in which individuals start to show symptoms, on average, in their 40s. It is caused by a dominant allele. Joe, a man in his 20s, just learned that his father has Huntington disease.
   a. What is the probability that Joe will also develop the disease?
   b. Joe and his new wife have been eager to start a family. What is the probability that their first child will eventually develop the disease?

2-27 Is the disease shown dominant or recessive? Why? Based on the limited information given in this pedigree, do you think the disease allele is rare or common in the population? Why?

2-28 Figure 2.20 on p. 30 shows the inheritance of Huntington disease in a family from a small, isolated village near Lake Maracaibo in Venezuela.
  a. Why could you not conclude definitively that the disease is the result of a dominant or a recessive allele solely by looking at the pedigree?
  b. Is there any information you could glean from the family's history that might imply the disease is due to a dominant allele?

2-29 The common grandfather of two first cousins has hereditary hemochromatosis, a recessive condition causing an abnormal buildup of iron in the body. Neither of the cousins has the disease nor do any of their relatives.
  a. If the first cousins mated with each other and had a child, what is the chance that the child would have hemochromatosis? Assume that the unrelated, unaffected parents of the cousins are not carriers.
  b. How would your calculation change if you knew that 1 out of every 10 unaffected people in the population (including the unrelated parents of these cousins) was a carrier for hemochromatosis?

2-30 Polycystic kidney disease is a dominant trait that causes the growth of numerous cysts in the kidneys. The condition eventually leads to kidney failure. A child with polycystic kidney disease is born to a couple, neither of whom shows the disease. What possibilities might explain this outcome?

2-31 Midphalangeal hair (hair on top of the middle segment of the fingers) is a common phenotype caused by a dominant allele *M*. Homozygotes for the recessive allele (*mm*) lack hair on the middle segment of their fingers. Among 1000 families in which both parents had midphalangeal hair, 1853 children showed the trait while 209 children did not. Explain this result.

2-32 People with nail-patella syndrome have poorly developed or absent kneecaps and nails. Individuals with alkaptonuria have arthritis as well as urine that darkens when exposed to air. Both nail-patella syndrome and alkaptonuria are rare phenotypes. In the following pedigree, vertical red lines indicate individuals with nail-patella syndrome, while horizontal green lines denote individuals with alkaptonuria.
  a. What are the most likely modes of inheritance of nail-patella syndrome and alkaptonuria? What genotypes can you ascribe to each of the individuals in the pedigree for both of these phenotypes?
  b. In a mating between IV-2 and IV-5, what is the chance that the child produced would have both nail-patella syndrome and alkaptonuria? Nail-patella syndrome alone? Alkaptonuria alone? Neither defect?

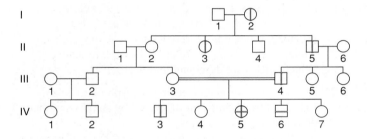

CHAPTER

3

Extensions to Mendel: Complexities in Relating Genotype to Phenotype

Coat color in deer is one of many traits in natural populations governed by the interactions of multiple genes with each other and the environment.

nlike the pea traits that Mendel examined, most human characteristics—height, hair color, skin color, voice quality, artistic or athletic ability, to name a few—do not fall neatly into just two opposing phenotypic categories. And because people cannot be simply divided into two contrasting camps for each trait (brown hair versus blond, high voice versus low), they seem to defy Mendelian analysis. The same can be said of traits expressed by many of the world's food crops; the size, shape, succulence, and nutrient content of the plants' edible parts vary over a wide range of values. Lentils (*Lens culinaris*) provide a graphic illustration of this variation. Lentils, a legume like peas and beans, are grown in many parts of the world as a rich source of both protein and carbohydrate. The mature plants set fruit in the form of diminutive pods that contain two small, round and flattened seeds, which can be ground into meal or used in soups, salads, and stews. These seeds come in an intriguing array of colors and patterns (Fig. 3.1), and since particular couplings of color and pattern are preferred in the cuisines of different cultures, there is con-

siderable commercial interest in producing specific seed coat combinations. But crosses between pure-breeding lines of lentils result in some startling surprises in relation to Mendel's laws. A cross between pure-breeding tan and pure-breeding gray parents, for example, yields an all brown F_1 generation; and when these novel hybrids are left to self-pollinate, the F_2 plants produce not only tan, gray, and brown lentils, but also green.

Beginning with the first decade of the twentieth century, geneticists subjected all kinds of puzzling variation in plants and animals to controlled breeding tests, using Mendel's 3:1 phenotypic ratio as a guideline in interpreting the results. If the traits under analysis behaved as predicted by Mendel's laws, it meant they were determined by a single gene with alternative alleles that show clear-cut dominance and recessiveness. Many traits, however, did not behave in this way. For some, there was no definitive dominance and recessiveness, or there were more than two alternative alleles in a particular cross. Other traits turned out to be **multifactorial,** that is, determined by two or more factors,

Figure 3.1 Some phenotypic variation poses a challenge to Mendelian analysis. In this array of red, green, and brown lentils, some of the seeds have speckled patterns, while others are clear. As we see in this chapter, combinations of alleles at two genes determine seed coat color, while combinations of the multiple alleles of a third gene produce the pattern.

As we look at the genetic factors that make it possible for plants and animals to generate the enormous amount of phenotypic variation seen in the natural world, we encounter the following general theme: In their attempt to sort things out, geneticists limit the number of variables under investigation at any one time. Mendel, for example, compared and mated pure-breeding, inbred strains of peas that differed from each other by one or a few traits. By establishing a genetic background that was uniform, or homogenized, for all other traits, he was able to visualize the actions of single genes in isolation. Similarly, twentieth-century geneticists have used inbred populations of fruit flies, mice, and other experimental organisms to study specific traits. Of course, geneticists cannot approach people in this way. Human populations are far from inbred, and if a phenotype arises from many genes that are not held constant, it may be difficult to see how a particular gene affects a particular trait. For this reason, until recently the genetic basis of much human variation remained a mystery. The advent of molecular biology in the 1970s, however, provided new tools that geneticists now use to unravel the genetics of complex human traits.

We divide our presentation of extensions to Mendelian analysis into two broad categories:

- Single-gene inheritance (a) in which pairs of alleles show deviations from complete dominance and recessiveness, (b) in which different forms of the gene are not limited to two alleles, and (c) where one gene may determine more than one trait.

- Multifactorial inheritance in which the phenotype arises from the interaction of one or more genes with the environment, chance, or each other.

including multiple genes interacting with each other or one or more genes interacting with the environment. The seed coat color of lentils is a multifactorial trait. Because such traits arise from an intricate network of interactions, they do not necessarily generate straightforward Mendelian phenotypic ratios. Nonetheless, simple extensions of Mendel's hypotheses that clarify the relationship between genotype and phenotype have allowed geneticists to explain all of the observed deviations without challenging Mendel's basic laws.

■ ■ ■

Extensions to Mendel for Single-Gene Inheritance

William Bateson, an early interpreter and defender of Mendel, who coined the terms "genetics," "allelomorph" (later shortened to "allele"), "homozygote," and "heterozygote," entreated the audience at a 1908 lecture: "Treasure your exceptions! . . . Keep them always un-covered and in sight. Exceptions are like the rough brickwork of a growing building which tells that there is more to come and shows where the next construction is to be." Consistent exceptions to simple Mendelian ratios revealed unexpected patterns of single-gene inheritance. By distilling the significance of these patterns, Bateson and other early geneticists extended the scope of Mendelian analysis and obtained a deeper understanding of the relationship between genotype and phenotype. We now look at the major extensions to Mendelian analysis elucidated over the last century.

Dominance Is Not Always Complete

A consistent working definition of dominance and recessiveness depends on the F_1 hybrids that arise from a mating between two pure-breeding lines. If a hybrid is identical to one parent for the trait under consideration, the allele carried by that parent is deemed dominant to the allele carried by the parent whose trait is not expressed in the hybrid. If, for example, a mating between a pure-breeding white line and a pure-breeding blue line produces F_1 hybrids that are white, the white allele of the gene for color is dominant to the blue allele. If the F_1 hybrids are blue, the blue allele is dominant to the white one (Fig. 3.2).

The determination of dominance and recessiveness does not imply anything about the actual functioning of the alleles. Although geneticists once believed that all dominant alleles were actually active in the organism, while recessive alleles were inactive, they now know of many cases where this generalization is not true. We discuss some of these cases later in the chapter.

Mendel described and relied on complete dominance in sorting out his ratios and laws, but it is not the only kind of dominance he observed. Figure 3.2 diagrams two situations in which neither allele of a gene is completely dominant. As the figure shows, crosses between true-breeding strains can produce hybrids with phenotypes that differ from both parents. We now explain how these phenotypes arise.

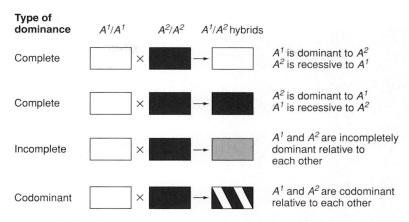

Figure 3.2 Different dominance relationships. The phenotype of the heterozygote defines the dominance relationship between two alleles of the same gene (here, A^1 and A^2). Dominance is complete when the hybrid resembles one of the two pure-breeding parents. Dominance is incomplete when the hybrid resembles neither parent; its novel phenotype is usually intermediate between the parental phenotypes. Codominance occurs when the hybrid shows the traits from both pure-breeding parents.

In Incomplete Dominance, the F_1 Hybrid Resembles Neither Pure-Breeding Parent

A cross between pure late-blooming and pure early-blooming pea plants results in an F_1 generation that blooms in between the two extremes. This is just one of many examples of **incomplete dominance,** in which the hybrid does not resemble either pure-breeding parent. F_1 hybrids that differ from both parents often express a phenotype that is intermediate between those of the pure-breeding parents. Thus, with incomplete dominance, neither parental allele is dominant or recessive to the other; both contribute to the F_1 phenotype. Mendel observed plants that bloomed midway between two extremes when he cultivated various types of pure-breeding peas for his hybridization studies, but he did not pursue the implications. Blooming time was not one of the seven characteristics he chose to analyze in detail, almost certainly because in peas, the time of bloom was not as clear-cut as seed shape or flower color.

In many plant species, flower color serves as a striking example of incomplete dominance. With the tubular flowers of four o'clocks or the double-lipped floret clusters of snapdragons, for instance, a cross between pure-breeding red-flowered parents and pure-breeding white yields hybrids with pink blossoms, as if a painter had mixed red and white pigments to get pink (Fig. 3.3a). If allowed to self-pollinate, the F_1 pink-blooming plants produce F_2 progeny bearing red, pink, and white flowers in a ratio of 1:2:1 (Fig. 3.3b). This is the familiar *genotypic* ratio of an ordinary single-gene F_1 self-cross. What is new is that because the heterozygotes look unlike either homozygote, the *phenotypic* ratios are an exact reflection of the genotypic ratios. The biochemical explanation for this type of incomplete dominance is that each allele of the gene under analysis specifies an alternative form of a protein molecule with an enzymatic role in pigment production. If the "white" allele does not give rise to a functional enzyme, no pigment appears. Thus, in snapdragons and four o'clocks, two "red" alleles per cell produce a double dose of a red-producing enzyme, which generates enough pigment to make the flowers

look fully red. In the heterozygote, one copy of the "red" allele per cell results in only enough pigment to make the flowers look pink. In the homozygote for the "white" allele, where there is no functional enzyme and thus no red pigment, the flowers appear white.

In Codominance, Alternative Traits Are Both Visible in the F_1 Hybrid

A cross between pure-breeding spotted lentils and pure-breeding dotted lentils produces heterozygotes that are both spotted and dotted (Fig. 3.4a). These F_1 hybrids illustrate a second significant departure from complete dominance. They look like both parents, which means that neither the "spotted" nor the "dotted" allele is dominant or recessive to the other. Since both traits show up equally in the heterozygote's phenotype, the alleles are termed **codominant.** Self-pollination of the spotted/dotted F_1 generation generates F_2 progeny in the ratio of 1 spotted : 2 spotted/dotted : 1 dotted. The Mendelian 1:2:1 ratio among these F_2 progeny establishes that the spotted and dotted traits are determined by alternative alleles of a single gene. Once again, because the heterozygotes can be distinguished from both homozygotes, the phenotypic and genotypic ratios coincide.

Figure 3.2 summarizes the differences between complete dominance, incomplete dominance, and codominance for phenotypes reflected in color variations. With complete dominance, F_1 progeny look like one of the true-breeding parents. With incomplete dominance, hybrids resemble neither of the parents and thus display neither pure-breeding trait. With codominance, the phenotypes of both pure-breeding lines show up simultaneously in the F_1 hybrid. Determinations of dominance relationships depend on what phenotype appears in the F_1 generation. Complete dominance, as we saw in Chapter 2, results in a 3:1 ratio of phenotypes in the F_2, while both incomplete dominance and codominance yield 1:2:1 F_2 ratios.

In humans, some of the complex membrane-anchored molecules that distinguish different types of red blood cells

(a) *Antirrhinum majus* **(snapdragons)**

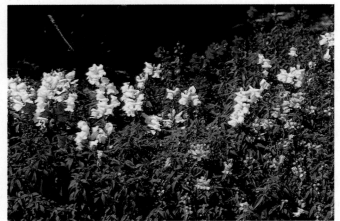

(b) **A Punnett square for incomplete dominance**

1 *AA* (red) : 2 *Aa* (pink) : 1 *aa* (white)

Figure 3.3 Pink flowers are the result of incomplete dominance. (a) Color differences in snapdragons reflect the activity of one pair of alleles. **(b)** The F₁ hybrids from a cross of pure-breeding red and white strains of snapdragons have pink blossoms, resembling neither parent. Flower colors in the F₂ appear in the ratio of 1 red : 2 pink : 1 white. This ratio signifies that the alleles of a single gene determine these three colors.

exhibit codominance. For example, one gene (*I*) with alleles *I^A* and *I^B* controls the presence of a sugar polymer that protrudes from the red blood cell membrane. The alternative alleles each encode a slightly different form of an enzyme that causes production of a slightly different form of the complex sugar. In heterozygous individuals, the red blood cells carry both the *I^A*-determined and the *I^B*-determined sugars on their surface, whereas the cells of homozygous individuals display the products of either *I^A* or *I^B* alone (Fig. 3.4b).

It often happens that alleles are codominant at the molecular level, in terms of the protein they determine, even if that codominance is not readily apparent at the level of the whole organism. An example is sickle-cell anemia. Homozygotes for the sickling allele have the disease; homozygotes for the nor-

(a) **Codominant lentil coat patterns**

1 *C^S C^S* (spotted) : 2 *C^S C^D* (spotted/dotted) : 1 *C^D C^D* (dotted)

(b) **Codominant blood group alleles**

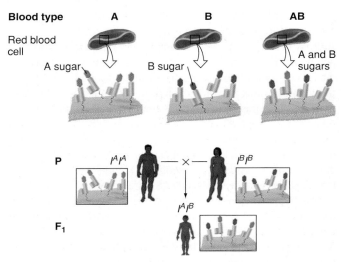

Figure 3.4 In codominance, F₁ hybrids display the traits of both parents. (a) A cross between pure-breeding spotted lentils and pure-breeding dotted lentils produces heterozygotes that are both spotted and dotted. Because each genotype has its own corresponding phenotype, the F₂ ratio is 1:2:1. **(b)** The *I^A* and *I^B* blood group alleles are codominant because the red blood cells of an *I^A I^B* heterozygote have both kinds of sugars at their surface.

mal allele do not. Heterozygotes are generally healthy under normal living conditions; thus the sickle-cell allele is recessive to the normal allele in its overall effect on well-being. But the red blood cells of a heterozygote carry both normal and sickling proteins; thus, the normal and sickling alleles are codominant at the molecular level.

This example illustrates that dominance relationships are defined in the context of the particular phenotype under analysis. When looked at from the point of view of different phenotypes, the same two alleles can display different dominance relationships. We return later to this important point.

Variations on Complete Dominance Do Not Negate Mendel's Law of Segregation

The dominance relations of a gene's alleles do not affect the alleles' transmission. Whether two alternative alleles of a single gene show complete dominance, incomplete dominance, or codominance depends on the kinds of proteins determined by the alleles and the biochemical function of those proteins in the cell. These same phenotypic dominance relations, however, have no bearing on the segregation of the alleles during gamete formation. As Mendel proposed, cells still carry two copies of each gene, and these copies—a pair of either similar or dissimilar alleles—segregate during gamete formation. Fertilization then restores two alleles to each cell without reference to whether the alleles are the same or different. Variations in dominance relations thus do not detract from Mendel's laws of segregation. Rather, they reflect differences in the way gene products control the production of phenotypes, adding a level of complexity to the task of sorting out the visible results of gene transmission and inferring genotype from phenotype.

A Gene May Have More Than Two Alleles

Mendel analyzed "either-or" traits controlled by genes with two alternative alleles, but for many traits there are more than two alternatives. Human blood types provide an example. If a person with blood type A mates with a person with blood type B, it is possible in some cases for the couple to have a child that is neither A nor B, but a third blood type called O. The reason? The gene for the ABO blood types has three alleles: I^A, I^B, and i (Fig. 3.5a). Allele I^A gives rise to blood type A by specifying an enzyme that adds sugar A; I^B results in blood type B by specifying an enzyme that adds sugar B; i does not produce a functional sugar-adding enzyme. Alleles I^A and I^B are both dominant to i, and blood type O is therefore a result of homozygosity for allele i.

Note in Fig. 3.5a that the A phenotype can arise from two genotypes, $I^A I^A$ or $I^A i$. The same is true for the B blood type, which can be produced by $I^B I^B$ or $I^B i$. But a combination of the two alleles $I^A I^B$ generates blood type AB.

We can draw several conclusions from these observations. First, as already stated, a given gene may have more than two alleles, or **multiple alleles;** in our example, the series of alleles is denoted I^A, I^B, i.

Second, although the ABO blood group gene has three alleles, each person carries only two of the alternatives—$I^A I^A$, $I^B I^B$, $I^A I^B$, $I^A i$, $I^B i$, or ii. There are thus six possible ABO genotypes. Because each individual carries no more than two alleles for each gene, no matter how many alleles there are in a series, Mendel's law of segregation remains intact, since in any sexually reproducing organism, the two alleles of a gene separate during gamete formation. Third, an allele is not inherently dominant or recessive; its dominance or recessiveness is always relative to a second allele. In other words, dominance relations are unique to a pair of alleles. In our example, I^A is completely dominant to i, but it is codominant with I^B. Given these dominance relations, the six genotypes possible with I^A, I^B, and i generate four different phenotypes: blood groups A, B, AB, and O. With

Genotypes	Corresponding Phenotypes: Type(s) of Molecule on Cell
$I^A I^A$ $I^A i$	A
$I^B I^B$ $I^B i$	B
$I^A I^B$	AB
ii	O

(b)

Blood Type	Antibodies in Serum
A	Antibodies against B
B	Antibodies against A
AB	No antibodies against A or B
O	Antibodies against A and B

(c)

Blood Type of Recipient	Donor Blood Type (Red Cells)			
	A	B	AB	O
A	+	−	−	+
B	−	+	−	+
AB	+	+	+	+
O	−	−	−	+

Figure 3.5 ABO blood types are determined by three alleles of one gene. (a) Six genotypes produce the four blood group phenotypes. **(b)** Blood serum contains antibodies against foreign red blood cell molecules. **(c)** If a recipient's serum has antibodies against the sugars on a donor's red blood cells, the blood types of recipient and donor are incompatible and coagulation of red blood cells will occur during transfusions. In this table, a plus (+) indicates compatibility and a minus (−) indicates incompatibility.

this background, you can understand how a type A and a type B parent could produce a type O child: The parents must be $I^A i$ and $I^B i$ heterozygotes, and the child receives an i allele from each parent.

An understanding of the genetics of the ABO system has had profound medical and legal repercussions. Matching ABO blood types is a prerequisite of successful blood transfusions, because people make antibodies to foreign blood cell molecules. A person whose cells carry only A molecules, for example, produces anti-B antibodies; B people manufacture anti-A antibodies; AB individuals make neither type of antibody; and O individuals produce both anti-A and anti-B antibodies (Fig. 3.5b). These anti-bodies cause coagulation of cells displaying the foreign molecules (Fig. 3.5c). As a result, people with blood type O have historically been known as universal donors because their red blood cells carry no surface molecules that will stimulate an antibody attack. In contrast, people with blood type AB are considered universal recipients, because they make neither anti-A nor anti-B antibodies, which, if present, would target the surface molecules of incoming blood cells.

Information about ABO blood types can also be used as legal evidence in court, to exclude the possibility of paternity or criminal guilt. In a paternity suit, for example, if the mother is type A and her child is type B, logic dictates that the I^B allele must have come from the father, whose genotype may be $I^A I^B$, $I^B I^B$, or $I^B i$. In 1944, the actress Joan Barry (phenotype A) sued Charlie Chaplin (phenotype O) for support of a child (phenotype B) whom she claimed he fathered. The scientific evidence contradicting her claim (the father had to carry an I^B allele and Chaplin was apparently ii) was admissible in court but did not convince the jury, and Chaplin had to pay. Today, with a public that is much more scientifically literate, it is likely that members of the jury would understand the scientific evidence, reject Barry's claim, and find Chaplin innocent.

Lentils offer another example of multiple alleles. A gene for seed coat pattern has five alleles: spotted, dotted, clear (pattern absent), and two types of marbled. Reciprocal crosses between pairs of pure-breeding lines of all patterns (marbled-1 × marbled-2, marbled-1 × spotted, marbled-2 × spotted, and so forth) have clarified the dominance relations of all possible pairs of the alleles to reveal a **dominance series** in which alleles are listed in order from most dominant to most recessive. For example, crosses of marbled-1 with marbled-2, or of marbled-1 with spotted or dotted or clear, produce the marbled-1 phenotype in the F_1 generation and a ratio of three marbled-1 to one of any of the other phenotypes in the F_2. This indicates that the marbled-1 allele is dominant to each of the other four alleles. Analogous crosses with the remaining four phenotypes reveal the dominance series shown in Fig. 3.6. Recall that dominance relations are meaningful only when comparing two alleles; an allele, such as marbled-2, can be recessive to a second allele (marbled-1) but dominant to a third and fourth (dotted and clear). The fact that all tested pairings of lentil seed coat pattern alleles yielded a 3:1 ratio in the F_2 generation (except for spotted × dotted, which yielded the 1:2:1 phenotypic ratio reflective of codominance) indicates that these lentil seed coat patterns are determined by different alleles of the same gene.

In some multiple allelic series, each allele is codominant with every other allele, and every distinct genotype therefore produces a distinct phenotype. This happens particularly with traits detectable only at the molecular level. An extreme example is the group of three major genes that encode a class of cell surface molecules in humans and other mammals known as **histocompatibility antigens.** Carried by all of the body's cells, except the red blood cells and sperm, histocompatibility antigens play a critical role in stimulating a proper immune response that destroys intruders (viral or bacterial, for example) while leaving the body's own tissues intact. Because each of the three major histocompatibility genes (called *HLA-A, HLA-B,* and *HLA-C* in humans) has between 20 and 100 alleles (and these are just the ones geneticists have already discovered), the number of possible allelic combinations creates a powerful potential for the phenotypic variation of cell surface molecules. Other than identical (that is, *monozygotic*) twins, no two people are likely to carry the same array of cell surface molecules.

Mutations Are the Source of New Alleles

How do the multiple alleles of an allelic series arise? The answer is that chance alterations of the genetic material, known

Parental Generation Parental seed coat pattern in cross Parent 1 × Parent 2	F₁ Generation F₁ phenotype	F₂ Generation Total F₂ frequencies and phenotypes		Apparent phenotypic ratio
marbled-1 × clear	marbled-1	798	296	3 : 1
marbled-2 × clear	marbled-2	123	46	3 : 1
spotted × clear	spotted	283	107	3 : 1
dotted × clear	dotted	1,706	522	3 : 1
marbled-1 × marbled-2	marbled-1	272	72	3 : 1
marbled-1 × spotted	marbled-1	499	147	3 : 1
marbled-1 × dotted	marbled-1	1,597	549	3 : 1
marbled-2 × dotted	marbled-2	182	70	3 : 1
spotted × dotted	spotted/dotted	168 339 157		1 : 2 : 1

Dominance series: marbled-1 > marbled-2 > spotted = dotted > clear

Figure 3.6   How to establish the dominance relations between multiple alleles. Pure-breeding lentils with different seed coat patterns are crossed in pairs, and the F_1 progeny are self-fertilized to produce an F_2 generation. The 3:1 or 1:2:1 F_2 monohybrid ratios from all of these crosses indicate that different alleles of a single gene determine all the traits. The phenotypes of the F_1 hybrids establish the dominance relationships between the alleles, as shown at the *bottom*. Note that spotted and dotted alleles are codominant to each other, but each is recessive to the marbled alleles and dominant to clear.

as **mutations,** arise spontaneously in nature. Once they occur in gamete-producing cells, they are faithfully inherited. Mutations that have phenotypic consequences can be counted, and such counting reveals that they occur at low frequency. The frequency of gametes carrying a mutation in a particular gene varies anywhere from 1 in 10,000 to 1 in 1,000,000. This range of rates exists because different genes have different mutation rates. Mutations make it possible to follow gene transmission. If, for example, a mutation specifies an alteration in an enzyme that normally produces yellow so that it now makes green, the new phenotype (green) will make it possible to recognize the new mutant allele. In fact, it takes at least two alleles, that is, some form of variation, to "see" the transmission of a gene. Thus, in segregation studies, geneticists can only analyze genes with variants; they have no way of following a gene that comes in only one form. If all peas were yellow, Mendel would not have been able to decipher the transmission patterns of the gene for the seed color trait. We discuss mutations in greater detail in Chapter 7.

The survival of a new allele within a population—that is, within a group of organisms of the same species that inhabit the same area—depends on many factors, particularly, its contribution to the reproductive success of the organism. Alleles that confer a survival advantage allow individuals carrying them to produce more offspring; these alleles will tend to increase in the population. Other alleles are deleterious and are lost from the population. Still other alleles are neutral; more than 99.9% of these are eventually lost from a large population because they provide no selective advantage, but occasionally, purely by chance, one may become established.

Since each organism carries two copies of every gene, you can calculate the number of copies of a gene in a given population by multiplying the number of individuals by 2. Each allele of the gene accounts for a percentage of the total number of gene copies, and that percentage is known as the **allele frequency.** An allele whose frequency is greater than 1% is by definition a **wild-type allele,** often designated by a superscript plus sign ($^+$). An allele with a frequency of less than 1% is considered a **mutant allele.** (A mutation is a newly arisen mutant allele.) In mice, for example, one of the main genes determining coat color is the *agouti* gene. The wild-type allele (*A*) produces fur with each hair having yellow and black bands that blend together from a distance to give the appearance of dark gray, or agouti. Researchers have identified in the laboratory 14 distinguishable mutant alleles for the *agouti* gene. One of these (*a^t*) is recessive to the wild type and gives rise to a black coat on the back and a yellow coat on the belly; another (*a*) is also recessive to *A* and produces a pure black coat (Fig. 3.7). In nature, wild-type agoutis (*AA*) survive to reproduce, while very few black-backed or pure black mutants (*a^t a^t* or *aa*) do so because their dark coat makes it hard for them to evade the eyes of predators. As a result, *A* is present at a frequency of much more than 99% and is thus the only wild-type allele in mice for the *agouti* gene. A gene with only one wild-type allele is **monomorphic.**

In contrast, some genes have more than one wild-type allele, which makes them **polymorphic.** For example, in the ABO blood type system, all three alleles—*I^A*, *I^B*, and *i*—have a frequency greater than 1%. They are thus all wild-type alleles and the gene is polymorphic.

A rather unusual mechanism leading to the proliferation of many different alleles occurs in the mating systems of wild species of tomatoes and petunias. Evolution of an "incompatibility" gene whose alleles determine acceptance or rejection of pollen has allowed these plants to prevent self-fertilization and

(a) *Mus musculus* **(house mouse) coat colors**

(b) Alleles of the *agouti* gene

Genotype	Phenotype
A–	agouti
a^t a^t	black/yellow
aa	black

(c) Evidence for a dominance series

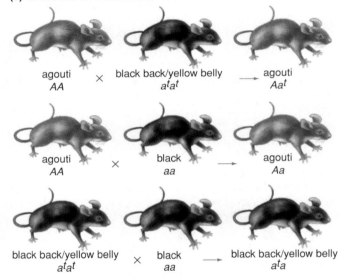

agouti
AA × black back/yellow belly
a^t a^t ⟶ agouti
Aa^t

agouti
AA × black
aa ⟶ agouti
Aa

black back/yellow belly
a^t a^t × black
aa ⟶ black back/yellow belly
a^t a

Dominance series: *A > a^t > a*

Figure 3.7 The mouse *agouti* gene: One wild-type allele, many mutant alleles. (a) Black-backed, yellow-bellied (*top left*); black (*top right*); and agouti (*bottom*) mice. **(b)** Genotypes and corresponding phenotypes for alleles of the *agouti* gene. **(c)** Crosses between pure-breeding lines make it possible to arrange these three alleles in a dominance series. Interbreeding of the F₁ hybrids (not shown) yields 3:1 phenotypic ratios of F₂ progeny, indicating that *A*, *a^t*, and *a* are in fact alleles of the same gene.

promote out-breeding. In this form of incompatibility, a plant cannot accept pollen carrying an allele identical to either of its own incompatibility alleles. If, for example, pollen carrying allele *S₁* of the incompatibility gene lands on the stigma of a plant that also carries *S₁* as one of its incompatibility alleles, a pollen tube will not grow (Fig. 3.8). As a result, every plant is heterozygous for the incompatibility gene (since the pollen grain and female reproductive organs needed to form the plant

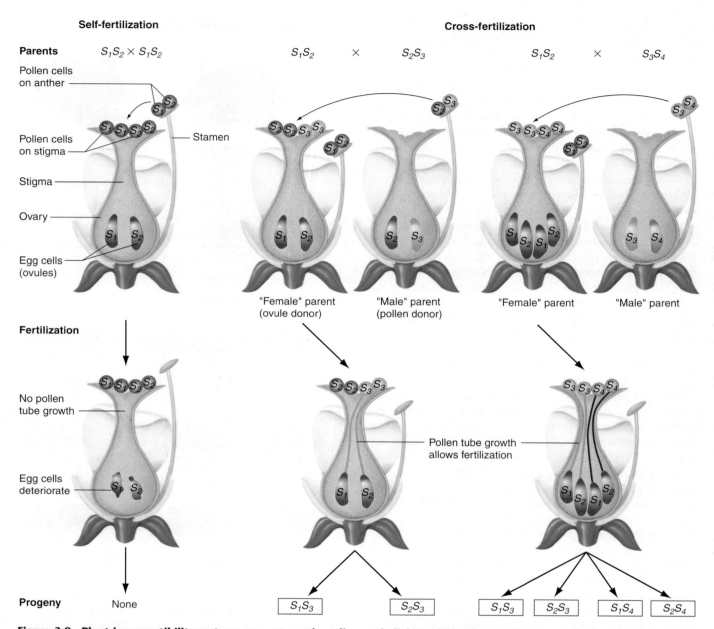

Figure 3.8 Plant incompatibility systems promote outbreeding and allele proliferation. A pollen grain carrying an allele of the self-incompatibility gene that is identical to either of the two alleles carried by a potential female parent is unable to grow a pollen tube; as a result, fertilization cannot take place. Because all the pollen grains produced by any one plant have one of the two alleles carried by the female reproductive parts of the same plant, self-fertilization is impossible.

cannot share alleles). Plants carrying rare alleles (that have arisen relatively recently by mutation and are not present in many other plants) will be able to send pollen to and receive pollen from most of the other plants in their population. In some species with this type of mating system, geneticists have detected as many as 92 alleles for the incompatibility gene, and there are probably many more alleles yet to be detected. Because this incompatibility mechanism encourages the proliferation of new mutants, there are numerous wild-type alleles as well as many mutants. This is an extreme case of multiple alleles, not seen with most genes.

One Gene May Contribute to Several Visible Characteristics

Mendel derived his laws from studies in which one gene determined one trait; but, always the careful observer, he himself noted possible departures. In listing the traits selected for his pea experiments, he remarked that specific seed coat colors are always associated with specific flower colors.

The phenomenon of a single gene determining a number of distinct and seemingly unrelated characteristics is known as **pleiotropy.** Since geneticists now know that each gene determines a specific protein and that each protein can have a cascade of effects on an organism, we can understand how pleiotropy arises. Among the aboriginal Maori people of New Zealand, for example, many of the men develop frequent respiratory problems and are also sterile. Researchers have found that the fault lies with the recessive allele of a single gene. The gene's normal dominant allele specifies a protein necessary for the action of cilia and flagella, both of which are hairlike structures extending from the surfaces of some cells. In men who are homozygous for the recessive allele, however, cilia that normally clear the airways fail to work effectively and flagella that normally propel sperm fail to do their job. Thus, one gene determines a protein that indirectly affects both respiratory function and reproduction.

Some Alleles May Cause Lethality

A significant variation of pleiotropy occurs in alleles that not only produce a visible phenotype but also affect viability. Mendel assumed that all genotypes are equally viable such that representatives of each genotype have an equal rate of survival. If this were not true, and a large percentage of, say, homozygotes for a particular allele died before germination or birth, you would not be able to count them after birth, and this would alter the 1:2:1 genotypic ratios and the 3:1 phenotypic ratios predicted for the F_2 generation.

Consider the inheritance of coat color in mice. As mentioned earlier, wild-type agouti (AA) animals have black and yellow striped hairs that appear dark gray to the eye. One of the 14 mutant alleles of the *agouti* gene gives rise to mice with a much lighter, almost yellow color. When inbred AA mice are mated to yellow mice, one always observes a 1:1 ratio of the two coat colors among the offspring (Fig. 3.9a). From this result, we can draw three conclusions: (1) All yellow mice must carry the *agouti* allele even though they do not express it; (2) yellow is therefore dominant to agouti; and (3) all yellow mice are heterozygotes. (Note again that dominance and recessiveness are defined in the context of each pair of alleles. Even though agouti (A) is dominant to the a^t and a mutations for black coat color, it can still be recessive to the yellow coat color allele.) If we designate the allele for yellow as A^y, the yellow mice in the preceding cross are A^yA heterozygotes, and the agoutis, AA homozygotes. So far, no surprises. But a mating of yellow to yellow produces a skewed phenotypic ratio of two yellow mice to one agouti (Fig. 3.9b). Among these progeny, matings between agouti mice show that the agoutis are all pure-breeding and therefore AA homozygotes as expected. There are, however, no pure-breeding yellow mice among the progeny. When the yellow mice are mated to each other, they unfailingly produce 2/3 yellow and 1/3 agouti offspring, a ratio of 2:1; they must therefore be heterozygotes. In short, one can never obtain pure-breeding yellow mice.

How can we explain this phenomenon? The Punnett square in Fig. 3.9b suggests one possibility: Two copies of the A^y allele prove fatal to the animal carrying them, whereas one copy of the allele produces a yellow coat. This means that the A^y allele affects two different traits: It is dominant to A in the determi-

(a) All yellow mice are heterozygotes.

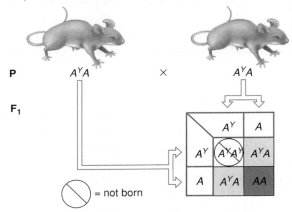

(b) Two copies of A^Y cause lethality.

Figure 3.9 A^Y**: A recessive lethal allele that also produces a dominant coat color phenotype. (a)** A cross between inbred agouti mice and yellow mice yields a 1:1 ratio of yellow to agouti progeny. The yellow mice are therefore A^YA heterozygotes, and for the trait of coat color, A^Y (for yellow) is dominant to A (for agouti). **(b)** Yellow mice do not breed true. In a yellow × yellow cross, the 2:1 ratio of yellow to agouti progeny indicates that the A^Y allele is a recessive lethal.

nation of coat color, but it is recessive to A in the production of lethality. An allele, such as A^y, that negatively affects the survival of a homozygote is known as a **recessive lethal allele.**

Because the A^y allele is dominant for yellow coat color, it is easy to detect carriers of this particular recessive lethal allele in mice, but such is not the case for the vast majority of recessive lethal mutations that do not simultaneously show a visible dominant phenotype for some other trait. Lethal mutations can arise in many different genes, and as a result, most animals, including humans, carry some recessive lethal mutations. Such mutations usually remain "silent," except in rare cases of homozygosity, which in people are often caused by consanguineous matings. If a mutation produces an allele that prevents production of a crucial molecule, homozygous individuals would not make any of the vital molecule and would not survive. Heterozygotes, by contrast, with only one copy of the deleterious mutation and one wild-type allele, would be able to produce 50% of the wild-type amount of the normal molecule; this is usually sufficient to sustain normal cellular processes such that life goes on.

TABLE 3.1 For Traits Determined by One Gene: Extensions to Mendel's Analysis Explain Alterations of the 3:1 Monohybrid Ratio

What Mendel Described	Extension	Extension's Effect on Heterozygous Phenotype	Extension's Effect on Ratios Resulting from an $F_1 \times F_1$ Cross
Complete dominance	Incomplete dominance Codominance	Unlike either homozygote	Phenotypes coincide with genotypes in a ratio of 1:2:1
Two alleles	Multiple alleles	Multiplicity of phenotypes	A series of 3:1 ratios
All alleles are equally viable	Recessive lethal alleles	No effect	2:1 instead of 3:1
One gene determines one trait	Pleiotropy: one gene influences several traits	Several traits affected in different ways, depending on dominance relations	Different ratios, depending on dominance relations for each affected trait

In the preceding discussion, we have described recessive alleles that result in the death of homozygotes prenatally, *in utero*. With some mutations, however, homozygotes may survive beyond germination or birth and then die later from the deleterious consequences of the genetic defect. An example is seen in human infants with Tay-Sachs disease. The seemingly normal newborns remain healthy for five to six months but then develop blindness, paralysis, mental retardation, and other symptoms of a deteriorating nervous system; the disease usually proves fatal by the age of six. Tay-Sachs disease results from the absence of an active lysosomal enzyme called hexosaminidase A, whose lack leads to the accumulation of a toxic waste product inside nerve cells. The approximate incidence of Tay-Sachs among live births is 1/35,000 worldwide, but it is 1/3000 among Jewish people of Eastern European descent. Reliable tests that detect carriers, in combination with genetic counseling and educational programs, have all but eliminated the disease in the United States.

Table 3.1 summarizes Mendel's basic assumptions about dominance, the number and viability of one gene's alleles, and the effects of each gene on phenotype, and then compares these assumptions with the extensions contributed by his twentieth-century successors. Through carefully controlled monohybrid crosses, these later geneticists analyzed the transmission patterns of the alleles of single genes, challenging and then confirming the law of segregation.

A Comprehensive Example: Sickle-Cell Anemia Illustrates Many Extensions to Mendel's Analysis of Single-Gene Inheritance

Sickle-cell anemia can serve as a comprehensive example of most of the points covered in this section on single-gene traits. The disease is the result of a faulty hemoglobin molecule. Hemoglobin is composed of two types of polypeptide chains, alpha (α) globin and beta (β) globin, each specified by a different gene: $Hb\alpha$ for α globin and $Hb\beta$ for β globin.

Normal red blood cells are packed full of millions upon millions of hemoglobin molecules, each of which picks up oxygen in the lungs and transports it to all the body's tissues.

Multiple Alleles
The β-globin gene has a normal wild-type allele ($Hb\beta^A$) that gives rise to fully functional β globin, as well as close to 400 mutant alleles that have been identified so far. The most common of these, $Hb\beta^S$, specifies an abnormal polypeptide that causes sickling among red blood cells (Fig. 3.10a). For simplicity, in the remainder of this discussion, we refer to the wild-type β-globin allele as A and the mutant sickling allele of the same $Hb\beta$ gene as S.

Pleiotropy
The S allele of the β-globin gene affects more than one trait (Fig. 3.10b). Hemoglobin molecules in the red blood cells of homozygous SS individuals behave aberrantly after releasing their oxygen. Instead of remaining soluble in the cytoplasm, they aggregate to form long fibers that deform the red blood cell from a normal biconcave disk to a sickle shape (see Fig. 3.10a). The deformed cells clog the small blood vessels, reducing oxygen flow to the tissues and giving rise to muscle cramps, shortness of breath, and fatigue. The sickled cells are also very fragile and easily broken. Consumption of fragmented cells by phagocytic white blood cells leads to a low red blood cell count, a condition called anemia. On the positive side, SS homozygotes are resistant to malaria, because although the organism that causes the disease, *Plasmodium falciparum*, can multiply with impunity in normal red blood cells, it cannot do so in cells that sickle. Infection by *P. falciparum* causes sickle-shaped cells to break down before the malaria organism has a chance to multiply.

Recessive Lethality
People who are homozygous for the recessive S allele often develop heart failure because of stress on the circulatory sys-

(a)

Phenotypes at Different Levels of Analysis	Normal AA	Carrier AS	Diseased SS	Dominance Relations at Each Level of Analysis
β-globin polypeptide production				*A* and *S* are codominant
Red blood cell shape at sea level	Normal	Normal	Sickled cells present	*A* is dominant *S* is recessive
Red blood cell concentration at sea level	Normal	Normal	Lower	
Red blood cell shape at high altitudes	Normal	Sickled cells present	Severe sickling	*A* and *S* show incomplete dominance
Red blood cell concentration at high altitudes	Normal	Lower	Very low, anemia	
Susceptibility to malaria	Normal susceptibility	Resistant	Resistant	*S* is dominant *A* is recessive

(b)

Figure 3.10 Pleiotropy of sickle-cell anemia: Dominance relations vary with the phenotype under consideration. (a) A normal red blood cell (*right*) is easy to distinguish from the sickled cell in the scanning electron micrograph at the *left*. **(b)** Different levels of analysis identify various phenotypes. Dominance relationships between the *S* and *A* alleles of the *Hbβ* gene vary with the phenotype and sometimes even change with the environment.

tem. Many sickle-cell sufferers die in childhood, adolescence, or early adulthood.

Different Dominance Relations

Comparisons of heterozygous carriers of the sickle-cell allele—individuals whose cells contain one *A* and one *S* allele—with homozygous *AA* (normal) and homozygous *SS* (diseased) individuals make it possible to distinguish different dominance relations for different phenotypic aspects of sickle-cell anemia (Fig. 3.10b). At the molecular level—the production of β globin—both alleles are expressed such that *A* and *S* are *codominant.* At the cellular level, in their effect on red blood cell shape, the *A* and *S* alleles show *incomplete dominance.* Although under normal oxygen conditions, the great majority of a heterozygote's red blood cells have the normal biconcave shape, when oxygen levels drop, sickling occurs in some cells. All *AS* cells, however, are resistant to malaria because like the *SS* cells described previously, they break down before the malarial organism has a chance to reproduce. Thus for the trait of resistance to malaria, the *S* allele is *dominant* to the *A* allele. But luckily for the heterozygote, for the phenotype of anemia, *S* is *recessive* to *A.* A corollary of this observation is that in its effect on general health under normal environmental conditions and its effect on red blood cell count, the *A* allele is *dominant* to *S.* Thus, for the β-globin gene, as for many other

genes, dominance and recessiveness are not an inherent quality of alleles in isolation; rather, they are specific to each pair of alleles and to the level of physiology at which the phenotype is examined.

In the 1940s, the incomplete dominance of the *A* and *S* alleles in the expression of red blood cell shape had significant repercussions for certain soldiers who fought in World War II. Aboard transport planes flying troops across the Pacific, several heterozygous carriers suffered sickling crises similar to those usually seen in *SS* homozygotes. The explanation is as follows. The heterozygous red blood cells of a carrier produce both normal and abnormal hemoglobin molecules. At sea level, these molecules together deliver sufficient oxygen, although less than the normal amount, to the body's tissues; but with a decrease in the amount of oxygen available at the high-flying altitudes, the hemoglobin picks up less oxygen, the rate of red blood cell sickling increases, and symptoms of the disease occur.

Extensions to Mendel for Multifactorial Inheritance

Although some traits are indeed determined by allelic variations of a single gene, the vast majority of common traits in all organisms are *multifactorial,* arising from the action of two or more genes, or from interactions between genes and the environment. In genetics, the term *environment* has an unusually broad meaning that encompasses all aspects of the outside world an organism comes into contact with. These include temperature, diet, and exercise as well as the uterine environment before birth (minute chemical differences, for example, can make a difference in development) and the psychological environment afterward (the amount of stress, for example, can affect various levels of phenotype).

In this section, we examine how geneticists again used breeding experiments and the guidelines of Mendelian ratios to analyze the complex network of interactions that give rise to multifactorial traits.

Two Genes Can Interact to Determine One Trait

Two genes can interact in several ways to determine a single trait, such as the color of a flower, a seed coat, a chicken's feathers, or a dog's fur, and each type of interaction produces its own signature of phenotypic ratios. Throughout our discussion of how two genes interact to affect one trait, we use big *A* and little *a* to represent alternative alleles of the first gene and big *B* and little *b* for those of the second gene.

Novel Phenotypes Can Emerge from the Combined Action of the Alleles of Two Genes

In the chapter opening we described a mating of tan and gray lentils that produced a uniformly brown F_1 generation, and then produced an F_2 generation containing lentils with brown, tan, gray, and green seed coats. An understanding of how this can happen emerges from experimental results demonstrating that the ratio of the four F_2 colors is 9 brown : 3 tan : 3 gray : 1 green (Fig. 3.11a). Recall from Chapter 2 that this is the same ratio Mendel observed in his analysis of the F_2 generations from dihybrid crosses following two independently assorting genes. In Mendel's studies, each of the four classes consisted of plants that expressed a combination of two unrelated traits. With lentils, however, we are looking at a single trait—seed coat color. The simplest explanation for the parallel ratios is that a combination of genotypes at two independently assorting genes interacts to produce the phenotype of seed coat color in lentils.

Results obtained from self-crosses with the various types of F_2 lentil plants support this explanation. Self-crosses of F_2 green individuals show that they are pure-breeding, producing an F_3 generation that is entirely green. Tans generate either all tan or tan plus green; grays similarly produce all gray or gray plus green; and browns either breed true, or instead they assort into two col-

(a) A dihybrid cross with lentil coat colors

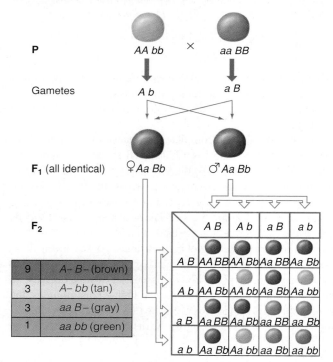

(b) Self-pollination of the F_2 to produce an F_3

Phenotypes of F_2 individual	Observed F_3 phenotypes	Expected proportion of F_2 population*
Green	Green	1/16
Tan	Tan	1/16
Tan	Tan, green	2/16
Gray	Gray, green	2/16
Gray	Gray	1/16
Brown	Brown	1/16
Brown	Brown, tan	2/16
Brown	Brown, gray	2/16
Brown	Brown, gray, tan, green	4/16

*This 1:1:2:2:1:1:2:2:4 F_2 genotypic ratio corresponds to a 9 brown:3 tan:3 gray:1 green F_2 phenotypic ratio.

(c) Sorting out the dominance relations by select crosses

Seed coat color of parents	F_2 phenotypes and frequencies	Ratio
Tan × green	231 tan, 85 green	3:1
Gray × green	2586 gray, 867 green	3:1
Brown × gray	964 brown, 312 gray	3:1
Brown × tan	255 brown, 76 tan	3:1
Brown × green	57 brown, 18 gray, 13 tan, 4 green	9:3:3:1

Figure 3.11 How two genes interact to produce new colors in lentils. (a) In a cross of pure-breeding tan and gray lentils, all the F_1 hybrids are brown, but four different phenotypes appear among the F_2 progeny. The 9:3:3:1 ratio of F_2 phenotypes suggests that seed coat color is determined by two independently assorting genes in a dihybrid cross. **(b)** Expected results of selfing individual F_2 plants of the indicated phenotypes to produce an F_3 generation, if seed coat color is determined by the interaction of two genes. The third column shows the proportion of the F_2 population that would be expected to produce the observed F_3 phenotypes. **(c)** Other two-generation crosses involving pure-breeding parental lines also support the two-gene hypothesis. In this table, the F_1 hybrid generation has been omitted.

ors (either tan plus brown or gray plus brown), or into all four colors (Fig. 3.11b). The two-gene hypothesis explains why there is

- only one green genotype: pure-breeding *aa bb,* but
- two types of tans: pure-breeding *AA bb* as well as tan- and green-producing *Aa bb,* and
- two types of grays: pure-breeding *aa BB* and gray- and green-producing *aa Bb,* yet
- four types of browns: true-breeding *AA BB,* brown- and tan-producing *AA Bb,* brown- and gray-producing *Aa BB,* and *Aa Bb* dihybrids that give rise to plants producing lentils of all four colors.

In short, for the two genes that determine seed coat color, both dominant alleles must be present to yield brown (*A– B–*); the dominant allele of one gene produces tan (*A– bb*); the dominant allele of the other specifies gray (*aa B–*); and the complete absence of dominant alleles (that is, the double recessive) yields green (*aa bb*). Thus, the four color phenotypes arise from four **genotypic classes,** with each class defined in terms of the presence or absence of the dominant alleles of two genes: (1) both present (*A– B–*), (2) one present (*A– bb*), (3) the other present (*aa B–*), and (4) neither present (*aa bb*). Note that the *A–* notation means that the second allele of this gene can be either *A* or *a,* while *B–* denotes a second allele of either *B* or *b.* Note also that only with a two-gene system in which the dominance and recessiveness of alleles at both genes is complete can the nine different genotypes of the F₂ generation be categorized into the four genotypic classes described. With incomplete dominance or codominance, the F₂ genotypes could not be grouped together in this simple way, as they would give rise to more than four phenotypes.

Further crosses between plants carrying lentils of different colors confirmed the two-gene hypothesis (review Fig. 3.11c). Thus, the 9:3:3:1 phenotypic ratio of brown to tan to gray to green in an F₂ descended from pure-breeding tan and pure-breeding gray lentils tells us not only that two genes assorting independently interact to produce the seed coat color, but also that each genotypic class (*A– B–, A– bb, aa B–,* and *aa bb*) determines a particular phenotype.

This is not always the case. In some two-gene interactions, the four F₂ genotypic classes defined by Mendel produce fewer than four observable phenotypes, because some of the phenotypes include two or more genotypic classes. For example, in the first decade of the twentieth century, William Bateson conducted a cross between two lines of pure-breeding white-flowered sweet peas (Fig. 3.12). Quite unexpectedly, all of the F₁ progeny were

(a) *Lathyrus odoratus* **(sweet peas)**

(b) A dihybrid cross involving complementary gene action

(c) A biochemical model for purple pigment production

Figure 3.12 Complementary gene action generates color in sweet peas. (a) White and purple sweet pea flowers. **(b)** The 9:7 ratio of purple to white F₂ plants indicates that at least one dominant allele for each gene is necessary for the development of purple color. **(c)** A possible biochemical explanation: Enzymes specified by the dominant alleles of the two genes may both be necessary for completion of a biochemical pathway for pigment production. The recessive alleles of both genes specify inactive enzymes.

purple. Self-pollination of these novel hybrids produced a ratio of 9 purple : 7 white in the F_2 generation. The explanation? Two genes work in tandem to produce purple sweet-pea flowers, and a dominant allele of both genes must be present to produce that color. A simple biochemical explanation for this type of **complementary gene action** is shown in Fig. 3.12c. Because it takes two enzymes catalyzing two separate biochemical reactions to change a colorless precursor into a colorful pigment, only the *A– B–* genotypic class, which produces active forms of both required enzymes, can generate colored flowers. The other three genotypic classes (*A– bb*, *aa B–*, and *aa bb*) become grouped together with respect to phenotype since they do not specify functional forms of one or the other requisite enzyme and thus give rise to no color, which is the same as white (review the Fast Forward box "Genes Encode Proteins" in Chapter 2). It is easy to see how the "7" part of the 9:7 ratio encompasses the 3:3:1 of the 9:3:3:1 ratio of two genes in action. The 9:7 ratio is the phenotypic signature of this type of complementary gene interaction in which the dominant alleles of two genes acting together (*A– B–*) produce color or some other trait, while the other three genotypic classes (*A– bb*, *aa B–*, and *aa bb*) do not.

In Epistasis, One Gene's Alleles Mask the Effects of Another Gene's Alleles

In some gene interactions, the four Mendelian genotypic classes produce fewer than four observable phenotypes because one gene masks the phenotypic effects of another. An example is seen in the sleek, short-haired coat of Labrador retrievers, which can be black, chocolate brown, or golden yellow. Which color shows up depends on the allelic combinations of two independently assorting coat color genes (Fig. 3.13a). The dominant *B* allele of the first gene determines black, while the recessive *bb* homozygote is brown. With the second gene, the dominant *E* allele has no visible effect on black or brown coat color, but a double dose of the recessive allele (*ee*) hides the effect of any combination of the black or brown alleles to yield gold. A gene interaction in which the effects of an allele at one gene hide the effects of alleles at another gene is known as **epistasis;** the allele that is doing the masking (in this case, the *e* allele of the *E* gene) is **epistatic** to the gene that is being masked (the *hypostatic gene*). In our example, where homozygosity for a recessive *e* allele of the second gene is required to hide the effects of another gene, the masking phenomenon is called **recessive epistasis** (because the allele causing the epistasis is recessive), and the recessive *ee* homozygote is considered epistatic to any allelic combination at the first gene.

Let's look at the phenomenon in greater detail. Crosses between pure-breeding black retrievers (*BB EE*) and one type of pure-breeding golden retriever (*bb ee*) create an F_1 generation of dihybrid black retrievers (*Bb Ee*). Crosses between these F_1 dihybrids produce an F_2 generation with nine black dogs (*B– E–*) for every three brown (*bb E–*) and four gold (*– – ee*) (Fig. 3.13a). Note that there are only three phenotypic classes

(a) A dihybrid cross showing recessive epistasis

(b) Molecular basis of the Bombay phenotype

Figure 3.13 Recessive epistasis: Coat color in Labrador retrievers and a rare human blood type. (a) Golden Labrador retrievers are homozygous for the recessive *e* allele, which masks the effects of the *B* or *b* alleles of a second coat color gene. In *E–* dogs, a *B–* genotype produces black and a *bb* genotype produces brown. **(b)** Homozygosity for the *h* Bombay allele is epistatic to the *I* gene determining ABO blood types. *hh* individuals fail to produce substance H, which is needed for the addition of A or B sugars at the surface of red blood cells.

because the two genotypic classes without a dominant *E* allele—the three *B– ee* and the one *bb ee*—combine to produce golden phenotypes. The telltale ratio of recessive epistasis in the F$_2$ generation is thus 9:3:4, with the 4 representing a combination of 3 (*B– ee*) + 1 (*bb ee*). Because the *ee* genotype completely masks the influence of the other gene for coat color, you cannot tell by looking at a golden Labrador what its genotype is for the black or brown (*B* or *b*) gene.

An understanding of recessive epistasis made it possible to resolve an intriguing puzzle in human genetics. In rare instances, two parents who appear to have blood type O, and thus genotype *ii*, produce a child who is either blood type A (genotype *I^{A}i*) or blood type B (genotype *I^{B}i*). This phenomenon occurs because an extremely rare trait, called the Bombay phenotype after its discovery in Bombay, India, superficially resembles blood type O. As Fig. 3.13b shows, the Bombay phenotype actually arises from homozygosity for a mutant recessive allele (*hh*) of a second gene that masks the effects of any ABO alleles that might be present.

Here's how it works at the molecular level. In the construction of the red blood cell surface molecules that determine blood type, type A individuals make an enzyme that adds polysaccharide A onto a base consisting of a sugar polymer known as substance H; type B individuals make an altered form of the enzyme that adds polysaccharide B onto the base; and type O

individuals make neither A-adding nor B-adding enzyme and thus have an exposed substance H in the membranes of their red blood cells. All people of A, B, or O phenotype carry at least one dominant wild-type *H* allele for the second gene and thus produce some substance H. In contrast, the rare Bombay-phenotype individuals, with genotype *hh* for the second gene, do not make substance H at all, so even if they make an enzyme that would add A or B to this polysaccharide base, they have nothing to add it onto; as a result, they appear to be type O. For this reason homozygosity for the recessive *h* allele of the H-substance gene masks the effects of the *ABO* gene, making the *hh* genotype epistatic to any combination of *I^A*, *I^B*, and *i* alleles. A person who carries *I^A*, *I^B*, or both *I^A* and *I^B* but is also an *hh* homozygote for the H-substance gene may appear to be type O, but he or she will be able to pass along an *I^A* or *I^B* allele in sperm or egg. The offspring receiving, let's say, an *I^A* allele for the ABO gene and a recessive *h* allele for the H-substance gene from its mother plus an *i* allele and a dominant *H* allele from its father would have blood type A (genotype *I^{A}i, Hh*), even though neither of its parents is phenotype A or AB.

Epistasis can also be caused by a dominant allele. In summer squash, two genes influence the color of the fruit (Fig. 3.14a). With one gene, the dominant allele (*A–*) determines yellow while homozygotes for the recessive allele (*aa*) are green. A second gene's dominant allele (*B–*) produces white, while *bb* fruit may be

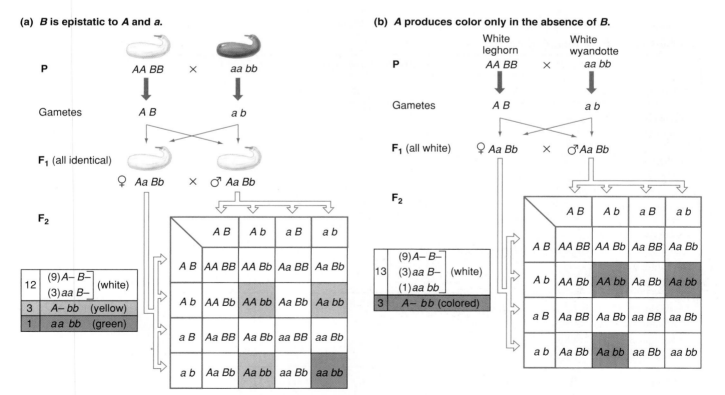

Figure 3.14   Dominant epistasis produces telltale phenotypic ratios of 12:3:1 or 13:3. (a) In summer squash, the dominant *B* allele causes white color and is sufficient to mask the effects of any combination of *A* and *a* alleles. As a result, yellow (*A–*) or green (*aa*) color is expressed only in *bb* individuals. **(b)** In the F$_2$ generation resulting from a dihybrid cross between white leghorn and white wyandotte chickens, the ratio of white birds to birds with color is 13:3. This is because at least one copy of *A* and the absence of *B* is needed to produce color.

either yellow or green, depending on the genotype of the first gene. In the interaction between these two genes, the presence of *B* hides the effects of either *A–* or *aa*, producing white fruit, and *B–* is thus epistatic to any genotype of the *Aa* gene. The recessive *b* allele has no effect on fruit color determined by the *Aa* gene. Epistasis in which the dominant allele of one gene hides the effects of another gene is called **dominant epistasis.** In a cross between white F₁ dihybrids (*Aa Bb*), the F₂ phenotypic ratio is 12 white : 3 yellow : 1 green (Fig. 3.14a). The "12" includes two genotypic classes: 9 *A– B–* and 3 *aa B–*. Another way of looking at this same phenomenon is that dominant epistasis restores the 3:1 ratio for the dominant epistatic phenotype (12 white) versus all other phenotypes (4 green plus yellow).

A variation of this ratio is seen in the feather color of certain chickens (Fig. 3.14b). White leghorns have a doubly dominant *AA BB* genotype for feather color; white wyandottes are homozygous recessive for both genes (*aa bb*). A cross between these two pure-breeding white strains produces an all-white dihybrid (*Aa Bb*) F₁ generation, but birds with color in their feathers appear in the F₂, and the ratio of white to colored is 13:3 (Fig. 3.14b). We can explain this ratio by assuming a kind of dominant epistasis in which *B* is epistatic to *A;* the *A* allele (in the absence of *B*) produces color; and the *a*, *B*, and *b* alleles produce no color. The interaction is characterized by a 13:3 ratio because the 9 *A– B–*, 3 *aa B–*, and 1 *aa bb* genotypic classes combine to produce only one phenotype: white.

So far we have seen that when two independently assorting genes interact to determine a trait, the 9:3:3:1 ratio of the four Mendelian genotypic classes in the F₂ generation can produce a variety of phenotypic ratios, depending on the nature of the gene interactions. The result may be four, three, or two phenotypes, composed of different combinations of the four genotypic classes. Table 3.2 summarizes some of the possibilities, correlating the phenotypic ratios with the genetic phenomena they reflect.

For Some Traits, Homozygosity for a Mutant Allele at Any One of Two or More Genes Produces the Phenotype

Close to 50 different genes have mutant alleles that can cause deafness in humans. This is because it takes many genes to generate the developmental pathway that brings about hearing, and a loss of function in any part of the pathway, for instance, in one small bone of the middle ear, can result in deafness. In other words, it takes a dominant wild-type allele at each of these 50 genes to produce normal hearing. Thus, deafness is a **heterogeneous trait:** A mutation at any one of a number of genes can give rise to the same phenotype.

It is not always possible to determine which of many different genes has mutated in a person who expresses a heterogeneous mutant phenotype. In the case of deafness, for example, it is usually not possible to discover whether a particular nonhearing

TABLE 3.2 **Summary of Discussed Gene Interactions**

| Gene Interaction | Example | F₂ Genotypic Ratios from an F₁ Dihybrid Cross | | | | F₂ Phenotypic Ratio |
		A– B–	A– bb	aa B–	aa bb	
None: Four distinct F₂ phenotypes	Lentil: seed coat color (see Fig. 3.11)	9	3	3	1	9:3:3:1
Complementary: One dominant allele of each of two genes is necessary to produce phenotype	Sweet pea: flower color (see Fig. 3.12b)	9	3	3	1	9:7
Recessive epistasis: Homozygous recessive of one gene masks both alleles of another gene	Retriever: coat color (see Fig. 3.13a)	9	3	3	1	9:3:4
Dominant epistasis I: Dominant allele of one gene hides effects of both alleles of another gene	Summer squash: color (see Fig. 3.14a)	9	3	3	1	12:3:1
Dominant epistasis II: Dominant allele of one gene hides effects of dominant allele of another gene	Chicken: feather color (see Fig. 3.14b)	9	3	3	1	13:3

man and a particular nonhearing woman carry mutations at the same gene, unless they have children together. If they have only children who can hear, they most likely carry mutations at two different genes, and the children carry one normal, wild-type allele at both of those genes (Fig. 3.15a). By contrast, if all of their children are deaf, it is likely that both parents are homozygous for a mutation in the same gene, and all of their children are also homozygous for this same mutation (Fig. 3.15b).

This method of discovering whether a particular phenotype arises from mutations in the same or separate genes is a naturally occurring version of an experimental genetic tool called the **complementation test.** Simply put, when what appears to be an identical *recessive* phenotype arises in two separate breeding lines, geneticists want to know whether mutations at the same gene are responsible for the phenotype in both lines. They answer this question by setting up a mating between individuals from the two lines in which the recessive mutations arose independently. If offspring receiving the two mutations—one from each parent—express the wild-type phenotype, complementation has occurred. The observation of complementation means that the original mutations affected two different genes, and for both genes, the normal allele from one parent can provide what the mutant allele of the same gene from the other parent cannot. Figure 3.15a illustrates one example of this phenomenon in humans. By contrast, if offspring receiving two recessive mutant alleles—again, one from each parent—express the mutant phenotype, complementation does not occur because the two mutations independently alter the same gene

(review Fig. 3.15b). Thus, the occurrence of complementation reveals genetic heterogeneity. Note that complementation tests cannot be used if either of the mutations is dominant to the wild type. Chapter 7 includes an in-depth discussion of complementation tests and their uses.

In medical genetics, where researchers and doctors are always working with undefined crosses, it is very important to know that a particular mutant phenotype can result from a mutation at one of several genes. The treatment of choice often depends on which gene is involved. For example, severe mental retardation in children is a heterogeneous trait that can be caused by many different factors, both genetic and environmental. One cause is homozygosity for a mutation that eliminates the activity of a gene encoding an enzyme known as phenylalanine hydroxylase. The enzyme normally converts the amino acid phenylalanine to tyrosine, another amino acid. Absence of the enzyme causes a buildup of phenylalanine, and after birth this buildup interferes with normal brain development, which in turn leads to mental retardation. Children born without the ability to make phenylalanine hydroxylase suffer from a genetic condition known as phenylketonuria, or PKU, which grows progressively worse in the early years of life. If a medical practitioner knows a newborn has the particular mutant genotype that gives rise to PKU, he or she can recommend the elimination of phenylalanine from the infant's diet. This dietary therapy can prevent mental retardation. Treating newborns affected by other causes of mental retardation in the same way would have no positive effect.

Figure 3.15 Genetic heterogeneity in humans: Mutations in many genes can cause deafness. Hearing requires the coordinated function of many complex structures within the ear. **(a)** Two deaf parents can have hearing offspring. This situation is an example of genetic complementation; it occurs if the nonhearing parents are homozygous for recessive mutations in different genes. **(b)** Two deaf parents may produce all deaf children. This happens when complementation does not occur because both parents carry mutations in the same gene.

In sum, there are several genetic variations on the theme of multifactorial traits: (1) genes can interact to generate novel phenotypes, (2) the dominant alleles of two interacting genes can both be necessary for the production of a particular phenotype, (3) one gene's alleles can mask the effects of another's, and (4) mutant alleles at one of two or more different genes can result in the same phenotype. In examining each of these categories, for the sake of simplicity, we have looked at examples in which one allele of each gene in a pair showed complete dominance over the other. But for any type of gene interaction, the alleles of one or both genes may exhibit incomplete dominance or codominance, and these possibilities increase the potential for phenotypic diversity. For example, Fig. 3.16 shows how incomplete dominance at both genes in a dihybrid cross generates additional phenotypic variation.

Although the possibilities for variation are manifold, none of the observed departures from Mendelian phenotypic ratios contradicts Mendel's genetic laws of segregation and independent assortment. The alleles of each gene still segregate as he proposed. Interactions between the alleles of many genes sim-

ply make it harder to unravel the complex relation of genotype to phenotype.

Breeding Studies Help Decide How a Trait Is Inherited

How do you know whether a particular trait is caused by the alleles of one gene or by two genes interacting in one of a number of possible ways? Breeding tests can usually resolve the issue. Phenotypic ratios diagnostic of a particular mode of inheritance (for instance, the 9:7 or 13:3 ratios indicating that two genes are interacting) can provide the first clues and suggest hypotheses. Further breeding studies can then show which hypothesis is correct. We have seen, for example, that yellow coat color in mice is determined by a dominant allele of the *agouti* gene, which also acts as a recessive lethal. We now look at two other mouse genes for coat color. Since we have already designated alleles of the *agouti* gene as *Aa*, we use *Bb* and *Cc* to designate the alleles of these additional genes.

A mating of one strain of pure-breeding white albino mice with pure-breeding brown results in black hybrids; and a cross between the black F$_1$ hybrids produces 90 black, 30 brown, and 40 albino offspring. What is the genetic constitution of these phenotypes? We could assume that we are seeing the 9:3:4 ratio of recessive epistasis and hypothesize that two genes, one epistatic to the other, interact to produce the three mouse phenotypes (Fig. 3.17a). But how do we know if this hypothesis is correct? We might also explain the data—160 progeny in a ratio of 90:30:40—by the activity of one gene (Fig. 3.17b). According to this one-gene hypothesis, albinos would be homozygotes for one allele (*bb*), brown mice would be homozygotes for a second allele (*BB*), and black mice would be heterozygotes (*Bb*) that have their own "intermediate" phenotype because *B* shows incomplete dominance over *b*. Under this system, a mating of black (*Bb*) to black (*Bb*) would be expected to produce 1 *BB* brown : 2 *Bb* black : 1 *bb* albino, or 40 brown : 80 black : 40 albino. Is it possible that the 30 brown, 90 black, and 40 albino mice actually counted were obtained from the inheritance of a single gene? Intuitively, the answer is yes: the ratios 40:80:40 and 30:90:40 do not seem that different. We know that if we flip a coin 100 times, it doesn't always come up 50 heads : 50 tails; sometimes it's 60:40 just by chance. So, how can we decide between the two-gene versus the one-gene model?

The answer is that we can use other types of crosses to verify or refute the hypotheses. For instance, if the one-gene hypothesis were correct, a mating of pure white F$_2$ albinos with pure-breeding brown mice similar to those of the parental generation would produce all black heterozygotes (brown [*BB*] × albino [*bb*] = all black [*Bb*]) (Fig. 3.17b). But if the two-gene hypothesis is correct, with recessive mutations at an albino gene (called *C*) epistatic to all expression from the *B* gene, different matings of pure-breeding brown (*bb CC*) with the F$_2$ albinos (*– – cc*) will give different results—all progeny are black; half are black and half brown; all are brown—depending on the albino's genotype at the *B* gene (see Fig. 3.17a). In fact, when the experiment is actually performed, the diversity of

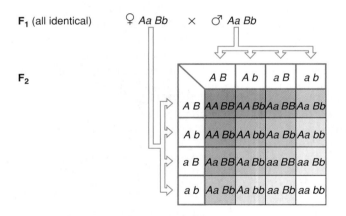

F$_1$ (all identical) ♀ *Aa Bb* × ♂ *Aa Bb*

F$_2$

		A B	*A b*	*a B*	*a b*
A B		*AA BB*	*AA Bb*	*Aa BB*	*Aa Bb*
A b		*AA Bb*	*AA bb*	*Aa Bb*	*Aa bb*
a B		*Aa BB*	*Aa Bb*	*aa BB*	*aa Bb*
a b		*Aa Bb*	*Aa bb*	*aa Bb*	*aa bb*

1	*AA BB*	purple shade 9
2	*AA Bb*	purple shade 8
2	*Aa BB*	purple shade 7
1	*AA bb*	purple shade 6
4	*Aa Bb*	purple shade 5
1	*aa BB*	purple shade 4
2	*Aa bb*	purple shade 3
2	*aa Bb*	purple shade 2
1	*aa bb*	purple shade 1 (white)

Figure 3.16 With incomplete dominance, the interaction of two genes can produce nine different phenotypes for a single trait. In this example, two genes produce purple pigments. Alleles *A* and *a* of the first gene exhibit incomplete dominance, as do alleles *B* and *b* of the second gene. Because the two alleles of each gene can combine to generate three different phenotypes, double heterozygotes have the potential for producing progeny of nine (3 × 3) different colors in a ratio of 1:2:2:1:4:1:2:2:1.

(a) Hypothesis 1 (two genes with recessive epistasis)

(b) Hypothesis 2 (one gene with incomplete dominance)

Figure 3.17 Specific breeding tests can help decide between hypotheses. Either of two hypotheses could explain the results of a cross tracking coat color in mice. **(a)** In one hypothesis, two genes interact with recessive epistasis to produce a 9:3:4 ratio. **(b)** In the other hypothesis, a single gene with incomplete dominance between the alleles generates the observed results. One way to decide between these models is to cross each of several albino F_2 mice with true-breeding brown mice. The two-gene model predicts several different outcomes depending on the $--cc$ albino's genotype at the B gene. The one-gene model predicts that all progeny of the crosses will be black.

results confirms the two-gene hypothesis. The comprehensive example on p. 67 outlines some of the details of the interactions of the three mouse genes for coat color.

With Humans, Pedigree Analysis Replaces Breeding Experiments

Breeding experiments cannot be applied to humans, for obvious reasons. But a careful examination of as many family pedigrees as possible can help elucidate the genetic basis of a particular condition. In a form of albinism known as ocular-cutaneous albinism (OCA), for example, people with the inherited condition have little or no pigment in their skin, hair, and eyes (Fig. 3.18a). The horizontal inheritance pattern seen in Fig. 3.18b suggests that OCA is determined by the recessive allele of one gene, with albino family members being homozygotes for that allele. But a 1952 paper on albinism reported a family in which two albino parents produced three normally pigmented children (Fig. 3.18c). How would you explain this phenomenon? The answer is that albinism is another example of heterogeneity: Mutant alleles at any one of several different genes can cause the condition. The reported mating was, in effect, an inadvertent complementation test, which showed that one parent was homozygous for an OCA-causing mutation in gene A, while the other parent was homozygous for an OCA-causing mutation in a different gene, B (compare with Fig. 3.15 on p. 59).

The Same Genotype Does Not Always Produce the Same Phenotype

In our discussion of gene interactions so far, we have looked at examples in which a genotype reliably fashions a particular phenotype. But this is not always what happens. Sometimes a genotype is not expressed at all; that is, even though the genotype is present, the expected phenotype does not appear. Other times, the trait caused by a genotype is expressed to varying degrees or in a variety of ways in different individuals. Factors

(a) Ocular-cutaneous albinism (OCA)

(b) OCA is recessive

(c) Complementation for albinism

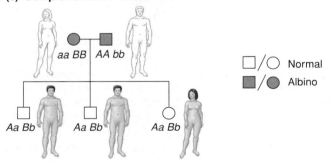

□/○ Normal
■/● Albino

Figure 3.18 Family pedigrees help unravel the genetic basis of albinism. (a) Tanzanian mother with one normally pigmented child and one albino offspring. **(b)** A pedigree following the inheritance of OCA in an inbred family indicates that the trait is recessive. **(c)** A family in which two albino parents have nonalbino children demonstrates that homozygosity for a recessive allele of either of two genes can cause OCA.

that alter the phenotypic expression of genotype include modifier genes, the environment (in the broadest sense, as defined earlier), and chance.

Phenotype Often Depends on Penetrance and Expressivity

Retinoblastoma, the most malignant form of eye cancer, arises from a dominant mutation of one gene, but only 75% of people who carry the mutant allele develop the disease. Geneticists use the term **penetrance** to describe how many members of a population with a particular genotype show the expected phenotype. Penetrance can be *complete* (100%), as in the traits that Mendel studied, or *incomplete,* as in retinoblastoma (see the Genetics and Society box "Disease Prevention Versus the Right to Privacy" on p. 64 for another example of incomplete penetrance). For retinoblastoma, the penetrance is 75%. In some people who show the trait, only one eye is affected, while in other individuals with the phenotype, both eyes are diseased. **Expressivity** refers to the degree or intensity with which a particular genotype is expressed in a phenotype. Expressivity can be *variable,* as in retinoblastoma (one or both eyes affected), or *unvarying,* as in pea color. As we will see, the incomplete penetrance and variable expressivity of retinoblastoma are the result of chance, but in other cases, it is modifier genes and/or the environment that causes such variations in the appearance of phenotype.

Modifier Genes Produce Secondary Effects on Phenotype

Not all genes that influence the appearance of a trait contribute equally to the phenotype. Major genes have a large influence, while **modifier genes** have a more subtle, secondary effect. Modifier genes alter the phenotypes produced by the alleles of other genes. There is no formal distinction between major and modifier genes. Rather, there is a continuum between the two, and the cutoff is arbitrary.

Modifier genes influence the length of a mouse's tail. The mutant *T* allele of the tail-length gene causes a shortening of the normally long wild-type tail. But not all mice carrying the *T* mutation have the same length tail. A comparison of several inbred lines points to modifier genes as the cause of this variable expressivity. In one inbred line, mice carrying the *T* mutation have tails that are approximately 75% as long as normal tails; in another inbred line, the tails are 50% normal length; and in a third line, the tails are only 10% as long as wild-type tails. Because all members of each inbred line grow the same length tail, no matter what the environment (for example, diet, cage temperature, or bedding), geneticists conclude it is genes and not the environment or chance that determines the length of a mutant mouse's tail. Different inbred lines most likely carry different alleles of the modifier genes that determine exactly how short the tail will be when the *T* mutation is present. Researchers have not yet identified the modifier genes or ascertained what they do, but the fact that all members of one inbred line differ in exactly the same way from all members of the second and third lines reinforces the interpretation that differences in *T*-produced tail length are caused by modifier genes.

The Environment Can Affect the Phenotypic Expression of a Genotype

Temperature is one element of the environment that can have a visible effect on phenotype. For example, temperature influences the unique coat color pattern of Siamese cats (Fig. 3.19). These domestic felines are homozygous for one of the multiple alleles of a gene that encodes an enzyme catalyzing the production of the dark pigment melanin. The form of the enzyme

Figure 3.19 In Siamese cats, temperature affects coat color. (a) A Siamese cat. **(b)** Melanin is produced only in the cooler extremities. This is because Siamese cats are homozygous for a mutation that renders an enzyme involved in melanin synthesis temperature sensitive. The mutant enzyme is active at lower temperatures but inactive at higher temperatures.

generated by the variant "*Siamese*" allele does not function at the cat's normal body temperature. It becomes active only at the lower temperatures found in the cat's extremities, where it promotes the production of melanin, which darkens the animal's ears, nose, paws, and tail. The enzyme is thus *temperature sensitive.* Under the normal environmental conditions in temperate climates, the Siamese phenotype does not vary much in expressivity from one cat to another, but one can imagine the expression of a very different phenotype—no dark extremities—in equatorial deserts, where the ambient temperature is at or above normal body temperature.

Temperature can also affect survivability. In one type of experimentally bred fruit fly (*Drosophila melanogaster*), some individuals develop and multiply normally at temperatures between 18°C and 29°C; but if the thermometer climbs beyond that cutoff for a short time, they become reversibly paralyzed, and if the temperature remains high for more than a few hours, they die. These insects carry a temperature-sensitive allele of the *shibire* gene, which encodes a protein essential for nerve cell transmission, among other functions. This type of allele is known as a **conditional lethal** because it is lethal only under certain conditions. The range of temperatures under which the insects remain viable is the **permissive** range; the lethal temperatures above that are **restrictive.** Thus, at one temperature the allele gives rise to a phenotype that is indistinguishable from the wild type, while at another temperature, the same allele generates a mutant phenotype (lethality). Flies with the wild-type *shibire* allele are viable even at the higher temperatures. The fact that some mutations are lethal only under certain conditions clearly illustrates that the environment can affect the penetrance of a phenotype. Since flies with temperature-sensitive *shibire* muta-

tions can survive only under carefully controlled environmental conditions, they do not survive in nature, where the mutation is more likely to be penetrant at some point in the life cycle.

Even in genetically normal individuals, exposure to chemicals or other environmental agents can have phenotypic consequences that are similar to those caused by mutant alleles of specific genes. A change in phenotype arising from such environmental agents that mimics the effects of a mutation in a gene is known as a **phenocopy.** Phenocopies are not heritable because they do not arise from a change in a gene. In humans, ingestion of the sedative thalidomide by pregnant women in the early 1960s produced a phenocopy of a rare dominant trait called *phocomelia.* By disrupting limb development in otherwise normal fetuses, the drug mimicked the effect of the phocomelia-causing mutation. When this became evident, thalidomide was withdrawn from the market.

Some types of environmental change may have a positive effect on an organism's survivability, as in the following example, where a straightforward application of medical science artificially reduces the penetrance of a mutant phenotype. As discussed earlier, children born with the genetically determined disease known as phenylketonuria, or PKU, will develop a range of neurological problems, including convulsive seizures and mental retardation, unless they are put on a special diet. Today, there is a reliable blood test for detecting the condition in newborns. Once a baby with PKU is identified, the doctor or other health-care practitioner prescribes a protective diet that excludes phenylalanine; the diet must also provide enough calories to prevent the infant's body from breaking down its own proteins and thereby releasing the damaging amino acid from within. Such dietary therapy—a simple change in the

Genetics and Society

Disease Prevention Versus the Right to Privacy

In one of the most extensive human pedigrees ever assembled, a team of researchers traced a familial pattern of blindness back through five centuries of related individuals to its origin in a couple who died in a small town in northwestern France in 1495. More than 30,000 French men and women alive in 1990 descended from that one fifteenth-century couple, and within this direct lineage resided close to half of all reported French cases of a particular blindness-producing disease, a form of hereditary juvenile glaucoma. The massive genealogic tree for the trait (when posted on the office wall, it was over 100 feet long) showed that the genetic defect follows a simple Mendelian pattern of transmission determined by the dominant allele of a single gene (Fig. A). The pedigree also showed that the dominant genetic defect displays incomplete penetrance: Not all people receiving the dominant allele from one or the other of their parents become blind; these sighted carriers may unknowingly pass the blindness-causing dominant allele to their children. Thus, with hereditary juvenile glaucoma, a parent, grandparent, or great-grandparent may possess the defective dominant allele but not show up as an affected patient in the family pedigree.

Unfortunately, people do not know they have the disease until their vision starts to deteriorate, and by that time, many of

their optic fibers have been irreversibly damaged and blindness is all but inevitable. Surprisingly, the existence of medical therapies that make it possible to arrest the nerve damage of hereditary juvenile glaucoma and prevent blindness created a quandary. Because treatment, to be effective, has to begin before symptoms of impending blindness show up, information in the extensive juvenile glaucoma pedigree could help doctors pinpoint people who are at risk, even though neither of their parents is blind. However, a French law protecting personal privacy forbids public circulation of the names in the pedigree. In fact, in the late 1980s, the researchers who compiled the massive family history wanted to give physicians the names of at-risk individuals living in their area; with such information the doctors could keep a close watch on certain patients and when necessary recommend drugs or surgery to prevent blindness. But interpreters of the law at that time maintained that if actual names were circulated, potential carriers of the disease might suffer discrimination in hiring or insurance. According to the law, individuals about whom information is collected must know how that information will be used, and at the time of the study, living participants were not told that their names would be given out to medical workers. As a result, the researchers could alert physicians to the hereditary nature and symptoms of the disease, but they could not give out names. And since only a long, arduous, expensive—and at the time infeasible—national screening program could identify affected individuals without those names, some sufferers of the disease would not receive treatment soon enough to prevent blindness.

France thus faced a serious ethical dilemma: On the one hand, giving out names could save perhaps thousands of people from blindness; on the other hand, laws designed to protect personal privacy and forestall discrimination precluded the dissemination of specific names. Until the legal problem was resolved, the only solution was a massive educational program to alert the general public to the problem so that concerned families could seek medical advice. Meanwhile, molecular geneticists were searching for the gene (or genes) that cause this particular form of juvenile glaucoma. Its (or their) identification could result in diagnostic tests based on the direct analysis of genotype (see the Fast Forward box "The Direct Analysis of Human Genotype" in Chapter 2). And these tests, if administered on a large enough scale with proper educational follow-up, could circumvent the problem of giving out names from the pedigree by delivering genetic information directly to the parents of newborns, in plenty of time for preventive therapy.

Male
Female
Blind without diagnosis
Glaucoma

Figure A   A pedigree showing the transmission of juvenile glaucoma. A small part of the genealogic tree: The vertical transmission pattern over seven generations shows that a dominant allele of a single gene causes juvenile glaucoma. The lack of glaucoma in V-2 followed by its reappearance in VI-2 reveals that the trait is incompletely penetrant. As a result, sighted heterozygotes may unknowingly pass the condition on to their children.

environment—although somewhat unpleasant because it is highly limiting, now enables many PKU infants to develop into healthy adults.

Finally, two of the top killer diseases in the United States—cardiovascular disease and lung cancer—also illustrate how the environment can alter phenotype by influencing both expressivity and penetrance. People may inherit a propensity to heart disease, but the environmental factors of diet and exercise contribute to the occurrence (penetrance) and seriousness (expressivity) of their condition. Similarly, some people are born genetically prone to lung cancer, but whether or not they get it (penetrance) is strongly determined by whether or not they choose to smoke.

Thus, various aspects of an organism's environment, including temperature, diet, and exercise, interact with its genotype to generate the functional phenotype, the ultimate combination of traits that determines what a plant or animal looks like and how it behaves.

Chance Can Affect Penetrance and Expressivity

In the retinoblastoma example described earlier, people are born with the disease mutation in one allele of the retinoblastoma gene in every cell. But whether or not they show the phenotype and whether they show it in one or both eyes depend on additional genetic events that occur randomly to alter the second allele of the gene in specific body cells. Examples of random events that can trigger the onset of the disease include exposure to cosmic rays that alter the genetic material in retinal cells or mistakes made by retinal cell machinery during cell division. Events of this type provide the second "hit"—a mutation in the second copy of the retinoblastoma gene—necessary to turn a normal retinal cell into a cancerous one. The phenotype of retinoblastoma thus results from a specific heritable mutation in a specific gene, but the incomplete penetrance and variable expressivity of the disease depend on chance genetic events that affect the other allele in certain cells.

By contributing to incomplete penetrance and variable expressivity, modifier genes, the environment, and chance give rise to phenotypic variation. Unlike dominant epistasis or recessive lethality, however, the probability of penetrance and the level of expressivity cannot be derived from the original Mendelian principles of segregation and independent assortment; they are determined empirically by observation and counting.

Even Continuous Variation Can Be Explained by Extensions to Mendelian Analysis

In Mendel's experiments, height in pea plants was determined by two segregating alleles of one gene (in the wild, it is determined by many genes, but in Mendel's inbred populations, the alleles of all these other genes were invariant). The phenotypes that resulted from these alternative alleles were clear-cut, either short or tall, and pea plant height was therefore known as a **discontinuous trait.** In contrast, because people do not produce inbred populations, height in humans is determined by segre-

gating alleles of many different genes whose interaction with each other and the environment produces phenotypes showing continuous variation; height in humans is thus an example of a **continuous trait.** Within most outbred human populations, individual heights vary over a range of values that when charted on a graph produce a bell curve (Fig. 3.20a). In fact, many human traits, including height, weight, and skin color, show continuous variation, rather than the clear-cut alternatives analyzed by Mendel. They also appear to blend and "unblend." Think for a moment of skin color. Children of marriages between people of African and Northern European descent, for example, often seem to be a blend of their parents' skin color. Matings of these F_1 individuals produce offspring with a wide range of skin pigmentation; a few may be as light as the original Northern European parent, a few as dark as the original African parent, but most will fall in a range between the two (Fig. 3.20b). For these reasons, early human geneticists were slow to accept Mendelian analysis. Because they were working with outbred populations, they found very few examples of "either-or" Mendelian traits in normal, healthy people.

By 1930, however, studies of corn and tobacco conclusively demonstrated that it is possible to provide a Mendelian

(a)

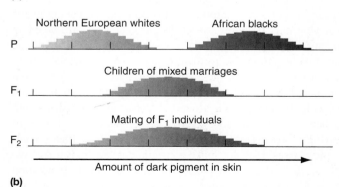

(b)

Figure 3.20 Continuous traits in humans. (a) Women runners at the start of a 5th Avenue mile race in New York City demonstrate that height is a trait showing continuous variation. **(b)** The skin color of most F_1 offspring is usually between the parental extremes, while the next (F_2) generation exhibits a broader distribution of continuous variation.

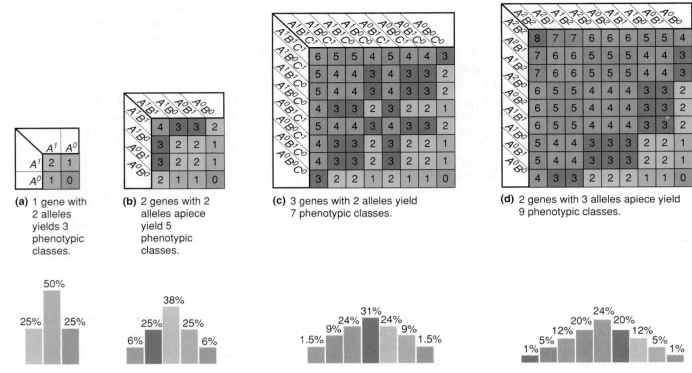

(a) 1 gene with 2 alleles yields 3 phenotypic classes.

(b) 2 genes with 2 alleles apiece yield 5 phenotypic classes.

(c) 3 genes with 2 alleles yield 7 phenotypic classes.

(d) 2 genes with 3 alleles apiece yield 9 phenotypic classes.

Figure 3.21 A Mendelian explanation of continuous variation. The more genes or alleles, the more possible phenotypic classes, and the greater the similarity to continuous variation. In these examples, several pairs of incompletely dominant alleles have additive effects. Percentages denote frequencies of each genotype expressed as fractions of the total population.

explanation of continuous variation by simply increasing the number of genes contributing to a phenotype. The more genes, the more phenotypic classes, and the more classes, the more the variation appears continuous. As a hypothetical example, consider a series of genes all affecting the height of pole beans. For each gene, there are two alleles, a *"0"* allele that contributes nothing to height and a *"1"* allele that increases the height of a plant by one unit. All alleles exhibit incomplete dominance relative to alternative alleles at the same gene. The phenotypes determined by all these genes are additive. What would be the result of a two-generation cross between pure-breeding plants carrying only *0* alleles at each height gene and pure-breeding plants carrying only *1* alleles at each height gene? If only one gene were responsible for height, and if environmental effects could be discounted, the F$_2$ population would be distributed among three classes: homozygous *0,0* plants with 0 height (they lie prostrate on the ground); heterozygous *0,1* plants with a height of 1; and homozygous *1,1* plants with a height of 2 (Fig. 3.21a). This distribution of heights over three phenotypic classes does not make a continuous curve. But for two genes, there will be five phenotypic classes in the F$_2$ generation (Fig. 3.21b); for three genes, seven classes (Fig. 3.21c); and for four genes, nine classes (not shown). The distributions produced by three and four genes begin to approach continuous variation, and if we add a small contribution from environmental variation, a smooth curve will appear. After all, we would expect bean plants to grow better in good soil, with ample sunlight and water. The environmental component effectively converts the stepped bar graph to a continuous curve by producing some variation in expressivity within each genotypic class. Moreover, additional variation might arise from more than two alleles at some genes (Fig. 3.21d), unequal contribution to the phenotype by the various genes involved (review Fig. 3.16 on p. 60), interactions with modifier genes, and chance. Thus, from what we now know about the relation between genotype and phenotype, it is possible to see how just a handful of genes that behave according to known Mendelian principles can easily generate continuous variation. In fact, studies suggest that most of the variation in skin color between people of Northern European and African descent (see Figure 3.20b) can be explained by the action of three to four genes.

Continuous traits (also called **quantitative traits**) vary quantitatively over a range of values and can usually be measured: the length of a tobacco flower in millimeters, the amount of milk produced by a cow per day in liters, or the height of a person in meters. They are usually **polygenic**—controlled by multiple genes—and show the additive effects of a large number of alleles, which creates an enormous potential for variation within a population. Polygenic traits are, by definition, multifactorial, but not all multifactorial traits are polygenic (some, for instance, are determined by the way the alleles of one gene interact with each other and the environment). We discuss the analysis and distribution of multifactorial traits in Chapter 20 on population genetics.

The Mouse's Coat Color: A Comprehensive Example of Multiple Alleles and Multifactorial Traits

Most field mice are a dark gray (agouti), but mice bred for specific mutations in the laboratory can be gray, tan, yellow, brown, black, or various combinations thereof. Here we look at the alleles of three of the genes that make such variation possible. Our review underscores how the various allelic interactions of just a handful of genes can produce an astonishing diversity of phenotypes.

Gene 1—Agouti or Other Color Patterns

The *agouti* gene determines the distribution of color on each hair and has multiple alleles. The wild-type allele A specifies the bands of yellow and black that give the agouti appearance; A^y gets rid of the black and thus produces solid yellow; a gets rid of the yellow and thus produces solid black; and a^t specifies black on the back and yellow on the stomach. The dominance series for this set of *agouti* gene alleles is: $A^y > A > a^t > a$. However, although A^y is dominant to all other alleles for coat color, it is recessive to all the others for lethality: A^yA^y homozygotes die before birth, while A^yA, A^ya^t, or $A^y a$ heterozygotes survive.

Gene 2—Black or Brown

A second gene specifies whether the dark color of each hair is black or brown. This gene has two alleles: B is dominant and designates black; b is recessive and generates brown. Since the A^y allele at the *agouti* gene completely eliminates the dark band of each hair, it acts in a dominant epistatic manner to the B gene. With all other *agouti* alleles, however, it is possible to distinguish the effects of the two different B alleles on phenotype. The $A– B–$ genotype gives rise to the wild-type agouti having black with yellow hairs. The $A– bb$ genotype generates a color referred to as cinnamon (with hairs having stripes of brown and yellow); $aa bb$ is all brown; and $a^ta^t bb$ is brown on the back and yellow on the stomach. A cross between two F_1 hybrid animals of genotype $A^ya Bb$ would yield an F_2 generation with yellow ($A^ya – –$), black ($aa B–$), and brown ($aa bb$) animals in a ratio of 8:3:1. This ratio reflects the dominant epistasis of A^y and the loss of a class of four ($A^yA^y – –$) due to prenatal lethality.

Gene 3—Albino or Pigmented

Like other mammals, mice have a third gene influencing coat color. A recessive allele (c) abolishes the function of the enzyme that leads to the formation of dark pigment melanin, making this allele epistatic to all other coat color genes. As a result, cc homozygotes are pure white, while $C–$ mice are agouti, black, brown, yellow, or yellow and black (or other colors and patterns), depending on what alleles they carry at the A and B genes, as well as at some 50 other genes known to play a role in determining the coat color of mice. Adding to the complex color potential are other alleles that geneticists have uncovered for the albino gene; these cause only a partial inactivation of the melanin-producing enzyme and thus have a partial epistatic effect on phenotype.

This comprehensive example of coat color in mice gives some idea of the potential for variation from just a few genes, some with multiple alleles. Amazingly, this is just the tip of the iceberg. When you realize that both mice and humans carry roughly 40,000–60,000 genes, the number of interactions that connect various alleles of these genes in the expression of phenotype is in the millions, if not the billions. The potential for variation and diversity among individuals is staggering indeed.

CONNECTIONS

Part of Mendel's genius was to look at the genetic basis of variation through a very narrow window, focusing his first glimpse of the mechanisms of inheritance on such fundamental phenomena that the import of his insights remains undiminished to this day. Mendel worked on just a handful of traits in inbred populations of one species, and for each trait, one gene with one completely dominant and one recessive allele determined two distinguishable, or discontinuous, phenotypes. Both the dominant and recessive alleles showed complete penetrance and negligible differences of expressivity. He knew from other examples that phenotype does not always reflect genotype, but he probably did not foresee how many complexities could cloud the correlation between the two.

In the first few decades of the twentieth century, many questioned the general applicability of Mendelian analysis, for it seemed to shed little light on the complex inheritance patterns of most plant and animal traits or on the mechanisms producing continuous variation. Simple embellishments, however, clarified the genetic basis of continuous variation and provided explanations for other apparent exceptions to Mendelian analysis. These embellishments included the ideas that dominance need not be complete, that one gene can have multiple alleles, that one gene can determine more than one trait, that several genes can contribute to the same trait, and that the expression of genes can be affected in a variety of ways by other genes, the environment, and chance. Each embellishment extends the range of Mendelian analysis and deepens our understanding of the genetic basis of variation. And no matter how broad the view, Mendel's basic conclusions—that genes are the discrete units of heredity and that the two similar or dissimilar copies of a gene segregate during gamete formation and randomly unite at fertilization—remain valid.

But what about the law of independent assortment? As it turns out, its application is not as universal as that of the law of segregation. Many genes do assort independently, but some do not; rather, they appear to be linked and transmitted together from generation to generation. An understanding of this fact emerged from studies that located Mendel's hereditary units, the genes, in specific cellular organelles, the chromosomes. In describing how researchers deduced that genes travel with chromosomes, Chapter 4 establishes the physical basis of inheritance, including segregation, and clarifies why some genes do, while others do not, assort independently.

ESSENTIAL CONCEPTS

1. The F_1 phenotype generated by each pair of alleles defines the dominance relationship between these alleles. One allele is not always completely dominant or completely recessive to another. With *incomplete dominance,* the F_1 hybrid phenotype resembles neither parent. With *codominance,* the F_1 hybrid phenotype contains observable components from both parents. Many allele pairs are codominant at the level of protein production.

2. One gene can contribute to multiple traits (*pleiotropy*); for such a gene, the dominance relation between any two alleles can vary according to the particular phenotype under consideration.

3. A single gene may have any number of alleles, each of which can cause the appearance of different phenotypes. New alleles arise by mutation. Alleles whose frequency is greater than 1% in a population are considered *wild types;* those whose frequency is below 1% are considered *mutants.* When two or more wild-type alleles exist for a gene, the gene is considered *polymorphic;* a gene with only one wild-type allele is *monomorphic.*

4. Two or more genes may interact in several ways to affect the production of a single trait. It is often possible to arrive at an understanding of these interactions by observing characteristic deviations from traditional Mendelian phenotypic ratios.

5. In *epistasis,* the action of an allele at one gene can hide traits normally caused by the expression of alleles at another gene. In *complementary gene action,* dominant alleles of two or more genes are required to generate a particular trait. In *heterogeneity,* mutant alleles at any one of two or more genes are sufficient to elicit a phenotype. The *complementation test* can reveal whether a particular phenotype arises from mutations in the same or separate genes.

6. In many cases, the route from genotype to phenotype can be modified by the environment, chance, or other genes. A phenotype shows *incomplete penetrance* when it is expressed in fewer than 100% of individuals with the same genotype. A phenotype shows *variable expressivity* when it is expressed at a quantitatively different level in different individuals with the same genotype.

7. A *continuous trait* can have any value of expression between two extremes. Most traits of this type are *polygenic,* that is, determined by the interactions of multiple genes.

SOCIAL AND ETHICAL ISSUES

1. John and Nancy's daughter is entering kindergarten and, as required by law, has to have a physical and blood test performed by their family doctor. The doctor knows that both John and Nancy have the AB blood type. But when results come back on their daughter's blood, the doctor discovers that her blood type is O. Neither parent is heterozygous for the Bombay allele. The only conclusion the doctor can reach from this result is that babies must have been switched at the hospital. The doctor is now in an awkward position. John and Nancy and their healthy daughter make a very happy family. If he explains to them the implication of his findings, their lives will be turned upside down. If not, he will be withholding personal information that one could argue rightly belongs to them. What would you do if you were the doctor?

2. A man who has committed several violent rapes is on trial, and his lawyer argues that the man cannot help himself because his behavior is the result of his genetic makeup. The man's father and brother also have a history of committing violent crimes. In defense of this position, the lawyer cites studies reported in 1993 on a small Dutch family indicating that some people inherit a propensity toward violent behavior. Affected individuals in this family have a mutation in the monoamine oxidase gene involved in the breakdown of neurotransmitters. Do you think the sentencing of the man on trial should depend in part on this type of scientific study? Our judicial system is based on the assumption that people act with free will and are therefore responsible for their actions. Are people with known genetic alterations

responsible for their actions? Should we make exceptions in the case of genetic predispositions? What role should society play in helping people whose behaviors may be influenced by the alleles they carry?

3. A teenage girl was diagnosed with chronic myelogenous leukemia, for which the treatment is a bone marrow transplant. A matching donor could not be found and her

parents decided to have another child, hoping that the younger sibling could serve as a donor of bone marrow for the older child. Is it ethical to conceive a child for the purpose of tissue or organ donation to a sibling? Consider situations in which the donor's life will be compromised by the donation as well as cases in which the donor's life will not be compromised.

S O L V E D   P R O B L E M S

I. Imagine you purchased an albino mouse (genotype cc) in a pet store. The c allele is epistatic to other coat color genes. How would you go about determining the genotype of this mouse at the brown locus? (In pigmented mice, BB and Bb are black, bb is brown.)

Answer

This problem requires an understanding of gene interactions, specifically epistasis. You have been placed in the role of experimenter and need to design crosses that will answer the question. To determine the alleles of the B gene present you need to eliminate the blocking action of the cc genotype. Because only the recessive c allele is epistatic, when a C allele is present, no epistasis will occur. To introduce a C allele during the mating, the test mouse you mate to your albino can have the genotype CC or Cc. (If the mouse is Cc, half of the progeny will be albino and will not contribute useful information, but the nonalbinos from this cross would be informative.) What alleles of the B gene should the test mouse carry? To make this decision, work through the expected results using each of the possible genotypes.

Test mouse genotype		Albino mouse	Expected progeny
BB	×	BB	all black
	×	Bb	all black
	×	bb	all black
Bb	×	BB	all black
	×	Bb	3/4 black, 1/4 brown
	×	bb	1/2 black, 1/2 brown
bb	×	BB	all black
	×	Bb	1/2 black, 1/2 brown
	×	bb	all brown

From these hypothetical crosses, you can see that a test mouse with either the Bb or bb genotype would yield distinct outcomes for each of the three possible albino mouse genotypes. However, a bb test mouse would be more useful and less ambiguous. First, it is easier to identify a mouse with the bb genotype because a brown mouse must have this homozygous recessive genotype. Second, the results are completely different for each of the three possible genotypes when you use the bb test

mouse. (In contrast, a Bb test mouse would yield both black and brown progeny whether the albino mouse was Bb or bb; the only distinguishing feature is the ratio.) *To determine the full genotype of the albino mouse, you should cross it to a brown mouse (which could be* CC bb *or* Cc bb*).*

II. In a particular kind of wildflower, the wild-type flower color is deep purple and the plants are true-breeding. In one true-breeding mutant stock, the flowers have a reduced pigmentation, resulting in a lavender color. In a different true-breeding mutant stock, the flowers have no pigmentation and are thus white. When a lavender-flowered plant from the first mutant stock was crossed to a white-flowered plant from the second mutant stock, all the F_1 plants had purple flowers. The F_1 plants were then allowed to self-fertilize to produce an F_2 generation. The 277 F_2 plants were 157 purple : 71 white : 49 lavender. Explain how flower color is inherited. Is this trait controlled by the alleles of a single gene? What kinds of progeny would be produced if lavender F_2 plants were allowed to self-fertilize?

Answer

Are there any modes of single-gene inheritance compatible with the data? The observations that the F_1 plants look different from either of their parents and that the F_2 generation is composed of plants with three different phenotypes exclude complete dominance. The ratio of the three phenotypes in the F_2 plants has some resemblance to the 1:2:1 ratio expected from codominance or incomplete dominance, but the results would then imply that purple plants must be heterozygotes. This conflicts with the information provided that purple plants are true-breeding.

Consider now the possibility that two genes are involved. From a cross between plants heterozygous for two genes (W and P), the F_2 generation would contain a 9:3:3:1 ratio of the genotypes $W- P-$, $W- pp$, $ww P-$, and $ww pp$ (where the dash indicates that the allele could be either a dominant or a recessive form). Are there any combinations of the 9:3:3:1 ratio that would be close to that seen in the F_2 generation in this example? The numbers seem close to a 9:4:3 ratio. What hypothesis would support combining two of the classes $(3 + 1)$? If *w* is epistatic to the *P* gene, then the *ww P–* and *ww pp* genotypic classes would have the same white phenotype. With this explanation, 1/3 of the F$_2$ lavender plants would be *WW pp,* and the remaining 2/3 would be *Ww pp.* Upon self-fertilization, *WW pp* plants would produce only lavender (*WW pp*) progeny, while *Ww pp* plants would produce a 3:1 ratio of lavender (*W– pp*) and white (*ww pp*) progeny.

III. Huntington disease (HD) is a rare dominant condition in humans that results in a slow but inexorable deterioration of the nervous system. HD shows what might be called "age-dependent penetrance," which is to say that the probability that a person with the HD genotype will express the phenotype varies with age. Assume that 50%

of those inheriting the *HD* allele will express the symptoms by age 40. Susan is a 35-year-old woman whose father has HD. She currently shows no symptoms. What is the probability that Susan will show symptoms in five years?

Answer

This problem involves probability and penetrance. Two conditions are necessary for Susan to show symptoms of the disease. There is a 1/2 (50%) chance that she inherited the mutant allele from her father and a 1/2 (50%) chance that she will express the phenotype by age 40. Since these are independent events, *the probability is the product of the individual probabilities, or 1/4.*

PROBLEMS

3-1 For each of the terms in the left column, choose the best matching phrase in the right column.

a. epistasis	1. one gene affecting more than one phenotype
b. modifier gene	2. the alleles of one gene mask the effects of alleles of another gene
c. conditional lethal	3. both parental phenotypes are expressed in the F$_1$ hybrids
d. permissive condition	4. a heritable change in a gene
e. reduced penetrance	5. cell surface molecules that are involved in the immune system and are highly variable
f. multifactorial trait	6. genes whose alleles alter phenotypes produced by the action of other genes
g. incomplete dominance	7. less than 100% of the individuals possessing a particular genotype express it in their phenotype
h. codominance	8. environmental conditions that allow conditional lethals to live
i. histocompatibility antigens	9. a trait produced by the interaction of alleles of at least two genes or from interactions between gene and environment
j. mutation	10. individuals with the same genotype have related phenotypes that vary in intensity
k. pleiotropy	11. a genotype that is lethal in some situations (for example, high temperature) but viable in others
l. variable expressivity	12. the heterozygote resembles neither homozygote

3-2 In four o'clocks, the allele for red flowers is incompletely dominant over the allele for white flowers, so that heterozygotes have pink flowers. What ratios of flower colors would you expect among the offspring of the following crosses: (a) pink × pink, (b) white × pink, (c) red × red, (d) red × pink, (e) white × white, and (f) red × white? If you specifically wanted to produce pink flowers, which of these crosses would be most efficient?

3-3 A cross between two plants that both have yellow flowers produces 80 offspring plants, of which 38 have yellow flowers, 22 have red flowers, and 20 have white flowers. If one assumes that this variation in color is due to inheritance at a single locus, what is the genotype associated with each flower color, and how can you describe the inheritance of flower color?

3-4 In radishes, color and shape are each controlled by a single locus with two incompletely dominant alleles. Color may be red (*RR*), purple (*Rr*), or white (*rr*) and shape can be long (*LL*), oval (*Ll*), or round (*ll*). What phenotypic classes and proportions would you expect among the offspring of a cross between two plants heterozygous at both loci?

3-5 A wild legume with white flowers and long pods is crossed to one with purple flowers and short pods. The F$_1$ offspring are allowed to self-fertilize, and the F$_2$ generation has 301 long purple, 99 short purple, 612 long pink, 195 short pink, 295 long white, and 98 short white. How are these traits being inherited?

3-6 Describe briefly:
   a. The genotype of a person who has sickle-cell anemia.
   b. The genotype of a person with a normal phenotype who has a child with sickle-cell anemia.
   c. The *total* number of different alleles of the β-globin gene that could be carried by five children with the same mother and father.

3-7 Assuming no involvement of the Bombay phenotype:
   a. If a girl has blood type O, what could be the genotypes and corresponding phenotypes of her parents?
   b. If a girl has blood type B and her mother has blood type A, what genotype(s) and corresponding phenotype(s) could the other parent have?

c. If a girl has blood type AB and her mother is also AB, what are the genotype(s) and corresponding phenotype(s) of any male who could *not* be her father?

3-8 There are several genes in humans in addition to the *ABO* gene that give rise to recognizable antigens on the surface of red blood cells. The *MN* and *Rh* genes are two examples. The *Rh* locus can contain either a positive or negative allele, with positive being dominant to negative. *M* and *N* are codominant alleles of the *MN* gene. The following chart shows several mothers and their children. For each mother–child pair, choose the father of the child from among the males in the right column, assuming one child per male.

	Mother	Child	Males
a.	O M Rh pos	B MN Rh neg	O M Rh neg
b.	B MN Rh neg	O N Rh neg	A M Rh pos
c.	O M Rh pos	A M Rh neg	O MN Rh pos
d.	AB N Rh neg	B MN Rh neg	B MN Rh pos

3-9 Alleles of the gene that determines seed coat patterns in lentils can be organized in a dominance series: marbled > spotted = dotted (codominant alleles) > clear. A lentil plant homozygous for the marbled seed coat pattern allele was crossed to one homozygous for the spotted pattern allele. In another cross, a homozygous dotted lentil plant was crossed to one homozygous for clear. An F_1 plant from the first cross was then mated to an F_1 plant from the second cross.
  a. What phenotypes in what proportions are expected from this mating between the two F_1 types?
  b. What are the expected phenotypes of the F_1 plants from the two original parental crosses?

3-10 In clover plants, the pattern on the leaves is determined by a single gene with multiple alleles that are related in a dominance series. Seven different alleles of this gene are known; an allele that determines the absence of a pattern is recessive to the other six alleles, each of which produces a different pattern. All heterozygous combinations of alleles show complete dominance.
  a. How many different kinds of leaf patterns (including the absence of a pattern) are possible in a population of clover plants in which all seven alleles are represented?
  b. What is the largest number of different genotypes that could be associated with any one phenotype? Is there any phenotype that could be represented by only a single genotype?
  c. In a particular field, you find that the large majority of clover plants lack a pattern on their leaves, even though you can identify a few plants representative of all possible pattern types. Explain this finding.

3-11 In a population of rabbits, you find three different coat color phenotypes: chinchilla (C), himalaya (H), and albino (A). To understand the inheritance of coat colors, you cross individual rabbits with each other and note the results in the following table.

Cross number	Parental phenotypes	Phenotypes of progeny
1	H × H	3/4 H : 1/4 A
2	H × A	1/2 H : 1/2 A
3	C × C	3/4 C : 1/4 H
4	C × H	all C
5	C × C	3/4 C : 1/4 A
6	H × A	all H
7	C × A	1/2 C : 1/2 A
8	A × A	all A
9	C × H	1/2 C : 1/2 H
10	C × H	1/2 C : 1/4 H : 1/4 A

  a. What can you conclude about the inheritance of coat color in this population of rabbits?
  b. Ascribe genotypes to the parents in each of the 10 crosses.
  c. What kinds of progeny would you expect, and in what proportions, if you crossed the chinchilla parents in crosses #9 and #10?

3-12 Some plant species have an incompatibility system different from that shown in Fig. 3.8. In this alternate kind of incompatibility, a mating cannot produce viable seeds if the male parent shares an incompatibility allele with the female parent. (Just as with the kind of incompatibility system shown in Fig. 3.8, this system also ensures that all plants are heterozygous for the incompatibility gene.) Five plants were isolated from a wild population of a species with this alternate type of incompatiblity. The results of the matings between each pair of the plants are given here (− means no seeds were produced; + means seeds were produced). How many different alleles of the incompatibility gene are present in this group of five plants? What are the genotypes of the plants?

	1	2	3	4	5
1	−	−	−	+	−
2		−	+	+	+
3			−	+	−
4				−	−
5					−

3-13 In mice, the A^y allele of the *agouti* gene is a recessive lethal allele, but it is dominant for yellow coat color. What phenotypes and ratios of offspring would you expect from the cross of a mouse heterozygous at the agouti locus (genotype $A^y A$) and also at the albino locus (Cc) to an albino mouse (cc) heterozygous at the agouti locus ($A^y A$)?

3-14 In a species of tropical fish, a colorful orange and black variety called montezuma occurs. When two montezumas are crossed, 2/3 of the progeny are montezuma and 1/3 are the wild-type, dark grayish green color. Montezuma is a single-gene trait and montezuma fish are never true-breeding.

a. Explain the inheritance pattern seen here and show how your explanation accounts for the phenotypic ratios given.

b. In this same species, the morphology of the dorsal fin is altered from normal to ruffled by homozygosity for a recessive allele designated *f*. What progeny would you expect to obtain, and in what proportions, from the cross of a montezuma fish homozygous for normal fins to a green, ruffled fish?

c. What phenotypic ratios of progeny would be expected from the crossing of two of the montezuma progeny from part b?

3-15 You picked up two mice (one female and one male) that had escaped from experimental cages in the animal facility. One mouse is yellow in color and the other is brown agouti. You know that this mouse colony has animals with different alleles at only three coat color genes: the agouti or nonagouti or yellow alleles of the *A* gene, the black or brown allele of the *B* gene, and the albino or nonalbino alleles of the *C* gene. However, you don't know which alleles of these genes are actually present in each of the animals that you've captured. To determine the genotypes, you breed them together. The first litter has only three pups. One is albino, one is brown (nonagouti), and the third is black agouti.

a. What alleles of the *A, B,* and *C* genes are present in the two mice you caught?

b. After raising several litters from these two parents, you have many offspring. How many different coat color phenotypes (in total) do you expect to see expressed in the population of offspring? What are the phenotypes?

3-16 A rooster with a particular comb morphology called walnut was crossed to a hen with a type of comb morphology known as single. The F_1 progeny all had walnut combs. When F_1 males and females were crossed to each other, 93 walnut and 11 single combs were seen among the F_2 progeny but there were also 29 birds with a new kind of comb called rose and 32 birds with another new comb type called pea.

a. Explain how comb morphology is inherited.

b. What progeny would result from crossing a homozygous rose-combed hen with a homozygous pea-combed rooster? What phenotypes and ratios would be seen in the F_2 progeny?

c. A particular walnut rooster was crossed to a pea hen, and the progeny consisted of 12 walnut, 11 pea, 3 rose, and 4 single chickens. What are the likely genotypes of the parents?

d. A different walnut rooster was crossed to a rose hen, and all the progeny were walnut. What are the possible genotypes of the parents?

3-17 A black mare was crossed to a chestnut stallion and produced a bay son and a bay daughter. The two offspring were mated to each other several times, and they produced offspring of four different coat colors: black, bay, chestnut, and liver. Crossing a liver grandson back to the black mare gave a black foal, and crossing a liver granddaughter back to the chestnut stallion gave a chestnut foal. Explain how coat color is being inherited in these horses.

3-18 Filled-in symbols in the pedigree below designate individuals suffering from deafness.

a. Study the pedigree, and explain how deafness is being inherited.

b. What is the genotype of the individuals in generation V? Why are they not affected?

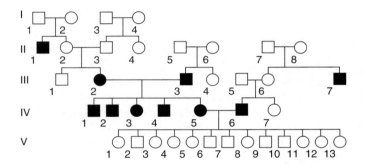

3-19 You do a cross between two true-breeding strains of zucchini. One has green fruit and the other has yellow fruit. The F_1 plants are all green, but when these are crossed, the F_2 plants consist of 9 green : 7 yellow.

a. Explain this result. What were the genotypes of the two parental strains?

b. Indicate the phenotypes, with frequencies, of the progeny of a testcross of the F_1 plants.

3-20 A student whose hobby was fishing pulled a carp out of Cayuga Lake that was very unusual: It had no scales on its body. She decided to investigate whether this strange nude phenotype had a genetic basis. She therefore obtained some inbred carp that were pure-breeding for the wild-type scale phenotype (body covered with scales in a regular pattern) and crossed them with her nude fish. To her surprise, the F_1 progeny consisted of wild-type fish and fish with a single linear row of scales on each side in a 1:1 ratio.

a. Can a single gene with two alleles account for this result? Why or why not?

b. To follow up on the first cross, the student allowed the linear fish from the F_1 generation to mate with each other. The progeny of this cross consisted of fish with four phenotypes: linear, wild type, nude, and scattered (the latter had a few scales scattered irregularly on the body). The ratio of these phenotypes was 6:3:2:1, respectively. How many genes appear to be involved in determining these phenotypes?

c. In parallel, the student allowed the phenotypically wild-type fish from the F_1 generation to mate with each other and observed, among their progeny, wild-type and scattered carp in a ratio of 3:1. How many genes with how many alleles appear to determine the difference between wild-type and scattered carp?

d. The student confirmed the conclusions of part c by crossing those scattered carp with her pure-breeding wild-type stock. Diagram the genotypes and phenotypes of the parental, F_1, and F_2 generations for this cross and indicate the ratios observed.

e. The student attempted to generate a true-breeding nude stock of fish by inbreeding. However, she found that this was impossible. Every time she crossed two nude fish, she found nude and scattered fish in the progeny, in a 2:1 ratio. (The scattered fish from these crosses bred true.) Diagram the phenotypes and genotypes of this gene in a nude × nude cross and explain the altered Mendelian ratio.

f. The student now felt she could explain all of her results. Diagram the genotypes in the linear × linear cross performed by the student (in part b). Show the genotypes of the four phenotypes observed among the progeny and explain the 6:3:2:1 ratio.

3-21 Explain the difference between epistasis and dominance. How many loci are involved in each case?

3-22 Two true-breeding white strains of the plant *Illegitimati noncarborundum* were mated and the F_1 progeny were all white. When the F_1 plants were allowed to self-fertilize, 126 white-flowered and 33 purple-flowered F_2 plants grew.

a. How could you describe inheritance of flower color? Describe how specific alleles influence each other and therefore affect phenotype.

b. A white F_2 plant is allowed to self-fertilize. Of the progeny, 3/4 are white-flowered and 1/4 are purple-flowered. What is the genotype of the white F_2 plant?

c. A purple F_2 plant is allowed to self-fertilize. Of the progeny, 3/4 are purple-flowered and 1/4 are white-flowered. What is the genotype of the purple F_2 plant?

d. Two white F_2 plants are crossed with each other. Of the progeny, 1/2 are purple-flowered and 1/2 are white-flowered. What are the genotypes of the two white F_2 plants?

3-23 "Secretors" (genotypes *SS* and *Ss*) secrete their A and B blood group antigens into their saliva and other body fluids, while "nonsecretors" (*ss*) do not. What would be the apparent phenotypic blood group proportions among the offspring of an $I^A I^B$ *Ss* woman and an $I^A I^A$ *Ss* man, if typing was done using saliva?

3-24 As you will learn in later chapters, duplication of genes is an important evolutionary mechanism. As a result, many cases are known in which a species has two or more nearly identical genes.

a. Suppose there are two genes, *A* and *B,* that specify production of the same enzyme. An abnormal phenotype results only if an individual does not make any of that enzyme. What ratio of normal versus abnormal progeny would result from a mating between two parents of genotype *Aa Bb,* where *A* and *B* represent alleles that specify production of the enzyme, while *a* and *b* are alleles that do not?

b. Suppose now that there are three genes specifying production of this enzyme, and again that a single functional allele is sufficient for a wild-type phenotype. What ratio of normal versus abnormal progeny would result from a mating between two triply heterozygous parents?

3-25 The following table shows the responses of blood samples from the individuals in the pedigree to anti-A and anti-B sera. A "+" in the anti-A row indicates that the red blood cells of that individual were clumped by anti-A serum and therefore the individual made A antigens, and a "−" indicates no clumping. The same notation is used to describe the test for the anti-B antigens.

a. Deduce the blood type of each individual from the data in the table.

b. Assign genotypes for the blood groups as accurately as you can from this data, explaining the pattern of inheritance shown in the pedigree. Assume that all genetic relationships are as presented in the pedigree (that is, there are no cases of false paternity).

	I-1	I-2	I-3	I-4	II-1	II-2	II-3	III-1	III-2
anti-A	+	+	−	+	−	−	+	+	−
anti-B	+	−	+	+	−	−	+	−	−

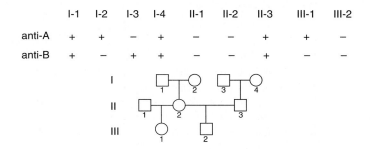

3-26 Normally, wild violets have yellow petals with dark brown markings and erect stems. Imagine you discover a plant with white petals, no markings, and prostrate stems. What experiment could you perform to determine whether the non-wild-type phenotypes are due to several different mutant genes or to the pleiotropic effects of alleles at a single locus? Explain how your experiment would settle the question.

3-27 Three genes in fruit flies affect a particular trait, and one dominant allele of *each* gene is necessary to get a wild-type phenotype.
   a. What phenotypic ratios would you predict among the progeny if you crossed triply heterozygous flies?
   b. You cross a particular wild-type male in succession with three tester strains. In the cross with one tester strain (*AA bb cc*), only 1/4 of the progeny are wild type. In the crosses involving the other two tester strains (*aa BB cc* and *aa bb CC*), half of the progeny are wild type. What is the genotype of the wild-type male?

3-28 Three different pure-breeding strains of corn that produce ears with white kernels were crossed to each other. In each case the F_1 plants were all red, while both red and white kernels were observed in the F_2 generation in a 9:7 ratio. These results are tabulated here.

	F_1	F_2
white-1 × white-2	red	9 red : 7 white
white-1 × white-3	red	9 red : 7 white
white-2 × white-3	red	9 red : 7 white

   a. How many genes are involved in determining kernel color in these three strains?
   b. Define your symbols and show the genotypes for the pure-breeding strains white-1, white-2, and white-3.
   c. Diagram the cross between white-1 and white-2, showing the genotypes and phenotypes of the F_1 and F_2 progeny. Explain the observed 9:7 ratio.

3-29 In foxgloves, there are three different petal phenotypes: white with red spots (WR), dark red (DR) and light red (LR). There are actually two different kinds of true-breeding WR strains (WR-1 and WR-2) that can be distinguished by two-generation intercrosses with true-breeding DR and LR strains:

Cross	Parental	F_1	F_2 WR	F_2 LR	F_2 DR
1	WR-1 × LR	all WR	480	39	119
2	WR-1 × DR	all WR	99	32	0
3	DR × LR	all DR	0	43	132
4	WR-2 × LR	all WR	193	64	0
5	WR-2 × DR	all WR	286	24	74

   a. What can you conclude about the inheritance of the petal phenotypes in foxgloves?
   b. Ascribe genotypes to the four true-breeding parental strains (WR-1, WR-2, DR, and LR).
   c. A WR plant from the F_2 generation of cross #1 is now crossed with an LR plant. Of 500 total progeny from this cross, there were 253 WR, 124 DR, and

123 LR plants. What are the genotypes of the parents in this WR × LR mating?

3-30 Figure 3.16 on p. 60 and Fig. 3.21b on p. 66 both show traits that are determined by two genes, each of which has two incompletely dominant alleles. But in Fig. 3.16 the gene interaction produces nine different phenotypes, while the situation depicted in Fig. 3.21b shows only five possible phenotypic classes. How can you explain this difference in the amount of phenotypic variation?

3-31 The garden flower *Salpiglossis sinuata* ("painted tongue") comes in many different colors. Several crosses are made between true-breeding parental strains to produce F_1 plants, which are in turn self-fertilized to produce F_2 progeny.

Parents	F_1 phenotypes	F_2 phenotypes
red × blue	all red	102 red, 33 blue
lavender × blue	all lavender	149 lavender, 51 blue
lavender × red	all bronze	84 bronze, 43 red, 41 lavender
red × yellow	all red	133 red, 58 yellow, 43 blue
yellow × blue	all lavender	183 lavender, 81 yellow, 59 blue

   a. State a hypothesis explaining the inheritance of flower color in painted tongues.
   b. Assign genotypes to the parents, F_1 progeny, and F_2 progeny for all five crosses.
   c. In a cross between true-breeding yellow and true-breeding lavender plants, all of the F_1 progeny are bronze. If you used these F_1 plants to produce an F_2 generation, what phenotypes in what ratios would you expect? Are there any genotypes that might produce a phenotype that you cannot predict from earlier experiments, and if so, how might this alter the phenotypic ratios among the F_2 progeny?

3-32 You have come into contact with two unrelated patients who express what you think to be a rare phenotype—a dark spot on the bottom of the foot. According to a medical source, this phenotype is seen in 1 in every 100,000 people in the population. The two patients give their family histories to you and you generate the pedigrees shown on the next page.
   a. Given that this trait is rare, do you think the inheritance is dominant or recessive? Are there any special conditions that appear to apply to the inheritance?
   b. Which nonexpressing members of these families must carry the mutant allele?
   c. If this trait is instead quite common in the population, what alternative explanation would you propose for the inheritance?
   d. Based on this new explanation (part c), which nonexpressing members of these families must have the genotype normally causing the trait?

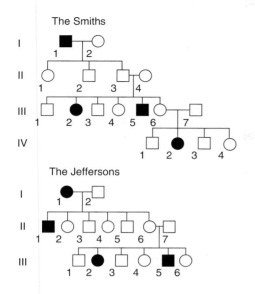

The Smiths

The Jeffersons

3-33 A married man and woman, both of whom are deaf, carry some mutant alleles (all recessive to wild type) in three different "hearing genes": *D1, D2,* and *D3*. Homozygosity for a mutant allele at any one of these three genes causes deafness. In addition, homozygosity for any two of the three genes together in the same genome will cause prenatal lethality (and spontaneous abortion) with a penetrance of 25%. Furthermore, homozygosity for the mutant alleles of all three genes will cause prenatal lethality with a penetrance of 75%. If the genotypes of the mother and father are as indicated here, what is the likelihood that a live-born child will be deaf?

Mother:	*D1/d1, D2/d2, d3/d3*
Father:	*d1/d1, D2/d2, D3/d3*

3-34 In a culture of fruit flies, matings between any two flies with hairy wings (wings abnormally containing additional small hairs along their edges) always produce both hairy-winged and normal-winged flies in a 2:1 ratio. You now take hairy-winged flies from this culture and cross them with four types of normal-winged flies; the results for each cross are shown in the following table. Assuming that there are only two possible alleles of the hairy-winged gene (one for hairy wings and one for normal wings), what can you say about the genotypes of the four types of normal-winged flies?

Type of normal-winged flies	Progeny obtained from cross with hairy-winged flies	
	Fraction with normal wings	Fraction with hairy wings
1	1/2	1/2
2	1	0
3	3/4	1/4
4	2/3	1/3

3-35 Familial hypercholesterolemia (FH) is an inherited trait in humans that results in higher than normal serum cholesterol levels (measured in milligrams of cholesterol per deciliter of blood [mg/dl]). People with serum cholesterol levels that are roughly twice normal have a 25 times higher frequency of heart attacks than unaffected individuals. People with serum cholesterol levels three or more times higher than normal have severely blocked arteries and almost always die before they reach the age of 20. The following pedigrees show the occurrence of FH in four Japanese families:

Family 1

Family 2

Family 3

Family 4

Key to serum cholesterol levels:

○ □ < 250 mg/dl

○ □ 250–500 mg/dl

● ■ > 500 mg/dl

a. What is the most likely mode of inheritance of FH based on this data? Are there any individuals in any of these pedigrees who do not fit your hypothesis? What special conditions might account for such individuals?

b. Why do individuals in the same phenotypic class (unfilled, yellow, or orange symbols) show such variation in their levels of serum cholesterol?

The Chromosome Theory of Inheritance

Each of these three human chromosomes carries
hundreds to thousands of genes.

Down syndrome was the first human genetic disorder attributable not to a gene mutation but to an abnormal number of chromosomes. In the spherical, membrane-bounded nuclei of plant and animal cells prepared for viewing under the microscope, *chromosomes* appear as brightly staining, threadlike, colored bodies. The nuclei of normal, healthy human cells carry 23 pairs of chromosomes for a total of 46. There are noticeable differences in size and shape among the 23 pairs, but within each pair, the two chromosomes appear to match exactly at this level of resolution. (The only exceptions are the male's sex chromosomes, designated X and Y, which constitute an unmatched pair.) Children born with Down syndrome have 47 chromosomes in each cell nucleus because they carry three, instead of the normal pair, of a very small chromosome referred to as number 21. The aberrant genotype, known as trisomy 21, gives rise to an abnormal phenotype, including a wide skull that is flatter than normal at the back, an unusually large tongue, learning disabilities caused by the abnormal development of the hippocampus and other parts of the brain, and a propensity to respiratory infections as well as heart disorders, rapid aging, and leukemia (Fig.4.1).

How can one extra copy of a chromosome that is itself of normal size and shape cause such wide-ranging phenotypic effects? The answer has two parts. First and foremost, chromosomes are the cellular organelles responsible for transmitting genetic information. That is, each chromosome carries a number of genes that interact with each other and those of other chromosomes to determine the development, behavior, and appearance of an individual. (Note that we use the term "organelle" in its most general sense to designate a large, complex structure of characteristic shape that performs essential functions within the cell, rather than in its more specific sense of an intracellular, membrane-bounded entity.) In this chapter, we describe how geneticists interpreted various bits of evidence to arrive at the idea that chromosomes are the carriers of genes, a proposal that became known as the **chromosome theory of inheritance.** The second part of the answer is that proper development depends not just on what type of genetic material is present but also on how much of it there is. Thus the mechanisms governing gene transmission during cell division must vigilantly maintain each cell's chromosome number.

Figure 4.1 Down syndrome: One extra chromosome 21 has widespread phenotypic consequences. Trisomy 21 usually causes changes in the physical characteristics of the head and body, as well as in the potential for learning. Many children with Down syndrome (such as the fifth grader in the multicolored, striped shirt at the center of the photograph) nonetheless participate fully in regular activities.

Proof that genes are located on and travel with chromosomes comes from both breeding experiments and the microscopic examination of cells. As we analyze the evidence, you will see that the behavior of chromosomes during one type of nuclear division called *meiosis* accounts for the segregation and independent assortment of genes proposed by Mendel. Meiosis figures prominently in the process by which most sexually reproducing organisms generate the gametes—eggs or sperm—that at fertilization unite to form the first cell of the next generation. This first cell is the fertilized egg, or *zygote.* Once the union of egg and sperm creates the zygote, a second kind of nuclear division, known as *mitosis,* occurs during the millions of cell divisions that propel development from a single cell to a complex multicellular organism. Mitosis is a conservative process that provides each of the many cells in an individual with the same number and types of chromosomes. Both meiosis and mitosis occur only in the eukaryotic cells of yeast and higher organisms; these cells with a nucleus are the focus of our discussion in this chapter.

The precise chromosome-parceling mechanisms of meiosis and mitosis are so crucial to the normal functioning of an organism that when the machinery does not function properly, errors in chromosome distribution can have dire repercussions on the individual's health and survival. Down syndrome, for example, is the result of a failure of chromosome segregation during meiosis. The meiotic error gives rise to an egg or sperm carrying an extra chromosome 21, which if incorporated in the zygote at fertilization, is passed on via mitosis to every cell of the developing embryo. Trisomy—three copies of a chromosome instead of two—can occur with other chromosomes, but in nearly all such cases, the condition is prenatally lethal and results in a miscarriage.

Two themes emerge in our discussion of meiosis and mitosis. First, it was direct microscopic observations of pairs of chromosomes segregating during gamete formation that led early twentieth-century investigators to recognize that *chromosome movements parallel the behavior of Mendel's genes and are likely to carry the genetic material.* This chromosome theory of inheritance was proposed in 1902 and confirmed in the following 15 years through elegant experiments performed mainly on the fruit fly *Drosophila melanogaster.* Second, by locating the genetic material in a cell organelle specialized for the storage and transmission of genetic information, the chromosome theory transformed the concept of a gene from an abstract particle to a physical reality—part of a chromosome that could be seen and manipulated.

Our discussion of the connection between Mendel's laws of heredity and the behavior of chromosomes during cell division examines:

- Observations and experiments that placed the hereditary material in the cell nucleus, specifically on the chromosomes.

- Mitosis, which distributes chromosomes equally to each of two daughter cells, ensuring that every cell in an organism carries the same set of chromosomes.

- Meiosis, which distributes one member of each chromosome pair to the cells that will become the gametes.

- Gametogenesis, a process including both meiotic and mitotic divisions, by which specialized germ cells differentiate into gametes.

- Validation of the chromosome theory of inheritance.

Chromosomes Contain the Genetic Material

One of the first questions asked at the birth of an infant—Is it a boy or a girl?—acknowledges that male and female are mutually exclusive characteristics like the yellow versus green of Mendel's peas. What's more, among humans and many other sexually reproducing species, there is a roughly 1:1 ratio of the two genders. In any randomly chosen human population, for example, approximately 50% are women and 50% men. Both males and females produce cells specialized for reproduction—sperm or eggs—that serve as a physical link to the next generation. In bridging the gap between generations, these gametes must each contribute half of the genetic material for making a normal, healthy son or daughter. Whatever part of the cell carries this material, its structure and function must be able to account for the either-or aspect of sex determination as well as the generally observed 1:1 ratio of men to women. As we will see, these two features of sex determination were among the earliest clues to the cellular basis of heredity.

Evidence That Genes Reside in the Nucleus

The nature of the link between sex and reproduction remained a mystery until Anton van Leeuwenhoek, one of the earliest and most astute of microscopists, discovered in 1667 that semen contains spermatozoa (literally "sperm animals"). He imagined that these microscopic creatures might enter the egg and somehow achieve fertilization, but it was not possible to confirm this hypothesis for another 200 years. Then, during a 20-year period starting in 1854 (about the same time Gregor Mendel was beginning his pea experiments), microscopists studying fertilization in frogs and sea urchins observed the union of male and female gametes and recorded the details of the process in a series of time-lapse drawings. These drawings, as well as later micrographs (photographs shot through a microscope), clearly show that egg and sperm nuclei are the only elements contributed equally by maternal and paternal gametes. This observation implies that something in the nucleus contains the hereditary material. In humans, the nucleus of a sperm is less than 1 millionth of a meter across at its widest point, and the nucleus of an egg, a little more than twice that size. It is indeed remarkable that the genetic link between generations is packaged within such an exceedingly small space.

Evidence That Genes Reside in Chromosomes

Further investigations, some dependent on technical innovations in microscopy, suggested that yet smaller, discrete structures within the nucleus are the repository of genetic information. In the 1880s, for example, a newly discovered combination of organic and inorganic dyes that stained the nucleus of a cell blue, the cytoplasm pink, and discrete nuclear structures brighter blue than their surroundings revealed the existence of the long, threadlike bodies within the nucleus that we call **chromosomes** (literally "colored bodies"). It was now possible to follow the movement of chromosomes during different kinds of cell division.

In embryonic cells, the brightly staining chromosomal threads split lengthwise in two just before cell division, and each of the two newly forming daughter cells receives one-half of every split thread. The kind of nuclear division followed by cell division that results in two daughter cells containing the same number and type of chromosomes as the original parent cell is called **mitosis** (from the Greek *mitos* meaning "thread" and *-osis* meaning "formation" or "increase"). In contrast to mitosis, in the cells that give rise to male and female gametes, a key event is the segregation of the chromosomes composing each pair such that the resulting gametes receive only one chromosome from each chromosome pair. The gametes thus have half the number of chromosomes as other cells within the same organism. The kind of nuclear division that generates egg or sperm cells containing half the number of chromosomes found in other cells within the same organism is called **meiosis** (from the Greek word for "diminution").

One Chromosome Pair Determines an Individual's Sex

In the first decade of the twentieth century, cytologists—scientists who use the microscope to study cell structure—realized that particular chromosomes carry the information for determining sex. In one study, Walter S. Sutton, a young American graduate student at Columbia University, obtained cells from the testes of the great lubber grasshopper (*Brachystola magna;* Fig. 4.2) and followed them through the meiotic divisions that produce sperm. After Sutton's original studies had been corrected for the existence of the very small Y chromosome (which he had missed), the following picture emerged. Prior to meiosis, precursor cells within the testes of a great lubber grasshopper contain 22 chromosomes in addition to an X chromosome and a Y chromosome—a total of 24 chromosomes. After meiosis, the sperm produced within these testes are of two equally prevalent types: one-half have a set of 11 chromosomes plus an X chromosome, while the other half have a set of 11 chromosomes plus a Y. By comparison, all of the eggs produced by females of the species carry an 11-plus-X set of chromosomes like the set found in the first class of sperm. When a sperm with an X chromosome fertilizes an egg, an XX female grasshopper results; when a Y-containing sperm fuses with an egg, an XY male develops. Sutton concluded that the X and the Y chromosomes determine sex.

Several researchers studying other organisms observed and argued the same point, and it soon became apparent that in many sexually reproducing multicellular species, two distinct chromosomes—known as the **sex chromosomes**—provide the basis of sex determination. One sex carries two copies of the same chromosome (a matching pair), while the other sex has one of each type of sex chromosome (an unmatched pair). The cells of normal human females, for example, contain 23 pairs of chromosomes. The two chromosomes of each pair, including the sex-determining X chromosomes, appear to be identical in size and shape. In males, however, there is one unmatched pair of chromosomes: the larger of these is the X; the smaller, the Y

Figure 4.2 The great lubber grasshopper. In this mating pair, the smaller male is astride the female.

(a)

(b)

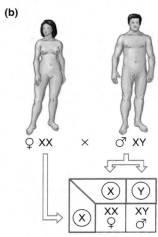

Figure 4.3 How the X and Y chromosomes determine sex in humans. (a) This colorized micrograph shows the human X chromosome on the *left* and the human Y on the *right*. **(b)** Children can receive only an X chromosome from their mother, but they can inherit either an X or a Y from their father.

mosome from every pair). As a result, at fertilization, each gamete would contribute one full set of chromosomes to the zygote. The fertilized egg, or zygote, would thus have two sets of chromosomes.

At Fertilization, Haploid Gametes Produce Diploid Zygotes

Sutton's studies confirmed this idea by showing that the chromosomes in a fertilized egg actually consist of two matching sets, one contributed by the maternal gamete, the other by the paternal gamete. The corresponding maternal and paternal chromosomes appear alike in size and shape (with the exception of the sex chromosomes in most species). While the zygote carries two complete chromosome sets, each gamete contains one-half the number of chromosomes in the zygote, or one complete set. Gametes and other cells that carry only a single set of chromosomes are called **haploid** (from the Greek word for "single"). Zygotes and other cells carrying two matching sets are **diploid** (from the Greek word for "double"). The number of chromosomes in a normal haploid cell is designated by the shorthand symbol n; the number of chromosomes in a normal diploid cell is then $2n$. Figure 4.4 shows diploid cells as well as the haploid gametes that arise from them in *Drosophila* where $2n = 8$ and $n = 4$. In humans, $2n = 46$; $n = 23$. You can

(Fig. 4.3a). Apart from this difference in sex chromosomes, the two sexes are not distinguishable at any other pair of chromosomes. Thus, geneticists can designate women as XX and men as XY and represent sexual reproduction as a simple cross between the two: XX × XY.

Now, if sex is an inherited trait subject to Mendel's rule of segregation—that is, if it is determined by a pair of sex chromosomes that separate to different cells during gamete formation—such a cross could account for both the mutual exclusion of genders and the near 1:1 ratio of males to females, which are among the most striking features of sex determination (Fig. 4.3b). And if chromosomes carry information defining the two contrasting sex phenotypes, it is not extravagant to infer that chromosomes also carry genetic information specifying other characteristics as well. If so, segregation of the chromosomes of every pair—sex and nonsex alike—during gamete formation would produce eggs and sperm each containing one full set of chromosomes (one chro-

Drosophila melanogaster

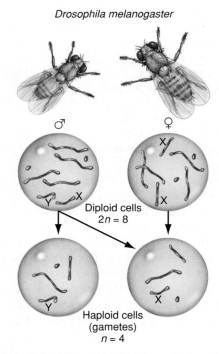

Figure 4.4 Diploid versus haploid: 2n versus n. Most body cells are diploid: They carry a maternal and paternal copy of each chromosome. Meiosis generates haploid gametes with only one copy of each chromosome. In *Drosophila,* diploid cells have eight chromosomes ($2n = 8$), while gametes have four chromosomes ($n = 4$). The fusion of gametes at fertilization yields a diploid zygote (*not shown*). Note that the chromosomes in this diagram are pictured before their replication.

see how the halving of chromosome number during meiosis and gamete formation, followed by the union of two gametes' chromosomes at fertilization, allows a constant 2*n* number of chromosomes to be maintained from generation to generation in all individuals of a species (except, of course, for those carrying rare abnormalities such as trisomy 21). The process of mitosis then ensures that all the cells of the developing individual have identical diploid chromosome sets.

The Number and Shape of Chromosomes Vary from Species to Species

Scientists analyze the chromosomal makeup of a cell by observing the nucleus when the chromosomes are most visible—at a specific moment in the cell cycle of growth and division, just before the nucleus divides. At this point, known as *metaphase* (described in detail later), individual chromosomes have duplicated and condensed from thin threads into compact rodlike structures, each one consisting of two identical halves known as **sister chromatids** attached to each other at a specific location called the **centromere** (Fig. 4.5). In **metacentric** chromosomes, the centromere is more or less in the middle; in **acrocentric** chromosomes, the centromere is very close to one end. When viewed under the microscope, some chromosomes appear to have their centromere right at one end, and geneticists originally coined the term "telocentric" to describe these chromosomes. Recent evidence from high-resolution cytological studies, however, indicates that all apparently telocentric chromosomes have at least a small region extending beyond the

centromere. As a result, even though it is not always possible to see this region under the microscope, cytogeneticists now refer to all chromosomes whose centromere appears near or at one end as acrocentric.

Cells in metaphase can be fixed and stained with one of several dyes that highlight the chromosomes and accentuate the centromeres. In each kind of chromosome, under the right conditions, the dyes also produce characteristic banding patterns of lighter and darker regions, making it possible to distinguish matching chromosome pairs. Chromosomes that match in size, shape, and banding are called **homologous chromosomes,** or **homologs.** The two homologs of each pair normally contain the same set of genes, although for some of those genes, they may carry different alleles. The differences between alleles are too small to show up in the microscope.

Figure 4.5 introduces a system of notation employed throughout the remainder of this book, using color to indicate degrees of relatedness between chromosomes. Thus, sister chromatids—the two daughter molecules of a duplicated chromosome that are joined by a single centromere—which are identical at the completion of replication, appear in the same shade of the same color. Homologous chromosomes, which carry the same genes but may vary in the identity of particular alleles, are pictured in different shades (light or dark) of the same color. *Nonhomologous chromosomes,* which carry completely unrelated sets of genetic information, appear in different colors.

To study the chromosomes of a single organism, geneticists cut images of the stained chromosomes from a micrograph and, by convention, arrange them in homologous pairs of decreasing size to produce a **karyotype.** Figure 4.6 shows the karyotype of a human male, with 46 chromosomes arranged in 22 matching pairs of nonsex chromosomes and one nonmatching pair of sex chromosomes. Despite this obvious difference in appearance, the X and Y chromosomes together constitute a pair. The 44

Figure 4.5 Metaphase chromosomes can be classified by centromere position. Before cell division, each chromosome replicates into two sister chromatids connected at a centromere. In highly condensed metaphase chromosomes, the centromere can appear near the middle (making it a metacentric chromosome), very near an end (forming an acrocentric chromosome), or anywhere in between. In a diploid cell, the chromosomes come in homologous pairs, with one chromosome in the pair from the mother and the other from the father.

Figure 4.6 Karyotype of a human male. Computer-enhanced photos of metaphase human chromosomes are paired and arranged in order of decreasing size. In a normal human male karyotype, there are 22 pairs of autosomes, as well as an X and a Y (2*n* = 46).

nonsex chromosomes are known as **autosomes.** As you can see, autosome number 21 is one of the smallest chromosomes. Interestingly, some dyes may reveal very slight differences in banding patterns between the maternally and paternally derived chromosomes of a homologous pair, and medical researchers can use this information to trace the origin of an extra chromosome, such as the extra 21 that causes Down syndrome. Data collected in studies of this type indicate that in 80% of cases, the third chromosome 21 comes from the egg; in 20%, from the sperm.

Through thousands of karyotypes, cytologists have verified that the cells of each species carry a distinctive diploid number of chromosomes. Among three species of fruit flies, for example, *Drosophila melanogaster* carries 8 (4 pairs of) chromosomes, *Drosophila obscura* carries 10 (5 pairs), and *Drosophila virilis,* 12 (6 pairs). Differences in the size, shape, and number of chromosomes reflect differences in the assembled genetic material that determines what each species looks like and how it functions. Mendel's peas contain 14 (7 pairs of) chromosomes in each diploid cell, macaroni wheat has 28 (14 pairs), giant sequoia trees 22 (11 pairs), goldfish 94 (47 pairs), dogs 78 (39 pairs), and people 46 (23 pairs). As these figures show, the number of chromosomes does not always correlate with the size or complexity of the organism.

There Is Variation Between Species in How Chromosomes Determine an Individual's Sex

As we have already seen, humans and other mammals have a pair of sex chromosomes that are identical in the XX female but different in the XY male. Several studies have shown that in humans, it is the presence or absence of the Y that actually makes the difference; that is, any person carrying a Y chromosome will look like a male. For example, rare humans with two X and one Y chromosomes (XXY) are males displaying certain abnormalities collectively called *Klinefelter syndrome.* Klinefelter males are typically tall, thin, and sterile, and they

may be mentally retarded. That these individuals are males shows that two X chromosomes are insufficient for female development in the presence of a Y. In contrast, humans carrying an X and no second sex chromosome (XO) are females with *Turner syndrome.* Turner females are usually sterile, lack secondary sexual characteristics such as pubic hair, are of short stature, and have folds of skin between their necks and shoulders (webbed necks). Even though these individuals have only one X chromosome, they develop as females because they have no Y chromosome.

Other species show variations on this XX = female / XY = male chromosomal strategy of sex determination. In fruit flies, for example, even though normal females are XX and normal males XY, it is ultimately the ratio of X chromosomes to autosomes (and not the presence or absence of the Y) that determines sex. In female *Drosophila,* the ratio is 1:1 (there are two X chromosomes and two copies of each autosome); in males, the ratio is 1:2 (there is one X chromosome but two copies of each autosome). Curiously, a rarely observed abnormal intermediate ratio of 2:3 produces intersex flies that display both male and female characteristics. Although the Y chromosome in *Drosophila* does not determine whether a fly looks like a male, it is necessary for male fertility; XO flies are sterile males. Table 4.1 compares how humans and *Drosophila* respond to unusual complements of sex chromosomes. Differences between the two species arise in part because the genes they carry on their sex chromosomes are not identical and in part because the strategies they use to deal with the presence of additional sex chromosomes are not the same. (The molecular mechanisms of sex determination of *Drosophila* are covered in detail in Chapter 17.)

The XX = female / XY = male strategy of sex determination is by no means universal. In some species of moths, for example, the females are XX but the males are XO. In birds and butterflies, it is the males that have the matching sex chromosomes, while the females have an unmatched set; in such

---

TABLE 4.1   Sex Determination in Fruit Flies and Humans

	Complement of Sex Chromosomes						
	XXX	XX	XXY	XO	XY	XYY	OY
Drosophila	Dies	Normal female	Normal female	Sterile male	Normal male	Normal male	Dies
Humans	Nearly normal female	Normal female	Klinefelter male (sterile); tall, thin	Turner female (sterile); webbed neck	Normal male	Normal or nearly normal male	Dies

Humans can tolerate extra X chromosomes better than *Drosophila* (compare the fate of XXX individuals). Lack of the X chromosome is lethal to both fruit flies and humans because this chromosome carries essential genes not found on other chromosomes. Additional Y chromosomes seem to have little effect in either species.

species, geneticists represent the sex chromosomes as ZZ in the male and ZW in the female. Yet other variations include the complicated sex-determination mechanisms of bees and wasps, in which females are diploid and males haploid, and the systems of certain fish, in which sex is determined by changes in the environment, such as fluctuations in temperature.

Despite these differences, it became clear that chromosomes carry the genetic information specifying sexual identity and probably many other characteristics as well. Sutton and other early adherents of the chromosome theory realized that because of this, the perpetuation of life itself depends on the proper distribution of chromosomes during cell division. In the next sections, as we examine step by step the microscopic maneuvers by which cells maintain their crucial chromosomal constancy, you will see that the behavior of chromosomes during mitosis and meiosis is exactly that expected of organelles carrying genes.

Mitosis Ensures That Every Cell in an Organism Carries the Same Chromosomes

The fertilized human egg is a single diploid cell that preserves its genetic identity unchanged through more than 100 generations of cells as it divides again and again to produce a full-term infant, ready to be born. As the newborn infant develops into a toddler, a teenager, and an adult, yet more cell divisions fuel continued growth and maturation. Mitosis, the nuclear division that apportions chromosomes in equal fashion to two daughter cells, is the cellular mechanism that preserves genetic information through all these generations of cells. In this section, we take a close look at how the nuclear division of mitosis fits into the overall scheme of cell growth and division.

If you were to peer through a microscope and follow the history of one cell through time, you would see that for much of your observation, the chromosomes resemble a mass of extremely fine tangled string—called **chromatin**—surrounded by the **nuclear envelope.** Each convoluted thread of chromatin is composed mainly of DNA (which carries the genetic information) and protein (which serves as a scaffold for packaging and managing that information, as described in Chapter 12). You would also be able to distinguish one or two darker areas of chromatin called **nucleoli** (singular, *nucleolus,* literally "small nucleus"; the function of these structures will be also be discussed in Chapter 12). Although the chromatin-laden nucleus appears quiescent during this period between cell divisions, it houses a great deal of invisible activity necessary for the growth and survival of the cell. One particularly important part of this activity is the accurate duplication of all the chromosomal material.

With continued vigilance, you would observe during a very short period in the life history of the cell a dramatic change in the nuclear landscape: The chromatin condenses into discrete threads and then each chromosome compacts even further into the twin rods clamped together at the centromere that can be identified in karyotype analysis (review Fig. 4.5 on p. 81). Each rod in a duo is called a **chromatid;** as described earlier, it is an exact duplicate of the other sister chromatid to which it is connected. Continued observation would reveal the doubled chromosomes beginning to jostle around and then lining up at the midplane of the cell. At this point, the sister chromatids comprising each chromosome separate to opposite poles of the now elongating cell, where they become identical sets of chromosomes. Each of the two identical sets eventually ends up enclosed in a separate nucleus in a separate cell. The two cells, known as *daughter cells,* are thus genetically identical.

The repeating pattern of cell growth (an increase in size) followed by division (the splitting of one cell into two) is called the **cell cycle** (Fig. 4.7). Only a small part of the cell cycle is spent in division; the period between divisions is called **interphase.**

During Interphase, Cells Grow and Replicate Their Chromosomes

Interphase consists of three parts: gap 1 (G_1), synthesis (**S**), and gap 2 (G_2) (Fig. 4.7). G_1 lasts from the birth of a new cell to the onset of chromosome replication; for the genetic material, it is

(a) The cell cycle

(b) Chromosomes replicate during S phase

Figure 4.7 The cell cycle: An alternation between interphase and mitosis. (a) Chromosomes replicate to form sister chromatids during synthesis (S phase); the sister chromatids segregate to daughter cells during mitosis (M phase). The gaps between the S and M phases are called the G_1 and G_2 phases. It is during the gaps that most cell growth takes place. **(b)** Interphase consists of the G_1, S, and G_2 phases together. In multicellular organisms, some terminally differentiated cells stop dividing and arrest in a so-called G_0 stage, as shown in part (a).

a period when the chromosomes are neither duplicating nor dividing. During this time, the cell is using the information from the genetic code to make and assemble the materials it needs to function normally and also achieves most of its growth. G_1 varies in length more than any other phase of the cell cycle. In rapidly dividing cells of the human embryo, for example, G_1 is as short as a few hours. In contrast, mature brain cells become arrested in a resting form of G_1 known as G_0 and do not normally divide again during a person's lifetime.

Synthesis (S) is the time when the cell duplicates its genetic material by synthesizing DNA. During duplication, each chromosome doubles to produce identical sister chromatids that will become visible when the chromosomes condense at the beginning of mitosis. The two sister chromatids remain joined to each other at the centromere.

Gap 2 (G_2) is the interval between chromosome duplication and the beginning of mitosis. During this time, the cell may grow (usually less than during G_1); it also synthesizes proteins that are essential to the subsequent steps of mitosis itself.

In addition, during interphase an array of fine microtubules crucial for many interphase processes becomes visible outside the nucleus. The microtubules radiate out into the cytoplasm from a single organizing center known as the **centrosome,** usually located near the nuclear envelope. In animal cells, the discernible core of each centrosome is a pair of small, darkly staining bodies called **centrioles** (Fig. 4.8a); the microtubule-organizing center of plants does not contain centrioles. During the S and G_2 stages of interphase, the centrosomes replicate, producing two centrosomes that remain in extremely close proximity.

During Mitosis (M Phase), Sister Chromatids Separate and Are Apportioned to Different Daughter Nuclei

Although the rigorously choreographed events of nuclear and cellular division occur as a dynamic and continuous process, for the sake of clarity, we take the traditional approach of analyzing the process in separate stages marked by visible cytological events. The artist's sketches in Fig. 4.8 illustrate these stages in

(a) Prophase: (1) Chromosomes condense and become visible; (2) centrosomes move apart toward opposite poles and generate new microtubules; (3) nucleoli begin to disappear.

(b) Prometaphase: (1) Nuclear envelope breaks down; (2) microtubules from the centrosomes invade the nucleus; (3) sister chromatids attach to microtubules from opposite centrosomes.

(c) Metaphase: Chromosomes align on the metaphase plate with sister chromatids facing opposite poles.

(d) Anaphase: (1) Centromeres divide; (2) the now separated sister chromatids move to opposite poles.

(e) Telophase: (1) Nuclear membranes and nucleoli re-form; (2) spindle fibers disappear; (3) chromosomes uncoil and become a tangle of chromatin.

(f) Cytokinesis: The cytoplasm divides, splitting the elongated parent cell into two daughter cells with identical nuclei.

Figure 4.8 Mitosis maintains the chromosome number of the parent cell nucleus in the two daughter nuclei. In the photomicrographs of newt lung cells, chromosomes are stained *blue* and microtubules appear either *green* or *yellow.* The artist's sketches illustrate the stages of mitosis in the nematode *Ascaris,* where 2n = 4.

the nematode *Ascaris,* whose diploid cells contain only four chromosomes (two pairs of homologous chromosomes).

Prophase: Chromosomes Condense (Fig. 4.8a)

During the whole of interphase, the cell nucleus remains intact and the chromosomes are indistinguishable aggregates of chromatin. At **prophase** (from the Greek *pro-* meaning "before"), the gradual emergence, or **condensation,** of individual chromosomes from the undifferentiated mass of chromatin marks the beginning of mitosis. Each emerging chromosome has already been duplicated during interphase and thus consists of sister chromatids attached at the centromere. At this stage in *Ascaris* cells, there are therefore four chromosomes with a total of eight chromatids. The progressive appearance of an array of individual chromosomes is a truly impressive event: Interphase DNA molecules as long as 3–4 cm condense into discrete chromosomes whose length is measured in microns (millionths of a meter). This is equivalent to compacting a 200 m length of thin string (as long as two football fields) into a cylinder 8 mm long and 1 mm wide. With condensation, the duplicated chromosomes convert from a metabolically active state to a condition suitable for subsequent transport to daughter cells. Another visible change in chromatin also takes place during prophase: The darkly staining nucleoli begin to break down and disappear.

In addition to these nuclear events, several important processes that characterize prophase occur outside the nucleus in the cytoplasm. The centrosomes, which replicated during interphase, now move apart and become clearly distinguishable as two separate entities in the light microscope. At the same time, the interphase scaffolding of long, stable microtubules disappears and is replaced by a set of microtubules that rapidly grow from and contract back toward their centrosomal organizing centers. The centrosomes continue to move apart, migrating around the nuclear envelope toward opposite ends of the nucleus, apparently propelled by forces exerted between interdigitated microtubules extending from both centrosomes.

Prometaphase: The Spindle Forms (Fig. 4.8b)

Prometaphase ("before middle stage") begins with the breakdown of the nuclear envelope, which allows microtubules extending from the two centrosomes to invade the nucleus. Chromosomes attach to these microtubules through the **kinetochore,** a structure in the centromere region of each chromatid that is specialized for conveyance. Each kinetochore contains proteins that act as molecular motors, enabling the chromosome to slide along the microtubule. When the kinetochore of a chromosome originally contacts a microtubule at prometaphase, the kinetochore-based motor moves the chromosome toward the centrosome from which that microtubule radiates. Microtubules growing from the two centrosomes randomly capture chromosomes by the kinetochore of one of the two sister chromatids. As a result, it is sometimes possible to observe groups of chromosomes congregating in the vicinity of each centrosome. In this early part of prometaphase, for each chromosome, one chromatid's kinetochore is attached to a microtubule but the sister chromatid's kinetochore remains unattached.

During prometaphase, three different types of microtubule fibers together form the **mitotic spindle;** all of these microtubules originate from the centrosomes, which function as the two "poles" of the spindle apparatus. Microtubules that extend between a centrosome and the kinetochore of a chromatid are called **kinetochore microtubules,** or **centromeric fibers.** Microtubules from each centrosome that are directed toward the middle of the cell are **polar microtubules;** polar microtubules originating in opposite centrosomes interdigitate near the cell's equator. Finally, there are short, unstable **astral microtubules** that extend out from the centrosome toward the cell's periphery.

Near the end of prometaphase, the kinetochore of each chromosome's previously unattached sister chromatid now associates with microtubules extending from the opposite centrosome. This event orients each chromosome such that one sister chromatid faces one pole of the cell, the other, the opposite pole. Experimental manipulation has shown that if both kinetochores become attached to microtubules from the same pole, the configuration is unstable; one of the kinetochores will repeatedly detach from the spindle until it associates with microtubules from the other pole. Thus, attachment of sister chromatids to opposite spindle poles is the only stable arrangement.

Metaphase: Chromosomes Align at the Cell's Equator (Fig. 4.8c)

During **metaphase** ("middle stage"), the connection of sister chromatids to opposite spindle poles sets in motion a series of jostling movements that cause the chromosomes to move toward an imaginary equator halfway between the two poles. The imaginary midline is called the **metaphase plate.** When the chromosomes are aligned along it, the forces pulling and pushing them toward or away from each pole are in a balanced equilibrium. As a result, any movement away from the metaphase plate is rapidly compensated by tension that restores the chromosome to its position equidistant between the poles.

Anaphase: Sister Chromatids Move to Opposite Spindle Poles (Fig. 4.8d)

The nearly simultaneous severing of the centromeric connection between the sister chromatids of all chromosomes indicates that **anaphase** (from the Greek *ana-* meaning "up" as in "up toward the poles") is underway. The separation of sister chromatids allows each chromatid to be pulled toward the spindle pole to which it is connected by its kinetochore microtubules; as the chromatid moves toward the pole, its kinetochore microtubules shorten. Since the arms of the chromatids lag behind the kinetochores, metacentric chromatids have a characteristic V shape during anaphase. The connection of sister chromatids to microtubules emanating from opposite spindle poles means that the genetic information migrating toward one pole is exactly the same as its counterpart moving toward the opposite pole.

Telophase: Identical Sets of Chromosomes Are Enclosed in Two Nuclei (Fig. 4.8e)

The final transformation of chromosomes and the nucleus during mitosis happens at **telophase** (from the Greek *telo-* meaning "end"). Telophase is like a rewind of prophase. The spindle fibers begin to disperse; a nuclear envelope forms around the group of chromatids at each pole; and one or more nucleoli reappears. The former chromatids now function as independent chromosomes, which decondense (uncoil) and dissolve into a tangled mass of chromatin. Mitosis, the division of one nucleus into two identical nuclei, is over.

Cytokinesis: The Cytoplasm Divides (Fig. 4.8f)

In the final stage of cell division, the daughter nuclei emerging at the end of telophase are packaged into two separate daughter cells. This final stage of division is called **cytokinesis** (literally "cell movement"). During cytokinesis, the elongated parent cell separates into two smaller independent daughter cells with identical nuclei. Cytokinesis usually begins during anaphase, but it is not completed until after telophase. The mechanism by which cells accomplish cytokinesis differs in animals and plants. In animal cells, cytoplasmic division depends on a **contractile ring** that pinches the cell into two approximately equal halves, similar to the way the pulling of a string closes the opening of a bag of marbles (Fig. 4.9a). Intriguingly, some types of molecules that form the contractile ring also participate in the mechanism responsible for muscle contraction. In plants, whose cells are surrounded by a rigid cell wall, a membrane-enclosed disk, known as the **cell plate,** forms inside the cell near the equator and then grows rapidly outward, thereby dividing the cell in two (Fig. 4.9b).

During cytokinesis, a large number of important organelles and other cellular components, including ribosomes, mitochondria, membranous structures such as Golgi bodies, and in plants, chloroplasts, must be parcelled out to the emerging daughter cells. The mechanism accomplishing this task does not appear to predetermine which organelle is destined for which daughter cell. Instead, because most cells contain many copies of these cytoplasmic structures, each new cell is bound to receive at least a few representatives of each component. This original complement of structures is enough to sustain the cell until synthetic activity can repopulate the cytoplasm with organelles.

Sometimes cytoplasmic division does not immediately follow nuclear division, and the result is a cell containing more than one nucleus. An animal cell with two or more nuclei is known as a **syncytium.** The early embryos of fruit flies are multinucleated syncytia (Fig. 4.10), as are the precursors of spermatozoa in humans and many other animals. A multinucleate plant tissue is called a **coenocyte;** coconut milk is a nutrient-rich food composed of coenocytes.

In summary, through mitosis plus cytokinesis, a parental cell generates two daughter cells that have the same number and kind of chromosomes as each other and as the original parental cell.

Regulatory Checkpoints Ensure Correct Chromosome Separation During Mitosis

The cell cycle is a complex series of coordinated events that proceed in sequence with maximum precision. In higher organisms, a cell's "decision" to divide depends on both intrinsic fac-

(a) Cytokinesis in an animal cell

Contractile ring

Cleavage furrow

(b) Cytokinesis in a plant cell

Cell plate

Figure 4.9 Cytokinesis: The cytoplasm divides, producing two daughter cells. (a) In this dividing frog zygote, the contractile ring at the cell's periphery has contracted to form a cleavage furrow visible from the exterior that will eventually pinch the cell in two. **(b)** In this dividing onion root cell, a cell plate that began forming near the equator of the cell expands to the periphery, separating the two daughter cells.

Figure 4.10 If cytokinesis does not follow mitosis, one cell may contain many nuclei. In fertilized *Drosophila* eggs, 13 rounds of mitosis take place without being followed by cytokinesis. The result is a single-celled syncytial embryo that contains several thousand nuclei. The photograph shows part of an embryo in which the nuclei are all dividing; chromosomes are in *red* and spindle fibers are in *green.* Nuclei at the *upper left* are in metaphase, while nuclei toward the *bottom right* are progressively later in anaphase. Membranes eventually grow around these nuclei, dividing the embryo into cells.

tors, such as conditions within the cell that register a sufficient size for division, and signals from the environment, such as hormonal cues or contacts with neighboring cells that encourage or restrain division. Once a cell has initiated events leading to division, usually during the G_1 period of interphase, everything else follows like clockwork. A number of checkpoints—moments at which the cell evaluates the results of previous steps—allow the precise, sequential coordination of cell-cycle events. Consequently, under normal circumstances, the chromosomes replicate before they condense, and the doubled chromosomes separate to opposite poles only after correct metaphase alignment of sister chromatids ensures equal distribution to the daughter nuclei (Fig. 4.11). In one illustration of the molecular basis of checkpoints, even a single kinetochore that has not attached to spindle fibers generates a molecular signal that prevents the sister chromatids of all chromosomes from separating at their centromeres. This signal makes the beginning of anaphase dependent on the prior proper alignment of all the chromosomes at metaphase. As a result of the multiple checkpoints operating during the cell cycle, each daughter cell reliably receives the right number of chromosomes.

Breakdown of the mitotic machinery can produce division mistakes that have crucial consequences for the cell. Improper chromosome segregation, for example, can cause serious malfunction or even the death of daughter cells. As the Fast Forward box "How Gene Mutations Cause Errors in Mitosis" explains, one source of improper segregation is gene mutations that disrupt mitotic structures, such as the spindle, kinetochores, or centrosomes. Other problems occur in cells where the normal restraints on cell division have broken down. Such cells may divide uncontrollably, leading to a tumor. We present the details of cell-cycle regulation, checkpoint controls, and cancer formation in Chapter 18.

Meiosis Produces Haploid Germ Cells, the Gametes

During the many rounds of cell division within an embryo, most cells either grow and divide via the mitotic cell cycle described earlier, or they stop growing and become arrested in G_0. These mitotically dividing and G_0-arrested cells are the so-called **somatic cells** whose descendants continue to make up the vast majority of each organism's tissues throughout the lifetime of the individual. Early in the embryonic development of animals, however, a group of cells is set aside for a different fate. These are the **germ cells:** cells destined for a specialized role in the production of gametes. Germ cells arise later in plants, during floral development instead of during embryogenesis. The germ cells become incorporated in the reproductive organs—ovaries and testes in animals; ovaries and anthers in flowering plants—where they ultimately undergo meiosis, the special two-part cell division that produces gametes (eggs and sperm or pollen) containing half the number of chromosomes as other body cells. Normal somatic cells have a finite lifetime; they undergo a certain number of mitotic divisions and then die. In contrast, the germ cells achieve a kind of immortality because, in giving rise to gametes, they transmit an individual's genes to the next generation.

The union of haploid gametes at fertilization yields diploid offspring that carry the combined genetic heritage of two parents. Sexual reproduction therefore requires the alternation of haploid and diploid generations. If gametes were diploid rather than haploid, the number of chromosomes would double in each successive generation such that in humans, for example, the children would have 92 chromosomes per cell, the grandchildren 184, and so on. Meiosis prevents this lethal, exponential accumulation of chromosomes. We now turn to a detailed examination of meiosis, focusing on how the first round of this two-part nuclear division gives rise to genetic variation as it halves a germ cell's chromosome number from two sets to one.

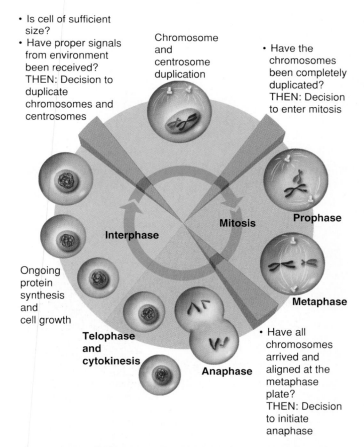

- Is cell of sufficient size?
- Have proper signals from environment been received? THEN: Decision to duplicate chromosomes and centrosomes

Chromosome and centrosome duplication

- Have the chromosomes been completely duplicated? THEN: Decision to enter mitosis

Interphase

Mitosis

Prophase

Metaphase

Ongoing protein synthesis and cell growth

Telophase and cytokinesis

Anaphase

- Have all chromosomes arrived and aligned at the metaphase plate? THEN: Decision to initiate anaphase

Figure 4.11 Checkpoints help regulate the cell cycle. Cellular checkpoints (*red wedges*) ensure that important events in the cell cycle occur in the proper sequence. At each checkpoint, the cell evaluates one or more prior events before deciding whether to proceed to the next step of the cell cycle. (For simplicity, we show here only two chromosomes per cell.)

Meiosis Consists of One Round of Chromosome Replication But Two Rounds of Nuclear Division

Unlike mitosis, meiosis consists of two successive nuclear divisions, logically named **division I of meiosis** and **division II of meiosis,** or simply **meiosis I** and **meiosis II.** With each round,

Fast Forward

How Gene Mutations Cause Errors in Mitosis

During each cell cycle, the chromosomes participate in a tightly patterned choreography that leads to duplication before condensation, midplane alignment before sister chromatid separation, and nuclear division before cytokinesis. Because of these sequential steps, synchronized in both time and space, the chromosomes convey a complete set of genes to each of two newly forming daughter cells. Not surprisingly, some of the very genes they carry encode proteins that shepherd them through the dance.

A variety of proteins, some assembled into structures such as centrosomes and microtubule fibers, make up the molecular machinery that helps coordinate the orderly progression of events in mitosis. Because a particular gene specifies each protein, we might predict that mutant alleles generating defects in particular proteins could disrupt the dance. Cells homozygous for a mutant allele might be unable to complete chromosome duplication or mitosis or cytokinesis because a required component of the molecular machinery is missing or unable to function. Experiments on organisms as disparate as yeast and fruit flies have borne out this prediction. Here we describe the effects of a mutation in one of the many *Drosophila* genes critical for proper chromosome segregation.

Although most mistakes in mitosis are eventually lethal to a multicellular organism, some mutant cells may manage to divide early in development, and when prepared for viewing under the microscope, these cells actually allow us to see the effects of defective mitosis. For example, Fig. A (*left panel*) shows the eight condensed metaphase chromosomes of a wild-type male fruit fly (*Drosophila melanogaster*): two pairs of large metacentric autosomes with the centromere in the center, a pair of dotlike autosomes that are so small it is not possible to see the centromere region, an acrocentric X chromosome with the centromere very close to one end, and a metacentric Y chromosome. Because most of the Y chromosome consists of a special form of chromatin known as heterochromatin, the two Y sister chromatids remain so tightly connected that they often appear as one. Figure B (*left panel*) shows the results of aberrant mitosis in an animal homozygous for a mutation in a gene that encodes a component of the chromosomal kinetochores. The mutation disrupted mitotic chromosome segregation during early development, producing cells with the wrong number of chromosomes. The problem in chromosome segregation revealed by the photo probably occurred during anaphase of the previous cell division. Figure A (*right panel*) shows a normal anaphase separation leading to the wild-type chromosome complement. Figure B (*right panel*) portrays an aberrant anaphase separation in a mutant animal that could lead to an abnormal chromosome complement similar to that depicted in the left panel of the same figure; you can see that many more chromatids are migrating to one spindle pole than to the other.

The smooth unfolding of each cell cycle depends on a diverse array of proteins. Some form the structures that transport chromosomes (like the kinetochores in this example) or that cleave cells in two; others help control when and where things happen. Particular genes specify each of the proteins active in mitosis and cytokinesis, and each protein makes a contribution to the coordinated events of the cell cycle. As a result, a mutation in any of a number of genes can disrupt the meticulously choreographed mitotic mechanism that, when functioning normally, creates generation upon generation of chromosomally identical diploid daughter cells.

X chromosome

Metaphase

Anaphase

Y chromosome

Figure A Metaphase and anaphase chromosomes in a wild-type male fruit fly.

Metaphase

Anaphase

Figure B Metaphase and anaphase chromosomes in a mutant fly. These cells are from a *Drosophila* male hemizygous for an X-linked mutation called *zw10*. The mutant metaphase cell (*left*) contains extra chromosomes as compared with the wild-type metaphase cell in Fig. A. In the mutant anaphase cell (*right*), more chromatids are moving toward one spindle pole than toward the other.

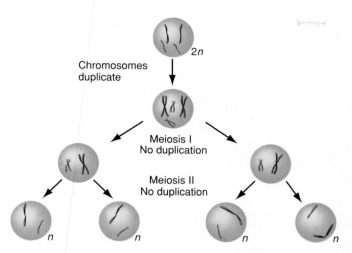

Figure 4.12   An overview of meiosis: The chromosomes replicate once, while the nuclei divide twice. In this figure, all four chromatids of each chromosome pair are shown in the same shade of the same color. Note that the chromosomes duplicate before meiosis I, but they do not duplicate between meiosis I and meiosis II.

the cell passes through a prophase, metaphase, anaphase, and telophase followed by cytokinesis. In meiosis I, the parent nucleus divides to form two daughter nuclei; in meiosis II, each of the two daughter nuclei divides, resulting in four nuclei (Fig. 4.12). These four nuclei—the final products of meiosis—become partitioned in four separate daughter cells because cytokinesis occurs after both rounds of division. The chromosomes duplicate at the start of meiosis I, but they do not duplicate again in meiosis II, which explains why the gametes generated by meiosis contain half the number of chromosomes found in other body cells. A close look at each round of meiotic division reveals the mechanisms by which each gamete comes to receive one full haploid set of chromosomes.

During Meiosis I, Homologous Chromosomes Pair, Exchange Parts, and Then Segregate from Each Other

The events of meiosis I are unique among nuclear divisions (Fig. 4.13, meiosis I). The process begins with the replication of chromosomes, after which each one consists of two sister chromatids. A key to understanding meiosis I is the observation that the centromeres joining these chromatids remain intact throughout the entire division, rather than splitting as in mitosis. As the division proceeds, homologous chromosomes align across the cellular equator to form a coupling that ensures proper chromosome segregation to separate nuclei. Moreover, during the time homologous chromosomes face each other across the equator, the maternal and paternal chromosomes of each homologous pair exchange parts, creating new combinations of alleles at different genes along the chromosomes. Afterward, the two homologous chromosomes, each consisting of two sister chromatids connected at a single, unsplit centromere, are pulled to opposite poles of the spindle. As a result,

it is homologous chromosomes (rather than sister chromatids as in mitosis) that segregate into different daughter cells at the conclusion of the first meiotic division. With this overview in mind, let us take a closer look at the specific events of meiosis I, bearing in mind that we analyze a dynamic, flowing sequence of cellular events by breaking it down somewhat arbitrarily into the easily pictured, traditional phases.

Prophase I: During This Longest, Most Complex Phase of Meiosis, Crossing-Over Occurs

Among the critical events of **prophase I** are the condensation of chromatin, the pairing of homologous chromosomes, and the reciprocal exchange of genetic information between these paired homologs (Fig. 4.13, meiosis I). These complicated events can take many days, months, or even years to complete. For example, in the female germ cells of several species, including our own, meiosis is suspended at prophase I until ovulation. This may sound surprising, but it will become clear when we discuss the details of egg formation later in this chapter.

Leptotene (from the Greek for "thin" and "delicate") is the first definable substage of prophase I, the time when the long, thin chromosomes begin to thicken (see Fig. 4.14a on p. 92 for a more detailed view). Each chromosome has already duplicated prior to prophase I (as in mitosis) and thus consists of two sister chromatids affixed at a centromere. At this point, however, these sister chromatids are so tightly bound together that they are not yet visible as separate entities.

Zygotene (from the Greek for "conjugation") begins as each chromosome seeks out its homologous partner and the matching chromosomes become zipped together in a process known as **synapsis**. The "zipper" itself is an elaborate protein structure called the **synaptonemal complex** that aligns the homologs with remarkable precision, juxtaposing the corresponding genetic regions of the chromosome pair (Fig. 4.14b).

Pachytene (from the Greek for "thick" or "fat") begins at the completion of synapsis when homologous chromosomes are united along their length. Each synapsed chromosome pair is known as a **bivalent** (because it encompasses two chromosomes), or a **tetrad** (because it contains four chromatids, which, as meiosis proceeds, will be parcelled out, one to each of the four products of meiosis). On one side of the bivalent is a maternally derived chromosome, on the other side a paternally derived one. Because X and Y chromosomes are not identical, they do not synapse completely; there is, however, a small region of similarity (or "homology") between the X and the Y chromosomes that allows for a limited amount of pairing.

During pachytene, structures called **recombination nodules** begin to appear along the synaptonemal complex, and an exchange of parts between nonsister (that is, between maternal and paternal) chromatids occurs at these nodules (see Fig. 4.14c on p. 92 for details). Such an exchange is known as **crossing-over;** it results in the **recombination** of genetic material. You can envision the outcome of the process by imagining that the two chromatids constituting a maternal chromosome are dark blue pieces of string, and the paternal chromatids are two light blue pieces of the same length. If you cut one dark

Meiosis: One diploid cell produces four haploid cells.

Meiosis I: A reduction division

Prophase I: Leptotene
1. Chromosomes thicken and become visible but the chromatids remain invisible.
2. Centrosomes begin to move towards opposite poles.

Prophase I: Zygotene
1. Homologous chromosomes enter *synapsis*.
2. The *synaptonemal complex* forms.

Prophase I: Pachytene
1. Synapsis is complete.
2. *Crossing-over*, genetic exchange between nonsister chromatids of a homologous pair, occurs.

Metaphase I
1. Tetrads line up along the *metaphase plate*.
2. Each chromosome of a homologous pair attaches to fibers from opposite poles.
3. Sister chromatids attach to fibers from the same pole.

Anaphase I
1. The centromere does not divide.
2. The chiasmata migrate off chromatid ends.
3. Homologous chromosomes move to opposite poles.

Meiosis II: An equational division

Prophase II
1. Chromosomes condense.
2. Centrioles move toward the poles.
3. The nuclear envelope breaks down at the end of prophase II (not shown).

Metaphase II
1. Chromosomes align at the metaphase plate.
2. Sister chromatids attach to spindle fibers from opposite poles.

Anaphase II
1. Centromeres divide and sister chromatids move to opposite poles.

Prophase I: Diplotene
1. Synaptonemal complex dissolves.
2. A *tetrad* of four chromatids is visible.
3. Crossover points appear as *chiasmata*, which hold nonsister chromatids together.
4. Meiotic arrest occurs at this time in many species.

Prophase I: Diakinesis
1. Chromatids thicken and shorten.
2. At the end of prophase I, the nuclear membrane (not shown earlier) breaks down and the spindle begins to form.

Telophase I
1. The nuclear envelope re-forms.
2. Resultant cells have half the number of chromosomes, each consisting of two sister chromatids.

Interkinesis
1. This is similar to interphase with one important exception: *No chromosomal duplication takes place.*
2. In some species, the chromosomes decondense; in others, they do not.

Telophase II
1. Chromosomes begin to uncoil.
2. Nuclear envelopes and nucleoli (not shown) re-form.

Cytokinesis
1. The cytoplasm divides, forming four new haploid cells.

Figure 4.13   To aid visualization of the chromosomes, the figure is simplified in two ways: (1) The nuclear envelope is not shown during prophase of either meiotic division. (2) The chromosomes are shown as fully condensed at zygotene; in reality the chromosomes continue to condense throughout prophase such that full condensation does not occur until diakinesis.

(a) Leptotene: Threadlike chromosomes begin to condense and thicken, becoming visible as discrete structures. Although the chromosomes have duplicated, the sister chromatids of each chromosome are not yet visible in the microscope.

(b) Zygotene: Chromosomes are clearly visible and begin active pairing with homologous chromosomes along the synaptonemal complex to form a bivalent, or tetrad.

(c) Pachytene: Full synapsis of homologues. Recombination nodules appear along the synaptonemal complex.

(d) Diplotene: Bivalent appears to pull apart slightly but remains connected at crossover sites, called chiasmata.

(e) Diakinesis: Further condensation of chromatids. Nonsister chromatids that have exchanged parts by crossing-over remain closely associated at chiasmata.

Figure 4.14   Prophase I of meiosis at very high magnification.

Figure 4.15   How crossing-over produces recombined chromosomes. Crossing-over and genetic recombination occur after the chromosomes have replicated into sister chromatids. Crossing-over involves cutting, at the same position, one chromatid from each chromatid pair. The subsequent reshuffling and splicing of the cut regions produces two recombinant chromatids of the original size.

blue and one light blue string at the same position and retie dark to light and light to dark, you still have two pieces of string of equal length, but they are recombined mosaics of the originals (Fig. 4.15). As a result of crossing-over, each homologous chromosome in a pair may no longer be of purely maternal or purely paternal origin; each one, however, is still a complete entity, with no gaps in its genetic information.

Diplotene (from the Greek for "twofold" or "double") is signaled by the gradual dissolution of the synaptonemal zipper complex and a slight separation of regions of the homologous chromosomes (see Fig. 4.14d). The aligned homologous chro-

mosomes of each bivalent nonetheless remain very tightly merged at intervals along their length called **chiasmata** (singular, *chiasma*), which represent the sites where crossing-over occurred.

Diakinesis (from the Greek for "double movement") is accompanied by further condensation of the chromatids. Because of this chromatid thickening and shortening, it can now clearly be seen that each tetrad consists of four separate chromatids, or viewed in another way, that the two homologous chromosomes of a bivalent are each composed of two sister chromatids held together at a centromere (see Fig. 4.14e). Nonsister chromatids that have undergone crossing-over remain closely associated at chiasmata. The end of diakinesis is analogous to the prometaphase of mitosis: The nuclear envelope breaks down and the microtubules of the spindle apparatus begin to form.

When Do Chromosome Pairing and Recombination Actually Take Place?   We have described the events of prophase I as they appear in the light microscope. But recent genetic experiments using the unicellular yeast *Saccharomyces cerevisiae* indicate that in this organism, chromosomes somehow begin to pair (or at least loosely align) during and possibly even prior to leptotene, well before synaptonemal complexes become visible during zygotene, and that recombination starts during leptotene or zygotene, before recombination nodules become visible during pachytene. Surprisingly, these experiments also show that recombination can occur even if a synaptonemal complex that tightly juxtaposes the homologous chromosomes does not form. Many investigators are now trying to understand how visible structures such as the synaptonemal complex, recombination nodules, and chiasmata contribute to chromosome pairing and the initiation and completion of genetic recombination.

Metaphase I: Homologous Chromosomes Attach to Spindle Fibers Growing from Opposite Poles (Fig. 4.13, Meiosis I)

There is an essential difference between the spindle formed during meiosis I and that formed during mitosis. As we have seen, during mitosis, each sister chromatid has a kinetochore that becomes attached to microtubules emanating from opposite spindle poles. In contrast, during meiosis I, the kinetochores of sister chromatids fuse so that each chromosome contains only a single functional kinetochore; there are thus no oppositely directed forces that could later pull the sister chromatids apart during the anaphase of meiosis I. Instead, during **metaphase I,** it is the kinetochores of homologous chromosomes that attach to microtubules from opposite spindle poles. As a result, in chromosomes aligned at the metaphase plate, the kinetochores of maternally and paternally derived chromosomes face opposite spindle poles, positioning the homologous chromosomes to which they are connected to move in opposite directions. Because the alignment and hookup of each bivalent is independent of every other bivalent's, the chromosomes facing each pole are a random mix of maternal and paternal origin.

Anaphase I: Homologous Chromosomes Move to Opposite Poles of the Spindle

At the onset of **anaphase I,** the chiasmata joining homologous chromosomes dissolve, which allows the maternal and paternal homologs to begin to move toward opposite spindle poles (see Fig. 4.13, meiosis I). Note that in the first meiotic division, the centromeres do not divide as they do in mitosis. Thus, from each homologous pair, one chromosome consisting of two sister chromatids joined at their centromere segregates to each spindle pole.

Recombination through crossing-over plays an important role in the proper segregation of homologous chromosomes during the first meiotic division. This is because chiasmata, in holding homologous chromosomes together, ensure that their kinetochores remain attached to opposite spindle poles throughout metaphase. When recombination does not occur within a bivalent, mistakes in hookup and conveyance may cause homologous chromosomes to move to the same pole instead of segregating to opposite poles. In some organisms, however, proper segregation of nonrecombinant chromosomes nonetheless occurs through other pairing processes. Investigators do not yet completely understand the nature of these processes and are currently evaluating several models to explain them.

Telophase I: Nuclear Envelopes Reform

The telophase of the first meiotic division, or **telophase I,** takes place when nuclear membranes begin to form around the chromosomes that have moved to the poles. Each of the incipient daughter nuclei contains one-half the number of chromosomes in the original parent nucleus, but each chromosome consists of two sister chromatids joined at the centromere (see Fig. 4.13, meiosis I).

In most species, cytokinesis follows telophase I, with daughter nuclei becoming enclosed in separate daughter cells. A short interphase then ensues. During this time, the chromosomes usually decondense, in which case they must recondense during the prophase of the subsequent second meiotic division. In some cases, however, the chromosomes simply stay condensed. Most importantly, there is no S phase during the interphase between meiosis I and meiosis II; that is, the chromosomes do not replicate during meiotic interphase. The relatively brief interphase between meiosis I and meiosis II is known as **interkinesis.**

During Meiosis II, Sister Chromatids Separate to Produce Haploid Gametes

The second meiotic division (meiosis II) proceeds in a fashion very similar to that of mitosis, but since the number of chromosomes in each dividing nucleus has already been reduced by half, the resulting daughter cells are haploid. The same process occurs in each of the two daughter cells generated by meiosis I, producing four haploid cells at the end of this second meiotic round (see Fig. 4.13, meiosis II).

Prophase II: The Chromosomes Condense as in the Prophase of Mitosis

If the chromosomes decondensed during the preceding interphase, they recondense during **prophase II.** At the end of prophase II, the nuclear envelope breaks down, and the spindle apparatus re-forms.

Metaphase II: Chromosomes Align at the Metaphase Plate

The kinetochores of sister chromatids attach to microtubule fibers emanating from opposite poles of the spindle apparatus, just as in mitotic metaphase. There are nonetheless two significant features of **metaphase II** that distinguish it from mitosis. First, the number of chromosomes is one-half that in mitotic metaphase of the same species. Second, in most chromosomes, the two sister chromatids are no longer strictly identical because of the recombination through crossing-over that occurred during meiosis I. The sister chromatids still contain the same genes, but they may carry different combinations of alleles.

Anaphase II: Sister Chromatids Separate to Opposite Spindle Poles

Just as in mitosis, severing of the centromeric connection between sister chromatids allows them to move toward opposite spindle poles during **anaphase II.**

Telophase II: Nuclear Membrane Formation, Followed by Cytokinesis, Creates Four Haploid Daughter Cells

Membranes form around each of four daughter nuclei in **telophase II,** and cytokinesis places each nucleus in a separate cell. The result is four haploid gametes.

A Summary of the Significant Events of Meiosis

During meiosis, duplication of chromosomes followed by two nuclear divisions reduces the chromosome number by half (review Fig. 4.13). This reduction in chromosome number is not haphazard because in metaphase I, homologous chromosomes pair up and then separate to different daughter cells. At the end of a normal meiosis I, each daughter cell contains one complete set of chromosomes, with each chromosome consisting of two attached sister chromatids. During meiosis II, these two sister chromatids separate to daughter nuclei.

Segregational errors during either meiotic division can lead to aberrations, such as trisomies, in the next generation. If, for example, the homologs of a chromosome pair do not segregate during meiosis I (a mistake known as **nondisjunction**), they may travel together to the same pole and eventually become part of the same gamete. At fertilization, such an error in the segregation of homologous chromosomes may result in any one of a large variety of possible trisomies. Most autosomal trisomies, as we already mentioned, are lethal *in utero;* one exception is trisomy 21, the genetic basis of Down syndrome. Like trisomy 21, extra sex chromosomes may also be nonlethal but cause a variety of mental and physical abnormalities, such as those seen in Klinefelter syndrome (see Table 4.1 on p. 82).

In contrast to rare mistakes in the segregation of one pair of chromosomes, some hybrid animals carry nonhomologous chromosomes that can never pair up and segregate properly. Figure 4.16 shows the two dissimilar sets of chromosomes carried by the diploid cells of a mule; the set inherited from the donkey father contains 31 chromosomes, while the set from the horse mother has 32 chromosomes. Viable gametes cannot form in these animals and the result is sterility.

Figure 4.16 Hybrid sterility: When chromosomes cannot pair during meiosis I, they will segregate improperly. The mating of a male donkey (*Equus asinus; green*) and a female horse (*Equus caballus; red*) produces a mule with 63 chromosomes. In this karyotype of a female mule, the first 13 donkey and horse chromosomes are homologous and pictured in pairs. Starting at chromosome 14, the donkey and horse chromosomes are numbered and pictured separately because they are too dissimilar to pair with each other during meiosis I.

Meiosis Contributes to Genetic Diversity

The wider the assortment of different gene combinations among members of a species, the greater the chance that at least some individuals will carry combinations of alleles that allow survival in a changing environment. Two aspects of meiosis contribute to genetic diversity in a population. First, since it is purely a matter of chance toward which pole the paternal and maternal homologs of each bivalent migrate during the first meiotic division, different gametes carry a different mix of maternal and paternal chromosomes. Figure 4.17a shows how two different patterns of homolog migration produce four different mixes of parental chromosomes in the gametes. The amount of potential variation generated by this random independent assortment increases with the number of chromosomes. In *Ascaris,* for example, where $n = 2$ (the chromosome complement shown in

Fig. 4.17a), the random assortment of homologs could produce only 2^2, or 4 types of gametes; in a person, however, where $n = 23$, this same mechanism alone could generate 2^{23}, or more than 8 million genetically different kinds of gametes. A second feature of meiosis—the reshuffling of genetic information through crossing-over during prophase I—ensures an even greater amount of genetic diversity in gametes. Because crossing-over recombines maternally and paternally derived genes, each chromosome in each different gamete produced by a person could consist of different combinations of maternal and paternal information (Fig. 4.17b). Of course, sexual reproduction adds yet another means of generating genetic diversity. At fertilization, any of a vast number of genetically diverse sperm can fertilize an egg with its own distinctive genetic constitu-

(a) Independent assortment **(b) Recombination**

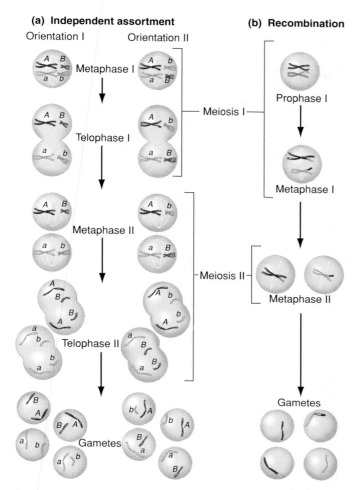

**Figure 4.17 How meiosis contributes to genetic diversity.
(a)** The variation resulting from the independent assortment of nonhomologous chromosomes increases with the number of chromosomes in the genome. **(b)** Crossing-over between homologous chromosomes ensures that each gamete produced by an individual is unique.

tion. It is thus not very surprising that, with the exception of identical twins, the 6 billion people in the world are all genetically unique.

Meiosis and Mitosis: A Comparison

Mitosis occurs in all types of eukaryotic cells (that is, cells with a membrane-bounded nucleus) and is a conservative mechanism that preserves the genetic status quo. Mitosis (followed by cytokinesis) produces growth by increasing the number of cells. It also promotes the continual replacement of roots, stems, and leaves in plants and the regeneration of blood cells, intestinal tissues, and skin in animals. Meiosis, on the other hand, occurs only in sexually reproducing organisms, in just a few specialized germ cells within the reproductive organs that produce haploid gametes. It is not a conservative mechanism; rather, the extensive combinatorial changes arising from meio-

sis are one source of the genetic variation that fuels evolution. Table 4.2 illustrates the significant contrasts between the two mechanisms of cell division.

Gametogenesis Requires Both Mitotic and Meiotic Divisions

In all sexually reproducing animals, the embryonic germ cells (collectively known as the **germ line**) undergo a series of mitotic divisions that yield a collection of specialized diploid cells, which subsequently divide by meiosis to produce haploid cells. As with other biological processes, there are variations on this general pattern. In some species, the haploid cells resulting from meiosis are the gametes themselves, while in other species, those cells must undergo a specific plan of differentiation to fulfill that function. Moreover, in certain organisms, the four haploid products of a single meiosis do not all become gametes. Thus gamete formation, or **gametogenesis,** gives rise to haploid gametes marked not only by the events of meiosis per se but also by cellular events that precede and follow meiosis. Nevertheless, although the specifics of gametogenesis vary widely from species to species as well as from sex to sex within the same species, the general pattern of mitotic divisions followed by a meiotic halving of chromosome number is nearly universal. Here we illustrate gametogenesis with a description of egg and sperm formation in humans. The details of gamete formation in several other organisms appear throughout the book in discussions of specific experimental studies as well as in the Genetic Portraits (Reference chapters A–E).

Oogenesis in Humans: Asymmetrical Meiotic Divisions Produce One Large Ovum

The end product of egg formation in humans is a large, nutrient-rich ovum whose stored resources can sustain the early embryo. The process, known as **oogenesis** (Fig. 4.18), begins when diploid germ cells in the ovary, called **oogonia** (singular, *oogonium*), multiply rapidly by mitosis and produce a large number of **primary oocytes,** which then undergo meiosis. For each primary oocyte, meiosis I results in the formation of two daughter cells that differ in size, so this division is asymmetric. The larger of these cells, the **secondary oocyte,** receives over 95% of the cytoplasm. The other small sister cell is known as the first **polar body.** During meiosis II, the secondary oocyte undergoes another asymmetrical division to produce a large haploid **ovum** and a small, haploid second polar body. The first polar body usually arrests its development and does not undergo the second meiotic division. However, the first polar body does divide in a small proportion of cases, producing two haploid polar bodies. The two (or rarely, three) small polar bodies apparently serve no function and disintegrate, leaving one large haploid ovum as the functional gamete. Thus, only one of the three (or rarely, four) products of a single meiosis serves as a female gamete. A normal human ovum carries 22 autosomes and an X sex chromosome.

TABLE 4.2 **Mitosis and Meiosis: A Comparison**

Mitosis	Meiosis
Occurs in somatic cells Haploid and diploid cells can undergo mitosis One round of division	Occurs in germ cells as part of the sexual cycle Two rounds of division, meiosis I and meiosis II Only diploid cells undergo meiosis

Mitosis is preceded by S phase (chromosome duplication).

Chromosomes duplicate prior to meiosis I but not before meiosis II.

Homologous chromosomes do not pair.

During prophase of meiosis I, homologous chromosomes pair (synapse) along their length.

Genetic exchange between homologous chromosomes is very rare.

Crossing-over occurs between homologous chromosomes during prophase of meiosis I.

Sister chromatids attach to spindle fibers from opposite poles during metaphase.

Homologous chromosomes (not sister chromatids) attach to spindle fibers from opposite poles during metaphase I.

The centromere splits at the beginning of anaphase.

The centromere does not split during meiosis I.

Sister chromatids attach to spindle fibers from opposite poles during metaphase II.

The centromere splits at the beginning of anaphase II.

Mitosis produces two new daughter cells, identical to each other and the original cell. Mitosis is thus genetically conservative.

Meiosis produces four haploid cells, one (egg) or all (sperm) of which can become gametes. None of these are identical to each other or to the original cell, because meiosis results in combinatorial change.

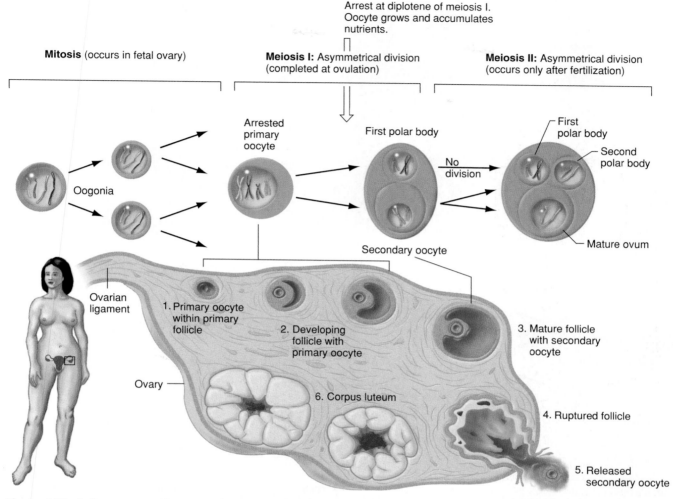

Figure 4.18 In humans, egg formation begins in the fetal ovaries and arrests during the prophase of meiosis I. Fetal ovaries contain about 500,000 primary oocytes arrested in the diplotene of meiosis I. If the egg released during a menstrual cycle is fertilized, meiosis is completed. Only one of the three (rarely, four) cells produced by meiosis serves as the functional gamete, or ovum.

Oogenesis begins in the fetus. By six months after conception, the fetal ovaries are fully formed and contain about half a million primary oocytes arrested in the diplotene of meiosis I. These cells, with their homologous chromosomes locked in synapsis, are the only oocytes the female will produce, so a girl is born with all the oocytes she will ever possess. From the onset of puberty, at about age 12, until menopause some 35–40 years later, most women release one primary oocyte each month (from alternate ovaries), amounting to roughly 480 oocytes released during the reproductive years. The remaining primary oocytes disintegrate during menopause. At ovulation, a released oocyte completes meiosis I and proceeds as far as the metaphase of meiosis II. If the oocyte is then fertilized, that is, penetrated by a sperm nucleus, it quickly completes meiosis II. The nuclei of the sperm and ovum then fuse to form the diploid nucleus of the zygote, and the zygote divides by mitosis to produce a functional embryo. Unfertilized oocytes exit the body during the menses stage of the menstrual cycle.

Interestingly, the long interval before completion of meiosis in oocytes released by women in their 30s, 40s, and 50s may contribute to an observed correlation between maternal age and meiotic segregational errors, including those that produce trisomies. Women in their mid-20s, for example, run a very small risk of trisomy 21; only 0.05% of children born to women of this age have this particular genetic defect. During the later childbearing years, however, the risk rapidly rises; at age 35 it is 0.9% of live births and at age 45, it is 3%. You would not expect this age-related increase in risk if meiosis were completed before the mother's birth.

Spermatogenesis in Humans: Symmetrical Meiotic Divisions Produce Four Sperm

The production of sperm, or **spermatogenesis** (Fig. 4.19), begins in the male testes in germ cells known as **spermatogonia.** Mitotic divisions of the spermatogonia produce many diploid cells, the **primary spermatocytes.** Unlike primary oocytes, primary spermatocytes undergo a symmetrical meiosis

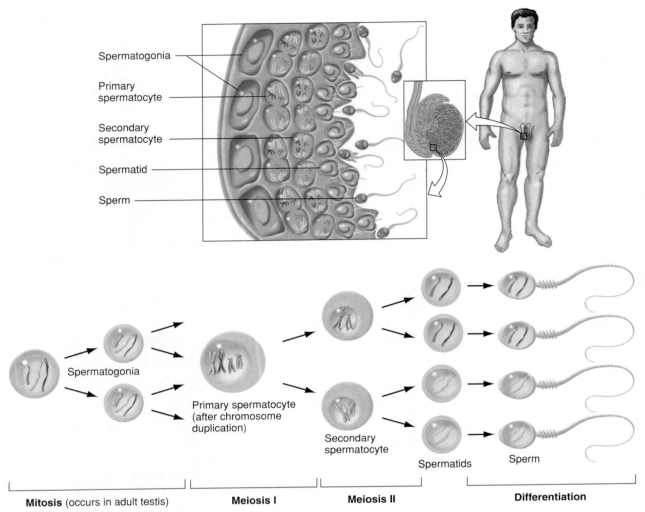

Figure 4.19   Human sperm form continuously in the testes after puberty. Spermatogonia are located near the exterior of seminiferous tubules in a human testis. Once they divide to produce the primary spermatocytes, the subsequent stages of spermatogenesis—meiotic divisions in the spermatocytes and maturation of spermatids into sperm—occur successively closer to the middle of the tubule. This allows for release of the mature sperm into the central lumen of the tubule for ejaculation.

I, producing two **secondary spermatocytes,** each of which undergoes a symmetrical meiosis II. At the conclusion of meiosis, each original primary spermatocyte thus yields four equivalent haploid **spermatids.** These spermatids then mature by developing a characteristic whiplike tail and by concentrating all their chromosomal material in a head, thereby becoming functional **sperm**—mobile microscopic missiles capable of delivering their nuclear payload to the ovum. A human sperm, much smaller than the ovum it will fertilize, contains 22 autosomes and *either* an X *or* a Y sex chromosome.

The timing of sperm production differs radically from that of egg formation. The meiotic divisions allowing conversion of primary spermatocytes to spermatids begin only at puberty, but meiosis then continues throughout a man's life. The entire process of spermatogenesis takes about 48–60 days: 16–20 for meiosis I, 16–20 for meiosis II, and 16–20 for the maturation of spermatids into fully functional sperm. Within each testis after puberty, millions of sperm are always in production, and a sin-

gle ejaculate can contain up to 300 million. Over a lifetime, a man can produce billions of sperm, almost equally divided between those bearing an X and those bearing a Y chromosome.

Validation of the Chromosome Theory

So far, we have presented two circumstantial lines of evidence in support of the chromosome theory of inheritance. First, the phenotype of sexual identity is associated with the inheritance of particular chromosomes. Second, the events of mitosis, meiosis, and gametogenesis ensure a constant number of chromosomes in the somatic cells of all members of a species over time; one would expect the genetic material to exhibit this kind of stability even in organisms with very different modes of reproduction. Final acceptance of the chromosome theory depended on researchers going beyond the circumstantial evidence to a rigorous demonstration of two key points: (1) that the inheritance of genes corresponds with the inheritance of chromosomes in every detail and (2) that the transmission of particular chromosomes coincides with the transmission of specific traits other than sex determination.

The Chromosome Theory Correlates Mendel's Laws with Chromosome Behavior During Meiosis

Walter Sutton first outlined the chromosome theory of inheritance in 1902–1903, building on the theoretical ideas and experimental results of Theodor Boveri in Germany, E. B. Wilson in New York (Sutton's mentor at Columbia and a preeminent cytologist who worked out the details of sex determination in several insect species), and others. In a 1902 paper, Sutton speculated that "the association of paternal and maternal chromosomes in pairs and their subsequent separation during the reducing division [that is, meiosis I] . . . may constitute the physical basis of the Mendelian law of heredity." In 1903, he suggested that chromosomes carry Mendel's hereditary units because:

1. Every cell contains two copies of each kind of chromosome and there are two copies of each kind of gene.

2. The chromosome complement, like Mendel's genes, appears unchanged as it is transmitted from parents to offspring through generations.

3. During meiosis, homologous chromosomes pair and then separate to different gametes just as the alternative alleles of each gene segregate to different gametes.

4. Maternal and paternal copies of each chromosome pair separate to opposite spindle poles without regard to the assortment of any other homologous chromosome pair, just as the alternative alleles of unrelated genes assort independently.

5. At fertilization, an egg's set of chromosomes unites with a randomly encountered sperm's set of chromosomes, just as alleles obtained from one parent unite at random with those from the other parent.

6. In all cells derived from the fertilized egg, one-half of the chromosomes and one-half of the genes are of maternal origin, the other half of paternal origin.

Figure 4.17 on p. 95 and Table 4.3 on p. 100 together show the intimate relationship between the chromosome theory of inheritance and Mendel's laws of segregation and independent assortment. The behavior of chromosomes parallels the behavior of genes. Walter Sutton's observation of these parallels led him to propose that chromosomes and genes are physically connected in some manner. Figure 4.17 follows different chromosomes carrying the alleles of two different genes through the process of meiosis. Meiosis ensures that each gamete will contain only a single chromatid of a bivalent, and thus only a single

allele of any gene on that chromatid. The independent behavior of two bivalents during meiosis means that the genes carried on different chromosomes will assort into gametes independently. Table 4.3 takes this demonstration one step further by illustrating what occurs when the gametes produced by this kind of meiotic chromosome behavior unite at random during fertilization. This table assigns the genes Mendel predicted for pea color and for pea shape to different (that is, nonhomologous) chromosomes. You can see that in this way, Mendel's laws of segregation and independent assortment result immediately from the chromosome theory of inheritance.

From a review of Fig. 4.17, you might wonder whether crossing-over abolishes the clear correspondence between Mendel's laws and the movement of chromosomes. The answer is no. Each chromatid of a homologous chromosome pair contains only one copy of a given gene, and only one chromatid from each pair of homologs is incorporated into each gamete. Because alternative alleles remain on different chromatids even after crossing-over has occurred, alternative alleles still segregate to different gametes as demanded by Mendel's first law. And because the orientation of nonhomologous chromosomes is completely random with respect to each other during both meiotic divisions, even if crossing-over occurs, the genes on different chromosomes assort independently, as demanded by Mendel's second law.

Specific Traits Are Transmitted with Specific Chromosomes

The fate of a theory depends in part on whether its predictions can be validated. Since genes determine traits, the prediction that chromosomes carry genes could be tested by breeding experiments that would show whether transmission of a specific chromosome coincides with transmission of a specific trait. Cytologists knew that one pair of chromosomes, the sex chromosomes, determines whether an individual is male or female. Would similar correlations exist for other traits?

In *Drosophila,* a Gene Determining Eye Color Resides on the X Chromosome

Thomas Hunt Morgan, an American experimental biologist with training in embryology, headed the research group whose findings eventually established a firm experimental base for the chromosome theory. Morgan chose to work with the fruit fly *Drosophila melanogaster* because it is extremely prolific and has a very short generation time, taking only 12 days for a fertilized egg to develop into a mature adult capable of producing hundreds of offspring. Morgan fed his flies mashed bananas and housed them in empty milk bottles capped with wads of cotton.

In 1910 a white-eyed male appeared among a large group of flies with brick-red eyes. A mutation had apparently altered a gene determining eye color, changing it from the normal wild-type allele specifying red to a new allele that produced white. When Morgan allowed the white-eyed male to mate with its red-eyed sisters, all the flies of the F_1 generation had red eyes; the red allele was clearly dominant to the white (Fig. 4.20, cross A).

TABLE 4.3   How the Chromosome Theory of Inheritance Explains Mendel's Laws

Sutton observed that within each somatic cell, the chromosomes are present in pairs. He also saw that each haploid gamete produced at the end of meiosis has one copy of each chromosome. At fertilization, the union of male and female gametes creates a zygote that again has two copies of each chromosome. This pattern of chromosome segregation during meiosis and the union of gametes at fertilization precisely mirrored what Mendel proposed for the alleles of a gene.

One parallel between the behavior of genes and the behavior of chromosomes allowed Sutton to explain Mendel's law of segregation, which as you will recall was based in large part on the results for the F_2 generation in a cross between F_1 hybrids. Sutton assumed that in the hybrids, the allele for round-seeded peas is found on one chromosome, and the allele for wrinkled peas is on the homologous chromosome. If this were true, then the mechanism of meiosis would produce two types of gametes, one carrying the round-allele chromosome, the other carrying the wrinkled-allele chromosome. At fertilization, each gamete has a 50% chance of uniting with a gamete carrying a round-allele chromosome and a 50% chance of uniting with a wrinkled-allele chromosome.

Four possible types of diploid zygotes are equally likely to result, but only one of these, in which both homologous chromosomes carry a wrinkled allele, will become wrinkled peas. This explains Mendel's 3:1 ratio of phenotypes in the F_2 generation.

Another parallel between the behavior of genes and the behavior of chromosomes allowed Sutton to rationalize Mendel's law of independent assortment as follows. One pair of homologous chromosomes carries the gene for seed texture. A second pair of homologous chromosomes carries the gene for seed color. Each homologous pair aligns at random at the

metaphase plate during meiosis I, orienting its two chromosomes toward opposite poles of the cell independently of all other homologous pairs. As a result, a dihybrid individual will produce four equally likely types of gametes with regard to the two traits in question. And as the Punnett square below affirms, this means that the assortment of traits carried by one pair of chromosomes is independent of the assortment of traits carried by all other homologous pairs.

As these figures demonstrate, Mendel's laws follow inexorably if (a) the two alleles of a gene are on homologous chromosomes, and (b) different genes are located on nonhomologous chromosomes. Sutton concluded that genes must therefore be carried by the chromosomes, and that the behaviors of chromosomes during meiosis that he observed in the microscope were the physical basis for the segregation of alleles and the independent assortment of genes.

Establishing a pattern of nomenclature for *Drosophila* geneticists, Morgan named the gene identified by the abnormal white eye color, the *white* gene, for the mutation that revealed its existence. The normal wild-type allele of the *white* gene, abbreviated w^+, is for brick-red eyes, while the counterpart mutant w

allele results in white eye color. The superscript $+$ signifies the wild type. By writing the gene name and abbreviation in lowercase, Morgan symbolized that the mutant w allele is recessive to the wild-type w^+. (If a mutation results in a dominant non-wildtype phenotype, the first letter of the gene name or of its abbre-

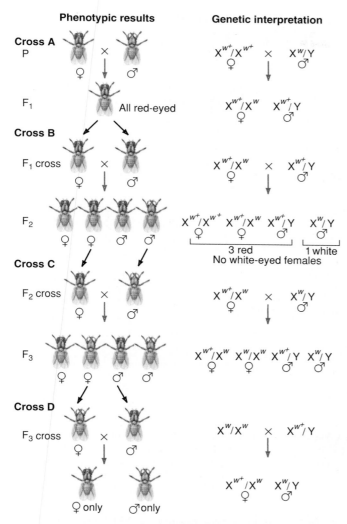

Phenotypic results

Genetic interpretation

Cross A
P

F$_1$ All red-eyed

Cross B

F$_1$ cross

F$_2$

3 red   1 white
No white-eyed females

Cross C

F$_2$ cross

F$_3$

Cross D

F$_3$ cross

♀ only   ♂only

Figure 4.20 A _Drosophila_ eye color gene is located on the X chromosome. X-linkage explains the inheritance of alleles of the _white_ gene.

viation is written in uppercase; thus the mutation known as _Bar_ eyes is dominant to the wild-type _Bar+_ allele.)

Morgan then crossed the red-eyed males of the F$_1$ generation with their red-eyed sisters (Fig. 4.20, cross B) and obtained an F$_2$ generation with the predicted 3:1 ratio of red to white eyes. But there was something askew in the pattern: Among the red-eyed offspring, there were two females for every one male, and all the white-eyed offspring were males. This result was surprisingly different from the observed transmission of Mendelian traits discussed in Chapters 2 and 3. In these fruit flies, the ratio of various phenotypes was not the same in male and female progeny. By mating F$_2$ red-eyed females with their white-eyed brothers (Fig. 4.20, cross C), Morgan obtained some females with white eyes, which then allowed him to mate a white-eyed female with a red-eyed wild-type male (Fig. 4.20, cross D). The result was exclusively red-eyed daughters and white-eyed sons. The pattern seen in cross D is known as **crisscross inheritance** because the males inherit their eye color

from their mothers, while the daughters inherit their eye color from their fathers. Note in Fig. 4.20 that the results of the reciprocal crosses red female × white male (cross A) and white female × red male (cross D) are not identical, again in contrast with Mendel's findings.

From the data, Morgan reasoned that the _white_ gene for eye color is **X linked,** that is, carried by the X chromosome (note that while symbols for genes and alleles are italicized, symbols for chromosomes are not). The Y chromosome carries no allele of this gene for eye color. Males, therefore, have only one copy of the gene, which they inherit from their mother along with their only X chromosome (their Y chromosome must come from their father). Thus, males are **hemizygous** for this eye color gene, because their diploid cells have _half_ the number of alleles carried by the female on her two X chromosomes.

If the single _white_ gene on the X chromosome of a male is the wild-type w^+ allele, he will have red eyes and a genotype that can be written X^{w+} / Y. (Here we designate the chromosome [X or Y] together with the allele it carries, to emphasize that certain genes are X linked. The slash [/], whose use is optional in this context, separates what is carried on the X chromosome [here, the w^+ allele for eye color] from what is carried on the Y chromosome [in this case, no allele for the gene in question]. With a pair of homologous X chromosomes, the symbol to the right of the slash [X^{w+} / X^w] would represent the second X and the allele it carries.) In contrast to an X^{w+}/ Y male, a hemizygous X^w/ Y male whose lone X carries the _w_ allele would have a phenotype of white eyes. Females with two X chromosomes can be one of three genotypes: X^w / X^w (white-eyed), X^w / X^{w+} (red-eyed because w^+ is dominant to _w_), or X^{w+} / X^{w+} (red-eyed).

As shown in Fig. 4.20, Morgan's assumption that the gene for eye color is X linked explains the results of his breeding experiments. Crisscross inheritance, for example, results because the only X chromosome in sons of a white-eyed mother (X^w / X^w) must carry the _w_ allele, so the sons will be white-eyed. In contrast, because daughters of a red-eyed (X^{w+} / Y) father must receive a w^+-bearing X chromosome from their father, they will have red eyes.

Analysis of Rare Mistakes in Meiosis Provided Further Support for the Chromosome Theory

Although Morgan's work strongly supported the hypothesis that the gene for eye color lies on the X chromosome, he himself continued to question the absolute validity of the chromosome theory until Calvin Bridges, one of his top students, found another key piece of evidence. Bridges repeated the cross Morgan had performed between white-eyed females and red-eyed males, but this time he did the experiment on a larger scale. As expected, the progeny of this cross consisted mostly of red-eyed females and white-eyed males. However, about 1 in every 2000 males had red eyes, and about the same small fraction of females had white eyes.

Bridges hypothesized that these exceptions arose through rare events in which the X chromosomes fail to separate during meiosis in females. He called such failures in chromosome segregation _nondisjunction_. As Fig. 4.21a shows, nondisjunction

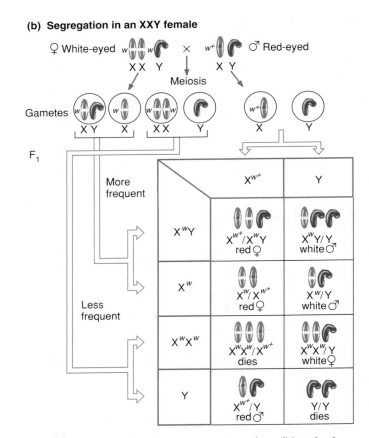

(a) Nondisjunction in an XX female

(b) Segregation in an XXY female

Figure 4.21 Nondisjunction: Rare mistakes in meiosis help confirm the chromosome theory. (a) Rare events of nondisjunction in an XX female produce XX and O eggs. The results of normal disjunction in the female are not shown. XO males are sterile because the missing Y chromosome is needed for male fertility in *Drosophila*. **(b)** In an XXY female, the three sex chromosomes can pair and segregate in two ways, producing progeny with unusual sex chromosome complements.

would result in some eggs with two X chromosomes and others with none. Fertilization of these chromosomally abnormal eggs could produce four types of zygotes: XXY (with two X chromosomes from the egg and a Y from the sperm), XXX (with two Xs from the egg and one X from the sperm), XO (with the lone sex chromosome from the sperm and no sex chromosome from the egg), and OY (with the only sex chromosome again coming from the sperm). When Bridges examined the sex chromosomes of the rare white-eyed females produced in his large-scale cross, he found that they were indeed XXY individuals who must have received two X chromosomes and with them two *w* alleles from their white-eyed X^w / X^w mothers. The exceptional red-eyed males emerging from the cross were XO; their eye color showed that they must have obtained their sole sex chromosome from their X^{w+} / Y fathers. In this study, transmission of the *white* gene alleles followed the predicted behavior of X chromosomes during rare meiotic mistakes, indicating that the X chromosome carries the gene for eye color. These results also suggested that zygotes with the two other abnormal sex chromosome karyotypes (XXX and OY) expected from nondisjunction in females die during embryonic development and thus produce no progeny.

Note that the sexual identity of XO males and XXY females, as well as the lack of XXX and OY adults, is consistent with the sex determination information presented in Table 4.1 on p. 82. OY flies do not survive because the X chromosome carries genes necessary for viability that are not found on the Y. Why XXX flies do not survive is not yet well understood. Researchers currently think that the flies die because of a lethal imbalance between proteins determined by autosomal genes and proteins determined by X-linked genes: (For further details of *Drosophila* sex determination and viability, see the comprehensive example in Chapter 17 on gene regulation in eukaryotes.)

Because the XXY white-eyed females have three sex chromosomes rather than the normal two, Bridges reasoned they would produce four kinds of eggs: XY and X, or XX and Y (Fig. 4.21b). You can visualize the formation of these four kinds of eggs by imagining that when the three chromosomes pair and disjoin during meiosis, two chromosomes must go to one pole and one chromosome to the other. With this kind of segregation, there are only two possible results; either one X and the Y go to one pole, and the second X to the other (yielding XY and X gametes), or the two Xs go to one pole and the

Y to the other (yielding XX and Y gametes). The first of these two scenarios occurs more often because it comes about when the two similar X chromosomes pair with each other, ensuring that they will go to opposite poles during the first meiotic division. The second, less likely possibility happens only if the two X chromosomes fail to pair with each other.

Bridges next predicted that fertilization of these four kinds of eggs by normal sperm would generate an array of sex chromosome karyotypes associated with specific eye color phenotypes in the progeny. Bridges verified all his predictions when he analyzed the eye color and sex chromosomes of a large number of offspring. For instance, he showed cytologically that all of the white-eyed females emerging from the cross in Fig. 4.21b had two X chromosomes and one Y chromosome, while one-half of the white-eyed males had a single X chromosome and two Y chromosomes. Bridges' painstaking observations provided compelling evidence that specific genes do in fact reside on specific chromosomes.

X- and Y-Linked Traits in Humans Are Identified by Pedigree Analysis

A person unable to tell red from green would find it nearly impossible to distinguish the rose, scarlet, and magenta in the flowers of a garden bouquet from the delicately variegated greens in their foliage, or to complete a complex electrical circuit by fastening red-clad metallic wires to red ones and green to green. Such a person has most likely inherited some form of red-green colorblindness, a recessive condition that runs in families and affects mostly males. Among Western Europeans in North America and Europe, 8% of men but only 0.44% of women have this vision defect. Figure 4.22 suggests to readers with normal color vision what people with red-green colorblindness see.

In 1911, E. B. Wilson, a contributor to the chromosome theory of inheritance, combined familiarity with studies of colorblindness and recent knowledge of sex determination by the X and Y chromosomes to make the first assignment of a human gene to a particular chromosome. The gene for red-green colorblindness, he said, lies on the X because the condition usually passes from a maternal grandfather through an unaffected carrier mother to roughly 50% of the grandsons.

Several years after Wilson made his colorblindness gene assignment, pedigree analysis established that various forms of hemophilia, or "bleeders disease" (in which the blood fails to clot properly), also result from mutations on the X chromosome that give rise to a relatively rare, recessive trait. In this context, rare means "infrequent in the population." The family histories under review, including one following the descendants of Queen Victoria of England (Fig. 4.23a), showed that most relatively rare X-linked traits exhibit five basic properties.

1. The trait appears in more males than females since a female has to receive two copies of the rare defective allele to show it, whereas a hemizygous male with only one copy will display the phenotype.

Figure 4.22 Red-green colorblindness is an X-linked recessive trait in humans. How the world looks to a person with either normal color vision (*top*) or a kind of red-green colorblindness known as deuteranopia (*bottom*).

2. The mutation and trait never pass from father to son because sons receive only a Y chromosome from their fathers.

3. An affected male, however, does pass the X-linked mutation to all his daughters, who serve as heterozygous carriers not manifesting the trait but passing it on to their offspring. One-half of the sons of these carrier females will inherit the defective allele and the trait.

4. Thus, the trait often skips a generation as the mutation passes from grandfather through a carrier daughter to grandson.

5. The only time the trait appears in successive generations is when a sister of an affected male is a carrier. If she is, one-half of her sons will show the trait.

Unlike colorblindness and hemophilia, some—although very few—of the known rare mutations on the X chromosome are dominant to the wild-type allele. With such dominant X-linked mutations, more females than males show the aberrant

(a) X-linked recessive: Hemophilia

(b) X-linked dominant: Hypophosphatemia

Figure 4.23   X-linked traits may be recessive or dominant.
(a) Pedigree showing inheritance of the recessive X-linked trait hemophilia in Queen Victoria's family. **(b)** Pedigree showing the inheritance of the dominant X-linked trait hypophosphatemia, commonly referred to as vitamin D–resistant rickets.

phenotype. This is because all of the daughters of an affected male but none of his sons will have the condition, while one-half of the sons and one-half of the daughters of an affected female will receive the dominant allele and therefore show the phenotype. Vitamin D–resistant rickets, or hypophosphatemia, is an example of an X-linked dominant trait. Figure 4.23b presents the pedigree of a family affected by this disease.

Theoretically, phenotypes caused by mutations on the Y chromosome should also be identifiable by pedigree analysis. Such traits would pass from an affected father to all of his sons, and from them to all future male descendants. Females would neither exhibit nor transmit a Y-linked phenotype. However, besides the determination of maleness itself, as well as a contribution to sperm formation and thus male fertility, no clear-cut Y-linked visible traits have turned up. The paucity of known Y-linked traits reflects the fact that the small Y chromosome contains very few genes. Indeed, one would expect the Y chromosome to have only a limited effect on phenotype because normal XX females do perfectly well without it.

Autosomal Genes Can Also Determine Differences Between the Sexes

Not all genes that have a differential effect in the two sexes reside on the X or Y chromosomes. Some autosomal genes govern traits that appear in one sex but not the other, or traits that are expressed differently in the two sexes. **Sex-limited traits** affect a structure or process that is found in one sex but not the other. Mutations in genes for sex-limited traits can influence only the phenotype of the sex that expresses those struc-

Figure 4.24   Male pattern baldness, a sex-influenced trait.
Apparently geneticists will not have therapies for male pattern baldness even by the twenty-third century.

tures or processes. A curious example of a sex-limited trait occurs in *Drosophila* males homozygous for an autosomal recessive mutation known as *stuck,* which affects the ability of mutant males to retract their penis and release the claspers by which they hold on to female genitalia during copulation. The mutant males have difficulty separating from females after mating. In extreme cases both individuals die, forever caught in their embrace. Because females lack penises and claspers, homozygous *stuck* mutant females can mate normally. **Sex-influenced traits** show up in both sexes, but expression of such traits may differ between the two sexes because of hormonal differences. Patterned baldness, a condition in which hair is lost prematurely from the top of the head but not from the sides (Fig. 4.24), is a sex-influenced trait in humans. Men heterozygous for the balding allele lose their hair while still in their 20s, whereas heterozygous women do not show any significant hair loss. In contrast, homozygotes in both sexes become bald (though the onset of baldness in homozygous women is usually much later in life than in homozygous men). This sex-influenced trait is thus dominant in men, recessive in women.

The Chromosome Theory Integrates Many Aspects of Gene Behavior

Mendel had assumed that genes are located in cells. The chromosome theory assigned the genes to a specific structure within cells and explained alternative alleles as physically matching parts of homologous chromosomes. In so doing, the theory provided an explanation of Mendel's laws. Because the mechanism of meiosis ensures that the matching parts of homologous chromosomes will segregate to different gametes (except in rare instances of nondisjunction), meiosis in specialized germ cells ensures that alleles segregate from each other as predicted by Mendel's first law. Because each homologous chromosome pair aligns independently of all others at meiosis I, genes carried on different chromosomes will assort independently, as predicted by Mendel's second law. The chromosome theory is also able to explain the creation of new alleles through mutation, a spontaneous change in a particular gene, that is, in a particular part of a chromosome. If a mutation occurs in the germ line, it can be transmitted to subsequent generations. Finally, through mitotic cell division in the embryo and after birth, each cell in a multicellular organism receives the same chromosomes—and thus the same maternal

and paternal alleles of each gene—as the zygote received from the egg and sperm at fertilization. This means that, with a few exceptions (to be discussed in later chapters), an individual's genome—the chromosomes and genes he or she carries—remains constant throughout life.

C O N N E C T I O N S

T. H. Morgan and his students, collectively known as the *Drosophila* group, acknowledged that Mendelian genetics could exist independently of chromosomes. "Why then, we are often asked, do you drag in the chromosomes? Our answer is that since the chromosomes furnish exactly the kind of mechanism that Mendelian laws call for; and since there is an ever-increasing body of information that points clearly to the chromosomes as the bearers of the Mendelian factors, it would be folly to close one's eyes to so patent a relation. Moreover, as biologists, we are interested in heredity not primarily as a mathematical formulation, but rather as a problem concerning the cell, the egg, and the sperm."

The *Drosophila* group went on to find several X-linked mutations in addition to white eyes. One made the body yellow

instead of brown, another shortened the wings, yet another made bent instead of straight body bristles. These findings raised several compelling questions. First, if the genes for all of these traits are physically linked together on the X chromosome, does this linkage affect their ability to assort independently, and if so, how? Second, does each gene have an exact chromosomal address, and if so, does this specific location in any way affect its transmission? In Chapter 5 we describe how the *Drosophila* group and others analyzed the transmission patterns of genes on the same chromosome in terms of known chromosome movements during meiosis, and then used the information obtained to localize genes at specific chromosomal positions.

E S S E N T I A L   C O N C E P T S

1. *Chromosomes* are cell organelles specialized for the storage and transmission of genetic material. Genes are located on chromosomes and travel with them during cell division and gamete formation.

2. In sexually reproducing organisms, *somatic cells* carry a precise number of homologous pairs of chromosomes, which is characteristic of the species. One chromosome of each pair is of maternal origin; the other, paternal.

3. During the first division of *meiosis,* homologous chromosomes in *germ cells* segregate from each other. As a result, each gamete receives one member of each matching pair, as predicted by Mendel's first law.

4. Also during the first meiotic division, the independent alignment of each pair of homologous chromosomes at the cellular midplane results in the independent

assortment of genes carried on different chromosomes, as predicted by Mendel's second law.

5. *Crossing-over* and the independent alignment of homologs during the first meiotic division generate diversity.

6. The second meiotic division generates gametes with a *haploid* number of chromosomes (n).

7. Fertilization—the union of egg and sperm—restores the *diploid* number of chromosomes ($2n$) to the zygote.

8. *Mitosis* underlies the growth and development of the individual. Through mitosis, diploid cells produce identical diploid progeny cells. During mitosis, the sister chromatids of every chromosome separate to each of two daughter cells; before the next cell division, the chromosomes again duplicate to form sister chromatids.

S O C I A L   A N D   E T H I C A L   I S S U E S

1. Gregg and Jennifer are in their late 20s and just had a baby born with *cri du chat* syndrome. The physical and mental manifestations are microcephaly (small head), mental retardation, and a catlike cry. This dominant condition is due to a loss of a portion of one copy of chromosome 5 and is readily detectable by karyotype analysis. The policy of the health organization to which

Gregg and Jennifer belong is to do routine amniocentesis on pregnant women over 35 years of age and on patients from a family with a history of a detectable genetic condition. (*Amniocentesis* is a technique in which doctors obtain fetal cells for genetic analysis by inserting a needle into the woman's abdomen and withdrawing a sample of amniotic fluid, which contains some cells shed by the fetus.) Jennifer fit into neither category and so was not tested. Jennifer and Gregg feel their doctor was remiss in not ordering the amniocentesis, and they want to bring a malpractice suit against the doctor and the health organization to which they belong. Is there just cause to bring this lawsuit? Should people have a say in which tests they receive? Should doctors be required to give all tests possible? What effect could this have on the cost of health insurance?

2. In 1965, a study on mentally subnormal males with violent tendencies in a Scottish institution was published, indicating that 7 of 197, or 3.5%, had the karyotype XYY. These individuals were, on average, 6 inches taller than the XY males in the institution. This percentage (3.5%) of XYY males in this institution seemed abnormally high when compared with that seen in the general population (about 0.1%). A large-scale study was initiated in 1968 in Boston to screen newborns for the XYY karyotype and to follow up on the development of these boys. In 1974, a group of scientists opposed this study, feeling that it was based on preliminary observations and could be damaging to the boys involved. A major concern was that the boys would be

unfairly branded as abnormal and violent. In addition, the parents had not been fully informed of the purpose of the testing. What is a fair way to approach the scientific question of the effects of an extra Y chromosome in humans? Can one separate social and cultural impacts from biological impacts of an extra Y chromosome?

3. Jack and Jessica are going to have a baby, and because Jessica is 38 years old, she had amniocentesis. The karyotype results indicate that the baby has three copies of chromosome 21 (trisomy 21 or Down syndrome). A genetic counselor and doctor have discussed with the couple the potential emotional and financial concerns in raising a Down syndrome child at home versus sending the child to an institution (many of which are state-supported). Jack and Jessica have decided to continue the pregnancy and raise their child at home. Given their decision to have the child knowing it will be affected, should they be eligible to receive state funds to defray the costs of special educational and therapeutic programs that they may need to raise their child at home? Should society pay for expenses accrued by children with genetic disorders if it is possible to screen for these disorders before birth?

S O L V E D   P R O B L E M S

I. In humans, chromosome 16 sometimes has a heavily staining area in the long arm near the centromere. This feature can be seen through the microscope but has no effect on the phenotype of the person carrying it. When such a "blob" exists on a particular copy of chromosome 16, it is a constant feature of that chromosome and is inherited. A couple conceived a child, but the fetus had multiple abnormalities and was miscarried. When the chromosomes of the fetus were studied, it was discovered that it was trisomic for chromosome 16, and that two of the three chromosome 16s had large blobs. Both chromosome 16 homologs in the mother lacked blobs, but the father was heterozygous for blobs. Which parent experienced nondisjunction, and in which meiotic division did it occur?

Answer

This problem requires an understanding of nondisjunction during meiosis. When individual chromosomes contain some distinguishing feature that allows one homolog to be distinguished from another, it is possible to follow the path of the two homologs through meiosis. In this case, since the fetus had two chromosome 16s with the blob, we can conclude that the extra chromosome came from the father (the only parent with a blobbed chromosome). In which meiotic division did the nondisjunction occur? When nondisjunction occurs during meiosis I, homologs fail

to segregate to opposite poles. If this occurred in the father, the chromosome with the blob and the normal chromosome 16 would segregate into the same cell (a secondary spermatocyte). After meiosis II, the gametes resulting from this cell would carry both types of chromosomes. If such sperm fertilized a normal egg, the zygote would have two copies of the normal chromosome 16 and one of the chromosome with a blob. On the other hand, if nondisjunction occurred during meiosis II in the father in a secondary spermatocyte containing the blobbed chromosome 16, sperm with two copies of the blob-marked chromosome would be produced. After fertilization with a normal egg, the result would be a zygote of the type seen in this spontaneous abortion. *Therefore the nondisjunction occurred in meiosis II in the father.*

II. (a) What sex ratio would you expect among the offspring of a cross between a normal male mouse and a female mouse heterozygous for a recessive X-linked lethal gene? (b) What would be the expected sex ratio among the offspring of a cross between a normal hen and a rooster heterozygous for a recessive Z-linked lethal?

Answer

This problem deals with sex-linked inheritance and sex determination.

a. Mice have a sex determination system of XX = female and XY = male. A normal male mouse ($X^R Y$) × a heterozygous female mouse (X^RX^r) would result in X^RX^R, X^RX^r, X^RY, and X^rY mice. The X^rY mice would die, so there would be a 2:1 ratio of females to males.

b. The sex determination system in birds is ZZ = male and ZW = female. A normal hen (Z^RW) × a heterozygous rooster (Z^RZ^r) would result in Z^RZ^R, Z^RZ^r, Z^RW, and Z^rW chickens. Because the Z^rW offspring do not live, the ratio of females to males would be 1:2.

III. A woman with normal color vision whose father was color-blind mates with a man with normal color vision.

a. What do you expect to see among their offspring?

b. What would you expect if it was the normal man's father who was color-blind?

Answer

This problem involves sex-linked inheritance.

a. The woman's father has a genotype of $X^{cb}Y$. Because the woman had to inherit an X from her father, she must have an X^{cb} chromosome, but because she has normal color vision, her other X chromosome must be X^{CB}. The man she mates with has normal color vision and therefore has an $X^{CB}Y$ genotype. Their children could equally probably be $X^{CB}X^{CB}$ (normal female), $X^{CB}X^{cb}$ (carrier female), $X^{CB}Y$ (normal male), or $X^{cb}Y$ (color-blind male).

b. If the man with normal color vision had a color-blind father, the X^{cb} chromosome would not have been passed on to him, since a man does not inherit an X chromosome from his father. The man has the genotype $X^{CB}Y$ and cannot pass on the color-blind allele.

PROBLEMS

4-1 Choose the best matching phrase in the right column for each of the terms in the left column.

a. meiosis	1. X and Y
b. gametes	2. chromosomes that do not differ between the sexes
c. karyotype	3. one of the two identical halves of a replicated chromosome
d. mitosis	4. microtubule organizing centers at the spindle poles
e. interphase	5. cells in the testes that undergo meiosis
f. syncytium	6. division of the cytoplasm
g. synapsis	7. haploid germ cells that unite at fertilization
h. sex chromosomes	8. an animal cell containing more than one nucleus
i. cytokinesis	9. pairing of homologous chromosomes
j. anaphase	10. one diploid cell gives rise to two diploid cells
k. chromatid	11. the array of chromosomes in a given cell
l. autosomes	12. the part of the cell cycle during which the chromosomes are not visible
m. centromere	13. one diploid cell gives rise to four haploid cells
n. centrosomes	14. cell produced by meiosis that does not become a gamete
o. polar body	15. the time during mitosis when sister chromatids separate
p. spermatocytes	16. connection between sister chromatids

4-2 One oak tree cell with 14 chromosomes undergoes mitosis.

a. How many daughter cells are formed, and what is the chromosome number in each cell?

b. If the same cell had undergone meiosis, how many cells would have resulted, and what would have been the chromosome number in each?

4-3 Humans have 46 chromosomes in each somatic cell.

a. How many chromosomes does a child receive from its father?

b. How many autosomes and how many sex chromosomes are present in each somatic cell?

c. How many chromosomes are present in a human ovum?

d. How many sex chromosomes are present in a human ovum?

4-4 In the moss *Polytrichum commune*, the haploid chromosome number is 7. A haploid male gamete fuses with a haploid female gamete to form a diploid cell that divides and develops into the multicellular sporophyte. Cells of the sporophyte then undergo meiosis to produce haploid cells called spores. What is the probability that an individual spore will contain a set of chromosomes all of which came from the male gamete? Assume no recombination.

4-5 A system of sex determination known as haplodiploidy is found in honeybees. Females are diploid and males (drones) are haploid. Male offspring result from the development of unfertilized eggs. Sperm are produced by mitosis in males and fertilize eggs in the females. Ivory eye is a recessive characteristic in honeybees; wild-type eyes are brown.

a. What progeny would result from an ivory-eyed queen and a brown-eyed drone? Give both genotype and phenotype for progeny produced from fertilized and non-fertilized eggs.

b. What would result from crossing a daughter from the mating in part a with a brown-eyed drone?

4-6 a. What are the four major stages of the cell cycle?
   b. Which stages are included in interphase?
   c. What events distinguish G_1, S, and G_2?

4-7 Indicate which of the cells illustrated matches each of the following stages of mitosis:
   a. anaphase
   b. prophase
   c. metaphase
   d. G_2
   e. telophase/cytokinesis

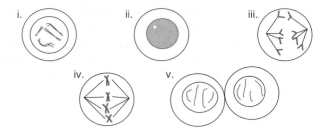

4-8 Somatic cells of chimpanzees contain 48 chromosomes.
   How many chromatids and chromosomes are present at: (a) anaphase of mitosis, (b) anaphase I of meiosis, (c) anaphase II of meiosis, (d) G_1 of mitosis, (e) G_2 of mitosis, (f) G_1 just prior to meiosis I, and (g) prophase of meiosis I?
   How many chromatids or chromosomes are present in: (h) an oogonial cell prior to S phase, (i) a spermatid, (j) a primary oocyte arrested prior to ovulation, (k) a secondary oocyte arrested prior to fertilization, (l) a second polar body, and (m) a chimpanzee sperm?

4-9 Complete the following statements using as many of the following terms as are appropriate: mitosis, meiosis I (first meiotic division), meiosis II (second meiotic division), and none (neither mitosis nor meiosis I nor meiosis II).
   a. The spindle apparatus is present in cells undergoing _____.
   b. Chromosome replication occurs just prior to _____.
   c. The cells resulting from _____ in a haploid cell have a ploidy of n.
   d. The cells resulting from _____ in a diploid cell have a ploidy of n.
   e. Homologous chromosome pairing regularly occurs during _____.
   f. Nonhomologous chromosome pairing regularly occurs during _____.
   g. Physical recombination leading to the production of recombinant progeny classes occurs during _____.
   h. Centromere division occurs during _____.
   i. Nonsister chromatids are found in the same cell during _____.

4-10 The five cells shown in figures a–e are all from the same individual. For each cell, indicate whether it is in mitosis, meiosis I, or meiosis II. What stage of cell division is represented in each case? What is n in this organism?

4-11 Females with Down syndrome are sometimes fertile. A 20-year-old woman has Down syndrome due to trisomy for chromosome 21. She wants to have a child. What is the probability that the child will have Down syndrome? How could this happen?

4-12 A person is simultaneously heterozygous for two autosomal genetic traits. One is a recessive condition for albinism (alleles A and a); this albinism gene is found near the centromere on the long arm of an acrocentric autosome. The other trait is the dominantly inherited Huntington disease (alleles HD and HD^+). The Huntington gene is located near the telomere of one of the arms of a metacentric autosome. Draw all copies of the two relevant chromosomes in this person as they would appear during metaphase of (a) mitosis, (b) meiosis I, and (c) meiosis II. In each figure, label the location on every chromatid of the alleles for these two genes, assuming that no recombination took place.

4-13 One of the first microscopic observations of chromosomes in cell division was published in 1905 by Nettie Stevens. Because it was hard to reproduce photographs at the time, she recorded these observations as *camera lucida* sketches. One such drawing, of a completely normal cell division in the mealworm *Tenebrio molitor,* is shown here. The techniques of the time were relatively unsophisticated by today's standards, and they did not allow her to resolve chromosomal structures that must have been present.

   a. Describe in as much detail as possible the kind of cell division and the stage of division depicted in the drawing.
   b. What chromosomal structure(s) cannot be resolved in the drawing?
   c. How many chromosomes are present in normal *Tenebrio molitor* gametes?

4-14 Some organisms, such as the single-celled alga *Chlamydomonas reinhardtii,* can reproduce sexually or asexually. Sexual reproduction is triggered by stressful conditions. In terms of the survival of the species, what could be an advantage obtained from sexual reproduction in these circumstances?

4-15 In humans:
  a. How many sperm develop from 100 primary spermatocytes?
  b. How many sperm develop from 100 secondary spermatocytes?
  c. How many sperm develop from 100 spermatids?
  d. How many ova develop from 100 primary oocytes?
  e. How many ova develop from 100 secondary oocytes?
  f. How many ova develop from 100 polar bodies?

4-16 Female mammals, including women, sometimes develop benign tumors called "ovarian teratomas" or "dermoid cysts" in their ovaries. Such a tumor begins when a primary oocyte escapes from its prophase I arrest and finishes meiosis I within the ovary. (Normally meiosis I does not finish until the primary oocyte is expelled from the ovary upon ovulation.) The secondary oocyte then develops as if it were an embryo, and it implants and develops within the follicle. Development is disorganized, however, and results in a tumor containing a wide variety of differentiated tissues, including teeth, hair, bone, muscle, nerve, and many others. If a dermoid cyst forms in a woman whose genotype is *Aa,* what are the possible genotypes of the cyst?

4-17 In a certain strain of turkeys, unfertilized eggs sometimes develop parthenogenetically to produce diploid offspring. (Females have ZW and males have ZZ sex chromosomes. Assume that WW cells are inviable.) What distribution of sexes would you expect to see among the parthenogenetic offspring according to each of the following models for how parthenogenesis occurs?
  a. The eggs develop without ever going through meiosis.
  b. The eggs go all the way through meiosis and then duplicate their chromosomes to become diploid.
  c. The eggs go through meiosis I, and the chromatids separate to create diploidy.
  d. The egg goes all the way through meiosis and then fuses at random with one of its three polar bodies (this assumes the first polar body goes through meiosis II).

4-18 Imagine you have two pure-breeding lines of canaries, one with yellow feathers and the other with brown feathers. In crosses between these two strains, yellow female × brown male gives only brown sons and daughters, while brown female × yellow male gives only brown sons and yellow daughters. Propose a hypothesis to explain these results.

4-19 In *Drosophila,* a cross was made between a yellow-bodied male with vestigial (not fully developed) wings and a wild-type female (brown body). The F₁ generation consisted of wild-type males and wild-type females. F₁ males and females were crossed and the F₂ progeny consisted of 16 yellow-bodied males with vestigial wings, 48 yellow-bodied males with normal wings, 15 males with brown bodies and vestigial wings, 49 wild-type males, 31 brown-bodied females with vestigial wings, and 97 wild-type females. Explain the inheritance of the two genes in question based on these results.

4-20 When Calvin Bridges observed a large number of offspring from a cross of white-eyed female *Drosophila* to red-eyed males, he observed very rare white-eyed females and red-eyed males among the offspring. He was able to show that these exceptions resulted from nondisjunction, such that the white-eyed females had received two Xs from the egg and a Y from the sperm, while the red-eyed males had received no sex chromosome from the egg and an X from the sperm. What progeny would have arisen from these same kinds of nondisjunctional events if they had occurred in the male parent? What would their eye colors have been?

4-21 Barred feather pattern is a Z-linked dominant trait in chickens. What offspring would you expect from (a) the cross of a barred hen to a nonbarred rooster? (b) the cross of an F₁ rooster from part a to one of his sisters?

4-22 Each of the four pedigrees shown here represents a human family within which a genetic disease is segregating. Affected individuals are indicated by filled-in symbols. One of the diseases is transmitted as an autosomal recessive condition, one as an X-linked recessive, one as an autosomal dominant, and one as an X-linked dominant. Assume all four traits are rare in the population.
  a. Indicate which pedigree represents which mode of inheritance, and explain how you know.
  b. For each pedigree, how would you advise the parents of the chance that their child (indicated by the hexagon shape) will have the condition?

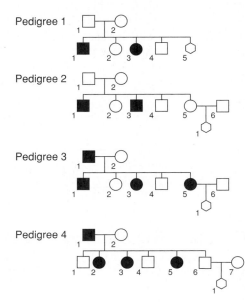

4-23 The pedigree shown here indicates the occurrence of albinism in a group of Hopi Indians, among whom the trait is unusually frequent. Assume that the trait is fully penetrant (all individuals with a genotype that could give rise to albinism will display this condition).
  a. Is albinism in this population caused by a recessive or a dominant allele?
  b. Is the gene sex-linked or autosomal?

What are the genotypes of the following individuals?
  c. individual I-1
  d. individual I-8
  e. individual I-9
  f. individual II-6
  g. individual II-8
  h. individual III-4

4-24 In 1995, doctors reported a Chinese family in which retinitis pigmentosa (progressive degeneration of the retina leading to blindness) affected only males. All six sons of affected males were affected, but all of the five daughters of affected males (and all of the children of these daughters) were unaffected.
  a. What is the likelihood that this form of retinitis pigmentosa is due to an autosomal mutation showing complete dominance?
  b. What other possibilities could explain the inheritance of retinitis pigmentosa in this family? Which of these possibilities do you think is more likely?

4-25 The pedigree at the bottom of this page shows five generations of a family that exhibits congenital hypertrichosis, a rare condition in which affected individuals are born with unusually abundant amounts of hair on their faces and upper bodies. The two small black dots in the pedigree indicate miscarriages.

  a. What can you conclude about the inheritance of hypertrichosis in this family, assuming complete penetrance of the trait?
  b. On what basis can you exclude other modes of inheritance?
  c. How many husbands did III-2 and III-9 have?

4-26 Consider the following pedigrees from human families containing a male with Klinefelter syndrome (a set of abnormalities seen in XXY individuals; indicated with shaded boxes). In each, *A* and *B* refer to codominant alleles of the X-linked *G6PD* gene. The *phenotypes* of each individual (A, B, or AB) are shown on the pedigree. Indicate if nondisjunction occurred in the mother or father of the son with Klinefelter syndrome for each of the three examples. Can you tell if the nondisjunction was in the first or second meiotic division?

4-27 The following is a pedigree of a family in which a rare form of colorblindness is found (filled-in symbols). Indicate as much as you can about the genotypes of all the individuals in the pedigree.

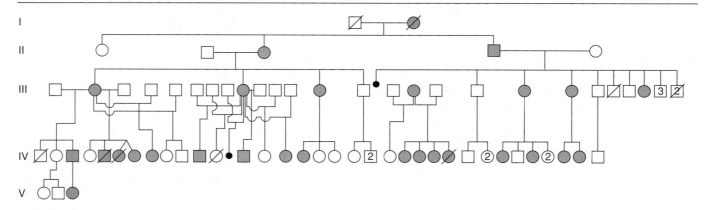

4-28 Duchenne muscular dystrophy (DMD) is caused by a relatively rare X-linked recessive allele. It causes progressive muscular wasting and usually leads to death before age 20.

   a. What is the probability that the first son of a woman whose brother is affected will be affected?

   b. What is the probability that the second son of a woman whose brother is affected will be affected, if her first son was affected?

   c. What is the probability that a child of an unaffected man whose brother is affected will be affected?

   d. An affected man mates with his unaffected first cousin; there is otherwise no history of DMD in this family. If the parents of this couple were sisters, what is the probability that the couple's first child will be an affected boy? An affected girl? An unaffected child?

   e. If the parents of the couple in part (d) were brother and sister, what is the probability that the couple's first child will be an affected boy? An affected girl? An unaffected child?

4-29 Several different antigens can be detected in blood tests. The following four traits were tested for each individual shown:

ABO type	(I^A and I^B codominant, i recessive)
Rh type	(Rh^+ dominant to Rh^-)
MN type	(M and N codominant)
$Xg^{(a)}$ type	($Xg^{(a+)}$ dominant to $Xg^{(a-)}$)

All of these blood type genes are autosomal, except for $Xg^{(a)}$, which is X linked.

Mother	AB	Rh^-	MN	$Xg^{(a+)}$
Daughter	A	Rh^+	MN	$Xg^{(a-)}$
Alleged father 1	AB	Rh^+	M	$Xg^{(a+)}$
Alleged father 2	A	Rh^-	N	$Xg^{(a-)}$
Alleged father 3	B	Rh^+	N	$Xg^{(a-)}$
Alleged father 4	O	Rh^-	MN	$Xg^{(a-)}$

   a. Which, if any, of the alleged fathers could be the real father?

   b. Would your answer to part (a) change if the daughter had Turner syndrome (the abnormal phenotype seen in XO individuals)? If so, how?

4-30 In *Drosophila,* the autosomal recessive *brown* eye color mutation displays interactions with both the X-linked recessive *vermilion* mutation and the autosomal recessive *scarlet* mutation. Flies homozygous for *brown* and simultaneously hemizygous or homozygous for *vermilion* have white eyes. Flies simultaneously homozygous for both the *brown* and *scarlet* mutations also have white eyes. Predict the F_1 and F_2 progeny of crossing the following true-breeding parents:

   a. vermilion females × brown males

   b. brown females × vermilion males

   c. scarlet females × brown males

   d. brown females × scarlet males

4-31 Scientists investigating xeroderma pigmentosum (XP; a human disease causing sensitivity to sunlight that can lead to skin cancer) independently identified two mice with this disease. Starting with the first mouse, they established a true-breeding line of sunlight-sensitive XP mice with short tails; starting with the other mouse, they established a true-breeding XP line with long tails. XP short-tailed females were crossed with XP long-tailed males. All the F_1 progeny (male and female) were found to be short-tailed and to have normal resistance to sunlight (that is, an XP^+ phenotype). The F_1 progeny were interbred to produce an F_2 generation.

Females:		Males:	
XP, short-tailed	22	XP, short-tailed	11
XP, long-tailed	0	XP, long-tailed	10
XP^+, short-tailed	28	XP^+, short-tailed	14
XP^+, long-tailed	0	XP^+, long-tailed	15

Explain the inheritance of both tail length and xeroderma pigmentosum in these mice.

4-32 In 1919, Calvin Bridges began studying an X-linked recessive mutation causing eosin-colored eyes in *Drosophila.* Within an otherwise true-breeding culture of eosin-eyed flies, he noticed rare variants that had much lighter cream-colored eyes. By intercrossing these variants, he was able to make a true-breeding cream-eyed stock. Bridges now crossed males from this cream-eyed stock with true-breeding wild-type females. All the F_1 progeny had red (wild-type) eyes. When F_1 flies were intercrossed, the F_2 progeny were: 104 females with red eyes, 52 males with red eyes, 44 males with eosin eyes, and 14 males with cream eyes. Assume this represents an 8:4:3:1 ratio.

   a. Formulate a hypothesis to explain the F_1 and F_2 results, assigning phenotypes to all possible genotypes.

   b. What do you predict in the F_1 and F_2 generations if the parental cross is between true-breeding eosin-eyed males and true-breeding cream-eyed females?

   c. What do you predict in the F_1 and F_2 generations if the parental cross is between true-breeding eosin-eyed females and true-breeding cream-eyed males?

4-33 As we learned in this chapter, the *white* mutation of *Drosophila* studied by Thomas Hunt Morgan is X linked and recessive to wild type. When true-breeding white-eyed males carrying this mutation were crossed with true-breeding purple-eyed females, all the F_1 progeny had wild-type (red) eyes. When the F_1 progeny were intercrossed, the F_2 progeny emerged in the ratio 3/8 wild-type females : 1/4 white-eyed males : 3/16 wild-type males : 1/8 purple-eyed females : 1/16 purple-eyed males.

   a. Formulate a hypothesis to explain the inheritance of these eye colors.

   b. Predict the F_1 and F_2 progeny if the parental cross was reversed (that is, if the parental cross was between true-breeding white-eyed females and true-breeding purple-eyed males).

4-34 The ancestry of a white female tiger bred in a city zoo is depicted in the following pedigree. White tigers are indicated with unshaded symbols. (As you can see, there was considerable inbreeding in this lineage. For example, Mohan was mated with his daughter.) In answering the following questions, assume that "white" is determined by allelic differences at a single gene and that the trait is fully penetrant. Explain your answers by citing the relevant information in the pedigree.

a. Could white coat color be caused by a Y-linked allele?

b. Could white coat color be caused by a dominant X-linked allele?

c. Could white coat color be caused by a dominant autosomal allele?

d. Could white coat color be caused by a recessive X-linked allele?

e. Could white coat color be caused by a recessive autosomal allele?

4-35 The Fast Forward box in this chapter describes mutations that interfere with chromosome segregation during mitosis in *Drosophila*. Researchers have also been interested in mutations that disrupt chromosome segregation during meiosis. A useful genetic marker for these studies is a special Y chromosome called $y^+ \cdot Y$, which contains a small bit of the X chromosome including the wild-type y^+ allele of the *yellow (y)* gene. The wild-type allele is dominant and produces a brown body color. The y allele yields yellow body color. How could you use the $y^+ \cdot Y$ chromosome to find mutants that have increased levels of nondisjunction during meiosis in *Drosophila* females?

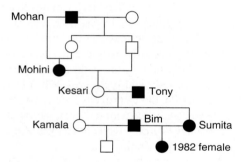

CHAPTER

5

Linkage, Recombination, and the Mapping of Genes on Chromosomes

Maps illustrate the spatial relationships of objects, such as the locations of subway
stations along subway lines. Genetic maps portray the positions of genes
along chromosomes.

In 1928, doctors completed a four-generation pedigree tracing two known X-linked traits: red-green colorblindness and hemophilia A (the more serious X-linked form of "bleeders disease"). The maternal grandfather of the family exhibited both traits, which means that his single X chromosome carried mutant alleles of the two corresponding genes. As expected, neither colorblindness nor hemophilia showed up in his sons and daughters, but two grandsons and one great-grandson inherited both of the X-linked conditions (Fig. 5.1a). The fact that none of the descendants of this family's maternal grandfather manifested one of the traits without the other suggests that the mutant alleles did not assort independently during meiosis. Instead they traveled together in the gametes forming one generation and then into the gametes forming the next generation, producing grandsons and great-grandsons with an X chromosome specifying both colorblindness and hemophilia. Genes that travel together more often than not exhibit **genetic linkage.**

In contrast, another pedigree following colorblindness and the slightly different B form of hemophilia, which also arises from

a mutation on the X chromosome, revealed a different inheritance pattern. A grandfather with hemophilia B and colorblindness had four grandsons, but only one of them exhibited both conditions. In this family, the genes for colorblindness and hemophilia appeared to assort independently, producing in the male progeny all four possible combinations of the two traits—normal vision and normal blood clotting, colorblindness and hemophilia, colorblindness and normal clotting, and normal vision and hemophilia—in approximately equal frequencies (Fig. 5.1b). Thus, even though the mutant alleles of the two genes were on the same X chromosome in the grandfather, they had to separate to give rise to grandsons III-2 and III-3. This separation of genes on the same chromosome is the result of **recombination,** the occurrence in progeny of new gene combinations not seen in previous generations. (Note that *recombinant progeny* can result in either of two ways: from the recombination of genes on the same chromosome during gamete formation, discussed in this chapter, or from the independent assortment of genes on nonhomologous chromosomes, previously described in Chapter 4.)

113

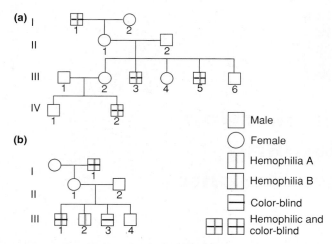

Figure 5.1 Pedigrees indicate that colorblindness and the two forms of hemophilia are X-linked traits. (a) Transmission of red-green colorblindness and hemophilia A. Colorblindness and hemophilia A travel together through the pedigree, indicating their genetic linkage. **(b)** Transmission of red-green colorblindness and hemophilia B. Even though the genes for both conditions are X linked, the mutant alleles are inherited together in only one of four grandsons in generation III. These pedigrees indicate that the gene for color-blindness is close to the hemophilia A gene, but far away from the hemophilia B gene, on the human X chromosome.

In this chapter, as we look at the tools geneticists devised to follow the transmission of genes linked on the same chromosome, you will see that recombination may separate those genes when homologous chromosomes exchange parts during meiosis. The farther apart two genes are, the greater the probability of separation through recombination. Extrapolating from this general rule, you can see that the gene for hemophilia A must be very close to the gene for red-green colorblindness, because, as Fig. 5.1a shows, the two rarely separate. By comparison, the gene for hemophilia B must lie far away from the colorblindness gene, because, as Fig. 5.1b indicates, new combinations of alleles of the two genes occur quite often. Geneticists can use data about how often genes separate during transmission to map the genes' relative locations on a chromosome. Such mapping is a key to sorting out and tracking down the components of complex genetic networks; it is also crucial to geneticists' ability to isolate and characterize genes at the molecular level.

Three general themes emerge from our discussion of genetic linkage and recombination. First, maps reflecting the frequency with which linked genes travel together helped validate the chromosome theory of inheritance by confirming that each chromosome carries many genes in a particular order. Second, the twin concepts of linkage and recombination help explain the genetic underpinnings of evolution. Linkage creates a potential for the simultaneous transmission of blocks of genes that function well together, while recombination produces a potential for gene reshuffling that may enhance the chances of survival under changing conditions. Third, chromosome maps led to further refinement in the concept of a gene. For Mendel, the gene was an abstract unit that controlled a trait and assorted independently of other such units. With the chromosome theory, the gene became part of a chromosome. With mapping, the gene acquired a chromosomal address—a precise location that became the focus of more accurate predictions about inheritance.

Our presentation of the transmission patterns of genes on the same chromosome describes:

■ Linkage and meiotic recombination: Genes linked on the same chromosome usually assort together, instead of independently; such linked genes may nonetheless become separated through recombination during meiosis.

■ Mapping: The frequency with which linked genes become separated through meiotic recombination generally reflects the physical distance between them; recombination data thus make it possible to determine the distance between genes along chromosomes.

■ Mitotic recombination: Rarely, recombination occurs during mitosis. In multicellular organisms, mitotic recombination can produce *genetic mosaicism* in which different cells have different genotypes.

Gene Linkage and Recombination

If people have roughly 40,000–60,000 genes but only 23 pairs of chromosomes, most human chromosomes must carry hundreds, if not thousands, of genes. By the beginning of the twenty-first century, with the completion of the first draft of the Human Genome Project, 698 human genes had been assigned to the X chromosome alone; and that number is likely to grow, at least slightly, as geneticists develop new techniques to analyze the X chromosome's DNA sequence. Recognition that many genes reside on each chromosome raises an important question. If genes on *different* chromosomes assort independently because of the independent segregation of homologs during meiosis, how do genes on the *same* chromosome assort?

Some Genes on the Same Chromosome Assort Together More Often Than Not

We begin our analysis with X-linked *Drosophila* genes because they were the first to be assigned to a specific chromosome and because it is easy to design and follow multigenerational crosses yielding large numbers of *Drosophila* progeny. As we outline various crosses, remember that females carry two alleles for each X-linked gene, while males carry only one. This, of course, is because each female has two X chromosomes, one from each parent, whereas each male has only one, the X inherited from his mother.

In Dihybrid Crosses, Departures from a 1:1:1:1 Ratio of F_1 Gametes Indicate That the Two Genes Are on the Same Chromosome

We look first at two X-linked genes that determine a fruit fly's eye color and body color. (Here and in subsequent discussion, lowercase italicized y and y^+ refer to alleles of the *yellow* gene for body color, while a capital Y refers to the Y chromosome, which does not carry genes for eye or body color.) In a cross between a female with mutant white eyes and a wild-type brown body ($w\,w\,y^+\,y^+$) and a male with wild-type red eyes and a mutant yellow body ($w^+\,y$ plus a Y chromosome, denoted as $w^+y\,/\,Y$), the F_1 offspring are evenly divided between brown-bodied females with normal red eyes ($w^+\,w\,y^+\,y$) and brown-bodied males with mutant white eyes ($w\,y^+\,/\,Y$) (Fig. 5.2). Note that the male progeny look like their mother because their phenotype directly reflects the genotype of the single X chromosome they received from her. The same is not true for the F_1 females, who received w and y^+ on the X from their mother and $w^+\,y$ on the X from their father. With two alleles for each X-linked gene, one derived from each parent, the dominance relations of each pair of alleles determine the female phenotype.

Now comes the significant cross for answering our question about the assortment of genes on the same chromosome. If these two *Drosophila* genes for eye and body color assort independently, as predicted by Mendel's second law, the F_1 females should make four kinds of gametes, with four different combinations of genes on the X chromosome—$w\,y^+$, $w^+\,y$, $w^+\,y^+$, and $w\,y$. These four types of gametes should occur with equal frequency, that is, in a ratio of 1:1:1:1. If it happens this way, approximately half of the gametes will be of the two **parental** types, carrying either the $w\,y^+$ allele combination seen in the original mother (the female of the P generation) or the $w^+\,y$ allele combination seen in the original father (the male of the P generation). The remaining half of the gametes will be of the two **recombinant** types, in which reshuffling has produced either $w^+\,y^+$ or $w\,y$ allele combinations not seen in the P generation parents of the F_1 females.

We can see if the 1:1:1:1 ratio of the four kinds of gametes actually materializes by counting the different types of male progeny in the F_2 generation, as these sons receive their only X-linked genes from their maternal gamete. The bottom part of Fig. 5.2 depicts the results of a breeding study that produced 9026 F_2 males. The relative numbers of the four X-linked gene combinations passed on by the F_1 females' gametes reflect a significant departure from the 1:1:1:1 ratio expected of independent assortment. By far the largest numbers of gametes carry the parental combinations $w\,y^+$ and w^+y. Of the total 9026 flies counted, 8897, or almost 99%, had these genotypes. In contrast, the new combinations w^+y^+ and $w\,y$ made up little more than

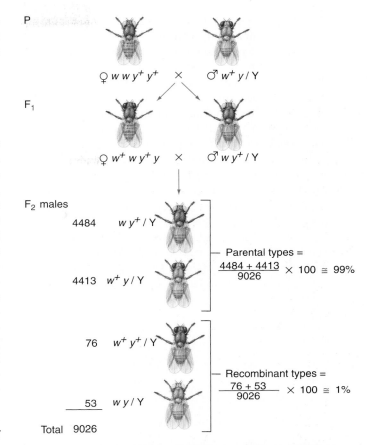

Figure 5.2 When genes are linked, parental combinations outnumber recombinant types. Doubly heterozygous $w^+\,w\,y^+\,y$ F_1 females produce four types of sons. Sons that look like the father ($w^+\,y\,/\,Y$) or mother ($w\,y^+\,/\,Y$) of the F_1 females are parental types. Other sons ($w^+y^+\,/\,Y$ or $w\,y\,/\,Y$) are recombinant types. For these two genes that are very close together on the *Drosophila* X chromosome, there are many more parental than recombinant types among the progeny.

1% of the total. We can explain this outcome in one of two ways. Either the $w\,y^+$ and $w^+\,y$ combinations are preferred because of some intrinsic chemical affinity between these particular alleles, or it is the parental combination of alleles the F_1 female receives from one or the other of her P generation parents that shows up most frequently.

A Preponderance of Parental Genotypes in the F_2 Generation Defines Linkage

The correct explanation of departures from a 1:1:1:1 ratio of F_1 gametes emerges from a second set of crosses (Fig. 5.3) in which the original parental generation consists of red-eyed, brown-bodied $w^+\,w^+\,y^+\,y^+$ females and white-eyed, yellow-bodied $w\,y\,/$ Y males, producing w^+y^+ and $w\,y$ gametes, respectively. All the F_1 females are still all $w^+\,w\,y^+\,y$ dihybrids, as in the first set of crosses. But what about the telltale F_2 males? What ratios of the four possible F_1 female gametes do they reflect?

This time, as Fig. 5.3 shows, $w^+ y$ and $w y^+$ are the recombinants that account for little more than 1% of the total, while $w y$ and $w^+ y^+$ are the parental combinations, which again add up to almost 99%. You can see that there is no preferred association of w^+ and y or of y^+ and w in this cross. Instead, a comparison of the two experiments illustrated in Figs. 5.2 and 5.3 shows that the observed frequencies of the various types of progeny depend on how the arrangement of alleles in the F_1 females originated. It is the **parental classes,** the combinations originally present in the P generation, that show up most frequently in the F_2 generation. The reshuffled **recombinant classes** occur less frequently. It is important to appreciate that the designation of "parental" and "recombinant" gametes or progeny of a doubly heterozygous F_1 female is operational, that is, determined by the particular set of alleles she receives from each of her parents. To keep track of parental allelic combinations, we can represent the genotype of the F_1 female in Fig. 5.3 as $w y / w^+ y^+$, with the alleles she received from different parents shown on different sides of the slash (/). As explained in Chapter 4, each side of the slash shows the alleles of one chromosome in a homologous pair.

When genes assort independently, the numbers of parental and recombinant F_2 progeny are equal, because a doubly heterozygous F_1 individual produces an equal number of all four types of gametes. By comparison, two genes are considered **linked** when the number of F_2 progeny with parental genotypes exceeds the number of F_2 progeny with recombinant genotypes. Instead of assorting independently, the genes behave as if they are connected to each other much of the time. The genes for eye and body color that reside on the X chromosome in *Drosophila* are an extreme illustration of the linkage concept. The two genes are so tightly coupled that the parental combinations of alleles— $w^+ y$ and $w y^+$ (in Fig. 5.2) or $w^+ y^+$ and $w y$ (in Fig. 5.3)—are reshuffled to form recombinants in only 1 out of every 100 gametes formed. In other words, the two parental allele combinations of these tightly linked genes are inherited together 99 times out of 100.

Percentages of Parental and Recombinant Classes Vary with the Gene Pair

Linkage is not always this tight. In *Drosophila,* a mutation for miniature wings (m) is also found on the X chromosome. A cross of red-eyed females with normal wings ($w^+ m^+ / w^+ m^+$) and white-eyed males with miniature wings ($w m$ / Y) yields an F_1 generation containing all red-eyed, normal-winged flies. The genotype of the F_1 females is $w^+ m^+ / w m$ (Fig. 5.4). Of the F_2 males, 67.2% are parental types ($w^+ m^+$ and $w m$), while the remaining 32.8% are recombinants ($w m^+$ and $w^+ m$). This preponderance of parental combinations among the F_2 genotypes reveals that the two genes are linked: The parental combinations of alleles travel together more often than not. But compared to the 99% linkage between the w and y genes for eye color and body color, the linkage of w to m is not that tight. The parental combinations for color and wing size are reshuffled in not 1 but instead in roughly 33 out of every 100 gametes.

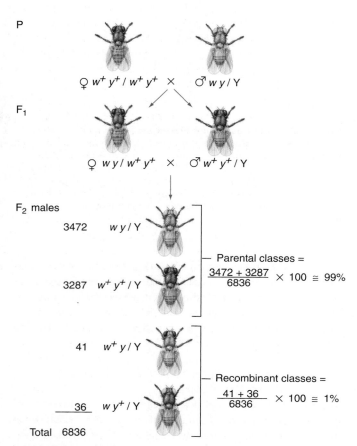

P

$♀\ w^+ y^+ / w^+ y^+ \times ♂ w y / Y$

F_1

$♀\ w y / w^+ y^+ \times ♂ w^+ y^+ / Y$

F_2 males

3472 $w y$ / Y

3287 $w^+ y^+$ / Y

Parental classes = $\dfrac{3472 + 3287}{6836} \times 100 \cong 99\%$

41   $w^+ y$ / Y

36   $w y^+$ / Y

Recombinant classes = $\dfrac{41 + 36}{6836} \times 100 \cong 1\%$

Total   6836

Figure 5.3   Designations of "parental" and "recombinant" relate to past history. The parental and recombinant classes here are the opposite of those in Fig. 5.2, because the F_1 females received different allelic combinations of the *white* and *yellow* genes from their parents. The percentages of recombinant and parental types among the progeny are essentially the same in both experiments, showing that the frequency of recombination is independent of the particular arrangement of alleles.

P $♀\ w^+ m^+ / w^+ m^+ \times ♂ w m$ / Y

F_1 $♀\ w^+ m^+ / w m \times w^+ m^+$ / Y

F_2 males

412 $w^+ m^+$ / Y ⎤ Parental   = $\dfrac{412 + 389}{1192} \times 100 = 67.2\%$
389 $w m$ / Y   ⎦ classes

206 $w^+ m$ / Y ⎤ Recombinant = $\dfrac{206 + 185}{1192} \times 100 = 32.8\%$
185 $w m^+$ / Y ⎦ classes

Total 1192

Figure 5.4   The recombination frequency depends on the gene pair. The recombination frequency (RF) for the w and m genes (32.8%) is very different from the RF for w and y (about 1%; see Figs. 5.2 and 5.3). The w and y genes are thus very close together, while the w and m genes are farther apart.

To determine whether two human genes on the X chromosome are linked, and if so, how tightly, geneticists analyze pedigrees similar to the one depicted in Fig. 5.1a. It takes many such pedigrees to develop a sample of adequate size. Such studies have shown that in families where a man of the P generation has both colorblindness and hemophilia A so that his daughters (the F_1) must carry an X chromosome with defects for both traits, roughly 97% of the daughters' sons (the F_2) are either affected by both conditions like their grandfather or show no phenotypic abnormality. The remaining 3% have *either* hemophilia *or* colorblindness, not both. These phenotypic ratios indicate that the genes underlying colorblindness and hemophilia A in humans are closely linked on the X chromosome. The reason no recombination showed up in the pedigree depicted in Fig. 5.1a is that the sample size was too small.

Autosomal Traits Also Exhibit Linkage

In the decades following the 1900 rediscovery of Mendel's laws, certain departures from the 9:3:3:1 ratio expected of two independently assorting non-sex-linked traits did not yield to analysis in terms of the gene interactions discussed in Chapter 3. T. H. Morgan and his colleagues realized that the concept of linkage might explain some of these anomalous results. Because all progeny receive two copies of genes carried by autosomes, it is harder to follow the transmission of non-sex-linked traits than those determined by the X chromosome. But a testcross in which a homozygous parent contributes only recessive alleles makes it possible to analyze the gene combinations transmitted by the gametes of the other, doubly heterozygous F_1 parent.

Fruit flies, for example, carry an autosomal gene for body color (in addition to the X-linked *y* gene); the wild type is once again brown, but a recessive mutation in this gene gives rise to black (*b*). A second autosomal gene helps determine the shape of a fruit fly's wing, with the wild type having straight edges and a recessive mutation (*c*) producing curves. Figure 5.5 depicts a cross between black-bodied females with straight wings

($b\ c^+\ /\ b\ c^+$) and brown-bodied males with curved wings ($b^+\ c\ /\ b^+\ c$). All the F_1 progeny are double heterozygotes ($b\ c^+\ /\ b^+\ c$) that are phenotypically wild type. In a testcross of the F_1 females with $b\ c\ /\ b\ c$ males, all of the offspring receive the recessive *b* and *c* alleles from their father. The phenotypes of the offspring thus indicate the kinds of gametes received from the mother. For example, a black fly with normal wings would be genotype $b\ c^+\ /\ b\ c$; since we know it received the *b c* combination from its father, it must have received *b* c^+ from its mother. As Fig. 5.5 shows, roughly 77% of the testcross progeny in one experiment received parental gene combinations (that is, allelic combinations transmitted into the F_1 females by the gametes of each of her parents), while the remaining 23% were recombinants. The parental classes thus outnumber the recombinant classes, and we can conclude that the autosomal genes for black body and curved wings are linked.

To summarize, many pairs of genes on both autosomes and sex chromosomes exhibit linkage; that is, they do not assort independently, but instead they are transmitted together more than 50% of the time. (Do not confuse the general concept of linkage with the idea that some genes are X linked, which simply means that they reside on the X chromosome.) For each pair of genes, a particular linkage percentage indicates how often parental combinations travel together. In the preceding examples, the autosomal genes for body color and wing shape were transmitted together 77% of the time, the X-chromosome genes for eye color and wing size were transmitted together 67.2% of the time, and the X-chromosome genes for eye and body color were transmitted together 99% of the time. Linkage is never 100%. No matter how tightly two genes are linked, if you observe enough individuals, you will find some recombinants.

The Chi Square Test Pinpoints the Probability That Experimental Results Are Evidence for Linkage

How do you know from a particular experiment whether two genes assort independently or are genetically linked? At first glance, this question should pose no problem. Discriminating between the two possibilities involves straightforward calculations based on assumptions well supported by observations. For independently assorting genes, a dihybrid F_1 female produces four types of gametes in equal numbers so that one-half of the F_2 progeny are of the parental classes and the other half of the recombinant classes. In contrast, for linked genes, the two types of parental classes by definition always outnumber the two types of recombinant classes in the F_2 generation.

The problem is that because real-world genetic transmission is based on chance events, in a particular study even unlinked, independently assorting genes can produce deviations from the 1:1:1:1 ratio, just as in 10 tosses of a coin, you may easily get 6 heads and 4 tails (rather than the predicted 5 and 5). Thus, if a breeding experiment analyzing the transmission of two genes shows a deviation from the equal ratios of parentals and recombinants expected of independent assortment, can we necessarily conclude the two genes are linked? Is it instead possible that the results represent a statistically acceptable chance fluctuation from the mean values expected of unlinked genes

Figure 5.5 Autosomal genes can also exhibit linkage. Genes for body color (*b*) and wing shape (*c*) are both autosomal. A testcross shows that the recombination frequency (RF) for this pair of genes is 23%. Because parentals outnumber recombinants, the *b* and *c* genes are genetically linked and must be on the same *Drosophila* autosome.

that assort independently? Such questions become more pressing in cases where linkage is not all that tight, so that even though the genes are linked, the percentage of recombinant classes approaches 50%.

To answer these kinds of questions, statisticians have devised a quantitative measure that indicates how often an experimentally observed deviation from the predictions of a particular hypothesis will occur solely by chance. This measure of the "goodness of fit" between observed and predicted results is a probability test known as the **chi square test.** The test is designed to account for the fact that the size of an experimental population (the "sample size") is an important component of statistical significance. To appreciate the role of sample size, it is useful to return to the proverbial coin toss before examining the details of the chi square test.

We have seen that in 10 tosses of a coin, because of chance, an outcome of 6 heads (60%) and 4 tails (40%) is not unexpected. In contrast, with 1000 tosses of the coin, a result of 600 heads (60%) and 400 tails (40%) would intuitively be highly unlikely. In the first case, a change in the results of one coin toss would alter the expected 5 heads and 5 tails to the observed 6 heads and 4 tails. In the second case, 100 tosses would have to change from tails to heads to generate the stated deviation from the predicted 500 heads and 500 tails. It is reasonable, even likely, that chance events could cause 1 deviation from the predicted number, but not 100. Two important concepts emerge from this simple presentation. First, a comparison of percentages or ratios alone will never allow you to determine whether or not *observed* data are significantly different from *predicted* values. Second, the absolute numbers obtained are important because they reflect the size of the experiment. The larger the sample size, the closer the observed percentages can be expected to match the values predicted by the experimental hypothesis, *if the hypothesis is correct.* The chi square test is therefore always calculated with numbers—actual data—and not percentages or proportions.

A critical prerequisite of the chi square test is the framing of a hypothesis that leads to clear-cut predictions. Although contemporary geneticists use the chi square test to interpret many kinds of genetic experiments, they use it most often to discover whether data obtained from breeding experiments provide evidence for or against the hypothesis that two genes are linked. But the problem with the general hypothesis that "genes *A* and *B* are linked" is that there is no precise prediction of what to expect in terms of breeding data. This is because, as we have seen, the frequency of recombination varies with each linked gene pair. In contrast, the alternative hypothesis "that genes *A* and *B* are *not* linked" gives rise to a precise prediction: that alleles at different genes will assort independently and produce 50% parental and 50% recombinant progeny. So, whenever a geneticist wants to determine whether two genes are linked, he or she actually tests whether the observed data are consistent with a **null hypothesis** of no linkage. If the chi square test shows that the observed data differ significantly from those expected with independent assortment, that is, they differ enough not to be reasonably attributable to chance alone, then the researcher can reject the null hypothesis of no linkage and accept the alternative of linkage between the two genes.

The box on the following page presents the general protocol of the chi square test. The final result of the calculations outlined in steps 1–5 is the determination of the numerical probability—the *p* value—that a particular set of observed experimental results represents a chance deviation from the values predicted by a particular hypothesis. If the probability is high, it is likely that the hypothesis being tested explains the data, and the observed deviation from expected results is considered *insignificant.* If the probability is very low, the observed deviation from expected results becomes *significant.* When this happens, it is unlikely that the hypothesis under consideration explains the data, and the hypothesis can be rejected.

Applying the Chi Square Test

Figure 5.6 depicts two sets of data obtained from testcross experiments asking whether genes *A* and *B* are linked. We first apply the chi square analysis to data accumulated in the first experiment. The total number of offspring is 50, of which 31 (that is, $17 + 14$) are observed to be parental types and 19 ($8 + 11$) recombinant types. Dividing 50 by 2, you get 25, the number of parental or recombinant offspring expected according to the null hypothesis of independent assortment (which predicts that parentals = recombinants). Now, considering first the parental types alone, you square the observed deviation from the expected value, and divide the result by the expected value. After doing the same for the recombinant types, you add the two quotients to obtain the value of chi square.

$$\chi^2 = \frac{(31 - 25)^2}{25} + \frac{(19 - 25)^2}{25} = 1.44 + 1.44 = 2.88$$

Cross: $A\,B\,/\,a\,b \;\times\; a\,b\,/\,a\,b$

Progeny	Experiment 1	Experiment 2
A B	17	34
a b	14	28
A b	8	16
A B	11	22
Total	50	100

Class	Observed / Expected		Observed / Expected	
Parentals	31	25	62	50
Recombination	19	25	38	50

Figure 5.6 Applying the chi square test to see if genes *A* and *B* are linked. The null hypothesis is that the two genes are unlinked. For experiment 1, $p > 0.05$, so it is not possible to reject the null hypothesis. For experiment 2, with a data set twice the size, $p < 0.05$. This is below the arbitrary boundary of significance used by most geneticists, which makes it possible to reject the null hypothesis and conclude with greater than 95% confidence that the genes are linked.

The Chi Square Test

The general protocol for using the chi square test and evaluating its results can be stated in a series of steps. Two preparatory steps precede the actual chi square calculation.

1. Use the data obtained from a breeding experiment to answer the following questions:
   a. What is the *total number* of offspring (events) analyzed?
   b. How many different *classes* of offspring (events) are there?
   c. In each class, what is the *number* of offspring (events) *observed?*

2. Calculate how many offspring (events) would be expected for each class if the null hypothesis (here, no linkage) were correct by multiplying the percentage predicted by the null hypothesis (here, 50% parentals and 50% recombinants) by the total number of offspring analyzed in the experiment. You are now ready for the chi square calculation.

3. To calculate chi square, begin with one class of offspring. Subtract the expected number from the observed number (which gives the deviation from the prediction for this class), square the result, and divide this amount by the expected number. Do this for all classes and sum the individual results (the deviation^2/expected) together. The final result is the chi square (χ^2) value. This step is summarized by the equation

   $$\chi^2 = \Sigma \, \frac{(Number\ observed\ -\ Number\ expected)^2}{Number\ expected}$$

   where Σ means "sum of all classes."

4. Compute the **degrees of freedom (df).** The df is a measure of the number of independently varying parameters in the experiment. For the types of experiments we are discussing, the number of degrees of freedom is one less than the number of classes. For example, if the total number of offspring in a testcross fall into four classes and you know the number present in any three, you can easily calculate the number in the fourth. Thus, if N = the number of classes, then the degrees of freedom (df) = $N - 1$. If there are 4 classes, there are only 3 df.

5. Use the chi square value together with the number of degrees of freedom to determine a ***p* value:** the probability that a deviation from the predicted numbers at least as large as that observed in the experiment will occur by chance. Although the *p* value is arrived at through a numerical analysis, geneticists routinely determine the value by a quick search through a table of critical χ^2 values, such as the one in Table 5.1 on page 120.

6. Evaluate the significance of the *p* value. You can think of the *p* value as the probability that the null hypothesis is true and the alternative hypothesis, in this case of linkage, is wrong. A value greater than 0.05 indicates that in more than 1 in 20 (or more than 5%) of the repetitions of an experiment of the same size, the observed data showing a deviation from the predictions of the null hypothesis could have been obtained by chance even if the null hypothesis were true; the data are therefore *not significant* for rejecting the null hypothesis and showing linkage. Statisticians have arbitrarily selected the 0.05 *p* value as the boundary between accepting and rejecting the null hypothesis. A *p* value of less than 0.05 means that you can consider the data showing deviation *significant* and reject the null hypothesis.

Since with two classes (parentals and recombinants), the number of degrees of freedom is 1, you scan the chi square table (see Table 5.1 on p. 120) for $\chi^2 = 2.88$ and df $= 1$, and you find by extrapolation that the corresponding *p* value is greater than 0.05 (roughly 0.09). From this, you can conclude that it is not possible to reject the null hypothesis on the basis of this experiment, which means that this data set is not sufficient to demonstrate linkage between *A* and *B*.

If you use the same strategy to calculate a *p* value for the data observed in the second experiment, where there are a total of 100 offspring and thus an expected number of 50 parentals and 50 recombinants, you get

$$\chi^2 = \frac{(62 - 50)^2}{50} + \frac{(38 - 50)^2}{50} = 2.88 + 2.88 = 5.76$$

The number of degrees of freedom (df) remains 1, so Table 5.1 arrives at a *p* value greater than 0.01 but less than 0.05. In this case, you can consider the difference between the observed and expected values to be significant. As a result you can reject the null hypothesis of independent assortment and conclude it is likely that genes *A* and *B* are linked.

A certain amount of subjectivity enters into any decision about the significance of a particular *p* value. As stated in the preceding box, statisticians have arbitrarily selected a *p* value of 0.05 as the boundary between significance and nonsignificance. Values lower than this indicate there would be less than 5 chances in 100 of obtaining the same results by random sampling if the null hypothesis were true. An extremely low *p* value thus suggests that the data shows major deviations from values predicted by the hypothesis; these values are significant enough to reject

TABLE 5.1 Critical Chi Square Values

Degrees of Freedom	p Values						
	Cannot Reject the Null Hypothesis				Null Hypothesis Rejected		
	0.99	0.90	0.50	0.10	0.05	0.01	0.001
	χ^2 calculations						
1	—	0.02	.45	2.71	3.84	6.64	10.83
2	0.02	0.21	1.39	4.61	5.99	9.21	13.82
3	0.11	0.58	2.37	6.25	7.81	11.35	16.27
4	0.30	1.06	3.36	7.78	9.49	13.28	18.47
5	0.55	1.61	4.35	9.24	11.07	15.09	20.52

χ^2 values that lie in the yellow-shaded region of this table allow you to reject the null hypothesis with > 95% confidence, and for recombination experiments, to postulate linkage.

the null hypothesis. The lower the p value, the clearer the case that the data cannot be regarded as chance deviations from the predictions of the hypothesis. For this reason, more conservative scientists often set the boundary of significance at $p = 0.01$, and would thus be less willing to reach a conclusion based on the results of experiment 2 in Fig. 5.6. In linkage studies, such a low p value means that it is highly unlikely (only 1 chance in 100) that the observed distribution of phenotypes arose from two unlinked genes, and we can conclude, with 99% confidence, that the genes are linked. On the other hand, higher p values (greater than 0.01 or 0.05, depending on the criterion used) do not necessarily mean that two genes are unlinked; it may mean only that the sample size is not large enough to provide an answer. With more data, the p value will normally rise if the null hypothesis of no linkage is correct and fall if there is, in fact, linkage.

Note that in Fig. 5.6 all of the numbers in the second set of data are simply double the numbers in the first set, with the percentages remaining the same. Thus, just by doubling the sample size from 50 to 100 individuals, it was possible to go from no significant difference to a significant difference between the observed and the expected values (using the $p = 0.05$ cutoff). In other words, the larger the sample size, the less the likelihood that a certain percentage deviation from expected results happened simply by chance. Bearing this in mind, you can see that it is not appropriate to use the chi square test when analyzing very small samples (of less than 10), which by their nature are usually insufficient to answer questions concerning linkage. This creates a problem for human geneticists, who cannot "construct" people as they can fruit flies and other experimental organisms nor develop large populations of inbred individuals. To achieve a reasonable sample size for linkage studies in humans, they must pool data from a large number of family pedigrees.

The subjectivity involved in determining the boundary of significance means that the chi square test *does not* prove linkage or its absence. What it *does* do is provide a quantitative measure of the likelihood that the data from an experiment can be explained by a particular hypothesis. The chi square analysis is thus a general statistical test for significance; it can be used with many different experimental designs and with hypotheses other than the absence of linkage. As long as it is possible to propose a hypothesis that leads to a predicted set of values for a defined set of data classes, you can readily determine whether or not the observed data are consistent with the hypothesis. If what you are looking for is any one of a range of outcomes that is not the outcome predicted by the hypothesis, that is, if the hypothesis serves as a "straw man," set up only to be knocked down, it is considered a null hypothesis.

When experiments lead to rejection of a null hypothesis, you may need to confirm the alternative. For instance, if you are analyzing the inheritance of two opposing traits on the basis of a null hypothesis that says the two phenotypes result from the segregation of two equally viable alleles of a single gene, you would expect a testcross between an F_1 heterozygote and a recessive homozygote to produce a 1:1 ratio of the two traits in the offspring. If instead, you observe a ratio of 6:4 and the chi square test produces a p value of 0.009, you can reject the null hypothesis. But you are still left with the question of what the absence of a 1:1 ratio means. There are actually two alternatives: Either the reverse of the null hypothesis is true and the two alleles of the single gene are not equally viable *or* more than one gene encodes the trait. The chi square test cannot tell you which possibility is correct, and you would have to study the matter further. The problems at the end of this chapter illustrate several applications of the chi square test that are pertinent to genetics.

Recombination Results When Crossing-Over During Meiosis Separates Linked Genes

It is easy to understand how genes that are physically connected on the same chromosome can be transmitted together and thus show genetic linkage. It is not as obvious why all linked genes always show some recombination in a sample population of sufficient size. Do the chromosomes participate in a physical process that gives rise to the reshuffling of linked genes that we call recombination? The answer to this question is of more than passing interest as it provides a basis for gauging relative distances between pairs of genes on a chromosome.

In 1909 the Belgian cytologist Frans Janssens described structures he had observed in the light microscope during prophase of the first meiotic division. He called these structures **chiasmata;** as described in Chapter 4, they seemed to represent regions in which nonsister chromatids of homologous chromosomes cross over each other (review Fig. 4.14 on p. 92). Making inferences from a combination of genetic and cytological data, Morgan suggested that the chiasmata observed through the light microscope were sites of chromosome breakage and exchange resulting in genetic recombination.

Reciprocal Exchanges Between Homologous Chromosomes Are the Physical Basis of Recombination

Morgan's idea that the physical breaking and rejoining of chromosomes during meiosis was the basis of genetic recombination

seemed reasonable. But although Janssens's chiasmata could be interpreted as signs of the process, before 1930 no one had produced visible evidence that crossing-over between homologous chromosomes actually occurs. The identification of **physical markers,** or cytologically visible abnormalities that make it possible to keep track of specific chromosome parts from one generation to the next, enabled researchers to turn the logical deductions about recombination into facts derived from experimental evidence. In 1931 Harriet Creighton and Barbara McClintock, who studied corn, and Curt Stern, who worked with *Drosophila,* published the results of experiments showing that genetic recombination depends on the reciprocal exchange of parts between maternal and paternal chromosomes. Stern, for example, bred female flies with two different X chromosomes, each containing a distinct physical marker. These same females were also doubly heterozygous for two X-linked **genetic markers**—genes that could serve as points of reference in determining whether particular progeny were the result of recombination.

Figure 5.7 diagrams the chromosomes of these heterozygous females. One X chromosome carried mutations producing carnation eyes (a dark ruby color, abbreviated *car*) that were kidney-shaped (*Bar*); in addition, this chromosome was marked physically by a visible discontinuity. The other X chromosome had wild-type alleles (+) for both the *car* and the *Bar* genes, and its physical marker consisted of part of the Y chromosome that had become connected to the X chromosome centromere.

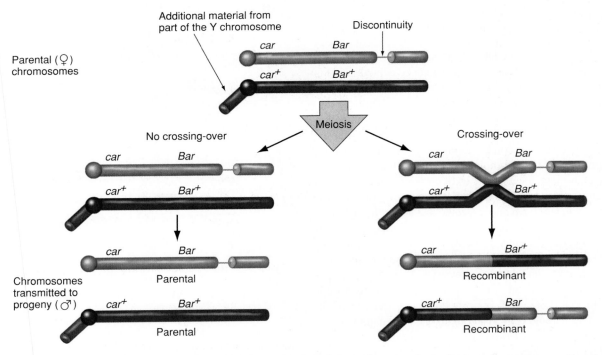

Figure 5.7 Evidence that recombination results from reciprocal exchanges between homologous chromosomes. Genetic recombination between the *car* and *Bar* genes on the *Drosophila* X chromosome is accompanied by the exchange of physical markers observable in the microscope. Note that this depiction of crossing-over is a simplification, as genetic recombination actually occurs after each chromosome has replicated into sister chromatids.

Figure 5.7 illustrates how the chromosomes in these *car Bar* / *car⁺ Bar⁺* females were transmitted to male progeny. According to the experimental results, all sons showing a phenotype determined by one or the other parental combination of genes (either *car Bar* or *car⁺ Bar⁺*) had an X chromosome that was structurally indistinguishable from one of the original X chromosomes in the mother. In recombinant sons, however, such as those that manifested carnation eye color and normal eye shape (*car Bar⁺* / Y), an identifiable exchange of the abnormal features marking the ends of homologous X chromosomes accompanied the recombination of genes. The evidence thus tied a particular instance of phenotypic recombination to the crossing-over of particular genes located in specifically marked parts of particular chromosomes. This was an elegant demonstration that genetic recombination is associated with the actual reciprocal exchange of segments between homologous chromosomes during meiosis.

Through the Light Microscope: Chiasmata Mark the Sites of Recombination

Figure 5.8 outlines what is currently known about the steps of recombination as they appear in chromosomes viewed through the light microscope. (As mentioned in Chapter 4, although this low-resolution view may not represent certain details of recombination with complete accuracy, it nonetheless provides a useful frame of reference.) In Fig. 5.8a, the two homologs of each chromosome pair have already replicated so that there are now two pairs of sister chromatids or a total of four chromatids within each bivalent. In Fig. 5.8b, the synaptonemal complex zips together homologous chromosome pairs along their length. The synaptonemal zipper aligns homologous regions of all four chromatids such that allelic DNA sequences are physically near each other (see Fig. 4.14b on p. 92 for a detailed depiction). This proximity facilitates crossing-over between homologous sequences. In Fig. 5.8c, the synaptonemal complex begins to disassemble. Although at least some steps of the recombination process occurred while the chromatids were zipped in synapsis, it is only now that the recombination event becomes apparent. As the zipper dissolves, homologous chromosomes remain attached at chiasmata, the actual sites of crossing-over. Visible in the light microscope, chiasmata indicate where chromatid sections have switched from one molecule to another. In Fig. 5.8d, during anaphase I, as the two homologues separate, starting at their centromeres, the ends of the two recombined chromatids pull free of their respective sister chromatids and the chiasmata shift from their original positions toward a chromosome end, or telomere. This movement of chiasmata is known as **terminalization.** In Fig. 5.8e and f, meiosis continues and eventually produces four haploid cells that contain one chromatid—now a chromosome—apiece. Homologous chromosomes have exchanged parts.

Recombination can also take place apart from meiosis. Indeed, as we explain near the end of this chapter, it sometimes, though rarely, occurs during mitosis. It also occurs with the circular chromosomes of prokaryotic organisms and with cellular organelles such as mitochondria and chloroplasts, which do not

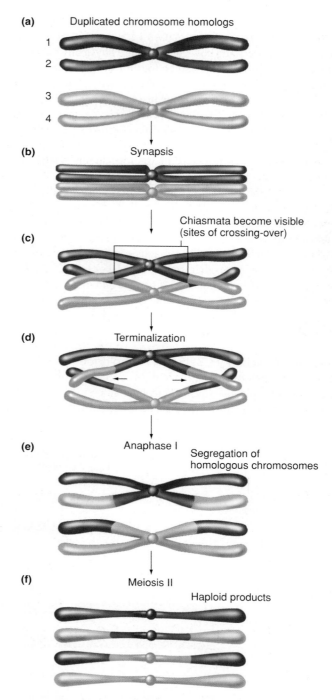

Figure 5.8 Recombination through the light microscope.
(a) A pair of duplicated homologous chromosomes very early in prophase of meiosis I. **(b)** During leptotene and zygotene of prophase I, the synaptonemal complex helps align corresponding regions of homologous chromosomes, allowing recombination. **(c)** As the synaptonemal complex disassembles during diplotene, homologous chromosomes remain attached at chiasmata. **(d)** and **(e)** The chiasmata terminalize (move toward the chromosome ends), allowing the recombined chromosomes to separate during anaphase and telophase. **(f)** The result of the process is recombinant gametes.

undergo meiosis and do not form chiasmata (see Chapters 14 and 15).

Recombination Frequencies for Pairs of Genes Reflect the Distances Between Them

Thomas Hunt Morgan's belief that chiasmata represent sites of physical crossing-over between chromosomes, and that such crossing-over may result in recombination, led him to the following logical deduction. He hypothesized that different gene pairs exhibit different linkage rates because genes are arranged in a line along a chromosome; and the closer together two genes are on the chromosome, the less their chance of being separated by an event that cuts and recombines the line of genes. To look at it another way, if we assume for the moment that chiasmata can form anywhere along a chromosome with equal likelihood, then the probability of a crossover occurring between two genes increases with the distance separating them. If this is so, then the frequency of genetic recombination also must increase with the distance between genes. To illustrate the point, imagine pinning to a wall 10 inches of ribbon with a line of tiny black dots along its length and then repeatedly throwing a dart to see where you will cut the ribbon. You would find that practically every throw of the dart separates a dot at one end of the ribbon from a dot at the other end, while few if any throws separate any two particular dots positioned right next to each other.

Alfred H. Sturtevant, one of Morgan's students, took this idea one step further, proposing that the percentage of recombination, or **recombination frequency (RF),** could be used as a gauge of the physical distance separating any two genes on the same chromosome. Sturtevant arbitrarily defined one RF percentage point as the unit of measure along a chromosome; later, another geneticist named the unit a **centimorgan (cM)** after T. H. Morgan. Mappers often refer to a centimorgan as a **map unit (m.u.).** Although the two terms are interchangeable, researchers prefer one or the other, depending on their experimental organism. *Drosophila* geneticists, for example, use map units while human geneticists use centimorgans. In Sturtevant's system, 1% RF = 1 cM = 1 m.u. A review of the two pairs of X-linked *Drosophila* genes we analyzed earlier shows how his proposal works. Because the X-linked genes for eye color (w) and body color (y) recombine in 1.1% of F_2 progeny, they are 1.1 m.u. apart (Fig. 5.9a). In contrast, the X-linked genes for eye color (w) and wing size (m) have a recombination frequency of 32.8 and are therefore 32.8 m.u. apart (Fig. 5.9b).

As a unit of measure, the map unit is simply an index of recombination probabilities assumed to reflect distances between genes. According to this index, the y and w genes are much closer together than the m and w genes. Geneticists have used this logic to map thousands of genetic markers to the chromosomes of *Drosophila,* building recombination maps step by step with closely linked markers. And as we see next, they have learned that markers very far apart on the same chromosome may appear unlinked even though their recombination distances relative to closely linked intervening markers confirm that they are indeed on the same chromosome.

Figure 5.9 Recombination frequencies are the basis of genetic maps. (a) 1.1 percent of the gametes produced by a female doubly heterozygous for the genes w and y are recombinant. The recombination frequency (RF) is thus 1.1%, and the genes are approximately 1.1 map units (m.u.) or 1 centimorgan (cM) apart. **(b)** The distance between the w and m genes is longer: 32.8 m.u. (or 32.8 cM).

Experimental Recombination Frequencies Between Two Genes Are Never Greater Than 50%

If the definition of linkage is that the proportion of recombinant classes is less than that of parental classes, a recombination frequency of less than 50% indicates linkage. But what can we conclude about the relative location of genes if there are roughly equal numbers of parental and recombinant progeny? And does it ever happen that recombinants are in the majority? To answer these questions, we analyze the results of two simple experiments.

Consider first a cross following two recessive mutations in *Drosophila*—rosy eye color (ry) and thick wing veins (tkv)—that are known to lie on different autosomes. A female with rosy eyes (ry/ry ; tkv^+/tkv^+) is mated to a male with thick wing veins (ry^+/ry^+ ; tkv/tkv); and the F_1 males and females are heterozygous for both traits (ry/ry^+; tkv/tkv^+). (Note that the semicolons separate genes on different chromosomes—in *Drosophila,* chromosomes 1 [the X], 2, 3, or 4—while the slashes separate alleles on the two homologs of a chromosome pair.) Figure 5.10a shows the distribution of progeny from a testcross between the dihybrid F_1 females (ry/ry^+ ; tkv/tkv^+) and males homozygous for both recessive traits (ry/ry; tkv/tkv). The result depicted is not surprising. Because the genes are not linked, they obey Mendel's second law of independent assortment and give rise to the four possible types of F_1 gametes with approximately equal frequency.

Compare this with a second cross following a different pair of recessive mutations known to lie at opposite ends of the same autosome: aristaless (al), which produces flies lacking part of the antennal structure, and blistered wings (bs). A testcross of doubly heterozygous F_1 females of genotype $al\ bs/al^+\ bs^+$ with doubly recessive $al\ bs/al\ bs$ males yielded 48.2% recombinant progeny ($al\ bs^+$ and $al^+\ bs$) in one experiment, for a recombination frequency of 48.2% (Fig. 5.10b). Thus, with genes located very far apart on the same chromosome, the recombination frequency obtained from a single crossover between the two is also approximately 50%.

Researchers have never observed statistically significant recombination frequencies between two genes greater than 50%, which means that in any cross or series of crosses, recombinants among the F_2 progeny are never in the majority. Genes

Figure 5.10 Unlinked genes show a recombination frequency of 50%. (a) The genes *ry* and *tkv* are genetically unlinked because they lie on different chromosomes. **(b)** The genes *al* and *bs* are genetically unlinked because they lie far apart on the same chromosome. As an exercise, use the chi square test to show that you cannot reject the null hypothesis of no linkage in either experiment.

on different chromosomes yield recombination frequencies of 50% as a result of independent assortment. Genes sufficiently far apart on the same chromosome show the same percentage. Later in the chapter we examine why 50% is the upper limit on recombination frequency between two genes, even for genes on the same chromosome that are separated by very large physical distances. For now, simply note that recombination frequencies near 50% suggest either that two genes are on different chromosomes or that they lie far apart on the same chromosome. The only way to tell whether the two genes are on the same chromosome is through a series of matings showing definite linkage with other genes that lie between them. In short, even though crosses between two genes lying very far apart on a chromosome may show no linkage at all (because recombinant and parental classes are equal), you can demonstrate that they are on the same chromosome if you can tie each of the widely separated genes to one or more common intermediaries.

Linkage and Recombination: A Summary

The important conclusions from the experimental data presented thus far in this chapter can be summarized as follows. Gene pairs that are close together on the same chromosome are linked and do not follow Mendel's law of independent assortment. Instead, alleles of linked genes are more or less tightly

coupled during transmission, leading to a preponderance of parental classes among the progeny of double heterozygotes. Linked alleles, however, can become separated through recombination, and the frequency with which this happens is different for each pair of genes. There is a correlation between recombination frequency and the distance separating two genes.

The mechanism of recombination is crossing-over: Some genes move from the maternal to the paternal chromosome and vice versa when homologs exchange parts during meiosis. Chiasmata are the visible signs of crossing-over, and if we assume that chiasmata can occur anywhere along a chromosome, the farther apart two genes are, the greater the opportunity for chiasmata to form between them. This explains why larger recombination frequencies reflect greater distances between genes. Recombination frequencies in pairwise crosses vary from 0% to 50%. Statistically significant values of less than 50% indicate that two genes are linked and must therefore be on the same chromosome. Two genes that show a recombination frequency of 50% are genetically unlinked, either because they are on different chromosomes *or* because they reside far apart on the same chromosome.

Knowledge of linkage and recombination paves the way for understanding how geneticists assign genes a relative chromosomal position by comparing the recombination frequencies of many gene pairs. The chromosomal position assigned to a gene is its **locus.** Each gene's locus is the same in all individuals of a species, and the process of determining that locus is known as **mapping.** As you will see, the results of mapping experiments consistently verify the idea that genes are arranged in a line along a chromosome.

Mapping: Locating Genes Along a Chromosome

Maps are images of the relative positions of objects in space. Whether depicting the floor plan of New York's Metropolitan Museum of Art, the layout of the Roman Forum, or the location of cities served by the railways of Europe, maps turn measurements (how far apart two rooms, ruins, or railway stops are from one another) into patterns of spatial relationships that add a new level of meaning to the original data of distances. Maps that assign genes to locations on particular chromosomes are no exception. By transforming genetic data into spatial arrangements, they sharpen our ability to predict the inheritance patterns of specific traits.

We have seen that recombination frequency (RF) is a measure of the distance separating two genes along a chromosome. We now examine how data from many crosses following two and three genes at a time can be compiled and compared to generate accurate, comprehensive gene/chromosome maps.

Two-Point Crosses: Comparisons Help Establish Relative Gene Positions

In his senior undergraduate thesis, Morgan's student A. H. Sturtevant asked whether data obtained from a large number of two-point crosses (crosses tracing two genes at a time) would support the idea that genes form a definite linear series along a chromosome. He began by looking at X-linked genes in *Drosophila.* Figure 5.11a lists his recombination data for several two-point crosses. Recall that the distance between two genes that yields 1% recombinant progeny—an RF of 1%—is 1 m.u. As an example of Sturtevant's reasoning, consider the three genes *w, y,* and *m.* If these genes are arranged in a line (instead of a more complicated branched structure, for example), then one of them must be in the middle, flanked on either side by the other two. The greatest genetic distance should separate the two genes on the outside, and this value should roughly equal the sum of the distances separating the middle gene from each outside gene. The data Sturtevant obtained are consistent with this idea, implying that *w* lies between *y* and *m* (Fig. 5.11b). Note that the left-to-right orientation of this map was selected at random; the map in Fig. 5.11b would be equally correct if it portrayed *y* on the right and *m* on the left.

Recombination Mapping Supports the Idea That Genes Are Arranged in a Line Along a Chromosome

By following exactly the same procedure for each set of three genes, Sturtevant established a self-consistent order for all the genes he investigated on *Drosophila*'s X chromosome (Fig. 5.11c; once again, the left-to-right arrangement is an arbitrary

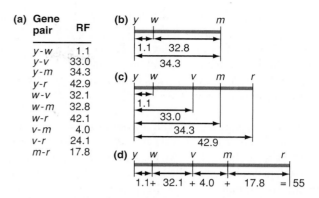

Figure 5.11 Mapping genes by comparisons of two-point crosses. (a) Sturtevant's data for the distances between pairs of X-linked genes in *Drosophila.* **(b)** Because the distance between *y* and *m* is greater than the distance between *w* and *m,* the order of genes must be *y-w-m.* **(c)** and **(d)** Maps for five genes on the *Drosophila* X chromosome. The left-to-right orientation is arbitrary. Note that the numerical position of the *r* gene depends on how it is calculated. The best genetic maps are obtained by summing many small intervening distances as in (d).

choice). By checking the data for every combination of three genes, you can assure yourself that this ordering makes sense. The fact that the recombination data yield a simple linear map of gene position supports the idea that genes reside in a unique linear order along a chromosome.

Two-Point Crosses Have Their Limitations

Though of great importance, the pairwise mapping of genes has several shortcomings that limit its usefulness. First, in crosses involving only two genes at a time, it may be difficult to determine gene order if some gene pairs lie very close together. For example, in mapping *y, w,* and *m,* 34.3 m.u. separate the outside genes *y* and *m,* while nearly as great a distance (32.8 m.u.) separates the middle *w* from the outside *m* (Fig. 5.11b). Before being able to conclude with any confidence that *y* and *m* are truly farther apart, that is, that the small difference between the values of 34.3 and 32.8 is not the result of sampling error, you would have to examine a very large number of flies and subject the data to a statistical test, such as the chi square test.

A second problem with Sturtevant's mapping procedure is that the actual distances in his map do not always add up, even approximately. As an example, suppose that the locus of the *y* gene at the far left of the map is regarded as position 0 (Fig. 5.11c). The *w* gene would then lie near position 1, and *m* would be located in the vicinity of 34 m.u. But what about the *r* gene, named for a mutation that produces rudimentary (very small) wings? Based solely on its distance from *y,* as inferred from the *y* ↔ *r* data in Fig. 5.11a, we would place it at position 42.9 (Fig. 5.11c). However, if we calculate its position as the sum of all intervening distances inferred from the data in Fig. 5.11a, that is, as the sum of *y* ↔ *w* plus *w* ↔ *v* plus *v* ↔ *m* plus *m* ↔ *r,* the locus of *r* becomes 1.1 + 32.1 + 4.0 + 17.8 = 55.0 (Fig. 5.11d). What explains this difference, and which of these two values is closer to the truth? Three-point crosses help provide some of the answers.

Three-Point Crosses: A Faster, More Accurate Way to Map Genes

The simultaneous analysis of three markers makes it possible to obtain enough information to position the three genes in relation to each other from just one set of crosses. To describe this procedure, we look at three genes linked on one of *Drosophila*'s autosomes.

A homozygous female with mutations for vestigial wings (*vg*), black body (*b*), and purple eye color (*pr*) was mated to a wild-type male (Fig. 5.12a). All the triply heterozygous F$_1$ progeny, both male and female, had normal phenotypes for the three characteristics, indicating that the mutations are autosomal recessive. In a testcross of the F$_1$ females with males having vestigial wings, black body, and purple eyes, the progeny were of eight different phenotypes reflecting eight different genotypes. The order in which the genes in each phenotypic class are listed in Fig. 5.12a is completely arbitrary; instead of *vg b pr*, one could write *b vg pr* or *vg pr b* to indicate the same genotype. This is because at the outset, we do not know the gene order. Deducing it is the goal of the mapping study. Thus, the way the data are tabulated does not necessarily indicate the actual order of genes along the chromosome.

In analyzing the data, we look at two genes at a time (recall that the recombination frequency is always a function of a pair of genes). For the pair *vg* and *b*, the parental combinations are *vg b* and *vg$^+$ b$^+$*; the nonparental recombinants are *vg b$^+$* and *vg$^+$ b*. To determine whether a particular class of progeny is parental or recombinant for *vg* and *b*, we do not care whether the flies are *pr* or *pr$^+$*. Thus, to the nearest tenth of a map unit, the *vg ↔ b* distance, calculated as the percentage of recombinants in the total number of progeny, is

$$\frac{252 + 241 + 131 + 118}{4197} \times 100$$

$$= 17.7 \text{ m.u. } (vg \leftrightarrow b \text{ distance})$$

Similarly, since recombinants for the *vg–pr* gene pair are *vg pr$^+$* and *vg$^+$ pr*, the interval between these two genes is

$$\frac{252 + 241 + 13 + 9}{4197} \times 100$$

$$= 12.3 \text{ m.u. } (vg \leftrightarrow pr \text{ distance})$$

while the distance separating the *b–pr* pair is

$$\frac{131 + 118 + 13 + 9}{4197} \times 100$$

$$= 6.4 \text{ m.u. } (b \leftrightarrow pr \text{ distance})$$

These recombination frequencies show that *vg* and *b* are separated by the largest distance (17.7 m.u., as compared with 12.3 and 6.4) and must therefore be the outside genes, flanking *pr* in the middle (Fig. 5.12b). But as with the X-linked *y* and *r* genes analyzed by Sturtevant, the distance separating the outside *vg* and *b* genes (17.7) does not equal the sum of the two intervening distances (12.3 + 6.4 = 18.7). In the next section we learn that the reason for this discrepancy is the rare occurrence of double crossovers.

Three-Point Crosses Allow Correction for Double Crossovers

Figure 5.13 depicts the homologous autosomes of the F$_1$ females that are heterozygous for the three genes *vg*, *pr*, and *b*. A close examination of the chromosomes reveals the kinds of crossovers that must have occurred to generate the classes and numbers of progeny observed. In this and subsequent figures, the chromosomes depicted are in late prophase/early metaphase of meiosis I, when there are four chromatids for each pair of homologous chromosomes. As we have suggested previously and demonstrate more rigorously later, prophase I is the stage at which recombination takes place. Note that we call the space between *vg* and *pr* "region 1" and the space between *pr* and *b* "region 2."

Recall that the progeny from a testcross between triply heterozygous F$_1$ females and males homozygous for the recessive allele of all three traits fall into eight groups (review Fig. 5.12). Flies in the two largest groups carry the same configurations of genes as did their grandparents of the P generation: *vg b pr* and *vg$^+$ b$^+$ pr$^+$*; they thus represent the parental classes (Fig. 5.13a). The next two groups—*vg$^+$ b pr* and *vg b$^+$ pr$^+$*—are composed

(a) Three-point cross results

P   ♀ *vg b pr / vg b pr*   ×   ♂ *vg$^+$ b$^+$ pr$^+$ / vg$^+$ b$^+$ pr$^+$*

F$_1$ (all identical)   *vg b pr / vg$^+$ b$^+$ pr$^+$*

Testcross ♀ *vg b pr / vg$^+$ b$^+$ pr$^+$*   ×   ♂ *vg b pr / vg b pr*

Testcross progeny			
1779	*vg b pr*	⎤	Parental combinations for
1654	*vg$^+$ b$^+$ pr$^+$*	⎦	all three genes
252	*vg$^+$ b pr*	⎤	Recombinants for *vg* relative to
241	*vg b$^+$ pr$^+$*	⎦	parental combinations for *b* and *pr*
131	*vg$^+$ b pr$^+$*	⎤	Recombinants for *b* relative to
118	*vg b$^+$ pr*	⎦	parental combinations for *vg* and *pr*
13	*vg b pr$^+$*	⎤	Recombinants for *pr* relative to
9	*vg$^+$ b$^+$ pr*	⎦	parental combinations for *vg* and *b*

4197

(b) Deduced genetic map

Figure 5.12 Analyzing the results of a three-point cross.
(a) Results from a three-point testcross of F$_1$ females simultaneously heterozygous for *vg*, *b*, and *pr*. **(b)** The gene in the middle must be *pr* because the longest distance is between the other two genes: *vg* and *b*. The most accurate map distances are calculated by summing shorter intervening distances, so 18.7 m.u. is a more accurate estimate of the genetic distance between *vg* and *b* than 17.7 m.u.

Figure 5.13 Inferring the location of a crossover event. Once you establish the order of genes involved in a three-point cross, it is easy to determine which crossover events gave rise to particular recombinant gametes. Note that double crossovers are needed to generate gametes in which the gene in the middle has recombined relative to the parental combinations for the genes at the ends.

of recombinants that must be the reciprocal products of a crossover in region 1 between *vg* and *pr* (Fig. 5.13b). Similarly the two groups containing *vg⁺ b pr⁺* and *vg b⁺ pr* flies must have resulted from recombination in region 2 between *pr* and *b* (Fig. 5.13c).

But what about the two smallest groups made up of rare *vg b pr⁺* and *vg⁺ b⁺ pr* recombinants? What kinds of chromosome exchange could account for them? Most likely, they result from two different crossover events occurring simultaneously, one in region 1, the other in region 2 (Fig. 5.13d). The gametes produced by such double crossovers still have the parental configuration for the outside genes *vg* and *b*, even though not one but two exchanges must have occurred.

Because of the existence of double crossovers, the *vg* ↔ *b* distance of 17.7 m.u. calculated in the previous section does not reflect all of the recombination events producing the gametes that gave rise to the observed progeny. To correct for this oversight, it is necessary to adjust the recombination frequency by adding the double crossovers twice, since each individual in the double crossover groups is the result of two exchanges between *vg* and *b*. The corrected distance is

$$\frac{252 + 241 + 131 + 118 + 13 + 13 + 9 + 9}{4197} \times 100$$

$$= 18.7 \text{ m.u.}$$

This value makes sense because you have accounted for all of the crossovers that occur in region 1 as well as all of the crossovers in region 2. As a result, the corrected value of 18.7 m.u. for the distance between *vg* and *b* is now exactly the same as the sum of the distances between *vg* and *pr* (region 1) and between *pr* and *b* (region 2).

As previously discussed, when Sturtevant originally mapped several X-linked genes in *Drosophila* by two-point crosses, the locus of the rudimentary wings (*r*) gene was ambiguous. A two-point cross involving *y* and *r* gave a recombination frequency of 42.9, but the sum of all the intervening distances was 55.0 (review Fig. 5.11 on p.125). This discrepancy occurred because the two-point cross (see Fig. 5.11c) ignored any double crossovers that might have occurred in the large interval between the *y* and *r* genes. The data summing the smaller intervening distances (see Fig. 5.11d) accounted for at least some of these double crossovers by catching recombinations of gene pairs between *y* and *r*. Moreover, each smaller distance is less likely to encompass a double crossover than a larger distance, so each number for a smaller distance is inherently more accurate. Note that even a three-point cross like the one for *vg, pr,* and *b* ignores the possibility of two recombination events taking place in, say, region 1. For greatest accuracy, it is always best to construct a map using many genes separated by relatively short distances.

Interference: The Number of Double Crossovers May Be Less Than Expected

In a three-point cross following three linked genes, of the eight possible genotypic classes, the two parental classes contain the largest number of progeny, while the two double recombinant classes, resulting from double crossovers, are always the smallest (see Fig. 5.12). We can understand why double-crossover progeny are the rarest by looking at the probability of their occurrence. If an exchange in region 1 of a chromosome does not affect the probability of an exchange in region 2, the probability that both will occur simultaneously is the product of their separate probabilities (recall the product rule in Chapter 2, p. 21). For example, if progeny resulting from recombination in region 1 alone account for 10% of the total progeny (that is, if region 1 is 10 m.u.), and progeny resulting from recombination in region 2 alone account for 20%, the probability of a double crossover (one event in region 1, the second in region 2) is 0.10 $\times$ 0.20 = 0.02, or 2%. This makes sense because the likelihood of two rare events occurring simultaneously is even less than that of either rare event occurring alone.

As we have seen, in a three-point cross of the $vg\ pr\ b$ trio of genes, the two classes of progeny arising from double crossovers contain the fewest progeny. The numerical frequencies of observed double crossovers, however, do not coincide with expectations derived from the law of the product. Let's look at the actual numbers. The probability of a single crossover between vg and pr is 0.123 (corresponding to 12.3 m.u.), and the probability of a single crossover between pr and b is 0.064 (6.4 m.u.). The product of these probabilities is

$$0.123 \times 0.064 = 0.0079 = 0.79\%$$

But the observed proportion of double crossovers (see Fig. 5.12) was

$$\frac{13 + 9}{4197} \times 100 = 0.52\%$$

The fact that the number of observed double crossovers is less than the number expected if the two exchanges are independent events suggests that the occurrence of one crossover reduces the likelihood that another crossover will occur in an adjacent part of the chromosome. This phenomenon—of crossovers not occurring independently—is called **chromosomal interference.** Interference may exist to ensure that every pair of homologous chromosomes undergoes at least one crossover event. The reasoning behind this hypothesis is straightforward: It is critical that every pair of homologous chromosomes sustain one or more crossover events because such events help the chromosomes orient properly at the metaphase plate during the first meiotic division. Indeed, homologous chromosome pairs without crossovers often segregate improperly. If only a limited number of crossovers can occur during each meiosis and interference lowers the number of crossovers on large chromosomes, then interference also raises the probability of crossovers occurring on small chromosomes. This would increase the likelihood that at least one crossover will take place on

every homologous pair. Though the molecular mechanism underlying interference is not yet clear, recent experiments suggest that interference is mediated by the synaptonemal complex.

Interference is not uniform and may vary even for different regions of the same chromosome. Investigators can obtain a quantitative measure of the amount of interference in different chromosomal intervals by first calculating a **coefficient of coincidence,** defined as the ratio between the actual frequency of double crossovers observed in an experiment and the number of double crossovers expected on the basis of independent probabilities.

$$\text{Coefficient of coincidence} = \frac{\text{frequency observed}}{\text{frequency expected}}$$

For the three-point cross involving vg, pr, and b, the coefficient of coincidence is

$$\frac{0.52}{0.79} = 0.66$$

Interference itself is defined as 1 − the coefficient of coincidence; in this case,

$$\text{Interference} = 1 - \text{coefficient of coincidence}$$

$$= 1 - 0.66 = 0.34$$

To understand the meaning of interference, it is helpful to contrast what happens when there is no interference with what happens when it is complete. If interference is 0, the frequency of observed double crossovers equals expectations, and crossovers in adjacent regions of a chromosome occur independently of each other. If interference is complete, that is, if interference = 1 because the coefficient of coincidence = 0, no double crossovers occur in the experimental progeny because one exchange effectively prevents another. As an example, in a particular three-point cross in mice, the recombination frequency for the pair of genes on the left (region 1) is 20, and for the pair of genes on the right (region 2), it is also 20. Without interference, the expected rate of double crossovers in this chromosomal interval is

$$0.20 \times 0.20 = 0.04, \text{ or } 4\%$$

but when investigators observed 1000 progeny of this cross, they found 0 double recombinants instead of the expected 40.

The Arrangement of Alleles in Double Recombinants Indicates the Relative Order of Three Genes

As we pointed out earlier, in three-point crosses of linked genes, the smallest classes of progeny are the two that contain double recombinants generated by double crossovers. It is possible to use the composition of alleles in these double crossover classes to determine which of the three genes lies in the middle, even without calculating any recombination frequencies. Consider again the progeny of a three-point testcross looking at the vg, pr, and b genes. The F_1 females are $vg\ pr\ b\ /\ vg^+\ pr^+\ b^+$. As Figs. 5.12 and 5.13d demonstrated, testcross progeny resulting from double crossovers in the trihybrid females of the F_1 generation received gametes from their mothers carrying the allelic combinations $vg\ pr^+\ b$ and $vg^+\ pr\ b^+$. In these individuals, the alleles of the vg and b genes retain their parental associations ($vg\ b$ and $vg^+\ b^+$), while the pr gene has recombined with respect to both the other genes ($pr\ b^+$ and $pr^+\ b;\ vg\ pr^+$ and $vg^+\ pr$). The same is true in all three-point crosses: In those gametes formed by double crossovers, the gene whose alleles have recombined relative to the parental configurations of the other two genes must be the one in the middle.

Comprehensive Example: Using a Three-Point Cross to Reanalyze Sturtevant's Map of the *Drosophila* X Chromosome

The technique of looking at double recombinants to discover which gene has recombined with respect to both other genes allows immediate clarification of gene order even in otherwise difficult cases. Consider the three X-linked genes y, w, and m that Sturtevant located in his original mapping experiment (see Fig. 5.11 on p. 125). Since the distance between y and m (34.3 m.u.) appeared slightly larger than the distance separating w and m (32.8 m.u.), he concluded that w was the gene in the middle. But because of the small difference between the two numbers, his conclusion was subject to questions of statistical significance. If, however, we look at a three-point cross following y, w, and m, these questions disappear.

Figure 5.14 tabulates the classes and numbers of male progeny arising from females heterozygous for the y, w, and m genes. Since these male progeny receive their only X chromosome from their mothers, their phenotypes directly indicate the gametes produced by the heterozygous females. In each row of the table, the genes appear in an arbitrary order which does not presuppose knowledge of the actual map. As you can see, the two classes of progeny listed at the top of the table outnumber the remaining six classes, which indicates that all three genes are linked to each other. Moreover, these largest groups, which are the parental classes, show that the two X chromosomes of the heterozygous females were $w^+\ y^+\ m$ and $w\ y\ m^+$.

Among the male progeny in Fig. 5.14, the two smallest classes, representing the double crossovers, have X chromosomes carrying $w^+\ y\ m^+$ and $w\ y^+\ m$ combinations, in which the w alleles are recombined relative to those of y and m. The w gene must therefore lie between y and m, verifying Sturtevant's original assessment.

To complete a map based on the $w\ y\ m$ three-point cross, you can calculate the interval between y and w (region 1)

$$\frac{49 + 41 + 1 + 2}{6823} \times 100 = 1.3\ \text{m.u.}$$

as well as the interval between w and m (region 2)

$$\frac{1203 + 1092 + 2 + 1}{6823} \times 100 = 33.7\ \text{m.u.}$$

The genetic distance separating y and m is the sum of

$$1.3 + 33.7 = 35.0\ \text{m.u.}$$

Figure 5.14 How three-point crosses verify Sturtevant's map. The parental classes correspond to the two X chromosomes in the F_1 female. The makeup of the double recombinant classes shows that w must be the gene in the middle.

Note that you could also calculate the distance between y and m directly by including double crossovers twice, to account for the total number of recombination events detected between these two genes.

$$\text{RF} = (1203 + 1092 + 49 + 41 + 2 + 2 + 1 + 1)/6823 \times 100 = 35.0\ \text{m.u.}$$

This method yields the same value as the sum of the two intervening distances (region 1 + region 2). Further calculations show that interference is considerable in this portion of the *Drosophila* X chromosome, at least as inferred from the set of data tabulated in Fig. 5.14. The percentage of observed double recombinants was

$$3/6823 = 0.00044, \text{ or } 0.044\%$$

(rounding to the nearest thousandth of a percent), while the percentage of double recombinants expected on the basis of independent probabilities by the law of the product is

$$0.013 \times 0.337 = 0.0044, \text{ or } 0.44\%$$

Thus, the coefficient of coincidence is

$$0.044/0.44 = 0.1$$

and the interference is

$$1 - 0.1 = 0.9$$

How Close Is the Correlation Between a Genetic Map and Physical Reality?

In 1911, when Sturtevant first devised his system of genetic mapping based on recombination frequencies, he acknowledged there was at the time no way to correlate a specific gene position with a physical part of the chromosome. Subsequent experiments, however, have shown that genetic maps do in fact accurately depict the order of genes along chromosomes. Thus, as

Fast Forward

Gene Mapping Leads to a Possible Cure for Cystic Fibrosis

For 40 years after the symptoms of cystic fibrosis were first described in 1938, there was no molecular clue—no visible chromosomal abnormality transmitted with the disease, no identifiable protein defect carried by affected individuals—suggesting the genetic cause of the disorder. As a result, there was no effective treatment for the 1 in 2000 Caucasian Americans born with the disease, most of whom died before they were 30. In the 1980s, however, geneticists were able to combine recently invented techniques for looking directly at DNA with maps constructed by linkage analysis to pinpoint a precise chromosomal position, or locus, for the cystic fibrosis gene. Knowledge of this locus made it possible to identify and clone the gene, discover the protein it encodes, understand the biochemical mechanisms disrupted by faulty alleles, and develop better therapies as well as a potential cure for this life-threatening condition.

The mappers of the cystic fibrosis gene faced an overwhelming task. They were searching for a gene that encoded an unknown protein, a gene that had not yet even been assigned to a chromosome. It could lie anywhere amid the 23 pairs of chromosomes in a human cell. Imagine looking for a close friend you lost track of years ago, who might now be anywhere in the world. You would first have to find ways to narrow the search to a particular continent (the equivalent of a specific chromosome in the gene mappers' search); then to a country (the long or short arm of the chromosome); next to the state or province, county, city, or town, and street (all increasingly narrow bands of the chromosome); and finally, to a house address (the locus itself). Here, we briefly summarize how researchers applied some of these steps in mapping the cystic fibrosis gene.

■ After a review of many family pedigrees containing first-cousin marriages had confirmed that cystic fibrosis is most likely determined by a single gene (the *CF* gene), investigators collected white blood cells from 47 families with two or more affected children, obtaining genetic data from 106 patients, 94 parents, and 44 unaffected siblings.

■ They next tried to discover if any other trait is reliably transmitted with the condition. Analyses of the easily obtainable serum enzyme paroxonase showed that its gene (known as the *PON* gene) is indeed linked to the *CF* gene. At first, this knowledge was not that helpful, because the *PON* gene had not yet been assigned to a chromosome.

■ Then, in the early 1980s, geneticists developed a large series of DNA markers, based on new techniques that enabled them to recognize variations in the genetic material. A **DNA marker** is a piece of DNA of known size, representing a specific locus, that comes in identifiable variations. These allelic variations segregate according to Mendel's laws, which means it is possible to follow their transmission as you would any gene's. For example, if one allelic form of DNA marker A cosegregates with the abnormal *CF* allele, while another form of marker A assorts with the normal copy of the gene, these identifiable variations in DNA serve as molecular landmarks located near the *CF* gene. Chapters 10 and 11 explain the discovery and use of DNA markers in greater detail; for now it is only important to know that they exist and can be identified.

we will see again and again in the many types of experiments presented throughout the remainder of this book, the *order of genes* revealed by mapping techniques corresponds to the order of those same genes along the DNA molecule of a chromosome.

In contrast, the *actual physical distances between genes*— that is, the amount of DNA separating them—does not always show a direct correspondence to genetic map distances. This is largely because the relationship between recombination frequency and physical distance along a chromosome is not simple. One complicating factor is the existence of double, triple, and even more crossovers. When genes are separated by 1 m.u. or less, double crossovers are not significant because the 0.01 × 0.01 = 0.0001 probability of their occurring is so small. But for genes separated by 20, 30, or 40 m.u., the probability of double crossovers skewing the data takes on greater significance. A second confounding factor is the 50% limit on the recombination frequency observable in a cross. This limit reduces the precision of RF as a measure of chromosomal distances. No matter how far apart two genes are on a long chromosome, they will never recombine more than 50% of the time.

Yet a third problem is that recombination is not uniform even over the length of a single chromosome: Certain "hotspots" are favored sites of recombination, while other areas—often in the vicinity of centromeres—are "recombination deserts" in which few crossovers ever take place.

Ever since Morgan, Sturtevant, and others began mapping, geneticists have generated mathematical equations called **mapping functions** to compensate for the inaccuracies inherent in relating recombination frequencies to physical distances. These equations generally make large corrections for RF values of widely separated genes, while barely changing the map distances separating genes that lie close together. This reflects the fact that multiple recombination events and the 50% limit on recombination do not confound the calculation of distances between closely linked genes. However, the corrections for large distances are at best imprecise, because mapping functions are based on simplifying assumptions (such as no interference) that rarely apply to the specific situation under consideration. Thus, the best way to create an accurate map is still by summing many smaller intervals, locating widely separated genes through link-

By 1986, linkage analyses of hundreds of DNA markers had shown that one marker, known as *D7S15*, is linked with both the *PON* gene and the *CF* gene. Once they had established the fact of linkage, researchers computed recombination frequencies and found that the distance from the DNA marker to the *CF* gene was 15 cM; from the DNA marker to the *PON* gene, 5 cM; and from *PON* to *CF*, 10 cM. After repeating their analyses with an ever-extending group of families, they concluded that the order of the three loci was *D7S15-PON-CF* (Fig. A). Since the *CF* gene could lie 15 cM in either of two directions from the DNA marker, the area under investigation was approximately 30 cM. And since the human genome consists of roughly 3000 cM, this step of linkage analysis narrowed the search to 1% of the human genome.

■ Next, the DNA marker *D7S15* was localized to the long arm of chromosome 7, which meant that the gene for cystic fibrosis also resides in that same region of chromosome 7. Researchers had now placed the *CF* gene in a certain country on a particular genetic continent.

■ Finally, investigators discovered linkage with several other markers on the long arm of chromosome 7. These included a DNA marker known as *J3.11*, a gene for a key molecule of the immune system (the β chain of the T-cell receptor), and an "oncogene" called *met* (a gene which has mutant alleles that contribute to cancer). Two of the markers turned out to be separated from the *CF* gene by a distance of only 1 cM. With all these markers, it became possible to place the *CF* gene in the middle third of chromosome 7's long arm, on band 31 (band 7q31, Fig. A). For families with at least one child who has cystic fibrosis, geneticists using DNA analyses of these closely linked markers could now identify carriers of an abnormal copy of the *CF* gene with substantial confidence.

Figure A How molecular markers helped locate the gene for cystic fibrosis *(CF)*.

By 1989, researchers had used this mapping information to identify and clone the *CF* gene on the basis of its location. And by 1992, they had shown it encodes a cell membrane protein that regulates the flow of chloride ions into and out of cells (review the Fast Forward box "Genes Encode Proteins" in Chapter 2). This knowledge has become the basis of new therapies to open up ion flow (older therapies treated the results of flow blockage with only minimal success), as well as gene therapies to introduce normal copies of the *CF* gene into the cells of CF patients. Although only in the early stages of development, such gene therapy holds out hope of a cure for cystic fibrosis.

---

age to common intermediaries. Because the most accurate genetic maps are drawn from a large series of genetic crosses, they are subject to continual refinement as more and more newly discovered genes are included.

Rates of recombination may differ from species to species. We know this because recent elucidation of the complete DNA sequences of several organisms' genomes has allowed investigators to compare the actual physical distances between genes (in base pairs of DNA) with genetic map distances. They found that in humans, a map unit corresponds on average to about 1 million base pairs. In yeast, however, where the rate of recombination per length of DNA is much higher than in humans, one map unit is approximately 2500 base pairs. Thus, although map units are useful for estimating distances between the genes of an organism, 1% RF can reflect very different expanses of DNA in different organisms. Recombination rates sometimes even vary between the two sexes of the same species. *Drosophila* provides an extreme example: No recombination occurs during meiosis in males. If you review the examples already discussed in this chapter, you will discover that they all measure recombination among the progeny of doubly heterozygous *Drosophila* females.

Multiple-Factor Crosses Help Establish Linkage Groups by Inference

Suppose you discovered a never-before-analyzed mouse trait—a bright pink tail—determined by the recessive mutation *e*. In a cross between agouti mice with pink tails (genotype *AA ee*) and black mice with wild-type tails (*aa EE*), the recombination frequency is 50%. How can you tell whether the *A* and *E* genes reside on separate chromosomes or far apart on the same chromosome? Like Sturtevant, you could conduct a series of crosses involving other genes (say, *B*, *C*, and *D*) that are known to reside on the same chromosome as the well-characterized *A* gene. If you can demonstrate that *A* is linked to *B*, *B* to *C*, *C* to *D*, and *D* to *E*, you can conclude that *A* and *E* are on the same chromosome. The Fast Forward box "Gene Mapping Leads to a Possible Cure for Cystic Fibrosis" describes how researchers used such linkage information to locate the gene for cystic fibrosis.

Figure 5.15 **_Drosophila melanogaster_ has four linkage groups.** A genetic map of the fruit fly, showing the position of many genes influencing visible phenotypes of body morphology, including those used as examples in this chapter (*highlighted in bold*). Because so many *Drosophila* genes have been mapped, each of the four chromosomes can be represented as a single linkage group.

Genes chained together by linkage relationships are known collectively as a **linkage group.** When enough genes have been assigned to a particular chromosome, the terms *chromosome* and *linkage group* become synonymous. And when the genetic map of a genome becomes so dense that it is possible to show that any gene on a chromosome is linked to another gene on the same chromosome, the number of linkage groups equals the number of pairs of homologous chromosomes in the species. Humans have 23 linkage groups, mice have 20, and fruit flies 4 (Fig. 5.15). The total genetic distance along a chromosome, which is obtained by adding many short distances between genes, may be much more than 50 m.u. For example, the two long *Drosophila* autosomes are both slightly more than 100 m.u. in length (Fig. 5.15), while the longest human chromosome is approximately 270 m.u. Recall, however, that even with the longest chromosomes, *pairwise* crosses between genes located at the two ends will not produce more than 50% recombinant progeny.

Tetrad Analysis in Fungi: A Powerful Tool for Mapping and Understanding the Mechanisms of Recombination

With *Drosophila,* mice, peas, people, and other diploid organisms, each individual represents only one of the four potential gametes generated by each parent in a single meiotic event. Thus, until now, our presentation of linkage, recombination, and mapping has depended on inferences derived from examining the phenotypes of diploid progeny resulting from random unions of random products of meiosis. For such diploid organisms, we do not know which, if any, of the parents' other progeny arose from gametes created in the same meiosis. Because of this limitation, the analysis of random products of meiosis in diploid organisms is based on statistical samplings of large populations.

In contrast, various species of fungi provide a unique opportunity for genetic analysis because they house all four haploid products of each meiosis in a sac called an **ascus** (plural, *asci*).

These haploid cells, or **ascospores** (also known as *haplospores*), can germinate and survive as viable haploid individuals that grow and perpetuate themselves by mitosis. The phenotype of such haploid fungi is a direct representation of their genotype, without complications of dominance. Figure 5.16 illustrates the life cycles of two fungal species that preserve their meiotic products in a sac. One, the normally unicellular baker's yeast (*Saccharomyces cerevisiae*), is sold in supermarkets and contributes to the texture, shape, and flavor of bread; it generates four ascospores with each meiosis. The other, *Neurospora crassa,* is a bread mold that renders the bread on which it grows inedible; it too generates four ascospores with each meiosis, but at the completion of meiosis, each of the four haploid ascospores immediately divides once by mitosis to yield four pairs, for a total of eight haploid cells. The two cells in each pair of *Neurospora* ascospores have the same genotype, because they arose from mitosis. Haploid cells of both yeast and *Neurospora* normally reproduce vegetatively (that is, asexually) by mitosis. However, sexual reproduction is possible because the haploid cells come in two mating types, and cells of opposite mating types can fuse to form a diploid zygote. In baker's yeast, these diploid cells are stable and can reproduce through successive mitotic cycles. Stress, such as that caused by a scarcity or lack of essential nutrients, induces the diploid cells of yeast to enter meiosis. In bread mold, the diploid zygote immediately undergoes meiosis, so the diploid state is only transient.

Mutations in haploid yeast cells affect many different traits, including the appearance of the cells and their ability to grow under particular conditions. For instance, yeast cells with the *his4* mutation are unable to grow in the absence of the amino acid histidine, while yeast with the *trp1* mutation cannot grow without an external source of the amino acid tryptophan. Geneticists who specialize in the study of yeast have devised a system of representing genes that is slightly different from the ones for *Drosophila* and mice. They use capital letters (*HIS4*) to designate dominant alleles and lowercase letters (*his4*) to represent recessives. For most of the yeast genes we will discuss, the wild-type alleles are dominant and may be represented by the alternative "+", while the symbol for the recessive alleles remains the lowercase abbreviation (*his4*).

After meiosis, the assemblage of four ascospores (or four pairs of ascospores) in a single ascus is called a **tetrad**. (Note that this is a second meaning for the term *tetrad*. In Chapter 4, a tetrad was the four homologous chromatids—two in each chromosome of a bivalent—synapsed during the prophase and metaphase of meiosis I. Here, it is the four products of a single meiosis held together in a sac. Since the four chromatids of a bivalent give rise to the four products of meiosis, the two meanings of tetrad refer to almost the same things.) In yeast, each tetrad is **unordered,** that is, the four meiotic products, known as spores, are arranged at random within the ascus. In *Neurospora crassa,* each tetrad is **ordered,** with the four pairs, or eight haplospores, arranged in a line. To analyze both unordered and ordered tetrads, researchers can release the spores

of each ascus, induce the haploid cells to germinate under appropriate conditions, and then analyze the genetic makeup of the resulting haploid cultures. The data they collect in this way enable them to identify the four products of a single meiosis and compare them with the four products of many other distinct meioses. Ordered tetrads offer another possibility. With the aid of a dissecting microscope, investigators can recover the ascospores in the order in which they occur within the ascus, and thereby obtain additional information that is useful for mapping. We look first at the analysis of randomly released spores, using the unordered tetrads of yeast as an example. We then describe the additional information that can be gleaned from the microanalysis of ordered tetrads, using *Neurospora* as our model organism.

Tetrads Can Be Characterized by the Number of Parental and Recombinant Spores They Contain

What kinds of tetrads arise when diploid yeast cells heterozygous for two genes on different chromosomes are induced to undergo meiosis? Consider a mating between a haploid strain of yeast of mating type *a,* carrying the *his4* mutation and the wild-type allele of the *TRP1* gene, and a strain of the opposite mating type α that has the genotype *HIS4 trp1.* The resulting *a /* α diploid cells are *his4 / HIS4 ; trp1 / TRP1,* as shown in Fig. 5.17a. When conditions promote meiosis, the two unlinked genes will assort independently to produce equal frequencies of two different kinds of tetrads. In one kind, all of the spores are parental in that the genotype of each spore is the same as one of the parents: *his4 TRP1* or *HIS4 trp1* (Fig. 5.17b). A tetrad that contains four parental class haploid cells is known as a **parental ditype (PD).** Note that *di-,* meaning two, indicates there are two possible parental combinations of alleles; the PD tetrad contains two of each combination. The second kind of tetrad, arising from the equally likely alternative distribution of chromosomes during meiosis, contains four recombinant spores: two *his4 trp1* and two *HIS4 TRP1* (Fig. 5.17c). This kind of tetrad is termed a **nonparental ditype (NPD),** because the two parental classes have recombined to form two reciprocal nonparental combinations of alleles.

A third kind of tetrad also appears when *his4 / HIS4 ; trp1 / TRP1* cells undergo meiosis. Called a **tetratype (T)** from the Greek word for "four," it carries four kinds of haploid cells: two different parental class spores (one *his4 TRP1* and one *HIS4 trp1*) and two different recombinants (one *his4 trp1* and one *HIS4 TRP1*). Tetratypes result from a crossover between one of the two genes and the centromere of the chromosome on which it is located (Fig. 5.17d).

Figure 5.17e displays the data from one experiment. Bear in mind that the column headings of PD, NPD, and T refer to tetrads (the group of four cells produced in meiosis) and not to individual haploid cells. Because the spores released from a yeast ascus are not arranged in any particular order, the order in which the spores are listed does not matter. The classification of a tetrad as PD, NPD, or T is based solely on the number of parental and recombinant spores found in the ascus.

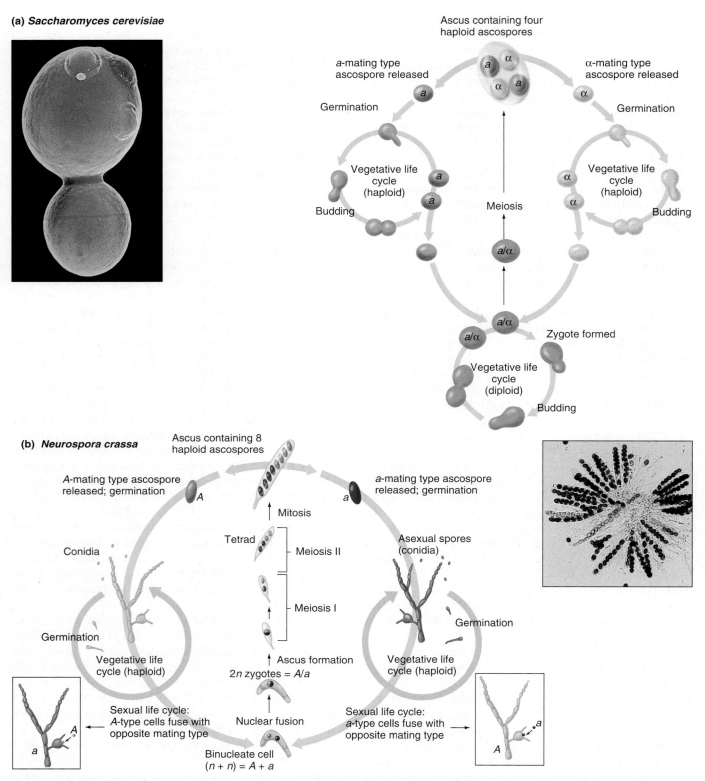

(a) *Saccharomyces cerevisiae*

(b) *Neurospora crassa*

Figure 5.16 The life cycles of the yeast *Saccharomyces cerevisiae* and the bread mold *Neurospora crassa*. Both *S. cerevisiae* and *N. crassa* have two mating types that can fuse to form diploid cells that undergo meiosis. **(a)** Yeast cells can grow vegetatively either as haploids or diploids. The products of meiosis in a diploid cell are four haploid ascospores that are arranged randomly in unordered yeast asci. **(b)** The diploid state in *Neurospora* exists only for a short period. Meiosis in *Neurospora* is followed by mitosis, to give eight haploid ascospores in the ascus. The ordered arrangement of spores in *Neurospora* asci reflects the geometry of the meiotic and mitotic spindles. The photograph showing a budding (mitotically dividing) yeast cell in part (a) is at much higher magnification that the photograph displaying *Neurospora* asci in part (b).

Figure 5.17 How meiosis can generate three kinds of tetrads when two genes are on different chromosomes. (a) Parental cross. **(b)** and **(c)** In the absence of recombination, the two equally likely alternative arrangements of two pairs of chromosomes yield either PD or NPD tetrads. T tetrads are made only if either gene recombines with respect to its corresponding centromere, as in **(d).** Numerical data in **(e)** show that the number of PD tetrads = the number of NPD tetrads when the two genes are unlinked.

When PD = NPD, Two Genes Are Unlinked

A cross following two unlinked genes must give equal numbers of individual parental and recombinant spores. This is simply another way of stating Mendel's second law of independent assortment, which predicts a 50% recombination frequency in such cases. Now, since T tetrads, regardless of their number, contain two recombinant and two nonrecombinant spores, and since all four spores in PD tetrads are parental, the only way 50% of the total progeny spores could be recombinant (as demanded by independent assortment) is if the number of NPDs (with four recombinant spores apiece) is roughly the same as the number of PDs (review Fig. 5.17). For this reason, if PD = NPD, the two genes must be unlinked, either because they reside on different chromosomes or because they lie very far apart on the same chromosome.

When PDs Significantly Outnumber NPDs, Two Genes Are Linked

The genetic definition of linkage is the emergence of more parental types than recombinants among the progeny of a doubly heterozygous parent. In the preceding section, we saw that tetratypes always contribute an equal number of parental and recombinant spores. Thus, in terms of tetrads, linkage exists only when PD >> NPD, that is, when the number of PD tetrads (carrying only parental type spores) substantially exceeds the number of NPD tetrads (containing only recombinants). By analyzing an actual cross involving linked genes, we can see how this follows from the events occurring during meiosis.

A haploid yeast strain containing the *arg3* and *ura2* mutations was mated to a wild-type *ARG3 URA2* haploid strain (Fig. 5.18). When the resultant *a*/α diploid was induced to sporulate (that is, undergo meiosis), the 200 tetrads produced had the distribution shown in Fig. 5.18. As you can see, the 127 PD tetrads far outnumber the 3 NPD tetrads, suggesting that the two genes are linked.

Figure 5.19 shows how we can explain the particular kinds of tetrads observed in terms of the various types of crossovers that could occur between the linked genes. If no crossing-over occurs between the two genes, the resulting tetrad must be PD; since none of the four chromatids participates in an exchange, all

P	arg3 ura2 (*a*-mating type)	×	ARG3 URA2 (α-mating type)	
Diploid cell		arg3 ura2 / ARG3 URA2		
		Meiosis		
Products of meiosis	**PD**	**NPD**		**T**
	arg3 ura2	arg3 URA2		arg3 ura2
	arg3 ura2	arg3 URA2		arg3 URA2
	ARG3 URA2	ARG3 ura2		ARG3 ura2
	ARG3 URA2	ARG3 ura2		ARG3 URA2
Number of tetrads	127	3		70

Figure 5.18 When Genes are Linked, PDs Exceed NPDs.

Duplication	Meiosis I	Meiosis II

(a) No crossing-over (NCO)

Parental ditype

(b) Single crossover (SCO)

Tetratype

(c) Double crossover (DCO) 2-strand

Parental ditype

(d) DCO 3-strand

Tetratype

(e) DCO 3-strand

Tetratype

(f) DCO 4-strand

Nonparental ditype

Figure 5.19 How crossovers between linked genes generate different tetrads. (a) PDs arise when there is no crossing-over. **(b)** Single crossovers between the two genes yield tetratypes. **(c)** to **(f)** Double crossovers between linked genes can generate PD, T, or NPD tetrads, depending on which chromatids participate in the crossovers.

of the products are of parental configuration (Fig. 5.19a). A single crossover between *ARG3* and *URA2* will generate a tetratype, containing four genetically different spores (Fig. 5.19b). But what about double crossovers? There are actually four different possibilities, depending on which chromatids participate, and each of the four should occur with equal frequency. A double crossover involving only two chromatids (that is, one where both crossovers affect the same two chromatids) produces only parental-type progeny, generating a PD tetrad (Fig. 5.19c). Three-strand double crossovers can occur in the two ways depicted in Fig. 5.19d and e; either way, a tetratype results. Finally, if all four strands take part in the two crossovers (one crossover involves two strands and the other crossover, the other two strands), all four progeny spores will be recombinant and the resulting tetrad is NPD (Fig. 5.19f). Thus, if two genes are linked, the only way to generate an NPD tetrad is through a four-strand double exchange. Meioses with crossovers generating such a specific kind of double recombination must be a lot rarer than no crossing-over or single crossovers, which produce PD and T tetrads, respectively. This explains why, if two genes are linked, PD must greatly exceed NPD.

Calculating the Recombination Frequency

Because we know that all of the spores in an NPD tetrad are recombinant, and half of the four spores in a tetratype are recombinant, we can say that

$$RF = \frac{NPD + 1/2T}{Total\ tetrads} \times 100$$

For the *ARG3 URA2* example in Fig. 5.18,

$$RF = \frac{3 + (1/2)(70)}{200} \times 100 = 19\ m.u.$$

It is reassuring that this formula gives the exact same result as calculating the RF as the percentage of individual recombinant spores. For example, the 200 tetrads analyzed in this experiment contain 800 (that is, 200×4) individual spores; each NPD ascus holds 4 recombinant ascospores, and each T tetrad contains 2 recombinants. Thus,

$$RF = \frac{(4 \times 3) + (2 \times 70)}{800} \times 100 = 19\ m.u.$$

Tetrad Analysis Confirms That Recombination Occurs at the Four-Strand Stage

Both T and NPD tetrads contain recombinant spores, and when tetrad analysis reveals linked genes, the T tetrads always outnumber the NPDs, as in the example we have been discussing. This makes sense, because all single and some double crossovers yield tetratypes, while only 1/4 of the rare double crossovers produce NPDs. The very low number of NPDs establishes that recombination occurs after the chromosomes have replicated, when there are four chromatids for each pair of homologs. If recombination took place before chromosome duplication, every single

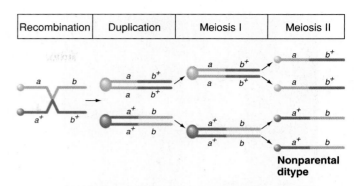

Figure 5.20 A mistaken model: Recombination before chromosome replication. If recombination occurred before the chromosomes duplicated, and if two genes were linked, most tetrads containing recombinant spores would be NPDs instead of Ts. Actual results, such as those shown in Fig. 5.18, show that the opposite is true.

crossover event would yield four recombinant chromatids and generate an NPD tetrad (Fig. 5.20). A model assuming that recombination occurs when there are two rather than four chromatids per pair of homologous chromosomes would thus predict more NPD than T tetrads, but experimental observations show just the opposite (compare Fig. 5.20 and Fig. 5.19).

The fact that recombination takes place after the chromosomes have replicated explains the 50% limit on recombination for genes on the same chromosome. Single crossovers between two genes generate T tetrads containing two out of four spores that are recombinant. Thus, even if one crossover occurred between two such genes in every meiosis, the observed recombination frequency would be 50%. The four kinds of double crossovers depicted in Fig. 5.19 yield either PD tetrads with 0/4 recombinants; T tetrads carrying 2/4 recombinants; differently derived T tetrads, still carrying 2/4 recombinants; and NPD tetrads with 4/4 recombinants. Because these four kinds of double crossovers almost always occur with equal frequency, no more than 50% of the progeny resulting from double (or, in fact, triple or more) crossovers can be recombinant.

Tetrad Analysis Also Demonstrates That Recombination Is Usually Reciprocal

Suppose you are following linked genes *A* and *B* in a cross between *A B* and *a b* strains of yeast. If the recombination that occurs during meiosis is reciprocal, every tetrad with recombinant progeny should contain equal numbers of both classes of recombinants. Observations have confirmed this prediction: Every T tetrad carries one *A b* and one *a B* spore, while every NPD tetrad contains two of each type of recombinant. We can thus conclude that meiotic recombination is almost always reciprocal, generating two homologous chromosomes that are inverted images of each other. There are, however, exceptions. Very rarely, a particular cross produces tetrads containing unequal numbers of reciprocal classes, and such tetrads cannot be classified as PD, NPD, or T. In these exceptional tetrads, the two input alleles of one of the genes, instead of segregating at a ratio of 2*A* : 2*a*, produce ratios of 1*A* : 3*a* or 3*A* : 1*a* or even 0*A* : 4*a* or 4*A* : 0*a* (Fig. 5.21).

Figure 5.21 In rare tetrads, the two alleles of a gene do not segregate 2:2. Researchers sporulated a *HIS4 / his4* diploid yeast strain and dissected the four haploid spores from three different tetrads. They then plated these spores on petri plates containing medium without histidine. Each row on the petri plate presents the four spores of a single tetrad. The top two rows show the normal 2:2 segregation of the two alleles of a single gene: two of the spores are *HIS4* and form colonies, whereas the other two spores are *his4* and cannot grow into colonies. The bottom row displays a rare tetrad with an unusual segregation of 3 *HIS4* : 1 *his4*.

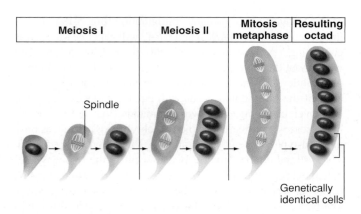

Figure 5.22 How ordered tetrads form. Spindles form parallel to the long axis of the growing *Neurospora* ascus, and the cells cannot slide around each other. The order of ascospores thus reflects meiotic spindle geometry. After meiosis, each haploid cell undergoes mitosis, producing an eight-cell ascus (an octad). The octad consists of four pairs of cells; the two cells of each pair are genetically identical.

In these same tetrads, markers such as *B/b* and *C/c* that flank the *A* or *a* allele on the same chromosome still segregate 2*B* : 2*b* and 2*C* : 2*c*. Moreover, careful phenotypic and genetic tests show that even when alleles do not segregate 2:2, only the original two input alleles occur in the progeny. Thus, recombination, no matter what ratios it generates, does not create new alleles. Geneticists believe that the unusual non-2:2 segregation ratios observed in rare instances result from molecular events at the site of recombination. We discuss these events at the molecular level in Chapter 6. For now, it is simply necessary to know that the unusual ratios exist but are quite rare.

Ordered Tetrads Help Locate Genes in Relation to the Centromere

Analyses of ordered tetrads, such as those produced by the bread mold *Neurospora crassa,* allow you to map the centromere of a chromosome relative to other genetic markers, information that you cannot normally obtain from unordered yeast tetrads. As described earlier, immediately after specialized haploid *Neurospora* cells of different mating types fuse at fertilization, the diploid zygote undergoes meiosis within the confines of a narrow ascus (review Fig. 5.16b on p. 134). At the completion of meiosis, each of the four haploid meiotic products divides once by mitosis, yielding an **octad** of eight haploid ascospores. Dissection of the ascus at this point allows one to determine the phenotype of each of the eight haploid cells.

The cross-sectional diameter of the ascus is so small that cells cannot slip past each other. Moreover, during each division after fertilization, the microtubule fibers of the spindle extend outward from the centrosomes parallel to the long axis of the ascus. These facts have two important repercussions. First,

when each of the four products of meiosis divides once by mitosis, the two genetically identical cells that result lie adjacent to each other (Fig. 5.22). Because of this feature, starting from either end of the ascus, you can count the octad of ascospores as four cell pairs and analyze it as a tetrad. And second, from the precise positioning of the four ascospore pairs within the ascus, you can infer the arrangement of the four chromatids of each homologous chromosome pair during the two meiotic divisions.

To understand the genetic consequences of the geometry of the ascospores, it is helpful to consider what kinds of tetrads you would expect from the segregation of two alleles of a single gene. The mutant *white-spore* allele (*ws*) alters ascospore color from wild-type black to white. In the absence of recombination, the two alleles (*ws*⁺ and *ws*) separate from each other at the first meiotic division because the centromeres to which they are attached separate at that stage. The second meiotic division and subsequent mitosis create asci in which the top four ascospores are of one genotype (for instance *ws*⁺) and the bottom four of the other (*ws*). Whether the top four are *ws*⁺ and the bottom four *ws,* or vice versa, depends on the random metaphase I orientation of the homologs that carry the gene relative to the long axis of the developing ascus.

The segregation of two alleles of a single gene at the first meiotic division is thus indicated by an ascus in which an imaginary line drawn between the fourth and the fifth ascospores of the octad cleanly separates haploid products bearing the two alleles. Such an ascus displays a **first-division segregation** pattern (Fig. 5.23a).

Suppose now that during meiosis I, a crossover occurs in a heterozygote between the *white-spore* gene and the centromere of the chromosome on which it travels. As Fig. 5.23b illustrates, this can lead to four equally possible ascospore arrangements, each one depending on a particular orientation of the four chro-

Meiosis I (first division)	Meiosis II (second division)	Mitosis	Segregation pattern of ascospores

Figure 5.23 Two segregation patterns in ordered asci. (a) In the absence of a crossover between a gene and its centromere, the two alleles of a gene will separate at the first meiotic division. The result is a first-division segregation pattern in which each allele appears in spores located on only one side of an imaginary line through the middle of the ascus. **(b)** A crossover between a gene and its centromere produces a second-division segregation pattern in which both alleles appear on the same side of the middle line (but in different spores).

matids during the two meiotic divisions. In all four cases, both *ws⁺* and *ws* spores are found on both sides of the imaginary line drawn between ascospores 4 and 5, because cells with only one kind of allele do not arise until the end of the second meiotic division. Octads carrying this configuration of spores display a **second-division segregation** pattern.

Since second-division segregation patterns result from meioses in which there has been a crossover between a gene and its centromere, the relative number of asci with this pattern can be used to determine the gene ↔ centromere distance. In an ascus showing second-division segregation, one-half of the ascospores are derived from chromatids that have exchanged parts, while the remaining half arise from chromatids that have not participated in crossovers leading to recombination. To calculate the distance between a gene and its centromere, you therefore simply divide the percentage of second-division segregation octads by 2. Geneticists use information about the location of centromeres to make more accurate genetic maps as well as to study the structure and function of centromeres.

A Numerical Example of Ordered-Tetrad Analysis
In one experiment, a *thr⁺arg⁺* wild-type strain of *Neurospora* was crossed with a *thr arg* double mutant. The *thr* mutants cannot grow in the absence of the amino acid threonine, while *arg* mutants cannot grow without a source of the amino acid arginine; cells carrying the wild-type alleles of both genes can grow

in medium that contains neither amino acid. From this cross, 105 octads, considered here as tetrads, were obtained. These tetrads were classified in seven different groups—A, B, C, D, E, F, and G—as shown in Fig. 5.24a. For each of the two genes, we can now find the distance between the gene and the centromere of the chromosome on which it is located.

To do this for the *thr* gene, we count the number of tetrads with a second-division segregation pattern for that gene. Drawing an imaginary line through the middle of the tetrads, we see that those in groups B, D, E, and G are the result of second-division segregations for *thr*, while the remainder show first-division patterns. The centromere ↔ *thr* distance is thus

$$\text{Percentage of second-division patterns} =$$

$$\frac{(1/2)(16 + 2 + 2 + 1)}{105} \times 100 = 10 \text{ m.u.}$$

Similarly, the second-division tetrads for the *arg* gene are in groups C, D, E, and G, so the distance between *arg* and its centromere is

$$\frac{(1/2)(11+2+2+1)}{105} \times 100 = 7.6 \text{ m.u.}$$

To ascertain whether the *thr* and *arg* genes are linked, we need to evaluate the seven tetrad groups in a different way, looking at the combinations of alleles for the two genes to see if the

(a) A *Neurospora* cross

Tetrad group	A	B	C	D	E	F	G
Segregation pattern	*thr arg*	*thr arg*	*thr arg*	*thr arg⁺*	*thr arg⁺*	*thr arg⁺*	*thr arg*
	thr arg	*thr⁺arg*	*thr arg⁺*	*thr⁺arg*	*thr⁺arg*	*thr arg⁺*	*thr⁺arg⁺*
	thr⁺arg⁺	*thr⁺arg⁺*	*thr⁺arg*	*thr⁺arg⁺*	*thr⁺arg*	*thr⁺arg*	*thr⁺arg⁺*
	thr⁺arg⁺	*thr arg⁺*	*thr⁺arg⁺*	*thr arg*	*thr arg⁺*	*thr⁺arg*	*thr arg*
Total in group	72	16	11	2	2	1	1

(b) Corresponding genetic map

arg ⟵—— 16.7 m.u. ——⟶ thr
7.6 m.u. ⟵—⟶ 10 m.u.

Figure 5.24 Genetic mapping by ordered-tetrad analysis: An example. (a) In ordered-tetrad analysis, tetrad classes are defined not only as PD, NPD, or T but also according to whether they show a first- or second-division segregation pattern. **(b)** Genetic map derived from the data in part (a). Ordered-tetrad analysis allows determination of the centromere's position as well as distances between genes.

tetrads in that group are PD, NPD, or T. We can then ask whether PD >> NPD. Referring again to Fig. 5.24, we find that groups A and G are PD, because all the ascospores show parental combinations, while groups E and F, with four recombinant spores, are NPD. PD is thus $72 + 1 = 73$, while NPD is $1 + 2 = 3$. From these data, we can conclude that the two genes are linked.

What is the map distance between *thr* and *arg?* For this calculation, we need to find the numbers of T and NPD tetrads. Tetratypes are found in groups B, C, and D, and we already know that groups E and F carry NPDs. Using the same formula for map distances as the one previously used for yeast,

$$RF = \frac{NPD + 1/2T}{Total\ tetrads} \times 100$$

we get

$$RF = \frac{3 + (1/2)(16 + 11 + 2)}{105} \times 100 = 16.7\ m.u.$$

Since the distance between *thr* and *arg* is larger than that separating either gene from the centromere, the centromere must lie between *thr* and *arg,* yielding the map in Fig. 5.24b. The distance between the two genes calculated by the formula (16.7 m.u.) is smaller than the sum of the two gene ↔ centromere distances $(10.0 + 7.6 = 17.6\ m.u.)$ because the formula does not account for all of the double crossovers. As always, calculating map positions for more genes with shorter distances between them produces the most accurate picture.

Mitotic Recombination Can Produce Genetic Mosaics

The recombination of genetic material is a critical feature of meiosis. The chiasmata that mark recombination sites keep homologous chromosomes together, allowing them to reach the metaphase plate and subsequently to segregate to opposite spindle poles during meiosis I. Moreover, recombination of genes during meiosis helps ensure the genetic diversity of the gametes. It is thus not surprising that eukaryotic organisms express a variety of enzymes (described in Chapter 6) that specifically initiate meiotic recombination. Recombination can also occur during mitosis. Unlike what happens in meiosis, however, mitotic crossovers are initiated by mistakes in chromosome replication or by chance exposures to radiation that break DNA molecules, rather than by a well-defined cellular program. As a result, mitotic recombination is a rare event, occurring no more frequently than once in a million somatic cell divisions. Nonetheless, the growth of a colony of yeast cells or the development of a complex multicellular organism involves so many cell divisions that geneticists can routinely detect these rare mitotic events.

Curt Stern, an early *Drosophila* geneticist, originally inferred the existence of mitotic recombination from observations of "twin spots" in a few fruit flies. **Twin spots** are adjacent islands of tissue that differ both from each other and from the tissue surrounding them. The distinctive patches arise from homozygous cells with a recessive phenotype growing amid a generally heterozygous cell population displaying the dominant phenotype. In *Drosophila,* the *yellow (y)* mutation changes body color from normal brown to yellow, while the *singed bristles (sn)* mutation causes body bristles to be short and curled rather than long and straight. Both of these genes are on the X chromosome. In 1936, Stern examined *Drosophila* females of genotype *y sn⁺ / y⁺ sn.* These heterozygotes were generally wild type in appearance, but Stern noticed that some flies carried patches of yellow body color, others had small areas of singed bristles, and still others displayed twin spots: adjacent patches of yellow cells and cells with singed bristles (Fig. 5.25). He assumed that mistakes in the mitotic divisions accompanying fly development could have led to these **mosaic** animals containing tissues of different genotypes. Individual yellow or singed patches could arise from chromosome loss or by mitotic nondisjunction.

Single yellow spot   Twin spot   Single singed spot

Figure 5.25 Twin spots: A form of genetic mosaicism. In a *y sn⁺ / y⁺ sn Drosophila* female, most of the body is wild type, but aberrant patches showing either yellow color or singed bristles sometimes occur. In some cases, yellow and singed patches are adjacent to each other, a configuration known as *twin spots.*

These errors in mitosis would yield XO cells containing only *y* (but not *y⁺*) or *sn* (but not *sn⁺*) alleles; such cells would show one of the recessive phenotypes.

The twin spots must have a different origin. Stern reasoned that they represented the reciprocal products of mitotic crossing-over between the *sn* gene and the centromere. The mechanism is as follows. During mitosis in a diploid cell, after chromosome duplication, homologous chromosomes occasionally—very occasionally—pair up with each other. While the chromosomes are paired, nonsister chromatids (that is, one chromatid from each of the two homologous chromosomes) can exchange parts by crossing-over. The pairing is transient, and the homologous chromosomes soon resume their independent positions on the mitotic metaphase plate. There, the two chromosomes can line up relative to each other in either of two ways (Fig. 5.26a). One of these orientations would yield two daughter cells that remain heterozygous for both genes and thus be indistinguishable from the surrounding wild-type cells. The other orientation, however, will generate two homozygous daughter cells, one *y sn⁺ / y sn⁺*, the other *y⁺sn / y⁺sn*. Since the two daughter cells would lie next to each other, subsequent mitotic divisions would produce adjacent patches of *y* and *sn* tissue (that is, twin spots). Note that if crossing-over occurs between *sn* and *y*, single spots of yellow tissue can form, but a reciprocal singed spot cannot be generated in this fashion (Fig. 5.26b).

Diploid yeast cells that are heterozygous for one or more genes exhibit mitotic recombination in the form of **sectors:** portions of a growing colony that have a different genotype than the

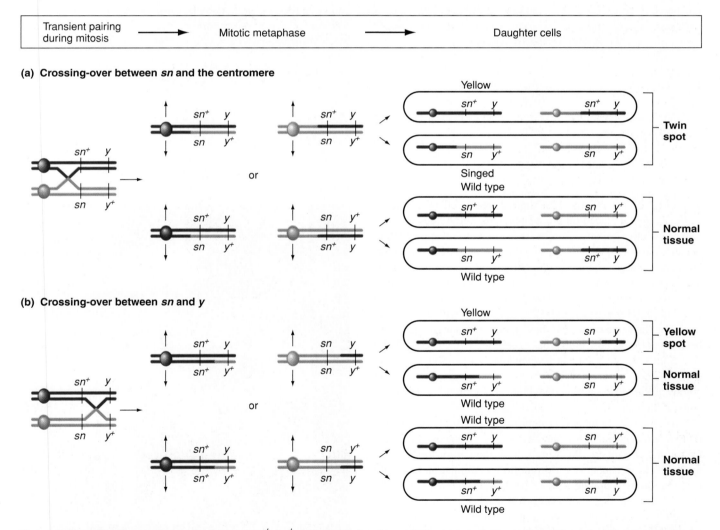

Figure 5.26 Mitotic crossing-over. **(a)** In a *y sn⁺ / y⁺ sn Drosophila* female, a mitotic crossover between the centromere and *sn* can produce two daughter cells, one homozygous for *y* and the other homozygous for *sn,* that can develop into adjacent aberrant patches (twin spots). This outcome depends on a particular distribution of chromatids at anaphase (*top*). If the chromatids are arranged in the equally likely opposite orientation, only phenotypically normal cells will result (*bottom*). **(b)** Crossovers between *sn* and *y* can generate single yellow patches. In contrast, a single mitotic crossover in these females cannot produce a single singed spot if the *sn* gene is closer to the centromere than the *y* gene. See if you can demonstrate this fact.

Genetics and Society

Mitotic Recombination and Cancer Formation

In humans, some tumors, such as those found in retinoblastoma, may arise as a result of mitotic recombination. Recall from our discussion of penetrance and expressivity in Chapter 3 that retinoblastoma is the most malignant form of eye cancer. In the United States, a genetic predisposition to the cancer is seen in about 1 in 20,000 live births. The retinoblastoma gene *(RB)* resides on chromosome 13, where the normal wild-type allele *(RB⁺)* encodes a protein that regulates retinal growth and differentiation. Cells in the eye need at least one copy of the normal wild-type allele to maintain control over cell division. A mutation in *RB⁺* leading to loss of normal gene function is designated *RB⁻*. If a cell loses both copies of *RB⁺*, it loses regulatory control, and a tumor results. The normal, wild-type *RB⁺* allele is thus known as a tumor-suppressor gene.

People with a genetic predisposition to retinoblastoma are born with only one functional copy of the normal *RB⁺* allele; their second chromosome 13 carries either a nonfunctional *RB⁻* allele or no *RB* gene at all. If a mutagen (such as radiation) or a mistake in gene replication or segregation destroys or removes the single remaining normal copy of the gene in a retinal cell in either eye, a retinoblastoma tumor will develop at that site. In one study of people with a genetic predisposition to retinoblastoma, cells taken from eye tumors were *RB⁻* homozygotes, while white blood cells from the same people were *RB⁺/RB⁻* heterozygotes. As Fig. A shows, mitotic recombination between the *RB* gene and the centromere of the chromosome carrying the gene provides one mechanism by which a cell in an *RB⁺/RB⁻* individual could become

remainder of the colony. If a diploid yeast cell of genotype *ADE2 / ade2* is placed on a petri plate, its mitotic descendents will grow into a colony. Usually, such colonies will appear white because the dominant wild-type *ADE2* allele specifies that color. However, many colonies will contain red sectors of diploid *ade2 / ade2* cells, which arose as a result of mitotic recombination events between the *ADE2* gene and its centromere (Fig. 5.27). (Homozygous *ADE2 / ADE2* cells will also be produced by the same event, but they cannot be distinguished from heterozygotes because both types of cells are white in color.) The size of the red sectors indicates when mitotic recombination took place. If they are large, it happened early in the growth of the colony, giving the resulting daughter cells a long time to proliferate; if they are small, it happened later.

Mitotic recombination is significant both as an experimental tool and because of the phenotypic consequences of particular mitotic crossovers. Problem 5-36 at the end of this chapter illustrates how geneticists use mitotic recombination to obtain information for mapping genes relative to each other and to the centromere. Mitotic crossing-over has also been of great value in the study of development because it can generate animals in which different cells have different genotypes (see Chapter 19). Finally, as the Genetics and Society box "Mitotic Recombination and Cancer Formation" explains, mitotic recombination can have major repercussions for human health.

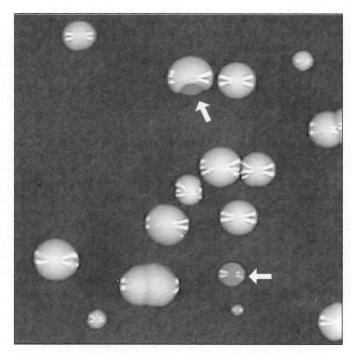

Figure 5.27   Mitotic recombination during the growth of diploid yeast colonies can create sectors. Arrows point to large, red *ade2 / ade2* sectors formed from *ADE2 / ade2* heterozygotes.

Transient pairing of homologous chromosomes 13 during mitosis	Mitotic metaphase	Daughter cells

Figure A   How mitotic crossing-over can contribute to cancer. Mitotic recombination during retinal growth in an RB^-/RB^+ heterozygote may produce an RB^-/RB^- daughter cell that lacks a functional retinoblastoma gene and thus divides out of control. The crossover must occur between the RB gene and its centromere. Only the arrangement of chromatids yielding this result is shown.

RB^-/RB^-. Once a homozygous RB^- cell is generated, it will divide uncontrollably, leading to tumor formation.

Only 40% of retinoblastoma cases follow the preceding scenario. The other 60% occur in people who are born with two normal copies of the RB gene. In such people, it takes two mutational events to cause the cancer. The first of these must convert an RB^+ allele to RB^-, while the second could be a mitotic recombination producing daughter cells that become cancerous because they are homozygous for the newly mutant, nonfunctional allele.

Interestingly, the role of mitotic recombination in the formation of retinoblastoma helps explain the incomplete penetrance and variable expressivity of the disease. People born as RB^+/RB^- heterozygotes may or may not develop the condition (incomplete penetrance). If, as usually happens, they do, they may have it in one or both eyes (variable expressivity). It all depends on whether and in what cells of the body mitotic recombination (or some other "homozygosing" event that affects chromosome 13) occurs.

C O N N E C T I O N S

Medical geneticists have used their understanding of linkage, recombination, and mapping to make sense of the pedigrees presented at the beginning of this chapter (see Fig. 5.1 on p. 114). The X-linked gene for red-green colorblindness must lie very close to the gene for hemophilia A because the two are tightly coupled. In fact, an examination of many pedigrees has shown that if a doubly heterozygous woman receives defective alleles of both genes from the same parent, as few as 3% of her sons will have only one of the conditions, as a result of recombination. This means that the genetic distance between the two genes is only 3 m.u. On the other hand, even though hemophilia B is also on the X chromosome, it lies far enough away from the red-green colorblindness locus that the two genes recombine relatively freely and thus, in a small sample, may appear genetically unlinked. The recombination distance separating the genes for colorblindness and hemophilia B is about 36 m.u. Pedigrees pointing to two different forms of hemophilia, one very closely linked to colorblindness, the other almost not linked at all, provided one of several indications that hemophilia is determined by more than one gene (Fig. 5.28).

Refining the human chromosome map poses a continuous challenge for medical geneticists. The newfound potential for finding and fitting more and more DNA markers into the map (review the Fast Forward box in this chapter) enormously improves the ability to assign precise locations to genes that cause disease. If several markers flank one such gene and it is possible

Hunter syndrome
Hemophilia B

Fragile X syndrome

Hemophilia A

G6PD deficiency: Favism
Drug-sensitive anemia
Chronic hemolytic anemia

Colorblindness (several forms)
Dyskeratosis congenita
Deafness with stapes fixation
TKCR syndrome

Adrenoleukodystrophy
Adrenomyeloneuropathy

Emery muscular dystrophy
SED tarda
Spastic paraplegia, X-linked

Figure 5.28   A genetic map of part of the human X chromosome.

to test a fetus for those markers, it is also possible to estimate the likelihood that the fetus has received the closely linked disease gene. DNA markers have also been of great importance in the construction of a physical map of the human genome, a process we discuss in Chapters 10 and 11.

The simultaneous occurrence of genetic linkage and recombination is universal among life-forms and presumably confers some advantage to a species beyond allowing geneticists to map its genes. Although definitive conclusions are beyond our reach, speculation about the possible advantages can help us understand why genomes are arranged as they are. Linkage provides the potential for transmitting favorable combinations of genes intact to successive generations, while recombination produces great flexibility in generating new combinations of alleles. Some new combinations may help a species adapt to changing environmental conditions, whereas the inheritance of successfully tested combinations can pre-serve what has worked in the past. Linkage and recombination thus help chromosomes function as organelles (in the broad sense) that not only transmit but also evolve genetic information.

In Chapters 2–5 of this book, we have examined how genes and chromosomes are transmitted. As important and useful as this knowledge is, it tells us very little about the structure and mode of action of the genetic material. In the next section (Chapters 6–8) we carry our analysis to the level of DNA, the actual molecule of heredity. In Chapter 6, we look at DNA structure and learn how the DNA molecule carries genetic information. In Chapter 7, we describe how geneticists defined the gene as a localized region of DNA containing many nucleotides that together encode the information to make a protein. In Chapter 8, we examine how the cellular machinery interprets the genetic information in genes to produce the multitude of phenotypes that make up an organism.

ESSENTIAL CONCEPTS

1. Gene pairs that are close together on the same chromosome are genetically linked because they are transmitted together more often than not. The hallmark of linkage is that among the progeny of double heterozygotes produced from testcrosses, the number of parental types is greater than the number of recombinant types.

2. The recombination frequencies of pairs of genes indicate how often two genes are transmitted together. For linked genes, the recombination frequency is less than 50%.

3. Gene pairs that assort independently exhibit a recombination frequency of 50%. This is because the number of parental types equals the number of recombinants. Genes may assort independently either because they are on different chromosomes or because they are far apart on the same chromosome.

4. Statistical analysis helps determine whether or not two genes assort independently. The probability value (p) calculated by the chi square test measures the likelihood that a particular set of data supports the null hypothesis of independent assortment, or no linkage. The lower the p value, the less likely is the null hypothesis, and the more likely the linkage. The chi square test can also be used to determine how well the outcomes of crosses fit other genetic hypotheses.

5. The greater the physical distance between linked genes, the higher the recombination frequency. However, recombination frequencies become more and more inaccurate as the distance between genes increases.

6. Recombination occurs because chromatids of homologous chromosomes exchange parts, that is, cross over, during the prophase of meiosis I, after the chromosomes have replicated.

7. Genetic maps are a visual representation of relative recombination frequencies. The greater the density of genes on the map (and thus the smaller the distance between the genes), the more accurate and useful the map becomes in predicting inheritance.

8. Organisms that retain all the products of one meiosis within an ascus reveal the relation between genetic recombination and the segregation of chromosomes during the two meiotic divisions. Organisms like *Neurospora* that produce ordered octads make it possible to locate a chromosome's centromere on the genetic map.

9. In diploid organisms heterozygous for two alleles of a gene, rare mitotic recombination between the gene and its centromere can produce *genetic mosaics* in which some cells are homozygous for one allele or the other.

SOCIAL AND ETHICAL ISSUES

1. Ophelia has been taking a genetics course and has learned about inherited predispositions to breast cancer. Realizing that many members of her family have had breast cancer, she wants to be tested to determine if there is an inherited form of the *BRCA1* gene that is responsible for the prevalence of cancer in her family. One of the very important individuals that will need to be tested is her great-aunt who had breast cancer many years ago.

Unfortunately, this aunt is showing early signs of senile dementia. The great-aunt has periods of lucidity when she seems to understand the need for testing and agrees to give a blood sample for the DNA testing. At other times, she is confused in her thinking overall and at these times refuses to have a sample taken for these purposes. She does routinely have blood drawn for other medical tests. Since she sometimes agrees to the DNA testing, is it ethical to do the test on the blood samples even though at other times she refuses?

2. Dean Hamer and colleagues from the National Institutes of Health in 1993 reported evidence that gene(s) in a particular region of the X chromosome correlated with homosexuality in males. To demonstrate this association, Hamer selected for his analysis those families in which there were two gay brothers and other gay males in the maternal lineage only. Is there scientific merit in preselecting your study group in this way? Why would a human geneticist feel it is necessary to study a subset of individuals?

3. In 1996, evidence was presented for linkage of a marker to a predisposition to prostate cancer. Using 66 families in which at least three males had prostate cancer, a genetic marker was found that is inherited with high frequency in those men with prostate cancer. Linked to this marker, somewhere in a region of about 10 million base pairs, is a gene that predisposes to prostate cancer. The estimate is that this genetic region is involved in about 3% of the *total* cases of prostate cancer. National newspapers announced these findings with the following headlines:

"Scientists Find Proof That Mutant Gene Can Increase Risk of Prostate Cancer"

"Prostate Cancer Gene Evidence Found"

"Scientists Zero in on Gene Tied to Prostate Cancer"

Do these headlines accurately represent the situation? Could they be misleading in any way? What would you use as a headline for an article on this study? What is the responsibility of the press in presenting new findings? Consider the responsibility of journalists/reporters to their employers (newspapers), their profession, and to the public.

S O L V E D   P R O B L E M S

I. The *Xg* locus on the human X chromosome has two alleles, a^+ and a. The a^+ allele causes the presence of the Xg surface antigen on red blood cells, while the recessive a allele does not allow antigen to appear. The *Xg* locus is 10 m.u. from the *Sts* locus. The *Sts* allele produces normal activity of the enzyme steroid sulfatase, while the recessive *sts* allele results in the lack of steroid sulfatase activity and the disease ichthyosis (scaly skin). A man with ichthyosis and no Xg antigen has a normal daughter with Xg antigen, who is expecting a child.

a. If the child is a son, what is the probability he will lack antigen and have ichthyosis?

b. What is the probability that a son would have both the antigen and ichthyosis?

c. If the child is a son with ichthyosis, what is the probability he will have Xg antigen?

Answer

a. This problem requires an understanding of how linkage affects the proportions of gametes. First designate the genotype of the individual in which recombination during meiosis affects the transmission of alleles: in this problem, the daughter. The X chromosome she inherited from her father (who had icthyosis and no Xg antigen) must be *sts a*. (No recombination could have separated the genes during meiosis in her father since he has only one X chromosome.) Because the daughter is normal and has the Xg antigen, her other X chromosome (inherited from her mother) must contain the *Sts* and a^+ alleles. Her X chromosomes can be diagrammed as:

$$\begin{array}{cc} sts & a \\ \hline \hline Sts & a^+ \end{array}$$

Because the *Sts* and *Xg* loci are 10 m.u. apart on the chromosome, there is a 10% recombination frequency. Ninety percent of the gametes will be parental: *sts a* or *Sts* a^+ (45% of each type) and 10% will be recombinant: *sts* a^+ or *Sts a* (5% of each type). The phenotype of a son directly reflects the genotype of the X chromosome from his mother. *Therefore, the probability that he will lack the Xg antigen and have icthyosis (genotype: sts a / Y) is 45/100.*

b. *The probability that he will have the antigen and ichthyosis (genotype: sts a^+ / Y) is 5/100.*

c. There are two classes of gametes containing the ichthyosis allele: *sts a* (45%) and *sts* a^+ (5%). If the total number of gametes is 100, then 50 will have the *sts* allele. Of those gametes, 5 (or 10%) will have the a^+ allele. *Therefore there is a 1/10 probability that a son with the* sts *allele will have the Xg antigen.*

II. *Drosophila* females of wild-type appearance but heterozygous for three autosomal genes are mated with males showing three autosomal recessive traits: glassy eyes, coal-colored bodies, and striped thoraxes. One thousand (1000) progeny of this cross are distributed in the following phenotypic classes:

Wild type	27
Striped thorax	11
Coal body	484
Glassy eye, coal body	8
Glassy eye, striped thorax	441
Glassy eye, coal body, striped thorax	29

a. Draw a genetic map based on this data.

b. Show the arrangement of alleles on the two homologous chromosomes in the parent females.

c. Normal-appearing males containing the same chromosomes as the parent females in the preceding cross are mated with females showing glassy eyes, coal-colored bodies, and striped thoraxes. Of 1000 progeny produced, indicate the numbers of the various phenotypic classes you would expect.

Answer

A logical, methodical way to approach a three-point cross is described here.

a. Designate the alleles:

t^+ = wild-type thorax	t = striped thorax
g^+ = wild-type eye	g = glassy eye
c^+ = wild-type body	c = coal-colored body

In solving a three-point cross, designate the types of events that gave rise to each group of individuals and the genotypes of the gametes obtained from their mother. (The paternal gametes contain only the recessive alleles of these genes [$t\ g\ c$]. They do not change the phenotype and can be ignored.)

Progeny	Number	Type of event	Genotype		
1. wild type	27	single crossover	t^+	g^+	c^+
2. striped thorax	11	single crossover	t	g^+	c^+
3. coal body	484	parental	t^+	g^+	c
4. glassy eye, coal body	8	single crossover	t^+	g	c
5. glassy eye, striped thorax	441	parental	t	g	c^+
6. glassy eye, coal body, striped thorax	29	single crossover	t	g	c

Picking out the parental classes is easy. If all the other classes are rare, the two most abundant categories are those gene combinations that have not undergone recombination. Then there should be two sets of two phenotypes that correspond to a single crossover event between the first and second genes, or between the second and third genes. Finally there should be a pair of classes containing small numbers that result from double crossovers. In this example, there are no flies in the double crossover classes, which would have been in the two missing phenotypic combinations: glassy eye and coal body, striped thorax.

Look at the most abundant classes to determine which alleles were on each chromosome in the female heterozygous parent. One parental class had the phenotype of coal body (484 flies), so one chromosome in the female must have contained the t^+, g^+, and c alleles. (Notice that we cannot yet say in what order these alleles are located on the chromosome.) The other parental class was glassy eye and striped thorax, corresponding to a chromosome with the t, g, and c^+ alleles.

To determine the order of the genes compare the $t^+\ g\ c^+$ double crossover class (not seen in the data) with the most similar parental class ($t\ g\ c^+$). The alleles of g and c retain their parental associations ($g\ c^+$) while the t gene has recombined with respect to both other genes in the double recombinant class. Thus the t gene is between g and c.

In order to complete the map, calculate the recombination frequencies between the center gene and each of the genes on the ends. For g and t the nonparental combinations of alleles are in classes 2 and 4, so RF = (11 + 8) / 1000 = 19 / 1000, or 1.9%. For t and c classes 1 and 6 are nonparental, so RF = (27 + 29) / 1000 = 56 / 1000, or 5.6%.

The genetic map is

b. The alleles on each chromosome were already determined (c, g^+, t^+ and c^+, g, t). Now that the order of loci has also been determined, the arrangement of the alleles can be indicated.

c. Males of the same genotype as the starting female ($c\ t^+\ g^+ / c^+\ t\ g$) could produce only two types of gametes: parental types $c\ t^+\ g^+$ and $c^+\ t\ g$ because there is no recombination in male *Drosophila*. The progeny expected from the mating with a homozygous recessive female are thus 500 coal body and 500 glassy eye, striped thorax flies.

III. The following asci were obtained in *Neurospora* when a wild-type strain ($ad^+ leu^+$) was crossed to a double mutant strain that cannot grow in the absence of adenine or leucine ($ad^- leu^-$). Only one member of each spore pair produced by the final mitosis is shown, since the two cells in a pair have the same genotype. Total asci = 120.

Spore pair			Ascus type		
1–2	$ad^+ leu^+$	$ad^+ leu^-$	$ad^+ leu^+$	$ad^+ leu^-$	$ad^- leu^+$
3–4	$ad^+ leu^+$	$ad^+ leu^-$	$ad^+ leu^-$	$ad^- leu^+$	$ad^+ leu^+$
5–6	$ad^- leu^-$	$ad^- leu^+$	$ad^- leu^+$	$ad^- leu^-$	$ad^- leu^-$
7–8	$ad^- leu^-$	$ad^- leu^+$	$ad^- leu^-$	$ad^+ leu^+$	$ad^+ leu^-$
# of asci	30	30	40	2	18

a. What genetic event causes the alleles of two genes to segregate to different cells at the second meiotic division, and when does this event occur?

b. Provide the best possible map for the two genes and their centromere(s).

Answer

This problem requires an understanding of tetrad analysis and the process (meiosis) that produces the patterns seen in ordered asci.

a. *A crossover between a gene and its centromere causes the segregation of alleles at the second meiotic division. The crossover event occurs during prophase of meiosis I.*

b. Using ordered tetrads you can determine whether two genes are linked, the distance between two genes, and the distance between each gene and its centromere. First designate the five classes of asci shown. The first class is a parental ditype (spores contain the same combinations of alleles as their parents); the second is a nonparental ditype; the last three are tetratypes. Next determine if these genes are linked. The number of PD = number of NPD, so the genes are not linked. When genes are unlinked, the tetratype asci are generated by a crossing-over event between a gene and its centromere. Looking at the *leu* gene, there is a second-division segregation pattern of that gene in the third and fourth asci types. Therefore, the percent of second-division segregation is

$$\frac{40+2}{120} \times 100 = 35\%$$

Because only half of the chromatids in the meioses that generated these tetratype asci were involved in the crossover, the map distance between *leu* and its centromere is 35/2, or 17.5 m.u. Asci of the fourth and fifth types show a second-division segregation pattern for the *ad* gene

$$\frac{2+18}{120} \times 100 = 16.6\%$$

Dividing 16.6% by 2 gives the recombination frequency and map distance of 8.3 m.u. *The map of these two genes is the following:*

PROBLEMS

5-1 Choose the phrase from the right column that best fits the term in the left column.

a. recombination	1. a statistical method for testing the fit between observed and expected results
b. linkage	2. an ascus containing spores of four different genotypes
c. chi square test	3. one crossover along a chromosome makes a second nearby crossover less likely
d. chiasma	4. when two loci recombine in less than 50% of gametes
e. tetratype	5. the relative chromosomal location of a gene
f. locus	6. the ratio of observed double crossovers to expected double crossovers
g. coefficient of coincidence	7. individual composed of cells with different genotypes
h. interference	8. formation of new genetic combinations by exchange of parts between homologs
i. parental ditype	9. when the two alleles of a gene are segregated into different cells at the first meiotic division
j. ascospores	10. an ascus containing only two nonrecombinant kinds of spores
k. first-division segregation	11. structure formed at the spot where crossing-over occurs between homologs
l. mosaic	12. fungal spores contained in a sac

5-2 Do the data that Mendel obtained fit his hypothesis? For example, Mendel obtained 315 yellow round, 101 yellow wrinkled, 108 green round, and 32 green wrinkled seeds from the selfing of *Yy Rr* individuals (a total of 556). His hypotheses of segregation and independent assortment predict a 9:3:3:1 ratio in this case. Use the chi square test to determine whether Mendel's data are significantly different from what he predicted. (The chi square test did not exist in Mendel's day, so he was not able to test his own data for goodness of fit to his hypotheses.)

5-3 Two genes control color in corn snakes as follows: *O– B–* snakes are brown, *O– bb* are orange, *oo B–* are black, and *oo bb* are albino. An orange snake was mated to a black snake, and a large number of F_1 progeny were obtained, all of which were brown. When the F_1 snakes were mated to one another, they produced 100 brown offspring, 25 orange, 22 black, and 13 albino.

a. What are the genotypes of the F_1 snakes?

b. What proportions of the different colors would have been expected among the F_2 snakes if the two loci assort independently?

c. Do the observed results differ significantly from what was expected, assuming independent assortment is occurring?

d. What is the probability that differences this great between observed and expected values would happen by chance?

5-4 A mouse from a true-breeding population with normal gait was crossed to a mouse displaying an odd gait called "dancing." The F_1 animals all showed normal gait.

a. If dancing is caused by homozygosity for the recessive allele of a single gene, what proportion of the F_2 mice should be dancers?

b. If mice must be homozygous for recessive alleles of both of two different genes to have the dancing phenotype, what proportion of the F_2 should be dancers if the two genes are unlinked?

c. When the F_2 mice were obtained, 42 normal and 8 dancers were seen. Use the chi square test to determine if these results better fit the one gene model from part (a) or the two gene model from part (b).

5-5 In *Drosophila,* males from a true-breeding stock with raspberry-colored eyes were mated to females from a true-breeding stock with sable-colored bodies. In the F_1 generation, all the females had wild-type eye and body color, while all the males had wild-type eye color but sable-colored bodies. When F_1 males and females were mated, the F_2 generation was composed of 216 females with wild-type eyes and bodies, 223 females with wild-type eyes and sable bodies, 191 males with wild-type eyes and sable bodies, 188 males with raspberry eyes and wild-type bodies, 23 males with wild-type eyes and bodies, and 27 males with raspberry eyes and sable bodies. Explain these results by diagramming the crosses, and calculate any relevant map distances.

5-6 Figure 5.6 on p. 118 applied the chi square method to test linkage between two genes by asking whether the observed numbers of parental and recombinant classes differed significantly from the expectation of independent assortment that parentals = recombinants. Another possible way to analyze the results from these same experiments is to ask whether the observed frequencies of the four genotypic classes (*A B, a b, A b,* and *a B*) can be explained by a null hypothesis predicting that they should appear in a 1:1:1:1 ratio.

a. What are the relative advantages and disadvantages of analyzing the data in these two different ways?

b. What is the null hypothesis in each case?

c. Which is a more sensitive test of linkage? (Analyze the data in Fig. 5.6 by the second method.)

d. How would both methods respond to a situation in which one allele of one of the genes causes reduced viability?

5-7 If the *a* and *b* loci are 40 cM apart, and an *AA BB* individual and an *aa bb* individual mate:

a. What gametes will the F_1 individuals produce, and in what proportions? What phenotypic classes in what proportions are expected in the F_2 generation (assuming complete dominance for both genes)?

b. If the original cross was *AA bb* × *aa BB,* what gametic proportions would emerge from the F_1? What would be the result in the F_2 generation?

5-8 *CC DD* and *cc dd* individuals were crossed to each other, and the F_1 generation was backcrossed to the *cc dd* parent; 903 *Cc Dd,* 897 *cc dd,* 98 *Cc dd,* and 102 *cc Dd* offspring resulted.

a. How far apart are the *c* and *d* loci?

b. What progeny and in what frequencies would you expect to result from testcrossing the F_1 generation from a *CC dd* × *cc DD* cross to *cc dd?*

5-9 If the *a* and *b* loci are 20 m.u. apart in humans, and an *A B / a b* woman mates with an *a b / a b* man, what is the probability that their first child will be *A b / a b?*

5-10 In a particular human family, John and his mother both have brachydactyly (a rare autosomal dominant causing short fingers). John's father has Huntington disease (another rare autosomal dominant). John's wife is phenotypically normal and is pregnant. Two-thirds of people who inherit the Huntington (*HD*) allele show symptoms by age 50, and John is 50 and has no symptoms. Brachydactyly is 90% penetrant.

a. What are the genotypes of John's parents?

b. What are the possible genotypes for John?

c. What is the probability the child will express both brachydactyly and Huntington disease by age 50 if the two genes are unlinked?

d. If these two loci are 20 m.u. apart, how will it change your answer to part (c)?

5-11 In mice, the autosomal locus coding for the β-globin chain of hemoglobin is 1 m.u. from the albino locus. Assume for the moment that the same is true in humans. The disease sickle-cell anemia is the result of homozygosity for a particular mutation in the β-globin gene.

a. A son is born to an albino man and a woman with sickle-cell anemia. What kinds of gametes will the son form, and in what proportions?

b. A daughter is born to a normal man and a woman who has both albinism and sickle-cell anemia. What kinds of gametes will the daughter form, and in what proportions?

c. If the son in part (a) grows up and marries the daughter in part (b), what is the probability that a child of theirs will be an albino with sickle-cell anemia?

5-12 In corn, the allele A allows the deposition of anthocyanin (blue) pigment in the kernels (seeds), while aa plants have yellow kernels. At a second gene, $W-$ produces smooth kernels, while ww kernels are wrinkled. A plant with blue smooth kernels was crossed to a plant with yellow wrinkled kernels. The F_1 generation consisted of 1447 blue smooth, 169 blue wrinkled, 186 yellow smooth, and 1510 yellow wrinkled.

a. Are the a and w loci linked? If so, how far apart are they?

b. What was the genotype of the blue smooth parent? Include the chromosome arrangement of alleles.

c. If a plant grown from a blue wrinkled F_1 seed is crossed to a plant grown from a yellow smooth F_1 seed, what kinds of kernels would be expected, and in what proportions?

5-13 Albino rabbits (lacking pigment) are homozygous for the recessive c allele (C allows pigment formation). Rabbits homozygous for the recessive b allele make brown pigment, while those with at least one copy of B make black pigment. True-breeding brown rabbits were crossed to albinos, which were BB. F_1 rabbits, which were all black, were crossed to the double recessive (bb cc). The progeny obtained were 34 black, 66 brown, and 100 albino.

a. What phenotypic proportions would have been expected if the b and c loci were unlinked?

b. How far apart are the two loci?

5-14 Cinnabar eyes (cn) and reduced bristles (rd) are autosomal recessive characters in *Drosophila*. A homozygous wild-type female was crossed to a reduced, cinnabar male, and the F_1 males were then crossed to the F_1 females to obtain the F_2. Of the 400 F_2 offspring obtained, 292 were wild type, 9 were cinnabar, 7 were reduced, and 92 were reduced, cinnabar. Explain these results and estimate the distance between the cn and rd loci.

5-15 A DNA variant has been found linked to a rare autosomal dominant disease in humans and can thus be used as a marker to follow inheritance of the disease allele. In an informative family (in which one parent is heterozygous for both the disease allele and the DNA marker in a known chromosomal arrangement of alleles, and his or her mate does not have the same alleles of the DNA variant), the reliability of such a marker as a predictor of the disease in a fetus is related to the map distance between the DNA marker and the gene causing the disease. Imagine that a man affected with the disease (genotype Dd) is heterozygous for the V^1 and V^2 forms of the DNA variant, with form V^1 on the same chromosome as the D allele and form V^2 on the same chromosome as d. His wife is V^3V^3 dd, where V^3 is another allele of the DNA marker. Typing of the fetus by amniocentesis reveals that the fetus has the V^2 and V^3 variants of the DNA marker. How likely is it that the fetus has inherited the disease allele D if the distance between the D locus and the marker locus is (a) 0 m.u., (b) 1 m.u., (c) 5 m.u., (d) 10 m.u., (e) 50 m.u.?

5-16 a. In *Drosophila*, crosses between F_1 heterozygotes of the form $A\ b\ /\ a\ B$ always yield the same ratio of phenotypes in the F_2 progeny regardless of the distance between the two genes (assuming complete dominance for both autosomal genes). What is this ratio? Would this also be the case if the F_1 heterozygotes were $A\ B\ /\ a\ b$?

b. If you intercrossed F_1 heterozygotes of the form $A\ b\ /\ a\ B$ in mice, the phenotypic ratio among the F_2 progeny would vary with the map distance between the two genes. Is there a simple way to estimate the map distance based on the frequencies of the F_2 phenotypes, assuming rates of recombination are equal in males and females? Could you estimate map distances in the same way if the mouse F_1 heterozygotes were $A\ B\ /\ a\ b$?

5-17 A true-breeding strain of Virginia tobacco has dominant alleles determining leaf morphology (M), leaf color (C), and leaf size (S). A Carolina strain is homozygous for the recessive alleles of these three genes. These genes are found on the same chromosome as follow:

An F_1 hybrid between the two strains is now backcrossed to the Carolina strain. Assuming no interference:

a. What proportion of the backcross progeny will resemble the Virginia strain for all three traits?

b. What proportion of the backcross progeny will resemble the Carolina strain for all three traits?

c. What proportion of the backcross progeny will have the leaf morphology and leaf size of the Virginia strain but the leaf color of the Carolina strain?

d. What proportion of the backcross progeny will have the leaf morphology and leaf color of the Virginia strain but the leaf size of the Carolina strain?

5-17 In the tubular flowers of foxgloves, wild-type coloration is red while a mutation called *white* produces white flowers. Another mutation, called *peloria*, causes the flowers at the apex of the stem to be huge. Yet another mutation, called *dwarf*, affects stem length. You cross a white-flowered plant (otherwise phenotypically wild type) to a plant that is dwarf and peloria but has wild-type red flower color. All of the F_1 plants are tall with white, normal-sized flowers. You cross an F_1 plant back to the dwarf and peloria parent, and you see the 543 progeny shown in the chart. (Only mutant traits are noted.)

dwarf, peloria	172
white	162
dwarf, peloria, white	56
wild type	48
dwarf, white	51
peloria	43
dwarf	6
peloria, white	5

a. Which alleles are dominant?
b. What were the genotypes of the parents in the original cross?
c. Draw a map showing the linkage relationships of these three loci.
d. Is there interference? If so, calculate the coefficient of coincidence and the interference value.

5-19 In *Drosophila,* three autosomal genes have the following map:

a. Provide the data, in terms of the expected number of flies in the following phenotypic classes, when $a^+ b^+ c^+ / a\,b\,c$ females are crossed to $a\,b\,c / a\,b\,c$ males. Assume 1000 flies were counted, and that there is no interference in this region.

a^+	b^+	c^+
a	b	c
a^+	b	c
a	b^+	c^+
a^+	b^+	c
a	b	c^+
a^+	b	c^+
a	b^+	c

b. If the cross were reversed, such that $a^+ b^+ c^+ / a\,b\,c$ males are crossed to $a\,b\,c / a\,b\,c$ females, how many flies would you expect in the same phenotypic classes?

5-20 A cross was performed between a haploid strain of yeast with the genotype $a\,f\,g$ and another haploid strain with the genotype $\alpha + +$ (*a* and α are mating types). The resulting diploid was sporulated, and a random sample of 101 of the resulting haploid spores was analyzed. The following genotypic frequencies were seen: 31 $\alpha + +$, 29 $a\,f\,g$, 13 $\alpha + g$, 14 $a\,f +$, 6 $a + g$, 6 $\alpha\,f +$, 1 $a + +$, and 1 $\alpha\,f\,g$.

a. Map the loci involved in the cross.
b. Assuming all three genes are on the same chromosome arm, is it possible that a particular ascus could contain an $\alpha\,f\,g$ spore but not an $a + +$ spore? If so, draw a meiosis that could generate such an ascus.

5-21 Male *Drosophila* expressing the recessive mutations *sc (scute), ec (echinus), cv (crossveinless),* and *b (black)* were crossed to phenotypically wild-type females, and the 3288 progeny listed here were obtained.

653	black, scute, echinus, crossveinless
670	scute, echinus, crossveinless
675	wild type
655	black
71	black, scute
73	scute
73	black, echinus, crossveinless
74	echinus, crossveinless
87	black, scute, echinus
84	scute, echinus
86	black, crossveinless
83	crossveinless
1	black, scute, crossveinless
1	scute, crossveinless
1	black, echinus
1	echinus

a. Diagram the genotype of the female parent.
b. Map these loci.
c. Is there evidence of interference? Justify your answer with numbers.

5-22 A snapdragon with pink petals, black anthers, and long stems was allowed to self-fertilize. From the resulting seeds, 650 adult plants were obtained. The phenotypes of these offspring are listed here.

78	red	long	tan
26	red	short	tan
44	red	long	black
15	red	short	black
39	pink	long	tan
13	pink	short	tan
204	pink	long	black
68	pink	short	black
5	white	long	tan
2	white	short	tan
117	white	long	black
39	white	short	black

a. Using *P* for one allele and *p* for the other, indicate how flower color is inherited.
b. What numbers of red : pink : white would have been expected among these 650 plants?

c. How are anther color and stem length inherited?
d. What was the genotype of the original plant?
e. Do any of the three genes show independent assortment?
f. For any genes that are linked, indicate the arrangements of the alleles on the homologous chromosomes in the original snapdragon, and estimate the distance between the genes.

5-23 The following list of four *Drosophila* mutations indicates the symbol for the mutation, the name of the gene, and the mutant phenotype:

Allele symbol	Gene name	Mutant phenotype
dwp	dwarp	small body, warped wings
rmp	rumpled	deranged bristles
pld	pallid	pale wings
rv	raven	dark eyes and bodies

You perform the following crosses with the indicated results:

Cross #1:
dwarp, rumpled females × pallid, raven males →
dwarp, rumpled males and wild-type females

Cross #2:
pallid, raven females × dwarp, rumpled males →
pallid, raven males and wild-type females

F_1 females from cross #1 were crossed to males from a true-breeding *dwarp rumpled pallid raven* stock. The 1000 progeny obtained were as follows:

pallid	3
pallid, raven	428
pallid, raven, rumpled	48
pallid, rumpled	23
dwarp, raven	22
dwarp, raven, rumpled	2
dwarp, rumpled	427
dwarp	47

Indicate the best map for these four genes, including all relevant data. Calculate interference values where appropriate.

5-24 Two crosses were made in *Neurospora* involving the mating type locus and either the *ad* or *p* genes. In both cases, the mating type locus (*A* or *a*) was one of the loci whose segregation was scored. One cross was *ad A* × *+ a* (cross a), and the other was *p A* × *+ a* (cross b). From cross a, 10 parental ditype, 9 nonparental ditype, and 1 tetratype asci were seen. From cross b, the results were 24 parental ditype, 3 nonparental ditype, and 27 tetratype asci.
a. What are the linkage relationships between the mating type locus and the other two loci?

b. Although these two crosses were performed in *Neurospora,* you cannot use the data given to calculate centromere-to-gene distances for any of these genes. Why not?

5-25 *Neurospora* of genotype *a + c* are crossed with *Neurospora* of genotype *+ b +*. The following tetrads are obtained (note that the genotype of the four spore *pairs* in an ascus are listed, rather than listing all eight spores):

a + c	a b c	+ + c	+ b c	a b +	a + c
a + c	a b c	a + c	a b c	a b +	a b c
+ b +	+ + +	+ b +	+ + +	+ + c	+ + +
+ b +	+ + +	a b +	a + +	+ + c	+ b +
137	141	26	25	2	3

a. In how many cells has meiosis occurred to yield this data?
b. Give the best genetic map to explain these results. Indicate all relevant genetic distances, both between genes and between each gene and its respective centromere.
c. Diagram a meiosis that could give rise to one of the three tetrads in the class at the far right in the list.

5-26 The *a, b,* and *c* loci are all on different chromosomes in yeast. When *a +* yeast were crossed to *+ b* yeast, and the resultant tetrads analyzed, it was found that the number of nonparental ditype tetrads was equal to the number of parental ditypes, but that there were no tetratype asci at all. On the other hand, many tetratype asci were seen in the tetrads formed after *a +* was crossed with *+ c,* and after *b +* was crossed with *+ c.* Explain these results.

5-27 A cross was performed between a yeast strain that requires methionine and lysine for growth (*met⁻ lys⁻*) and another yeast strain, which is *met⁺ lys⁺*. One hundred asci were dissected, and colonies were grown from the four spores in each ascus. Cells from these colonies were tested for their ability to grow on petri plates containing either minimal medium (min), min + lysine (lys), min + methionine (met), or min + lys + met. The asci could be divided into two groups based on this analysis:

Group 1: In 89 asci, cells from two of the four spore colonies could grow on all four kinds of media, while the other two spore colonies could grow only on min + lys + met.

Group 2: In 11 asci, cells from one of the four spore colonies could grow on all four kinds of petri plates. Cells from a second one of the four spore colonies could grow only on min + lys plates and on min + lys + met plates. Cells from a third of the four spore colonies could only grow on min + met plates and on min + lys +

met. Cells from the remaining colony could only grow on min + lys + met.

  a. What are the genotypes of each of the spores within the two types of asci?

  b. Are the *lys* and *met* genes linked? If so, what is the map distance between them?

  c. If you could extend this analysis to many more asci, you would eventually find some asci with a different pattern. For these asci, describe the phenotypes of the four spores. List these phenotypes as the ability of dissected spores to form colonies on the four kinds of petri plates.

5-28 Two genes are located on the same chromosome as follows:

A haploid cross of the form *C D* × *c d* is made.

  a. What proportions of PD, NPD, and T tetrads would you expect if this cross was made between strains of *Saccharomyces cerevisiae* and the interference in this region = 1?

  b. If the interference in this region = 0?

  c. What kinds of tetrads, and in what proportions, would you expect if this cross was made between strains of *Neurospora crassa* and the interference in this region = 1? (Consider not only whether a tetrad is PD, NPD, or T but also whether the tetrad shows first or second division segregation for each gene.)

  d. If the interference in this region = 0?

5-29 Indicate the percentage of tetrads that would have 0, 1, 2, 3, or 4 viable spores after *Saccharomyces cerevisiae* *a* / α diploids of the following genotypes are sporulated:

  a. A true-breeding wild-type strain (with no mutations in any gene essential for viability).

  b. A strain heterozygous for a null (completely inactivating) mutation in a single essential gene.

For the remaining parts of this problem, consider crosses between yeast strains of the form *a* × *b*, where *a* and *b* are both temperature-sensitive mutations in different essential genes. The cross is conducted under permissive (low-temperature) conditions. Indicate the percentage of tetrads that would have 0, 1, 2, 3, or 4 viable spores subsequently measured under restrictive (high-temperature) conditions.

  c. *a* and *b* are unlinked, and both are 0 m.u. from their respective centromeres.

  d. *a* and *b* are unlinked. *a* is 0 m.u. from its centromere, while *b* is 10 m.u. from its centromere.

  e. *a* and *b* are 0 m.u. apart.

  f. *a* and *b* are 10 m.u. apart. Assume all crossovers between *a* and *b* are SCOs (single crossovers).

  g. In part (f), if a four-strand DCO (double crossover) occurred between *a* and *b,* how many of the spores in the resulting tetrad would be viable at high temperature?

5-29 A research group has selected three independent *trp⁻* haploid strains of *Neurospora*, each of which cannot grow in the absence of the amino acid tryptophan. They first mated these three strains with a wild-type strain of opposite mating type, and then they analyzed the resultant octads. For all three matings, two of the four spore pairs in every octad could grow on minimal medium (that is, in the absence of tryptophan), while the other two spore pairs were unable to grow on this minimal medium.

  a. What can you conclude from this result?

In the matings of mutant strains 1 and 2 with wild type, one of the two topmost pairs in some octads had spores that could grow on minimal medium while the other of the two topmost pairs in the same octads had spores that could not grow on minimal medium. In the mating of mutant strain 3 with wild type, either all the spores in the two topmost pairs could grow on minimal medium or all could not grow on minimal medium.

  b. What can you conclude from this result?

The researchers next prepared two separate cultures of each mutant strain; one of these cultures was of mating type *A* and the other of mating type *a*. They mated these strains in pairwise fashion, dissected the resultant octads, and determined how many of the individual spores could grow on minimal medium. The results are shown here.

Mating	% of octads with *x* number of spores viable on minimal medium				
	0	2	4	6	8
1 × 2	78	22	0	0	0
1 × 3	46	6	48	0	0
2 × 3	42	16	42	0	0

  c. For each of the three matings in the table, how many of the 100 octads are PD? NPD? T?

  d. Draw a genetic map explaining all of the preceding data. Assume that the sample sizes are sufficiently small that none of the octads are the result of double crossovers.

  e. Although this problem describes crosses in *Neurospora*, it does not help in this particular case to present the matings in the table as ordered octads. Why not?

  f. Why in this particular problem can you obtain gene–centromere distances from the crosses in the table, even though the data are not presented as ordered octads?

5-31 A yeast strain that cannot grow in the absence of the amino acid histidine *(his⁻)* is mated with a yeast strain that cannot grow in the absence of the amino acid lysine *(lys⁻)*. Among the 400 unordered tetrads resulting from this mating, 233 were PD, 11 were NPD, and 156 were T.

a. What types of spores are in the PD, NPD, and T tetrads?

b. What is the distance in map units between the *his* and *lys* genes?

c. Assuming that none of these tetrads was caused by more than two crossovers between the genes, how can you estimate the number of meioses that generated these 400 tetrads in which zero, one, or two crossovers took place?

d. Based on your answer to part (c), what is the mean number of crossovers per meiosis in the region between the two genes?

e. The equation RF = 100 × (NPD + 1/2T) / total tetrads accounts for some, but not all, double crossovers between two genes. Which double crossovers are missed? Can you extrapolate from your answer to part (d) to obtain a more accurate equation for calculating map distances between two genes from the results of tetrad analysis?

f. Using your corrected equation from part (e), what is a more accurate measurement of the distance in map units between the *his* and *lys* genes?

5-32 Write the number of *different kinds* of phenotypes excluding gender you would see among a large number of progeny from an F₁ mating between individuals of identical genotype that are heterozygous for one or two genes (that is, *Aa* or *Aa Bb*) as indicated. No gene interactions means that the phenotype determined by one gene is not influenced by the genotype of the other gene.

a. One gene; *A* completely dominant to *a*.

b. One gene; *A* and *a* codominant.

c. One gene; *A* incompletely dominant to *a*.

d. Two unlinked genes; no gene interactions; *A* completely dominant to *a,* and *B* completely dominant to *b*.

e. Two genes, 10 m.u. apart; no gene interactions; *A* completely dominant to *a,* and *B* completely dominant to *b*.

f. Two unlinked genes; no gene interactions; *A* and *a* codominant, and *B* incompletely dominant to *b*.

g. Two genes, 10 m.u. apart; *A* completely dominant to *a,* and *B* completely dominant to *b;* and with recessive epistasis between the genes.

h. Two unlinked duplicated genes (that is, *A* and *B* perform the same function); *A* and *B* completely dominant to *a* and *b,* respectively.

i. Two genes, 0 m.u. apart; no gene interactions; *A* completely dominant to *a,* and *B* completely dominant to *b*. (There are two possible answers.)

5-33 A single yeast cell placed on a solid agar will divide mitotically to produce a colony of about 10^7 cells. A haploid yeast cell that has a mutation in the *ade2* gene will produce a red colony; an *ade2⁺* colony will be white. Some of the colonies formed from diploid yeast cells with a genotype of *ade2⁺ / ade2⁻* will contain sectors of red within a white colony.

a. How would you explain these sectors?

b. Although many different white colonies are roughly the same size, the red sectors within some of the white colonies vary markedly in size. Why? Do you expect more of the red sectors to be relatively large or relatively small?

5-34 A diploid strain of yeast has a wild-type phenotype but the following genotype:

a, b, c, d, and *e* all represent recessive alleles that yield a visible phenotype, and *leth* represents a recessive lethal mutation. All genes are on the same chromosome, and *a* is very tightly linked to its centromere (indicated by dark circle). Which of the following phenotypes could be found in sectors resulting from mitotic recombination in this cell? (1) *a;* (2) *b;* (3) *c;* (4) *d;* (5) *e;* (6) *b e;* (7) *c d;* (8) *c d e;* (9) *d e;* (10) *a b.* Assume that double mitotic crossovers are too rare to be observed.

5-35 In *Drosophila,* the *yellow (y)* gene is near the end of the acrocentric X chromosome, while the *singed (sn)* gene is located near the middle of the X chromosome. On the wings of female flies of genotype *y sn / y⁺ sn⁺,* you can very rarely find patches of yellow tissue within which a small subset of cells also have singed bristles.

a. How can you explain this phenomenon?

b. Would you find similar patches on the wings of females having the genotype *y⁺ sn / y sn⁺?*

5-36 Neurofibromas are tumors of the skin that can arise when a skin cell that is originally *NF1⁺ / NF1⁻* loses the *NF1⁺* allele. This wild-type allele encodes a functional tumor suppressor protein, while the *NF1⁻* allele encodes a nonfunctional protein.

A patient of genotype *NF1⁺ / NF1⁻* has 20 independent tumors in different areas of the skin. Samples are taken of normal, noncancerous cells from this patient, as well as of cells from each of the 20 tumors. Extracts of these samples are analyzed by a technique called gel electrophoresis that can detect variant forms of four different proteins (A, B, C, and D) all encoded by genes that lie on the same autosome as *NF1.* Each protein has a slow (S) and a fast (F) form that are encoded by different alleles (for example, A^S and A^F). In the extract of normal tissue, slow and fast variants of all four proteins are found. In the extracts of the tumors, 12 had only the fast variants of proteins A and D but both the fast and slow variants of proteins B and C; 6 had only the fast variant of protein A but both the fast and slow variants of proteins B, C, and D; and the remaining 2 tumor extracts had only the fast variant of protein A, only the slow variant of protein B, the fast and slow variants of protein C, and only the fast variant of protein D.

a. What kind of genetic event described in this chapter could cause all 20 tumors, assuming that all the tumors are produced by the same mechanism?

b. Draw a genetic map describing this data, assuming that this small sample represents all the types of tumors that could be formed by the same mechanism in this patient. Show which alleles of which genes lie on the two homologous chromosomes. Indicate all relative distances that can be estimated.

c. Another mechanism that can lead to neurofibromas in this patient is a mitotic error producing cells with 45 rather than the normal 46 chromosomes. How can this mechanism cause tumors? How do you know, just from the results described, that none of these 20 tumors are formed by such mitotic errors?

d. Can you think of any other type of error that could produce the results described?

6

DNA: How the Molecule of Heredity Carries, Replicates, and Recombines Information

The double-helical structure of DNA provides an explanation for the
accurate transmission of genetic information from generation
to generation over billions of years.

For nearly 4 billion years the double-stranded DNA molecule
has served as the almost universal bearer of genetic information; 3.7 billion years ago the earliest bacterial cells incorporated it
into their chromosomes, and about 2 billion years ago, when the
eukaryotic precursors of plants, animals, and fungi evolved from
these simple cells, their chromosomes also carried a DNA molecule. Since that time, the hardware—the structure of the molecule
itself—has not changed. In contrast, evolution has honed and
vastly expanded the software—the programs of genetic information that the molecule stores, expresses, and transmits from one
generation to the next.

Under special conditions of little or no oxygen, DNA can withstand a wide range of temperature, pressure, and humidity and remain relatively intact for hundreds, thousands, even tens of
thousands of years. Molecular sleuths of the 1980s and 1990s have
retrieved the evidence: 100-year-old DNA from preserved tissue of
the quagga (a partially striped horselike creature that became extinct in the final years of the nineteenth century); 8000-year-old
DNA from human skulls found in the swamps of Florida; and

30,000-year-old DNA from a Neanderthal skeleton (Fig. 6.1).
Surprisingly, this ancient DNA still carries readable sequences—
shards of decipherable information that become veritable time
machines for the direct viewing of genes in long-vanished organisms and species. Comparisons with equivalent expanses of modern DNA make it possible to identify the precise mutations that
have fueled evolution. Analyses of Neanderthal and human DNA,
for example, have helped anthropologists settle a long-running
debate about the relation of the two species; the genetic evidence
shows that Neanderthals and early humans probably did not in-
termix and that *Homo sapiens,* upon their arrival in Europe some
30,000 years ago, most likely displaced Neanderthals over a period
of time.

Francis Crick, codiscoverer of DNA's double helical structure
and a leading theoretician of molecular biology, has written that
"almost all aspects of life are engineered at the molecular level,
and without understanding molecules, we can only have a very
sketchy understanding of life itself." In Chapters 1–5 we examined how Mendel and his successors used data from breeding

155

(a)

(b)

(c)

Figure 6.1   Ancient DNA still carries information. Molecular biologists have successfully extracted DNA from **(a)** a 100-year-old quagga; **(b)** an 8000-year-old human skull; **(c)** a Neanderthal skeleton excavated in 1856 in Germany's Neander Valley whose DNA was analyzed in the late 1990s. These findings attest to the chemical stability of DNA, the molecule of inheritance.

experiments to deduce the presence and activity of genes, analyze the effect of alternative alleles on phenotype, and tie genes to the movements of chromosomes in time and space. With this approach, it is possible to predict the outcomes of genetic crosses in the absence of detailed knowledge of the molecules and biochemical reactions that underlie the events observed. But without understanding what genes are made of, it is impossible to decipher the exact mechanisms by which they determine phenotypes, transmit instructions between generations, and evolve new information. For this reason, we shift our perspective in this chapter to an examination of DNA, the molecule that composes the genetic material.

As we extend our analysis to the molecular level, two general themes emerge. First, DNA's genetic functions flow directly from its molecular structure—the way its atoms are arranged in space. Second, most of DNA's genetic functions depend on spe-

cialized proteins that interact with the molecule of heredity. One such protein, for example, binds to DNA to begin the process of replication.

In our discussion of the structure and function of DNA, we describe:

- How investigators pinpointed DNA as the genetic material.

- The elegant Watson-Crick model of DNA structure.

- How DNA structure provides for the storage of genetic information.

- How DNA structure gives rise to the semiconservative mode of molecular replication.

- How DNA structure promotes the recombination of genetic information.

Experiments Designate DNA As the Genetic Material

Though accepted today as fact, it was not always obvious that DNA is the genetic material. It took a cohesive pattern of results from experiments performed over more than 50 years to convince the scientific community that DNA is the molecule of heredity. We now present key pieces of the evidence.

Chemical Characterization Localizes DNA in the Chromosomes

In 1869 Friedrich Miescher extracted a weakly acidic, phosphorus- rich material from the nuclei of human white blood cells and named it "nuclein." It was unlike any previously reported chemical compound, and its major component turned out to be DNA, although it also contained some contaminants. The full chemical name of DNA is **<u>d</u>eoxyribo<u>n</u>ucleic <u>a</u>cid,** reflect-

ing three characteristics of the substance: one of its constituents is a sugar known as deoxyribose; it is found mainly in cell nuclei; and it is acidic.

After purifying DNA from the nuclein and treating it with various reagents, researchers established that DNA is composed of four different subunits linked in a long chain (Fig. 6.2). The four different subunits belong to a class of compounds known as **nucleotides,** the bonds joining one nucleotide to another are covalent **phosphodiester bonds,** and the linked chain of repeating subunits is a type of **polymer.** A procedure first reported in 1923 made it possible to discover where in the cell DNA resides. Named the Feulgen reaction after its designer, the procedure relies on a chemical called the Schiff reagent, which stains DNA red. In a preparation of stained cells, the chromosomes redden, while other areas of the cell remain relatively colorless. The reaction shows that DNA is localized almost exclusively within specific nuclear organelles—the chromosomes.

Figure 6.2 The chemical composition of DNA. A single strand of a DNA molecule consists of a chain of nucleotide subunits (*colored boxes*). Each nucleotide is made of the sugar deoxyribose connected to an inorganic phosphate group and to one of four nitrogenous bases. The phosphodiester bonds that link the nucleotide subunits to each other attach the phosphate group of one nucleotide to the deoxyribose sugar of the preceding nucleotide.

The finding that DNA is a component of chromosomes does not prove that the molecule has anything to do with genes. Eukaryotic chromosomes also contain an even greater amount of protein by weight. Because proteins are built of 20 different amino acids whereas DNA carries just four different subunits, many researchers thought proteins had greater potential for diversity and were better suited to serve as the genetic material. These same scientists assumed that even though DNA was an important part of chromosome structure, it was too simple to specify the complexity of genes.

Bacterial Transformation Implicates DNA As the Substance of Genes

Several studies supported the idea that DNA would be the genetic material. The most important of these used single-celled bacteria as their experimental organism. Bacteria carry their genetic material in a single circular chromosome that lies in the nucleoid region of the cell without being enclosed in a nuclear membrane. The single circular chromosome (which makes the cells haploid) and lack of a nuclear membrane (which makes them *prokaryotes,* meaning "before a true nucleus") are two features that distinguish bacteria from the eukaryotic cells of higher organisms. Another feature is their method of cell division. With only one chromosome, bacteria do not undergo meiosis to produce germ cells, and they do not apportion their replicated chromosomes to daughter cells by mitosis; rather they divide by a process known as binary fission. Even with these acknowledged differences, at least some investigators in the first half of the twentieth century believed that the genetic material of bacteria was similar to that found in eukaryotic organisms.

As with any species, one prerequisite of genetic studies in bacteria is the detection of alternative forms of a trait within a population. In a 1923 study of *Streptococcus pneumoniae* bacteria grown in laboratory media, Frederick Griffith distinguished two bacterial forms: smooth (S) and rough (R). S is the wild type; a mutation in S gives rise to R. From observation and biochemical analysis, Griffith determined that S forms appear smooth because they synthesize a polysaccharide capsule that surrounds pairs of cells. R forms, which arise spontaneously as mutants of S, are unable to make the capsular polysaccharide and as a result, their colonies appear to have a rough surface (Fig. 6.3a). We now know that the R form lacks an enzyme necessary for synthesis of the capsular polysaccharide. Because the polysaccharide capsule helps protect the bacteria from an animal's immune response, the S bacteria are virulent and kill most laboratory animals exposed to them (Fig. 6.3b.1); by contrast, the R forms fail to cause infection (Fig. 6.3b.2). In humans, the virulent S forms of *S. pneumoniae* can cause pneumonia.

Transformation

In 1928, Griffith published the astonishing finding that genetic information from dead bacterial cells could somehow be transmitted to live cells. He was working with two kinds of bacteria— live R forms and heat-killed S forms. Neither the heat-killed S forms nor the live R forms produced infection when injected into laboratory mice (Fig. 6.3b.2 and 3); but a mixture of the two killed the animals (Fig. 6.3b.4). Bacteria recovered from the blood of the dead animals were living S forms (Fig. 6.3b.4). The ability of a substance to change the genetic characteristics of an organism is known as **transformation.** Something from the heat-killed S bacteria must have transformed the living R bacteria into S. This transformation was permanent and most likely genetic, because all future generations of the bacteria grown in culture were the S form.

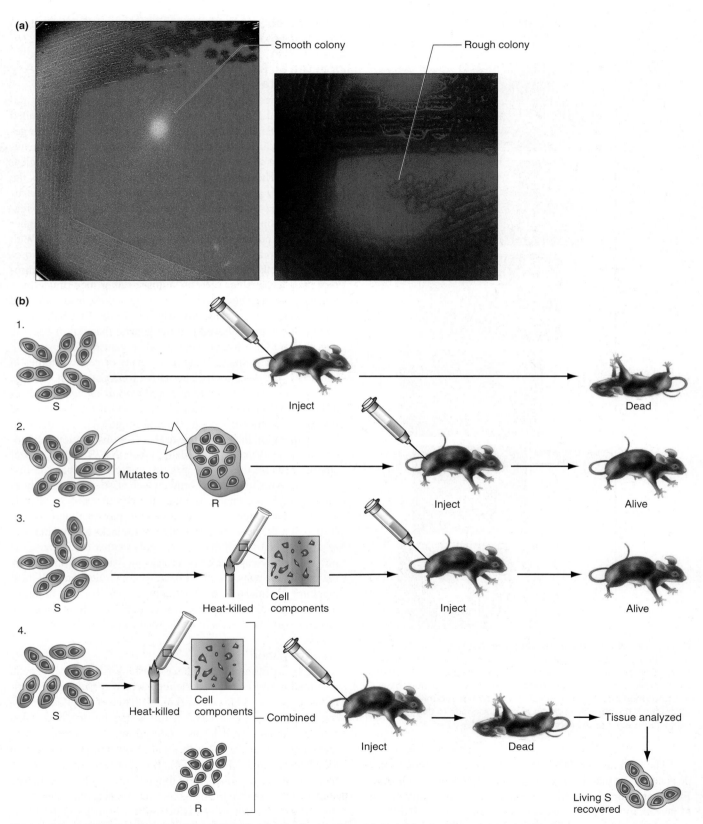

Figure 6.3 Griffith's demonstration of bacterial transformation. (a) Smooth (S) and rough (R) colonies of *S. pneumoniae.* **(b)** Griffith's experiment: (1) S bacteria are virulent and can cause lethal infections when injected into mice. (2) Injections of R mutants by themselves do not cause infections that kill mice. (3) Similarly, injections of heat-killed S bacteria do not cause lethal infections. (4) Lethal infection does result, however, from injections of live R bacteria mixed with heat-killed S strains; the blood of the dead host mouse contains living S-type bacteria.

Bacterial Transformation Is Caused by DNA

By 1929, two other laboratories had repeated these results, and in 1931, investigators in Oswald T. Avery's laboratory found they could achieve transformation without using any animals at all, simply by growing R-form bacteria in medium in the presence of components from dead S forms (Fig. 6.4a). Avery then embarked on a quest that would remain the focus of his work for almost 15 years: "Try to find in that complex mixture, the active principle!" In other words, try to identify the heritable substance in the bacterial extract that induces transformation from harmless R bacteria to pathogenic S bacteria. Avery dubbed the substance he was searching for the "transforming principle" and spent many years trying to purify it sufficiently to be able to identify it unambiguously. He and his coworkers eventually prepared a tangible, active transforming principle. In the final part of their procedure, a long whitish wisp materialized from ice-cold alcohol solution and wound around the glass stirring rod to form a fibrous wad of nearly pure principle (Fig. 6.4b).

Once purified, the transforming principle had to be characterized. In 1944 Avery and two coworkers, Colin MacLeod and Maclyn McCarty, published the cumulative findings of experiments designed to determine the transforming principle's chemical composition (Fig. 6.4c). In these experiments, the purified transforming principle was active at the extraordinarily high dilution of 1 part in 600 million. Although the preparation was almost pure DNA, the investigators nevertheless exposed it to various enzymes to see if some molecule other than DNA could cause transformation. Enzymes that degraded RNA, protein, or polysaccharide had no effect on the transforming principle, but an enzyme that degrades DNA completely destroyed its activity. The tentative published conclusion was that the transforming principle appeared to be DNA. In a personal letter to his brother, Avery went one step further and confided that the transforming principle "may be a gene."

Despite the paper's abundance of concrete evidence, many within the scientific community still resisted the idea that DNA

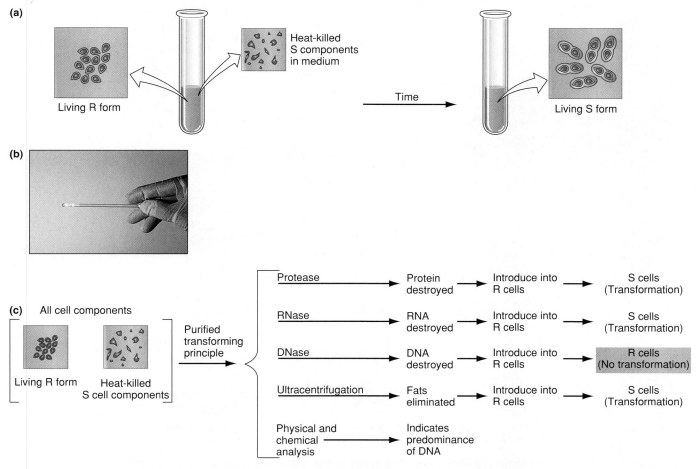

Figure 6.4 The transforming principle is DNA: Experimental confirmation. (a) Bacterial transformation occurs in culture medium containing the remnants of heat-killed S bacteria. This indicates that some "transforming principle" from the heat-killed S bacteria is taken up by the live R bacteria, converting (transforming) them into virulent S strains. **(b)** A wad of purified DNA from heat-killed S-type *S. pneumoniae* wound around a glass stirring rod. Small amounts of this preparation are highly efficient in causing bacterial transformation. **(c)** Chemical fractionation of the transforming principle. Treatment of purified DNA with a DNA-degrading enzyme destroys its ability to cause bacterial transformation, while treatment with enzymes that destroy other macromolecules has no effect on the transforming principle. In contrast to the purified DNA shown in part (b), highly purified preparations of protein, lipids, or carbohydrates do not contain active transforming principle.

is the molecule of heredity. They argued that perhaps Avery's results reflected the activity of contaminants; or perhaps genetic transformation was not happening at all, and instead, the transforming principle was somehow triggering a physiological switch in the transformed bacteria. Unconvinced for the moment, these scientists remained attached to the idea that proteins are the prime candidates for the genetic material.

Convincing Evidence That Genes Are DNA: The Molecule Carries the Information Required for the Replication of Bacterial Viruses

Not everyone shared this skepticism. Alfred Hershey and Martha Chase anticipated that they could assess the relative importance of DNA and protein in gene transmission by infecting bacterial cells with viruses called **phages,** short for **bacterio-**

phages (literally "bacteria eaters"). Viruses are the simplest of organisms. By structure and function, they fall somewhere between living cells capable of reproducing themselves and macromolecules such as proteins. Because viruses depend on their host cell for most of the machinery required for growth and replication, they can be very small indeed and contain very few genes. For many kinds of phage, each particle consists of roughly equal weights of protein and DNA (Fig. 6.5a). These phage particles can reproduce themselves only after infecting a bacterial cell. Thirty minutes after infection, the cell bursts and hundreds of newly made phages spill out (Fig. 6.5b). The question is: What substance directs the production of the new phage particles—DNA or protein?

With the invention of the electron microscope in 1939, it became possible to see individual phages, and surprisingly, electron micrographs revealed that the entire phage does not en-

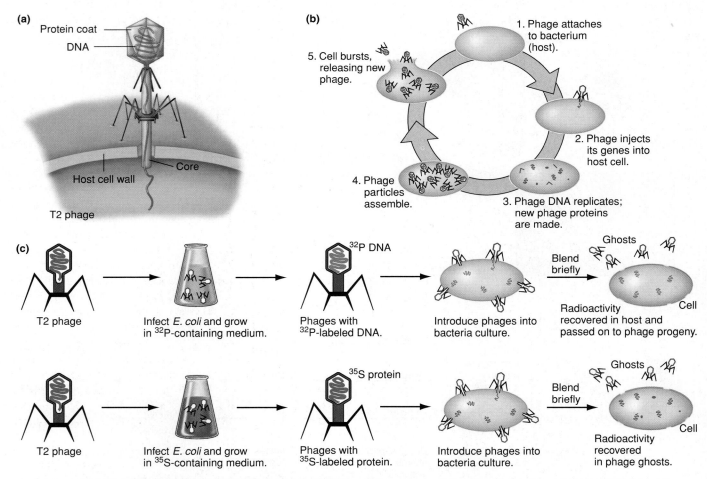

Figure 6.5 Experiments with viruses provide convincing evidence that genes are made of DNA. **(a)** and **(b)** Bacteriophage T2 structure and life cycle. The phage particle consists of DNA contained within a protein coat. The virus attaches to the bacterial host cell and injects its genes (the DNA) through the bacterial cell wall into the host cell cytoplasm. Inside the host cell, these genes direct the formation of new phage DNA and proteins, which assemble into progeny phages that are released into the environment when the cell bursts. **(c)** The Hershey-Chase Waring blender experiment. T2 bacteriophage particles either with ^{32}P-labeled DNA or with ^{35}S-labeled proteins were used to infect bacterial cells. After a short incubation, Hershey and Chase shook the cultures in a Waring blender and spun the samples in a centrifuge to separate the empty viral ghosts from the heavier infected cells. Most of the ^{35}S-labeled proteins remained with the ghosts, while most of the ^{32}P-labeled DNA was found in the sediment with the T2 gene-containing infected cells.

ter the bacterium it infects. Instead, a viral shell—called a *ghost*—remains attached to the outer surface of the bacterial cell wall. Because the empty phage coat remains outside the bacterial cell, one investigator likened phage particles to tiny syringes that bind to the cell surface and inject transforming principle into the host cell.

In their famous Waring blender experiment of 1952, Alfred Hershey and Martha Chase tested the idea that the ghost left on the cell wall is composed of protein while the injected transforming principle consists of DNA (Fig. 6.5c). A type of phage known as T2 served as their experimental organism. They grew two separate sets of T2 in bacteria maintained in two different culture media, one infused with radioactively labeled phosphorus (^{32}P), the other with radioactively labeled sulfur (^{35}S). Since proteins incorporate sulfur but no phosphorus and DNA contains phosphorus but no sulfur, phages grown on ^{35}S would have radioactively labeled protein while particles grown on ^{32}P would have radioactive DNA. The radioactive tags would enable the investigators to determine the location of each material when the phages infected fresh cultures of bacterial cells.

After exposing one culture of bacteria to ^{32}P-labeled phage and another culture to ^{35}S-labeled phage, Hershey and Chase used a Waring blender to disrupt each culture, effectively separating the viral ghosts from the bacteria harboring the viral genes. Centrifugation of the cultures then separated the heavier infected cells, which ended up in a pellet, from the lighter phage ghosts, which remained suspended in the supernatant. Most of the radioactive ^{32}P (in DNA) went to the pellet, while most of the radioactive ^{35}S (in protein) remained in the supernatant. This confirmed that the extracellular ghosts were indeed mostly protein while the injected viral material specifying production of more phages was mostly DNA. Bacteria containing the radiolabeled phage DNA behaved just as in a normal phage infection, producing and disgorging hundreds of progeny particles. From these observations, Hershey and Chase concluded that phage genes are made of DNA.

The Hershey-Chase experiment, although less rigorous than the Avery project, had an enormous impact. In the minds of many investigators, it confirmed Avery's results and extended them to viral particles. The spotlight was now on DNA.

The Watson-Crick Model: DNA Is a Double Helix

Under appropriate conditions, DNA can align in fibers to produce an ordered structure. And just as a crystal chandelier scatters light to produce a distinctive pattern on the wall, DNA fibers scatter X rays to produce a characteristic diffraction pattern (Fig. 6.6). A knowledgeable X-ray crystallographer can interpret DNA's diffraction pattern to elucidate selected aspects of the molecule's three-dimensional structure. When in the spring of 1951 the 23-year-old James Watson learned that the genetic material could project a diffraction pattern, he realized that DNA "must have a regular structure that could be solved in a straightforward fashion."

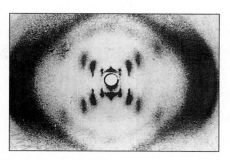

Figure 6.6   X-ray diffraction patterns reflect the helical structure of DNA. Photograph of an X-ray diffraction pattern produced by oriented DNA fibers, taken by Rosalind Franklin and Maurice Wilkins in late 1952. The crosswise pattern of X-ray reflections indicates that DNA is helical.

In this section, we analyze DNA's three-dimensional structure, looking first at significant details of the nucleotide building blocks; then at how those subunits are linked together in a polynucleotide chain; and finally, at how two chains associate to form a double helix.

Nucleotides Are the Basic Building Blocks of DNA

DNA is a long polymer composed of subunits known as *nucleotides*. Each nucleotide consists of a deoxyribose sugar, a phosphate, and a nitrogenous base. Detailed chemical knowledge of these constituents and the way they combine played an important role in Watson and Crick's model building.

Figure 6.7 depicts the chemical composition of deoxyribose, phosphate, and the four nitrogenous bases; how these components come together to form a nucleotide; and how phosphodiester bonds link the nucleotides in a chain. Each individual carbon or nitrogen atom in the central ring structure of a nitrogenous base is assigned a number: from 1–9 for purines, and 1–6 for pyrimidines. The carbon atoms of the deoxyribose sugar are distinguished from atoms within the nucleotide base by the use of primed numbers from $1'–5'$. Covalent attachment of a base to the $1'$ carbon of deoxyribose forms a nucleoside. The addition of a phosphate group to the $5'$ carbon forms a complete nucleotide. As Fig. 6.7 shows, a DNA chain composed of many nucleotides has **polarity:** an overall direction. Phosphodiester bonds always form a covalent link between the $3'$ carbon of one nucleoside and the $5'$ carbon of the following nucleoside. The consistent orientation of the nucleotide building blocks gives a chain overall direction, such that the two ends of a single chain are chemically distinct.

At the $5'$ end, the sugar of the terminal nucleotide has a free $5'$ carbon atom, free in the sense that it is not linked to another nucleotide. Depending on how the DNA is synthesized or isolated, the $5'$ carbon of the nucleotide at the $5'$ end may carry either a hydroxyl or a phosphate group. At the other—$3'$—end of the chain, it is the $3'$ carbon of the final nucleotide that is free. Along the chain between the two ends, this $5'$-to-$3'$ polarity is conserved from nucleotide to nucleotide. By convention, a DNA chain is described in terms of its bases, written with the $5'$-to-$3'$ direction

A detailed look at DNA's chemical constituents.

(a) The separate entities

1. Deoxyribose sugar

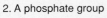
Ribose

2. A phosphate group

3. Four nitrogenous bases

Purines

Adenine **(A)**

Guanine **(G)**

Pyrimidines

Thymine **(T)**

Cytosine **(C)**

(b) Assembly into a nucleotide

1. Attachment of base to sugar

Nucleoside

2. Addition of phosphate

Purine nucleotide

Pyrimidine nucleotide

(c) Nucleotides linked in a directional chain

5' end

T

Phosphodiester bond

A

C

Phosphodiester bond

G

3' end

going from left to right (unless otherwise noted). The chain depicted in Fig. 6.7c, for instance, would be TACG.

A Directional Base Sequence Can Carry Information

Information can be encoded only in a sequence of symbols whose order varies according to the message to be conveyed. Without this sequence variation, there is no potential for carrying information. Because DNA's sugar-phosphate backbone is chemically identical for every nucleotide in a DNA chain, the only difference between nucleotides is in the identity of the nitrogenous base. Thus, if DNA carries genetic information, that information must consist of variations in the sequence of the A, G, T, and C bases. The information constructed from the 4-letter language of DNA bases is analogous to the information built from the 26-letter alphabet of English or French or Italian. Just as you can combine the 26 letters of the alphabet in different ways to generate the words of a book, so too different combinations of the four bases in very long sequences of nucleotides can encode the information for constructing an organism.

The Double Helix Contains Two Antiparallel Chains That Associate by Complementary Base Pairing

Watson and Crick's discovery of the structure of the DNA molecule ranks with Darwin's theory of evolution by natural selection and Mendel's laws of inheritance in its contribution to our understanding of biological phenomena. The Watson-Crick structure, first embodied in a model that superficially resembled the Tinker Toys of preschool children, was based on an understanding and interpretation of all the chemical and physical data available at the time. Watson and Crick published their findings in the scientific journal *Nature* in April 1953.

The Physical Data: X-Ray Diffraction Patterns Indicate Molecular Parameters

The diffraction patterns of oriented DNA fibers do not, on their own, contain sufficient information to reveal structure. For instance, the number of diffraction spots, whose intensities and positions constitute the X-ray data (review Fig. 6.6), is considerably less than the number of unknown coordinates of all the atoms in an oriented DNA molecule. Nevertheless, the photographs do reveal a wealth of structural information to the trained eye. Excellent X-ray images produced by Rosalind Franklin and Maurice Wilkins showed that the molecule is spiral-shaped, or helical; the spacing between repeating units along the axis of the helix is 3.4 Å; the helix undergoes one complete turn every 34 Å; the diameter of the molecule is 20 Å. Given this diameter, the molecule must consist of more than one polynucleotide chain.

Key Chemical Data Show That Nucleotides Exhibit Complementarity

If DNA contains more than one chain of nucleotides, what forces hold these chains together? Erwin Chargaff obtained data on the nucleotide composition of DNA from various sources that provided an important clue. Despite large variations between species in the relative amounts of the bases, in every individual, the ratio of A to T is always about 1:1, and the ratio of G to C is the same (Table 6.1). Watson grasped that the 1:1 ratios of A to

TABLE 6.1 **Chargaff's Data on Nucleotide Base Composition in the DNA of Various Organisms**

| Organism | Percentage of Base in DNA | | | | Ratios | |
	A	T	G	C	A:T	G:C
Staphylococcus afermentams	12.8	12.9	36.9	37.5	0.99	0.99
Escherichia coli	26.0	23.9	24.9	25.2	1.09	0.99
Yeast	31.3	32.9	18.7	17.1	0.95	1.09
*Caenorhabditis elegans**	31.2	29.1	19.3	20.5	1.07	0.96
*Arabadopsis thaliana**	29.1	29.7	20.5	20.7	0.98	0.99
Drosophila melanogaster	27.3	27.6	22.5	22.5	0.99	1.00
Honeybee	34.4	33.0	16.2	16.4	1.04	0.99
Mus musculus (mouse)	29.2	29.4	21.7	19.7	0.99	1.10
Human (liver)	30.7	31.2	19.3	18.8	0.98	1.03

*Data for *C. elegans* and *A. thaliana* is based on that for close relative organisms.
Note that even though the level of any one nucleotide is different in different organisms, the amount of A always approximately equals the amount of T, and the level of G is always similar to that of C. Moreover, as you can calculate for yourself, the total amount of purines (A plus G) nearly always equals the total amount of pyrimidines (C plus T).

The structure of DNA: A double helix composed of antiparallel chains associated by complementary base pairing.

(a) In a leap of imagination, Watson and Crick took the known facts about DNA's chemical composition and physical arrangement in space and constructed a wire-frame model that not only united the evidence but also served as a basis for explaining the molecule's function.

(b) In the model, two DNA chains spiral around an axis with the sugar-phosphate backbones on the outside and pairs of bases (one from each chain) meeting in the middle. Although both chains wind around the helix axis in a right-handed sense, chemically one of them runs 5' to 3' upward, while the other runs in the opposite direction of 5' to 3' downward. In short, the *two chains are antiparallel*. The base pairs are essentially flat and perpendicular to the helix axis, and the planes of the sugars are roughly perpendicular to the base pairs. As the two chains spiral about the helix axis, they wrap around each other once every 10 base pairs, or once every 34 Å. The result is a double helix that looks like a twisted ladder with the two spiraling structural members composed of sugar-phosphate backbones and the rungs consisting of base pairs.

(c) In a space-filling representation of the model, the overall shape is that of a cylinder with a diameter of 20 Å whose axis is the axis of the double helix. The backbones spiral around the axis like threads on a screw, but because there are two backbones, there are two threads, and these two threads are vertically displaced from each other. This displacement of the backbones generates two grooves, one much wider than the other, that also spiral around the helix axis. Biochemists refer to the wider groove as the **major groove** and the narrower one as the **minor groove.**

The *two chains of the double helix are held together by hydrogen bonds between complementary base pairs*, A–T and G–C (see Fig. 6.9). Since the overall shapes of the two base pairs are quite similar, either pair can fit into the structure at each position along the DNA. Moreover, each base pair can be accommodated in the structure in two ways that are the reverse of each other: an A purine may be on strand 1 with its corresponding T pyrimidine on strand 2 or the T pyrimidine may be on strand 1 and the A purine on strand 2. The same is true of G and C base pairs. With complementary base pairing, the sugar-phosphate backbone remains in a relatively regular conformation along the entire length of both DNA chains, and the broad outlines of that conformation are independent of base sequence.

(d) Interestingly, within the double-helical structure, the spatial requirements of the base pairs are satisfied if and only if each pair consists of one small pyrimidine and one large purine, and even then, only for the particular pairing of A–T and G–C. Pyrimidine–pyrimidine pairs are too small for the structure, and purine–purine pairs are too large. In addition, A–C and G–T pairs do not fit well together; that is, they do not easily form hydrogen bonds. Complementary base pairing is thus a logical outgrowth of the molecule's steric

(a)

(d)

Pyrimidine–pyrimidine

Purine–purine

Purine–pyrimidine

├── 20 Å diameter ──┤

requirements. Although any one nucleotide pair forms only two or three hydrogen bonds, the sum of these connections between successive base pairs in a long DNA molecule composed of thousands or millions of nucleotides is one basis of the molecule's great chemical stability.

(b)

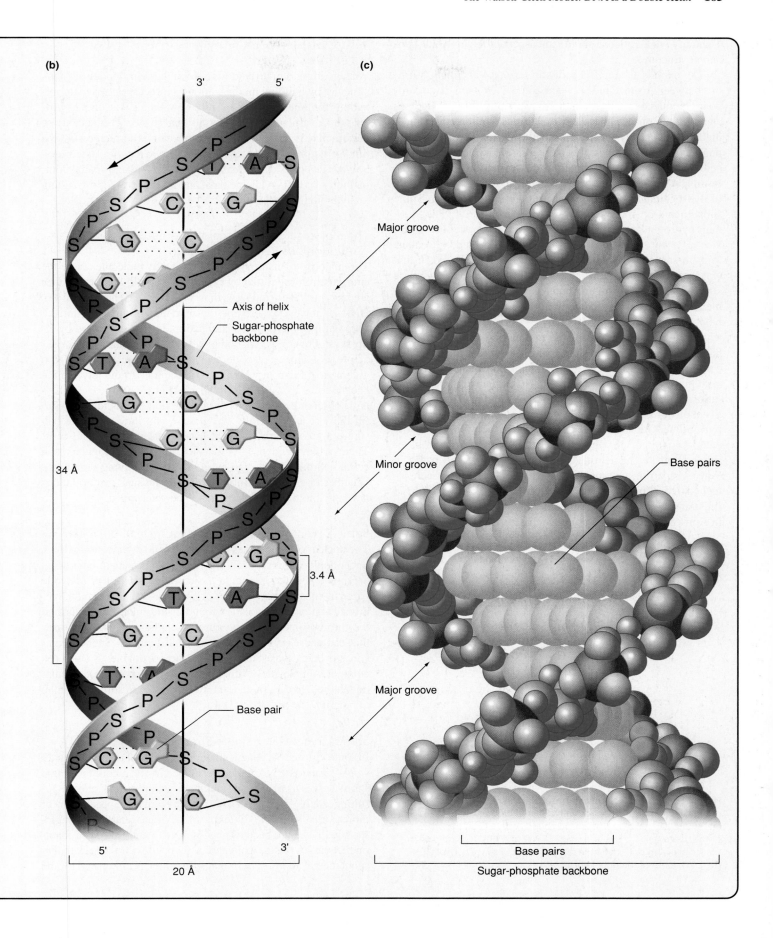

3'   5'

Axis of helix

Sugar-phosphate
backbone

34 Å

3.4 Å

Base pair

5'   3'

20 Å

(c)

Major groove

Minor groove

Base pairs

Major groove

Base pairs

Sugar-phosphate backbone

T and of G to C reflect a significant aspect of the molecule's inherent structure.

To explain Chargaff's ratios in terms of chemical affinities between A and T and between G and C, Watson made cardboard cutouts of the bases in their proper physiological shapes and tried to match these up in various combinations, like pieces in a jigsaw puzzle. He knew that the substituents on purines and pyrimidines play a crucial role in molecular interactions as they can participate in the formation of **hydrogen bonds:** weak electrostatic bonds that result in a partial sharing of hydrogen atoms between reacting groups (Fig. 6.9). Some substituents on the nitrogenous bases provide, or "donate," their hydrogen atoms for bonding, while others are able to create, or "accept," bond formation with the donated hydrogens. Watson saw that A and T could be paired together such that two hydrogen bonds formed between them. If G and C were similarly paired, hydrogen bonds could also easily connect the nucleotides carrying these two bases. (Watson originally posited two hydrogen bonds between G and C, but there are actually three.) Remarkably, the two pairs—A–T and G–C—had essentially the same shape. This meant that the two pairs could fit in any order between two sugar-phosphate backbones without distorting the structure. It also explained the Chargaff ratios—always equal amounts of A and T and of G and C. Note that both of these base pairs consist of one purine and one pyrimidine. Crick connected the chemical facts with the X-ray data, recognizing that because of the geometry of the base-sugar bonds in nucleotides, the orientation of the bases in Watson's pairing scheme could only arise if the bases were attached to backbones running in opposite directions. Figure 6.8 illustrates and explains the model Watson and Crick proposed in April 1953: DNA as a double helix.

Figure 6.9 Complementary base pairing. An A on one strand can form two hydrogen bonds with a T on the other strand. A G on one strand can form three hydrogen bonds with a C on the other strand. Note that the size and shape of A–T and of G–C base pairs are similar, allowing both to fill the same amount of space between the two backbones of the double helix.

The Double Helix May Assume Alternative Forms

Watson and Crick arrived at the double helix model of DNA structure by building models, not by a direct structural determination from the data alone. And even though Watson has written that "a structure this pretty just had to exist," the beauty of the structure is not necessarily evidence of its correctness. At the time of its presentation, the strongest evidence for its correctness was its physical plausibility, its chemical and spatial compatibility with all available data, and its capacity for explaining many biological phenomena.

The Biological Significance of Alternative Overall Configurations Remains Unknown

The majority of naturally occurring DNA molecules have the configuration suggested by Watson and Crick. Such molecules are known as **B-form** DNA; they spiral to the right (Fig. 6.10a). DNA is, however, more polymorphic than originally assumed. Some of the small double-stranded molecules examined by X-ray crystallography look completely different. One type, for example, contains nucleotide sequences that cause the DNA to assume a **Z form** in which the helix spirals to the left and the backbone takes on a zigzag shape (Fig. 6.10b). Researchers have observed many kinds of unusual non-B structures *in vitro* (in the test tube, literally "in glass"), and they speculate that some of these might occur at least transiently in living cells. There is some evidence, for instance, that Z DNA might exist in certain chromosomal regions *in vivo* (in the living organism). Whether or not the Z form and other unusual conformations have any biological role remains to be determined.

Some DNA Molecules Are Circular Instead of Linear

The nuclear chromosomes of all eukaryotic organisms are long, linear double helixes, but some chromosomes are circular (Fig. 6.11a and b). These include the chromosomes of prokaryotic bacteria, the chromosomes of organelles such as the mitochondria and chloroplasts that are found inside eukaryotic cells, and the chromosomes of some viruses, including the papovaviruses that can cause cancers in animals and humans. Such circular chromosomes consist of covalently closed, double-stranded circular DNA molecules. Although neither strand of these circular double helixes has an end, the two strands are still antiparallel in polarity.

Some Viruses Carry Single-Stranded DNA

In some viruses, the genetic material consists of relatively small, single-stranded DNA molecules. Once inside a cell, the single strand serves as a mold for making a second strand, and the resulting double-stranded DNA then governs the production of more virus particles. Examples of viruses carrying single-stranded DNA are bacteriophages φX174 and M13, and mammalian parvoviruses, which are associated with fetal death and spontaneous abortion in humans. In both φX174 and M13, the

(a)

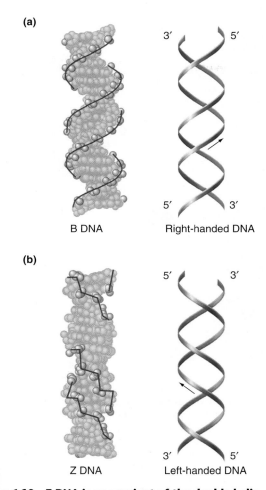

B DNA   Right-handed DNA

(b)

Z DNA   Left-handed DNA

Figure 6.10   Z DNA is one variant of the double helix.
(a) Normal Watson-Crick B-form DNA forms a right-handed helix with a smooth backbone. **(b)** Z-form DNA is left-handed and has an irregular backbone.

single DNA strand is in the form of a covalently closed circle; in the parvoviruses, it is linear (Fig. 6.11c and d).

Alternative B and Z configurations; circularization of the molecule; and single strands that are converted to double helices before replication and expression—these are minor variations on the double-helical theme. Despite such experimentally determined departures of detail, the Watson-Crick double helix remains *the* model for thinking about DNA structure. This model describes those features of the molecule that have not changed throughout billions of years of evolutionary history.

DNA Structure Is the Foundation of Genetic Function

Without sophisticated techniques for determining base sequence, one cannot distinguish bacterial DNA from human DNA. This is because all DNA molecules have the same general chemical properties and physical structure. Proteins, by comparison, are a much more diverse group of molecules with a much greater complexity of structure and function. In his account of the discovery of the double helix, Crick referred to this difference when he said that "DNA is, at bottom, a much less sophisticated molecule than a highly evolved protein and for this reason reveals its secrets more easily."

There are four basic DNA "secrets," embodied in four questions:

1. How does the molecule carry information?

2. How is that information copied for transmission to future generations?

3. What mechanisms allow the information to change?

4. How does the information govern the expression of phenotype?

The double-helical structure of DNA provides a potential solution to each of these questions, endowing the molecule with the capacity to carry out all the critical functions required of the genetic material.

(a)

(b)

(c)

(d)

Figure 6.11   DNA molecules may be linear or circular, double-stranded or single-stranded. These electron micrographs of naturally occurring DNA molecules show **(a)** a fragment of a long, linear double-stranded human chromosome, **(b)** a circular double-stranded papovavirus chromosome, **(c)** a linear single-stranded parvovirus chromosome, and **(d)** a circular single-stranded bacteriophage M13 chromosome.

In the remainder of this chapter, we describe how DNA's structure enables it to carry genetic information, replicate that information with great fidelity, and reorganize the information through recombination. How the information changes through mutation and how the information determines phenotype are the subjects of Chapters 7 and 8.

DNA Stores Information in the Sequence of Its Bases

The information content of DNA resides in the sequence of its bases. The four bases in each chain are like the letters of an alphabet; they may follow each other in any order, and different sequences spell out different "words." Each "word" has its own meaning, that is, its own effect on phenotype. AGTCAT, for example, means one thing, while CTAGGT means another. Although DNA has only four different letters, or building blocks, the potential for different combinations and thus different sets of information in a long chain of nucleotides is staggering. Some human chromosomes, for example, contain chains that are 250 million nucleotides long; because the different bases may follow each other in any order, such chains contain $4^{250,000,000}$ (which translates to 1 followed by 150,515,000 zeros) potential nucleotide sequences.

Much of DNA's Sequence-Specific Information Is Accessible Only When the Double Helix Is Unwound

The unwinding of a DNA molecule exposes a single file of bases on each of two strands (Fig. 6.12a). Proteins "read" the information in DNA and carry out its instructions by binding to a specific sequence or by synthesizing a stretch of RNA or DNA complementary to a specific sequence (Fig. 6.12b).

Some Genetic Information Is Accessible Even in Intact, Double-Stranded DNA Molecules

This information emerges in part from differences between the four bases that appear in the major and minor grooves and in part from conformational irregularities in the sugar-phosphate backbone. Within the grooves, certain atoms at the periphery of the bases are exposed, and particularly in the major groove, these atoms may provide chemical information. Such information is in the form of spatial patterns of hydrogen bond donors and acceptors as well as in the distinct patterns of particular substituent shapes. Proteins can access this information to "sense" the base sequence in a stretch of DNA without disassembling the double helix (Fig. 6.12b.2). The proteins that help turn genes on and off make use of these subtle conformational differences. Nevertheless, the amount of information exposed in either groove or on the backbone as distinct shapes is relatively limited.

Some proteins can recognize specific base sequences in double-stranded DNA. The Fast Forward box "Restriction Enzymes Recognize Specific Base Sequences in DNA" ex-

Figure 6.12   DNA stores information in the sequence of its bases. (a) A partially unwound DNA double helix. Note that different structural information is available in the double-stranded and unwound regions of the molecule. **(b.1)** Unwinding of the helix allows the enzyme DNA polymerase to copy a strand of DNA. **(b.2)** Some proteins can bind to specific sequences within double-helical DNA. Here, a protein called GCN4 interacts closely with features of a particular sequence of bases that are accessible to the protein in the major groove of the double helix.

plains how bacteria use enzymatic proteins of this type to stave off viral infection and how geneticists use these same enzymes to cut DNA at particular sites.

A Few Viruses Use RNA As the Repository of Genetic Information

In all cellular forms of life and many viruses, DNA carries the genetic information. Prokaryotes such as *Escherichia coli* bacteria carry their DNA in a double-stranded, covalently closed circular chromosome. Eukaryotic cells package their DNA in two or more double-stranded linear chromosomes. DNA viruses carry it in small molecules that are single- or double-stranded, circular, or linear.

By contrast, some viruses, including those that cause polio and AIDS, use RNA as their genetic material (Fig. 6.13). There are three major chemical differences between RNA and DNA. First, RNA takes its name from the sugar ribose, which it incorporates instead of the deoxyribose found in DNA (Fig. 6.13a.1). Second, RNA usually contains the base uracil (U) instead of the base thymine (T); U, like T, base pairs with A (Fig. 6.13a.3).

Figure 6.13   RNA: Chemical constituents and complex folding pattern. (a) and **(b)** Each ribonucleotide contains the sugar ribose, an inorganic phosphate group, and a nitrogenous base. RNA contains the pyrimidine uracil (U) instead of the thymine (T) found in DNA. **(c)** Phosphodiester bonds join ribonucleotides into an RNA chain. Most RNA molecules are single-stranded but are sufficiently flexible so that some regions can fold back and form base pairs with other parts of the same molecule.

Finally, most RNA molecules are single-stranded and contain far fewer nucleotides than the very long DNA molecules found in nuclear chromosomes. Some completely double-stranded RNA molecules do nonetheless exist, although they are rare. Even within a single-stranded RNA molecule, however, if folding brings two oppositely oriented regions that carry complementary nucleotide sequences alongside each other, they can form a double-stranded, base-paired stretch within the molecule. This means that, compared to the relatively simple, double-helical shape of a DNA molecule, many RNAs have a complicated structure of short double-stranded segments interspersed with single-stranded loops (Fig. 6.13c).

RNA has the same ability to carry information in the sequence of its bases but is much less stable than DNA. In addition to serving as the genetic material for an array of viruses, RNA fulfills several vital functions in all cells. For example, it participates in gene expression and protein synthesis. It also plays a significant role in DNA replication, which we now describe.

DNA Replication: Copying Genetic Information for Transmission to the Next Generation

In one of the most famous sentences of the scientific literature, Watson and Crick wrote at the end of their 1953 paper proposing the double helix model: "It has not escaped our notice that the specific pairing we have postulated immediately suggests a possible copying mechanism for the genetic material." This copying, as we saw in Chapter 4, precedes the transmission of chromosomes from one generation to the next via meiosis, and it is also the basis of the chromosome duplication prior to each mitosis that allows all the daughter cells of a developing organism to receive a complete copy of the genetic information.

Complementary Base Pairing Produces Semiconservative Replication: An Overview

In the process of replication postulated by Watson and Crick, the double helix unwinds to expose the bases in each strand of

Fast Forward

Restriction Enzymes Recognize Specific Base Sequences in DNA

In many types of bacteria, the unwelcome arrival of viral DNA mobilizes minute molecular weapons known as **restriction enzymes.** Each enzyme has the twofold ability to recognize a specific sequence of four to six base pairs anywhere within a DNA molecule and to sever a covalent bond in the sugar-phosphate backbone at a particular position within or near that sequence on each strand. When a bacterium calls up its reserve of restriction enzymes at the first sign of invasion, the ensuing shredding and dicing of selected stretches of viral DNA incapacitates the virus's genetic material and thereby restricts infection.

Since the early 1970s, geneticists have isolated more than 300 types of restriction enzymes and named them for the bacterial species in which they orginate. *Eco*RI, for instance, comes from *E. coli; Hpa*I is one of several restriction enzymes found in *Haemophilus parainfluenzae; Hind*III derives from *Haemophilus influenzae.* Each enzyme recognizes a different base sequence and cuts the DNA strand at a precise spot in relation to that sequence. *Eco*RI recognizes the sequence 5′. . .GAATTC. . .3′ and cleaves between the G and the first A. *Hpa*I recognizes the sequence 5′. . .GTTAAC. . .3′ and makes a cut

between the second T and the first A. *Hind*III sees the sequence 5′. . .AAGCTT. . .3′ and shears between the two As. The DNA of a bacteriophage called lambda (λ) carries the GAATTC sequence recognized by *Eco*RI in five separate places; the enzyme thus cuts the linear lambda DNA at five points, cutting it into six pieces. The DNA of a phage known as ϕX174, however, contains no *Eco*RI recognition sequences. By comparison, the *Hpa*I enzyme can cut λ DNA in 13 places and ϕX174 DNA in 4, while *Hind*III can cut λ DNA in 6 places and ϕX174 in 10.

How, you may wonder, do bacteria keep their arsenal of restriction enzymes from attacking their own DNA? The answer is a group of companion enzymes known as *methylases,* which attach methyl groups (–CH$_3$) in a precisely defined way. With every cycle of DNA replication, the methylating enzymes recognize the exact same sites that would be recognized by the cell's restriction enzymes, "see" a methyl group on the template strand at those sites, and add a methyl group at the exact same sites on the newly forming daughter strand. This site-specific methylation of every newly forming strand shields the bacterial DNA from recognition by the

Figure A   Three restriction enzymes in action. The restriction enzymes *Eco*RI, *Hpa*I, and *Hind*III recognize different six-base-pair-long symmetrical sequences in double-stranded DNA molecules. Each enzyme severs the phosphodiester bonds between the same two adjacent nucleotides on each DNA strand. If these bonds are in the middle of the six base recognition site (as with *Hpa*I), the resultant DNA fragments will have blunt ends. If the bonds are offset from the center of the recognition site (as with *Eco*RI and *Hind*III), the products of cleavage will have sticky ends. Note that any sticky end produced by cleavage with a particular restriction enzyme is complementary in sequence to any other sticky end made by the same enzyme.

1. *Eco*RI cuts DNA.

*Eco*RI

Plasmid

Insulin gene

Human DNA

2. Complementary sticky ends exposed.

Sticky ends

Sticky ends

3. Sticky ends form base pairs with each other.

4. Ligase seals breaks in DNA backbones.

5. Recombinant plasmid inserted into bacterial cell.

Bacterial chromosome

6. Population of bacterial cells grown containing recombinant plasmid.

Figure B   One use of recombinant DNA technology: Harnessing bacteria to copy the human insulin gene.
Geneticists cleave both a small circular chromosome called a plasmid and human chromosomal DNA with the *Eco*RI restriction enzyme. This cleavage converts the plasmid to a linear piece of DNA with *Eco*RI sticky ends, while converting the human DNA into thousands of fragments *Eco*RI sticky ends; one of these human chromosomal fragments contains the insulin gene. When these pieces of DNA are mixed together, they can adhere to each other in various combinations because of the complementarity of their sticky ends; the enzyme DNA ligase will seal these fragments together by forming the proper phosphodiester bonds. One possible outcome of this process is the creation of a recombinant DNA molecule in which the human insulin gene is spliced into the circular plasmid. Any *E. coli* cells transformed with this recombinant plasmid will become miniature factories for the synthesis of insulin.

cell's own restriction enzymes. The methylases recognize only single-stranded restriction sites that form opposite an already methylated mirror image. Because viral DNA enters the bacterial cell as a double-stranded helix that is not methylated at the restriction enzyme's recognition sites, the methylases do not protect it.

Figure A illustrates the results of three restriction enzymes in action. Note that each recognition sequence in double-stranded DNA is symmetrical; that is, the base sequences on the two strands are

identical when each is read in the 5'-to-3' direction. Thus, each time an enzyme recognizes a short 5'-to-3' sequence on one strand, it finds the exact same sequence in the 5'-to-3' direction of the complementary antiparallel strand. In other words, viewed in the 5'-to-3' direction, the double-stranded recognition sequence is palindromic; like the phrase "TAHITI HAT" or the number 1881, it reads the same backward and forward (although the analogy is not exact because the language of DNA is double-stranded, while English follows only a single line at a time). Note also that the cuts made by restriction enzymes leave two types of DNA ends. *Hpa*I cleaves right in the middle of the six-base sequence, making a smooth cut straight through both chains of the double helix to create what geneticists call "blunt" ends. In contrast, *Eco*RI and *Hind*III do not cleave the two strands of the double helix at the same spot in the middle of a recognition sequence, and as a result, the cuts they make are offset in opposite directions on the two strands. When the weak hydrogen bonds between the strands dissociate, these cuts leave short, protruding single-stranded flaps known as **sticky, or cohesive, ends.** Like a tiny finger of velcro, each flap can stick to—that is, re-form hydrogen bonds with—a complementary sequence protruding from the end of another piece of DNA.

Even though restriction enzymes evolved in bacteria as a means of protection against viruses, given the appropriate base sequence, they can cut the DNA of any organism. And as Fig. B shows, any two DNA molecules with complementary protruding sequences can join together, no matter how unrelated the rest of their base sequences may be. The cutting and splicing together of DNA from two origins creates a **recombinant DNA** molecule.

In the mid-1970s, geneticists took advantage of the activity of restriction enzymes to splice together DNA from any two organisms. In their hands, the enzymes served as precision scissors that, in effect, revolutionized the study of life and gave birth to recombinant DNA technology. Although the sticky ends created by restriction enzymes enable two unrelated DNA molecules to come together by base pairing, another enzyme, known as DNA ligase, is required to stabilize the recombinant molecule. The ligase seals the breaks in the backbones of both strands.

Figure B illustrates one application of recombinant DNA technology: the splicing of the human gene for insulin into a small circle of DNA known as a plasmid, which can replicate inside a bacterial cell. Here is how it works. *Eco*RI is added to solutions of plasmids and human genomic DNA, where it cleaves the DNA molecules. The cleavage converts the circular plasmids to linear DNAs with *Eco*RI sticky ends; it also fragments each copy of the human genomic DNA into hundreds of thousands of pieces, all of which terminate with *Eco*RI sticky ends. When the solutions are then mixed together, the different fragments can adhere to each other in any combination, because of the complementarity of their sticky ends. In one combination, a human genomic fragment containing the insulin gene will re-create a circular DNA molecule after adhering to the two ends of a linearized plasmid. And just as restriction enzymes operate as scissors, DNA ligase acts as a glue that seals the breaks in the DNA backbone by forming new phosphodiester bonds. Investigators can treat bacteria with the recombinant plasmids containing the insulin gene exactly as Avery treated bacteria with his "transforming" DNA. The recombinant DNA molecules will enter some cells. When the bacteria copy their own chromosome in preparation for cell division, they will also make copies of any resident plasmids along with all the genes the plasmids contain. In the illustrated example, the plasmid carrying the gene for insulin also carries sequences that can direct the expression of an inserted gene into protein. As the bacterial culture grows, so does the number of plasmids carrying sequences that direct the expression of the human insulin gene into protein. Eventually a population of bacteria grows up in which every cell not only contains a copy of the human gene but also makes the insulin encoded by that gene.

Techniques for designing and constructing recombinant DNA molecules and for harnessing bacteria to produce large quantities of a particular gene and its protein product hold great promise for medicine, agriculture, and industry. Researchers are currently using the technology to try to develop, among many other projects, a relatively inexpensive nose spray able to kill the viruses that cause the common cold, drugs to help fight cancer and AIDS, corn containing proteins with the nutritional value of beef proteins, and a process for making automobile fuel from discarded corn stalks. These and other recombinant DNA possibilities depend on the use of restriction enzymes to recognize and cut specific base sequences.

DNA. Each of the two separated strands then acts as a **template,** or molecular mold, for the synthesis of a new second strand (Fig. 6.14). The new strand forms as complementary bases align opposite the exposed bases on the parent strand. That is, an A at one position on the original strand signals the addition of a T at

the corresponding position on the newly forming strand; a T on the original signifies addition of an A; similarly G calls for C, and C calls for G. This type of base pairing, which determines the nucleotide sequence of the new strand, is known as **complementary base pairing.** Once the appropriate base has aligned opposite and formed hydrogen bonds with its complement, enzymes join the base's nucleotide to the preceding nucleotide by a phosphodiester bond, eventually linking a whole new line of nucleotides into a continuous strand. This mechanism of complementary base pairing followed by the coupling of successive nucleotides yields two "daughter" double helixes that each contain one of the original DNA strands intact (that is, "conserved") and one completely new strand (Fig. 6.15a). For this reason, such a pattern of double helix duplication is called **semiconservative replication:** a copying in which one strand is conserved from the parent molecule and the other is newly synthesized.

Watson and Crick's proposal is not the only replication mechanism imaginable. Figure 6.15b and c illustrate two possible alternatives. With *conservative* replication, one of the two "daughter" double helixes would consist entirely of original DNA strands, while the other helix would consist of two newly synthesized strands. With *dispersive* replication, both "daughter" double helixes would carry blocks of original DNA interspersed with blocks of newly synthesized material. These alternatives are less satisfactory than semiconservative replication because they do not immediately suggest a mechanism for copying the information in the sequence of bases, and they do not explain the research data (presented later) as well.

Experimental Proof of Semiconservative Replication

In 1958 Matthew Meselson and Franklin Stahl performed an experiment that confirmed the semiconservative nature of DNA replication (Fig. 6.17, p. 176). The experiment depended on being able to differentiate preexisting "parental" DNA from newly synthesized daughter DNA. To accomplish this, they controlled the isotopic composition of the nucleotides incorporated in the newly forming daughter strands as follows. They grew *E. coli* bacteria for many generations on media in which all the nitro-

1. Original double helix.

2. Strands separate.

3. Complementary bases align opposite templates.

4. Enzymes link sugar-phosphate elements of aligned nucleotides into a continuous new strand.

Template

Template

Daughter helices

Templates

New strands

Figure 6.14 The model of DNA replication postulated by Watson and Crick. Unwinding of the double helix allows each of the two strands to serve as a template for the synthesis of two new strands by complementary base pairing. The end result: A single double helix becomes two identical daughter double helixes.

(a) Semiconservative (b) Conservative   (c) Dispersive

Parent DNA

First-generation daughter DNA

Second-generation daughter DNA

Figure 6.15   Three possible models of DNA replication. DNA from the original double helix is *blue;* newly made DNA is *purple.*
(a) Semiconservative replication (the Watson-Crick model).
(b) Conservative replication: The parental double helix remains intact; both strands of the daughter double helix are newly synthesized.
(c) Dispersive replication: At completion, both strands of both double helices contain both original and newly synthesized material.

gen was the normal isotope ^{14}N; these cultures served as a control. They grew other cultures of *E. coli* for many generations on media in which the only source of nitrogen was the heavy isotope ^{15}N. After several generations of growth on heavy-isotope medium, essentially all the nitrogen atoms in the DNA of these bacterial cells were labeled with (that is, contained) ^{15}N. The cells in some of these cultures were then transferred to new medium in which all the nitrogen was ^{14}N. Any DNA synthesized after the transfer would contain the lighter isotope.

Meselson and Stahl isolated DNA from cells grown in the different nitrogen-isotope cultures and then subjected these DNA samples to *equilibrium density gradient centrifugation,* an analytic technique they had just developed. In a test tube, they dissolved the DNA in a solution of the dense salt cesium chloride (CsCl) and spun these solutions at very high speed (about 50,000 revolutions per minute) in an ultracentrifuge. Over a period of two to three days, the centrifugal force (roughly 250,000 times the force of gravity) causes the formation of a stable gradient of CsCl concentrations, with the highest concentration, and thus highest CsCl density, at the bottom of the tube. The DNA in the tube forms a sharply delineated band at a position where its own density equals that of the CsCl. Since DNA con-

taining ^{15}N is denser than DNA containing ^{14}N, pure ^{15}N DNA will form a band lower, that is, closer to the bottom of the tube, than pure ^{14}N DNA (Fig. 6.17).

As Fig. 6.16 shows, when cells with pure ^{15}N DNA were transferred into ^{14}N medium and allowed to divide once, DNA from the resultant first-generation cells formed a band at a density intermediate between that of pure ^{15}N DNA and that of pure ^{14}N DNA. A logical inference is that the DNA in these cells contains equal amounts of the two isotopes. This finding invalidates the "conservative" model, which predicts the appearance of bands reflecting only pure ^{14}N and pure ^{15}N with no intermediary band. In contrast, DNA extracted from second-generation cells that had undergone a second round of division in the ^{14}N medium produced two observable bands, one at the density corresponding to equal amounts of ^{15}N and ^{14}N, the other at the density of pure ^{14}N. These observations invalidate the "dispersed" model, which predicts a single band between the two bands of the original generation.

Meselson and Stahl's observations are consistent only with semiconservative replication: In the first generation after transfer from the ^{15}N to the ^{14}N medium, one of the two strands in every daughter DNA molecule carries the heavy isotope label; the other, newly synthesized strand carries the lighter ^{14}N isotope. The band at a density intermediate between that of ^{15}N DNA and ^{14}N DNA represents this isotopic hybrid. In the second generation after transfer, half of the DNA molecules have one ^{15}N strand and one ^{14}N strand, while the remaining half carry two ^{14}N strands. The two observable bands—one at the hybrid position, the other at the pure ^{14}N position—reflect this mix.

By confirming the predictions of semiconservative replication, the Meselson-Stahl experiment disproved the conservative and dispersive alternatives. We now know that the semiconservative replication of DNA is nearly universal. To understand material presented in the remainder of this book, it is important to understand precisely how semiconservative replication relates to the structure of chromosomes in eukaryotic cells during the mitotic cell cycle (review Fig. 4.7). Early in interphase, each eukaryotic chromosome contains a single continuous linear double helix of DNA. Later, during the S-phase portion of interphase, the cell replicates the double helix semiconservatively; after this semiconservative replication, each chromosome is composed of two sister chromatids joined at the centromere. Each sister chromatid is a double helix of DNA, with one strand of parental DNA and one strand of newly synthesized DNA. At the conclusion of mitosis, each of the two daughter cells receives one sister chromatid from every chromosome in the cell. This process preserves chromosome number and identity during mitotic cell division because the two sister chromatids are identical in base sequence to each other and to the original parental chromosome.

The Molecular Mechanism of Replication: Doubling the Double Helix

Although Watson and Crick's model for semiconservative replication, depicted in Fig. 6.15a, is a simple concept to grasp, the

(a) **Initiation: Preparing the double helix for complementary base pairing.** A prerequisite of DNA replication is the unwinding of a portion of the double helix. Such unwinding exposes the bases in each DNA strand, making it possible for them to pair with complementary nucleotides. Initiation begins with the unwinding of the double helix at a particular short sequence of nucleotides known as the **origin of replication**. Each circular *E. coli* chromosome has a single origin of replication. Several proteins bind to the origin, forming a stable complex in which a small region of DNA is unwound and the two complementary strands

are separated. The first of the proteins to recognize and bind to the origin of replication is called the *initiator protein*. A DNA-bound initiator attracts an enzyme called DNA helicase, which catalyzes the localized unwinding of the double helix. The opening up of a region of DNA creates two Y-shaped areas, one at either end of the unwound area, or **replication bubble.** Each Y is called a **replication fork** and consists of the two unwound DNA strands branching out into unpaired (but complementary) single strands. These single strands will serve as **templates**—molecular molds—for fashioning new strands of DNA. The molecule is now ready for replication.

Actual formation of new DNA strands depends on the action of a complex enzyme known as **DNA polymerase III,** which adds nucleotides, one after the other, to the end of a growing DNA strand. DNA polymerase operates according to three strict rules: First, it can copy only DNA that is unwound and maintained in the single-stranded state; second, it adds nucleotides only to the end of an existing chain (that is, it cannot establish the first link in the chain); and third, it functions in only one—the 5'-to-3'—direction. The requirement for a single-stranded template is satisfied by formation of the replication bubble. The requirement for an already existing chain end, however, means that something else must start the about-to-be-constructed chain. That "something else" is RNA. Construction of a very short new strand consisting of a few nucleotides of RNA provides an "end" to which DNA polymerase can add the nucleotides of DNA. Since this short stretch of RNA prepares the ground for DNA polymerase activity, it is called an RNA **primer.** An enzyme called primase synthesizes the RNA primer at the replication fork, where the primer then base pairs with the single-stranded DNA template. With the double helix unwound and the primer in place, DNA replication can proceed. The third characteristic of DNA polymerase activity—one way only—determines some of the special features of subsequent steps.

(b) **Elongation: Connecting the correct sequence of nucleotides into a continuous new strand of DNA.** Elongation—the linking together of appropriately aligned nucleotide subunits into a continuous new strand of DNA—is the heart of replication. We have seen that the lineup of bases is determined by complementary base pairing with the template strand. An A on the template strand will form hydrogen bonds only with a T on the forming strand, and a C on the template will pair only with a G. Thus, the order of bases in the template specifies the order of bases in the newly forming strand. Once

formation of phosphodiester bonds is known as **polymerization.** The DNA polymerase III enzyme first joins the correctly paired nucleotide to the 3′ hydroxyl end of the RNA primer, and then it continues to add the appropriate nucleotides to the 3′ end of the growing chain. As a result, the DNA strand under construction

complementary base pairing has determined the next nucleotide to be added to the newly forming strand, DNA polymerase III catalyzes the actual joining of this properly positioned nucleotide to the preceding nucleotide. The linkage of subunits through the

grows in the 5′-to-3′ direction. The fact that the new strand is antiparallel to the template strand means that the DNA polymerase molecule actually moves along that template strand in the 3′-to-5′ direction.

As DNA replication proceeds, helicase progressively unwinds the double helix to expose successive series of nucleotides in their unpaired, single-stranded state. DNA polymerase III can then move in the same direction as the fork to synthesize one of the two new chains under construction at the fork. The enzyme encounters no problems in the polymerization of this chain—called the **leading strand**—because it can add nucleotides continuously to the growing 3′ end as soon as the unraveling fork exposes the corresponding bases on the template strand. The movement of the replication fork, however, presents problems for the synthesis of the second new DNA chain: the **lagging strand.** The polarity of the lagging strand is opposite that of the leading strand, yet as we have seen, DNA polymerase functions only in the 5′-to-3′ direction. This means that in synthesizing the lagging strand, the polymerase must travel in a direction opposite to that of the replication fork. How can the polymerase link the nucleotides of the lagging strand if it cannot approach that strand's recently unwound template at the replication fork?

The answer is that the lagging strand is synthesized *discontinuously* as small fragments of about 1000 bases called **Okazaki fragments** (after two of their discoverers Reiji and Tuneko Okazaki).

DNA polymerase III still synthesizes these small fragments in the normal 5′-to-3′ direction, but because the enzyme can add nucleotides only to the 3′ end of an existing strand, each Okazaki fragment is initiated by the synthesis of a short RNA primer. The primase enzyme catalyzes formation of the RNA primer for each upcoming Okazaki fragment as soon as the replication fork has progressed a sufficient distance along the DNA. Polymerase then adds nucleotides to this new primer, creating an Okazaki fragment that extends as far as the 5′ end of the primer of the previously synthesized fragment. Finally, DNA polymerase I and other enzymes replace the RNA primer of the previously made Okazaki fragment with DNA, and an enzyme known as DNA ligase covalently joins successive Okazaki fragments into a continuous strand of DNA. With the completion of both leading and lagging strands, DNA replication is complete.

After replacement of RNA primers with DNA bases, DNA ligase joins Okazaki fragments into a continuous strand.

^{15}N cells transferred to^{14}N medium

Control: *E. coli* grown for many generations in ^{14}N medium.

1. *E. coli* grown for many generations in ^{15}N medium.

2. Cells replicate once to produce first generation of daughter cells.

30 minutes

3. Cells replicate a second time to produce a second generation of daughter cells.

Extract DNA from cells.

Extract DNA from cells.

Extract DNA from cells.

Extract DNA from cells.

Centrifuge

Centrifuge

Centrifuge

Centrifuge

DNA bands in cesium chloride gradient

^{14}N ^{14}N

^{15}N ^{15}N

^{15}N ^{14}N

^{14}N ^{14}N
^{15}N ^{14}N

Results confirm prediction of semiconservative replication.

Figure 6.17 How the Meselson-Stahl experiment confirmed semiconservative replication. (1) *E. coli* cells were grown in heavy ^{15}N medium. **(2)** and **(3)** Some of these cells were transferred to ^{14}N medium and allowed to divide either once or twice. When DNA from each of these sets of cells was prepared and centrifuged in a cesium chloride gradient, the density of the extracted DNA conformed to the predictions of the semiconservative mode of replication, as shown at the *bottom* of the figure, where *blue* indicates heavy original DNA and *purple* depicts light, newly synthesized DNA. The results are inconsistent with the conservative and dispersive models for DNA replication (compare with Fig. 6.15b and c).

process by which it occurs is quite complicated. Replication does not happen spontaneously any time a mixture of DNA and nucleotides is present. Rather, it occurs at a precise moment in the cell cycle, depends on a network of interacting regulatory elements, requires considerable input of energy, and involves a complex array of the cell's molecular machinery, including a variety of enzymes.

Here we present highlights of replication's two basic steps: **initiation,** during which proteins open up the double helix and prepare it for complementary base pairing, and **elongation,** during which proteins connect the correct sequence of nucleotides into a continuous new strand of DNA. Figure 6.16, illustrates the molecular mechanism of replication using *E. coli* bacteria as the model organism. Although the bacterial chromosome is circular, Fig. 6.16 presents a blowup of only a very small segment of that chromosome, which appears linear in the drawing. Biochemists, notably Nobel laureate Arthur Kornberg and members of his laboratory, deduced the steps by which each bacterial cell copies its single circular chromosome from biochemical studies in which

they purified individual components of the replication machinery. Remarkably, they were eventually able to mix these purified molecules together in a test tube and observe DNA replication proceeding outside the cell.

The Mechanics of DNA Replication at the Chromosomal Level

DNA replication, which depends in part on DNA polymerase, is complicated by the way the polymerase enzyme functions. All known DNA polymerases can create new strands in only one direction: 5′ to 3′. DNA polymerase can thus lengthen existing DNA chains only by adding nucleotides to the 3′ hydroxy group of the DNA strand. As a result, when the double helix unwinds during replication, one newly synthesized strand (the *leading strand*) can grow continuously into the opening Y-shaped area, but the other new strand (the *lagging strand*) can come into existence only as a series of smaller Okazaki fragments (named for their discoverers). These fragments must be joined together at a second stage of the process.

As Fig. 6.16 shows, DNA replication depends on the coordinated activity of many different proteins, including two different DNA polymerases called pol I and pol III (*pol* is short for polymerase). Pol III plays the major role in producing the new strands of complementary DNA, while pol I fills in the gaps between newly synthesized Okazaki segments. Other enzymes contribute to the initiation process: DNA helicase unwinds the double helix. A special group of single-stranded binding proteins keep the DNA helix open. An enzyme called primase creates RNA primers to initiate DNA synthesis. The ligase enzyme welds together Okazaki fragments.

It took many years for biochemists and geneticists to discover how the tight collaboration of many proteins drives the intricate mechanism of DNA replication. Today they believe that programmed molecular interactions of this kind underlie most of the biochemical processes that occur in cells. In these processes, a group of proteins, each performing a specialized function, like the workers on an assembly line, cooperate in the manufacture of complex macromolecules.

Recall that the origin of replication has two forks (Fig. 6.18). As a result, replication is generally **bidirectional,** with the replication forks moving in opposite directions (Fig. 6.18d). At each fork, polymerase copies both template strands, one in a continuous fashion, the other discontinuously as Okazaki fragments. As replication proceeds, DNA helicase progressively unwinds DNA ahead of each fork, while other proteins separate and stabilize the two single strands. Behind each fork, as elongation replicates chains complementary to both templates, double helixes form spontaneously. In the circular *E. coli* chromosome, there is only one origin of replication. When its two forks, moving in opposite directions, meet at a designated *termination region* about halfway around the circle from the origin of replication, replication is complete (Fig. 6.18 d–f).

Not surprisingly, local unwinding of the double helix at a replication fork affects the chromosome as a whole. In *E. coli*, the unwinding of a section of a covalently closed circular chromosome overwinds and distorts the rest of the molecule (Fig. 6.18a and b). Overwinding reduces the number of helical turns to less than the one every 10.5 nucleotides characteristic of B-form DNA. The chromosome accommodates the strain of distortion by twisting back upon itself. You can envision the effect by imagining a coiled telephone cord that overwinds and bunches up with use. The additional twisting of the DNA molecule is called **supercoiling.** Movement of the replication fork causes more and more supercoiling. This cumulative supercoiling, if left unchecked, would wind the chromosome up so tight that it would impede the progress of the replication fork. A group of enzymes known as **DNA topoisomerases** help relax the supercoils by nicking one or both strands of the DNA (Fig. 6.18c). Just as a telephone cord freed at the receiver end can unwind and restore its normal coiling pattern, the DNA strands, after nicking, can rotate relative to each other and thereby restore the normal coiling density of one helical turn per 10.5 nucleotide pairs. The activity of topoisomerases allows replication to proceed through the entire chromosome by preventing supercoils from accumulating in front of the replication fork.

(a) Original double helix.

Origin of replication

Termination region

(b) Unwinding distorts molecule.

Newly replicated DNA

Replication forks

Overwound, supercoiled region

Unreplicated DNA

(c) Topoisomerase relaxes supercoils by nicking, unwinding, and suturing the DNA.

1. Topoisomerase in position to cut DNA
2. DNA cut by topoisomerase
3. Cut strands rotate to unwind
4. Cut ends of strands rejoined by ligase

(d) Replication is bidirectional.

Termination region

(e) Replication is complete when replication forks meet at the termination region.

Termination region

(f) Topoisomerases separate entwined daughter chromosomes, yielding two daughter molecules.

Figure 6.18   The bidirectional replication of a circular bacterial chromosome: An overview. (a) and **(b)** Replication proceeds in two directions from a single origin of replication, creating two replication forks that move in opposite directions around the circle. Local unwinding of DNA at the replication forks creates supercoiled twists in the DNA in front of the replication fork. **(c)** The action of topoisomerase enzymes that nick the DNA helps reduce this supercoiling. **(d)** and **(e)** When the two replication forks meet at the replication termination region, the entire chromosome has been copied. **(f)** Topoisomerase enzymes separate the two daughter chromosomes, which are usually entwined.

Replication of the original double helix sometimes produces intertwined daughter molecules whose clean separation also depends on topoisomerase activity.

In the much larger, linear chromosomes of eukaryotic cells, bidirectional replication proceeds roughly as just described but from many origins of replication. The multiple origins ensure that copying is completed within the time allotted (that is, within the S period of the cell cycle). Because of the three rules governing DNA polymerase activity (see Fig. 6.16a), replication of the very ends of linear chromosomes is problematic. But eukaryotic chromosomes have evolved specialized termination structures known as **telomeres,** which ensure the maintenance and accurate replication of the two ends of each linear chromosome. (Chapter 12 presents the details of eukaryotic chromosome replication.)

Throughout replication in both circular and linear chromosomes, the disentangling of parent strands and respiraling of daughter helixes occurs continuously as enzymes, including helicases and topoisomerases, wind tight, unwind, and rewind the double helix before and after the passage of polymerase, in part by nicking and rotating the two strands in relation to each other. At completion, replication has produced two identical double-stranded daughter DNA molecules.

Cells Must Ensure the Accuracy of Their Genetic Information—Before, During, and After Replication

DNA is the sole repository of the vast amount of information required to specify the structure and function of most organisms. In some species, this information may lie in storage for many years and undergo replication many times before it is called on to generate progeny. During this time, the organism must protect the integrity of the information, for even the most minor change—the simple substitution of one base for another—can have disastrous consequences, such as the production of severe genetic disease or even death. Each organism ensures the informational fidelity of its DNA in three important ways.

Redundancy

Since either strand of the double helix can specify the sequence of the other, each double-stranded DNA molecule is informationally redundant. Built into the structure of the molecule itself, this redundancy provides a basis for the repair of errors arising either from chemical alterations sustained during storage or from malfunctions of the replication machinery.

Enzymes That Repair Chemical Damage to DNA

The cell has an array of enzymes devoted to the repair of nearly every imaginable type of chemical damage. We describe how these enzymes carry out their corrections in Chapter 7. The evolution of sophisticated enzymatic mechanisms for detecting and correcting DNA damage underlines the importance of maintaining the fidelity of genetic information.

The Remarkable Precision of the Cellular Replication Machinery

Evolution has perfected the cellular machinery for DNA replication to the point where errors that could creep in during the copying of the information are exceedingly rare. For example, DNA polymerase has acquired the ability to remove unmatched nucleotides from a new strand of DNA; as a result, the enzyme attaches a free nucleotide to a growing leading or lagging strand only if the nucleotide's base is correctly paired with its complementary base on the parent strand. We examine the molecular mechanisms of this type of DNA repair in Chapter 7.

All of these safeguards help ensure that the information content of DNA will be transmitted intact from generation to generation of cells and organisms. However, as we see next, new combinations of existing information arise naturally as a result of recombination.

Recombination Reshuffles the Information Content of DNA

Mutation, the ultimate source of all new alleles, is a relatively rare phenomenon at any one point on a chromosome. The most important mechanism for generating genomic diversity in sexually reproducing species is the production of new combinations of *already existing alleles.* This type of diversity increases the chances that at least some offspring of a mating pair will be able to survive and reproduce in a changing environment.

New combinations of already existing alleles arise from two different meiotic events: crossing-over, in which two homologous chromosomes exchange parts; and independent assortment, in which each pair of homologous chromosomes segregates free from the influence of other pairs, via random spindle attachment (see Chapter 4). Independent assortment can produce gametes carrying new allelic combinations of genes on different chromosomes; but for genes on the same chromosome, independent assortment alone will only conserve the existing combinations of alleles. Crossing-over, however, can generate new allelic combinations of linked genes. The evolution of crossing-over thus compensated for a significant disadvantage of linkage, which would otherwise lock in only existing combinations of alleles on the same chromosome.

Historically, geneticists have used the term "recombination" to indicate the production of new combinations of alleles by any means, including independent assortment. In the remainder of this chapter, we use **recombination** to mean the generation of new allelic combinations—new combinations of nucleotides along a single DNA molecule—through genetic exchange between homologous chromosomes. In our discussion, we refer to the products of recombination as **recombinants:** chromosomes that carry a mix of alleles derived from different homologs.

As we examine recombination at the molecular level, we look first at experiments demonstrating that crossing-over occurs and then at the molecular details of a crossover event.

During Recombination, DNA Molecules Break and Rejoin

Indirect Physical Evidence of Breaking and Rejoining

When viewed through the light microscope, recombinant chromosomes bearing physical markers appear to result from two homologous chromosomes breaking and exchanging parts as they rejoin (see Figs. 5.7 and 5.8 in Chapter 5). Since the recombined chromosomes, like all other chromosomes, are composed of one long DNA molecule, a logical expectation is that they should show some physical signs of this breakage and rejoining at the molecular level. To evaluate this hypothesis, researchers selected a bacterial virus, **lambda,** as their model organism. Lambda had a distinct experimental advantage for this particular study: It is about half DNA, so the density of the whole virus reflects the density of its DNA.

The experimental approach was similar in principle to the one in which Meselson and Stahl monitored a change in DNA density to follow DNA replication, only in this case, the researchers used the change in DNA density to look at recombination (Fig. 6.19). They grew two bacterial viruses that were genetically marked to keep track of recombination, one in medium with a heavy isotope, the other in medium with a light isotope. They then infected the same bacterial cell with the two viruses under conditions that permitted little if any viral replication. With this type of coinfection, recombination could occur between "heavy" and "light" viral DNA molecules. After allowing time for recombination and the repackaging of viral DNA into virus particles, the experimenters isolated the viruses released from the lysed cells and analyzed them on a density gradient. Those viruses that had not participated in recombination formed bands in two distinct positions, one heavy and one light, as expected. Those viruses that had undergone recombination, however, migrated to intermediate densities, which corresponded to the position of the recombination event. If the recombinant derived most of its alleles and hence most of its chromosome from a "heavy" DNA molecule, its density was skewed toward the gradient's heavy region; by comparison, if it derived most of its alleles and chromosome from a "light" DNA molecule, it had a density skewed toward the light region of the gradient. These experimental results demonstrated that recombination at the molecular level results from the breakage and rejoining of DNA molecules.

Extreme Magnification Would Reveal Heteroduplexes That Mark Spots of Recombination

Recall that chiasmata, which are visible in the light microscope, indicate where chromatids from homologous chromosomes have crossed over, or exchanged parts (see Fig. 5.8). A 100,000-fold magnification of the actual site of recombination within a DNA molecule would reveal the breakage, exchange, and rejoining that constitute the molecular mechanism of crossing-over according to the lambda study. Although it is not yet possible to magnify DNA to the point of distinguishing base sequences under the microscope, a variety of molecular and genetic procedures enable geneticists to make deductions equivalent to such a 100,000-fold magnification. The data obtained provide the following two clues about the mechanism of recombination. First, the products of recombination are almost always in exact register, with not a single base pair lost or gained. Geneticists originally deduced this from observing that recombination usually does not cause mutations; today we know this to be true from analyses of DNA sequence (which we discuss in Chapter 9). Second, the two strands of a recombinant DNA molecule do not break and rejoin at the same location on the double helix. Instead, the break points on each strand can be offset from each other by hundreds or even thousands of base pairs. The segment of the DNA molecule located between the two break points is called a **heteroduplex region** (from the Greek *hetero* meaning "other" or "different") (Fig. 6.20). This name applies not only because one strand of the double helix in this region is of maternal origin, while the other is paternal, but also because the pairing of maternal and paternal strands may produce mismatches in which bases are not complementary. Such mismatches occur at sites in the sequence that differ between the maternal and paternal homologs involved in the recombination event. The base sequences of homologs differ by 0.1% or less among individuals within any species. Within a heteroduplex, these mismatches prevent proper pairing at the mismatched base pairs, but they do not prevent double helix formation along the neighboring complementary nucleotides. Mismatched heteroduplex molecules do not persist for long. The same DNA repair enzymes that operate to correct mismatches during replication can move in to resolve them during recombination. The outcome of the repair enzymes' work depends on which strand they correct. For example, a repaired G–T mismatch could become either G–C or A–T.

The heteroduplex region of a DNA molecule that has undergone crossing-over has one break point on each strand of the

Figure 6.19 DNA molecules break and rejoin during recombination: The experimental evidence. Matthew Meselson and Jean Weigle infected *E. coli* cells with two different genetically marked strains of bacteriophage lambda previously grown in the presence of heavy (^{13}C and ^{15}N) or light (^{12}C and ^{14}N) isotopes of carbon and nitrogen. They then spun the progeny bacteriophages released from the cells on a CsCl density gradient. The position of phages in the gradients reflects the relative density of their DNA. The genetic recombinants had densities intermediate between the heavy and light parents.

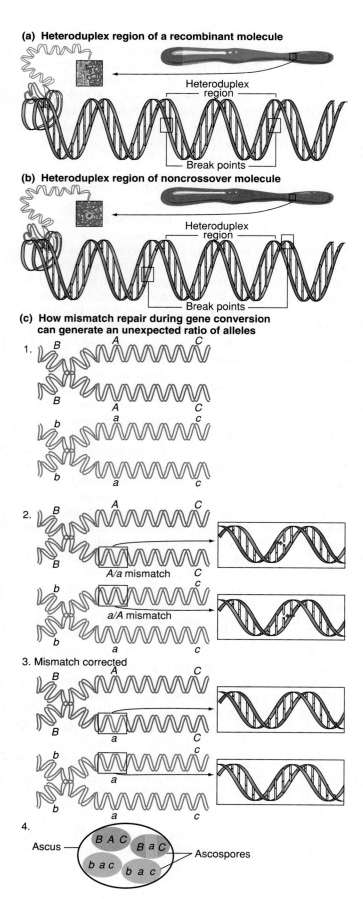

(a) Heteroduplex region of a recombinant molecule

Heteroduplex region

Break points

(b) Heteroduplex region of noncrossover molecule

Heteroduplex region

Break points

(c) How mismatch repair during gene conversion can generate an unexpected ratio of alleles

1. B A C
   B A C
   b a c
   b a c

2. B A C
   B A/a mismatch C
   b a/A mismatch c
   b a c

3. Mismatch corrected
   B A C
   B a C
   b c
   b a c

4. Ascus — B A C B a C b a c b a c — Ascospores

Figure 6.20 Heteroduplex regions occur at sites of genetic exchange. (a) A heteroduplex region separates portions of a chromosome derived from alternative parental homologs after crossing-over. **(b)** A heteroduplex region left behind after an aborted crossover attempt: Sequences from the same parental molecule are found on both sides of the heteroduplex region. The heteroduplexes depicted in (a) and (b) are thought to be two alternative products of the same molecular intermediate (see Fig. 6.22). **(c.1–c.3)** Mismatch repair of heteroduplex regions such as those in (a) or (b) can **(c)** generate gene conversion observed as deviations from the expected 2:2 segregation of parental alleles.

double helix (Fig. 6.20a). Beyond the heteroduplex region, *both* strands of one DNA molecule have been replaced by both strands of its homolog. There is, however, an alternative type of heteroduplex region in which the initiating and resolving cuts are on the same DNA strand (Fig. 6.20b). With this type of heteroduplex, only one short segment of one strand has traded places with one short segment of a homologous nonsister strand. Like the first type of heteroduplex, a short heteroduplex arising from a single-strand exchange may also contain a mismatch.

In either type of heteroduplex, mismatch repair may alter one allele to another. If, for example, the original homologs carried the *A* allele in one segment of two sister chromatids and the *a* allele in the corresponding segment of the other pair of sister chromatids, the *A:a* ratio of alleles in these original homologs would be 2:2. Mismatch repair might change that *A:a* allele ratio from 2:2 to 3:1 (that is, three *A* alleles for every one *a* allele) or 1:3 (one *A* allele for every three *a* alleles; Fig. 6.20c). Any deviation from the expected 2:2 segregation of parental alleles is known as **gene conversion,** because one allele (such as *A*) has been converted to the other (in this case *a*).

Although the unusual ratios resulting from gene conversion occur in many types of organisms, geneticists have studied them most intensively in yeast, where tetrad analysis makes it possible to follow all four meiotic products from a single cell (Fig. 6.21). Interestingly, observations in yeast indicate that gene conversion is associated with crossing-over about 50% of the time, but it is an isolated event not associated with a crossover between flanking markers the other 50% of the time. As we see later, both outcomes derive from the same proposed molecular intermediate, which may or may not lead to a crossover.

A Molecular Model of Crossing-Over

A variety of experimental observations provide the framework for a detailed model of crossing-over during meiosis. First, tetrad analysis shows that only two of the four meiotic products from a single cell are affected by any individual recombination event. One member of each pair of sister chromatids remains unchanged (Fig. 6.21). This provides evidence that recombination occurs during meiotic prophase, after completion of DNA

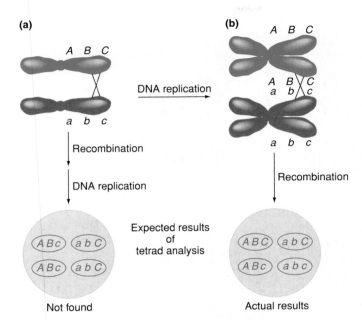

(a)

A B C

a b c

DNA replication →

Recombination ↓

DNA replication ↓

(ABc) (abC)

(ABc) (abC)

Not found

(b)

A B C

A B C
a b c

a b c

Recombination ↓

Expected results of tetrad analysis

(ABC) (abC)

(ABc) (abc)

Actual results

Figure 6.21 When does crossing-over occur? (a) If crossing-over occurs before DNA replication, all four meiotic products would carry the recombinant chromosome. This is not observed. **(b)** If crossing-over occurs after DNA replication, only two meiotic products will carry recombinant chromosomes. This is, in fact, observed.

replication. The observation that recombination occurs only between homologous regions and is highly accurate, that is, in exact register, suggests an important role for base pairing between complementary strands derived from the two homologs. The observation that crossover sites are often associated with heteroduplex regions further supports the role of base pairing in the recombination process; it also implies that the process is initiated by single-strand exchange between nonsister chromatids. Finally, the observation of heteroduplex regions associated with gene conversion in the absence of crossing-over indicates that not all recombination events lead to crossovers.

The current molecular model for meiotic recombination derives almost entirely from results obtained in experiments on yeast. Researchers have found, however, that the protein Spo11, which plays a crucial role in initiating meiotic recombination in yeast, is also essential for meiotic recombination in other fungi, nematodes, plants, fruit flies, and mammals. This finding suggests that the mechanism of recombination presented in detail in Fig. 6.22—and known as the "double-strand-break model"—has been conserved throughout the evolution of eukaryotes. In the figure, we focus on the two nonsister chromatids involved in a single recombination event and show the two nonrecombinant chromatids only at the beginning of the process. These two non-

recombinant chromatids, depicted in the outside positions in Fig. 6.22, step 1, remain unchanged throughout recombination.

Only cells undergoing meiosis express the Spo11 protein, which is responsible for a rate of meiotic recombination several orders of magnitude higher than that found in mitotically dividing cells. Meiotic recombination begins when Spo11 makes a double-strand break in one of the four chromatids. In yeast, where meiotic double-strand breaks have been mapped, it is clear that Spo11 has a preference for some genomic sequences over others, resulting in "hot spots" for crossing over.

Unlike meiotic cells, mitotic cells do not usually initiate recombination in response to an instruction from the normal cell-cycle program; instead, recombination in mitotic cells is a consequence of environmental damage to the DNA. X rays and ultraviolet light, for example, can cause either double-strand breaks or single-strand nicks. The cell's enzymatic machinery works to repair the damaged DNA site. Since the physical and chemical insults that damage DNA occur at random along the molecule, they result in a random distribution of reparative recombination events. Moreover, since nonmeiotic recombination depends on DNA damage, the amount of such recombination is low in the absence of damage and increases in proportion to the amount of damage sustained by the cell.

The double-strand-break repair model of meiotic recombination was proposed in 1983, well before the direct observation of any recombination intermediates. Since that time, scientists have seen—at the molecular level—the formation of double-strand breaks, the resection of those breaks to produce 3′ single-strand tails, and intermediate recombination structures in which single strands from two homologs have invaded each other. The double-strand-break repair model for recombination has become established because it explains much of the data so far obtained from genetic and molecular studies as well as the five properties of recombination deduced from breeding experiments:

1. Homologs physically break, exchange parts, and rejoin.

2. Breakage and repair create reciprocal products of recombination.

3. Recombination events can occur anywhere along the DNA molecule.

4. Precision in the exchange—no gain or loss of nucleotide pairs—prevents mutations from occurring during the process.

5. Gene conversion—in which a small segment of information from one homologous chromosome transfers to the other—can give rise to an unequal yield of two different alleles. Fifty percent of gene conversion events are associated with crossing-over between flanking markers, but an equal 50% are not associated with crossover events.

A model of recombination at the molecular level.

Step 1 Double-strand break formation. During meiotic prophase, the meiosis-specific Spo11 protein makes a double-strand break on one of the chromatids by breaking the phosphodiester bonds between adjacent nucleotides on both strands of the DNA.

Chromatids followed in figure

Spo11

Step 2 Resection. The 5′ ends on each side of the break are degraded to produce two 3′ single-stranded tails.

3′ single-stranded tails

Step 3 First strand invasion. One single-stranded tail is recognized and bound by an enzyme that also binds to a double helix in the immediate vicinity. In *E. coli*, the enzyme that simultaneously binds to the 3′ tail and the double helix is called *Rec*A. It plays a major role in the ensuing steps of the process, although many other enzymes collaborate with it. Their combined efforts open up the *Rec*A-bound double helix, promoting its invasion by the single displaced tail from the other duplex. *Rec*A then moves along the double helix, prying it open in front and releasing it to snap shut behind. With *Rec*A as its guide, the invading strand scans the base sequence it passes in the momentarily unwound stretches of DNA duplex. As soon as it finds a complementary sequence of sufficient length, it becomes immobilized by dozens of hydrogen bonds and forms a stable heteroduplex. Meanwhile, the strand displaced by the invading tail forms a D-loop (for displacement loop), which is stabilized by binding of the single-strand-binding (SSB) protein that played a similar role in DNA replication. D-loops have been observed in electron micrographs. In eukaryotes, strand invasion is mediated by a *Rec*A-like meiosis-specific protein, Dmc1.

First strand invasion

D-loop

Single-strand-binding protein

Step 4 Formation of a double Holliday junction. New DNA synthesis (indicated by *dotted string*) to the invading 3′ tail enlarging the D-loop until the single-stranded bases on the displaced strand can form a complementary base pair with the 3′ tail on the nonsister chromatid. New DNA synthesis from this tail re-creates the DNA duplex on the bottom chromatid. The 5′ end on the right side of the break is then connected to the 3′ end of the invading strand. The resulting X structures are called Holliday junctions after Robin Holliday, the scientist who first proposed them.

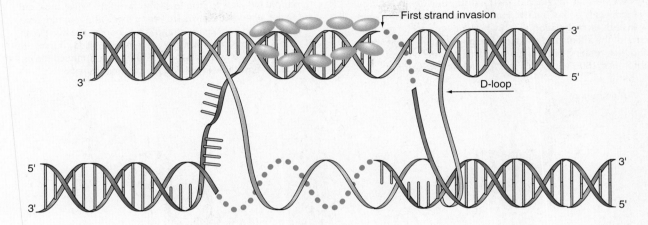

Step 5 Branch migration. The next step, branch migration, results from the tendency of both invading strands to "zip up" by base pairing along the length of their newly formed complementary strands. The DNA double helices unwind in front of this double zipping action and two newly created heteroduplex molecules rewind behind it. Branch migration thus lengthens in the heteroduplex region of both DNA molecules from tens of base pairs to hundreds or thousands. Because the two invading strands began their scanning from complementary bases at slightly different points on the homologous chromatids, branch migration produces two heteroduplex regions that are somewhat different in length.

Step 6 The Holliday intermediate. For meiosis to proceed, the two interlocked nonsister chromatids must disengage. There are two equally likely paths to such a resolution of crossing-over. To understand these alternative resolutions, it is helpful to modify the way we view the interlocked intermediate structure. In this figure we show only one of the two Holliday intermediates associated with each recombination shift. By pushing out each of the four arms of the interlocked structure into an X pattern shown here and then rotating one set of arms from the same original chromatid 180°, you obtain the "isomerized cross-strand exchange configuration" pictured in step 7, commonly referred to as the "Holliday intermediate." It is important to realize that this is simply a different way of looking at the structure for pedagogical purposes. In reality, there is no preferred conformation of chromatid arms relative to each other in this small, localized region. Rather, the arms are free to move about at random, constrained only by the strands that connect the two DNA molecules to each other. The view of the Holliday intermediate, however, clearly reveals that the four single-stranded regions all play an equal role in holding the structure together.

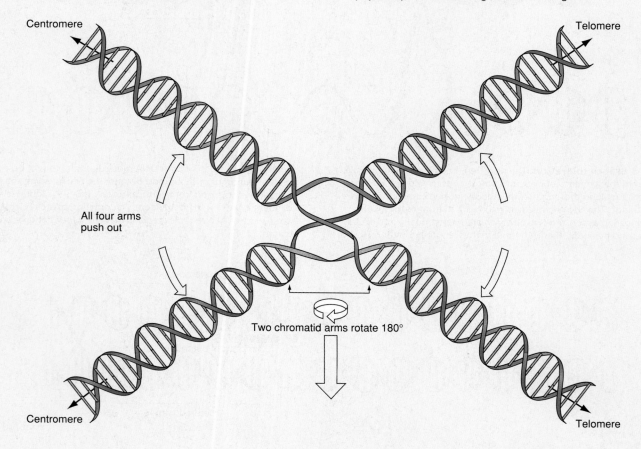

Centromere

Telomere

All four arms push out

Two chromatid arms rotate 180°

Centromere

Telomere

Step 7 Alternative resolutions. If endonucleases make a horizontal cut (as in this illustration) across a Holliday intermediate, the freed centromeric and telomeric strands of both homolog 1 and homolog 2 can become ligated. If the endonucleases make a vertical cut across a strand from homolog 1 and homolog 2, the newly freed strand from the centromeric arm of homolog 1 can now be ligated to the freed strand from

the telomeric arm of homolog 2. Likewise, the telomeric strand from homolog 1 can now be ligated to the centromeric strand from homolog 2. This leads to crossing-over between two homologs. However, the resolution of the second Holliday intermediate will determine whether an actual crossing-over event is consummated, as detailed in step 8.

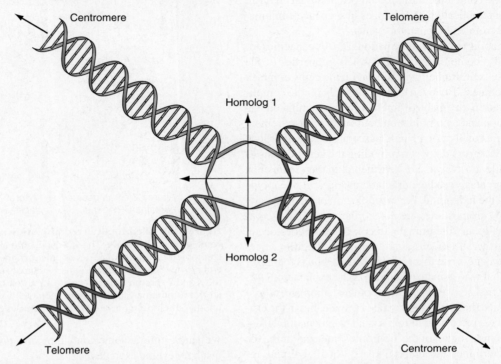

Step 8 Probability of crossover occurring. Because there are two Holliday junctions, both must be resolved. Resolution of both Holliday junctions in the same plane results in a

noncrossover chromatid. For a crossover to occur, the two Holliday junctions must be resolved in opposite planes.

Resolution of Intermediate 1	Resolution of Intermediate 2	RESULT
Horizontal	Horizontal	Gene conversion
Horizontal	Vertical	Crossing over
Vertical	Horizontal	Crossing over
Vertical	Vertical	Gene conversion

CONNECTIONS

The Watson-Crick model for the structure of DNA, the single most important biological discovery of the twentieth century, clarified how the genetic material fulfills its primary function of carrying information: Each long, linear or circular molecule carries one of a vast number of potential arrangements of the four nucleotide building blocks (A, T, G, and C). The model also suggested how base complementarity could provide a mechanism for both accurate replication and changes in sequence combinations that arise from recombination events.

Unlike its ability to carry information, DNA's capacities for replication and recombination are not solely properties of the DNA molecule itself. Rather they depend on the cell's complex enzymatic machinery. But even though they rely on the complicated orchestration of many different proteins acting on the DNA, replication and recombination both occur with extremely high fidelity—normally not a single base pair is gained or lost. Rarely, however, errors do occur, providing the genetic basis of evolution. While most errors are detrimental to the organism, a very small percentage produce dramatic changes in phenotype without killing the individual. For example, although most parts of the X and Y chromosomes are not similar enough to recombine, occasionally an "illegitimate" recombination does occur. Depending on the site of crossing-over, such illegitimate recombination may give rise to an XY individual who is female or an XX individual who is male (Fig. 6.23). The explanation is as follows. In the first six weeks of development, a human embryo has the potential to become either male or female, but in the critical seventh week, information from a small segment of DNA—the sex-determining region of the Y chromosome containing the *SRY* gene—determines the embryo's sex. An illegitimate recombination between the X and the Y that shifts the *SRY* gene from the Y to the X chromosome creates a Y chromosome lacking *SRY* and an X chromosome with *SRY*. Fertilization of eggs by sperm with a Y chromosome lacking *SRY* generates XY individuals that develop as females; fertilization of eggs by sperm

Figure 6.23 Illegitimate recombination may produce an XY female or an XX male. The *SRY* gene, normally located on the Y chromosome, dictates male development; in the absence of *SRY*, an embryo develops as a female. Rare, illegitimate recombination between the X and Y chromosome can create a Y chromosome without *SRY* and an X chromosome with *SRY*. Fertilization with gametes containing these unusual chromosomes will produce XY females or XX males.

with an X chromosome containing *SRY* produces XX individuals that develop as males.

How do genes such as *SRY* produce their phenotypic effects? We begin to answer this question in Chapter 7, where we describe how geneticists learned that a gene is a specific sequence of nucleotides in a discrete region of DNA that encodes a particular protein.

ESSENTIAL CONCEPTS

1. DNA is the nearly universal genetic material. Experiments showing that DNA causes bacterial transformation and is the agent of virus production in phage-infected bacteria demonstrated this fact.

2. According to the Watson-Crick model, proposed in 1953 and confirmed in the succeeding decades, the DNA molecule is a double helix composed of two antiparallel strands of nucleotides; each nucleotide consists of one of four nitrogenous bases (A, T, G, or C), a deoxyribose sugar, and a phosphate. An A on one strand pairs with a T on the other, and a G pairs with a C.

3. DNA carries information in the sequence of its bases, which may follow one another in any order.

4. The DNA molecule reproduces by semiconservative replication. In this type of replication, the two DNA strands separate and the cellular machinery then synthesizes a complementary strand for each. By producing exact copies of the base sequence information in DNA, semiconservative replication allows life to reproduce itself.

5. Recombination arises from a highly accurate cellular mechanism that includes the base pairing of homologous strands of nonsister chromatids. Recombination generates new combinations of alleles.

SOCIAL AND ETHICAL ISSUES

1. In 1986, a published paper from a highly respected cancer research laboratory was retracted because of questions about the validity of some of its data. The research was in a very specialized area of immunology and involved a collaboration between a cellular immunology laboratory and a molecular immunology laboratory. Is the head scientist of each laboratory responsible for understanding and evaluating the details of each part of a published work, including research in specialized areas outside his or her own area of expertise? Because federal grant money was used to fund the research, the National Institutes of Health (NIH) and Congress set up a special committee to determine if there had been scientific misconduct. The scientists under investigation felt they were considered guilty until proven innocent. What is a good system for monitoring scientific research? Are scientists the best watchdogs of themselves or should there be a nonscientist oversight committee? Would an oversight committee understand the research and the research community well enough to monitor activities of the scientific community? (Epilogue: The scientists were exonerated of any wrongdoing.)

2. In 1990, researchers reported the recovery and amplification of a small fragment of DNA from a fossilized leaf that was 20 million years old. (Subsequent studies suggested that all samples of purported million-year old DNA may actually be the result of unintentional contamination from contemporary organisms.) Nevertheless, the recovery of small bits of ancient DNA inspired Michael Crichton to write the best-selling novel

Jurassic Park. In his story, researchers re-create dinosaurs by isolating and piecing together bits of DNA. The necessity of recovering enough intact long stretches of dinosaur DNA make this a highly improbable scenario, yet the movie portrayed this work with an aura of authenticity. Does a science fiction writer have a responsibility to indicate what is truly possible, and where the line between science and fiction is likely to lie? Does this become more of an imperative when a book is made into a movie that reaches a mass audience? Can the reality of what is currently possible be conveyed within the framework of science fiction?

3. Debra is a new graduate student who just joined a research laboratory. She will be following up on work done by a graduate student, Dan, who completed his Ph.D. degree and left the laboratory to do a postdoctoral fellowship at another university. When Debra begins doing the experiments needed to complete a joint publication with Dan and the head of the laboratory, she finds she cannot repeat some of Dan's experiments. She feels the protocols and experimental data he left behind were incomplete. When she contacts Dan to get more information, he is extremely defensive and reluctant to send her information that would enable her to repeat his work. He feels his analysis was complete and doesn't need to be repeated. Debra starts to suspect that Dan fabricated some of the data. What should she do? Should she confront Dan? Should she go to the head of the lab with the problem? Should she try to work out the protocols herself and not raise the issue with anyone?

SOLVED PROBLEMS

```
5' TAAGCGTAACCCGCTAA  CGTATGCGAAC  GGGTCCTATTAACGTGCGTACAC 3'

3' ATTCGCATTGGGCGATT  GCATACGCTTG  CCCAGGATAATTGCACGCATGTG 5'
```

I. Imagine that the double-stranded DNA molecule shown here was broken at the sites indicated by spaces in the sequence, and that before the breaks were repaired, the DNA fragment between the breaks was reversed. What would be the base sequence of the repaired molecule? Explain your reasoning.

Answer

To answer this question, you need to keep in mind the polarity of the DNA strands involved.

The top strand has the polarity left to right of 5' to 3'. The reversed region must be rejoined with the same

polarity. Label the polarity of the strands within the inverting region. To have a 5'-to-3' polarity maintained on the top strand, the *fragment that is reversed must be flipped over,* so the strand that was formerly on the bottom is now on top (see below).

II. A new virus has recently been discovered that infects human lymphocytes. The virus can be grown in the laboratory using cultured lymphocytes as host cells. Design an experiment using a radioactive label that would tell you if the virus contains DNA or RNA.

```
5'                    3'5'          3'5'                  3'
TAAGCGTAACCCGCTAAGTTCGCATACGGGGTCCTATTAACGTGCGTACAC

ATTCGCATTGGGCGATTCAAGCGTATGCCCCAGGATAATTGCACGCATGTG
3'                    5'3'          5'3'                  5'
```

Answer

Use your knowledge of the differences between DNA and RNA to answer this question. RNA contains uracil instead of the thymine found in DNA. *You could set up one culture in which you add radioactive uracil to the media and a second one in which you add radioactive thymine to the culture.* After the viruses have infected cells and produced more new viruses, collect the newly synthesized virus. Determine which culture produced radioactive viruses. If the virus contains RNA, the collected virus grown in media containing radioactive uracil will be radioactive, but the virus grown in radioactive thymine will not be radioactive. If the virus contains DNA, the collected virus from the culture containing radioactive thymine will be radioactive, but the virus from the radioactive uracil culture will not. (You might also consider using radioactively labeled ribose or deoxyribose to differentiate between an RNA- and DNA-containing virus. Technically this does not work as well, because the radioactive sugars are processed by cells before they become incorporated into nucleic acid, thereby obscuring the results.)

III. If you expose a culture of human tissue culture cells (for example, HeLa cells) to ^{3}H-thymidine during S phase, how would the radioactivity be distributed over a pair of homologous chromosomes at metaphase? Would the radioactivity be in (a) one chromatid of one homolog, (b) both chromatids of one homolog, (c) one chromatid each of both homologs, (d) both chromatids of both homologs, or (e) some other pattern? Choose the correct answer and explain your reasoning.

Answer

This problem requires application of your knowledge of the molecular structure and replication of DNA and how it relates to chromatids and homologs. DNA replication occurs during S phase, so the ^{3}H-thymidine would be incorporated into the new DNA strands. A chromatid is a replicated DNA molecule, and each new DNA molecule contains one new strand of DNA (semiconservative replication). *The radioactivity would be in both chromatids of both homologs (d).*

PROBLEMS

6-1 For each of the terms in the left column, choose the best matching phrase in the right column.

a. transformation	1. the strand that is synthesized discontinuously during replication
b. bacteriophage	2. the sugar within the nucleotide subunits of DNA
c. pyrimidine	3. a nitrogenous base containing a double ring
d. deoxyribose	4. noncovalent bonds that hold the two strands of the double helix together
e. hydrogen bonds	5. Meselson and Stahl experiment
f. complementary bases	6. Griffith experiment
g. origin	7. structures at ends of eukaryotic chromosomes
h. Okazaki fragments	8. two nitrogenous bases that can pair via hydrogen bonds
i. purine	9. a nitrogenous base containing a single ring
j. topoisomerases	10. a short sequence of bases where unwinding of the double helix for replication begins
k. semiconservative replication	11. a virus that infects bacteria
l. lagging strand	12. short DNA fragments formed by discontinuous replication of one of the strands
m. telomeres	13. enzymes involved in controlling DNA supercoiling

6-2 Griffith, in his 1928 experiments, demonstrated that bacterial strains could be genetically transformed. The evidence that DNA was the "transforming principle" responsible for this phenomenon came later. What was the key experiment that Avery, MacCleod, and McCarty performed to prove that DNA was responsible for the genetic change from rough cells into smooth cells?

6-3 During bacterial transformation, DNA that enters a cell is not an intact chromosome but consists of randomly generated fragments of chromosomal DNA. In a transformation where the donor DNA was from a bacterial strain that was a^+ b^+ c^+ and the recipient was a^- b^- c^-, 55% of the cells that became a^+ were also transformed to c^+, but only 2% of the a^+ cells were b^+. Is gene b or c closer to gene a?

6-4 Nitrogen and carbon are more abundant in proteins than sulfur. Why did Hershey and Chase use radioactive sulfur instead of nitrogen and carbon to label the protein portion of their bacteriophages in their experiments to determine whether parental protein or parental DNA is necessary for progeny phage production?

6-5 Imagine you have three test tubes containing identical solutions of purified, double-stranded human DNA. You expose the DNA in tube 1 to an agent that breaks the sugar-phosphate (phosphodiester) bonds. You expose the DNA in tube 2 to an agent that breaks the bonds that attach the bases to the sugars. You expose the DNA in tube 3 to an agent that breaks the hydrogen bonds. After treatment, how would the structures of the molecules in the three tubes differ?

6-6 What information about the structure of DNA was obtained from X-ray crystallographic data?

6-7 If 30% of the bases in human DNA are A, (a) what percentage are C? (b) what percentage are T? and (c) what percentage are G?

6-8 Which of the following statements are true about double-stranded DNA?

  a. A + C = T + G

  b. A + G = C + T

  c. A + T = G + C

  d. A/G = C/T

  e. A/G = T/C

  f. (A 1 G)/T = 1

6-9 DNA in cells is most often double-stranded, but viruses contain either double-stranded or single-stranded DNA. How could you determine if the nucleic acid in a DNA virus is single-stranded or double-stranded?

6-10 When a double-stranded DNA molecule is exposed to high temperature, the two strands separate and the molecule loses its helical form. We say the DNA has been denatured. (Denaturation also occurs when DNA is exposed to acid or alkaline solutions.)

  a. Regions of the DNA that contain many A–T base pairs are the first to become denatured as the temperature of a DNA solution is raised. Thinking about the chemical structure of the DNA molecule, why do you think the A–T-rich regions denature first?

  b. If the temperature is lowered, the original DNA strands can reanneal or renature. In addition to the full double-stranded molecules, some molecules of the type shown here are seen when the molecules are examined under the electron microscope. How can you explain these structures?

6-11 An RNA virus that infects plant cells is copied into a DNA molecule once it enters the plant cell. What would be the sequence of bases in the first strand of DNA made complementary to the section of viral RNA shown here?

5′ CCCUUGGAACUACAAAGCCGAGAUUAA 3′

6-12 Bacterial transformation and bacteriophage labeling experiments proved that DNA was the hereditary material in bacteria and in DNA-containing viruses. Some viruses do not contain DNA but have RNA inside the phage particle. An example is the tobacco mosaic virus (TMV) that infects tobacco plants, causing lesions in the leaves. Two different variants of TMV exist that have different forms of a particular protein in the virus particle that can be distinguished. It is possible to reconstitute TMV *in vitro* (in the test tube) by mixing purified proteins and RNA. The reconstituted virus can then be used to infect the host plant cells and produce a new generation of viruses. Design an experiment to show that RNA acts as the hereditary material in TMV.

6-13 The bases of one of the strands of DNA in a region where DNA replication begins are shown here. What is the sequence of the primer that is synthesized complementary to the bases in bold? (Indicate the 5′ and 3′ ends of the sequence.)

5′ AGGCCTCG**AATTCGTATAGCTTTCAGAAA** 3′

6-14 If you expose human tissue culture cells (for example, HeLa cells) to ^{3}H-thymidine just as they enter S phase, then wash this material off the cells and let them go through a second S phase before looking at the chromosomes, how would you expect the ^{3}H to be distributed over a pair of homologous chromosomes? (Ignore the effect recombination could have on this outcome.) Would the radioactivity be in (a) one chromatid of one homolog, (b) both chromatids of one homolog, (c) one chromatid each of both homologs, (d) both chromatids of both homologs, or (e) some other pattern? Choose the correct answer and explain your reasoning. (This problem extends the analysis begun in solved problem III.)

6-15 When Meselson and Stahl grew *E. coli* in ^{15}N medium for many generations, and then transferred to ^{14}N medium for one generation, they found that the bacterial DNA banded at a density intermediate between that of pure ^{15}N DNA and pure ^{14}N DNA following equilibrium density centrifugation. When they allowed the bacteria to replicate one additional time in ^{14}N medium, they observed that half of the DNA remained at the intermediate density, while the other half banded at the density of pure ^{14}N DNA. What would they have seen after an additional generation of growth in ^{14}N medium? After two additional generations?

6-16 Replicating structures in DNA can be observed in the electron microscope. Regions being replicated appear as bubbles.

  a. Assuming bidirectional replication, how many origins of replication are active in this DNA molecule?

  b. How many replication forks are present?

  c. Assuming that all replication forks move at the same speed, which origin of replication was activated last?

6-17 Draw a bidirectional replication fork and label the origin of replication, the leading strands, lagging strands, and the 5′ and 3′ ends of all strands shown in your diagram.

6-18 Diagram replication occurring at the end of a double-stranded linear chromosome. Show the leading and lagging strands with their primers. (Indicate the 5′ and 3′ ends of the strands.) What difficulty is encountered in producing copies of both DNA strands at the end of a chromosome?

6-19 Indicate the role of each of the following in DNA replication: (a) topoisomerase, (b) helicase, (c) primase, and (d) ligase.

6-20 Researchers have discovered that during replication of the circular DNA chromosome of the animal virus SV40, the two newly completed daughter double helices are intertwined. What would have to happen for the circles to come apart?

6-21 In yeast, gene conversion occurs equally frequently with recombination of genetic markers flanking the region of gene conversion and without it. Why is this so?

6-22 Imagine that you have done a cross between two strains of yeast, one of which has the genotype $A\ B\ C$ and the other $a\ b\ c$, where the letters refer to three rather closely linked genes in the order given. You examine many tetrads resulting from this cross, and you find two that do not contain the expected two B and two b spores. In tetrad I, the spores are $A\ B\ C$, $A\ B\ C$, $a\ B\ c$, and $a\ b\ c$. In tetrad II, the spores are $A\ B\ C$, $A\ b\ c$, $a\ b\ C$, and $a\ b\ c$. How have these unusual tetrads arisen?

6-23 What properties would you expect of a strain of *E. coli* that has a mutant allele (null or nonfunctional) of the *recA* gene? Explain.

6-24 From a cross between $e^+ f^+ g^+$ and $e^- f^- g^-$ strains of *Neurospora*, recombination between these linked genes resulted in a few octads containing the following ordered set of spores:

$$e^+\ f^+\ g^+$$
$$e^+\ f^+\ g^+$$
$$e^+\ f^-\ g^+$$
$$e^+\ f^-\ g^+$$
$$e^-\ f^-\ g^-$$
$$e^-\ f^-\ g^-$$
$$e^-\ f^-\ g^-$$
$$e^-\ f^-\ g^-$$

a. Where was recombination initiated?
b. Where did the resolving cut get made?
c. Why do you end up with $2:6\ f^+:f^-$ but $4:4\ e^+:e^-$?

REFERENCES

Avery, O. T., MacLeod, C. M., and McCarty, M. (1944). Studies on the chemical nature of the substance inducing transformation of pneumococcal types. *J. Exp. Med.* 98:451–60.

Griffith, F. (1928). The significance of pneuomococcal types. *J. Hyg.* 27:113–59.

Hershey, A. D., and Chase, M. (1952). Independent functions of viral protein and nucleic acid in growth of bacteriophage. *J. Gen. Physiol.* 36:39–56.

Higuchi, R., Bowman, B., Freiberger, M., Ryder, O. A., and Wilson, A. C. (November 15–21, 1984). DNA sequences from the quagga, an extinct member of the horse family. *Nature* 312(5991):282–84.

Meselson, M., and Stahl, F. W. (1958). The replication of DNA in *E. coli. Proc. Nat. Acad. Sci. USA* 44:671–82.

Ovchinnikov, I. V., Gotherstrom, A., Romanova, G. P., Kharitonov, V. M., Liden, K., and Goodwin, W. (March 30, 2000). Molecular analysis of Neanderthal DNA from the northern Caucasus. *Nature* 404(6777):490–93.

Szostak, J. W., Orr-Weaver, T. L., Rothstein, R. J., and Stahl, F. W. (May 1983). The double-strand-break repair model for recombination. *Cell* 33(1):25–35.

Watson, J. D., and Crick, F. H. C. (1953). A structure for deoxyribose nucleic acid. *Nature* 171:737–38.

CHAPTER

7

Anatomy and Function of a Gene: Dissection Through Mutation

A scale played on a piano keyboard and a gene on a chromosome are both a series
of simple, linear elements (keys or nucleotide pairs) that produce information.
A wrong note or an altered nucleotide pair calls attention to the
structure of the musical scale or the gene.

Human chromosome 3 consists of approximately 220 million base pairs and carries 1000–2000 genes. Somewhere on the long arm of the chromosome resides the gene for rhodopsin, a light-sensitive protein active in the rod cells of our retinas. Under the microscope, a lone chromosome 3 released from the nucleus of a human cell looks like a long, undifferentiated looping string of DNA (Fig. 7.1). At very high magnification—unfortunately higher than that obtainable with today's microscopes—this genetic twine would appear as a double helix composed of two antiparallel strands of nucleotides held together by the hydrogen bonds of complementary base pairs. But which nucleotides within the scramble make up the rhodopsin gene, and what does the gene look like? Knowledge that DNA is the molecule of heredity does not in itself answer the question "What is a gene?" From Mendel on, geneticists who delineated genes through breeding studies and pedigree analyses considered a gene to be a unit of information that affects phenotype; but the string of DNA in chromosome 3 does not resolve into 1000–2000 observable units. Detailed studies of mutations helped resolve this dilemma.

The rhodopsin gene determines perception of low-intensity light. People who carry the normal, wild-type allele of the gene see well in a dimly lit room and on the road at night. One simple change—a mutation—in the rhodopsin gene, however, diminishes light perception just enough to lead to night blindness; other alterations in the gene cause the destruction of rod cells, resulting in total blindness. Medical researchers have so far identified more than 30 mutations in the rhodopsin gene that affect vision. How can one gene sustain so many different mutations? And how can those mutations have such different phenotypic effects?

In this chapter, we describe how geneticists examined the relationship between mutations, genes, and phenotype as they tried to understand, at the molecular level, what genes are and how they function. The researchers considered **mutations** to be heritable changes in base sequence that affect phenotype. From ingenious experiments engineered to reveal how different mutations can occur in different parts of the same gene, and how various mutations in the same gene can have different phenotypic effects, the researchers inferred that physically, a **gene** is a specific

Figure 7.1 The DNA of each human chromosome contains hundreds to thousands of genes. The DNA of this human chromosome has been spread out and magnified 50,000X. Even at this magnification, there are no topological signs that reveal where along the DNA the genes reside.

These findings provided fresh insight into gene function, clarifying that the nucleotide sequence of a gene specifies the amino acids that must be strung together to make a protein. The order of amino acids determines each protein's three-dimensional structure—how it folds in space—which in turn, determines how the protein functions in the cell.

One general theme emerges from our discussion. Knowledge of what genes are and how they work deepens our understanding of Mendelian genetics by providing a biochemical explanation for how genotype influences phenotype. One mutation in the rhodopsin gene, for example, causes the substitution of one particular amino acid for another in the construction of the rhodopsin protein, and this specific substitution changes the protein's ability to absorb photons and thus a person's ability to perceive light.

In our presentation of how genes govern the relationship between mutation and phenotype, we examine:

■ What mutations are; how often they occur; some of the events that cause them; and how they affect the survival of an individual and the evolution of species.

■ What mutations tell us about gene structure: A gene is a specific sequence of nucleotide pairs in a discrete region of DNA that acts as a unit of function, usually by encoding the instructions for making a particular protein.

■ What mutations tell us about gene function: Genes encode proteins by directing the correct assembly of the amino acids that compose each polypeptide; one or more polypeptides compose a protein.

■ How genotype correlates with phenotype: Phenotype depends on the structure and amount of protein produced; mutations that alter a gene's instructions for the corresponding sequence of amino acids alter protein structure and protein function and, as a result, phenotype.

protein-encoding segment of DNA composed of a distinctive set of nucleotide pairs (also referred to as base pairs) in a discrete region of a chromosome. From these same studies, they concluded that a gene is not simply a bead on a string, changeable only as a whole and only in one way, as some had believed. Rather, genes are divisible, and each gene's subunits—the individual nucleotide pairs—can mutate independently and recombine with each other.

Mutations: Primary Tools of Genetic Analysis

Mutations Are Heritable Changes in Base Sequences That Modify the Information Content of DNA

We saw in Chapter 3 that any allele existing at a frequency of more than 1% in a natural population under study is known as a **wild-type allele.** Many experimental geneticists, however, have a slightly different definition of a "wild-type allele," considering it to be the one allele that dictates the most commonly found phenotype in a population; with this definition, all other alleles—no matter what their frequency in the population—would be mutant alleles. Under either definition, a mutation that changes a wild-type allele of a gene to a different allele is called a **forward mutation.** The resulting novel mutant allele can be either recessive or dominant to the original wild type. Geneticists diagram forward mutations as $A^+ \rightarrow a$ when the mutation is recessive and as $b^+ \rightarrow B$ when the mutation is dominant. Mutations can also cause a novel mutant allele to revert back to wild type ($a \rightarrow A^+$, or $B \rightarrow b^+$) in a process known as **reverse mutation,** or **reversion.** In this chapter, we designate most wild-type alleles, whether recessive or dominant, with a "+".

Mendel originally defined genes by the visible phenotypic effects—yellow or green, round or wrinkled—of their alternative alleles. In fact, the only way he knew that genes existed at all was because alternative alleles for seven particular pea genes had arisen through forward mutations in the genetic material. Close to a century later, knowledge of DNA structure clarified that such mutations are heritable changes in DNA base sequence. DNA thus carries the potential for genetic change in the same place it carries genetic information—in the sequence of its bases.

One Way Geneticists Classify Mutations Is by Their Effect on the DNA Molecule

A **substitution** occurs when a base at a certain position in one strand of the DNA molecule is replaced by one of the other three bases (Fig. 7.2a); during DNA replication, a base substitution in one strand will cause a new base pair to appear in the daughter molecule generated from that strand. Substitutions can be subdivided into *transitions,* in which one purine (A or G) replaces the other purine, or one pyrimidine (C or T) replaces the other,

and *transversions,* in which a purine changes to a pyrimidine or vice versa. Other types of mutations produce more complicated rearrangements of DNA sequence. A **deletion** occurs when a block of one or more nucleotide pairs is lost from a DNA molecule; an **insertion** is just the reverse—the addition of one or more nucleotide pairs (Fig. 7.2b and c). Deletions and insertions can be as small as a single base pair or as large as megabases (that is, millions of base pairs). Researchers can see the larger changes under the microscope, in the observation of karyotypes. Even more complex mutations include **inversions,** 180° rotations of a segment of the DNA molecule (Fig. 7.2d), and **reciprocal translocations,** in which parts of two nonhomologous chromosomes change places (Fig. 7.2e). Large-scale DNA rearrangements, including megabase deletions and insertions as well as inversions and translocations, cause major genetic reorganizations that can change either the order of genes along a chromosome or the number of chromosomes in an organism. We discuss these **chromosomal rearrangements,** which affect many genes at a time, in Chapter 13. In this chapter, we focus on mutations that alter only one gene at a time.

Only a small fraction of the mutations in a genome actually alter the nucleotide sequences of genes in a way that affects gene function. By changing one allele to another, these mutations modify the structure or amount of a gene's protein product, and the modification in protein structure or amount influences phenotype. All other mutations either alter genes in a way that does not affect their function, or they change the DNA between genes. We discuss mutations without observable phenotypic consequences in Chapters 10 and 11; such mutations are very useful for mapping genes and tracking differences between individuals. In the remainder of this chapter, we focus on those mutations that have an impact on gene function and thereby influence phenotype.

Spontaneous Mutations Influencing Phenotypes Occur at a Very Low Rate

Mutations that modify gene function happen so infrequently that geneticists must examine a very large number of individuals from a formerly homogeneous population to detect the new phenotypes that reflect these mutations. In one ongoing study, dedicated investigators have monitored the coat colors of millions of specially bred mice and discovered that on average, a single gene mutates to a recessive allele at the rate of 1×10^{-5} per gene per generation (Fig. 7.3). In the study, roughly 11 out of every 1 million gametes produced in a generation have a mutation in a given gene. Studies of several other organisms have yielded similar results: an average spontaneous rate of $2–12 \times 10^{-6}$ mutations per gene per generation.

Looking at the mutation rate from a different perspective, you could ask how many mutations there might be in the genes of an individual. To find out, you would simply multiply the rate of $2–12 \times 10^{-6}$ mutations per gene times 50,000, a generous current estimate of the number of genes in the human genome, to obtain an answer of between 0.1–0.6 mutations per haploid

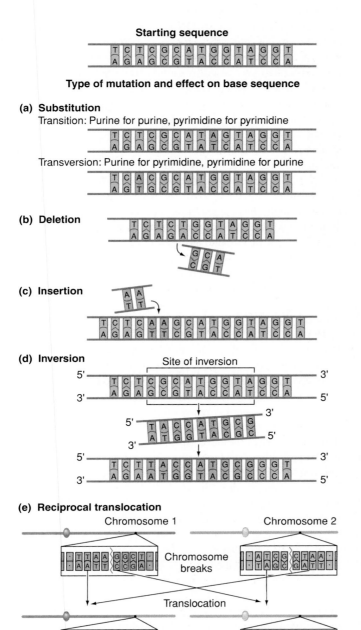

Figure 7.2   Mutations classified by their effect on DNA.

a

b

Locus[a]	Number of gametes tested	Number of mutations	Mutation rate ($\times 10^{-6}$)
a^- (albino)	67,395	3	44.5
b^- (brown)	919,699	3	3.3
c^- (nonagouti)	150,391	5	33.2
d^- (dilute)	839,447	10	11.9
ln^- (leaden)	243,444	4	16.4
	2,220,376	25	11.2 (average)

[a] Mutation is from wild type to the recessive allele shown.

Figure 7.3 Rates of spontaneous mutation. (a) Mutant mouse coat colors: albino (*left*), brown (*right*). **(b)** Mutation rates from wild type to recessive alleles for five coat color genes. Mice from highly inbred wild-type strains were mated with mice homozygous for recessive coat color alleles. Progeny with mutant coat colors indicated the presence of recessive mutations in gametes produced by the inbred mice.

genome. This very rough calculation would mean that, on average, 1 new mutation affecting phenotype could arise in every 2–10 human gametes.

The Mutation Rate Varies from Gene to Gene

Although the average mutation rate per gene is $2–12 \times 10^{-6}$, this number masks considerable variation in the mutation rates for different genes. Experiments with many organisms show that mutation rates range from less than 10^{-8} to more than 10^{-3} per gene per generation. Variation in the mutation rate of different genes within the same organism reflects differences in gene size (larger genes are larger targets that sustain more mutations) as well as differences in the susceptibility of particular phenotypes to mutations in the genes that govern them. In addition, the average rate of gene mutation varies somewhat among species because of differences in certain aspects of DNA metabolism.

Forward Mutations Usually Occur More Often Than Reverse Mutations

In the mouse coat color study, when researchers allowed brother and sister mice homozygous for a recessive mutant allele of one of the five mutant coat color genes to mate with each other, they could estimate the rate of reversion by examining the F_1 offspring. Any progeny expressing the dominant wild-type phenotype for a particular coat color, of necessity, carried a gene that

had sustained a reverse mutation. Calculations based on observations of several million F_1 progeny revealed a reverse mutation rate ranging from $0–2.5 \times 10^{-6}$ per gene per generation; the rate of reversion varied somewhat from gene to gene. In this study, then, the rate of reversion was significantly lower than the rate of forward mutation, most likely because there are many ways to disrupt gene function, but there are only a few ways to restore function once it has been disrupted. The conclusion that the rate of reversion is significantly lower than the rate of forward mutation holds true for most types of mutation. In one extreme example, deletions of more than a few nucleotide pairs can never revert, because DNA information that has disappeared from the genome cannot spontaneously reappear.

There is one rare exception to the very low rate of reversion. Forward mutations caused by the insertion into a chromosome of **transposable elements (TEs)** sometimes revert at a high rate. This is because TEs are DNA elements that rapidly move around the genome, sometimes by excision from one site and insertion at another. A clean excision of a TE that had caused a forward mutation would produce a reversion to wild type, because the excision would restore the normal DNA sequence of the gene (see p. 199 of this chapter for more on transposable elements).

Although all these estimates of mutation rates are extremely rough, they nonetheless support three general conclusions: (1) Mutations affecting phenotype occur very rarely; (2) different genes mutate at different rates; and (3) the rate of forward mutation (a disruption of gene function) is almost always higher than the rate of reversion (which requires a restoration of function to a previously disrupted gene).

Spontaneous Mutations Arise from Many Kinds of Random Events

Because spontaneous mutations affecting a gene occur so infrequently, it is very difficult to study the events that produce them. To overcome this problem, researchers turned to bacteria as the experimental organisms of choice. With these single-celled microbes, it is easy to grow many millions of individuals and then search rapidly through enormous populations to find the few that carry a novel mutation. In one study, investigators spread wild-type bacteria on the surface of agar containing sufficient nutrients for growth as well as a large amount of a bacteria-killing substance, such as an antibiotic or a bacteriophage. While most of the bacterial cells died, a few showed resistance to the bactericidal substance and continued to grow and divide. The descendants of a single resistant bacterium, produced by many rounds of binary fission, formed a mound of genetically identical cells called a **colony.**

The few bactericide-resistant colonies that appeared presented a puzzle. Had the cells in the colonies somehow altered their internal biochemistry to produce a life-saving response to the antibiotic or bacteriophage? Or did they carry heritable mutations conferring resistance to the bactericide? And if they did carry mutations, did those mutations arise by chance from random spontaneous events that take place continuously, even in the absence of a bactericidal substance, or did they only arise in

(a) Two hypotheses for the origin of bactericide resistance

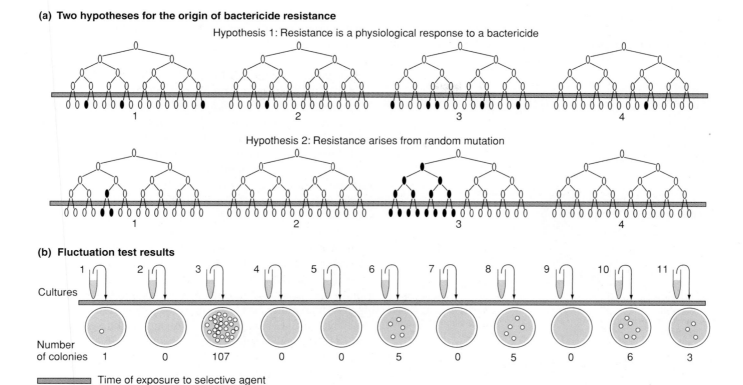

Hypothesis 1: Resistance is a physiological response to a bactericide

Hypothesis 2: Resistance arises from random mutation

(b) Fluctuation test results

Cultures

Number of colonies

Time of exposure to selective agent

Figure 7.4   The Luria-Delbrück fluctuation experiment. (a) Hypothesis 1: If resistance arises only after exposure to a bactericide, all bacterial cultures of equal size should produce roughly the same number of resistant colonies. Hypothesis 2: If random mutations conferring resistance arise before exposure to bactericide, the number of resistant colonies in different cultures should vary (fluctuate) widely. **(b)** Actual results showing large fluctuations suggest that mutations in bacteria occur as spontaneous mistakes independent of exposure to a selective agent.

response to environmental signals (in this case, the addition of the bactericide) that elicit a compensating change in genetic information?

Mutations Are Chance Occurrences Modifying the Genome at Random

In 1943 Salvador Luria and Max Delbrück devised an experiment to examine the origin of bacterial resistance (Fig. 7.4). According to their reasoning, if bacteriophage-resistant colonies arise in direct response to infection by bacteriophages, separate suspensions of bacteria containing equal numbers of cells will generate similar, small numbers of resistant colonies when spread in separate petri plates on nutrient agar suffused with phages. By contrast, if resistance arises from mutations that occur spontaneously even when the phages are not present, different liquid cultures, when spread on separate petri plates, will generate very different numbers of resistant colonies. This is because the mutation conferring resistance can, in theory, arise at any time during the growth of the culture. If it happens early, the cell in which it occurs will produce many mutant progeny prior to petri plating; if it happens later, there will be far fewer mutant progeny when the time for plating arrives. After plating, these numerical differences will show up as fluctuations in the numbers of resistant colonies growing in the different petri plates.

The results of this **fluctuation test** were clear: Most plates supported zero to a few resistant colonies, but a few harbored hundreds of resistant colonies. From this observation of a substantial fluctuation in the number of resistant colonies in different petri plates, Luria and Delbrück concluded that bacterial resistance arises from mutations that exist before exposure to bacteriophage. After exposure, the bactericide in the petri plate becomes a selective agent that kills off nonresistant cells, allowing only the preexisting resistant ones to survive. Figure 7.5 illustrates how researchers used another technique, known as *replica plating,* to demonstrate even more directly that the mutations conferring bacterial resistance occur before the cells "see" the bactericide that selects for their resistance.

These key experiments showed that bacterial resistance to phages and other bactericides is the result of mutations. And these mutations do not arise in particular genes as a directed response to environmental change; instead they occur spontaneously as a result of random processes that can happen at any time and hit the genome at any place. Once such random changes occur, however, they usually remain stable. If the resistant mutants of the Luria-Delbrück experiment, for example, were grown for many generations in medium that did not contain bacteriophages, they would nevertheless remain resistant to this bactericidal virus.

(a) The replica plating technique

1. Invert master plate; pressing against velvet surface leaves an imprint of colonies. Save plate.

2. Invert second plate (replica plate); pressing against velvet surface picks up colony imprint.

Master plate
No penicillin in medium

Penicillin in medium

Velvet

3. Incubate plate.

S = penicillin-sensitive bacteria
R = penicillin-resistant bacteria

Replica plate

4. Only penicillin-resistant colonies grow. Compare with position of colonies on original plate.

(b) Mutations occur prior to penicillin exposure

10^7 colonies of penicillin-sensitive bacteria

Make three replica plates. Incubate to allow penicillin-resistant colonies to grow.

Master plate
No penicillin in medium

Penicillin in medium

Velvet

Penicillin in medium

Penicillin in medium
Penicillin-resistant colonies grow in the same position on all three plates.

Figure 7.5 Replica plating verifies that bacterial resistance is the result of preexisting mutations. (a) Pressing a *master plate* onto a velvet surface transfers some cells from each bacterial colony onto the velvet. Pressing a *replica plate* onto the velvet then transfers some cells from each colony onto the replica plate. Investigators track which colonies on the master plate are able to grow on the replica plate (here, only penicillin-resistant ones). **(b)** A large number of colonies on a master plate without penicillin are sequentially transferred to three replica plates with penicillin. Resistant colonies grow in the same positions on all three replicas, showing that some colonies on the master plate had multiple resistant cells before exposure to the antibiotic.

We now describe some of the many kinds of random events that cause mutations, and we discuss how cells cope with the damage.

Hydrolysis, Radiation, Ultraviolet Light, and Oxidation Can Alter the Information Stored in DNA

Chemical and physical assaults on DNA are quite frequent. Geneticists estimate, for example, that the hydrolysis of a purine base, A or G, from the deoxyribose-phosphate backbone occurs 1000 times an hour in every human cell. This kind of DNA alteration is called **depurination** (Fig. 7.6a). Because the resulting *apurinic site* cannot specify a complementary base, the DNA replication process sometimes introduces a random base opposite the apurinic site, causing a mutation in the newly synthesized complementary strand three-quarters of the time. Another naturally occurring process that may modify DNA's information content is **deamination:** the removal of an amino (–NH₂) group. Deamination can change cytosine to uracil (U), the nitrogenous base found in RNA but not in DNA. Because U pairs with A rather than G, deamination followed by replication may alter a C–G base pair to a T–A pair in future generations of DNA molecules (Fig. 7.6b); such a C–G to T–A change is a transition mutation. Other assaults include naturally occurring radiation such as cosmic rays and X rays, which break the sugar-phosphate backbone (Fig. 7.6c); ultraviolet light, which causes adjacent thymine residues to become chemically linked into thymine–thymine dimers (Fig. 7.6d); and oxidative damage to any of the four bases (Fig. 7.6e). All of these changes alter the information content of the DNA molecule.

Cells have evolved many mechanisms for dealing with these problems. In one, specialized proteins patrol the DNA for irregularities, detecting and repairing chemical changes that disrupt the normal Watson-Crick base pairing. When they find such a change, they cut on either side of the damage and remove the intervening bases; DNA polymerase then fills in the gap by complementary base pairing with the information in the undamaged strand (Fig. 7.7a). This repair mechanism is known as **excision repair** since it excises the damage. Excision repair uses the redundancy of information in the double helix as the basis for filling in the excised information. The significance of this correction device is evident from the genetic disease *xeroderma pigmentosum*. People with the disease lack the enzymatic ability to recognize and excise thymine–thymine dimers produced by ultraviolet light. Unless they avoid all exposure to sunlight, their inability to correct damaged DNA results in a whole series of mutations that eventually leads to skin cancer (Fig. 7.7b). Even in people with normal genes for excision enzymes, excision repair, though doggedly persistent, is not 100% effective. As a result, a small number of the changes caused by chemical assaults, ultraviolet light, and radiation slip through to become established mutations transmitted from one generation to the next.

Mistakes During DNA Replication Can Also Alter Genetic Information

If the cellular machinery for some reason incorporates the wrong base during replication, for instance, a C opposite an A instead of

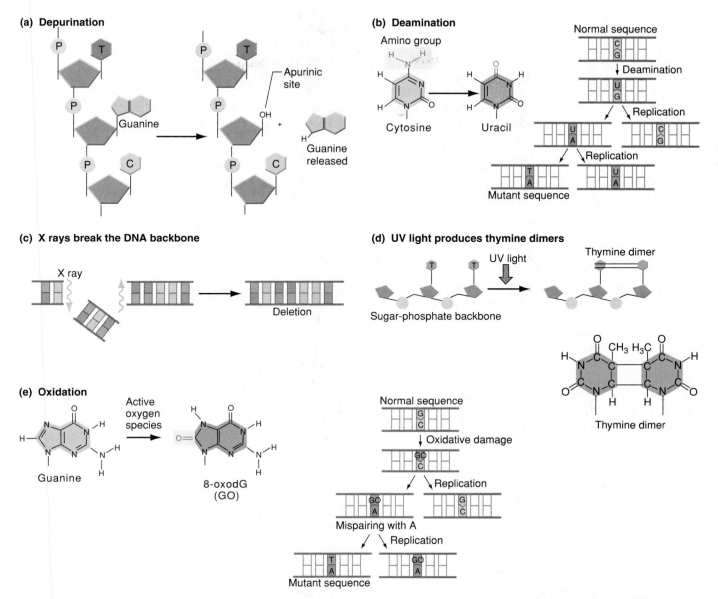

Figure 7.6 How natural processes can change the information stored in DNA. (a) In depurination, the hydrolysis of A or G bases leaves a DNA strand with an unspecified base. **(b)** In deamination, the removal of an amino group from C initiates a process that eventually (after DNA replication) causes a transition. **(c)** X rays break the sugar-phosphate backbone and thereby split a DNA molecule into smaller pieces, which may be spliced back together improperly. **(d)** Ultraviolet (UV) radiation causes adjacent Ts to form dimers, which can disrupt the readout of genetic information. **(e)** Irradiation causes the formation of *free radicals* (such as oxygen molecules with an unpaired electron) that can damage individual bases. Here, the pairing of the altered base GO with A creates a transversion that changes a G–C base pair to T–A.

the expected T, then during the next replication cycle, one of the daughter DNAs will have the normal A–T base pair while the other will have a mutant G–C. Careful measurements of the fidelity of replication *in vivo,* in both bacteria and human cells, show that such errors are exceedingly rare, occurring less than once in every 10^9 base pairs. That is equivalent to typing this entire book 1000 times while making only one typing error. Considering the complexities of helix unwinding, base pairing, and polymerization, this level of accuracy is amazing. How do cells achieve it?

The replication machinery minimizes errors through successive stages of correction. In the test tube, DNA polymerases replicate DNA with an error rate of about one mistake in every 10^6 bases copied. This rate is about 1000-fold worse than that achieved by the cell. Even so, it is impressively low and is only attained because polymerase molecules provide, along with their polymerization function, a proofreading/editing function in the form of a nuclease that is activated whenever the polymerase makes a mistake. This nuclease portion of the polymerase molecule, called the *3'-to-5' exonuclease,* recognizes a mispaired

(a) Excision repair

1. Exposure to UV light.

2. Thymine dimer forms.

3. Endonuclease nicks strand containing dimer.

4. Damaged fragment is released from DNA.

5. DNA polymerase fills in the gap with new DNA (yellow).

6. DNA ligase seals the repaired strand.

(b) Xeroderma pigmentosum

Figure 7.7 Excision repair removes damaged DNA and fills in the gap. (a) Excision repair enzymes release damaged regions of DNA. DNA polymerases and DNA ligase then resynthesize the missing information (*yellow*) and reseal the now-corrected strand. **(b)** Skin lesions in a xeroderma pigmentosum patient, who lacks a critical excision repair enzyme.

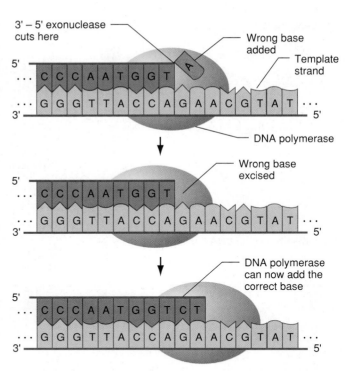

Figure 7.8 DNA polymerase's proofreading function. If DNA polymerase mistakenly adds an incorrect nucleotide at the 3'-end of the strand it is synthesizing, the enzyme's 3'-to-5' exonuclease activity will remove this nucleotide, giving the enzyme a second chance to add the correct nucleotide.

base and excises it, allowing the polymerase to copy the nucleotide correctly on the next try (Fig. 7.8). Without its nuclease portion, DNA polymerase would have an error rate of one mistake in every 10^4 bases copied, so its editing function improves the fidelity of replication 100-fold. DNA polymerase *in vivo* is part of a replication system including many other proteins that collectively improve on the error rate another 10-fold, bringing it to within about 100-fold of the fidelity attained by the cell.

The 100-fold higher accuracy of the cell depends on a backup system that notices and corrects residual errors in the newly replicated DNA. Active only *after* polymerization, the backup system has a difficult problem to solve. Suppose that a

G–C pair were copied to produce two daughter molecules, one of which has the correct G–C base pair, while the other has an incorrect G–T. The backup system can recognize the incorrectly matched G–T base pair because the improper base pairing distorts the double helix. But how does the system know whether to correct the pair to a G–C or to an A–T?

Bacteria solve this problem by placing a distinguishing mark on the parental DNA strands at specific places: Everywhere the sequence GATC occurs, an enzyme system puts a methyl group on the A (Fig. 7.9a). Shortly after replication, the old template strand bears the methyl mark, while the new daughter strand is as yet unmarked. Whenever the enzymatic system detects an improper pair, it excises a few nucleotides around the mispaired nucleotides from the new, unmethylated strand and resynthesizes the information using the old, methylated strand as a template (Fig. 7.9b–d). Later, with the completion of replication and repair, enzymes mark the new strand as well, so that its parental origin will be evident in the next round of replication. This repair system is known as **methyl-directed mismatch repair.** Like excision repair, it uses the redundancy of DNA's double strands as the basis for restoring information. Unlike excision repair, which recognizes and corrects irregularities caused by chemical or physical assaults, methyl-directed mismatch repair recognizes base pairs mismatched during replication. Such mismatches show

(a) Parental strands are marked with methyl groups.

(b) Enzyme system recognizes mismatch in replicated DNA.

(c) DNA on unmarked new strand is excised.

(d) Repair and methylation of newly synthesized DNA strand.

Figure 7.9 Methyl-directed mismatch repair corrects mistakes in replication. (a) GATC sequences on parental strands (*blue*) are marked with a methyl group, but the GATCs on newly synthesized strands (*purple*) do not immediately receive a methyl group. If a replication error produces a mismatch **(b)**, repair enzymes remove nucleotides from the mismatched region of the unmarked, newly synthesized strand **(c)**. Just as in excision repair (see Fig. 7.7a), DNA polymerase and DNA ligase restore the missing information and reseal the corrected strand **(d)**.

up as bulges and hollows in the newly replicated double-stranded DNA.

Eukaryotic cells also have a mismatch correction system, but it is not yet known how they distinguish template from newly replicated strands. Researchers have nevertheless discovered that mismatch repair in humans has medical repercus-

sions. Defects in mismatch repair enzymes are associated with colon and other forms of cancer.

Unequal Crossing-Over and Transposons Cause DNA Changes That Are Not Susceptible to Excision or Mismatch Repair

Some mutations arise from events other than chemical and physical assaults or replication errors. Significant among the mechanisms giving rise to such mutations is erroneous recombination. For example, in **unequal crossing-over,** which occurs during meiosis, one homologous chromosome ends up with a duplication (a kind of insertion), while the other homolog sustains a deletion. As Fig. 7.10a shows, some forms of red-green colorblindness arise from deletions and duplications in the genes that enable us to perceive red and green wavelengths of light; these reciprocal informational changes are the result of unequal crossing-over. Another notable mechanism for altering DNA sequence involves the units of DNA known as *transposable elements (TEs).* TEs move from place to place within the genome, sometimes causing a change in DNA function when they insert themselves into a new chromosomal location. As mentioned earlier, because certain types of TEs move around the genome via a mechanism that excises them from their original location (Fig. 7.10b), some TE-induced mutations have a high rate of reversion. Chapter 13 discusses additional genetic consequences of TE behavior.

The Instability of Certain Trinucleotide Repeats Causes Mutations

In 1992 a group of molecular geneticists discovered an unusual and completely unexpected type of mutation in humans: the excessive amplification of a CGG base triplet normally repeated only a few to 50 times in succession. If, for example, a normal allele of a gene carries 5 consecutive repetitions of the base triplet CGG (that is, CGGCGGCGGCGGCGG on one strand), an abnormal allele resulting from mutation could carry 200 repeats in a row. Further investigations revealed that repeats of several trinucleotides—CAG, CTG, and GAA, in addition to CGG—can be unstable such that the number of repeats often increases or decreases in different cells of the same individual. Instability can also occur during the production of gametes, resulting in changes in repeat number from one generation to the next. The expansion and contraction of trinucleotide repeats has now been found not only in humans but in many other species as well.

The rules governing trinucleotide repeat instability appear to be quite complicated, but one general feature is that the larger the number of repeats at a particular location, the higher the probability that expansion and contraction will occur. Usually, tracts with less than 30–50 repetitions of a triplet change in size only very infrequently, and the mutations that do occur cause only small variations in the repeat number. Larger tracts involving hundreds of repeats change in size more frequently, and they result in more variation in the number of repetitions. Researchers have not yet determined the precise mechanism of

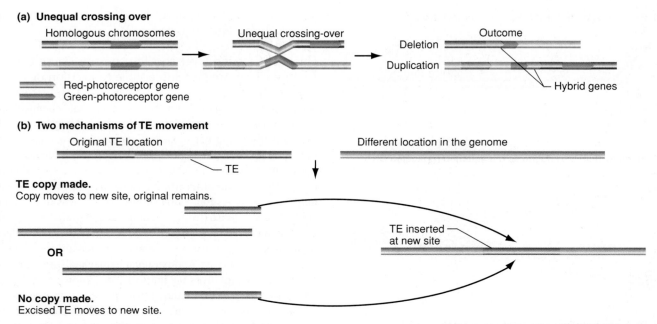

Figure 7.10   How unequal crossing-over and the movement of transposable elements (TEs) change DNA's information content.
(a) If two nearby regions contain a similar DNA sequence, the two homologous chromosomes may pair out of register during meiosis and produce gametes with either a deletion or a reciprocal duplication of the intervening region. Colorblindness in humans can result from unequal crossing-over between the nearby and highly similar genes for red and green photoreceptors. **(b)** TEs move around the genome. Some TEs copy themselves before moving, while others are excised from their original positions during the transposition process. Insertion of a TE into a gene often has phenotypic consequences.

triplet repeat amplification. One possibility is that regions with long trinucleotide repeats form unusual DNA structures that are hard to replicate because they force the copying machinery to slip off, then hop back on, slip off, then hop back on when it gets to this sequence. Such stopping and starting may produce a replication "stutter" that causes synthesis of the same triplet repeat over and over again, expanding the number of copies. This type of error could conversely shrink the size of the trinucleotide repeat tract if, after slipping off, the replication machinery restarts copying at a repeat further down the template sequence. Whatever the mechanism, mutations of long trinucleotide stretches occur quite often, suggesting that the enzymes for excision or mismatch repair are not very efficient at repairing them.

The expansion of trinucleotide repeats is at the root of one of the most common forms of human mental retardation as well as many other disorders of the nervous system. The Genetics and Society box "Amplified Trinucleotide Repeats May Have Medical Consequences" discusses the fascinating implications of this type of DNA-sequence instability for human genetics and medicine.

Mutagens Induce Mutations

Mutations make genetic analysis possible, but most mutations appear spontaneously at such a low rate that researchers have looked for controlled ways to increase their occurrence. H. J.

Muller, an original member of T. H. Morgan's *Drosophila* group, first showed that exposure to a dose of X rays higher than the naturally occurring level increases the mutation rate in fruit flies (Fig. 7.11). He exposed male *Drosophila* to large doses of X rays and then mated these males with females that had one X chromosome containing an easy-to-recognize dominant mutation causing Bar eyes. This X chromosome (called a *balancer*) also carried chromosomal rearrangements known as inversions that prevented it from crossing-over with other X chromosomes. (Chapter 13 explains the details of this phenomenon.) Some of the F$_1$ daughters of this mating were heterozygotes carrying a mutagenized X from their father and a *Bar*-marked X from their mother. If X rays induced a recessive lethal mutation anywhere on the paternally derived X chromosome, then these F$_1$ females would be unable to produce non-Bar-eyed sons. Thus simply by noting the presence or absence of non-Bar-eyed sons, Muller could establish whether a mutation had occurred in any of the more than 1000 genes on the X chromosome that are essential to *Drosophila* viability. He concluded that the greater the X-ray dose, the greater the frequency of recessive lethal mutations.

Any physical or chemical agent that raises the frequency of mutations above the spontaneous rate is called a **mutagen.** Researchers use many different mutagens to produce mutations for study. With the Watson-Crick model of DNA structure as a guide, they can understand the action of most mutagens at the molecular level. The X rays used by Muller to

Figure 7.11 Exposure to X rays increases the mutation rate in *Drosophila*. F_1 females are constructed that have an irradiated paternal X chromosome (*red line*), and a *Bar*-marked "balancer" maternal X chromosome (*wavy blue line*). These chromosomes cannot recombine because the balancer chromosome has multiple inversions (as explained in Chapter 13). If the paternal chromosome has an X-ray-induced recessive lethal mutation (*m*), the F_1 female can produce only Bar-eyed sons. If this X chromosome has no such mutation, she will produce both Bar-eyed and non-Bar-eyed sons.

induce mutations on the X chromosome, for example, can break the sugar-phosphate backbones of DNA strands, sometimes at the same position on the two strands of the double helix. Multiple double-strand breaks produce DNA fragmentation, and the improper "stitching back together" of the fragments can cause inversions, deletions, or other rearrangements (see Fig. 7.6c on p. 197). Another example of a molecular mechanism of mutagenesis involves mutagens known as **base analogs,** which are so similar in chemical structure to the normal nitrogenous bases that the replication machinery can incorporate them into DNA in place of the normal bases (Fig. 7.12a on pp. 204–205). Since a base analog may have pairing properties different from those of the base it replaces, it can cause base substitutions on the complementary strand synthesized in the next round of DNA replication. Other chemical mutagens generate substitutions by directly altering a base's chemical structure and properties (Fig. 7.12b). Again, the effects of these changes become fixed in the genome when the altered base causes incorporation of an incorrect complementary base during a subsequent round of replication. Yet another class of chemical mutagens consists of compounds known as **intercalators:** flat, planar molecules that can sandwich themselves between successive base pairs and disrupt the machinery for replication, recombination, or repair (Fig. 7.12c). The disruption may eventually generate deletions or insertions of a single base pair.

Thus, even though the cells of *E. coli, Drosophila,* mice, humans, and all other organisms have mechanisms that keep mutations to a minimum, allowing the cells to achieve a maximum of fidelity in the use and transmission of their genes, some changes in DNA information content do occur. The very low level of spontaneous mutation is the best most cells can achieve under normal conditions, and exposure to mutagens raises that rate.

Impact: Mutations Have Consequences for the Evolution of Species and the Survival of Organisms

"The capacity to blunder slightly is the real marvel of DNA. Without this special attribute, we would still be anaerobic bacteria and there would be no music." In these two sentences, the eminent medical scientist and self-appointed "biology watcher" Lewis Thomas acknowledges that changes in DNA are behind the phenotypic variations that are the raw material on which natural selection has acted for billions of years to drive evolution. The wide-ranging variation in the genetic makeup of the human population—and other populations as well—is, in fact, the result of a balance between the continuous introduction of new mutations, the loss of deleterious mutations because of the selective disadvantage they impose on the individuals that carry them, and the increase in frequency of rare mutations that provide a selective advantage to the individuals carrying them (or that spread through a population by other means; see Chapter 20 for the details of population genetics).

In sexually reproducing multicellular organisms, only germ-line mutations that can be passed on to the next generation play a role in the evolution of a species. Nevertheless, mutations in somatic cells can still have an impact on the well-being and survival of individuals. Somatic mutations in genes that help regulate the cell cycle may, for example, lead to cancer. The U.S. Food and Drug Administration tries to identify potential cancer-causing

Genetics and Society

Amplified Trinucleotide Repeats May Have Medical Consequences

Expansions of the base triplet CGG cause a heritable disorder known as *fragile X syndrome*. Adults affected by this syndrome manifest several physical anomalies, including an unusually large head, long face, large ears, and in men, large testicles. They also exhibit moderate to severe mental retardation. Fragile X syndrome has been found in men and women of all races and ethnic backgrounds. The fragile X mutation is, in fact, a leading genetic cause of mental retardation worldwide, second only to the trisomy 21 that results in Down syndrome.

Specially prepared karyotypes of cells from people with fragile X symptoms reveal a slightly constricted, so-called fragile site near the tip of the long arm of the X chromosome (Fig. A). In some pictures of these karyotypes, the X has actually broken at this site, releasing a small piece containing the end of the chromosome (not shown). The long tracts of CGG trinucleotides, which make up the fragile X mutation, apparently produce a localized constricted region that breaks easily. Geneticists named the fragile X disorder for this specific pinpoint of fragility more than 20 years before they identified the mutation that gives rise to it.

The gene in which the fragile X mutation occurs is called *FMR-1* (for <u>f</u>ragile-X-associated <u>m</u>ental <u>r</u>etardation). Near one end of the gene, different people carry a different number of repeats of the sequence CGG, and geneticists now have the molecular tools to quantify these differences. Normal alleles contain 5–54 of these triplet repeats, while the *FMR-1* gene in people with fragile X syndrome contains 200–4000 repeats of the exact same triplet (Fig. B.1). The rest of the gene's base sequence is the same in both normal and abnormal alleles.

The triplet repeat mutation that underlies fragile X syndrome has a surprising transmission feature. Alleles with a full-blown mutation are foreshadowed by *premutation alleles* that carry an intermediate number of repeats—more than 50 but fewer than 200 (Fig. B.1). Premutation alleles do not themselves generate fragile

Figure A   A karyotype reveals a fragile X chromosome. The fragile X site is seen on the bottom of both chromatids of the X chromosome at the *right*.

X symptoms in most carriers, but they show significant instability and thus forecast the risk of genetic disease in a carrier's progeny. The greater the number of repeats in a premutation allele, the higher the risk of disease in that person's children. For example, if a woman carries a premutation allele with 60 CGG repeats, 17% of her offspring run the risk of exhibiting fragile X syndrome. If she carries a premutation allele with 90 repeats, close to 50% of her offspring will show symptoms. In short, the change that produces a premutation allele increases the likelihood that the *FMR-1* gene will incur more mutations; and the larger the original number of

agents (known as carcinogens) by using the Ames test to screen for chemicals that cause mutations in bacterial cells (Fig. 7.13 on p. 206). This test asks whether a particular chemical can induce histidine$^+$ (*his$^+$*) revertants of a special histidine$^-$ (*his$^-$*) mutant strain of the bacterium *Salmonella typhimurium*. The *his$^+$* revertants can synthesize all the histidine they need from simple compounds in their environment, whereas the original *his$^-$* mutants cannot make histidine, so they can survive only if histidine is supplied in the environment. The advantage of the test is that only revertants can grow on petri plates that do not contain any histidine,

so it is possible to examine large numbers of cells from an originally *his$^-$* culture to find the rare *his$^+$* revertants induced by the chemical in question. To increase the sensitivity of mutation detection, the *his$^-$* strain used in the Ames test system contains a second mutation inactivating the excision repair system, which prevents the ready repair of mutations, and a third mutation causing defects in the cell wall, which allows chemicals easier access to the cell interior.

Since most agents that cause mutations in bacteria should also damage the DNA of higher eukaryotic organisms, any mu-

(1) Effect of (CGG) repeat number

$(CGG)_{<50}$

Wild-type alleles

$(CGG)_{50-200}$

Premutation alleles

$(CGG)_{>200}$

Disease-causing alleles

(2) A fragile X pedigree

22/29   82   29/80

22/83   22/90   ~500

>200   >200

☐ Unaffected
■ Affected
■ Heterozygous or hemizygous for premutation allele

Figure B   Amplification of CGG triplet repeats correlates with the fragile X syndrome. (1) *FMR-1* genes in unaffected people generally have fewer than 50 CGG repeats. Unstable premutation alleles have between 50 and 200 repeats. Disease-causing alleles have more than 200 CGG repeats, and some have up to 4000. (2) A fragile X pedigree showing the number of CGG repeats in different individuals. Fragile X syndrome patients are almost always the progeny of mothers who carried premutation alleles.

causing Huntington disease consist of an expanded CAG trinucleotide repeat within the Huntington chorea gene; but in this case, the amplification occurs most often when the gene is transmitted by the father. The mutations resulting in Huntington disease may thus occur during spermatogenesis. Several other genetic disorders of the nervous system also arise from fluctuating numbers of triplet repeats. Why these repeats are so unstable and prone to expansion (and occasionally shrinkage) is an area of active research.

The CGG trinucleotide repeat expansion underlying fragile X syndrome has interesting implications for genetic counseling. There are thousands of possible alleles of the *FMR-1* gene, ranging from the smallest normal allele isolated to date, with 5 triplet repeats, to the largest abnormal allele so far isolated, with roughly 4000 repeats. This enormous number of alleles gives rise to a slow gradation of normal-to-abnormal phenotypes, with the abnormal phenotypes progressing from mild retardation without other evidence of disease to severe retardation combined with physical anomalies. The relation between genotype and phenotype is clear at both ends of the triplet-repeat spectrum: Individuals whose alleles contain less than 55 repeats are normal, while people with an allele carrying more than 200 repeats are almost always moderately to severely retarded. With an intermediate number of repeats, however, expression of the mental retardation phenotype is highly variable, depending to an unknown degree on chance, the environment, and modifier genes.

This range of variable expressivity leads to an ethical dilemma: Where should medical geneticists draw the line in their assessment of risk? Prospective parents with a history of mental retardation on either side of the family may consult with a counselor to determine their options. The counselor would first test the parents for premutation alleles to establish the possible involvement of fragile X syndrome. He or she would also want to analyze the fetal cells directly by amniocentesis, to determine whether the fetus carries an expanded number of CGG repeats in its *FMR-1* gene. If the results indicate the presence of an allele in the middle range of triplet repeats, the counselor will have to acknowledge that this is a grey area in which prediction can only be imperfect. The prospective parents' difficult decision of whether or not to continue the pregnancy will then rest on the very shaky ground of an inconclusive, overall evaluation of risk.

additional CGG repeats, the larger the number of subsequent additions. Interestingly, the expansion of *FMR-1* premutation alleles has some as-yet-unexplained relation to the parental origin of the repeats. Whereas most male carriers transmit their *FMR-1* allele with only a small change in the number of repeats, many women with premutation alleles bear children with 250–4000 CGG repeats in their *FMR-1* gene (Fig. B.2). One possible explanation is that whatever conditions generate fragile X mutations occur most readily during oogenesis.

This same transmission pattern does not hold for all triplet repeat mutations. In 1993, geneticists found that the mutations

tagen that increases the rate of mutation in bacteria might be expected to cause cancer in people and other mammals. Mammals, however, have complicated metabolic processes capable of inactivating some otherwise hazardous chemicals. Other biochemical events in mammals can create a mutagenic substance from some otherwise nonhazardous chemicals. To simulate the action of mammalian metabolism, toxicologists often add a solution of rat liver enzymes to the chemical under analysis by the Ames test (Fig. 7.13). Because this simulation is not perfect, Food and Drug Administration agents ultimately assess whether

bacterial mutagens identified by the Ames test can cause cancer in rodents when included in their diet.

In summary, random errors that change the information content of DNA are the ultimate source of variation within and between species. But even though chance mutations provide the raw material of evolution, most changes in genes are deleterious to an organism or its progeny. Cells have evolved several enzymatic systems that either prevent mutations from forming or eliminate those that do. These safeguards enable organisms to keep mutations to a minimum.

Type of mutagen	Chemical action of mutagen
(a) Replace a base: Base analogs have a chemical structure almost identical to that of a DNA base.	5-Bromouracil–normal state, behaves like thymine Adenine 5-Bromouracil–rare state, behaves like cytosine Guanine 5-Bromouracil: almost identical to thymine. Normally pairs with A; in transient state, pairs with G.

(b) Alter base structure and properties:

Hydroxylating agents: add a hydroxyl (–OH) group

Cytosine → N-4-Hydroxycytosine (C*) Adenine

Hydroxylamine adds –OH to cytosine; with the –OH, hydroxylated C now pairs with A instead of G.

Alkylating agents: add ethyl (–CH₂–CH₃) or methyl (–CH₃) groups

Guanine → O-6-Ethylguanine (G*) Thymine

Ethylmethane sulfonate adds an ethyl group to guanine or thymine. Modified G pairs with T above, and modified T pairs with G (not shown).

Deaminating agents: remove amine (–NH₂) groups

Cytosine → Uracil Adenine

Adenine → Hypoxanthine Cytosine

Nitrous acid modifies cytosine to uracil, which pairs with A instead of G; modifies adenine to hypoxanthine, a base that pairs with C instead of T.

(c) Insert between bases: Intercalating agents

Proflavin   Intercalated proflavin molecules

Proflavin intercalates into the double helix. This disrupts DNA metabolism, eventually resulting in deletion or addition of a base pair.

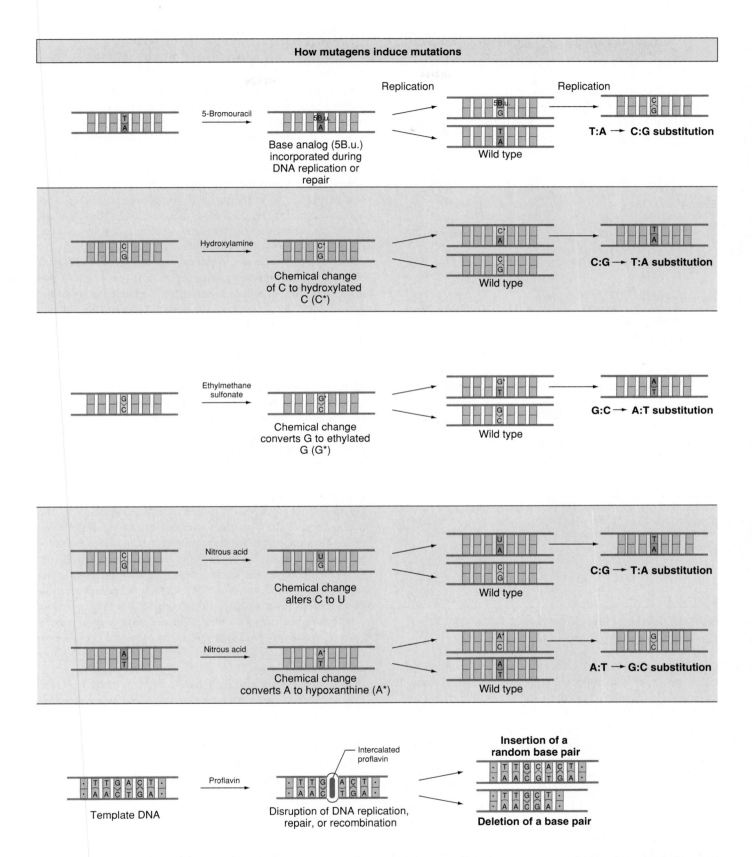

Figure 7.12 How mutagens alter DNA. (a) Base analogs incorporated into DNA may pair aberrantly, which allows the addition of incorrect nucleotides to the opposite strand during replication. **(b)** Some mutagens alter the structure of bases such that they pair inappropriately in the next round of replication. **(c)** Intercalating agents are roughly the same size and shape as a base pair of the double helix. Their incorporation into DNA produces insertions or deletions of single base pairs by unknown mechanisms.

Figure 7.13 The Ames test identifies potential carcinogens through their mutagenicity. A compound to be tested is mixed with cells of a *his⁻* strain of *Salmonella typhimurium,* as well as with a solution of rat liver enzymes (which can sometimes convert a harmless compound into a mutagen). The mixture is plated on a petri plate without histidine, so that only *his⁺* revertants are able to grow. If this plate (*left*) has more *his⁺* revertants than a control plate without the compound (*right*), the compound is considered mutagenic and a potential carcinogen.

What Mutations Tell Us About Gene Structure

One way to learn how DNA sequences along a chromosome constitute genes is to collect a large series of mutations in the same gene and analyze how these mutations are arranged with respect to each other. For this approach to be successful, it is necessary to establish that various mutations are, in fact, in the same gene. This is not a trivial exercise, as illustrated by the following situation.

Early *Drosophila* geneticists identified a large number of X-linked recessive mutations affecting the normally red wild-type eye color (Fig. 7.14). The first of these to be discovered produced the famous white eyes studied by Morgan's group. Other mutations caused a whole palette of hues to appear in the eyes: darkened shades such as garnet and ruby; bright colors such as vermilion, cherry, and coral; and lighter pigmentations known as apricot, buff, and carnation. Were these mutations multiple alleles of a single gene, or did they affect more than one gene?

Complementation Testing Reveals Whether Two Mutations Are in the Same or Different Genes

Researchers commonly define a gene as a functional unit that directs the appearance of a molecular product that in turn contributes to a particular phenotype. They can then use this definition to determine whether two mutations are in the same or different genes. If two homologous chromosomes in an individual each carries a mutation recessive to wild type, a normal phenotype will result if the mutations are in different genes. The normal phenotype occurs because the dominant wild-type alleles on each of the two homologs make up for, or **complement,** the defect in the other chromosome, generating enough of both gene products to yield a normal phenotype (Fig. 7.15a, *left*). By contrast, if the recessive mutations on the two homologous chromosomes are in the same gene, no wild-type allele of that gene exists in the individual and neither mutated copy of the gene will be able to perform the normal function. As a result, there will be no complementation and no normal gene product, and a mutant phenotype will appear (Fig. 7.15a, *right*). Ironically, a collection of mutations that do *not* complement each other is known as a **complementation group.** Geneticists often use "complementation group" as a synonym for "gene" because the mutations in a complementation group all affect the same unit of function, thus, the same gene.

A simple test based on the idea of a gene as a unit of function can determine whether or not two mutations are alleles of the same gene. You simply examine the phenotype of a heterozygous individual in which one homolog of a particular chromosome carries one of the recessive mutations and the other homolog carries the other recessive mutation. If the phenotype is wild type, the mutations cannot be in the same gene. This technique for determining whether two homologous chromosomes carry nonallelic mutations of different genes that complement each other or allelic mutations of the same gene that do not complement each other is known as **complementation testing.** For example, because a female *Drosophila* heterozygous for garnet and ruby (*garnet ruby⁺/ garnet⁺ ruby*) has wild-type brick-red eyes, it is possible to conclude that the mutations causing garnet and ruby colors are in different genes.

Complementation testing has in fact shown that garnet, ruby, vermilion, and carnation pigmentation are governed by separate genes. But chromosomes carrying mutations yielding white, cherry, coral, apricot, and buff phenotypes fail to complement each other. These mutations therefore make up different alleles of a single gene. *Drosophila* geneticists named this gene the *white,* or *w,* gene after the first mutation observed; they designate the wild-type allele as *w⁺* and the various mutations as *w¹* (the original white mutation discovered by T. H. Morgan, often simply designated as *w*), *w^{cherry}, w^{coral}, w^{apricot},* and *w^{buff}.* As an example, the eyes of a *w¹/w^{apricot}* female are a dilute apricot color. Figure 7.15b illustrates how researchers collate data from many complementation tests in a **complementation table.** Such a table helps visualize the relationships among a large group of mutants.

Figure 7.14 ***Drosophila* eye color mutations produce a variety of phenotypes.** Flies carrying different X-linked eye color mutations; a wild-type eye is at the *far right*.

(a) Complementation testing

| | **Complementation** | | **No complementation** | |

| | Defective gene | Functional gene | Defective gene | Functional gene |

Maternal chromosome

m₁
G   R

m₁
G   R

| | Functional gene | Defective gene | Defective gene | Functional gene |

Paternal chromosome

G   R
m₂

G   R
m₂

Conclusion: m₁ and m₂ are in different genes.
m₁/m₂ has wild-type phenotype because one chromosome supplies gene *G* function while the other supplies gene *R* function.

Conclusion: m₁ and m₂ are in the same gene.
m₁/m₂ has mutant phenotype because organism has no gene *G* function.

(b) A complementation table: X-linked eye color mutations in *Drosophila*

Mutation	white	garnet	ruby	vermilion	cherry	coral	apricot	buff	carnation
white	−	+	+	+	−	−	−	−	+
garnet		−	+	+	+	+	+	+	+
ruby			−	+	+	+	+	+	+
vermilion				−	+	+	+	+	+
cherry					−	−	−	−	+
coral						−	−	−	+
apricot							−	−	+
buff								−	+
carnation									−

(c) Genetic map: X-linked eye color mutations in *Drosophila*

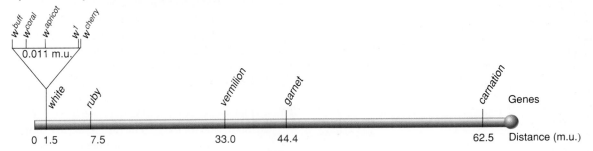

w^buff w^coral w^apricot w^1 w^cherry

0.011 m.u.

white   ruby   vermilion   garnet   carnation   Genes

0 1.5   7.5   33.0   44.4   62.5   Distance (m.u.)

Figure 7.15   Complementation testing of *Drosophila* eye color mutations. (a) A heterozygote has one mutation (m₁) on one chromosome and a different mutation (m₂) on the homologous chromosome. If the mutations are in different genes, the heterozygote will be wild type; the mutations complement each other (*left*). If both mutations affect the same gene, the phenotype will be mutant; the mutations do not complement each other (*right*). Complementation testing makes sense only when both mutations are recessive to wild type. **(b)** This complementation table reveals five complementation groups (five different genes) for eye color. Several mutations are alleles of one gene (*white*). **(c)** Recombination mapping shows that mutations in different genes are often far apart, while different mutations in the same gene are very close together.

In *Drosophila*, mutations in the *w* gene map in the same region of the X chromosome, while mutations in other eye color genes lie elsewhere on the chromosome (Fig. 7.15c). This result suggests that genes are not disjointed entities with parts spread out from one end of a chromosome to another; each gene, in fact, occupies only a relatively small, discrete area of a chromosome. Studies defining genes at the molecular level have shown that most genes consist of 1000–20,000 contiguous base pairs (bp). In humans, among the shortest genes are the roughly 500-bp-long genes that govern the production of histone proteins, while the longest gene so far identified is the Duchenne muscular dystrophy (*DMD*) gene, which has a length of more than 2 million nucleotide pairs. All known human genes fall somewhere between these extremes. To put these figures in perspective, an average human chromosome is approximately 130 million base pairs in length.

Although complementation testing makes it possible to distinguish mutations in different genes from mutations in the same gene, it does not clarify how the structure of a gene can accommodate different mutations and how these different mutations can alter phenotype in different ways. Does each mutation change the whole gene at a single stroke in a particular way, or does it change only a specific part of a gene, while other mutations alter other parts?

A Gene Is a Linear Sequence of Nucleotide Pairs That Can Mutate Independently and Recombine with Each Other

The American geneticist Seymour Benzer used recombination analysis to show that two different mutations that did not complement each other and were therefore known to be in the same gene can in fact change different parts of that gene. He reasoned that if recombination can occur not only between genes but within a gene as well, crossovers between homologous chromosomes carrying different mutations known to lie in the same gene could in theory generate a wild-type allele (Fig. 7.16). Because

Figure 7.16 How recombination within a gene could generate a wild-type allele. Suppose a gene, indicated by the region between brackets, is composed of many sites that can mutate independently. Recombination between mutations m₁ and m₂ at different sites in the same gene produces a wild-type allele and a reciprocal allele containing both mutations.

mutations affecting a single gene lie very close together on a chromosome, it is necessary to examine a very large number of progeny to see even one crossover event between them. The resolution of the experimental system must thus be extremely high, allowing rapid detection of rare genetic events. For his experimental organism, Benzer chose bacteriophage T4, a DNA virus that infects *Escherichia coli* cells (Fig. 7.17a.1). Because each T4 phage that infects a bacterium generates 100–1000 phage progeny in less than an hour, Benzer could easily produce enough rare recombinants for his analysis (Fig. 7.17a.2). Moreover, by exploiting a peculiarity of certain T4 mutations, he devised conditions that allowed only recombinant phages, and not parental phages, to proliferate.

The Experimental System: Analyzing Multiple *rII⁻* Mutations of Bacteriophage T4

Even though bacteriophages are too small to be seen without the aid of an electron microscope, they leave a trail of activity detectable with the unaided eye (Fig. 7.17a.3). To observe this trail, researchers mix a population of bacteriophage particles with a much larger number of bacteria and then pour this mixture onto a petri plate, where the cells are immobilized in a nutrient agar. If a single phage infects a single bacterial cell somewhere on this so-called **lawn** of bacteria, the cell produces and releases progeny viral particles that diffuse away to infect adjacent bacteria, which in turn produce and release yet more phage progeny. With each release of virus particles, the bacterial host cell dies. Thus, several cycles of phage infection, replication, and release produce a circular cleared area in the plate, called a **plaque**, devoid of living bacterial cells. The rest of the petri plate surface is covered by an opalescent lawn of living bacteria. Most plaques contain from 1 million to 10 million viral progeny of the single bacteriophage that originally infected a cell in that position on the petri plate. Sequential dilution of phage-containing solutions makes it possible to measure the number of phages in a particular plaque and arrive at a countable number of viral particles (Fig. 7.17a.4).

In the mid-1950s through the early 1960s when Benzer was looking for genetic traits associated with bacteriophage T4, he found mutants that, when added to a lawn of *E. coli* B strain bacteria, produced larger plaques with sharper, more clearly rounded edges than those produced by the wild-type bacteriophage (Fig. 7.17b). Because these changes in plaque morphology seemed to result from the abnormally rapid lysis of the host bacteria, Benzer named the mutations giving rise to this phenotype *r* for "rapid lysis." Many *r* mutations map to a region of the T4 chromosome known as the *rII* region; these are called *rII⁻* mutations. An additional property of *rII⁻* mutations makes them ideal for the genetic **fine structure mapping** (the mapping of mutations within a gene) undertaken by Benzer. Wild-type *rII⁺* bacteriophages form plaques of normal shape and size on cells of both the *E. coli* B strain and a strain known as *E. coli* K(λ). The *rII⁻* mutants, however, have an altered host range: They cannot form plaques with *E. coli* K(λ) cells, although as we have seen, they produce large, unusually distinct plaques with *E. coli* B cells (Fig. 7.17b). The reason that *rII⁻* mutants are unable to infect cells of the K(λ) strain was not clear to Benzer, but their inability to do so enabled him to develop an extremely simple and effective test for *rII$^+$* gene function that he could use to understand gene structure.

The *rII* Region Has Two Genes

Before he could check whether two mutations in the same gene could recombine, Benzer had to be sure he was really looking at two mutations in the same gene. To verify this, he performed customized complementation tests tailored to two significant characteristics of bacteriophage T4: They are haploid (that is, each phage carries a single T4 chromosome), and they can replicate only in a host bacterium. Because T4 phages are haploid, Benzer needed to ensure that two T4 chromosomes entered the same bacterial cell in order to test for complementation between the mutations. In his complementation tests, he simultaneously infected *E. coli* K(λ) cells with two types of T4 chromosomes—one carried one *rII$^-$* mutation, the other carried a different *rII$^-$* mutation—and then looked for cell lysis (Fig. 7.17c). To ensure that the two kinds of phages would infect almost every bacterial cell, he added many more phages of each type than there were bacteria. If the two *rII$^-$* mutations were in different genes, each of the mutant T4 chromosomes would supply one wild-type *rII$^+$* gene function, making up for the lack of that function in the other chromosome and resulting in lysis. On the other hand, if the two *rII$^-$* mutations were in the same gene, no plaques would appear, because neither mutant chromosome would be able to supply the missing function.

Benzer had to satisfy one final experimental requirement: For the complementation test to be meaningful, he had to make sure that the two *rII$^-$* mutations under analysis were each recessive to wild type and did not interact with each other to produce an *rII$^-$* phenotype dominant to wild type. He checked these points by a control experiment in which he placed the two *rII$^-$* mutations on the same chromosome and then simultaneously infected *E. coli* K(λ) with these double *rII$^-$* mutants and with wild-type phages (Fig. 7.17c). If the mutations were recessive and did not interact with each other, the cells would lyse, in which case the complementation test would be interpretable.

The significant distinction between the actual complementation test and the control experiment is in the placement of the two *rII$^-$* mutations. In the complementation test, one *rII$^-$* mutation is on one chromosome, while the other *rII$^-$* mutation is on the other chromosome; two mutations arranged in this way are said to be in the *trans* configuration. In the control experiment, the two mutations are on the same chromosome, in the so-called *cis* configuration. The complete test, including the complementation test and the control experiment, is known as a *cis-trans* test. Benzer called any complementation group identified by the cis-trans test a **cistron,** and some geneticists still use the term "cistron" as a synonym for "gene."

Tests of many different pairs of *rII$^-$* mutations showed that they fall into two complementation groups: *rIIA* and *rIIB*. With this knowledge, Benzer could look for two mutations in the same gene and then see if they ever recombine to produce wild-type progeny.

A Gene Is Composed of Mutable Subunits That Can Recombine

When Benzer infected *E. coli* B strain bacteria with a mixture of phages carrying different mutations in the same gene (*rIIA$_1$* and *rIIA$_2$*, for example), he did observe the appearance of *rII$^+$* progeny (Fig. 7.17d). He knew these wild-type progeny resulted from recombination and not reverse mutations because the frequencies of the *rII$^+$* phage particles he observed were much higher than the frequencies of *rII$^+$* revertants seen among progeny produced by infecting B strain bacteria with either mutant alone. On the basis of these observations, he drew three conclusions about gene structure: (1) A gene consists of different parts that can each mutate; (2) recombination between different mutable sites in the same gene can generate a normal, wild-type allele; and (3) a gene performs its normal function only if all of its components are wild type.

From what we now know about the molecular structure of DNA, this all makes perfect sense. The separate mutable sites are different nucleotides within the gene, and recombination can occur between nucleotides within a gene as well as between genes.

But how are the nucleotide pairs that make up a gene arranged—in a continuous row or dispersed in precise patterns around the genome? And do the various mutations that affect gene function alter many different nucleotides or only a small subset within each gene?

A Gene Is a Discrete Linear Set of Nucleotide Pairs

To answer these questions about the arrangement of nucleotides in a gene, Benzer eventually obtained thousands of spontaneous and mutagen-induced *rII$^-$* mutations that he mapped with respect to each other. To map the location of a thousand mutants through comparisons of all possible two-point crosses, he would have had to set up a million ($10^3 \times 10^3$) matings. But by taking advantage of deletion mutations, he could obtain the same information with far fewer crosses.

Using Deletions to Map Mutations and Define a Gene

As we have seen, deletions are mutations that remove contiguous nucleotide pairs along a DNA molecule. In crosses between bacteriophages carrying a mutation and bacteriophages carrying deletions of the corresponding region, no wild-type recombinant progeny can arise, because neither chromosome carries the proper information at the location of the mutation. However, if the mutation lies outside the region deleted from the homologous chromosome, wild-type progeny can appear (Fig. 7.18a). This is true whether the mutation is a **point mutation,** that is, a mutation of one nucleotide, or is itself a deletion. Crosses between any uncharacterized mutation and a known deletion thus immediately reveal whether or not the mutation resides in the region deleted from the other phage chromosome. This method of deletion mapping provides a rapid way of finding the general location of a mutation. Using a series of overlapping deletions, Benzer divided the *rII* region into a series of intervals. He could then assign any point mutation to an interval by observing

How Benzer Analyzed the *rII* Genes of Bacteriophage T4.

(a.1)

Head
Tail
Tail fibers
100 nm

Viral chromosome
Sheath
Tail fibers

(a.2)

1. Phage injects its DNA into host cell.

Host chromosome

2. Phage proteins synthesized; DNA replicated. Host chromosome degraded.

4. Lysis of host cell.

3. Assembly of phages within host cell.

(a.3)

(a.4)

Pipette out
0.01 ml   0.01 ml   0.1 ml

0.1 ml

Add plating bacteria

1 ml 1 ml   1 ml   1 ml
Concentrated solution of bacteriophages
Tubes containing medium without phage

25 plaques

(a) Working with bacteriophage T4

1. Bacteriophage T4 (at a magnification of approximately 100,000×) and in an artist's rendering. The viral chromosome is contained within a protein head. Other proteinaceous parts of the phage particle include the tail fibers, which help the phage attach to host cells, and the sheath, a conduit for injecting the phage chromosome into the host cell.

2. The lytic cycle of bacteriophage T4. A single phage particle infects a host cell; the phage DNA replicates and directs the synthesis of viral protein components using the machinery of the host cell; the new DNA and protein components assemble into new bacteriophage particles. Eventual lysis of the host cell releases up to 1000 progeny bacteriophages into the environment.

3. Clear plaques of bacteriophages in a lawn of bacterial cells. A mixture of bacteriophages and a large number of bacteria are poured onto the agar surface of a petri plate. Uninfected bacterial cells grow, producing an opalescent lawn. A bacterial cell infected

by even a single bacteriophage will lyse and release progeny bacteriophages, which can infect adjacent bacteria. Several cycles of infection result in a plaque: a circular cleared area containing millions of bacteriophages genetically identical to the one that originally infected the bacterial cell.

4. Counting bacteriophages by serial dilution. A small sample of a concentrated solution of bacteriophages is transferred to a test tube containing fresh medium, and a small sample of this dilution is transferred to another tube of fresh medium. Successive repeats of this process increase the degree of dilution. A sample of the final dilution, when mixed with bacteria and poured on the agar of a petri plate, yields a countable number of plaques from which it is possible to extrapolate back and calculate the number of bacteriophage particles in the starting solution. The original 1 ml of solution in this illustration contained roughly 2.5×10^7 bacteriophages.

(b) Phenotypic properties of *rII⁻* mutants of bacteriophage T4

1. *rII⁻* mutants, when plated on *E. coli* B cells, produce plaques that are larger and more distinct (with sharper edges) than plaques formed by *rII⁺* wild-type phage.

2. *rII⁻* mutants are particularly useful for looking at rare recombination events, because they have an altered host range. In contrast to *rII⁺* wild-type phages, *rII⁻* mutants cannot form plaques in lawns of *E. coli* strain K(λ) host bacteria.

(b.1)

rII⁺
rII⁻
rII⁺

(b.2)

T4 strain	*E. coli* strain	
	B	K(λ)
rII⁻	Large, distinct	No plaques
rII⁺	Small, fuzzy	Small, fuzzy

(c.1) Complementation test
(*trans* configuration)

rllA nonfunctional

rllB functional

rllA functional

rllB functional

No complementation
- no cell lysis
- no phage progeny

Complementation
- cell lysis
- phage progeny

(c) A customized complementation test between *rll⁻* mutants of bacteriophage T4

1. *E. coli* K(λ) cells are simultaneously infected with an excess of two different *rll⁻* mutants (m$_1$ and m$_2$). Inside the cell, the two mutations will be in *trans;* that is, they lie on different chromosomes. If the two mutations are in the same gene, they will affect the same function and cannot complement each other, so no progeny phages will be produced. If the two mutations are in different genes (*rllA* and *rllB*), they will complement each other, leading to progeny phage production and cell lysis.

(c.2) Control
(*cis* configuration)

If mutations are recessive, cell lysis.

If mutations are dominant, no cell lysis.

━━ Gene *rllA*　　━━ Gene *rllB*

2. An important control for this complementation test is the simultaneous infection of *E. coli* K(λ) bacteria with a wild-type T4 strain and a T4 strain containing both m$_1$ and m$_2$. Inside the infected cells, the two mutations will be in *cis;* that is, they lie on the same chromosome. Release of phage progeny shows that both mutations are recessive to wild type and that there is no interaction between the mutations that prevents the cells from producing progeny phages. Complementation tests are meaningful only if the two mutations tested are both recessive to wild type.

(d.1) Recombination test

Recombination

rll⁺
wild type

rllA$_1$ + *rllA$_2$*
double mutant

Forms plaques on *E. coli* K(λ)

No plaques on *E. coli* K(λ)

(d) Detecting recombination between two mutations in the same gene

1. *E. coli* B cells are simultaneously infected with a large excess of two different *rllA⁻* mutants (*rllA$_1$* and *rllA$_2$*). If no recombination between the two *rllA⁻* mutations takes place, progeny phages will carry either of the original mutations and will be phenotypically *rll⁻*. If recombination between the two mutations occurs, one of the products will be an *rll⁺* recombinant, while the reciprocal product will be a double mutant chromosome containing both *rllA$_1$* and *rllA$_2$*. When the phage progeny subsequently infect *E. coli* K(λ) bacteria, only *rll⁺* recombinants will be able to form plaques.

(d.2) Control

No plaques on
E. coli K(λ)

2. As a control, *E. coli* B cells are infected with a large amount of only one kind of mutant (*rllA$_1$* or *rllA$_2$*). The only *rll⁺* phages that can result are revertants of either mutation. This control experiment shows that such revertants are extremely rare and can be ignored among the *rll⁺* progeny made in the recombination experiment at the *left* in part d.1. Even if the two *rllA⁻* mutations are in adjacent base pairs, the number of *rll⁺* recombinants obtained is more than 100 times higher than the number of *rll⁺* revertants the cells infected by a single mutant could produce.

(a) Using deletions for rapid mapping

(b) Portion of the *rIIA* deletion map at increasing resolutions

(c) Fine structure of the *rII* region

Figure 7.18 Fine structure mapping of the bacteriophage T4 *rII* genes. (a) A phage cross between a point mutation and a deletion removing the DNA at the position of the mutation cannot yield wild-type recombinants. The same is true if two different deletion mutations overlap each other. **(b)** Large deletions divide the *rII* locus into regions; finer deletions divide each region into subsections. Point mutations, such as 271 (*in red at bottom*), map to region 3 if they do not recombine with deletions PT1, PB242, or A105 but do recombine with deletion 638 (*top*). Point mutations can be mapped to subsections of region 3 using other deletions (*middle*). Recombination tests map point mutations in the same subregion (*bottom*). Point mutations 201 and 155 cannot recombine to yield wild-type recombinants because they affect the same nucleotide pair. **(c)** Benzer's fine structure map. Hot spots are locations with many independent mutations that cannot recombine with each other.

whether or not it recombined to give *rII*⁺ progeny when crossed with the series of deletions (Fig. 7.18b).

Benzer mapped 1612 spontaneous point mutations and several deletions in the *rII* locus of bacteriophage T4 through recombination analysis. He first used recombination to determine the relationship between the deletions. He next found the approximate location of individual point mutations by observing which deletions could recombine with each mutant to yield wild-type progeny. He then performed recombination tests between all point mutations known to lie in the same small region

of the chromosome. These results produced a map of the "fine structure" of the region (Fig. 7.18c).

From the observation that the number of mutable sites in the *rII* region is very close to the number of nucleotides estimated to be in this region, Benzer inferred that a mutation can arise from the change of a single nucleotide, and that recombination can occur between adjacent nucleotide pairs. From the observation that mutations within the *rII* region form a self-consistent, linear recombination map, he concluded that a gene is composed of a continuous linear sequence of nucleotide pairs within the DNA. And from observations that the positions of mutations in the *rIIA* gene did not overlap those of the *rIIB* gene, he inferred that the nucleotide sequences composing those two genes are separate and distinct. A *gene* is thus a linear set of nucleotide pairs, located within a discrete region of a chromosome, that serves as a unit of function.

Hot Spots Are More Prone to Mutation

Some sites within a gene spontaneously mutate more frequently than others and as a result are known as **hot spots** (Fig. 7.18c). The existence of hot spots suggests that certain nucleotides can be altered more readily than others. Treatment with mutagens also turns up hot spots, but because mutagens have specificities for particular nucleotides, the highly mutable sites that turn up with various mutagens are often at different positions in a gene than the hot spots resulting from spontaneous mutation.

Nucleotides are chemically the same whether they lie within a gene or in the DNA between genes, and as Benzer's experiments show, the molecular machinery responsible for mutation and recombination does not discriminate between those nucleotides that are *intragenic* (within a gene) and those that are *intergenic* (between genes). The main distinction between DNA within and DNA outside a gene is that the array of nucleotides composing a gene has evolved a function that determines phenotype. We now describe how geneticists discovered what that function is.

What Mutations Tell Us About Gene Function

In the early 1900s, the British physician Dr. Archibald Garrod showed that a human genetic disorder known as *alkaptonuria* is determined by the recessive allele of an autosomal gene. Garrod analyzed family pedigrees and performed biochemical analyses on family members with and without the trait. The urine of people with alkaptonuria turns black on exposure to air. Garrod found that a substance known as homogentisic acid, which blackens upon contact with oxygen, accumulates in the urine of alkaptonuria patients. These same people excrete all of the homogentisic acid they ingest, while people not suffering from the condition do not have any homogentisic acid in their urine even after ingesting the substance. From these observations, he concluded that people with alkaptonuria are incapable of metabolizing homogentisic acid to the breakdown products generated by normal individuals (Fig. 7.19). Because many biochemical reactions within the cells of organisms are catalyzed by enzymes, Garrod hypothesized that lack of the enzyme that breaks down homogentisic acid is the cause of alkaptonuria; in the absence of this enzyme, the acid accumulates and causes the urine to turn black on contact with oxygen. He called this condition an "inborn error of metabolism."

Garrod studied several other inborn errors of metabolism and suggested that all arose from mutations that prevented a particular gene from producing an enzyme required for a specific

Figure 7.19 Alkaptonuria: An inborn error of metabolism. The biochemical pathway in humans that degrades phenylalanine and tyrosine via homogentisic acid (HA). In alkaptonuria patients, the enzyme HA hydroxylase is not functional so it does not catalyze the conversion of HA to maleylacetoacetic acid. As a result, HA, which oxidizes to a black compound, accumulates in the urine.

biochemical reaction. In today's terminology, the wild-type allele of the gene would allow production of functional enzyme (in the case of alkaptonuria, the enzyme is homogentisic acid oxidase), whereas the mutant allele would not. Because the single wild-type allele in heterozygotes generates sufficient enzyme to prevent accumulation of homogentisic acid and thus alkaptonuria, the mutant allele is recessive.

The One Gene, One Enzyme Hypothesis: A Gene Contains the Information for Producing a Specific Enzyme

George Beadle and Edward Tatum carried out a series of experiments on the bread mold *Neurospora crassa* (whose life cycle was described in Chapter 5) during the 1940s that demonstrated a direct relation between genes and the enzymes that catalyze specific biochemical reactions. Their strategy was simple. They first isolated a number of mutations that disrupted synthesis of the amino acid arginine, a compound needed for *Neurospora* growth. They next hypothesized that different mutations blocked different steps in a particular **biochemical pathway:** the orderly series of reactions that allows *Neurospora* to obtain simple molecules from the environment and convert them step by step into successively more complicated molecules culminating in arginine. A study showing that mutations can block discrete biochemical steps in the pathway resulting in arginine would constitute strong evidence that different genes control the appearance of different enzymes, each of which catalyzes a specific reaction in the biochemical pathway.

The Experimental Evidence for "One Gene, One Enzyme"

Figure 7.20a illustrates the experiments Beadle and Tatum performed to test their hypothesis. They first obtained a set of mutagen-induced mutations that prevented *Neurospora* from synthesizing arginine. Cells with any one of these mutations were unable to make arginine and could therefore grow on a minimal medium containing salt and sugar only if it had been supplemented with arginine. A mutant microorganism that can grow on minimal medium only if it has been supplemented with one or more growth factors not required by wild-type strains is known as an **auxotroph.** The cells just mentioned were arginine auxotrophs. (In contrast, a wild-type cell that can synthesize a particular growth factor and thus grow in its absence on minimal medium is a **prototroph** for that factor. In a more general meaning, *prototroph* refers to a wild-type cell that can synthesize all required growth factors and thus grow on minimal medium alone.) Recombination analyses located the auxotrophic arginine-

blocking mutations in four distinct regions of the genome, and complementation tests showed that each of the four regions correlated with a different complementation group. On the basis of these results, Beadle and Tatum concluded that at least four genes support the biochemical pathway for arginine synthesis. They named the four genes *ARG-E, ARG-F, ARG-G,* and *ARG-H.*

They next asked whether any of the mutant *Neurospora* strains could grow in minimal medium supplemented with any of a number of compounds that are intermediates in the biochemical pathway leading to arginine, instead of with arginine itself. If a mutant grew on medium supplemented in such a way, it would indicate that *Neurospora* is able to convert the intermediate compound into arginine. After finding three such intermediates—ornithine, citrulline, and argininosuccinate—Beadle and Tatum compiled a table describing which arginine auxotrophic mutants were able to grow on minimal medium supplemented with each of the intermediates (Fig. 7.20b).

(a) Isolation of arginine auxotrophs

(b) Growth response if nutrient is added to minimal medium

Mutant strain	Nothing	Supplements			
		Ornithine	Citrulline	Arginino-succinate	Arginine
Wildtype: *Arg⁺*	+	+	+	+	+
Arg-E⁻	−	+	+	+	+
Arg-F⁻	−	−	+	+	+
Arg-G⁻	−	−	−	+	+
Arg-H⁻	−	−	−	−	+

(c) Inferred biochemical pathway

Gene:

ARG-E   ARG-F   ARG-G   ARG-H

Enzymes:

Acetylornithinase   Ornithine transcarbamylase   Argininosuccinate synthetase   Argininosuccinate lyase

Reactions:

N-Acetylornithine ⟶ Ornithine ⟶ Citrulline ⟶ Argininosuccinate ⟶ Arginine

Carbamyl phosphate   Aspartate

Figure 7.20 Beadle and Tatum's experiment supporting the "one gene, one enzyme" hypothesis. (a) Beadle and Tatum mated an X-ray-mutagenized strain of *Neurospora* with another strain, and they isolated haploid ascospores that grew on complete medium. Cultures that failed to grow on minimal medium were nutritional mutants. Nutritional mutants that could grow on minimal medium plus arginine were *arg⁻* auxotrophs. **(b)** The ability of wild-type and mutant strains to grow on minimal medium supplemented with intermediates in the arginine pathway. **(c)** Each of the four *ARG* genes encodes an enzyme needed to convert one intermediate to the next in the pathway.

Interpretation of Results: Genes Encode Enzymes

On the basis of these results, Beadle and Tatum proposed a model of how *Neurospora* cells synthesize arginine (Fig. 7.20c). In the linear progression of biochemical reactions by which a cell constructs arginine from the constituents of minimal medium, each intermediate is both the product of one step and the substrate for the next. Each reaction in the precisely ordered sequence is catalyzed by a specific enzyme, and the presence of each enzyme depends on one of the four *ARG* genes. A mutation in one gene blocks the pathway at a particular step because the cell lacks the corresponding enzyme; this lack prevents the cell from growing in the absence of arginine. Supplementing the medium with any intermediate that occurs beyond the blocked reaction will restore growth, because the organism has all the remaining enzymes required to convert the intermediate to arginine. Supplementation with an intermediate that occurs before the missing enzyme will not work because the cell is unable to convert the intermediate into arginine.

By inference, because each mutation abolishes the cell's ability to make an enzyme capable of catalyzing a certain reaction, each gene controls the synthesis or activity of an enzyme, or as stated by Beadle and Tatum: one gene, one enzyme. Of course, the gene and the enzyme are not the same thing; rather, the sequence of nucleotides in a gene contains information that somehow encodes the structure of an enzyme molecule.

While the analysis of the arginine pathway studied by Beadle and Tatum was straightforward, studies of biochemical pathways are not always so easy to interpret, because some biochemical pathways are not linear progressions of stepwise reactions. For example, a branching pathway occurs if different enzymes act on the same intermediate to convert it into two different end products. If the cell requires both of these end products for growth, a mutation in a gene encoding any of the enzymes required to synthesize the intermediate would make the cell dependent on the addition to minimal medium of both end products. A second possibility is that a cell might employ either of two independent, parallel pathways to synthesize a needed end product. In such a case, a mutation in a gene encoding an enzyme in one of the pathways would be without effect. Only a cell with mutations for genes specifying enzymes in both pathways would display an aberrant phenotype.

Even with nonlinear progressions such as these, careful genetic analysis can reveal the nature of the biochemical pathway on the basis of Beadle and Tatum's insight that genes encode proteins.

Genes Specify the Identity and Order of Amino Acids in Polypeptide Chains

Although the one gene, one enzyme hypothesis was a critical advance in our understanding of how genes influence phenotype, it is an oversimplification. Not all genes govern the construction of enzymes active in biochemical pathways. Enzymes are one class of the molecules known as proteins, and cells contain many different kinds of protein, only some of which behave as enzymes. Among the other types are proteins that provide shape and rigidity to a cell, proteins that transport molecules in and out of cells, proteins that help fold DNA into chromosomes, and proteins that act as hormonal messengers. Genes direct the synthesis of all proteins, enzymes and nonenzymes alike. Moreover, as we see next, genes actually determine the construction of polypeptides, and because some proteins are composed of more than one type of polypeptide, more than one gene determines their construction.

Proteins Are Linear Polymers of Amino Acids Linked by Peptide Bonds

To understand how the information in a gene specifies the production of a particular protein, it is necessary to review the chemical composition of proteins. Proteins are polymers composed of building blocks known as **amino acids.** Cells use mainly 20 different amino acids to synthesize the proteins they need. All of these amino acids have certain basic features, encapsulated by the formula NH_2–CHR–$COOH$ (Fig. 7.21a). The –$COOH$ component, also known as *carboxylic acid,* is, as the name implies, acidic; the –NH_2 component, also known as an *amino group,* is basic. The R refers to side chains that distinguish each of the 20 amino acids (Fig. 7.21b). An R group can be as simple as a hydrogen atom (in the amino acid glycine), or it can be as complex as a benzene ring (in phenylalanine). Some side chains are relatively neutral and nonreactive, others are acidic, and still others are basic.

During protein synthesis, a cell's protein-building machinery links amino acids by constructing covalent **peptide bonds** that join the –$COOH$ group of one amino acid to the –NH_2 group of the next (Fig. 7.21c). A pair of amino acids connected in this fashion is a **dipeptide;** several amino acids linked together constitute an **oligopeptide.** The amino-acid chains that make up proteins contain hundreds to thousands of amino acids joined by peptide bonds and are known as **polypeptides.** Proteins are thus linear polymers of amino acids. Like the chains of nucleotides in DNA, polypeptides have a chemical polarity. One end of a polypeptide is called the **N terminus** because it contains a free amino group that is not connected to any other amino acid. The other end of the polypeptide chain is the **C terminus,** because it contains a free carboxylic acid group.

The Primary Business of Most Genes Is to Specify the Amino-Acid Sequence of a Polypeptide

Each protein is composed of a unique sequence of amino acids. In fact, the chemical properties that enable structural proteins to give a cell its shape, or enzymes to catalyze specific reactions, or hormones to act as messengers are a direct consequence of the identity, number, and linear order of amino acids in the protein.

If genes encode proteins, then at least some mutations could be changes in a gene that alter the proper sequence of amino acids in the protein encoded by that gene. In the mid-1950s, Vernon Ingram began to establish what kinds of changes particular mutations cause in the corresponding protein. Using recently developed techniques for determining the sequence of amino acids in a protein, he compared the amino-acid sequence of the normal adult form of hemoglobin (HbA) with that of

(a) Generic amino acid structure

Amino (–NH$_2$) group

CHR group

Carboxyl (–COOH) group

(c) Peptide bond formation

AA1   AA2   AA3

N terminus   C terminus

Peptide bonds

(b) Amino acids with nonpolar R groups

Amino acids with uncharged polar R groups

Amino acids with basic R groups

Amino acids with acidic R groups

Figure 7.21   Proteins are chains of amino acids linked by peptide bonds. (a) Amino acids contain a basic amino group (–NH$_2$), an acidic carboxylic acid group (–COOH), and a ⟩CHR moiety, where R stands for one of the 20 different side chains. **(b)** Table of amino acids commonly found in proteins, arranged according to the properties of their R groups. **(c)** One molecule of water is lost when a covalent amide linkage (a peptide bond) is formed between the –COOH of one amino acid and the –NH$_2$ of the next amino acid. Polypeptides such as the tripeptide shown here have polarity; they extend from an N terminus (with a free amino group) to a C terminus (with a free carboxylic acid group).

(a) From mutation to phenotype

Normal individual

Sickle-cell individual

N ... N ...

Valine, Histidine, Leucine, Threonine, Proline, Glutamic acid

Valine, Histidine, Leucine, Threonine, Proline, Valine, Glutamic acid

1. The polypeptide: the β chain of hemoglobin

Glutamic acid

Valine

2. The protein: (made of two α and two β chains)

Free proteins

Long fibers

3. Red blood cell making thousands of hemoglobin molecules

Disk-shaped

Sickle-shaped

(b) Sickle-cell anemia is pleiotrophic

Sickling of red blood cells

Rapid destruction of sickle cells	Clumping of cells; interference with circulation	Accumulation of red blood cells in spleen
Anemia	Local failures in blood supply	Enlargement and damage to spleen
Fatigue, heart damage, overactivity of bone marrow	Damage to heart, kidney, muscle/joints, brain, lung, gastrointestinal tract	

(c) β-chain substitutions/variants

Amino-acid position

	1	2	3 ···	6	7 ···	26 ···	63 ···	125 ···	146
Normal (HbA)	Val	His	Leu	Glu	Glu	Glu	His	Val	Glu	His
HbS	Val	His	Leu	Val	Glu	Glu	His	Val	Glu	His
HbC	Val	His	Leu	Lys	Glu	Glu	His	Val	Glu	His
HbG San Jose	Val	His	Leu	Glu	Gly	Glu	His	Val	Glu	His
HbE	Val	His	Leu	Glu	Glu	Lys	His	Val	Glu	His
HbM Saskatoon	Val	His	Leu	Glu	Glu	Glu	Tyr	Val	Glu	His
Hb Zurich	Val	His	Leu	Glu	Glu	Glu	Arg	Val	Glu	His
HbM Milwaukee 1	Val	His	Leu	Glu	Glu	Glu	His	Glu	Glu	His
HbDβ Punjab	Val	His	Leu	Glu	Glu	Glu	His	Val	Gln	His

Figure 7.22   The molecular basis of sickle-cell and other anemias. (a) Substitution of glutamic acid with valine at the sixth amino acid from the N terminus affects the three-dimensional structure of the β chain of hemoglobin. Hemoglobins incorporating the mutant β chain form aggregates that cause red blood cells to sickle. **(b)** Red blood cell sickling has many phenotypic effects. **(c)** Other mutations in the β-chain gene also cause anemias.

hemoglobin in the bloodstream of people homozygous for the mutation that causes sickle-cell anemia (HbS). Remarkably, he found only a single amino-acid difference between the wild-type and mutant proteins (Fig. 7.22a). The sixth amino acid from the N terminus of one of the two polypeptide chains making up hemoglobin was glutamic acid in normal individuals but valine in sickle-cell patients. Ingram thus established that a mutation substituting one amino acid for another had the power to change the structure and function of hemoglobin and thereby alter the phenotype from normal to sickle-cell anemia (Fig. 7.22b). We now know that the glutamic acid–to-valine change affects the solubility of hemoglobin within the red blood cell. The hemoglobin molecule consists of two alpha (α) and two beta (β) chains complexed in a tetramer of four chains. At low concentrations of oxygen, the less soluble sickle-cell form of hemoglobin aggregates into long chains that deform the red blood cell (review Fig. 7.22a).

Because people suffering from a variety of inherited anemias also had defective hemoglobin molecules, Ingram and other geneticists were able to determine how a large number of different mutations affected the amino-acid sequence of hemoglobin (Fig. 7.22c). Most of the altered hemoglobins had a change in only one amino acid. In different patients, the alter-

ation was generally in different amino acids, but occasionally, two independent mutations resulted in different substitutions for the same amino acid. Geneticists use the term **missense mutation** to describe a genetic alteration that causes the substitution of one amino acid for another.

The Sequence of Amino Acids in Polypeptides Determines a Protein's Three-Dimensional Shape and Thus Its Function

Despite the uniform nature of protein construction—a line of amino acids joined by peptide bonds—each type of polypeptide folds into a unique three-dimensional shape. The linear sequence of amino acids within a polypeptide is its **primary structure.** Each unique primary structure places constraints on how a chain can arrange itself in three-dimensional space. Because the R groups distinguishing the 20 amino acids have dissimilar chemical properties, some amino acids form hydrogen bonds or electrostatic bonds when brought into proximity with other amino acids. Nonpolar amino acids, for example, may become associated with each other by hydrophobic interactions that "hide" them from water; while two cysteine amino acids can form covalent disulfide bridges (–S–S–) through the oxidation of their –SH groups. All of these interactions

(a) Interactions determining polypeptide structure

(b) 1° structure

(c) 2° structures

α helix

β-pleated sheets

(d) 3° structure

Myoglobin

Figure 7.23 Levels of polypeptide structure. (a) Covalent and noncovalent interactions determine the structure of a polypeptide. **(b)** A polypeptide's primary (1°) structure is its amino-acid sequence. **(c)** Localized regions form secondary (2°) structures such as α helices and β-pleated sheets. **(d)** The tertiary (3°) structure is the complete three-dimensional arrangement of a polypeptide. In this portrait of myoglobin, the iron-containing heme group, which carries oxygen, is in *red,* while the polypeptide itself is in *green.*

(Fig. 7.23a) help stabilize the polypeptide in a specific three-dimensional conformation. The primary structure (Fig. 7.23b) determines three-dimensional shape by generating localized regions with a characteristic geometry known as **secondary structure** (Fig. 7.23c), as well as the other folds and twists that together with the secondary structure produce the ultimate three-dimensional **tertiary structure** of the entire polypeptide (Fig. 7.23d). Normal tertiary structure—the way a long chain of amino acids naturally folds in three-dimensional space under physiological conditions—is known as a polypeptide's *native configuration.* Various forces, including hydrogen bonds, electrostatic bonds, hydrophobic interactions, and disulfide bridges, help stabilize the native configuration.

It is worth repeating that primary structure—the sequence of amino acids in a polypeptide—directly determines secondary and tertiary structures because inherent in the linear sequence of amino acids is the information required for the chain to fold into its native configuration. Several examples illustrate this fact. For instance, protein biologists know that a polypeptide chain is unable to form a particular kind of secondary structure called an α helix in the vicinity of a proline. On the basis of this and other structural information they have devised computer programs that examine the primary sequence of a protein for amino acids that favor or disrupt α-helix formation. They can use these programs

to predict the α-helical composition of a polypeptide in its native state with considerable accuracy. In a different example, many proteins unfold, or become **denatured,** when exposed to urea and mercaptoethanol or to increasing heat or pH because these treatments disrupt the interactions between amino acids that normally stabilize the secondary and tertiary structures. When conditions return to normal, many proteins spontaneously refold into their native configuration without help from other agents. No other information beyond the primary structure is needed to achieve the proper three-dimensional shape of such proteins.

Some Proteins Consist of More than One Polypeptide

Certain proteins, such as the rhodopsin that promotes black-and-white vision, consist of one polypeptide. Many others, however, such as the lens crystallin protein, which provides rigidity and transparency to the lenses of our eyes, or the hemoglobin molecule that transports oxygen to and from our tissues, are composed of two or more polypeptide chains that aggregate in a specific way (Fig. 7.24a). The individual polypeptides in an aggregate are known as *subunits* and the complex of subunits is often referred to as a *multimer.* The three-dimensional configuration of subunits in a multimer is a complex protein's **quaternary structure.** The same forces that stabilize the native form of a polypeptide (that is, hydrogen bonds, electrostatic bonds,

(a) A multimer with identical subunits

β2 lens crystallin

Two identical subunits

β2 lens crystallin gene

(b) A multimer with nonidentical subunits

Hemoglobin

Two α subunits   Two β subunits

*Hb*α gene   *Hb*β gene

(c) One polypeptide in different proteins

IL-4 Receptor   IL-2 Receptor   IL-7 Receptor

IL-4   IL-2   IL-2Rα   IL-7

IL-4R   γ   IL-2Rβ   γ   IL-7R   γ

γ   Gamma-chain subunit

Defective gamma chain → XSCID

(d) Microtubules: large assemblies of subunits

α tubulin
β tubulin — Tubulin dimer

Assembly of microtubules: mitotic metaphase

Microtubule

Chromosomes aligned on spindle apparatus

Disassembly of microtubules: mitotic telophase

Spindle apparatus breaks down

Figure 7.24 Multimeric proteins. (a) β2 lens crystallin contains two copies of one kind of subunit; the subunits are all the product of a single gene. The peptide backbones of the two subunits are shown in different shades of *purple*. **(b)** Hemoglobin is composed of two different kinds of subunits, each encoded by a different gene. **(c)** Three distinct protein receptors for the immune-system molecules called interleukins (ILs; *purple*). All contain a gamma (γ) chain (*yellow*), plus other receptor-specific polypeptides (*green*). A mutant γ chain blocks the function of all three receptors, leading to XSCID, a complete failure of the immune system. **(d)** One α-tubulin and one β-tubulin polypeptide associate to form a tubulin dimer. Many tubulin dimers form a single microtubule. The mitotic spindle is an assembly of many microtubules.

hydrophobic interactions, and disulfide bridges) also contribute to the maintenance of quaternary structure. As Fig. 7.24a shows, in some multimers, the two or more interacting subunits are identical polypeptides. These identical chains are encoded by one gene. In other multimers, by contrast, more than one kind of polypeptide makes up the protein (Fig. 7.24b). The different polypeptides in these multimers are encoded by different genes.

Alterations in just one kind of subunit, caused by a mutation in a single gene, can affect the function of a multimer. The adult hemoglobin molecule, for example, consists of two α and two β subunits, with each type of subunit determined by a different gene—one for the α chain and one for the β chain. A mutation in the *Hbβ* gene resulting in an amino-acid switch at position 6 in the *β* chain causes sickle-cell anemia. Similarly, if several multimeric proteins share a common subunit, a single

mutation in the gene encoding that subunit may affect all the proteins simultaneously. An example is an X-linked mutation in mice and humans that incapacitates several different proteins all known as interleukin (IL) receptors. Since all of these receptors are essential to the normal function of immune-system cells that fight infection and generate immunity, this one mutation causes the life-threatening condition known as X-linked severe combined immune deficiency (XSCID; Fig. 7.24c).

The polypeptides of complex proteins can assemble into extremely large structures capable of changing with the needs of the cell. For example, the microtubules that make up the spindle during mitosis are gigantic assemblages of mainly two polypeptides: α tubulin and β tubulin (Fig. 7.24d). The cell can assemble these subunits into very long hollow tubes that grow or shrink as needed at different stages of the cell cycle.

Fast Forward

Using Mutagenesis to Look at Biological Processes

Geneticists can use mutations to dissect complicated biological processes into their protein components. To determine the specific, dedicated role of each protein, they introduce mutations into the genes encoding the protein. The mutations knock out, or delete, functional protein either by preventing protein production altogether or by altering it such that the resulting protein is nonfunctional. The researchers then observe what happens when the cell or organism attempts to perform the biological process without the deleted protein.

In the 1960s, Robert Edgar set out to delineate the function of the proteins determined by all the genes in the T4 bacteriophage genome. As we have seen, the T4 virus infects *Escherichia coli* bacteria. After a single viral particle infects a bacterium, the host cell stops producing bacterial proteins and becomes a factory for making only viral proteins. Thirty minutes after infection, the bacterial cell lyses, releasing 100 new viral particles. The head of each particle carries a DNA genome 200,000 base pairs in length that encodes at least 120 genes.

Edgar's experimental design was to obtain many different mutant bacteriophages, each containing a mutation that inactivates one of the genes essential for viral reproduction. By analyzing what went wrong with each type of mutant during the infective cycle, he would learn something about the function of each of the proteins produced by the T4 genome.

There was just one barrier to implementing this plan. A mutation that prevents viral reproduction by definition makes the virus unable to reproduce and therefore unavailable for experimental growth and study. The solution to this dilemma came with the discovery of *conditional lethal mutants:* microbes or other organisms (in this case, viral particles) carrying mutations that are lethal to the organism under one condition but not another. One type of conditional lethal mutant used by Edgar was temperature sensitive; that is, the mutant T4 phage could reproduce at low, but not at high, temperatures. The temperature-sensitive mutations producing these mutants changed one amino acid in a polypeptide such that the protein was stable and functional at a low temperature but became unstable and non-functional at a higher temperature. Temperature-sensitive muta-

tions can occur in almost any gene. Edgar isolated thousands of conditional lethal bacteriophage T4 mutants, and using complementation studies, he discovered that they fall into 65 complementation groups. These complementation groups defined 65 genes whose function is required for bacteriophage replication.

Edgar next studied the consequences of infecting bacterial cells under *restrictive conditions,* that is, under conditions in which the mutant protein could not function. For the temperature-sensitive mutants, the restrictive condition was high temperature. He found that mutations in 17 genes prevented viral DNA replication and concluded that these 17 genes contribute to that process. Mutations in most of the other 48 genes did not impede viral DNA replication but were necessary for the construction of complete viral particles. Electron microscopy showed that mutations in these 48 genes caused the accumulation of partially constructed viral particles. Edgar assumed that these partial particles were normal intermediates in the virus construction pathway, and on the basis of this assumption, he used the incomplete particles to plot the path of viral assembly. As Fig. A illustrates, three subassembly lines—one for the tail, one for the head, and one for the tail fibers—come together during the assembly of the viral product. Once the heads are completed and filled with DNA, they attach to the tails, after which attachment of the fibers completes particle construction. It would have been very difficult to discern this trilateral assembly pathway by any means other than mutagenesis-driven genetic dissection.

Between 1990 and 1995, molecular geneticists determined the complete DNA sequence of the T4 genome, and then using the genetic code dictionary (described in Chapter 8), they translated that sequence into coding regions for proteins. In addition to the 65 genes identified by Edgar, another 55 genes became evident from the sequence. Edgar did not find these genes because they are not essential to viral reproduction under the conditions used in the laboratory. The previously unidentified genes most likely play important roles in the T4 life cycle outside the laboratory, perhaps when the virus infects hosts other than the *E. coli* strain normally used in the laboratory or when the virus grows under different environmental conditions and is competing with other viruses.

Summation: Each Gene Directs the Synthesis of a Particular Polypeptide

Because more than one gene governs the production of some multimeric proteins, and because not all proteins are enzymes, the "one gene, one enzyme" hypothesis is not broad enough to define gene function. A more accurate statement is "one gene,

one polypeptide": Each gene governs the construction of a particular polypeptide. As we will see in Chapter 8, even this reformulation does not encompass the function of all genes, as a few genes in all organisms do not determine the construction of proteins; instead, they encode RNAs that are not translated into polypeptides.

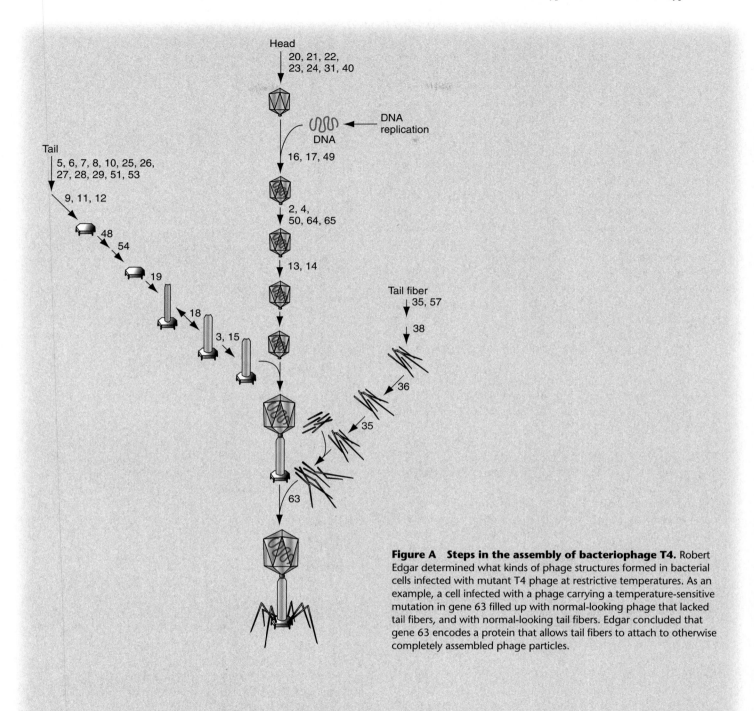

Figure A Steps in the assembly of bacteriophage T4. Robert Edgar determined what kinds of phage structures formed in bacterial cells infected with mutant T4 phage at restrictive temperatures. As an example, a cell infected with a phage carrying a temperature-sensitive mutation in gene 63 filled up with normal-looking phage that lacked tail fibers, and with normal-looking tail fibers. Edgar concluded that gene 63 encodes a protein that allows tail fibers to attach to otherwise completely assembled phage particles.

How Genotype Correlates with Phenotype

Cells contain a staggering variety of proteins, each of which performs a specific developmental or biochemical role (see the Fast Forward box "Using Mutagenesis to Look at Biological Processes"). Knowledge that genes encode proteins by determining the sequence of amino acids in a polypeptide enabled geneticists to begin to understand how genes control complex phenotypes, and how gene function determines the recessiveness or dominance of an allele.

Mutations can affect phenotype by altering the amino-acid composition of a protein or by altering the amount of normal protein produced. For missense mutations that alter a protein's amino-acid sequence, it is possible to explain the differing phenotypic effects of mutations in the same gene by assuming that each amino acid in a polypeptide has its own specific effect on protein structure. Thus changes in different types of amino acids at different positions in a polypeptide chain will have different effects on protein function that will in turn influence phenotype in different ways.

Most proteins have an active site that carries out a particular task while other parts of the protein support the shape and position of that site. Mutations that change the identity of amino acids at the active site may have more serious consequences than those that affect amino acids outside the active site. Some kinds of amino-acid substitution, such as replacement of an amino acid having a basic side chain with an amino acid having an acidic side chain, would be more likely to compromise protein function than substitutions that retain the chemical characteristics of the original amino acid.

Some mutations do not affect the amino-acid composition of a polypeptide but still generate an abnormal phenotype. As we see in Chapter 8, such mutations change the amount of normal polypeptide produced by generating altered DNA sequences that disrupt or abolish the genetic processes responsible for decoding a gene into a polypeptide.

Dominance Relations Between Alleles Depend on the Relation Between Protein Function and Phenotype

The one gene, one polypeptide hypothesis helps account for the dominance relationships between the two alleles of a gene in a diploid organism. The theory is simple: Phenotype depends on protein activity, and protein activity varies with the allele; an alternate allele can change the amino-acid sequence or the amount of polypeptide produced. Applications of the theory, however, take some unexpected turns, because the amount of protein required to generate a particular phenotype varies from protein to protein, and thus the gene activity required to produce a normal phenotype varies from gene to gene.

Alleles That Produce Nonfunctional Proteins Are Usually Recessive

Mutations that abolish the function of a protein encoded by the wild-type allele are known as **null mutations,** or **null alleles** (Fig. 7.25). Such mutations either prevent synthesis of the protein or promote synthesis of a protein incapable of carrying out any function. For example, a deletion of an entire gene would by definition be a null allele. In an A^+/a heterozygote, in which allele a is recessive to wild-type allele A^+, the A^+ allele would generate functional protein, while the null a allele would not. If the amount of protein produced by the single A^+ allele (usually, though not always, half the amount produced in an A^+/A^+ cell) is above the threshold amount sufficient to fulfill the normal biochemical requirements of the cell, then the phenotype of the A^+/a heterozygote will be wild type. For the large number of

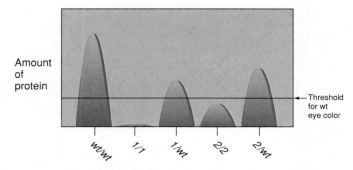

Figure 7.25 Why some mutant alleles are recessive. Researchers subjected fly extracts to rocket immunoelectrophoresis; rocket size reflects the amount of a particular enzyme called xanthine dehydrogenase. Flies need only 10% of the enzyme produced in wild-type strains (*wt/wt*) to have normal eye color. Null allele 1 and hypomorphic allele 2 are recessive to wild type because *1/wt* or *2/wt* heterozygotes have more than the low threshold amount of enzyme needed for normal eye color.

genes that function in this way, it is clear that A^+/A^+ cells actually make more than twice as much of the protein needed for the normal phenotype.

A **hypomorphic mutation** produces either much less of a protein or a protein with very weak but detectable function (Fig. 7.25). In a B^+/b heterozygote, where b is a hypomorphic allele recessive to wild-type allele B^+, the amount of protein activity will be somewhat greater than half the amount in a B^+/B^+ cell. Usually, this is enough activity to fulfill the normal biochemical requirements of the cell. Most hypomorphic mutations are detectable only in homozygotes, and only if the reduction in protein amount or function is sufficient to cause an abnormal phenotype.

Incomplete Dominance Can Arise When the Phenotype Varies in Proportion to the Amount of Functional Protein

Some combinations of alleles generate phenotypes that vary continuously with the amount of functional gene product. For example, mutations in a single pigment-producing gene can generate a red-to-white spectrum of flower colors, with the white resulting from the absence of an enzyme in a biochemical pathway (Fig. 7.26). Consider three alleles of the gene encoding this enzyme: R^+ specifies a high, wild-type amount of the enzyme; r^{50} generates half the normal amount of the same enzyme (or the full amount of an altered form that has half the normal level of activity); and r^0 is a null allele. R^+/r^0 heterozygotes produce pink flowers whose color is halfway between red and white because one-half the R^+/R^+ level of enzyme activity is not enough to generate a full red. Combining R^+ or r^0 with the r^{50} allele produces pigmentation intermediate between red and pink or between pink and white.

Dominance Can Reflect Several Kinds of Phenomena

With phenotypes that are exquisitely sensitive to the amount of functional protein produced, even a relatively small change of

Figure 7.26   When a phenotype varies continuously with levels of protein function, incomplete dominance results.

twofold or less can cause a switch between distinct phenotypes. For example, a heterozygote for a null or hypomorphic mutation that generates only half the normal amount of functional gene product may look completely different from the wild type. The *T* locus in mice has just such a mutation, with an easy-to-visualize dominant phenotype (Fig. 7.27a). Mice require the wild-type protein product of the *T*-locus gene during embryogenesis for the normal development of the posterior portion of the spinal cord and tail. Embryos heterozygous for a null mutation at the *T* locus produce only half the normal amount of the *T*-determined protein and mature into viable offspring that are normal in all respects except for the absence of the distal two-thirds of their tail. The severely shortened tail reflects the embryo's sensitivity to the level of *T*-gene product available during morphogenesis; half the normal amount of T protein is below the threshold needed for normal development. Geneticists sometimes use the term **haploinsufficiency** to describe situations in which one wild-type allele does not provide enough of

(a) Haploinsufficiency

(b) Dominant negative mutations

Functional enzyme	Nonfunctional enzyme			
$d^+d^+d^+d^+$	$d^+d^+d^+D$	$d^+d^+Dd^+$	$d^+Dd^+d^+$	$Dd^+d^+d^+$
	d^+d^+DD	d^+DDd^+	d^+Dd^+D	Dd^+Dd^+
	Dd^+d^+D	DDd^+d^+	d^+DDD	Dd^+DD
	DDd^+D	$DDDd^+$	$DDDD$	

D = dominant mutant subunit
d^+ = wild-type subunit

(c) *Kinky:* **A dominant negative mutation**

(d)   A result of ectopic expression

Figure 7.27   Why some mutant alleles are dominant.
(a) Mice heterozygous for a null mutation of the *T* locus have short tails because one-half the normal amount of T-locus protein is insufficient to make a wild-type tail. The unusual *t* allele enhances the phenotype caused by the dominant null *T* allele. **(b)** With proteins composed of four subunits encoded by the same gene, a dominant negative mutant may inactivate 15 out of every 16 multimers. **(c)** The *Kinky* allele in mice is a dominant negative mutation that causes a kink in the tail. **(d)** A neomorphic dominant mutation in the fly *Antennapedia* gene causes ectopic expression of a leg-determining gene in structures that normally produce antennae; the photo shows two legs growing out of the head.

a gene product. Only a minority of phenotypes are so sensitive to the amount of a particular protein. Thus, as described earlier, null and hypomorphic alleles usually produce phenotypes that are recessive to wild type.

Another mechanism that can result in dominance is the extremely rare **hypermorphic mutation** that produces an allele generating either more protein than the wild-type allele or the same amount of a more efficient protein. If excess protein activity alters phenotype, the hypermorphic allele is dominant. A hypermorphic mutation in the rhodopsin gene produces a rhodopsin protein that is activated whether or not light is present, resulting in constant, low-level stimulation of rhodopsin in the photoreceptor cells that detect black and white. These cells, known as rod cells, function primarily at night. People with the mutation can still see in bright daylight, but they have congenital night blindness. The blindness probably arises because the constant rhodopsin stimulation prevents adaptation of the rod cells to the very low light intensities present at night.

In yet another mechanism leading to dominance, some alleles of genes encode subunits of multimers that block the activity of the subunits produced by normal alleles; such blocking alleles are called **dominant negative alleles.** Consider, for example, a gene encoding a polypeptide that associates with three other identical polypeptides in a four-subunit enzyme. All four subunits are products of the same gene. If a dominant mutant allele D directs the synthesis of a polypeptide that can still assemble into aggregates but whose presence in the multimer—even as one subunit out of four—abolishes enzyme function, the chance of a heterozygote producing a multimer composed solely of functional wild-type d^+ subunits is 1 in 16: $(1/2)^4 = 1/16 = 6.25\%$ (Fig. 7.27b). As a result, total enzyme activity in D/d^+ heterozygotes is far less than that seen in wild-type d^+/d^+ homozygotes. Dominant negative mutations can also affect subunits in multimers composed of more than one type of polypeptide. The *Kinky* allele at the *fused* locus in mice is an example of such a dominant negative mutation (Fig. 7.27c).

Finally, a very rare class of dominant alleles arises from **neomorphic mutations** that generate a novel phenotype. Some neomorphic mutations produce proteins with a new function, while others cause genes to produce the normal protein but at an inappropriate time or place. A striking example of inappropriate protein production is the *Drosophila* gene *Antennapedia,* active during embryonic and larval stages. Normally, the gene makes its protein product in tissues destined to become legs; the protein ensures that these tissues develop into legs and not, for example, head structures such as antennae. Dominant mutations of the gene cause production of the protein in the head region of the animal, where the *Antennapedia* gene is not normally active. Here, the misplaced protein causes tissues that would normally develop into antennae to develop into legs (Fig. 7.27d). Production of a protein outside of its normal place or time is called **ectopic expression.**

In summary, mutations alter phenotypes by altering the structure of the protein produced or the amount or place of protein production. Some mutations cause a change in amino-acid sequence that partially blocks protein function. Other mutations cause a change in amino-acid sequence that completely disrupts protein function. Still other mutations prevent a gene from making protein in the first place, while yet others reduce (or rarely, increase) the amount of protein made. Dominance relations between the wild-type and mutant alleles of genes in diploid organisms depend on how drastically a mutation influences protein production and how thoroughly phenotype depends on the normal wild-type amount of the protein.

How Gene Mutations Affect Light-Receiving Proteins and Vision: A Comprehensive Example

Researchers first described anomalies of color perception in humans close to 200 years ago. Since that time, they have discovered a large number of mutations that modify human vision. By examining the phenotype associated with each mutation and then looking directly at the DNA alterations inherited with the mutation, they have learned a great deal about the genes influencing human visual perception and the function of the proteins they encode.

Several attributes of human subjects facilitate the experimental analysis of vision. First, people can recognize and describe variations in the way they see, from trivial differences in what the color red looks like, to not seeing any difference between red and green, to not seeing any color at all. Second, the highly developed science of psychophysics provides sensitive, noninvasive tests for accurately defining and comparing phenotypes. One diagnostic test, for example, is based on the fact that people perceive each color as a mixture of three different wavelengths of light—red, green, and blue—and can adjust ratios of red, green, and blue light of different intensities to match an arbitrarily chosen fourth wavelength such as yellow. The mixture of wavelengths does not combine to form the fourth wavelength; it just appears that way to the eye. A person with normal vision, for instance, will select a well-defined proportion of red and green lights to match a particular yellow, but a person who can't tell red from green will permit any proportion of these two color lights to make the same match. Finally, since inherited variations in the visual system rarely affect fecundity or longevity in modern human societies, mutations generating many of the new alleles that change visual perception remain in a population over time.

The Cellular and Molecular Basis of Vision
The Cells

People perceive light through neurons in the retina at the back of the eye (Fig. 7.28a). These neurons are of two types: rods and cones. The rods, which make up 95% of all light-receiving neurons, are stimulated by weak light over a range of wavelengths. At higher light intensities, the rods become saturated and no longer send meaningful information to the brain. This is when the cones take over, processing wavelengths of bright light that enable us to see color. The cones come in three forms—one specializes in the reception of red light, a second in the reception of green, and a third in the reception of blue. For each photoreceptor cell, the act of reception consists of absorbing photons from light of a particular wavelength, transducing information about the number and energy of those photons to electrical signals, and transmitting the signals via the optic nerve to the brain.

Four Genes, Four Polypeptides

The protein that receives photons and triggers the processing of information in rod cells is rhodopsin. It consists of a single polypeptide chain containing 348 amino acids that snakes back and forth across the cell membrane (Fig. 7.28b). One lysine within the chain associates with retinal, a carotenoid pigment molecule that actually absorbs photons. The amino acids in the vicinity of the retinal constitute rhodopsin's active site; by positioning the retinal in a particular way, they determine its response to light. Each rod cell contains approximately 100 million molecules of rhodopsin in its specialized membrane. As we have seen, the gene governing the production of rhodopsin is on chromosome 3.

The protein that receives and initiates the processing of photons in the blue cones is a relative of rhodopsin, also consisting of a single polypeptide chain containing 348 amino acids, and also encompassing one molecule of retinal. Slightly less than half of the 348 amino acids in the blue-receiving protein are the same as those found in rhodopsin; the rest are different and account for the specialized light-receiving ability of the protein (Fig. 7.28b). The gene for the blue protein is on chromosome 7.

Similarly related to rhodopsin are the red- and green-receiving proteins in the red and green cones. These are also single polypeptides associated with retinal and embedded in the cell membrane, although they are both slightly larger at 364 amino acids in length (Fig. 7.28b). Like the blue protein, the red and green proteins differ from rhodopsin in nearly half of their amino acids; they differ from each other in only four amino acids out of every hundred. Even these small differences, however, are sufficient to differentiate the light sensitivities of the two types of cones and confer on them distinct spectral sensitivities. The genes for the red and green proteins both reside on the X chromosome in a tandem head-to-tail arrangement. Most individuals have one red gene and one to three green genes on their X chromosomes (Fig. 7.28c).

The Rhodopsin Gene Family Evolved by Duplication and Divergence

The similarity in structure and function between the four rhodopsin proteins suggests that the genes encoding these polypeptides arose by duplication of an original photoreceptor gene and then divergence through the accumulation of many mutations. Many of the mutations that promoted the ability to see color must have provided advantages that evolutionary forces selected for over a period of millions of years. The red and green genes are the most similar, differing by less than five nucleotides out of every hundred. This suggests that they diverged from each other only in the relatively recent evolutionary past. The

(a) Photoreceptor-containing cells

(b) Photoreceptor proteins

Rhodopsin protein

Blue-receiving protein

Green-receiving protein

Red-receiving protein

(c) Red/green pigment genes

X chromosomes from normal individuals:

(d) Evolution of visual pigment genes

Primordial gene

Red gene   Green gene   Blue gene   Rhodopsin gene

Figure 7.28 The cellular and molecular basis of vision.
(a) Rod and cone cells in the retina carry membrane-bound photoreceptors. **(b)** The photoreceptor in rod cells is rhodopsin. The blue-, green-, and red-receiving proteins in cone cells are related to rhodopsin. **(c)** One red photoreceptor gene and one to three green photoreceptor genes are clustered on the X chromosome. **(d)** The genes for rhodopsin and the three color receptors probably evolved from a primordial photoreceptor gene through three gene duplication events followed by divergence of the duplicated copies.

less pronounced amino-acid similarity of the red or green proteins with the blue protein, and the even lower relatedness between rhodopsin and any color photoreceptor, reflect earlier duplication and divergence events (Fig. 7.28d).

How Mutations in the Rhodopsin Gene Family Influence the Way We See

Many Amino-Acid Substitutions in Rhodopsin Result in Partial or Complete Blindness

At least 29 different single nucleotide substitutions in the rhodopsin gene cause an autosomal dominant vision disorder known as *retinitis pigmentosa* that begins with an early loss of rod function followed by a slow progressive degeneration of the peripheral retina. Figure 7.29a shows the location of the amino acids affected by these mutations. These amino-acid changes result in abnormal rhodopsin proteins that either do not fold properly, or once folded are unstable. While normal rhodopsin is an essential structural element of rod cell membranes, these nonfunctional mutant proteins are retained in the body of the cell, where they remain unavailable for insertion into the membrane. Rod cells that cannot incorporate enough rhodopsin into their membrane eventually die. Depending on how many rod cells die, partial or complete blindness ensues.

Other mutations in the rhodopsin gene cause the far less serious condition of night blindness (Fig. 7.29a). These hypomorphic mutations change the protein's amino-acid sequence so that the threshold of stimulation required to trigger the vision cascade increases. With the change, very dim light is no longer enough to initiate vision.

Mutations in the Cone-Cell Pigment Genes Alter Color Vision in Predictable Ways

Vision problems caused by mutations in the cone-cell pigment genes are less severe than vision problems caused by similar defects in the rod cells' rhodopsin genes, most likely because the rods make up 95% of a person's light-receiving neurons, while the cones comprise only about 5%. Some mutations in the blue gene on chromosome 7 cause *tritanopia,* a defect in the ability to discriminate between colors that differ only in the amount of blue light they contain (Figs. 7.29b and 7.30). Mutations in the red gene on the X chromosome can modify or abolish red protein function and as a result, the red cone cells' sensitivity to light. For example, a change at position 203 in the red-receiving protein from cysteine to arginine disrupts one of the disulfide bonds required to support the protein's tertiary structure (see Fig. 7.29c). Without that bond, the protein cannot stably maintain its native configuration, and a person with the mutation has red colorblindness. In other mutations affecting the red gene, small deletions from neighboring DNA that does not specify amino acids reduce or eliminate gene function, probably by incapacitating control elements needed for gene activity (see Fig. 7.29c). People whose cells carry one of these small deletions are partially or completely red color-blind.

Figure 7.29 How mutations modulate light and color perception. (a) Amino-acid substitutions (*black dots*) that disrupt rhodopsin's three-dimensional structure result in retinitis pigmentosum. Other substitutions diminishing rhodopsin's sensitivity to light cause night blindness. **(b)** Substitutions in the blue pigment can produce tritanopia (blue colorblindness). **(c)** Red colorblindness results from particular mutations that destabilize the red photoreceptor (*left*), or by the deletion of DNA regions that control the expression of the red gene (*right*). **(d)** Unequal crossing-over between the red and green genes can alter their arrangement and create hybrid photoreceptor proteins. **(e)** Deletions that remove a region controlling expression of both the red and green genes prevent the synthesis of both photoreceptors, resulting in blue cone monochromacy.

Unequal Crossing-Over Between the Red and Green Genes Produces Most Variations in Red-Green Perception

People with normal color vision have a single red gene; some of these normal individuals also have a single adjacent green gene, while others have two to five green genes. The red and green genes are 96% identical in DNA sequence; the different green genes, 99.9% identical. The proximity and high degree of homology make these genes unusually prone to unequal crossing-over. A variety of unequal recombination events produce DNA containing no red gene, no green gene, various combinations of

Figure 7.30 **How the world looks to a person with tritanopia.** Compare with Fig. 4.22 on p. 103.

green genes, or hybrid red-green genes (see Fig. 7.29d). These different DNA combinations account for the large majority of the known aberrations in red-green color perception, with the

remaining abnormalities stemming from point mutations, as described earlier. Because the accurate perception of red and green depends on the differing ratios of red and green light processed, people with no red or no green gene perceive red and green as the same color (see Fig. 4.22 on p. 103).

Some Mutations Knock Out the Ability to See Both Red and Green

Geneticists have detected seven different deletions that produce an X-linked trait in which green and red cones do not function. The trait is called *blue cone monochromacy,* because people with the condition see only blue-related colors in bright light. Characterization at the DNA level has shown that all seven deletions overlap in a 600 bp segment outside the coding region of the red and green genes (see Fig. 7.29e). It is likely that this 600 bp region serves as a long-range control element necessary for the activity of all the genes in the red-green array.

In short, we see the way we do in part because four genes direct the production of four polypeptides in the rod and cone cells of our retina. Mutations that alter those polypeptides or their amounts change our perception of light or color.

C O N N E C T I O N S

Careful studies of mutations showed that genes are linear arrays of mutable elements that direct the assembly of amino acids in a polypeptide. The mutable elements are the nucleotide building blocks of DNA.

Biologists call the parallel between the sequence of nucleotides in a gene and the order of amino acids in a polypeptide

colinearity. In the next chapter, we see that colinearity arises from base pairing, a genetic code, specific enzymes, and macromolecular assemblies like ribosomes that guide the flow of information from DNA through RNA to protein. How cells accomplish the DNA-directed assembly of polypeptides is the subject of Chapter 8.

E S S E N T I A L   C O N C E P T S

1. Mutations are alterations in the nucleotide sequence of the DNA molecule that occur by chance and modify the genome at random. Once they occur, mutations can be transmitted from generation to generation when DNA replicates.
   a. Mutations that affect phenotype occur naturally at a very low rate, which varies from gene to gene. Forward mutations usually occur more often than reversions.
   b. The agents of spontaneously occurring mutations include chemical hydrolysis, radiation such as cosmic rays and ultraviolet light, and mistakes during DNA replication. Mutagens raise the frequency of mutation above the spontaneous rate. Cells have evolved enzyme systems that minimize mutations.

   c. Mutations are the raw material of evolution. The action of natural selection on heritable mutations is a major agent of evolution.

2. What mutations tell us about gene structure:
   a. Mutations within the same gene usually fail to complement each other. The concept of a complementation group thus defines the gene as a unit of function.
   b. A gene is composed of a linear sequence of nucleotide pairs in a discrete, localized region of a chromosome. Recombination can occur within a gene, even between adjacent nucleotide pairs.

3. What mutations tell us about gene function:
   a. The function of most genes is to specify the linear sequence of amino acids in a particular polypeptide.

b. The sequence of amino acids in a polypeptide determines the polypeptide's three-dimensional structure, which in turn determines its function. Each protein consists of one, two, or more polypeptides. One gene encodes one polypeptide. Proteins composed of two or more different subunits are encoded by two or more genes.

4. How genotype correlates with phenotype:
   a. The functions of proteins produced within an organism determine phenotype.
   b. Some mutations modify phenotype by altering the sequence of amino acids and thus the function of a protein. Other mutations do not affect a polypeptide's amino-acid sequence but instead affect the amount, time, or place of protein production.
   c. Dominance relations between alleles depend on the relation between protein function and phenotype. Recessive alleles often produce nonfunctional proteins. Incomplete dominance arises when phenotype varies with the amount of functional protein produced. Dominant alleles reflect either situations in which small perturbations in the amount of protein produced disrupt normal phenotypes or other rare events that affect protein structure or synthesis.

SOCIAL AND ETHICAL ISSUES

1. Sean was diagnosed with AIDS in 1986. In 1985, the drug AZT (azidothymidine) had been shown to be effective in inhibiting replication of the HIV-1 virus, which causes AIDS, and was thus a potential therapeutic agent. Sean read about research indicating this effect and wanted to begin taking the drug as soon as possible. However, the standard procedure for testing new drugs involves extensive clinical trials that can take years and require frequent monitoring. One way Sean could get the drug before it was commercially available was to volunteer for a drug-testing trial. However, Sean lives in a small community far from the research universities where the trials were taking place. Is it fair that those people who do not live near a major research facility are unable to get drugs that are in clinical trials? Does an individual have a right to demand that the medical community get the drug for them? Should physicians be allowed to prescribe drugs for AIDS patients before the drugs are approved, if the patients would like to try them?

2. The American Cancer Society (ACS) is a nonprofit organization that funds much cancer research, treatment, and public education. In 1996, the ACS decided to offer exclusive endorsements to *Nicoderm* nicotine patches, which are used to combat addiction to smoking. The ACS logo would appear on the packaging of the patches and the ACS would, in return, receive a sum of money that it would use to support its cancer-fighting programs. Is this an appropriate activity for a charitable organization such as the ACS? Could these endorsements be damaging to the reputation of the organization? What criteria could or should the ACS use in deciding whose product it will endorse?

3. Chemicals that are mutagenic are identified by tests such as the Ames test, which measures the level of mutagenesis in bacteria. The susceptibility of humans to mutagenic chemicals may vary depending on the genetic makeup of the individual. The dose that affects one person may be different from that which affects another. However, there are few, if any, reliable tests that determine a person's level of susceptibility. If this is true, is it a good idea to translate the results of the Ames test of mutability in bacteria to a prediction of carcinogenicity in humans? Often reports of Ames test results on a chemical make newspaper headlines. Is this a useful and honest way to report findings that could affect human health or do people need to consider other variables to make an informed decision?

SOLVED PROBLEMS

I. Mutations can often be reverted to wild type by treatment with mutagens. The type of mutagen that will reverse a mutation gives us information about the nature of the original mutation. The mutagen EMS almost exclusively causes transitions; proflavin is an intercalating agent that causes insertion or deletion of a base; ultraviolet (UV) light causes single-base substitutions. Cultures of several *E. coli* *met*$^-$ mutants were treated with three mutagens separately and spread onto a plate lacking methionine to look for revertants. (In the chart $-$ indicates that no colonies grew, and $+$ indicates that some *met*$^+$ revertant colonies grew.)

Mutant number	Mutagen treatment		
	EMS	Proflavin	UV light
1	+	−	+
2	−	+	−
3	−	−	−
4	−	−	+

a. Given the results, what can you say about the nature of the original mutation in each of the strains?

b. Experimental controls are designed to eliminate possible explanations for the results, thereby ensuring that data are interpretable. In the experiment described, we scored the presence or absence of colonies. How do we know if colonies that appear on plates are mutagen-induced revertants? What else could they be? What control would enable us to be confident of our revertant analysis?

Answer

To answer this question, you need to understand the concepts of mutation and reversion.

a. Mutation 1 is reverted by the mutagen that causes transitions, *so mutation 1 must have been a transition.* Consistent with this conclusion is the fact the UV light can also revert the mutation and the intercalating agent proflavin does not cause reversion. *Mutation 2 is reverted by proflavin and therefore must be either an insertion or a deletion of a base.* The other two mutagens do not revert mutation 2. Mutation 3 is not reverted by any of these mutagenic agents. It is therefore not a single-base substitution, a single-base insertion, or a single-base deletion. *Mutation 3 could be a deletion of several bases or an inversion.* Mutation 4 is reverted by UV light so it is a single-base change, but it is not a transition, since EMS did not revert the mutation. *Mutation 4 must be a transversion.*

b. *The colonies on the plates could arise by spontaneous reversion of the mutation.* Spontaneous reversion should occur with lower frequency than mutagen-induced reversion. The important control here is to *spread each mutant culture without any mutagen treatment onto selective media to assess the level of spontaneous reversion.*

II. Imagine that 10 independently isolated recessive lethal mutations (l^1, l^2, l^3, etc.) map to chromosome 7 in mice. Complementation testing was done by mating all pairwise combinations of heterozygotes bearing these lethal mutations. Absence of complementation was scored by examining pregnant females for dead fetuses. A + in the chart means that the two lethals complemented, and dead embryos were not found. A − indicates that dead embryos were found, at the rate of about one in four conceptions. (The crosses between heterozygous mice would be expected to yield the homozygous recessive showing the lethal phenotype in 1/4 of the embryos.) The lethal mutation in the parental heterozygotes for each cross are listed across the top and down the left side of the chart

(that is, l^1 indicates a heterozygote in which one chromosome bears the l^1 mutation and the homologous chromosome is wild type).

	l^1	l^2	l^3	l^4	l^5	l^6	l^7	l^8	l^9	l^{10}
l^1	−	+	+	+	+	−	−	+	+	+
l^2		−	+	+	+	+	+	+	+	−
l^3			−	−	−	+	+	−	−	+
l^4				−	−	+	+	−	−	+
l^5					−	+	+	−	−	+
l^6						−	−	+	+	+
l^7							−	+	+	+
l^8								−	−	+
l^9									−	+
l^{10}										−

How many genes do the 10 lethal mutations represent? What are the complementation groups?

Answer

This problem involves the application of the complementation concept to a set of data. There are two ways to analyze these results. You can focus on the mutations that do complement each other, conclude that they are in different genes, and begin to create a list of mutations in separate genes. Alternatively, you can focus on mutations that do not complement each other and therefore are alleles of the same genes. The latter approach is more efficient when several mutations are involved. For example, l^1 does not complement l^6 and l^7. These three alleles are in one complementation group. l^2 does not complement l^{10}; they are in a second complementation group. l^3 does not complement l^4, l^5, l^8, or l^9, so they form a third complementation group. *There are three complementation groups.* (Note also that for each mutant, the cross between individuals carrying the same alleles resulted in no complementation, because the homozygous recessive lethal was generated.) *The three complementation groups consist of: (1) l^1, l^6, l^7; (2) l^2, l^{10}; (3) l^3, l^4, l^5, l^8, l^9.*

III. W, X, and Y are the intermediates (in that order) in a biochemical pathway whose product is Z. $Z^−$ mutants are found in five different complementation groups. *Z1* mutants will grow on Y or Z but not W or X. *Z2* mutants will grow on X, Y, or Z. *Z3* mutants will only grow on Z. *Z4* mutants will grow on Y or Z. Finally, *Z5* mutants will grow on W, X, Y, or Z.

a. Order the five complementation groups in terms of the steps they block.

b. What does this genetic information reveal about the nature of the enzyme that carries out the conversion of X to Y?

Answer

This problem requires that you understand complementation and the connection between genes and enzymes in a biochemical pathway.

a. A biochemical pathway represents an ordered set of reactions that must occur to produce a product. This problem gives the order of intermediates in a pathway for producing product Z. The lack of any enzyme along the way will cause the phenotype of Z^- but the block can occur at different places along the pathway. If the mutant grows when given an intermediate compound, the enzymatic (and hence gene) defect must be before production of that intermediate compound. The *Z1* mutants that grow on Y or Z (but not on W or X) must have a defect in the enzyme that produces Y. *Z2* mutants have a defect prior to X; *Z3* mutants have a defect prior to Z; *Z4*

mutants have a defect prior to Y; *Z5* have a defect prior to W. *The five complementation groups can be placed in order of activity within the biochemical pathway as follows:*

$$\text{Z5} \quad \xrightarrow{\quad} \text{W} \quad \xrightarrow{\text{Z2}} \text{X} \quad \xrightarrow{\text{Z1, Z4}} \text{Y} \quad \xrightarrow{\text{Z3}} \text{Z}$$

b. Mutants *Z1* and *Z4* affect the same step, but because they are in different complementation groups we know they are in different genes. *Mutations Z1 and Z4 are probably in genes that encode subunits of a multisubunit enzyme that carries out the conversion of X to Y.* Alternatively, there could be a currently unknown additional intermediate step between X and Y.

PROBLEMS

7-1 The following is a list of mutational changes. For each of the specific mutations described, indicate which of the following terms could apply, either as a description of the mutation or as a possible cause. More than one term from the right column can apply to each statement in the left column.

1. an A–T base pair in the wild-type gene is changed to a G–C pair
2. an A–T base pair is changed to a T–A pair
3. the sequence AAGCTTATCG is changed to AAGCTATCG
4. the sequence AAGCTTATCG is changed to AAGCTTTATCG
5. the sequence AACGTTATCG is changed to AATGTTATCG
6. the sequence AACGTCACACACACATCG is changed to AACGTCACATCG
7. the gene map in a given chromosome arm is changed from *bog-rad-fox1-fox2-try-duf* (where *fox1* and *fox2* are highly homologous, recently diverged genes) to *bog-rad-fox1-fox3-fox2-try-duf* (where *fox3* is a new gene with one end similar to *fox1* and the other similar to *fox2*)
8. the gene map in a chromosome is changed from *bog-rad-fox1-fox2-try-duf* to *bog-rad-fox2-fox1-try-duf*
9. the gene map in a given chromosome is changed from *bog-rad-fox1-fox2-try-duf* to *bog-rad-fox1-mel-qui-txu-sqm*

a. transition
b. base substitution
c. transversion
d. inversion
e. translocation
f. deletion
g. insertion
h. deamination
i. X-ray irradiation
j. intercalator
k. unequal crossing-over

7-2 The DNA sequence of a gene from three independently isolated mutants is given here. Using this information, what is the sequence of the wild-type gene in this region?

mutant 1   ACCGTAATCGACTGGTAAACTTTGCGCG
mutant 2   ACCGTAGTCGACCGGTAAACTTTGCGCG
mutant 3   ACCGTAGTCGACTGGTTAACTTTGCGCG

7-3 Over a period of several years, a large hospital kept track of the number of births of babies displaying the trait achondroplasia. Achondroplasia is a very rare autosomal dominant condition resulting in dwarfism with abnormal body proportions. After 120,000 births, it was noted that there had been 27 babies born with achondroplasia. One physician was interested in determining how many of these dwarf babies result from new mutations and whether the apparent mutation rate in his area was higher than normal. He looked up the families of the 27 dwarf births and discovered that 4 of the dwarf babies had a dwarf parent. What is the apparent mutation rate of the achondroplasia gene in this population? Is it unusually high or low?

7-4 Among mammals, measurements of the rate of generation of autosomal recessive mutations have been made almost exclusively in mice, while many measurements of the rate of generation of dominant mutations have been made both in mice and in humans. Why do you think there has been this difference?

7-5 In a genetics lab, Kim and Maria infected a sample from an *E. coli* culture with a particular virulent bacteriophage. They noticed that most of the cells were lysed, but a few survived. The survival rate in their sample was about 1×10^{-4}. Kim was sure the bacteriophage induced the resistance in the cells, while Maria thought that resistant mutants probably already existed in the sample of cells they used. Earlier, for a different experiment, they had spread a dilute suspension of *E. coli* onto solid medium in a large petri dish, and, after seeing that about 10^5 colonies were growing up, they had replica-plated that plate onto three other plates. Kim and Maria decided to use these plates to test their theories. They pipette a suspension of the bacteriophage onto each of the three replica plates.

What should they see if Kim is right? What should they see if Maria is right?

7-6 Suppose you wanted to study genes controlling the structure of bacterial cell surfaces. You decide to start by isolating bacterial mutants that are resistant to infection by a bacteriophage that binds to the cell surface. The selection procedure is simple: Spread cells from a culture of sensitive bacteria on a petri plate, expose them to a high concentration of phages, and pick the bacterial colonies that grow. To set up the selection you could (1) spread cells from a single liquid culture of sensitive bacteria on many different plates and pick every resistant colony *or* (2) start many different cultures, each grown from a single colony of sensitive bacteria, spread one plate from each culture, and then pick a single mutant from each plate. Which method would ensure that you are isolating many independent mutations?

7-7 A wild-type male *Drosophila* was exposed to a large dose of X rays and was then mated to an unirradiated female, one of whose X chromosomes carried both a dominant mutation for the trait *Bar* eyes and several inversions. Many F₁ females from this mating were recovered who had the *Bar,* multiply inverted X chromosome from their mother, and an irradiated X chromosome from their fathers. (The inversions ensure that viable offspring of these F₁ females will not have recombinant X chromosomes, as explained in Chapter 13.) After mating to normal males, most F₁ females produced *Bar* and wild-type sons in equal proportions. There were three exceptional F₁ females, however. Female A produced as many sons as daughters but half of the sons had *Bar* eyes, and the other half had white eyes. Female B produced half as many sons as daughters, and all of the sons had *Bar* eyes. Female C produced 75% as many sons as daughters. Of these sons, 2/3 had *Bar* eyes and 1/3 had wild-type eyes. Explain the results obtained with each exceptional F₁ female.

7-8 A wild-type *Drosophila* female was mated to a wild-type male that had been exposed to X rays. One of the F₁ females was then mated with a male that had the following recessive markers on the X chromosome: *yellow* body (*y*), *crossveinless* wings (*cv*), *cut* wings (*ct*), *singed* bristles (*sn*), and *miniature* wings (*m*). These markers are known to map in the order *y−cv−ct−sn−m*. The progeny of this second mating were unusual in two respects. First, there were twice as many females as males. Second, while all of the males were wild type in phenotype, 1/2 of the females were wild type and the other 1/2 exhibited the *ct* and *sn* phenotypes.
a. What did the X rays do to the irradiated male?
b. Draw the X-chromosome pair present in a progeny female fly produced by the second mating that was phenotypically *ct* and *sn*.
c. If the *ct* and *sn* female fly whose chromosomes were drawn in part (b) was then crossed to a wild-type male,

what phenotypic classes would you expect to find among the progeny males?

7-9 When the *his⁻ Salmonella* strain used in the Ames test is exposed to substance X, no *his⁺* revertants are seen. If, however, rat liver supernatant is added to the cells along with substance X, revertants do occur. Is substance X a potential carcinogen for human cells? Explain.

7-10 When a particular mutagen identified by the Ames test is injected into mice, it causes the appearance of many tumors, showing that this substance is carcinogenic. When cells from these tumors are injected into other mice not exposed to the mutagen, almost all of the new mice develop tumors. However, when mice carrying mutagen-induced tumors are mated to unexposed mice, virtually all of the progeny are tumor free. Why can the tumor be transferred horizontally (by injecting cells) but not vertically (from one generation to the next)?

7-11 When 1 million cells of a culture of haploid yeast carrying a *met⁻* auxotrophic mutation were plated on petri plates lacking methionine (met), five colonies grew. You would expect cells in which the original *met⁻* mutation was reversed (by a base change back to the original sequence) would grow on the media lacking methionine, but some of these apparent reversions could be due to a mutation in a different gene that somehow suppresses the original *met⁻* mutations. How would you be able to determine if the mutations in your five colonies were due either to a precise reversion of the original *met⁻* mutation or to the generation of a suppressor mutation in a gene on another chromosome?

7-12 Imagine that you caught a female albino mouse in your kitchen and decided to keep it for a pet. A few months later, while vacationing in Guam, you caught a male albino mouse and decided to take it home for some interesting genetic experiments. The question you are wondering about is whether the two mice are both albino due to mutations in the same gene. What could you do to find out the answer to this question? Assume that both mutations are recessive.

7-13 In a haploid yeast strain, eight recessive mutations were found that resulted in a requirement for the amino acid lysine. All the mutations were found to revert at a frequency of about 1×10^{-6}, except mutations 5 and 6, which did not revert. Matings were made between *a* and α cells carrying these mutations. The ability of the resultant diploid strains to grow on minimal medium in the absence of lysine is shown in the chart on the next page. (+ means growth and − means no growth.)
a. How many complementation groups were revealed by this data? Which point mutations are found within which complementation groups?

	1	2	3	4	5	6	7	8
1	−	+	+	+	+	−	+	−
2	+	−	+	+	+	+	+	+
3	+	+	−	−	−	−	−	+
4	+	+	−	−	−	−	−	+
5	+	+	−	−	−	−	−	+
6	−	+	−	−	−	−	−	−
7	+	+	−	−	−	−	−	+
8	−	+	+	+	+	−	+	−

b. The same diploid strains are now induced to undergo sporulation. The vast majority of resultant spores cannot form colonies when plated on minimal medium minus lysine (auxotrophic spores). However, particular diploids can produce rare spores that do form colonies when plated on minimal medium minus lysine (prototrophic spores). The following table shows whether (+) or not (−) any prototrophic spores are formed upon sporulation of the various diploid cells.

	1	2	3	4	5	6	7	8
1	−	+	+	+	+	−	+	+
2	+	−	+	+	+	+	+	+
3	+	+	−	+	−	+	+	+
4	+	+	+	−	−	−	+	+
5	+	+	−	−	−	−	+	+
6	−	+	+	−	−	−	+	+
7	+	+	+	+	+	+	−	+
8	+	+	+	+	+	+	+	−

When prototrophic spores occur during sporulation of the diploids just discussed, what ratio of auxotrophic to prototrophic spores would you generally expect to see in any tetrad containing such a prototrophic spore? Explain the ratio you expect.

c. Using the data from all parts of this question, draw the best map of the eight lysine mutations under study. Show the extent of any deletions involved, and indicate the boundaries of the various complementation groups.

7-14 a. You have a test tube containing 5 ml of a solution of bacteriophage, and you would like to estimate the number of bacteriophage in the tube. Assuming the tube actually contains a total of 15 billion bacteriophage, design a serial dilution experiment that would allow you to estimate this number. Ideally, the final plaque-containing plates you count should contain more than 10 and less than 1000 plaques.

b. When you count bacteriophage by the serial dilution method as in part (a), you are assuming a *plating efficiency* of 100%: that is, the number of plaques on the petri plate exactly represents the number of bacteriophage you mixed with the plating bacteria. Is there any way to test the possibility that only a certain percentage of bacteriophage particles are able to form plaques (so that the plating efficiency would be less than 100%)? Conversely, why is it fair to assume that any plaques are initiated by one rather than multiple bacteriophage particles?

7-15 a. Seymour Benzer's fine structure analysis of the *rII* region of bacteriophage T4 depended in large part on deletion analysis as shown in Fig. 7.18a on p. 212. But to perform such deletion analysis, Benzer had to know which *rII⁻* bacteriophage strains were deletions and which were point mutations. How do you think he was able to distinguish *rII⁻* deletions from point mutations?

b. Benzer concluded that recombination can occur between adjacent nucleotide pairs, even within the same gene. How was he able to make this statement? At the time, Benzer had two relevant pieces of information: (i) the total length in μm of the bacteriophage T4 chromosome (measured in the electron microscope) and (ii) many mutations in many bacteriophage T4 genes, including *rIIA* and *rIIB*.

c. Figure 7.18c on p. 212 shows Benzer's fine structure map of point mutations in the *rII* region. A key feature of this map is the existence of "hot spots," which Benzer interpreted as nucleotide pairs that were particularly susceptible to mutation. How could Benzer say that all the independent mutations in a hot spot were due to mutations of the same nucleotide pair?

7-16 You found five T4 *rII⁻* mutants that will not grow on *E. coli* K(λ). You mixed together all possible combinations of two mutants (as indicated in the chart below), added the mixtures to *E. coli* K(λ), and scored for the ability of the mixtures to grow and make plaques (indicated as a + in the chart).

	1	2	3	4	5
1	−	+	+	−	+
2		−	−	+	−
3			−	+	−
4				−	+
5					−

a. How many genes were identified by this analysis?

b. Which mutants belong to the same complementation groups?

7-17 Nine *rII⁻* mutants of bacteriophage T4 were used in pairwise infections of *E. coli* K(λ) hosts. Six of the mutations in these phages are point mutations; the other three are deletions. The ability of the doubly infected cells to produce progeny phages in large numbers is scored in the chart at the top of the next page.

	1	2	3	4	5	6	7	8	9
1	−	−	+	+	−	−	−	+	+
2		−	+	+	−	−	−	+	+
3			−	−	+	−	+	−	−
4				−	+	−	+	−	−
5					−	−	−	+	+
6						−	−	−	−
7							−	+	+
8								−	−
9									−

The same nine mutants were then used in pairwise infections of *E. coli* B hosts. The production of progeny phage that can subsequently lyse *E. coli* K(λ) hosts is now scored. In the table, 0 means the progeny do not produce any plaques on *E. coli* K(λ) cells; − means that only a very few progeny phages produce plaques; and + means that many progeny produce plaques (more than 10 times as many as in the − cases).

	1	2	3	4	5	6	7	8	9
1	−	+	+	+	+	−	−	+	+
2		−	+	+	+	+	−	+	+
3			0	−	+	0	+	+	−
4				−	+	−	+	+	+
5					−	+	−	+	+
6						0	0	−	+
7							0	+	+
8								−	+
9									−

a. Which of the mutants are the three deletions? What criteria did you use to reach your conclusion?
b. If you know that mutation 9 is in the *rIIB* gene, draw the best genetic map possible to explain the data, including the positions of all point mutations and the extent of the three deletions.
c. There should be one uncertainty remaining in your answer to part (b). How could you resolve this uncertainty?

7-18 The *rosy (ry)* gene of *Drosophila* encodes an enzyme called xanthine dehydrogenase. Flies homozygous for *ry* mutations exhibit a rosy eye color. Heterozygous females were made that had ry^{41} *Sb* on one homolog and *Ly* ry^{564} on the other homolog, where ry^{41} and ry^{564} are two independently isolated alleles of *ry*. *Ly* (*Lyra* [narrow] wings) and *Sb* (*Stubble* [short] bristles) are dominant markers to the left and right of *ry*, respectively. These females are now mated to males homozygous for ry^{41}. Out of 100,000 progeny, 8 have wild-type eyes, *Lyra* wings, and *Stubble* bristles, while the remainder have rosy eyes.
a. What is the order of these two *ry* mutations relative to the flanking genes *Ly* and *Sb*?
b. What is the genetic distance separating ry^{41} and ry^{564}?

7-19 The pathway for arginine biosynthesis in *Neurospora crassa* involves several enzymes that produce a series of intermediates.

$$\text{N-acetylornithine} \xrightarrow{argE} \text{ornithine} \xrightarrow{argF} \text{citrulline} \xrightarrow{argG} \text{argininosuccinate} \xrightarrow{argH} \text{arginine}$$

a. If you did a cross between *argE*⁻ and *argH*⁻ *Neurospora* strains, what would be the distribution of Arg⁺ and Arg⁻ spores within parental ditype and nonparental ditype asci? Give the spore types in the order in which they would appear in the ascus.
b. For each of the spores in your answer to part (a), what nutrients could you supply in the media to get spore growth?

7-20 In corn snakes, the wild-type color is brown. One autosomal recessive mutation causes the snake to be orange, and another causes the snake to be black. An orange snake was crossed to a black one, and the F_1 offspring were all brown. Assume that all relevant genes are unlinked.
a. Indicate what phenotypes and ratios you would expect in the F_2 generation of this cross if there is one pigment pathway, with orange and black being different intermediates on the way to brown.
b. Indicate what phenotypes and ratios you would expect in the F_2 generation if orange pigment is a product of one pathway, black pigment is the product of another pathway, and brown is the effect of mixing the two pigments in the skin of the snake.

7-21 In a certain species of flowering plants with a diploid genome, four enzymes are involved in the generation of flower color. The genes encoding these four enzymes are on different chromosomes. The biochemical pathway involved is as follows; the figure shows that either of two different enzymes is sufficient to convert a blue pigment into a purple pigment.

white $\longrightarrow$ green $\longrightarrow$ blue $\rightrightarrows$ purple

A true-breeding green-flowered plant is mated with a true-breeding blue-flowered plant. All of the plants in the resultant F_1 generation have purple flowers. F_1 plants are allowed to self-fertilize, yielding an F_2 generation. Show genotypes for P, F_1, and F_2 plants, and indicate which genes specify which biochemical steps. Determine the fraction of F_2 plants with the following phenotypes: white flowers, green flowers, blue flowers, and purple flowers. Assume the green-flowered parent is mutant in only a single step of the pathway.

7-22 Null mutations are valuable genetic resources because they allow a researcher to determine what happens to an organism in the complete absence of a particular protein. However, it is often not a trivial matter to determine whether a mutation represents the null state of the gene.

a. Geneticists sometimes use the following test for the "nullness" of an allele in a diploid organism: If the abnormal phenotype seen in a homozygote for the allele is identical to that seen in a heterozygote where one chromosome carries the allele in question and the homologous chromosome is known to be completely deleted for the gene, then the allele is null. What is the underlying rationale for this test? What limitations might there be in interpreting such a result?

b. Can you think of other methods to determine whether an allele represents the null state of a particular gene?

7-23 In each of the following cross schemes, two true-breeding plant strains are crossed to make F_1 plants, all of which have purple flowers. The F_1 plants are then self-fertilized to produce F_2 progeny as shown here.

Cross	Parents	F_1	F_2
1	blue × white	all purple	9 purple : 4 white : 3 blue
2	white × white	all purple	9 purple : 7 white
3	red × blue	all purple	9 purple : 3 red : 3 blue : 1 white
4	purple × purple	all purple	15 purple : 1 white

a. For each cross, explain the inheritance of flower color.

b. For each cross, show a possible biochemical pathway that could explain the data.

c. Which of these crosses is compatible with an underlying biochemical pathway involving only a single step that is catalyzed by an enzyme with two dissimilar subunits, both of which are required for enzyme activity?

d. For each of the four crosses, what would you expect in the F_1 and F_2 generations if all relevant genes were tightly linked?

7-24 The intermediates A, B, C, D, E, and F all occur in the same biochemical pathway. G is the product of the pathway, and mutants 1 through 7 are all G^-, meaning that they cannot produce substance G. The following table shows which intermediates will promote growth in each of the mutants. Arrange the intermediates in order of their occurrence in the pathway, and indicate the step in the pathway at which each mutant strain is blocked. A + in the table indicates that the strain will grow if given that substance, an O means lack of growth.

	Supplements						
Mutant	A	B	C	D	E	F	G
1	+	+	+	+	+	O	+
2	O	O	O	O	O	O	+
3	O	+	+	O	+	O	+
4	O	+	O	O	+	O	+
5	+	+	+	O	+	O	+
6	+	+	+	+	+	+	+
7	O	O	O	O	+	O	+

7-25 The following noncomplementing *E. coli* mutants were tested for growth on four known precursors of thymine, A–D.

	Precursor/product				
Mutant	A	B	C	D	Thymine
9	+	−	+	−	+
10	−	−	+	−	+
14	+	+	+	−	+
18	+	+	+	+	+
21	−	−	−	−	+

a. Show a simple linear biosynthetic pathway of the four precursors and the end product, thymine. Indicate which step is blocked by each of the five mutations.

b. What precursor would accumulate in the following double mutants: 9 and 10? 10 and 14?

7-26 The pathways for the biosynthesis of the amino acids glutamine (gln) and proline (pro) involve one or more common intermediates. Auxotrophic yeast mutants numbered 1–7 are isolated that require either glutamine or proline or both amino acids for their growth, as shown in the table below (+ means growth; − no growth). These mutants are also tested for their ability to grow on the intermediates A–E. What is the order of these intermediates in the glutamine and proline pathways, and at which point in the pathway is each mutant blocked?

Mutant	A	B	C	D	E	gln	pro	gln+ pro
1	+	−	−	−	+	−	+	+
2	−	−	−	−	−	−	+	+
3	−	−	+	−	−	−	−	+
4	−	−	−	−	−	+	−	+
5	−	−	+	+	−	−	−	+
6	+	−	−	−	−	−	+	+
7	−	+	−	−	−	+	−	+

7-27 Antibodies were made that recognize six proteins that are part of a complex inside the *Caenorhabditis elegans* one-cell embryo. The mother produces proteins that are believed to assemble stepwise into a structure in the egg, beginning at the embryo's inner surface. The antibodies were used to detect the protein location in embryos produced by mutant mothers (who are homozygous recessive for the gene[s] encoding each protein). The *C. elegans* mothers are self-fertilizing hermaphrodites so no wild-type copy of a gene will be introduced during fertilization. In the table, * means the protein was present and at the embryo surface, − means that the protein was not present, and + means that the protein was present but not at the embryo surface. Assume all mutations prevent production of the corresponding protein.

Mutant in gene for protein	Protein production and location					
	A	B	C	D	E	F
A	−	+	*	+	*	+
B	*	−	*	*	*	*
C	*	+	−	+	*	+
D	*	+	*	−	*	+
E	+	+	+	+	−	+
F	*	+	*	*	*	−

Complete the following figure, showing the construction of the hypothetical protein complex, by writing the letter of the proper protein in each circle. The two proteins marked with arrowheads can assemble into the complex independently of each other, but both are needed for the addition of subsequent proteins to the complex.

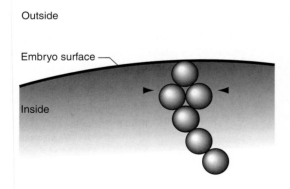

7-28 In 1952, an article in the *British Medical Journal* reported interesting differences in the behavior of blood plasma obtained from several individuals who suffered from X-linked recessive hemophilia. When mixed together, the cell-free blood plasma from certain combinations of individuals could form clots in the test tube. For example, the following table shows whether (+) or not (−) clots could form in various combinations of plasma from four individuals with hemophilia:

1 and 1	−	2 and 3	+
1 and 2	−	2 and 4	+
1 and 3	+	3 and 3	−
1 and 4	+	3 and 4	−
2 and 2	−	4 and 4	−

What do these data tell you about the inheritance of hemophilia in these individuals? Do these data allow you to exclude any models for the biochemical pathway governing blood clotting?

7-29 Mutations in an autosomal gene in humans cause a form of hemophilia called von Willebrand disease (vWD). This gene specifies a blood plasma protein cleverly called von Willebrand factor (vWF). vWF stabilizes factor VIII, a blood plasma protein specified by the wild-type hemophilia A gene. Factor VIII is needed to form blood clots. Thus, factor VIII is rapidly destroyed in the absence of vWF.

Which of the following might successfully be employed in the treatment of bleeding episodes in hemophiliac patients? Would the treatments work immediately or only after some delay needed for protein synthesis? Would the treatments have only a short-term or a prolonged effect? Assume that all mutations are null and that the plasma is cell-free.

a. transfusion of plasma from normal blood into a vWD patient
b. transfusion of plasma from a vWD patient into a different vWD patient
c. transfusion of plasma from a hemophilia A patient into a vWD patient
d. transfusion of plasma from normal blood into a hemophilia A patient
e. transfusion of plasma from a vWD patient into a hemophilia A patient
f. transfusion of plasma from a hemophilia A patient into a different hemophilia A patient
g. injection of purified vWF into a vWD patient
h. injection of purified vWF into a hemophilia A patient
i. injection of purified factor VIII into a vWD patient
j. injection of purified factor VIII into a hemophilia A patient

7-30 Adult hemoglobin is a multimeric protein with four polypeptides, two of which are α globin and two of which are β globin.

    a. How many genes are needed to define the structure of the hemoglobin protein?

    b. If a person is heterozygous for wild-type alleles and alleles that would yield amino-acid substitution variants for both α globin and β globin, how many different kinds of hemoglobin protein would be found in the person's red blood cells and in what proportion? Assume all alleles are expressed at the same level.

7-31 In addition to the predominant adult hemoglobin, HbA, which contains two α-globin chains and two β-globin chains ($\alpha_2\beta_2$), there is a minor hemoglobin, HbA$_2$, composed of two α and two δ chains ($\alpha_2\delta_2$). The β- and δ-globin genes are arranged in tandem and are highly homologous. Draw the chromosomes that would result from an event of unequal crossing-over between the β and δ genes.

7-32 a. What are the differences between null, hypomorphic, hypermorphic, dominant negative, and neomorphic mutations?

    b. For each of these kinds of mutations, would you predict they would be dominant or recessive to a wild-type allele in producing a mutant phenotype?

7-33 The α- and β-thalassemias are blood diseases that can arise from deletions in the α- or β-globin genes or from mutations that lead to underexpression of the α or β globins. In this problem, assume that contributions of two alleles are additive. Persons who have less than 35% of normal levels of β globin are defined as having clinical β-thalassemia. Alice is clinically affected, with only 20% of normal levels of β globin. Her father (Bill) has 70%, her mother (Carol) has 50%, her two brothers (David and Earl) have 50% and 70%, and her sister (Fran) has 100% of normal β globin.

    a. What are the genotypes of Bill and Carol?

    b. Define the levels of expression associated with each allele in each parent.

    c. If Bill and Carol have another baby, what is the probability it will have clinical β-thalassemia?

    d. What is the probability this next baby will have normal levels of β globin?

7-34 Most mammals, including "New World" primates such as marmosets (a kind of monkey), are *dichromats:* they have only two kinds of rhodopsin-related color receptors. "Old World" primates such as humans and gorillas are *trichromats* with three kinds of color receptors. Primates diverged from other mammals roughly 65 million years ago (Myr), while Old World and New World primates diverged from each other roughly 35 Myr.

    a. Using this information, define on Figure 7.28d the time span of any events that can be dated.

    b. Some New World monkeys have an autosomal color receptor gene and a single X-linked color receptor gene. The X-linked gene has three alleles, each of which encodes a photoreceptor that responds to light of a different wavelength (all three wavelengths are different from that recognized by the autosomal color receptor). How is color vision inherited in these monkeys?

    c. About 95% of all light-receiving neurons in humans and other mammals are rod cells containing rhodopsin, a pigment that responds to low level light of many wavelengths. The remaining 5% of light-receiving neurons are cone cells with pigments that respond to light of specific wavelengths of high intensity. What does this suggest about the lifestyle of the earliest mammals?

7-35 This problem refers to Figure A in the Fast Forward box on pp. 220–221. For each part below, describe what structures Robert Edgar would have seen in the electron microscope if he examined extracts of *E. coli* cells infected with the indicated temperature-sensitive mutant strains of bacteriophage T4 under restrictive conditions.

    a. A strain with a mutation in gene 19

    b. A strain with a mutation in gene 16

    c. Simultaneous infection with two mutant strains, one in gene 13 and the other in gene 14. The polypeptides produced by genes 13 and 14 associate with each other to form a multimeric protein that governs one step of phage head assembly (see Fig. A on p. 221).

    d. A strain whose genome contains mutations in both genes 15 and 35

8

Gene Expression: The Flow of Genetic Information from DNA via RNA to Protein

The ability of an aminoacyl-tRNA synthetase (*red*) to recognize a particular tRNA (*blue*) and couple it to its corresponding amino acid (*not shown*) is central to the molecular machinery that converts the language of nucleic acids into the language of proteins.

A dedicated effort to determine the complete nucleotide sequence of the haploid genome in a variety of organisms has been underway since 1990. This massive endeavor has been more successful than many scientists thought possible. By 2001, the sequence of As, Ts, Gs, and Cs in the genomes of more than 20 different species, including the bacterium *Escherichia coli*, the yeast *Saccharomyces cerevisiae*, the fruit fly *Drosophila melanogaster*, the nematode *Caenorhabditis elegans*, the plant *Arabidopsis thaliana*, and humans (*Homo sapiens*), had already been deciphered. With this sequence information in hand, geneticists can consult the *genetic code*—the nearly universal dictionary equating nucleotide sequence with amino-acid sequence—to decide what parts of a genome are likely to be genes. They can also identify genes through matches with nucleotide sequences already known to encode proteins in other organisms. As a result, modern geneticists can use the complete nucleotide sequence of a genome to predict the total number of genes in an organism, and by extension, to discover the number and amino-acid sequences of all the polypeptides that determine phenotype. Knowledge of DNA

sequence thus opens up powerful new possibilities for understanding an organism's growth and development at the molecular level.

Studies of the tiny nematode *Caenorhabditis elegans* illustrate the kinds of insights researchers can gain from this DNA-sequence-based approach. *C. elegans* is a roundworm 1 mm in length that lives in soils throughout the world (Fig. 8.1). Feeding on bacteria, it grows from fertilized egg to adult—either hermaphrodite or male—in just three days. Each hermaphrodite produces between 250 and 1000 progeny. Because of its small size, short life cycle, and capacity for prolific reproduction, *C. elegans* is an ideal subject for genetic analysis.

The haploid genome of *C. elegans* contains 100 million base pairs distributed among six chromosomes. In 1998, an international team of investigators in the United States and Britain reported the sequencing and preliminary analysis of virtually the entire *C. elegans* genome. Using their knowledge of the concepts explored in this chapter, they found that the nematode genome carries about 19,000 genes. Interestingly, roughly 15% of the genes recognized

Figure 8.1   C. elegans: An ideal subject for genetic analysis. Micrograph of several adult worms.

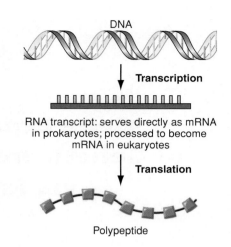

Figure 8.2   Gene expression: The flow of genetic information from DNA via RNA to protein. In transcription, the enzyme RNA polymerase copies DNA to produce an RNA transcript. In translation, the cellular machinery uses instructions in mRNA to synthesize a polypeptide, following the rules of the genetic code.

as having a known function encode molecules that play some role in **gene expression:** the process by which cells convert DNA sequence information to RNA and then decode the RNA information to the amino-acid sequence of a polypeptide (Fig. 8.2). The fact that 15% of nematode genes encode components of gene expression suggests the importance of the process to the life of the organism. About 2800 of the genes in the nematode either generate the machinery that enables genes to be interpreted as proteins, or they directly regulate this machinery to determine when, where, and how much of a protein will be made. In this chapter, we describe the cellular mechanisms that carry out gene expression. As intricate as some of the details may appear, the general scheme of gene expression is elegant and straightforward: *Within each cell, genetic information flows from DNA to RNA to protein.* In 1957 Francis Crick proposed that genetic information moves in only one direction and named his concept of a one-way molecular transfer the "Central Dogma" of molecular biology. As Crick explained, "Once 'information' has passed into protein, it cannot get out again."

Inside most cells, as the Central Dogma suggests, genetic information flows from one class of molecule to another in two distinct stages (Fig. 8.2). If you think of genes as instructions written in the language of nucleic acids, the cellular machinery first transcribes a set of instructions written in the DNA dialect to the same instructions written in the RNA dialect. The conversion of DNA-encoded information to its RNA-encoded equivalent is known as **transcription.** The product of transcription is a **transcript:** a molecule of **messenger RNA (mRNA)** in prokaryotes, a molecule of RNA that undergoes processing to become an mRNA in eukaryotes. In the second stage of gene expression, the cellular machinery translates the mRNA to its polypeptide equivalent in the language of amino acids. This decoding of nucleotide information to a sequence of amino acids is known as **translation.** It takes place on molecular workbenches called **ribosomes,** which are composed of proteins and ribosomal RNAs (rRNAs); and it depends on the universal dictionary known as the **genetic code,** which defines each amino acid in terms of specific sequences of

three nucleotides. It also depends on **transfer RNAs (tRNAs),** small RNA adaptor molecules that place specific amino acids at the correct position in a growing polypeptide chain. The tRNAs can bring amino acids to the right place on the translational machinery because tRNAs and mRNAs have complementary nucleotides that can base pair with each other.

Four general themes emerge from our discussion of gene expression. First, the pairing of complementary bases figures prominently in the precise transfer of information from DNA to RNA and from RNA to polypeptide. Second, the polarities of DNA, RNA, and protein molecules help guide the mechanisms of gene expression: the 3'-to-5' transcription of a template DNA strand yields a polar mRNA that builds from its 5' to its 3' end; the 5'-to-3' translation of this mRNA yields a polar protein running from its amino terminal to its carboxyl terminal. Third, like DNA replication and recombination (previously discussed in Chapter 6), gene expression requires an input of energy and the participation of several specific proteins and macromolecular assemblies like ribosomes at different points in the process. Fourth, since the accurate one-way flow of genetic information determines protein structure, mutations that change this information or obstruct its flow can have dramatic effects on phenotype.

As we examine how cells use the sequence information contained in DNA to construct proteins, we present:

- The genetic code: How triplets of the four nucleotides unambiguously specify 20 amino acids, making it possible to translate information from a nucleotide chain to a sequence of amino acids.

- Transcription: How RNA polymerase, guided by base pairing, synthesizes a single-stranded mRNA copy of a gene's DNA template.

- Translation: How base pairing between mRNA and tRNAs directs the assembly of a polypeptide on the ribosome.

- Significant differences in gene expression between prokaryotes and eukaryotes.

- A comprehensive example of gene expression in *C. elegans*.

- How mutations affect gene information and expression.

■ ■ ■

The Genetic Code: How Precise Groupings of the Four Nucleotides Specify 20 Amino Acids

A code is a system of symbols that equates information in one language with information in another. A useful analogy for the genetic code is the Morse code, which uses dots and dashes to transmit messages over radio or telegraph wires. Various groupings of the dot-dash symbols represent the 26 letters of the English alphabet. Because there are many more letters than the two symbols (dot or dash), groups of one, two, three, or four dots or dashes in various combinations represent individual letters. For example, the symbol for C is dash dot dash dot (- · - ·), the symbol for O is dash dash dash (- - -), D is dash dot dot (- · ·), and E is a single dot (·). Because anywhere from one to four symbols specify each letter, the Morse code requires a symbol for "pause" (in practice, a short interval of time) to signify where one letter ends and the next begins.

In the Genetic Code, a Triplet Codon Represents Each Amino Acid

The language of nucleic acids is written in four nucleotides—A, G, C, and T in the DNA dialect; A, G, C, and U in the RNA dialect—while the language of proteins is written in amino acids. To understand how the sequence of bases in DNA or RNA encodes the order of amino acids in a polypeptide chain, it is essential to know how many distinct amino acids there are. Watson and Crick produced the now accepted list of the 20 amino acids that are genetically encoded by DNA or RNA sequence over lunch one day at a local pub. They created the list by analyzing the amino-acid sequence of a variety of naturally occurring polypeptides. Amino acids that are present in only a small number of proteins or in only certain tissues or organisms did not qualify as standard building blocks; Crick and Watson correctly assumed that such amino acids arise when proteins undergo modification after their synthesis. By contrast, amino acids that are present in most, though not necessarily all, proteins made the list. The question then became: How can four nucleotides encode 20 amino acids?

Just as the Morse code conveys information through different groupings of dots and dashes, the four nucleotides encode 20 amino acids through specific groupings of A, G, C, and T or A, G, C, and U. Researchers initially arrived at the number of letters per grouping by deductive reasoning and later confirmed their guess by experiment. They reasoned that if only one nucleotide represented an amino acid, there would be information for only four amino acids: A would encode one amino acid;

G, a second amino acid; C, a third; and T, a fourth. If two nucleotides represented each amino acid, there would be $4^2 = 16$ possible combinations of couplets.

Of course, if the code consisted of groups containing one *or* two nucleotides, it would have $4 + 16 = 20$ groups and could account for all the amino acids, but there would be nothing left over to signify the pause required to denote where one group ends and the next begins. Groups of three nucleotides in a row would provide $4^3 = 64$ different triplet combinations, more than enough to code for all the amino acids. If the code consisted of doublets and triplets, a signal denoting pause would once again be necessary. But a triplets-only code would require no symbol for "pause" if the mechanism for counting to three and distinguishing among successive triplets were very reliable.

Although this kind of reasoning—explaining the unknown in terms of the known by looking for the simplest possible explanation—generates a theory, it does not prove it. As it turned out, however, the experiments described later in this chapter did indeed demonstrate that groups of three nucleotides represent all 20 amino acids. Each nucleotide triplet is called a **codon.** Each codon, designated by the bases defining its three nucleotides, specifies one amino acid. For example, GAA is a codon for glutamic acid (Glu), and GUU is a codon for valine (Val). Because the code comes into play only during the translation part of gene expression, that is, during the decoding of messenger RNA to polypeptide, geneticists usually present the code in the RNA dialect of A, G, C, and U, as depicted in Fig. 8.3. When speaking of genes, they can substitute T for U to show the same code in the DNA dialect.

If you knew the sequence of nucleotides in a gene or its transcript as well as the sequence of amino acids in the corresponding polypeptide, you could deduce the genetic code without understanding how the cellular machinery uses the code to translate from nucleotides to amino acids. Although techniques for determining both nucleotide and amino-acid sequence are available today, this was not true when researchers were trying to crack the genetic code in the 1950s and 1960s. At that time, they could establish a polypeptide's amino-acid sequence, but not the nucleotide sequence of DNA or RNA. Because of their inability to read nucleotide sequence, they used an assortment of genetic and biochemical techniques to fathom the code. They began by examining how different mutations in a single gene affected the amino-acid sequence of the gene's polypeptide product. In this way, they were able to use the abnormal (specific mutations) to understand the normal (the general relationship between genes and polypeptides).

Second letter

		U	C	A	G	
First letter	U	UUU UUC } Phe UUA UUG } Leu	UCU UCC UCA UCG } Ser	UAU UAC } Tyr UAA Stop UAG Stop	UGU UGC } Cys UGA Stop UGG Trp	U C A G
	C	CUU CUC CUA CUG } Leu	CCU CCC CCA CCG } Pro	CAU CAC } His CAA CAG } Gln	CGU CGC CGA CGG } Arg	U C A G
	A	AUU AUC } Ile AUA AUG Met	ACU ACC ACA ACG } Thr	AAU AAC } Asn AAA AAG } Lys	AGU AGC } Ser AGA AGG } Arg	U C A G
	G	GUU GUC GUA GUG } Val	GCU GCC GCA GCG } Ala	GAU GAC } Asp GAA GAG } Glu	GGU GGC GGA GGG } Gly	U C A G

Third letter

Figure 8.3 The genetic code: 61 codons represent the 20 amino acids, while 3 codons signify stop. To read the code, find the first letter in the *left column,* the second letter along the *top,* and the third letter in the *right column;* this reading corresponds to the 5′-to-3′ direction along the mRNA.

A Gene's Nucleotide Sequence Is Colinear with the Amino-Acid Sequence of the Encoded Polypeptide

We have seen that DNA is a linear molecule with base pairs following one another down the intertwined chains. Proteins, by contrast, have complicated three-dimensional structures. Even so, if unfolded and stretched out from N terminus to C terminus, proteins have a one-dimensional, linear structure—a specific sequence of amino acids. If the information in a gene and its corresponding protein are colinear, the consecutive order of bases in the DNA from the beginning to the end of the gene would stipulate the consecutive order of amino acids from one end to the other of the outstretched protein. Note that this hypothesized relationship implies that both a gene and its protein product have definite polarities with an invariant relation to each other.

Charles Yanofsky, in studying the *E. coli* gene for a subunit of the enzyme tryptophan synthetase during the 1960s, was the first to compare maps of mutations within a gene to the particular amino-acid substitutions that resulted. He began by generating a large number of *trp⁻* auxotrophic mutants that carried mutations in the *trpA* gene for the tryptophan synthetase subunit. He next made a fine structure recombinational map of these mutations; and then he purified and determined the amino-acid sequence of the mutant tryptophan synthetase subunits. As Fig. 8.4a illustrates, Yanofsky's data showed that the order of

mutations mapped within the DNA of the gene by recombination was indeed colinear with the positions of the amino-acid substitutions occurring in the resulting mutant proteins. In spite of this colinearity in order, distances on the genetic map (measured in map units) do not exactly reflect the number of amino acids between the amino-acid substitutions. This is because recombination as seen on this very high resolution map does not occur with an equal probability at every base pair within the gene.

Nonoverlapping Codons Are Set in a Reading Frame

By carefully examining the results of his analysis, Yanofsky, in addition to confirming the existence of colinearity, deduced key features of codons and helped establish many parameters of the genetic code relating nucleotides to amino acids.

A Codon Is Composed of More Than One Nucleotide

Yanofsky observed that different **point mutations** (changes in only one nucleotide pair) may affect the same amino acid. In one example shown in Fig. 8.4a, mutation #23 changed the glycine (Gly) at position 211 of the wild-type polypeptide chain to arginine (Arg), while mutation #46 yielded glutamic acid (Glu) at the same position. In another example, mutation #78 changed the glycine at position 234 to cysteine (Cys), while mutation #58 produced aspartic acid (Asp) at the same position. In both cases, Yanofsky also found that recombination could occur between the two mutations that changed the identity of the same amino acid, and such recombination would produce a wild-type tryptophan synthetase gene (Fig. 8.4b). Because the smallest unit of recombination is the base pair, two mutations capable of recombination—in this case, in the same codon because they affect the same amino acid—must be in different (although nearby) nucleotides. Thus, a codon contains more than one nucleotide.

Each Nucleotide Is Part of Only a Single Codon

As Fig. 8.4a illustrates, each of the point mutations in the tryptophan synthetase gene characterized by Yanofsky alters the identity of only a single amino acid. This is also true of the point mutations examined in many other genes, such as the human genes for rhodopsin and hemoglobin (see Chapter 7). Since point mutations change only a single nucleotide pair and most point mutations affect only a single amino acid in a polypeptide, each nucleotide in a gene must influence the identity of only a single amino acid. If, on the contrary, a nucleotide were part of more than one codon, a mutation in that nucleotide would affect more than one amino acid.

A Codon Is Composed of Three Nucleotides, and the Starting Point of Each Gene Establishes the Reading Frame

Although the most efficient code that would allow four nucleotides to specify 20 amino acids requires three nucleotides per codon, more complicated scenarios are possible. But in 1955

(a) Colinearity of genes and proteins

(b) Recombination within a codon

Figure 8.4 Mutations in a gene are colinear with the sequence of amino acids in the encoded polypeptide. (a) The relationship between the genetic map of *E. coli*'s *trpA* gene and the positions of amino-acid substitutions in mutant tryptophan synthetase proteins. Each mutation changes only a single amino acid, so each nucleotide is part of only a single codon. **(b)** Codons must include two or more base pairs. When two mutant strains with different amino acids at the same position were crossed, recombination could produce a wild-type allele. Thus, the two point mutations must be in different nucleotide pairs.

Francis Crick and Sydney Brenner obtained convincing evidence for the triplet nature of the genetic code in studies of mutations in the bacteriophage T4 *rIIB* gene (Fig. 8.5). They induced the mutations with proflavin, an intercalating mutagen that can insert itself between the paired bases stacked in the center of the DNA molecule (Fig. 8.5a). Their assumption was that proflavin would act like other mutagens, causing single-base substitutions. If this were true, it would be possible to generate revertants through treatment with other mutagens that might restore the wild-type DNA sequence. Surprisingly, genes with proflavin-induced mutations did not revert to wild type upon treatment with other mutagens known to cause nucleotide substitutions. Only further exposure to proflavin caused proflavin-induced mutations to revert to wild type (Fig. 8.5b). Crick and Brenner had to explain this observation before they could proceed with their phage experiments. With keen insight, they correctly guessed that proflavin does not cause base substitutions; instead, it causes insertions or deletions. This hypothesis explained why base-substituting mutagens could not revert proflavin-induced mutations; it was also consistent with the structure of proflavin. By intercalating between base pairs, proflavin would distort the double helix and thus interfere with the action of enzymes that function in the repair, replication, or recombination of DNA, eventually causing the deletion or addi-

tion of one or more nucleotide pairs to the DNA molecule (review Fig. 7.12c on pp. 204–205).

Crick and Brenner began their experiments with a particular proflavin-induced *rIIB⁻* mutation they called FC0. They next treated this mutant strain with more proflavin to isolate an *rIIB⁺* revertant (see Fig. 8.5b). By recombining this revertant with wild-type bacteriophage T4, Crick and Brenner were able to show that the revertant's chromosome actually contained two different *rIIB⁻* mutations (Fig. 8.5c). One was the original FC0 mutation; the other was the newly induced FC7. Either mutation by itself yields a mutant phenotype, but their simultaneous occurrence in the same gene yielded an *rIIB⁺* phenotype. Crick and Brenner reasoned that if the first mutation was the deletion of a single base pair, represented by the symbol (−), then the counteracting mutation must be the insertion of a base pair, represented as (+). The restoration of gene function by one mutation canceling another in the same gene is known as **intragenic suppression.** On the basis of this reasoning, they went on to establish T4 strains with different numbers of (+) and (−) mutations in the same chromosome. Figure 8.5d tabulates the phenotypes associated with each combination of proflavin-induced mutations.

In analyzing the data, Crick and Brenner assumed that each codon is a trio of nucleotides, and for each gene there is a

(a) The mutagen proflavin can insert between two base pairs.

Proflavin

Molecule of proflavin inserted between stacked base pairs

(b) Consequences of exposure to proflavin.

rIIB⁺ wild-type

Exposure to proflavin
FC0

rIIB⁻

Exposure to proflavin
FC0   FC7

rIIB⁺ revertant

Original   Second
mutation   mutation

(c) rIIB⁺ revertant X wild type yields rIIB⁻ recombinants.

FC0   FC7

rIIB⁻   FC0   rIIB⁻   FC7

(d) Different sets of mutations generate either a mutant or a normal phenotype.

Proflavin-induced mutations (+) insertion (−) deletion	Phenotype
− or +	Mutant
− − or + +	Mutant
− − − − or − − − − − or + + + + or + + + + +	Mutant
− +	Wild type
− − − or − − − − − or + + + or + + + + + +	Wild type

Figure 8.5   Studies of frameshift mutations in the bacteriophage T4 rIIB gene showed that codons consist of three nucleotides. (a) The mutagen proflavin slips between adjacent base pairs, eventually causing a deletion or insertion. **(b)** Treatment with proflavin produces a mutation at one site (FC0). A second proflavin exposure results in a second mutation (FC7) within the same gene, which suppresses FC0. **(c)** When the revertant is crossed with a wild-type strain, crossing-over separates the two rIIB⁻ mutations FC0 and FC7. The reversion to an rIIB⁺ phenotype was thus the result of intragenic suppression. **(d)** Evidence for a triplet code.

single starting point. This starting point establishes a **reading frame:** the partitioning of groups of three nucleotides such that the sequential interpretation of each succeeding triplet generates the correct order of amino acids in the resulting polypeptide chain. If codons are read in order from a fixed starting point, one mutation will counteract another if the two are equivalent mutations of opposite signs; in such a case, each insertion compensates for each deletion, and this counterbalancing restores the reading frame (Fig. 8.6a). The gene would only regain its wild-type activity, however, if the portion of the polypeptide encoded between the two mutations of opposite sign is not required for protein function, because in the double mutant, this region would have an improper amino-acid sequence. Similarly, if a gene sustains three or multiples of three changes of the same sign, the encoded polypeptide can still function, because the mutations do not alter the reading frame for the majority of amino acids (Fig. 8.6b). The resulting polypeptide will, however, have one extra or one fewer amino acid than normal (designated by three plus signs (+) or three minus signs (−), respectively), and the region encoded by the part of the gene between the first and the last mutations will not contain the correct amino acids.

By contrast, a single nucleotide inserted into or deleted from a gene alters the reading frame, and thereby it affects the identity of not only one amino acid but of all other amino acids beyond the point of alteration (Fig. 8.6c). Changes that alter the grouping of nucleotides into codons are called **frameshift mutations:** they shift the reading frame for all codons beyond the point of insertion or deletion, almost always abolishing the function of the polypeptide product.

A review of the evidence tabulated in Fig. 8.5d supports all these points. A single (−) or a single (+) mutation destroyed the function of the rIIB gene and produced an rIIB⁻ phage. Similarly, any gene with two base changes of the same sign (− − or + +), or with four or five insertions or deletions of the same sign (for example, + + + +), also generated a mutant phenotype. However, genes containing three or multiples of three mutations of the same sign (for example, + + + or − − − − − −), as well as genes containing a (+ −) pair of mutations, generated rIIB⁺ wild-type individuals. In these last examples, intragenic suppression allowed restitution of the reading frame and thereby restored the lost or aberrant genetic function produced by other frameshift mutations in the gene.

correct triplet
incorrect triplet

(a) Intragenic suppression: 2 mutations of opposite sign.

Single base insertion (+) and single base deletion (−)

(b) Intragenic suppression: 3 mutations of the same sign.

Three single base deletions (− − −)

Three single base insertions (+ + +)

(c) Some frameshift mutations.

Single base deletion (−)

Single base insertion (+)

Figure 8.6 Codons consist of three nucleotides read in a defined reading frame. The phenotypic effects of proflavin-induced frameshift mutations depend on whether the reading frame is restored and whether the part of the gene with an altered reading frame specifies an essential or nonessential region of the polypeptide.

Most Amino Acids Are Specified by More Than One Codon

As Fig. 8.6a illustrates, intragenic suppression occurs only if, in the region between two frameshift mutations of opposite sign, a gene still dictates the appearance of amino acids, even if these amino acids are not the same as those appearing in the normal protein. If the frameshifted part of the gene encodes instructions to stop protein synthesis, by introducing, for example, a triplet of nucleotides that does not correspond to any amino acid, then wild-type polypeptide production will not continue.

This is because polypeptide synthesis would stop before the compensating mutation could reestablish the correct reading frame.

The fact that intragenic suppression occurs as often as it does suggests that the code includes more than one codon for some amino acids. Recall that there are 20 common amino acids but $4^3 = 64$ different combinations of three nucleotides. If each amino acid corresponded to only a single codon, there would be $64 - 20 = 44$ possible triplets not encoding an amino acid. These noncoding triplets would act as "stop" signals and prevent further polypeptide synthesis. If this happened, more than half of all frameshift mutations (44/64) would cause protein synthesis to stop at the first codon after the mutation, and the chances of extending the protein each amino acid farther down the chain would diminish exponentially. As a result, intragenic suppression would rarely occur. However, we have seen that many frameshift mutations of one sign can be offset by mutations of the other sign. The distances between these mutations, estimated by recombination frequencies, are in some cases large enough to code for more than 50 amino acids, which would be possible only if most of the 64 possible triplet codons specified amino acids. Thus, the data of Crick and Brenner provide strong support for the idea that the genetic code is **degenerate:** Two or more nucleotide triplets specify most of the 20 amino acids (see the genetic code in Fig. 8.3 on p. 240).

Cracking the Code: Biochemical Manipulations Revealed Which Codons Represent Which Amino Acids

Although the genetic experiments just described enabled remarkably prescient insights about the nature of the genetic code, they did not make it possible to assign particular codons to their corresponding amino acids. This awaited the discovery of messenger RNA and the development of techniques for synthesizing simple messenger RNA molecules that researchers could use to manufacture simple proteins in the test tube.

The Discovery of Messenger RNAs, Molecules for Transporting Genetic Information

In the 1950s researchers exposed eukaryotic cells to amino acids tagged with radioactivity and observed that protein synthesis incorporating the radioactive amino acids into polypeptides takes place in the cytoplasm, even though the genes for those polypeptides are sequestered in the cell nucleus. From this discovery, they deduced the existence of an intermediate molecule, made in the nucleus and capable of transporting DNA sequence information to the cytoplasm, where it can direct protein synthesis. RNA was a prime candidate for this intermediary information-carrying molecule. Because of RNA's potential for base pairing with a strand of DNA, one could imagine the cellular machinery copying a strand of DNA into a complementary strand of RNA in a manner analogous to the DNA-to-DNA copying of DNA replication. Subsequent studies in eukaryotes on the incorporation of radioactive uracil (a base found only in

RNA) into molecules of RNA showed that although the molecules are synthesized in the nucleus, at least some of them migrate to the cytoplasm. Among those RNA molecules that migrate to the cytoplasm are the messenger RNAs, or mRNAs, depicted in Fig. 8.2 on p. 238. They arise in the nucleus from the transcription of DNA sequence information through base pairing and then move (after processing) to the cytoplasm, where they determine the proper order of amino acids during protein synthesis.

Using Synthetic mRNAs and *In Vitro* Translation to Discover Which Codons Designate Which Amino Acids

Knowledge of mRNA served as the framework for two experimental breakthroughs that led to the deciphering of the genetic code. In the first, biochemists obtained cellular extracts that, with the addition of mRNA, synthesized polypeptides in a test tube. They called these extracts "*in vitro* translational systems." The second breakthrough was the development of techniques enabling the synthesis of artificial mRNAs containing only a few codons of known composition. When added to *in vitro* translational systems, these simple, synthetic mRNAs directed the formation of very simple polypeptides.

In 1961, Marshall Nirenberg and Heinrich Matthaei added a synthetic poly-U (5′ . . . UUUUUUUUUUUU . . . 3′) mRNA to a cell-free translational system derived from *E. coli*. With the poly-U mRNA, phenylalanine (Phe) was the only amino acid incorporated into the resulting polypeptide (Fig. 8.7a). Because UUU is the only possible triplet in poly-U, UUU must be a codon for phenylalanine. In a similar fashion, Nirenberg and Matthaei showed that CCC encodes proline (Pro), AAA is a codon for lysine (Lys), and GGG encodes glycine (Gly) (Fig. 8.7b).

The chemist Har Gobind Khorana later made mRNAs with repeating dinucleotides, such as poly-UC (5′ . . . UCUCUCUC . . . 3′), repeating trinucleotides, such as poly-UUC, and repeating tetranucleotides, such as poly-UAUC, and used them to direct the synthesis of slightly more complex polypeptides. As Fig. 8.7b shows, his results limited the coding possibilities, but some ambiguities remained. For example, poly-UC encodes the polypeptide N . . . Ser-Leu-Ser-Leu-Ser-Leu . . . C in which serine and leucine alternate with each other. Although the mRNA contains only two different codons (5′ UCU 3′ and 5′ CUC 3′), it is not obvious which corresponds to serine and which to leucine.

Nirenberg and Philip Leder resolved these ambiguities in 1965 with experiments in which they added short, synthetic mRNAs only three nucleotides in length to an *in vitro* translational system containing 1 radioactive amino acid and 19 unlabeled amino acids, all attached to tRNA molecules. They then poured through a filter the mixture of synthetic mRNAs and translational systems containing a tRNA-attached, radioactively labeled amino acid (Fig. 8.8). tRNAs carrying an amino acid normally go right through a filter. If, however, a tRNA carrying an amino acid binds to a ribosome, it will stick in the filter, because this larger complex of

(a) Poly-U mRNA encodes polyphenylalanine.

Analyze radioactive polypeptides synthesized

Synthetic mRNA

In vitro translational system plus radioactive amino acids

(b) Analyzing the coding possibilities.

Synthetic mRNA	Polypeptides synthesized
	Polypeptides with one amino acid
poly-U UUUU ...	Phe-Phe-Phe ...
poly-C CCCC ...	Pro-Pro-Pro ...
poly-A AAAA ...	Lys-Lys-Lys ...
poly-G GGGG ...	Gly-Gly-Gly ...
Repeating dinucleotides	**Polypeptides with alternating amino acids**
poly-UC UCUC ...	Ser-Leu-Ser-Leu ...
poly-AG AGAG ...	Arg-Glu-Arg-Glu ...
poly-UG UGUG ...	Cys-Val-Cys-Val ...
poly-AC ACAC ...	Thr-His-Thr-His ...
Repeating trinucleotides	**Three polypeptides each with one amino acid**
poly-UUC UUCUUCUUC ...	Phe-Phe.... and Ser-Ser.... and Leu-Leu....
poly-AAG AAGAAGAAG ...	Lys-Lys.... and Arg-Arg.... and Glu-Glu....
poly-UUG UUGUUGUUG ...	Leu-Leu.... and Cys-Cys.... and Val-Val....
poly-UAC UACUACUAC ...	Tyr-Tyr.... and Thr-Thr.... and Leu-Leu....
Repeating tetranucleotides	**Polypeptides with repeating units of four amino acids**
poly-UAUC UAUCUAUC ...	Tyr-Leu-Ser-Ile-Tyr-Leu-Ser-Ile...
poly-UUAC UUACUUAC ...	Leu-Leu-Thr-Tyr-Leu-Leu-Thr-Tyr...
poly-GUAA GUAAGUAA ...	none
poly-GAUA GAUAGAUA ...	none

Figure 8.7   How geneticists used synthetic mRNAs to limit the coding possibilities. (a) Poly-U mRNA generates a poly-phenylalanine polypeptide. **(b)** Polydi-, polytri-, and polytetra-nucleotides encode simple polypeptides. Some synthetic mRNAs, such as poly-GUAA, contain stop codons in all three reading frames and thus specify the construction only of short peptides.

ribosome, amino-acid-carrying tRNA, and small mRNA cannot pass through the filter. Nirenberg and Leder could thus use this approach to see which small mRNA caused the entrapment of which radioactively labeled amino acid in the filter. For example, they knew from Khorana's earlier work that CUC encoded either serine or leucine. When they added the synthetic triplet CUC to an *in vitro* system where the radioactive amino acid was serine, this tRNA-attached amino acid passed through the filter, and the filter thus emitted no radiation. But when they added the same triplet to a system where the radioactive amino acid was leucine, the filter lit up with radioactivity, indicating that the radioactively tagged leucine attached to a tRNA had bound to the ribosome-mRNA complex and gotten stuck in the filter. CUC thus encodes leucine, not serine. Nirenberg and Leder used this technique to determine all the codon–amino acid correspondences shown in the genetic code (see Fig. 8.3 on p. 240).

Figure 8.8 Cracking the genetic code with mini-mRNAs.
Nirenberg and Leder added trinucleotides of known sequence, in combination with tRNAs charged with a radioactive amino acid, to an *in vitro* extract containing ribosomes. If the trinucleotide specified the radioactive amino acid, the amino-acid-bearing tRNA formed a complex with the ribosomes and thus became trapped when the solution was passed through a filter. The experiments shown here indicate that the codon CUC specifies leucine, not serine.

The 5′-to-3′ Direction in mRNA Corresponds to the N-Terminal-to-C-Terminal Direction in the Polypeptide

In studies using synthetic mRNAs, when investigators added the six-nucleotide-long 5′ AAAUUU 3′ to an *in vitro* translational system, the product N-Lys-Phe-C emerged, but no N-Phe-Lys-C appeared. Since AAA is the codon for lysine and UUU is the codon for phenylalanine, this means that the codon closest to the 5′ end of the mRNA encoded the amino acid closest to the N terminus of the corresponding polypeptide. Similarly, the codon nearest the 3′ end of the mRNA encoded the amino acid nearest the C terminus of the resulting polypeptide.

To understand how the polarities of the macromolecules participating in gene expression relate to each other, remember that although the gene is a segment of a DNA double helix, only one of the two strands serves as a template for the mRNA. This strand is known as the **template strand.** The other strand is the **RNA-like strand,** because it has the same polarity and sequence (written in the DNA dialect) as the RNA. Figure 8.9 diagrams the respective polarities of a gene's DNA, the mRNA transcript of that DNA, and the resulting polypeptide.

Nonsense Codons Cause Termination of a Polypeptide Chain

Although most of the simple, repetitive RNAs synthesized by Khorana were very long and thus generated very long polypeptides, a few did not. These RNAs had signals that stopped con-

Figure 8.9 Correlation of polarities in DNA, mRNA, and polypeptide. The template strand of DNA is complementary to both the RNA-like DNA strand and the mRNA. The 5′-to-3′ direction in an mRNA corresponds to the N-terminus-to-C-terminus direction in the polypeptide.

struction of a polypeptide chain. As it turned out, three different triplets—UAA, UAG, and UGA—do not correspond to any of the amino acids. When these codons appear in frame, translation stops. As an example of how investigators established this fact, consider the case of poly-GUAA (review Fig. 8.7b). This mRNA will not generate a long polypeptide because in all possible reading frames, it contains the **stop codon** UAA.

The three stop codons that terminate translation are also known as **nonsense codons.** For historical reasons, researchers often refer to UAA as the *ocher* codon, UAG as the *amber* codon, and UGA as the *opal* codon. The historical basis of this nomenclature is the last name of one of the early investigators—Bernstein—which means "amber" in German; ocher and opal derive from their similarity with amber as semiprecious materials.

The Genetic Code: A Summary

From the experiments just described, researchers learned that the genetic code is a complete, unabridged dictionary equating the 4-letter language of the nucleic acids with the 20-letter language of the proteins. The following list summarizes the code's main features:

1. The code consists of *triplet codons,* each of which specifies an amino acid. As written in Fig. 8.3 on p. 240, the code shows the 5′-to-3′ sequence of the three nucleotides in each mRNA codon; that is, the first nucleotide depicted is at the 5′ end of the codon.

2. The codons are *nonoverlapping.* In the mRNA sequence 5′ GAAGUUGAA 3′, for example, the first three nucleotides (GAA) form one codon; nucleotides 4 through 6 (GUU) form the second; and nucleotides 7 through 9 (GAA), the third. Each nucleotide is part of only one codon.

3. The code includes three *stop,* or *nonsense, codons:* UAA, UAG, and UGA. These codons do not encode an amino acid and thus terminate translation.

4. The code is *degenerate,* which means that in many cases more than one codon specifies the same amino acid.

Despite its "degeneracy," the code is unambiguous, because each codon specifies only one amino acid.

5. In reading the transcript of a gene, the cellular machinery scans a single strand of mRNA from a fixed starting point that establishes a *reading frame*. As we see later, the nucleotide triplet AUG, which specifies the amino acid methionine, serves in certain contexts as the **initiation codon,** marking the precise spot in the nucleotide sequence of an mRNA where the code for a particular polypeptide begins. The position of the initiation codon specifies the reading frame for the rest of the sequence by determining how the nucleotides become grouped into triplets. Note that without a set reading frame established by an initiation codon, there would be three possible ways to read a piece of mRNA, depending on whether you begin with the first, second, or third base in the sequence.

6. Moving from the 5′ to the 3′ end of an mRNA, each successive codon is sequentially interpreted into an amino acid, starting at the N terminus and moving toward the C terminus of the resulting polypeptide.

7. Mutations may modify the message encoded in a sequence of nucleotides in three ways. *Frameshift mutations* are nucleotide insertions or deletions that alter the genetic instructions for polypeptide construction by changing the reading frame. *Missense mutations* change a codon for one amino acid to a codon for a different amino acid. *Nonsense mutations* change a codon for an amino acid to a stop codon.

Using Genetics to Verify the Code

The experiments that cracked the genetic code by assigning codons to amino acids were all *in vitro* studies using cell-free extracts and synthetic mRNAs. A logical question thus arose: Do living cells construct polypeptides according to the same rules? Early evidence that they do came from studies analyzing how mutations actually affect the amino-acid composition of the polypeptides encoded by a gene. Most mutagens change a single nucleotide in a codon. As a result, most missense mutations that change the identity of a single amino acid should be single-nucleotide substitutions, and analyses of these substitutions should conform to the code. Yanofsky, for example, found two *trp⁻* auxotrophic mutations in the *E. coli* tryptophan synthetase gene that produced two different amino acids (arginine, or Arg, and glutamic acid, or Glu) at the same position—amino acid 211—in the polypeptide chain (Fig. 8.10a). According to the code, both of these mutations could have resulted from single-base changes in the GGA codon that normally inserts glycine (Gly) at position 211.

Even more telling were the *trp⁺* revertants of these mutations subsequently isolated by Yanofsky. As Fig. 8.10a illustrates, single-base substitutions could also explain the amino-acid changes in these revertants. Note that some of these substitutions restore Gly to position 211 of the polypeptide,

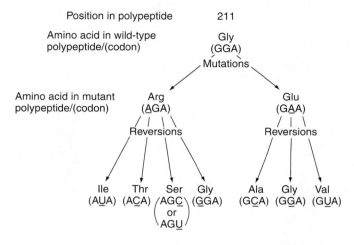

(a) Altered amino acids in *trp⁻* mutations and *trp⁺* revertants

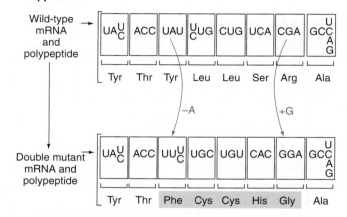

(b) Amino acid alterations that accompany intragenic suppression

Figure 8.10 Experimental verification of the genetic code. **(a)** Single-base substitutions can explain the amino-acid substitutions of *trp⁻* mutations and *trp⁺* revertants. **(b)** The genetic code predicts the amino-acid alterations (*yellow*) that would arise from single-base-pair deletions and suppressing insertions.

while others place amino acids such as Ile, Thr, Ser, Ala, or Val at this site in the tryptophan synthetase molecule. In relation to Gly, these other amino acids constitute conservative substitutions because their presence at position 211 in the polypeptide chain is compatible with (that is, largely conserves) the enzyme's function.

Yanofsky obtained better evidence yet that cells use the genetic code *in vivo* by analyzing proflavin-induced frameshift mutations of the tryptophan synthetase gene (Fig. 8.10b). He first treated populations of *E. coli* with proflavin to produce *trp⁻* mutants. Subsequent treatment of these mutants with more proflavin generated some *trp⁺* revertants among the progeny. The most likely explanation for the revertants was that their tryptophan synthetase gene carried both a single-base-pair deletion and a single-base-pair insertion (− +). Upon determining the amino-acid sequences of the tryptophan synthetase enzymes made by the revertant strains, Yanofsky found that he could use the genetic code to predict the precise amino-acid alterations that had occurred by assuming the revertants had a specific single-base-pair insertion and a specific single-base-pair deletion.

Yanofsky's results helped confirm not only amino-acid codon assignments but other parameters of the code as well. His interpretations make sense only if codons do not overlap and are read from a fixed starting point with no pauses or commas separating the adjacent triplets.

In Chapter 9, we describe molecular techniques developed since the 1970s that make it possible to read the exact nucleotide sequence of a piece of DNA. In studies using these techniques, geneticists have determined the DNA sequences of genes encoding proteins of known amino-acid sequence. These nucleotide sequences verify that the codons assigned to each amino acid by *in vitro* translational systems incorporating synthetic mRNAs are in fact used for translation *in vivo*.

The Genetic Code Is Almost, But Not Quite, Universal

We now know that virtually all cells alive today—from the prokaryotic cells of *E. coli* bacteria through the range of eukaryotic cells in yeast, *C. elegans,* peas, lentils, and people— use the same basic genetic code. One early indication of this uniformity was that a translational system derived from one organism could use the mRNA from another organism to convert genetic information to the encoded protein. Rabbit hemoglobin mRNA, for example, when injected into frog eggs or added to cell-free extracts from wheat germ, directs the synthesis of rabbit hemoglobin proteins. More recently, comparisons of DNA and protein sequences have revealed a perfect correspondence according to the genetic code between codons and amino acids in almost all organisms examined.

The universality of the code is an indication that it evolved very early in the history of life. Once it emerged, it remained constant over billions of years, in part because evolving organisms would have little tolerance for change. A single change in the genetic code could disrupt the production of hundreds or thousands of proteins in a cell—from the DNA polymerase that is essential for replication to the RNA polymerase that is required for gene expression to the tubulin proteins that compose the mitotic spindle—and would therefore be lethal.

Researchers were thus quite amazed to observe a few exceptions to the universality of the code. In some species of the single-celled eukaryotic protozoans known as ciliates, the codons UAA and UAG, which are nonsense codons in most organisms, specify the amino acid glutamine; in other ciliates, UGA, the third stop codon in most organisms, specifies cysteine. These ciliates use the remaining nonsense codons as stop codons. Other systematic changes in the genetic code exist in mitochondria, the semiautonomous, self-reproducing organelles within eukaryotic cells that are the sites of ATP formation. Each mitochondrion has its own chromosomes and its own apparatus

for gene expression (which we describe in detail in Chapter 15). In the mitochondria of yeast, CUA specifies threonine instead of leucine. It may be that ciliates and mitochondria tolerated these changes in the genetic code because the changes affected very few proteins. For instance, the nonsense codon UGA might have found only infrequent use in one kind of primitive ciliate, so its switch to a "sense" codon would not have made a tremendous difference in protein production. Similarly, mitochondria might have survived a few changes in the code because they synthesize only a handful of proteins.

The experimental evidence presented so far helped define a nearly universal genetic code. But although cracking the code made it possible to understand the broad outlines of information flow between gene and protein, it did not explain exactly how the cellular machinery accomplishes gene expression. This is our focus as we present the details of transcription and translation.

Transcription: RNA Polymerase Synthesizes a Single-Stranded RNA Copy of a Gene

Transcription is the process by which the polymerization of ribonucleotides guided by complementary base pairing produces an RNA transcript of a gene. The template for the RNA transcript is one strand of that portion of the DNA double helix that composes the gene.

Details of the Process

Figure 8.11 depicts the basic components of transcription and illustrates key events in the process as it occurs in the bacterium *E. coli.* Although certain details may vary slightly in other organisms, the following four main points always hold true: (1) The enzyme **RNA polymerase** catalyzes transcription. (2) DNA sequences near the beginning of genes, called **promoters,** signal RNA polymerase where to begin transcription. As seen in Fig. 8.12 on p. 250, most bacterial gene promotors share two short regions in which the promoters have almost identical nucleotide sequence. The nucleotides at these positions have evolved substantial similarity in different promoters because they are the sites at which RNA polymerase makes particularly strong contact with the promoters. (3) RNA polymerase adds nucleotides to the growing RNA polymer in the 5′- to-3′ direction. (4) Sequences in the RNA products, known as **terminators,** tell RNA polymerase where to stop transcription.

As you examine Fig. 8.11 to see how components of the cellular machinery work together to begin, carry out, and end transcription, bear in mind that a gene consists of two antiparallel strands of DNA, as mentioned earlier. One—the *RNA-like strand*—has the same polarity and sequence (except for T instead of U) as the emerging RNA transcript. The second—the *template strand*—has the opposite polarity and a complementary sequence that enables it to serve as the template for making the RNA transcript. When geneticists refer to the sequence of a gene, they usually mean the sequence of the RNA-like strand.

Transcription: How cells copy the DNA of a gene into the RNA of a transcript.

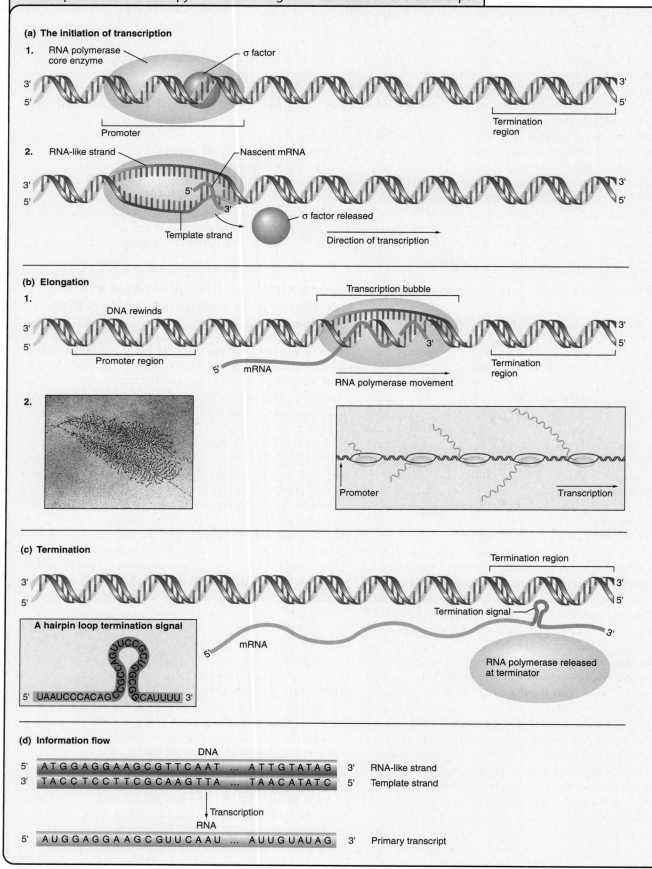

(a) The initiation of transcription

1. RNA polymerase core enzyme
   σ factor
   3'
   5'
   Promoter
   Termination region

2. RNA-like strand
   Nascent mRNA
   3'
   5'
   5'
   3'
   Template strand
   σ factor released
   Direction of transcription

(b) Elongation

1. Transcription bubble
   DNA rewinds
   3'
   5'
   Promoter region
   5'   mRNA   3'
   RNA polymerase movement
   Termination region

2.
   Promoter
   Transcription

(c) Termination

Termination region

3'
5'
Termination signal

A hairpin loop termination signal
   GUUCCGC
   CCGCCAG   GCG
   GCGC
   5' UAAUCCCACAG   GCAUUUU 3'

mRNA
5'
3'

RNA polymerase released at terminator

(d) Information flow

DNA

5' ATGGAGGAAGCGTTCAAT ... ATTGTATAG 3'   RNA-like strand
3' TACCTCCTTCGCAAGTTA ... TAACATATC 5'   Template strand

↓ Transcription

RNA

5' AUGGAGGAAGCGUUCAAU ... AUUGUAUAG 3'   Primary transcript

(a) The Initiation of Transcription

1. *RNA polymerase binds to double-stranded DNA at the beginning of the gene to be copied.* RNA polymerase recognizes and binds to **promoters,** specialized DNA sequences near the beginning of a gene where transcription will start. Although the promoters of different genes vary substantially in size and sequence, all promoters contain two characteristic short sequences of 6–10 nucleotide pairs that help bind RNA polymerase. These short sequences are nearly identical in different promoters (Fig. 8.12). In bacteria, the complete RNA polymerase (the *holoenzyme*) consists of a core enzyme, plus a σ subunit involved only in initiation. The σ subunit reduces RNA polymerase's general affinity for DNA but simultaneously increases RNA polymerase's affinity for the promoter. As a result, the RNA polymerase holoenzyme can hone in on a promoter and bind tightly to it, forming a so-called *closed promoter complex.*

2. *After binding to the promoter, RNA polymerase unwinds part of the double helix, exposing unpaired bases on the strand that will serve as the template.* The complex formed between the RNA polymerase holoenzyme and an unwound promoter is called an *open promoter complex.* The enzyme identifies the template strand and chooses the two nucleotides with which to initiate copying. Guided by base pairing with these two nucleotides, RNA polymerase aligns the first two ribonucleotides of the new RNA, which will be at the 5′ end of the final RNA product. The end of the gene transcribed into the 5′ end of the mRNA is often called the *5′ end of the gene.* The mRNA grows in the 5′-to-3′ direction because RNA polymerase adds nucleotides to the 3′ end of the growing RNA chain. After aligning the first two ribonucleotides, RNA polymerase catalyzes the formation of a phosphodiester bond between them, resulting in a diribonucleotide. Soon thereafter, the RNA polymerase releases the σ subunit. This release marks the end of initiation.

(b) Elongation: Constructing an RNA Copy of the Gene

1. *When the σ subunit separates from the RNA polymerase, the enzyme loses its enhanced affinity for the promoter sequence and regains its strong generalized affinity for any DNA.* These changes enable the core enzyme to leave the promoter yet remain bound to the gene. The core enzyme now moves along the chromosome, unwinding the double helix to expose the next single-stranded region of the template. The enzyme extends the RNA by linking to the 3′ end of the growing chain a ribonucleotide positioned by complementarity with the template strand. As the enzyme extends the mRNA in the 5′-to-3′ direction, it moves in the antiparallel 3′-to-5′ direction along the DNA template strand.

The region of DNA unwound by RNA polymerase is called the **transcription bubble.** Within the bubble, the nascent RNA chain remains base paired with the DNA template, forming a DNA-RNA hybrid. However, in those parts of the gene behind the bubble which have already been transcribed, the DNA double helix reforms, displacing the RNA which hangs out of the transcription complex as a single strand with a free 5′ end.

2. *Once an RNA polymerase has moved off the promoter, other RNA polymerase molecules can move in to initiate transcription.* If the promoter is very strong, that is, if it can rapidly attract RNA polymerase, the gene can undergo transcription by many RNA polymerases simultaneously. Here we show an electron micrograph and an artist's interpretation of simultaneous transcription by several RNA polymerases. As you can see, there is a range of RNAs in production; those in the early steps of elongation are shorter than those in later steps of elongation. The promoter for this gene thus lies very close to where the shortest RNA is emerging from the DNA.

Geneticists often use the direction traveled by RNA polymerase as a reference when discussing various features within a gene. If for example, you started at the 5′ end of a gene at point A and moved along the gene in the same direction as RNA polymerase to point B, you would be traveling in the **downstream** direction. If by contrast, you started at point B and moved in the opposite direction from RNA polymerase to point A, you would be traveling in the **upstream** direction.

(c) Termination: The End of Transcription

RNA sequences that signal the end of transcription are known as **terminators.** There are two types of terminators: *intrinsic terminators,* which cause the RNA polymerase core enzyme to terminate transcription on its own, and *extrinsic terminators,* which require proteins other than RNA polymerase—particularly a polypeptide known as *rho*—to bring about termination. All terminators, whether intrinsic or extrinsic, are specific sequences in the mRNA; they are transcribed from specific DNA regions and they signal the termination of transcription. Terminators often form **hairpin loops** in which nucleotides within the mRNA pair with nearby complementary nucleotides. Upon termination, RNA polymerase and a completed RNA chain are both released from the DNA.

(d) The Product of Transcription Is a Single-Stranded Primary Transcript

The RNA produced by the action of RNA polymerase on a gene is a single strand of nucleotides known as a *primary transcript.* The bases in the primary transcript are complementary to the bases between the initiation and termination sites in the template strand of the gene. The nucleotides carrying these bases include groupings for a start codon, codons specifying all the amino acids in the polypeptide to be built, and a stop codon.

(a) Transcription initiation signals in bacteria

(b) Strong *E. coli* promoters

Figure 8.12 The promoters of 10 different bacterial genes. Only the sequence of the RNA-like strand is shown; numbering starts at the first transcribed nucleotide (+1) **(a)** Most promoters are upstream of (that is, before) the start point of transcription. **(b)** All promoters in *E. coli* share two different short stretches of nucleotides (*yellow*) that are essential for recognition of the promoter by RNA polymerase. The most common nucleotides at each position in each stretch constitute a *consensus sequence;* invariant nucleotides are in *bold.*

The Genetics and Society box "HIV and Reverse Transcription: An Unusual DNA Polymerase Gives the AIDS Virus an Evolutionary Edge" on pp. 252–253 describes how the AIDS virus uses an uncommon variation of transcription, known as **reverse transcription,** to construct a double strand of DNA from an RNA template.

As Fig. 8.11d illustrates, the result of transcription is a single strand of RNA known as a **primary transcript.** In prokaryotic organisms, the RNA produced by transcription is the actual messenger RNA that guides protein synthesis. In eukaryotic organisms, by contrast, most primary transcripts undergo *processing* in the nucleus before they migrate to the cytoplasm to direct protein synthesis. This processing has played a fundamental role in the evolution of complex organisms.

In Eukaryotes, RNA Processing After Transcription Produces a Mature mRNA

Some RNA processing in eukaryotes modifies only the 5′ or 3′ ends of the primary transcript, leaving untouched the information content of the rest of the mRNA. Other processing deletes blocks of information from the middle of the primary transcript so that the content of the mature mRNA is related, but not identical, to the complete set of DNA nucleotide pairs in the original gene.

Adding a Methylated Cap at the 5′ End and a Poly-A Tail at the 3′ End

The nucleotide at the 5′ end of a eukaryotic mRNA is a G in reverse orientation from the rest of the molecule; it is connected through a triphosphate linkage to the first nucleotide in the pri-

mary transcript. This backward G is not transcribed from the DNA. Instead, a special *capping enzyme* adds it to the primary transcript after polymerization of the transcript's first few nucleotides. Enzymes known as *methyl transferases* then add methyl ($-CH_3$) groups to the backward G and to one or more of the succeeding nucleotides in the RNA, forming a so-called **methylated cap** (Fig. 8.13). The methylated cap is critical for efficient translation of the mRNA into protein, even though it does not help specify an amino acid.

Like the 5′ methylated cap, the 3′ end of most eukaryotic mRNAs is not encoded directly by the gene. In a large majority of eukaryotic mRNAs, the 3′ end consists of 100–200 As, referred to as a **poly-A tail** (Fig. 8.14). Addition of the tail is a two-step process. First, ribonuclease cleaves the primary transcript to form a new 3′ end; cleavage depends on the sequence AAUAAA, which is found in poly-A-containing mRNAs 11–30 nucleotides upstream of the position where the tail is added. Next the enzyme *poly-A polymerase* adds As onto the 3′ end exposed by cleavage. The actual function of the poly-A tail remains a matter of some controversy, but most investigators believe it may have three functions: aiding mRNA transport out of the nucleus into the cytoplasm; stabilizing mRNAs in the cytoplasm so that they can serve as messages for a longer time; and increasing the efficiency of the initial steps of translation.

RNA Splicing Removes Introns from the Primary Transcript

Another kind of RNA processing became apparent in the late 1970s, after researchers had developed techniques that enabled them to analyze nucleotide sequences in both DNA and RNA.

Figure 8.13 Structure of the methylated cap at the 5′ end of eukaryotic mRNAs. Capping enzyme connects a backward G to the first nucleotide of the primary transcript through a triphosphate linkage. Methyl transferase enzymes then add methyl groups to this G and to one or two of the nucleotides first transcribed from the DNA template.

Figure 8.14 How RNA processing adds a tail to the 3′ end of eukaryotic mRNAs. A ribonuclease recognizes AAUAAA in a particular context of the primary transcript and cleaves the transcript 11–30 nucleotides downstream to create a new 3′ end. The enzyme poly-A polymerase then adds 100–200 As onto this new 3′ end.

Using these techniques, which we describe in Chapter 9, they began to compare eukaryotic genes with the mRNAs derived from them. Their expectation was that just as in prokaryotes, the DNA nucleotide sequence of a gene's RNA-like strand would be identical to the RNA nucleotide sequence of the messenger RNA (with the exception of U replacing T in the RNA).

Surprisingly, they found that the DNA nucleotide sequences of many eukaryotic genes are much longer than their corresponding mRNAs, suggesting that RNA transcripts, in addition to receiving a methylated cap and a poly-A tail, undergo extensive internal processing.

An extreme example of the length difference between primary transcript and mRNA is seen in the human gene for dystrophin (Fig. 8.15 on p. 254). Abnormalities in the dystrophin gene underlie the genetic disorder of Duchenne muscular dystrophy (DMD). The dystrophin gene is 2.5 million nucleotides—or 2500 kilobases (kb)—long, whereas the corresponding mRNA is roughly 14,000 nucleotides, or 14 kb, in length. Obviously the gene contains DNA sequences that are not present in the mature mRNA. Those regions of the gene that do end up in the mature mRNA are scattered throughout the 2500 kb of DNA.

Sequences found in both a gene's DNA and the mature messenger RNA are called **exons** (for "expressed regions"). The *DMD* gene has more than 80 exons. The sequences found in the DNA of the gene but not in the mature mRNA are known as **introns** (for "intervening regions"). Introns interrupt, or separate, the exon sequences that actually end up in the mature mRNA. Introns can occur anywhere within a gene: between nucleotides that eventually form part of a single codon in the mature mRNA; between nucleotides encoding adjacent codons in the message; or within regions transcribed into parts of the mRNA that are not translated into protein. All of a gene's nucleotides that are transcribed into mRNA codons must be located in exons, but the three nucleotides of a single codon are often present in two different (but successive) exons. Finally,

HIV and Reverse Transcription: An Unusual DNA Polymerase
Gives the AIDs Virus an Evolutionary Edge

The AIDS-causing human immunodeficiency virus (HIV) is the most intensively analyzed virus in history. From laboratory and clinical studies spanning more than a decade, researchers have learned that each viral particle is a rough-edged sphere consisting of an outer envelope enclosing a protein matrix, which in turn surrounds a cut-off cone-shaped core (Fig. A). Within the core lies an enzyme-studded genome: two identical single strands of RNA associated with many molecules of an unusual DNA polymerase known as **reverse transcriptase.**

During infection, the AIDS virus binds to and injects its cone-shaped core into cells of the human immune system (Fig. B). It next uses reverse transcriptase to copy its RNA genome into double-stranded DNA molecules in the cytoplasm of the host cell. The double helixes then travel to the nucleus where another enzyme inserts them into a host chromosome. Once integrated into a host-cell chromosome, the viral genome can do one of two things. It can commandeer the host cell's protein synthesis machinery to make hundreds of new viral particles that bud off from the parent cell, taking with them part of the cell membrane and sometimes resulting in the host cell's death. Alternately, it can lie latent inside the host chromosome, which then copies and transmits the viral genome to two new cells with each cell division.

The events of this life cycle make HIV a **retrovirus:** an RNA virus that after infecting a host cell copies its own single strands of RNA into double helixes of DNA that a viral enzyme then integrates

into a host chromosome. RNA viruses that are not retroviruses simply infect a host cell and then use the cellular machinery to make more of themselves, often killing the host cell in the process. The viruses that cause hepatitis A, the common cold, and rabies are this type of RNA virus. Unlike retroviruses, they are not transmitted by cell division to a geometrically growing number of new cells.

Reverse transcription, the foundation of the retroviral life cycle, is inconsistent with the one-way, DNA-to-RNA-to-protein flow of genetic information described in this chapter. Because it was so unexpected, the phenomenon of reverse transcription encountered great resistance in the scientific community when first reported by Howard Temin of the University of Wisconsin and David Baltimore, then of MIT. It is now an established fact. In the normal order of molecular genetics, DNA polymerases construct DNA polymers from a DNA template, replicating a cell's double-helical genome for transmission to daughter cells during the cell cycle; and RNA polymerases construct RNA polymers from a DNA template, copying the information of genes into an RNA transcript that is then translated

HIV viral particle

Core
Protein matrix
RNA
Reverse transcriptase
Bilipid outer layer

Figure A   Structure of the AIDS virus.

3. DNA copy of virus genome enters nucleus.

4. DNA copy of virus genome integrates into host chromosome.

Host DNA

2. Core disintegrates, releasing RNA. Reverse transcriptase produces DNA from viral RNA genome.

5. Transcription of integrated virus makes viral RNA genome.

1. Virus particles attach to host cell membrane.

Host cell

6. Core forms; new virus particles bud from host cell.

Figure B   Life cycle of the AIDS virus.

some exons or parts of exons near the beginning and end of a gene may not carry amino-acid-specifying codons. These exons instead encode untranslated regions at the 5′ and 3′ ends of the mRNA that play important roles in regulating the efficiency of mRNA translation into protein.

Exons vary in size from 50 to a few thousand nucleotide pairs (in the *DMD* gene, the mean exon length is 200 bp), while

introns range from 50 to over 100,000 nucleotide pairs (in the *DMD* gene, the mean intron length is 35 kb, but one intron is an amazing 400 kb long). The greater size variation seen in introns compared to exons reflects the fact that introns do not appear in mature mRNAs and thus do not have to encode polypeptides. Because of this, there are fewer restraints on the sequence and number of nucleotides within introns. Geneticists into protein. Reverse transcriptase, by comparison, is a remarkable DNA polymerase that can construct a DNA polymer from either an RNA or a DNA template.

In addition to its comprehensive copying abilities, reverse transcriptase has another feature not seen in most DNA polymerases: inaccuracy. As we saw in Chapter 6, normal DNA polymerases replicate DNA with an error rate of one mistake in every million nucleotides copied. Reverse transcriptase, however, introduces one mutation in every 5000 incorporated nucleotides.

HIV uses this capacity for mutation, in combination with its ability to integrate its genome into the chromosomes of immune-system cells, to gain a tactical advantage over the immune response of its host organism. Cells of the immune system seek to overcome an HIV invasion by multiplying in response to the proliferating viral particles. The numbers are staggering. Each day of infection in every patient, from 100 million to a billion HIV particles are released from infected immune-system cells. As long as the immune system is strong enough to withstand the assault, it may respond by producing as many as 2 billion new cells daily. Many of these new cells produce antibodies targeted against proteins on the surface of the virus. But just when an immune response to a particular HIV protein wipes out those viral particles carrying the targeted protein, virions incorporating new forms of the protein resistant to the current immune response make their appearance. After many years of this complex chase, capture, and destruction by the immune system, the changeable virus outruns the host's immune response and gains the upper hand. Thus, the intrinsic infidelity of HIV's reverse transcriptase, by enhancing the virus's ability to compete in the evolutionary marketplace, increases its threat to human life and health.

The inherent mutability arising from reverse transcriptase's error-prone copying has undermined two therapeutic approaches that have helped control infections other than AIDS: drugs and vaccines. Some of the antiviral drugs approved in the United States for treatment of HIV infection—AZT (zidovudine), ddC (dideoxycytidine), and ddI (dideoxyinosine)—block viral replication by interfering with the action of reverse transcriptase. Each drug is similar to one of the four nucleotides, and when reverse transcriptase incorporates one of the drug molecules rather than a genuine nucleotide into a growing DNA polymer, the enzyme cannot extend the chain any further. However, the drugs are toxic at high doses and thus can be administered only at low doses that do not destroy all viral particles. Because of this limitation and the virus's high rate of mutation, mutant reverse transcriptases soon appear that work even in the presence of the drugs.

Similarly, researchers are having trouble developing safe, effective vaccines. Since HIV infects cells of the immune system and a vaccine works by stimulating immune-system cells to multiply, some of the vaccines tested so far actually increase the activity of the virus; others have only a weak effect on viral replication. Moreover, if it were possible to produce a vaccine that could generate a massive immune response against one, two, or even several HIV proteins at a time, such a vaccine might be effective for only a short while—until enough mutations built up to make the virus resistant.

For these reasons, the AIDS virus will most likely not succumb entirely to drugs or vaccines that target proteins active at various stages of its life cycle, although combinations of these therapeutic tools will remain an important part of the medical arsenal for prolonging an AIDS patient's life. In 1996, for example, medical researchers found that a therapeutic "cocktail" including at least one anti-reverse-transcriptase drug and a relatively new kind of drug known as a protease inhibitor (which blocks the enzymes that cleave a long, inactive polyprotein into shorter, functional viral proteins) could reduce the viral load of some very sick AIDS patients to undetectable levels, thereby relieving their disease symptoms. One year later, however, a clinical study revealed that for slightly more than 50% of patients receiving the drug cocktail in a San Francisco hospital, the treatment lost its effectiveness after six months. For the roughly 50% of patients for whom the cocktail was still effective, the new therapy offers some hope for controlling AIDS unless and until the virus in them evolves resistance to the new treatment.

Thus, after more than a decade of cooperative research and clinical experience, scientists know a great deal about the structure and life cycle of the AIDS virus, but they have been unable to develop drugs and vaccines that stop or even significantly slow its spread for more than a short while. In the mid-1990s, epidemiologists estimated that new infections of the AIDS virus occurred once every 15 seconds. By the year 2000, that rate of infection added up to nearly 35 million HIV-infected people worldwide.

A self-preserving capacity for mutation, perpetuated by reverse transcriptase, is surely one of the main reasons for HIV's success. Ironically, it may also provide a basis for its subjugation. Researchers are studying what happens when the virus increases its mutational load. Their reasoning is as follows: If reverse transcriptase's error rate determines the size and integrity of the viral population in a host organism, greatly accelerated mutagenesis might push the virus beyond the error threshold that allows it to function. In other words, too much mutation might destroy the virus's infectivity, virulence, or capacity to reproduce. If geneticists could figure out how to make this happen, they might be able to give the human immune system the advantage it needs to overrun the virus.

know very little about the rules that determine the number, size, and position of introns in different genes. In humans, although a small percentage of genes do not contain introns, the large majority of genes interrupt their coding sequences anywhere from 1–60 times.

How do cells make a mature mRNA from a gene whose coding sequences are interrupted by introns? The answer is that cells first make a primary transcript containing all of a gene's introns and exons, and then they remove the introns from the primary transcript by **RNA splicing,** the process that deletes introns and joins together successive exons to form a mature mRNA consisting only of exons (Fig. 8.15). Because the first and last exons of the primary transcript become the 5′ and 3′ ends of the mRNA, while all intervening introns are spliced

Splicing removes introns from a primary transcript.

Figure 8.15   The human dystrophin gene: An extreme example of RNA splicing. Though the dystrophin gene is 2500 kb (or 2.5 Mb) long, the dystrophin mRNA is only 14 kb long. More than 80 introns are removed from the 2500 kb primary transcript to produce the mature mRNA (which is not drawn to scale).

(a) Short sequences dictate where splicing occurs.

(b) Two sequential cuts remove the intron.

Figure 8.16   How RNA processing splices out introns and joins adjacent exons. (a) Three short sequences within the primary transcript determine the specificity of splicing. (1) The splice-donor site occurs where the 3′ end of an exon abuts the 5′ end of an intron. In most splice-donor sites, a GU dinucleotide (*arrows*) that begins the intron is flanked on either side by a few purines (Pu; that is, A or G). (2) The splice-acceptor site is at the 3′ end of the intron where it joins with the next exon. The final nucleotides of the intron are always AG (*arrows*) preceded by 12–14 pyrimidines (Py; that is, C or U). (3) The branch site, which is located within the intron about 30 nucleotides upstream of the splice acceptor, must include an A (*arrow*) and is usually rich in pyrimidines. **(b)** Two sequential cuts, the first at the splice-donor site and the second at the splice-acceptor site, remove the intron (which is subsequently degraded), allowing precise splicing of adjacent exons.

out, a gene must have one more exon than introns. To construct the mature mRNA, splicing must be remarkably precise. For example, if an intron lies within a codon, splicing must remove the intron and reconstitute the codon without disrupting the reading frame of the mRNA.

Figure 8.16 illustrates the details of RNA splicing. Three types of short sequences within the primary transcript—**splice donors, splice acceptors,** and **branch sites**—help ensure the specificity of splicing. These sites make it possible to sever the connections between an intron and the exons that precede and follow it, and then to join the formerly separated exons. The mechanism of splicing involves two sequential cuts in the primary transcript. The first cut is at the splice-donor site, at the 5′ end of the intron. After this first cut, the new 5′ end of the intron attaches, via a novel 2′–5′ phosphodiester bond, to an A at the branch site located within the intron, forming a so-called *lariat structure.* The second cut is at the splice-acceptor site, at the 3′ end of the intron; this cut removes the intron. The discarded intron is degraded, and the precise splicing of adjacent exons completes the process of intron removal.

Splicing normally requires a complicated intranuclear machine called the **spliceosome,** which ensures that all of the splicing reactions take place in concert (Fig. 8.17). The spliceosome consists of four subunits known as small nuclear ribonucleoproteins, or snRNPs (pronounced "snurps"). Each snRNP

contains one or two small nuclear RNAs (snRNAs) 100–300 nucleotides long, associated with proteins in a discrete particle. Certain snRNAs can base pair with the splice donor and splice acceptor sequences in the primary transcript, so these snRNAs are particularly important in bringing together the two exons that flank an intron. Given the complexities of spliceosome structure, it is remarkable that a few primary transcripts can splice themselves without the aid of a spliceosome or any additional factor. These rare primary transcripts function as **ribozymes:** RNA molecules that can act as enzymes and catalyze a specific biochemical reaction.

It might seem strange that eukaryotic genes incorporate DNA sequences that are spliced out of the mRNA before translation and thus do not encode amino acids. No one knows exactly why introns exist. One hypothesis proposes that they make it possible to assemble genes from various exon building blocks, which encode modules of protein function. This type of assembly would allow the shuffling of exons to make new genes, a process that appears to have played a key role in the evolution of complex organisms. The exon-as-module proposal is attractive because it is easy to understand the selective advantage of the potential for exon shuffling. Nevertheless, it remains a hypothesis without proof. There is no hard evidence for or against the hypothesis, and introns may have become established through means that scientists have yet to imagine.

Alternative Splicing Produces Different mRNAs from the Same Primary Transcript

Normally, RNA splicing joins together the splice donor and splice acceptor at the opposite ends of an intron, resulting in removal of the intron and fusion of two successive—and now adjacent—exons. For some genes, however, RNA splicing during development is regulated so that at certain times or in certain tissues, some splicing signals may be ignored. As an example, splicing may occur between the splice donor site of one intron and the splice acceptor site of a different intron downstream. Such **alternative splicing** produces different mRNA molecules that may encode related proteins with different—though partially overlapping—amino-acid sequences. In effect then, alternative splicing can tailor the nucleotide sequence of a primary transcript to produce more than one kind of polypeptide.

In mammals, alternative splicing of the gene encoding the antibody heavy chain determines whether the antibody proteins become embedded in the membrane of the B lymphocyte that makes them or are instead secreted into the blood. The antibody heavy-chain gene has eight exons and seven introns; exon number 6 has a splice-donor site within it. To make the membrane-bound antibody, all exons except for the right-hand part of number 6 are joined to create an mRNA encoding a hydrophobic (water-hating), lipid-loving C terminus (Fig. 8.18a). For the secreted antibody, only the first six exons (including the right part of 6) are spliced together to make an mRNA encoding a heavy chain with a hydrophilic (water-loving) C terminus. The alternate forms of RNA splicing thus allow physiological compartmentalization of the antibody molecule. Membrane-bound antibody molecules serve as receptors for foreign molecules

Spliceosome components

Five snRNAs
(small nuclear RNAs) + ~50 proteins

Four snRNPs (small nuclear ribonucleic particles), which assemble into a spliceosome

Proteins

snRNA

Figure 8.17   Splicing is catalyzed by the spliceosome. (*Top*) The spliceosome is assembled from four snRNP subunits, each of which contains one or two snRNAs and several proteins. (*Bottom*) A view of three spliceosomes in the electron microscope.

(a) Alternative splicing produces two different mRNAs from the same gene.

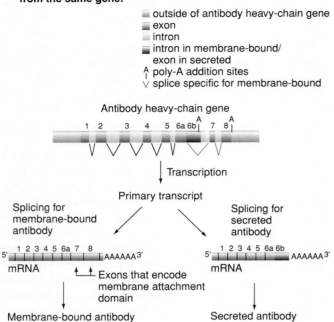

(b) Trans-splicing combines exons from different genes.

Figure 8.18 Different mRNAs can be produced from the same primary transcript. (a) Alternative splicing of the primary transcript for the antibody heavy chain produces mRNAs that encode different kinds of antibody proteins. **(b)** Rare trans-splicing events combine exons from different genes into one mature mRNA.

(termed *antigens*); the binding of antigen triggers proliferation and differentiation of the B lymphocyte into an antibody-secreting factory. The secreted antibodies in the blood can interact directly with antigens, including portions of viruses and bacteria, targeting them for destruction.

A rare and unusual strategy of alternative splicing, seen in *C. elegans* and a few other eukaryotes, is **trans-splicing,** in which the spliceosome joins an exon of one gene with an exon of another gene (Fig. 8.18b). Special nucleotide sequences in the RNAs make trans-splicing possible.

In summary, RNA processing may add a methylated cap or a poly-A tail, delete introns from a primary transcript,

determine which exons from a primary transcript are joined together to form a mature mRNA, and even join the exons of different genes.

Translation: Base Pairing Between mRNA and tRNAs Directs Assembly of a Polypeptide on the Ribosome

Virtually all cells—both prokaryotic and eukaryotic—use the same basic genetic code to translate the sequence of nucleotides in a messenger RNA to the sequence of amino acids in the corresponding polypeptide. This process of translation takes place on ribosomes that coordinate the movements of transfer RNAs carrying specific amino acids with the genetic instructions of an mRNA. As we examine the cell's translation machinery, we first describe the structure and function of tRNAs and ribosomes, and then we explain how these components interact during translation.

Transfer RNAs Mediate the Translation of mRNA Codons to Amino Acids

There is no obvious chemical similarity or affinity between the nucleotide triplets of mRNA codons and the amino acids they specify. Rather, *transfer RNAs (tRNAs)* serve as adaptor molecules that mediate the transfer of information from nucleic acid to protein.

tRNAs Carry an Anticodon at One End and an Amino Acid at the Other

tRNAs are short, single-stranded RNA molecules 74–95 nucleotides in length. Several of the nucleotides in tRNAs contain modified bases produced by chemical alterations of the principal A, G, C, and U nucleotides (Fig. 8.19a). Each tRNA carries one particular amino acid, and cells must have at least one tRNA for each of the 20 amino acids specified by the genetic code. The name of a tRNA reflects the amino acid it carries. For example, tRNA^Gly carries the amino acid glycine.

As Fig. 8.19b shows, it is possible to consider the structure of a tRNA molecule on three levels. (1) The nucleotide sequence of a tRNA constitutes the primary structure. (2) Short complementary regions within a tRNA's single strand can form base pairs with each other to create a characteristic cloverleaf shape; this is the tRNA's secondary structure. (3) Folding in three-dimensional space creates a tertiary structure that looks like a compact letter L. At one end of the L, the tRNA carries an **anticodon:** three nucleotides complementary to an mRNA codon specifying the amino acid carried by the tRNA (Fig. 8.19b). The anticodon never forms base pairs with other regions of the tRNA; it is always available for base pairing with its complementary mRNA codon. As with other complementary base sequences, during pairing at the ribosome, the strands of anticodon and codon run antiparallel to each other. If, for example, the anticodon is 3′ CCU 5′, the complementary mRNA codon is 5′ GGA 3′, specifying the amino acid glycine.

At the other end of the L—where the 5′ and 3′ ends of the tRNA strand are found—the tRNA connects to the amino acid specified by the codon. Enzymes known as **aminoacyl-tRNA synthetases** are the catalysts of this attachment (Fig. 8.20). These enzymes are extraordinarily specific, recognizing unique features of a particular tRNA—despite its general structural similarities with all other tRNAs—while also recognizing the corresponding amino acid (see the opening photo of this chapter on p. 237). Aminoacyl-tRNA synthetases are, in fact, the only molecules that read the languages of both nucleic acid and protein. They are thus the actual molecular translators. There is at least one aminoacyl-tRNA synthetase for each of the 20 amino acids, and like tRNA, each synthetase functions with only one amino acid. Figure 8.20 shows the two-step process that establishes the covalent bond between an amino acid and the 3′ end of its corresponding tRNA. A tRNA covalently coupled to its amino acid is called a **charged tRNA.**

Base Pairing Between an mRNA Codon and a tRNA Anticodon Directs Amino-Acid Incorporation into a Growing Polypeptide

While attachment of the appropriate amino acid charges a tRNA, the amino acid itself does not play a significant role in determining where it becomes incorporated in a growing polypeptide chain. It is the specific interaction between a tRNA's anticodon and an mRNA's codon that makes that decision. A simple experiment illustrates this point (Fig. 8.21). Researchers can subject a charged tRNA to chemical treatments that, without altering the structure of the tRNA, change the amino acid it carries. One treatment replaces the cysteine carried by tRNA^{Cys} with alanine. When investigators then add the tRNA^{Cys} charged with alanine to a cell-free translational system, the system incorporates *alanine* into the growing polypeptide wherever the mRNA contains a cysteine codon complementary to the anticodon of the tRNA^{Cys}.

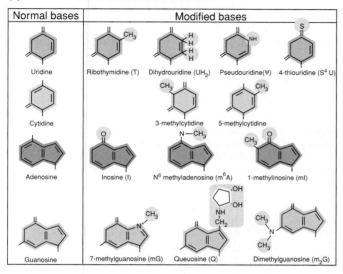

(a) Some tRNAs contain modified bases.

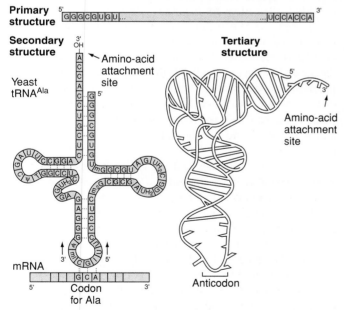

(b) Each tRNA has a primary, secondary, and tertiary structure.

Figure 8.19 tRNAs mediate the transfer of information from nucleic acid to protein. (a) Many tRNAs contain modified bases produced by chemical alterations of A, G, C, and U. **(b)** The primary structures of tRNA molecules fold to form characteristic secondary and tertiary structures. The anticodon and the amino-acid attachment site are at opposite ends of the "L" formed by a tRNA.

Figure 8.20 Aminoacyl-tRNA synthetases catalyze the attachment of tRNAs to their corresponding amino acids. The aminoacyl tRNA synthetase first uses the energy generated by ATP hydrolysis to activate the amino acid, forming an AMP-amino acid. The enzyme then transfers the carboxyl group of the amino acid from AMP to the hydroxyl (–OH) group of the ribose at the 3′ end of the tRNA, producing a charged tRNA.

Figure 8.21 Base pairing between an mRNA codon and a tRNA anticodon determines where an amino acid becomes incorporated into a growing polypeptide. A tRNA with an anticodon for cysteine, but carrying the amino acid alanine, adds alanine whenever the mRNA codon for cysteine appears.

Figure 8.22 Wobble: Some tRNAs recognize more than one codon for the amino acid they carry. (a) The G at the 5′ end of the anticodon shown here can pair with either U or C at the 3′ end of the codon. **(b)** The chart shows the pairing possibilities for other nucleotides at the 5′ end of an anticodon; I = inosine.

Wobble: Some tRNAs Can Recognize More Than One Codon

Although there is at least one kind of tRNA for each of the 20 amino acids, cells do not necessarily carry tRNAs with anticodons complementary to all of the 61 possible codon triplets in the degenerate genetic code. *E. coli*, for example, makes 79 different tRNAs containing 42 different anticodons. Although several of the 79 tRNAs in this collection obviously have the same anticodon, 61 − 42 = 19 of 61 potential anticodons are not represented. Thus 19 mRNA codons will not find a complementary anticodon in the *E. coli* collection of tRNAs. Another bacterial species, *Mycoplasma capricolum*, makes 30 different tRNAs carrying 29 different anticodons. In this species, only two tRNAs have the same anticodon, but 61 − 29 = 32 mRNA codons will not find a complementary anticodon in this collection of tRNAs. Although it is understandable that different species assemble their repertoire of tRNAs in different ways, how can an organism construct proper polypeptides if some of the codons in its mRNAs cannot locate tRNAs with complementary anticodons?

The answer is that some tRNAs can recognize more than one codon for the amino acid with which they are charged. That is, the anticodons of these tRNAs can interact with more than one codon for the same amino acid. Although researchers do not fully understand this "promiscuous" base pairing between codons and anticodons, Francis Crick spelled out a few of the rules that govern it. The rules are consistent with the genetic code. By analyzing the genetic code, Crick concluded that the 3′ nucleotide in many codons adds nothing to the specificity of the codon. For example, 5′ GGU 3′, 5′ GGC 3′, 5′ GGA 3′, and 5′ GGG 3′ all encode glycine (review Fig. 8.3 on p. 240). It does not matter whether the 3′ nucleotide is U, C, A, or G as long as the first two letters are GG. The same is true for other amino acids encoded by four different codons, such as valine, where the first two bases must be GU, but the third base can be U, C, A, or G. For amino acids specified by two different codons, the first two bases of the codon are, once again, always the same, while the third base must be either one of the two

purines (A or G) or one of the two pyrimidines (U or C). Thus, 5′ CAA 3′ and 5′ CAG 3′ are both codons for glutamine; 5′ CAU 3′ and 5′ CAC 3′ are both codons for histidine. If Pu stands for either purine and Py stands for either pyrimidine, then CAPu represents the codons for glutamine, while CAPy represents the codons for histidine.

In fact, the 5′ nucleotide of a tRNA's anticodon can often pair with more than one kind of nucleotide in the 3′ position of an mRNA's codon. (Recall that after base pairing, the bases in the anticodon run antiparallel to the bases in the codon.) A single tRNA charged with a particular amino acid can thus recognize several or even all of the codons for that amino acid. This flexibility in base pairing between the 3′ nucleotide in the codon and the 5′ nucleotide in the anticodon is known as **wobble** (Fig. 8.22a).

Crick's "wobble rules," shown in Fig. 8.22b, delimit what kind of flexibility in base pairing is consistent with the genetic code. For example, methionine (Met) is specified by a single codon (5′ AUG 3′). As a result, Met-specific tRNAs must have a C at the 5′ end of their anticodons (5′ CAU 3′), because this is the only nucleotide at that position that can base pair only with the G at the 3′ end of the Met codon. By contrast, a single isoleucine-specific tRNA with the modified nucleotide inosine (I) at the 5′ position of the anticodon can recognize all three codons (5′ AUU 3′, 5′ AUC 3′, and 5′ AUA 3′) for isoleucine.

Ribosomes Are the Sites of Polypeptide Synthesis

Ribosomes facilitate polypeptide synthesis in various ways. First, they recognize mRNA features that signal the start of translation. Second, they help ensure accurate interpretation of the genetic code by stabilizing the interactions between tRNAs and mRNAs; without a ribosome, codon-anticodon recognition, mediated by only three base pairs (one of which may wobble), would be extremely weak. Third, they supply the enzymatic activity that links the amino acids in a growing polypeptide chain. Fourth, by moving 5′ to 3′ along an mRNA molecule, they expose the mRNA codons in sequence, ensuring the linear addition of amino acids. Finally, ribosomes help end polypeptide synthesis by dissociating from both the mRNA directing polypeptide construction and the polypeptide product itself.

Ribosomes Are Complex Structures Composed of RNA and Protein

In *E. coli,* ribosomes consist of 3 different **ribosomal RNAs (rRNAs)** and 52 different ribosomal proteins (Fig. 8.23a). These components associate to form two different ribosomal subunits called the 30S subunit and the 50S subunit (with S designating a coefficient of sedimentation related to the size and shape of the subunit; the 30S subunit is smaller than the 50S subunit). Before translation begins, the two subunits exist as separate entities in the cytoplasm. Soon after the start of translation, they come together to reconstitute a complete ribosome. Eukaryotic ribosomes have more components than their prokaryotic counterparts, but they still consist of two dissociable subunits.

(a) A ribosome has two subunits composed of RNA and protein.

Complete ribosomes	Subunits	Nucleotides	Proteins
Prokaryotic			
	50S	23S RNA 3000 nucleotides	31
70S	30S	16S RNA 1700 nucleotides 5S RNA 120 nucleotides	21
Eukaryotic			
	60S	28S RNA 5000 nucleotides 5.8S RNA 160 nucleotides 5S RNA 120 nucleotides	~ 45
80S	40S	18S RNA 2000 nucleotides	~ 33

(b) Different parts of a ribosome have different functions.

Peptidyl transferase
Peptidyl (P) site
Aminoacyl (A) site
Large subunit
Small subunit

Figure 8.23 The ribosome: Site of polypeptide synthesis.
(a) A ribosome has two subunits, each composed of rRNA and various proteins. **(b)** The small subunit initially binds to mRNA. The large subunit contributes the enzyme peptidyl transferase, which catalyzes the formation of peptide bonds in the growing polypeptide chain. The two subunits together form the A and P tRNA binding sites.

Different Parts of the Ribosome Have Different Functions

The small 30S subunit, for example, is the part of the ribosome that initially binds to mRNA. The larger 50S subunit contributes an enzyme known as **peptidyl transferase,** which catalyzes formation of the peptide bonds joining adjacent amino acids (Fig. 8.23b). Both the small and the large subunits contribute to two distinct areas known as the **peptidyl (or P) site** and the **aminoacyl (or A) site,** which simultaneously bind to two different tRNA molecules. Finally, other regions of the ribosome distributed over the two subunits serve as points of contact for some of the additional proteins that play a role in translation.

The Mechanism of Translation

Translation consists of an **initiation** phase that sets the stage for polypeptide synthesis; **elongation,** during which amino acids are added to a growing polypeptide; and a **termination** phase that brings polypeptide synthesis to a halt and enables the ribosome to release a completed chain of amino acids. Figure 8.24 illustrates the details of the process, focusing on translation as it occurs in bacterial cells. As you examine the figure, note the following points about the flow of information during translation. First, the first codon to be translated—the initiation codon—is an AUG set in a special context at the 5′ end of the gene's reading frame (which is *not* directly at the 5′ end of the mRNA). Second, special initiating tRNAs carrying a modified form of methionine called formylmethionine (fMet) recognize the initiation codon. Thus, fMet is the first amino acid incorporated into all bacterial polypeptides and defines their N termini. Third, the ribosome moves along the mRNA in the 5′-to-3′ direction, revealing successive codons in a stepwise fashion. Fourth, at each step of translation, the polypeptide grows by the addition of the next amino acid in the chain to its C terminus. Finally, translation terminates when the ribosome reaches a UAA, UAG, or UGA nonsense codon at the 3′ end of the gene's reading frame. These points explain the biochemical basis of colinearity; that is, the correspondence between the 5′-to-3′ direction in the mRNA and the N-terminus-to-C-terminus direction in the resulting polypeptide.

During elongation, the translation machinery adds about 2–15 amino acids per second to the growing chain. The speed is higher in prokaryotes and lower in eukaryotes. At these rates, construction of an average size 300-amino-acid polypeptide (from an average-length mRNA that is somewhat longer than 1000 nucleotides) could take as little as 20 seconds or as long as 2.5 minutes.

Processing After Translation Can Change a Polypeptide's Structure

Protein structure is not irrevocably fixed at the completion of translation. Several different processes may subsequently modify a polypeptide's structure. Cleavage may remove amino acids, such as the N-terminal fMet, from a polypeptide (Fig. 8.25a), or it may generate several smaller polypeptides from one larger product of translation (Fig. 8.25b). In the latter case, the larger polypeptide made immediately after translation but

Translation: How cells use mRNA and charged tRNAs to assemble a polypeptide on the ribosome.

Initiation phase

Prokaryotic

Eukaryotic

(a) Initiation: Setting the stage for polypeptide synthesis The first three nucleotides of an mRNA do not serve as the first codon to be translated into an amino acid. Instead, a special signal indicates where along the mRNA translation should begin. In prokaryotes, this signal is called the **ribosome binding site,** and it has two important elements. The first is a short sequence of six nucleotides—usually 5' . . . AGGAGG . . . 3'—named the **Shine-Dalgarno box** after its discoverers.

The second element in an mRNA's ribosome binding site is the triplet 5' AUG 3', which serves as the initiation codon. A special initiator tRNA, whose 5' CAU 3' anticodon is complementary to AUG, recognizes an AUG preceded by the Shine-Dalgarno box of a ribosome binding site. The initiator tRNA carries **N-formylmethionine (fMet),** a modified methionine whose amino end is blocked by a formyl group. The specialized fMet tRNA functions only at an initiation site. An AUG codon

located within an mRNA's reading frame is recognized by a different tRNA that is charged with an unmodified methionine. This tRNA cannot start translation.

During initiation, the 3' end of the rRNA in the 30S ribosomal subunit binds to the mRNA's Shine-Dalgarno box (*not shown*), the fMet tRNA binds to the mRNA's initiation codon, and a large 50S ribosomal subunit associates with the small subunit to round out the ribosome. At the end of initiation, the fMet tRNA sits in the P site of the completed ribosome. Proteins known as **initiation factors** (*not shown*) play a transient role in the initiation process.

In eukaryotes, the small ribosomal subunit binds first to the methylated cap at the 5' end of the mature mRNA. It then migrates to the initiation site—usually the first AUG it encounters as it scans the mRNA in the 5'-to-3' direction. The initiator tRNA in eukaryotes carries unmodified methionine (Met) instead of fMet.

Elongation

Direction of ribosome movement

(b) Elongation: The addition of amino acids to a growing polypeptide

With the mRNA bound to the complete two-subunit ribosome and with the initiating tRNA in the P site, the elongation of the polypeptide begins. A group of proteins known as **elongation factors** (*not shown*) usher the appropriate tRNA into the A site of the ribosome. The anticodon of this charged tRNA must recognize the next codon in the mRNA. The ribosome holds the initiating tRNA at its P site and the second tRNA at its A site so that peptidyl transferase can catalyze formation of a peptide bond between the amino acids carried by the two tRNAs. This bond-forming reaction connects the fMet at the P site to the amino acid carried by the tRNA at the A site; it also disconnects fMet from the initiating tRNA. As a result, the tRNA at the A site now carries two amino acids. The N terminus of this dipeptide is fMet; the C terminus is the second amino acid, whose carboxyl group remains covalently linked to its tRNA.

Once formation of the first peptide bond causes the initiating tRNA in the P site to release its amino acid, the ribosome moves, exposing the next mRNA codon. Like the first steps of elongation, the ribosome's movement requires the help of elongation factors and an input of energy. As the ribosome moves, the initiating tRNA in the P site, which no longer carries an amino acid, dissociates from the ribosome, and the other tRNA carrying the dipeptide shifts from the A site to the P site.

The empty A site now receives another tRNA, whose identity is determined by the next codon in the mRNA. Peptidyl transferase then catalyzes formation of a second peptide bond, generating a chain of three amino acids. This tripeptide is connected at its C terminus to the tRNA currently in the A site; the fMet at its N terminus hangs free. With each subsequent round of ribosome movement and peptide bond formation, the peptide chain grows one amino acid longer.

Because the elongation machinery adds amino acids to the C terminus of the lengthening polypeptide, polypeptide synthesis proceeds from the N terminus to the C terminus. As a result, fMet in prokaryotes (Met in eukaryotes), the first amino acid in the growing chain, will be the N-terminal amino acid of all finished polypeptides that do not undergo protein processing (see main text). Moreover, because the nucleotide triplets encoding amino acids progressively closer to a polypeptide's C terminus are found successively farther from the 5' end of the mRNA, the ribosome must move along the mRNA in the 5'-to-3' direction.

The movement of ribosomes along the mRNA has important implications. Once a ribosome has moved far enough away from the mRNA's ribosome binding site, that site becomes accessible to other ribosomes. In fact, several ribosomes can work on the same mRNA at one time. A complex of several ribosomes translating from the same mRNA is called a **polyribosome.** This complex allows the simultaneous synthesis of many copies of a polypeptide from a single mRNA.

Polyribosome

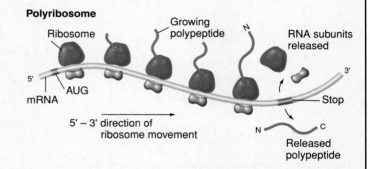

(c) Termination: The ribosome releases the completed polypeptide

No normal tRNAs carry anticodons complementary to the three nonsense (stop) codons UAG, UAA, and UGA. Thus, when movement of the ribosome brings a nonsense codon into the ribosome's A site, no tRNAs can bind to that codon through complementary base pairing. Instead, proteins called **release factors** recognize the termination codons and bring polypeptide synthesis to a halt. During termination, three events must occur: the tRNA specifying the C-terminal amino acid releases the completed polypeptide, the same tRNA as well as the mRNA separate from the ribosome, and the ribosome dissociates into its large and small subunits.

(a) Cleavage may remove an amino acid.

Enzyme removes fMet

(b) Cleavage may split a polyprotein.

Polyprotein

Cleavage

Multiple smaller polypeptides

(c) Addition of chemical constituents may modify a protein.

Serine

A phosphate group is added to serine

P   Phosphorylation

Figure 8.25 Posttranslational processing can modify polypeptide structure. Cleavage may remove an amino acid from the N-terminus end of a polypeptide **(a)** or split a larger *polyprotein* into two or more smaller functional proteins **(b)**. **(c)** Chemical reactions may add a phosphate or other functional group to an amino acid in the polypeptide.

before cleavage into smaller polypeptides is often called a **polyprotein.** The addition of chemical constituents, such as phosphate groups, methyl groups, or even carbohydrates, to specific amino acids may also modify a polypeptide after translation (Fig. 8.25c). Such additions are known as **posttranslational modifications.** Posttranslational changes to a protein can be very important: for example, the biochemical function of many enzymes directly depends on the addition (or sometimes removal) of phosphate groups.

There Are Significant Differences in Gene Expression Between Prokaryotes and Eukaryotes

In Eukaryotes, the Nuclear Membrane Prevents the Coupling of Transcription and Translation

In *E. coli* and other prokaryotes, transcription takes place in an open intracellular space undivided by a nuclear membrane; translation occurs in the same open space and is sometimes coupled directly with transcription (Table 8.1). This coupling is possible because transcription extends mRNAs in the same 5'-to-3' direction as the ribosome moves along the mRNA. As a result, ribosomes can begin to translate a partial mRNA that the RNA polymerase is still in the process of transcribing from the DNA. The coupling of transcription and translation has significant consequences for the regulation of gene expression in prokaryotes. For example, in an important regulatory mechanism called *attenuation,* which we describe in Chapter 16, the rate of translation of some mRNAs directly determines the rate at which the corresponding genes are transcribed into these mRNAs. Such coupling cannot occur in eukaryotes because the nuclear envelope physically separates the sites of transcription and RNA processing in the nucleus from the site of translation in the cytoplasm. As a result, translation in eukaryotes can only affect the rate at which genes are transcribed in more indirect ways.

The Initiation of Translation Also Differs Between Prokaryotes and Eukaryotes

In prokaryotes, translation begins at a ribosome binding site on the mRNA, which is defined by a short, characteristic sequence of nucleotides called a *Shine-Dalgarno box* adjacent to an initiating AUG codon (see Fig. 8.24a). There is nothing to prevent an mRNA from having more than one ribosome binding site, and in fact, many prokaryotic messages are **polycistronic:** they contain the information of several genes (sometimes referred to as *cistrons*), each of which can be translated independently starting at its own ribosome binding site (Table 8.1). In eukaryotes, by contrast, the small ribosomal subunit first binds to the methylated cap at the 5' end of the mature mRNA and then migrates to the initiation site. This is almost always the first AUG codon encountered by the ribosomal subunit as it moves along, or scans, the mRNA in the 5'-to-3' direction (see Fig. 8.24a and Table 8.1). The mRNA region between the 5' cap and the initiation codon is sometimes referred to as either the **5'-untranslated region (5' UTR)** or the **5'-untranslated leader.** Because of this scanning mechanism, initiation in eukaryotes takes place at only a single site on the mRNA, and each mRNA contains the information for translating only a single kind of polypeptide.

Another translational difference between prokaryotes and eukaryotes is in the composition of the initiating tRNA. In prokaryotes, it carries a modified form of methionine known as formylmethionine, while in eukaryotes it carries an unmodified methionine (see Table 8.1). This difference means that immediately after translation, prokaryotic polypeptides all have fMet at their N termini, while eukaryotic polypeptides have Met at that same location. Posttranslational cleavage events in both prokaryotes and eukaryotes often create mature proteins that no longer have fMet or Met at their N termini (Fig. 8.25a).

Eukaryotic mRNAs Require More Processing Than Prokaryotic mRNAs

Table 8.1 reviews other important differences in gene structure and expression between prokaryotes and eukaryotes. In particular, introns interrupt eukaryotic, but not prokaryotic, genes

TABLE 8.1 Differences Between Prokaryotes and Eukaryotes in the Details of Gene Expression

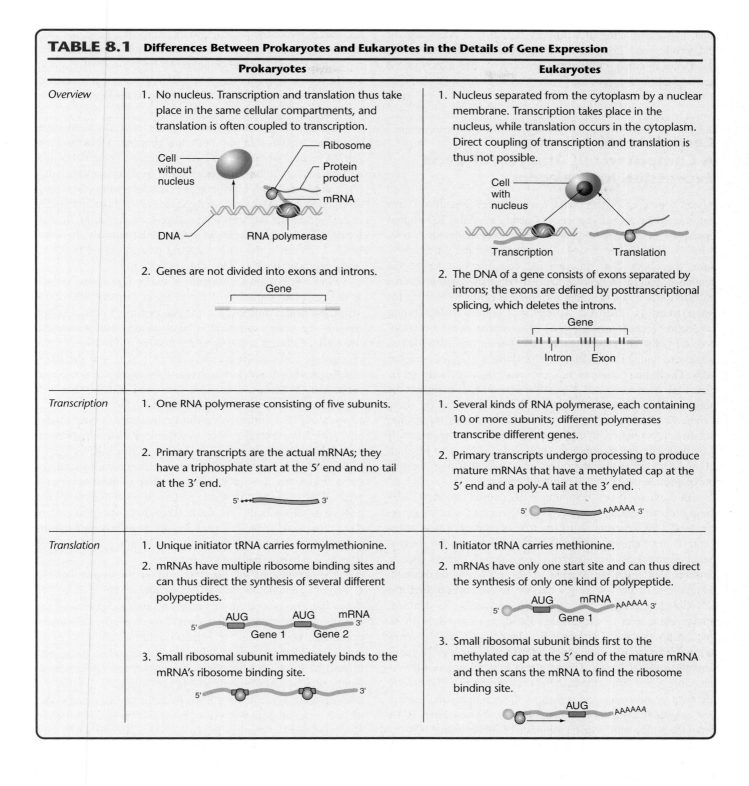

	Prokaryotes	**Eukaryotes**
Overview	1. No nucleus. Transcription and translation thus take place in the same cellular compartments, and translation is often coupled to transcription.	1. Nucleus separated from the cytoplasm by a nuclear membrane. Transcription takes place in the nucleus, while translation occurs in the cytoplasm. Direct coupling of transcription and translation is thus not possible.
	2. Genes are not divided into exons and introns.	2. The DNA of a gene consists of exons separated by introns; the exons are defined by posttranscriptional splicing, which deletes the introns.
Transcription	1. One RNA polymerase consisting of five subunits.	1. Several kinds of RNA polymerase, each containing 10 or more subunits; different polymerases transcribe different genes.
	2. Primary transcripts are the actual mRNAs; they have a triphosphate start at the 5′ end and no tail at the 3′ end.	2. Primary transcripts undergo processing to produce mature mRNAs that have a methylated cap at the 5′ end and a poly-A tail at the 3′ end.
Translation	1. Unique initiator tRNA carries formylmethionine.	1. Initiator tRNA carries methionine.
	2. mRNAs have multiple ribosome binding sites and can thus direct the synthesis of several different polypeptides.	2. mRNAs have only one start site and can thus direct the synthesis of only one kind of polypeptide.
	3. Small ribosomal subunit immediately binds to the mRNA's ribosome binding site.	3. Small ribosomal subunit binds first to the methylated cap at the 5′ end of the mature mRNA and then scans the mRNA to find the ribosome binding site.

such that the splicing of a primary transcript is necessary for eukaryotic gene expression. Other types of RNA processing that occur in eukaryotes but not prokaryotes add a methylated cap and a poly-A tail to the 5′ and 3′ ends of the mRNAs.

Comprehensive Example: A Computerized Analysis of Gene Expression in *C. elegans*

At the beginning of this chapter we saw that geneticists have determined the precise sequence of nearly all of the 100 million base pairs in the haploid genome of the tiny nematode *C. elegans*. Using their knowledge of gene structure and gene expression, they have also programmed computers to locate the sequences within the genome likely to be genes. Their programs include instructions to search for possible exons by looking for **open reading frames:** strings of amino-acid-encoding nucleotide triplets uninterrupted by in-frame nonsense (stop) codons. Other algorithms ignore potential introns, identified as sequences lying between likely splice-donor and splice-acceptor sites. Once the computer has retrieved regions likely to be genes, the researchers ask it to use the genetic code to project the amino-acid sequences of the polypeptides encoded by these genes. Finally, they scan computerized databases for similar amino-acid sequences in the polypeptides of other organisms. If they find a similar sequence in a polypeptide of known function in another organism, they can conclude that the worm's polypeptide probably has a parallel function.

In this way, investigators have discovered that the *C. elegans* genome contains roughly 19,000 genes, of which approximately 15% encode components of the worm's gene-expression machinery. Many of these gene-expression components are proteins. For example, more than 60 genes encode proteins that function as parts of the ribosome workbench, while more than 300 genes encode **transcription factors:** DNA-binding proteins that regulate transcription. By contrast, a large contingent of expression-related genes produce RNAs that are not translated into protein. There are 659 tRNA genes in the *C. elegans* genome, about 100 rRNA genes, and 72 genes for spliceosomal RNAs. The relatively high numbers of RNA-encoding genes reflect the fact that the genome contains several identical or near-identical copies of these untranslated genes. For example, even though there are 72 spliceosomal RNA genes, there are only 5 different kinds of spliceosomal RNAs.

Computerized predictions based on genomic DNA sequences alone are valuable but not infallible tools. Computer programs are currently very good at predicting the introns and exons and the primary amino-acid sequence of genes encoding proteins that are well conserved in evolution. But certain details of the transcription and translation of these genes cannot be established without isolating and characterizing their corresponding mRNAs. For example, although the computer can accurately locate the protein-coding exons of a gene, the gene may contain additional exons and introns at its 5′ or 3′ ends that are more difficult for the computer to find. Similarly, without biochemical analysis of the gene's RNA products, it is not possible to know whether alternative splicing of the gene's primary transcript produces different mRNAs.

Using techniques described in the next three chapters, researchers have obtained both the genomic DNA and the mRNA sequences for many *C. elegans* genes. These data allow an examination of the structure of these genes in nucleotide-by-nucleotide detail. One of these genes encodes a particular type of collagen protein. This single-polypeptide protein is a component of the hard cuticle that surrounds and protects the worm. Related forms of collagen occur in all multicellular animals. In vertebrates, collagen is the most abundant protein, found in bones, teeth, cartilage, tendons, and other tissues.

Figure 8.26 shows a diagram of the collagen gene as well as the complete sequence of the gene, its primary RNA transcript, the mature mRNA, and the polypeptide product. As you can see, the gene's structural features include three exons and two introns, as well as the signals that allow transcription, RNA processing, and translation into collagen. Note that the ATG-initiated reading frame for the protein begins only in the second exon. This is because the entire first exon and the first four nucleotides of the second exon correspond to the 5′-untranslated region (5′-UTR). Similarly, the third exon contains both amino-acid-specifying codons as well as sequences transcribed into an untranslated region near the 3′ end of the mature mRNA (the 3′-UTR) just upstream of the poly-A tail. The general structure of the collagen gene is similar to the structure of most eukaryotic genes. This is because the basic pattern of gene expression has remained substantially the same throughout evolution, even though the details, such as gene length, exon number, and the spacing or size of the untranslated 5′ and 3′ ends, vary from gene to gene and from organism to organism.

The sequencing of the 100 million nucleotides in the *C. elegans* genome not only led to the identification of 19,000 genes but also helped reveal some uncommon features in the way the worm expresses its genes. Like bacteria, *C. elegans* transcribes some groups of adjacent genes as one long polycistronic primary transcript; it is one of the very few eukaryotic organisms in which researchers have observed this predominantly prokaryotic phenomenon. In rare instances, worms use trans-splicing to create an mRNA from the primary transcripts of two different genes (review Fig. 8.18b on p. 256); before observing trans-splicing in the nematode, researchers had seen it mainly in trypanosomes, the single-celled protozoans that cause sleeping sickness. The introns that *C. elegans* deletes through RNA splicing are fewer and smaller than the introns found in most other animal genes, and they are unusually rich in As and Ts. Researchers will be able to apply these insights clarifying *C. elegans'* mechanisms of gene expression to studies of the worm's growth and development (see the genetic portrait of *C. elegans* in Reference C).

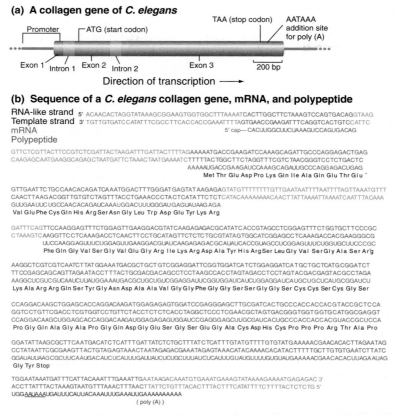

Figure 8.26 **Expression of a C. elegans gene for collagen. (a)** Landmarks in the collagen gene. **(b)** Comparison of the sequence of the collagen gene's DNA with the sequence of nucleotides in the mature mRNA (*purple*) pinpoints the start of transcription, the location of exons (*red*) and introns (*green*), and the position of the AAUAAA poly-A addition signal (*underlined in purple*). Translation of the mRNA according to the genetic code determines the amino acids of the protein product.

How Mutations Affect Gene Expression

We have seen that the information in DNA is the starting point of gene expression. The cell transcribes that information into mRNA and then translates the mRNA information into protein. Mutations that alter the nucleotide pairs of DNA may modify any of the steps or products of gene expression.

Mutations in a Gene's Coding Sequence Can Alter the Gene Product

Because of the nature of the genetic code, mutations in a gene's amino-acid-encoding exons generate a range of repercussions (Fig. 8.27a).

Silent Mutations Do Not Alter the Amino Acid Specified

One consequence of the code's degeneracy is that some mutations, known as **silent mutations,** direct the inclusion of the same amino acid as the unmutated DNA and therefore have no effect on the amino-acid composition of the encoded polypep-

tide or on phenotype. The majority of silent mutations change the third nucleotide of a codon, the position at which most codons for the same amino acid differ. For example, a change from GCA to GCC in a codon would still yield alanine in the protein product.

Missense Mutations Replace One Amino Acid with Another

Missense mutations that cause substitution in the polypeptide of an amino acid with chemical properties similar to the one it replaces may have little or no effect on protein function. Such substitutions are *conservative*. For example, a mutation that alters a GAC codon for aspartic acid to a GAG codon for glutamic acid is a conservative substitution because both amino acids have acidic R groups. By contrast, *nonconservative* missense mutations that cause substitution of an amino acid with very different properties are likely to have more noticeable consequences. A change of the same GAC codon for aspartic acid to GCC, a codon for alanine (an amino acid with an uncharged, nonpolar R group), is an example of a nonconservative substitution. The effect on phenotype of any missense mutation is

(a) Types of mutation in a gene's coding sequence

Wild-type mRNA	5' GCU GGA GCA CCA GGA CAA GAU GGA 3'
Wild-type polypeptide	N Ala Gly Ala Pro Gly Gln Asp Gly C

Silent mutation

GCU GGA GCC CCA GGA CAA GAU GGA
Ala Gly Ala Pro Gly Gln Asp Gly

Missense mutation

GCU GGA GCA CCA AGA CAA GAU GGA
Ala Gly Ala Pro Arg Gln Asp Gly

Nonsense mutation

GCU GGA GCA CCA GGA UAA GAU GGA
Ala Gly Ala Pro Gly Stop

Frameshift mutation

GCU GGA GCC ACC AGG ACA AGA UGG A
Ala Gly Ala Thr Arg Thr Arg Trp

(b) Sites of mutation outside the coding sequence that can disrupt gene expression

Ribosome binding site or 5'-untranslated leader
Sites needed for splicing
Promoter
In frame stop codon (TAG)
Transcription termination
Exon   Intron   Exon

Figure 8.27 How mutations in a gene can affect its expression. (a) Mutations in a gene's coding sequences. *Silent mutations* are not expressed because they do not alter the protein's primary structure. *Missense mutations* replace one amino acid with another. *Nonsense mutations* shorten a polypeptide by replacing a codon with a stop signal. *Frameshift mutations* result in a change in reading frame downstream of the addition or deletion. **(b)** Mutations in sites outside of the coding region can also disrupt gene expression.

difficult to predict since it depends on how a particular amino-acid substitution changes a protein's structure and function.

Nonsense Mutations Change an Amino-Acid-Specifying Codon to a Stop Codon

Nonsense mutations therefore result in the production of proteins smaller than those encoded by wild-type alleles of the same gene. The shorter, *truncated proteins* lack all amino acids between the amino acid encoded by the mutant codon and the C terminus of the normal polypeptide. The mutant polypeptide will be unable to function if it requires the missing amino acids for its activity.

Frameshift Mutations Result from the Insertion or Deletion of Nucleotides Within the Coding Sequence

As we have seen, if the number of extra or missing nucleotides is not divisible by 3, the insertion or deletion will skew the reading frame downstream of the mutation. As a result, **frameshift mutations** cause unrelated amino acids to appear in place of amino acids critical to protein function, destroying or diminishing polypeptide function.

Mutations in a Gene Outside the Coding Sequence Can Also Alter Gene Expression

Mutations that produce a variant phenotype are not restricted to alterations in codons. Since gene expression depends on several signals other than the actual coding sequence, changes in any of these critical signals can disrupt the process (see Fig. 8.27b).

We have seen that promoters and termination signals in the DNA of a gene instruct RNA polymerase where to start and stop transcription. Changes in the sequence of a promoter that make it hard or impossible for RNA polymerase to recognize the site diminish or prevent transcription. Mutations in a termination signal can diminish the amount of mRNA produced and thus the amount of gene product.

In eukaryotes, most primary transcripts have splice-acceptor sites, splice-donor sites, and branch sites that allow splicing to join exons together with precision in the mature mRNA. Changes in a splice-acceptor or -donor site can obstruct splicing, resulting in no mature mRNA and thus no normal polypeptide.

Mature mRNAs have ribosome binding sites and in-frame stop codons indicating where translation should start and stop. Mutations affecting a ribosome binding site would lower the affinity of the mRNA for the small ribosomal subunit; since this affinity helps determine the efficiency of translation, such mutations are likely to diminish the amount of polypeptide product. Mutations in a stop codon would produce longer than normal proteins that might be unstable or nonfunctional.

Mutations in Genes Encoding the Molecules That Implement Expression May Affect Transcription, mRNA Splicing, or Translation

Gene expression depends on an astonishing number and variety of macromolecules (Table 8.2). A separate gene encodes the subunits of each macromolecule. The genes for all the proteins are transcribed and translated the same as any other gene. The genes for all the rRNAs, tRNAs, and snRNAs are transcribed but *not* translated. Many mutations in these genes have a dramatic effect on phenotype.

Mutations Altering Genes Encoding Proteins or RNAs Involved in Gene Expression Are Usually Lethal

This is because such mutations adversely affect the synthesis of all proteins in a cell. Even a 50% reduction in the amount of some of the proteins enumerated in Table 8.2 can have severe repercussions. In *Drosophila,* for example, null mutations in many of the genes encoding the various ribosomal proteins are lethal when homozygous. This same mutation in a heterozygote causes a dominant *Minute* phenotype in which the slow growth of cells delays the fly's development.

Mutations in tRNA Genes Can Suppress Mutations in Protein-Coding Genes

If more than one gene encoded a molecule with the same role in gene expression, a mutation in one of these genes would not be lethal and might even be useful. Bacterial geneticists have

TABLE 8.2	The Cellular Components of Gene Expression
Function	**Cellular Components**
*Transcription**	Core RNA polymerase
	Sigma subunit
	Rho factor
Splicing and RNA Processing	snRNAs
	Protein components of spliceosomes
	Additional splicing factors
	Capping enzyme
	Methyl transferases
	Poly-A polymerase
Translation	mRNAs
	tRNAs
	Aminoacyl-tRNA synthetases
	rRNAs
	Protein components of ribosomes
	Translation factors
Protein Processing	Deformylases
	Amino peptidases
	Proteases
	Methylases
	Hydroxylases
	Glycosylases
	Kinases
	Phosphatases

*For simplicity, we list here only proteins from prokaryotic organisms involved in transcription. The cellular components needed for transcription in eukaryotic organisms are more complex; for example, eukaryotes have three different kinds of RNA polymerase, each made of numerous subunits.

Figure 8.28 Nonsense suppression. (a) A nonsense mutation that generates a stop codon causes production of a truncated, nonfunctional polypeptide. **(b)** A second, nonsense-suppressing mutation in a tRNA gene causes addition of an amino acid in response to the stop codon, allowing production of a full-length polypeptide.

found, for example, that mutations in certain tRNA genes can suppress the effect of a nonsense mutation in other genes. The tRNA-gene mutations that have this effect give rise to **nonsense suppressor tRNAs.** Consider, for instance, an otherwise wild-type *E. coli* population with an in-frame UAG nonsense mutation in the tryptophan synthetase gene. All cells in this population make a truncated, nonfunctional form of the tryptophan synthetase enzyme and are thus tryptophan auxotrophs (*trp⁻*) unable to synthesize tryptophan (Fig. 8.28a). Subsequent exposure of these auxotrophs to mutagens, however, generates some *trp⁺* cells that carry two mutations: one is the original

tryptophan synthetase nonsense mutation, the second is a mutation in the gene that encodes a tRNA for the amino acid tyrosine. Evidently, the mutation in the tRNA gene suppresses the effect of the nonsense mutation, restoring the function of the tryptophan synthetase gene. As Fig. 8.28b illustrates, the basis of this nonsense suppression is that the tRNA^{Tyr} mutation changes an anticodon that recognizes the codon for tyrosine to an anticodon complementary to the UAG stop codon. The mutant tRNA can therefore insert tyrosine into the polypeptide at the position of the in-frame UAG nonsense mutation, allowing the cell to make at least some full-length enzyme. Similarly, mutations in the anticodons of other tRNA genes can suppress UGA or UAA nonsense mutations.

Cells with a nonsense-suppressing mutation in a tRNA gene can only survive if two conditions coexist with the mutation. First, the cell must have other tRNAs that recognize the same codon as the suppressing tRNA recognized before mutation altered its anticodon. Without such tRNAs, the cell has no way to insert the proper amino acid in response to that codon (in our example, the codon for tyrosine). Second, the suppressing tRNA must have only a weak affinity for the stop codons normally found at the ends of mRNA coding regions. If this were not the case, the suppressing tRNA would wreak havoc in the cell, producing a whole array of aberrant polypeptides that are longer than normal. One way cells guard against this possibility is that for many genes, termination depends on two stop codons in a row. Because a suppressing tRNA's chance of inserting an amino acid at both of these codons is very low, only a small number of extended proteins arise.

CONNECTIONS

The promoter, a particular region of the DNA double helix, signals the starting point of gene expression. Beginning at this site, RNA polymerase unwinds the DNA and transcribes the simple linear sequence of nucleotides of the gene's template strand into a complementary messenger RNA containing all the information required for construction of a complex polypeptide. The cell then uses the genetic code to translate the mRNA nucleotide sequence into a colinear chain of amino acids whose sequence dictates its three-dimensional folding into a functional protein. Interestingly, the DNA within a cell is much more stable than either the RNA or proteins constructed from it. The half-life of most prokaryotic mRNAs is measured in minutes, while that of eukaryotic mRNAs is measured in hours. The half-life of many proteins is about a day. DNA is so stable it remains intact even after the organism dies.

Our knowledge of gene expression enables us to redefine the concept of a gene. A gene is not simply the DNA that is transcribed into the mRNA codons specifying the amino acids of a particular polypeptide. Rather, *a gene is all the DNA sequences needed for expression of the gene into a polypeptide product.* A gene therefore includes the promoter sequences that govern where transcription begins and, at the opposite end, signals for the termination of transcription. A gene also includes sequences dictating where translation starts and stops; such information is an important part of the mRNA. In addition to all of these features, eukaryotic genes contain introns that are spliced out of the primary transcript to make the mature mRNA. Because of introns, most eukaryotic genes are much larger than prokaryotic genes.

Even with introns, a single gene carries only a very small percentage of the nucleotide pairs in the chromosomes that make up a genome. The average gene in *C. elegans* is about 4000 nucleotide pairs in length, and there are roughly 19,000 genes. The worm's haploid genome, however, contains approximately 100 million nucleotide pairs distributed among six chromosomes containing an average of 16–17 million nucleotide pairs apiece. In humans, where genes tend to have more introns, the average gene is 16,000 nucleotide pairs in length, and there are 40,000–60,000 of them. But the haploid human genome has roughly 3 billion (3,000,000,000) nucleotide pairs distributed among 23 chromosomes containing an average of 130 million nucleotide pairs apiece. How can geneticists sort through these enormous haystacks of genetic information to find the pinpoints of interest: the DNA of a single gene or the small changes within that gene that alter a healthy phenotype to disease?

In Chapters 9–11, we describe how researchers analyze the mass of genetic information in the chromosomes of a genome as they try to discover what parts of the DNA are genes and how those genes influence phenotype. They begin their analysis by breaking the DNA into pieces of manageable size, making many copies of those pieces to obtain enough material for study, and characterizing the pieces down to the level of nucleotide sequence. They then try to reconstruct the DNA sequence of an entire genome by determining the spatial relationship between the many pieces. Finally, they use the knowledge they have obtained to examine the genomic variations that make individuals unique.

ESSENTIAL CONCEPTS

1. *Gene expression* is the process by which cells convert the DNA sequence of a gene to the RNA sequence of a transcript, and then decode the RNA sequence to the amino-acid sequence of a polypeptide.

2. The nearly universal *genetic code* consists of 64 *codons,* each one composed of three nucleotides. Of these codons, 61 specify amino acids, while 3—UAA, UAG, and UGA—are *nonsense* or *stop codons* that do not specify an amino acid.
   a. The code is *degenerate:* More than one codon specifies every amino acid except methionine and tryptophan.
   b. AUG in the context of a ribosome binding site is the *initiation codon;* it establishes a *reading frame* that determines the grouping of nucleotides into triplet codons.
   c. The code is *nonoverlapping.* Within a reading frame, the first three nucleotides constitute one codon; the next three, the second codon; and so forth.

3. Gene expression based on the genetic code produces the *colinearity* of a gene's nucleotide sequence and a protein's sequence of amino acids.

4. *Transcription* is the first stage of gene expression. During transcription, *RNA polymerase* synthesizes a

single-stranded *primary transcript* from a DNA template.

   a. RNA polymerase initiates transcription by binding to the *promoter* sequence of the DNA and unwinding the double helix to expose bases for pairing.

   b. RNA polymerase extends the RNA in the 5′-to-3′ direction by catalyzing formation of phosphodiester bonds between successively aligned nucleotides.

   c. *Terminator* sequences in the RNA cause RNA polymerase to dissociate from the DNA.

   d. In prokaryotes, the primary transcript is the *mRNA* that guides polypeptide synthesis.

5. In eukaryotes, *RNA processing* after transcription produces a mature mRNA that travels from the nucleus to the cytoplasm to direct polypeptide synthesis.

   a. RNA processing adds a methylated cap to the 5′ end and a poly-A tail to the 3′ end of the eukaryotic mRNA.

   b. The *spliceosome* removes *introns* from the primary transcript and precisely splices together the remaining *exons. Alternative splicing* makes it possible to produce different mRNAs from the same primary transcript.

6. *Translation* is the stage of gene expression when the cell synthesizes protein according to instructions in the mRNA.

   a. *tRNAs* carry amino acids to the translation machinery. *Aminoacyl-tRNA synthetases* connect amino acids to their corresponding tRNAs. Each tRNA molecule has an *anticodon* complementary to the mRNA codon specifying the amino acid it carries. Because of *wobble,* some tRNA anticodons recognize more than one mRNA codon.

   b. Translation occurs on complex molecular machines called *ribosomes*. Ribosomes have two binding sites for tRNAs—P and A—and an enzyme known as *peptidyl transferase* that catalyzes formation of a peptide bond between amino acids carried by the tRNAs at these two sites.

   c. Initiation: To start translation, part of the ribosome binds to a *ribosome binding site* on the mRNA, which includes the AUG initiation codon. Special initiating tRNAs with codons complementary to AUG carry the amino acid fMet in prokaryotes or

Met in eukaryotes to the ribosomal P site. This amino acid will become the N terminus of the growing polypeptide.

   d. Elongation: When the carboxyl group of the amino acid connected to a tRNA at the ribosome's P site becomes attached through a peptide bond to the amino acid carried by the tRNA at the A site, the ribosome travels three nucleotides toward the 3′ end of the mRNA. This movement exposes the next codon and allows the next round of amino-acid addition. This mechanism of translation ensures that the 5′-to-3′ direction in the mRNA corresponds to the N-terminus-to-C-terminus direction in the polypeptide under construction.

   e. Termination: When the ribosome encounters in-frame nonsense (stop) codons, it ends translation by releasing the mRNA and disconnecting the complete polypeptide from the tRNA.

7. *Posttranslational processing* may alter a polypeptide by adding or removing chemical constituents to or from particular amino acids, or by cleaving the polypeptide into smaller molecules.

8. Mutations affect gene expression in several ways.

   a. Mutations in a gene may modify the message encoded in a sequence of nucleotides. *Silent mutations* usually change the third letter of a codon and have no effect on polypeptide production. *Missense mutations* change the codon for one amino acid to the codon for another amino acid, and thereby direct incorporation of a different amino acid. *Nonsense mutations* change a codon for an amino acid to a stop codon, causing synthesis of a truncated polypeptide. *Frameshift mutations* change the reading frame of a gene, altering the identity of amino acids downstream of the mutation.

   b. Mutations outside of coding sequences that alter signals required for transcription, mRNA splicing, or translation can also disrupt gene expression.

   c. Mutations in genes encoding molecules of the gene-expression machinery are often lethal. Among the exceptions to this rule are mutations in tRNA genes that suppress mutations in polypeptide-encoding genes.

S O C I A L   A N D   E T H I C A L   I S S U E S

1. Scientists have been exploring the possibility that chemicals similar in structure to human hormones (hormone mimics) affect physiological processes such as fertility, cancer development, and intelligence. Hormone mimics are found in several synthetic compounds such as pesticides, drugs, and plastics that have become increasingly widespread in our environment. The evidence that these chemicals are harmful is still viewed by most scientists as preliminary. Chemical companies that produce hormone-mimic pesticides have a major stake in these studies. What is a company's responsibility to society when this type of

controversy appears? Should it immediately stop production of a potentially harmful chemical even if this stoppage leads to a large loss of jobs and the potential loss of some crops? When should a chemical be banned and what criteria should be used in calling for a ban?

2. An ethnobotanist who works with native peoples of Peru on medicinal plants of the rain forest approached a large pharmaceutical corporation with his newly acquired knowledge. He hoped to get some benefits for the native Peruvians as compensation for conveying knowledge on medicinal plants to the company. Scientists and executives in the company found several potential plant-based drugs using information from the indigenous peoples. However, when they tried expressing the plant genes to produce the drugs they found that their yield was low. As an alternative, chemists in the company began to synthesize the compounds that had been identified in some of the medicinal plants. Because the chemists are synthesizing the compounds, the company says it owes neither the native peoples nor the country of Peru anything. Is this a responsible policy for the company? Consider the responsibility the company has to the ethnobotanist, to its stockholders, to the native peoples, and to Peru.

3. In 1992, the biotechnology company Calgene applied for approval of their rot-resistant tomato, the Flavr Savr® tomato. They had engineered the tomato plant by introducing a copy of the gene encoding a softening enzyme downstream of, and in reverse orientation to, a strong promoter active in the fruit. When the engineered gene is transcribed, an "antisense RNA," complementary in sequence to the softening enzyme mRNA, is produced. The antisense RNA binds to the normal mRNA (sense RNA) and blocks translation of the softening enzyme so the tomatoes stay ripe longer. The FDA approved the sale of the tomato without any additional labeling because the DNA that had been introduced into the tomato was tomato DNA. (The FDA's policy on bioengineered foods is to label food as genetically engineered only when a new substance has been introduced that could cause an allergy or when there has been a change in the food's nutritional value.) Several consumer groups were upset by the FDA's action and wanted to see labeling about genetic modification appear on the tomatoes in the stores. The protest significantly stalled the sale of these tomatoes. Was the consumer group's action, which blocked an advance that people might have wanted, warranted? Are these groups important watchdogs of everyone's well-being or are they hindering progress?

SOLVED PROBLEMS

I. A geneticist examined the amino-acid sequence of a particular protein in a variety of mutants in *E. coli*. The amino acid in position 40 in the normal enzyme is glycine. The following table shows the substitutions the geneticist found at amino-acid position 40 in six mutant forms of the enzyme.

mutant 1	cysteine
mutant 2	valine
mutant 3	serine
mutant 4	aspartic acid
mutant 5	arginine
mutant 6	alanine

Determine the nature of the base substitution that must have occurred in the DNA in each case. Which of these mutants would be capable of recombination with mutant 1 to form a wild-type gene?

Answer

To determine the base substitutions, use the genetic code table (see Fig. 8.3). The original amino acid was glycine, which can be encoded by GGU, GGC, GGA, or GGC. Mutant 1 results in a cysteine at position 40; Cys codons are either UGU or UGC. A change in the base pair in the

DNA encoding the first position in the codon (a G–C to T–A transversion) must have occurred, and the original glycine codon must therefore have been either GGU or GGC. Valine (in mutant 2) is encoded by GUN (with N representing any one of the four bases), but assuming that the mutation is a single base change, the Val codon must be either GUU or GUC. The change must have been a G–C to T–A transversion in the DNA for the second position of the codon. To get from glycine to serine (mutant 3) with only one base change, the GGU or GGC would be changed to AGU or AGC, respectively. There was a transition (G–C to A–T) at the first position. Aspartic acid (mutant 4) is encoded by GAU or GAC, so the DNA of mutant 4 is the result of a G–C to A–T transition at position 2. Arginine (mutant 5) is encoded by CGN, so the DNA of mutant 5 must have undergone a G–C to C–G transversion at position 1. Finally, alanine (mutant 6) is encoded by GCN, so the DNA of mutant 6 must have undergone a G–C to C–G transversion at position 2. Mutants 2, 4, and 6 affect a base pair different from that affected by mutant 1, so they could recombine with mutant 1.

In summary, the sequence of nucleotides on the RNA-like strand of the wild-type and mutant genes at this position must be:

wild type	5′ G G T/C 3′
mutant 1	5′ T G T/C 3′
mutant 2	5′ G T T/C 3′
mutant 3	5′ A G T/C 3′
mutant 4	5′ G A T/C 3′
mutant 5	5′ C G T/C 3′
mutant 6	5′ G C T/C 3′

II. The double-stranded circular DNA molecule that forms the genome of the SV40 virus can be denatured into single-stranded DNA molecules. Because the base composition of the two strands differs, the strands can be separated on the basis of their density into two strands designated W(atson) and C(rick). When each of the purified preparations of the single strands was mixed with mRNA from cells infected with the virus, hybrids were formed between the RNA and DNA. Closer analysis of these hybridizations showed that RNAs that hybridized with the W preparation were different from RNAs that hybridized with the C preparation. What does this tell you about the transcription templates for the different classes of RNAs?

Answer

An understanding of transcription and the polarity of DNA strands in the double helix are needed to answer this question. *Some genes use one strand of the DNA as a template; others use the opposite strand as a template.* Because of the different polarities of the DNA strands, one set of genes would be transcribed in a clockwise direction

on the circular DNA (using say the W strand as the template), and the other set would be transcribed in a counterclockwise direction (with the C strand as template).

III. Geneticists interested in human hemoglobins have found a very large number of mutant forms. Some of these mutant proteins are of normal size, with amino-acid substitutions, while others are short, due to deletions or nonsense mutations. The first extralong example was named Hb Constant Spring, in which the β globin has several extra amino acids attached at the C-terminal end. What is a plausible explanation for its origin? Is it likely that Hb Constant Spring arose from failure to splice out an intron?

Answer

An understanding of the principles of translation and RNA splicing are needed to answer this question. Because there is an extension on the C-terminal end of the protein, *the mutation probably affected the termination (nonsense) codon rather than affecting splicing of the RNA.* This could have been a base change or a frameshift or a deletion that altered or removed the termination codon. The information in the mRNA beyond the normal stop codon would be translated until another stop codon in the mRNA was reached. A splicing defect could explain Hb Constant Spring only in the more unlikely case that an incorrectly spliced mRNA would encode a protein much longer than normal.

P R O B L E M S

8-1 For each of the terms in the left column, choose the best matching phrase in the right column.

a. codon

b. colinearity

c. reading frame

d. frameshift mutation

e. degeneracy of the genetic code

f. nonsense codon

g. initiation codon

h. template strand

i. RNA-like strand

j. intron

1. removing base sequences corresponding to introns from the primary transcript

2. UAA, UGA, or UAG

3. the strand of DNA that has the same base sequence as the primary transcript

4. a transfer RNA molecule to which the appropriate amino acid has been attached

5. a group of three mRNA bases signifying one amino acid

6. most amino acids are not specified by a single codon

7. using the information in the nucleotide sequence of a strand of DNA to specify the nucleotide sequence of a strand of RNA

8. the grouping of mRNA bases in threes to be read as codons

9. AUG in a particular context

10. the linear sequence of amino acids in the polypeptide corresponds to the linear sequence of nucleotide pairs in the gene

k. RNA splicing

l. transcription

m. translation

n. alternative splicing

o. charged tRNA

p. reverse transcription

11. produces different mature mRNAs from the same primary transcript

12. addition or deletion of a number of base pairs other than three into the coding sequence

13. a sequence of base pairs within a gene that is not represented by any bases in the mature mRNA

14. the strand of DNA having the base sequence complementary to that of the primary transcript

15. using the information encoded in the nucleotide sequence of an mRNA molecule to specify the amino-acid sequence of a polypeptide molecule

16. copying RNA into DNA

8-2 Match the hypothesis from the left column to the observation from the right column that gave rise to it.

a. existence of an intermediate messenger between DNA and protein

1. two mutations affecting the same amino acid can recombine to give wild type

b. the genetic code is nonoverlapping

c. the codon is more than one nucleotide

d. the genetic code is based on triplets of bases

e. stop codons exist and terminate translation

f. the amino-acid sequence of a protein depends on the base sequence of an mRNA

2. one or two base deletions (or insertions) in a gene disrupt its function; three base deletions (or insertions) are often compatible with function

3. artificial messages containing certain codons produced shorter proteins than messages not containing those codons

4. protein synthesis occurs in the cytoplasm, while DNA resides in the nucleus

5. artificial messages with different base sequences gave rise to different proteins in an *in vitro* translation system

6. single base substitutions affect only one amino acid in the protein chain

8-3 How would the artificial mRNA 5′. . . GUGUGUGU. . . 3′ be read according to each of the following models for the genetic code?
a. two base, not overlapping
b. two base, overlapping
c. three base, not overlapping
d. three base, overlapping
e. four base, not overlapping

8-4 An example of a portion of the T4 *rIIB* gene in which Crick and Brenner had recombined one + and one − mutation is shown here. (The RNA-like strand of the DNA is shown.)

```
wild type    5′ AAA AGT CCA TCA CTT AAT GCC 3′
mutant       5′ AAA GTC CAT CAC TTA ATG GCC 3′
```

a. Where are the + and − mutations in the mutant DNA?
b. What alterations in amino acids occurred in this double mutant, which produces wild-type plaques?
c. How can you explain the fact that amino acids are different in the double mutant compared to the wild-type sequence, yet the phage is wild type?

8-5 In the *HbS* allele (sickle-cell allele) of the human β-globin gene, the sixth amino acid in the β-globin chain is changed from glutamic acid to valine. In *HbC* the sixth amino acid in β globin is changed from glutamic acid to lysine. What would be the order of these two mutations within the map of the β-globin gene?

8-6 The amino-acid sequence of part of a protein has been determined:

N . . . Gly Ala Pro Arg Lys . . . C

A mutation has been induced in the gene encoding this protein using the mutagen proflavin. The resulting mutant protein can be purified and its amino-acid sequence determined. The amino-acid sequence of the mutant protein is exactly the same as the amino-acid sequence of the wild-type protein from the N terminus of the protein

to the glycine in the preceding sequence. Starting with this glycine, the sequence of amino acids is changed to the following:

N . . . Gly His Gln Gly Lys . . . C

Using the amino-acid sequences, one can determine the sequence of 14 nucleotides from the wild-type gene encoding this protein. What is this sequence?

8-7 When the artificial mRNA 5′. . .UCUCUCUC. . . 3′ was added to an *in vitro* protein synthesis system, investigators found that proteins composed of alternating leucine and serine were made. What experiments were done to determine whether leucine was specified by CUC and serine by UCU, or vice versa?

8-8 The following diagram describes the mRNA sequence of part of the *A* gene and the beginning of the *B* gene of phage φX174. In this phage, there are some genes that are read in overlapping reading frames. For example, the code for the *A* gene is used for part of the *B* gene but the reading frame is displaced by one base. Shown here is the single mRNA with the codons for proteins A and B indicated.

aa	5 6 7 8 9 10 11 12 13 14 15 16
A	AlaLysGluTrpAsnAsnSerLeuLysThrLysLeu
mRNA	GCUAAAGAAUGGAACAACUCACUAAAAACCAAGCUG
B	MetGluGlnLeuThrLysAsnGlnAla
aa	1 2 3 4 5 6 7 8 9

Given the following amino-acid changes, indicate the base change that occurred in the mRNA and the consequences for the other protein sequence.
a. Asn at position 10 in protein A is changed to Tyr.
b. Leu at position 12 in protein A is changed to Pro.
c. Gln at position 8 in protein B is changed to Leu.
d. The occurrence of overlapping reading frames is very rare in nature. When it does occur, the extent of the overlap is not very long. Why do you think this is the case?

8-9 The coding sequence for gene *F* is read from left to right on the figure below. The coding sequence for gene *G* is read from right to left. Which strand of DNA (top or bottom) serves as the template for transcription of each gene?

8-10 Locate as accurately as possible those of the listed items which are shown on the following figure. Some items are not shown. (a) 5′ end of DNA template strand; (b) 3′ end of mRNA; (c) ribosome; (d) promoter; (e) codon; (f) an amino acid; (g) DNA polymerase; (h) 5′ UTR; (i) centromere; (j) intron; (k) anticodon; (l) N terminus; (m) 5′ end of charged tRNA; (n) RNA polymerase; (o) 3′ end of uncharged tRNA; (p) a nucleotide; (q) mRNA cap; (r) peptide bond; (s) P site; (t) aminoacyl-tRNA synthetase; (u) hydrogen bond; (v) exon; (w) 5′ AUG 3′; (x) potential "wobble" interaction.

8-11 Concerning the figure for the previous problem (8-10):
a. Which process is being represented?
b. What is the next building block to be added to the growing chain in the figure? To what end of the growing chain will this building block be added? How many building blocks will there be in the chain when it is completed?
c. What other building blocks have a known identity?
d. What details could you add to this figure that would be different in a eukaryotic cell versus a prokaryotic cell?

8-12 If you mixed the mRNA of a human gene with the genomic DNA for the same gene and allowed the RNA and DNA to form a hybrid, what would you be likely to see in the electron microscope? Your figure should include hybridization involving both DNA strands (template and RNA-like) as well as the mRNA.

8-13 How many possible open reading frames (frames without stop codons) are there that extend through the following sequence?

```
5′...CTTACAGTTTATTGATACGGAGAAGG...3′
3′...GAATGTCAAATAACTATGCCTCTTCC...5′
```

8-14 In prokaryotes, a search for genes in a DNA sequence involves scanning the DNA sequence for long open reading frames (that is, reading frames uninterrupted by stop codons). What problem can you see with this approach in eukaryotes?

8-15 The sequence of a segment of mRNA, beginning with the initiation codon, is given here, along with the corresponding sequences from several mutant strains.

Normal	AUGACACAUCGAGGGGUGGUAAACCCUAAG...
Mutant 1	AUGACACAUCCAGGGGUGGUAAACCCUAAG...
Mutant 2	AUGACACGAGGGGUGGUAAACCCUAAG...
Mutant 3	AUGACGCAUCGAGGGGUGGUAAACCCUAAG...
Mutant 4	AUGACACAUCGAGGGGUUGGUAAACCCUAAG...
Mutant 5	AUGACACAUUGAGGGGUGGUAAACCCUAAG...
Mutant 6	AUGACAUUUACCACCCCUCGAUGCCCUAAG...

a. Indicate the type of mutation present in each and translate the mutated portion of the sequence into an amino-acid sequence in each case.
b. Which of the mutations could be reverted by treatment with EMS (ethylmethane sulfonate; see Fig. 7.12 on pp. 204–205)? with proflavin?

8-16 The sequence of a complete eukaryotic gene encoding the small protein Met Tyr Arg Gly Ala is shown here. All of the written sequences on the template strand are transcribed into RNA.

```
5′ CCCCTATGCCCCCCTGGGGGAGGATCAAAACACTTACCTGTACATGGC 3′
3′ GGGGATACGGGGGGACCCCCTCCTAGTTTTGTGAATGGACATGTACCG 5′
```

a. Which strand is the template strand? Which direction (right-to-left or left-to-right) does RNA polymerase move along the template as it transcribes this gene?
b. What is the sequence of the nucleotides in the processed mRNA molecule for this gene? Indicate the 5′ and 3′ polarity of this mRNA.
c. A single base mutation in the gene results in synthesis of the peptide Met Tyr Thr. What is the sequence of nucleotides making up the mRNA produced by this mutant gene?

8-17 Describe the steps in transcription and translation that require complementary base pairing.

8-18 The yeast gene encoding a protein found in the mitotic spindle was cloned by a laboratory studying mitosis. The gene encodes a protein of 477 amino acids.
a. What is the minimum length in nucleotides of the protein-coding part of this yeast gene?
b. A partial sequence of one DNA strand in an exon containing the middle of the coding region of the yeast gene is given here. What is the sequence of nucleotides of the mRNA in this region of the gene? Show the 5′ and 3′ directionality of your strand.

```
5′ GTAAGTTAACTTTCGACTAGTCCAGGGT 3′
```

c. What is the sequence of amino acids in this part of the yeast mitotic spindle protein?

8-19 A particular protein has the amino-acid sequence

N . . . Ala-Pro-His-Trp-Arg-Lys-Gly-Val-Thr . . . C

within its primary structure. A geneticist studying mutations affecting this protein discovered that several of the mutants produced shortened protein molecules that terminated within this region. In one of them, the His became the terminal amino acid.
a. What DNA single-base change(s) would cause the protein to terminate at the His residue?
b. What other potential sites do you see in the DNA sequence encoding this protein where mutation of a single base pair would cause premature termination of translation?

8-20 In studying normal and mutant forms of a particular human enzyme, a geneticist came across a particularly interesting mutant form of the enzyme. The normal enzyme is 227 amino acids long, but the mutant form was 312 amino acids long, having that extra 85 amino acids as a block in the middle of the normal sequence. The inserted amino acids do not correspond in any way to the normal protein sequence. What are possible explanations for this phenomenon? How would you distinguish among them?

8-21 A mutant *B. adonis* bacterium has a nonsense suppressor tRNA that inserts glutamine (Gln) to match a UAG (but not other nonsense) codons.
   a. What is the anticodon of the suppressing tRNA? Indicate the 5′ and 3′ ends.
   b. What is the sequence of the template strand of the wild-type tRNA^Gln^-encoding gene that was altered to produce the suppressor, assuming that only a single-base-pair alteration was involved?
   c. What is the *minimum* number of *tRNA^Gln^* genes that could be present in a wild-type *B. adonis* cell? Describe the corresponding anticodons.

8-22 You are studying mutations in a bacterial gene that codes for an enzyme whose amino-acid sequence is known. In the wild-type protein, proline is the fifth amino acid from the amino terminal end. In one of your mutants with nonfunctional enzyme, you find a serine at position number 5. You subject this mutant to further mutagenesis and recover three different strains. Strain A has a proline at position number 5 and acts just like wild type. Strain B has tryptophan at position number 5 and also acts like wild type. Strain C has no detectable enzyme function at any temperature, and you can't recover any protein that resembles the enzyme. You mutagenize strain C and recover a strain (C-1) that has enzyme function. The second mutation in C-1 responsible for the recovery of enzyme function does not map at the enzyme locus.
   a. What is the nucleotide sequence in both strands of the wild-type gene at this location?
   b. Why does strain B have a wild-type phenotype? Why does the original mutant with serine at position 5 lack function?
   c. What is the nature of the mutation in strain C?
   d. What is the second mutation that arose in C-1?

8-23 Another class of suppressor mutations, not described in the chapter, are mutations that suppress missense mutations.
   a. Why would bacterial strains carrying such missense suppressor mutations generally grow more slowly than strains carrying nonsense suppressor mutations?
   b. What other kinds of mutations can you imagine in genes encoding components needed for gene expression that would suppress a missense mutation in a protein-coding gene?

8-24 Yet another class of suppressor mutations not described in the chapter are mutations in tRNA genes that can suppress frameshift mutations. What would have to be true about a tRNA that could suppress a frameshift mutation involving the insertion of a single base pair?

8-25 There is at least one nonsense suppressing tRNA known that can suppress more than one type of nonsense codon.
   a. What is the anticodon of such a suppressing tRNA?
   b. What stop codons would it suppress?
   c. What are the amino acids most likely to be carried by this nonsense suppressing tRNA?

8-26 An investigator was interested in studying UAG nonsense suppressor mutations in bacteria. In one species of bacteria, she was able to select two different mutants of this type, one in the *tRNA^Tyr^* gene and the other in the *tRNA^Gln^* gene, but in a second species she was not able to obtain any such nonsense suppressor mutations, even after very extensive effort. What could explain the difference between the two species?

8-27 Do you think each of the following types of mutations would have very severe effects, mild effects, or no effect at all?
   a. Nonsense mutations occurring in the sequences encoding amino acids near the N terminus of the protein
   b. Nonsense mutations occurring in the sequences encoding amino acids near the C terminus of the protein
   c. Frameshift mutations occurring in the sequences encoding amino acids near the N terminus of the protein
   d. Frameshift mutations occurring in the sequences encoding amino acids near the C terminus of the protein
   e. Silent mutations
   f. Conservative missense mutations
   g. Nonconservative missense mutations affecting the active site of the protein
   h. Nonconservative missense mutations not in the active site of the protein

8-28 Using recombinant DNA techniques (which will be described in the next chapter), it is possible to take the DNA of a gene from any source and place it on a chromosome in the nucleus of a yeast cell. When you take the DNA for a human gene and put it into a yeast cell chromosome, the altered yeast cell can make the human protein. But when you remove the DNA for a gene normally present on yeast mitochondrial chromosomes and put it on a yeast chromosome in the nucleus, the yeast cell cannot synthesize the correct protein, even though the gene comes from the same organism. Explain. What would you need to do to ensure that such a yeast cell could make the correct protein?

8-29 The following is a list of mutations that have been discovered in a gene that has more than 60 exons and encodes a very large protein of 2532 amino acids. Indicate whether or not each mutation could cause a detectable change in the size or the amount of mRNA and/or a detectable change in the size or the amount of the protein product. (Detectable changes in size or amount must be greater than 1% of normal values.) What kind of change would you predict?

a. Lys576Val (changes amino acid 576 from lysine into valine)
b. Lys576Arg
c. AAG576AAA (changes codon 576 from AAG to AAA)
d. AAG576UAG
e. Met1Arg (there are at least two possible scenarios for this mutation)
f. promoter mutation
g. one base-pair insertion into codon 1841
h. deletion of codon 779
i. IVS18DS, G–A, + 1 (this mutation changes the first nucleotide in the eighteenth intron of the gene, causing exon 18 to be spliced to exon 20, thus skipping exon 19)
j. deletion of the poly-A addition site
k. G-to-A substitution in the 5′ UTR
l. insertion of 1000 base pairs into the sixth intron (this particular insertion does not alter splicing)

8-30 Considering further the mutations described in the previous problem (8-29):
a. Which of the mutations could be null mutations?
b. Which of the mutations would be most likely to result in an allele that is recessive to wild type?
c. Which of the mutations could result in an allele dominant to wild type? What mechanism(s) could explain this dominance?

8-31 Arrange the following list of eukaryotic gene elements in the order they would appear in the genome and in the direction traveled by RNA polymerase along the gene. Assume the gene's single intron interrupts the open reading frame. Note that some of these names are abbreviated and thus do not distinguish between elements in DNA versus RNA. For example, "splice-donor site" is an abbreviation for "DNA sequences transcribed into the splice-donor site" because splicing takes place on the gene's RNA transcript, not on the gene itself. Geneticists often use this kind of shorthand for simplicity, even though it is imprecise. (a) splice-donor site; (b) 3′ UTR; (c) promoter; (d) stop codon; (e) nucleotide to which methylated cap is added; (f) initiation codon; (g) transcription terminator; (h) splice-acceptor site; (i) 5′ UTR; (j) poly-A addition site; (k) splice branch site.

8-32 Concerning the list of eukaryotic gene elements in the previous problem (8-31):
a. Which of the element names in the list are abbreviated? (That is, which of these elements actually occur in the gene's primary transcript or mRNA rather than in the gene itself?)
b. Which of the elements in the list are found partly or completely in the first exon of this gene (or the RNA transcribed from this exon)? In the intron? In the second exon?

9

Deconstructing the Genome: DNA at High Resolution

Colonies of bacterial cell clones containing recombinant DNA molecules.

The vivid red color of our blood stems from its life-sustaining ability to carry oxygen. This ability, in turn, derives from billions of red blood cells suspended in proteinaceous solution, each one packed with close to 280 million copies of the protein pigment known as hemoglobin (Fig. 9.1a). The hemoglobin picks up oxygen in the lungs and transports it to tissues throughout the body, where, after its release, the oxygen contributes to a multitude of metabolic reactions. A normal adult hemoglobin molecule consists of four polypeptide chains, two alpha (α) and two beta (β) globins, each surrounding an iron-containing small molecular structure known as a heme group (Fig. 9.1b). The iron atom within the heme sustains a reversible interaction with oxygen, binding it firmly enough to hold it on the trip from lungs to body tissue but loosely enough to release it where needed. The intricately folded α and β chains protect the iron-containing hemes from molecules in the cell's interior. Each hemoglobin molecule can carry up to four oxygen atoms, one per heme; and it is these oxygenated hemes that impart a scarlet hue to the pigment molecules and thus to the blood cells that carry them.

Interestingly, the genetically determined molecular composition of hemoglobin changes several times during human development, enabling the molecule to adapt its oxygen-transport function to the varying environments of the embryo, fetus, newborn, and adult (Fig. 9.1c). In the first five weeks after conception, the red blood cells carry *embryonic hemoglobin,* which consists of two alphalike zeta (ζ) chains and two betalike epsilon (ϵ) chains. Thereafter, throughout the rest of gestation, the cells contain *fetal hemoglobin,* composed of two bona fide α chains and two β-like gamma (γ) chains. Then, shortly before birth, production of *adult hemoglobin,* composed of two α and two β chains, begins to climb. By the time an infant reaches three months of age, almost all of his or her hemoglobin is of the adult type.

The various forms of hemoglobin evolved in response to the different environments that exist at different stages of human development. The early embryo, which is not yet associated with a fully functional placenta, has the least access to oxygen in the maternal circulation. The fetus, which is associated with a fully functional placenta, has access to the gases dissolved in the maternal

(a)

Figure 9.1 Hemoglobin is composed of four polypeptide chains that change during development. (a) Scanning electron micrograph of adult human red blood cells loaded with one kind of protein—hemoglobin. **(b)** Adult hemoglobin consists almost entirely of two α and two β polypeptide chains. **(c)** The hemoglobin, carried by red blood cells, switches slowly during human development from an embryonic form containing two α-like ξ chains and two β-like ε chains, to a fetal form containing two α chains and two β-like γ chains, and finally to the adult form containing two α and two β chains. In a small percentage of adult hemoglobin molecules, a β-like δ chain replaces the actual β chain.

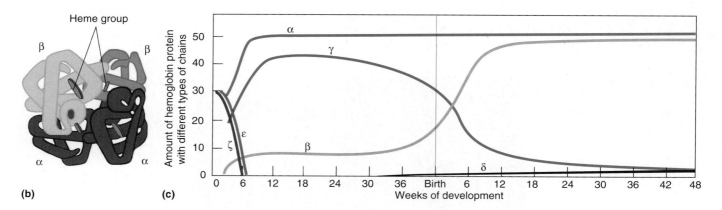

(b)   **(c)**

bloodstream, but the amount of oxygen in the maternal circulation is less than that found in the air that enters our lungs after birth. Both embryonic and fetal hemoglobin evolved to bind oxygen more tightly than adult hemoglobin would at the lower levels of oxygen found in the embryonic and fetal environments. All the hemoglobins readily release their oxygen to cells, which have an even lower level of oxygen than any source of the gas. After birth, when oxygen is abundantly available in the lungs, adult hemoglobin, with its more relaxed kinetics of oxygen binding, allows for the most efficient pickup and delivery of the vital gas.

What structure and arrangement of hemoglobin genes facilitate the crucial switches in the type of protein produced? And what can go wrong with these genes to generate hemoglobin disorders? Such disorders are the most common genetic diseases in the world and include sickle-cell anemia, which arises from an altered β chain, and thalassemia, which results from decreases in the amount of either α- or β-chain production.

To answer these questions, researchers must have a way of looking at the hemoglobin genes of individuals with normal and abnormal phenotypes. But these genes lie buried in a diploid human genome containing 6 billion base pairs distributed among 46 different strings of DNA (the chromosomes) that range in size from 60 million to 360 million base pairs each. In this chapter, we describe the powerful tools of modern molecular analysis that medical researchers now use to search through these enormously long strings of information for genes that may be only several thousand base pairs in length. Ideally suited to the analysis of complex biological systems, these tools take advantage of

enzymes and biochemical reactions that occur naturally within the simplest life-forms, bacterial cells. Researchers refer to the whole kit of modern tools and reactions as **recombinant DNA technology.** The most important enzymes in this tool kit are ones that operate on the DNA molecule itself: restriction enzymes that cut DNA, ligases that join two molecules together, and polymerases that synthesize new DNA strands. The most important uncatalyzed biochemical reaction is **hybridization:** the binding together of two DNA strands with complementary nucleotide sequences. Hybridization results from the natural propensity of complementary single-stranded molecules to form stable double helixes. A final tool in the recombinant DNA kit is the natural process of DNA replication, which occurs within growing colonies of cells.

Developed during a technological revolution that began in the mid-1970s, recombinant DNA technology makes it possible to characterize DNA molecules directly, rather than indirectly through the phenotypes they produce. Geneticists use recombinant DNA techniques to gather information unobtainable in any other way, or to analyze the results of breeding and cytological studies with greater speed and accuracy than ever before.

Two general themes recur in our discussion. First, the new molecular tools of genetic analysis emerged from a knowledge of DNA structure and function. For example, complementary base pairing between similar or identical stretches of DNA is the basis of hybridization techniques that use labeled DNA probes to identify related DNA molecules. Second, the speed, sensitivity, and accuracy of the new tools make it possible to answer questions that were impossible to resolve just a short time ago. In one recent study, for instance, researchers used a protocol for making many copies of a specific DNA segment to trace the activity of the AIDS virus and found that, contrary to prior belief, the virus, after infecting a person, does not become latent inside all the cells it enters; it remains active in certain cells of the lymph nodes and adenoids.

As we illustrate the power of the modern tools for analyzing a seemingly simple biological system—the hemoglobins—that turned out to be surprisingly complex, we describe how researchers use recombinant DNA technology to carry out five basic operations:

- Cut the enormously long strings of DNA into much smaller fragments with scissorslike *restriction enzymes;* and separate the small fragments according to size through *gel electrophoresis.*

- Isolate, amplify, and purify the fragments through *molecular cloning.*

- Use purified DNA fragments as hybridization *probes* to identify the presence of similar sequences in libraries of clones or in complex mixtures of DNA or RNA molecules.

- Rapidly isolate and amplify previously defined genomic or mRNA sequences from new individuals or cell sources through the *polymerase chain reaction (PCR).*

- Determine the precise *sequence* of bases within isolated DNA fragments.

Fragmenting Complex Genomes into Bite-Size Pieces for Analysis

Every intact diploid human body cell, including the precursors of red blood cells, carries two nearly identical sets of 3 billion base pairs of information that, when unwound, extend 2 m in length and contain two copies of 40,000–60,000 genes. If you could enlarge the cell nucleus to the size of a basketball, the unwound DNA would have the diameter of a fishing line and a length of 200 km. This is much too much material and information to study as a whole. To reduce its complexity, researchers first cut the genome into "bite-size" pieces.

Restriction Enzymes Fragment the Genome at Specific Sites

Researchers use restriction enzymes to cut the DNA released from the nuclei of cells at specific sites. These well-defined cuts generate fragments suitable for manipulation and characterization. A **restriction enzyme** recognizes a specific sequence of bases anywhere within the genome and then severs two covalent bonds (one in each strand) in the sugar-phosphate backbone at particular positions within or near that sequence. The fragments generated by restriction enzymes are referred to as **restriction fragments,** and the act of cutting is often called **digestion.**

Restriction enzymes originate in and can be purified from bacterial cells. In these prokaryotic cells, the enzymes protect against invasion by viruses through the digestion of viral DNA. Bacteria shield their own DNA from digestion by indigenous restriction enzymes through the selective addition of methyl groups ($-CH_3$) to the restriction recognition sites in their DNA. In the test tube, restriction enzymes from bacteria recognize target sequences of 4–8 bp in DNA isolated from any other organism, and cut the DNA at or near these sites. Table 9.1 lists the names, recognition sequences, and microbial origins of just 10 of the more than 100 commonly used restriction enzymes.

TABLE 9.1   Ten Commonly Used Restriction Enzymes

Enzyme	Sequence of Recognition Site	Microbial Origin
*Taq*I	5' T C G A 3' 3' A G C T 5'	*Thermus aquaticus* YTI
*Rsa*I	5' G T A C 3' 3' C A T G 5'	*Rhodopseudomonas sphaeroides*
*Sau*3AI	5' G A T C 3' 3' C T A G 5'	*Staphylococcus aureus* 3A
*Eco*RI	5' G A A T T C 3' 3' C T T A A G 5'	*Escherichia coli*
*Bam*HI	5' G G A T C C 3' 3' C C T A G G 5'	*Bacillus amyloliquefaciens* H.
*Hind*III	5' A A G C T T 3' 3' T T C G A A 5'	*Haemophilus influenzae*
*Kpn*I	5' G G T A C C 3' 3' C C A T G G 5'	*Klebsiella pneumoniae* OK8
*Cla*I	5' A T C G A T 3' 3' T A G C T A 5'	*Caryophanon latum*
*Bss*HII	5' G C G C G C 3' 3' C G C G C G 5'	*Bacillus stearothermophilus*
*Not*I	5' G C G G C C G C 3' 3' C G C C G G C G 5'	*Nocardia otitidiscaviarum*

Genetics and Society

Serendipity in Science: The Discovery of Restriction Enzymes

Most of the tools and techniques for cloning and analyzing DNA fragments emerged from studies of bacteria and the viruses that infect them. Molecular biologists had observed, for example, that viruses able to grow abundantly on one strain of bacteria grew poorly on a closely related strain of the same bacteria. While examining the mechanisms of this discrepancy, they discovered restriction enzymes.

To follow the story, one must know that researchers compare rates of viral proliferation in terms of *plating efficiency:* the fraction of viral particles that enter and replicate inside a host bacterial cell, causing the cell to lyse and release viral progeny, which go on to infect and replicate inside neighboring cells, which lyse and release further virus particles, which continue this process. When a petri plate is coated with a continuous "lawn" of bacterial cells, an active viral infection can be observed as a visibly cleared spot, or plaque, where bacteria have been eliminated (see Fig. 7.18). The plating efficiency of lambda virus grown on *E. coli* C is nearly 1.0. This means that 100 original virus particles will cause close to 100 plaques on a lawn of *E. coli* C bacteria.

The plating efficiency of the same virus grown on *E. coli* K12 is only 1 in 10^4 or 0.0001. The ability of a bacterial strain to prevent the replication of an infecting virus, in this case the growth of lambda on *E. coli* K12, is called **restriction.**

Interestingly, restriction is rarely absolute. Although lambda virus grown on *E. coli* K12 produces almost no progeny (viruses infect cells but can't replicate inside them), a few viral particles do manage to proliferate. If their progeny are then tested on *E. coli* K12, the plating efficiency is nearly 1.0. The phenomenon in which growth on a restricting host modifies a virus so that succeeding generations grow more efficiently on that same host is known as **modification.**

What mechanisms account for restriction and modification? Studies following viral DNA after bacterial infection found that during restriction, the viral DNA is broken into pieces and degraded (Fig. A). When the enzyme responsible for the initial breakage was isolated, it was found to be an endonuclease, an enzyme that breaks the phosphodiester bonds in the viral DNA molecule, usually making double-strand cuts at a specific sequence in the viral chromosome. Because this breakage restricts the biological activity of the viral DNA, researchers called the enzymes that accomplish it *restriction enzymes.* Subsequent studies showed that the small percentage of viral DNA that escapes digestion and goes on to generate new viral particles has been modified by the addition of methyl groups during its replication in the host cell. Researchers named the enzymes that add methyl groups to specific DNA sequences **modification enzymes.**

Figure A **Operation of the restriction enzyme/modification system in nature.** *E. coli* C does not have a functional restriction enzyme/modification system and is susceptible to infection by the lambda phage. In contrast, *E. coli* K12 generally resists infection by the viral particles produced during an *E. coli* C-phage infection. This is because *E. coli* K12 makes several restriction enzymes, including *Eco*RI, which cut the lambda DNA molecule before it can express itself. However, in rare K12 cells, the lambda DNA replicates before restriction enzymes can attack it, and the newly replicated lambda DNA is then modified by an enzyme that protects its recognition sites from attack by the host cell's restriction enzymes. This lambda DNA can now generate phage particles, which eventually destroy the bacterial cell.

In nature, pairs of restriction-modification enzymes protect bacterial cells from foreign DNA. Unmodified DNA of any type falls prey to digestion by restriction enzymes. But as described in the Fast Forward box in Chapter 6 on pp. 170–172, a bacterial cell's own DNA is newly methylated and protected from the same digestive fate by modification enzymes that operate with each round of replication. Modification enzymes also ensure the survival of those few viral DNA molecules that happen to evade digestion. These viral DNA molecules, modified by methylation, then use the host cell's machinery to produce new viral particles.

Biologists have identified a large number of restriction-modification systems in a variety of bacterial strains. Purification of the systems has yielded a mainstay of recombinant DNA technology: the battery of restriction enzymes used to cut DNA *in vitro* for cloning, mapping, and ligation (see Table 9.1).

This example of serendipity in science sheds some light on the debate between administrators who distribute and oversee research funding and scientists who carry out the research. Microbial investigators did not set out to find restriction enzymes; they could not have known that these enzymes would be one of their finds. Rather, they sought to understand the mechanisms by which viruses infect and proliferate in bacteria. Along the way, they discovered restriction enzymes and how they work. Their insights have had broad applications in genetic engineering and biotechnology. DNA cloning, for example, has produced, among many other medically useful products, human insulin for diabetics, which provokes far fewer allergic reactions than the pig insulin that preceded it; tissue plasminogen activator (TPA), a protein that helps dissolve blood clots; and gamma interferon, a protein produced by the human immune system that protects laboratory-grown human cells from infection by herpes and hepatitis viruses. The politicians and administrators in charge of allocating funds often want to direct research spending to urgent health or agricultural problems, assuming that much of the basic research upon which such studies depend has been completed. The scientists in charge of laboratory research call for a broad distribution of funds to all projects investigating interesting biological phenomena, arguing that we still know relatively little about molecular biology and that the insights necessary for understanding disease may emerge from studies on any organism. The validity of both views suggests the need for a balanced approach to the funding of research activities.

For the majority of these enzymes, the recognition site contains 4–6 base pairs and exhibits a kind of palindromic symmetry in which the base sequences of each of the two DNA strands are identical when read in the 5′-to-3′ direction. Because of this, base pairs on either side of a central line of symmetry are mirror images of each other. Each enzyme always cuts at the same place relative to its specific recognition sequence, and most enzymes make their cuts in one of two ways: either straight through both DNA strands right at the line of symmetry to produce fragments with **blunt ends,** or displaced equally in opposite directions from the line of symmetry by one or more bases to generate fragments with single-stranded ends (Fig. 9.2). Geneticists often refer to these protruding single strands as **sticky ends.** They are considered "sticky" because they are free to base pair with a complementary sequence from the DNA of *any* organism cut by the same restriction enzyme. (The Genetics and Society box on p. 280 of this chapter and the Fast Forward box on p. 170 of Chapter 6 contain more information about restriction enzymes.)

Different Restriction Enzymes Produce Fragments of Different Lengths

For each restriction enzyme, it is possible to calculate the average length of the fragments the enzyme will generate and then use that information to estimate the approximate number and distribution of recognition sites in a genome. The estimate depends on two simplifying assumptions: First, that each of the four bases occurs in equal proportions such that a genome is composed of 25% A, 25% T, 25% G, and 25% C; second, that the bases are randomly distributed in the DNA sequence. Although these assumptions are never precisely valid, they enable us to determine the average distance between recogni-

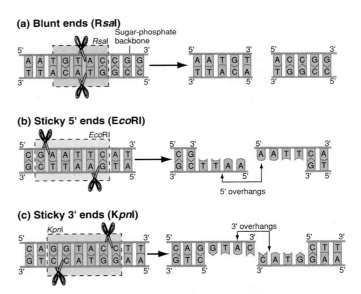

Figure 9.2 Restriction enzymes cut DNA molecules at specific locations to produce restriction fragments with either blunt or sticky ends. (a) The restriction enzyme *Rsa*I produces blunt-ended restriction fragments. **(b)** *Eco*RI produces sticky ends with a 5′ overhang. **(c)** *Kpn*I produces sticky ends with a 3′ overhang.

tion sites of any length by the general formula 4^n, where n is the number of bases in the site (Fig. 9.3).

The Size of the Recognition Site Is the Primary Determinant of Fragment Length

According to the 4^n formula, *Rsa*I, which recognizes the four-base-sequence GTAC, will cut on average once every 4^4, or

(a) 1. Probability that a four-base recognition site will be found in a genome =

$$1/4 \times 1/4 \times 1/4 \times 1/4 = 1/256$$

2. Probability that a six-base recognition site will be found =

$$1/4 \times 1/4 \times 1/4 \times 1/4 \times 1/4 \times 1/4 = 1/4,096$$

(b) Intact Human DNA

1. Four-base

*Rsa*I sites

2. Six-base

*Eco*RI sites

3. Eight-base

*Not*I sites

Figure 9.3 The number of base pairs in a restriction enzyme recognition site determines the average distance between restriction sites in a genome and thus the size of fragments produced. *Rsa*I recognizes and cuts at a 4 bp site, *Eco*RI cuts at a 6 bp site, and *Not*I cuts at an 8 bp site. The actual distances between restriction sites in the genome are variable.

every 256 base pairs (bp), creating fragments averaging 256 bp in length. By comparison, the enzyme *Eco*RI, which recognizes the six-base-sequence GAATTC, will cut on average once every 4^6, or 4096 bp; since 1000 base pairs = 1 kilobase pair, researchers often round off this large number to roughly 4.1 kilobase pairs, abbreviated 4.1 kb. Similarly, an enzyme such as *Not*I, which recognizes the eight bases GCGGCCGC, will cut on average every 4^8 bp, or every 65.5 kb. Note, however, that because the actual distances between restriction sites for any enzyme vary considerably, very few of the fragments produced by the three enzymes mentioned here will be precisely 65.5 kb, 4.1 kb, or 256 bp in length.

The Timing of Exposure to a Restriction Enzyme Helps Determine Fragment Size

Geneticists often need to produce DNA fragments of a particular length—larger ones to study the organization of a chromosomal region, smaller ones to examine a whole gene, and ones that are smaller still for DNA sequence analysis (that is, for the determination of the precise order of bases in a DNA fragment).

Intact human DNA

*Eco*RI sites

Enzyme cuts at one random site in five.

Figure 9.4 Partial digests. By reducing the time allotted for the reaction, you can ensure that an enzyme actually cuts only a subset of the total number of recognition sites within a DNA sample. This diagram shows three copies of the same DNA sequence. The chosen reaction time allowed for the cutting of only 20% of all the *Eco*RI sites. Since the cutting is at random, the particular 20% of sites at which cuts occur is different, even on the identical DNA molecules shown here.

If their goal is 4 kb fragments, they have a range of six-base-cutter enzymes to choose from. By exposing the DNA to a six-base cutter for a long enough time, they give the restriction enzyme ample opportunity for digestion. The result is a **complete digest** in which the DNA has been cut at every one of the recognition sites it contains. But what if the researchers need 20 kb fragments—a length in between that produced by a six-base cutter and an eight-base cutter? The answer is a **partial digest:** a cutting obtained by controlling the amount of enzyme or the amount of time the DNA is exposed to the restriction enzyme. With *Eco*RI, for example, exposing the DNA just long enough for the enzyme to cut on average only one of five target sites would produce fragments averaging 20 kb in length (Fig. 9.4). Researchers determine the correct timing by trial and error. Complete or partial digestion with different enzymes produces fragments of average lengths that are appropriate for different experimental uses.

Different Restriction Enzymes Produce Different Numbers of Fragments from the Same Genome

We have seen that the four-base cutter *Rsa*I cuts the genome on average every 4^4 (256) bp. If you exposed the haploid human genome with its 3 billion bp to *Rsa*I for a sufficient time under appropriate conditions, you would ensure that all of the recognition sites in the genome that can be cleaved will be cleaved, and you would get

$$\frac{3,000,000,000 bp}{\sim 256 bp} = \begin{array}{l} \sim 12,000,000 \text{ fragments that are} \\ \sim 256 \text{ bp in average length} \end{array}$$

By comparison, the six-base cutter *Eco*RI cuts the DNA on average every 4^6 (4096) bp, or every 4.1 kb. If you exposed the haploid human genome with its 3 billion bp, or 3 million kb, to *Eco*RI in the proper way, you would get

$$\frac{3,000,000,000 bp}{\sim 4100 bp} = \begin{array}{l} \sim 700,000 \text{ fragments that are} \\ \sim 4.1 \text{ kb in average length} \end{array}$$

And if you exposed the same haploid human genome to the eight-base cutter *Not*I, which cuts on average every 4^8 (65,536) bp, or 65.5 kb, you would obtain

$$\frac{3,000,000,000 bp}{\sim 65,500 bp} = \begin{array}{l} \sim 46,000 \text{ fragments that are} \\ \sim 65.5 \text{ kb in average length} \end{array}$$

Clearly, the larger the recognition site, the smaller the number of fragments generated by enzymatic digestion.

Restriction enzymes were first used to study the very small genomes of viruses like bacteriophage lambda (λ), whose genome has a length of approximately 48.5 kb, and the animal tumor virus SV40, whose genome has a length of 5.2 kb. We now know that the six-base cutter *Eco*RI digests lambda DNA into 5 fragments, and the four-base cutter *Rsa*I digests SV40 into 12 fragments. But when molecular biologists first used restriction enzymes to digest these viral genomes, they also needed a tool that could distinguish the different fragments in a genome from each other and determine their sizes. That tool was gel electrophoresis.

Gel Electrophoresis Distinguishes DNA Fragments According to Size

Electrophoresis is the movement of charged molecules in an electric field. Biologists use it to separate many different types of molecules, for example, DNA of one length from DNA of other lengths, DNA from protein, or one kind of protein from another. In this discussion, we focus on its application to the separation of DNA fragments of varying length in a gel (Fig. 9.5). To carry out such a separation, you place a solution of DNA molecules into indentations called wells at one end of a porous gel-like matrix. When you then place the gel in a buffered aqueous solution and set up an electric field between electrodes affixed at either end, the electric field causes all charged molecules in the wells to migrate in the direction of the electrode having an opposite charge. Since all of the phosphate groups in the backbone of DNA carry a net negative charge in a solution near neutral pH, DNA molecules are pulled through a gel toward the positive electrode.

Several variables determine the rate at which DNA molecules (or any other molecules) move during electrophoresis. These variables are the strength of the electric field applied across the gel, the composition of the gel, the charge per unit volume of molecule (known as charge density), and the physical size of the molecule. The only one of these variables that actually differs among any set of linear DNA fragments migrating in a particular gel is size. This is because all molecules placed in a well are subjected to the same electric field and the same gel matrix; and all DNA molecules have the same charge density (because the charge of all nucleotide pairs is nearly identical). As a result, only differences in size cause different linear DNA molecules to migrate at different speeds during electrophoresis.

With linear DNA molecules, differences in size are proportional to differences in length: the longer the molecule, the larger the volume it will occupy as a random coil. The larger the volume a molecule occupies, the less likely it is to find a pore in the matrix big enough to squeeze through and the more often it will bump into the matrix. And the more often the molecule bumps into the matrix, the lower its rate of migration (also referred to as its mobility). With this background, you can follow the steps of Fig. 9.5a to determine the length of the restriction fragments in the DNA under analysis.

When electrophoresis is completed, the gel is incubated with a fluorescent DNA-binding dye called ethidium bromide. After the unbound dye has been washed away, it is easy to visualize the DNA by placing the gel under an ultraviolet light. The actual size of restriction fragments observed on gels is always determined by comparison to migration distances of known *marker fragments* that are subjected to electrophoresis in an adjacent lane of the gel.

DNA molecules range in size from small fragments of less than 10 bp to whole human chromosomes that have an average length of 130,000,000 bp. No one sizing procedure has the capacity to separate molecules throughout this enormous range. To detect DNA molecules in different size ranges, researchers use a variety of protocols based mainly on two kinds of gels: polyacrylamide (formed by covalent bonding between acrylamide monomers), which is good for distinguishing smaller DNA fragments; and agarose (formed by the noncovalent association of agarose polymers), which is suitable for looking at larger fragments. Figure 9.5b illustrates these differences.

Restriction Maps Provide a Rough Roadmap of Virus Genomes and Other Purified DNA Fragments

Researchers can use restriction enzymes not only as molecular scissors to create DNA fragments of different sizes but also as an analytic tool to create maps of viral genomes and other purified DNA fragments. These maps, called *restriction maps,* show the relative order and distances between multiple restriction sites, which thus act as landmarks along a DNA molecule.

There are numerous approaches to the derivation of a restriction map. One of the most commonly used methods involves digestion with multiple restriction enzymes—alone or mixed together—followed by gel electrophoresis to visualize the fragments produced. If the relative arrangement of sites for the various restriction enzymes employed does not create too many fragments, the data obtained can provide enough information to piece together a map showing the position of each restriction site. Figure 9.6 shows how a process of elimination allows you to infer the arrangement of restriction sites consistent with the results of digestion, using either of two enzymes alone or both enzymes simultaneously. This is one way to make a restriction map.

Today, molecular biologists can use automated DNA sequencing (along with computer analysis) to rapidly obtain detailed information on the size and sequence of any DNA molecule in the size range of SV40 or bacteriophage λ; we describe such sequencing techniques later in this chapter. Although the sequence information obtained in this way can immediately pinpoint restriction sites for enzymes whose recognition sequences are known (see Table 9.1), many investigators still

Gel electrophoresis separates DNA molecules according to size.

1. Attach a comb to a clear acrylic plate with clamps.

2. Pour heated agarose into plate. Allow to cool.

3. Remove comb. Shallow wells are left in gel.

4. Place gel in buffered solution in the gel box. Set up electrodes and current source.

Size marker well

5. Place solution of DNA fragments of known size into the size marker well. Place solution of unknown sized fragments into adjacent well. Turn on current; allow fragments to migrate for 1–20 hours.

Unknown-sized-fragment well

Direction of fragment migration

6. Remove gel from gel box. Incubate with ethidium bromide, then wash to remove excess dye.

Size markers | Unknown-sized DNA fragments

kb
12
8
4
2
1

7. Expose gel to UV light. DNA molecules will appear as red bands. A photo of the bands will provide a black-and-white image.

(a) Preparing the gel. To prepare an agarose gel containing wells for samples, you follow the illustrated steps. You then place the prepared gel on a base inside a gel tank that contains a buffered solution. Into one well, you place a solution containing DNA fragments of unknown size. Into another well you put a second sample carrying DNA "size markers." These markers are DNA fragments whose sizes have been calibrated exactly through complete sequence determination. You now turn on the electric current, allow the fragments to migrate for 1–20 hours, and then remove the gel from the electrophoresis chamber and incubate it with ethidium bromide, a fluorescent dye and intercalator that will bind tightly to any DNA fragments in the gel. After intercalation has occurred, immersion of the gel in water will wash away any unbound dye molecules. Then, with exposure to ultraviolet light, the bound dye will absorb photons in the UV range of the electromagnetic spectrum and give off photons in the visible red range. When this happens, all the DNA molecules will

appear as red bands, and a snapshot will show the relative positions to which they have migrated in the gel.

To determine the length of a DNA fragment, you chart the mobility of the band composed of that fragment relative to the migration of the size marker bands in the adjacent gel lane.

(b) Different types of gels separate different-sized DNA molecules. It is possible to separate DNA molecules in gels composed of two different types of chemical matrices: polyacrylamide and agarose. The average size of pores in polyacrylamide is much smaller than the average pores in agarose. Polyacrylamide can separate DNA molecules in the range of 10–500 nucleotides in length; agarose separates only larger molecules. The polyacrylamide gel on the *left* was used to separate restriction fragments formed by the digestion of eight different DNA samples with an enzyme that recognizes a four-base sequence. The fragments range in size from 90–340 bases in length. The agarose gel on the *right* was used to separate restriction fragments in a series of samples digested with *Eco*RI. The first lane contains size markers that range in size from 2.2–8.3 kb. The other lanes contain digested whole genomic DNA from different mice.

carry out restriction mapping (by methods such as the one shown in Fig. 9.6) as an inexpensive yet effective first step in the characterization of novel DNA molecules.

Cloning Fragments of DNA

While restriction enzyme digestion and gel electrophoresis provide a means for analyzing simple DNA molecules, the genomes of animals, plants, and even microorganisms are far too large to be analyzed in this way. For example, the *E. coli* genome is approximately 4,200 kb or 4.2 megabases (Mb) in length. An *Eco*RI digestion of this genome would produce approximately 1000 fragments, and as we calculated earlier, the same digestion of human DNA would produce 700,000 different fragments. If you subjected these complex mixtures of DNA fragments to gel electrophoresis, all you would see at the end is a smear rather than discrete bands (review Fig. 9.5b). To study any one fragment within this complex mixture, it is necessary first to purify it away from all the other fragments and then to

amplify it, that is, make many identical copies of the single purified molecule. Researchers can then apply various biochemical techniques—including restriction mapping and DNA sequencing—to the analysis of the isolated DNA fragment.

Scientists now use two strategies to accomplish the purification and amplification of individual fragments: *molecular cloning,* which replicates individual fragments of previously uncharacterized DNA; and the *polymerase chain reaction* (or *PCR*), which can purify and amplify a previously sequenced genomic region (or a transcribed version of it) from any source much more rapidly than cloning. Here we present the protocol for the molecular cloning of DNA. Later in the chapter, we describe PCR.

Molecular cloning is the process that takes a complex mixture of restriction fragments and uses living cells to purify and make many exact replicas of just one fragment at a time. It consists of two basic steps. In the first, DNA fragments that fall within a specified range of sizes are inserted into specialized chromosome-like carriers called *vectors,* which ensure the transport, replication, and purification of individual inserts. In

Figure 9.6 How to infer a restriction map from the sizes of restriction fragments produced by two restriction enzymes. Divide a purified preparation of cloned DNA into three aliquots; expose the first aliquot to *Eco*RI, the second to *Bam*HI, and the third to both enzymes. Now separate by gel electrophoresis the restriction fragments that result from each digestion and determine their sizes in relation to defined markers separated on the same gel. Finally, use a process of elimination to derive the only arrangement that can account for the results obtained with all three samples. To make sure that you understand how the restriction map at the bottom of the figure is the only arrangement that can account for all the data, explain how the pattern of restriction fragments in all three lanes of the gel is consistent with the map.

the second step, the combined vector-insert molecules are transported into living cells, and the cells make many copies of these molecules. Since all the copies are identical, the group of replicated DNA molecules is known as a **DNA clone.** A DNA clone is any amplified set of purified DNA molecules; it can consist of vectors alone, insert fragments alone, or a molecular combination of the two. DNA clones may be purified for immediate study or stored within cells as collections of clones known as *libraries* for future analysis. We now describe each step of molecular cloning.

Cloning Step 1: Ligation of Fragments to Cloning Vectors Creates Recombinant DNA Molecules

On their own, restriction fragments cannot reproduce themselves in a cell. To make replication possible, it is necessary to splice each fragment to a **vector:** a specialized DNA sequence that can enter a living cell, signal its presence to an investigator by conferring a detectable property on the host cell, and provide a means of replication for itself and the foreign DNA inserted into it. A vector must also possess distinguishing physical traits, such as size or shape, by which it can be purified away from the host cell's genome. There are several types of vectors (Table 9.2), and each one behaves as a chromosome capable of accepting foreign DNA inserts and replicating independently of the host cell's genome. The cutting and splicing together of vector and inserted fragment—DNA from two different origins—creates a **recombinant DNA molecule.**

Sticky Ends Facilitate Recombinant DNA Fabrication

Two characteristics of single-stranded, or "sticky," ends provide a basis for the efficient production of a vector-insert recombinant: The ends are available for base pairing, and no matter what the origin of the DNA (bacterial or human, for example), sticky ends produced with the same enzyme are complementary in sequence. You simply cut the vector with the same restriction enzyme used to generate the fragment of genomic DNA, and mix the digested vector and genomic DNAs together in the presence of DNA ligase. You then allow time for the base pairing of complementary sticky ends from the DNA of two different origins and for the ligase to stabilize the molecule by sealing the breaks in the sugar-phosphate backbones. Certain laboratory "tricks" (discussed later) help prevent two or more genomic fragments from joining with each other rather than with vectors. It is also possible to ligate restriction fragments having blunt ends onto vectors that also have blunt ends, although without the sticky ends, the ligation process is much less efficient.

There Are Several Types of Vectors

Available vectors differ from one another in biological properties, carrying capacity, and the type of host they can infect. The simplest vectors are minute circles of double-stranded DNA known as **plasmids** that can gain admission to and replicate in the cytoplasm of many kinds of bacterial cells, independently of the bacterial chromosomes (Fig. 9.7). The most useful plasmids contain several recognition sites, one for each of several different restriction enzymes; for example, one *Eco*RI site, one *Hpa*I site, and so forth. This provides flexibility in the choice of enzymes that can be used to digest the DNA containing the fragment, or fragments, of interest. Exposure to any one of these restriction enzymes opens up the vector at the corresponding recognition site, allowing the insertion of a foreign DNA fragment, without at the same time splitting the plasmid into many pieces and thereby destroying its continuity and integrity (Fig. 9.7).

A plasmid carrying a foreign insert is known as a *recombinant plasmid.* Each plasmid vector also carries an origin of replication and a gene for resistance to a specific antibiotic. The origin of replication enables it to replicate independently inside a bacterium. The gene for antibiotic resistance signals its presence by conferring on the host cell the ability to survive in a medium containing a specific antibiotic; the resistance gene thereby enables experimenters to select for

TABLE 9.2 **Various Vectors and the Size of the Inserts They Carry**

Vector	Form of Vector	Host	Typical Carrying Capacity (Size of Insert Accepted)	Major Uses
Plasmid	Double-stranded circular DNA	E.coli	Up to 15 kb	cDNA libraries; subcloning
Bacteriophage lambda	Virus (linear DNA)	E.coli	Up to 25 kb	Genomic and cDNA libraries
Cosmid	Double-stranded circular DNA	E.coli	30–45 kb	Genomic libraries
Bacteriophage P1	Virus (circular DNA)	E.coli	70–90 kb	Genomic libraries
BAC	Bacterial artificial chromosome	E.coli	100–500 kb	Genomic libraries
YAC	Yeast artificial chromosome	Yeast	250–2000 kb (2 megabases)	Genomic libraries

propagation only those bacterial cells that contain a plasmid (Fig. 9.8). Antibiotic resistance genes and other vector genes that make it possible to pick out cells harboring a particular DNA molecule are called **selectable markers.** Plasmids fulfill the final requirement for vectors—ease of purification—because they can be purified away from the genomic DNA of the bacterial host by several techniques that take advantage of size and other differences, as described later. Plasmid vector restriction sites useful for cloning are ones that do not interrupt either the vector origin of replication or the coding region of the selectable marker.

Common plasmid vectors are 2–4 kb in length and capable of carrying up to 15 kb of foreign DNA. In some experimental situations, however, it is necessary to isolate DNA fragments that are larger than this limit. A variety of larger-capacity vectors have been developed to accommodate different sizes of DNA fragments. Table 9.2 lists the properties of a few types of vectors and the host cells in which they grow.

Several types of vectors have been constructed from the genomes of naturally occurring viruses such as bacteriophage lambda (λ). Phage λ is a double-stranded DNA virus that infects *E. coli*. As mentioned earlier, each λ chromosome is 48.5 kb long; it can be engineered to receive central inserts of up to 25 kb in length that replace nonessential viral sequences. The remaining viral sequences on both sides of the insert represent vector arms containing genes required to form whole virus particles; these particles infect host cells with extremely high rates of efficiency and then multiply voraciously. Foreign DNA inserted between the two arms of a λ vector will be packaged into the λ virus particles and amplified along with them inside the host cells.

The largest-capacity vectors are *artificial chromosomes:* recombinant DNA molecules formed by combining chromosomal replication and segregation elements with a DNA insert. Such vectors contain the crucial chromosomal elements that allow them to replicate within a particular host. A bacterial arti-

ficial chromosome (BAC) can accommodate a DNA insert of 150 kb; a yeast artificial chromosome can accommodate a DNA insert of 2000 kb.

The utility of these different cloning vectors becomes apparent when you consider the structure of the β-globin locus. Situated on chromosome 11, the entire locus, including multiple β-globin-like genes and all the regulatory information for proper expression of these genes, spans 70 kb of DNA. Only a few vectors (including BACs and YACs) can ferry this much material. By contrast, the collection of exons and introns that constitute a single β-globin gene is only 1.4 kb in length and can thus fit easily into a plasmid. Most other genes are larger than individual β-globin genes, in part because they have greater coding capacity and many more introns; as a result, to be cloned as a single insert, they require a λ, BAC, or occasionally even a YAC vector.

Cloning Step 2: Host Cells Take up and Amplify Vector-Insert Recombinants

Although each type of vector functions in a slightly different way and enters a specific kind of host, the general scheme of entering a host cell and taking advantage of the cellular environment to replicate itself is the same for all. We divide our discussion of this step of cloning into three parts: getting foreign DNA into the host cell; selecting cells that have received a DNA molecule; and distinguishing insert-containing recombinant molecules from vectors without inserts. Figure 9.8 illustrates the three-part process with a plasmid vector containing an origin of replication, the gene for resistance to ampicillin (*amp*^{r}), and the *E. coli lacZ* gene, which encodes the enzyme β-galactosidase. By constructing the vector with a common restriction site like *Eco*RI right in the middle of the *lacZ* gene, researchers can insert foreign DNA into the gene at that location and then use the disruption of *lacZ* gene function to distinguish insert-containing recombinant molecules from vectors without inserts (as described in the caption and later in the text).

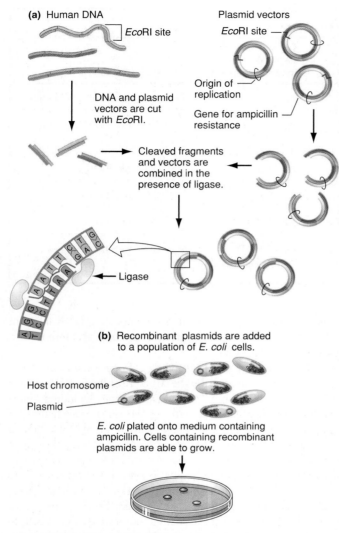

(a) Human DNA

EcoRI site

Plasmid vectors

EcoRI site

Origin of replication

Gene for ampicillin resistance

DNA and plasmid vectors are cut with EcoRI.

Cleaved fragments and vectors are combined in the presence of ligase.

Ligase

(b) Recombinant plasmids are added to a population of E. coli cells.

Host chromosome

Plasmid

E. coli plated onto medium containing ampicillin. Cells containing recombinant plasmids are able to grow.

Figure 9.7 Creating recombinant DNA molecules with plasmid vectors. DNA from human tissue is cut with EcoRI to reduce the genome into a mixture of EcoRI restriction fragments. A plasmid vector is also cut with EcoRI. The plasmid has a single EcoRI recognition site, and digestion with the enzyme converts it into a single linear molecule. The cut genomic DNA and the cut plasmid vectors are mixed together in the presence of the enzyme ligase, which sutures the two types of molecules to each other to form a circular recombinant DNA molecule.

Many of the plasmid vectors used today incorporate most if not all of the features depicted in Fig. 9.8.

Transformation: Vectors Carry Insert DNA into Cells

Transformation, we saw in Chapter 6, is the process by which a cell or organism takes up a foreign DNA molecule, changing the genetic characteristics of that cell or organism. What we now describe is similar to what Avery and his colleagues did in the transformation experiments that determined DNA was the molecule of heredity (see p. 159 of Chapter 6), but the method

outlined here is more efficient. First, recombinant DNA molecules are added to a suspension of specially prepared *E. coli*. Under conditions favoring entry, such as suspension of the bacterial cells in a cold $CaCl_2$ solution or treatment of the solution with high-voltage electric shock (a technique known as *electroporation*), the plasmids will enter about 1 in 1000 cells (Fig. 9.8a). These protocols increase the permeability of the bacterial cell membrane, in essence, punching temporary holes through which the DNA gains entry. The probability that any one plasmid will enter any one cell is so low (0.001) that the probability of simultaneous entry of two plasmids into a single cell is insignificant ($0.001 \times 0.001 = 0.000001$).

How Do You Know Which Cells Have Been Transformed?

To identify the 0.1% of cells housing a plasmid, you decant the bacteria-plasmid mixture onto a plate containing agar, nutrients, and ampicillin. Only cells transformed by a plasmid providing resistance to ampicillin will be able to grow and multiply in the presence of the antibiotic. The plasmid's origin of replication enables it to replicate in the bacterial cell independently of the bacterial chromosome; in fact, most plasmids replicate so well that a single bacterial cell may end up with hundreds of identical copies of the same plasmid molecule. Each viable plasmid-containing bacterial cell will multiply to produce a colony of tens of millions of genetically identical cells. The colony as a whole is considered a **cellular clone.** Such clones show up on the agar plate as spots about 1 mm in diameter, easily big enough to see yet small enough for many distinct ones to appear on a plate (Fig. 9.8b, see also chapter opening photo). The millions of identical plasmid molecules contained within a colony together make up a **DNA clone.** They can be purified away from other cellular material as described below. The process through which a particular plasmid molecule is amplified to millions of copies inside the multiplying cells of a colony derived from a single bacterial progenitor is called **molecular cloning.**

How Do You Know Whether the Plasmids Inside Bacterial Cells Contain an Insert?

If prepared under proper conditions (including, for example, a controlled ratio of fragments to vectors during the ligation procedure), most plasmids will contain an insert. Some plasmids, however, will slip through without one. Figure 9.8c, shows how the system we are discussing distinguishes cells with just vectors from cells with vectors containing inserts. The medium on which the transformed, ampicillin-resistant bacteria grow contains, in addition to nutrients and ampicillin, a chemical compound known as X-Gal. This compound serves as a substrate for the reaction catalyzed by the intact β-galactosidase enzyme (the protein encoded by the *lacZ* gene); one product of the reaction is a new, blue-colored chemical. Cells containing vectors without inserts turn blue because they carry the original intact β-galactosidase gene. Cells containing plasmids with inserts remain

Figure 9.8 How to identify transformed bacterial cells containing plasmids with DNA inserts. Under proper conditions (including, for example, exposure to phosphatase and a controlled ratio of fragments to vectors), most plasmids will accept an insert. Some, however, will slip through without one. To identify such interlopers, plasmid vectors are constructed so that they contain the *lacZ* gene with a restriction site right in the middle of the gene. If the vector reanneals to itself without inclusion of an insert, the *lacZ* gene will remain uninterrupted; if it accepts an insert, the gene will be interrupted. **(a)** Transformation: When added to a culture of bacteria, plasmids enter about 1 in 1000 cells. **(b)** Only cells transformed by a plasmid carrying a gene for ampicillin resistance will grow. **(c)** Cells containing vectors that have reannealed to themselves without the inclusion of an insert will express the uninterrupted *lacZ* gene that they carry. The polypeptide product of the gene is β-galactosidase. Reaction of this enzyme with a substrate known as X-Gal produces a molecule that turns the cell blue. By contrast, if the opened ends of the plasmid vector have reannealed to a fragment of foreign DNA, this insert will interrupt and inactivate the *lacZ* gene on the plasmid, rendering it unable to produce a functional polypeptide. As a result of this process, called *insertional inactivation,* any cells containing recombinant plasmids will not generate active β-galactosidase and will therefore not turn blue. It is thus possible to visually distinguish colorless colonies that contain inserts from blue colonies that do not.

colorless, because the interrupted *lacZ* gene does not allow production of functional β-galactosidase enzyme.

To Purify Cloned DNA, You Separate Recombinant Plasmid Vector from Host DNA, and DNA Insert from Vector

To obtain a test tube filled with copies of just the cloned DNA insert, you first purify the cloned vector-insert recombinant molecule away from the host cell chromosome and other cellular components. You then purify the DNA insert away from vector DNA. The purification of cloned DNA away from the host chromosome depends on the unique physical characteristics that distinguish the two types of DNA. The cloned molecule is always much smaller and sometimes different in form (circular versus linear) from the host chromosome. Figure 9.9a illustrates the protocol that exploits these physical differences to separate cloned plasmids from bacterial chromosomes.

To separate genomic inserts from plasmid vectors, you simply cut the purified vector-insert molecules with the same restriction enzyme that you used to make the recombinants in the first place. You then subject the resulting mixture of insert DNA disengaged from vector DNA to gel electrophoresis (Fig. 9.9b). At the completion of electrophoresis, the region of the gel that contains the insert is cut out and the DNA within it purified. Molecular cloning ultimately yields a large amount of purified fragments—billions of identical copies of the original small piece of genomic DNA—that are now ready for study.

Libraries Are Collections of Cloned Fragments

Moving step by step from the DNA of any organism to a single purified DNA fragment is a long and tedious process. Fortunately, scientists do not have to return to step 1 every time they need to purify a new genomic fragment from the

Figure 9.9 Purifying cloned DNA. (a) Separation of plasmid from the bacterial chromosome and other cellular components. In the plasmid-*E. coli* system, the bacterial chromosome is about 4000 kb long, while the recombinant plasmid is from 4–20 kb in length. Cell lysis causes the large bacterial chromosome to fragment into random linear fragments while the small circular plasmid molecules remain intact. As supercoiled plasmid DNA intercalates ethidium bromide, it becomes negatively supercoiled, making it denser than chromosomal DNA. The buoyant density of the plasmids is distinguishable from that of the chromosomes when the two are added to a cesium chloride (CsCl) gradient saturated with ethidium bromide. The band containing plasmid is extracted and purified. **(b)** Separation of the genomic DNA insert from vector DNA. The purified plasmids are digested with the same enzyme that was used to construct the recombinant molecule in the first place, releasing vector and insert fragments that differ in size. With gel electrophoresis, the two types of fragments can be separated from each other, and the band containing the insert can be cut out and purified. **(c)** Subcloning: From YAC clones to lambda clones to plasmids. The large genomes of higher eukaryotes (such as those of mice and humans) can be readily divided into a few thousand YAC clones that each contain 1 Mb of genomic DNA. To study gene-size regions from a particular YAC, however, it is convenient to subclone 20 kb fragments into a lambda vector. To analyze particular regions of a gene, it is useful to perform another round of subcloning, placing fragments that are 2 kb or smaller into a plasmid vector.

same organism. Instead, they can build a **genomic library:** a storable collection of cellular clones that contains copies of every sequence in the whole genome inserted into a suitable vector. Like traditional book libraries, genomic libraries store large amounts of information for retrieval upon request. They make it possible to start a new cloning project at an advanced stage, when the initial cloning step has already been completed and the only difficult task left is to determine which of the many clones in a library contains the DNA sequence of interest. Once the correct cellular clone is identified, it can be amplified to yield a large amount of the desired genomic fragment.

How to Compile a Genomic Library

If you digested the genome of a single cell with a restriction enzyme and ligated every fragment to a vector with 100% efficiency, and then transformed all of these recombinant DNA molecules into host cells with 100% efficiency, the resulting set of clones would represent the entire genome in a fragmented form. A hypothetical collection of cellular clones that includes one copy—and one copy only—of every sequence in the entire genome would be a *complete genomic library.*

How many clones will be present in this hypothetical library? If you started with the 3,000,000 kb of DNA from a haploid human sperm and reliably cut it into a series of 150 kb restriction fragments, you would generate 3,000,000/150 = 20,000 genomic fragments. If you placed each and every one of these fragments into BAC cloning vectors that were then transformed into *E. coli* host cells, you would create a perfect library of 20,000 clones that collectively carry every locus in the genome. The number of clones in this perfect library defines a **genomic equivalent.** For any library, to find the number of clones that constitute one genomic equivalent, you simply divide the length of the genome (here, 3,000,000 kb) by the average size of the inserts carried by the library's vector (in this case, 150 kb).

In real life, it is impossible to obtain a perfect library. Each step of cloning is far from 100% efficient, and the DNA of a single cell does not supply sufficient raw material for the process. Researchers must thus harvest DNA from the millions of cells in a particular tissue or organism. If you make a genomic library with this DNA by collecting only one genomic equivalent (20,000 clones for a human library in BAC vectors), the uncertainties of random sampling of DNA fragments would mean that some human DNA fragments would appear more than once, while others might not be present at all. To increase the chance that a library contains at least one clone with a particular genomic fragment, the number of clones in the library must exceed one genomic equivalent. Including four to five genomic equivalents produces an average of four to five clones for each locus, and a 95% probability that any individual locus is present at least once.

For many types of molecular genetics experiments, it is advantageous to use a vector like a BAC that allows the cloning of large fragments of DNA. This is because the number of clones in a genomic equivalent is much smaller than in libraries constructed with a vector that accepts only smaller fragments, so investigators have to search through fewer clones to find a particular clone of interest. For example, a genomic equivalent of BAC clones with 150 kb inserts contains only 20,000 clones, while a genomic equivalent of plasmid clones with 10 kb inserts would contain 300,000 clones.

But BACs have a disadvantage: It takes a lot more work to characterize the larger inserts they carry. Researchers must often **subclone** smaller restriction fragments from a BAC clone to look more closely at a particular region of genomic DNA (Fig. 9.9c). Sometimes, it is useful to reduce complexity one step at a time, subcloning a BAC DNA fragment into a lambda

vector and then subcloning fragments of the lambda clone into a plasmid vector. This process of **subcloning** establishes smaller and smaller DNA clones that are enriched for the sequences of interest.

cDNA Libraries Carry Information from the RNA Transcripts Present in a Particular Tissue

Often, only the information in a gene's coding sequence is of experimental interest, and it would be advantageous to limit analysis to the gene's exons without having to determine the structure of the introns as well. Because coding sequences account for a very small percentage of genomic DNA in higher eukaryotes, however, it is inefficient to look for them in genomic libraries. The solution is to generate **cDNA libraries,** which store sequences copied into DNA from all the RNA transcripts present in a particular cell type, tissue, or organ. Because they are obtained from RNA transcripts, these sequences carry only exon information. As we saw in Chapter 8, exons contain all the protein-coding information of a gene, as well as sequences that will appear in the 5′ and 3′ untranslated regions at the ends of corresponding mRNAs.

To produce DNA clones from mRNA sequences, researchers rely on a series of *in vitro* reactions that mimic several stages in the life cycle of viruses known as **retroviruses.** Retroviruses, which include among their ranks the HIV virus that causes AIDS, carry their genetic information in molecules of RNA. As part of their gene-transmission kit, retroviruses also contain the unusual enzyme known as **RNA-dependent DNA polymerase,** or simply **reverse transcriptase** (review the Genetics and Society box in Chapter 8, p. 252). After infecting a cell, a retrovirus uses reverse transcriptase to copy its single strand of RNA into a mirror-image-like strand of complementary DNA, often abbreviated as **cDNA.** The reverse transcriptase, which can also function as a DNA-dependent DNA polymerase, then makes a second strand of DNA complementary to this first cDNA strand (and equivalent in sequence to the original RNA template). Finally, this double-stranded copy of the retroviral RNA chromosome integrates into the host cell's genome. Although the designation cDNA originally meant a single strand of DNA complementary to an RNA molecule, it now refers to any DNA—single- or double-stranded—derived from RNA. In 1975, researchers began to use purified reverse transcriptase to convert mRNA sequences from any tissue in any organism into cDNA sequences for cloning.

Suppose you were interested in studying the structure of a mutant β-globin protein. You have already analyzed hemoglobin obtained from a patient carrying this mutation and found that the alteration affects the amino-acid structure of the protein itself and not its regulation, so you now need only look at the sequence of the mutant gene's coding region to understand the primary genetic defect. To establish a library enriched for the mutant gene sequence and lacking all the extraneous information from genomic introns and nontranscribed regions, you would first obtain mRNA from the cytoplasm of the patient's red blood cell precursors (Fig. 9.10a). About 80% of the total

(a) Red blood cell precursors

Release mRNA from cytoplasm and purify.

(b) Add oligo(dT) primer. Treat with reverse transcriptase in presence of four nucleotides.

Reverse transcriptase

(c) Denature cDNA-mRNA hybrids and digest mRNA with RNase. cDNA acts as template for synthesis of second cDNA strand in the presence of four nucleotides and DNA polymerase.

S1 nuclease cuts hairpin loop   DNA polymerase

(d) Insert cDNA into vector.

Figure 9.10 Converting RNA transcripts to cDNA. (a) Obtain mRNA from red blood cell precursors (which still contain a nucleus). **(b)** Add oligo(dT) primers to initiate reverse transcription of cDNA from the mRNA template. The result is a hybrid cDNA-mRNA molecule. **(c)** Heat the mixture to separate mRNA and cDNA strands, and then eliminate the mRNA transcript through digestion with RNase. The 3′ end of the cDNA strands will loop around and bind by chance to complementary nucleotides within the same strand, forming the primer necessary to initiate DNA polymerization. Add DNA polymerase and the four nucleotides to produce a second cDNA strand complementary to the first. After the reaction is completed, the enzyme S1 nuclease is used to cleave the "hairpin loop" at one end. **(d)** Insert the newly created double-stranded DNA molecule into a vector for cloning.

mRNA in these red blood cells is from the α- and β-hemoglobin genes, so the mRNA preparation contains a much higher proportion of the sequence corresponding to the β-globin gene than do the genomic sequences found in a cell's nuclear DNA.

The addition of reverse transcriptase to the total mRNA preparation—as well as ample amounts of the four deoxyribose nucleotide triphosphates, and primers to initiate synthesis—will generate single-stranded cDNA bound to the mRNA template (Fig. 9.10b). The primers used in this reaction would be oligo(dT)—single-stranded fragments of DNA containing about 20 Ts in a row—that can bind through hybridization to the poly-A tail at the 3′ end of eukaryotic mRNAs and initiate polymerization of the first cDNA strand. Upon exposure to high temperature, the mRNA-cDNA hybrids separate, or denature, into single strands. The addition of an RNase enzyme that digests the original RNA strands leaves intact single strands of cDNA. Most of these fold back on themselves at their 3′ end to form transient hairpin loops via base pairing with random complementary nucleotides in nearby sequences in the same strand. These hairpin loops will serve as primers for synthesis of the second DNA strand. Now the addition of DNA polymerase, in the presence of the requisite nucleotide triphosphates, initiates the production of a second cDNA strand from the just synthesized single-stranded cDNA template (Fig. 9.10c).

After using restriction enzymes and ligase to insert the double-stranded cDNA into a suitable vector (Fig. 9.10d) and then transforming the vector-insert recombinants into appropriate host cells, you would have a library of double-stranded cDNA fragments, with the cDNA fragments in each individual clone corresponding to an mRNA molecule in the red blood cells that served as your sample. In contrast to a genomic library, this cDNA library does not contain all the DNA sequences in the genome. Instead, it includes only the exons from that part of the genome that the red blood precursors were actively transcribing for translation into protein. Since fully spliced RNA transcripts are on average a few kilobases in length, the cDNA fragments constructed in their image fall within that size range as well. The percentage of clones in the library with cDNA specific for the β-globin protein will reflect the fraction of red blood cell mRNAs that are β-globin messages. For genes expressed infrequently or in very few tissues, you will have to screen many clones of a cDNA library to find the gene of interest. For highly expressed genes, such as the β-globin gene, you will have to screen only a few clones in a red blood cell precursor library.

Genomic Versus cDNA Libraries

Figure 9.11 compares genomic and cDNA libraries. The main advantage of genomic libraries is that the genomic clones within them represent all regions of DNA equally and show what the intact genome looks like in the region of each clone. The chief advantage of cDNA libraries is that the cDNA clones reveal which parts of the genome contain the information used in making proteins in specific tissues, as determined from the prevalence of the mRNAs for the genes involved. To gain as much information as possible about a gene's structure and function, researchers rely on both types of libraries.

Random 100 kb genomic region

Gene *A*
expressed only
in brain

Gene *B*
expressed in
all tissues

Gene *C*
expressed only
in liver

Clones from a genomic library with 20 kb inserts that are homologous
to this region

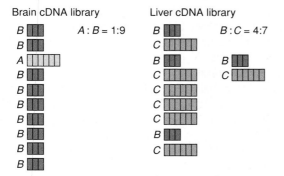

Contains part of gene *A*

Contains parts of genes *B* and *C*

Contains all of gene *C*

Contains only last exon of gene *A*

Clones from cDNA libraries

Brain cDNA library

A : *B* = 1:9

Liver cDNA library

B : *C* = 4:7

Figure 9.11   A comparison of genomic and cDNA libraries.
Every tissue in a multicellular organism can generate the same genomic
library, and the DNA fragments in that library collectively carry all the
DNA of the genome. Individual genomic fragments may contain no
gene, one gene, more than or less than one gene, introns that are
transcribed into RNA but spliced out before translation, regulatory
sequences that are not transcribed at all, or any combination of these
possibilities. On average, the clones of a genomic library represent every
locus an equal number of times. By contrast, every tissue in a
multicellular organism generates a different cDNA library. Clones of this
library represent only the fraction of the genome that is being actively
transcribed. Within a cDNA library derived from any single tissue type,
the frequency with which particular fragments appear is proportional to
the level of the corresponding mRNA in that tissue. Thus, the differential
expression of genes within different tissues can be ascertained by the
numbers of clones of each type that show up in each library.

So far, we have presented molecular cloning technology as
a powerful set of tools used by scientists engaged in basic
research to analyze genes and their products. This powerful
technology also makes it possible to go beyond analysis to the
production of new medicines and other products of commercial
value. When molecular cloning technology is used for commercial applications, it is often referred to as *biotechnology.*

Expression Vectors Provide a Means for Producing Large Amounts of a Specific Polypeptide

The modern biotechnology industry sprang into existence in
the late 1970s. It was based on a specialized cloning technol-

ogy that allows the large-scale production of well-defined
polypeptides with important medical or commercial value.
The production process begins with the ligation of a DNA
fragment containing just the coding region of the desired
polypeptide to a specialized **expression vector** that contains
a *promoter* and other regulatory sequences recognized by the
transcriptional apparatus of the host cell (Fig. 9.12). The first
expression vectors were plasmids that contained an *E. coli*
promoter. Among the first human genes to be cloned into a
plasmid expression vector was the gene for insulin. In bacterial cells transformed with an expression vector-insert construct, *E. coli* RNA polymerase recognizes the promoter in the
vector DNA and proceeds to transcribe the insulin-coding
region contained in the insert. The *E. coli* ribosomes then
translate the insulin mRNA to human insulin (Fig. 9.12a).
Purification from a large culture of *E. coli* cells prepares the
insulin for commercial use (see Fig. 9.12b).

There is one significant drawback to the use of bacterial
cells as hosts to produce human proteins. Human and other
eukaryotic polypeptides produced in bacterial cells are not
properly folded into a functional three-dimensional structure,
because the bacterial cytoplasmic environment is not compatible with eukaryotic folding processes. To overcome this
problem, scientists developed specialized expression vectors
that can both replicate and transcribe insert sequences inside
the eukaryotic cells of yeast. Sometimes though, even the
yeast cells do not achieve proper polypeptide folding. For
these stubborn cases, scientists developed expression vectors
that are specific for cultured mammalian cells. There are
trade-offs as biotechnologists move from bacterial cells to
yeast cells to mammalian cells as factories for producing
human proteins. With bacteria, it is possible to obtain large
quantities of proteins most efficiently and cheaply, but in
many cases, the proteins are not fully functional. Proteins
made in mammalian cells are almost always fully functional,
but the process is much more expensive and has a lower yield.
Yeast is in the middle on both accounts. Biotechnologists
must balance efficiency and functionality in their choice of an
appropriate host cell.

It is possible to produce an entire cDNA library in an
expression vector. Such a library is called an *expression library.*
Most of the clones in an expression library will be nonproduc-
tive, either because the insert contains a coding sequence in the
"wrong" direction relative to the promoter or because the
cDNA is an incomplete copy of the coding sequence.
Nevertheless, if the library is produced in an appropriate manner, a proportion of the clones will produce functional protein.
As a result, instead of using a DNA probe to locate a specific
cDNA fragment, you can use a fluorescently labeled antibody
that binds to the protein product of the gene you wish to clone.
To find the gene sequence producing the protein, you simply
make a replica of the host cell colonies, lyse these replicas, and
expose their contents to the protein-specific antibody labeled
with fluorescence. The antibody, by interacting with the protein, will identify the cell colony containing the cDNA
sequence responsible for the protein (Fig. 9.12b).

(a) An expression vector allows production of specific polypeptide

Bacterial cell

(b) Screening for insulin gene expression

Overlay plate with nitrocellulose paper. Pick up cells.

Lyse cell. Treat with NaOH. Proteins adhere to paper.

Incubate paper in solution of labeled insulin antibody. Antibodies will bind to insulin protein.

Wash filter. Expose to UV light and identify flourescent spots. Compare with original plate in order to find bacterial clone containing human insulin gene.

Figure 9.12 An expression vector can be used to produce a desired polypeptide. (a) A cDNA fragment containing just the coding region of insulin is ligated to a plasmid vector containing an *E. coli* promoter next to the restriction site used to cut it open. The recombinant DNA molecule is transformed into *E. coli* cells, where it is recognized by RNA polymerase and transcribed into insulin mRNA. When the host-cell translational machinery sees the transcripts, it translates them into insulin protein. **(b)** Screening an expression library with an antibody probe. Clones of an expression library are partially transferred to a nitrocellulose paper disk. The disk is removed from the plate and exposed to a solution that lyses the cells and releases the proteins, which are attached to the disk through baking. The disk is then exposed to a solution containing a fluorescently labeled antibody to insulin. Antibody molecules bind to protein within spots formed by clones expressing the insulin gene. After washing away unbound antibody, exposure to UV light allows the identification of colonies expressing the gene product recognized by the antibody. The recombinant DNA clones in these colonies must carry the DNA sequence that encodes the protein recognized by the antibody.

Hybridization Is Used to Identify Similar DNA Sequences

While you can "screen" an expression library with an antibody that recognizes the protein product of the gene that you want to clone, the DNA inserts in all other types of libraries do not signal their presence in such a straightforward fashion. So, once you have collected the hundreds of thousands of human DNA fragments in a genomic or a nonexpressing cDNA library, how do you find the gene you wish to study, the proverbial "needle in the haystack"? For example, how would you go about finding a genomic clone that contained just a portion of the β-globin gene? One way is to take advantage of hybridization—the natural propensity of complementary single-stranded molecules to form stable double helixes. If a clone containing any portion of a β-globin sequence is already available—perhaps as a previously obtained cDNA clone—it can be denatured (separated into single strands), linked with a radioactive or fluorescent tag, and then used to probe a whole genome library that is spread out as a series of colonies on one or more petri plates. The tagged DNA probe will hybridize with denatured DNA from the genomic clones that contain a complementary sequence. After nonhybridizing probe is washed away, only the tiny number of β-globin-containing clones (among the hundreds of thousands in the library) are tagged by virtue of their hybridization to the probe. Individual cellular clones can then be retrieved from the library and put into culture to produce larger amounts of material in preparation for recovery of the purified DNA insert.

Hybridization has a single critical requirement: The region of complementarity between two single strands must be sufficiently long and accurate to produce a large enough number of hydrogen bonds to generate a cohesive force. Accuracy refers to the percentage of bases within the complementary regions that are actual complements of each other (C–G or A–T). If the overall region is at least several hundred bases long, strands that show only 80% complementarity can still form hybrids. The cohesive force formed by adding together large numbers of hydrogen bonds will counteract the thermal forces that tend to disrupt the double helix. Individual hydrogen bonds are very weak, and the two (A:T) or three (G:C) hydrogen bonds that form between individual pairs of bases are far too weak to prevent thermal forces from pulling them apart. But if two single strands form hydrogen bonds between 25 or more contiguous base pairs, the combined force is sufficient to counteract the thermal forces that operate even at room temperature. Hybridization can occur between any two single strands of nucleic acid: DNA/DNA, DNA/RNA, or RNA/RNA. In this section, we focus on DNA/DNA hybridization.

We now describe how to screen a library with a DNA probe.

Preparing the Library

Before you can use a DNA probe to screen a library, you must distribute the library's clones on petri plates and place a piece of nitrocellulose paper over each plate, allowing some, but not all, of the cells from each colony to transfer to the nitrocellulose disk replica (Fig. 9.13). You now lyse the bacteria on the disk with an

Master plate containing genomic
library of mouse clones.

Overlay a nitrocellulose disk
to make a replica of the plate.

Nitrocellulose
disk

Remove disk from plate and lyse cells on
it and denature DNA with NaOH. Bake
and treat with UV light to bind DNA strands
to disk.

Disk replica

Labeled
human
cystic fibrosis
sequences

Add labeled probe.
Colonies with complementary DNA
sequences hybridize to probe and restrain it.

Human probe

Mouse gene

Original plate

Wash disk, expose to X-ray film.

Compare with original plate to locate bacterial
clone with desired genomic fragment.

Figure 9.13 Screening a library of clones by hybridization to a labeled probe. To determine which clone in a library has a DNA fragment of interest, you place a wet nitrocellulose paper disk onto the plate containing cellular clones and then remove it. This procedure allows most cells within the original colonies on the plate to remain in their original positions but transfers a small portion of each colony to the disk. Treating the disk under special conditions causes the cells to lyse and release their DNA; the DNA is then denatured with NaOH and attached to the disk through baking and exposure to UV light. Next, a labeled DNA probe in solution is applied directly to the disk; the solution is left on the disk long enough to promote hybridization between the probe and any complementary strands of DNA that may be present within one or more spots that correspond to one or more of the clones. The disk is removed and the nonhybridizing probe is washed away. If the probe is labeled with a radioactive isotope, a piece of X-ray film is placed up against the disk and exposed to record the location of spots that are radioactive because of hybridization to the probe. This process is referred to as *autoradiography*. By comparing the location of spots on the film with the original bacterial plate, it is possible to identify the clone or clones that contain DNA fragments of interest.

alkaline solution, which both releases the DNA from the cells and denatures it into single strands. Baking the disk or treating it with ultraviolet (UV) light will attach the single-stranded DNA

molecules from the cells of each colony to the filter at the confined position corresponding to that clone's location in the petri plate. Some of the single-stranded DNAs attached to the nitrocellulose paper will be from the recombinant plasmids that replicated in the cells making up the colony. The library's DNA is now ready for screening with DNA probes.

Constructing the DNA Probes

DNA probes are purified fragments of single-stranded DNA 25 to several thousand nucleotides in length that are subsequently labeled with a radioactive isotope (typically ^{32}P) or a fluorescent dye. DNA probes can be produced from previously cloned fragments of DNA, purified fragments of DNA amplifed by PCR (described in the next section), or short single strands of chemically synthesized DNA.

In chemical synthesis, an automated DNA synthesizer adds specified nucleotides, one at a time, through chemical reactions, to a growing DNA strand (Fig. 9.14a). Modern synthesizers can produce specific sequences that range in size from 25–100 nucleotides. An investigator can instruct the computer controlling the DNA synthesizer to construct a particular sequence of As, Ts, Cs, and Gs. Within a few hours, the machine produces the desired short DNA chains, which are known as **oligonucleotides.**

To produce labeled probes during chemical synthesis, scientists provide the synthesizer with nucleotides that are prelabeled. To fabricate labeled probes from previously purifed DNA fragments or clones, they use DNA polymerase in combination with prelabeled nucleotides.

With the availability of oligonucleotide synthesis, it is possible to generate probes indirectly from a polypeptide sequence whose corresponding gene coding sequence is unknown, rather than directly from a known DNA sequence. The process by which one turns a polypeptide sequence into a set of corresponding DNA sequences is known as *reverse translation* (Fig. 9.14b). To perform a reverse translation, an investigator first translates the amino-acid sequence of a protein into a DNA sequence via the genetic code dictionary. Recall, however, that the genetic code is "degenerate," with more than one codon for many individual amino acids. Without knowing the coding DNA sequence, it is often impossible to predict which of several codons is actually present in the genome. To simplify the task, investigators choose peptide sequences containing amino acids encoded by as few potential codons as possible. They must then synthesize a mixture of oligonucleotides containing all possible combinations of codons for each amino acid. This is no problem for an automated DNA synthesizer: An investigator can direct the machine to add in a defined mixture of nucleotides (A and G, for example) at each ambiguous position in the oligonucleotide. With this indirect method of obtaining a DNA probe, researchers can locate and clone genes even if they have only partial coding information based on the proteins the genes encode.

Screening the Library

To find the hemoglobin genes in a genomic library that has been replicated onto nitrocellulose paper disks, you simply expose the library's DNA to a solution of α- and β-globin

(a)

(b) Synthesizing DNA probes based on reverse translation

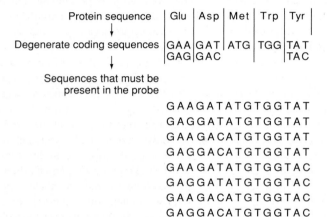

Protein sequence	Glu	Asp	Met	Trp	Tyr
↓					
Degenerate coding sequences	GAA GAG	GAT GAC	ATG	TGG	TAT TAC
↓					

Sequences that must be
present in the probe

GAAGATATGTGGTAT
GAGGATATGTGGTAT
GAAGACATGTGGTAT
GAGGACATGTGGTAT
GAAGATATGTGGTAC
GAGGATATGTGGTAC
GAAGACATGTGGTAC
GAGGACATGTGGTAC

Figure 9.14 How to make oligonucleotide probes for screening a library. (a) A DNA synthesizer is a machine that automates the addition through chemical reactions of specified nucleotides to the growing DNA chains, known as oligonucleotides. Four of the five smaller bottles at the *top* contain solutions of each of the four different nucleotide triphosphates (A, T, C, and G), and the fifth bottle contains a mixture of the four nucleotides for situations where an investigator wants to create an oligonucleotide with a random base in that position. The other bottles contain various reagents used in the various chemical reactions that the synthesizer carries out. **(b)** Reverse translation. An amino-acid sequence can be "reverse translated" into a degenerate DNA sequence, which can be programmed into a DNA synthesizer to create a set of oligonucleotides that must include the one present in the actual genomic DNA.

probes labeled with radioactivity or fluorescence (see Fig. 9.13). After a sufficient period of incubation with radioactively labeled probe, you wash away all probe molecules that have not hybridized to complementary DNA strands stuck to the disk, place a large piece of X-ray film in direct contact with the paper disk to record the radioactive emissions from the hybridizing probe, and then develop the X-ray film to see the dark spots representing the sites of hybridization. An X-ray film that has been processed in this way to identify the locations of hybridizing radioactive probes is called an **autoradiograph,** and the process of creating an autoradiograph is called **autoradiography.** If you use fluorescence-labeled probes, you can see the hybridizing clones directly by shining ultraviolet light on the disks. The spots revealed by hybridization correspond to genomic clones containing sequences complementary to the α- and β-globin probes. Knowing the position on the disk that

gave rise to these spots, you can return to the original master petri plates and pick out the corresponding colonies containing the cloned inserts of interest. You can then harvest and purify the cloned DNA identified by the globin probes and proceed with further analysis. The DNA clones recovered with this screening procedure would lead to the discovery that the human genome carries one gene for β globin, but two similar α-globin genes back to back, named α1 and α2.

Of great importance to geneticists is the fact that hybridization can occur between single strands that are not completely complementary, including related sequences from different species. In general, two single DNA strands that are longer than 50–100 bp will hybridize so long as the extent of their complementarity is more than 80%, even though mismatches may appear throughout the resulting hybrid molecule. Imperfect hybrids are less stable than perfect ones, but geneticists can exploit this difference in stability to evaluate the similarity between molecules from two different sources. Hybridization, for example, occurs between the mouse and human genes for the cystic fibrosis protein. Researchers can thus use the human genes to identify and isolate the corresponding mouse sequences, and then use these sequences to develop a mouse containing a defective cystic fibrosis gene. Such a mouse provides a model for cystic fibrosis in a species that, unlike humans, is amenable to experimental analysis.

Gel Electrophoresis Combined with Hybridization Provides a Tool for Mapping DNA Fragments

We have just seen how researchers use hybridization to screen a library of thousands of clones for particular clones carrying DNA sequences complementary to specific probes. Hybridization can also provide information about sequences related to a gene in a cloned DNA fragment or a whole genome. The protocol for accomplishing this task combines gel electrophoresis (review discussion on p. 283) with the hybridization of DNA probes to DNA targets immobilized within nitrocellulose paper.

Suppose you had a clone of a gene called *H2K* from the mouse major histocompatibility complex (or MHC). The *H2K* gene is known to play a critical role in the body's ability to mount an immune response to foreign cells. You want to know whether there are other genes in the mouse genome that are similar to *H2K* and, by extrapolation, also play a role in the immune response. To get an estimate of the number of *H2K*-like genes that exist in the genome, you could turn to a hybridization technique called the **Southern blot,** named for Edward Southern, the British scientist who developed it. Figure 9.15 illustrates the details of the technique.

Southern blotting can identify individual *H2K*-like DNA sequences within the uncloned expanse of DNA present in a mammalian genome. Cutting the total genomic DNA with *Eco*RI produces about 700,000 fragments. When you separate these fragments by gel electrophoresis and stain them with ethidium bromide, all you see is a smear, because it is impossible to distinguish 700,000 fragments spread over a distance of some 10 cm (Fig. 9.15). But you can still blot the smear of fragments to a nitrocellulose filter paper and probe the resulting Southern blot with a labeled *H2K* clone, which picks out the bands containing the *H2K*-like gene sequences. The result shown in Fig. 9.15 is a pattern of approximately two dozen fragments that constitute a series of related genes within the mouse MHC. The Southern blot thus makes it possible to start with a very complex mixture and find the small number of fragments among hundreds of thousands that are related to your original clone.

Southern blotting can also determine the location of one cloned sequence (such as a 1.4 kb human β-globin cDNA sequence from a plasmid vector) within a larger cloned sequence (such as 30 kb cosmid containing the β-globin locus and surrounding genome). Suppose you want to use these two clones to discover the location of the gene within the cosmid as well as to learn which parts of the full gene are exons and which parts are introns. To answer both of these questions in a very straightforward fashion you would turn to Southern blotting. First you would use gel electrophoresis of restriction-enzyme digested cosmid DNA, and staining with ethidium bromide, to construct a restriction map of the cosmid. Next, you would transfer the restriction fragments from the gel onto a filter paper, and probe the filter by hybridization to the labeled β-globin cDNA clone. An autoradiograph will then show the precise cosmid restriction fragments that carry coding regions of the β-globin gene. With very high resolution restriction maps of cosmid subclones and high-resolution gel electrophoresis of small restriction fragments (produced with four-cutter restriction enzymes), it would be possible to distinguish restriction fragments containing exons from the in-between fragments that contain the introns of the β-globin gene.

The Polymerase Chain Reaction Provides a Rapid Method for Isolating DNA Fragments

Genes are rare targets in a complex genome: the β-globin gene, for example, spans only about 1400 of the 3,000,000,000 nucleotide pairs in the haploid human genome. Cloning overcomes the problem of studying such rarities by amplifying large amounts of a specific DNA fragment in isolation. But cloning is a tedious, labor-intensive process. Once a sequence is known, or even partially known, molecular biologists now use an alternative method to recover versions of the same sequence from any source material, including related sequences from different species, from different individuals within the same species, and from sequences in RNA transcripts or whole genomes. This alternative method is called the **polymerase chain reaction,** or **PCR.** First developed in 1985, PCR is faster, less expensive, and more flexible in application than cloning. From a complex mixture of DNA—like that present in a person's blood sample—PCR can isolate a purified DNA fragment in just a few hours.

PCR is also extremely efficient. We have seen that in creating a genomic or cDNA library, it is necessary to use a large number of cells from one or more tissues as the source of DNA

or mRNA. By contrast, the single copy of a genome present in one sperm cell, or the minute amount of severely degraded DNA recovered from the bone marrow of a 30,000-year-old Neanderthal skeleton, provides enough material for PCR to make 100 million or more copies of a target DNA sequence in an afternoon.

How PCR Achieves the Exponential Accumulation of Target DNA

The polymerase chain reaction is a kind of reiterative loop in which an operation is repeatedly applied to the products of earlier rounds of the same operation. You can liken it to the operation of an imaginary generously paying automatic slot machine. You start the machine by inserting a quarter, at which point the handle cranks, and the machine pays out two quarters; it then reinserts those two coins, cranks, and produces four quarters; reinserts the four, cranks, and spits out eight coins, and so on. By the twenty-second round, this fantasy machine delivers more than 4 million quarters.

The PCR operation brings together and exploits the method of DNA hybridization described earlier in this chapter and the essential features of DNA replication described in Chapter 6. Once a specific genomic region (which may be anywhere from a few dozen base pairs to 25 kb in length) has been chosen for amplification, an investigator uses knowledge of the sequence (obtained from previous sequencing based on the method described later in this chapter) to synthesize two oligonucleotides that correspond to the two ends of the target region. One oligonucleotide is complementary to one strand of DNA at one end of the region; the other oligonucleotide is complementary to the other strand at the other end of the region. The process of amplification is initiated by the hybridization of these oligonucleotides to denatured DNA molecules within the sample. The oligonucleotides act as primers directing DNA polymerase to create new strands of DNA complementary to both strands between the two primed sections (Fig. 9.16).

This initial replication is followed by subsequent rounds in which both the starting DNA and the copies synthesized in previous steps become templates for further replication, resulting in an exponential increase by doubling the number of copies of the replicated region with each step. Figure 9.16 diagrams the steps of the PCR operation, showing how you could use it to obtain many copies of a small portion of the β-globin gene for further study.

PCR Products Can Be Used Just Like Cloned Restriction Fragments

When properly executed, PCR provides all the highly enriched DNA you could want for unambiguous analyses of many types. PCR products can be labeled to produce hybridization probes for detecting related clones within libraries and related sequences within whole genomes spread out on Southern blots. PCR products can be sequenced (as described in the next section) to determine the exact genetic information they contain. Because the products are obtained without cloning, it is possible to amplify and learn the sequence of a specific DNA segment in

Feature Figure 9.15

Southern blot analysis.

Genomic DNA was purified from the tissues of seven mice and each sample was subjected to digestion with the restriction enzymes *Eco*RI. Digested samples were separated by electrophesis in an agarose gel as illustrated in Fig. 9.5.

Stain with ethidium bromide to visualize total genomic DNA under UV illumination.

Next you place the gel in a strongly alkaline solution that causes the DNA to denature (that is, to separate into single stands) and then in a neutralizing solution. You now cover the gel containing the separated DNA restriction fragments with a piece of nitrocellulose filter paper. On top of the filter paper, you place a stack of paper towels, and beneath the gel, a sponge saturated with buffer. Within this setup, the paper towels act as a blotter, pulling liquid from the buffer-saturated sponge, through the gel, and into the towels themselves. As the

liquid is drawn out of the gel, so too is the DNA. But the large DNA molecules do not pass through the filter into the paper towels. Instead, they become trapped in the nitrocellulose at points directly above their locations in the gel, forming a Southern blot, which consists of the nitrocellulose filter containing DNA fragments in a pattern that is a replica of their migration pattern in the gel.

Buffer solution

Sponge

Gel

Nitrocellulose paper

Stacked paper towels

The Southern blot is removed from the blotting apparatus, incubated with NaOH to denature the transfered DNA, and then baked and exposed to UV to attach the single-stranded DNA to the blot.

High temperature

UV exposure

The blot is incubated with radioactive probe for a mouse major histocompatibility gene *H2K*.

Blot is removed,

washed, and exposed to X-ray film.

Autoradiograph

A B C D E F G

15.0 —

9.1 —

5.3 —

3.3 —

2.0 —

1
2a
2b
3
5
7
8
9
10
11
12
13
14
15
16

In each genomic DNA sample, the *H2K* probe hybridizes to all 20–30 related major histocompatability genes present within the mouse genome.

Polymerase chain reaction.

Suppose you are a physician and wish to understand the molecular details of the β-globin gene mutation that causes the expression of a novel form of anemia in one of your patients. Although the sequence of the normal β-globin gene as well as the sequence of many of its mutant alleles is known, you suspect that your patient may carry a new, as-yet-uncharacterized β-globin allele of unknown sequence. To characterize the potentially novel allele, you could turn to PCR. You begin by preparing a small amount of genomic DNA from skin, blood, semen, or other tissue that is easy to obtain from your patient suffering from the novel anemia. You then synthesize

two specific oligonucleotide primers, each a short single-stranded chain of 16–26 nucleotides, whose sequence is dictated by the already known sequence of the wild-type β-globin allele. One of these oligonucleotides is complementary to one strand of a part of the β-globin gene to the left of the region you believe might contain the "target sequence," in this case, the mutation; the other oligonucleotide is complementary to the opposite DNA strand from a part of the β-globin gene to the right of the region of interest. As you will see, the target DNA amplified by PCR is that stretch of the genome lying between the two primers.

Next put the patient's genomic DNA in a test tube along with the specially prepared primers, a solution of the four deoxynucleotides, and *Taq* DNA polymerase, a specialized polymerase obtained from *Thermus aquaticus* bacteria living in hot springs. This specialized DNA polymerase remains active at the high temperatures employed

during the PCR protocol, which is not true of polymerases from organisms that live under more moderate conditions. Now place the test tube with these components in a machine called a thermal cycler, which repeatedly changes the temperature of incubation according to a preset program.

In a typical program, the cycler heats the solution to 94°C for 5 minutes. At this temperature, the target DNA separates into single strands. The temperature next lowers to 50–60°C for 2 minutes to allow the primers to base pair with complementary sequences in the single-stranded genomic DNA. Specifics of both temperature and timing within these ranges depend on the length and GC:AT ratio of

the primer sequences. The thermal cycler then raises the temperature to 72°C, the temperature at which the *Taq* polymerase functions best. Holding the temperature at 72°C for 2–5 minutes (depending on the length of the target sequence) allows DNA synthesis to proceed. At the end of this period, with the completion of DNA synthesis, the first round of PCR is over.

To start the next round, the cycler again raises the temperature to 94°C, but this time for only about 20 seconds, to denature the short stretches of DNA consisting of one of the original strands of genomic DNA and a newly synthesized complementary strand initiated by a primer. These short single strands become the templates for the second round of replication because the synthesized primers are able to anneal to them.

Denature strands   Base pairing of primers   Elongation of primers

The machine repeats the cycle again and again, generating an exponential increase in the amount of target sequence: 22 repetitions produce over a million copies of the target sequence; 32 repetitions over a billion. The length of the accumulating DNA strands becomes fixed at the length of the DNA between the 5' ends of the two primers, as shown. This is because, beginning with round 3, the 5' end of a majority of templates is defined by a primer that has been incorporated in one strand of a PCR product.

a very short time. In fact, in checking for a particular hemoglobin mutation, one could start with a blood sample and determine a DNA sequence within two days.

As an analytic tool, PCR has several advantages over cloning. First, it provides the ultimate in sensitivity: The minimum input for the polymerase chain reaction to proceed is a single DNA molecule. Second, as we have seen, it is very fast, requiring no more than a few hours to generate enough amplified DNA for analysis; by comparison, the multiple steps of cloning usually require at least a week to complete. Third, in the sometimes highly competitive research world, it is an agent of democracy: Once the base sequence of the oligonucleotide primers that define a particular target sequence appears in print, anyone with the relatively small amount of funds needed to synthesize or buy these primers can reproduce the reaction. PCR is nevertheless unsuitable in certain situations. Because the protocol only copies DNA fragments up to 25 kb in length, it cannot amplify larger regions of interest. And because the synthesis of PCR primers depends on sequence information from the target region, the protocol cannot serve as the starting point for the analysis of genes or genomic regions that have not yet been cloned and sequenced.

PCR Has Many Uses

PCR is one of the most powerful techniques in molecular biology. Its originator, Kary Mullis, received the 1993 Nobel Prize in chemistry for his 1985 invention of this tool for genetic analysis. By making it possible to amplify a specific sequence directly from complex DNAs (either genomic or cDNA), PCR has made molecular analysis an essential component of genotype detection and gene mapping; we describe its applications in these areas in Chapters 10 and 11. By making it possible to analyze traces of partially degraded DNA from ancient samples (based on related sequences from living organisms) and to compare homologous sequences from a variety of sources, PCR has revolutionized evolutionary studies, enabling researchers to analyze sequences from both living and extinct organisms and to determine the relatedness between these organisms with greater accuracy than ever before (Chapter 21 discusses evolution at the molecular level). By making it possible to analyze the same sequence in hundreds or thousands of individual samples from humans, other animals, or plants without having to make hundreds or thousands of libraries, PCR has facilitated the study of gene diversity at the nucleotide level in populations and greatly simplified the process of monitoring genetic changes in a group over time (see Chapter 20 for the details of population genetics). Finally, PCR has helped bring molecular genetics to many fields outside of traditional genetics. The following example of its use in diagnosing infectious disease provides an inkling of its potential impact on medicine.

AIDS, like other viral diseases, though not inherited, is in one sense a genetic disease, because it is caused by the activity of foreign DNA inside a subgroup of somatic cells. HIV, the virus associated with AIDS, gains entry to a person's body through the bloodstream or lymphatic system, then docks at specific membrane receptors on a few types of white blood cells, fuses with the cell membrane, and releases its RNA chromosome along with several copies of reverse transcriptase into the cell (see the Genetics and Society box on p. 252 in Chapter 8). Once inside the cell, the reverse transcriptase copies the RNA to cDNA. The double-stranded DNA copy of the viral genome then integrates itself into the host genome where, known as a *provirus,* it can lie latent for up to 10 years or become active at any time. When activated, it directs the cellular machinery to make more viral particles.

Standard tests for HIV detect antibodies to the virus, but it may take several months for the antibodies produced by an infected person's immune system to reach levels that are measurable in the blood. Then, in another few months, when ongoing viral activity inside many types of circulating white blood cells subsides, most of the antibodies may disappear from the circulation. This is because once the viral particles have entered the latent state, they are literally in hiding and able to avoid detection by the immune system.

With PCR, it is possible to detect small amounts of virus circulating in the blood or lymph very soon after infection, before antibody production is in full swing. PCR can also detect viral DNA incorporated in the genome of any cell, picking up as little as 1–10 copies of viral DNA per million cells. Thus, with PCR, it becomes possible to confirm and begin treating HIV infection during the critical period before antibodies reach measurable levels. It also becomes possible to follow the progress of each person's HIV infection and tailor therapies accordingly, using a large dose of certain drugs to combat a large amount of viral activity but small doses of perhaps other drugs to prevent a small number of cells from emerging from latency.

In summary, once the base sequence of a particular locus has been determined, it becomes possible to develop a PCR assay capable of picking out any allelic version of that locus that is present in any DNA sample. This is because if the primer templates find a match, PCR will amplify any sequence between the primers, even those with a slightly different order of bases. Automation of the PCR procedure makes it possible to screen thousands of samples in a short time.

DNA Sequence Analysis

The definitive description of any DNA molecule—from a short fragment in a DNA clone to a whole human chromosome—is the sequence of its bases. The DNA sequence of a gene provides a staggering amount of practical information. Restriction enzyme recognition sites useful for manipulating a gene are immediately visible from the DNA sequence. The codons in open reading frames allow scientists to predict the amino-acid sequence of the protein product of the gene. Once the protein's amino-acid sequence is known, computer programs can be instructed to model the three-dimensional shape of the protein, thereby projecting something about its function. Comparison of genomic and cDNA sequences immediately shows how a gene is divided into exons and introns and may suggest whether alternative splicing of the gene's primary transcript occurs. Compilations of the DNA sequences of many genes whose expression is turned on by the same environmental cues, as well as the sequences of the genes' surrounding regions, can supply clues about the DNA signals that determine this kind of gene regulation. Even an exploration of the DNA sequences between genes can provide important information about the evolution of genomes.

No matter how large the ultimate goal—up to and including the determination of the sequence of the whole human genome—all sequencing projects depend on the protocol described next. It allows the sequence determination of 800 bases at a time from a DNA fragment that has been isolated and amplified by cloning or PCR. For larger sequencing projects, the protocol is repeated on hundreds, thousands, or even millions of individually purified DNA fragments. The final extended sequence is then pieced together like a puzzle according to computer programs that search for overlaps in the sequence information generated by each application of the protocol.

The protocol currently used for determining DNA sequence derives from one of two sequencing technologies developed in the 1970s. One technology, developed by Alan Maxam and Walter Gilbert, is based on the chemical cleavage of DNA molecules at specific nucleotide types. The second technology, developed by Fred Sanger, is based on the enzymatic extension of DNA strands to a defined terminating base. Maxam, Gilbert, and Sanger all won the Nobel Prize for their contribution to DNA sequencing technology. Their techniques have a similar accuracy, which approaches 99.9%. But only the Sanger technique is readily amenable to automation, and it is only with automation that DNA sequencing reaches its full potential.

General Principles of the Procedure

There are two steps to the Sanger method of sequencing, whose object is to reveal the order of base pairs in an isolated DNA molecule. The first step is the generation of a complete series of single-stranded subfragments complementary to a portion of the DNA template under analysis. (Although both strands of a DNA fragment are present in a typical DNA sample, only one is used as a template for sequencing.) Each subfragment differs in length by a single nucleotide from the preceding and succeeding fragments; and the graduated set of fragments is known as a *nested array*. A critical feature of the subfragments is that each one is distinguishable according to its terminal 3′ base. Thus, each subfragment has two defining attributes—relative length and one of four possible terminating nucleotides. In the second step of the sequencing process, biologists analyze the mixture of DNA subfragments through polyacrylamide gel electrophoresis, under conditions that allow the separation of DNA molecules differing in length by only a single nucleotide.

The original Sanger sequencing procedure (illustrated in Fig. 9.17) begins with the denaturing into single strands of the DNA to be sequenced. The single strands are then mixed in solution with DNA polymerase, the four deoxynucleotide triphosphates (dATP, dCTP, dGTP, and dTTP), and a radioactively labeled oligonucleotide primer complementary to DNA adjacent to the 3′ end of the template strand under analysis. (Typically, sequencing is performed on purified DNA clones or amplified PCR products. For sequencing DNA clones, the vector-insert recombinant molecule is retained intact, and a primer is used that is complementary to the vector sequence immediately adjacent to the insert DNA. For sequencing PCR products, the PCR primer can also serve as the sequencing primer.) The solution is next divided into four aliquots. To each one, an invesitgator adds a small amount of a single type of deoxynucleotide triphosphate lacking the 3′-hydroxyl group that is critical for the formation of the phosophodiester bonds that lead to chain extension (review Fig. 6.7); this nucleotide analogue is called a dideoxynucleotide and it comes in four forms: **ddTTP, ddATP, ddGTP,** or **ddCTP** (abbreviated from here on as ddT, ddA, ddG, and ddC).

In each sample reaction tube, the oligonucleotide primer will hybridize at the same location on the template DNA strand. As a primer, it will supply a free 3′ end for DNA chain extension by DNA polymerase. The polymerase will add nucleotides to the growing strand that are complementary to the nucleotides of the sample's template strand (that is, the actual DNA strand under analysis). The addition of nucleotides will continue until, by chance, a dideoxynucleotide is incorporated instead of a normal nucleotide. The absence of a 3′-hydroxyl group in the dideoxynucleotide prevents the DNA polymerase from forming a phosphodiester bond with any other nucleotide, ending the polymerization for that new strand of DNA. Next, after allowing enough time for the polymerization of all molecules to reach completion, an investigator separates the templates from the newly synthesized strands by denaturing the DNA at high temperature. Each sample tube now holds a whole collection of single-stranded radioactive DNA chains as well as the nonradioactive single strands of the template DNA. The lengths of the radioactive chains reflect the distance from the 5′ end of the oligonucleotide primer to the position in the sequence at which the specific dideoxynucleotide present in that particular tube was incorporated into the growing chain. (Recall that construction of this chain, which is complementary to the template DNA under analysis, was initiated by a radioactive primer.)

The samples in the four tubes are now electrophoresed in adjacent lanes on a polyacrylamide gel, and the gel is subjected to autoradiography. Because the template strands are not labeled, they do not show up on the autoradiograph. The investigator reads out the sequence of the radioactive strand by starting at the bottom and moving up the autoradiograph, determining which lane carries each subsequent band in the ascending series, as shown in Fig. 9.17. Recall that as you ascend the radiograph, each step pictures a chain that is one nucleotide longer than the chain of the step below. Once the sequence of the newly synthesized DNA is known, it is a simple matter to convert this sequence (on paper or by computer) into the complementary sequence of the template strand under analysis.

Sanger sequencing.

Begin by mixing the purified, denatured DNA with a labeled oligonucleotide primer that hybridizes to a particular complementary site on one strand of the cloned insert.

Next divide the mixture of cloned DNA and labeled primer into four aliquots and add a special preparation of nucleotides along with DNA polymerase to catalyze the polymerization of nucleotides from

the primer. The special nucleotide preparation is spiked with a chain-terminating dideoxynucleoside analog of only one of the four bases. When an analog nucleoside combines with a phosphate, the result is an analog nucleotide. One aliquot, for example, receives nucleotides A, T, C, and G and an analog of A; a second aliquot receives A, T, C, and G and an analog of T; and so forth.

Because the analogs have no oxygen at the 3' position in the sugar, they prevent the further addition of nucleotides to the strand and thus terminate a growing chain wherever they become incorporated in place of an actual nucleotide. The aliquot that has the dideoxy

form of thymidine, for example, will generate a population of DNA molecules that terminate at each of the thymidines in the original template strand under analysis.

Then use electrophoresis on a polyacrylamide gel to separate the fragments in each of the four aliquots and arrange them by size. The resolution of the gel is such that you can distinguish DNA molecules that differ in length by only a single base. The appearance of a DNA fragment of a particular length demonstrates the presence of a particular nucleotide at that position in the strand.

Gel analysis of fragments

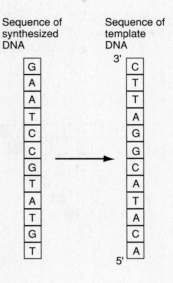

Suppose, for example, that the aliquot polymerized in the presence of dideoxythymidine shows fragments 32, 35, and 39 bases in length. These fragments indicate that thymidine is present at those positions in the strand of nucleotides. In practice, one does not independently determine the exact lengths of each fragment.

Instead, one starts at the bottom of the gel, looks at which of the four lanes has a band in it, records that base, then moves up one position and determines which lane has the next band, and so on. In this way, it is possible to read several hundred bases from a single set of reactions.

To automate the DNA sequencing process, molecular geneticists changed the method of labeling the newly formed complementary DNA strands. Instead of placing a single radioactive label on the primer oligonucleotide, they labeled each of the four chain-terminating dideoxynucleotides with a different color fluorescent dye. As a result, instead of four separate reactions, as in the original procedure, all four dideoxynucleotides could be combined in a single reaction mixture that could be analyzed in a single lane on a gel (Fig. 9.18). A DNA sequencing machine follows the DNA chains of each length in the ascending series through a special detector that can distin-

guish the different colors associated with each terminating dideoxynucleotide. Thus, in each lane of a gel, it is possible to run a different DNA fragment for complete sequence analysis. Modern DNA sequencing machines can run 100 or more samples simultaneously.

Sequencing Long Regions of DNA

DNA sequences are a form of digital information that is easy to record and analyze with computers. When a project calls for the determination of a long DNA sequence, such as a plasmid containing a 10 kb insert or even a 150 kb BAC clone, multiple

(a) Automated sequencing

1. Generate nested array of fragments; each with a fluorescent label corresponding to the terminating 3' base.

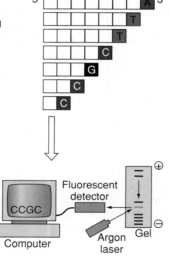

2. Fragments separated by electrophoresis in a single vertical gel lane.

3. As migrating fragments pass through the scanning laser, they fluoresce. A fluorescent detector records the color order of the passing bands. That order is translated into sequence data by a computer.

(b)

(c)

CTNGCTTTGGAGAAAGGCTCCATTGNCAATCAAGACACACAGAGGTGTCCTCTTTTTCCCCTGGTCAGCGNCCAGGTACATNGCACCAAGGCTGCGTAGTGAACTTGNCACCAGNCCATGGAC
CTatGCTTTGGAGAAAGGCTCCATTGgCAATCAAGACACACAGAGGTGTCCTCTTTTTCcCCTGGTCAGCGaCCAGGTACATgGCACCAAGGCTGCGTAGTGAACTTGcCACCAGcCCATGGAC

Figure 9.18   Automated sequencing. (a) For automated sequencing, the Sanger protocol is performed with all four individually labeled terminating nucleotides present in a single reaction. At completion of the reaction, DNA fragments terminating at every base in the sequence will be present and color-coded according to the identity of the terminating nucleotide. Separation by gel electrophoresis is next. A laser and accompanying detector are positioned near the bottom of the polyacrylamide gel; as each fragment moves past the laser, the beam determines the color of the fluorescent label on the terminal base and records that color in a computer. **(b)** Output from automated sequencing machine. Each lane displays the sequence obtained with a separate DNA sample and primer. **(c)** The computer reads the sequence of the strand complementary to the template strand from right to left, which corresponds to the 5′-to-3′ direction. The machine then automatically generates the top, complementary sequence, recording any ambiguity in the base call as an "N." A technician can resolve most such ambiguities by direct examination of the fluorescence signals. If the technician concludes with high certainty that a particular N is, for example, the base G, he or she replaces that N with a "g" in the bottom sequence.

rounds of sequencing must be performed to obtain data that the computer can combine into a single final sequence.

There are two basic strategies for sequencing the relatively large DNA inserts of a clone: primer walking and shotgun sequencing.

Primer walking begins with the use of two vector-based primers, flanking the two ends of an insert. Each primer makes it possible to sequence into one strand of the unknown insert for 500–800 nucleotides. The information obtained at the end of these sequences can be used to program an automated sequencer to synthesize new primers for the next round of sequencing. An investigator repeats this process until he or she has sequenced both strands of the entire insert. Although the two strands are complementary to each other, and thus define each other entirely, it is useful to have two independently obtained sequences to confirm the accuracy of the data obtained.

The primer-walking protocol is a reasonable approach to obtaining a sequence for a relatively small clone of perhaps 5 kb in length. But for a 150 kb BAC clone, under ideal conditions, primer walking would require 300 sequential rounds of DNA sequencing and primer synthesis. This could take a year to complete.

Shotgun sequencing provides a much faster way to determine the DNA sequence of a larger clone. To carry out shotgun sequencing, you begin by fragmenting the BAC clone, or other DNA starting material, at random into a large number of smaller pieces. Partial digestion of the DNA with a four-base cutter or through intense shearing with high-frequency, high-intensity sound waves (Fig. 9.19) will generate these smaller fragments. You can then clone and sequence them using an oligonucleotide primer complementary to a vector sequence directly adjacent to the unknown insert. If you ascertain the sequences of a sufficient number of random fragments (the number is determined by the total size of the original clone or genome), the likelihood that you have obtained complete coverage of the original DNA fragment is high. You now enter the sequences into a computer programmed to look for sequence overlaps. If the number of sequenced fragments is large enough, it is possible to find overlaps of sequences extending across the entire original clone. Any gaps can be filled in by the primer-walking protocol described previously.

The shotgun approach thus relies on redundancy—you have to gather sequence information on three to four times the actual number of base pairs in the original clone to make it work. Even so, it is an efficient method of sequencing. The downside of shotgun sequencing is that only large laboratories equipped with many automated DNA synthesizers able to operate simultaneously can carry it out.

Understanding the Genes for Hemoglobin: A Comprehensive Example

Geneticists have used the tools of recombinant DNA technology to analyze the clusters of related genes that make up the α- and β-globin loci. They have isolated mRNA from red blood cell precursors and with the use of reverse transcriptase, produced cDNA libraries. They easily probed these libraries for α-globin and β-globin cDNA clones because 80% of the transcripts within red blood cell precursors code for these polypeptides, and the proportion of α-globin and β-globin cDNA clones in the libraries was equivalently high. They have sequenced individual cDNA clones to identify those that carried coding sequences for either α-globin or β-globin polypeptides. They have used PCR to amplify for sequencing the globin genes present in many individuals, both normal and diseased.

Geneticists have probed genomic libraries with the α- and β-globin cDNA clones and identified genomic clones carrying globin genes and their surrounding regions. They have also used cDNA clones to probe Southern blots containing restriction fragments of the genomic clones to determine the number and location of coding sequences within each genomic locus. Finally, they have sequenced the entire chromosomal regions that contain the α-globin or β-globin genes.

Fundamental insights from these studies have helped explain how the linear information of DNA encompasses all the

Original fragment

Random cut with four-base recognition restriction enzyme

Clone fragments

Sequence from vector-based primer

Primer — — Chain extension

Enter data into computer; look for overlaps.

Individual sequences

Sequences derived from overlapping fragments

Figure 9.19 Sequencing a long DNA molecule. In shotgun sequencing, the long DNA molecule is reduced into a large number of smaller fragments that are cloned and individually sequenced. The individual sequences are aligned by computer to generate a single continuous sequence that spans the entire length of the original molecule.

instructions for the development of the hemoglobin system, including the changes in globin expression during normal development. The studies have also clarified how the globin genes evolved, and how a large number of different mutations produce the phenotypic permutations that give rise to a range of globin-related disorders.

The Genes Encoding Hemoglobin Occur in Two Clusters on Two Separate Chromosomes

The α-globin gene cluster contains three functional genes and spans about 28 kb on chromosome 16 (Fig. 9.20a). All the genes in the α-gene cluster point in the same direction; that is, they all use the same strand of DNA as the template for transcription. Moving in the 5′-to-3′ direction along the RNA-like strand, the α or α-like genes appear in the order ζ, α2, α1. The genes in the β-gene cluster, like those in the α-gene cluster, all point in the same direction. The β-globin gene cluster covers 50 kb on chromosome 11 and contains five functional genes in the order ε, Gγ, Aγ, δ, and β (Fig. 9.20b). Geneticists refer to the chromosomal region carrying all of the α-globin-like genes

as the **α-globin locus** and the region containing the β-globin-like genes as the **β-globin locus.** Note that the term **locus** signifies a location on a chromosome; that location may be as small as a single nucleotide or as large as a cluster of related genes.

The Succession of Genes in Each Cluster Correlates with the Sequence of Their Expression During Development

The linear organization of the genes in the α- and β-gene clusters reflects the order in which they are expressed during development. For the α-like chains, that temporal order is ζ during the first five weeks of embryonic life, followed by the two α chains during fetal and adult life. For the β-like chains, the order is ε during the first five weeks of embryonic life; then the two γ chains during fetal life; and finally, within a few months of birth, mostly β but also some δ chains (see Fig. 9.1 on p. 278 and Fig. 9.20).

The fact that the order of genes on the chromosomes reflects the order of their expression during development suggests that whatever mechanism turns these genes on and off takes advantage of their relative positions. We now understand what that mechanism is: A *locus control region* (or *LCR*) associated with specialized DNA binding proteins at the 5′ end of each locus works its way down the locus, bending the chromatin back on itself to turn genes on and off in order. We describe this regulatory mechanism in more detail on p. 594 of Chapter 17.

One consequence of a master regulatory element that controls an entire gene complex is seen in a rare medical condition with a surprising prognosis. In some adults, the red blood cell precursors express neither β or δ chains. Although this should be a lethal situation, these adults remain healthy. Cloning and sequence analysis of the β-globin locus from affected adults show that they have a deletion extending across the β and δ genes. Because of this deletion, the master regulatory control can't switch at birth, as it normally would, from γ-globin production to β- and δ-globin production (Fig. 9.20c). People with this rare condition continue to produce large amounts of fetal γ globin throughout adulthood, and that γ globin is sufficient to maintain a near-normal level of health.

A Variety of Mutations Account for the Diverse Symptoms of Globin-Related Diseases

By comparing DNA sequences from affected individuals with those from normal individuals, researchers have learned that there are two general classes of disorders arising from alterations in the hemoglobin genes. In one class, mutations change the amino-acid sequence and thus the three-dimensional structure of the α- or β-globin chain, and these structural changes result in an altered protein whose malfunction causes the destruction of red blood cells. Diseases of this type are known as hemolytic anemias (Fig. 9.21a). An example is sickle-cell anemia, caused by an A-to-T substitution in the sixth codon of the β-globin chain. This simple change in DNA sequence alters

(a) α-globin genes on human chromosome 16

kb — 30 — 20 — 10 — 0

LCR   ζ   ψζ ψα1 α2 α1

(b) β-globin genes on human chromosome 11

LCR ψβ2 ε   Gγ Aγ ψβ1   δ   β

kb — 60 — 50 — 40 — 30 — 20 — 10 — 0

(c) Mutant deleted form of β-globin locus

LCR ψβ2 ε   Gγ Aγ ψβ1   δ   β

β-δ deletion

■ Expressed genes   ■ Pseudogenes

Figure 9.20 The genes for the polypeptide components of human hemoglobin are located in two genomic clusters on two different chromosomes. (a) Schematic representation of the α-globin gene cluster on chromosome 16. The cluster contains three functional genes and two pseudogenes (designated by the symbol ψ); pseudogenes are sequences that look like but do not function as genes. **(b)** Schematic representation of the β-globin gene cluster on chromosome 11; this cluster has five functional genes and two pseudogenes. Upstream from both the α- and β-globin gene clusters lies the locus control region (LCR), which interacts with and activates sequential genes in each complex during human development. **(c)** In this example of a mutant chromosome, the adult β-globin genes have been deleted; as a result, the LCR cannot switch from activating the fetal genes to activating the adult genes, and the fetal genes remain active in the adult.

(a.1) Major types of structural variants causing hemolytic anemias

Name	Molecular basis of mutation	Change in polypeptide	Pathophysiological effect of mutation	Inheritance
HbS	Single nucleotide substitution	β 6 Glu ↓ Val	Deoxygenated HbS polymerizes → sickle cells → vascular occlusion and hemolysis	Autosomal Recessive
HbC	Single nucleotide substitution	β 6 Glu ↓ Lys	Oxygenated HbC tends to crystallize → less deformable cells → mild hemolysis; the disease in HbS:HbC compounds is like mild sickle-cell anemia	Autosomal Recessive
Hb Hammersmith	Single nucleotide substitution	β 42 Phe ↓ Ser	An unstable Hb → Hb precipitation → hemolysis; also low O_2 affinity	Autosomal Dominant

(b.1) Clinical results of various α-thalassemia genotypes

Clinical condition	Genotype		Number of functional α genes	α-chain production
Normal		αα/αα	4	100%
Silent carrier		αα/α–	3	75%
Heterozygous α-thalassemia— mild anemia	or	α–/α– or αα/– –	2	50%
HbH (β₄) disease— moderately severe anemia		α–/– –	1	25%
Homozygous α-thalassemia— lethal		– –/– –	0	0%

(a.2) Basis of sickle-cell anemia

β6 triplet codon →
Amino-acid replacement →
Hemoglobin variant →

GAG
↓ Substitution
GTG

β6 glutamic acid
↓
β6 valine
↓
HbS

In oxygenated blood / In deoxygenated blood

Vaso-occlusion

(a.3)

(b.2) A β-thalassemia patient makes only α globin, not β globin.

α   β

Four α subunits combine to make abnormal hemoglobin.

Abnormal hemoglobin molecules clump together, altering shape of red blood cells. Abnormal cells carry reduced amounts of oxygen.

To compensate for reduced oxygen level, medullary cavities of bones enlarge to produce more red blood cells.

Spleen enlarges to remove excessive number of abnormal red blood cells.

Too few red blood cells in circulation result in anemia.

(b.3)

Figure 9.21 Mutations in the DNA for hemoglobin produce two classes of disease. (a.1) The major types of hemoglobin variants causing hemolytic anemias. **(a.2)** The basis of sickle-cell anemia. **(a.3)** Sickling red blood cells appear as crescents among more rounded nonsickling cells. **(b.1)** Thalassemias associated with deletions in the α-globin polypeptide. **(b.2)** The physiological basis of β-thalassemia major. **(b.3)** Child suffering from β-thalassemia major.

the sixth amino acid in the chain from glutamic acid to valine, which, in turn, modifies the form and function of the affected hemoglobin molecules. Red blood cells carrying these altered molecules often have abnormal shapes that cause them to block blood vessels or be degraded.

The second major class of hemoglobin-related genetic diseases arises from DNA mutations that reduce or eliminate the production of one of the two globin polypeptides. The disease state resulting from such mutations is known as thalassemia, from the Greek words *thalassa* meaning "sea" and *emia* meaning "blood"; the name arose from the observation that a relatively high rate of this blood disease occurs among people who live near the Mediterranean Sea. Several different types of mutation can cause thalassemia, including those that delete an entire α or β gene, those that alter the sequence in regions that are outside the gene but necessary for its regulation, or those that alter the sequence within the gene such that no protein can be produced. The consequence of these changes in DNA sequence is the total absence or a deficient amount of one or the other of the normal hemoglobin chains. Because there are two α genes (α1 and α2) that see roughly equal expression beginning a few weeks after conception, individuals carrying deletions within the α-globin locus may be missing anywhere from one to four copies of the α-globin genes (Fig. 9.21b). A person lacking only one would be a heterozygote for the deletion of one of two α genes; a person missing all four would be a homozygote for deletions of both α genes. The range of mutational possibilities explains the range of phenotypes seen in α-thalassemia. Individuals missing only one of four possible copies of the α genes are normal; those lacking two of the four have a mild anemia, and those without all four die before birth. The fact that the α chains are expressed early in fetal life explains why the α-thalassemias are detrimental *in utero*. By contrast, β-thalassemia major, the disease occurring in people who are homozygotes for most deletions of the single β gene, also usually results in death, but not until early childhood.

Comparisons of the altered DNA sequences from affected individuals with wild-type sequences from normal individuals have helped illuminate the sequences necessary for normal hemoglobin expression. In some β-thalassemia patients, for example, disease symptoms arise from the alteration of a few nucleotides adjacent to the 5′ end of the coding region for the β chain. Data of this type have defined sequences that are important for expression of the β-globin locus. One such segment is the TATA box, a sequence found in many eukaryotic promoters (Fig. 9.22a; see p. 585 in Chapter 17 for a more detailed discussion). In other thalassemia patients, the entire α-globin locus and adjacent regulatory segments, including the TATA box, are intact, but a mutation has altered the LCR found far to the 5′ side of all the α-like genes. This LCR is necessary for a high level of tissue-specific expression of all α-like genes in red blood cell progenitors (Fig. 9.22b). Mutations in the TATA box or the locus control region, depending on how disruptive they are, produce α- or β-thalassemias of varying severity.

Figure 9.22 Mutations in globin regulatory regions can affect globin gene activity and result in thalassemia.
(a) Mutations in the TATA box associated with the β-globin gene can eliminate transcription and cause β-thalassemia. **(b)** A locus control region is present 25–50 kb upstream of the α-globin gene cluster. The function of the LCR is to open up the chromatin domain associated with the complete cluster of α-globin genes. In contrast, the regulatory regions next to each particular gene fine-tune gene expression during development. Mutations in the LCR can prevent expression of all the α-globin genes, resulting in severe α-thalassemia.

The α- and β-Globin Loci House Multiple Genes That Evolved from One Ancestral Gene

From sequencing the DNA of all the human globin genes, researchers can see that the genes form a closely related group, or family, of genes that evolved by duplication and divergence from one ancestral gene (Fig. 9.23). All gene sequences evolve continuously. The two separate products of a duplication,

Figure 9.23 Evolution of the globin gene family. Duplication of an ancestral gene followed by divergence of the separate duplication products established the α- and β-globin lineages. Further rounds of duplication and divergence within the separate lineages generated the two sets of genes and pseudogenes of the globin gene family.

which start out identical, eventually diverge because their sequences evolve independently of each other. The members of a gene family may be grouped together on one chromosome (like the very closely related β-globin genes) or dispersed on different chromosomes (like the less closely related α- and β-globin genes). All the β-like genes are exactly the same length and have two introns at exactly the same positions. All the α-like genes also have two introns at exactly the same positions, but these positions are different from those of the β genes. The sequences of all the β-like genes are more similar to each other than they are to the α-like sequences, and vice versa. These comparisons suggest that a single ancestral globin gene duplicated, and one copy moved to another chromosome. With time, one of the two gene copies gave rise to the α lineage, the other to the β lineage. Each lineage then underwent further duplications to generate the present array of three α-like and five β-like genes in humans.

Interestingly, the duplications also produced genes that eventually lost the ability to function. Molecular geneticists made this last deduction from data showing two additional α-like sequences within the α locus and one β-like sequence within the β locus that no longer have the capacity for proper expression. The reading frames are interrupted by frameshifts, missense mutations, and nonsense codons, while regions needed to control the expression of the genes have lost key DNA signals. Sequences that look like, but do not function as, a gene are known as **pseudogenes;** they occur throughout all higher eukaryotic genomes.

CONNECTIONS

The tools of recombinant DNA technology grew out of an understanding of the DNA molecule and its interaction with the enzymes that operate on DNA in normal cells. Geneticists use the tools singly or in combination to look at DNA directly. Through cloning, hybridization, PCR, and sequencing, they have been able to isolate the genes that encode, for example, the hemoglobin proteins; identify sequences near the genes that regulate their expression; determine the complete nucleotide sequence of each gene; and discover the changes in sequence produced by the hundreds of mutations that affect hemoglobin production. The results give a fascinating and detailed picture of how the nucleotides along a DNA molecule determine protein structure and function and how mutations in sequence produce far-ranging and varied effects on human health. The methods of classical genetics that we examined in Chapters 2–5 complement those of recombinant DNA technology to produce an integrated picture of genes and genomes at many levels.

In Chapter 10, we describe how the use of recombinant DNA technology has expanded from the analysis of single genes and gene complexes to the sequencing and examination of whole genomes. Through the automation of sequencing and high-powered computer analysis of the data, scientific teams have determined the DNA sequence of the entire human genome and the genomes of many other organisms as well. DNA arrays, another recently developed technology, allow scientists to follow the patterns of expression from all the genes in the genome as a tissue develops normally or undergoes carcinogenic transformation. With these new technologies, geneticists are entering an era where they will be able to look at all the parts of a biological system simultaneously. This new capability for global analysis will allow them to better understand the normal processes of differentiation and development and how these processes can go awry.

ESSENTIAL CONCEPTS

1. An intact eukaryotic genome is too complex for most types of analysis. Geneticists have appropriated the enzymes that normally operate on the DNA molecule inside a bacterial cell and used them in the test tube to create the tools of recombinant DNA technology. *Restriction enzymes* cut DNA into pieces of a defined size, *ligase* splices the pieces together, *DNA polymerase* makes DNA copies, and *reverse transcriptase* copies RNA into DNA.

2. Restriction enzymes cut DNA into restriction fragments at specific sites that are 4–8 bp in length. Enzymes that recognize a 4 bp site digest DNA into fragments that average 256 bp in length; enzymes that recognize 6 bp digest DNA into fragments that average 4 kb in length; enzymes that recognize 8 bp digest DNA into fragments that average 64 kb in length. *Complete digestion* at every site recognized by a particular enzyme can occur under optimal reaction conditions. Under less than optimal conditions (of temperature, enzyme concentration, or duration of reaction), a *partial digestion* occurs in which only a random fraction of restriction sites are digested. Partial digestion enables investigators to obtain a sample of DNA fragments

having an average size in-between those obtainable by complete digestion with enzymes that recognize 4, 6, or 8 bp sites. Most restriction enzymes produce DNA molecules with *sticky ends,* but some produce DNA molecules with *blunt ends.*

3. Gel electrophoresis provides a method for separating DNA fragments according to their size. When biologists subject a viral genome, plasmid, or small chromosome to restriction digestion and gel electrophoresis, they can observe the resulting DNA fragments by ethidium bromide staining and size the fragments by comparing their migration within the gel with the migration of known marker fragments.

4. Cloning DNA fragments:
   a. Restriction fragments and cloning vectors with matching sticky ends can be spliced together to produce *recombinant DNA molecules.* A *cloning vector* is a DNA sequence that can enter a host cell, make its presence known, and provide a means of replicating and purifying both itself and any DNA to which it is spliced. Different types of vectors carry different-size inserts, which can serve as the basis for different levels of analysis.
   b. Once inside a living cell, vector-insert recombinants are replicated during each cell cycle (just as the cell's own chromosomes are) in a process known as cloning. The millions of cells arising from a single cell by consecutive divisions make up a *cellular clone.* The vector-insert recombinant molecules inside the cells of a clone, often referred to as *DNA clones,* can be purified by procedures that separate recombinant molecules from host DNA. The insert can then be cut away from the vector by restriction enzymes and purified.
   c. *Genomic libraries* are random collections of vector-insert recombinants containing the DNA fragments of a given species. The most useful libraries carry four to five *genomic equivalents. cDNA libraries* carry DNA copies of the RNA transcripts produced in a particular tissue. The clones in a cDNA library represent only that part of the genome transcribed and spliced into mRNA in the cells of the specific tissue, organ, or organism.

5. *Hybridization* is the natural propensity of complementary DNA strands to form stable double helixes. Hybridization makes it possible to use previously purified DNA fragments as labeled probes. Biologists use such probes to identify clones containing identical or similar sequences within genomic or cDNA libraries. Hybridization can also be used in combination with gel electrophoresis as part of an analytic technique called *Southern blotting.* Southern blot hybridization allows an investigator to determine the numbers and positions of complementary sequences within isolated DNA fragments or whole genomes of any complexity.

6. The *polymerase chain reaction* (or *PCR*) is a method for the rapid purification and amplification of a single DNA fragment from a complex mixture such as the whole human genome. The DNA fragment to be amplified is defined by a pair of oligonucleotide primers complementary to either end on opposite strands. The PCR procedure operates through a reiterative loop that amplifies the sequence between the primers in an exponential manner. PCR is used in place of cloning to purify DNA fragments whenever sequence information for primers is already available.

7. *Sequencing* provides the ultimate description of a cloned fragment. Automation has increased the speed and scope of sequencing.

S O C I A L   A N D   E T H I C A L   I S S U E S

1. A grocery chain in the Midwest has been receiving milk from the same dairy supplier for many years. Recently, that supplier, as well as many of the dairy farmers in the Midwest, have begun to give their cows bovine somatotropin hormone (BST), a peptide that is mass-produced using a cloned *BST* gene. Treatment with this growth hormone results in a 10% to 25% increase in milk production. Because milk production per cow is increased, the price of the milk is lower. The peptide is injected directly into cows and is undetectable in the milk. Cows treated with the hormone do have an increased incidence of the relatively common disease mastitis, which is treated with antibiotics. A consequence of increased mastitis is that small amounts of antibiotics may be more often found in the milk. Some people worry that this exposure to antibiotics could cause an increase in antibiotic-resistant bacteria in humans, while many experts argue that this risk is negligible. The grocery chain would like to carry this milk because it has a lower price. Does it have a responsibility to customers to inform them about the conditions under which the milk was produced? What proportion of customers at a typical

grocery store are likely to have the knowledge base required to understand the process sufficiently to make an informed decision as to whether to accept milk from BST-injected cows? There are several ways to deal with complicated scientific information that could potentially affect society. Which do you think makes the most sense? (a) Consumers should always be provided with the scientific information even if they are unable to evaluate its significance. (b) An independent commission should evaluate the scientific information and inform consumers of its significance. (c) The government should be trusted to protect the health and well-being of its citizens.

2. Kerry Mullis, while working at Cetus Corporation, developed the polymerase chain reaction technique, one of the most powerful new tools of molecular biology. As an employee of the company, he had previously relinquished much of his right to royalties from the patent and received a compensation that is small relative to the value of the technique he developed. Should an employee be required to give most of the patent rights to the company? How should these rights be handled in a

research university? If the idea belonged to one person, should that person alone profit? Is it possible to determine accurately the relative contributions of individual researchers and of the institutions with which they are associated to the development of new ideas that become commercially viable?

3. A new biotechnology company that has an interest in cloning genes involved in the immune response has approached a scientist at the university and wants to support some of her research. A stipulation of receiving the money is that the research cannot be published in journals until papers have been filed to cover any potential patent. A postdoctoral fellow who has been doing much of the research on this project is currently looking for a job and wants to present this research in his seminars at job interviews. His boss has warned him that this may go against the scientist's contract with the company. What should the postdoctoral fellow do? Is industry support of research in the university compatible with academic life in the university?

S O L V E D   P R O B L E M S

I. The following map of the plasmid cloning vector pBR322 shows the locations of the ampicillin (amp) and tetracycline (tet) resistance genes as well as two unique restriction enzyme recognition sites, one for *Eco*RI and one for *Bam*HI. You digested this plasmid vector with both *Eco*RI and *Bam*HI enzymes and purified the large *Eco*RI-*Bam*HI vector fragment. You also digested the cellular DNA that you want to insert into the vector with both *Eco*RI and *Bam*HI. After mixing the plasmid vector and the fragments together and ligating, you transformed an ampicillin-sensitive strain of *E. coli* and selected for ampicillin-resistant colonies. If you test all of your selected ampicillin-resistant transformants for tetracycline resistance, what result do you expect and why?

EcoRI

*Bam*HI

amp

tet

Answer

This problem requires an understanding of vectors and the process of combining DNAs using sticky ends generated by restriction enzymes.

The plasmid must be circular to replicate in *E. coli* and in this case a circular molecule will be formed only if the insert fragment joins with the cut vector DNA. The cut vector will not be able to re-ligate without an inserted fragment because the *Bam*HI and *Eco*RI sticky ends are not complementary and cannot base pair. All ampicillin-resistant colonies therefore contain a *Bam*HI-*Eco*RI fragment ligated to the *Bam*HI-*Eco*RI sites of the vector. Fragments cloned at the *Bam*HI-*Eco*RI site interrupt and therefore inactivate the tetracycline resistance gene. *All ampicillin-resistant clones will be tetracycline sensitive.*

II. The gene for the human peptide hormone somatostatin (encoding nine amino acids) is completely contained on an *Eco*RI (5′ G ^A A T T C 3′) fragment, which can be cut out of the larger fragment shown here. (The ^ symbol indicates the site where the sugar-phosphate backbone is cut by the restriction enzyme.)

5′ GCCG&AATTCGATCCTATCAACACGAAGTGAAAGTCTTACAACCCATG&AATTCGATTCG 3′

3′ CGGCTTAAGCTAGGATAGTTGTGCTTCACTTTCAGAATGTTGGGTACTTAAGCTAAGC 5′

a. What is the amino-acid sequence of human somatostatin?

b. Indicate the direction of transcription of this gene.

c. The first step in synthesizing large amounts of human somatostatin for pharmacological treatments involves constructing a so-called fusion gene. In this fusion construct, the N terminus of the protein encoded by the fusion gene consists of the N-terminal half of the *lacZ* gene (encoding β-galactosidase), while the remainder of the product of the fusion gene is human somatostatin. A family of three plasmid vectors for the construction of such a fusion gene has been created. All of these vectors have an ampicillin resistance gene and part of the *lacZ* gene encoding the first 583 amino acids of the β-galactosidase protein. The *Eco*RI fragment (that is, the fragment produced by cutting with *Eco*RI) containing human somatostatin can be inserted into the single *Eco*RI restriction site on the vectors. The sequence of three vectors in the vicinity of the *Eco*RI site is shown here. The numbers refer to amino acids in the β-galactosidase protein with the N-terminal amino acid being number 1. The DNA sequence presented is the same as that of the *lacZ* mRNA (with Ts replacing the Us found in RNA). In which of these three vectors must the *Eco*RI fragment containing human somatostatin be inserted to generate a fusion protein with an N-terminal region from β-galactosidase and a C-terminal region from human somatostatin?

```
          582 583
          Gly  Asn                    EcoRI
pWR590-1  5′ GGCAACCGGGCGAGCTCGAATTCG

          582 583
          Gly  Asn                    EcoRI
pWR590-2  5′ GGCAACCCGGGCGAGCTCGAATTC

          582 583
          Gly  Asn               EcoRI
pWR590-3  5′ GGCAACGGGGCAGCTCGAATTCGA
```

Answer

This problem requires an understanding of the sticky ends formed by restriction enzyme digestion and the requirement of appropriate reading frames for the production of proteins.

a. The only complete open reading frame (ATG start codon to a stop codon) is found on the bottom strand (underlined on the following figure). *The amino-acid sequence is met-gly-cys-lys-thr-phe-thr-ser-cys.*

b. Based on the amino-acid sequence determined for part *a, the gene must be transcribed from right to left.*

c. The cut site for *Eco*RI is after the G at the 5′ end of the *Eco*RI recognition sequence on each strand. For each of the three vectors, the cut will be shifted relative to the reading frame of the *lacZ* gene by one base. The *Eco*RI fragment containing somatostatin can ligate to the vector in two possible orientations, but since we know the sequence on the bottom strand codes for the protein, a fusion protein will be produced only if the fragment is inserted with that coding sequence on the same strand as the vector coding sequences. Consider only this orientation to determine which vector will produce the fusion protein. The *Eco*RI fragment to be inserted into the vector next to the *lacZ* gene has five nucleotides that precede the first codon of the somatostatin gene (see following figure) and therefore requires one more nucleotide to match the reading frame of the vector. For pWR590-1, the cut results in an in-frame end; pWR590-2 has one base extra beyond the reading frame; pWR590-3 has a two-base extension past the reading frame. *The EcoRI fragment must be inserted in the vector pWR590-2 to get somatostatin protein produced.*

```
5′ GCCGAATTCGATCCTATCAACACGAAGTGAAAGTCTTACAACCCATGAATTCGATTCG 3′
3′ CGGCTTAAGCTAGGATAGTTGTGCTTCACTTTCAGAATGTTGGGTACTTAAGCTAAGC 5′
```

III. Imagine you have cloned a 14.7 kb piece of DNA, which contains restriction sites as shown here.

Numbers under the segments represent the sizes of the regions in kilobases (kb). You have labeled the left end of the molecule with ^{32}P.

a. What radioactive bands would you expect to see following electrophoresis if you did a complete digestion with *Bam*HI? *Eco*RI? *Hin*dIII?

b. What radioactive bands would you expect to see following electrophoresis if you did a partial digestion with *Bam*HI? *Eco*RI? *Hin*dIII?

Answer

This problem deals with partial and complete digests and radioactive labeling of fragments.

a. Only the left-most fragment would be seen after complete digestion with any of the three enzymes since only the left end contains radioactivity. *Radioactive bands seen after digestion with* Bam*HI: 2 kb;* Eco*RI: 3.4 kb; and* Hin*dIII: 6.9 kb.*

b. Partial digestions result in all possible combinations of adjacent fragments but again only those containing the left end of the fragment will be radioactive. For example, partial digestion with *Bam*HI results in the following fragments (asterisks indicate fragments that are radioactive). *Bam*HI: 2*, 7.9, 1.2, 9.9*, 12.7, and 14.7*. *The radioactive fragments are* Bam*HI: 2, 9.9, 14.7 kb;* Eco*RI: 3.4, 10.7, 14.7 kb; and* Hin*dIII: 6.9, 14.7 kb.*

PROBLEMS

9-1 Match each of the terms in the left column to the best-fitting phrase from the right column.

a. oligonucleotide

b. vector

c. sticky ends

d. recombinant DNA

e. reverse translation

f. genomic library

g. genomic equivalent

h. cDNA

i. PCR

j. hybridization

1. a DNA molecule used for transporting, replicating, and purifying a DNA fragment
2. a collection of the DNA fragments of a given species, inserted into a vector
3. DNA copied from RNA by reverse transcriptase
4. stable binding of single-stranded DNA molecules to each other
5. efficient and rapid technique for amplifying the number of copies of a DNA fragment
6. computational method for determining the possible sequence of base pairs associated with a particular region of a polypeptide
7. contains genetic material from two different organisms
8. the number of DNA fragments that are sufficient in aggregate length to contain the entire genome of a specified organism
9. short single-stranded sequences found at the ends of many restriction fragments
10. a short DNA fragment that can be synthesized by a machine

9-2 Approximately how many restriction fragments would result from the complete digestion of the human genome (3×10^9 bases) with the following restriction enzymes? (The recognition sequence for each enzyme is given in parentheses.)
a. *Sau*3A (∧GATC)
b. *Bam*HI (G∧GATCC)
c. *Sfi*I (GGCCNNNN∧NGGCC)

9-3 What purpose do selectable markers serve in vectors?

9-4 What is an artificial chromosome, and what purpose does it serve for molecular geneticists?

9-5 A plasmid vector pBS281 is cleaved by the enzyme *Bam*HI (G∧GATCC), which recognizes only one site in the DNA molecule. Human DNA is partially digested with the enzyme *Mbo*I (∧GATC), which recognizes many sites in human DNA. These two digested DNAs are now ligated together. Consider only those molecules in which the pBS281 DNA has been joined with a fragment of human DNA. Answer the following questions concerning the junction between the two different kinds of DNA.

a. What proportion of the junctions between pBS281 and all possible human DNA fragments can be cleaved with *Mbo*I?

b. What proportion of the junctions between pBS281 and all possible human DNA fragments can be cleaved with *Bam*HI?

c. What proportion of the junctions between pBS281 and all possible human DNA fragments can be cleaved with *Xor*II (C∧GATCG)?

d. What proportion of the junctions between pBS281 and all possible human DNA fragments can be cleaved with *Eco*RII (Pu Pu A ∧ T Py Py)? (Pu and Py stand for purine and pyrimidine, respectively.)

e. What proportion of all possible junctions that can be cleaved with *Bam*HI will result from cases in which the cleavage site in human DNA was not a *Bam*HI site in the human chromosome?

9-6 Consider four different kinds of human libraries: a genomic library, a brain cDNA library, a liver cDNA library, and a unique genomic sequence (no repetitive DNA) library.

a. Assuming inserts of approximately equal size, which would contain the greatest number of different clones?

b. Would you expect any of these not to overlap the others at all in terms of the sequences it contains? Explain.

c. How do these four libraries differ in terms of the starting material for constructing the clones in the library?

9-7 As a molecular biologist and horticulturist specializing in snapdragons, you have decided that you need to make a genomic library to characterize the flower color genes of snapdragons.

a. How many genomic equivalents would you like to have represented in your library to be 95% confident of having a clone containing each gene in your library?

b. How do you determine the number of clones that should be isolated and screened to guarantee this number of genomic equivalents?

9-8 Imagine that you are a molecular geneticist studying a particular gene in which mutations cause a serious human disease. The gene, including its flanking regulatory sequences, spans 200 kb of DNA. The distance from the first to the last coding base is 140 kb, which is divided among 10 exons and 9 introns. The exons contain a total of 38.7 kb, and the introns contain 101.3 kb of DNA. You would like to obtain the following for your work: (a) an intact clone of the whole gene, including flanking sequences; (b) a clone containing the entire coding sequences but no noncoding sequences; and (c) a clone of

exon 3, which is the site of the most common disease-causing mutation in this gene. Based on the size of your inserts, what vectors would you use to obtain each of these clones and why?

9-9 Listed here are steps you would take to clone the *idiosynchratase* gene from the hoot owl. Put these into their appropriate order (there are multiple correct answers).

a. Mix together BAC vector DNA and hoot owl DNA with ligase.

b. Expose nitrocellulose paper disks to UV radiation and baking.

c. Extract genomic DNA from hoot owl cells.

d. Perform autoradiography.

e. Produce a labeled DNA probe to the *idiosynchratase* gene.

f. Completely digest BAC vector DNA with the restriction enzyme *Hind*III.

g. Place nitrocellulose paper disks onto the agar surface to transfer colonies.

h. Incubate DNA probe with nitrocellulose paper disks.

i. Distribute bacteria onto a petri plate containing agar and nutrients and allow growth into colonies.

j. Partially digest hoot owl genomic DNA with the restriction enzyme *Hind*III.

k. Transform bacteria.

9-10 Expression vectors containing sequences from both the animal virus SV40 and a bacterial plasmid can be used to produce polypeptide products in both *E. coli* and animal cells.

a. Give the general characteristics that are necessary for this type of plasmid to be used as a vector in two very different types of cells.

b. What is required for *E. coli* expression of a human protein?

9-11 Why do longer DNA molecules move more slowly than shorter ones during electrophoresis?

9-12 You have a circular plasmid containing 9 kb of DNA, and you wish to map its *Eco*RI and *Bam*HI sites. When you digest the plasmid with *Eco*RI and run the resulting DNA on a gel, you observe a single band at 9 kb. You get the same result when you digest the DNA with *Bam*HI. When you digest with a mixture of both enzymes, you observe two bands, one 6 kb and the other 3 kb in size. Explain these results. Draw a map of the restriction sites.

9-13 A 49 bp *Eco*RI fragment containing the somatostatin gene was inserted into the vector pWR590 shown here. The sequence of the inserted fragment is:

```
5′ AATTCGATCCTATCAACACGAAGTGAAAGTCTTACAACCCATGAATTCG   3′
3′   GCTAGGATAGTTGTGCTTCACTTTCAGAATGTTGGGTACTTAAGCTTAA 5′
```

Distances between adjacent restriction sites are shown. What are the patterns of restriction digests with *Eco*RI (G∧AATTC) or with *Mbo*I (∧GATC) before and after cloning the somatostatin gene into the vector? (E = *Eco*RI, M = *Mbo*I)

9-14 The *Notch* gene, involved in *Drosophila* development, is contained within a restriction fragment of *Drosophila* genomic DNA produced by cleavage with the enzyme *Sal*I. The restriction map of this *Drosophila* fragment for several enzymes (*Sal*I, *Pst*I, and *Xho*I) is shown here; numbers indicate the distances between adjacent restriction sites. This fragment is cloned by sticky-end ligation into the single *Sal*I site of a bacterial plasmid vector that is 5.2 kb long. The plasmid vector has no restriction sites for *Pst*I or *Xho*I enzymes.

P = *Pst*I; S = *Sal*I; X = *Xho*I

Make a sketch of the expected patterns seen after agarose gel electrophoresis and staining of a *Sal*I digest (alone), of a *Pst*I digest (alone), of a *Xho*I digest (alone), of the plasmid containing vector and *Drosophila* fragment. Indicate the fragment sizes in kilobases.

9-15 The linear bacteriophage λ genomic DNA has at each end a single-strand extension of 20 bases. (These are "sticky ends" but are not, in this case, produced by restriction enzyme digestion.) These sticky ends can be ligated to form a circular piece of DNA. In a series of separate tubes, either the linear or circular forms of the DNA are digested to completion with *Eco*RI, *Bam*HI, or a mixture of the two enzymes. The results are shown here.

a. Which of the samples (A or B) represents the circular form of the DNA molecule?

b. What is the total length of the linear form of the DNA molecule?

c. What is the total length of the circular form of the DNA molecule?

d. Draw a restriction map of the linear form of the DNA molecule. Label all restriction enzyme sites as *Eco*RI or *Bam*HI.

9-16 The following fragments were found after digestion of a circular plasmid with restriction enzymes as noted. Draw a restriction map of the plasmid.

*Eco*RI: 7.0 kb
*Sal*I: 7.0 kb
*Hin*dIII: 4.0, 2.0, 1.0 kb
*Sal*I + *Hin*dIII: 2.5, 2.0, 1.5, 1.0 kb
*Eco*RI + *Hin*dIII: 4.0, 2.0, 0.6, 0.4 kb
*Eco*RI + *Sal*I: 2.9, 4.3 kb

9-17 You need to create a restriction map of a 7 kb circular plasmid you are studying. The following figure shows the results of digesting the plasmid DNA with various restriction endonucleases. The figure represents a photo of a gel on which the DNA products of the digestions were subjected to electrophoresis. The numbers across the top of the blot represent the lanes. The DNAs in each lane were treated in the following way: lane 1: *Eco*RI; lane 2: *Hin*dIII; lane 3: *Pst*I; lane 4: *Bam*HI; lane 5: *Eco*RI + *Hin*dIII; lane 6: *Bam*HI + *Hin*dIII; lane 7: *Bam*HI + *Eco*RI; lane 8: *Pst*I + *Eco*RI; and lane 9: *Pst*I + *Bam*HI. The numbers down the left edge of the gel represent the molecular sizes of the various bands, in kilobases. Draw a map of the plasmid indicating restriction sites and their spatial relation to each other.

9-18 You have purified an interesting protein from mouse brains that influences the ability of animals to taste sweet substances. You name the protein SWT. You now want to study the gene that encodes this protein, which you've determined is a single polypeptide monomer.

  a. Assuming that you have unlimited quantities of purified SWT, describe two approaches that you could use to obtain a cDNA clone of the coding sequence for this polypeptide.

You have been successful in cloning the full-length cDNA of the SWT transcript as a 2 kb insert in a 4 kb plasmid vector. You use the insert from this cDNA clone to probe a genomic library of BAC clones produced from an inbred (totally homozygous) mouse liver, and you recover a SWT-gene-containing clone with a 150 kb genomic insert. You now take purified DNA from (1) mouse liver, (2) the BAC clone containing the SWT gene, and (3) the SWT cDNA clone and digest all three samples with the restriction enzyme (*Not*I), which has an 8 bp recognition site. The three samples of digested DNA are subjected to gel electrophoresis. After the completion of electrophoresis, the gel is stained with ethidium bromide.

  b. Compare the patterns of staining that you are likely to see in the different lanes of the gel.

You now subject the same three DNA samples (genomic, BAC, and plasmid) to digestion with an enzyme (*Rsa*I) that recognizes a 4 bp recognition site and then perform gel electrophoresis and ethidium bromide staining.

  c. Compare the staining patterns that you are likely to see in the different lanes of the gel.

You now subject both gels to Southern blot analysis with a radioactive probe made from the purified insert of the cDNA clone.

  d. Compare the results that you are likely to see in the different lanes of both gels.

9-19 A fragment of human DNA with an *Eco*RI site at one end and a *Hind*III site at the other end is labeled with ^{32}P at the *Eco*RI end. This molecule is now partially digested with either restriction enzyme *Hha*I or *Sau*3A, and the partial digests are fractionated on an agarose gel. An autoradiograph of this gel is presented here.

  a. How large is the *Eco*RI to *Hind*III fragment in base pairs?
  b. Draw a map of this fragment showing the locations of the *Hha*I and *Sau*3A recognition sites and the distances of sites from one end.
  c. The same fragment, labeled with ^{32}P in the same fashion, is now digested to completion with *Hha*I, and the products are separated on an agarose gel. Sketch the expected pattern for the ethidium bromide stained gel and autoradiography of this gel.

9-20 a. Given the following restriction map of a cloned 10 kb piece of DNA, what size fragments would you see after digesting this linear DNA fragment with each of the enzymes or combinations of enzymes listed: (1) *Eco*RI, (2) *Bam*HI, (3) *Eco*RI + *Hind*III, (4) *Bam*HI + *Pst*I, and (5) *Eco*RI + *Bam*HI.
  b. What fragments in the last three double digests would hybridize with a probe made from the 4 kb *Bam*HI fragment?

9-21 Professor Ozone is hard at work trying to produce huge quantities of the Ozonase enzyme to reduce smog levels in his home town of Los Angeles. The protein is produced naturally in small quantities by an elusive bacterium called Phew. Professor Ozone has already determined the amino-acid sequence of the protein and now he wants to clone the gene that encodes the protein. Describe two different approaches that he could use to create and identify clones of the gene. Describe how he should go about producing large quantities of the enzyme.

9-22 a. If you are presented with the following sequencing gel pattern, what can you say about the sequence of the template strand used in these sequencing reactions?

    b. If the template for sequencing is the strand that resembles the mRNA, is this portion of the genome likely to be within a coding region? Explain your answer.

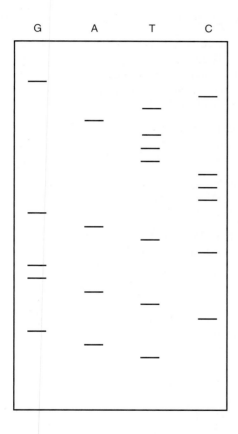

9-23 You have isolated a *trp⁻ E. coli* mutant and to study the nature of the mutation, you isolate DNA from this mutant and determine the sequence of the DNA in the *trp* gene. After carrying out a set of sequencing reactions, you obtain the following results with electrophoresis and autoradiography. The wild-type mRNA sequence is

5′ UAGAUAAGGAAUGUAAGAUAUUACUGAGAUUUAAC 3′

What would be the mRNA sequence of the mutant? Where does it differ from the wild-type sequence?

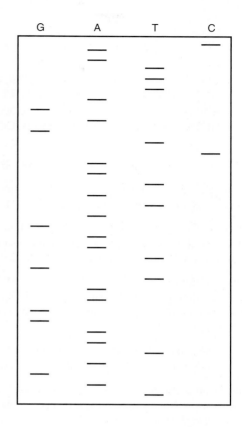

9-24 You read the following sequence directly from a gel.

5′ TCTAGCCTGAACTAATGC 3′

    a. Make a drawing that reproduces the gel pattern from which this sequence was read.

    b. Assuming this sequence is from an exon in the middle of a gene, does this newly synthesized strand or the template strand have the same sequence as the mRNA for the gene (except that Ts are present instead of Us)? Justify your answer.

    c. Using the genetic code table, give the amino-acid sequence of the hexapeptide (six amino acids) translated from the 18-base message. Indicate which is the amino terminal end of the peptide.

9-25 Which of the following set(s) of primers could you use to amplify the target DNA sequence indicated by the dots?

```
5′ GGCTAAGATCTGAATTTTCCGAG . . . TTGGGCAATAATGTAGCGCCTT 3′
3′ CCGATTCTAGACTTAAAAGGCTC . . . AACCCGTTATTACATCGCGGAA 5′
```

   a. 5′ GGAAAATTCAGATCTTAG 3′; 5′ TGGGCAATAATGTAGCG 3′
   b. 5′ GCTAAGATCTGAATTTTC 3′; 3′ ACCCGTTATTACATCGC 5′
   c. 3′ GATTCTAGACTTAAAGGC 5′; 3′ ACCCGTTATTACATCGC 5′
   d. 5′ GCTAAGATCTGAATTTTC 3′; 5′ TGGGCAATAATGTAGCG 3′

REFERENCES

Baltimore, D. (1970). RNA-dependent DNA polymerase in virions of RNA tumor viruses. *Nature 226*:1209–11.

Cohen, S. N., Chang, A. C., Boyer, H. W., and Helling, R. B. (1973). Construction of biologically functional bacterial plasmids in vitro. *Proc Natl Acad Sci U S A. 70*:3240–44.

Maniatis, T., Hardison, R. C., Lacy, E., Lauer, J., O'Connell, C., Quon, D., Sim, G. K., and Efstratiadis, A. (1978). The isolation of structural genes from libraries of eukaryotic DNA. *Cell 2*:687–701.

Maxam, A. M., and Gilbert, W. (1977). A new method for sequencing DNA. *Proc Natl Acad Sci U S A. 74*:560–64.

Nathans, D., and Smith, H. O. (1975). Restriction endonucleases in the analysis and restructuring of DNA molecules. *Ann. Rev. Biochem. 44*:273–93.

Saiki, R. K., Scharf, S., Faloona, F., Mullis, K. B., Horn, G. T., Erlich, H. A., and Arnheim, N. (1985). Enzymatic amplification of beta-globin genomic sequences and restriction site analysis for diagnosis of sickle cell anemia. *Science 230*:1350–54.

Sanger F., and Coulson, A. R. (1975). A rapid method for determining sequences in DNA by primed synthesis with DNA polymerase. *J Mol Biol. 94*:441–48.

Southern, E. M. (1975). Detection of specific sequences among DNA fragments separated by gel electrophoresis. *J Mol Biol. 98*:503–17.

Temin, H. M., and Mizutani, S. (1970). RNA-dependent DNA polymerase in virions of Rous sarcoma virus. *Nature 226*:1211–13.

10

Reconstructing the Genome Through Genetic and Molecular Analysis

The human genome, present in the nucleus of each cell, contains the instructions for transforming a single fertilized egg cell into an adult with 10^{14} cells. Each genome has ~3 billion letters of the DNA language divided among 24 distinct chromosomes ranging in size from 45 million to 280 million letters. The Human Genome Project undertook the challenging task of deciphering this book of life.

Since the mid-19th Century, three advances have radically transformed the field of genetics: Mendel's discovery of fundamental principles in the 1860s, Watson and Crick's elucidation of DNA structure in 1953, and the Human Genome Project from 1990 to the present. In previous chapters we examined Mendel's ideas and impact, as well as the contributions of Watson and Crick. In this chapter, we discuss the Human Genome Project and the field of *genomics* that it spawned.

The **Human Genome Project** was initiated to sequence and analyze the human genome in conjunction with the genomes of several model organisms. A **genome** is the entire collection of chromosomes present in the nucleus of each cell of an individual organism. The human genome, for example, contains 24 different chromosomes (22 autosomes and the X and Y sex chromosomes). These 24 strings of Gs, Cs, Ts, and As contain approximately 3 billion nucleotides and range in size from 45 million to 279 million bp. **Genomics,** the study of whole genomes, is a branch of biology dedicated to the development and application of more effective

mapping, sequencing, and computational tools. Genomicists use large-scale, or *global,* molecular techniques for linkage analysis, physical mapping, and the sequencing of genomes to generate vast amounts of data, which they then analyze by computer. Sophisticated programs enable genomicists to predict the existence and, in some cases, the general functions of previously undefined genes. Their molecular biologist colleagues can then confirm the computer-generated projections through biochemical studies and genetic engineering in model organisms. **Model organisms,** often used in genomic analysis, have many genetic mechanisms and cellular pathways in common with each other and with humans. Unlike humans, however, they are amenable to classical breeding experiments and direct experimental manipulation of the genome. The fruit fly studied by T. H. Morgan and his group, the nematode worm introduced on p. 237 in Chapter 8, and the common house mouse are three examples of model organisms. For more about model organisms, see pp. 651–653 in Chapter 19 as well as References A–E on pp. 739–865.

The history of the Human Genome Project provides a fascinating glimpse into how genomics and the study of model organisms have changed the practice of genetics. In the spring of 1985, the first meeting on the project took place in Santa Cruz, California. The chancellor of the University of California at Santa Cruz had assembled 12 biologists of diverse backgrounds to explore the idea of starting an institute to sequence the human genome. After two days of vigorous discussion, the 12 biologists concluded that it would, indeed, be possible to develop the technology required to accomplish this then seemingly impossible objective. However, the group was split on whether it would be a good idea scientifically. Two aspects of their discussion were striking. First, the concept of the Human Genome Project introduced the idea of discovery science—a new scientific approach to biology. In **discovery science,** one seeks to identify all the elements of a biological system—for instance, the complete sequences of the 24 chromosomes that contain the 3 billion nucleotides of the human genome—and place them in a database to enrich the infrastructure of biology. Researchers doing discovery science "discover" the data and make it available for biological analysis. Discovery science stands in contrast to hypothesis-driven approaches to biology in which one asks questions and seeks experimental verification of possible answers. Second, the Human Genome Project required the development of high-throughput automated DNA sequencing technology as well as the computational tools necessary for capturing, storing, and analyzing the vast amounts of cloning, mapping, and sequence data associated with obtaining genome sequences. These technological and computational developments would require integrating the disciplines of engineering, computer science, physics, and mathematics with biology, leading to a cross-disciplinary approach that would eventually transform modern biology and genetics.

Not surprisingly, most biologists initially viewed the Human Genome Project with skepticism, for several reasons. They thought the project would not be very interesting, since many geneticists claimed that only about 2% of the genome represented genes; the remaining 98%, they argued, was "junk." In the mid-1980s, most biologists also believed that the Human Genome Project was not really a scientific endeavor because it was not hypothesis-driven. Many did not understand how the discovery approach to determining the sequence of the human genome would revolutionize the power and potential of genetic and other biological studies. Finally, some viewed the HGP as big science that would inappropriately compete for funds with the more fruitful and productive small, hypothesis-driven science. Reasonable answers existed for each of these concerns. We now understand, for example, that some of the DNA sequences originally considered to be "junk" may play an important role in resculpting the structure of the genome, and that the sequences can occasionally evolve into genes.

In 1988 the National Academy of Sciences appointed a committee (half proponents and half opponents) to consider the scientific merits of the Human Genome Project. After a year of vigorous debate and analysis, the committee unanimously endorsed the project. This was a major turning point in its acceptance by the community of biologists and geneticists. The government-funded Human Genome Project began in 1990 with a projected 15-year time scale and a $3 billion budget for completing the human genome sequence. Through periodic reassessments, the Human Genome Project came to have six distinct objectives:

1. to generate physical, genetic, and sequence maps of the human genome;

2. to sequence the genomes of a variety of model organisms;

3. to develop improved technologies for mapping and sequencing;

4. to develop computational tools for capturing, storing, analyzing, displaying, and distributing map and sequence information;

5. to sequence EST fragments of cDNAs (*expressed sequence tags,* or *ESTs,* are single-sequence runs [~600 bp] on the cDNA inserts) and eventually full-length cDNAs encoding the expressed mRNAs in different cell types of humans and mice;

6. to consider the ethical, social, and legal challenges posed by genomic information.

Biology's first large-scale project is close to achieving many of its goals. A rough sequence draft of the human genome was completed in February 2001 (Fig. 10.1); in this "draft" the sequence did not yet have an appropriate level of accuracy (an error rate of 1/10,000), and it had some gaps. The final sequence was completed in 2003, two years ahead of the originally proposed 2005 finish date. The early finish was, in part, catalyzed by the 1998 promise of Celera, a private company, to complete a draft of the genome in just three years, employing a novel sequencing strategy. The Celera challenge prompted the federally supported genome effort to move its timetable for sequencing a draft of the human genome ahead by several years. Indeed, this private/public competition benefited biology and genetics through the earlier availability of the human genome sequence.

In addition to the human genome sequence, draft or finished genome sequences existed for eight model organisms by 2002 (see Table 10.1 on p. 324): *Escherichia coli,* representing prokaryotes; *Saccharomyces cerevisiae,* or the bread yeast, representing single-celled eukaryotes; *Drosophila melanogaster,* the fruit fly, and *Caenorhabditis elegans,* the nematode, together representing multicellular animals of moderate complexity; *Arabidopsis thaliana,* a mustard weed, representing simple plants; *Oryza sativa,* or rice, representing a more complex plant; *Fugu rubripes,* the Japanese puffer fish, representing fish and a very compact veretebrate genome; and *Mus musculus,* the house mouse, representing the mammalian class of animals, which includes humans. Researchers chose to analyze these models in conjunction with the human genome because the organisms have increasingly complex genomes that are amenable to genetic analysis and that have already been extensively mapped and studied. By comparing the DNA of model organisms to that of humans, researchers seek to uncover genes and other critical DNA elements that are conserved across evolutionary lines. The Human Genome Project catalyzed efforts to sequence these myriad model organism genomes as well

Figure 10.1 Covers of the February 2001 issues of *Nature* and *Science*. *Nature* published a draft of the human genome sequence obtained by the public effort; *Science* published a draft produced by private efforts.

as more than 100 additional microbial and parasitic genomes. Such sequencing projects have revolutionized the study of microbiology as well as that of genetics, molecular biology, and plant development.

In other accomplishments, production of a human genomewide genetic map has made it possible to identify more than 1000 genes causing different human diseases. The identification of genes predisposing to specific diseases creates new opportunities for diagnosis and therapy as well as for understanding the biology of the disease. More than 4 million human and 2 million mouse *expressed sequence tags (ESTs)* have been sequenced; and projects are underway to sequence most mouse and human full-length cDNA sequences. These EST and cDNA sequences are playing an important role in defining gene locations, alternative RNA splicing patterns, and exon/intron boundaries in genomic sequences.

The Human Genome Project has also catalyzed the development of high-throughput automated DNA sequencing instruments (for example, a 96-capillary automated DNA sequencer has

a 2000-fold throughput increase over the original sequencing prototype instrument), given rise to a variety of novel strategies for mapping and sequencing genomes, and supported the development of computational tools for dealing with the ensuing explosion of genomic information.

Yet another significant aspect of the project is that it is the first scientific project to include in its statement of purpose consideration of the social, ethical, and legal challenges emerging from its projected results; for example, the issue of genetic privacy. At the suggestion of James Watson, the first director of the Human Genome Project, 3% to 5% of the project's research funds were dedicated to the study of ethical, legal, and social implications of mapping the genome—ELSI is the acronym for this arm of the project. A critical concept is to educate society about the opportunities and challenges of the new genetics so that public and private figures alike can make wise decisions about resource allocations and the rules and regulations governing applications of the new technologies.

Completion of the Human Genome Project is moving biology into a postgenome sequencing era where researchers can use global measurements to analyze complex systems. The ultimate goal of genomic sequencing and genetic analysis is to view the genome of any organism as a whole, both to make a connection between every gene and its specific function and to understand how all the genes fit together on the chromosomes. With this knowledge for all genes in the human genome, biologists may one day understand the molecular basis of every genetic disease, as well as the networks of molecular interactions that bring each person to life in a unique way.

As we describe the tools and potential applications of genomics, we encounter a compelling general theme: The social and personal repercussions of the information gained from genomic analysis are generating new areas of biological concern that require close attention as the new knowledge unfolds.

In this chapter, we examine:

- Challenges and strategies of genome analysis, including genome size; features to be analyzed; problems with DNA variations, or polymorphisms; and the development of three types of whole-genome maps—high-density linkage maps, long-range physical maps, and complete sequence maps.

- Major insights emerging from complete and nearly complete genome sequences, including the number and types of genes, the extent of repeated sequences, genome organization and structure, evolution by lateral gene transfer, and areas of future investigation.

- High throughput tools for analyzing genomes and their protein products, including DNA sequencers, DNA arrays, and mass spectrometers.

- Two paradigm changes propelled by the availability of whole-genome sequences and the new tools of genomic analysis: *systems biology,* a change in the way biology can be done; and *predictive and preventive medicine,* changes in the ways medicine will be practiced.

Analyses of Genomes

The Genomes of Living Organisms Vary Enormously in Size

The genomes of microbes and eukaryotes range from the 700,000 base pairs (700 kb) in a single microbial chromosome to more than 3 billion base pairs (3 gigabase pairs, or 3 Gb) distributed among the 5–96 chromosomes of various mammals. Table 10.1 gives the genome sizes of representative microbes, plants, and animals. One amoeba, *Amoeba dubia,* has a genome of more than 600 billion base pairs, and wheat has a genome of 15 billion base pairs. To put these numbers in perspective, the human genome is 200 times larger than the yeast genome and 200 times smaller than the genome of *Amoeba dubia.* Thus the information content of a genome is not necessarily proportional to its size. The large size of some genomes presents fascinating challenges for their ultimate characterization and analysis.

Genomicists Look at Two Basic Features of Genomes: Sequence and Polymorphisms

Two primary goals of genomic researchers are to determine the sequences of each of the chromosomes in the genome and to identify as many as possible of the **polymorphisms,** or allelic variants, present throughout the genome.

There Are Several Major Challenges to Achieving These Two Objectives

In their quest to sequence and analyze genomic DNA, genomicists must confront various issues.

Questions About Sequencing. We've seen that the lengths of chromosomes range from 700 kb to more than 500 million base pairs (500 Mb), and each chromosome is a single continuous double-stranded string of bases. Current technology allows one to sequence, on average, 600 bp at a time (these 600 bp are known as a *read*). One question is: How does one sequence a

TABLE 10.1 **A Comparison of the Developmental Complexity and Genome Features* of Model Organisms**

Organism		Developmental Complexity	Genome Size* (Megabases)	Predicted Genes	# of Genes per Million bp Sequenced	Date Genome Finished
Type	**Species**					
Bacterium	*Escherichia coli*	1-cell prokaryote	4.64	4200	905	1997
Archaebacterium	*Halobacterium sp. NRC-1*	1-cell	2.57	2630 (2411)	1023	2000
Yeast	*Saccharomyces cerevisiae*	1-cell eukaryote	12.07	5800	483	1996
Worm	*Caenorhabditis elegans*	~1000 cells	100	19,099	197	1998
Fly	*Drosophila melanogaster*	~50,000 cells	180	13,601	117	2000
Mustard weed	*Arabidopsis thaliana*	10^{10} cells	125	25,000	221	2000
Pufferfish (draft)	*Fugu rubripes*	10^{12} cells	380	38,000	118	2002
Rice (draft)	*Oryza sativa*	5×10^{10} cells	Ind/jap 466/420	46,000–56,000/ 32,000–50,000	127–155/ 82–128	2002
Mouse (unfinished)	*Mus musculus*	10^{11} cells	3200	40,000	8–11	2002
Human (draft)	*Homo sapiens*	10^{14} cells	3200	40,000–60,000	9–12	2002

*Haploid genome size, including heterochromatin.

Note: For the sequenced genomes of model organisms: Gene numbers are taken from the original sequence publications; most numbers have since changed slightly and different sources give different estimates, depending on protocols. For rice, two different strains have been sequenced: *Oryza sativa* L. ssp. *Japonica* and *Oryza sativa* L. ssp. *Indica.*

500 Mb chromosome 600 bp at a time? Moreover, the DNA sequencing error rate is about 1% per sequence read (that is, per 600 bp), and this raises a second question: How accurate should a genome sequence be? It must be accurate enough to readily delineate important chromosomal features, for example, exons and genes. Most geneticists believe that error rates between one error per 1000 bp (1/1000) and one error per 10,000 bp (1/10,000) are acceptable. The Human Genome Project has selected an error rate of less than 1/10,000 for its finished sequence.

How Does One Distinguish Sequence Errors from Polymorphisms? Higher organisms are diploid; therefore, each genome contains maternal and paternal complements of chromosomes with their attendant polymorphisms. The rate of polymorphisms in humans is about 1 in 500 bp. How does one distinguish these polymorphisms from DNA sequence errors?

Repeated Sequences May Be Hard to Place. The genomes of higher organisms are replete with *repeated sequences*. These repetitive sequences range from tens of base pairs, as in simple sequence repeats like (AT)n, to replicated chromosomal blocks that can span hundreds of kilobases (see pp. 376–378 in Chapter 11 as well as pp. 251–253 of the Genetics and Society box in Chapter 8 for more details on DNA repeats). Repetitive sequences may be tandemly arrayed (that is, positioned side by side), or they may be scattered across the genome. All repeated sequences are identical when they arise, but they diverge as they age. They are thus evolutionary signposts that reveal how much time has elapsed since a duplication event occurred. Young repeats, of recent origin, are very similar, if not identical; older repeats, of more ancient origin, may be quite diverged in sequence. It is the young, very similar repeats longer than the 600 bp reads that constitute a challenge for current approaches to sequencing. If, for example, a 600 bp read falls in a repeat that has 10,000 copies dispersed around the genome, how do you know which of the repeats you are looking at and where in a particular chromosome it belongs?

Unclonable DNA Cannot Be Sequenced. As we saw on pp. 286–287 in Chapter 9, the first step of any initial sequencing is cloning. But, a significant fraction of genomic DNA, often referred to as **heterochromatic DNA,** is not clonable. Hetero-chromatic DNA may constitute up to 30% or more of the genome (see Table 10.1), and many believe it contains few genes. How do we know what heterochromatic regions contain if we cannot clone and sequence them?

Given these challenges, how do genome analysts proceed?

A Divide and Conquer Strategy Makes It Possible to Meet Most of the Challenges If one can sequence only 600 bp at a time, a divide and conquer strategy is the solution for sequencing multimegabase chromosomes. With this approach, the chromosomes are broken into small overlapping pieces, and all pieces are cloned. After the 600 bp at both ends of the clones are sequenced, the clones, whose ends overlap, are reassembled into their original chromosomal strings. Figure 10.2 outlines this strategy. The overlapping ends of the cloned DNA inserts are necessary for the final assembly, or joining, of the sequence information (Fig. 10.2b) as there is no way to assemble or determine the order of adjacent fragments with no sequence overlap (Fig. 10.2a). The sequencing of 600 bp at both ends of each DNA insert provides 1200 bp of sequence per clone analyzed.

To achieve an error rate of 1/10,000, each DNA sequence region must be sequenced multiple times from independently derived clones. Trial and error has shown that the sequencing of every chromosomal region from 10 independent inserts can generate an error rate of less the 1/10,000. This tactic is termed **10-fold sequence coverage.** Ten-fold coverage readily permits the random errors (seen on 1 in 10 of the sequence fragments) to be distinguished from the polymorphisms (seen on 5 in 10 sequence fragments). With 10-fold coverage, the 600 bp reads are assembled into individual DNA strings computationally by alignment of the overlapping sequences from each insert.

A critical question in the divide and conquer strategy is whether to generate the overlapping fragments for sequence analyses directly from the genome (a whole-genome approach) or to generate them from large insert clones that researchers have previously ordered in a physical map (a hierarchal approach). We consider this issue when we describe the three types of chromosomal maps—linkage, physical, and sequence—that view the genome at differing levels of resolution. First, however, we provide a brief overview of basic analytic techniques.

Four Relatively Simple Techniques Make Genome Characterization Possible

Cloning, hybridization, PCR amplification, and sequencing (described in detail on pp. 285–307 in Chapter 9), as well as several computational tools, lie at the heart of the various mapping and sequencing strategies for genome analysis.

(a) Adjacent fragments
↑ Failure to assemble because there is no overlap.
| Sequence both ends of fragments.
↑ Clone.
| Break into adjacent fragments.

(b) Overlapping fragments
| Break into overlapping fragments.
| Clone.
| Sequence both ends of fragment.
↓ Reassemble string by sequence overlap.

Figure 10.2 The divide and conquer strategy for DNA sequencing. The *arrows* indicate sequence determined from each end of the clone. The *vertical bars* indicate overlapping sequences between two clones, which make sequence assembly possible. **(a)** The original DNA string is cut into adjacent fragments that cannot be assembled. **(b)** The original DNA string is cut into overlapping fragments that can be assembled.

Review of Techniques for Mapping and Sequencing

Cloning. Fragments of genomic DNA ranging in size from 500 bp to more than 1,000,000 bp may be cloned into a variety of vectors (review Table 9.2). The entire collection of cloned DNA fragments for a particular DNA source is termed a **library**. A library may have low coverage of a particular genome with each DNA sequence region on average represented just one time or less, or it may have high coverage with each sequence, represented on average 10 times or more.

Hybridization. Researchers use hybridization to determine the location of particular DNA sequences in a library of fragments. Recall that hybridization depends on the molecular complementarity of a target DNA sequence with a DNA fragment from the library. A hybridization protocol is carried out after some type of fragment separation, such as the plating out of individual insert colonies on a filter. Through hybridization with specific, defined fragments, the clones containing a particular gene or chromosomal region can be readily identified for mapping or sequencing.

PCR Amplification. Using the polymerase chain reaction (PCR), researchers can directly amplify particular regions of the genome up to a millionfold or more. This amplification requires that DNA sequences on either side of the region to be amplified are known so that PCR primers may be selected. Regions ranging from 1 bp to more than 20 kb may be amplified by PCR. A critical point for genomewide analyses is that the pair of PCR primer sequences must be unique (not in a repetitive region). If they are not, the PCR protocol will amplify multiple regions of the genome.

DNA Sequencing. An automated DNA sequencer employing the Sanger sequencing method can determine DNA sequences about 600 bp in length from cloned DNA fragments.

Computational Tools. Four types of computational tools are important to genomics: programs for identifying the sequence matches between a particular DNA fragment and a large population of previously sequenced fragments; programs for identifying the overlaps of individual sequences and assembling them into strings of bases; programs for estimating the error rates in different DNA sequences; and programs for identifying genes in chromosomal sequences.

We now describe how researchers use these basic tools in various approaches to the mapping and sequencing of genomes.

Large-Scale Maps Serve As Guides to Whole Genomes

Chromosomal maps show the location of genes, genetic markers, centromeres, telomeres, and other points of interest along the chromosome of an organism. In Chapters 5 and 9, we described techniques for mapping a small number of loci in a relatively small region of a genome. Here we examine how genomicists expand on these techniques to produce genetic, physical, and sequence maps for a whole genome. Geneticists use the terms "linkage" and "genetic" interchangeably when talking about maps made through linkage analyses of genetic relationships. In the rest of this chapter, as in Chapter 11, we use the term "linkage."

The analysis of the human genome focused initially on a genomewide linkage map, then a physical map, and finally a sequence map. These maps were successively integrated into each other. The linkage map was produced by a global genomewide approach; the physical map by a hierarchical divide and conquer approach; and the sequence map by both strategies.

In Chapter 11, we include a detailed discussion of how geneticists use linkage maps to identify genes predisposing to disease and to analyze complex traits. Here we consider how researchers generate and employ dense genomewide linkage maps.

High-Density Linkage Maps: Computerized Analyses of Transmission Data Position Unlimited Numbers of Markers in Relation to Each Other

With the traditional linkage analysis described in Chapter 5, geneticists map only a handful of loci in any one cross because easy-to-follow phenotypes, such as those of color and gross morphology, tend to interfere with each other's expression. In addition, the number of good phenotypic markers is relatively limited (compared, for example, to the number of genes). Moreover, a single phenotype may be encoded by many genes. For example, 34 different genes cause mice to have white spots. Each white-spot gene maps apart from the others, but in a two-point cross following, say, black coat color and white spots, it is impossible to tell which one of the 34 white-spot genes is contributing to the phenotype observed. This problem is particularly acute in human genetics since it is not possible to breed people with combinations of mutations appropriate for linkage analysis. As a result, with loci defined solely by variant phenotypes, the only way to construct an all-encompassing linkage map is through the tedious process of accumulating and pooling information from many two- and three-point crosses.

Geneticists have solved the problems of limited assayable phenotypic traits and traits encoded by multiple genes by using some of the recently developed tools for the direct detection of genotype (described in detail in Chapter 11). With these tools, they can follow the segregation of a very large number of loci in a single experimental plant or animal cross or in a human pedigree, and they use the transmission data to build a high-density linkage map. Such a high-density map shows the relative positions of thousands to tens of thousands of closely spaced (that is, densely packed) DNA markers. (Recall from the Fast Forward box on pp. 130–131 in Chapter 5 that a DNA marker is a DNA fragment of known size and location that comes in identifiable variants.) Since Chapter 11 covers linkage maps in detail, we now describe only the large-scale maps built from genomewide identification and typing of markers.

The Making of Large-Scale Linkage Maps

The human genome contains millions of DNA polymorphisms scattered across the different chromosomes. Two types of DNA polymorphisms are commonly employed for large-scale linkage mapping: *single nucleotide polymorphisms (SNPs)* in which, for example, the single nucleotide T in one sequence is replaced by a G in the corresponding sequence; and *simple sequence repeats (SSRs)* in which a set of two to five or so nucleotides is repeated 4–50 or more times (CGTCGTCGTCGT. . .; Fig. 10.3). SSRs have a significant tendency to expand or contract during chromosomal replication, and these processes generate length polymorphisms. (Note that another term for SSRs is *microsatellites*. In the rest of this chapter, we use SSRs to denote simple sequence repeats, which are useful in genomic analysis. In Chapter 11, where we discuss how to distinguish individual genomes, we use the term "microsatellites" and contrast with larger-unit "minisatellites"; the patterns of both help identify individual genomes.)

SNPs and SSRs may lie within a gene, where they can cause alterations in the resulting gene function ranging from profound to no effect. Hence, these types of polymorphisms may have a profound effect on phenotype or little to no effect at all. Most polymorphisms seem to have little effect on the organism.

Both SNPs and SSRs, which come in identifiable variants, can serve as DNA markers. In making genomewide linkage maps, researchers must be able to rapidly identify and order these DNA markers across the chromosomes. They must also be able to determine the genotypes of large numbers of ordered polymorphisms in populations ranging from hundreds to thousands of individuals.

Genomewide Identification of Genetic Markers

The initial human (and mouse) genetic maps produced by the Human Genome Project employed the highly polymorphic simple sequence repeats (SSRs). As mentioned previously, these tandemly repeated di-, tri-, tetra-, or pentanucleotides readily expand or contract the numbers of their repeats in different individuals, thus generating size polymorphisms. Hybridization probes of simple sequence repeats, for example (AT)n, can identify clone inserts with these same repeat sequences in a genomic library. These clones can then be sequenced to obtain unique 5′ and 3′ adjacent sequences for PCR primers. Recall that PCR primer sequences must be unique in the genome; if they are not, a polymerase chain reaction using them could simultaneously amplify several similar regions, with confusing results. After the identification of hundreds to thousands of SSRs, researchers use recombination analyses to map them in families. They then order these genetic markers across the genome to generate global linkage maps.

Recently, the DNA sequence analysis of *orthologous gene* regions and cDNA clones from different individuals has identified millions of potential SNPs. The newly identified SNPs are being verified by suitable recombinational analyses in family pedigrees.

Orthologous genes, defined by their sequence similarities, are genes in two different species that arose from the same gene in the two species' common ancestor. Orthologous genes are part of the more general category of **homologous genes,** which are genes with enough sequence similarity to be evolutionarily related somewhere back in evolutionary history. Orthologous genes are homologous, but homologous genes do not have to be orthologous. They could, for example, be *paralogous.* **Paralogous genes** arise by duplication (sometimes multiple duplications) within the same species, often within the same chromosome; paralogous genes often constitute a multigene family. The differences between paralogous, orthologous, and homologous genes are differences of degree and timing. *Homologous* is a general term that includes all genes with sufficient sequence similarity to have evolved from a common ancestor situated anywhere in evolutionary time. With *orthologous* genes, the common ancestor is the immediate ancestor to the two species. The duplication events that create *paralogous* genes take place in the same species. These three terms have become important to genomicists as they refine their analyses of various genomes.

Today, genome analysts can determine the order of SNPs and SSRs across the genome by sequence comparisons with the human genome sequence. For example, the current human and mouse linkage maps contain more than 20,000 and 10,000 SSRs, respectively, and more than 4 million and 10,000 potential SNPs, respectively. With these markers, it will be possible to construct very dense maps. The idea is to have a map of ordered genetic markers dense enough to permit the identification of the gene or genes encoding polymorphic phenotypic traits, for example, by linkage analyses in families with the variant gene (or genes). There is an argument among geneticists about how dense this genetic map has to be to identify one or more genes that, for example, predispose to a particular disease. Some argue that thousands of markers genomewide are adequate; others argue that hundreds of thousands of markers are necessary; still others argue that it may be impossible to map complex traits encoded by multiple genes (see Chapter 11). Until recently, genomicists used approximately 500 human SSRs evenly spaced across the genome for linkage mapping—which provided on average 1 SSR for every 6 Mb. The identification of SNPs alone has given them more than 500,000 DNA markers across the genome, a 1000-fold increase in marker density.

Single nucleotide polymorphism (SNP)	...GCAA T TCCCGATT...
	...GCAA G TCCCGATT...
Simple sequence repeat (SSR)	...GCATTATATATATATC...
	...GCATTATAT[ ]C...

Figure 10.3 Two common types of polymorphisms employed for genetic mapping. SNPs occur at rates of 1/500 to 1/1000 across the human genome. The SSR illustrated here is a dinucleotide (A T) repeat, whose alleles exhibit the (A T)$_5$ and (A T)$_2$ alternatives. SSRs occur, on average, every 20–40 kb across the genome.

(a) PCR amplification

PCR primer 1

Individual 1   GCAGCAAT(AT)$_{25}$ GGTAAAAC
Individual 2   Unknown
Individual 3   GCAGCAAT(AT)$_5$ GGTAAAAC

PCR primer 2

(b) Size separation

Individuals

1   2   3

Electrophoresis

(A T)$_{25}$

(A T)$_5$

Figure 10.4 The two-stage assay for simple sequence repeats (SSRs). Individuals 1 and 3 are homozygous for the (A T)$_{25}$ and (A T)$_5$ alleles, respectively. Individual 2 has an unknown genotype. In the first stage, PCR primers 1 and 2 amplify the SSRs directly from genomic DNA. In the second stage, electrophoresis separates the alleles by size. Individual 2 turns out to be a heterozygote with both the (A T)$_{25}$ and (A T)$_5$ alleles.

Genomewide Typing of Genetic Markers

Researchers generally genotype SSRs by examining the polymorphic size differences of allelic PCR products via gel electrophoresis (Fig. 10.4). Robots and 384-well PCR machines, together with the automated loading of four-color, 96-capillary DNA sequencing machines, allow thousands of SSRs to be analyzed per day per machine. The typing of SNPs can also be done by the DNA sequence analysis of PCR products, although it is now more often carried out by specialized assays employing oligonucleotide arrays or mass spectrometry, which we discuss on p. 351 of this chapter.

Genomewide linkage maps provide DNA sequences ordered along the chromosomes as well as hybridization probes. Investigators can use both the ordered sequences and the probes to localize DNA clone inserts and DNA sequences along the individual chromosomes in the making of large-scale, or long-range, physical and sequence maps.

Long-Range Physical Maps: Karyotypes and Genomic Libraries Provide the Basis for Positioning Markers on Chromosomes

A *physical map* is a constellation of overlapping DNA fragments that are ordered and oriented and span each of the chromosomes in a genome. Physical maps, the molecular counterparts of linkage maps, are based on the direct analysis of genomic DNA. Such maps chart the actual number of base pairs (bp), kilobases (kb), or megabases (Mb) that define and separate a locus, or site, from its neighbors in a particular region of a particular chromosome. Linkage and physical maps can be calibrated against one another. For example, in humans, 1 cM ≈ 1 Mb, whereas in mice, 1 cM ≈ 2 Mb.

As we saw on pp. 350–351 in Chapter 9, geneticists build relatively short-range physical maps by exposing DNA clones to two or more restriction enzymes and comparing the fragments produced in this way; they then use hybridization probes to ascribe a precise position to the genes and markers included in the mapped clone. The major difference between the physical maps discussed earlier and those developed by genomicists is one of scale. A map of a cosmid clone, for example, would span 30,000 bp of genomic DNA. Here we learn how molecular geneticists map a human chromosome averaging 130,000,000 bp in length (approximately 4300 times longer than a single cosmid).

In developing a genomewide physical map, the length and stability of the clone inserts are significant when generating the fragments to be used. The inserts should be as long as possible; and they should be very stable (that is, they should not delete or rearrange during vector propagation). Two types of vectors that accept larger clone inserts used to construct physical maps are yeast artificial chromosomes, or YACs (insert size ranges from 100–1,000,000 Mb), and bacterial artificial chromosomes, or BACs (ranging from 50–300 kb). Because BAC inserts are more stable and simpler to purify from their host DNA, they have become the insert of choice for physical mapping.

Another consideration in the development of genomewide physical maps is how to determine the order of the clones across the genome. As noted earlier, the cloned DNA inserts must overlap at their ends. These insert overlaps can be produced by random shearing or partial restriction enzyme digests of genomic DNA before cloning. There are two general approaches to assigning insert location—a top-down approach in which the insert is hybridized against a karyotype of the entire genome, and a bottom-up approach in which the overlapping sequences of tens of thousands of clones are determined by restriction mapping or the analysis of sequence tagged sites. **Sequence tagged sites (STSs)** are short DNA sequences that are readily located and amplified by PCR and that identify unique physical genomic locations. To cover the entire human genome with BACs containing inserts averaging 200 kb, more than 15,000 BACs must be analyzed (3×10^9 bp/genome/2×10^5 bp/BAC = 1.5×10^4 BACs/genome).

The FISH Protocol Locates Cloned Loci—Genes or Markers—Directly on the Chromosomes of a Karyotype—A Top-Down Approach

We saw in Chapter 4 that the chromosomes of actively dividing cells in the metaphase of mitosis, when stained with a Giemsa dye and viewed in the light microscope, show an identifiable series of dark and light regions referred to as bands and interbands (Fig. 4.6 and 10.5a). (The dark regions are A–T-rich and gene-poor, while the light bands are G–C-rich and gene-rich.) The number, intensity, and width of each band and interband are highly reproducible characteristics that a skilled cytogeneticist can use to distinguish each pair of homologs from all other chromosomes in a cell. The visual description of a complete set of chromosomes in one cell of an organism is a *karyotype*. To fa-

Figure 10.5   The human karyotype: Banding distinguishes the chromosomes. (a) Photograph of a complete set of human chromosomes at metaphase. Staining with Giemsa dye accentuates the bands and interbands. **(b)** Idiograms for the complete set of human chromosomes. An idiogram is an idealized diagram of the banding pattern associated with a stained chromosome. **(c)** Chromosome 7 at three different levels of banding resolution. As staining techniques improve, it becomes possible to resolve what previously appeared as a single band into a series of bands and interbands, producing more and more bands along each chromosome. Thus, at one resolution, 7q31 appears as one band. At a slightly higher resolution, 7q31 becomes two bands (7q31.1 and 7q31.3) flanking an interband (7q31.2); and at an even higher resolution, 7q31.3 itself appears as two bands (7q31.31 and 7q31.33) and an interband (7q31.32).

cilitate the presentation and comparison of data from different karyotypes in the same and diverse species, investigators convert the light and dark bands actually observed under the microscope into black-and-white diagrams of the chromosomes (Fig. 10.5a and b). In these diagrams, the autosomes are numbered in order of descending length. In humans, we have seen that chromosome 1 is the longest autosome and chromosome 22 is the shortest. The X and Y sex chromosomes are intermediate in size and small, respectively. The shorter arm of each chromosome is designated "p" (for *petit*) and the longer arm is "q" (for following *p*). Karyotypers number each band and interband, starting at the centromere and moving out along each arm toward the telomere.

How do researchers use karyotypes in making maps? They can, of course, locate the centromere and telomeres of each chromosome through microscopic observation. But what about the loci that lie between those points? A protocol known as **fluorescent *in situ* hybridization,** or **FISH,** makes it possible to map any cloned locus to a site on a metaphase chromosome.

Figure 10.6 illustrates the steps of the FISH protocol. In the final step, sites of hybridization appear as spots of illumination on the chromosome under the microscope.

The FISH analysis has several advantages over linkage analysis for the mapping of a newly cloned locus to a particular chromosomal site. First, all clones can be mapped by the FISH protocol, whereas only those clones that detect polymorphisms can be mapped by linkage analysis. Second, researchers can perform a FISH analysis on any cloned locus in isolation, whereas linkage mapping requires the analysis of one locus in relation to another. Third, FISH requires only a single sample on a microscope slide, whereas linkage analysis requires genotype information from a large cohort of individuals. For this reason, human geneticists often use the FISH protocol as a rapid way to obtain the general chromosomal position of a newly cloned locus. Finally, the FISH protocol enables direct comparisons between the map positions of cloned loci and the aberrations (such as deletions or translocations) that are visible in the chromosomes of individuals with unusual chromosomes. The main disadvantage of

Figure 10.6   The FISH protocol. (a) The technique. (1) First, drop cells arrested in the metaphase stage of the cell cycle onto a microscope slide. The force of the droplet hitting the slide causes the cells to burst open with the chromosomes spread apart. (2) Next fix the chromosomes and gently denature the DNA within them such that the overall chromosomal structure is maintained even though each DNA double helix opens up at numerous points. (3) Label a DNA probe with a fluorescent dye, add it to the slide, incubate the probe with the slide long enough for hybridization to occur, and wash away unhybridized probe. (4) Now place the slide under a special microscope that focuses ultraviolet (UV) light on the chromosomes. The UV light causes the bound probe to fluoresce in the visible range of the spectrum. You can view the fluorescence through the eyepiece and photograph it. **(b)** A fluorescence micrograph. Photograph of a baby hamster kidney cell subjected to FISH analysis. It shows the microtubular structure.

FISH is that its revolving power is much less than that possible with linkage analysis. The metaphase bands and interbands used as guideposts in FISH protocols contain the equivalent of 4–8 Mb or 4–8 cM of chromosomal material. In contrast, we have seen that linkage analysis can resolve map positions down to a fraction of a centimorgan.

Karyotypic analysis thus enables the generation of low-resolution physical maps that locate cloned genes and markers on particular parts of chromosomes. We now describe how geneticists use the cloning techniques described in Chapter 9 to expand these maps with a much finer resolution of loci.

The Use of DNA Hybridization and Restriction Mapping to Order and Analyze the Clones of a Genomic Library—A Bottom-up Approach

It is possible to build a continuous high-resolution map across an entire chromosome by piecing together cloned fragments in a one-dimensional jigsaw puzzle. Figure 10.7 illustrates the basic steps of this kind of mapping: identifying an ordered series of genomic clones that overlap each other from one end of the chromosome to the other; analyzing each clone separately to obtain as much information as possible about its restriction sites and gene locations; aligning the overlapping ends of clones, using information from the restriction maps; localizing sequence tagged sites; and combining the physical information from each clone into a physical map of the entire chromosome. Computer programs facilitate the assembly of the large amount of data.

How do researchers achieve the first step of piecing together an ordered series of overlapping DNA clones? As we saw in Chapter 9, a genomic library represents all the DNA of a particular cell incorporated into clones reproduced by cell colonies that are arranged at random on a plate. If the research goal is to identify a gene of interest, this randomness does not matter; so long as there are enough clones in the library, one or more will carry the desired gene, and hybridization probes can isolate it from the pack. But what if the goal is to identify all the overlapping clones in an ordered set, as shown in Fig. 10.7? In such

(a) Identify an ordered series of overlapping genomic clones.

(b) Analyze each clone for restriction sites and gene locations.

(c) Create maps of overlapping genomic clones.

(d) Combine information into a single continuous physical map that spans the length of the chromosome.

Figure 10.7 Building a whole chromosome physical map. **(a)** To produce a whole chromosome physical map, you first order a set of overlapping genomic clones that extend from one end of the chromosome to the other. Subsequent figures describe various methods of obtaining this ordered set of clones. **(b)** You next map the restriction sites of each clone in the set through restriction analysis, and analyze individual restriction fragments in other ways, such as Northern blot analysis, to identify transcription units. **(c)** Computers overlay the different types of maps for each clone onto the overlapping clones to obtain a continuous map. **(d)** The result is a single continuous map extending the length of the chromosome.

a case, the significant feature of each clone is its chromosomal location, rather than its genetic function.

There are two approaches to developing an ordered set of clones: one uses linkage maps to derive physical maps, the other relies exclusively on physical mapping techniques. The two approaches are complementary. In the mapping of chromosomes, researchers use both.

The approach based on linkage mapping begins with a high-density linkage map of a selected chromosome containing evenly spaced markers no more than 1 cM apart. To generate an ordered set of clones across the chromosome, you use each marker as a hybridization probe to retrieve from the library clones carrying a complementary sequence (Fig. 10.8a.1). The clones retrieved by each marker probe are, of course, much larger than the marker itself. Consider, for example, using 6000 markers evenly spaced across the human genome's 3000 Mb to screen a YAC library of the human genome. The markers will lie approximately 0.5 cM from each other; in the human genome, 0.5 cM is the equivalent of 500 kb. Since each YAC in the library carries a genomic insert roughly 1 Mb in length, there is a good chance that a YAC clone picked up by screening the

Assigning chromosomal positions: A top-down approach

Use M1 as hybridization probe to screen YAC library. Use M2 as hybridization probe to screen YAC library.

2. Do yM1 and yM2 overlap?

(b) Data generated by screening a YAC library with seven polymorphic markers

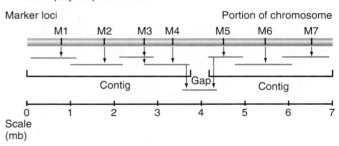

(c) Complete contig spanning region between M1 and M7

Figure 10.8 Using a high-density linkage map to build an overlapping set of genomic clones. (a) Marker probes retrieve and identify clones that overlap. (1) Markers that map near each other (M1 and M2) serve as probes to screen a YAC library and recover genomic clones. (2) Testing of the clones recovered by nearby markers for overlap consists of using the end fragments from each clone to probe the other. A positive hybridization signal indicates that the end fragment of one clone is present in the other clone. **(b)** Hypothetical data obtained from screening a YAC library with seven closely linked markers. The seven clones recovered fall into two separate overlapping sets, or contigs. The contig data show there is an uncloned genomic "gap" between markers 4 and 5. This gap can be filled by using the end fragments from each of these clones to rescreen the YAC library. **(c)** With the addition of the new clone, a single contig extends across the genomic region between markers M1 and M7.

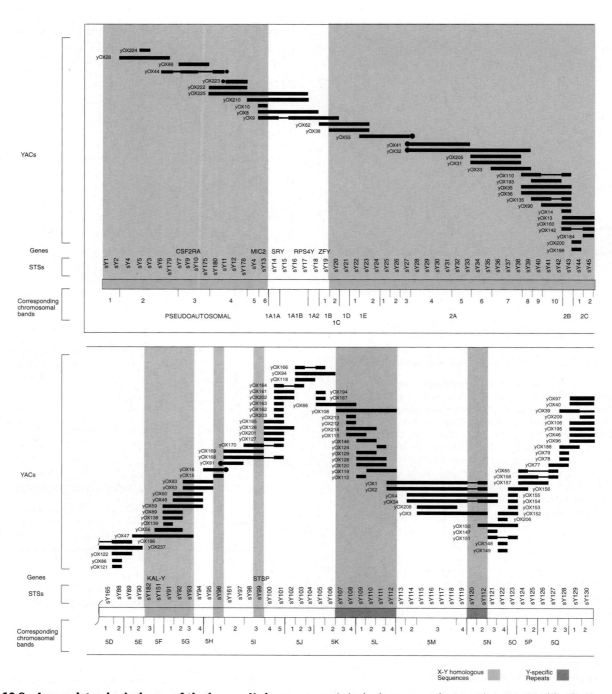

Figure 10.9 A complete physical map of the human Y chromosome. A single-chromosome-long contig was created for the Y chromosome. The contig consists of the series of overlapping YAC clones shown. The STS markers used to establish the contig are indicated by name just above the chromosome idiogram. Gene locations are indicated by name above the markers.

library with one marker (say, marker one, or M1) will overlap a YAC clone recovered by screening the same library with a neighboring marker (M2). You can use cross-hybridization between one YAC clone and the next to demonstrate an overlap between the two YAC clones (Fig. 10.8a.2). Figure 10.8b shows hypothetical data generated by screening a YAC library with seven polymorphic markers (M1–M7). The YACs recovered on screening with M1, M2, M3, and M4 contain regions of overlap, as do the YACs recovered on screening with M5, M6, and

M7. A set of two or more partially overlapping cloned DNA fragments that together cover an uninterrupted stretch of the genome form a **contig** (from the word *contiguous*). The YACs defined by M1 through M4 make up one contig, while the YACs defined by M5 through M7 form a second contig. Each of these contigs is roughly 3–4 Mb in length, while the YACs of which they are composed are about 1 Mb long.

The ultimate goal of high-resolution physical mapping is the generation of one large contig for each chromosome in a

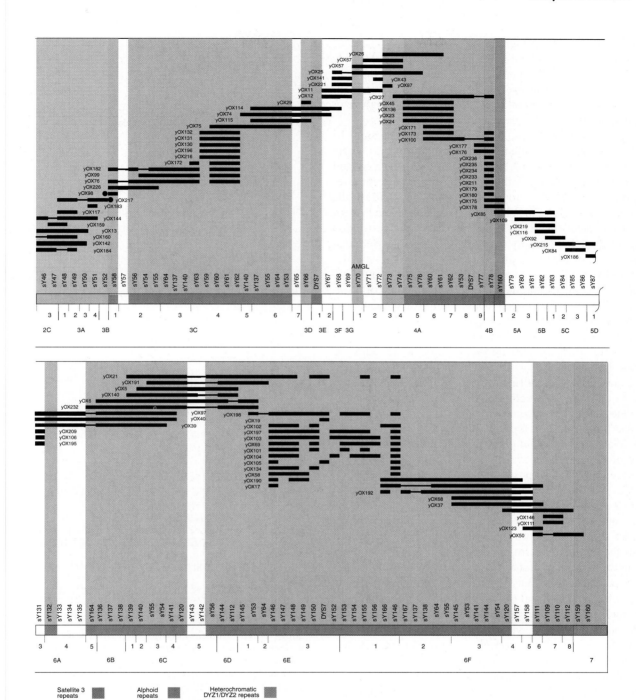

Satellite 3 repeats · Alphoid repeats · Heterochromatic DYZ1/DYZ2 repeats

Figure 10.9 Continued

genome. But because of the somewhat variable spacing between linkage markers, as well as variation in the length of cloned YAC fragments, "gaps" between smaller contigs arise when the distance between two markers is by chance too large to allow recovery of overlapping YACs. The gap between M4 and M5 in Fig. 10.8b is an example. The greater the density of molecular markers, the smaller the number of gaps. To fill in the gap separating two contigs, one can use small fragments from the ends of the contigs as probes to rescreen the YAC library for one or more additional clones that may bridge the gap and thereby convert

two smaller contigs into one larger contig. This method of using the ends of unconnected contigs as probes to retrieve clones that extend into an unmapped region is known as **chromosome walking.** The critical feature of any contig is that through the generation of physical maps for each of the clones it contains, it is possible to piece together a physical map that spans the entire contig (Fig. 10.8c). In 1993, researchers constructed a complete physical map of the human Y chromosome using 207 molecular markers to recover 196 overlapping YAC clones that composed a contig spanning 60 Mb of genomic DNA (Fig. 10.9).

In contrast to using a high-density linkage map as the starting point for deriving a long-range physical map, it is possible to generate a whole-genome physical map by physical techniques alone. These exclusively physical approaches do not depend on linkage information derived from pedigree analysis or the initial positioning of molecular markers.

In one such approach, researchers create a cosmid library containing individual clones covering the whole genome of interest (Fig. 10.10a). They examine each cosmid with molecular techniques that yield a unique pattern of restriction fragment sizes known as a **DNA fingerprint** of the genomic region within the clone (Fig. 10.10b). (Note that the term "fingerprint" denotes a particular pattern of restriction fragment sizes. It can apply to any length of DNA. Here that pattern defines a very small genomic region in a specific clone. In Chapter 11, we use the same term for the pattern of restriction fragments that define the whole genome of a single individual.) Researchers then program a computer to look for overlaps between the fingerprints (Fig. 10.10c) and use the overlaps to arrange the cosmids into 24 chromosomal contigs (Fig. 10.10d). This bottom-up approach breaks the genome into a random jumble of thousands of clones and then generates sufficient fingerprinting information for each clone to allow reconstruction of the small fragments into a complete set of chromosomes.

The most common type of fingerprint is a list of fragment sizes obtained from digesting the cloned DNA with a series of restriction enzymes (review Fig. 10.10b). Each cloned region, when exposed to enough different restriction enzymes, becomes defined by a unique set of fragments of different sizes. When the computer has stored data for a sufficient number of clones (usually tens of thousands), it searches through the entire set, testing each clone against every other clone in the set, for evidence of fingerprint overlaps between pairs of clones (Fig. 10.10c). The premise behind this strategy is that two truly overlapping genomic clones will share a subset of matching restriction fragments (from their region of overlap), whereas two unrelated clones would be unlikely to have the same number of matching fragments. With the fingerprinting of more and more clones, the gaps between contigs fill in, resulting in longer and longer contigs (Fig. 10.10d). Yeast geneticists used this bottom-up approach to generate a single contig spanning the entire length of yeast chromosome I.

Using PCR Amplification of Sequence Tagged Sites (STSs) to Order Clones—Another Bottom-up Approach

A second strategy for determining clone insert overlaps is denoted the *sequence tagged site approach*. The idea is very simple. One identifies a unique sequence in the genome and synthesizes PCR primers to amplify a portion of it. With the synthesis of these primers, the unique sequence becomes a **sequence tagged site (STS).** All STSs have a unique chromosomal address; some STSs are polymorphic, others are not. STSs may come from the ends of clone inserts (for example, BACs),

Bottom-up approach

(a) Digest chromosomes with restriction enzyme. Clone fragments into a genomic cosmid library.

(b) Fingerprint each cosmid for the pattern of restriction fragment sites.

(c) Feed fingerprint data into computer programmed to look for overlaps.

(d) Arrange overlapping cosmid inserts into 24 chromosomal contigs.

Contig

Figure 10.10 Building a whole chromosome physical map of overlapping genomic clones without linkage information. (a) The starting material is a complete library of unordered genomic clones. **(b)** Each clone is subjected to fingerprint analysis. The fingerprints are in the form of gel patterns of restriction fragments. **(c)** The fingerprint information from many clones is fed into a computer, which looks for evidence of overlaps between fingerprints from different clones. **(d)** With the analysis of a sufficient number of clones, the computer can generate long contigs of clones extending across each chromosome. Gaps between contigs can be filled in by the method described in Fig. 10.9c.

from randomly generated genome sequences (for example, 2 kb fragments of randomly sheared genomic DNA), from cDNA clones, or from SSRs and SNPs. It is possible to test for an STS using its paired PCR primers in, for example, 20,000 different BAC clones. The test will identify all clones that contain the

Figure 10.11   A hypothetical physical map generated by the analysis of sequence tagged sites (STS). Each STS represents a unique segment of the genome that has been amplified by a pair of PCR primers. Here, eight BAC clones are interrogated with seven STS markers. This process can unambiguously determine the physical order of the BAC clones in this example.

STS's sequence. The fact that the sequence is unique in the genome means that the identified clones must overlap. You can repeat this testing process for as many STSs as are necessary to create a physical map (Fig. 10.11). The SSR and SNP DNA markers described previously are polymorphic STSs that can also be used to order clones in the physical map. Ordering by this means permits an integration of the genomewide linkage and physical maps.

Once a combination of approaches to generating whole chromosome contigs has produced an ordered array of overlapping clones for all the chromosomes in the genome, researchers can use subcloning, restriction mapping, and hybridization (as described in Chapter 9) to study the clones of each contig in greater detail. In this way, they can locate and analyze genes and markers of interest. They can then have a computer combine all the information for all the clones into a detailed physical map showing a variety of landmarks across the contigs of all the chromosomes in the genome.

A Sequence Map Is The Highest-Resolution Genomic Map

Sequence maps show the order of nucleotides in a cloned piece of DNA. The goal of the Human Genome Project is the determination of the complete nucleotide sequence for every chromosome in the genome (of humans and model organisms). Two basic strategies have been employed for sequencing the human genome: the hierarchical shotgun approach and the whole-genome shotgun approach. The term **shotgun** means that the overlapping insert fragments to be sequenced have been randomly generated either from large insert clones (BACs) or from the whole genome by shearing with sound (sonication) or partial digestion with restriction enzymes.

The Hierarchical Shotgun Sequencing Strategy

The publicly funded effort to obtain a draft sequence of the human genome employed the hierarchical shotgun strategy (Fig. 10.12). In this multistep approach, researchers first generate a genomic BAC library, develop a map of overlapping BAC clones across the genome, and then select a set of minimally overlapping BACs across the genome. The final, minimally overlapping set of BACs is called a **minimal tiling path** of BAC clones. To sequence each BAC clone on the minimal tiling path, a researcher randomly shears it into ~2 kb fragments, clones these fragments in plasmids, and then sequences both ends of a sufficient number of plasmid inserts to generate 10-fold coverage across the BAC insert. If the BAC insert is 200 kb and 1200 bp (or 1.2 kb) are sequenced for each plasmid insert (600 bp from each end), the sequencing of each BAC would require the sequencing of 200 kb/BAC $\times$ 10/1.2 kb/plasmid = 1.667×10^3 plasmids/BAC. One advantage of this divide and conquer strategy is that you need to assemble only about 1700 plasmid insert sequences for each BAC. The data from the linkage maps (showing SSRs or SNPs) and the physical maps (showing gene locations, STSs, and restriction maps) are also used to help assemble the sequence maps of the individual chromosomes. The hierarchical shotgun strategy requires a fair amount of front-end work to create the physical map of BAC clones and to generate a plasmid library for each of the 20,000 or so BAC clones to be sequenced.

The Whole-Genome Shotgun Sequencing Strategy

The private company Celera employed the whole-genome shotgun strategy to obtain its draft sequence of the human genome. In this approach, the whole-genome DNA is randomly sheared three times, first to construct a plasmid library with ~2 kb inserts, second to generate a plasmid library with ~10 kb inserts, and third to produce a BAC library of ~200 kb inserts (Fig. 10.13). In theory, it should be possible to sequence the 2 kb and 10 kb inserts to attain approximately 6-fold and 3-fold coverage, respectively (3×10^9 [the genome] $\times$ 6/1200 [bp sequenced per insert] = 15×10^6 plasmid clones; and $3 \times 10^9 \times 3/1200 = 7.5 \times 10^6$ plasmid clones, respectively). One could then sequence the insert ends from the 200 kb BAC library to a 1-fold coverage ($3 \times 10^9 \times 1/1200 = 2.5 \times 10^6$ BACs). The 6-fold plus 3-fold plus 1-fold coverages would amount to a 10-fold coverage. (In the Celera draft, the total coverage was about 6-fold). A genomewide shotgun computer program would then assemble all these sequences into the chromosomal strings.

The whole-genome shotgun strategy has several advantages: It does not require the construction of a physical map. It

Hierarchical shotgun sequencing

Genomic DNA

BAC library

Organized
mapped large
clone contigs

BAC to be
sequenced
from minimum
tiling path

Shotgun
clones

Shotgun   ...ACCGTAAATGGGCTGATCATGCTTAAA
sequence   TGATCATGCTTAAACCCTGTGCATCCTACTG...

Assembly   ...ACCGTAAATGGGCTGATCATGCTTAAACCCTGTGCATCCTACTG...

Figure 10.12   Idealized representation of the hierarchical shotgun sequencing strategy. (a) Construct a library by fragmenting the target genome and cloning it into a large fragment cloning vector, in this case, BAC vectors. **(b)** Organize the genomic DNA fragments represented in the library into a physical map, and select and sequence the BAC clones on a minimum tiling path by the random shotgun strategy. The minimum tiling path is indicated by the darker clones. **(c)** Assemble the cloned shotgun sequences to reconstruct the sequence of the genome.

depends on the construction of only one BAC and two plasmid libraries. It overcomes the problem posed by repeat sequences scattered throughout the genome because paired end sequences from clones of three insert sizes, 2 kb, 10 kb, and 200 kb, make it possible to bridge most lengths of repetitive sequences. It relies on only a single highly automated and very mature technology—DNA sequencing. The Celera effort did in the end incorporate data from the public effort into its own data. Since then, however, Celera has sequenced and assembled a draft mouse genome sequence using the whole-genome shotgun method without the use of data from other linkage, physical, or sequence maps.

Two limitations for all large-scale sequencing efforts are the fact that some genomic sequences cannot be cloned (for example, sequences in the darkly staining, more condensed *heterochromatin* described on pp. 429–432 in Chapter 12), and some sequences rearrange or sustain deletions when cloned (for example, some tandemly arrayed repeats).

In the future, large genome sequencing efforts will probably use a combination of hierarchical and whole-genome shot-

gun strategies. The public effort in sequencing the mouse genome is pioneering such a two-prong approach.

The Sequencing of the Human Genome

Most of the sequencing of the draft of the human genome took place in the last year or so of the project. There were three reasons for this. First, DNA sequencing instruments had evolved to run 96 samples (in 96 capillaries) simultaneously and to complete 6 runs in 24 hours. In theory, each instrument could thus sequence 345,600 bp per day (600 bp/capillary × 96 capillaries × 6 runs/day). Second, an automated factory-like production line generated sufficient DNA fragments to supply the high-throughput sequencers on a daily basis. Finally, several very large sequencing centers emerged with 100–300 sequencing instruments a piece. The theoretical throughput of a 300-instrument center is 103,680,000 bp per day (345,600 × 300). Hence, such a center, in principle, could provide a 10-fold sequence coverage of the human genome in about 30 days (3×10^9 bp in the human genome $\times 10/1.037 \times 10^8$). In practice, the throughput of sequencing machines is considerably lower—perhaps 10-fold or more—

Figure 10.13 Hypothetical whole-genome shotgun sequencing strategy. Only three libraries of differing-sized fragments, here 2 kb, 10 kb, and 200 kb, need to be constructed. The big challenge is to assemble these sequences in the presence of large numbers of repeats.

because of limitations on providing adequate insert DNA for sequencing, experimental errors, machine failures, poor-quality data, and failures in data capture.

The Integration of Linkage, Physical, and Sequence Maps

Linkage, physical, and sequence maps all locate sequence loci along the chromosomes of the genome. The SSR and SNP DNA markers are both amenable to PCR amplification. Thus they can be readily integrated into the physical map by PCR analysis across all of the insert clones of the physical map. Hence, the linkage map can be fully integrated into the physical map. The physical map employs PCR-based STS markers. The SSR, SNP, and STS markers, all amplifiable by PCR, have unique primer-detected sequences of 20 bp or more that can readily be placed on the sequence map—indeed, the orientation and spacing of the paired PCR primer sequences provides a dual check that the marker assignments for the PCR primers are correct. The inte-

gration of the linkage, physical, and sequence maps provides a check on the correct order of each map against the other two.

Finding Genes in a Sequenced Genome

There are many approaches to the identification of genes in chromosomal sequences. First, computational similarity analyses can match EST (expressed sequence tag) or cDNA sequences against their chromosomal counterparts. Recall that such sequences represent transcribed genes. Researchers can use a full-length cDNA sequence to delineate the exons and introns of a gene within the genome sequence. Researchers can also use sufficiently similar EST or cDNA sequences from other animals to identify exons. The similarity programs have associated statistical analyses that determine the significance of various levels of sequence match.

Second, relatively successful computational gene-finding programs are available. These programs analyze the specific features of genes (start and termination sites, intron/exon junctions). They also take advantage of the fact that different animals (and plants and microbes) show individual preferences in the codons used for particular amino acids; these codon preferences can be integrated into the gene-finding programs. The gene-finding programs are 70% to 90% successful in identifying some but not necessarily all the genes or exons in an extended sequence.

Finally, researchers can compare complete genome sequences from appropriately diverged organisms. For the comparison to work, the organisms' genomes should have diverged to the point that the nonfunctional regions of the genome are significantly different while the functional regions remain relatively conserved. Functional chromosomal regions include exons and transcription factor binding sites (in control regions, discussed in detail on pp. 592–594 in Chapter 17). In comparisons of the orthologous chromosomal sequences of two organisms, conserved blocks 25 bp to several kilobases in length are automatically candidates for exons or control regions. Comparative genomic analysis is particularly powerful when many of the genes have been defined in one of the two organisms.

Researchers can determine whether hypothetical genes are real genes by scanning across many different cell and tissue types with appropriate PCR primers to identify the genes' expressed mRNA.

The Human Genome Project Has Changed the Practice of Biology, Genetics, and Genomics

The draft of the human genome sequence, in conjunction with the complete genome sequences of model organisms, has profoundly changed biology, genetics, and genomics. The complete genome sequences have provided a genetic parts list for each organism through the identification of most of its genes as well as the adjacent sequences controlling gene expression.

The genome sequencing achievements to date will expedite future gene-finding and gene-function analyses in several ways. First, the identification of genes in one organism's genome facilitates identification by sequence homology of genes in a second organism. If the second organism's genome sequence is available,

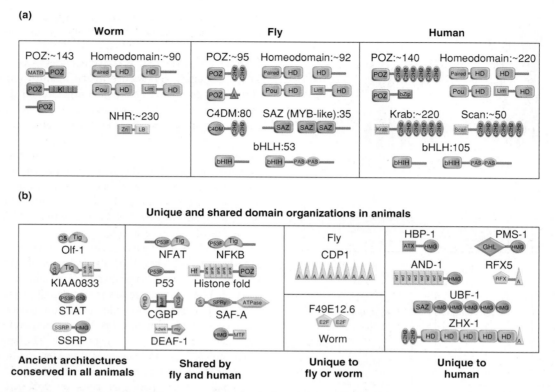

Figure 10.14 How the domains and architectures of transcription factors have expanded in specific lineages. (a) Specific families of transcription factors have expanded in the worm (nematode), fly (*Drosophila*), and human proteomes. The proteome is the collection of all proteins present in an organism or individual cell type. The diagram shows the approximate numbers of each domain identified in each of the three species. Some domains appear in different animal lineages; others are lineage-specific. **(b)** Samples of transcription factor architectures found: in all animals (ancient architectures), in only fruit flies and humans, and uniquely in one lineage. The capital-letter designations indicate something about the different domain structures or functions. For example, HD indicates a <u>ho</u>meo<u>d</u>omain that can bind to the 3' end of a gene's DNA and help regulate transcription.

its genes can be identified by a computer search. If the second organism's genome is not available and the two genomes are sufficiently similar, PCR primers from the first organism's genome (if sufficiently similar to the second organism's sequences) can be used to identify genes in the second organism's genome.

Second, the comparisons of all the genes within and between organisms has shown that many of the genes are *paralogous* or *orthologous* to one another (review definitions on p. 327 of this chapter). Hence, knowledge of how a gene functions in one organism may help us understand its function in a second organism.

Third, the complete genome sequences have revealed that genes often encode one or more **protein domains,** or discrete functional units. The gene regions that encode protein domains are like the cars of a train. Each train is composed of many different cars, and each kind of car (engine, flat car, dining car, caboose) has a discrete function. Different trains may carry different combinations of cars and thus execute different functions. Similarly, many genes are composed of distinct regions that encode discrete protein domains. Each gene may express different domain-encoding regions by the shuffling, addition, or deletion of exons. Through these genetic mechanisms, protein architecture, defined by combinations of domains, may change with evolution. Figure 10.14, which shows a series of domains

associated with different *transcription factors,* illustrates the idea of different domains and the shifting of domain architecture across the species. Recall from p. 249 in Chapter 8 that transcription factors are proteins that bind to DNA promoters and other control regions; the DNA binding of transcription factors influences the temporal and spatial patterns of gene expression as well as its amplitude. Differing domains enable transcription factors to bind with different DNA sequences and also to interact uniquely with specific protein cofactors. Just as the makeup of a train's cars provides insights into what the train does, so too the specific gene-encoded domains of a protein are associated with particular functions and provide insights into what the protein does. Biologists may guess at the function of a new protein (or the gene that encodes it) by analogy, after searching for the protein's sequence in a database of all known domains.

Fourth, the genome sequence has established ready access to the identification of human polymorphisms, and the genomewide genetic map (using SSRs and SNPs as markers) has made it possible to correlate some of the many DNA polymorphisms and their associated genes with predispositions to disease and interesting aspects of physiology.

Finally, comparisons between sequenced genomes will more readily allow researchers to assemble the sequence frag-

Human chromosomes

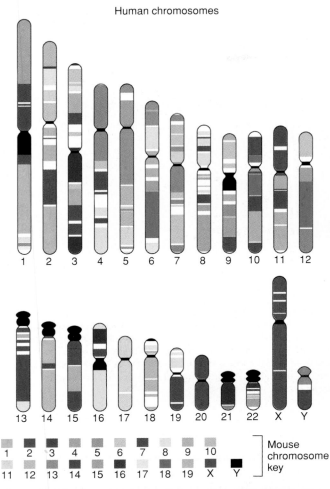

Figure 10.15 Conserved segments or syntenic blocks in the human and mouse genomes. Human chromosomes, with segments containing at least two genes whose order is conserved in the mouse genome; the segments appear as *color blocks.* Each color corresponds to a particular mouse chromosome. Centromeres; subcentromeric heterochromatin of chromosomes 1, 9, and 16; and the repetitive short arms of 13, 14, 15, 21, and 22 are in *black.*

ments of an appropriately related, newly sequenced genome into its chromosomal strings. For example, the mouse and human genomes, which diverged 85 million years ago, exhibit striking similarities in genes as well as their orthologous order. Indeed, the mouse and human genomes exhibit approximately 180 homologous blocks of chromosomal sequence, ranging in size from 24 kb to 90.5 Mb—for an average of 17.6 Mb (Fig. 10.15). Such blocks of linked loci are called **syntenic blocks.** The orders of these blocks, as determined from their chromosomal linkage relationships, are totally different in the two organisms. It is as if one genome had been cut into 180 pieces of varying size and then randomly assembled into the other genome. Within each of the human syntenic blocks is information useful for assembling the corresponding homologous mouse sequences and vice versa. *Conserved synteny,* in which the same two or more loci are linked in different species, also exists between the human and puffer fish genomes, which diverged more than 400 million

years ago. In this case, though, the syntenic blocks are relatively small—averaging about 10 kb in length.

Thus, as these few examples show, successful sequence analysis of the human and model organism genomes has changed the strategies for biology, genomics, and medicine.

Major Insights from the Human and Model Organism Genome Sequences

In addition to promoting new analytical approaches, the draft sequence of the human genome has provided striking new insights into the architecture, gene organization, structural elements, and the evolution of chromosomes. We now describe some of the main lessons and surprises.

There Are Approximately 40,000 Human Genes

The first surprise to emerge from the human genome sequence was the discovery of just 30,000 genes, a much lower number than expected. A back-of-the-envelope calculation (with no experimental support) done at the initiation of the Human Genome Project had suggested that there are perhaps 100,000 human genes—approximately 1 gene every 30 kb.

For a variety of reasons, it appears likely that the final gene number will fall around 40,000. First, not all of the genome has been accurately sequenced; hence, some genes have yet to be counted. Second, small genes are difficult to identify. For example, genomic researchers have recently discovered many small functional RNAs (about 25 bp long) in a variety of organisms. These small RNAs, which appear to play a regulatory role in controlling the expression levels of complementary mRNA, could number in the thousands. In addition, there are many other types of genes expressed only at the RNA level that are difficult to identify; and geneticists estimate that there are hundreds to thousands of additional genes encoding small peptides that are also difficult to identify. Finally, some genes are rarely expressed and do not reflect normal codon usage patterns. Such genes would not be detected by ESTs or common gene-finding programs.

Genes Encode Either Noncoding RNAs or Proteins

Noncoding RNA Genes

The noncoding RNAs, probably representing more than 2000 genes, fall into four main categories:

1. transfer RNAs (tRNAs), which are the adaptors that translate the triplet code of RNA into the amino-acid sequence of proteins;

2. ribosomal RNAs (rRNAs), which constitute fundamental components of the ribosome—the protein-synthesizing molecular machine;

3. small nucleolar RNAs (snoRNAs), which are required for rRNA processing and base modification in the nucleolus (an area of the genome that is rich in RNA);

Figure 10.16 **Examples of domain accretion in chromatin proteins.** Domain accretions in various lineages are shown using schematic representations of domain architectures (not to scale). *Asterisks* indicate the mobile domains that have participated in the accretion. Species in which a domain architecture has been identified are indicated *above* the diagrams (Y, yeast; W, worm; F, fly; V, vertebrate). Protein names are *below* the diagrams. The domains are SET, a chromatin protein methyltransferase domain; SW12, a superfamily II helicase/ATPase domain; Sa, sant domain; Br, bromo domain; Ch, chromodomain; C, a cysteine triad motif associated with the Msl-2 and SET domains; a, AT hook motif; EP1/EP2, enhancer of polycomb domains 1 and 2; Znf, zinc finger; sja, SET-JOR-associated domain; Me, DNA methylase/Hrx-associated DNA-binding zinc finger; Ba, bromo-associated homology motif. **(a–c)** Different examples of accretion.

4. small nuclear RNAs (snRNAs), which are critical components of spliceosomes (the molecular machines that splice out introns from pre-mRNAs in the nucleus).

There are also many additional noncoding RNAs expressed at lower levels.

Protein-Coding Genes Generate the Proteome

From the draft sequence of the human genome, investigators predicted more than 30,000 protein-coding genes. The collective translation of the 30,000 predicted genes into proteins is termed the **proteome.** The human proteome contains *homologs*—in this context, proteins related by evolutionary descent—for 61% of the fruit fly proteome, 43% of the worm proteome, and 46% of the yeast proteome. Proteomic analyses have revealed that there are about 1200 gene families containing two or more members, 92 (7%) of which are vertebrate-specific; these vertebrate-specific families encode proteins active in immunity, other types of defense, and the nervous system. Analysts have also shown that there has been a striking increase in the complexity of the proteome from yeast to humans. This increase in complexity has come about in five ways:

1. *More genes.* The human genome has more genes than the genomes of the simpler model organisms (see Table 10.1).

2. *The shuffling, increase, or decrease of functional modules.* Many proteins contain discrete functional domains that, as described earlier, can be joined in many different orders by addition, deletion, or shuffling (Fig. 10.16). The number and order of a protein's domains constitute its **domain architecture.** Although the human organism has only a modest increase in the total number of protein domains over the model organisms, it has evolved many new gene arrangements that alter domain architecture (Fig. 10.17).

3. *More paralogs.* Many human gene families have significantly expanded their numbers of paralogs (Table 10.2). For example, humans have about 1000 olfactory genes whereas fish have perhaps 60.

4. *Alternative RNA splicing.* Different pathways of RNA splicing may significantly expand the numbers of proteins. For example, the three human neurexin genes, which encode molecules that guide the appropriate connections of neurons, or nerve cells, may collectively generate more than 2000 splice variants. Humans exhibit significantly more alternative splicing than their model organism counterparts. Key questions include how many of the splice variants encode distinct functional proteins (rather than proteins with the same function) and whether

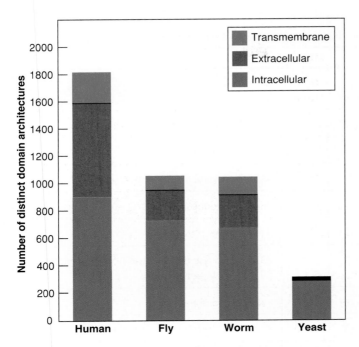

Figure 10.17 Number of distinct domain architectures in the four eukaryotic genomes. The number of architectures is split into three cellular environments: intracellular, extracellular, and transmembrane.

different variants represent different addresses for telling neurons where to go.

5. *Chemical modification of proteins.* Human proteins may be modified by more than 400 different chemical reactions, each capable of altering the proteins' functions. Thus, the typical human cell might have perhaps 20,000 different types of mRNAs and perhaps 200,000 different proteins. Humans can make more protein modifications than their simple model organism counterparts.

By all of these criteria, the human proteome is considerably more complex than its counterparts in the fly, worm, or yeast.

Repeat Sequences Constitute More Than 50% of the Human Genome

For the purposes of genomic analysis, repeat sequences fall into five classes: (1) transposon-derived repeats scattered across the genome, which constitute more than 45% of the genome; (2) inactive copies of protein-coding genes and small structural RNAs known as *processed pseudogenes* (because they lack introns as well as regulatory sequences); processed pseudogenes arise by retroposition, or the copying of RNA into DNA, followed by the random integration of this cDNA into chromosomes; (3) simple sequence repeats of short multimers two to

five nucleotides in length, such as (AA)n, (AT)n, or (GCC)n (which make up about 3% of the genome); (4) segmental duplications of 10–300 kb copied from one region to another region of the genome (more than 5% of the genome); and (5) blocks of repeated sequences at the centromeres, telomeres, and other chromosomal features.

As mentioned on p. 194 of Chapter 7, *transposons* are units of DNA that move from place to place in the genome. Transposon-derived repeats, which we have seen constitute more than 45% of human DNA, fall into four classes: long interspersed elements (LINEs), short interspersed elements (SINEs), long terminal repeat retroposons, and DNA transposons. The first three transpose through RNA intermediates and the fourth transposes directly as DNA. Figure 10.18 shows the general properties of these repeats. Some transposon-derived repeats have all the machinery required for independent transposition and are termed **autonomous;** others are nonautonomous—they employ the transposition machinery of the autonomous elements. Transposon-derived repeats, which were once considered *junk DNA,* have given rise to at least 47 different human genes. Moreover, they constitute a rich paleontological record of evolutionary events such as mutation and selection because researchers can recognize cohorts of repeats born at the same time and follow their fates in the genomes of different species. Finally, transposon-driven repeats have reshaped the genome by facilitating chromosomal rearrangements and translocations. (See pp. 461–464 of Chapter 13 and pp. 715–717 of Chapter 21 for a more detailed discussion of transposons and other types of repeats.)

The Genome Contains Distinct Types of Gene Organization

The organization of genes is strikingly different across the genome. Examples of different types of gene organization include gene families, gene-rich regions, and gene-poor deserts. A fundamental question is whether these different organizations have biological meaning.

Gene Families

Closely related genes that are members of multigene families abound throughout the genome. Such families may be clustered together on one chromosome or dispersed on several chromosomes. Examples of gene families include the genes that encode histones, hemoglobins, immunoglobins, actins, collagens, and heat-shock proteins. We described the hemoglobin-encoding globin gene family on pp. 310–311 of Chapter 9. Here we take a brief look at the human olfactory multigene family, which has about 1000 members. The olfactory genes, which encode olfactory receptor molecules, evolved in two ways. In the first, one gene underwent multiple duplications to create roughly 20 copies, or a family of approximately 20 paralogs. These family members, which were originally identical, then diverged to paralogs that were quite distinct from one another. Next, a massive

TABLE 10.2 The Most Populous Gene Families in the Human Proteome and Other Species

Human		Fly		Worm		Yeast		Mustard weed		
No. of Genes	Rank	No. of Genes	Rank	No. of Genes	Rank	No. of Genes	Rank	No. of Genes	Rank	
765	(1)	140	(9)	64	(34)	0	(na)	0	(na)	Immunoglobulin domain
706	(2)	357	(1)	151	(10)	48	(7)	115	(20)	C2H2 zinc finger
575	(3)	319	(2)	437	(2)	121	(1)	1049	(1)	Eukaryotic protein kinase
569	(4)	97	(14)	358	(3)	0	(na)	16	(84)	Rhodopsin-like GPCR superfamily
433	(5)	198	(4)	183	(7)	97	(2)	331	(5)	P-loop motif
350	(6)	10	(65)	50	(41)	6	(36)	80	(35)	Reverse transcriptase (RNA-dependent DNA polymerase)
300	(7)	157	(6)	96	(21)	54	(6)	255	(8)	rrm domain
277	(8)	162	(5)	102	(19)	91	(3)	210	(10)	G-protein β WD-40 repeats
276	(9)	105	(13)	107	(17)	19	(23)	120	(18)	Ankyrin repeat
267	(10)	148	(7)	109	(15)	9	(33)	118	(19)	Homeobox domain
252	(11)	77	(22)	71	(31)	27	(17)	27	(73)	PH domain
242	(12)	111	(12)	81	(25)	15	(27)	167	(12)	EF-hand family
222	(13)	81	(20)	113	(14)	0	(na)	17	(83)	EGF-like domain
215	(14)	72	(23)	62	(35)	25	(18)	3	(97)	SH3 domain
210	(15)	114	(11)	126	(12)	85	(12)	379	(4)	RING finger
188	(16)	115	(10)	54	(38)	7	(35)	392	(2)	Leucine-rich repeat
171	(17)	0	(na)	0	(na)	0	(na)	0	(na)	KRAB box
165	(18)	63	(27)	51	(40)	2	(40)	4	(96)	Fibronectin type III domain
162	(19)	70	(24)	66	(33)	2	(40)	15	(85)	PDZ domain
155	(20)	87	(17)	78	(27)	79	(4)	148	(13)	Helicase C-terminal domain

TABLE 10.2   (Continued)

No. of Genes	Rank	No. of Genes	Rank	No. of Genes	Rank	No. of Genes	Rank	No. of Genes	Rank	
Human		**Fly**		**Worm**		**Yeast**		**Mustard weed**		
150	(21)	86	(18)	46	(43)	36	(11)	125	(17)	TPR repeat
133	(22)	65	(26)	99	(20)	2	(40)	31	(69)	Ion transport protein
131	(23)	84	(19)	41	(46)	7	(35)	106	(24)	Helix–loop–helix DNA-binding domain
123	(24)	43	(34)	36	(49)	9	(33)	82	(34)	C2 domain
119	(25)	4	(71)	22	(63)	1	(41)	2	(98)	SH2 domain
118	(26)	210	(3)	12	(73)	1	(41)	15	(85)	Serine protease, trypsin family
114	(27)	19	(56)	16	(69)	0	(na)	0	(na)	Cadherin domain
113	(28)	78	(21)	117	(13)	1	(41)	54	(50)	BTB/POZ domain
112	(29)	35	(40)	108	(16)	12	(30)	21	(79)	Tyrosine-specific protein phosphatase and dual specificity protein phosphatase family
106	(30)	18	(57)	169	(9)	0	(na)	5	(95)	Collagen triple helix repeat
94	(31)	141	(8)	134	(11)	40	(10)	194	(11)	Esterase/lipase/thioesterase
89	(32)	38	(38)	18	(67)	8	(34)	18	(82)	HMG1/2 (high mobility group) box
87	(33)	56	(29)	92	(22)	8	(34)	12	(88)	Neutral zinc metallopeptidase
84	(34)	37	(39)	24	(61)	16	(26)	71	(39)	PHD-finger
83	(35)	32	(43)	24	(61)	1	(41)	14	(86)	Cation channels (non-ligand gated)
81	(36)	38	(38)	36	(49)	4	(38)	8	(92)	LIM domain
81	(36)	8	(67)	45	(44)	3	(39)	17	(83)	VWA domain
80	(37)	33	(42)	30	(55)	3	(39)	18	(82)	Calponin homology domain
77	(38)	20	(55)	20	(65)	10	(32)	9	(91)	RhoGAP domain

Forty most populous gene families found in the human proteome compared with equivalent numbers from other species. na, not applicable (used when there are no proteins in an organism in that family).

Figure 10.18 Almost all transposable elements in mammals fall into one of four classes of interspersed repeats (as described in the text).

duplication event created 30 different families from the original 20-paralog family. In this massive event, the entire family, or portions of it, duplicated and translocated to about 30 sites, or clusters, around the genome. Each of the 30 sites contains all or a portion of the 20 or so paralogs of the original family (Fig. 10.19). Interestingly, the equivalent paralogs in each of the 30 clusters, for example paralog 3 at all 30 sites, are more closely related to each other than to other paralogs, say 2 or 7, in the same cluster. In the second mode of evolution, a single olfactory gene recently duplicated to generate a gene family of mostly similar paralogs. This is the class I family depicted in Fig. 10.19.

Perhaps the expansion of olfactory receptor genes reflects an evolutionary response to the selective pressures for a more acute and discriminating sense of smell in some mammals. Clearly, gene families have a variety of mechanisms for generating paralogs. About half of the human olfactory receptor genes are pseudogenes. Why some gene families remain as single clusters and why others are scattered across the genome is a fascinating question. Another is, Why are there so many pseudogenes in some gene families? We discuss the evolution of multigene families in greater detail on pp. 718–722 of Chapter 21.

Gene-Rich Regions

Some chromosomal regions are densely packed with genes. For example, the 700 kb class III region of the major histocompatibility complex on chromosome 6 contains 60 genes encoding many diverse functions (Fig. 10.20). Only one is a pseudogene. This is the most gene-rich region of the human genome; 70% of the DNA in this region is transcribed. Moreover, the region has a high G C content, 54% versus a genomewide average of 45%. High G C content is also seen in other gene-rich areas. Why are these genes so densely packed? Is there a functional explanation or is it a reflection of the chance events that shape chromosomal architecture?

Gene Deserts

The current human genome draft contains 82 gene deserts, which are chromosomal regions containing a megabase or more of DNA with no identifiable genes. Deserts span 144 Mb, or 3% of the genome. The largest desert is 4.1 Mb in length.

One explanation for the existence of gene deserts is that they contain genes that are difficult to identify. An example is the class of genes called *big genes*. A big gene is a single gene whose nuclear transcript spans 500 kb or more of chromosomal DNA. The largest of the big genes is the gene for dystrophin, which spans 2.3 Mb. One hundred twenty-four big genes encompassing 112 Mb of chromosomal DNA have been defined to date. Interestingly, many big genes have modest-sized mRNAs; the exons encoding these RNAs typically encompass about 1% of the total chromosomal gene region in which they occur. This means the exons are widely scattered across large regions composed mainly of introns and, hence, are extremely difficult to find by computational methods alone. Full-length cDNA sequences are often essential for the delineation of exon/intron structures in big genes. But because big genes are synthesized very slowly, their synthesis cannot be completed in rapidly dividing cells. It is in neurons, which do not divide, that a large fraction of the big genes are expressed.

Once again, several intriguing questions arise: Do gene-rich or gene-poor areas have functional significance or do these different chromosomal organizations represent random fluctuations in the evolutionary events mediating chromosomal evolution and organization? Similarly, do big genes provide some selective advantage, possibly functional and/or regulatory, or does their existence reflect chance events that reshaped the chromosomal architecture at the genomic level?

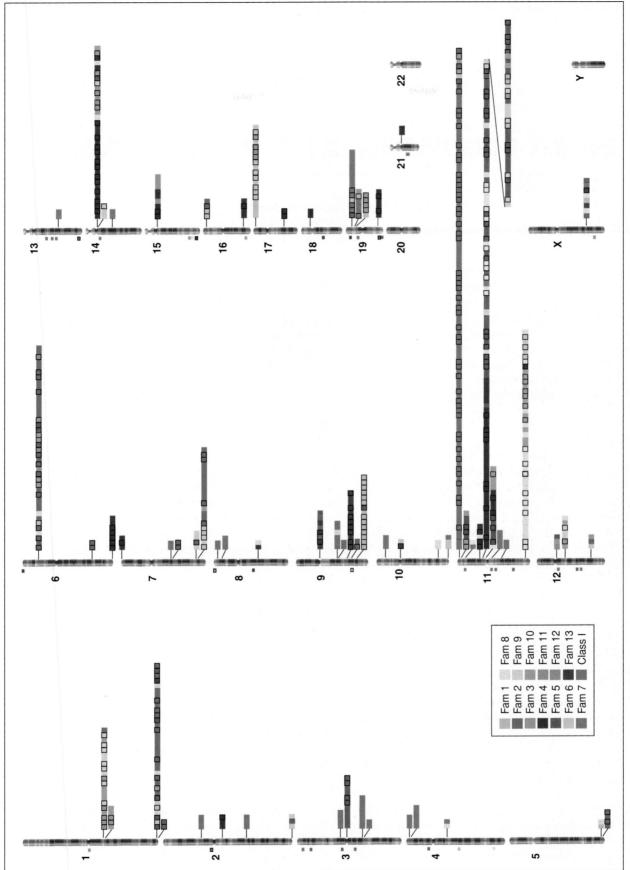

Figure 10.19 Olfactory receptor (OR) gene families. ORs are depicted as *squares*, colored by family (*see key*). All class I families are colored equally. Unclassified ORs are indicated in *light gray*. *Framed squares* denote intact or functional genes. ORs to the *left* of each chromosome indicate single genes, and those to the *right* are in clusters of two or more. The largest cluster on chromosome 11 is shown split for convenience.

Figure 10.20 Class III region of the human major histocompatibility complex. This region contains 60 genes and is 700 kb in length.

Combinatorial Strategies May Amplify Genetic Information and Generate Diversity

Combinatorial amplification results from the potential for combining a set of basic elements in many different ways. A simple slot machine, for example, may contain 3 wheels, each carrying 7 different symbols; from its 21 basic elements (7 + 7 + 7), it can generate 343 different combinations (7^3, or $7 \times 7 \times 7$). In biology, combinatorial amplification occurs at both the DNA and RNA levels.

At the DNA Level

Antibody and T-cell receptor genes, for example, are encoded by a multiplicity of gene segments (Fig. 10.21). The human T-cell receptor family has 45 functional variable (V) gene segments, two functional diversity (D) gene segments, and 11 functional joining (J) gene segments. In an individual T cell, any D element may first join to any J element by deletion of the intervening DNA. This joined DJ element may, in turn, join to any V element—once again by deletion of the intervening DNA—to generate a complete V-D-J gene. This combinatorial process can generate 990 different V-D-J genes ($45 \times 2 \times 11 = 990$), although in a given T cell, only one such functional rearrangement occurs. Thus, from 58 gene elements (45 + 2 + 11), a combinatorial joining mechanism can generate 990 V-D-J genes.

T cells, whose receptors are capable of interacting with foreign molecular structures (called antigens), are driven by contact with such antigens to divide and expand their numbers 1000-fold or more. This antigen-triggered expansion by mitosis to a clone of genetically identical cells descended from a common ancestor amplifies the useful combinatorial information and is a key part of every immune response. The particular combinatorial gene arrangements in a few of the original population of T cells produce T-cell receptors more precisely fitted to the antigen. These receptors will then bind the antigen, and such binding will trigger the clonal expansion of the cells that carry the tightly fitting receptors; as a result, the specificity and strength of the immune response will increase.

Figure 10.21 A schematic diagram of the human β T-cell receptor gene family. The variable (antigen recognition) gene is composed of three elements: V, D, and J. During T-cell development any D may join with any J. Any V may join with any DJ.

At the RNA Level

The splicing together of RNA exons in different orders is another way in which combinatorial strategies can increase information and generate diversity. Further diversity results from the initiation of transcription at distinct promoter regions, which create transcripts with different numbers of exons. The three neurexin genes illustrate both of these combinatorial RNA strategies (Fig. 10.22). Each neurexin gene contains two promoter regions (producing α and β mRNAs) and five sites at which alternative splicing can occur. Together, these three genes can probably generate more than 2000 alternatively spliced forms of mRNA. A key question is, do these forms have distinct functions and if so, how many? Or are they simply the random results of the splicing process? Certainly alternative mRNA splicing will generate distinct functional gene products for some genes, but we have yet to learn the extent to which a single gene may generate functionally distinct alternative mRNA forms.

Evolution May Occur by the Lateral Transfer of Genes from One Organism to Another

One of the fundamental dictates of evolutionary theory has been that evolution occurs in a vertical manner; that is, evolutionary changes occur over time as parents pass their rearranged and mutated genomes to their progeny. However, initial analyses of a draft of the human genome have suggested that more than 200 human genes may have arisen by direct, or *lateral, transfer* from other organisms such as bacteria. **Lateral gene transfer** is the direct transfer of genes from one species into the germ line of another species. Despite some dispute over whether all of the 200 human genes mentioned really represent lateral gene transfers, it appears that lateral gene transfer can and does occur, al-β-neurexin 3

α-neurexin 3

Neurexin statistics			
Gene	Length in human	Length in pufferfish	Number of potential alternative splice variants in human
NRXN1	1112 kb	>163 kb	292
NRXN2	117 kb	unknown	194
NRXN3	1692 kb	>181 kb	1764

Figure 10.22 The organization of the three neurexin genes. Each gene has two promoters (α and β) to initiate mRNA synthesis and five sites at which alternative RNA splicing can occur. The *pink rectangles* indicate exons affected by alternative splicing. Numbers at the top of the figure designate exons.

though we do not yet understand how. Knowledge of the potential for lateral transfers opens up the possibility that one organism may evolve by using the genes of another organism. This possibility is obviously limited to the exchange of genetic information between organisms that have access to one another's DNA (for example, bacteria in the human intestines).

The possibility of lateral gene transfer suggests that, in principle, the evolutionary potential of one species is enormously enhanced by access to successful evolutionary strategies of other species. It also adds fuel to the debate over genetic engineering, providing a counterpoint to the argument that humans are genetically unique.

Males Appear to Have More Than a Twofold Increased Rate of Mutation in Meiosis over Females

Comparisons of orthologous repeat sequences on the X and Y chromosomes have shown that the rate of mutation in meiosis in males is more than twice as high as the rate in females. These comparisons take into account the fact all Y chromosomes reside in males and X chromosomes spend twice as much time in females as in males. Presumably, the same differential mutation mechanism operates on the autosomes, although it is not possible to measure its operation in the same way there. The tentative conclusion is that the majority of all human mutations probably occurs in males. Since these mutations can generate defects as well as diversity, males may give rise to more defects but also more of the diversity that is the basis of evolution.

The Different Human Races Appear to Have Very Few Uniquely Distinguishing Genes

The human population encompasses a number of different races—Caucasian, Mongoloid, Negroid (or Black), American Indian, and so forth—each of which is distinguishable by certain phenotypic features such as skin color, skin folds at the corners of the eyes, etc. Genome centers in Europe, Japan, China, and North America have sequenced significant portions of the genome from at least three different races. From these sequences, it appears that the range of polymorphisms within a particular race (for example, Black) can be much greater than the range between any two individuals in different races. Indeed, it is clear that very few genes will turn out to be race-specific. This underscores the basic genetic similarity of all humans. Genetically, humans are a single race.

The Sequences of the Human and Model Organism Genomes Reaffirm That All Living Organisms Evolved from a Common Ancestor

The availability of complete genomic sequences has created an opportunity to analyze large numbers of gene products (mRNAs, proteins) in the context of specific cell types or in the context of the changing patterns of gene expression during the activation of developmental or physiological pathways.

Preliminary analyses support the conclusion of earlier molecular studies that the basic cellular mechanisms of all living organisms (for example, cell cycle, fundamental features of metabolism) have remarkably similar genetic components. This observation, in turn, supports the idea that life evolved once, and we are all descendants of that fortuitous event. The similarity of basic genetic components also reaffirms that the analysis of appropriate biological systems in model organisms can provide fundamental insights into how the corresponding human systems function.

In the Future, Other Features of Chromosomes Will Become Increasingly Important

An understanding of several as yet little understood chromosomal features will become important to study in a global manner as we move into the postgenomic (that is, the post-genome-sequencing) era and develop more effective tools for their characterization. One significant feature is the *chemical modification of bases*. A common modification of DNA is the methylation of Cs in the promoter regions of genes. This post-DNA-synthesis event may suppress the readout of individual genes and thereby alter the expression of developmental programs or physiological responses. Other base modifications may also be important.

A second feature is the *interaction of various proteins with the chromosomes*. These interactions can modify the three-dimensional structure of chromosomes and, in so doing, alter their ability to express genes in some regions. For example, the histone proteins catalyze the formation of DNA packaging units known as nucleosomes. Specific DNA-histone interactions thus provide chromosomes with a higher-order structure that allows them to compact; and more compact chromosomal structures are generally incapable of expressing genes (see pp. 418–424 of Chapter 12 for a more thorough discussion of chromosome structure).

A third feature is the *three-dimensional structure of chromosomes in the nucleus.* Such three-dimensional structure may help determine the interactions of specific chromosomal regions with particular regions of the nuclear envelope. The three-dimensional structures of chromosomes are also likely to influence the readouts of the developmental and physiological programs encoded in the genome.

We need to develop far more effective tools for looking at chromosomal chemical modifications, chromosomal interactions with proteins and other macromolecules, and the formation and influence of three-dimensional chromosomal structures.

Table 10.3 lists a variety of web sites where you can directly view detailed information on some of the preceding in-

sights into genetics and biology as well as data about other aspects of the Human Genome Project.

High-Throughput Genomic and Proteomic Platforms Permit the Global Analysis of Gene Products

The Human Genome Project has given us the complete sequences of a variety of genomes. The next step is to carry out global analyses of this basic data to gain a more complete understanding of how cells and larger biological systems use genetic information to create a whole organism.

The need for global analysis has driven the development of powerful high-throughput platforms that enable researchers to generate the information necessary to carry out large-scale, automated searches of global databases. In this context, a *platform* denotes all the components needed for an automated acquisition of a set of data. A high-throughput genomic platform might include automated instrumentation to prepare DNA fragments for sequencing, an automated sequencer to carry out the sequencing, and a set of computational tools to capture and store the information obtained from sequencing.

We now describe some of the main instruments genomicists have developed for large-scale analyses. In our discussion of these tools, we use *genomics* to mean the global analysis of chromosomal and RNA features (such as sequence, genetic markers, mRNAs) and *proteomics* to mean the global analysis of complex protein mixtures.

TABLE 10.3 **Sources of Publicly Available Sequence Data and Other Relevant Genomic Information**

http://genome ucsc.edu/
University of Santa Cruz
Contains the updated assembly of the human genome sequence.

http://www.ensembl.org
EBI/Sanger Centre
Allows access to DNA and protein sequences with automatic baseline annotation.

http://www.ncbi.nlm.nih.gov/genome/guide/
NCBI
Views of chromosomes and maps and loci with links to other NCBI resources

http://compbio.ornl.gov/channel/index.html
Oak Ridge National Laboratory
Provides Java viewers for human genome data.

http://hgrep.ims.u-tokyo.ac.jp/
RIKEN and the University of Tokyo
Gives an overview of the entire human genome structure.

http://www.ncbi.nlm.nih.gov/Omim/
Oneline Mendelian Inheritance in Man
Contains information about human genes and disease.

http://www.nhgri.nih.gov/ELSI/and
http://www.ornl.gov/hgmis/elsi/elsi.html
NHGRI and DOE
Contains information, links, and articles on a wide range of social, ethical, and legal issues.

High-Throughput Instruments

Three high-throughput technologies have played a key role in genomic and proteomic platforms: the DNA sequencer, DNA arrays, and the mass spectrometer.

The DNA Sequencer

The automated DNA sequencer employs the dideoxy (Sanger) sequencing strategy described in detail on pp. 303–306 of Chapter 9. The automated format is a three-step process (Fig. 10.23), which we briefly summarize. Step 1: DNA polymerase synthesizes a complementary DNA strand from a primer sequence across the unknown DNA insert to be sequenced. The unknown sequences serve as the template strands for the primer-initiated synthesis of complementary strands. The four chain-terminating dideoxy bases are used in separate reactions at specific concentrations such that roughly 0.5% of the growing chains are stopped at each G, C, A, and T across the sequence. The primers for each of the four reactions are labeled with one of four fluorescent dyes. Hence, the A-terminating fragments are labeled with one dye and the T-, G-, and C-terminating fragments are each labeled with a second, third, and fourth dyes. Step 2: The four different sets of fragments are pooled and then separated by a type of electrophoresis that separates fragments differing by one base. Each of the separated fragments is color-coded with a distinct fluorescent dye, depending on the terminal base in the frag-

Figure 10.23 Sanger sequencing scheme. (a) Each reaction contains the unknown template DNA cloned in a single-strand sequencing vector, a primer oligonucleotide complementary to a region of the vector near the 3' end of the unknown DNA, a DNA polymerase, all four deoxy nucleotide triphosphates (dNTPs), and one dideoxynucleotide triphosphate (ddNTP). The detection label (·) is shown attached to the 5' end of the primer. **(b)** Schematic depictions of the gel electrophoresis detection system used in DNA sequencing. Real-time detection of bands from four sequencing reactions, each using a different-colored label, run in a single gel track. **(c)** Data from DNA sequencing experiments. Superimposed four-color traces obtained in a sequencing experiment using four-color DNA sequencing on a DNA sequencer. Data shown correspond to the same region of the vector as that in (a). **(d)** Commercial implementation employing fluorescence scanning of slab electrophoresis gel.

ment. Step 3: The specific fluorescence of each fragment identifies its terminal nucleotide. For example, all fragments ending in A will fluorescence red, while those ending in T will fluoresce yellow.

The sequencing instruments described earlier can analyze 96 or, more recently, 384 DNA fragments at a time. Moreover, several large sequencing centers have created sequencing production lines in which the production of DNA fragments for sequencing has been semiautomated, and the capture and quality assessment of the DNA sequence data have been fully automated. The most recent 96-capillary instruments can, in principle, sequence more than 1,500,000 bp in 24 hours.

DNA Arrays

A DNA array is a large set of DNA fragments displayed on a solid support. To make a DNA array, researchers either synthesize DNA fragments encoding portions of individual genes at se-

lected sites on a solid support or "spot" (that is, deposit) presynthesized DNA fragments at precise locations on the solid support (Fig. 10.24). They then hybridize a complex mixture of fluorescent or radio-labeled DNA or RNA fragments (for example, mRNAs or cDNAs) with the array of anchored fragments to analyze in a quantitative manner the levels of expression of individual genes represented on the array. In this manner, they can analyze tens of thousands of gene-expression patterns.

There are three types of DNA arrays.

Macroarrays. After the creation of a cDNA library, which is then copied on agarose and placed in microtiter wells, an arrayer robot can pick up DNA from each of 96 or 384 wells and spot it on the nylon membranes forming an array. The robot can individually spot up to 20,000 clones from a cDNA library on the nylon membranes from the microtiter wells. The complex cDNA mixture to be analyzed against the array is radio-labeled,

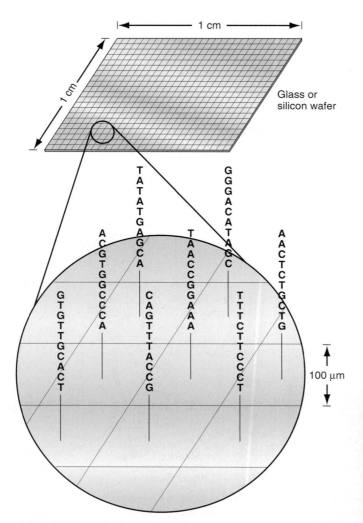

Figure 10.24   An oligonucleotide array. The oligonucleotides are covalently linked to the glass or silicon. Typically, the oligonucleotides range in size from 20–60 bp.

and any hybridization is detected by autoradiography or with a phosphoimager (an instrument for detecting radiolabels). These microtiter plate-based arrays are termed **macroarrays.** Macroarrays, which are relatively inexpensive, are useful for preliminary studies of gene expression in organisms or cell types where only a small fraction of the genes have been sequenced or characterized.

Microarrays.   PCR amplified gene fragments are spotted on glass slides. A single slide can hold roughly 5,000–40,000 PCR products. The mixture of mRNAs or cDNAs to be analyzed is labeled with fluorescent dyes and hybridized. The level of fluorescence in a single DNA spot is proportional to the level of gene expression. These small glass-slide arrays are termed **microarrays.** A researcher can use microarrays to simultaneously analyze two different cDNA (or RNA) samples taken from one kind of cell in two different states or from two cell types. To distinguish the samples, each is labeled with a different-colored dye (Fig. 10.25a). It is possible to measure the levels of the two dyes

(a)

(b)

Figure 10.25   Two-color DNA microarrays. (a) Two separate cDNA samples, one from normal yeast, the other from a mutated yeast, are labeled with red and green fluorescent dyes and hybridized to a PCR microarray. When both labeled messages hybridize to a PCR product, a *yellow* color is produced. The *red* hybridizations represent mRNAs (cDNAs) overexpressed in the mutant, and the *green* hybridizations represent mRNAs overexpressed in the normal yeast. **(b)** The entire yeast genome of ~6200 genes is represented on this DNA microarray. The array is hybridized with two different populations of labeled mRNA.

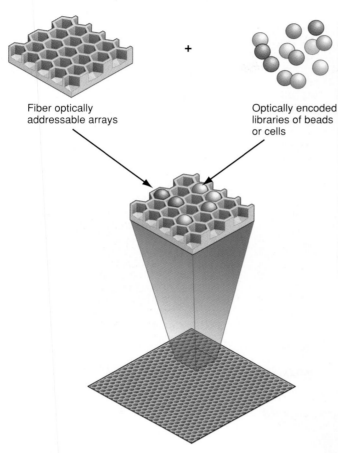

Figure 10.26   An optical fiber approach to DNA array analyses (see the text for details).

Figure 10.27   Mass spectrometer. The mass spectrometer consists of three components: (1) an ionization source, (2) a mass analyzer, and (3) a detector. The text explains the function of each component.

optical fiber 1 mm in diameter (Fig. 10.26). Fifty thousand beads, each with a different 25-mer, can be placed in these wells and used to interrogate a target sample, either for SNPs or patterns of gene expression. With this approach, it is easy to extend (scale up) the number of measurements that can be made. Since current instruments employ 96 optical fibers, they can carry out close to 5 million measurements per cycle of analysis (50,000, or $5 \times 10^4 \times 96 = 4,700,000$). The objective of all global technologies is high-throughput (the processing of a large amount of information) and low cost per unit of information analyzed.

The Mass Spectrometer

The mass spectrometer is an instrument that measures the masses of interesting molecules. For our purposes, such molecules include proteins, peptides, and oligonucleotides. The mass spectrometer requires that the molecules to be analyzed can be ionized and transferred to a vacuum. The spectrometer determines the masses by measuring the molecules' migration rates in an electric field (small masses migrate more rapidly than large masses). Mass spectrometers consist of three components: (1) a source of ionization, (2) a mass analyzer, and (3) a detector (Fig. 10.27). The ionization source ionizes the peptides or oligonucleotides; the mass analyzer separates the ionized fragments according to their mass-to-charge ratios (m/z); and the detector measures the separation times or distances, which researchers can use to determine the fragment mass.

Commercial mass spectrometers use different combinations of ionization sources and mass analyzers. These instruments can accurately measure the masses of peptides, small proteins, or oligonucleotides, and they sequence peptides by further fragmentation and mass measurement of the derivative peptides. An instrument known as the *ion trap tandem spectrometer* has the ability to separate different-sized fragments in the first mass spectrometer, select fragments for further fragmentation by collision with ions, and then measure the masses of the resulting peptide fragments in the second mass spectrometer. In this manner, it can sequence peptides (and oligonucleotides). The mass spectrometer is thus a powerful instrument for identifying and sequencing peptide fragments from proteins in complex mixtures and for SNP typing.

Global Genomic Strategies and High-Throughput Platforms

The instruments just described are the analytic centers of high-throughput platforms. As mentioned, such platforms include the

for each spot on the array to determine patterns of gene expression in the two different cell states or cell types (Fig. 10.25b). The differences in gene expression can be detected only if they vary by a factor of two or more. This technology is relatively inexpensive compared to the third type of array.

Oligonucleotide Arrays.   This third type of array involves the synthesis or spotting of oligonucleotides at densities of 100,000 or more on glass slides. An **oligonucleotide** is a short segment of DNA or RNA that is chemically synthesized to precise specifications. Oligonucleotides are usually 20–60 nucleotides in length.

The synthesized oligonucleotide arrays have two significant advantages over their macro- and microarray counterparts: First, they can detect SNPs and distinguish closely related gene products (genes differing by one or a few nucleotides) or alternative RNA splicing products. Second, they do not require keeping track of a large library of clones or PCR products. A variation of the oligonucleotide array uses presynthesized fragments 50–70 nucleotides in length for each gene. These 50- to 70-nucleotide sequences can be spotted on glass slides.

One new approach to scaling up the density of oligonucleotide arrays comes from the application of optic fiber technology. It is possible to create 50,000 wells in the end of an

20 cycles of proprietary
base-by-base sequencing

1 million beads

1 million beads arranged in a monolayer

Each of a million beads
provides a 20-base-pair
signature sequence

	Signature sequence	Number of beads
1	GATCAATCGGACTTGTCGAA	2
2	GATCGTGCATCAGCAGTACT	53
3	GATCCGATACAGCTTTGGGC	212
4	GATCTATGGGTATAGTCGAG	349
5	GATCCAGCGTTTGGTGCTTG	417
6	GATCCCAGCAAGATAACAGC	561
7	GATCCTCCTGTCTCACATGT	672
8	GATCACTTCTCTCATTAACA	702
9	GATCTACAGAACTCGGTGAG	814
.		
.		
30,285	GATCGGACCGAATCGACTAT	2,935

Figure 10.28 Lynx Therapeutics sequencing strategy of multiple parallel signature sequencing (MPSS). The text describes how this recently introduced technique for quantitatively defining the composition of a cell's transcriptome is used.

automated preparation of the samples to be analyzed, the analytic instruments themselves, and automated software for capturing and assessing the quality of the global data sets gathered. We now describe how genomic researchers use the tools in selected platforms.

Analysis of Genomic and RNA Sequences

The platform of choice for genomic and EST or full-length RNA sequencing is the automated DNA sequencer.

Quantitative Analysis of mRNA Levels. Researchers employ three types of DNA sequencing to determine the quantities of different mRNAs in a mixture. They have used the automated DNA sequencer, for example, to generate more than 4 million human ESTs from hundreds of different cell types. More recently, they have developed a technique termed **serial analysis of gene expression,** or **SAGE.** With SAGE, they can synthesize small cDNA tags about 15 bp in length from the 3′ ends of mRNA and link these small tags together into 600 bp DNA that can then be sequenced. In principle, each 600 bp DNA string contains 15 bp tags from 40 different genes ($15 \times 40 = 600$).

The sequence analyses of thousands of these linked 15 bp tags permit a quantitative estimate of the mRNAs present in the original cell type.

Finally, the very recent massively parallel signature sequence technique (MPSS) appears to be the most powerful approach to quantitatively defining the transcriptomes of individual cell types (Fig. 10.28). A **transcriptome** is the population of mRNAs expressed in a single cell or cell type. MPSS is the only technique that readily (and cost effectively) allows the routine identification of most of a cell's rarely expressed mRNAs. In this context, "rarely expressed" means that there are fewer than 10 copies per cell. The MPSS approach could, for example, take a cDNA library of 1 million clones and amplify each clone with a unique PCR tag so that the products from each clone can be attached to a single nylon bead (Fig. 10.28). The million beads are displayed in a flow cell and then the million different sequences are simultaneously sequenced with fluorescent reporter groups to generate 1 million sequence tags of 17–20 bp. If the genome of the organism whose cell type was used to make the cDNA library has been sequenced (and, hence, all or most of the genes have been identified), then the MPSS technique can provide a precise quantitative measure of most of the mRNA types in the cell down to those from genes expressed in only a single copy. This sensitivity is significant because many biologically important genes are expressed at very low levels (less than 50 copies per cell). The typical mammalian cell has perhaps 300,000 transcripts representing 20,000 different mRNAs. Ninety percent of the mRNAs are expressed at levels of less than 100 copies per cell.

The DNA array approaches, particularly the microarray and oligonucleotide arrays, also permit a semiquantitative analysis of patterns of gene expression. The two-color microarray analyses have been particularly useful in comparing different tissue or cell types (for example, normal versus cancerous). Most of the time, however, the DNA arrays do not accurately detect mRNAs expressed at less than 50 copies per cell.

Analysis of SNPs.   SNPs can be analyzed by oligonucleotide arrays with the use of molecular hybridization, or they can be analyzed by mass spectrometry. There are a variety of different approaches for distinguishing SNPs by mass spectrometry. All involve the analysis of oligonucleotide fragments that reflect the differences in the masses of the single nucleotide polymorphisms. Some of the high-throughput platforms such as the optical fiber array will be able to analyze 1 million or more SNPs per day.

Global Proteomic Strategies and High-Throughput Platforms

The analysis of proteomes is more challenging than the analysis of genomes for two reasons. First, the range of protein expression in cells is enormous: from 1 copy to 10^6 copies/cell. Since there is no protein equivalent of PCR, the tools of protein analysis need to function across concentration differences spanning six orders of magnitude. Second, proteins in complex mixtures have many different features that need to be identified and/or characterized. These include the gene encoding each protein; levels of expression in particular cell types, which change across developmental stages or physiological responses; covalent modifications (for example, phosphorylation, glycosylation); interactions with other proteins, macromolecules, or small molecules; compartmentalization within the cell; state of activation; half-lives; three-dimensional structure; and the relationship between structure and function. In sum, proteins are dynamic molecules constantly changing in concentration, interactions, cellular location, chemical modifications, and so forth. The ability to analyze all of these features is important to an understanding of how proteins function in the systems in which they participate.

The Identification of Proteins in Complex Mixtures

Proteins encoded by genomes where most of the genes have been sequenced and annotated can be identified as follows. The cells of a particular type are purified. The proteins are then extracted from the cells and digested with trypsin, a protease that cleaves at arginine and lysine residues. These peptides are then partially fractionated on a reverse phase column

(a column separating peptides based on their ability to repel water, or hydrophobicity). Successive aliquots, or portions, from the column are introduced into the mass spectrometer to provide mass measurements of the partially fractionated tryptic peptides. Alternatively, proteins can be separated by two-dimensional gel electrophoresis (separation by size in one dimension and by charge in the second) and protein spots can be cut out, digested with trypsin, and the resulting peptides injected into the mass spectrometer (Fig. 10.29). In every yeast cell, for example, there are potentially 6000 proteins and 350,000 tryptic peptides. Most of these peptides exhibit unique masses that can be placed in a database. The mass of each of the peptides analyzed in the mass spectrometer can then be compared by a software program against the database of genomewide peptide masses to identify the gene that encodes the protein from which the tryptic peptide is derived. A single mass spectrometry run can assign hundreds of peptide fragments to the genes that encode them.

The Quantification of Changes in Protein Concentration in Different Cell or Tissue States

A technique called *isotope analysis,* which employs prepared reagents known as *isoptope-coded affinity tags (ICATs),* permits researchers to analyze the changing patterns of protein expression in two different cellular states. The ICAT reagent has three components: (1) a biotin tag (biotin is a molecule that binds tightly to avidin molecules, and this binding ability provides a means for purifying proteins or peptides of interest); (2) a linker to which eight hydrogens or eight deuteriums can be attached to create light (hydrogen) or heavy (deuterium) chemical isotope forms (differing by 8 mass units); and (3) a chemical group that reacts with cysteine amino acids and will thus attach the ICAT reagent to all cysteines in a protein or peptide (Fig. 10.30a).

In one example of this isotope analysis technique, a normal cell is labeled with the light isotope reagent and its cancerous counterpart with the heavy one (Fig. 10.30b). Equal quantities of these cells are then mixed, their proteins are purified and digested with trypsin, and the cysteine-ICAT-labeled peptides are purified by affinity chromatography (that is, poured through a tube filled with avidin known as an avidin column; the avidin will bind all biotin-labeled peptides, and all other peptides will flow through the column and out). The biotin-labeled peptides are then fractionated on a reverse phase column, and successive aliquots of the biotin-labeled fractionated peptides are analyzed in two ways in the mass spectrometer. First, as shown in Fig. 10.30b, the area under the curves of the isotopic pairs of peptides (hydrogen/deuterium) is determined. These areas are proportional to the concentrations of peptides (and by extension, the proteins) expressed in the two cell types. After further fragmentation in the mass spectrometer, the peptide pairs are analyzed to determine their amino-acid sequences. With these sequences in hand, it is possible to determine the gene that encodes the protein from the database table of the masses of tryptic fragments. The ICAT technique is thus similar to the two-color microarray studies: the ICAT quantitates proteins, the microarrays quantitate mRNAs, from two different cellular states.

Figure 10.29   A strategy using two-dimensional gels and mass spectrometry to identify the genes encoding proteins in complex mixtures. A two-dimensional gel separates proteins by size and charge. Individual spots can be eluted from the gel, digested with trypsin to produce peptides, and analyzed in a tandem mass spectrometer (MS/MS). The first mass spectrometer (Q1) separates the peptide fragments (MS base peak); then one fragment (*orange*) is moved to the collision cell (Q2) and fragmented into components, which can be analyzed in the second mass spectrometer (Q3). Researchers can search the mass spectrum obtained in this way against a database of theoretical fragments to identify peptides and proteins and, by extrapolation, the genes that encode them.

Work is now underway to develop high-throughput platforms for measuring many of the other features of proteins (such as phosphorylation and activation).

Identification of Protein-Protein Interactions

There are two commonly employed strategies for characterizing protein-protein interactions: affinity capture combined with mass spectrometry and yeast two-hybrid interactions.

Affinity Capture/Mass Spectrometry. In this method, an antibody or other specifically interacting molecule is used to "pull down" one protein from a complex mixture (Fig. 10.31). The antibody or other specifically interacting molecule is called a capture molecule. This affinity capture procedure will also pull down all proteins associated with the first protein. These complexes of pulled-down proteins can then be identified in the

mass spectrometer. One global approach to analyzing protein-protein interactions is to attach a nucleic-acid sequence encoding a good affinity tag (for example, protein A) to the 5′ or 3′ end of a gene. When this tagged gene is integrated into the genome of the organism, the protein it encodes can be pulled down as just described. It should be possible, for example, to so modify all 6200 yeast genes and construct 6200 strains, each with a different protein A tag to a different gene. Subsequent pull-down experiments with each protein could shed light on the global network of all (most) protein-protein interactions.

Yeast Two-Hybrid Interactions. The yeast two-hybrid approach depends on the fact that the galactose (Gal 4) transcription factor is composed of two domains—a DNA-binding domain and an activator domain—and the Gal 4 molecule functions only when the two domains are adjacent to one another

Figure 10.30 The isotope-coded affinity tag approach to the quantitation of complex protein mixtures from two different states. (a) The isotope-coded affinity tag reagent (described in the text). **(b)** The strategy for labeling the proteins of two cell types with the light and heavy reagents (also described in the text).

(Fig. 10.32). In this approach, a gene of interest is subcloned into a yeast-expression vector as a fusion to the DNA-binding domain of Gal 4, which is under the control of a promoter that is always turned on in yeast cells. These yeast cells are then mated (fused) to a library of yeast cells, which, in theory, each contain the fusion of 1 of the 6200 yeast genes to the Gal 4 activator domain. After this mating, the two chimeras are expressed in the same yeast cell. If the Gal 4 DNA-binding domain comes into contact with the activation domain because of a physical interaction between the two chimeras, the resulting complex will reconstitute the activity of the Gal 4 transcription factor and thus activate genes under Gal 4 control. If, for example, Gal 4 controls the gene responsible for histidine synthesis, then histidine will be synthesized when (and only when) Gal 4 is activated. Thus, fused clones (known as *transformants*) are screened for their ability to grow when there is no histidine in their media (under such conditions, the clones must synthesize the compound to survive). Yeast clones that survive can be examined to determine which fused proteins are interacting. A

gene synthesizing a colored reporter group can also be put under Gal 4 control to obtain a visual indication of the interactions of the fused protein.

The yeast two-hybrid approach has been used to define an extensive map of protein-protein interactions in yeast (~4000 proteins with 7500 interactions). This network of protein interactions is termed an *interactome*. Researchers are now applying this approach to the analysis of protein networks in other organisms.

Uses of Genomic and Proteomic High-Throughput Platforms

For each interesting organism, geneticists would like to use the high-throughput platforms described to sequence the genome, define quantitatively the transcriptomes and proteomes in each cell type, delineate the nature of proteome interactions in various cell types, and analyze each of the other general features of the proteome (such as location of proteins, state and level of their activation, half-life, three-dimensional structure). They

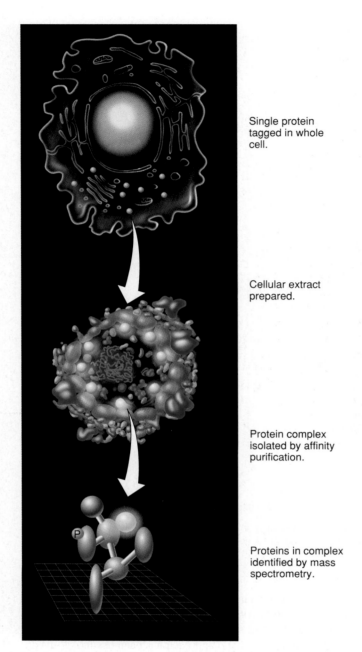

Single protein tagged in whole cell.

Cellular extract prepared.

Protein complex isolated by affinity purification.

Proteins in complex identified by mass spectrometry.

Figure 10.31 How to use affinity purification and mass spectrometry to identify protein interactions. First, use genetic engineering to label the problem of interest (see text). Next, prepare a cellular extract and isolate the protein complex by affinity purification. Finally, identify the interacting proteins in the complex by mass spectrometry.

(a) Proteins interacting (B/ORF)

(b) Proteins failing to interact

Figure 10.32 The yeast two-hybrid approach to analyzing protein interactions. (a) *DBD* indicates the Gal4 DNA-binding domain. *B* indicates the protein of interest whose interacting partners are to be identified; B is termed the **Bait.** *AD* depicts the activator domain of Gal4. *ORF* indicates the open reading frames whose proteins are to be tested against the Bait for interactions. A DBD/B expression vector is placed in one set of yeast cells, and the AD/ORF expression vector is placed in a second yeast strain. It is possible to construct ~6200 AD/ORF strains. The DBD/B and AD/ORF strains are fused. If the B and ORF proteins interact, the Gal4 transcription factor is able to turn on expression of the gene (*HIS*) for histidine synthesis. The *HIS*-expressing strains are then selected (as described in the text). **(b)** Strains in which the proteins fail to interact do not turn on the *HIS* gene and as a result, they die under the conditions used to select the *HIS*-expressing strains.

would also like to have strains of an organism that have resulted from different genetic manipulations, for example, separate strains with a different gene destroyed or knocked out. In some organisms, global studies are underway to obtain many mutations of each gene (for example, by chemical mutagenesis). These are important because different mutations in the same gene can produce different levels of activity that would not be seen in all-or-none knockout mutations. Geneticists would also

like to be able to express each protein from an organism in order to study it individually or in combination with other proteins. In yeast, many of these global discovery projects are quite advanced; in other organisms, such programs are just beginning.

The high-throughput platforms can also be applied to genetic or environmental experimental manipulations of organisms or cells. Transcriptome or proteome analyses of the knockout strains and chemical mutants described would be examples. Researchers could also use these tools to compare normal and disease states. For example, microarray analyses of several types of cancer (leukemia, prostate, breast, melanoma) and the corresponding normal tissues have revealed markers that allow phenotypically similar cancers to be separated into distinct groups. This presumably indicates that they are different cancers with different prognoses and possibly requiring different treatments. The knockout strains and global protein interaction data from yeast have been used to study a variety of biological systems in yeast, including the cell cycle. Similar types of studies are underway in higher organisms.

The availability of complete genome sequences and high-throughput platforms has catalyzed the beginning of two distinct paradigm changes: systems biology and predictive/preventive medicine. We use the word *paradigm* to denote the philosophical and theoretical framework of a scientific discipline within which researchers formulate hypotheses and perform experiments to test them.

Systems Biology

A **biological system** is a collection of elements that carry out a specific biological task. **Systems biology** is the global study of multiple components of biological systems and their simultaneous interactions. Systems biology is beginning to complement the more classical hypothesis-driven studies analyzing one gene or one protein at a time. It is made possible by the following consequences of the Human Genome Project:

- the sequencing of the human and model organisms genomes to provide a genetics parts list of all the genes and predictions of the proteins they encode;

- the development of high-throughput platforms, which pioneered the acquisition of global data sets and thus laid the basis for discovery science;

- the development of powerful computational tools that make it possible to acquire, store, analyze, integrate, display, and model biological information;

- the use of model organisms, which allow scientists to study the genetic and environmental *perturbations* (experimental manipulations) of biological systems to obtain additional global data sets;

- the use of comparative genomics to determine the logic of life for individual organisms and how that logic has changed in different evolutionary lineages.

In a biological system, all the elements work together to give rise to emergent properties. For example, the cells and molecules of the immune system work together to generate immunity (the ability to recognize and destroy foreign pathogens) and tolerance (the ability to distinguish self from nonself and not react against the self). The interacting neurons and neurotransmitters of the brain give rise to the emergent properties of memory, consciousness, and the ability to learn. Systems biology uses experimental manipulations (perturbations) in an attempt to understand how all of the elements of a system interact. It is also interested in snapshots of how biological information changes across developmental time or physiological responses. Ultimately, a systems biologist would like to understand how the elements of a system give rise to its emergent properties.

The systems biology approach incorporates the following six steps:

1. Formulate a computer-based model for the system based on the available biological data and current understanding.

2. Employ discovery science to define as many of the system's elements as possible (for example, sequence the genome to define all the genes, sequence the mRNAs in

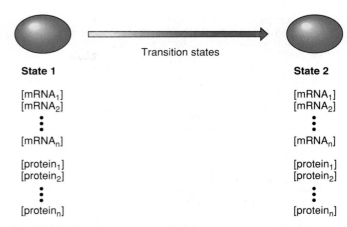

Figure 10.33 A global analysis of the mRNAs and proteins from two different states of perturbation. This is a key step in systems biology that employs high throughput platforms for the analysis of mRNAs and proteins (see text).

relevant cell types (or tissues) to define the transcriptomes, perform assays to define the proteomes in relevant cell types, define the protein-protein and protein-DNA interactions in relevant cell types).

3. Perturb the system either genetically (by knocking out, or destroying, genes, for example) or environmentally (by adding nutrients or changing the temperature or other conditions) and measure changes in the global relationships of all the elements, one to another (Fig. 10.33).

4. Integrate the different levels of biological information gathered from the perturbations (for example, DNA sequence, mRNA levels, protein levels, protein-protein interactions characterized, protein-DNA interactions analyzed, *biomodules* defined, networks of biomodules within the cell surveyed) and compare these integrated data against predictions of the model. A **biomodule** is simply a group of proteins that carry out a biological function; a biomodule can itself be a system or several biomodules can in combination make up a system.

5. Formulate hypotheses to explain disparities between the experimental data and the model, and use these hypotheses as the basis for a second round of global perturbations.

6. Refine the model after integrating data from the preceding step and repeat steps 3 through 6 until model and experiment are in accord with one another. The refined model should predict not only the behavior of the system given any perturbation but also its emergent properties.

- Well-studied pathway involving carbon utilization in yeast
- ~ 9 genes involved in specific processing of galactose sugar: includes structural genes (e.g., enzymes) and control genes (e.g., transcription factors)

- Enzymes are transcriptionally up-regulated 1000X when cells are stimulated by galactose

Figure 10.34 The yeast galactose-utilization system. This schematic depicts the chemical reactions that convert galactose to glucose-6 phosphate and indicates the enzymes that catalyze these reactions.

Researchers have successfully used this general systems approach to look at how yeast, the simplest eukaryotic cell, turns various genes on and off in the utilization of galactose. The galactose-utilization system (or galactose biomodule) in yeast converts the sugar galactose to a second sugar glucose-6 phosphate, which is a major component in one carbohydrate pathway for energy production. The initial model illustrated in Fig. 10.34 summarizes more than 30 years of galactose-utilization analysis in which researchers looked at one gene or one protein at a time.

The galactose-utilization system has nine elements: four enzymes, which catalyze the necessary chemical-modification reactions; a transporter molecule, which transports galactose across the yeast membrane and also sets the activity state of the galactose system such that when galatose is present, the system is turned on, and when galactose is absent, the system is shut down; and four transcription factors, which turn the system on and off.

To analyze the galactose-utilization system, researchers carried out three types of global experiments. They first constructed nine genetically perturbed yeast strains; in each strain, one of the nine genes encoding the protein elements of the system had been destroyed (knocked out). The investigators then used global microarrays with all 6200 yeast genes represented to analyze the nine knockout strains as well as the wild-type strain in both the presence and the absence of galactose. These analyses represented 20 perturbations (9 knockout + 1 wild type = 10 yeast strains × 2 system states).

Several interesting observations emerged from these experiments. First, more than eight different gene-expression patterns were recorded, which suggested that the galactose system has additional control features not captured in the original model (Fig. 10.34). In two instances, researchers formulated new hypotheses, generated new perturbations (double knockouts), and carried out a second round of global microarray analyses, which afforded new insights on the regulation of galactose utilization that could be incorporated into the original model (Fig. 10.35). These fundamental new insights helped explain the operation of the galactose system. Second, the expression patterns of 997 yeast genes changed in 1 or more of the 20 perturbations. These 997 genes could be clustered into 16 groups, each exhibiting similar patterns of gene-expression change across the perturbations. Strikingly, some of the genes within each cluster were known to function in different systems in the yeast cell (for example, during the cell cycle, in amino-acid synthesis, in other aspects of carbohydrate metabolism). The investigators hypothesized that each of these systems was directly or indirectly connected to the galactose-utilization system such that perturbations in the galactose system caused changes in the others. A second round of analyses using protein-protein and protein-DNA interactions confirmed this hypothesis.

Databases cataloging more than 7000 protein-protein interactions and 500 protein-DNA (transcription factor/promoter region) interactions exist for yeast. Investigators correlated these interactions with the 997 genes perturbed in the knockout experiments. They then constructed a graph or model of the integrated mRNA-expression changes, the protein-protein and protein-DNA interactions (Fig. 10.36). These data supported the hypothesis that interactions occur between the galactose system and many other metabolic and cellular systems. This finding, in turn, pointed to additional global analyses—for example, to define in more detail how the different systems communicate with one another.

Finally, the researchers used the ICAT technology to compare the concentrations of various proteins in two types of wild-type yeast cells: those with the galactose system running and those with the system shut off. They examined about 300 proteins, 30 of which exhibited significant changes in concentration; significantly 15 of these 30 had no changes in their mRNA concentrations. One conclusion is that for these proteins, the regulation of concentration must operate at a posttranscriptional level.

The galactose-utilization experiments show that in studying biological systems, it is important to analyze many types of data and integrate the various levels of biological information obtained. In the yeast studies investigators looked at mRNA levels, protein levels, protein-protein interactions, and protein-DNA interactions. The experiments also illustrate how useful global, iterative, hypothesis-driven studies and model refinement are in the examination of biological systems.

Systems biology is at an early stage of development and faces many challenges. These include the need to develop

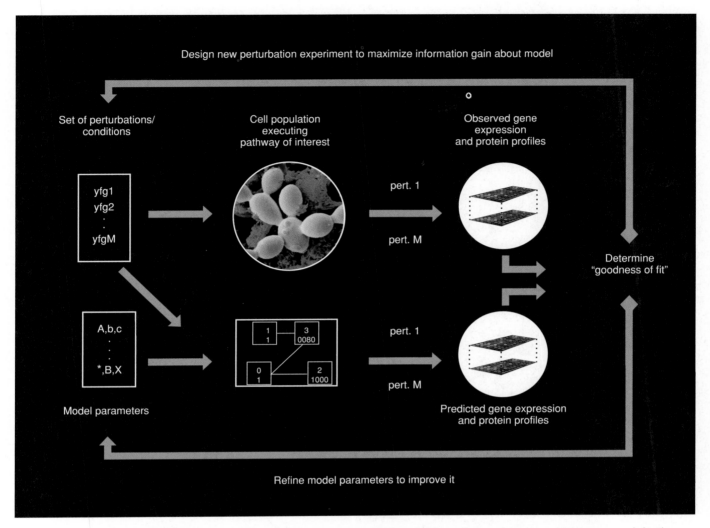

Figure 10.35 Diagram of the iterative nature of the systems approach to biology. To start the systems biology approach for the galactose-utilization system (see text), a preliminary model was constructed from 30 years of prior genetic, molecular, and cellular studies. Then the nine genes in the system were knocked out (perturbed) and each knockout was tested for global mRNA changes with the system on (galactose present) and off (galactose absent). The experiment observations failed to match the predictions from the model in eight cases. For two of these observations, hypotheses were formulated to explain the discrepancies and additional perturbations were carried out to discriminate between the possibilities. The model was then refined in accordance with the new data. This process can be iterated as many times as necessary to bring experiment and theory into accord with one another. Other perturbations used the ICAT approach to study protein-level changes in different perturbations (see text). Also, the mRNA and protein concentration data were integrated with protein-protein and DNA-protein interaction data to get a more complete picture of the system (see text). These integrative, iterative, and model refining processes are integral components of systems biology.

■ more effective platforms offering higher throughput, better-quality data, greater sensitivity, and lower cost;

■ more powerful computational and mathematical tools for storing, integrating, graphically displaying, analyzing, and mathematically modeling biological systems;

■ the tools for single-molecule and single-cell analyses;

■ better tools for carrying out global genetic and environmental perturbations of model organisms.

Predictive/Preventive Medicine

The second paradigm change to emerge from the Human Genome Project is the potential for predictive/preventive medicine. The project has provided access to the DNA polymorphisms underlying human variability. Although most polymorphisms fall outside of genes and thus do little to change human phenotypes, some are responsible for differences in normal physiology (tall or short, thin or fat), and others predispose to disease. With the potential for carrying out dense genomewide genetic analyses,

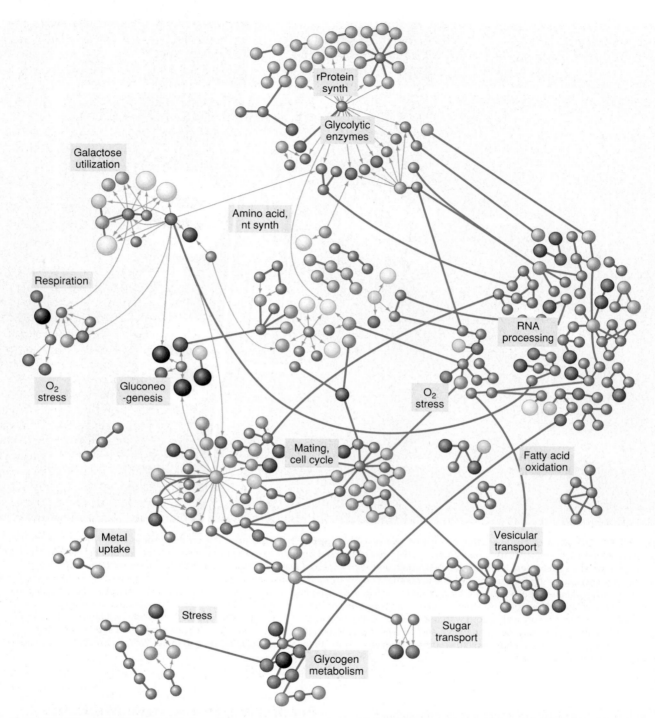

Figure 10.36   How genetic perturbation of the galactose-utilization system in yeast (the knockout of individual genes) affects the network of interactions with other metabolic and functional systems. This network was developed by combining the clusters of mRNAs defined by the knockout perturbation experiments and the protein-protein and protein-DNA interaction data. The *yellow arrows* indicate protein-DNA interactions (transcription factor activity) and the *blue bars* indicate protein-protein interactions. The *red circle* indicates where the *Gal4* gene has been knocked out. A *gray scale* indicates levels of mRNA expression: *black* = high levels; *white* = low levels.

clinical researchers will in the next 10–15 years identify hundreds of genes that predispose to disease (mostly late-onset diseases such as cancer, cardiovascular disease, immunological diseases, neurological diseases, and metabolic diseases). For example, the breast cancer 1 gene (*BRCA1*), when present in women in a single defective copy, predisposes 70% of them to breast cancer by the time they are 60 years old. Why only 70%? Either there are environmental factors that operate in concert with the defective gene, or there are other modifying, disease-predisposing genes that are present only 70% of the time. In either case, a prediction can be made about the future likelihood of disease for individuals carrying the defective gene. In time, physicians will be able to scan the genomes of the young and provide a probabilistic projection of what the future may hold with regard to a wide variety of diseases. Of course, being able to predict a disease without being able to cure or prevent it leaves physicians in a very uncomfortable position. As the techniques of systems biology mature, scientists will be able to understand defective genes in the context of the biological systems in which they operate, and they will learn how to circumvent the limitations of these defective genes—with, for example, novel drugs, environmental controls (diet), or other approaches such as stem cell transplants or gene therapy. The idea will be to design preventive measures to avoid or greatly delay the onset of the disease. In time, these preventive measures could extend the productive and creative life span of individuals by 10–30 years.

Predictive/Preventive Medicine Raises Challenging Social, Ethical, and Legal Issues for Which There Are No Simple Solutions

The preceding view of medicine in the twenty-first century raises a host of pressing questions about the privacy of genetic information, limitations on the use of genetic testing, the patenting of DNA sequences, society's view of older people, the training of physicians, and the extent to which the human genetic engineer should seek to engineer himself or herself.

We discuss the privacy issue—the need to ensure that genetic information is confidential, to be given out only at the discretion of the tested individual—in the Genetics and Society boxes in Chapters 2 and 3. In the Genetics and Society box in Chapter 2 and in the Genetics and Society box "Social and Ethical Issues Surrounding Preimplantation Embryo Diagnosis" in Chapter 11, we outline considerations that make it necessary to establish guidelines for genetic screening; among those considerations are who should be tested and whether testing should be carried out if no therapies or preventive measures exist to help an individual with a genetic predisposition. Most geneticists emphasize the need for educating the general public about the benefits and drawbacks of testing. Meanwhile, the U.S. Congress has passed a law ensuring the privacy of medical records. The Genetics and Society box on pp. 362–363 of this chapter discusses the "Patentability of DNA."

With predictive/preventive medicine, many people could remain productive and creative into their nineties. Our society will thus have to rethink its philosophies of retirement, social security, access to medical care, and related topics.

The physicians trained today will be practicing predictive/preventive medicine in 20 years. How should their training change to prepare them for the new ways of thinking about medicine? In the same vein, how can we educate society about the very different nature of predictive/preventive medicine?

Finally, what about the social and ethical concerns raised by genetic engineering, that is, the ability to alter specific genes of an organism in specific ways? Many of the preventative or remedial measures constitute **somatic gene therapy** in which medical practitioners compensate for a faulty gene by inserting a replacement gene into the affected tissue where the gene is expressed. Somatic gene therapy causes biochemical and physiological changes in the genetically modified tissue or tissues that die with the individual. A potential therapy for cystic fibrosis, for example, consists in inserting a wild-type *CFTR* gene into lung cells. This type of genetic engineering is not different in kind from drug therapies aimed at correcting a particular physiological deficiency (for example, the insulin injections aimed at treating diabetes). The alterations resulting from somatic gene therapy affect only the somatic cells of the individual undergoing the therapy and cannot be transmitted to offspring.

Most of the controversy surrounding genetic engineering stems from the potential for **germ-line gene therapy:** modifications of the human germ line. Germ-line gene therapy produces changes in germ cells that are passed on to progeny. Ethical concerns focus on what is and is not appropriate. Should, for example, a couple be able to eliminate a cancer-predisposing gene in their unborn child? Should they be allowed to alter the child's potential for obesity or longevity? Should they have the option of choosing the child's eye color?

Many geneticists and bioethicists support the idea of somatic gene therapy but oppose germ-line therapy. At the same time, they urge serious and open discussion of the issues before perfection of the technology overtakes our ability to control its use.

Genetics and Society

The Patentability of DNA

Some people argue that genetic information, the naturally occurring raw material of life's evolution, is a common heritage that belongs to everyone. Yet since the mid-1970s, universities and biotechnology companies have sought patents on specific DNA sequences—human, bacterial, and plasmid. To understand this apparent paradox, we take a brief look at the history and purpose of U.S. patent law and examine its application to various types of DNA sequences.

Congress passed the first patent act, penned by Thomas Jefferson, in 1790. That same year the U.S. Patent Office awarded the first patent to Samuel Hopkins of Vermont for the making of potash, a potassium-based compound used to produce fertilizer and soap. One hundred and ninety years later, in 1980, the patent office granted two patents for specific uses of genes. The first was to a microbiologist at General Electric for a genetically engineered oil-eating bacterium that could help clean up spills; the second was to Stanford University and the University of California at San Francisco (UCSF) for the basic gene-splicing and gene-cloning technology invented by researchers at those institutions.

Patent examiners evaluate the patentability of a product or process by three criteria: Is it *new?* Is it *nonobvious?* Is it *useful?* In the DNA arena, the courts have made the following interpretations of patent law. Raw materials of nature, such as wild-type DNA in a living organism, are not novel and thus are not patentable; modified products, such as bacterial DNA altered by a synthetic mutation or human DNA in a mouse genome, are novel and thus eligible for patents by the novelty criterion. DNA-based processes that produce a novel material, such as clones of a gene (consisting of a DNA construct in a vector), are also patentable. Publication in a scientific journal makes an item, such as a DNA sequence, obvious, thus unpatentable per se; but a specific use of a published sequence may be nonobvious and thus patentable. A well-defined use might be a particular test for a particular genetic aberration or a particular process for the manufacture of a particular therapeutic agent.

Since the mid-1980s, patent examiners have used these guidelines to grant DNA-related patents on a case-by-case basis. The cDNA for insulin integrated into a bacterial plasmid has received a patent; this cDNA does not occur naturally in significant quantities, but it is fabricated through a human-devised protocol using reverse transcriptase. By contrast, human cDNA fragments isolated at random from brain cells did not qualify for a patent, in part because the potential use of these fragments could not be defined without identification of the complete cDNAs and the proteins they encode. In another example, researchers at the University of Utah obtained a patent on *BRCA1* sequences to be used as a test for predisposition to breast cancer. These researchers mapped and sequenced the gene and incorporated that knowledge into the design of a socially useful test; thus their application passes muster. To date, the processing of DNA-related patent applications has taken 8–12 years.

The rationale for granting patents is to encourage innovation—the invention of useful contributions to society—by providing a time-limited monopoly to protect an invention from imitation. This commercial protection is given in exchange for the complete disclosure of the information related to a product or process. In fact, in pursuing a patent, a company protects its interests by making available as much information as possible about the modified product or process. Patents protect use for profit; they do not interfere with research or other noncommercial uses of the information in the patent. Institutions, companies, or individuals holding patents can profit from their inventions themselves or license the right to profit from them to another party. Stanford and UCSF, for example, originally granted many nonexclusive licenses to companies

C O N N E C T I O N S

One of the triumphs of modern genetics has been the determination of the complete sequence of the human genome and the genomes of many model genomes. With powerful new "shotgun" strategies for sequencing large genomes, genomicists have sequenced more than 100 different microbe, plant, and animal genomes. The genome projects now underway in laboratories around the world will eventually add hundreds of additional genomes to this list.

The Human Genome Project, which supported the development of high-throughput DNA sequencing platforms, has catalyzed the emergence of high-throughput platforms for analyzing gene expression and protein-protein interaction. Based on these tools, analyses of complete genome sequences have provided insights into the architecture and evolution of genomes and their corresponding proteomes. The genetics parts lists made available by the genome sequences are transforming using recombinant DNA technology in the production of various agents for a fee of $10,000 a year. In the first four years of its patent, Stanford received $3 million in licensing fees and patent royalties.

Different countries apply the basic tenets of patent law in different ways. In the United States, the purified form of a gene or protein is patentable because genes and proteins do not exist in nature in purified form. In England, a naturally occurring gene sequence is not patentable no matter what form it is in. And in France, the code on intellectual property declares unpatentable "the human body, its elements and products as well as knowledge of the partial or total structure of a human gene."

There is currently healthy debate over the application of patent criteria originally developed for mechanical and chemical inventions to DNA sequences and various life-forms. The debate focuses on several areas of concern.

Openness Versus Secrecy

How does the patent process affect the open exchange of ideas, which is a prerequisite of basic research? A company putting a great deal of money into the research and development of what it hopes will be a patentable gene therapy may withhold publication of its data or publish only partial results until its patent application is in the pipeline. For example, in the 1980s, before the discovery of the cystic fibrosis gene, one gene diagnostics company published an article about markers for the *CF* gene but, in an effort to protect their work, did not include the fact that those markers were on chromosome 7. As it happens, other groups subsequently found closer markers and then the gene. Still the question remains: Do commercial considerations interfere with the free exchange of ideas and results about genomes? If so, should something be done about this?

Profitability Versus the Social Good

Who has control over genetic data, deciding what applications to pursue with commercial funding? What criteria do they use? Most companies consider potential profitability as the basis for pursuing research and development. Some people wonder whether they should also factor into the equation concerns about serving the poor (who cannot pay much for drugs or therapies) and the relatively few patients suffering from uncommon diseases.

Development and Funding of Socially Useful Research Applications

Since companies must pour large sums of money into developing DNA-based drugs or therapies, testing them on large numbers of people, and bringing them to market, many researchers—at both universities and commercial enterprises—maintain that the survival of biotechnology companies, particularly small ones, depends on patent protection. Without the possibility of a patent, they argue, companies may not be able to afford to use knowledge of, for example, the *CF* gene and the transmembrane conductance regulator it encodes to develop therapies that may ease and extend the lives of thousands of CF sufferers. Development of drugs or therapies that cost a great deal of money, researchers maintain, simply will not occur without patents. On the other hand, many of the patents providing the basis for development and marketing are on inventions made in university laboratories; and the investigators in some of these laboratories are nonprofit recipients of federal grants that are ultimately funded by taxpayers. Is it proper for companies to use these inventions as the basis for financial gain? Most government agencies allow such patents because they themselves do not have the funds to develop research ideas for the marketplace. Furthermore, nearly all of the major technological advances of the late twentieth century in computers and electronics were based to one degree or another on research funded by federal money. We do not question the right of computer companies to profit off extensions of this research.

Thus, the availability of purified DNA, especially specific stretches of the human genome, raises several issues that the framers of patent law could not foresee. Vigorous public debate will help scientists, businesspeople, and lawmakers distill some of the answers. One radical and perhaps impractical proposal to deal with the legitimate concerns on both sides of the public-funding-versus-private-profit issue suggests creation of an international corporation that would hold patents on all human cDNA. This corporation would license the patents for development at auction and reinvest the proceeds from licensing fees in basic research.

---

the practice of biology and medicine. The integration of knowledge about genes, their products, and their regulatory networks has led to a new paradigm of systems biology.

In this chapter we focused on the human genome in a generic sense, as the common framework of genetic information that defines the human species. Indeed, every individual person carries a diploid genome that is 99.9% identical to that carried by every other individual. But the flip side of a 99.9% identity is a 0.1% nonidentity that distinguishes people (other than identical twins) from each other. This 0.1% difference translates into 6 million DNA sequence differences, or polymorphisms, which are responsible for all of the inherited ways in which individuals differ from one another. Of critical interest to medical researchers are the specific DNA polymorphisms that either directly alter or indirectly mark the genes that cause or predispose to disease. In Chapter 11, on the direct detection of individual genotypes, we discuss how researchers identify and use DNA polymorphisms to uncover disease-causing genes and other genes of interest in humans and other species. The discussion includes an examination of complex genetic traits, which arise from interactions among multiple genes and the environment.

ESSENTIAL CONCEPTS

1. Large-scale maps are the foundation of genomics.
   a. *High-density linkage or genetic maps* developed by computer analysis of raw genotyping data show the relative positions of closely spaced DNA markers. High-throughput platforms have been developed for the rapid analysis of large numbers of genetic markers (millions per day). Two types of commonly used genetic markers are single nucleotide polymorphisms (SNPs) and simple sequence repeats (SSRs), or microsatellites.
   b. Long-range *physical maps* chart the features of chromosomes.
      1. Fluorescent *in situ* hybridization, or *FISH,* locates a cloned locus to a particular band on a particular chromosome.
      2. DNA hybridization and fingerprinting order the clones of a genomic library into overlapping clusters that span the chromosomes of the species. One approach to assigning chromosomal positions uses the markers of a high-density linkage map and sequence tagged sites (STSs) as hybridization probes to obtain overlapping, large-insert clones that span the chromosomes. Gaps between the clusters of clones (contigs) can be filled in by surveying new large insert libraries with STS markers from either contig end. Another approach does not rely on linkage mapping and genetic markers. Instead, physical analysis begins with the complete set of individual clones in a whole-genome library. Researchers establish fingerprints for each of the clones and make a computer search to determine overlaps in fingerprints that indicate overlaps in the corresponding clones. With a large enough number of clones, it is possible to form contigs that extend across every chromosome.
      3. Researchers use restriction mapping and hybridization probes to characterize the fine details of each clone in a contig.
      4. Computers combine the information for individual clones into detailed whole chromosome physical maps.
   c. Long-range sequence maps, compiled from the sequences of subclones, provide a readout of every nucleotide in each chromosome. The subclones are derived either from previously mapped large insert clones (hierarchical shotgun approach) or directly from the genome (whole-genome shotgun approach).
   d. Linkage, physical, and sequence maps can be readily integrated because the markers for the linkage and physical maps (SSRs, SNPs, STSs) use sequence-based PCR primers that generate unique molecular addresses in the genome and thus can be specifically localized on the genome.

2. A variety of major insights have emerged from analyses of human and model organism genomes.
   a. The number of human genes, approximately 40,000 is surprisingly low.
   b. Genes fall into two major classes: noncoding RNA genes, generally representing RNA molecular machines; and protein-coding genes encoding the protein structural components and molecular machines. The collection of proteins in a particular cell is a *proteome,* and the collection of mRNAs, a *transcriptome.* Transcriptomes and proteomes are dynamic and change throughout the time spans of development and physiological responses. Proteins are often composed of one or more domains that encode discrete functions.
   c. Repeat sequences constitute more than 50% of the human genome. Some repeat sequences have evolved to become genes or control sequences. Others may catalyze chromosomal rearrangements and translocations.
   d. Evolution can occur by the lateral transfer of genes from one organism to another. Humans, for example, may have 200 or more genes derived recently from bacteria.
   e. The different human races have very few, if any, uniquely distinguishing genes. Genetically, there appears to be a single race of humans.
   f. The sequences of microbes, plants, and animals all employ the same genetic code and show a remarkable similarity among many basic biological systems. This affirms the idea that we all descended from a single common ancestor.

3. The Human Genome Project has catalyzed the development of the DNA sequencer, DNA arrays, and the use of the mass spectrometer for genomics and proteomics. These instruments have been integrated into high-throughput platforms for DNA sequencing and genotyping, as well as for mRNA and protein analysis.

4. The Human Genome Project has spawned paradigm changes in our approaches to biology and medicine. Systems approaches, in which researchers study all the elements of a system, complement the more classical biological approaches, in which researchers analyze one gene or protein at a time. Predictive medicine will allow researchers to correlate polymorphisms with disease predisposition and eventually formulate probabilistic health-history predictions for each individual. Preventive medicine will place the defective genes in their biological systems and develop ways to avoid their limitations with, for example, drugs, diet, gene therapy, or stem cell therapy. The potential of predictive/preventive medicine raises social, ethical, and legal questions for which there are no easy answers.

SOCIAL AND ETHICAL ISSUES

1. Several companies were formed in the 1990s with the purpose of sequencing portions of the human genome and using this information to devise new drugs. These companies sought to patent each bit of DNA sequence that they decoded even if they did not know whether it encoded a gene product. Their justification was to ensure that any use of the sequence would yield payback for their research. Do you think patents should be granted for each partial gene sequence that has been determined? Do you think patents should be granted for entire gene sequences even if we do not know what the gene does? What criteria do you think are compatible with facilitating private-industry drug development, encouraging research, and protecting intellectual property?

2. Researchers who have similar interests and complementary expertise often collaborate on projects.

The Johnston lab has recently started a collaboration with a lab in San Francisco to use a unique tissue culture cell line that they (the San Francisco lab) had established. At a scientific meeting, a graduate student from the San Francisco lab heard Marcus Johnston give a seminar talk in which he discussed future work and described this cell line, but he did not mention where it came from. Johnston also did not describe it accurately. The graduate student was very upset with this but then thought perhaps she was being too sensitive and when the work was done and publication occurred, appropriate credit would be given to the San Francisco lab. Should the student speak up at the end of the talk to correct misinformation? Should she approach Johnston later? Should she wait and talk with her advisor first? Should she do nothing and assume the misinformation will be corrected by someone else?

SOLVED PROBLEMS

I. A physical map of overlapping clones (a contig) was available for an area of a human chromosome containing three genes (and part of a fourth gene) that are transcribed within ovarian tumor tissue. The restriction map of the region is shown; tick marks above the line indicate *Bam*H1 restriction sites, while those below the line indicate *Xho*I sites. Sizes of DNA fragments between adjacent restriction sites are given in kilobases. Individual restriction fragments were purified, made radioactive, and used as probes for Northern blots of poly-A-containing RNA derived from ovarian tumors. In a Northern blot, mRNAs are analyzed just as DNA fragments are in a Southern blot (see Chapter 9). The poly-A-containing mixture of mRNAs from a particular cell type is subjected to electrophoresis—separating the mRNAs according to size (smaller mRNA migrate more rapidly than larger ones). The mRNAs are then transferred from the gel to a nitrocellulose filter and a radio-labeled probe for a particular gene (or gene fragment) is hybridized against the mRNAs on the filter. If size standards are also run, the presence and size of particular mRNAs can be established. The resulting autoradiograms are presented below the restriction map. Using these data, characterize the four genes within the contig in the following ways:

a. What is the length of the mRNA for each of the three complete genes?

b. What is the minimum length of the primary transcript for the largest of these RNAs?

c. What is the minimum number of exons for each gene?

d. What is the minimum number of introns for each gene?

Answer

This problem requires an understanding of primary and processed transcripts and the analysis of RNAs using Northern hybridization.

a. The bands on the Northern blots that hybridize with the probes represent the mRNAs from this region. The three mRNAs corresponding to the genes in this region are *1.1, 2.4,* and *6.8 kb* in length. The 4.3 kb transcript comes from a gene that is only partly contained in the DNA used as a probe since the one DNA fragment that hybridizes is only 2.0 kb in length.

b. The bands on the Northern blot represent processed transcripts (introns have been removed). Primary mRNAs are made by copying from contiguous DNA sequences, including those regions that will be removed by splicing. The minimum length of a primary transcript is based on the sizes of the restriction fragments that hybridize with the RNAs and any intervening fragments. *For the 6.8 kb largest mRNA, the minimum size is 2.8 + 1.6 + 1.4 + 3.1 + 2.6 + 2.9 + 2.2 + 0.5 + 4.4 kb, or 21.5 kb.* We cannot say from these data whether transcription begins before the 2.8 kb fragment or extends beyond the 4.4 kb fragment.

c. and d. The minimum number of exons is determined by counting the number of hybridizing fragments or groups of contiguous fragments. This is a minimum number because any of the fragments could contain more than one exon. The nonhybridizing fragments that separate these must contain introns.

Transcript	Minimum number of exons	Minimum number of introns
1.1 kb	one exon	none
2.4 kb	two exons	one
6.8 kb	four exons	three

II. Reverse transcriptase, the enzyme used to synthesize cDNA starting with mRNA as a template, often falls off the template before completely copying the mRNA. When screening for a cDNA clone of a gene, it is therefore not uncommon to isolate partial cDNA clones. What comparison could you make experimentally that would indicate if you had isolated a plasmid clone containing a partial cDNA?

Answer

For this problem, consider the two alternatives: you have the complete cDNA clone or you have a partial cDNA clone. What would be the differences that you could detect experimentally? The cDNA must be as long as its corresponding mRNA to be full length. You need to find out what the full length of the message is to do the comparison. The way to *measure the length of mRNA is to run a Northern gel and hybridize with the cDNA as a probe.* (Another alternative would be to hybridize a fragment from near the 5′ end of the gene to your clone. This would require that you have a fragment that you know is from the 5′ (end of the gene.)

PROBLEMS

10-1 For each of the terms in the left column, choose the best matching phrase in the right column.

a. sequence tagged sites (STSs)	1. a map showing the order of cloned inserts of DNA
b. whole-genome shotgun sequencing	2. fragments of DNA on a chip
c. DNA array	3. the presence in the germ line of one organism of a gene from a second organism
d. alternative RNA splicing	4. a genome sequencing strategy that avoids physical mapping
e. lateral gene transfer	5. a polymorphic single-base site
f. isotope-coded affinity tagging	6. the joining together of the exons of a gene in different combinations
g. physical map	7. a map of genetic markers that are separated by less than 1 cm
h. SNP	8. a global method for quantitating protein levels in two different cell types
i. high-density linkage map	9. unique DNA sequences that serve as molecular markers

10-2 Give two different reasons for the much higher ratio of total DNA to DNA that encodes proteins in the human genome compared to bacterial genomes.

10-3 Using a simple sequence repeat (SSR) DNA probe, you have isolated several clones from a library. You would like to identify unique sequences adjacent to these repeat DNAs found throughout the genome to establish some SSR-based STSs.
 a. How could you determine if there are unique sequences within the clones you have isolated?
 b. The most useful STSs are those associated with SSRs that are polymorphic in the population (that is, have sequence variants). How could you determine if an STS is associated with a polymorphic SSR?

10-4 What are the advantages of the FISH protocol over linkage mapping for the initial characterization of the chromosomal location of a gene?

10-5 A clear limitation to gene mapping in humans is that family sizes are small, so it is very difficult to collate enough data to get accurate recombination frequencies. A technique that circumvents this problem begins with the purification of DNA from single sperm cells. (Remember that recombination occurs during meiosis. Analysis of a population of sperm provides a large data set for linkage studies.) The DNA from single sperm cells can be used for polymerase chain reaction (PCR) studies. The PCR reactions can be run such that a PCR primer will not hybridize to DNA if there is a single-nucleotide difference between the sequence of the primer and the template DNA to be analyzed. A PCR product will be formed only if both primers hybridize to the template.

A series of three pairs of PCR primers were synthesized according to DNA sequences from human genomic clones. The alpha primers amplify a fragment 100 bp long; the beta primers produce a 200 bp product; the delta primers produce a 400 bp product. Twenty-five sperm obtained from one man were subjected to PCR analysis with these three pairs of primers simultaneously. Refer to the following data set.
 a. Is this man homozygous for any nucleotide polymorphism detected by this method? If so, which one?
 b. Determine the genetic map information you can from this data.

400 bp →

200 bp →

100 bp →

```
  1  2  3  4  5  6  7  8  9 10 11 12 13 14 15 16 17 18 19 20 21 22 23 24 25
```

Sperm number

10-6 The map of human chromosomes in males and females indicates that the centimorgan distances between genes is different in the two sexes.
   a. What could account for this difference in distances?
   b. Do you think the physical distances between genes in males and females would differ also? Why or why not?

10-7 A human genomic library was made by partially cutting human genomic DNA with the enzyme *Sau*3A (recognition site: 5′ ^ G A T C 3′) and then ligating the resultant fragments into the *Bam*HI site (5′ G ^ G A T C C 3′) of the plasmid vector pCU999 shown here. This plasmid contains a "polylinker" in which the *Bam*HI site (B) is flanked on one side by a site for *Sal*I (S) and on the other side by a site for *Not*I (N). The distance between the *Sal*I and *Not*I sites in the vector is about 20 bp.
   a. One particular clone from this library (clone 1) was isolated as part of a chromosome walk and was then cut by *Not*I (N), *Sal*I (S), and by *Not* I + *Sal* I (N + S). The fragments were separated by electrophoresis on an agarose gel and stained with ethidium bromide. A Southern bl0.367ot of this DNA was made, and the filter was probed with radioactive DNA made from the plasmid vector alone. The results are shown here. (Assume you cannot see a signal on the autoradiograph if the amount of homology with the probe is less than 50 bp.) Show the positions of all *Sal*I and *Not*I restriction sites in the genomic DNA part of clone 1, and indicate the distance between adjacent sites.

S   B   N

pCU999
2.5 kb

Restriction Digests of Clone 1

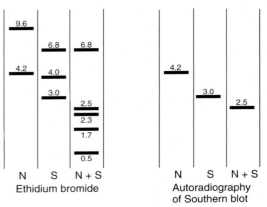

N	S	N + S		N	S	N + S
Ethidium bromide   Autoradiography
                             of Southern blot

   b. A clone with an overlapping fragment of human genomic DNA (clone 2) was isolated from the same genomic library. It was digested with the same enzymes as before and electrophoresed and transferred to nitrocellulose paper as before. This time, the Southern blot was probed with radioactive clone 1 (genomic DNA + vector DNA). The results appear here. Draw a map of clone 2, showing the positions of all *Sal*I and *Not*I restriction sites in the genomic DNA part of clone 2, indicating the distance between adjacent sites.

Restriction Digests of Clone 2

N   S   N + S
Ethidium bromide

N   S   N + S
Autoradiography
of Southern blot

c. Whole-genomic human DNA was now cut by *Sal*I, and identical blots of this digest were probed with clone 1 or clone 2. The results of this whole-genome Southern blot are shown next. The position of six *Sal*I sites in this region of the human genome can be mapped using the preceding data. Draw a *Sal*I map of this region.

Whole-Genome Southern Blot

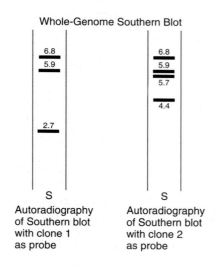

S
Autoradiography
of Southern blot
with clone 1
as probe

S
Autoradiography
of Southern blot
with clone 2
as probe

10-8 To make a set of clones more suitable for the analysis of DNA sequence, a series of cosmid clones was prepared by digesting a YAC and subcloning the resulting restriction fragments. The restriction patterns of the inserted fragments are shown next. Arrange these four cosmids into a physical map, showing the order of the clones and the overlap between them.

10-9 In an alternative approach to that described in problem 10-8, a cosmid library was probed with an insert from a YAC and 22 cosmids were obtained. All but two of the cosmids could be arranged as an overlapping group (a contig). This set of cosmids covered the length of the YAC insert. What explanation can you offer for the two cosmids that did not fit into this contig? Why did these hybridize with the YAC probe?

10-10 Would it be possible to construct a complete physical map of the human genome (assuming a 3 Gb size) with 15,000 BAC clones each with a 200 kb insert? Why or why not?

10-11 List three independent techniques you could use to identify DNA sequences encoding human genes within a cloned genomic region.

10-12 What are two potential difficulties with the whole-genome shotgun sequencing strategy?

10-13 A restriction map of part of the *Drosophila* X chromosome has been determined; it is presented here. Messenger RNA from *Drosophila* adults was purified, fractionated on agarose gels according to size, and then transferred to nitrocellulose filters. Radioactive recombinant DNAs corresponding to fragments A–G were hybridized to identical copies of the nitrocellulose filters containing the fractionated mRNAs. Using the map information, indicate the region of the DNA that must be transcribed to form three primary transcripts, which are then processed into the three mRNAs observed. Indicate where exons and introns occur in relation to the restriction fragments.

10-14 To study only the genes expressed in certain tissues, cDNA libraries are made starting with messenger RNA from a specific tissue. A mouse cDNA library was constructed as follows: cDNA complementary to the RNAs from the pituitary gland was prepared and then digested with the restriction enzyme *Eco*RI (E). The DNA fragments generated in this manner were inserted into the single *Eco*RI site of the plasmid vector pBR319 (see the map of this vector). A probe of a growth factor gene found in *Drosophila* was used to isolate the recombinant plasmid called pcDNA1. A clone was also isolated from a mouse genomic library that had been prepared by digesting mouse DNA with *Eco*RI and ligating the fragments into the pBR319 plasmid vector. The plasmid that hybridized with the probe is called pGEN1.

pcDNA1 and pGEN1 were subjected to digestion with combinations of the three restriction enzymes: *Bam*HI (B), *Hin*dIII (H), and *Pst*I (P). (There are no *Hin*dIII sites in the vector.) Ethidium-bromide-stained patterns of the agarose gels are presented.
a. How large is the cDNA insert in pcDNA1?
b. How large is the genomic DNA insert in pGEN1?
c. Make a map of pcDNA1 and pGEN1 indicating the positions of all B, H, and P sites and the distances in kilobases between all adjacent sites.
d. How many introns are indicated by this data, and how long is each intron? Show the position of this (these) intron(s) on the appropriate map.
e. Could you have learned this information by DNA sequencing of the genomic clone? How would you have used the sequence information?

10-15 Using a cDNA library, you isolated two different cDNA clones that hybridized with your probe for a nerve growth factor. The beginning and ending sequences of the clones are the same but the middle sequence is different. How can you explain the different cDNAs?

10-16 What sequence information about a gene is lacking in a cDNA library?

10-17 An interesting phenomenon found in vertebrate DNA is the existence of pseudogenes, nonfunctional copies of a gene found elsewhere in the genome. Pseudogenes appear to be double-stranded DNA copies of mature mRNA inserted into the chromosome. What sequence information would be a clue(s) that the source of these pseudogenes is cDNA?

10-18 Discuss three observations that suggest not all repeat sequences are junk.

10-19 With new information from the Human Genome Project, many new genes will be identified for which the function is not known.
a. What features of the DNA sequence might help you determine the function of a newly identified gene?
b. What are two other types of analysis that would help you learn more about a new gene?

10-20 a. If you found a zinc-finger domain (which facilitates DNA binding) in a newly identified gene, what hypothesis would you make about the gene's function?
b. In another gene, what would a high percentage of similarity throughout the gene with a previously identified gene in the same organism suggest about the origin of the gene?

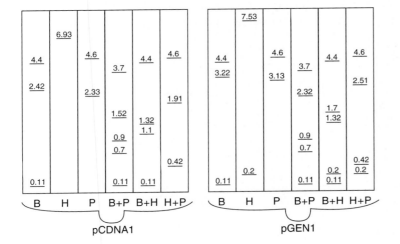

10-21 From an expression library, you isolated a clone and made an antibody to the protein produced by the cloned gene. When the antibody was used in an *in situ* hybridization experiment using a mouse embryo, the protein localized to the brain. Northern analysis showed that there were three mRNA transcripts present in the brain. Two of these mRNAs were also present in tissues other than the brain. Propose a hypothesis to explain the presence of RNAs in many tissues but protein only in the brain.

10-22 List an advantage of each of the following model organisms for the analysis of gene function. Give an example of the type of gene you might choose to analyze in each of these model organisms.
a. yeast (*S. cerevisiae*)
b. nematodes (*C. elegans*)
c. mouse (*M. musculus*)

10-23 The following factors pose technical challenges for the design of gene therapy experiments. Suggest ways to get around these problems.
a. large gene size in eukaryotes (up to 100 kb of DNA in the genome)
b. the need to get genes expressed in specific tissues

10-24 Researchers carried out the following experiments on the galactose utilization system in yeast. (1) They grew wild-type yeast in the presence and absence of galactose (that is, with the gal system running and shut down, respectively). (2) They used microarrays to quantitate all mRNA expression differences between these two states. (3) They used the ICAT approach to quantitate protein expression differences in 300 proteins in these two states. About 30 of 300 proteins changed expression levels under these conditions. However, for 15 of these 30 proteins, there was no change in their corresponding mRNA levels between the two states. Suggest two explanations for these observations.

10-25 It is often difficult to find genome-unique PCR primers in certain regions of the genome. Offer two explanations.

10-26 Complete genome sequences indicate that the human genome has roughly 40,000–60,000 genes, and the worm (nematode) genome has 19,000 genes. Explain how the human genome can encode a creature enormously more complex than the worm with only one-third more genes in its genome.

REFERENCES

Golub, T. R., Slonim, D. K., Tamayo, P., Huard, C., Gaasenbeck, M., Mesivov, J. P., Coller, H., Loh, M. L., Downing, J. R., Caligiuri, M. A., Bloomfield, G. D., and Lander, E. S. (1999). Molecular Classification of Cancer: Class Discovery and Class Prediction by Gene Expression Monitoring. *Science* 286:531–37.

Glusman, G., Rowen, L., Lee, I., Boysen, C., Roach, J. C., Smit, A. F., Wang, K., Koop, B. F., and Hood, L. (2001). Comparative Genomics of the Human and Mouse T Cell Receptor Loci. *Immunity* 15:337–49.

Glusman, G., Yanai, I., Rubin, I., and Lancet, D. (2001). The Complete Human Olfactory Subgenome. *Genome Research* 11:685–702.

Gygi, S. P., Rist, B., Gerber, S. A., Turecek, F., Gelb, M. H., and Aebersold, R. (1999). Quantitative Analysis of Complex Protein Mixtures Using Isotope-Coded Affinity Tags. *Nature Biotechnology* 17:994–99.

The Human Genome. (2001). *Nature* 409. This issue contains the draft sequence generated by the public International Human Genome Sequencing Consortium.

The Human Genome. (2001). *Science* 291. This issue contains the draft sequence generated by Celera.

Ideker, T., Galitski, T., and Hood, L. (2001). A New Approach to Decoding Life: Systems Biology. *Annual Review of Genomics and Human Genetics* 2:343–72.

Lander, E. S. (1999). Array of Hope. *Nature Genetics* 21:1–4. This issue contains a wide variety of articles on the DNA array technology and its applications.

Schwikowski, B., Uetz, P., and Fields, S. (2001). A Network of Protein-Protein Interactions in Yeast. *Nature Biotechnology* 18:1257–61.

The Direct Detection of Genotype Distinguishes Individual Genomes

An eighteenth-century painting depicting the 64 offspring of the ninth Sultan of Yogyakarta and his 13 wives. In this family tree, boys are depicted as fruit, girls as leaves. The two fruits emerging from a single stem on the *right* indicate twins. The painting is currently located in the Sultan's Palace inside the central walled-in city (Kraton) of Yogyakarta (on the island of Java in present-day Indonesia).

A couple whose firstborn suffers from cystic fibrosis learns from the medical diagnosis of that child's symptoms that both parents are carriers of a recessive disease-producing mutation. Together they run a 25% risk of transmitting the life-threatening hereditary condition to a second child. In the early 1990s, one such couple did not want to take that chance and, after genetic counseling, decided to try an experimental protocol: *in vitro* fertilization combined with the direct detection *in vitro* of the embryo's genotype, before its placement in the mother's womb.

The procedure took less than a week. At the start, a team of medical workers, including an obstetrician and a reproductive biologist, obtained 10 eggs from the woman and fertilized them with sperm from the man. Three days later, after the fertilized eggs had undergone several mitotic divisions to generate embryos with 6–10 cells, a research assistant used micropipettes to remove one cell from each embryo (Fig. 11.1). Since the natural formation of healthy identical twins results from embryos being split into two groups of cells at this stage, the removal of a single cell would not prevent normal development.

Next, the research assistant used PCR (see p. 285 of Chapter 9) to amplify from each isolated cell a specific DNA segment of chromosome 7 containing the site of the most common mutation within the gene responsible for cystic fibrosis (the *CFTR* gene). Previous cloning and sequencing of the gene made it possible to obtain primers bracketing this segment. Since the embryonic cells were diploid, they each supplied two alleles of the gene, one from the mother and one from the father.

To *genotype,* that is, detect which alleles were present in the material amplified from each cell, clinicians divided each cell's PCR product into two equal portions, or aliquots. They exposed one aliquot to an allele-specific hybridization probe for the normal *CFTR* allele, the other to an allele-specific probe for the mutant allele. If only the first aliquot from a sample produced a signal, the embryo represented by the sample was a homozygote for the normal allele. If both aliquots were positive for the probe, the embryo was a heterozygote. And if only the second aliquot glowed in response to the probe, the embryo carried two copies of the disease allele.

371

(a)

Ovary

1. Ripe eggs are retrieved from the ovary with a syringe.

2. Extracted eggs are fertilized with sperm.

3. At 6–10-cell stage, one cell is removed from each of six viable embryos.

4. In each of six isolated cells, site of common mutation in *CF* gene is amplified with PCR.

Primers

Known site of mutation

Amplified products

5. Divide PCR product into two portions. Denature. Apply one dot of each sample onto nitrocellulose filter.

6. Hybridize for normal *CF* allele.

7. Hybridize with ASO for mutant *CF* allele.

(b) Diagnosis

	Cell 1	Cell 2*	Cell 3	Cell 4*	Cell 5	Cell 6
Normal ASO	○	●	○	●	●	○
Mutant ASO	●	○	●	●	○	●

* Cells from embryos later transplanted into uterus

Figure 11.1 Preimplantation embryo diagnosis. Plucking one cell from an eight-cell embryo for the direct detection of genotype.

It took about 8 hours to determine the genotype of each isolated cell and, by inference, of each embryo. On the same day that the embryos were biopsied and genotyped, the doctor, in consultation with the parents, selected for placement in the mother's womb two embryos of known genotype. Of the two embryos they chose, one was a heterozygous carrier, the other a homozygote for the normal allele of the *CFTR* gene. The use of two to four embryos improves the chances of at least one implantation and is a part of most *in vitro* fertilization procedures. The selection of a carrier means that other embryos homozygous for the normal allele were most likely defective in some other way. As it happened, the embryo carrying two normal alleles of the *CFTR* gene implanted into the woman's uterus; the carrier embryo did not.

With this approach, the woman was able to embark on a pregnancy free of genetic concern, knowing that the embryo developing within her womb carried two copies of the normal allele for the *CFTR* gene; as a result, this child, as well as its children, would be free of cystic fibrosis. Nine months later, the mother gave birth to a healthy 7 lb 3 oz baby girl. When evaluated by a pediatrician at her four-week checkup, the infant daughter was found to be completely normal in both physical and mental development. At the same time, testing at an independent laboratory confirmed that neither of her homologous chromosome 7s carried the *CFTR* mutation.

The couple in this report worked with university doctors and laboratories whose funding and operations are approved by review boards that monitor social and ethical as well as scientific issues. The Genetics and Society box "Social and Ethical Issues Surrounding Preimplantation Embryo Diagnosis" on pp. 374–375 discusses several of these issues.

In this chapter, we examine how geneticists use an array of molecular tools to detect DNA differences among individuals. Once cloning and sequencing have identified a DNA difference, less cumbersome protocols can use this information for the rapid detection of genotype in many individuals—human, animal, or plant. The techniques are sensitive enough to operate on single hair follicles and even single cells from human embryos.

The ability to detect genotype directly at the DNA level has far-reaching consequences. It provides geneticists with tools for mapping and cloning the genes responsible for human diseases that, like the cystic fibrosis gene, were previously defined only by their effect on phenotype. Once the genes are cloned, DNA genotyping can help predict the possibility of future disease—*in vitro, in utero,* or after birth—or reveal the presence of a silent recessive disease allele in a carrier who shows no evidence of disease. The DNA-based technology is also the foundation of DNA fingerprinting, a method for evaluating the most prevalent DNA differences among individuals. DNA fingerprinting can provide legally acceptable evidence for or against a match between a suspect's genome and that of biological material (such as blood, semen, or even a few hairs) left behind at the scene of a crime.

Two general themes emerge from our discussion. First, the ability to distinguish genotypic differences of all kinds extends our concept of a locus and the alleles that define it. Mendel, Morgan, and other geneticists working before the advent of molecular technology identified a locus indirectly, inferring its existence from the observation of alternative phenotypes generated by alternative alleles of a gene; in this early era of genetic analysis, loci and genes were considered equivalents. By contrast, modern geneticists, able to look at genotype directly, can pick out genotypic differences in both genes and noncoding DNA regions on the basis of changes in DNA sequence alone. As a result, a **locus** is now a designated location anywhere on chromosome; it can range in size from a single base pair to more than a megabase pair. A **gene** is simply a coding locus. An **allele** is one of two or more alternative DNA sequences found at a locus—either coding or noncoding—in different homologs of the same chromosome, in a single individual, or in different individuals. While different alleles of a gene and its associated regulatory sequences may affect the function of the gene's RNA or polypeptide product with consequent effects on overall phenotype, alleles at noncoding, nonregulatory loci have no effect on phenotype. We will see that researchers can use DNA variations in noncoding regions as allelic markers to identify, locate, isolate, and follow the transmission of nearby genes.

The second general theme is that techniques for the direct detection of genotype provide access to genetic details about human individuals never before available. This potential for gathering and storing genetic data raises questions about the information so obtained: Who should have access to it? And how should it be used?

Our presentation of strategies for distinguishing genomes describes:

- Four classes of DNA variation that serve as the basis for the direct detection of genotype.

- Genotyping protocols that use hybridization, gel electrophoresis, PCR, microarrays, and sequencing (described in previous chapters) to distinguish alleles of the four types of polymorphic loci. These protocols include DNA fingerprinting, which produces a quick snapshot of a combination of genomic features that can identify an individual.

- Positional cloning: the application of direct genotype detection to linkage analysis, with the ultimate goal of cloning the genes behind any phenotypic differences among people, animals, or plants.

- Ways to look at the genes behind the complex inheritance patterns arising from incomplete penetrance, variable expressivity, and other phenomena that confound straightforward linkage analyses.

- The use of haplotype association studies to map disease loci in humans and other organisms. This novel strategy does not depend on the genetic pattern of allele transmission from parent to child; instead it relies on chromosomal evolutionary history.

Genetics and Society

Social and Ethical Issues Surrounding Preimplantation Embryo Diagnosis

■ In 1990, Germany passed a law that prohibits preimplantation embryo testing for any purpose.

■ A 1993 report from Canada's Commission on Reproductive Technologies warns against allowing market forces to determine the use of reproductive technologies. It also calls for creation of a permanent regulatory and licensing body to govern all aspects of the new reproductive practices, including sperm banks and *in vitro* fertilization.

■ In 1994, France and Norway passed legislation that limits genetic testing to situations in which the results are medically therapeutic, and the legislation authorizes governmental bodies to establish the criteria for defining "therapeutic" in this context. These laws prohibit the use of genetic testing for sex selection and normal trait enhancement.

■ In 2001, the governing board of the American Society for Reproductive Medicine (ASRM)—the major organization representing fertility doctors in the United States—decided that preimplantation sex selection is acceptable when parents already have a child of one sex and want to have another child of the opposite sex. The U.S. government has no stated position on preimplantation embryo diagnosis or other reproductive technologies. It views these practices as affairs of family life, and like marriage, divorce, and adoption, it

considers their regulation to be the purview of state law. However, most states have no specific laws either.

■ In 2002, U.S. women could sell their eggs to the highest bidder in 48 states. Advertisements in some college newspapers offered up to $50,000 for eggs from women who met highly selective criteria of height, appearance, SAT scores, and grades. In all countries of Western Europe, as well as Australia and Canada, egg selling is strictly forbidden.

This range of responses to the issues generated by the new reproductive technologies shows a diversity of approaches based in part on national culture and history. It also reflects international apprehension about the potential for misuse and abuse of the new technologies. Here are some of the main concerns.

When Should the Tests Be Used?

The couple in our opening story whose firstborn suffered from cystic fibrosis faced a medical problem. Preimplantation diagnosis could help them have a second child unaffected by the disease. With no cure at present for CF and no therapy that allows CF-affected people to look forward to a life of normal length, this is an example of medically therapeutic testing. Governmental and professional committees in most industrialized countries permit preimplantation diagnosis for this purpose, but in Germany and Japan, it is banned. Even in countries where preimplantation selection against disease is permitted, there is vocal opposition from a signif-

---

DNA Variation Is Multifaceted and Widespread

Members of the Same Species Show Enormous Sequence Variation in Their Genomes

If molecular geneticists were to clone and sequence the same 250 kb region of genomic DNA encompassing the cystic fibrosis (*CFTR*) gene on chromosome 7 from the two homologs of most healthy people, they would find about one difference in every 1000 bp, or 250 DNA differences in all (Fig. 11.2). With so many differences between these allelic sequences, you might suspect that one allele is wild type and the other mutant.

But a geneticist would tell you that none of the observed differences is likely to have any effect on the actual function of the *CFTR* gene; both homologs have wild-type activity. This is because only a small fraction of the total DNA stored in our chromosomes actually encodes or regulates the expression of

Two cystic fibrosis (*CFTR*) alleles from two healthy individuals

Figure 11.2 Base-pair differences between DNA cloned from the cystic fibrosis locus of two healthy individuals. The differences highlighted between the two *CFTR* alleles map to a single intron and have no effect on the function of the gene.

icant minority of people who hold a religious belief that all human embryos—even at the earliest stages of development—have a right to life. Others object to preimplantation diagnosis because they think people shouldn't have a right to interfere with the natural process of allele segregation. While geneticists view this process as fundamentally random, some people believe that it is guided in a purposeful way by a supernatural force or by "nature."

How Should the Tests Be Carried Out?

The couple in our opening story began by consulting a genetic counselor and then worked with medical practitioners associated with a university laboratory. Most geneticists agree that counseling before a procedure should foster an open discussion of all the issues (including the possibility that, in the case of cystic fibrosis, the tests might give false negatives) and that long-term follow-up should be part of the process. Preimplantation testing itself, like other forms of genetic testing, should be carried out by highly trained personnel in licensed laboratories. These accredited laboratories operate according to professional standards and have scientific and ethical review boards that monitor all work.

Who Should Have Access to the Technology?

The combination of *in vitro* fertilization and preimplantation testing cost $6000–$10,000 in 2002. Should the government provide tests for people who cannot afford them? How should society decide this issue? (A related discussion of access to medical technology appears in the Genetics and Society box on pp. 34–35 of Chapter 2.)

Should Parents Have the Right to Make Any Genetic Decision?

If, for instance, parents decide to have a child affected by a genetic disease, should they bear all financial responsibility for the child's care, or does society have an obligation to bear the cost of such medical treatment? On the other hand, how should physicians handle a request from prospective parents who wish to select *against* alleles responsible for minor diseases like myopia or late-onset diseases like Huntington disease or Alzheimer's disease? What about selection *for* alleles that provide a child with a relative advantage like complete resistance to infection by HIV (which is inherited naturally by ~10% of individuals in some populations).

Who Should Have Access to Test Results?

Just the parents? The parents and eventually the child? The parents, the child, and certain community institutions, such as schools? Some combination of these plus commercial enterprises such as insurance companies and places of employment? (We discuss this same question of privacy in relation to other types of genetic testing in the Genetics and Society boxes in Chapters 2 and 3.)

What Constitutes a Person?

Cultural and religious beliefs, rather than scientific knowledge, are the basis for answers to this question. Some people see preimplantation diagnosis as an alternative to abortion that allows a couple to make a decision before pregnancy begins. Others argue that even at the eight-cell stage, a microscopic preimplantation embryo is the equivalent of a human being, and rejection of an embryo is the equivalent of killing a human being. The difference between these two positions is in part the result of different religious beliefs about the moral significance of embryonic cells.

Although there are no simple solutions to these complex issues, geneticists around the globe agree on the need for continuous open discussion and tight oversight of the development of the new reproductive technologies.

---

proteins. Within the 250 kb of the *CFTR* gene, only about 4.5 kb (slightly less than 2%) consists of protein-coding exons or regions that play a role in gene regulation. When mutations occur in these coding or regulatory sequences, they are rarely passed on to future generations, because their effect on gene function is almost always negative. As a consequence, the vast majority of the gene's DNA variation that remains in a population is in noncoding, nonregulatory regions.

Geneticists consider variations at any position in the genome as alternative alleles of a particular chromosomal location, or *locus*. Before the development of recombinant DNA technology, the identification of a locus depended on mutations that gave rise to alternative phenotypes; investigators would conduct breeding experiments and follow these alternative phenotypes by observation. Today, any variation in DNA sequence, once discovered, produces a locus that can be looked at directly, even if it does not affect phenotype.

When two or more alleles exist at a DNA locus, the locus is considered **polymorphic,** and the variations themselves are called **DNA polymorphisms.** When a particular polymorphic DNA locus is useful for mapping studies, disease diagnosis, or any other analytical task, the locus is called a **DNA marker;** like a mile marker on a highway, it marks a particular point in the genome. If that position has no known function, the site is an **anonymous locus.**

Studies of numerous animal species indicate that the average frequency of polymorphism between any two homologous chromosomes is just as high as that found within the cystic fibrosis locus. The data are remarkably consistent. In all species evaluated, a sequence difference between homologous chromosomes from the same individual or from two different individuals appears on average once in every 1000 bp. Thus a comparison of the 3×10^9 bp of DNA in two haploid human genomes (for instance, those in any two sperm or egg cells) would exhibit 3 million allelic differences.

In itself, this is an astonishing level of variation. But comparison of a second pair of haploid human genomes would also reveal differences at 3 million loci, and only some of these differences would be the same as those found in the first comparison; others would be at new loci. Indeed, the more human genomes you look at, the more polymorphic loci you will find. The total number of polymorphic human loci may be 100 million or more. This is an enormous reservoir of potential DNA markers.

Geneticists Categorize DNA Polymorphisms in Four Different Classes

For the purposes of genotyping, DNA polymorphisms can be placed into one of four classes based on their impact on DNA sequence, their frequency in the genome, their stability, and their role in genetic analysis (Table 11.1). All four types of variation are found throughout the genomes of all multicellular eukaryotic organisms.

Single Nucleotide Polymorphisms (SNPs)

The simplest, most prevalent, and most generally useful class of DNA polymorphisms arise from single base-pair substitutions. Such base-pair changes, which can be triggered by mutagenic chemicals or mistakes in DNA replication, are referred to as **single nucleotide polymorphisms,** or **SNPs.** SNPs account for the vast majority of the total variation that we just discussed. Nearly all SNPs are biallelic (that is, they have only two alleles). The ratio of the two alleles among the chromosomes in a population can range from less than 1:100 to 50:50.

Sequencing the same region of the genome from several individuals can identify large numbers of SNPs. By December 2001, the international SNP Consortium—a nonprofit organization coordinating the efforts of both public and commercial research centers—and private biotechnology companies had identified and mapped over 5 million human SNPs.

Although SNPs in coding sequences can alter the amino-acid sequence of a gene product and have a direct impact on phenotype, the vast majority of SNPs occur at anonymous loci. There is no evolutionary advantage or disadvantage to mutations at these noncoding, nonregulatory loci. For this reason, and because the probability that any particular base in the human genome will mutate is so low (1×10^{-9} per generation), scientists believe that each anonymous SNP represents a single mutational event. This means that if you and a friend share an allele at an anonymous SNP locus, you both got that allele from the same ancestor (who may have lived thousands or even hundreds of thousands of years ago). The fact that every random pair of human beings on the planet share many unlinked SNP alleles indicates an entangled recent common ancestry for all people.

We will see that even though anonymous locus SNPs do not have a direct effect on phenotype, some lie so close to a disease gene or other genes influencing significant phenotypic differences (such as positive or negative responses to a particular medication) that they can serve as **DNA markers:** specific DNA loci with identifiable variations. Medical researchers can use such markers to identify and follow phenotypic differences in groups of people.

Microsatellites

The genomes of humans and other complex organisms are loaded with loci defined by repeated sequences. The repeating unit can be as short as a single base pair or as long as tens of kilobases. The number of repeats can vary between two and thousands. Here, we discuss the category of repeat sequences known as **microsatellites,** which are DNA elements composed of one-, two-, or three-base sequences repeated in tandem 15–100 times. Examples are AAAAAAAAAAAAAAA or CACACACACACACACACACACA. In the mammalian genome, the CA-repeat microsatellite occurs once in every 30,000 bp. Microsatellites arise spontaneously from random events that initially produce a short repeated sequence with four to five repeat units. Once a short microsatellite mutates into existence, however, it can expand into a longer sequence by the process shown in Fig. 11.3.

TABLE 11.1 Classes of DNA Polymorphisms

Class	Size of Locus	Number of Alleles	Number of Loci in Population	Rate of Mutation	Use	Method of Detection
SNP	Single base pair	2	100 million	10^{-9}	Linkage mapping	PCR followed by ASO hybridization or primer extension
Microsatellite	30–300 bp	2–10	200,000	10^{-3}	Linkage mapping	PCR and gel electrophoresis
Multilocus Minisatellite	1–20 kb	2–10	30,000	10^{-3}	DNA fingerprinting	Southern blot and hybridization
Small Changes in DNA Content (deletions and duplications)	1–100 bp	2	N/A	$<10^{-9}$	Linkage mapping	PCR and gel electrophoresis

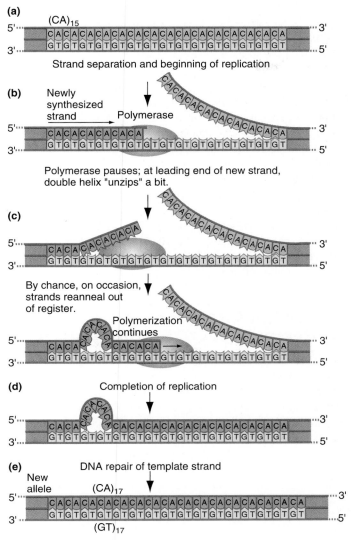

(a) $(CA)_{15}$

Strand separation and beginning of replication

(b) Newly synthesized strand → Polymerase

Polymerase pauses; at leading end of new strand, double helix "unzips" a bit.

(c)

By chance, on occasion, strands reanneal out of register.

Polymerization continues

(d) Completion of replication

(e) DNA repair of template strand

New allele $(CA)_{17}$

$(GT)_{17}$

Figure 11.3 Microsatellites are highly polymorphic because of their potential for faulty replication. (a) A microsatellite consisting of 15 tandem repeats of the CA dinucleotide sequence. **(b)** During replication, the strands of the DNA molecule separate and DNA polymerase moves in the 5′-to-3′ direction along each strand. **(c)** Throughout the replication process, pauses can occur if the required nucleotides are, by chance, not in the vicinity of the polymerase. During a pause, the newly synthesized strand may "unzip" or "breathe" apart from the template strand over a length of 10–20 bases. **(d)** When the required nucleotides become available, the newly synthesized strand reanneals to the template and acts as a primer for further replication. After the reannealing in a microsatellite, the new strand may be out of register such that the polymerase begins by adding one or more nucleotides across from a part of the template strand that has already been replicated. **(e)** The resulting DNA molecule will have one or more identical repeats in the newly synthesized strand that were not present in the template strand. DNA repair processes then adjust the template strand to make it the same length as the newly synthesized strand.

Unlike SNPs—which are biallelic and do not change after the mutational event that gave rise to them—individual microsatellite loci often mutate into multiple alleles. Research shows that faulty DNA replication is the main mutational mechanism (Fig. 11.3). Because the same short homologous unit (CA, for example) is repeated over and over again, DNA polymerase may develop a stutter during replication, that is, it may slip and make a second copy of the same dinucleotide, or skip over a dinucleotide. Microsatellites are thus highly polymorphic in the number of repeats they carry, with many alleles distinguishable at each microsatellite locus.

New alleles arise at microsatellite loci at an average rate of 10^{-3} per locus per gamete (that is, one in every thousand gametes). This frequency is much greater than the single nucleotide mutation rate of 10^{-9} and results in a large amount of microsatellite variation among unrelated individuals within a population. At the same time, the rate of microsatellite mutation is low enough that changes usually do not occur within a few generations of even a large family; because of this, microsatellites can serve as relatively stable, highly polymorphic DNA markers in linkage studies of human families, other animals, and plants.

Minisatellites

Minisatellites, a second important category of genomic DNA repeats, are larger than microsatellites. The repeating units that compose them are 20–100 bp long, and each unit is repeated up to thousands of time per locus. This gives each minisatellite locus a total length of 0.5–20 kb (Fig. 11.4). Like microsatellites,

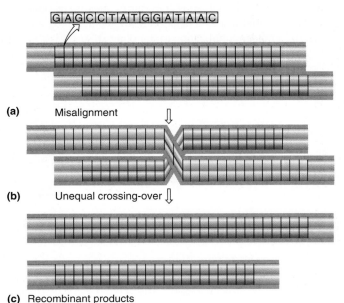

GAGCCTATGGATAAC

(a) Misalignment

(b) Unequal crossing-over

(c) Recombinant products

Figure 11.4 Minisatellites are highly polymorphic because of their potential for misalignment and unequal crossing-over during meiosis. Minisatellites are composed of relatively long tandem repeating units of identical sequence. **(a)** Misalignment and **(b)** unequal crossing-over produce **(c)** recombinant products that contain different numbers of repeating units than either parental locus; each new recombinant product is a new allele.

minisatellites arise from random events and are dispersed throughout the genomes of all vertebrates.

Minisatellite loci tend to be highly polymorphic because it is easy for homologous chromosomes with sections of many short repeats to line up out of register along the repeated units during synapsis (Fig. 11.4). Any crossing-over within the mis-aligned region will be unequal and produce reciprocal homologs that contain either duplications or deletions of one or more of the repeat units. The result is two new alleles—one with fewer, the other with a greater number of units than those on the parental chromosomes.

For geneticists, the most useful minisatellites loci can cross-hybridize to between 5–10 loci distributed throughout the genome. Using several unrelated minisatellite sequences as hybridization probes can provide an overview of the whole genome. The frequency of such multilocus minisatellites is about 1 per 100,000 bp, for a total of ~30,000 in the whole human genome.

Deletions, Duplications, and Insertions at Nonrepeat Loci

The DNA changes in this broad category are the result of muta-genic events that expand or contract the length of a nonrepeti-tive DNA locus (which excludes micro- and minisatellites) by deleting, duplicating, or inserting one or more base pairs. Deletions, duplications, and insertions can range in size from a few base pairs to multiple megabases. Chapter 13 examines the larger scale chromosomal alterations in genomic content.

Small deletions and duplications—from a few base pairs to a kilobase in length—most often arise from unequal crossing-over between nonhomologous sites during meiotic recombina-tion. The material lost as a deletion from one homolog will be gained as a tandem duplication to the other (Fig. 11.5a). Small

insertions ranging in size from hundreds to thousands of bases can also be caused by transposable elements that integrate at random into the genome (Fig. 11.5b), as described in detail in Chapter 13.

Deletions, duplications, and insertions of nonrepeated genomic sequences are much less common than SNPs, microsatellites, and minisatellites, but as we see in the next sec-tion, they are easy to analyze, and some are associated with par-ticular types of disease mutations.

SNPs, microsatellites, minisatellites, and deletions, inser-tions, or duplications at nonrepeat loci provide the basis for detecting genotypic differences among individuals. Once com-parative DNA sequencing or other analytical studies have identified a DNA variation or set of variations, researchers can use faster, less cumbersome tools to follow the transmis-sion of specific variants in families and populations. We now describe the primary techniques for the direct detection of DNA polymorphisms.

Detecting DNA Genotypes of Different Types of Polymorphisms

One general approach to determining an individual's genotype at a particular polymorphic locus is to extract genomic DNA from the individual, use that DNA to obtain clones of the locus, and then subject those clones to DNA sequencing. But as we saw in Chapter 9, this approach is labor-intensive, time-consuming, and expensive. And you can never be certain that you have sequenced both copies of the locus in a person or other organism unless you sequence multiple clones of the same region and eventually see two different DNA sequences, which cannot happen with a homozygote. Thus, cloning fol-

(a) Unequal crossing-over

(b) Transposon insertion

Figure 11.5 Deletion, duplications, and insertions cause variation in genomic content. (a) Unequal crossing-over occurs between different loci on homologous chromosomes. One product has a deletion over the region between the crossover sites; the second product has a duplication over the same region. **(b)** A transposon can insert into a chromosome and cause an insertion. If the insertion occurs within a gene, the gene will no longer function.

lowed by sequencing is impractical for genotyping large numbers of loci in large numbers of individuals.

Practical protocols for large-scale DNA genotyping must be inexpensive, rapid, accurate, and easy to perform on hundreds or thousands of samples. Automated DNA genotyping methods that can analyze hundreds of thousands of samples are needed for both large-scale clinical diagnoses and the mapping of complex human traits.

We now describe a variety of genotyping methods suitable for different types of polymorphisms.

There Are Three Approaches to the Direct Detection of SNPs

Southern Blots or PCR Can Detect SNPs That Alter Restriction Sites

A small proportion of SNPs by chance eliminate or create a restriction site recognized by a restriction enzyme. When this happens, researchers can use the restriction enzyme to distinguish between the two alleles. Consider, for example, the SNP shown in Fig. 11.6. A single base pair change from A:T in allele 1 to G:C in allele 2 determines the presence or absence of an *Eco*RI restriction site.

Southern Blot Analysis. Until recently, a Southern blot (see pp. 296–299 in Chapter 9) was the tool of choice for detecting a SNP-caused restriction site polymorphism. In the protocol, genomic DNA from the test sample is treated with *Eco*RI, and the resulting restriction fragments are separated by gel electrophoresis and then transferred to a filter paper. The resulting Southern blot is hybridized with a DNA probe obtained from the region between the polymorphic restriction site and an adjacent nonpolymorphic restriction site.

The length of the genomic restriction fragment that hybridizes to the probe reveals which version of the polymorphic restriction site is present. The probe in Fig. 11.6 detects a 3 kb restriction fragment in DNA with SNP allele 1 and a 5 kb restriction fragment in DNA with SNP allele 2. Because the different SNP alleles, when analyzed in this way, change the size of the hybridizing restriction fragment detected on the Southern blot, this type of polymorphism is called a *restriction fragment length polymorphism*, or *RFLP*.

PCR Analysis. Once the sequences on both sides of a restriction site polymorphism are known, genotypers can detect the polymorphism by a PCR-based protocol that is much quicker and cheaper, and requires much less material than Southern blot analysis. There are three steps to this protocol: (1) amplification by PCR of a several hundred base-pair region encompassing the SNP; (2) exposure of the PCR products to the appropriate restriction enzyme; and (3) evaluation of the samples by gel electrophoresis and ethidium bromide staining, followed by a reading of the size of the DNA fragments off the gel (Fig. 11.7).

We illustrate this experimental approach with a solution to the real-life problem of detecting the mutation at the β-globin locus that is responsible for sickle-cell anemia. Sickle-cell ane-

(a)

(b)

Figure 11.6 SNP-caused restriction fragment length polymorphisms can be detected by Southern blot analysis. **(a)** A SNP can create a restriction site polymorphism at an *Eco*RI site. The two SNP alleles will produce different-sized restriction fragments that can be detected with a DNA probe from a region between the polymorphic site and an adjacent site. **(b)** Southern blot analysis and hybridization with the probe can distinguish among the three possible genotypes at this SNP locus.

mia occurs, as we have seen, when a person carries two copies of a β-globin gene with a single-base substitution that replaces an A with a T and changes the encoded amino acid from glutamic acid to valine. The normal allele is called *A,* and the sickle-cell allele *S.* Since the sickle-cell mutation also by chance destroys the recognition site of the restriction enzyme *Mst*II (Fig. 11.7a), it is possible to use PCR and restriction enzyme digestion to detect the mutant allele.

Suppose a carrier couple (both of genotype *AS*) have a child with sickle-cell anemia (genotype *SS*) and want to know the genotype of the fetus they have recently conceived.

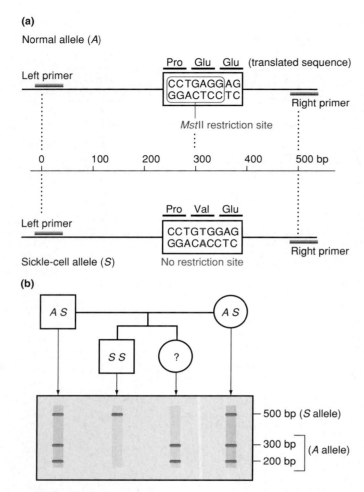

(a)

Normal allele (A)

Pro Glu Glu (translated sequence)

Left primer

CCTGAGG AG
GGACTCC TC

Right primer

*Mst*II restriction site

0   100   200   300   400   500 bp

Pro Val Glu

Left primer

CCTGTGGAG
GGACACCTC

Right primer

Sickle-cell allele (S)   No restriction site

(b)

A S   A S

S S   ?

500 bp (S allele)

300 bp
200 bp   } (A allele)

Figure 11.7 Sickle-cell genotypes can be detected rapidly with a PCR-based protocol (a) The normal (A) and sickle-cell (S) alleles at the β-globin locus differ by a single base-pair substitution that changes glutamic acid to valine in the protein product. The base-pair change also eliminates the restriction site *Mst*II. **(b)** PCR amplification of the region containing this SNP, with the primers shown, produces a 500 bp product. Exposure of the normal PCR product to *Mst*II digests this DNA fragment into two smaller fragments of 200 and 300 bp in size; exposure of the mutant PCR product to the restriction enzyme has no effect. Gel electrophoresis and ethidium bromide staining distinguish the three possible genotypes. Results shown are for a hypothetical family in which both parents are carriers, the first child is afflicted with the disease, and the genotype of the fetus is initially unknown. The results of this analysis show that the fetus will be homozygous for the normal A allele.

Through amniocentesis, their doctor recovers fetal cells from the pregnant woman's womb. The doctor subjects the genomic DNA in this sample, as well as in samples from both parents and the first child, to PCR amplification with primers complementary to sequences on either side of the sickle-cell mutation (Fig. 11.7a). The doctor then mixes the restriction enzyme *Mst*II with the PCR products and separates the resulting DNA fragments according to size by gel electrophoresis. It is easy to distinguish DNA from the normal allele—which is digested into two fragments by *Mst*II—from the indigestible DNA of the

mutant allele. Figure 11.7b shows the results: The fetus is *AA*, so the younger sibling will neither have sickle-cell anemia nor carry the sickle-cell trait.

Allele-Specific Oligonucleotide (ASO) Hybridization Can Detect All SNPs

Unfortunately for genotypers, most differences resulting from SNPs do not alter restriction sites. Fortunately, however, there are two general methods for detecting all SNPs. Here we describe a protocol that exploits differences in hybridization between a probe and two target sequences that differ by a single nucleotide.

A SNP is the proverbial needle in a haystack—a single base pair among, for example, the 3 billion in the human genome. The first step in nearly all methods for SNP genotype detection is the amplification by PCR of a short sequence—a few hundred base pairs of DNA—surrounding the SNP. With this step, the SNP becomes one nucleotide in several hundred, rather than one in several billion, and is much easier to detect.

How Allele-Specific-Oligonucleotide Hybridization Detects SNPs. It is only with very short probes—oligonucleotides containing around 20 bases—that single-base changes provide a large enough difference to be readily detected. This is because for very small DNA molecules—those composed of no more than 50 bp—the length of the molecule itself helps determine whether the double helix remains intact or falls apart. The effective length and therefore the strength of the hydrogen-bond forces holding together the double helix of a short-probe/short-target DNA hybrid depends on the longest stretch that does not contain any mismatches. When the two strands do not match exactly, there may not be enough weak hydrogen bonds in a row to hold them together.

If, for example, a 21-base probe hybridizes to a target strand that differs at a single base in the middle of the sequence, the effective length of the resulting double-stranded hybrids is only 10 bp. Since a 10 bp hybrid is significantly less stable than a 21 bp hybrid, one can devise hybridization conditions (by choosing the right temperatures) under which the perfect 21 bp hybrids will remain intact while the imperfect hybrids will not (Fig. 11.8a). By comparison, molecules longer than 50 bp can maintain their double helix conformation even with intermittent mismatches. Once a critical number of hydrogen bonds required for double helix stability is achieved, any further increase in the number of hydrogen bonds makes no difference (Fig. 11.8b).

Short oligonucleotides of 18–24 bases that hybridize to only one of the two alleles at a SNP locus under appropriate conditions are known as **allele-specific oligonucleotides,** or **ASOs** (Fig. 11.9a). Figure 11.9b–e shows the results of an ASO-based test to distinguish the three possible genotypes at an SNP locus.

The power of the PCR/ASO hybridization protocol is its simplicity. It does not require the running of a gel, and it can be automated if the necessary robotic resources are available. The disadvantage is that the reaction conditions needed for the opti-

Figure 11.8   Short hybridization probes can distinguish single-base mismatches. (a) Researchers allow hybridization between a short 21-base probe and two different target sequences. **(1)** A perfect match between probe and target extends across all 21 bases. When the temperature rises, this hybrid has enough hydrogen bonds to remain intact. **(2)** With a single-base mismatch in the middle of the probe, the effective length of the probe-target hybrid is only 10 bases. When the temperature rises, this hybrid does not have enough hydrogen bonds to remain intact, and it falls apart. **(b)** Researchers allow hybridization to occur with probes of 50 bases and 100 bases. **(1)** A perfect match between a 50-base probe and its target bases achieves a maximum of stability such that any extension in the length of the match has no significant effect on the temperature at which the hybrid falls apart. **(2)** Thus, it is not easy to distinguish a 100 bp hybrid with one mismatched base from a 100 bp hybrid with a perfect match.

mal distinction between two SNP alleles are difficult to predict. Some SNPs are so sensitive to conditions that slight alterations in temperature can lead to false positives or negatives in allele detection.

We now describe a significant clinical application of ASO analysis.

Large-Scale Multiplex ASO Analysis with Microarrays Can Be Used for the Clinical Detection of *BRCA1* Mutations. Breast cancer, like cystic fibrosis and many other common diseases, arises from any one of hundreds of different mutations in the same gene—some known, some still undefined. Thus, while a positive result in a screen for one particular mutation, for example, in the *BRCA1* gene, can identify a disease genotype, a negative result doesn't rule out the possibility of other disease-causing mutations. The only sure way to discover whether a person carries any of the many possible mutations is to sequence the entire gene from that individual.

The gel-based sequencing protocol described in Chapter 9 is too expensive for this type of routine clinical analysis. A more rapid and efficient approach relies on SNP analysis with ASOs. Figure 11.10 illustrates how to use this whole-gene mutation-detection protocol with the *BRCA1* gene. The *BRCA1* coding region is 5500 bp long. To detect a mutation

anywhere in the sequence, for each position, you design four ASOs with identical sequences flanking the position but with a different one of the four bases at the position itself. After repeating this process for all 5500 positions, you have 22,000 different ASOs. You now place each of these on the surface of a microarray.

In our discussion of ASOs, DNA from the genomic samples was attached to a surface and labeled probes were presented in the solution for hybridization. By comparison, in this version of the protocol, it is the probes that are attached to the surface (of the microarray) and the sample that is labeled.

After PCR amplification of the *BRCA1* coding region, the sample is labeled with a fluorescent dye and applied to the microarray under conditions that optimize hybridization only between perfectly complementary sequences. An automated analysis of fluorescence at each ASO now determines which base is at each position in the *BRCA1* coding sequence. Figure 11.10 compares the results from a normal individual carrying two wild-type alleles and a person heterozygous for a single base mutation.

The construction of a microarray chip and determination of the optimal conditions for the analysis of a particular genomic region is time-consuming and expensive. But once the chip is designed and the conditions determined, each use of the

Figure 11.10   The clinical diagnosis of *BRCA1* mutations depends on sequencing with allele-specific oligonucleotides on microarrays. A small section of a microarray with ASOs for each possible base between positions 2420 and 2440 of the *BRCA1* coding region. Each column has four rectangles that separately contain ASOs differing only at the nucleotide position under analysis. *BRCA1* DNA from any one allele can only be complementary to one of the four ASOs in each column, but DNA from individuals heterozygous for a single nucleotide mutation will hybridize to two ASOs in a particular column. Automated analysis determines which one or two of the four ASOs in each column display the highest level of hybridization. The readout for each base position is shown below the corresponding column. The *top* microarray has been hybridized to amplified genomic *BRCA1* DNA from an individual with a normal homozygous genotype; the *bottom* microarray shows hybridization to DNA from a heterozygote. The lower microarray shows evidence of heterozygosity for a mutation from T–C at position 2430.

Figure 11.9   ASOs can determine genotype at any SNP locus.
(a) The first step is the synthesis of two ASOs—oligonucleotides that are complementary to the short sequence surrounding a defined SNP and differ only at the SNP itself. One ASO is complementary to one allele, and the second ASO is complementary to the other allele at a unique SNP locus. **(b)** Genomic DNA samples can have only one of three genotypes at this SNP locus. **(c)** After amplification of individual samples by PCR, the amplified DNA of each sample is divided into two aliquots, which are denatured and blotted onto different spots on a filter paper. **(d)** One aliquot from each sample is hybridized to the wild-type ASO; the other aliquot is hybridized to the sickle-cell ASO. **(e)** Autoradiography indicates the genotype of each individual.

technology is relatively inexpensive. This approach is thus most appropriate for detecting mutations at loci like *BRCA1*, which occur often enough in humans to warrant whole population screening for their presence.

Single Nucleotide Primer Extension Can Also Detect SNPs

An alternative to the ASO detection of SNP genotypes is called *DNA-polymerase-assisted single nucleotide primer extension*, or just *primer extension*. It is simpler and more accurate for all loci than the PCR/ASO method.

The primer extension protocol illustrated in Fig. 11.11 takes advantage of the high fidelity with which DNA polymerase attaches only template-complementary nucleotides to a growing DNA strand during DNA replication. As in all methods of SNP detection, the first step is to amplify by PCR the locus of interest. The amplified DNA is denatured into single strands and then mixed with an oligonucleotide primer that is complementary to a short sequence of DNA directly adjacent to the polymorphic nucleotide. The primer will hybridize to the same single strand of DNA from both SNP alleles (since they are identical in the adjacent region).

DNA polymerase is now added to the reaction along with a dideoxynucleotide type (ddCTP in the figure) that is complementary to one of the two SNP alleles. If the 13-base primer binds to a DNA strand from allele 1 of this SNP, DNA poly-

Figure 11.11 **SNP detection with DNA polymerase-assisted single nucleotide primer extension.** **(a)** The DNA sequence surrounding a SNP locus is amplified by PCR from a heterozygous genomic sample. The two PCR products differ only at the single base pair that defines the SNP locus. **(b)** Denaturation produces four types of single-stranded molecules. **(c)** A 13-nucleotide-long primer complementary to the sequence adjacent to the polymorphic base pair forms hybrids with DNA from both alleles. **(d)** With the addition of DNA polymerase and ddCTP, only primers bound to allele 1 can be extended to a length of 14 nucleotides (nt). In a separate experiment with ddATP, only allele 2 can be extended. **(e)** Mass spectrometry can determine which oligonucleotides (1 and/or 2) are present, which indicates the genotype of the genomic sample.

merase will extend the primer with a ddCTP, since it is complementary to the G nucleotide at the next position on the template. However, the enzyme cannot extend the primer bound to allele 2 because ddCTP is not the proper nucleotide.

When the reaction is completed, samples homozygous for allele 1 will contain oligonucleotides that have all been extended to 14 bases; samples homozygous for allele 2 will have only unextended 13-base oligonucleotides; and heterozygous samples will have an equal proportion of 13- and 14-base oligonucleotides. To replicate results, a second reaction can be carried out with just the dideoxynucleotide complementary to the base in the second SNP allele. The results should be the reverse of those produced in reaction 1 (Fig. 11.11d).

You now need a technique to distinguish oligonucleotides 13 bp long from those 14 bp in length. You could use gel electrophoresis, but another method called "Time-of-Flight mass spectrometry" is much more rapid and accurate. In it, minute quantities of the reaction products are vaporized and subjected to an electric field in a small vacuum chamber. The electric field moves negatively charged oligonucleotides toward a detector. In the field, lighter molecules travel faster than heavier molecules. The detector records the time that it takes for molecules to reach it. Oligonucleotides that differ by a single nucleotide are easily separated by this protocol.

The mass spectrometry detection system is rapid and easily automated. Researchers can genotype up to 10 loci simultaneously by combining primers of different lengths for each locus.

How to Detect Alleles That Change the Length of a Locus

The SNP category of DNA polymorphisms is defined by alternative versions of particular DNA sequences; as a result, all SNP detection methods are based on the ability to detect nucleotide variation. In contrast, the other three categories of DNA polymorphisms—microsatellites, minisatellites, and deletions/duplications/insertions—are all associated with variations in the actual size of the locus. Alleles that differ in size from one another can be directly and easily distinguished by gel electrophoresis.

When the size difference between alleles is less than 5 kb, genotyping can be readily and rapidly accomplished as follows.

You begin by using a pair of primers complementary to sequences on either side of the actual polymorphism to amplify by PCR the locus from an individual's DNA. You then subject the PCR products to gel electrophoresis to separate DNA fragments according to their size. After staining with ethidium bromide, each allele shows up as a specific band. All possible homozygous and heterozygous genotypes can be distinguished in this way (Fig. 11.12).

A single researcher can use this protocol to genotype hundreds of samples in a single day, without any specialized equipment (other than the apparatus necessary for PCR and gel electrophoresis). The protocol is also amenable to automation using fluorescently tagged primers and electrophoresis in the same apparatus that is used for automated DNA sequencing (described on pp. 302–307 in Chapter 9). In Chapter 13, we present an in-depth look at the general category of deletions, duplications, and insertions. Here we focus our attention on the special categories of microsatellites and minisatellites, which are both defined by polymorphic differences in numbers of repeating elements.

A Direct Look at Microsatellites

Microsatellites are by far the most prominent DNA loci defined by alleles that differ in size. The most useful microsatellites consist of 2 or 3 bp units repeated 15–100 times in a row. These microsatellites are highly polymorphic, with multiple alleles that differ in 2 or 3 bp unit increments. All of the multiple alleles can be detected as different-sized PCR products (Fig. 11.12).

Microsatellites have played a crucial role in the development of linkage maps across the genomes of mice, humans, and other vertebrates. The reasons for their widespread use include their frequent appearance in all vertebrate genomes, the ease with which they can be identified by hybridization to simple sequence probes, and their extensive polymorphism.

A small number of genes naturally contain microsatellite sequences with triplet repeat units within their coding regions. The propensity of the microsatellite sequences to change in size from one generation to the next can produce mutant alleles with drastic effects on phenotype. One example of a disease in this category is Huntington disease (HD).

Huntington disease is transmitted as an autosomal dominant mutation. Over 30,000 Americans show one or more symptoms of the disease—involuntary, jerky movements; unsteady gait; mood swings; personality changes; slurred speech; impaired judgment. An additional 150,000 have an affected parent, which gives them a 50:50 chance of expressing the dominant condition themselves as they age. Although symptoms usually show up between the ages of 30 and 50, the first signs of the disease have appeared in people as young as 2

(a) Determine sequences flanking microsatellites.

(b) Amplify alleles by PCR.

(c) Analyze PCR products by gel electrophoresis and staining.

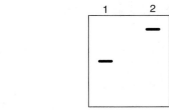

(d) Example of population with three alleles.

Six diploid genotypes are present in this population.

Figure 11.12 Detection of microsatellite polymorphisms by PCR and gel electrophoresis. (a) Microsatellite alleles differ in length. Left and right primers are devised based on sequences that flank the microsatellite locus. **(b)** Genomic DNA is amplified by PCR with primers specific for the microsatellite locus. **(c)** Gel electrophoresis and ethidium bromide staining distinguish the alleles from each other. **(d)** Microsatellites are often highly polymorphic with many different alleles present in a population. With just three alleles, there are six possible genotypes. With N alleles, there will be $\frac{N}{2}$ $(N + 1)$ genotypes.

and as old as 83. Some people with a family history of HD would like to know their genotype before deciding whether to have a family.

In 1993, after 10 years of intensive research, investigators identified and cloned the *HD* gene. With the gene in hand, they were able to uncover the unusual mutation that causes the disease (Fig. 11.13a). Unlike the vast majority of disease mutations, which result from base-pair changes or the elimination of genetic information, HD is caused by too much genetic information: an expansion of a CAG trinucleotide repeat in the coding sequence, which translates into a string of glutamine amino acids.

It is possible to detect *HD* alleles directly with the same size-based PCR procedure used to detect other microsatellite alleles. The normal allele contains up to 34 repeats while disease-causing alleles carry 42 or more. In general, the greater the number of repeats, the earlier the age of disease onset (Fig. 11.13b). Those who inherit a disease allele invariably get the disease if they live long enough. Thus, although expressivity, which depends on the number of triplet repeats, is variable, penetrance is complete. Several other diseases caused by triplet

repeat expansion have been uncovered, including a variety of neurological disorders, and the fragile X syndrome described in the Genetics and Society box on pp. 202–203 in Chapter 7.

A Direct Look at Minisatellites and DNA Fingerprinting

Minisatellites are another type of locus where alleles are distinguished by size, rather than by nucleotide sequence. We have seen that minisatellite alleles vary in length from 0.5–20 kb. Many minisatellite loci are amenable to the PCR analysis described for microsatellites. But the real power of minisatellites lies in the fact that particular minisatellite sequences often occur at a small number of different genomic loci. With restriction enzyme digestion, gel electrophoresis, and Southern blot hybridization using a cross-hybridizing minisatellite probe, researchers can look simultaneously at allelic variation at these multiple unlinked loci (Fig. 11.14). (This strategy wouldn't work for microsatellites because their core mono-, di-, or trinucleotide sequences are each present thousands of times in the genome.)

Because minisatellites provide a generalized picture of the whole genome rather than of a single locus, and because they are highly polymorphic, they are ideally suited for comparing the overall genetic relationship between two individuals or between an individual and a DNA sample recovered from the scene of a crime.

How many loci would you have to examine to be certain that two DNA samples come from the same individual (or identical twins) and no one else? A simple calculation shows that the probability of two unrelated individuals having identical genotypes at a locus with two equally prevalent alleles is 37.5%—quite a high probability. However, the chance that the same two individuals will be identical at 10 such loci, all unlinked, is only 0.375^{10}, or 0.005%—quite a low probability. The result of 0.005% means there is 1 chance in 20,000 that the two will by chance have the same genotype at 10 unlinked loci. By extension, if you simultaneously detect genotype at 24 separate two-allele loci, the chance of two individuals being the same at all 24 drops to 0.375^{24}, or 1 in 17 billion. Since the total human population is less than 8 billion, there is virtually no chance that two individuals (who are not identical twins) would have the same genotype at all 24 loci. In short, a relatively small number of loci are sufficient to produce a genotype pattern that, like a traditional fingerprint, will be unique for each individual (or pair of identical twins) within the species.

DNA Fingerprinting Defined. In 1985, Alec Jeffreys and coworkers made two key findings. Each minisatellite locus is highly polymorphic; and most minisatellites occur at multiple (usually between 2 and 50) sites scattered around the genome. As a result, they realized, minisatellite probes would be perfect reagents for obtaining a **DNA fingerprint:** a pattern produced by the simultaneous detection of genotype at a group of unlinked, highly polymorphic loci.

Fingerprinting protocols use restriction enzymes and Southern blotting to detect length polymorphisms at minisatellite

(a) Basic structure of the *HD* gene

CAG CAG CAG CAG CAG CAG CAG CAG CAG CAG CAG CAG CAG CAG CAG
Each triplet encodes glutamine.

(b) Some alleles at the *HD* locus

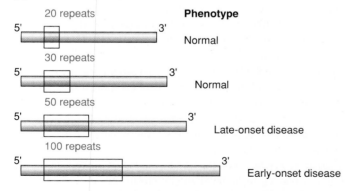

Figure 11.13 Mutations at the Huntington disease locus are caused by expansion of a triplet repeat microsatellite in a coding region. (a) Near the 5′ end of the coding region is a repeating triplet sequence that codes for a string of glutamines. **(b)** Different alleles at the HD locus have different numbers of repeating units. Fewer than 34 repeats gives a normal phenotype. As the number of repeats increases beyond 42, the onset of the disease is earlier.

(a) Digest DNA with restriction enzyme that does not cut inside microsatellite.

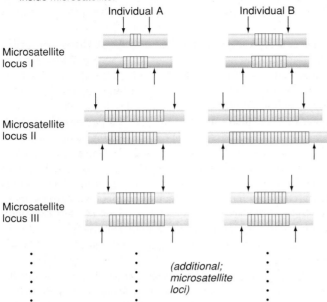

Individual A   Individual B

Microsatellite locus I

Microsatellite locus II

Microsatellite locus III

(additional; microsatellite loci)

(b) Run DNA samples on a gel. Perform Southern blotting. Hybridize with probe containing microsatellite sequence.

Individuals
A   B   C   D

Figure 11.14 Minisatellite analysis provides a broad comparison of whole genomes. (a) Two individuals each carry two alleles at three loci containing the same minisatellite sequence. Additional minisatellite loci are not shown. The *arrows* indicate a restriction site recognized by a particular restriction enzyme. Notice that minisatellite lengths are different both among alleles at a locus and among different loci. **(b)** After restriction enzyme digestion, gel electrophoresis, Southern blotting, and hybridization to the minisatellite probe, researchers can obtain an autoradiograph of the type shown here for four individuals. The darkness (or intensity) of each band depends on the degree of cross-hybridization.

loci. The simplest DNA fingerprint reveals which alleles an individual carries for one family of minisatellites. As you saw in Fig. 11.14, to obtain such a fingerprint, you simply digest a sample of genomic DNA with a restriction enzyme that does not cut within the minisatellite itself but does cut within closely flanking sites. You then run the restriction fragments on a gel, perform a Southern blot (as described on pp. 298–299), and probe the blot with a labeled DNA fragment derived from the minisatellite. The result is the actual DNA fingerprint, which consists of an autoradiograph displaying the pattern of different-sized fragments representing all the loci of that minisatellite in that genome. The most useful minisatellite families have 10–20 members per genome. This range of numbers is small enough to allow the resolution of all the loci as individual bands on an autoradiograph, but it is large enough to provide true fingerprint information. If one fingerprint is not sufficient to resolve the relationship between two different DNA samples, investigators can always obtain data from two, three, or even more minisatellite families.

Figure 11.15 illustrates an interesting example of the utility of DNA fingerprinting. In 1997, scientists from the Roslin Institute in Scotland announced that they had cloned a sheep by injecting a diploid nucleus from an adult udder cell (grown in culture) into an unfertilized egg whose own genetic material had been removed. Initially, many scientists were skeptical of this result and thought that "Dolly" might actually be the result of a fertilization between some contaminating sperm and the egg. If this were the case, Dolly's genome would be unique. Instead, the results of the fingerprint analysis shown in Fig. 11.15 demonstrated that Dolly's DNA fingerprint was identical to that of the adult udder cell used to clone her. This established beyond a doubt that her genome was indeed a clone of that cell's DNA.

DNA Fingerprints Can Identify Individuals and Determine Parentage. DNA fingerprints are a powerful tool for forensic analysis. Prosecutors and defense attorneys alike use them to show the likelihood of a suspect's presence at the scene of a crime or to prove the innocence of someone falsely accused. In one case, a man arrested on rape charges in 1981 had blood type alleles that matched those of cells in the semen found on the victim. The crude blood-typing tests available at that time did not have the resolving power to prove that the semen definitely came from the accused. But after the victim picked the accused out of a lineup, the prosecution used the test results as evidence to help obtain a conviction. Eleven years later, a defense lawyer filed an appeal based on the finding that DNA fingerprints of the semen obtained from the victim and from cells of the convicted man were different. With this evidence, the court reversed its decision and in 1993, set the convicted man free. Since that time, over 100 other men unjustly convicted of rape have been released from jail thanks to DNA fingerprint analysis.

Also in 1993, the courts for the first time accepted plant DNA as evidence in a murder trial. The defendant, accused of killing a woman whose body was found abandoned in the

1 2 U C D 3 4 5 6 7 8 9 10 11 12

kb

—12

—10

— 8

— 6

— 4

Figure 11.15 DNA fingerprint analysis confirmed that Dolly was cloned from an adult udder cell. Genomic DNA samples were prepared from the donor udder cells (U), from the cell culture prepared from the udder cells (C), from Dolly's blood cells (D), and from control sheep 1–12. These DNAs were digested with the restriction enzyme *Mbo*I and subjected to gel electrophoresis and Southern blot analysis with a minisatellite probe. The DNA fingerprints of the 12 control sheep are all different from each other and from the cells involved in the Dolly experiment. Dolly's DNA fingerprint is identical to the fingerprints of both the udder cells and the derived cell culture. This result provides very strong evidence that Dolly is a clone of the ewe that donated the udder cells.

Arizona desert, owned a pickup truck. When police searched the back of the truck, they found a few seed pods from a Palo Verde tree. The county sheriff's department asked a molecular geneticist if there was any way to match the pods to an individual tree at the scene of the crime. DNA fingerprinting was used to compare the pods recovered from the bed of the pickup truck with a variety of Palo Verde trees. Only one match was found, and it was to the tree located at the site of the crime. Although this evidence does not prove the defendant was at the scene of the crime, it strongly suggests that the defendant's truck was there.

More recently, DNA fingerprints demonstrated that skeletal remains unearthed at Ekaterinburg in the Ural Mountains of Russia belonged to Czar Nicholas II and his family, who were murdered in 1918 during the Bolshevik revolution. Geneticists established the relationship by comparing DNA from the excavated bones with samples obtained from a number of living relatives of the Romanov family, including Prince Philip, Duke of Edinburgh. This information disproved the claim of Anna Anderson that she was the Grand Duchess Anastasia; in three independent analyses, her DNA (obtained from hair and from biopsy samples removed during an examination for cancer years before her death in 1984) did not match that of members of the Romanov line—living or dead.

When researchers examine groups of polymorphic DNA loci to obtain DNA fingerprints that can help determine the genetic relatedness of DNA samples, the chromosomal location of each locus is unimportant to the analysis. But a knowledge of the chromosomal location of individual DNA markers can be of great use to geneticists who want to determine the chromosomal location of genes responsible for disease phenotypes and other inherited differences among people, animals, and plants. The chromosomal positions of DNA markers linked to genes defined by phenotype alone can serve as a basis for cloning those genes in a process called *positional cloning*.

Positional Cloning: From DNA Markers to Gene Clones

In Rare Cases, It Is Possible to Move from a Disease Phenotype to the Causative Gene Without Linkage Analysis

Medical researchers learned from the analysis of pedigrees like the one shown in Fig. 11.16a that hemophilia A is an X-linked recessive trait governed by a single gene. The main symptom of uncontrollable bleeding in affected individuals prompted geneticists to make an educated guess at the biochemical function of the responsible gene. The function of the wild-type hemophilia A gene, they proposed, is production of a normal clotting factor; mutations that inactivate this factor produce hemophilia A.

Once molecular investigators worked out the details of the blood-clotting cascade (Fig. 11.16b), they could look for clotting factors in normal individuals that were absent in hemophiliacs (Fig. 11.16c). In this way, they identified a protein known as Factor VIII. They next determined the amino-acid sequence of Factor VIII and used the genetic code to predict the nucleotide sequence of the corresponding gene. This information allowed them to develop a degenerate oligonucleotide probe that could identify clones of the Factor VIII gene within genomic libraries (Fig. 11.16d). When they sequenced this gene from people suffering from hemophilia A, they found mutations with an absolute correlation to the disease phenotype and thereby verified the gene as the causative agent of the disease.

Figure 11.16 How geneticists identified and cloned the hemophilia A gene. **(a)** A pedigree of the royal family descended from Queen Victoria. This family tree uses the standard pedigree symbols. **(b)** The blood-clotting cascade. Vessel damage induces a cascade of enzymatic events that convert inactive factors to active factors. The cascade results in the transformation of fibrinogen to fibrin and the formation of a clot. **(c)** Blood tests can determine whether an active form of each factor involved in the clotting cascade is present. The results of such analyses show that many hemophiliacs, such as those found in Queen Victoria's pedigree, lack an active Factor VIII in their blood. **(d)** Researchers purified Factor VIII, determined its amino-acid sequence, used this information to infer all possible degenerate coding sequences, constructed oligonucleotides for a region with minimal degeneracy, probed a genomic library with these oligonucleotides, and obtained genomic clones of the *Factor VIII* gene. They sequenced these clones to determine the gene's actual coding sequence.

Of the thousands of genes responsible for known human genetic diseases, only a tiny number can be identified in the manner just outlined. Much more often, even with a clearly defined mutant phenotype, it is hard to make an educated guess about the protein altered by the disease allele.

Cystic fibrosis, for example, is a recessive autosomal genetic condition inherited by 1 child in every 2500 born from two parents of European descent. But because the trait is recessive, the frequency of unaffected carriers in the population is much greater, about 1 in every 36 people. Many carriers come from families where the disease has never appeared, and so the first birth of a child with the disease can come as a complete shock. Children with the disease have a variety of symptoms arising from abnormally viscous secretions in the lungs, pancreas, sweat glands, and several other tissues. Malfunction of the pancreas leads to defective secretion of pancreatic enzymes necessary for normal digestion of food. Hyperviscous secretions in the lungs cause severe infections and breathing difficulties. Even with modern medical treatments that combat some symptoms of the disease, most cystic fibrosis patients die before the age of 30.

Unfortunately, the gross symptoms of cystic fibrosis did not provide insight into the underlying molecular cause of the disease. Hundreds of proteins contribute to the process of cell secretion, and most of these were still unidentified in the 1980s. Without a way to determine which one is defective in cystic fibrosis, investigators had no simple way to work their way from gene function to the amino-acid sequence of a particular protein and from there to the protein's DNA coding sequence.

In Positional Cloning, Linkage Analysis with DNA Markers Helps Identify Disease Genes

The dilemma for characterizing the genes associated with most human hereditary diseases thus becomes: How can you use a phenotype that is not yet associated with a detectable molecular defect as a tool for discovering the gene sequence responsible for the phenotype? The answer is by combining linkage analysis (as described in Chapter 5) with the use of DNA markers described earlier in this chapter to localize the human disease gene to a specific region of chromosomal DNA. Then other techniques can determine which gene—among the small number in this region—contains mutations that correlate with the disease phenotype. This entire protocol is called **positional cloning.**

We saw in Chapter 5 that a simple two-point cross can demonstrate linkage if the two loci under analysis lie close enough together on the same chromosome that the rate of recombination is significantly less than the 50% expected with independent assortment. We also saw that the frequency of recombination between the two loci provides a direct measure of the distance separating them, as recorded in centimorgans (cM), or map units (m.u.). (Geneticists studying humans, mice, and other mammals use the centimorgan unit of measure, which we adopt in this chapter.) Finally, we learned that it is possible to integrate multiple pairs of linked loci into a "linkage group" by performing many different two- or three-point crosses with overlapping sets of loci. The **linkage maps** constructed from these crosses depict the distances between loci as well as the order in which they occur on a chromosome.

With the use of anonymous DNA markers, rather than markers defined by phenotype, there is no limit to the number of loci that can be mapped in a single cross or extended human family. In place of a traditional three-point cross, it becomes possible to perform linkage analysis combining hundreds of DNA loci with the disease locus of interest. If genetic linkage can be demonstrated between a disease trait and one or more previously mapped DNA markers, then the gene responsible for the trait must lie in the same subchromosomal region as those DNA markers.

We now describe the entire process of positional cloning: the use of linkage analysis to map the disease locus to a small chromosomal region; the determination of all possible "candidate" genes within this region; and the identification of the candidate gene responsible for the disease (Fig. 11.17).

Step 1 of Positional Cloning: Correlating Phenotypic Transmission with One Area of the Genome

Discovery of a DNA marker that shows linkage to the disease locus is the first goal of positional cloning. For traits expressed in plants or small animals, it is a simple matter to set up a single testcross for the production of hundreds of offspring that can be easily analyzed to identify the map position of the trait in question. For human traits such as disease phenotypes, breeding is not a possibility. Instead, researchers try to find many different extended families—each with a large number of children—in which some individuals express the mutant phenotype and others do not. The Genetics and Society box "Using Human Pedigrees and LOD Scores to Calculate the Probability That Two Loci Are Linked" describes the mathematics used to demonstrate linkage between a marker and an abnormal trait in human pedigrees.

The mapping of a human disease locus begins with the genotyping of all members of the disease-carrying families for a series of DNA markers ideally spaced at even intervals of 10 cM on each chromosome. If the distance between markers is greater, the chances of demonstrating linkage between a marker and the disease locus are reduced; if the distance is shorter, investigators would have to do many more genotyping assays without much gain in statistical power.

Since the human genome is approximately 3000 cM long as measured in map distances, you would need approximately 300 markers to map the genome at 10 cM intervals. Because a rough draft of the human genome has been sequenced, you already know the chromosomal positions of these DNA markers. The only unknown map position is that of the disease locus. You can thus view a 301-point linkage analysis as 300 separate two-point crosses, each one a test for linkage between an individual DNA marker and the disease locus. If the number of genotyped individuals is sufficiently large, the disease locus must show linkage to one or more of these markers. Finding linkage to at least one marker of known position will place the disease locus in a particular subregion of a particular chromosome (Fig. 11.17a).

Figure 11.17 Positional cloning: From phenotype to chromosomal location to guilty gene. (a) Diagram of a human chromosome with four markers—M1, M2, M3, and M4—used in the linkage analysis of a disease phenotype. Each marker provides "linkage coverage" of a portion of the chromosome. The region of coverage for each marker is the region within which linkage to a disease locus could be uncovered. In the hypothetical study illustrated here, linkage has been detected to adjacent markers M1 and M2. This suggests that the gene responsible for the disease lies between those markers. **(b)** With this information, an investigator could type additional markers that lie between M1 and M2 to position the disease locus with higher resolution. **(c)** Looking for candidate genes. Analysis of the region between recombination sites that define the smallest area within which the disease locus can lie should reveal the presence of candidate genes. **(d)** Finding the correct candidate through comparisons of the structure and expression of each candidate gene in many diseased and nondiseased individuals. A correlation between a mutant structure or expression for a particular candidate gene and the disease phenotype can provide evidence that a particular gene is responsible for the disease phenotype.

Higher-Resolution Mapping with Additional DNA Markers. The next step is to increase the resolution of the linkage map in the area of the gene. Researchers accomplish this by choosing additional DNA markers that lie closer together in the vicinity of the disease locus. If, for example, the first stage of linkage analysis were to place a disease gene somewhere between markers M1 and M2 on the 175-cM-long chromosome shown in Fig. 11.17, you could query an electronic repository for additional markers between the M1 and M2 positions on the chromosome and then genotype each marker in all members of the pedigree used in the initial link-

age analysis (Fig. 11.17b). The goal of this second stage is to identify DNA markers so tightly linked to the disease locus that they show less than 1% recombination in all individuals tested. If you could map the disease locus between two markers separated by 1 cM, the gene would lie within a delineated region of less than 1000 kb, since the human genome contains 3000 Mb of DNA and a total-linkage distance of 3000 cM.

In 1984, the Huntington disease (*HD*) locus became the first human disease gene to be successfully mapped by positional cloning. Figure 11.18 shows the five-generation, 104-member family pedigree used to demonstrate linkage between a previously mapped DNA marker named G8 and the *HD* locus. Preliminary linkage between the *HD* locus and the G8 marker placed the disease gene on human chromosome 4. Further linkage analysis then narrowed down the map position of *HD* to less than 1000 kb.

Step 2 of Positional Cloning: Identifying Candidate Genes

Once the locus of a gene for a particular disease has been narrowed down to a 1000 kb genomic region, researchers catalogue and investigate all the coding sequences within that region. Current estimates of gene number in the human genome center around 50,000. If these genes were spaced evenly throughout the genome, there would be an average density of 50,000 genes/3000 Mb = 17 genes per 1000 kb.

Cataloguing All the Genes in the Identified Region. As we saw in Chapter 10, there are several ways to identify coding regions within the human genome. One approach depends on the different forces of natural selection that act upon coding regions versus nonfunctional sequences. Though mutations can occur at any nucleotide in the genome, mutations in genes and their regulatory regions are usually not tolerated. As a result, these sequences evolve much more slowly than nonfunctional sequences. In the 60 million years that mice and humans have been diverging from a common ancestor, nonfunctional sequences in both species have accumulated so many mutations that they no longer show any homology to each other. In contrast, coding sequences in humans almost always have a conserved homolog in mice, and often, the mouse and human coding sequences will hybridize with each other. Now that both the human and mouse genomes are completely sequenced, cross-hybridization experiments are no longer necessary. Instead, a specialized computer algorithm can be used as a form of "electronic hybridization" to identify all DNA regions conserved between the two species (Fig. 11.19).

A second independent method of gene identification relies on a different type of computer algorithm that detects exonlike coding regions based on the presence of an open reading frame, a knowledge of codon usage, and the structure of splice sites. A DNA sequence that looks like a coding region and has been conserved across distant species has a high probability of representing a gene.

A third independent approach for identifying genes is through their appearance in one or more EST clones. As we

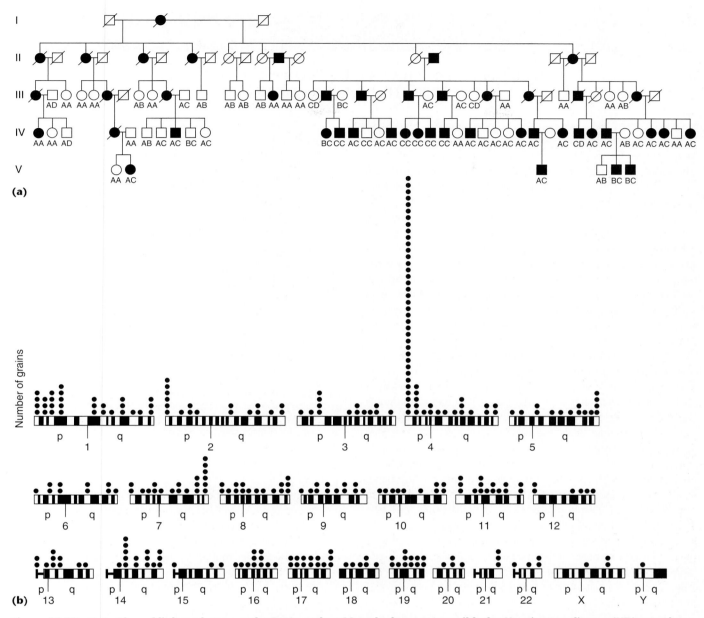

Figure 11.18   Detection of linkage between the DNA marker G8 and a locus responsible for Huntington disease (HD) was the first step in the cloning of the *HD* gene. (a) Portion of a large Venezuelan pedigree affected by HD, or Huntington disease. For living members of the pedigree, alleles at the G8 marker locus are indicated (A, B, C, and D). It is easy to see the cotransmission of marker alleles with the mutant and wild-type alleles at the *HD* locus. **(b)** Pedigree analysis shows that the *HD* locus is within 5 cM of the G8 marker. Thus, karyotypic localization of the G8 marker can provide a chromosomal location for the *HD* gene. Researchers carried out a karyotypic localization analysis by a precursor to the FISH technique in which radioactive probe was hybridized to metaphase chromosomes. Investigators counted the number of grains caused by radioactive decay in each chromosomal region. The results of this localization procedure appear in the figure as *dots* above an idiogram of the human karyotype. The tip of the small arm of chromosome 4 has many more grains than any other chromosomal region. This is the location of the G8 marker and, by inference, the *HD* locus. (All the other grains represent background noise.)

saw in Chapter 10 (p. 337), hundreds of thousands of random EST clones have been sequenced from cDNA libraries compiled from many different human tissues, and these sequences have been entered in databases. Since cDNAs correspond to RNA transcripts, the presence of a sequence in an EST means that the sequence must be derived from a gene.

Gene Expression Patterns Can Pinpoint Candidate Genes. How can you tell which of the 17 or so protein-coding genes in a region of interest might be the one responsible for a particular trait? The expression pattern of these genes—that is, the level of gene transcription in different tissues at different stages of development—can provide clues. For example, the gene

Genetics and Society

Using Human Pedigrees and LOD Scores to Calculate the Probability That Two Loci Are Linked

A geneticist cannot obtain a yes or no answer to the question of whether two loci are linked from pedigree data alone. All he or she can do is calculate the probability with which a particular set of data is likely to occur in the absence or presence of linkage between the two loci.

With experimental organisms, the statistical tool for such linkage analysis is the chi square (χ^2) test (see pp. 118–120 in Chapter 5). This test makes it possible to compare the numbers of recombinant and nonrecombinant chromosomes observed in the offspring of a planned cross and transform the data into linkage probabilities. The χ^2 test, however, is not applicable to human pedigree analysis. Because human matings are not planned experiments, linkage analysis in humans requires specialized statistical tools. We now describe those tools.

Consider a simple two-generation human pedigree following the unusual phenotype of six fingers on each hand in a family with two parents and eight children (Fig. A). The phenotype is caused by a completely penetrant, dominant mutation at a single locus known as *SF* (for *six fingers*). The father expresses the unusual phenotype, while the mother does not. The fact that some of the children do not have six fingers on each hand indicates that the father is a heterozygote for the *SF* mutation with an *SF* / + genotype. The mother, with only five fingers on each hand, is homozygous wild type (+ / +) at the *SF* locus. Like the father, the six-fingered offspring are *SF* / + heterozygotes; like the mother, the normal offspring are + / + homozygotes.

Suppose that you want to evaluate the possibility that the *SF* gene is linked to a polymorphic marker (*M*) with alleles 1 and 2. A single-base-pair difference, detectable by PCR-based protocols, distinguishes the two alleles. If PCR-based analysis demonstrates that the father is heterozygous *M1* / *M2* while the mother is homozygous *M1* / *M1* (as shown in the electrophoresis results depicted below the pedigree in Fig. A), the pedigree would represent a situation analogous to an experimental testcross in which one parent is heterozygous at both loci under analysis (*SF* / +; *M1* / *M2*); the

other parent is homozygous at both loci (+ / +; *M1* / *M1*); and the segregation of alleles from each paternal locus to each offspring is distinguishable with absolute confidence.

If it turns out that the *SF* and the *M* loci are linked, it is possible that the father's mutant *SF* allele is on the same homolog as his *M2* allele, in which case his genotype would be $\frac{SF \leftrightarrow M2}{+ \; \leftrightarrow M1}$, but it is equally possible that his *SF* allele is on the same homolog as his *M1* allele, in which case his genotype would be $\frac{SF \leftrightarrow M1}{+ \; \leftrightarrow M2}$.

Each of these two possible allele associations, or *haplotype configurations*, is referred to as a **phase of linkage.**

(a)

Unusual phenotype (six fingers per hand)   Normal phenotype (five fingers per hand)

M1
M2

Genotype for marker

● ○ Female
■ □ Male
● ■ Unusual phenotype
○ □ Normal phenotype

(b)

Odds value (ratios)	LOD score *(Z)*
0.01 (1:100)	-2
0.1 (1:10)	-1
1 (1:1)	0
10 (10:1)	1
100 (100:1)	2
1000 (1000:1)	3

Figure A   Using human pedigrees and LOD scores. (a) Two-generation pedigree showing transmission of the hypothetical autosomal dominant phenotype of six fingers and alleles at the *M* marker. **(b)** Relationship between odds values and LOD scores.

responsible for the neurodegenerative phenotype of Huntington disease must be expressed in the brain, while a gene responsible for breast cancer must be expressed in breast tissue. This criterion is by no means fail-safe, but knowledge of expression patterns often allows a researcher to eliminate most of the genes in the mapped region. The remaining genes are considered "candidates" for the disease locus.

There are two main ways to obtain information about gene-expression patterns. First, in well-studied organisms like humans, the information may already be available in public

databases of EST sequences. For example, if a particular gene is represented in many random ESTs from a lung cDNA library, but is never found in similar numbers of random clones from cDNA libraries from other tissues, the gene is probably expressed uniquely in lung tissue.

A second way to determine gene expression patterns is to perform a Northern blot. Recall from p. 296 in Chapter 9 that in Southern blotting, researchers separate DNA fragments by electrophoresis, blot the separated fragments to a special filter paper, expose the filter to radioactive probes that hybridize to

In experimental testcrosses, the phase of linkage is known at the outset because the haplotypes of the grandparents are known. But since human pedigrees are not "set up," the phase of linkage in a pedigree's first-generation parents is unknown. This makes it impossible to determine which inherited allelic combinations are recombinant and which are nonrecombinant. How then is it possible to use pedigree data to calculate the probability of linkage?

Geneticists Use Odds of Linkage and LOD Scores to Discover Evidence for or against Linkage

With just a cursory look at the pedigree in Fig. A, it is clear that in seven out of eight children, the paternal *M2* allele has been inherited with the wild-type phenotype, the paternal *M1* allele with the six-finger phenotype. Intuitively, these results seem more likely if the *M* marker and the *SF* loci are linked than if they are not. (Don't be alarmed by the fact that the wild-type mother is homozygous for the *M1* allele. Since the marker and the disease locus can recombine, different individuals can have different haplotype associations of alleles at linked loci.)

How can you convert this hunch into a numerical estimate of the probability of linkage? The answer is with a statistical approach known as **LOD score analysis.** The "LOD" of this type of analysis is an acronym for "logarithm of the odds"; the odds are the likelihood that linkage exists relative to the likelihood that it does not. Although gamblers speak of odds in terms of ratios, such as 3:2 or 10:1, for mathematical calculations, it is convenient to convert these ratios, through division, to real numerical values. Thus, 3:2 = (3 divided by 2) = 1.5; 10:1 = (10 divided by 1) = 10. If the calculated odds value is greater than 1.0, the likelihood of linkage is greater than the likelihood of no linkage. With an odds value of less than 1.0, the opposite is true.

Using a probability analysis that we will not describe here, it is possible to calculate the odds value for the pedigree in Fig. A as 6.3, a result that supports the hunch of linkage. But, just because the odds favor linkage does not mean that this is the correct interpretation of the data. To be on the safe side, the human genetics community has set an odds value of 1000 (in other words, 1 chance in 1001 of no linkage) as the minimal cutoff for accepting odds calculations as evidence of linkage between two loci.

Because data from a single human pedigree rarely generate a sufficiently high odds value, it helps to obtain data from multiple *SF*-segregating families. But how many families does it take to make the numbers large enough, and how does one combine data from different families? You cannot simply add up the recombinant and nonrecombinant chromosomes (since we don't know with certainty which are which) across all families, as one would for a testcross because the phase of linkage may be different in different families. Instead, you must use the product rule to combine probabilities derived from many independent families into a single grand probability for or against linkage.

To do this, you convert the odds for each family into a fraction in which the probability of linkage *P*(L) is the numerator and the probability of no linkage *P*(NL) is the denominator. The odds calculated for each family would then be *P*(L) / *P*(NL), and the combined odds calculated for, say, three independent families would be:

$$\left[\text{Odds} = \frac{P(\text{L})}{P(\text{NL})} = \frac{P1(\text{L})}{P1(\text{NL})} \times \frac{P2(\text{L})}{P2(\text{NL})} \times \frac{P3(\text{L})}{P3(\text{NL})} \right]$$

If a geneticist obtained an odds value of 6.3 with a single pedigree, he or she would want to figure out how many additional families of similar size would have to be analyzed to reach the 1000 cutoff. If the geneticist assumes that the odds for each family will be ~6.3, he or she can multiply 6.3 × 6.3 × 6.3 . . . until crossing the 1000 threshold. But it is more convenient to work with experimental values that can be added—rather than multiplied together—to get a composite value. The mathematical trick for converting multiplication to addition consists in converting odds values into LOD scores through the common logarithmic function based on the following mathematical principle: log (*A* × *B* × *C*) = (log *A*) + (log *B*) + (log *C*).

The table in Fig. A shows the LOD scores that correspond to six key odds values. The cutoff value of 1000 converts to a LOD score of 3.0. The critical feature of LOD scores is that values obtained from multiple families for the same two loci can be summed together to produce a single composite LOD score. As we see in the Fig. A table, the odds value of 6.3, obtained from the simple *SF* pedigree, converts to a LOD score of 0.8. It would take four families with a similar LOD score to push the composite score above the 3.0 cutoff: 0.8 + 0.8 + 0.8 + 0.8 = 3.2.

If the two loci under consideration are not linked, the assumption that additional families will have similar LOD scores would turn out to be wrong. Instead, there would be some negative LOD scores that when added to the initial positive number would give a negative total, supporting a hypothesis of no linkage.

---

complementary sequences in the blot, and finally, observe a band on an autoradiograph if the probe has hybridized to a stretch of DNA on the filter. By comparison, in Northern blotting, it is the RNA transcripts in the cells of a particular tissue that are separated by gel electrophoresis. The ensuing blotting, probing, and autoradiography are the same as for the Southern blotting (Fig. 11.20a).

It is possible to infer two types of information from Northern blot data. First, since RNA samples are not enzymatically treated before electrophoresis, the position of bands on the blot provides a direct measure of RNA size; and from the size of an RNA, you can estimate the transcript's coding capacity and thence the size of the protein it encodes. Second, Northern blot analysis of RNA samples from many different tissues enables one to determine which specific tissue a gene is expressed in as well as the relative levels of its expression in all cells where transcription is occurring.

Investigators can use Northern blots to identify the various transcription units in a large cloned genomic region. To accomplish this, they divide the genomic region into many smaller

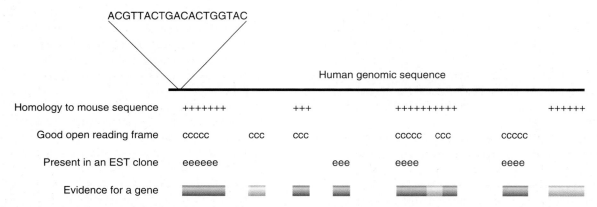

Figure 11.19 Computational analysis of genomic sequence can uncover genes. Genomic sequences can be analyzed computationally in two ways. First, each small region of the human genome can be compared with the whole mouse genome for the presence of homologous sequences. In this hypothetical example, human sequences that exhibit homology to mouse sequences are indicated as +++. Second, each region can be analyzed for the presence of an open reading frame containing codons typical of most human genes, bounded by splice-site sequences. Regions that pass this test are indicated as *ccc*. Sequences that pass both tests are considered very likely to represent coding regions. Sequences that pass just one test are possible coding regions. Incontrovertible evidence for a transcribed sequence comes when it is found to be present in an EST clone. *Red boxes* indicate sequences that are definitely or probably transcribed; *orange boxes* show possible coding regions.

fragments and label each fragment for use as a Northern blot probe. Genomic fragments that do not include coding regions will not light up bands in any RNA sample; fragments from genes expressed in all tissues will light up a band in all RNA samples; and fragments from genes expressed only in specific tissues will light up a band in some RNA samples.

Figure 11.20b shows how Northern blot analysis supported the hypothesis that the Y-linked *SRY* gene is responsible for testes differentiation. When a fragment of this conserved gene is used to probe a Northern blot with RNA samples from several different tissues, a 1.1 kb RNA transcript is found in the testes, but not in the ovaries, lungs, or kidneys. This finding makes the *SRY* gene an excellent candidate for testes-determining factor (TDF). However, there is one more requirement for clinching the case, which we describe in the next section.

Step 3 of Positional Cloning: Finding the Gene Responsible for the Phenotype

At this point in positional cloning, you have a set of candidate genes. They all map to the small region of the genome where the trait locus must lie, and they are transcribed in tissues consistent with your expectations based on the disease phenotype. How do you determine which one of these candidates is the gene responsible for the trait? Even if you have just one candidate, how do you confirm that it is the correct one and that you have not missed other candidate genes? You must compare groups of individuals with normal and abnormal phenotypes. If the DNA sequence or transcription of the candidate gene is consistently altered in all individuals with the abnormal phenotype, the evidence is very good that your candidate is in fact the gene responsible.

Expression Patterns. Many genetic diseases are caused by mutations that alter the pattern of gene expression in the mutant tissue. Mutations in regulatory regions can prevent transcrip-

tion, and others at splice junctions can disrupt splicing and lead to aberrant mRNAs and polypeptide synthesis. Other diseases are caused by nonsense mutations in the coding region. These mutations allow the production of properly sized mRNAs but prevent formation of the polypeptide product. Thus, a first step in vetting a candidate gene is to compare both its RNA and polypeptide expression in individuals who express and do not express the trait. RNA expression can be assayed with a Northern blot or by PCR amplification of cDNA with primers specific for the candidate transcript. Polypeptide expression can be assayed with an antibody specific for the wild-type protein. If a candidate gene is not expressed, is underexpressed, is overexpressed, or produces a smaller or larger transcript or polypeptide—or no transcript or polypeptide at all—in all phenotypically abnormal individuals, there is a good chance that it is responsible for the trait.

Sequence Differences. Some mutations, such as the one that causes sickle-cell anemia, produce their effect by substituting one amino acid in the polypeptide for another. The best way to detect subtle missense mutations of this type is by sequencing the coding region of the candidate gene from phenotypically normal and abnormal individuals. If all the phenotypically abnormal people have mutations in the candidate gene that are not present in the control group, this is powerful evidence that the particular candidate and the disease locus are one and the same.

The Ultimate Proof: Transgenic Modification of Phenotype. Incontrovertible evidence that a particular candidate gene is responsible for a particular phenotype must ultimately be obtained in an animal model where it is possible to induce a specific change in the gene and confirm that this change causes the predicted phenotype. The technology used to accomplish this goal is called **transgenic technology.**

(a) 1.

2. Load RNA samples in wells of a gel.

3. Separate RNA samples by gel electrophoresis. Blot onto filter. Expose filter to labeled hybridization probe.

Figure 11.20 Northern blots: Snapshots of gene expression. (a) The protocol. **(1)** Purify RNA from each tissue to be examined for expression of the gene under investigation; here, you are looking at the *SRY* candidate for the testes-determining factor, so it is important to analyze testicular tissue as well as several control tissues (ovary, lung, and kidney in this experiment). **(2)** Make an agarose gel, load each of the four RNA samples into a different well, and load a fifth well with RNA size markers. Now subject the gel to an electric current that causes the RNA in each sample to migrate along a lane toward the bottom of the gel. The mobility of each RNA transcript in a sample depends on its size: smaller RNAs move faster, while larger RNAs migrate more slowly. When the smallest RNAs reach the bottom of the gel, turn off the current. Staining the RNAs in each lane would produce a smear reflecting the presence of so many RNAs of different sizes that they cannot be resolved from each other. **(3)** Blot the RNA within the gel and fix it to a filter so that each RNA molecule retains its position relative to all the other molecules. Expose the filter to labeled probe and allow the label to hybridize for several hours. **(4)** Wash away unhybridized probe. Place the filter on a film for autoradiography. Develop the film. You will see bands only in those lanes containing a tissue where the gene represented by the probe has been expressed. **(b)** Northern blot results obtained using an *SRY* clone as a probe. The results show that *SRY* is expressed in the testes, but not the ovary, lung, or kidney. This result makes *SRY* a good candidate for the *TDF* locus. In a control experiment, researchers probed an identical blot with the same RNA samples using a clone containing the actin gene. As expected, a band of the same size appears in every lane (*shown below the* SRY *results*). This control demonstrates the integrity of the RNA samples used in this study.

The term **transgene** describes any piece of foreign DNA that researchers have inserted into an organism's genome through experimental manipulation of early-stage embryos or germ cells. An individual carrying a transgene is known as a *transgenic* plant or animal. In 1981, researchers developed a

general method for transferring foreign DNA of any type (naturally occurring or chemically synthesized) into the germ line of any mammal. They accomplished the transfer by injecting cloned DNA molecules through very fine pipettes into the haploid egg and sperm nuclei (known as pronuclei) present in

Transgenic analysis

Foreign DNA injected into male pronucleus of newly fertilized egg. → Injected eggs surgically implanted into uterus of "foster" mother and allowed to develop. → Mice are born with foreign DNA in every cell nucleus.

Figure 11.21 Transgenic analysis can prove the equivalence of a candidate gene and a trait locus. To demonstrate that *SRY* is the same as the *TDF* locus, researchers injected a clone containing the *SRY* gene into mouse embryos with two X chromosomes. The transgenic mice that resulted from this protocol had a penis. This observation proves that *SRY* is the equivalent of *TDF*.

newly fertilized eggs (Fig. 11.21). They then placed the injected eggs into the uterus of a "foster" mother and allowed them to develop. Up to 50% of the mice born to these foster mothers had integrated the injected DNA into their chromosomes; these transgenic mice later passed the transgene on through their germ cells to their own offspring.

Researchers have successfully inserted transgenes into the genomes of cows, pigs, goats, sheep, monkeys, and other mammals. The only limitation of current protocols is that the site of transgene integration into the host genome cannot be predetermined and is, instead, random.

Transgenic technology was used to confirm the function of the candidate gene *SRY* as the testes-determining factor (*TDF*), which we described earlier. Two to five copies of the complete *SRY* gene were injected into a group of fertilized eggs from a mating between two normal parents. The embryos were placed into foster mothers. Two to three weeks after birth, the mice

that developed were checked by karyotypic analysis (to determine the combination of sex chromosomes, either XX or XY), by Southern blot analysis (to see whether the transgene had integrated into the genome), and by visual examination (to check for the presence of a male sex organ). The identification of several transgenic XX animals with a penis (Fig. 11.21) proved beyond a doubt that the *SRY* coding region is equivalent to the *TDF* gene.

Comprehensive Example: Positional Cloning of the Cystic Fibrosis Gene Leads to a Potential Therapy

Figure 11.22 reviews the steps and benefits of positional cloning as medical geneticists used it to identify the cystic fibrosis (*CF*) gene. In 1985 researchers uncovered linkage between the *CF* locus and a marker called met, which placed the *CF* gene in the middle of human chromosome 7. Further linkage analysis carried out on a large number of families each having two or more members affected by CF showed that the met locus is 2 cM from the *CF* gene. Next, the investigators identified two additional markers—XV-2c and KM-19—that displayed no recombination with the disease gene and were thus much closer to it. They then cloned the genomic region between these two markers and identified four candidate genes within it (Fig. 11.22a).

As discussed earlier, children afflicted with cystic fibrosis have a variety of symptoms arising from abnormally viscous secretions in the lungs, pancreas, and sweat glands. Northern blot analyses of the four candidate genes revealed that only one, subsequently called *CFTR* (for cystic fibrosis transmembrane conductance regulator), was expressed in all of these tissues and no others (Fig. 11.22b). When medical geneticists examined the *CFTR* gene in many CF-affected people, they found that in every patient, small deletions or substitutions had destroyed *CFTR* function by either eliminating polypeptide production or disrupting protein structure and function (Fig. 11.22c and d).

The evidence pointing to the *CFTR* gene as the causative agent of cystic fibrosis is very strong because the disease affects so many people, all of whom show mutations in the gene. If a disease is not as prevalent, it may be difficult to accumulate enough evidence to be as confident that a suspected candidate gene is really the right one. To resolve this dilemma, scientists rely on the fact that most human genes are conserved in the mouse genome. If the candidate gene is present in the mouse genome, they can create mutations of this gene in mice and see whether the mutations produce the same phenotype as that observed in disease-afflicted people.

We have seen that one method for altering the genomes of experimental animals is the injection of transgenic DNA into the pronucleus of the one-cell embryo. Unfortunately, the DNA injection method does not work to alter existing genes in the genome; it only adds new genes. But with cystic fibrosis and many other human diseases, the phenotype arises from mutations that eliminate gene function. How can scientists eliminate, or "knock out," the function of the homologous gene in mice?

Figure 11.22   Positional cloning of the cystic fibrosis gene: A review. **(a)** Linkage analysis places the *CF* locus on chromosome 7. This idiogram of chromosome 7 shows the positions of markers—met, XV-2c, KM-19, and J3.11—that define the 1 cM region within which the *CF* locus must lie. Four candidate genes were identified within this region. **(b)** Northern blot analysis reveals that only one of the candidate genes is expressed in the lungs and pancreas, tissues affected by cystic fibrosis. This candidate (now known as *CFTR*) is not expressed in the brain, liver, or testes, all tissues that are not affected in diseased individuals. **(c)** Every CF patient has a mutated allele of the *CFTR* gene on both of her or his chromosome 7 homologs. The locations and numbers of mutations uncovered at different positions within the *CFTR* gene are indicated under the diagram of the gene shown schematically divided into its 24 exons. **(d)** The structure of the wild-type polypeptide schematically divided into its different domains. CFTR is a membrane protein. TMD-1 and TMD-2 are transmembrane domains. The other domains lie on the cytoplasmic side of the properly positioned protein and are involved in intracellular signaling.

The answer is through another kind of genome-altering technology called **targeted mutagenesis.** This technology is more technically demanding and complex than nuclear injection. Its application for the production of mutations in the mouse *CFTR* gene is described in detail in Reference E, the genetic portrait of the mouse as an experimental organism (pp. 845–865). Here we provide a brief summary of the protocol. Essentially, researchers introduce DNA molecules containing a mutant nonfunctional copy of the mouse *CFTR* gene into normal mouse embryonic cells growing in culture. In rare cells (less than one in a million), a molecule containing the mutant sequence undergoes a double recombination event with the homologous wild-type gene within the cell's chromosomes. Sophisticated techniques of selection and screening can identify these rare cells, which contain a mutant *CFTR* gene in place of a wild-type copy. The mutant cells are introduced into normal mouse embryos, where they become incorporated into the germ line of the developing animals. Gametes created from such germ-line cells produce offspring referred to as **knockout mice** because they carry an induced mutation in a targeted gene. When animals heterozygous for the *CFTR* knockout were bred together, they produced progeny homozygous for the *CFTR* mutation. These offspring exhibited the disease symptoms characteristic of cystic fibrosis in people.

Medical scientists now use the ASO-based genotyping protocols to determine *CFTR* genotypes for families concerned about the disease (review Fig. 11.9). These techniques can be applied to adults (to see if they are carriers), to fetuses (to establish what the disease status of the newborn will be), or to preimplantation embryos (to select nonafflicted ones for initiating a pregnancy). Medical researchers also use cells containing specialized clones of the gene to synthesize the wild-type protein

and learn how its absence or inactivation causes disease symptoms. The Fast Forward box on p. 32 in Chapter 2 summarizes our current understanding of the CFTR polypeptide: It is a cell membrane protein that regulates the passage of chloride ions from and into both cells and the extracellular environment of tissues like the lungs.

This understanding of gene function suggests a potential therapy: inserting a copy of the wild-type *CFTR* gene into a small fraction of a CF patient's lung cells. Cells that express the inserted, wild-type gene may be able to produce enough functional protein to eliminate the buildup of fluids in the lungs, which is the leading cause of death from CF. Clinical trials to test the safety and efficacy of such a therapy are in progress in both mice and humans.

Genetic Dissection of Complex Traits

In humans, only a small fraction of disease traits follow the simple Mendelian pattern of single-gene inheritance seen in cystic fibrosis, Huntington disease, and hemophilia A. Most common characteristics of human appearance, such as height, skin color, the shape of the face, hair type (curly, wavy, or straight), and many essential measures of human physiology important to human health (including blood pressure, cholesterol level, basal metabolism, and susceptibility to infectious diseases) have a more complex pattern of inheritance.

We now review the chief causes of complex inheritance and describe the problems they pose for the linkage mapping and positional cloning protocols described so far for single-gene traits (Table 11.2). We then explain how researchers adapt the procedures for analyzing single-gene traits to the more dif-

TABLE 11.2 Complexities That Alter Traditional Mendelian Ratios

Category of Complexity	Problem in Observed Relationship Between Genotype and Phenotype	Changes Necessitated in Mapping Strategy
Incomplete penetrance	Disease genotype can occur in an individual who does not express the disease phenotype	Eliminate nondiseased individuals from analysis
Phenocopy	Disease phenotype can be expressed by an individual who does not have the disease genotype	Program computer to flag unexpected cases of supposed double recombination events that flank a putative disease locus
Genetic heterogeneity	In different families, different disease genotypes are responsible for the same disease phenotype	Divide complete set of disease-transmitting families into subgroups (based on various parameters such as average age of onset) and perform linkage analysis separately on each subgroup
Polygenic determination	Mutant alleles at more than one locus influence expression of the disease phenotype in a single individual	Program computer to search for complex patterns of association between the disease trait and multiple loci

ficult task of identifying, mapping, and characterizing the genes that contribute to complex traits.

How the Causes of Complex Inheritance Patterns Confound Linkage Mapping and Thus Positional Cloning

Incomplete Penetrance: When a Mutant Genotype Does Not Always Cause a Mutant Phenotype

Unlike the link between smoking and lung cancer, no environmental factor has yet been connected with breast cancer. Rather, the higher incidence of breast cancer in women whose mothers or aunts have the malignancy suggests that a strong genetic factor predisposes some women to this condition. Medical investigators used a positional cloning protocol to map and clone the *BRCA1* (BReast CAncer 1) gene, one of several genes that can cause breast cancer. Significantly, only 66% of women who carry a mutant allele at the *BRCA1* locus develop breast cancer by the age of 55. As seen in the first pedigree in Fig. 11.23, it is possible for a mother to carry a mutant *BRCA1* allele and remain disease-free, while her daughter becomes afflicted with the disease. Thus, although the mutant *BRCA1* allele predisposes a woman to breast cancer, it does not guarantee that the disease phenotype will occur; that is, it is not completely penetrant. By comparison, a disease such as sickle-cell anemia, in which a mutant genotype always causes a mutant phenotype, is completely penetrant.

The causes of incomplete penetrance vary from trait to trait and from individual to individual. With breast cancer, it seems that chance plays the largest role in determining which predisposed individuals get the disease—through the accumulation of secondary somatic mutations. With heart disease, the individ-

ual's environment—especially diet and amount of exercise—plays a large role in determining whether a predisposing genotype results in a mutant phenotype, and if so, at what age.

Incomplete penetrance hampers linkage mapping and positional cloning for one main reason: Individuals who do not express a mutant phenotype may nevertheless carry a mutant genotype. The simplest solution to this problem is to exclude all nondiseased individuals from the analysis. With age-dependent traits like breast cancer and Huntington disease, such exclusion has meant that in disease-carrying families, the majority of children and adults under the age of 40 could not be included in the analysis. As a result, many more families were required for the studies that led to the mapping and cloning of the genes associated with both diseases.

Variable Expressivity: When Expression of a Mutant Trait Differs from Person to Person

The variation may be in age of onset, phenotypic severity, or any other measurable parameter. Variable expressivity does not normally interfere with genetic analysis, because geneticists can use any degree of mutant phenotype as evidence for the presence of a mutant allele.

Phenocopy: When the Disease Phenotype Is Not Caused by Any Inherited Predisposing Mutation

The observation that 3% of women who do not carry a mutation at the *BRCA1* locus or have any family history of the disease still develop breast cancer by age 55 suggests that the disease can arise entirely from one or more somatic mutations in the breast cells themselves. This form of the disease is considered a **phenocopy** because it is indistinguishable from the inherited form of the disease yet is not caused by an inherited

Figure 11.23 Genetic heterogeneity: Mutations at different loci can give rise to the same disease phenotype. Both pedigrees in this figure show evidence of the transmission of a dominant mutation with incomplete penetrance that causes breast cancer. Linkage analysis shows that the mutation in the first pedigree resides on chromosome 17, whereas the mutation in the second pedigree is on chromosome 13. The first family is segregating a mutant *BRCA1* allele, while the second family is segregating a mutant *BRCA2* allele.

mutant genotype. The percentage of women who develop phenocopy breast cancer rises to 8% by age 80.

We have seen that researchers focus on families with a history of disease to map predisposing alleles. If a small but significant fraction of women who develop the disease carry wild-type alleles, the correlation between the inheritance of the disease locus (or a locus-linked marker) and expression of the disease will diminish. Phenocopies thus make it more difficult to map disease-causing loci.

Genetic Heterogeneity: When Mutations at More Than One Locus Cause the Same Phenotype

Sometimes it is possible to use sophisticated diagnostic techniques to separate what appears to be a single disease into a set of related diseases caused by mutations in different genes. For example, researchers can distinguish insulin-dependent from insulin-independent diabetes on the basis of their different physiological origins. However, even when the limit of disease subdivision has been reached, what appears to be a homogeneous phenotype may still arise from genetic heterogeneity. The seemingly simple disease of thalassemia is a case in point. Mutations in either the α-globin or the β-globin gene can cause the same phenotype: severe reduction or elimination of the functional hemoglobin molecules produced in red blood cell precursors.

Genetic heterogeneity complicates attempts to map disease-causing loci in the following way. Although individual human families usually segregate only a single mutation responsible for a rare disease, most families do not have enough members to provide sufficient data for determining linkage (see the Genetics and Society box "Using Human Pedigrees and LOD Scores to Calculate the Probability That Two Loci Are Linked"). For this reason, linkage studies in humans almost always combine data from multiple families. But if a disease is heterogeneous, a marker linked to the disease locus in one family may assort independently from a different disease locus in a second family. When data from the two families are combined, the calculated probability of linkage between the marker and the initial disease locus will drop below that obtained with the first family alone.

Genetic heterogeneity is suspected whenever a comprehensive analysis of many families, each with several affected members, fails to map a locus responsible for the disease trait. In such a case, investigators try to divide the complete set of disease-transmitting families into subsets—based on any of several parameters—and then combine only the families in each subset for linkage analysis. For example, when researchers combined the data from a large set of breast-cancer-prone families, they found no evidence of linkage to any marker. But when they selected a subset of the families in which the average age of disease onset was less than 47 years, they obtained strong evidence for a disease locus on chromosome 17, named *BRCA1* (Fig. 11.23). It is now clear that mutations in *BRCA1* cause the onset of breast cancer at an earlier age than predisposing alleles at other loci. Classification of families into early-onset and late-onset groups may be helpful with any trait showing age-dependent expression. With traits for which classification by age of onset fails to produce evidence of linkage with a disease locus, classification by other variables, such as severity of the expressed phenotype, may prove helpful.

Once researchers have identified a first locus responsible for a disease, they can use DNA markers at that locus to determine whether it is responsible for the disease in other families with a disease history. A process of elimination may identify a subset of families that must inherit the disease because of a predisposing allele at a different locus or loci. This type of testing and elimination identified a group of families in which *BRCA1* could not be the locus predisposing women to breast cancer. The combined data from these non-*BRCA1* families revealed a second breast cancer locus, on chromosome 13, named *BRCA2* (Fig. 11.23). Mutations at *BRCA1* or *BRCA2* account for many inherited breast cancers, but not all. As of 2002, there was still at least one, and probably more, breast cancer loci yet to be discovered.

Polygenic Inheritance: When Two or More Genes Interact in the Expression of a Phenotype

So far, we have examined ways in which diseases caused by mutations in a single gene are associated with complex patterns of inheritance. But as we saw in Chapter 3, many traits arise from the interaction between two or more genes. Some such polygenic traits are discrete: They either show up or they don't. The occurrence of a heart attack, or myocardial infarction, is a discrete polygenic trait. Other polygenic traits are quantitative: They vary over a continuous range of measurement, from one extreme through the normal range to the opposite extreme. Blood sugar levels, cholesterol levels, and depression are examples of quantitative traits. Loci that control the expression of such quantitative traits are known as **quantitative trait loci,** or **QTLs.** Although extreme values of QTL expression are considered abnormal, the border between normal and abnormal is arbitrary.

There is virtually an unlimited number of transmission patterns for polygenic traits. A completely penetrant discrete trait may require mutations at multiple loci to cause the abnormal phenotype. With other discrete polygenic traits, penetrance may increase as the number of mutant loci increases. With quantitative polygenic traits, the measured degree of expression (expressivity) may vary with the number of mutant loci present in the individual, or with the degree to which different mutations at the same locus alter the level of polypeptide production.

Many other factors can complicate the analysis of polygenic traits. Some members of a set of interacting polygenic loci may make a disproportionately large (or small) contribution to the penetrance or expressivity of the trait. Mutations at some loci in a set may be recessive, while mutations at other loci are dominant or codominant. Some traits may arise from a mixture of polygenic and heterogeneous components. For example, one form of a disease may be caused by mutations at loci A, B, C, and D; a second form, from mutations at B and E; and a third form, from a single mutation at F. The more complex the inheritance pattern of a trait, the more difficult it is to identify the loci involved.

How Geneticists Move from Complex Traits to Sets of Contributing Loci

The mouse is a model mammalian organism in which genetic analysis is a powerful tool for analyzing traits too complex for dissection with human pedigrees. Many of the numerous genetically homogeneous inbred mouse lines express a variety of traits that correspond to similar traits in humans. The analysis of two contrasting genotypes represented by two parental inbred mouse strains is advantageous because it eliminates genetic heterogeneity. Since in a cross between two inbred strains, there are only two alternative alleles at every marker and disease locus, such a cross is the equivalent of a family

pedigree documenting several hundred children all born to the same two parents. Another advantage of the mouse is that the possibility of maintaining all animals under identical conditions greatly reduces the impact of environmental differences.

Human geneticists are not the only researchers interested in the dissection of complex traits. Such analysis is also important in agriculture. For example, dairy and crop farmers are interested in milk production in cows and the taste, texture, and shape of fruits. Methods for dissecting complex traits are basically the same in all sexually reproducing experimental organisms. Here we examine the genetics of pulp content in tomatoes, but you could perform analogous experiments to look at complex traits in mice, fruit flies, or nematodes.

The pulp of a tomato is the succulent, edible solid part of the fruit—everything but the water, seeds, and skin. Manufacturers of canned tomatoes, tomato sauce, and tomato paste look for fruit with a high pulp content because it produces the highest product yield per weight of starting material (Fig. 11.24a). A strain of tomato plants with a maximum of pulp content, however, might

Figure 11.24 Using linkage analysis to identify loci that contribute to the expression of complex traits, such as pulp content in tomatoes. (a) Tomatoes with an ideal amount of pulp (*right*) and with too little pulp (left).
(b) Researchers use a two-generation cross to map loci that contribute to the amount of pulp in individual tomatoes. The first-generation cross is between parental lines that differ greatly in pulp content. The F₁ plants are all genetically identical, but their gametes contain recombinant chromosomes that assort to different offspring in the F₂ generation. The researchers correlate expression of the trait in the F₂ offspring with transmission of particular chromosomal regions. They then use statistical analysis to identify significant correlations between chromosomal regions and expression of the trait.

(a)

Too little pulp   Ideal pulp

(b)

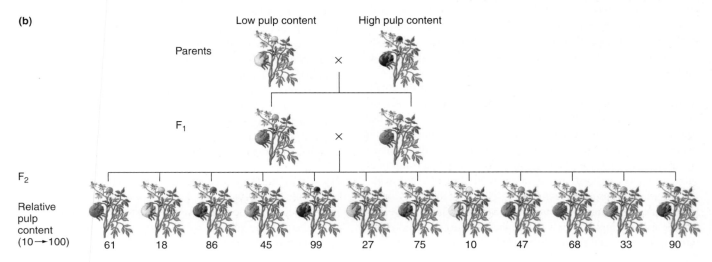

not have an ideal taste or texture. If you could identify DNA markers closely linked to loci that contribute to pulp content, you could cross an ideal pulp strain with other strains ideal for taste and texture, and then follow the markers to identify F_2 plants that had received the ideal pulp alleles and were otherwise genetically the same as parents with the best taste and texture.

The first step in such an analysis is the identification of two inbred strains with extreme, reproducible differences in the expression of the trait of interest.

To find markers linked to loci responsible for high pulp content, you cross plants from the high- and low-pulp strains and then cross the identical F_1 hybrids with each other to generate several hundred F_2 offspring. Because of independent assortment and recombination in the F_1 hybrids, the F_2 generation will have a range of genotype combinations and thus exhibit a range of "pulp" phenotypes (Fig. 11.24b). Some plants will have a pulp content similar to one parent's; others will have a pulp content similar to the other parent's. With most plants, the pulp content will lie somewhere between the two parental extremes. You now determine the genotype at polymorphic DNA markers spaced evenly across the genome, at 20 cM intervals, in each F_2 plant, and analyze the data for each marker in all F_2 plants. In your analysis, you look for correlations between marker genotype and pulp phenotype.

For example, if a marker is closely linked to a semidominant gene with a large influence on pulp content, most of the F_2 plants with high pulp content will be homozygous at this marker for the allele from the high-pulp parental strain, and most of the F_2 plants with low pulp content will be homozygous at this marker for the allele from the low-pulp parental strain. In contrast, markers not linked to a pulp-content locus would show no correlation between marker genotype and plant pulp phenotype.

You could use this basic process to find all tomato loci with a significant effect on pulp content, and then perform further analyses to understand the interactions between the various loci. When it is possible to map the genes contributing to every major trait of interest to tomato growers and processors, it will be possible, through further breeding and selection of individuals carrying the appropriate combination of marker alleles, to custom design a tomato plant possessing only the most desirable traits.

Eventually, the techniques of positional cloning described earlier will be used to clone the desirable genes themselves. When this happens, agricultural researchers will be able to use the tools of recombinant DNA technology to improve upon available characteristics and then insert the relevant modified genes directly into the plant genome. By combining traditional genetic analysis, positional cloning, recombinant DNA technology, and transgenic technology, scientists hope to bring forth a new generation of crops that can grow in hostile environments with higher yields and higher nutritional value. Such crops will help feed the world's population.

Haplotype Association Studies for High-Resolution Mapping in Humans

Linkage analysis in humans is not as simple as it is in animals or plants for which researchers design specific crosses to generate large numbers of offspring. In humans, traditional linkage analysis can only be done in retrospect with families that already exist. And because most human parents have few children, it is extremely difficult to generate high-resolution linkage maps, especially for diseases with complex inheritance patterns.

To overcome these limitations, geneticists developed a novel genetic approach to mapping disease loci that looks at the ancestry of chromosomes in present-day populations. This genetic approach is called *haplotype association analysis.*

Haplotypes Are Sets of Closely Linked Alleles
A contraction of the phrase "haploid genotype," the term **haplotype** refers to a specific combination of two or more DNA marker alleles situated close together on the same DNA molecule, that is, on the same homolog. As SNPs are the most prominent class of polymorphic DNA markers, they are also the most frequent markers in haplotypes.

Haplotypes come into existence by chance when a series of closely linked mutations accumulate one by one over time in the surviving descendants of an original ancestral chromosome (Fig. 11.25). The genetic distance over which the multiple loci of a haplotype extend must be short enough that the alleles are likely to remain associated with each other over many generations. In practice, haplotypes can extend in length from 1–100 kb. In humans, a 100 kb genomic distance is roughly equivalent to 0.1 cM, which means that the chance of such a haplotype being severed by a recombination event is 1 in a 1000. Recombination within a 1 kb haplotype will occur at the even lower frequency of 1 in 100,000 generations.

The formation of a SNP at any particular base pair is an extremely rare event (with a probability of $\sim 10^{-9}$ per gamete). The probability of sequentially forming multiple SNPs is the product of the probability of generating each individual SNP. Thus, a particular haplotype containing four or more SNPs can arise only once in human history (the chance of generating the same haplotype a second time is less than 1 in 10^{36} for four SNPs, which is 10^{26} times greater than the whole human population). Clearly, all individuals who carry a particular haplotype composed of four or more loci must have inherited it from the same ancestor, who may have lived hundreds or even thousands of years ago. Haplotype analysis thus provides a powerful tool for identifying ancient genetic relationships among people who have no recent ancestor in common.

Ancient Disease Loci Are Associated with Particular Haplotypes
Haplotype association analysis is based on two facts. First, a particular disease phenotype in many people from the same population is often the result of a single mutational event that

(a) Formation of haplotypes over evolutionary time

(b) Comparison of two final haplotypes with a compilation of five SNPs spread over 5000 base pairs

Figure 11.25 Haplotypes are formed by sequential mutations in a small genomic region. (a) Two pathways of divergence from an original 5 kb chromosomal region (*displayed in the middle of the two divergence pathways*). Along one pathway, two mutations accumulate; along the second pathway, three mutations accumulate. **(b)** The surviving descendant chromosomes represent two distinct haplotypes that differ at five separate SNP loci.

occurred in a long-forgotten common ancestor who lived and died centuries ago. Second, the disease mutation occurs in a haplotype of very closely linked SNPs that is passed down from generation to generation along with the disease mutation itself. While the mutation may be a single-base-change needle in the 10^9-bp-genomic haystack, its associated haplotype can magnify the needle 100,000-fold. That is, instead of looking for a single base change, you can follow a number of nearby, closely associated polymorphisms spread over an area of 100 kb.

To perform haplotype association analysis, you start with a population of people that has been genetically isolated for a long time, such as native Icelanders (who are descendants of Vikings from 1000 years ago) or the Amish community of central Pennsylvania (who descended from a small group of eighteenth-century Swiss immigrants). You collect DNA samples from a subgroup that expresses the disease of interest, such as schizophrenia, or obesity, and choose an equal number of DNA samples from random, unrelated people who do not exhibit the trait. Now you genotype each individual in both subgroups for haplotypes (groups of SNPs) throughout the entire genome. The vast majority of haplotypes will have no association with the

locus responsible for the disease; the frequency of appearance of such haplotypes will be the same in the disease group and the test group. You hope to find a haplotype present at a very high frequency in the disease group and a much lower frequency in the control group (Fig. 11.26). Alleles at separate loci—in this case, haplotype marker alleles and the disease allele—that are associated with each other at a frequency significantly higher than that expected by chance, are said to be in **linkage disequilibrium.** A finding of linkage disequilibrium not only means that the two (or more) loci are linked to each other, it also means that the disease mutation arose on a chromosome that already carried the SNP alleles in the haplotype.

When genetic mapping by association works, it resolves much more than traditional pedigree analysis. This is because for many inherited diseases, recombination events over the numerous generations between the mutated ancestor and the current population have narrowed to 100 kb or less the DNA markers that remain part of the disease-associated haplotype. Most genomic regions of this size contain fewer than three candidate genes whose contribution to the disease can be evaluated by positional cloning.

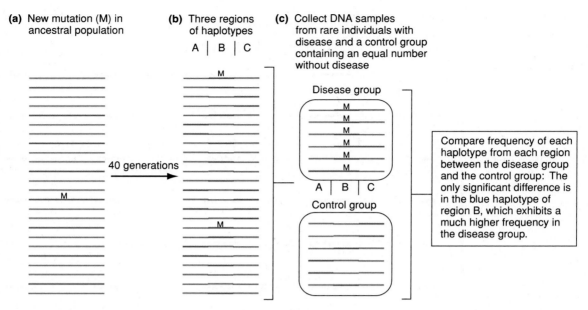

(a) New mutation (M) in ancestral population

(b) Three regions of haplotypes

A | B | C

(c) Collect DNA samples from rare individuals with disease and a control group containing an equal number without disease

40 generations

Disease group

A | B | C

Control group

Compare frequency of each haplotype from each region between the disease group and the control group: The only significant difference is in the blue haplotype of region B, which exhibits a much higher frequency in the disease group.

Figure 11.26 Haplotype association allows high-resolution gene mapping. (a) Representation of the same subchromosomal region in different individuals within an ancestral population that lived several thousand years ago. Lines of the same color have the same set of alleles because of common descent. The disease mutation (M) occurred on one ancestral chromosome. **(b)** After 40 generations, the original subchromosomal region has broken apart through recombination into three smaller regions (*1 on the left, 2 in the middle, and 3 on the right*). Each smaller region still occurs in three different haplotypes. **(c)** You now collect DNA samples from a number of rare, unrelated people with the disease and from an equal number of people without the disease, and compare the frequency of each haplotype from regions A, B, and C in the disease and control groups. Genotyping of the samples will enable you to compare haplotype frequencies in the two groups. Haplotypes of regions not closely linked to the disease locus will occur with the same frequency in both groups. The only significant difference will be in the one haplotype that encompasses the disease locus (the *blue* haplotype of region B); it has a much higher frequency (here 100%) in the disease group than in the control group.

CONNECTIONS

In Chapter 9, we described the new tools of recombinant DNA technology that enable geneticists to look directly at the DNA sequence in individual chromosomes. In Chapter 10, we explained how high-speed automated machines incorporate these tools to generate huge amounts of sequence information; analysis of these data by sophisticated computer algorithms produces whole-genome sequence maps showing the locations of functional units, including the coding sequences of genes. In this chapter, we examined how researchers apply the tools of recombinant DNA technology and genomics to the direct analysis of genotype at the DNA level. Health professionals can use direct genotyping to diagnose hereditary disease; forensic experts can use it to determine the identity and degree of relatedness of DNA samples. Finally, researchers can use direct genotyping to identify and clone the genes responsible for any inherited trait that differs in its expression among individuals within a population. With these new approaches, the human species has become a superb system for genetic analysis.

Will this enormous capacity for obtaining genetic information provide a complete understanding of life? Far from it. Biologists will need much more than a knowledge of the instructions specified by nucleotide sequence to comprehend the networks of interactions that bring an organism to life. This is because chromosomes are not simply repositories of genetic information. Composed of many protein molecules in addition to DNA, they are complex organelles that package and manage the availability of DNA. The activity of large chromosomal regions is controlled not just by local DNA sequence but by the way the DNA is packaged. Some regions, for example, are marked in a way that irreversibly extinguishes all activity, including gene expression. Moreover, chromosomes have a hierarchic organization that enables them to go through a cycle of condensation and decondensation with each cell cycle, and they contain specialized regions that dictate how they duplicate and segregate during nuclear division. In Chapter 12, we examine how proteins interact with DNA to generate a chromosome's functional complexity.

ESSENTIAL CONCEPTS

1. Using the tools of recombinant DNA technology to compare the DNA sequences of homologous chromosomal regions carried by different members of a species, researchers have detected enormous variation in nearly all animals and plants.

   a. The ability to distinguish genotypic differences of all kinds extends our concept of a locus and the alleles that define it. A *locus* is a designated location anywhere on a chromosome; it can range in size from a single base pair to more than a megabase pair. A *gene* is a coding locus. A DNA locus without any apparent function is an *anonymous locus*.

   b. When two or more alleles exist at a DNA locus, the locus is *polymorphic* and the variations themselves are *DNA polymorphisms*. Polymorphic DNA loci that are useful for genetic studies are known as *DNA markers*.

   c. There are four classes of DNA polymorphisms: single nucleotide polymorphisms, or SNPs; microsatellites; minisatellites; and deletions, duplications, and insertions in nonrepeat loci.

2. Once a DNA polymorphism at a particular locus has been uncovered, researchers can use a variety of methods to determine the genotype at that locus in any individual.

   a. There are three ways to detect SNPs.

   (1) A small proportion of SNPs, by chance, eliminate or create a restriction site. PCR provides a rapid, cost-effective method for their detection.

   (2) A general method for detecting SNPs exploits differences in hybridization between a probe and two target sequences in the center of the region of hybridization that differ by a single nucleotide. Probes developed for this purpose are called allele-specific oligonucleotides, or ASOs.

   (3) Primer extension analysis takes advantage of the high fidelity with which DNA polymerase attaches complementary nucleotides to a growing DNA strand during DNA replication.

   b. Polymorphisms in microsatellite and minisatellite loci, and small deletions, duplications, and insertions in nonrepeat loci, change the size of a locus and are easy to detect by PCR.

   (1) Microsatellite variants are detectable in this way as are microsatellite-related disease loci like the one responsible for HD.

   (2) Minisatellite alleles can also be detected by PCR. Simultaneous Southern blot analysis of cross-hybridizing minisatellite loci can identify multiple unlinked, polymorphic loci around the genome. The result of this analysis is called a DNA fingerprint. DNA fingerprinting provides a quick snapshot of genomic features that can identify an individual or reveal the relationship between two samples from the same population. DNA fingerprints are a powerful tool for forensic analysis.

3. Positional cloning identifies the genes responsible for traits whose molecular cause is unknown.

   a. To localize a trait-affecting gene to a specific region of chromosomal DNA, researchers combine formal linkage analysis with the use of DNA markers.

   b. Once they have narrowed the location of the gene to a small enough genomic region, researchers catalogue all possible candidate genes that map to that region, using computational tools that search for coding sequences within genome databases and sequences that are conserved between distantly related species. They can also search for EST sequences in clones derived from cDNA libraries, and they can use Northern blots to identify gene sequences that are transcribed into RNAs in tissues affected by the trait.

   c. To identify the one candidate gene that is responsible for the trait, researchers compare groups of phenotypically normal and abnormal individuals. A finding that the DNA sequence or transcription of the candidate gene is altered in all individuals exhibiting the mutant trait is strong evidence that the candidate gene is responsible for the trait.

   d. Incontrovertible evidence that a candidate gene causes a particular phenotype is best obtained in an experimental animal model where it is possible to induce a specific genetic change and demonstrate the appearance of a predicted phenotype. Two technologies can accomplish this goal: nuclear injection of DNA into embryos to produce transgenic mice, and targeted mutagenesis.

4. Most genetically determined trait variation among individuals of a species is inherited in a complex manner.

   a. With incomplete penetrance, a mutant genotype does not always cause a mutant phenotype.

   b. Mutant traits that arise in the absence of a mutant genotype are considered *phenocopies*.

   c. Mutant traits caused by mutations at any one of two or more alternative loci show *genetic heterogeneity*.

   d. A phenotype controlled by alleles at multiple loci is a *polygenic* trait.

e. If different combinations of alleles cause quantifiable differences in a trait, the trait is a *quantitative trait,* and the loci involved are *quantitative trait loci,* or *QTLs.*

f. The contemporary method of dissecting complex traits is basically the same in all sexually reproducing experimental organisms.

5. Haplotype association analysis is a novel genetic approach to mapping disease loci that is not dependent on the analysis of genetic transmission of alleles from parent to child.

a. The term *haplotype* refers to a specific combination of two or more DNA markers situated close together on the same chromosomal homolog. All individuals who carry a particular haplotype must have inherited it from the same ancestor.

b. Expression of a specific disease phenotype in many people from the same population is often the result of an ancient mutational event in a haplotype of very closely linked SNPs that has been passed intact from generation to generation.

c. When alleles at separate loci are associated with each other at a significantly higher frequency than is expected by chance, they are in *linkage disequilibrium.* A finding of linkage disequilibrium is strong evidence that a locus lies in a small, defined chromosomal region.

S O C I A L   A N D   E T H I C A L   I S S U E S

1. DNA fingerprinting has been accepted as evidence in many recent court trials. As a consequence, jurors and judges who do not have scientific training are being asked to evaluate the results of such tests, including how carefully a laboratory performed the test and the statistical relevance of the matches between sets of data. Is it fair to try someone on the basis of this kind of genetic evidence using a panel of jurors who do not have a prior understanding of the basic principles behind the technique? Can you think of ways to improve the use of these data in criminal cases?

2. Bob and Karen have a child who was born with Tay-Sachs, a recessive genetic disorder. When Karen became pregnant again, genetic tests were done on all family members, including amniocentesis on the fetus. Results showed that the fetus is unaffected, Karen is a carrier of a mutated allele, but Bob is not a carrier. These results strongly suggest that Bob could not be the father of the first child. It is possible, but unlikely, that a germ-line mutation arose in Bob's cells. Additional genetic tests could settle the paternity issue. Who should be told? The mother only? Both mother and father? Neither?

3. Mary is a 25-year-old woman who has a good job with a company where she has room to advance in her career. When she accepted her job at the age of 22, she signed a contract that stipulated that she would tell her supervisor of any change that occurred in her health status. A footnote in the contract stated explicitly that a change in health status was defined not only as a symptomatic change but also as a prognostic change. This footnote was meant to include persons who tested positive for HIV even when they did not yet exhibit the symptoms of AIDS. It seemed clear to Mary that the purpose of this stipulation was to provide the company with grounds for dismissal. Three years after the beginning of Mary's employment, Mary's 56-year-old mother Elizabeth was diagnosed with breast cancer. Because several other women in the family had breast cancer, there is a high probability that there is an inherited predisposition to breast cancer in the family. Tests indicated that her mother does carry a mutant allele of the *BRCA1* gene, and therefore, Mary has a 50% chance of having a *BRCA1* allele that predisposes her to breast cancer. The counselor encouraged Mary to take the test since there are preventative measures that can be taken if she finds out she is positive. But, if she takes the test and the result shows that she carries the *BRCA1* mutation, she believes that she will have to tell her boss and possibly lose her job. On the other hand, if the test shows that she doesn't have the mutation, a large load will be taken off her mind as she realizes her risk of disease is no more than that of other women without the *BRCA1* mutation. What would you do if you were Mary?

4. A couple who are both dwarfs want to have a child who is a dwarf. Both the parents have achondroplasia, the dominant inherited form of dwarfism. (The parents are heterozygous for the achondroplasia allele. Homozygous individuals die before age one of severe skeletal abnormalities.) The parents have told the genetic counselor at the testing clinic that they will abort a fetus if it *does not* carry the dwarf allele. Should the clinic perform the test for the achrondroplasia gene, knowing that the parents will choose to abort a healthy fetus simply because it will not be a dwarf?

5. Josephine read in the newspaper that an individual's propensity to engage in risk-taking behaviors is influenced by genetic makeup. She would like her children to take a more aggressive or assertive approach to new situations than she ever did. She thinks that the possibility of germ cell gene therapy (altering DNA of the

egg or sperm) to modify behavioral genes such as this would be very attractive and would like to see development of the technology to do this type of manipulation. Is it ethical to change the genetic makeup of the germ cells? Do humans have the right to change traits in their children? in themselves? What is the difference between controlling a child's genes and controlling a child's environment when the goal of each form of control is to maximize a child's chances of success? Would your answer be different if it were possible to introduce an allele that caused resistance to HIV?

S O L V E D   P R O B L E M S

I. The figure shows the pedigree of a family in which a completely penetrant, autosomal dominant disease is transmitted through two generations, together with a corresponding Southern blot with individual pedigree samples digested with *Eco*RI and probed with a DNA fragment that detects a restriction fragment length polymorphism (RFLP). Do the data suggest the existence of genetic linkage between the RFLP locus and the disease locus? If so, what is the estimated genetic distance between the two loci?

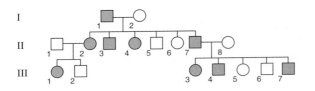

Answer

To solve this problem you need to understand how DNA polymorphisms can be followed through a pedigree, and how they can be tested for linkage to a locus defined by phenotype alone. First examine the Southern blot pattern to determine what the forms of the DNA polymorphism are. The two segregating DNA alleles in this pedigree are represented by RFLPs having sizes of 8 kb and 7 kb. Some individuals are heterozygous with both restriction fragments, and some are homozygous with just the 8 kb fragment or the 7 kb fragment alone.

When two parents have one DNA allele in common, but are different at the second allele, it is possible for a child to inherit the common allele from either or both parents. If the child is homozygous for the allele, then he must have received it from both parents. But, if the child has just one copy of the common allele, exclusion analysis can be used to determine which parent had to be the one that transmitted it. For example, children II-5, II-6, and II-7 have a 7 kb allele that could only have come from their father (I-1), because their mother doesn't carry this allele. By exclusion, their second allele—which is

8 kb—must have come from their mother, even though it is present in the genomes of both parents.

In the second generation set of siblings (II-2 through II-7), inheritance of the paternal 8 kb allele correlates with

inheritance of the disease allele in five of six children: II-2, II-3, and II-4 inherit the paternal 8 kb allele along with the disease, and II-5 and II-6 do not receive the paternal 8 kb allele and do not exhibit the disease. The remaining child (II-7) exhibits the disease but inherited the 7 kb allele from his father. There are two possible explanations for this discrepant individual. First, the RFLP locus could be unlinked to the disease locus, and the six out of seven transmission correlation could be a chance event. Second, the loci could indeed be linked with the II-7 child representing a recombination event that brings the disease locus onto the same chromosome as the 7 kb allele.

To distinguish between these possibilities, you need to examine transmission to the third-generation children using the same logic but with different facts. First, in the third-generation family on the left, the diseased parent (II-2) is homozygous for the 8 kb RFLP allele and thus, no useful data on linkage can be obtained from her children III-1 and III-2. But useful data can be obtained from the third-generation family on the right. If your hypothesis of linkage is correct, it should now be the 7 kb allele that is transmitted in correlation with the disease allele. Indeed, III-3, III-4, and III-7 receive a paternal 7 kb allele together with the disease allele, and III-5 and III-6 do not receive either the 7 kb allele or the disease allele.

When linkage data are combined from the entire pedigree, you find that there are 11 informative offspring. *In 10 of 11 cases, alleles at the disease locus show cotransmission with alleles at the RFLP locus.* This is evidence for linkage between the two loci. Further studies with additional families would be required to confirm this linkage.

II. A clear limitation to gene mapping in humans is that family sizes are small, so it is very difficult to collect enough data to get accurate recombination frequencies. A technique that circumvents this problem begins with the purification of DNA from single sperm cells. (Remember that recombination occurs during meiosis. Analysis of individual sperm in a large population can provide a large data set for linkage studies.) The DNA from single sperm cells can be used for SNP studies. Four pairs of primers were used for PCR amplification of four defined SNP loci from one man's somatic cells and from 21 single sperm that he provided for this research. Each of these primer pairs amplifies a different SNP locus referred to as A, B, C, and D. The four pairs of PCR primers were used simultaneously on each sample of DNA. Each of the amplified DNAs was divided into eight aliquots (identical subsamples), and these aliquots were denatured and spotted onto eight nitrocellulose membrane strips (vertically as shown in the figure). Each of these strips was then hybridized with a different ASO (allele-specific oligonucleotide). There are two different ASOs for each SNP. For example, ASOs named A1 and A2 detect different alleles at SNP locus A. A black spot indicates that the amplified DNA hybridized to the ASO probe.

a. Based on the results shown, which SNP loci could be X-linked?

b. Which SNP loci could be on the Y chromosome?

c. Which SNP loci must be autosomal and homozygous?

d. Which SNP loci must be autosomal and heterozygous?

e. Do any SNP loci appear to be linked to each other?

f. Ignoring the results from sperm number 21, what is the distance between the two linked SNP loci?

g. How could you map the genomic region defined by SNP locus A?

h. What event could have given rise to sperm number 21?

Answer

For this problem, you need to understand how ASOs detect SNP alleles and the advantages and limitations of SNPs as DNA marker loci. An advantage of SNP analysis by PCR is that the technique is so sensitive that the single alleles present within individual sperm cells can be assayed. An ASO result is either positive or negative. If the result is positive (as indicated by a black dot of hybridization in this example), a tested somatic cell sample can be either homozygous or heterozygous for the corresponding ASO allele; a positive ASO result by itself doesn't distinguish between these possibilities. If the result is negative, the sample doesn't contain the ASO allele under analysis, but nothing can be said about the alleles (if any) that are present at the tested locus.

a. Half of the sperm cells will not have an X chromosome and would not be expected to show a positive result with any ASO for any X-linked SNP locus. The other half of the sperm cell will carry the same SNP allele. *Gene C shows this type of pattern.*

b. Similarly, a gene on the Y chromosome would be found in only half the sperm. Again *gene C is a candidate for a Y linked gene.*

c. If an individual is homozygous at an autosomal SNP locus, all the sperm from that individual will show hybridization to one ASO for that locus and not any other. *Locus A appears to be homozygous and autosomal.*

d. At a heterozygous SNP locus, one ASO will hybridize to approximately one-half of the sperm samples, and a second ASO will hybridize to those samples that do not hybridize to the first ASO. *SNP loci B and D show this type of pattern.*

e. Alleles at linked loci will segregate together more than 50% of the time and would therefore end up in the same sperm. Alleles B1 and D2 are transmitted together more often than not, and the reciprocal alleles B2 and D1 are also transmitted together more often than not. *This result suggests that loci B and D are linked.*

f. Sperm 3, 9, and 18 show evidence of recombination between alleles at the B and D loci. Three out of 20, or 15%, are recombinant. *The distance between the B and D loci is therefore 15 cM.*

g. Since SNP locus A is homozygous in this individual, it can't be mapped in a linkage analysis. But, SNPs are found at an approximate rate of one in a thousand base pairs. *By sequencing several kilobases of genomic DNA around SNP locus A from this individual, you could identify a nearby SNP locus that is heterozygous and that could be mapped in a linkage analysis.*

h. *Sperm sample number 21 could accidentally have two sperm cells rather than one. It is also possible that a single sperm in this sample has accidently received two copies of the chromosome that carries loci B and D through meiotic nondisjunction.*

ASO:	Gene A		Gene B		Gene C		Gene D	
	A1	A2	B1	B2	C1	C2	D1	D2
Somatic cell	●		●	●		●	●	●
Sperm number								
1	●		●					●
2	●			●			●	
3	●			●		●	●	
4	●			●		●		●
5	●			●				●
6	●		●					●
7	●			●		●	●	
8	●			●			●	
9	●			●		●		●
10	●		●					●
11	●		●					●
12	●			●		●		●
13	●			●				●
14	●		●			●	●	
15	●		●			●		●
16	●			●		●	●	
17	●		●					●
18	●			●		●	●	
19	●			●			●	
20	●		●			●	●	
21	●		●	●			●	●

P R O B L E M S

11-1 Choose the phrase from the right column that best fits the term in the left column.

a.	DNA polymorphism	1.	DNA element composed of tandemly repeated identical sequences
b.	haplotype	2.	two different nucleotides appear at the same position in genomic DNA from different individuals
c.	RFLP	3.	combination of alleles at multiple tightly linked loci that are transmitted together over many generations
d.	ASO	4.	location on a chromosome
e.	SNP	5.	a DNA sequence that occurs in two or more variant forms
f.	DNA fingerprinting	6.	a short oligonucleotide probe that will hybridize to only one allele at a chosen SNP locus
g.	minisatellite	7.	detection of genotype at a number of unlinked highly polymorphic loci using one probe
h.	locus	8.	variation in the length of a restriction fragment detected by a particular probe due to nucleotide changes at a restriction site

11-2 Imagine that you have identified a SNP marker that lies 1 cM away from a locus causing a rare hereditary autosomal dominant disease. You test additional nearby markers and find one that shows no recombination with the disease locus in the one large family that you have used for your linkage analysis. Furthermore, you discover that all afflicted individuals have a G base at this SNP on their mutant chromosomes, while all wild-type chromosomes have a T base at this SNP. You would like to think that you have discovered the disease locus and the causative mutation but realize you need to consider other possibilities.
   a. What is another possible interpretation of the results?
   b. How would you go about obtaining additional genetic information that could support or eliminate your hypothesis that the base-pair difference is responsible for the disease?

11-3 Mutations at microsatellite loci occur at a frequency of 1×10^{-3}, which is much higher than base substitutions at other loci.
   a. What is the nature of microsatellite polymorphisms?
   b. By what mechanism are these polymorphisms generated?

   c. Minisatellites also mutate at a relatively high frequency. Do these mutations occur by the same or a different mechanism?

11-4 Restriction site polymorphisms can be detected by Southern blot analysis or by PCR analysis. What are the similarities and differences between the two approaches? Which one is more efficient?

11-5 Each of the following reagents can be used to detect SNP polymorphisms. Where is the polymorphism located in relation to the probe or primer DNA sequence used in each of these techniques?
   a. allele-specific oligonucleotide (ASO)
   b. primer-extension oligonucleotide
   c. RFLP probe

11-6 The ASO technique was used to determine the genotypes of 10 family members with regard to sickle-cell anemia, as shown here. Each pair of dots represents the results of ASO analysis for the DNA from one person. The upper row represents hybridization with the normal oligonucleotide, and the lower row represents the results of hybridization using the mutant oligonucleotide. The three replications of the assay were incubated at 100°C (upper set), 90°C (middle set), and 80°C (lower set).
   a. Why do the three replications of the same sample set look different?
   b. What are the genotypes of the individuals?

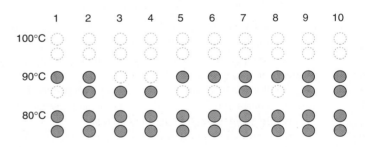

11-7 Given the two allelic sequences shown here and the site at which a single-base polymorphism occurs (underlined in the sequence), what sequences would you use as oligonucleotide probes for ASO analysis of genotype by hybridization? (Assume that ASO probes are usually 19 bp in length.)

```
allele 1:
5' GGCATTGCATGCTAACCCTATAAATGCGCTAGGCGTAGTTAGCTGGGAATAAAAAGCT 3'
allele 2:
5' GGCATTGCATGCTAACCCTATAAATGGGCTAGGCGTAGTTAGCTGGGAATAAAAAGCT 3'
```

11-8 Approximately 3% of the population carries a mutant allele at the *CFTR* gene responsible for cystic fibrosis. New disease-causing mutations at this locus arise at a frequency of 1 in 10^4 gametes. A genetic counselor is examining a family in which both parents are known to be carriers for a *CFTR* mutation. Their first child was born with the disease, and the parents have come to the counselor to assess whether the new fetus inside the mother is also diseased, is a carrier, or is completely wild type at the *CF* locus. DNA samples from each family member and the fetus are tested by PCR and gel electrophoresis for a microsatellite marker within one of the *CFTR* gene's introns. The following results are obtained:

```
        Dad    Mom    First   Fetus
                      child
      ┌──────────────────────────────┐
      │ ▬                     ▬        │
      │        ▬                       │
      │               ▬       ▬       │
      │ ▬             ▬               │
      │                       ▬       │
      └──────────────────────────────┘
```

a. What is the probability that the child who will develop from this fetus will exhibit the disease?
b. If this child grows up and gets married, what is the probability that one of her children will be afflicted with the disease?

11-9 When a researcher begins to choose DNA markers for linkage analysis of a disease trait in a particular family, what is the first criterion used in this choice?

11-10 Mice can be genetically engineered to express a hereditary disease that afflicts people. Strains of these mutant mice can provide biomedical scientists with a model for testing possible therapies.
a. What is the best strategy for creating a mouse model for a particular form of hypercholesterolemia that results from a mutant human gene that overexpresses a cholesterol-forming enzyme?
b. What is the best strategy for creating a mouse model for hemophilia?

11-11 The police discovered the body of a woman who had been brutally beaten and raped while working late in her office one evening. They suspected that one of her seven coworkers might be responsible for the crime. To test this possibility, they recovered semen from her vagina and used it to prepare a DNA sample. They also took DNA samples from each of her coworkers. All eight samples were subjected to DNA fingerprinting analysis based on restriction digestion, gel electrophoresis, Southern blotting, and probing with a minisatellite. The results are shown in the photo shown here.

1   2   3   4   5   6   7   Crime
                                sample

Is any one of her coworkers likely to be the perpetrator of the crime? Which one? How likely is it that this particular person is the perpetrator rather than another person unknown to the authorities (provide a general estimate)?

11-12 DNA fingerprinting can be used to settle cases concerning paternity. In the DNA fingerprint shown, the mother's DNA sample is in lane 1, the daughter is in lane 2, and two samples of men that could be the father are in lanes in 3 and 4. Can you determine from this data if one of the men must be the father?

sites surrounding the gene vary among individuals, producing at least three RFLPs in the population: the sizes of the RFLP alleles are 13, 10, and 7 kb. Shown here are two small pedigrees of families in which individuals with the disease are shaded in black. Below each pedigree symbol is the corresponding result obtained from the DNA sample in a Southern blot analysis with the FM probe.

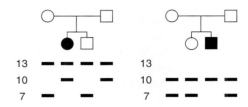

a. Which restriction fragment is associated with the disease mutation in the father shown in the left pedigree?

b. Which restriction fragment is associated with the disease mutation in the mother shown in the right pedigree?

c. If the male child from the left pedigree marries the female child from the right pedigree, what is the probability that their child will be diseased?

d. What is the probability their child will be a carrier?

11-16 The pedigrees indicated here were obtained with three unrelated families with members that express the same completely penetrant disease caused by a dominant mutation that is linked at a distance of 10 cM from a marker locus with three alleles numbered 1, 2, and 3. The marker alleles present within each live genotype are indicated below the pedigree symbol. The phenotypes of the newly born labeled individuals—A, B, C, and D—are unknown. What is the probability of disease expression in each of these individuals?

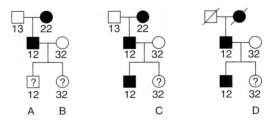

11-13 Angela and George have one child, and Angela has sickle-cell anemia. They want to have more children but do not want any of them to have to suffer with this disease. They also do not want to be in a position of having to abort a fetus, so they elect to have *in vitro* fertilization and embryo screening. Briefly list the steps they must take to accomplish this goal.

11-14 Experiments to find the gene, or genes, responsible for an inherited human disease can be conducted in humans or mice. What are the advantages and disadvantages of each system?

11-15 A relatively frequent, completely penetrant recessive disease known as the foul mouth syndrome (FM) has been found to be due to a variety of mutations in the *FM* gene, which has recently been cloned. Analysis of Southern blots of human DNA cleaved with the enzyme *Hpa*I and probed with a radioactively labeled fragment of the *FM* gene has revealed that *Hpa*I does not cleave within the gene itself. However, the positions of *Hpa*I

11-17 Arrange the following statements in the appropriate order for performing the primer extension protocol to detect genotype at a SNP locus. Assume that you will first prepare a common reagent needed to genotype any sample.
  a. Denature DNA sample.
  b. Add in DNA polymerase and one type of dideoxynucleotide triphosphate.
  c. Construct an oligonucleotide primer from the sequence adjacent to the SNP.
  d. Hybridize oligonucleotide primer with amplified DNA sample.
  e. Obtain genomic DNA from the person who you want to genotype.
  f. Analyze products by mass spectrometry.
  g. PCR amplify genomic region around SNP locus.

11-18 A rare human disease leads to overgrowth of the heart without any other effect on the afflicted individual. Linkage analysis with DNA markers has been used to map the disease locus to a small chromosomal region. This region has been divided into five DNA fragments (named A through E) that are each labeled and used to probe Northern blots containing a single lane of RNA from one of three tissues—liver, heart, and muscle—taken from nondiseased cadavers. The results are shown here.

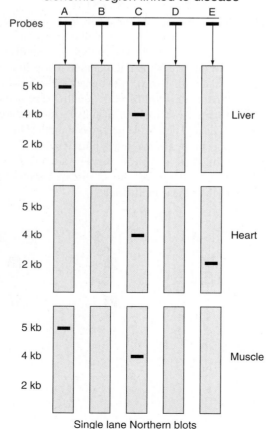

Genomic region linked to disease

Single lane Northern blots

  a. Which of the five DNA fragments is likely to contain a gene?
  b. How many genes have been identified in this region by Northern blot analysis?
  c. Is it possible that there are more genes in this region than those detected here?
  d. Which of the five DNA fragments could possibly contain the gene responsible for this disease?
  e. Which fragment is most likely to contain the gene?
  f. What is your next step in testing this candidate gene as the causative factor for the disease?

11-19 You have decided to study a disease trait that is very rare. You have found an extended family in which a number of nonsibling individuals express the disease. The pedigree is as follows:

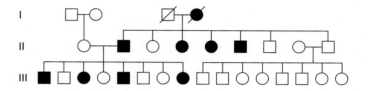

  a. What is the most likely genetic basis for the disease?
  b. Are there any individuals in the pedigree who *must* carry the disease mutation even though they don't express the disease trait? If so, list those individuals according to their generation number (in Roman numerals) and their number within the generation (counting from left to right across the entire pedigree).

11-20 You have decided to study another disease trait that is very rare. You have searched far and wide to come up with an extended family in which a number of nonsibling individuals express the disease. The pedigree is as follows:

  a. What is the most likely genetic basis for the disease?
  b. Are there any individuals in the pedigree who *must* carry the disease mutation even though they don't express the disease trait? If so, list those individuals according to their generation number (in Roman numerals) and their number within the generation (counting from left to right across the entire pedigree).

11-21 You have decided to study another disease trait that is very rare. You have searched far and wide to come up with an extended family in which a number of nonsibling individuals express the disease. One pair of fraternal twins are indicated with lines that join together at a vertex.

The pedigree is as follows:

a. What is the most likely genetic basis for the disease?

b. Are there any individuals in the pedigree who *must* carry the disease mutation even though they don't express the disease trait? If so, list those individuals according to their generation number (in Roman numerals) and their number within the generation (counting from left to right across the entire pedigree).

11-22 You have decided to study another disease trait that is very rare. You have found an extended family in which a number of nonsibling individuals express the disease. One pair of identical twins are indicated with a horizontal line that joins both symbols together and with descent lines that join together at a vertex.

The pedigree is as follows:

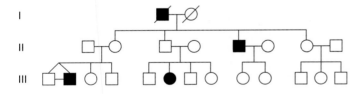

a. What is the most likely genetic basis for the disease?

b. Are there any individuals in the pedigree who *must* carry the disease mutation even though they don't express the disease trait? If so, list those individuals according to their generation number (in Roman numerals) and their number within the generation (counting from left to right across the entire pedigree).

11-23 List three independent conceptual approaches for finding genes within a large cloned and sequenced genomic region.

11-24 The next disease that you decide to tackle is Pinocchio syndrome, which causes the noses of afflicted individuals to grow larger when they tell a lie. You discover a family in which this disease is segregating as indicated in the following pedigree. You have reason to believe that a SNP locus called SNP1 is linked to the Pinocchio gene, and you test each individual in the family with ASOs that recognize allele 1 or allele 2 at this SNP locus with the genotyping results shown here.

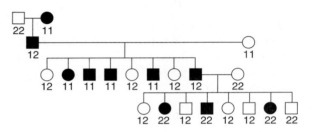

a. What is the most likely genetic basis for Pinocchio syndrome?

b. Is it likely or unlikely that the Pinocchio locus is linked to the SNP1 locus?

c. Sequence analysis shows the SNP1 locus actually resides in the middle of a coding region. How likely is it that this coding region is equivalent to the Pinocchio locus?

11-25 Among the most prevalent diseases that afflict human beings is heart disease, which can have a severe impact on quality of life as well as result in premature death. While heart disease mostly afflicts those who are older, 1% or 2% of people in their thirties, and even their twenties, suffer from this disease. There are genetic and environmental components to this disease. Use this information to answer the following questions.

a. What strategy might you use to choose families to participate in a linkage study of heart disease-causing genes?

b. Once you have cloned a gene that you believe plays a role in heart disease, how would you confirm this role?

R E F E R E N C E S

Handyside, A. H., Kontogianni, E. H., Hardy, K., and Winston, R. M. L. (1990). Pregnancies from biopsied human preimplantation embryos sexed by Y-specific DNA amplification. *Nature 344*:768–70.

Jeffreys, A. J., Wilson, V., and Thein, S. L. (1985). Individual-specific "fingerprints" of human DNA. *Nature 316*:76–79.

Koopman, P., Gubbay, J., Vivian, N., Goodfellow, P., and Lovell-Badge, R. (1991). Male development of chromosomally female mice transgenic for Sry. *Nature 351*:117–21.

Kwok, P. Y. (2001). Genetic association by whole-genome analysis? *Science 294*:1669–70.

Li, J., Butler, J. M., Tan, Y., Lin, H., Royer, S., Ohler, L., Shaler, T. A., Hunter, J. M., Pollart, D. J., Monforte, J. A., et al. (1999). Single nucleotide polymorphism determination using primer extension and time-of-flight mass spectrometry. *Electrophoresis 20*:1258–65.

Saiki, R. K., Bugawan, T. L., Horn, G. T., Mullis, K. B., and Erlich, H. A. (1986). Analysis of enzymatically amplified beta-globin and HLA-DQ alpha DNA with allele-specific oligonucleotide probes. *Nature 324*:163–66.

Signer, E. N., Dubrova, Y. E., Jeffreys, A. J., Wilde, C., Finch, L. M., Wells, M., and Peaker, M. (1998). DNA fingerprinting Dolly. *Nature 394*:329–30.

Syvanen, A. C. (2001). Accessing genetic variation: genotyping single nucleotide polymorphisms. *Nat Rev Genet 2*:930–42.

Weber, J. L., and May, P. E. (1989). Abundant class of human DNA polymorphisms which can be typed using the polymerase chain reaction. *Am. J. Hum. Genet. 44*:388–96.

Witte, J. S., Elston, R. C., and Schork, N. J. (1996). Genetic dissection of complex traits. *Nature Genetics 12*: 355–56.

The Eukaryotic Chromosome: An Organelle for Packaging and Managing DNA

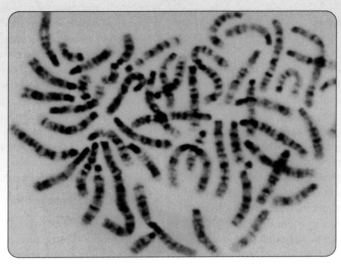

Mouse mitotic metaphase chromosomes stained for karyotype analysis
(magnification 600X).

To build an artificial chromosome able to function as part of a cellular genome, what chromosomal components would you need to assemble? Three sets of observations can help answer this question. In Chapter 4 you saw that when chromosomes segregate during mitosis or meiosis, some of the cell's newly elongated spindle fibers attach to the centromere, the pinched-in portion of the chromosome viewed under the microscope (Fig. 12.1). A centromere is therefore essential to proper chromosome segregation. Second, scientists have found that the ends of linear chromosomes remain intact even though linear pieces of DNA, when introduced into a cell, are susceptible to degradation by nucleases. A search for the chromosomal element that prevents degradation revealed that specific repeated DNA sequences and special proteins compose the tips of linear chromosomes. Known as **telomeres,** these protective DNA-protein caps are another critical chromosome component. Finally, you know that DNA must replicate before cell division so that each daughter cell receives a complete copy of the genomic DNA. To ensure that your artificial chromosome replicates properly, you must include at least one origin of replication.

In the 1980s, molecular geneticists took these three key chromosomal elements—centromeres, telomeres, and origins of replication—and, using raw materials from the yeast *Saccharomyces cerevisiae,* constructed the first artificial chromosome, known appropriately as a yeast artificial chromosome, or YAC. Single-celled yeast was the organism of choice because it is easy to manipulate, its genetic machinery resembles that found in the cells of higher organisms, and it is so far the only eukaryotic organism whose origins of replication, centromeres, and telomeres have been defined as discrete, small segments of DNA. Yeast cells have 16 chromosomes that range in length from 235,000 to more than 2 million base pairs. Within each chromosome, the functional elements, such as centromeres, are only 100–1000 bp long. In contrast, human chromosomes have an average length of 100 million base pairs and a centromere may span 1 million base pairs.

Manipulation of the YAC construction process has produced insights into chromosome function. Plasmids containing only origins of replication but no centromere or telomere replicate but do not segregate properly. Plasmids with origins of replication and a

415

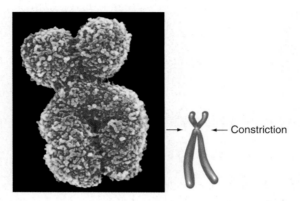

Figure 12.1 A chromosome's centromere looks like a constriction under the microscope. Color-enhanced scanning electron micrograph of a human metaphase chromosome and interpretive drawing show that the centromere corresponds to the major constriction of mitotic chromosomes.

centromere but no telomeres replicate and segregate fairly well if they are circular; if they are linear, they fragment and eventually become lost from the cell. Small plasmids carrying all three elements replicate and segregate as linear chromosomes, but unlike natural yeast chromosomes, they do not segregate accurately; instead, they segregate at random. Studies aimed at understanding this phenomenon revealed that a minimum length of 100,000–150,000 bp is essential to normal chromosome function. YACs carrying 11,000 bp show segregation errors in 50% of cell divisions. YACs containing 55,000 bp show segregation errors in 1.5% of cell divisions. With artificial chromosomes extended to more than 100,000 bp, the rate of segregation error falls to 0.3%. Although the longer artificial chromosomes function well, their frequency of segregation error is still 200 times greater than that seen with natural yeast chromosomes of normal size, indicating that some subtle aspects of chromosome structure and function remain to be discovered.

The successful construction and analysis of YACs confirms that three noncoding elements are essential to chromosome function: centromeres, telomeres, and origins of replication. Observations of YAC behavior show there is a minimum length required for accu-

rate chromosome segregation. Geneticists can use YACs for further studies of mitosis and meiosis. They can also use them as vectors for cloning large segments of DNA, such as those used in the Human Genome Project. Recall from Chapter 9 that cosmids can accept at most 40,000 bp of foreign DNA, while YACs can accommodate up to 1 million base pairs.

In this chapter, we examine the structure and function of significant features of the eukaryotic chromosome. Our discussion reveals that the chromosome is a dynamic organelle for the packaging, replication, segregation, and expression of the information in a single long molecule of DNA. (As noted earlier in the book, we consider chromosomes "organelles" because they are highly organized cellular structures that have important functions and are transmitted to daughter cells at cell division. This contrasts with the usage of many biologists who consider an entity to be an organelle only if it is surrounded by a membrane.) Each chromosome consists of one DNA molecule combined with a variety of proteins. Flexible DNA-protein interactions condense the chromosome for segregation during mitosis and decondense it for replication or gene expression during interphase. Specific sequences in the chromosomal DNA dictate where spindle attachment occurs for proper segregation; others determine where replication begins.

One general theme emerges from our discussion. Chromosomes have a versatile, modular structure for packaging DNA that supports a remarkable flexibility of form and function.

Our presentation of the chromosome as an organelle for packaging and managing genetic information describes:

- The components: one long DNA molecule and many kinds of proteins.

- Chromosome structure: how variable DNA-protein interactions create reversible levels of compaction.

- Specialized elements that ensure the accurate duplication and segregation of chromosomes: telomeres, origins of replication, and centromeres.

- How chromosomal packaging influences gene activity: Decondensation precedes gene expression; extreme condensation silences genes.

■ ■ ■

The Components of Eukaryotic Chromosomes: DNA, Histones, and Nonhistone Proteins

When viewed under the light microscope, chromosomes seem to change shape, character, and position as they pass through the cell cycle. During interphase, they look like a tangled mass of grainy string scattered throughout the nucleus. By metaphase of mitosis, they appear as a set number of paired bars facing each other across the cellular midplane. In this section, we describe what chromosomes are made of; then, in succeeding sections,

we explain how these chromosomal components associate, dissociate, and reassociate to produce the observed metamorphoses of structure.

Each Chromosome Packages a Single Long Molecule of DNA

Researchers learned from chemical and physical analyses that each chromosome within a cell nucleus contains one long linear molecule of DNA. In one study, they placed chromosomal DNA between two cylinders, stretched the DNA by rotating one of the cylinders, and measured the DNA's rate of recoil. Shorter molecules recoil faster than longer ones. When they applied this measure to the DNA in a *Drosophila* chromosome, the length of the DNA molecule was sufficient to account for all the DNA in a single chromosome. It is also possible to separate the DNA molecules of a relatively small genome by size using pulse-field gel electrophoresis, in which timed pulses of current are applied along two alternating paths through the gel. For the yeast genome, the number and sizes of DNA molecules separated using this technique correspond to the number and sizes expected if each chromosome contains a single DNA molecule.

The Protein Components of Eukaryotic Chromosomes: Histones and Nonhistone Proteins

By itself, DNA does not have the ability to fold up small enough to fit in the cell nucleus. For sufficient compaction, it depends on interactions with two categories of proteins: histones and nonhistone chromosomal proteins. **Chromatin** is the generic term for any complex of DNA and protein found in a cell's nucleus. Chromosomes are the separate pieces of chromatin that behave as a unit during cell division. Chromatin is the same chemical substance that Miescher extracted from the nuclei of white blood cells and named *nuclein* in 1869. Although chromatin is roughly 1/3 DNA, 1/3 histones, and 1/3 nonhistones by weight, it may also contain traces of RNA. Because these RNA bits result mainly from gene transcription and are probably unrelated to chromatin structure, we do not include RNA in our discussion of chromatin components.

Histone Proteins Abound in the Chromatin of All Eukaryotic Cells

Discovered in 1884, **histones** are relatively small proteins with a preponderance of the basic, positively charged amino acids lysine and arginine. The histones' strong positive charge enables them to bind to and neutralize the negatively charged DNA throughout the chromatin. Histones make up half of all chromatin protein by weight and are classified into five types of molecules: H1, H2A, H2B, H3, and H4 (Table 12.1). The last four types—H2A, H2B, H3, and H4—form the core of the most rudimentary DNA packaging unit—the **nucleosome**—and are therefore referred to as **core histones.** (We examine the role of these histones in nucleosome structure later.) All five types of histones appear throughout the chromatin of nearly all diploid eukaryotic cells. Histone synthesis occurs during S phase of the cell cycle, when chromosome replication requires more histones for the packaging of newly made DNA. Special regulatory mechanisms tightly correlate DNA and histone synthesis so that both occur at the appropriate time. During protein synthesis, specific modifications occur on amino acids within the histones. The acetylation of (addition of acetyl groups to) lysines as well as various methylations may be important for the functioning and assembly of chromatin.

Interestingly, there is great similarity among particular histone proteins in different organisms. In the H4 proteins of pea plants and calves, for example, all but two of the sequence of 102 amino acids are identical. That histones have changed so little throughout evolution underscores the importance of their contribution to chromatin structure.

Nonhistone Proteins Are a Heterogeneous Group Named by Default

Fully half of the mass of protein in the chromatin of most eukaryotic cells is not composed of histones. Rather, it consists of hundreds or even thousands of different kinds of nonhistone proteins, depending on the organism. The chromatin of a diploid genome contains from 200–2,000,000 molecules of each kind of nonhistone protein. Not surprisingly, this large variety of proteins fulfills many different functions, only a few of which have been defined to date. Some nonhistone proteins play a purely structural role, helping to package DNA into structures distinct from the histone-containing nucleosomes. The proteins that form the structural backbone, or *scaffold,* of the chromosome fall in this category (Fig. 12.2a). Others, such as DNA polymerase, are active in replication. Still others are active in chromosome

TABLE 12.1	**Some Molecular Characteristics of the Histone Proteins Found in Calf Thymus***		
Histone	**Basic/Acidic Amino Acids**	**Molecular Weight**	**Total Amino Acid Residues**
H1	5.4	23,000	224
H2A	1.4	13,960	129
H2B	1.7	13,774	125
H3	1.8	15,273	135
H4	2.5	11,236	102

*Note the small size of the histones; in addition, their predominantly basic amino-acid constitution ensures that they will be positively charged under physiological conditions.

(a)

(b)

Figure 12.2 Nonhistone proteins have diverse functions.
(a) Some nonhistone proteins form the chromosome scaffold. When the human chromosome in this picture was gently treated with detergents to remove the histones and some of the nonhistone proteins, a dark scaffold composed of some of the remaining nonhistone proteins, and in the shape of the two sister chromatids, became visible. Loops of DNA freed by the detergent treatment surround the scaffold. **(b)** Some nonhistone proteins power chromosome movements along the spindle during cell division. In this figure, chromosomes are stained in *blue* and a nonhistone protein known as CENP-E is stained in *red.* CENP-E is located at the sister kinetochores of each duplicated chromosome and appears to play a major role in moving separated sister chromatids toward the spindle poles during anaphase.

segregation; for example, the motor proteins of kinetochores help move chromosomes along the spindle apparatus and thus expedite the transport of chromosomes from parent to daughter cells during mitosis and meiosis (Fig. 12.2b). By far the largest class of nonhistone proteins foster or regulate transcription and RNA processing during gene expression. Mammals carry 5000–10,000 different proteins of this kind. By interacting with DNA, these proteins influence when, where, and at what rate genes give rise to their protein products.

Unlike the distribution of histones, which are found in similar amounts in all cells of an organism and are dispersed relatively evenly throughout the chromosomes, the distribution of nonhistone proteins is uneven. Nonhistone proteins appear in different amounts and in different proportions in different tissues within the same organism. The difference between universal and sporadic distribution no doubt reflects the functional differences between the two categories of proteins. Whereas histones have the singular task of interacting with DNA to form universal units of compaction, the nonhistone proteins fulfill diverse functions specific to different tissues (for example, tissue-

specific transcription factors), to different moments in the life cycle (for example, proteins of the synaptonemal complex), or to different parts of a chromosome (for example, proteins involved in centromere function).

In summary, chromatin is roughly 1/3 DNA, 1/3 histones, and 1/3 nonhistone proteins. We now examine how the components of chromatin associate to package long strands of DNA into chromosomes that can condense or decondense as circumstances require.

Chromosome Structure: Variable DNA-Protein Interactions Create Reversible Levels of Compaction

Stretched out in a thin, straight thread, the DNA of a single human cell would be 6 feet in length. This is, of course, much longer than the microscopic cell itself, whose dimensions are measured in fractions of millimeters. Several levels of compaction enable the DNA to fit inside the cell (Table 12.2). First,

TABLE 12.2 **The Different Levels of Chromosome Compaction**

Mechanism	Status	What It Accomplishes
Nucleosome	Confirmed by crystal structure	Condenses naked DNA 7-fold to a 100 Å fiber
Supercoiling	Hypothetical model (the 300 Å fiber predicted by the model has been seen in the electron microscope)	Causes additional 6-fold compaction, achieving a 40- to 50-fold condensation relative to naked DNA
Radial Loops—Scaffold	Hypothetical model (preliminary experimental support exists for this model)	Through progressive compaction of 300 Å fiber, condenses DNA to rodlike mitotic chromosome that is 10,000 times more compact than naked DNA

the winding of DNA around histones forms small nucleosomes. Next, tight coiling gathers the DNA with nucleosomes together into higher-order structures. Other levels of compaction, which researchers do not yet understand, produce the metaphase chromosomes observable in the microscope.

The Nucleosome: The Fundamental Unit of Chromosomal Packaging Arises from DNA's Association with Histones

The electron micrograph of chromatin in Fig. 12.3a shows long, nub-studded fibers bursting from the nucleus of a chick red blood cell. The chromatin fibers resemble beads on a string, with the beads having a diameter of about 100 Å and the string a diameter of about 20 Å (1 Å = 10^{-10}m = 0.1 nm). The 20 Å string is DNA. Figure 12.3b illustrates how DNA wraps around histone cores to form the chromatin fiber's observed beads-on-a-string structure. Each bead is a nucleosome containing roughly 160 bp of DNA wrapped around a core composed of eight histones—two each of H2A, H2B, H3, and H4, arranged as shown in the figure. The 160 bp of DNA wrap twice around this core octamer. An additional 40 bp form **linker DNA,** which connects one nucleosome with the next. Histone H1 lies outside the core, apparently associating with DNA where the DNA enters and leaves the nucleosome. When investigators use specific reagents to remove H1 from the chromatin, some DNA unwinds from each nucleosome, but the nucleosomes do not fall apart; about 140 bp remain wrapped around each core.

Researchers can crystallize the nucleosome cores and subject the crystals to X-ray diffraction analysis. The pictures obtained confirm the nucleosome structure just described and also indicate that the DNA does not coil smoothly around the histone core (Fig. 12.3c). Instead, it bends sharply at some positions and barely at all at others. Because the sharp bending may occur only with some DNA sequences and not others, base sequence helps dictate preferred nucleosome positions along the DNA. It is not yet known exactly what elements of base sequence dictate nucleosome interaction. The bending required for a nucleosome is close to the limit possible for B DNA. The stiffer forms of A DNA and Z DNA cannot bend sufficiently to form nucleosomes. Although DNA sequence and B-form structure help de-

termine where nucleosomes occur, other as yet undetermined factors also contribute to nucleosome formation.

The spacing and structure of nucleosomes affect genetic function. The nucleosomes of each chromosome are not evenly spaced, but they do have a well-defined arrangement along the chromatin. This arrangement is transmitted with high fidelity from parent to daughter cells. The spacing of nucleosomes along the chromosome is critical because DNA in the regions between nucleosomes is readily available for interactions with proteins that initiate expression, replication, and further compaction. The way in which DNA is wound around a nucleosome also plays a role in determining whether and how certain proteins interact with specific DNA sequences. This is because some DNA sequences in the nucleosome, despite their proximity to the histone core, can still be recognized by nonhistone binding proteins.

Packaging into nucleosomes condenses naked DNA about sevenfold. With this condensation, the 2 m of DNA in a diploid human genome shortens to approximately 0.25 m (a little less than a foot) in length. This is still much too long to fit in the nucleus of even the largest body cell, and additional compaction is required.

Models of Higher-Level Packaging Seek to Explain the Extreme Compaction of Chromosomes at Mitosis

The details of chromosomal condensation beyond the nucleosome remain unknown, but researchers have proposed several models to explain the different levels of compaction (see Table 12.2).

Formation of the 300 Å Fiber Through Supercoiling

One model of additional compaction beyond nucleosomal winding proposes that the 100 Å nucleosomal chromatin supercoils into a 300 Å superhelix, achieving a further sixfold chromatin condensation. Support for this model comes in part from electron microscope images of 300 Å fibers that contain about six nucleosomes per turn (Fig. 12.4a). Whereas the 100 Å fiber is one nucleosome in width, the 300 Å fiber looks three beads wide. Removal of some H1 from a 300 Å fiber causes it to

Figure 12.3 Nucleosomes: The basic building blocks of chromatin. (a) In the electron microscope, nucleosomes look like beads on a string. **(b)** Nucleosome structure: The DNA in each nucleosome wraps twice around a nucleosome core. An additional molecule of each histone (H2A, H2B, H3, and H4) in the nucleosome core is hidden in this view. Histone H1 associates with the DNA as it enters and leaves the nucleosome. **(c)** The structure of the nucleosome as determined by X-ray crystallography: In this overhead view of the core particle, you can see that the DNA (*orange*) actually bends sharply at several places as it wraps around the core histone octamer (*blue* and *turquoise*).

unwind to 100 Å. Adding back the H1 reinstates the 300 Å fiber. Although electron microscopists can actually see the 300 Å fiber, they do not know exactly how it forms. Higher levels of compaction are even less well understood.

The Radial Loop-Scaffold Model Seeks to Explain Compaction of the 300 Å Fiber

This model proposes that certain nonhistone proteins, including topoisomerase II, bind to chromatin every 60–100 kb and tether the supercoiled, nucleosome-studded 300 Å fiber into structural loops (Fig. 12.4b). Evidence that nonhistone proteins fasten these loops comes from chemical manipulations in which the removal of histones does not cause the chromatin to unfold completely. Other nonhistone proteins may, in turn, gather the loops into daisylike rosettes (Fig. 12.4c); and additional nonhistone proteins may then compress the rosette centers into a compact bundle. A range of nonhistones thus forms the condensation scaffold depicted in Fig. 12.2a. This proposal of looping and gathering is known as the **radial loop-scaffold model** of compaction. To visualize how it achieves condensation, imagine a long piece

of string. To shorten it, you knot it at intervals to form loops separated by straight stretches; the knots are at the base of each loop. To shorten the string still further, you clip together sets of knots. Finally, you pin together all the clips. In this image, the knots, clips, and pins function as the condensation scaffold.

The radial loop-scaffold model of chromosome packaging offers a simple explanation of progressive chromosome compaction from interphase to metaphase chromosomes. At interphase, the nucleosome-studded chromatin forms many structural loops, which are anchored together in rosettes in some areas. This initial looping and gathering compresses the genetic material sufficiently to fit into the nucleus, where it appears as a mass of tangled string. As the chromosomes enter prophase of mitosis, looping and gathering increase and bundling through protein cross-ties begins. By metaphase, the height of looping, gathering, and bundling achieves a 250-fold compaction of the roughly 40-fold-compacted 300 Å fiber, giving rise to the highly condensed, rodlike shapes we refer to as mitotic chromosomes.

Several pieces of biochemical and micrographic evidence support the radial loop-scaffold model. For example, metaphase

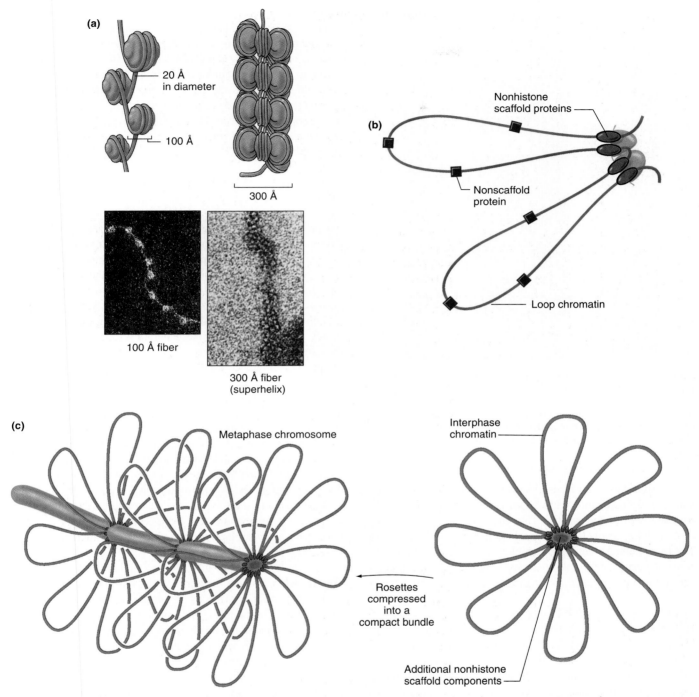

Figure 12.4 Models of higher-level packaging. (a) Electron micrographs contrasting the 100 Å fiber (*left*) with the 300 Å fiber (*right*). The line drawings show the probable arrangement of nucleosomes (with *green cores*) in these structures. **(b)** The radial loop-scaffold model for yet higher levels of compaction. According to this model, the 300 Å fiber is first drawn into loops, each including 60–100 kb of DNA, that are tethered at their bases by nonhistone scaffold proteins (*brown* and *orange*) including topoisomerase II (*brown*). **(c)** Additional nonhistone proteins might gather several loops together into daisylike rosettes and then compress the rosette centers into a compact bundle.

chromosomes from which experimenters have extracted all the histones still maintain their familiar X-like shapes (see Fig. 12.2). Moreover, electron micrographs of whole-mounted mitotic chromosomes show loops of chromatin at the periphery of the chromosomes (Fig. 12.5). And recent analyses of DNA indicate that special, irregularly spaced repetitive base sequences associate with nonhistone proteins to define the chromatin loops. These stretches of DNA are known as **scaffold-associated regions,** or **SARs.** Found at the base of the chromatin loops, SARs are most likely the sites at which the DNA is anchored to the condensation scaffold.

Figure 12.5 Experimental support for the radial loop-scaffold model. A close-up of the image in Fig. 12.2, this electron micrograph shows long DNA loops emanating from the protein scaffold at the *bottom* of the picture. The two ends of each DNA loop appear to attach to adjacent locations in the protein scaffold. Note that there are only loops—not ends—at the *top* of the photo.

Despite bits and pieces of experimental evidence, studies that directly confirm or reject the radial loop-scaffold model have not yet been completed. Thus the loops and scaffold concept of higher-level chromatin packaging remains a hypothesis. The hypothetical status of this higher-level compaction model contrasts sharply with nucleosomes, which are entities that investigators have isolated, crystallized, and analyzed in detail.

A Closer Look at Karyotypes: Fully Compacted Metaphase Chromosomes Have Unique, Reproducible Banding Patterns That Identify Them

We have seen that different levels of packaging compact the DNA in human metaphase chromosomes 10,000-fold (see Table 12.2). With this amount of compaction, the centromere region and telomeres of each chromosome become visible. We have also seen (in Chapter 4) that various staining techniques reveal a characteristic banding pattern for each metaphase chromosome, establishing a karyotype. In G-banding, for instance, chromosomes are first gently heated and then exposed to Giemsa stain; this DNA dye preferentially darkens certain regions to produce alternating dark and light "G bands." Each G band is a very large segment of DNA from 1–10 Mb in length, containing many loops. On analysis with low-resolution techniques, a human karyotype contains approximately 300 dark G bands and light

interbands. High-resolution G-banding techniques enable cytologists to subdivide the G bands into smaller bands and interbands that picture chromosomes at even higher resolution (Fig. 12.6a). There are nearly 2000 identifiable G bands and interbands in the chromosomes of the standard high-resolution human karyotype. G-banding serves as the basis for most karyotyping because G-band preparations are very stable and require only a good light microscope for detection. The karyotypes described in Chapter 4 present a cell's mitotic metaphase chromosomes by size, shape, and banding patterns.

Banding Patterns Are Highly Reproducible, Although No One Knows for Sure What They Represent

Most molecular geneticists think the bands produced by staining probably do *not* embody differences in base composition over long distances but more likely reflect an uneven packaging of loops determined in some way by the spacing and density of short repetitive DNA sequences. Although the detailed biochemical basis of banding is not yet understood, every time a chromosome replicates, whatever underlies its banding pattern is faithfully reproduced. The fact that banding patterns are so highly reproducible from one generation to the next indicates they are an intrinsic property of each chromosome, determined by the DNA sequence itself.

Researchers Can Use Bands to Locate Genes

Geneticists can designate the chromosomal location of a gene by describing its position in relation to the bands on the p (short) or q (long) arm of a particular chromosome. For this purpose, the p and q arms are subdivided into regions, and within each region, the dark and light bands are numbered consecutively. The X-linked genes for colorblindness, for example, reside at q27-qter, which means they are located on the X chromosome's long (q) arm somewhere between the beginning of the seventh band in the second region and the end of the telomere (terminus, or ter; Fig. 12.6b).

Geneticists Can Use Banding Patterns to Analyze Chromosomal Differences Between Species

In all placental mammals, the diploid genome carries roughly 6 billion base pairs, but this amazingly similar amount of DNA is packaged into different numbers of chromosomes with different banding patterns. Deer mice have 4 chromosomes in the nucleus of each body cell; cats have 38; house mice have 40; rabbits 44; humans 46; dogs have 48; cattle have 60; horses have 64; and hippopotamuses have 96.

High-resolution banding techniques, which divide chromosomes into so many stripes that their diagrams look a little like the universal bar codes on grocery-store items, enable geneticists to compare the banding patterns of different species. The similarities between closely related species are striking. We know, for example, that humans have 46 chromosomes. Their nearest primate relatives, the great apes (chimpanzees, gorillas, and orangutans) carry 48 chromosomes per diploid cell. As Fig. 12.6c shows, a single event—the fusion of two acrocentric chromosomes in the great apes to form metacentric chromo-

Figure 12.6 Chromosomes can be characterized by their banding patterns. (a) The karyotype of a human female examined by high-resolution G-banding techniques reveals approximately 1000 bands and interbands in the haploid complement of chromosomes. **(b)** Genes for colorblindness in humans have been localized to a small region near the tip of the long arm of the X chromosome. **(c)** A comparison of banding patterns in the two largest human chromosomes and their counterparts in the great apes.

some 2 in humans—can account for the numeric difference. Note that the banding patterns of the two great ape chromosomes that later fused are identical to those of human chromosome 2, except in the region of fusion and at the chromosome tips. Human banding patterns are almost identical to chimpanzees' for 13 chromosomes, to gorillas for 9, and to orangutans for 8. In the rest of the chromosomes, almost all the same bands are present, but they are distributed slightly differently.

Banding Pattern Differences Between Individuals Can Reveal the Cause of Genetic Disease

Karyotypes are a powerful tool for research and genetic counseling. As we have seen, medical geneticists use them to diagnose abnormalities related to chromosome number. Down syndrome, for example, is the result of three chromosome 21s. Some genetic disorders result from very slight additions or deletions of genetic material somewhere within the normal number

of chromosomes. With high-resolution G-banding techniques, clinicians can uncover some of these discrepancies by comparing the chromosomes of a person suffering from the genetic disorder with those of a standard karyotype.

A striking example of how genetic counselors exploit the phenomenon of exact and highly reproducible chromosomal packaging for clinical purposes is seen in the case history of a young boy identified in the medical literature as BB. With no known family history of genetic disorders, BB suffered from defects in four X-linked traits: Duchenne muscular dystrophy; chronic granulomatous disease, which impairs the infection-fighting ability of white blood cells; the rare McLeod's blood type, which results in bouts of anemia; and retinitis pigmentosa, in which a deterioration of the retina eventually causes blindness. The simultaneous occurrence of four X-linked conditions is so rare that one doctor asked medical geneticists to scrutinize the boy's X chromosome for the cause of the problems. High-resolution banding revealed the answer—deletion of a small white band sandwiched between two larger dark bands, which removed the four genes in question.

In summary, the DNA-protein interactions that define chromatin structure generate several levels of packaging: tightly wound nucleosomes, some form of supercoiling, and variously gathered larger loops. The flexible but reproducible structure of chromatin confers on each chromosome a structural versatility that enables it to adapt to the widely different functions of duplication, segregation, and the regulation of gene expression. In the following sections, we examine how chromosome structure helps determine those functions.

Specialized Chromosomal Elements Ensure Accurate Replication and Segregation of Chromosomes

The Accurate Duplication of Chromosomal DNA Depends on Origins of Replication and Telomeres

As the chromosomes decondense for copying during replication, special DNA sequences that do not encode proteins regulate the timing and accuracy of the process. Some of these sequences serve as origins of replication that signal where and when the DNA double helix opens up to form replication forks; others function as telomeres that protect the ends of individual chromosomes from progressive decay.

There Are Many Origins of Replication

During replication the enzyme DNA polymerase assembles a new string of nucleotides according to a DNA template, linking about 50 nucleotides per second in a typical human cell. At this rate and with only one origin of replication, it would take the polymerase about 800 hours, a little more than a month, to copy the 130 million base pairs in an average human chromosome.

But the length of the cell cycle in actively dividing human tissues is much shorter, some 24 hours, and S phase (the period of DNA replication) occupies only about a third of this time. Eukaryotic chromosomes meet these time constraints through multiple origins of replication that can function simultaneously (Fig. 12.7a). Most mammalian cells carry 10,000 such origins strategically positioned among the chromosomes. As we saw in Chapter 6, each origin of replication binds proteins that unwind the two strands of the double helix, separating them to produce two mirror-image replication forks. Replication then proceeds in two directions (bidirectionally), going one way at one fork and the opposite way at the other, until the forks run into adjacent forks. As replication opens up a chromosome's DNA, a replication bubble becomes visible in the electron microscope, and with many origins, many bubbles appear. The DNA running both ways from one origin of replication to the endpoints, where it merges with DNA from adjoining replication forks, is called a **replication unit,** or **replicon.** As yet unidentified controls tie the number of active origins to the length of S phase. In *Drosophila,* for example, early embryonic cells replicate their DNA in less than 10 minutes. To complete S phase in this short a time, their chromosomes use many more origins of replication than are active later in development when S phase is 6–10 times longer. Thus, all origins of replication are not necessarily active during all the mitotic divisions that create an organism.

The 10,000 origins of replication scattered throughout the chromatin of each mammalian cell nucleus are separated from each other by 30–300 kb of DNA, which suggests that there is at least one origin of replication per loop of DNA. Origins of replication in yeast (known as autonomously replicating sequences, or ARSs) can be isolated by their ability to permit replication of plasmids in yeast cells. ARS origins are capable of binding to the enzymes that initiate replication, and they consist of an A–T-rich region of DNA adjacent to special flanking sequences (Fig. 12.7b). By digesting interphase chromatin with DNase I, an enzyme that fragments the chromatin only at points

(a)

5. kb

(b)

Figure 12.7   Eukaryotic chromosomes have multiple origins of replication. (a) Electron micrograph and diagrammatic interpretation, showing a region of replicating DNA from a *Drosophila* embryo. Many origins of replication are active at the same time, creating multiple replicons.
(b) Structure of the yeast origin of replication ARSI. The *orange-boxed* sequence is the A–T-rich consensus region found in all ARS elements. The *blue boxes* are the flanking sequences close to the ARSI consensus region that promote function.

Consensus region

5' •••CAAATTTCGTCAAAAATGCTAAGAAATAGGTTATTACTTTTATTTAAGTATTGTTTGTGCCTTTTGAAAAGCAAGCATAAAAGATCTAAACATAAAATCTGTAAAATAAC•••3'
3' •••GTTTAAAGCAGTTTTTACGATTCTTTATCCAATAATGAAAATAAATTCATACAAACACGGAAAACTTTTCGTTCGTATTTTCTAGATTTGTATTTTAGACATTTTATTG•••5'

where the DNA is not protected inside a nucleosome, investigators have determined that origins of replication are accessible regions of DNA devoid of nucleosomes.

Telomeres Preserve the Integrity of Linear Chromosomes

The linear chromosomes of eukaryotic cells terminate at both ends in protective caps called telomeres (Fig. 12.8). Composed of DNA associated with proteins, these caps contain no genes but are nevertheless crucial in preserving the structural integrity of each chromosome. Chromosomes unprotected by telomeres fuse end to end, producing entities with two centromeres. During anaphase of mitosis, if the two centromeres are pulled in opposite directions, the DNA between them will rupture, resulting in broken chromosomes that segregate poorly and eventually disappear from the cell.

Cells must thus preserve their telomeres to maintain the normal genetic complement on which their viability depends. But the replication of telomeres is problematic. As we saw in Chapter 6, DNA polymerase, a key component of the replication machinery, functions only in the 5′-to-3′ direction and can add nucleotides only to the 3′ end of an existing chain. With this constraint, the enzyme cannot possibly replicate some of the nucleotides at the 5′ ends of the two DNA strands (one of which is in the telomere at one end of the chromosome, the other of which is in the telomere at the other end). In short, DNA polymerase can reconstruct the 3′ end of each newly made DNA strand in a chromosome, but not the 5′ end (Fig. 12.9). If left to its own devices, the enzyme would fail to fill in an RNA primer's length of nucleotides at the 5′ end of every new chromosomal strand with

each cell cycle. As a result, the chromosomes in successive generations of cells would become shorter and shorter, losing crucial genes as their DNA diminished.

Telomeres and an enzyme called **telomerase** provide a countermeasure to these problematic quirks of DNA polymerase. Telomeres consist of special repetitive DNA sequences. Human telomeres are composed of the base sequence TTAGGG repeated 250–1500 times (see Fig. 12.8). The number of repeats varies with the cell type. Sperm have the longest telomeres. The same exact TTAGGG sequence occurs in the telomeres of all mammals as well as in birds, reptiles, amphibians, bony fish, and many plant species. Some much more distantly related organisms also have repeats in their telomeres but with slightly different sequences. For example, the telomeric repeat in the chromosomes of the ciliate *Tetrahymena* is TTGGGG. The close conservation of these repeated sequences across phyla suggests that they perform a vital function that emerged in the earliest stages of the evolutionary line leading to eukaryotic organisms, long before dinosaurs roamed the earth. The special repetitive sequences of telomeres contain no protein-encoding genes. In fact, telomeres prevent the transcription of genes brought into their vicinity.

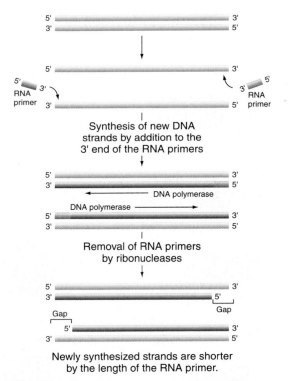

Synthesis of new DNA strands by addition to the 3′ end of the RNA primers

Removal of RNA primers by ribonucleases

Newly synthesized strands are shorter by the length of the RNA primer.

Figure 12.9 DNA polymerase cannot reconstruct the 5′ end of a DNA strand. Even if an RNA primer at the 5′ end can begin synthesis of a new strand, a gap will remain when ribonucleases eventually remove the primer. The requirement of DNA polymerase for a primer on which to continue polymerization means that the enzyme cannot fill this gap, so newly synthesized strands would always be shorter than parental strands if DNA polymerase were the only player in the production of new ends. In this figure, both parental DNA strands are in *light blue,* the two newly synthesized strands in *dark blue,* and the RNA primers in *red.*

Figure 12.8 Telomeres protect the ends of eukaryotic chromosomes. Human telomeres light up in yellow upon *in situ* hybridization with fluorescent probes (FISH) that recognize the base sequence TTAGGG. The telomeres of humans and many other species contain many repeats of this 6 bp motif.

The telomere DNA helps maintain and replicate chromosome ends by binding two types of proteins: protective proteins and telomerase. The bound protective proteins, which recognize the single-stranded TTAGGG sequences at the very ends of a chromosome, shield these ends. When the proteins are dislodged, the telomere attracts telomerase, which can also bind to the single-stranded TTAGGG sequence. The bound enzyme extends the telomere, roughly restoring it to its original length.

Telomerase is an unusual enzyme consisting of protein in association with RNA. Because of this mix, it is called a *ribonucleoprotein*. The RNA portion of the enzyme contains CCCUAA repeats that are complementary to the TTAGGG repeats in telomeres, and they serve as a template for adding new TTAGGG repeats to the end of the telomere (Fig. 12.10). In many cells, including the perpetually reproducing cells of yeast and the germ cells of humans, some kind of feedback mechanism appears to maintain the optimal number of repeats at the telomeres. In human somatic cells where the gene for telomerase is silenced, the lack of telomerase activity results in the progressive shortening of telomeres as cells divide. Many studies and observations have shown that telomeres are critical to chromosome function. In addition to preventing a chromosomal shortening during replication that could dismantle vital genes nucleotide by nucleotide, the telomeres maintain the integrity of the chromosomal ends. Broken chromosomes that lack telomeres are recognized as defective by the cellular DNA repair machinery, which often remedies the situation by putting the broken ends back together, restoring the telomeres. Sometimes, however, the unprotected, broken, nontelomeric ends are subject to inappropriate repair resulting in chromosome fusion, or they may attract enzymes that degrade the chromosome entirely. Both fusion and degradation disrupt chromosome number and function. Thus, even though they normally carry no genes, telomeres contain information essential to the duplication, segregation, and stability of chromosomes.

Does Telomerase Activity Forestall Aging?

The activity of telomerase in normal yeast cells ensures the full reconstruction of each chromosome's ends with each DNA replication. In studies where researchers deleted the yeast gene for telomerase, the telomeres shortened at the rate of about 3 bp per generation, and after significant loss of telomeric length, the chromosomes began to break, and the yeast cells died. Telomerase activity, it seems, endows normal yeast cells with the potential for immortality; given the proper conditions and continual telomere reconstruction, the cells can reproduce forever.

In humans, the telomerase gene is part of every cell's genome. Germ-line cells, which maintain their chromosomal ends through repeated rounds of DNA replication, express the gene, as do some stem cells; but many normal somatic cells, which have a finite life span, do not. In these differentiated somatic cells, the telomeres shorten slightly with each cell division. This shortening helps determine how many times a particular cell is able to divide. In culture, most somatic cells, after dividing for 30 or 40 or 50 generations, show signs of senescence and then die. Some researchers who study aging in humans have suggested that it may be possible to forestall aging by slowing cellular senescence through the activation of telomerase expression. They reason that when cells of the immune system, for instance, lose a certain telomeric length, they divide less often, and as a result, the immune system gradually loses its capacity for protection. If it were possible to preserve the telomeres, this loss of cells and immune function might not occur.

As with many aspects of human biology, a telomerase-based boost to longevity might not be the simple solution it appears at first glance, because it could make cells susceptible to cancerous proliferation. Many human tumor cells that become immortal exhibit telomerase activity. In such cells, the abnormal activation of the telomerase gene seems to bestow the capacity to divide indefinitely. Cells isolated from human ovarian tumors, for example, express the telomerase enzyme and maintain stable telomeres; cells from normal ovarian tissue do not. Oncologists hypothesize from these and other observations that expression of telomerase in cancerous human cells may keep those cells from losing their telomeres and eventually dying, and thereby perpetuate tumors. Some scientists even speculate that the genetic program silencing the telomerase gene in the somatic cells of differentiated human tissues may have evolved as a defense against cancer. Ironically, if this were true, the same mechanism that helps humans resist cancer blocks their biological potential for immortality.

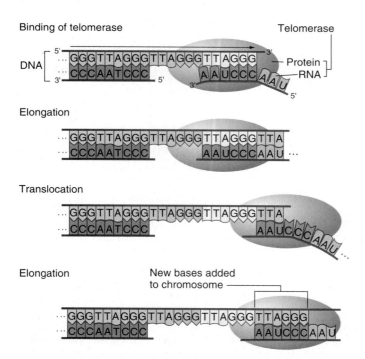

Figure 12.10 How telomerase extends telomeres. Telomerase binds to the ends of chromosomes because of complementarity between the CCCUAA repeats of telomerase RNA (*orange*) and the TTAGGG repeats of telomeres. Telomerase RNA CCCUAA repeats serve as templates for adding TTAGGG repeats to the ends of telomeres. After a telomere has acquired a new repeat, the telomerase enzyme moves (translocates) to the newly synthesized end, allowing additional rounds of telomere elongation.

Proteins Are an Essential Part of Reproducing Chromatin Structure

DNA replication is only one step in chromatin duplication. The complex process also includes the synthesis and incorporation of histone and nonhistone proteins to regenerate tissue-specific chromatin structure. Researchers speculate that the process works something like this. Before DNA synthesis can take place, the chromatin fiber must unwind. Next, as DNA replication proceeds, newly formed DNA must associate with histones, either preexisting histones or recently synthesized histones that have just made their way to the nucleus. The synthesis and transport of histones must be tightly coordinated with DNA synthesis since the nascent DNA becomes incorporated into nucleosomes within minutes of its formation. Proteins that mediate assembly of nucleosomes have been identified in several organisms, including yeast and humans. Finally, the nucleosomal DNA must interact in specific ways with a variety of proteins to compact in the same pattern as before. An exception to the exact replication of compaction patterns occurs in differentiating cells. Changes in available nuclear proteins produce slightly different folding patterns that promote the expression of different genes. Studies with mammalian cells have shown that some hormones can induce changes in gene expression if and only if they are present during chromatin replication.

The Segregation of Condensed Chromosomes Depends on Centromeres

When cell nuclei divide at mitosis or meiosis II, the two chromatids of each replicated chromosome must separate from one another and segregate such that each daughter cell receives one and only one chromatid from each chromosome. At meiosis I, homologous chromosomes must pair and segregate such that each daughter cell receives one and only one chromosome from each homologous pair. The centromeres of eukaryotic chromosomes ensure this precise distribution during different kinds of cell division by serving as segregation centers.

Centromeres Appear As Constrictions in the Chromosomes

These centromeric constrictions arise because centromeres are contained within blocks of repetitive, simple noncoding sequences, known as **satellite DNAs,** which have a very different chromatin structure and different higher-order packaging than other chromosomal regions (see Fig. 12.1). There are many different kinds of satellite DNA, each consisting of short sequences 5–300 bp long repeated in tandem thousands or millions of times to form large arrays. The predominant human satellite, "alphoid DNA," is a noncoding sequence 171 bp in length; it is present in a block of tandem repeats extending over a megabase of DNA in the centromeric region of each chromosome. Various human centromeres also contain several satellites unrelated to alphoid, which give their centromeric regions a complex structure. Although most satellite sequences lie in centromeric regions, some satellites are found outside the centromere on the chromosome arms.

The centromere can occur almost anywhere on a chromosome, except at the very ends (which must, instead, be telomeres). As we have seen, in a metacentric chromosome, the centromere is at or near the middle, while in an acrocentric chromosome, it is near one end.

Centromeres Have Two Functions That Ensure Proper Segregation

First, they are the sites that hold sister chromatids together—up to the beginning of anaphase in mitosis and up to anaphase II in meiosis. Researchers do not yet know how centromeres hold sister chromatids together. One model suggests that the DNA of sister chromatids somehow intertwines; another model proposes that proteins located at the centromere interconnect the two sister chromatids. *Drosophila* researchers have recently isolated a centromeric protein that offers support for the second model; found on meiotic chromosomes prior to anaphase II, the protein helps keep sister chromatids together until this time.

In addition to holding sister chromatids together, centromeres contribute to proper chromosome segregation through elaboration of a **kinetochore:** a specialized structure composed of DNA and proteins that is the site at which chromosomes attach to the spindle fibers (Fig. 12.11a). Some of the kinetochore proteins are motor proteins that help power chromosome movement during mitosis and meiosis. During mitosis, a kinetochore develops late in prophase on each sister chromatid, at the part of the centromere that faces one or the other cellular pole. By prometaphase, the kinetochores on the two sister chromatids attach to spindle fibers emanating from centrosomes at opposite poles of the cell. Although it is not yet clear what ensures this bipolar attachment, it appears that kinetochores somehow measure the tension arising when sister chromatids that are connected through their centromere are pulled in opposite directions. At the beginning of anaphase, the centromere "splits," freeing the sister chromatids to migrate toward opposite poles, with the assistance of the motor proteins in the kinetochore (review Fig. 4.8). During meiosis, controls on centromere splitting make sure that the sister chromatids of each homolog migrate together to the same pole during meiosis I but to opposite poles during meiosis II (review Fig. 4.13).

Using Centromere Function to Analyze Centromere Structure

Investigators can exploit the centromere's role in chromosome segregation to isolate and then analyze the specific chromosomal regions that make up a centromere. If removal of a DNA sequence disrupts chromosome segregation and reinsertion of that same sequence restores stable transmission, the sequence must be part of the centromere. In the yeast *S. cerevisiae*, centromeres consist of two highly conserved nucleotide sequences, each only 10–15 bp long, separated by approximately 90 bp of A–T-rich DNA (Fig. 12.11b). This means that a short stretch of roughly 120 nucleotides is sufficient to specify a centromere. The centromere sequences of different yeast chromosomes are so closely related that the centromere of one chromosome can substitute for that of another. This indicates that while all centromeres play the same role in chromosome segregation, they do not help distinguish one chromosome from another.

(a)

(b)

Figure 12.11 Centromere structure and function.
(a) Structure of centromeres in higher organisms. Centromeres hold sister chromatids together and contain information for the construction of a kinetochore, the structure that allows the chromosome to bind to spindle fibers. In this micrograph, an anaphase chromatid is migrating toward the spindle pole via microtubules attached to the kinetochore. The kinetochore is organized into a complicated "trilaminar plate." **(b)** Structure and DNA sequence organization of yeast centromeres.

The centromeres of higher eukaryotic organisms are much larger and more complex than those of yeast. In these multicellular organisms, the centromeres lie buried in a considerable amount of darkly staining, highly condensed chromatin, which makes it difficult to discover which specific DNA sequences are critical to centromere function. Although repetitive satellite DNA sequences in the vicinity of the centromeric constriction probably play some role in chromosome segregation, some geneticists believe that role is a secondary one and that, as in yeast, longer conserved DNA sequences determine the centromere's primary contribution to chromosome segregation. The kinetochores in higher eukaryotes attach to many spindle fibers instead of just one, as in yeast. Researchers think that these complex kinetochores are likely to consist of repeating structural subunits, with each subunit responsible for attachment to one fiber.

How Chromosomal Packaging Influences Gene Activity

Origins of replication, telomeres, and centromeres are specialized chromosomal elements composed of specific noncoding DNA sequences packaged in specific ways for proper function. Chromosomal packaging also affects the function of coding sequences. We now examine how different types of chromosomal packaging influence gene expression.

Controlled Decompaction Precedes Gene Expression

Cells express their genes mainly during interphase when the chromosomes have decondensed, or decompacted. Even the relatively decompacted interphase chromatin, however, requires further unwinding to open up the higher orders of packaging and expose the DNA inside nucleosomes for transcription.

Boundary Elements Delimit Areas of Decompaction

Molecular geneticists do not yet understand the mechanism that releases the rosettes, loops, and superhelixes to expose a coding sequence for transcription, but whatever it is, it removes the proteins that gather the chromosomal loops into higher-order structures such as rosettes, and it also unfastens the cross-ties that anchor the chromosome loops in the protein scaffold. Some sort of *boundary elements,* or *insulators,* prevent the decompaction from spreading beyond a certain point.

Nucleosomes in the Decompacted Area Unwind to Allow the Initiation of Transcription

RNA polymerase, the enzyme that transcribes genes, is so large compared to a supercoil of nucleosomes in the 300 Å solenoid fiber that it cannot reach all of the sequences it needs to copy. To provide the RNA polymerase with access to the DNA, some loosening of individual nucleosomes must occur at the beginning of genes. Some of the nonhistone proteins known as transcription factors carry out this task by unwinding nucleosomes and dislodging histones from DNA at the 5′ end of genes, where transcription will begin. The nucleosome-free regions of the chromosome become sufficiently decompacted that they are open to interactions with RNA polymerase and other transcription factors, which can now recognize the promoter and initiate gene expression.

Researchers use DNase I, an enzyme that breaks DNA into fragments by splitting the phosphodiester bonds between nucleotides, as a tool for analyzing the patterns of chromatin compaction. DNA in highly compacted chromatin is relatively impervious to DNase I because it is shielded by the structure of the nucleosome (Fig. 12.12a). Decondensed chromatin is much more vulnerable to fragmentation. Fragmentation by DNase I is reflected in the generation of shorter DNA fragments. Small amounts of DNase I digest DNA at those sites that are most vulnerable; such sites are not protected by proteins or do not reside

Figure 12.12 DNA compaction and transcription. (a) The sensitivity of DNA to DNase I measures levels of chromosome compaction. **(b)** Location of DH sites in short gene-containing segments of chromosomes. Most, but not all, DH sites correspond to the 5′ end of genes (*arrows*). **(c)** One model of how RNA polymerase transcribes through nucleosomes: As the enzyme moves, it partially unwinds the nucleosomes. After the RNA polymerase passes through a section, the DNA sequences can loop around and interact with the histones.

in compacted chromatin. Treatment with very small amounts of DNase I reveals that even highly compacted chromatin is punctuated with **DNase hypersensitive (DH) sites.** Most (though not all) DH sites are at the 5′ ends of genes to be transcribed by the cell (Fig. 12.12b). This is evidence that the unwinding of these 5′ regions is a prerequisite of transcription.

Interestingly, it appears that once transcription begins, the transcribed part of most genes does not become entirely free of nucleosomes. However, although the body of the gene does not become DNase hypersensitive, it is still more sensitive to DNase I than is the highly condensed chromatin. Some researchers speculate that as RNA polymerase copies DNA, the histones that remain bound to the DNA somehow step around the transcribing enzyme (Fig. 12.12c).

By Influencing Gene Expression, Chromatin Packaging Affects Tissue Differentiation

Significantly, the number and distribution of DNase hypersensitive sites vary from tissue to tissue. In one type of white blood cell, for example, DH sites expose the 5′ ends of the genes for making antibodies. In young red blood cells, it is the 5′ ends of the gene for hemoglobin synthesis. In muscle cells, it is the genes for myoglobin, myosin, actin, and dystrophin; in nerve cells, the genes for neurotransmitters; and so forth. Thus, by the time an individual matures, because of slight differences in packaging, different areas of chromatin unwind for expression during interphase in different types of cells.

At present, the mechanism that selectively alters certain areas of chromatin for specific gene activity remains a black, or at least a gray, box. Molecular geneticists know what happens—chromatin unwinds at points of gene expression and it is mediated by a variety of molecular interactions (for example, the enzyme topoisomerase II can dislodge histone H1). But they are only beginning to discover how many different proteins contribute to the complex network of molecular interactions and what signals coordinate their activity. One thing is clear, however. Many DNA sequences, including a variety of repetitive A–T-rich regions to which different proteins bind, collaborate to ensure the cell-to-cell transmission of particular chromosomal folding patterns. These patterns produce multiple levels of packaging that function as multiple points of control over the accessibility of genes. In each cell, this transmitted chromatin structure allows the decompaction leading to expression of relatively few genes, while repressing all the others.

Extreme Condensation Silences Expression

In cells stained with certain DNA-binding chemicals, a small proportion of chromosomal regions appear much darker than others when viewed under the light microscope. Geneticists call these darker regions **heterochromatin;** they refer to the contrasting lighter regions as **euchromatin.** The distinction between euchromatin and heterochromatin also shows up in the electron microscope, where the heterochromatin appears much more condensed than the euchromatin. Microscopists first identified dark-staining heterochromatin in the decondensed chromatin of interphase cells, where it tends to localize at the periphery of the nucleus (Fig. 12.13a). Even highly compacted metaphase chromosomes show the differential staining of heterochromatin versus euchromatin (Fig. 12.13b). Most of the heterochromatin in highly condensed chromosomes is found in regions flanking the centromere, but in some animals, heterochromatin forms in other regions of the chromosomes. In *Drosophila* the entire Y chromosome, and in humans most of the Y chromosome, is heterochromatic. Chromosomal regions that remain condensed in heterochromatin at most times in all cells are known as **constitutive heterochromatin.**

Autoradiography reveals that cells actively expressing their housekeeping and specialty genes incorporate radioactive RNA precursors into RNA almost exclusively in regions of euchromatin (see Fig. 12.13a). This observation indicates that euchromatin contains most of the sites of transcription and thus almost all of the genes. By contrast, heterochromatin appears to be transcriptionally inactive for the most part, probably because it is so tightly packaged that the enzymes required for transcription of the few genes it contains cannot access the correct DNA sequences.

(a)

Nuclear envelope ⎯⎯⎯⎯⎯

Peripheral heterochromatin ⎯

(b)

Figure 12.13   Heterochromatin versus euchromatin. (a) In this image of decondensed interphase chromatin, areas containing heterochromatin are bleached white and tend to be packed just inside the nuclear envelope, while ribonucleoprotein containing nuclear domains remain darker and granular. The cell from which this thin section was taken was pulse-labeled with [³H]uridine to mark sites of RNA synthesis (*dark spots*), most of which occur in euchromatin. **(b)** In this image, human metaphase chromosomes were stained by a special C-banding technique that darkens the constitutive heterochromatin, most of which localizes to regions surrounding the centromere.

Two specialized phenomena—position effect variegation in *Drosophila* and Barr bodies in mammalian females—clearly illustrate the correlation between heterochromatin formation and a loss of gene activity.

Position Effect Variegation in *Drosophila:* Moving a Gene Near Heterochromatin Prevents Its Expression

The *white⁺* (*w⁺*) gene in *Drosophila* is normally located near the telomere of the X chromosome, in a region of relatively decondensed euchromatin. When a chromosomal rearrangement such as an inversion or a translocation places the gene next to highly compacted heterochromatin near the centromere, the gene's expression may cease. Such rearrangements silence *w⁺* gene expression in some cells and not others, producing **position-effect variegation.** In flies carrying the wild-type *w⁺* allele, cells in the eye with an active *w⁺* gene are red, while cells with an inactive *w⁺* gene are white. Apparently, when normally euchromatic genes come into the vicinity of heterochromatin, the heterochromatin can spread into the euchromatic regions, shutting off gene expression in those cells where the heterochromatic "invasion" takes place. In such a situation, the DNA of the gene has not been altered, but the relocation has altered the gene's packaging in some cells. The phenomenon of position effect variegation thus reflects the existence of **facultative heterochromatin:** regions of chromosomes (or even whole chromosomes) that are heterochromatic in some cells and euchromatic in other cells of the same organism.

Position-effect variegation of red and white eye color in *Drosophila* produces eyes that are a mosaic of red and white patches of different sizes (Fig. 12.14a). The position and size of the patches vary from eye to eye. Such variation suggests that the "decision" determining whether heterochromatin spreads to the *w⁺* gene in a particular cell is the result of a random process. Because patches composed of many adjacent cells have the same color, the decision must be made early in the development of the eye. Once made, the decision determining whether the white gene will be on or off is transmitted to all the cell's mitotic descendants. These descendants occupy a particular region of the eye, forming patches of red or white cells respectively.

One interesting property revealed by position-effect variegation is that heterochromatin can spread over more than 1000 kb of previously euchromatic chromatin. For example, some rearrangements that bring the *w⁺* gene near heterochromatin also place the *roughest⁺* gene in the same vicinity, although a little farther away from the centromeric heterochromatin (Fig. 12.14b). The wild-type *roughest⁺* (*rst⁺*) gene normally produces a smooth eye surface. In flies carrying the rearrangements, however, some large white-colored patches contain within them smaller rough-surfaced patches, indicating that the heterochromatin inactivated first the *w⁺* and then the *rst⁺* gene. Red-colored, rough-surfaced patches never form, which means that the heterochromatin does not skip over genes as it spreads along the chromosome.

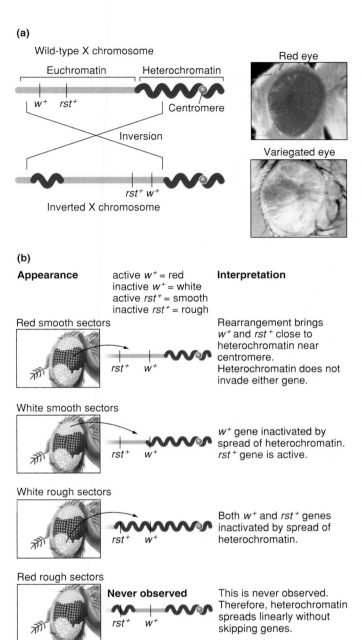

(a)

Wild-type X chromosome

Euchromatin   Heterochromatin

w^+ rst^+ Centromere

Inversion

Inverted X chromosome

rst^+ w^+

Red eye

Variegated eye

(b)

Appearance   active w^+ = red   **Interpretation**
                      inactive w^+ = white
                      active rst^+ = smooth
                      inactive rst^+ = rough

Red smooth sectors

rst^+ w^+

Rearrangement brings w^+ and rst^+ close to heterochromatin near centromere. Heterochromatin does not invade either gene.

White smooth sectors

rst^+ w^+

w^+ gene inactivated by spread of heterochromatin. rst^+ gene is active.

White rough sectors

rst^+ w^+

Both w^+ and rst^+ genes inactivated by spread of heterochromatin.

Red rough sectors

Never observed

rst^+ w^+

This is never observed. Therefore, heterochromatin spreads linearly without skipping genes.

Figure 12.14 Position-effect variegation in *Drosophila* is a phenotypic effect of facultative heterochromatin. (a) When the w^+ eye color gene (which specifies red color) is brought near an area of heterochromatin through a chromosomal rearrangement such as an inversion, the eyes of the fly can become variegated, with some red cells and some white cells. **(b)** A model for position-effect variegation of the w^+ and *roughest* (rst^+) genes postulates that heterochromatin can spread from its normal location surrounding the centromere to nearby genes, causing their inactivation. Cells in which w^+ is active are red; those in which the gene is inactive are white. Cells in which rst^+ is active have a smooth surface; those in which this gene is inactive have a rough surface. Heterochromatin can spread different distances in different cells, but it usually does not skip genes.

Geneticists currently know very little about how heterochromatin invades euchromatin, what determines the extent of heterochromatin invasion in a particular cell, or how that cell's descendants inherit its specific heterochromatin pattern. They can, nevertheless, use the phenomenon of position-effect variegation to identify the molecules involved in heterochromatin formation. In one procedure, they obtained mutations that either enhance the amount of variegation produced by genes positioned near heterochromatin or diminish the amount of variegation. Enhancement reflects gene inactivation in more cells; diminishment reflects gene inactivation in fewer cells. The researchers later isolated by molecular cloning several of the genes that had mutated, and they raised antibodies against the mutant protein products of these genes. In this way, they discovered that at least some of the genes influencing heterochromatin formation encode proteins that localize selectively to the heterochromatin. The researchers hypothesize that if these proteins migrate from established heterochromatin to nearby DNA sequences, position-effect variegation results.

Barr Bodies Are Another Example of the Correlation Between Heterochromatin and a Decrease in Genetic Activity

In both fruit flies and mammals, normal males have one copy of the X chromosome, while females have two. The two sexes, however, require equal amounts of most proteins encoded by genes on the X chromosome. To compensate for the discrepancy in the dose of X-linked genes, male *Drosophila* double the rate at which they express the genes on their single X. Mammals have a different control mechanism for dosage compensation: the random inactivation of all but one X chromosome in each of the female's somatic cells. The inactive X chromosomes are observable in interphase cells as darkly stained heterochromatin masses. Geneticists call these densely staining X chromosomes **Barr bodies** after Murray Barr, the cytologist who discovered them.

The inactive X chromosome in mammalian females is another example of facultative heterochromatin. Here, a whole X chromosome becomes completely heterochromatic in some cells while remaining euchromatic in others. Most genes on the X chromosome are available for transcription only in cells where the chromosome is euchromatic; very few genes are available for transcription when the X becomes heterochromatic.

An XX female has one Barr body in each somatic cell. The X chromosome that remains genetically active in these cells decondenses and stains as expected during interphase. XY male cells do not contain Barr bodies. As a result, normal male and female mammals have the same number of active X chromosomes. Females with XXX or XXXX karyotypes can survive because they have two or three Barr bodies and only a single active X. Barr bodies are not restricted to females, but their presence does require more than one X chromosome per cell. Cells from XXY Klinefelter males also contain a Barr body.

The "decision" determining which X chromosome in each cell becomes a Barr body occurs at random in the early stages of development and is inherited by the descendants of each cell. In humans, for example, two weeks after fertilization, when an XX female embryo consists of 500–1000 cells, one of the X chromosomes in each cell condenses to a Barr body. Each embryonic cell "decides" independently which X it will be. In some cells, it is the X inherited from the mother; in others, it is the X inherited from the father. Once the determination is made, it is clonally perpetuated so that all of the thousands or millions of cells descended by mitosis from a particular embryonic cell condense the same X chromosome to a Barr body and thereby inactivate it. Female mammals are thus a mosaic of cells containing either a maternally or a paternally derived inactivated X chromosome. In an individual female that is heterozygous for an X-linked gene, some cells will express one allele, while other cells will express the alternative. In females heterozygous for an X-linked mutation that would be lethal in a male, the relation between the two populations of activated X cells can make the difference between life and death. If the X chromosome carrying the wild-type allele is active in a high enough proportion of cells where expression of the gene is required, the individual will survive.

About a dozen genes on the X chromosomes of both mice and humans escape Barr body inactivation, most likely because they are in small regions that do not become completely heterochromatic. Most of these genes remain active on both X chromosomes in a female. One, however, is inactive on the euchromatic X but active on the heterochromatic X, and it may contribute to the mechanism that causes X inactivation. This gene is referred to as *Xist* for X inactivation specific transcript. An X chromosome that does not contain *Xist* cannot be inactivated. Deletion of the *Xist* gene abolishes a chromosome's capacity for X inactivation. Cells carrying one X chromosome that lacks the *Xist* gene must inactivate the other X. The specific function of the *Xist* gene is to produce an unusually large *cis*-acting RNA transcript. Studies using fluorescent molecular probes show that this *cis*-acting *Xist* RNA, unlike most transcripts, never leaves the nucleus. One model for how the *Xist* RNA produces inactivation proposes that the unusually large *cis*-acting RNA binds to the X chromosome that produces it, perhaps in association with other proteins; this binding makes most genes on that chromosome inaccessible to transcription factors.

Although position-effect variegation and formation of Barr bodies are specialized phenomena, they epitomize the relationship between chromatin structure and gene function in all eukaryotic chromosomes. The compaction of chromosomes, which prepares the sister chromatids of each duplicated chromosome for accurate segregation and transport through mitosis and meiosis, makes DNA inaccessible to the proteins that initiate gene expression. Such compaction reaches its height during metaphase and anaphase, curtailing the transcription of most genes at those phases of the cell cycle. Decompaction, by contrast, precedes gene expression. The opening up and closing off of chromosomal areas depends on a dynamic network of DNA-protein interactions, which modify the density of chromatin packaging to appropriate levels for genetic function and passage through the cell cycle. Once established, the chromatin structure that determines a gene's expression is stably transmitted to daughter cells.

Unusual Chromosome Structures Clarify the Correlation Between Chromosome Packaging and Gene Function

The chromosomes of some highly specialized cells present an unusual appearance, as do the regions of interphase chromosomes in normal cells that carry genes for ribosomal RNA (rRNA) synthesis. When viewed in the light microscope, for instance, chromosomes in the salivary gland cells of fruit fly larvae look like thickened, striped lengths of rope radiating from a darkened central core. Most cells in interphase of the cell cycle exhibit a large, well-defined sphere within the nucleus. Detailed studies of these unusual forms have clarified how chromosome structure affects gene function.

Polytene Chromosomes Magnify Patterns of Compaction and Tie Them to Gene Expression

Within the salivary gland cells of *Drosophila* larvae, the interphase chromosomes go through 10 rounds of replication without ever entering mitosis. As a result, the sister chromatids never separate and each chromosome consists of 2^{10} (1024) double helices. In addition, because homologous chromosomes in the somatic cells of *Drosophila* remain tightly paired throughout interphase, pairs of homologs form a cable of double thickness containing 2048 double helices of DNA (1024 from each homolog). These giant chromosomes consisting of many identical chromatids lying in parallel register are called **polytene chromosomes** (Fig. 12.15a). When stained and viewed in the light microscope, *Drosophila* polytene chromosomes have an irregular fine-grain banding pattern in which denser dark bands alternate with lighter interbands. Each chromosome has a distinctive banding pattern different from that of all other chromosomes. Microscopic examination supports the idea that the chromatin of each dark band is 10 times more condensed than the chromatin in the lighter interbands (Fig. 12.15b).

The centromeres of polytene chromosomes fuse to form a dense heterochromatic mass known as the **chromocenter.** The chromocenter locks the chromosomes into a starfishlike formation in which the undulating chromosome arms radiate from a central point (see Fig. 12.15a). The DNA in the chromocenter undergoes fewer duplications than the DNA in the remainder of the chromosome. As a result, whereas the normal somatic chromosomes in *Drosophila* have almost equal amounts of heterochromatin and euchromatin, polytene chromosomes have relatively little heterochromatin by mass.

Geneticists once thought there might be a one-to-one relationship between the bands of polytene chromosomes and genes, but studies using the molecular techniques described in

Figure 12.15 Polytene chromosomes in the salivary glands of *Drosophila* larvae. (a) A low-resolution photo (*left*) showing all the polytene chromosomes in a larval salivary gland cell. Above the photo is a picture of mitotic chromosomes of a diploid cell at the same scale. On the *right* is a higher-resolution image revealing the irregular fine-grain pattern of bands and interbands and a chromosome puff: a highly transcribed, decondensed region of a polytene chromosome. Below is a drawing of the banding pattern seen in polytene chromosomes. Note the chromocenter, where the centromeres of the chromosomes fuse. Although normal chromosomes are about equally divided between euchromatin and heterochromatin, the heterochromatin at the chromocenter is only a small part of a polytene chromosome because it is underreplicated. **(b)** *Left:* a very-high-resolution electron micrograph of a small section of a *Drosophila* polytene chromosome; *right:* a hypothetical model showing how the chromatin of the bands is more condensed than the chromatin of the interbands. **(c)** *In situ* hybridization of a probe containing the *white* gene to a single band (3C2) near the tip of the *Drosophila* X chromosome.

Chapters 9 and 10 show that this is not so. Some bands contain more than one gene; others do not seem to carry any. However, even though the 1:1 ratio of bands to genes has turned out to be too simplistic, most geneticists still believe that the division into bands has some functional significance. One possibility is that the bands represent units of transcriptional regulation containing genes activated at the same time.

The large size and precisely reproducible banding patterns of polytene chromosomes make them an invaluable research tool. Because the large number of bands reliably subdivides each chromosome into very small segments, geneticists can use them as a physical guide to mapping. The chromosomes of *Drosophila,* for example, collectively carry about 5000 bands that range in size from 3 kb to more than 150 kb. Investigators designate these bands by numbers and letters of the alphabet. The familiar *white* gene is located in band 3C2 on the X chromosome. In mapping, the banding pattern of polytene chromosomes provides extremely high resolution for *in situ* hybridization tests (Fig. 12.15c). A particular fragment of DNA can be localized to a 30–100 kb region of a polytene chromosome. By contrast, *in situ* hybridization with normal metaphase chromosomes locates a fragment in a much larger 5–10 Mb region.

Polytene chromosomes have also provided a basis for studying the relation between chromosome structure and gene activity. For example, when a gene in a salivary gland cell undergoes a very high rate of transcription, the band containing that gene decondenses and that region of the polytene chromosome swells to form a large, diffuse structure called a **chromosomal puff** (see Fig. 12.15a). The correlation between puffs and regions of high transcriptional activity is yet another example of the general rule that chromatin must unwind to allow gene expression. The pattern of puffing indicates which genes are being expressed in the salivary gland cells at a specific time. At an early stage in the development of salivary glands, for example, one band puffs while at a later stage, this puff has disappeared and other bands are forming puffs. Researchers have recently learned that the early puffing band contains a gene encoding a protein that helps activate transcription of genes in the later puffing bands.

Decondensed Chromatin in the Nucleolus of Interphase Cells Contains rRNA Genes Actively Undergoing Transcription

The most obvious nuclear feature of nearly every interphase eukaryotic cell viewed through the light microscope is the large sphere-shaped organelle known as the **nucleolus.** Within the nucleolus, long loops of DNA from several chromosomes each carry a cluster of rRNA genes; together the genes make up the **nucleolar organizer** (Fig. 12.16a). The nucleolus also houses growing rRNA transcripts and completed transcripts from arrays of rRNA genes repeated in tandem that look like felled Christmas trees or feathers (Fig. 12.16b). Like the chromosomal puffs of polytene chromosomes, the decompacted DNA in the nucleolus is a sign of high levels of transcription.

Figure 12.16   The nucleolus. (a) A nucleolus can contain hundreds of copies of rRNA genes (collectively known as the nucleolus organizer); the genes are organized as a series of transcription units separated by nontranscribed spacer regions. **(b)** The nucleolus is the site of rRNA gene transcription. To generate this image, a region of chromatin from the nucleolus was spread for visualization in the electron microscope. Newly transcribed rRNAs appear as short, wispy strands emanating from the DNA; the longer strands are closer to completion. Particles near the end of each RNA transcript are thought to be partially assembled ribosomal subunits. Each transcription unit of rRNA genes generates a large precursor RNA that corresponds to one of the "Christmas trees branches," or "feathers" in this micrograph. Processing of these large precursor RNAs produces the 18S, 5.8S, and 28S rRNAs of the ribosomal subunit.

CONNECTIONS

Eukaryotic chromosomes package and manage the genetic information in DNA through a modular chromatin design whose flexibility allows back-and-forth shifts between different levels of organization. These reversible changes in chromatin structure reliably sustain a variety of chromosome functions, producing selective unwinding for gene expression, universal unwinding for replication, and coordinated compaction for segregation and transport. Histones and nonhistone proteins provide the framework for chromatin and help regulate changes in chromosome structure and function. Noncoding sequences that specify the origins of replication, centromeres, and telomeres are essential to chromosome duplication, segregation, and integrity.

Although the faithful function, replication, and transmission of chromosomes underlies the perpetuation of life within each species, chromosomal changes do occur. We have already described two mechanisms of change: mutation of individual nucleotides (Chapter 7) and homologous recombination, which exchanges bases between homologs (Chapters 4, 5, and 6). In the next chapter, we examine broader chromosomal rearrangements that produce different numbers of chromosomes, reshuffle genes between nonhomologous chromosomes, and reorganize the genes of a single chromosome. These large-scale modifications, by altering the genetic content of a genome, provide some of the important variations that fuel evolution.

E S S E N T I A L   C O N C E P T S

1. Each chromosome consists of one long molecule of DNA compacted by *histone* and *nonhistone* proteins. The five types of histones—H1, H2A, H2B, H3, and H4—are essential to the establishment of generalized chromosome structure. There are more than 1000 kinds of nonhistone proteins: Some are structural; others are active in replication and segregation; but most help determine the time and place of gene expression.

2. DNA-protein interactions create reversible levels of compaction.
   a. The naked DNA wraps around core histones to form *nucleosomes,* which are secured by H1.
   b. Models of higher-level compaction suggest that some sort of *supercoiling* condenses the nucleosomal fiber to a wider fiber. Nonhistone proteins then anchor this fiber to form *loops* and higher levels of compaction (perhaps in the form of *rosettes* and bunches of rosettes). In metaphase chromosomes, higher levels of folding compact DNA 10,000-fold.
   c. In fully compacted metaphase chromosomes, the *centromere* and *telomeres* become visible under the microscope. Giemsa staining of metaphase chromosomes reveals highly reproducible banding patterns that researchers can use to locate genes, analyze chromosomal differences between species, and diagnose some genetic diseases.

3. Specialized elements are required for normal chromosome function.

   a. *Origins of replication* are sites accessible for the binding of proteins that initiate replication. In eukaryotic chromosomes, many origins of replication ensure timely DNA replication.
   b. Telomeres, composed of repetitive base sequences, protect the ends of chromosomes, ensuring their integrity. The enzyme *telomerase* helps reconstruct the complete telomere with each cell division.
   c. Centromeres, which appear as constrictions in metaphase chromosomes, ensure proper segregation by holding sister chromatids together and by elaborating kinetochores, which properly attach sister chromatids to spindle fibers for mitosis and meiosis.

4. Chromosomal packaging influences gene activity.
   a. Conditions that decondense selected areas of chromatin precede and facilitate gene expression. Puffs in *Drosophila* polytene chromosomes and the *nucleoli* in most interphase cells contain decondensed chromatin that is highly transcribed. Boundary elements called insulators delimit these areas of decompaction. Nucleosomes in the decompacted areas unwind to allow the initiation of transcription.
   b. Extreme condensation silences gene expression. Extremely condensed chromosomal areas appear as darkly staining *heterochromatin* under the microscope. Position-effect variegation in *Drosophila* and Barr bodies in mammals are examples of the correlation between heterochromatin formation and a loss of gene activity.

S O C I A L   A N D   E T H I C A L   I S S U E S

1. Recent evidence has shown that the ends of chromosomes (telomeres) get shorter as cells age because of a lack of the enzyme telomerase which maintains the ends and length of chromosomes. In contrast, cancer cells that are "immortal" and divide rapidly have telomerase activity. Whether the lack of telomeres in older cells is a cause or a result of aging is still under investigation. A proposal that included experimentation on telomerase has been submitted to the National Institutes of Health. A stated goal of the proposal was exploration of ways to extend the human life span. Is this a valid way to spend federal research monies given that many commercial companies are already exploring ways to use these findings to extend the human life span? What are the consequences for society if people live longer? What do you think of the goal of a longer life?

2. A city school board put forward a proposal to screen school children with learning disabilities for the fragile X mutation (see the Genetics and Society box "Amplified Trinucleotide Repeats May Have Medical Consequences"

in Chapter 7). Fragile X is the most common form of inherited mental retardation and is caused by an unusual expansion of a triplet nucleotide repeat in a gene on the X chromosome. (The triplet expansion is extreme enough to be seen as a cytological abnormality as well as by DNA sequence analysis.) The extent of mental impairment and other abnormal phenotypes varies with the size of the repeat. The proposal was well intended; the school board hoped to use the screening to identify students with potential disabilities and give them extra help and consideration. However, many parents and professionals opposed the testing. How could selection for testing harm a child? How could identification of fragile X harm a child? What are the benefits of such testing?

3. With available data on the DNA sequence of the human genome, what has changed about information that a researcher might choose to share with other researchers? For example, a scientist has worked for two years on identifying a particular human disease gene recently determined the 10 Mb region of a chromosome where the

gene is located. The next steps are to use genomic sequence information to identify potential gene candidates and determine which of the candidates is altered by disease-causing mutations. The scientist hesitates to talk about his finding because he thinks many other researchers will immediately use the data and design PCR primers to investigate the potential genes and mutations in affected individuals. Does he have a right to withhold the information until he has completed his analysis of mutants? Does the answer to this question depend on whether his work is supported by public (government) funding agencies or by a private biotechnology company?

S O L V E D   P R O B L E M S

I. One can construct YACs that range in size from 15 kb to 1 Mb. Based on DNA length, what level of chromosome compaction would you predict for a YAC of 50 kb compared with a YAC of 500 kb?

Answer

To answer this question, you need to apply information about the amount of DNA needed to get different types of chromosome condensation.

The 500 kb YAC would probably be more condensed than the 50 kb YAC based on its larger size. The DNA of both YACs would be wound around histones to form the nucleosome structure (160 bp around the core histones plus 40 bp in linker region). That DNA would be further compacted into 300 Å fibers that contain six nucleosomes per turn. *The 500 kb YAC would be compacted at a higher level of order, presumably in radial loops that occur every 60–100 kb in the chromosome, but the 50 kb YAC is not large enough to be packaged in this way.*

II. To clone and express genes in yeast, researchers constructed recombinant plasmids in the 1970s. Most of these early yeast plasmids did not contain an origin of replication and therefore were not maintained as autonomous plasmids separate from the chromosomes. If one of these plasmids integrated into the chromosome by recombination, it was replicated as part of the chromosome and was stably inherited from one division to the next. Later, when yeast origins of replication (ARS elements) were characterized, researchers constructed autonomous yeast plasmids that could be maintained and transmitted independently of the chromosome. Describe how you would distinguish a plasmid integrated into a chromosome from an autonomous plasmid, using hybridization techniques.

Answer

This question requires an understanding of the use of hybridization (that is, the source of DNA and the probe used) and the interpretation of results. Think about differences between the two plasmid states (integrated and autonomous) and how the differences might be visualized using hybridization.

If plasmid DNA is used as a probe in a hybridization with a blot of restriction enzyme-digested DNA from cells containing integrated plasmids and from cells containing autonomous plasmids, the hybridizing bands will be different (see the diagram). In the DNA sample containing the autonomous plasmid, the probe will hybridize to band(s) that together equal the total size of the original plasmid. If DNA from cells containing the integrated plasmid is digested with a restriction enzyme that cuts within the plasmid, two new fragments appear that contain both plasmid DNA and the chromosomal DNA on either side of the integration site in the chromosome. In the example in the figure, a 4 kb plasmid containing a *Eco*RI site recombined with the chromosome at a site within a 2.5 kb *Eco*RI chromosomal fragment. The 2.2 kb and 4.3 kb fragments are joint fragments containing both chromosomal DNA and plasmid DNA. Both bands will hybridize with the probe.

III. Mouse geneticist Mary Lyon proposed in 1961 that all but one copy of the X chromosome were inactivated in mammals.

a. What cytological finding supports the Lyon hypothesis?

b. What is a genetic result that supports the Lyon hypothesis?

Answer

This question requires an understanding of the experimental observations on X inactivation.

a. Microscopic examination of cells produces cytological evidence. *The number of Barr bodies seen in the cells of individuals with different numbers of X chromosomes supports the hypothesis.* For example, cells from XX females have one Barr body, whereas cells from an XXX female have two Barr bodies.

b. *Examination of the phenotype of cells from a female heterozygous for two different alleles of an X-linked gene produces genetic evidence that supports the Lyon hypothesis.* In these females, some cells have the phenotype associated with one allele, while other cells have the phenotype associated with the other allele.

P R O B L E M S

12-1 For each of the terms in the left column, choose the best matching phrase in the right column.

a. telomere	1. site at the base of chromatin loops that anchors loops to the scaffold
b. G bands	2. origin of replication in yeast
c. polytene chromosomes	3. repetitive DNA found near the centromere in higher eukaryotes
d. kinetochore	4. specialized structure at the end of a linear chromosome
e. nucleosome	5. giant chromosomes in *Drosophila* salivary glands
f. ARS	6. complexes of DNA and protein in the eukaryotic nucleus
g. satellite DNA	7. small basic proteins that bind to DNA and form the core of the nucleosome
h. chromatin	8. complex of DNA and proteins where spindle fibers attach to a chromosome
i. SAR	9. beadlike structure consisting of DNA wound around histone proteins
j. histones	10. regions of a chromosome that are distinguished by staining differences

12-2 The 16 chromosomes of the yeast *S. cerevisiae* can be separated by size using pulse-field gel electrophoresis. You have just isolated a clone of a yeast gene and want to know on which chromosome the gene is located. How could you use pulse-field gel electrophoresis to find this out?

12-3 What difference is there in the compaction of chromosomes during metaphase and interphase?

12-4 What is the role of the core histones in compaction compared to the role of histone H1?

12-5 The histone proteins H3 and H4 are modified in predictable and consistent ways that are conserved across species. One of the modifications is addition of an acetyl group to the twelfth lysine in the H4 protein. If you were a geneticist working in yeast and had a clone of the H4 gene, what could you do to test whether the acetylation at this specific lysine was necessary for the functioning of chromatin?

12-6 Many proteins other than histones are found associated with chromosomes. What roles do these nonhistone proteins play?

12-7 The mitotic cell divisions in the early embryo of *D. melanogaster* occur very rapidly (every 8 minutes).

a. If there was one bidirectional origin in the middle of each chromosome, how many nucleotides would DNA polymerase have to add per second to replicate all the DNA in the longest chromosome (66 Mb) during the 8-minute early embryonic cell cycles? (Assume that replication occurs during the entire cell division cycle.)

b. In fact, many origins of replication are active on each chromosome during the early embryonic divisions and are spaced approximately 7 kb apart. Calculate the average rate (per second) with which DNA polymerase adds complementary nucleotides to a growing chain in the early *Drosophila* embryo.

12-8 The extreme ends of chromosomes (telomeres) contain repeated DNA sequences (for example, 5′ TTAGGG 3′ in humans). If a blot of genomic DNA was hybridized using a probe of 3′ AATCCC 5′, all end fragments would hybridize with the probe.

a. If you wanted to examine an end of one specific chromosome by hybridization, what would you have to use as a probe?

b. When a probe specific for one chromosome end is used in Southern hybridization, a blurred band appears (see the figure). Why is the hybridizing band not a distinct sharp band as seen with probes that hybridize with internal sequences on the chromosome?

Size standard   Genomic DNA

12-9 The enzyme telomerase consists of protein and an RNA containing a template sequence that directs the addition of an end sequence appropriate for the species. Telomere sequences (TTGGGG) from the ciliated protozoan *Tetrahymena* were cloned onto the ends of a linear YAC, which was then transformed into yeast. The YAC survived as a linear piece of DNA but the YAC now had TGGTGG sequences at the very ends in addition to TTGGGG. Why do you think these sequences were added?

12-10 Chromosome assembly factor (CAF-1) is a complex of proteins that was identified biochemically in extracts from human cells. The sequence of amino acids in the proteins was identified.
a. How could you use this data to look for homologous genes in yeast?
b. As a geneticist, why would you find it advantageous to identify the yeast genes to further your understanding of chromatin assembly?

12-11 A number of yeast-derived elements were added to the bacterial plasmid pBR322. Yeast that require uracil for growth (Ura⁻ cells) were transformed with these modified plasmids and Ura⁺ colonies were selected by growth in media lacking uracil. For plasmids containing each of the elements listed, indicate whether you expect the plasmid to integrate into a chromosome or be maintained separately as a plasmid. If the plasmid is maintained autonomously, is it stably inherited by all of the daughter cells of subsequent generations when you no longer select for Ura⁺ cells (grow the yeast in media containing uracil)?
a. URA⁺ gene
b. URA⁺ gene, ARS
c. URA⁺ gene, ARS, CEN (centromere)

12-12 A protein called CBF1 was identified in yeast as a centromere binding protein. You want to know if similar proteins are present in human cells.
a. Starting with a human cDNA expression library from actively dividing cells, how could you isolate a clone containing a human gene similar to the yeast gene?
b. Imagine you obtained a clone containing a human gene similar to the yeast gene. How could you test to see if the protein encoded by the human gene is associated with a human centromere region?
c. If you had a haploid yeast strain with a temperature-sensitive mutation in the CBF1 gene, what phenotype(s) could you imagine when the temperature is raised to the nonpermissive temperature?

12-13 A DNA fragment containing yeast centromere DNA was cloned into a TRP ARS plasmid, YRp7, causing the plasmid to become mitotically very stable (that is, the plasmid was transmitted during mitotic divisions to each daughter cell). The assay for mitotic stability consists of growing a transformed cell without selection for the plasmid and determining the number of Trp⁺ colonies remaining after 20 generations of growth under conditions that are not selective for the plasmid. To identify the region of the cloned fragment that contained centromere DNA, you cut the initial fragment into smaller pieces, reclone those pieces into YRp7, and test for mitotic stability. Based on the map that follows and results of the mitotic stability assay, where is the centromeric DNA located?

Results of Mitotic Stability Assay

Plasmid DNA	Percentage of Trp⁺ colonies after 20 generations
YRp7	0.9
YRp7 + 5.5 kb *Bam*HI (B)	68
YRp7 + 3.5 kb *Bam*HI-*Hin*dIII (H)	0.5
YRp7 + 2.0 kb *Bam*HI-*Hin*dIII	80
YRp7 + 0.6 kb *Sau*3A (S)	76
YRp7 + 1.0 kb *Hin*dIII-*Sau*3A	0.7

12-14 Another vector system for cloning large fragments of DNA that investigators developed for genome analysis uses a derivative of the baculovirus that infects arthropods. One of the advantages of these BAC vectors compared to YACs is that DNA cloned into these vectors is less likely to get rearranged. Clones of a human *Sfi*I fragment were isolated from BAC- and YAC-based libraries using the same probe but the *Hin*dIII digests of these clones were very different from each other. How could you determine which of these two clones, if either, has the same order and size as the *Hin*dIII fragments present in the human genome?

12-15 If a sequence that acts as a boundary for the controlled unfolding of chromatin is removed, what effect could this have on genes in adjacent regions?

12-16 Some genes are highly transcribed in cells of one tissue type but not in other tissues. Based on the hypothesis presented in this chapter, which of the patterns of DNase digestion shown here in the vicinity of gene *X* would you expect to see in DNA isolated from cells in which gene *X* is highly expressed? (Vertical lines indicate sites where DNase I digests the DNA.)

a. Cell type I

___|_____|_____|___

b. Cell type II

__||_|_____|||__|_____||_|_|_|_|___|_|_||_

12-17 Give examples of constitutive and facultative heterochromatin in
a. *Drosophila*
b. humans

12-18 *Drosophila* geneticists have isolated many mutations that modify position-effect variegation. Dominant *suppressors of variegation* [*Su(var)s*] cause less frequent inactivation of genes brought near heterochromatin by chromosome rearrangements, while dominant *enhancers of variegation* [*E(var)s*] cause more frequent inactivation of such genes.
a. What effects would each of these two kinds of mutations have on position-effect variegation of the *white* gene in *Drosophila.*
b. Assuming that these *Su(var)* and *E(var)* mutations are loss of function (null) alleles in the corresponding genes, what kinds of proteins do you think these genes encode?

12-19 How many Barr bodies are present in humans with the following karyotype?
a. an XX female
b. an XY male
c. an XX male (known as an exceptional male*)
d. an XXY male
e. an XXXX female
f. an XO female

*In an exceptional XX male, one of the two X chromosomes has essentially all of the genes normally found on the X chromosome plus a small region from the Y chromosome that carries the testis-determining factor required for male development.

12-20 A pair of twin sisters were believed to be identical until one was diagnosed with Duchenne muscular dystrophy, an X-linked trait. (Her sister did not have the disease.) Does this finding mean they are definitely not identical twins (derived from fertilization of one egg)? Why or why not?

12-21 Females with the genotype $X^{CB} X^{cb}$ are rarely color-blind although some have only partial color vision. Speculate on why this is true.

12-22 What are two characteristics of polytene chromosomes that make them useful for cytogenetics?

REFERENCES

de Lange, Titia. (2001). Telomere capping—One strand fits all. *Science* 292:1075–76.

Fangman, W. L., and Brewer, B. J. (1992). A question of time. *Cell* 71:363–66.

Mirkovich, J., Mirault, M.-E., and Laemmli, U. K. (1984). Organization of the higher-order chromatin loop: Specific DNA attachment sites on nuclear scaffold. *Cell* 39:223–32.

Murray, A. W., and Szostak, J. W. (1983). Construction of artificial chromosomes in yeast. *Nature* 305:189–93.

13

Chromosomal Rearrangements and Changes in Chromosome Number Reshape Eukaryotic Genomes

Chromosomal rearrangements can be mapped with high precision on *Drosophila* polytene chromosomes.
The *red arrow* points to a very large transposable element that has inserted into one chromosome but
is not present in the paired homologous wild-type chromosome (*black arrow*).

Studies comparing the genome of the laboratory mouse (*Mus musculus*) with that of humans reveal an evolutionary paradox. At the DNA level, there is a close similarity of nucleotide sequence across hundreds of thousands of base pairs; but at the chromosomal level, normal mouse and human karyotypes bear little resemblance to each other. Researchers have, for example, sequenced regions encompassing more than 2000 kb of mouse and human DNA containing a complex of genes that encode proteins known as T-cell receptors. Key components of the mammalian immune system, these receptors become embedded in the membrane of the immune system's T cells, where they recognize foreign molecules and microorganisms and orchestrate the immune responses that dispose of them. Comparisons of the corresponding mouse and human regions show that the nucleotide sequences of the T-cell receptor genes are similar (though not identical) in the two species, as are the order of the genes and the relative positions of a variety of noncoding sequences (of unknown function) along the chromosome. Comparisons of mouse and human Giemsa-stained karyotypes, however, reveal no conservation of banding patterns between the 20 mouse and 23 human chromosomes.

Data for resolving this apparent paradox come from linkage maps, a level of genetic analysis that is intermediate between DNA sequencing and karyotyping. Because virtually all genes cloned from the mouse genome are conserved in the human genome and vice versa, it is possible to use the techniques described in Chapters 5 and 10 to construct linkage maps for the two genomes from the same set of markers. The maps become a resource for assessing the similarity of linkage groups. If the order of genes in one species were unrelated to the order of genes in the other, identical linkage relationships would occur by chance and appear only rarely. By contrast, if gene order were strictly conserved, it would be possible to align the whole-chromosome linkage maps of both species and see an exact correspondence. Actual comparisons of the mouse and human linkage maps reveal a picture of intermediate conservation. Genes closely linked in *M. musculus* tend to be closely linked in humans, but genes that are less tightly linked in one species tend not to be linked at all in the other. From careful

441

Figure 13.1 Comparing the mouse and human genomes.
Mouse chromosome 1 contains large blocks of sequences found on human chromosomes 1, 2, 6, and 18; mouse chromosome 3 includes large fragments of DNA that are located on human chromosomes 1, 3, 4, and 8.

analysis of these linkage patterns, geneticists learned that it is possible to "reconstruct" the mouse genome by breaking the human genome into approximately 170 fragments, each an average length of 17.6 Mb, and pasting these fragments together in a different order (Fig. 13.1). Since a 17.6 Mb fragment would occupy no more than one or two bands of a stained chromosome, this level of conservation is not visible in karyotypes. It does, however, show up in the sequence of a smaller genomic region, such as that encoding the T-cell receptors.

These findings contribute to our understanding of how complex life-forms evolved. Although mice and humans diverged from a common ancestor about 65 million years ago, the DNA sequence in many regions of the two genomes is very similar. It is thus possible to hypothesize that the mouse and human genomes evolved through a series of approximately 170 reshaping events during which the chromosomes broke apart and the resulting fragments resealed end to end in novel ways. After each event, the newly rearranged chromosomes somehow became fixed in the genome of the emerging species. Both nucleotide sequence differences and differences in genome organization thus contribute to dissimilarities between the species.

In this chapter, we examine two types of events that reshape genomes: (1) **rearrangements,** which reorganize the DNA sequences within one or more chromosomes, and (2) **changes in chromosome number** involving losses or gains of entire chromosomes or sets of chromosomes (Table 13.1). Rearrangements

and changes in chromosome number may affect gene activity or gene transmission by altering the position, order, or number of genes in a cell. Such alterations often, but not always, lead to a genetic imbalance that is harmful to the organism or its progeny.

Two general themes emerge from our discussion. First, karyotypes generally remain constant within a species, not because rearrangements and changes in chromosome number occur infrequently (they are, in fact, quite common), but because the genetic instabilities and imbalances produced by such changes usually place individual cells or organisms and their progeny at a selective disadvantage. Second, despite selection against chromosomal variations, related species almost always have different karyotypes, with closely related species (such as chimpanzees and humans) diverging by only a few rearrangements and more distantly related species (such as mice and humans) diverging by a larger number of rearrangements. These observations suggest there is some correlation between karyotypic rearrangements and the evolution of new species.

In our examination of events that reshape eukaryotic genomes, we discuss:

■ Rearrangements of DNA sequences within and between chromosomes: deletions, duplications, inversions, translocations, and movements of transposable elements.

■ Changes in chromosome number: aneuploidy, including monosomy and trisomy; monoploidy; and polyploidy.

TABLE 13.1   Chromosomal Rearrangements and Changes in Chromosome Number

Chromosomal Rearrangements

	Before	After
Deletion: Removal of a segment of DNA	1 2 3 4 5 6 7 8	1 2 3 5 6 7 8
Duplication: Increase in the number of copies of a chromosomal region	1 2 3 4 5 6 7 8	1 2 3 2 3 4 5 6 7 8
Inversion: Half-circle rotation of a chromosomal region	1 2 3 4 5 6 7 8 (180° rotation)	1 4 3 2 5 6 7 8
Translocations:		
Nonreciprocal: Unequal exchanges between nonhomologous chromosomes	1 2 3 4 5 6 7 8 / 12 13 14 15 16 17 18	12 13 4 5 6 7 8 / 14 15 16 17 18
Reciprocal: Parts of two nonhomologous chromosomes trade places	1 2 3 4 5 6 7 8 / 12 13 14 15 16 17 18	12 13 14 15 5 6 7 8 / 1 2 3 4 16 17 18
Transposition: Movement of short DNA segments from one position in the genome to another	1 2 3 4 5 6 7 8	1 2 4 5 6 3 7 8

Changes in Chromosome Number

Euploidy: Cells that contain only complete sets of chromosomes

   Diploidy (2*x*): Two copies of each homolog

   Monoploidy (*x*): One copy of each homolog

Polyploidy: More than the normal diploid number of chromosome sets

   Triploidy (3*x*): Three copies of each homolog

   Tetraploidy (4*x*): Four copies of each homolog

Aneuploidy: Loss or gain of one or more chromosomes producing a chromosome number that is not an exact multiple of the haploid number

   Monosomy (2*n* − 1)

   Trisomy (2*n* + 1)

   Tetrasomy (2*n* + 2)

Rearrangements of DNA Sequences Within Chromosomes

All chromosomal rearrangements, no matter how small, alter DNA sequence. Some do so by removing or adding base pairs. Others relocate chromosomal regions without changing the number of base pairs they contain. This chapter focuses on heritable rearrangements that can be transmitted through the germ-line from one generation to the next, but it also explains that the genomes of somatic cells can undergo changes in nucleotide number or order. For example, the Fast Forward box on p. 445–447—"Development of the Immune System Depends on Programmed DNA Rearrangements"—describes how the normal development of the human immune system depends on noninherited, programmed rearrangements of the genome in somatic cells.

Deletions Remove Material from the Genome

We saw in Chapter 7 that **deletions** remove one or more contiguous base pairs of DNA from a chromosome. They may arise from errors in replication, from faulty meiotic or mitotic recombination, and from exposure to X rays or other chromosome-damaging agents that break the DNA backbone (Fig. 13.2a). Here we use the symbol *Del* to designate a chromosome that has sustained a deletion. However, many geneticists, particularly those working on *Drosophila*, prefer the term *deficiency* (abbreviated as *Df*) to deletion.

Small deletions often affect only one gene, while large deletions can generate chromosomes lacking tens or even hundreds of genes. In higher organisms, geneticists usually find it difficult to distinguish small deletions affecting only one gene from point mutations; they can resolve such distinctions only through analysis of the DNA itself. For example, deletions can result in smaller restriction fragments or polymerase chain reaction (PCR) products, whereas most point mutations would not cause such changes (Fig. 13.2b). Larger deletions are sometimes identifiable because they affect the expression of two or more adjacent genes. Very large deletions are visible at the relatively low resolution of a karyotype, showing up as the loss of one or more bands from a chromosome.

Homozygosity for a Deletion Is Often, But Not Always, Lethal

Because many of the genes in a genome are essential to an individual's survival, homozygotes (*Del/Del*) or hemizygotes (*Del/Y*) for most deletion-bearing chromosomes do not survive. In rare cases where the deleted chromosomal region is devoid of genes essential for viability, however, a deletion hemi- or homozygote may survive. For example, *Drosophila* males hemizygous for an 80 kb deletion including the *white* (*w*) gene survive perfectly well in the laboratory; lacking the *w*+ allele required for red eye pigmentation, they have white eyes.

(a) DNA breakage may cause deletions

(b) Detecting deletions using PCR

Figure 13.2 Deletions: Origin and detection. (a) If a chromosome sustains two double-strand breaks, a deletion will result if the chromosomal fragments are not properly religated. **(b)** One way to detect deletions is by PCR. The two PCR primers shown will amplify a larger PCR product from wild-type DNA than from DNA with a deletion.

Heterozygosity for a Deletion Is Often Detrimental

Usually, the only way an organism can survive a deletion of more than a few genes is if it carries a nondeleted wild-type homolog of the deleted chromosome. Such a *Del/+* individual is known as a *deletion heterozygote*. Nonetheless, the missing segment cannot be too large, as heterozygosity for very large deletions is almost always lethal. Even small deletions can be harmful in heterozygotes. Newborn humans heterozygous for a relatively small deletion from the short arm of chromosome 5 have *cri du chat* syndrome (from the French for "cry of the cat"), so named because the symptoms include an abnormal cry reminiscent of a mewing kitten. The syndrome also leads to mental retardation.

Why should heterozygosity for a deletion have harmful consequences when the *Del/+* individual has at least one wild-type copy of all of its genes? The answer is that changes in **gene dosage**—the number of times a given gene is present in the cell nucleus—can create a **genetic imbalance.** This imbalance in gene dosage alters the amount of a particular protein relative to all other proteins, and this alteration in the relative amounts of protein can have a variety of phenotypic effects, depending on

Fast Forward

Development of the Immune System Depends on Programmed DNA Rearrangements

The human immune system is a marvel of specificity and diversity, including as it does close to a trillion lymphocytes of more than a billion different varieties. The lymphocytes, one category of white blood cell, come in two types: B cells, which make antibodies, also called *immunoglobulins* (*Igs*); and T cells, some of which coordinate and control immune responses while others directly dispatch infectious agents. Both B cells and T cells respond to microorganisms by proliferating, communicating with other immune-system cells, and gearing up to produce molecular missiles (such as antibodies and interferon) that subdue the invaders. To simplify our discussion, we focus on the maturation of B cells and the genetics of antibody formation, but variations of the mechanisms we describe also apply to the synthesis of receptor molecules critical for T cell function.

The Clonal Expansion of Specific B Cells Generates a Specific Antibody Response

The B cells of the human immune system produce antibodies of close to a billion different binding specificities. Each B cell, however, carries membrane-bound antibodies of only one specificity, which can bind to bacterial or viral proteins (called *antigens* in the context of immune responses) of complementary specificity. The binding of antibody to antigen causes the B cell carrying the membrane-bound antibodies to proliferate. The resulting clone of B cells matures generation by generation to a population containing two types of differentiated cells: dedicated antibody producers called *plasma cells* that every second secrete 1000 antibodies of the same binding specificity as the original membrane-bound antibodies, and *memory cells* that circulate in the blood and lymph ready to subdue in subsequent encounters antigens similar to those that caused the initial clonal expansion.

One intriguing question about antibody responses is: How can a genome containing only 40,000–60,000 (4–6×10^4) genes encode a billion (10^9) different types of antibodies? The answer is that programmed gene rearrangements, in conjunction with somatic mutations and the diverse pairing of polypeptides of different sizes, can generate roughly a billion binding specificities from a much smaller number of genes. To understand the mechanism of this diversity, it is necessary to know how antibodies are constructed and how B cells come to express the antibody-encoding genes determining specific antigen-binding sites.

The Genetics of Antibody Formation Produce Specificity and Diversity

All antibody molecules consist of a single or multiple copies of the same basic molecular unit. That unit is composed of two pairs of polypeptides: two identical light and two identical heavy chains (Fig. A). Each light and each heavy chain has a constant (C) domain and a variable (V) domain. The C domain of the heavy chain

Figure A   How antibody specificity emerges from molecular structure. Two heavy chains and two light chains held together by disulfide (–S–S–) bonds form the basic unit of an antibody molecule. Both heavy and light chains have variable (V) domains near their N termini, which associate to form the antigen-binding site. "Hypervariable" stretches of amino acids within the V domains vary extensively between antibody molecules. The remainder of each chain is composed of a C (constant) domain; that of the heavy chain has several subdomains (C_{H1}, hinge, C_{H2}, and C_{H3}).

determines whether the antibody falls into one of five major classes (designated IgM, IgG, IgE, IgD, and IgA), which influence where and how an antibody functions. For example, IgM antibodies form early in an immune response and are anchored in the B-cell membrane, where they help initiate an immune response; IgG antibodies emerge later in a response and are secreted into the blood serum, where they target antigens for destruction. The C domains of the light and heavy chains are not involved in determining the specificity of antibodies. That function falls to the V domains. The variable domains of light and heavy chains come together to form the antigen-binding site, which defines an antibody's specificity.

The DNA for all domains of the heavy chain resides on chromosome 14 (Fig. B). This heavy-chain gene region consists of more than 100 V-encoding segments, each preceded by a promoter, several D (for diversity) segments, several J (for joining) segments, and nine C-encoding segments preceded by an enhancer (a short DNA segment that aids in the initiation of transcription by interacting with the promoter; see Chapter 17 for details). In all germ-line cells and in most somatic cells, including the cells destined to become B lymphocytes, these various gene segments lie far apart on the chromosome. During B-cell development,

Heavy-chain gene region

Figure B **The heavy-chain gene region on chromosome 14.** The DNA of germ-line cells (as well as all non-antibody-producing cells) contains more than 100 V_H segments, about 20 D segments, 6 J_H segments, and 9 C_H segments (*top*). Each V_H and C_H segment is composed of two or more exons, as seen in the alternate view of the same DNA on the next line. In B cells, somatic rearrangements bring together random, individual V_H, D, and J_H segments. The promoter of the selected V_H segment is now activated by the enhancer adjacent to C_μ, allowing transcription of the newly constructed heavy-chain gene into a primary transcript, which is subsequently spliced into a mature mRNA. The μ heavy chain translated from this mRNA is the type of heavy chain found in IgM antibodies. Later in B-cell development, other rearrangements (*not shown*) connect the same V-D-J variable region to other C_H segments such as C_δ, allowing the synthesis of other antibody classes.

however, somatic rearrangements juxtapose random, individual V, D, and J segments together to form the particular variable region that will be transcribed. These rearrangements also place the newly formed variable region next to a C segment and its enhancer, and they further bring the promoter and enhancer into proximity, allowing transcription of the heavy-chain gene. RNA splicing removes the introns from the primary transcript, making a mature mRNA encoding a complete heavy-chain polypeptide.

The somatic rearrangements that shuffle the same deck of V, D, J, and C segments at random in each B cell of each individual permit expression of one, and only one, specific heavy chain. Without the rearrangements, antibody gene expression cannot occur. Random somatic rearrangements also generate the actual genes that will be expressed as light chains. The somatic rearrangements allowing the expression of antibodies thus generate enormous diversity of binding sites through the random selection and recombination of gene elements.

Several other mechanisms add to this diversity. First there is the imprecise joining of each gene's DNA elements, perpetrated by cutting and splicing enzymes that sometimes trim DNA from or add nucleotides to the junctions of the segments they join. Next, random somatic mutations in a rearranged gene's V region increase the variation of the antibody's V domain. Finally, in every

B cell, two copies of a specific H chain that emerged from random DNA rearrangements combine with two copies of a specific L chain that emerged from random DNA rearrangements to create molecules with a specific, possibly unique, binding site. Each B cell will display antibodies of only this specificity in its membrane, and it will manufacture large quantities of these same antibodies when stimulated by antigen. The fact that any light chain can pair with any heavy chain exponentially increases the potential diversity of antibody types. For example, if there were 10^4 different light chains and 10^5 different heavy chains, there would be 10^9 possible combinations of the two.

Interestingly, the somatic rearrangements that lead to antibody expression may also be responsible for the phenomenon of "allelic exclusion" in which a cell expresses either the paternal or maternal allele of an antibody gene—but not both. One would expect each B cell to express several types of antibodies—some bearing a paternal H and L chain, some a maternal H and L chain, and some a mix of the two. But this rarely, if ever, happens. The "one cell, one antibody" phenomenon prevails; each B cell makes only one kind of H chain and one kind of L chain at a time. Immunogeneticists believe that allelic exclusion is one consequence of the rearrangement process. In B cells in which a successful rearrangement has occurred, the rearrangement apparatus shuts down, precluding the expression of more than one allele.

Mistakes by the Enzymes That Carry Out Antibody Gene Rearrangements Can Lead to Cancer

RagI and RagII are enzymes that interact with DNA sequences in antibody genes to help catalyze the rearrangements required for gene expression. In carrying out their rearrangement activities, however, the enzymes sometimes make a mistake that results in a reciprocal translocation between human chromosomes 8 and 14. After this translocation, the enhancer of the chromosome 14 heavy-chain gene lies in the vicinity of the unrelated *c-myc* gene from chromosome 8. Under normal circumstances, *c-myc* gener-

ates a transcription factor that turns on other genes active in cell division, at the appropriate time and rate in the cell cycle. The translocated antibody-gene enhancer accelerates expression of *c-myc*, causing B cells containing the translocation to divide out of control. This uncontrolled B-cell division leads to a cancer known as Burkitt's lymphoma (Fig. C).

Thus, although programmed gene rearrangements contribute to the normal development of a healthy immune system, misfiring of the rearrangement mechanism can promote disease.

Chapter 21 describes the evolution of the gene families that encode antibodies and other immune system proteins.

Figure C Misguided translocations can lead to Burkitt's lymphoma. In DNA from this Burkitt's lymphoma patient, translocations bring transcription of the *c-myc* gene under the control of the enhancer adjacent to C_μ. As a result, B cells produce abnormally high levels of the c-myc protein. Apparently, the RagI and RagII enzymes have mistakenly connected a J_H segment to the *c-myc* gene from chromosome 8, instead of to a heavy-chain D segment as would normally occur.

how the proteins function and how critical the maintenance of a precise ratio of proteins is to the survival of the organism. For some rare genes, the normal diploid level of gene expression is essential to individual survival; fewer than two copies of such a gene results in lethality. In *Drosophila*, a single dose of the locus known as *Triplolethal* (Tpl^+) is lethal in an otherwise diploid individual. For certain other genes, the phenotypic consequences of a decrease in gene dosage are noticeable but not catastrophic. For example, *Drosophila* containing only one copy of the wild-type *Notch* gene have visible wing abnormalities but otherwise seem to function normally (Fig. 13.3). In contrast with these unusual examples, diminishing the dosage of most genes produces no obvious change in phenotype. There is a catch, however. Although a single dose of any one gene may not cause substantial harm to the individual, the genetic imbalance resulting from a single dose of many genes at the same time can be lethal. Humans, for example, cannot survive, even as heterozygotes, with deletions that remove more than about 3% of any part of their haploid genome.

There is another answer to the question of why heterozygosity for a deletion can be harmful. If a somatic cell heterozygous for a deletion sustains a mutation in the remaining wild-type region of the homologous chromosome, the cell may become unable to make the protein encoded by a gene in the deleted region. If this protein helps control cell division, the cell may divide out of control and generate a tumor. Thus, individuals born heterozygous for certain deletions have a greatly increased risk of losing both copies of certain genes and developing cancer. One case in point is retinoblastoma (RB), the most malignant form of eye cancer, which was previously introduced in Chapter 5 (pp. 142–143). Karyotypes of normal, noncancerous tissues from many people suffering from retinoblastoma reveal heterozygosity for deletions on chromosome 13. Cells from the retinal tumors of these same patients have a mutation in the remaining copy of the *RB* gene on the non-deleted chromosome 13. Chapter 18, "Cell-Cycle Regulation and the Genetics of Cancer," explains in detail how deletion of certain chromosomal regions greatly increases the risk of cancer,

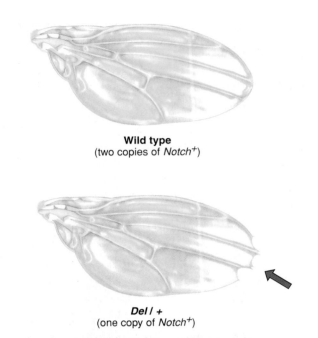

Wild type
(two copies of *Notch⁺*)

Del / +
(one copy of *Notch⁺*)

Figure 13.3 Heterozygosity for deletions may have phenotypic consequences. Flies carrying only one copy of the *Notch⁺* gene instead of the normal two copies have abnormal wings.

(a) Formation of a deletion loop

Normal homolog

Homolog with a deletion

(b) A deletion loop in *Drosophila* polytene chromosomes

Figure 13.4 Deletion loops form in the chromosomes of deletion heterozygotes. (a) During prophase of meiosis I, the undeleted region of the normal chromosome has nothing with which to pair and thus forms a deletion loop. No recombination can occur between genes within the deletion loop. In this simplified figure, each line represents two chromatids. **(b)** Deletion loops also form in the polytene chromosomes of *Drosophila*. The *thick arrow* points to the wild-type chromosome in this *Del* / + animal; the corresponding region is missing from the *Del* homolog (*thin arrow*).

and how researchers have used this knowledge to locate and clone some of the genes whose mutant forms cause cancer.

Heterozygosity for Deletions Affects Mapping Distances

Because recombination between maternal and paternal homologs can occur only at regions of similarity, map distances derived from genetic recombination frequencies in deletion heterozygotes will be aberrant. For example, no recombination is possible between genes *C, D,* and *E* in Fig. 13.4a because the DNA in this region of the normal, nondeleted chromosome has nothing with which to recombine. In fact, during the pairing of homologs in prophase of meiosis I, the "orphaned" region of the nondeleted chromosome forms a **deletion loop**—an unpaired bulge of the normal chromosome that corresponds to the area deleted from the other homolog. The progeny of a *Del* / + heterozygote will always inherit the markers in a deletion loop (*C, D,* and *E* in Fig. 13.4a) as a unit. As a result, these genes cannot be separated by recombination, and the map distances between them, as determined by the phenotypic classes in the progeny of a *Del* / + individual, will be zero. In addition, the genetic distance between loci on either side of the deletion (such as between markers *B* and *F* in Fig. 13.4a) will be shorter than expected since fewer crossovers can occur between them.

Because homologous chromosomes pair in the polytene chromosomes of *Drosophila,* as they do at prophase I of meiosis in any diploid organism, deletion loops are also visible in the polytene chromosomes of a deletion heterozygote (Fig. 13.4b). The high resolution of the banding patterns in these unusual chromosomes makes it possible to pinpoint the region of the deletion in relation to missing bands. By contrast, deletion loops

are not present in human karyotypes because homologous metaphase chromosomes do not pair with each other during mitosis. Geneticists can still determine the extent of a deletion in these chromosomes by observing which bands are missing from the *Del* chromosome, as described later.

Deletions in Heterozygotes Can "Uncover" Genes

A deletion heterozygote is, in effect, a hemizygote for genes on the normal, nondeleted chromosome that are missing from the deleted chromosome. If the normal chromosome carries a mutant recessive allele of one of these genes, the individual will exhibit the mutant phenotype. This phenomenon is sometimes called **pseudodominance.** In *Drosophila,* for example, the *scarlet* (*st*) eye color mutation is recessive to wild type. However, an animal heterozygous for the *st* mutation and a deletion that removes the *scarlet* gene (*st* / *Del*) will have bright scarlet eyes, rather than wild-type, dark red eyes. In these circumstances, the deletion "uncovers" (that is, reveals) the phenotype of the recessive mutation (Fig. 13.5). Geneticists can look for recessive phenotypes to determine whether a deletion has removed a particular gene. If the phenotype of a recessive-

Figure 13.5 In deletion heterozygotes, pseudodominance shows that a deletion has removed a particular gene. A fly of genotype *st/Del* (where *Del* is a deletion of the region in which the *scarlet* [*st*] gene normally lies) displays the recessive scarlet eye color. The deletion has thus "uncovered" the scarlet mutation.

Figure 13.6 Using deletions to assign genes to bands on *Drosophila* polytene chromosomes. *Red bars* show the bands removed by various deletions; for example, *Df 258-45* eliminates bands 3B3–3C3. Complementation experiments determined whether these deletions uncovered the *white* (*w*), *roughest* (*rst*), or *facet* (*fa*) genes. For instance, *w/Df 258-45* females have white eyes, so the *w* gene is removed by this deletion. The *w* gene must lie within bands 3C2–3 (*green*) because that is the region common to the deletions that uncover *w*. Similarly, the *rst* gene must be in bands 3C5–6 (*yellow*), and the *fa* gene in band 3C7 (*blue*).

allele/ deletion heterozygote is mutant, the deletion has uncovered the mutated locus; the gene thus lies inside the region of deletion. If the trait determined by the gene is wild type in these heterozygotes, the deletion has not uncovered the recessive allele, and the gene must lie outside the deleted region. You can consider this experiment as a complementation test between the mutation and the deletion: the uncovering of a mutant recessive phenotype demonstrates a lack of complementation because neither chromosome can supply wild-type gene function.

Using Deletions to Locate Genes

Geneticists can use deletions that alter chromosomal banding patterns to map genes relative to specific regions of metaphase or polytene chromosomes. A deletion that results in the loss of one or more bands from a chromosome and also uncovers the recessive mutation of a particular gene places that gene within the missing chromosomal segment.

The greater the number of distinguishable bands in a chromosome, the greater the accuracy of gene localization by this strategy. For this reason, the polytene chromosomes of *Drosophila,* which provide more than a 100-fold increase over metaphase chromosomes in the resolution of bands and interbands, are a prized mapping resource. If researchers find that a small deletion removing only a few polytene chromosome bands uncovers a gene, or that several overlapping larger deletions affect the same gene, they can assign the gene to one or a small number of bands, often representing less than 100 kb of DNA. Figure 13.6 shows how geneticists used this strategy to assign three genes to regions containing only one or two polytene chromosome bands on the *Drosophila* X chromosome.

Geneticists can use deletions analyzed at even higher levels of resolution to help locate genes on cloned fragments of DNA. They must first determine whether a particular deletion uncovers a recessive allele of the gene of interest and then ascertain which DNA sequences are removed by the deletion. Several molecular techniques can show whether a particular DNA sequence is part of a deletion. Suppose you are trying to determine whether a small segment of the *Drosophila* X chromosome in the vicinity of the *white* gene has been deleted. You could use purified DNA fragments as probes for *in situ* hybridization to polytene chromosomes prepared from female flies heterozygous for various deletions in this region of their X chromosomes. If a probe hybridizes to a *Del* chromosome,

the deletion has not completely removed that particular fragment of DNA; lack of a hybridization signal on a *Del* chromosome, however, indicates that the fragment has been deleted (Fig. 13.7a).

Southern blot analysis of genomic DNAs from individuals with deleted chromosomes could supply similar information (Fig. 13.7b). You would hybridize various probles to blots containing genomic DNA from a deletion heterozygote, and compare the results with those obtained from normal DNA (Fig. 13.7c). In such a comparison, the deletion heterozygote would show a 50% decrease in band signal intensity if the region containing the probe were deleted. Moreover, if one of the breakpoints of a deletion occurs within a restriction fragment homologous to the probe, a genomic Southern blot could reveal this by generating bands of unexpected sizes. This combination of genetic and molecular information can help pinpoint the DNA corresponding to genes of interest.

Duplications Add Material to the Genome

Duplications increase the number of copies of a particular chromosomal region. In **tandem** duplications, repeats of a region lie adjacent to each other, either in the same order or in reverse order (Fig. 13.8a). In **nontandem** (or *dispersed*) duplications, the two or more copies of a region are not adjacent to each other and may lie far apart on the same chromosome or on different chromosomes. Duplications arise by chromosomal breakage and faulty repair, unequal crossing-over, or errors in DNA replication

(a) Characterizing deletions with *in situ* hybridization.

(b) Probing the location of a deletion breakpoint.

(c) Southern blot results

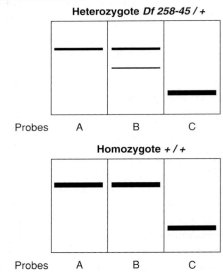

Figure 13.7 Using deletions to locate genes at the molecular level. (a) A labeled probe hybridizes to the wild-type chromosome but not to the *Df 258-45* deletion chromosome. The *Df 258-45* deletion chromosome thus lacks DNA homologous to the probe. **(b)** Probes A, B, and C, prepared from cloned genomic DNA, were hybridized to Southern blots of restriction enzyme (RE)-cut whole genomic DNA from either a *Df 258-45/+* heterozygote or a wild-type (*+/+*) control. **(c)** Probe A sequences are deleted, while probe C sequences are not. A novel band of lower intensity appears when *Df 258-45/+* DNA is hybridized with probe B, so the right breakpoint of *Df 258-45* is within the probe B sequence.

(Fig. 13.8b). In this book, we use *Dp* as the symbol for a chromosome carrying a duplication.

Most duplications have no obvious phenotypic consequences and can be detected only by cytological or molecular means. Sufficiently large duplications, for example, show up as repeated bands in metaphase or polytene chromosomes. During the prophase of meiosis I in heterozygotes for such duplications (*Dp/+*), the repeated bands form a **duplication loop**—a bulge

in the *Dp*-bearing chromosome that has no similar region with which to pair in the unduplicated normal homologous chromosome. Duplication loops can occur in several alternative configurations (Fig. 13.8c). Such loops also form in the polytene chromosomes of *Drosophila* duplication heterozygotes, where the pattern of the bands in the duplication loops is a repeat of that seen in the other copy of the same region elsewhere on the chromosome.

(a) Types of duplications

Tandem duplications

Normal chromosome   A B C D E F G

Same order   A B C B C D E F G

Reverse order   A B C C B D E F G

Nontandem (dispersed) duplications

Same order   A B C D E F B C G

Reverse order   A B C D E F C B G

(b) Chromosome breakage can produce duplications

X rays break one chromosome in two places

Nontandem duplication

X rays break homologous chromosome in one place

(c) Different kinds of duplication loops

Duplicated chromosome

Normal chromosome

Figure 13.8 Duplications: Structure, origin, and detection.
(a) In tandem duplications, the repeated regions lie adjacent to each other in the same or in reverse order. In nontandem duplications, the two copies of the same region are separated. **(b)** In one scenario for duplication formation, X rays break one chromosome twice and its homolog once. A fragment of the first chromosome can then insert elsewhere on its homolog, producing a nontandem duplication. **(c)** Duplication loops form when chromosomes pair in duplication heterozygotes (*Dp/+*). During prophase I, the duplication loop can assume different configurations that maximize the pairing of related regions. A single line represents two chromatids in this simplified diagram.

Duplications Can Affect Phenotype

Although duplications are much less likely to affect phenotype than are deletions of comparable size, some duplications do have phenotypic consequences for visible traits or for survival. Geneticists can use such phenotypes to identify individuals whose genomes contain the duplication. Duplications can produce a novel phenotype either by increasing the number of copies of a particular gene or set of genes, or by placing the genes bordering the duplication in a new chromosomal environment that alters their expression. These phenotypic consequences often arise even in duplication heterozygotes (*Dp/+*). For example, *Drosophila* heterozygous for a duplication including the *Notch*⁺ gene have abnormal wings that signal the three copies of *Notch*⁺ (Fig. 13.9a); we have already seen that *Del/+* flies with only one copy of the *Notch*⁺ gene have a different kind of wing abnormality (review Fig. 13.3 on p. 448). In another example from *Drosophila,* the locus known as *Triplolethal* (*Tpl*⁺) is lethal when present in one or three doses in an otherwise diploid individual (Fig. 13.9b). Thus heterozygotes for a *Tpl* deletion (*Del/+*) or for a *Tpl*⁺ duplication (*Dp/+*) do not survive. Heterozygotes carrying one homolog deleted for the locus and the other homolog duplicated for the locus (*Del/Dp*) are viable because they have two copies of *Tpl*⁺. Organisms are usually not so sensitive to additional copies of a gene; but just as for large deletions, imbalances for the many genes included in a very large duplication have additive deleterious effects that jeopardize survival. In humans, heterozygosity for duplications covering more than 5% of the haploid genome is most often lethal.

(a) Duplication heterozygosity can cause visible phenotypes.

Wild-type wing: two copies of *Notch*⁺ gene   Three copies of *Notch*⁺ *gene*   Aberrant wing veins

(b) For rare genes, survival requires exactly two copies.

Tpl⁺ / *Tpl*⁺	+ / +	Living fly (two copies *Tpl*⁺)
Del / +		Lethal (one copy *Tpl*⁺)
Dp / +		Lethal (three copies *Tpl*⁺)
Dp / *Del*		Living fly (two copies *Tpl*⁺)

Figure 13.9 The phenotypic consequences of duplications.
(a) Duplication heterozygotes (*Dp/+*) have three copies of genes contained in the duplication. Flies with three copies of the *Notch*⁺ gene have aberrant wing veins. This phenotype differs from that caused by only one copy of *Notch*⁺ (see Fig. 13.3). **(b)** In *Drosophila,* three copies or one copy of *Tpl*⁺ are lethal.

Unequal Crossing-Over Between Duplications Increases or Decreases Gene Copy Number

In individuals homozygous for a tandem duplication (*Dp/Dp*), homologs carrying the duplications occasionally pair out of register during meiosis. **Unequal crossing-over,** that is, recombination resulting from such out-of-register pairing, generates gametes containing increases to three and reciprocal decreases to one in the number of copies of the duplicated region. In *Drosophila,* tandem duplication of several polytene bands near the X chromosome centromere produces the Bar phenotype of kidney-shaped eyes. *Drosophila* females homozygous for the Bar eye duplication produce mostly Bar eye progeny. Some progeny, however, have wild-type eyes, while other progeny have double-Bar eyes that are even smaller than Bar eyes (Fig. 13.10). The genetic explanation is that flies with wild-type eyes carry X chromosomes containing only one copy of the region in question, flies with Bar eyes have X chromosomes containing two copies of the region, and flies with double-Bar eyes have X chromosomes carrying three copies. Unequal crossing-over in females homozygous for double-Bar chromosomes can yield progeny with even more extreme phenotypes associated with four or five copies of the duplicated region. Duplications in homozygotes thus allow for the expansion and contraction of the number of copies of a chromosomal region from one generation to the next.

How Duplications and Deletions Affect Phenotype and Evolution: A Summary

Both duplications and deletions alter the number of genes on a chromosome and as a result may affect the phenotypes of heterozygotes. Heterozygosity for deletions or duplications produces one or three copies of a gene in an otherwise diploid organism. These changes in gene dosage create an imbalance in gene products that can alter visible phenotypes; for a very few genes, the alterations in dosage are lethal to the organism. The deleterious effects of genetic imbalance are generally additive; such that heterozygotes for very large deletions or duplications of virtually any chromosomal region cannot survive. Both duplications and deletions can also alter phenotype by placing a gene in a new chromosomal location that modifies its expression. As a simple example, a deletion or duplication may relocate a gene close to a regulatory DNA sequence that will inappropriately suppress or enhance the transcription of the relocated gene.

Finally, deletions (by removing material from the genome) and duplications (by adding material to the genome) function as engines driving the evolution of the genome. Indeed, the duplication of chromosomal segments followed by the expansion or contraction of the number of copies of the duplicated regions has played a significant role in the evolution of present-day genomes. Families of tandemly repeated genes, such as those for hemoglobin and color perception, arose in this way (see the comprehensive examples in Chapters 7 and 9). We discuss the molecular basis of this evolutionary mechanism in more detail in Chapter 21.

Inversions Reorganize the DNA Sequence of a Chromosome

The half-circle rotation of a chromosomal region known as an **inversion (*In*)** can occur when radiation produces two double-strand breaks in a chromosome's DNA. The breaks release a middle fragment, which may turn 180° before religation to the flanking chromosomal regions, resulting in an inversion (Fig. 13.11a). Inversions may also result from rare crossovers between related DNA sequences present in two positions on the same chromosome in inverted orientation (Fig. 13.11b), or they may arise by the action of transposable genetic elements (discussed later). Inversions that include the centromere are **pericentric,** while inversions that exclude the centromere are **paracentric** (see Fig. 13.11a).

Most inversions do not result in an abnormal phenotype, because even though they alter the order of genes along the chromosome, they do not add or remove DNA and therefore do not change the identity or number of genes. Geneticists can detect some inversions that do not affect phenotype, especially those that cause cytologically visible changes in banding patterns, or those that suppress recombination in heterozygotes (as described later) and thereby change the expected results of linkage analysis. In natural populations, however, many inversions that do not affect phenotype go undetected.

If one end of an inversion lies within the DNA of a gene (Fig. 13.11c), a novel phenotype can occur. Inversion following an intragenic break separates the two parts of the gene, relocat-

Figure 13.10 Unequal crossing-over can increase or decrease copy number. Duplication of the X chromosome polytene region 16A causes Bar eyes. Unequal pairing and crossing-over during meiosis in females homozygous for this duplication produce chromosomes that have either one copy of region 16A (conferring normal eyes) or three copies of 16A (causing the more abnormal double-Bar eyes).

ing one part to a distant region of the chromosome, while leaving the other part at its original site. Such a split disrupts the gene's function. If that function is essential to viability, the inversion acts as a recessive lethal mutation, and homozygotes for the inversion will not survive.

Inversions can also produce unusual phenotypes by moving genes residing near the inversion breakpoints to chromosomal environments that alter their normal expression. For example, mutations in the *Antennapedia* gene of *Drosophila* that transform antennae into legs (review Fig. 7.27 on p. 223) are inversions that place the gene in a new regulatory environment, next

to sequences that cause it to be transcribed in tissues where it would normally remain unexpressed. Inversions that reposition genes normally found in a chromosome's euchromatin to a position near a region of heterochromatin can also produce an unusual phenotype; spreading of the heterochromatin may inactivate the gene in some cells, leading to position-effect variegation, as discussed in Chapter 12 (see particularly Fig. 12.14 on p. 431).

Heterozygosity for Inversions Reduces the Number of Recombinant Progeny

Individuals heterozygous for an inversion (*In / +*) are *inversion heterozygotes*. In such individuals, when the chromosome carrying the inversion pairs with its homolog at meiosis, formation of an **inversion loop** allows the tightest possible alignment of homologous regions. In an inversion loop, one chromosomal region rotates to conform to the similar region in the other homolog (Fig. 13.12). Crossing-over within an inversion loop produces aberrant recombinant chromatids whether the inversion is pericentric or paracentric.

If the inversion is pericentric and a single crossover occurs within the inversion loop, each recombinant chromatid will have a single centromere—the normal number—but will carry a duplication of one region and a deletion of a different region (Fig. 13.13a). Gametes carrying these recombinant chromatids will have an abnormal dosage of some genes. After fertilization, zygotes created by the union of these abnormal gametes with normal gametes are likely to die because of genetic imbalance.

If the inversion is paracentric and a single crossover occurs within the inversion loop, the recombinant chromatids will be unbalanced not only in gene dosage but also in centromere number (Fig. 13.13b). One crossover product will be an **acentric fragment** lacking a centromere; while the reciprocal crossover product will be a **dicentric chromatid** with two centromeres. Because the acentric fragment without a centromere cannot attach to the spindle apparatus during the first meiotic division, the cell cannot package it into either of the daughter nuclei; as a result, this chromosome is lost and will not be included in a gamete. By contrast, at anaphase of meiosis I, opposing spindle forces pull the dicentric chromatid toward both spindle poles at the same time with such strength that the dicentric chromatid breaks at random positions along the chromosome. These broken chromosome fragments are deleted for many of their genes. This loss of the acentric fragment and breakage of the dicentric chromatid results in genetically unbalanced gametes, which at fertilization will produce lethally unbalanced zygotes that cannot develop beyond the earliest stages of embryonic development. Consequently, no recombinant progeny resulting from a crossover in a paracentric inversion loop survive. Any surviving progeny are nonrecombinants.

In summary, whether an inversion is pericentric or paracentric, crossing-over within the inversion loop of an inversion heterozygote has the same effect: formation of recombinant gametes that after fertilization prevent the zygote from developing. Since only gametes containing chromosomes that did not recombine within the inversion loop can yield viable progeny,

(a) Chromosome breakage can produce inversions.

(b) Intrachromosomal recombination can also cause inversions.

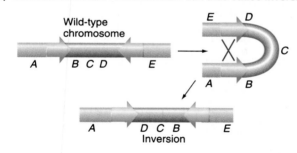

(c) Inversions can disrupt gene function.

Figure 13.11 Inversions: Origins, types, and phenotypic effects. (a) Inversions can arise when chromosome breakage produces a DNA segment that rotates 180° before it reattaches to the chromosome. When the rotated segment includes the centromere, the inversion is *pericentric;* when the rotated segment does not include the centromere, the inversion is *paracentric*. **(b)** If a chromosome has two copies of a sequence in reverse orientation, rare intrachromosomal recombination between the two copies can give rise to an inversion. **(c)** An inversion can affect phenotype if it disrupts a gene. Here, the inversion *In(1)y⁴* inactivates the *y* (*yellow*) gene by dividing it in two, thereby altering the fly's body color.

Figure 13.12   Inversion loops form in inversion heterozygotes. To maximize pairing during prophase of meiosis I in an inversion heterozygote (*In/+*), homologous regions form an inversion loop. (*Top*) Simplified diagram in which one line represents a pair of sister chromatids. (*Bottom*) Electron micrograph of an inversion loop during meiosis I in an *In/+* mouse.

Figure 13.13   Why inversion heterozygotes produce few if any recombinant progeny. Throughout this figure, each line represents one chromatid, and different shades of *green* indicate the two homologous chromosomes. **(a)** The chromatids formed by recombination within the inversion loop of a pericentric inversion heterozygote are genetically unbalanced. Note that the duplicated or deleted genes in these recombinant chromatids lie outside the inversion loop. **(b)** The chromatids formed by recombination within the inversion loop of a paracentric inversion heterozygote are not only genetically unbalanced but also contain two or no centromeres, instead of the normal one.

inversions act as **crossover suppressors.** This does not mean that crossovers do not occur within inversion loops, but simply that there are no recombinants among the viable progeny of an inversion heterozygote.

Geneticists use crossover suppression to create **balancer chromosomes,** which contain multiple, overlapping inversions (both pericentric and paracentric), as well as a marker mutation that produces a visible dominant phenotype (Fig. 13.14). The viable progeny of a *balancer/+* heterozygote will receive either the balancer or the chromosome of normal order (*+*), but they cannot inherit a recombinant chromosome containing parts of both. Researchers can distinguish these two types of viable progeny by the presence or absence of the dominant marker phenotype. Geneticists often generate balancer heterozygotes to ensure that a chromosome of normal order, along with any mutations of interest it may carry, is transmitted to the next genera-

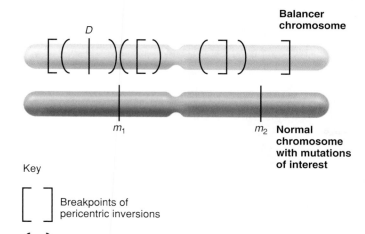

Key

Breakpoints of
pericentric inversions

Breakpoints of
paracentric inversions

Figure 13.14 Balancer chromosomes are useful tools for genetic analysis. Balancer chromosomes carry both a dominant marker D as well as inversions (*brackets*) that prevent the balancer chromosome from recombining with an experimental chromosome carrying mutations of interest (m_1 and m_2). A parent heterozygous for the balancer and experimental chromosomes will transmit either the balancer or the experimental chromosome, but not a recombinant chromosome, to its surviving progeny.

tion unchanged by recombination. To help create genetic stocks, the marker in most balancer chromosomes not only causes a dominant visible phenotype, but it also acts as a recessive lethal mutation that prevents the survival of balancer chromosome homozygotes. The *Drosophila* portrait (Reference D) discusses this and other significant uses of balancer chromosomes in genetic analysis on pp. 815–816.

Translocations Attach Part of One Chromosome to Another Chromosome

Translocations are large-scale mutations in which part of one chromosome becomes attached to a nonhomologous chromosome, or in which parts of two different chromosomes trade places. This second type of translocation is known as a **reciprocal translocation** (Fig. 13.15a). It results when two breaks, one in each of two chromosomes, yield DNA fragments that do not religate to their chromosome of origin; rather, they switch places and become attached to the other chromosome. Depending on the positions of the breaks and the sizes of the exchanged fragments, the translocated chromosomes may be so different from the original chromosomes that the translocation is visible in a cytological examination (Fig. 13.15b).

Robertsonian translocations are an important type of cytologically visible reciprocal translocations that arise from breaks at or near the centromeres of two acrocentric chromosomes (Fig. 13.16). The reciprocal exchange of broken parts generates one large metacentric chromosome and one very small chromosome containing few, if any, genes. This tiny chromosome may subsequently be lost from the organism. Robertsonian

(a) **Two chromosome breaks can produce a reciprocal translocation.**

(b) **Chromosome painting reveals a reciprocal translocation.**

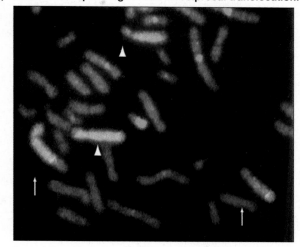

Figure 13.15 Reciprocal translocations are exchanges between nonhomologous chromosomes. (a) In a reciprocal translocation, the region gained by one chromosome is the region lost by the other chromosome. **(b)** Karyotype of a human genome containing a translocation. Researchers used a "chromosome-painting" hybridization technique that renders chromosomes in different colors. The two translocated chromosomes are stained both red *and* green (*arrows*). Two normal, non-translocated chromosomes are stained entirely red *or* entirely green (*arrowheads*), indicating that this person is heterozygous for the translocation.

A Robertsonian translocation

Figure 13.16 Robertsonian translocations can reshape genomes. In a Robertsonian translocation, reciprocal exchanges between two acrocentric chromosomes generate a large metacentric chromosome and a very small chromosome. The latter may carry so few genes that it can be lost without ill effect.

(a) **Leukemia patients have too many white blood cells.**

Normal

Leukemic

(b) **The genetic basis for chronic myelogenous leukemia.**

Figure 13.17 How a reciprocal translocation helps cause one kind of leukemia. (a) Uncontrolled divisions of large, dark-staining white blood cells in a leukemia patient (*right*) produce a higher ratio of white to red blood cells than that in a normal individual (*left*). **(b)** A reciprocal translocation between chromosomes 9 and 22 contributes to chronic myelogenous leukemia. This rearrangement makes an abnormal hybrid gene composed of part of the *c-abl* gene and part of the *bcr* gene. The hybrid gene encodes an abnormal fused protein that disrupts controls on cell division. *Black arrows* indicate PCR primers that will generate a PCR product only from DNA containing the hybrid gene.

translocations are named after W. R. B. Robertson, who in 1911 was the first to suggest that during evolution, metacentric chromosomes may arise from the fusion of two acrocentrics.

Most individuals bearing reciprocal translocations are phenotypically normal because they have neither lost nor gained genetic material. As with inversions, however, if one of the translocation breakpoints occurs within a gene, that gene's function may change or be destroyed. Or if the translocation places a gene normally found in the euchromatin of one chromosome near the heterochromatin of the other chromosome, normal expression of the gene may cease in some cells, giving rise to position-effect variegation (see p. 430 in Chapter 12).

Several kinds of cancer are associated with translocations in somatic cells. In normal cells, genes known as *protooncogenes* help control cell division. Translocations that relocate these genes can turn them into tumor-producing *oncogenes* whose protein products have an altered structure or level of expression that leads to runaway cell division. For example, in almost all patients with

chronic myelogenous leukemia, a type of cancer caused by overproduction of certain white blood cells, the leukemic cells have a reciprocal translocation between chromosomes 9 and 22 (Fig. 13.17). The breakpoint in chromosome 9 occurs within an intron of a protooncogene called *c-abl;* the breakpoint in chromosome 22 occurs within an intron of the *bcr* gene. After the translocation, parts of the two genes are adjacent to one another. During transcription, the RNA-producing machinery runs these two genes together, creating a long primary transcript. After splicing, the mRNA is translated into a fused protein in which 25 amino acids at the N terminus of the *c-abl*-determined protein are replaced by about 600 amino acids from the *bcr*-determined protein. The activity of this fused protein releases the normal controls on cell division, leading to leukemia. (See the Fast Forward box in this chapter for another example of a translocation-induced cancer called Burkitt's lymphoma.)

Medical practitioners can exploit the rearrangement of DNA sequences that accompany cancer-related translocations for diagnostic and therapeutic purposes. To confirm a diagnosis of myelogenous leukemia, for example, they first obtain a blood sample from the patient, and they then use a pair of PCR primers derived from opposite sides of the breakpoint—one synthesized from the appropriate part of chromosome 22, the other from chromosome 9—to carry out a PCR on DNA from the blood cells. The PCR will amplify the region between the primers only if the DNA sample contains the translocation (Fig. 13.17b). To monitor the effects of chemotherapy, they again obtain a blood sample and extract genomic DNA from the white blood cells. If the sample contains even a few malignant cells, a PCR test with the same two primers will amplify the DNA translocation from those cells, indicating the need for more therapy. PCR thus becomes a sensitive assay for this type of leukemic cell.

Pharmaceutical researchers have recently exploited their understanding of the molecular nature of the translocation underlying chronic myelogenous leukemia to achieve a stunning breakthrough in the treatment of this cancer. The protein encoded by *c-abl* is a *protein tyrosine kinase,* an enzyme that adds phosphate groups to tyrosine amino acids on other proteins. This enzyme is an essential part of the set of signals that dictate cell growth and division. Normal cells closely regulate the activity of the c-abl protein, blocking its function most of the time but activating it in response to stimulation by growth factors in the environment. By contrast, the fused protein encoded by *bcr/c-abl* in cells carrying the translocation is not amenable to regulation. It is always active, even in the absence of growth factor, and this leads to runaway cell division. Pharmaceutical companies have developed a drug called Gleevec® that specifically inhibits the enzymatic activity of the protein tyrosine kinase encoded by *bcr/c-abl.* In clinical trials, 98 percent of participants experienced a complete disappearance of leukemic blood cells and the return of normal white cells. This drug is now the standard treatment for chronic myelogenous leukemia and is a model for new types of cancer treatments that home in on cancer cells without hurting healthy ones.

Heterozygosity for Translocations Diminishes Fertility and Results in Pseudolinkage

Translocations, like inversions, produce no significant genetic consequences in homozygotes if the breakpoints do not interfere with gene function. During meiosis in a translocation homozygote, chromosomes segregate normally according to Mendelian principles (Fig. 13.18a). Even though the genes have been rearranged, both haploid sets of chromosomes in the individual have the same rearrangement. As a result, all chromosomes will find a single partner with which to pair at meiosis, and there will be no deleterious consequences for the progeny.

In translocation heterozygotes, however, certain patterns of chromosome segregation during meiosis produce genetically unbalanced gametes that at fertilization become deleterious to the zygote. In a translocation heterozygote, the two haploid sets of chromosomes do not carry the same arrangement of genetic information. As a result, during prophase of the first meiotic division, the translocated chromosomes and their normal homologs

assume a crosslike configuration in which four chromosomes, rather than the normal two, pair to achieve a maximum of synapsis between similar regions (Fig. 13.18b). To keep track of the four chromosomes participating in this crosslike structure, we denote the chromosomes carrying translocated material with a *T* and the chromosomes with a normal order of genes with an *N*. Chromosomes *N1* and *T1* have homologous centromeres found in wild type on chromosome 1; *N2* and *T2* have centromeres found in wild type on chromosome 2.

During anaphase of meiosis I, the mechanisms that attach the spindle to the chromosomes in this crosslike configuration still usually ensure the disjunction of homologous centromeres, bringing homologous chromosomes to opposite spindle poles (that is, *T1* and *N1* go to opposite poles, as do *T2* and *N2*). Depending on the arrangement of the four chromosomes on the metaphase plate, this normal disjunction of homologs produces one of two equally likely patterns of segregation (Fig. 13.18c). In the **alternate segregation pattern,** the two translocation chromosomes (*T1* and *T2*) go to one pole, while the two normal chromosomes (*N1* and *N2*) move to the opposite pole. Both kinds of gametes resulting from this segregation (*T1, T2* and *N1, N2*) carry the correct haploid number of genes; and the zygotes formed by union of these gametes with a normal gamete will be viable. By contrast, in the **adjacent-1 segregation pattern,** homologous centromeres disjoin so that *T1* and *N2* go to one pole, while the *N1* and *T2* go to the opposite pole. As a result, each gamete contains a large duplication (of the region found in both the normal and the translocated chromosome in that gamete) and a correspondingly large deletion (of the region found in neither of the chromosomes in that gamete), which make them genetically unbalanced. Zygotes formed by union of these gametes with a normal gamete are usually not viable.

Because of the unusual cruciform pairing configuration in translocation heterozygotes, nondisjunction of homologous centromeres occurs at a measurable but low rate. This nondisjunction produces an **adjacent-2 segregation pattern** in which the homologous centromeres *N1* and *T1* go to the same spindle pole while the homologous centromeres *N2* and *T2* go to the other spindle pole (Fig. 13.18c). The resulting genetic imbalances are lethal after fertilization to the zygotes containing them.

Thus, of all the gametes generated by translocation heterozygotes, only those arising from alternate segregation, which is slightly less than half the total, can produce viable progeny when crossed with individuals who do not carry the translocation. As a result, the fertility of most translocation heterozygotes, that is, their capacity for generating viable offspring, is diminished by at least 50%. This condition is known as **semisterility.** Corn plants illustrate the correlation between translocation heterozygosity and semisterility: The demise of genetically unbalanced ovules produces gaps in the ear where kernels would normally appear (Fig. 13.18d); in addition, genetically unbalanced pollen grains are abnormally small (not shown).

The semisterility of translocation heterozygotes undermines the potential of genes on the two translocated chromosomes to assort independently. Mendel's second law requires that all gametes

(a) Segregation in a translocation homozygote

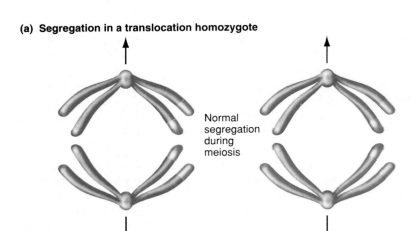

Normal
segregation
during
meiosis

(b) Chromosome pairing in a translocation heterozygote

(c) Segregation in a translocation heterozygote

Segregation pattern	Alternate		Adjacent - 1		Adjacent - 2 (less frequent)	
	Balanced N1 + N2	Balanced T1 + T2	Unbalanced T1 + N2	Unbalanced N1 + T2	Unbalanced N1 + T1	Unbalanced N2 + T2
Gametes	a b c d e f	p q r s t u	A B C D S T U	P Q R E F	a b c d e f	A B C D S T U	p q r s t u	P Q R E F
Type of progeny when mated with normal *abcdefpqrstu* homozygote	*abcdef pqrstu*	*ABCDEF PQRSTU*	None surviving	None surviving	None surviving	None surviving

(d) Semisterility in corn

Figure 13.18   The meiotic segregation of chromosomes that have sustained reciprocal translocations. In all parts of this figure, each bar or line represents one chromatid. **(a)** In a translocation homozygote (*T/T*), chromosomes segregate normally during meiosis I. **(b)** In a translocation heterozygote (*T/+*), the four relevant chromosomes assume a cruciform configuration to maximize pairing. The alleles of genes on chromosomes in the original order (*N1* and *N2*) are shown in lowercase; the alleles of these genes on the translocated chromosomes (*T1* and *T2*) are in uppercase letters. **(c)** There are three possible segregation patterns in a translocation heterozygote from the cruciform configuration shown in part (b). Only the alternate segregation pattern gives rise to balanced gametes. As a result, genes that were originally on two different chromosomes will behave as if they are genetically linked. **(d)** This semisterile ear of corn comes from a plant heterozygous for a reciprocal translocation. It has fewer kernels than normal because unbalanced ovules are aborted.

resulting from both possible metaphase alignments of two chromosomal pairs produce viable progeny. But as we have seen, in a translocation heterozygote, only the alternate segregation pattern yields viable progeny in outcrosses; the equally likely adjacent-1 pattern and the rare adjacent-2 pattern do not. Because of this, genes near the translocation breakpoints on the nonhomologous chromosomes participating in a reciprocal translocation exhibit **pseudolinkage:** They behave as if they are linked.

Figure 13.18c illustrates why pseudolinkage occurs in a translocation heterozygote. In the figure, lowercase *a b c d e f* represent the alleles of genes present on normal chromosome 1

(*N1*) and *p q r s t u* are the alleles of genes on a nonhomologous normal chromosome 2 (*N2*). The alleles of these genes on the translocated chromosomes *T1* and *T2* are in uppercase. In the absence of recombination, Mendel's law of independent assortment would predict that genes on two different chromosomes will appear in four types of gametes in equal frequencies; for example, *a p, A P, a P,* and *A p.* But alternate segregation, the only pattern that can give rise to viable progeny, produces only *a p* and *A P* gametes. Thus, in translocation heterozygotes such as these, the genes on the two nonhomologous chromosomes act as if they are linked to each other.

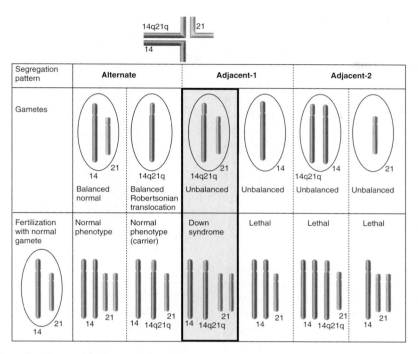

Figure 13.19   How translocation Down syndrome arises. In heterozygotes for a translocation involving chromosome 21, such as 14q21q (a Robertsonian translocation between chromosomes 21 and 14), adjacent-1 segregation can produce gametes with two copies of part of chromosome 21. If such a gamete unites with a normal gamete, the resulting zygote will have three copies of part of chromosome 21. Depending on which region of chromosome 21 is present in three copies, this tripling may cause Down syndrome.

Translocations Can Sometimes Help Map Important Genes

In humans, approximately 1 of every 500 individuals is heterozygous for some kind of translocation. While most such people are phenotypically normal, their fertility is diminished because many of the zygotes they produce abort spontaneously. As we have seen, this semisterility results from genetic imbalances associated with gametes formed by adjacent-1 or adjacent-2 segregation patterns. But such genetic imbalances are not inevitably lethal to the zygotes. If the duplicated or deleted regions are very small, the imbalanced gametes generated by these modes of segregation may produce children.

An important example of this phenomenon is seen among individuals heterozygous for certain reciprocal translocations involving chromosome 21. These people are phenotypically normal but produce some gametes from the adjacent-1 segregation pattern that have two copies of a part of chromosome 21 near the tip of its long arm (Fig. 13.19). At fertilization, if a gamete with the duplication unites with a normal gamete, the resulting child will have three copies of this region of chromosome 21. A few individuals affected by Down syndrome have, in this way, inherited a third copy of only a small part of chromosome 21. These individuals with **translocation Down syndrome** provide evidence that the entirety of chromosome 21 need not be present in three copies to generate the phenotype.

Geneticists are now mapping the chromosome 21 regions duplicated in translocation Down syndrome patients to find the one or more genes responsible for the syndrome. Although chromosome 21 is the smallest human autosome, it nevertheless contains an estimated 500 genes, most of them in the 43 million base pairs of its long arm. The mapping of genes relative to the breakpoints of one or more such translocations considerably simplifies the task of identifying those genes that in triplicate produce the symptoms of Down syndrome. One way to locate which parts of chromosome 21 are responsible for Down syndrome is to obtain cloned chromosome 21 sequences from the Human Genome Project and then use these clones as FISH (fluorescence *in situ* hybridization, described on pp. 329–330 in Chapter 10) probes for the genome of the translocation Down syndrome patient. If the probe lights up the translocation chromosome as well as the two normal copies of chromosome 21 (Fig. 13.19), it identifies a region of the genome that is of potential importance to the syndrome.

Translocations and Inversions Have Several Effects in Common: A Summary

Reciprocal translocations, like inversions, bring together pieces of DNA that were not adjacent before the rearrangement, without altering the amount of DNA in the genome. Both reciprocal translocations and inversions may affect gene function and as a result phenotype if one of the breakpoints of the rearrangement is within a gene, or if the rearrangement places a gene in a chromosomal environment that modifies its expression. Homozygosity or heterozygosity for translocations and inversions that do not affect gene function is usually without obvious phenotypic effect. However, heterozygosity for both types of rearrangements can produce genetically imbalanced gametes that are deleterious to the zygotes they help create. Inversion heterozygotes produce genetically imbalanced gametes from crossing-over within an inversion loop; this results in crossover suppression. Translocation heterozygotes produce genetically imbalanced gametes from two of three possible meiotic segregation patterns; this leads to semi-sterility and pseudolinkage. Finally, translocations and inversions, by reducing production of viable progeny from heterozygotes, can be catalysts of speciation, as we explain later in this chapter's comprehensive example.

Transposable Genetic Elements Move from Place to Place in the Genome

Large deletions and duplications, as well as inversions and translocations, are major chromosomal reorganizations visible at the relatively low resolution of a karyotype. Small deletions and duplications are lesser chromosomal reorganizations that reshape genomes without any visible effect on karyotype. Another type of cytologically invisible sequence rearrangement with a significant genomic impact is **transposition:** the movement of small segments of DNA—entities known as **transposable elements (TEs)**—from one position in the genome to another.

Marcus Rhoades in the 1930s and Barbara McClintock in the 1950s inferred the existence of TEs from intricate genetic studies of corn. At first, the scientific community did not appreciate the importance of their work because their findings did not support the conclusion from classical recombination mapping that genes are located at fixed positions on chromosomes. Once the cloning of TEs made it possible to study them in detail, geneticists not only acknowledged their existence, but they discovered TEs in the genomes of virtually all organisms, from bacteria to humans. In 1983 Barbara McClintock received the Nobel Prize for her insightful studies on movable genetic elements (Fig. 13.20).

Copia is a transposable element in *Drosophila.* If you examined the polytene chromosomes from two strains of flies isolated from different geographic locations, you would find in general that the chromosomes appear identical. A probe derived from the *white* gene for eye color, for example, would hybridize to a single site near the tip of the X chromosome in both strains (review Fig. 12.15 on p. 433). However, a probe including the *copia* TE would hybridize to 30–50 sites scattered throughout the genome, and the overall positions of *in situ* hybridization would not be the same in the two strains. Some sites would be identical in the two polytene sets, but others would be different (Fig. 13.21). These observations suggest that since the time the strains were separated geographically, the *copia* sequences have moved around (transposed) in different ways in the two genomes even though most genes have remained in a fixed position.

Any segment of DNA that evolves the ability to move from place to place within a genome is by definition a transposable element, regardless of its origin or function. TEs need not be se-

Figure 13.20 Barbara McClintock: Discoverer of transposable elements.

Figure 13.21 Transposable elements (TEs) can move to many locations in a genome. A probe for the *copia* TE hybridizes to multiple sites (*black bands* superimposed over the *blue* chromosomes) that differ in two different fly strains.

quences that do something for the organism; indeed, many scientists regard them primarily as "selfish" parasitic entities carrying only information that allows their self-perpetuation. Some TEs, however, appear to have evolved functions that help their host. In one interesting example, TEs maintain the length of *Drosophila* chromosomes. *Drosophila* telomeres, in contrast to those of most organisms, do not contain TTAGGG repeats that are extendable by the telomerase enzyme (see pp. 425–426 in Chapter 12). Certain TEs, however, combat the shortening of chromosome ends that accompanies every cycle of replication by jumping with high frequency into DNA very near chromosome ends. As a result, chromosome size stays relatively constant.

Most transposable elements in nature range from 50 bp to approximately 10,000 bp (10 kb) in length. A particular TE can be present in a genome anywhere from one to hundreds of thousands of times. *Drosophila melanogaster,* for example, harbors approximately 80 different TEs, each an average of 5 kb in length, and each present an average of 50 times. These TEs con-

(a) *Alu* **SINEs in the human genome.**

0.28 kb *Alu* units: ~300,000 found
dispersed throughout human genome
at ~10 kb intervals

kb 0   10   20   30   40   50

(b) TEs cause mottling in corn.

Figure 13.22 TEs in human and corn genomes. (a) The human genome carries about 300,000 copies of the 0.28 kb *Alu* retroposon, the major human SINE; these are spaced, on average, about 10 kb apart. The longer LINE elements are, on average, spaced further apart (*not shown*). **(b)** Movements of a transposon mottles corn kernels when the transposon jumps into or out of genes that influence pigmentation.

stitute $80 \times 50 \times 5 = 20,000$ kb, or roughly 12.5% of the 160,000 kb *Drosophila* genome. Mammals carry two major classes of TEs: **LINEs,** or long interspersed elements; and **SINEs,** or short interspersed elements. The human genome contains approximately 20,000 copies of the main human LINE—*L1*—which is up to 6.4 kb in length. The human genome also carries 300,000 copies of the main human SINE—*Alu*—which is 0.28 kb in length (Fig. 13.22a). These two TEs thus constitute roughly 7% of the 3,000,000 kb human genome. Because some TEs exist in only one or a few closely related species, it is probable that some elements arise and then disappear rather frequently over evolutionary time. Chapter 21 describes the evolutionary origins of LINEs and SINEs.

Classification of TEs on the basis of how they move around the genome distinguishes two groups. **Retroposons** transpose via reverse transcription of an RNA intermediate. The *Drosophila copia* elements and the human SINEs and LINEs just described are retroposons. **Transposons** move their DNA directly without the requirement of an RNA intermediate. The genetic elements in corn responsible for mottling the kernels are transposons (Fig. 13.22b). Some biologists use the term "transposon" in the broader sense to refer to all TEs. In this book, we reserve it for the direct-movement class of genetic elements, and we use "transposable elements (TEs)" to indicate all DNA segments that move about in the genome, regardless of the mechanism.

Retroposons: Transcription Generates an RNA That Encodes a Reverse-Transcriptase-Like Enzyme

This enzyme with reverse-transcriptase activity, like the reverse transcriptase made by the AIDS-causing virus HIV described in the Genetics and Society box on pp. 252–253 of Chapter 8, can copy RNA into a single strand of cDNA and then use that single DNA strand as a template for producing double-stranded cDNA. Many retroposons also encode polypeptides other than reverse transcriptase.

Some retroposons have a poly-A tail at the 3' end of the RNA-like DNA strand, a configuration reminiscent of mRNA molecules (Fig. 13.23a). Other retroposons end in *long terminal repeats* (*LTRs*), nucleotide sequences repeated in the same orientation at both ends of the element (Fig. 13.23a). The structure of this second type of retroposon is similar to the integrated DNA copies of RNA tumor viruses (known as retroviruses), suggesting that retroviruses evolved from this kind of retroposon or vice versa. In support of this notion, researchers sometimes find retroposon transcripts enclosed in viruslike particles.

The structural parallels between retroposons, mRNAs, and retroviruses, as well as the fact that retroposons encode a reverse-transcriptase-like enzyme, prompted investigators to ask whether retroposons move around the genome via an RNA intermediate. Experiments in yeast helped confirm that they do. In one study, a copy of the *Ty1* retroposon found on a yeast plasmid contained an intron in one of its genes; after transposition into the yeast chromosome, however, the intron was not there (Fig. 13.23b). Since removal of introns occurs only during mRNA processing, researchers concluded that the *Ty1* retroposon passes through an RNA intermediate during transposition.

The mechanisms by which various retroposons move around the genome resemble each other in general outline but differ in detail. Figure 13.23c outlines what is known of the process for the better understood LTR-containing retroposons. As the figure illustrates, one outcome of transposition via an RNA intermediate is that the original copy of the retroposon remains in place while the new copy inserts in another location. With this mode of transmission, the number of copies can increase rapidly with time. Human LINEs and SINEs, for example, occur in tens of thousands or even hundreds of thousands of copies within the genome. Other retroposons, however, such as the *copia* elements found in *Drosophila,* do not proliferate so profusely and exist in much more moderate copy numbers of 30–50. Currently unknown mechanisms may account for these differences by regulating the rate of retroposon transcription or by limiting the number of copies through selection at the level of the whole organism.

Transposons Encode Transposase Enzymes That Catalyze the Events of Transposition

A hallmark of transposons—TEs whose movement does not involve an RNA intermediate—is that their ends are inverted repeats of each other, that is, a sequence of base pairs at one end is present in mirror image at the other end (Fig. 13.24a). The inverted repeat is usually 10–200 bp long.

Figure 13.23 Retroposons: Structure and movement. (a) Some retroposons have a poly-A tail at the 3′ end of the RNA-like DNA strand (*top*); others are flanked on both sides by long terminal repeats (LTRs; *bottom*). **(b)** Researchers constructed a plasmid bearing a *Ty1* retroposon that contained an intron. When this plasmid was transformed into yeast cells, researchers could isolate new insertions of *Ty1* into yeast genomic DNA. The newly inserted *Ty1* did not have the intron, which implies that transposition involves splicing of a primary transcript to form an intronless mRNA. **(c)** The reverse-transcriptase-like enzyme synthesizes double-stranded retroposon cDNA in a series of steps. Insertion of this double-stranded cDNA into a new genomic location (*blue*) involves a staggered cleavage of the target site that leaves "sticky ends"; polymerization to fill in the sticky ends produces two copies of the 5 bp target site.

DNA between the transposon's inverted repeats commonly contains a gene encoding a transposase, a protein that catalyzes transposition. The exact sequence of the steps resulting in transposition is not yet known, but as Fig. 13.24a illustrates, the process includes excision of the transposon from its original genomic position and integration into a new location. The double-stranded breaks at the transposon's excision and integration sites are repaired in different ways in different cases. Figure 13.24b shows two of the possibilities so far observed. In *Drosophila,* after excision of a transposon known as a *P element,* DNA exonucleases first widen the resulting gap and then

repair it using either a sister chromatid or a homologous chromosome as a template. If the template contains the *P* element and DNA replication is completely accurate, repair will restore a *P* element to the position from which it was excised; this will make it appear as if the *P* element remained at its original location during transposition (Fig. 13.24b, *left*). If the template does not contain a *P* element, the transposon will be lost from the original site after transposition (Fig. 13.24b, *right*).

Some strains of *D. melanogaster* are called "P strains" because they harbor many copies of the *P* element; "M strains" of the same species do not carry the *P* element at all. Virtually all

(a) Transposon structure

Transposase enzyme recognizes inverted
repeats and excises the transposon.

Genomic DNA
at original
insertion site

5'
GGTGTACAAGT
3'
CCACATGTTCA
Transposase gene
ACTTGTACACC
TGAACATGTGG
3'
5'

Inverted repeats (IR)

Transposon

(b) How *P* element transposons move

Transposition of *P* element to new location

5' TTACGAAT ... TTACGAAT 3'
3' AATGCTTA ... AATGCTTA 5'

P element in original
genomic position

P element
excised

Excision of *P* element leaves
a gap at its original location

P element

5' TCAATGAT ... TCAATGAT 3'
3' AGTTACTA ... AGTTACTA 5'

Transposase

Exonucleases widen the gap

Repair of gap using a sister chromatid or homologous chromosome containing a *P* element

Repair of gap using a homologous
chromosome lacking a *P* element

Transposon remains in original position

Transposon no longer
at original position

Figure 13.24 Transposons: Structure and movement. (a) Most transposons contain inverted repeats at their ends and encode a transposase enzyme that recognizes these inverted repeats. The transposase cuts at the borders between the transposon and adjacent genomic DNA, and it also helps the excised transposon integrate at a new site. **(b)** Transposase-catalyzed integration of *P* elements creates a duplication of 8 bp present at the new target site. A gap remains when transposons are excised from their original position. After exonucleases widen the gap, cells repair the gap using related DNA sequences as templates. Depending on whether the template contains or lacks a *P* element, the transposon will appear to remain or to be excised from its original location during transposition.

commonly used laboratory flies are M strains, while many flies isolated from natural populations since 1950 are P strains. Since Thomas Hunt Morgan and coworkers in the early part of the twentieth century isolated the flies that have proliferated into most current laboratory strains, these observations suggest that P elements did not enter *D. melanogaster* genomes until around 1950. The prevalence of P elements in many contemporary natural populations attests to the rapidity with which transposable elements can spread once they enter a species' genome.

Interestingly, the mating of male flies from P strains with females from M strains causes a phenomenon called **hybrid dysgenesis**. This creates a series of defects including sterility of offspring, mutation, and chromosome breakage. One of the more interesting effects of hybrid dysgenesis is to promote the movement of P elements to new positions in the genome. Because elevated levels of transposition can foster many kinds of genetic changes (see below), some geneticists speculate that hybrid dysgenesis-like events involving various transposons in different species had a strong impact on evolution. The *Drosophila* portrait in Reference D on pp. 816–818 provides more information on the molecular mechanisms underlying hybrid dysgenesis and the ways in which fly geneticists use this phenomenon to introduce new genes into *Drosophila*.

Genomes Often Contain Defective Copies of Transposable Elements

Many copies of TEs sustain deletions either as a result of the transposition process itself (for example, incomplete reverse transcription of a retroposon RNA) or as a result of events following transposition (for example, faulty repair of a site from which a P element was earlier excised). If a deletion removes the promoter needed for transcription of a retroposon, that copy of the element cannot generate the RNA intermediate for future movements. If the deletion removes one of the inverted repeats at one end of a transposon, transposase will be unable to catalyze transposition of that element. Such deletions create defective TEs unable to transpose again. Most SINEs and LINEs in the human genome are defective in this way. Other types of deletions create defective elements that are unable to move on their own, but they can move if nondefective copies of the element elsewhere in the genome supply the deleted function. For example, a deletion inactivating the reverse transcriptase gene in a retroposon or the transposase gene in a transposon would "ground" that copy of the element at one genomic location if it is the only source of the essential enzyme in the genome. If reverse transcriptase or transposase were provided by other copies of the same element in the genome, however, the defective copy could move. Defective TEs that require the activity of nondeleted copies of the same TE for movement are called **nonautonomous elements;** the nondeleted copies that can move by themselves are **autonomous elements.**

Transposable Elements Can Generate Mutations in Adjacent Genes

Insertion of a TE near or within a gene can affect gene expression and alter phenotype. We now know that the wrinkled pea

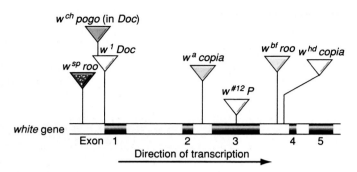

Figure 13.25 TEs can cause mutations on insertion into a gene. Many spontaneous mutations in the *white* gene of *Drosophila* arise from insertions of TEs such as *copia, roo, pogo,* or *Doc.* The consequences for the eye color phenotype (indicated by the color in the *triangles*) depend on the element involved and where in the *white* gene it inserts.

mutation first studied by Mendel resulted from insertion of a TE into the gene for starch-branching enzyme I. In *Drosophila,* a large percentage of spontaneous mutations, including the w^1 mutation discovered by T. H. Morgan in 1910, are caused by insertion of TEs (Fig. 13.25). Surprisingly, in light of the large numbers of LINEs and SINEs in human genomes, only a handful of mutant human phenotypes are known to result from insertion of TEs. Among these is a B-type hemophilia caused by *Alu* insertion into a gene encoding clotting factor IX; recall that *Alu* is the main human SINE.

A TE's effect on a gene depends on what the element is and where within or near the gene it inserts (Fig. 13.25). If an element lands within a protein-coding exon, the additional DNA may shift the reading frame or supply an in-frame stop codon that truncates the polypeptide. If the element falls in an intron, it could diminish the efficiency of splicing. Some of these inefficient splicing events might completely remove the element from the gene's primary transcript; this would still allow some—but less than normal—synthesis of functional polypeptide. TEs that land within exons or introns may also provide a transcription stop signal that prevents transcription of gene sequences downstream of the insertion site. Finally, insertions into regions that regulate transcription, such as promoters, can influence the amount of gene product made in particular tissues at particular times during development. Some transposons insert preferentially into the upstream regulatory regions of genes, and some even prefer specific types of genes, such as tRNA genes.

Transposable Genetic Elements Can Also Generate Chromosomal Rearrangements

Retroposons and transposons can trigger spontaneous chromosomal rearrangements other than transpositions in several ways. Sometimes, deletion or duplication of chromosomal material adjacent to the transposon occurs as a mistake during the transposition event itself. In another mechanism, if two copies of the same TE occupy nearby but not identical sites in homologous chromosomes, the two copies of the TE in heterozygotes carry-

(a) Unequal crossing-over between TEs.

Transposable element

Duplication

Deletion

(b) Two transposons can form a large, composite transposon.

IR w^+ IR

IR IR

Composite transposon can move to new location.

Figure 13.26 How TEs generate chromosomal rearrangements and relocate genes. (a) If a TE (*pink*) is found in slightly different locations on homologous chromosomes (here on opposite sides of segment *B*), unequal crossing-over between the TEs will produce reciprocal deletions and duplications of the intervening region. **(b)** If two copies of the same transposon are nearby on the same chromosome, transposase can recognize the outermost inverted repeats (IRs), creating a composite transposon that allows intervening genes such as *w⁺* (*red*) to jump to new locations.

ing both types of homolog may pair with each other and cross over (Fig. 13.26a). The recombination resulting from this unequal crossover would produce one chromosome deleted for the region between the two TEs and a reciprocal homolog with a tandem duplication of the same region. The duplication associated with the *Bar* mutation in *Drosophila* (review Fig. 13.10 on p. 452) probably arose in this way.

Transposition Can Relocate Genes

When two copies of a transposon occur in nearby but not identical locations on the same chromosome, the inverted repeats of the transposons are positioned such that an inverted version of the sequence at the 5' end of the copy on the left will exist at the 3' end of the copy to its right (Fig. 13.26b). If transposase acts on this pair of inverted repeats during transposition, it allows the entire region between them to move as one giant transposon, mobilizing and relocating any genes the region contains. Some composite transposons, such as that pictured in the figure at the beginning of this chapter on p. 441, carry as much as 400 kb of DNA. In prokaryotes, the capacity of two TEs to relocate the intervening genes helps mediate the transfer of drug resistance between different strains or species of bacteria, as will be discussed in Chapter 14.

Summary of Transposition

Although pieces of DNA that can move from place to place in the genome might enhance or diminish a genome's ability to respond to a changing environment, it is usually helpful to think of TEs as segments of "selfish" DNA that exist for their own sake. The movement of TEs has three main genetic consequences.

1. The insertion of both retroposons and transposons can mutate a gene; precise removal of the element from the gene then causes a reversion to wild type.

2. Side effects of the transposition process itself or of unequal crossing-over between copies of the same TE can help generate chromosomal rearrangements.

3. Recognition by transposase of two nearby transposons on the same chromosome can relocate the intervening genes.

Because of these genetic consequences, TEs make a major contribution to the evolution of genomes.

Rearrangements and Evolution: A Speculative Comprehensive Example

We saw at the beginning of this chapter that roughly 170 chromosomal rearrangements could reshape the human genome to a form that resembles the mouse genome. This suggests two things. First, although most chromosomal variations, including single-base changes and chromosomal rearrangements, are deleterious to an organism or its progeny, a few are either neutral or provide an advantage for survival and manage to become fixed in a population. Second, some rearrangements contribute to the processes underlying speciation. We still do not know enough of the details to describe the precise rearrangements that separate the mouse and human genomes. It is nonetheless useful to consider here some general examples of how chromosomal rearrangements might contribute to evolution.

Deletions A small deletion that moves a coding sequence of one gene next to a promoter or other regulatory element of an adjacent gene may, rarely, allow expression of a protein at a novel time in development or in a novel tissue. If the new time or place of expression is advantageous to the organism, it might become established in the genome.

Duplications An organism cannot normally tolerate mutations in a gene essential to its survival, but duplication would provide two copies of the gene. If one copy remained intact to perform the essential function, the other would be free to evolve a new function. The genomes of most higher plants and animals, in fact, contain many **gene families**—sets of closely related genes with slightly different functions that most likely arose from a succession of gene duplication events. In vertebrates, some *multigene families* have hundreds of members.

Inversions Suppose one region of a chromosome has three mutations that together greatly enhance the reproductive fitness of the organism. In heterozygotes where one homolog carries the mutations and the other does not, recombination could undo the beneficial linkage. If, however, the three mutations are part of an inversion, crossover suppression will ensure that they remain together as they spread through the population.

Translocations Suppose that a flood or other natural disaster separates two small populations of a species. After the separation, a translocation occurs and becomes fixed in one of the populations, so that now one population is homozygous for the normal gene order and the other is homozygous for the translocated gene order. If the two populations subsequently resume contact, heterozygotes formed by matings between individuals from each population would be semisterile, providing a partial breeding barrier between the two populations. New mutations favoring matings between homozygotes (for example, a behavioral mutation that makes one population seek mates in the spring and the other seek mates in the fall) would minimize the risk of producing semisterile progeny. The rapid spread of such mutations as a result of improved fertility would increase the breeding barrier between the populations and thus contribute to their eventual separation into different species.

Transpositions Movement of TEs may cause novel mutations, a small proportion of which might be selected for because they are advantageous to the organism. TEs can also help generate potentially useful duplications and inversions.

Because of these possibilities, many geneticists speculate that chromosomal rearrangements and transpositions are, like point mutations, important instruments of evolution.

Changes in Chromosome Number

We have seen that in peas, *Drosophila,* and humans, normal diploid individuals carry a $2n$ complement of chromosomes, where n is the number of chromosomes in the gametes. All the chromosomes in the haploid gametes of these diploid organisms are different from one another. In this section, we examine two types of departure from chromosomal diploidy found in eukaryotes: (1) aberrations in usually diploid species that generate cells or individuals whose genomes contain one to a few chromosomes more or less than the normal $2n$, for example, $2n + 1$ or $2n - 1$, and (2) species whose genomes contain complete but nondiploid sets of chromosomes, for example, $3n$ or $4n$.

The Loss or Gain of One or More Chromosomes Results in Aneuploidy

Individuals whose chromosome number is not an exact multiple of the haploid number (n) for the species are **aneuploids** (review Table 13.1 on p. 443). Individuals lacking one chromosome from the diploid number ($2n - 1$) are **monosomic,** while individuals having one chromosome in addition to the normal diploid set ($2n + 1$) are **trisomic.** Organisms with four copies of a particular chromosome ($2n + 2$) are **tetrasomic.**

Autosomal Aneuploidy Is Harmful to the Organism

Monosomy, trisomy, and other forms of aneuploidy create a genetic imbalance that is usually deleterious to the organism. In humans, monosomy for any autosome is generally lethal, but medical geneticists have reported a few cases of monosomy for

chromosome 21, one of the smallest human chromosomes. Although born with severe multiple abnormalities, these monosomic individuals survived beyond birth. Similarly, trisomies involving a human autosome are also highly deleterious. Individuals with trisomies for larger chromosomes, such as 1 and 2, are almost always aborted spontaneously early in pregnancy. Trisomy 18 causes Edwards syndrome, and trisomy 13 causes Patau syndrome; both phenotypes include gross developmental abnormalities that result in early death.

The most frequently observed human autosomal trisomy, trisomy 21, results in Down syndrome. As one of the shortest human autosomes, chromosome 21 contains about 1.5% of the DNA in the human genome. Although there is considerable phenotypic variation among Down syndrome individuals, traits such as mental retardation and skeletal abnormalities are usually associated with the condition. Many Down syndrome babies die in their first year after birth from heart defects and increased susceptibility to infection. We saw earlier (in the discussion of translocations) that some people with Down syndrome have three copies of only part of, rather than the entire, chromosome 21. It is thus probable that genetic imbalance for only one or a few genes may be a sufficient cause of the condition.

Humans Tolerate X Chromosome Aneuploidy Because X Inactivation Compensates for Dosage

Although the X chromosome is one of the longest human chromosomes and contains 5% of the DNA in the genome, individuals with X chromosome aneuploidy, such as XXY males, XO females, and XXX females, survive quite well compared with aneuploids for the larger autosomes. The explanation for this tolerance of X-chromosome aneuploidy is that X-chromosome inactivation equalizes the expression of most X-linked genes in individuals with different numbers of X chromosomes. As we saw in Chapter 12, X-chromosome inactivation represses expression of most genes on all but one X chromosome in a cell. As a result, even if the number of X chromosomes varies, the amount of protein generated by most X-linked genes remains constant. Human X-chromosome aneuploidies are nonetheless not without consequence. XXY men have Klinefelter syndrome, and XO women have Turner syndrome. The aneuploid individuals affected by these syndromes are usually infertile and display skeletal abnormalities, leading in the men to unusually long limbs and in the women to unusually short stature.

If X inactivation were 100% effective, we would not expect to see even the relatively minor abnormalities of Klinefelter syndrome, because the number of functional X chromosomes— one—would be the same as in normal individuals. One explanation is that during X inactivation, several genes near the telomere and centromere of the short arm of the human X chromosome escape inactivation and thus remain active. As a result, XXY males make twice the amount of protein encoded by these few genes as XY males (Fig. 13.27).

The reverse of X inactivation is *X reactivation;* it occurs in the oogonia, the female germ-line cells which develop into the oocytes that undergo meiosis (review Fig. 4.18 on p. 97). Reactivation of the previously inactivated X chromosomes in

XY
Normal ♂
X   Y

XX ♀
Normal
X   X

XXY
Klinefelter ♂
X   X   Y

▪ Active genes

Some X genes expressed
at twice the level of normal
males

Figure 13.27 Why aneuploidy for the X chromosome can have phenotypic consequences. X chromosome inactivation does not affect all genes on the X chromosome. As a result, in XXY Klinefelter males, a few X chromosome genes are expressed inappropriately at twice their normal level.

the oogonia ensures that every mature ovum (the gamete) receives an active X. If X reactivation did not occur, half of a woman's eggs (those with inactive X chromosomes) would be incapable of supporting development after fertilization. The phenomenon of X reactivation in the oogonia might help explain the infertility of women with Turner syndrome. With X reactivation, oogonia in normal XX females have two functional doses of X chromosome genes; but the corresponding cells in XO Turner women have only one dose of the same genes and may thus undergo defective oogenesis.

Aneuploidy Results from Meiotic Nondisjunction

How does aneuploidy arise? Mistakes in chromosome segregation during meiosis produce aneuploids of different types, depending on when the mistakes occur. If homologous chromosomes do not separate (that is, do not disjoin) during the first meiotic division, two of the resulting haploid gametes will carry both homologs, and two will carry neither. Union of these gametes with normal gametes will produce aneuploid zygotes, half monosomic, half trisomic (Fig. 13.28a, *left*). By contrast, if meiotic nondisjunction occurs during meiosis II, only two of the four resulting gametes will be aneuploid (Fig. 13.28a, *right*). Abnormal $n + 1$ gametes resulting from nondisjunction in a cell that is heterozygous for alleles on the nondisjoining chromosome will be heterozygous if the nondisjunction happens in the first meiotic division, but they will be homozygous if the nondisjunction takes place in the second meiotic division. (We assume here that no recombination has occurred between the heterozygous gene in question and the centromere, as would be the case for genes closely linked to the centromere.) It is possible to use this distinction to determine when a particular nondisjunction occurred (review Fig. 13.28a). The nondisjunction events that give rise to Down syndrome, for example, occur much more frequently in mothers (90%) than in fathers (10%). Interestingly, in women such nondisjunction events occur more often during the first meiotic division (about 75% of the time) than during the second. By contrast, when the nondisjunction event leading to Down syndrome takes place in men, the reverse is true.

Recently obtained data show that many meiotic nondisjunction events in humans result from problems in meiotic recombination. By tracking DNA markers, clinical investigators can establish whether recombination took place anywhere along chromosome 21 during meioses that created $n + 1$ gametes. In approximately one-half of Down syndrome cases caused by nondisjunction during the first meiotic division in the mother (that is, in about 35% of all Down syndrome cases), no recombination occurred between the homologous chromosome 21s in the defective meioses. This result makes sense because chiasmata, the structures associated with crossing-over, hold the maternal and paternal homologous chromosomes together in a bivalent at the metaphase plate of the first meiotic division (review Feature Figure 4.13 on pp. 90–91). In the absence of recombination and thus of chiasmata, there is no mechanism to ensure that the maternal and paternal chromosomes will go to opposite poles at anaphase I. The increase in the frequency of Down syndrome children associated with increasing maternal age (see Chapter 4, p. 97) may therefore reflect a decline in the effectiveness of the mother's machinery for meiotic recombination.

If an aneuploid individual survives and is fertile, the incidence of aneuploidy among his or her offspring will generally be extremely high. This is because half of the gametes produced by meiosis in a monosomic individual lack the chromosome in question, while half of the gametes produced in a trisomic individual have an additional copy of the chromosome (Fig. 13.28b).

Aneuploid Mosaics Can Result from Mitotic Nondisjunction or Chromosome Loss

As a zygote divides many times to become a fully formed organism, mistakes in chromosome segregation during the mitotic divisions accompanying this development may, in rare instances, augment or diminish the complement of chromosomes in certain cells. In **mitotic nondisjunction,** the failure of two sister chromatids to separate during mitotic anaphase generates reciprocal trisomic and monosomic daughter cells (Fig. 13.29a). Other types of mistakes, such as a lagging chromatid not pulled to either spindle pole at mitotic anaphase, result in a **chromosome loss** that produces one monosomic and one diploid daughter cell (Fig. 13.29b).

Aneuploid cells arising from either mitotic nondisjunction or chromosome loss may survive and undergo further rounds of cell division, producing clones of cells with an abnormal chromosome count. Nondisjunction or chromosome loss occurring early in development will generate larger aneuploid clones than the same events occurring later in development. The side-by-side existence of aneuploid and normal tissues results in a **mosaic** organism whose phenotype depends on what tissue bears the aneuploidy, the number of aneuploid cells, and the specific genes on the aneuploid chromosome. Many examples of mosaicism involve the sex chromosomes. If an XX *Drosophila* female loses one of the X chromosomes during the first mitotic division after fertilization, the result is a **gynandromorph** composed of equal parts male and female tissue (Fig. 13.29c).

Interestingly, in humans many Turner syndrome females are mosaics carrying some XX cells and some XO cells. These

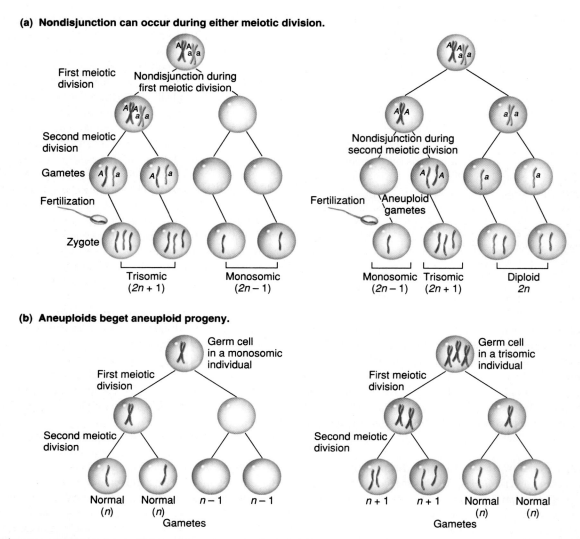

(a) Nondisjunction can occur during either meiotic division.

(b) Aneuploids beget aneuploid progeny.

Figure 13.28 Aneuploidy is caused by problems in meiotic chromosome segregation. (a) If trisomic progeny inherit two different alleles (*A* and *a*) of a centromere-linked gene from one parent, the nondisjunction occurred in meiosis I (*left*). If the two alleles inherited from one parent are the same (*A* and *A*; or *a* and *a*), the nondisjunction occurred during meiosis II (*right*). **(b)** Because aneuploids carry chromosomes that have no homolog with which to pair, aneuploid individuals frequently produce aneuploid progeny.

individuals began their development as XX zygotes, but with the loss of an X chromosome during the embryo's early mitotic divisions, they acquired a clone of XO cells. Similar mosaicism involving the autosomes also occurs. For example, physicians have recorded several cases of mild Down syndrome arising from mosaicism for trisomy 21. In people with Turner or Down mosaicism, the existence of some normal tissue appears to ameliorate the condition, with the individual's phenotype depending on the particular distribution of diploid versus aneuploid cells.

Some Euploid Species Contain Complete But Nondiploid Sets of Chromosomes

In contrast to aneuploids, **euploid** cells contain only complete sets of chromosomes. Most euploid species are diploid, but some euploid species are **polyploids** that carry three or more

complete sets of chromosomes (see Table 13.1 on p. 443). When speaking of polyploids, geneticists use the symbol *x* to indicate the **basic chromosome number,** that is, the number of different chromosomes that make up a single complete set. Triploid species, which have three complete sets of chromosomes are then 3*x*; tetraploid species with four complete sets of chromosomes are 4*x*; and so forth. For diploid species, *x* is identical to *n*—the number of chromosomes in the gametes—because each gamete contains a single complete set of chromosomes. This identity of *x* = *n* does not hold for polyploid species, as the following example illustrates. Commercially grown bread wheat has a total of 42 chromosomes: 6 nearly (but not wholly) identical sets each containing 7 different chromosomes. Bread wheat is thus a hexaploid with a basic number of *x* = 7 and 6*x* = 42. But each triploid gamete has one-half the total number

(a) Mitotic nondisjunction

Diploid cell at metaphase

Mitosis

Nondisjunction of one set
of sister chromatids at anaphase
produces aneuploid daughter cells.

Trisomic cell   Monosomic cell

(b) Mitotic chromosome loss

Diploid cell at metaphase

Mitosis

Lagging chromatid at
anaphase is lost. Aneuploid
daughter cell produced.

Monosomic cell   Normal diploid cell

(c) A gynandromorph

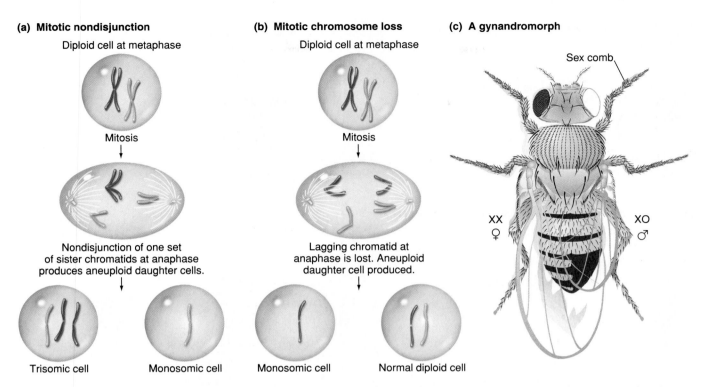

Sex comb

XX
♀

XO
♂

Figure 13.29   Mistakes during mitosis can generate clones of aneuploid cells. Mitotic nondisjunction **(a)** or chromosome loss during mitosis **(b)** can create monosomic or trisomic cells that can divide to produce aneuploid clones. **(c)** If an X chromosome is lost during the first mitotic division of an XX *Drosophila* zygote, one daughter cell will be XX (female) while the other will be XO (male). Such an embryo will grow into a gynandromorph. Here, the zygote was *w⁺ m⁺ / w m,* so the XX half of the fly (*left*) has red eyes and normal wings; loss of the *w⁺ m⁺* X chromosome gives the XO half of the fly (*right*) white eyes (*w*), miniature wings (*m*), and a male-specific sex comb on the front leg.

of chromosomes, so $n = 21$. Thus, for bread wheat, x and n are not the same. Another form of euploidy, in addition to polyploidy, exists in **monoploid** (x) organisms, which have only one set of chromosomes.

Monoploidy and polyploidy are rarely observed in animals. Among the few examples of monoploidy are some species of ants and bees in which the males are monoploid, while the females are diploid. Males of these species develop *parthenogenetically* from unfertilized eggs, and these males produce gametes through a modified mitosis. Polyploidy in animals normally exists only in species with unusual reproductive cycles, such as hermaphroditic earthworms, which carry both male and female reproductive organs, and goldfish, which are parthenogenetically tetraploid species. In *Drosophila,* it is possible, under special circumstances, to produce triploid and tetraploid females, but never males. In humans, polyploidy is always lethal, usually resulting in spontaneous abortion during the first trimester of pregnancy.

Monoploid Organisms Contain a Single Copy of Each Chromosome and Are Usually Infertile

Botanists can produce monoploid plants experimentally by special treatment of germ cells from diploid species that have completed meiosis and would normally develop into pollen. (Note

that monoploid plants obtained in this manner can also be considered haploids because $x = n$.) The treated cells divide into a mass of tissue known as an *embryoid*. Subsequent exposure to plant hormones enables the embryoid to develop into a plant (Fig. 13.30a). Monoploid plants may also arise from rare spontaneous events in a large natural population. Most monoploid plants, no matter how they originate, are infertile. Since the chromosomes have no homologs with which to pair during meiosis I, they are distributed at random to the two spindle poles during this division. Rarely do all chromosomes go to the same pole, and if they do not, the resulting gametes are defective as they lack one or more chromosomes. The greater the number of chromosomes in the genome, the lower the likelihood of producing a gamete containing all of them.

Despite such gamete-generating problems, monoploid plants and tissues are of great value to plant breeders. They make it possible to visualize normally recessive traits directly, without crosses to achieve homozygosity. Plant researchers can also introduce mutations into individual monoploid cells, select for desirable phenotypes, such as resistance to herbicides, and use hormone treatments to grow the selected cells into monoploid plants (Fig. 13.30b). They can then convert monoploids of their choice into homozygous diploid plants by treating tissue with *colchicine,* an alkaloid drug obtained from the autumn

(a) How to create a monoploid plant

(b) Using monoploloid plants to select for herbicide resistance

(c) Using colchicine to double chromosome numbers

Figure 13.30 The creation and use of monoploid plants. (a) Under certain conditions, haploid pollen grains can grow into haploid embryoids. When treated with plant hormones, haploid embryoids grow into monoploid plants. **(b)** Researchers select monoploid cells for recessive traits such as herbicide resistance. They then grow the selected cells into a resistant embryoid, which (with hormone treatment) eventually becomes a mature, resistant monoploid plant. Treatment with colchicine doubles the chromosome number, creating diploid cells that can be grown in culture with hormones to make a homozygous resistant diploid plant. **(c)** Colchicine treatment prevents formation of the mitotic spindle and also blocks cytokinesis, generating cells with twice the number of chromosomes. *Blue, red,* and *green* colors denote nonhomologous chromosomes.

Figure 13.31 Many polyploid plants are larger than their diploid counterparts. A comparison of octaploid (*left*) and diploid (*right*) strawberries.

(a) Formation of a triploid organism

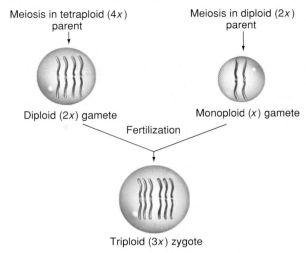

Meiosis in tetraploid (4*x*) parent

Meiosis in diploid (2*x*) parent

Diploid (2*x*) gamete

Monoploid (*x*) gamete

Fertilization

Triploid (3*x*) zygote

(b) Meiosis in a triploid organism

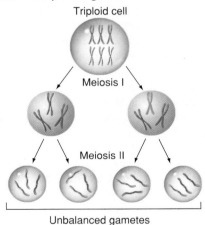

Triploid cell

Meiosis I

Meiosis II

Unbalanced gametes

Figure 13.32 The genetics of triploidy. (a) Production of a triploid (*x* = 3) from fertilization of a monoploid gamete by a diploid gamete. Nonhomologous chromosomes are either *blue* or *red*. **(b)** Meiosis in a triploid produces unbalanced gametes because meiosis I produces two daughter cells with unequal numbers of any one type of chromosome. If *x* is large, balanced gametes with equal numbers of all the chromosomes are very rare.

crocus. By binding to tubulin—the major protein component of the spindle—colchicine prevents formation of the spindle apparatus. In cells without a spindle, the sister chromatids cannot segregate after the centromere splits, so that there is often a doubling of the chromosome set following treatment with colchicine (Fig. 13.30c). The resulting diploid cells can be grown into diploid plants that will express the desired phenotype and produce fertile gametes.

Polyploidy Has Accompanied the Evolution of Many Cultivated Plant Species

Roughly one out of every three known species of flowering plants is a polyploid, and because polyploidy often increases plant size and vigor, many polyploid plants with edible parts have been selected for agricultural cultivation. Most commercially grown alfalfa, coffee, and peanuts are tetraploids (4*x*). MacIntosh apple and Bartlett pear trees that produce giant fruits are also tetraploids. Commercially grown strawberries are octaploids (8*x*) (Fig. 13.31). The evolutionary success of polyploid plant species may stem from the fact that polyploidy, like gene duplication, provides additional copies of genes; while one copy continues to perform the original function, the others can evolve new functions. However, as we now show, the consequences of polyploidy depend in large part on whether the number of chromosome sets is even or odd.

Triploids Are Almost Always Sterile

Triploids (3*x*) result from the union of monoploid (*x*) and diploid (2*x*) gametes (Fig. 13.32a). The diploid gametes may be the products of meiosis in tetraploid (4*x*) germ cells, or they may be the products of rare spindle or cytokinesis failures during meiosis in a diploid.

Sexual reproduction in triploid organisms is extremely inefficient because meiosis produces mostly unbalanced gametes. During the first meiotic division in a triploid germ cell, three sets of chromosomes must segregate into two daughter cells; regardless of how the chromosomes align in pairs, there is no way to ensure that the resulting gametes obtain a complete, balanced *x* or 2*x* complement of chromosomes. In most cases, at the end of anaphase I, two chromosomes of any one type move to one pole, while the remaining chromosome of the same type moves to the opposite pole. The products of such a meiosis have two copies of some chromosomes and one copy of others (Fig. 13.32b). If the number of chromosomes in the basic set is large, the chance of obtaining any balanced gametes at all is remote. Thus, fertilization with gametes from triploid individuals does not produce many viable offspring.

It is possible to propagate some triploid species, such as bananas and watermelons, through asexual reproduction. The fruits of triploid plants are seedless because the unbalanced gametes do not function properly in fertilization or, if fertilization occurs, the resultant zygote is so genetically unbalanced that it cannot develop. Either way, no seeds form. Like triploids, all polyploids with odd numbers of chromosome sets (such as $5x$ or $7x$) are sterile because they cannot reliably produce balanced gametes.

Tetraploids Are Often the Source of New Species

During mitosis, if the chromosomes in a diploid ($2x$) tissue fail to separate after replication, the resulting daughter cells will be tetraploid ($4x$; Fig. 13.33a). If such tetraploid cells arise in reproductive tissue, subsequent meioses will produce diploid gametes. Rare unions between diploid gametes produce tetraploid organisms. Self-fertilization of a newly created tetraploid organism will produce an entirely new species, because crosses between the tetraploid and the original diploid organism will produce infertile triploids (review Fig. 13.32a). Tetraploids made in this fashion are **autopolyploids,** a kind of polyploid that derives all its chromosome sets from the same species.

Maintenance of a tetraploid species depends on the production of gametes with balanced sets of chromosomes. Most successful tetraploids have evolved mechanisms ensuring that the four copies of each group of homologs pair two by two to form two **bivalents**—pairs of synapsed homologous chromosomes (Fig. 13.33b). Because the chromosomes in each bivalent become attached to opposite spindle poles during meiosis I, meiosis regularly produces gametes carrying two complete sets of chromosomes. The mechanism requiring that each chromosome pair with only a single homolog suppresses other pairing possibilities, such as a 3:1, which cannot guarantee equivalent chromosome segregation.

Tetraploids, with four copies of every gene, generate unusual Mendelian ratios. For example, even if there are only two alleles of a gene (say, A and a), five different genotypes are possible: $A A A A, A A A a, A A a a, A a a a, a a a a$. If the phenotype depends on the dosage of A, then five phenotypes, each corresponding to one of the genotypes, will appear. The segregation of alleles during meiosis in a tetraploid is similarly complex. Consider an $A A a a$ heterozygote in which the A gene is closely linked to the centromere, and the A allele is completely dominant. What are the chances of obtaining progeny with the recessive phenotype, generated by only the $a a a a$ genotype? As Fig. 13.33c illustrates, if during meiosis I, the four chromosomes carrying the gene align at random in bivalents along the metaphase plate, the expected ratio of gametes is $2 A A : 8 A a : 2 a a = 1 A A : 4 A a : 1 a a$. The chance of obtaining $a a a a$ progeny during self-fertilization is thus $1/6 \times 1/6 = 1/36$. In other words, since A is completely dominant, the ratio of dominant to recessive phenotypes, determined by the ratio of A- - - to $a a a a$ genotypes is 35:1. The ratios will be different if the gene is not closely linked to the centromere or if the dominance relationship between the alleles is not so simple.

New levels of polyploidy can arise from the doubling of a polyploid genome. Such doubling occurs on rare occasions in nature; it also results from controlled treatment with colchicine or other drugs that disrupt the mitotic spindle. The doubling of a tetraploid genome yields an octaploid ($8x$). These higher-level polyploids created by successive rounds of genome doubling are autopolyploids because all of their chromosomes derive from a single species.

Some Polyploids Have Agriculturally Desirable Traits Derived from Two Species

Polyploidy can also arise from crosses between members of two species, even if they have different numbers of chromosomes. Hybrids in which the chromosome sets come from two or more distinct, though related, species are known as **allopolyploids.** In crosses between octaploids and tetraploids, for example, fertilization unites tetraploid and diploid gametes to produce hexaploid progeny. Fertile allopolyploids arise only rarely, under special conditions, because chromosomes from the two species differ in shape, size, and number, so they cannot easily pair with each other. The resulting irregular segregation creates genetically unbalanced gametes such that the hybrid progeny will be sterile. Chromosomal doubling in germ cells, however, can restore fertility by creating a pairing partner for each chromosome. Organisms produced in this manner are termed **amphidiploids** if the two parental species were diploids; they contain two diploid genomes, each one derived from a different parent. As the following illustrations show, it is hard to predict the characteristics of an amphidiploid or other allopolyploids.

A cross between cabbages and radishes, for example, leads to the production of amphidiploids known as *Raphanobrassica*. The gametes of both parental species contain 9 chromosomes; the sterile F_1 hybrids have 18 chromosomes, none of which has a homolog. Chromosome doubling in the germ cells after treatment with colchicine, followed by union of two of the resulting gametes, produces a new species: a fertile *Raphanobrassica* amphidiploid carrying 36 chromosomes—a full complement of 18 (9 pairs) derived from cabbages and a full complement of 18 (9 pairs) derived from radishes. Unfortunately, this amphidiploid has the roots of a cabbage plant and leaves resembling those of a radish, so it is not agriculturally useful.

By contrast, crosses between tetraploid (or hexaploid) wheat and diploid rye have led to the creation of several allopolyploid hybrids with agriculturally desirable traits from both species (Fig. 13.34 on p. 474). Some of the hybrids combine the high yields of wheat with rye's ability to adapt to unfavorable environments. Others combine wheat's high level of protein with rye's high level of lysine; wheat protein does not contain very much of this amino acid, an essential ingredient in the human diet. The various hybrids between wheat and rye form a new crop known as *Triticale*. Some triticale strains produce nutritious grains that already appear in breads sold in health food stores. Plant breeders are currently assessing the usefulness of various triticale strains for large-scale agriculture.

(a) Generation of tetraploid (4x) cells

(b) Pairing of chromosomes as bivalents

(c) Gametes formed by A A a a tetraploids

Chromosomes	Pairing		Gametes produced by random spindle attachment
1. ─●─ A ─ 2. ─●─ A ─	1 ↑A 3 ↑a 2 ─ ↓A 4 ─ ↓a **or**	1 ↑A 4 ↑a 2 ─ ↓A 3 ─ ↓a	1 + 3 A a 1 + 4 A a **or** 2 + 4 A a 2 + 3 A a
3. ─●─ a ─ 4. ─●─ a ─	1 ↑A 2 ↑A 3 ─ ↓a 4 ─ ↓a **or**	1 ↑A 4 ↑a 3 ─ ↓a 2 ─ ↓A	1 + 2 A A 1 + 4 A a **or** 3 + 4 a a 2 + 3 A a
	1 ↑A 2 ↑A 4 ─ ↓a 3 ─ ↓a **or**	1 ↑A 3 ↑a 4 ─ ↓a 2 ─ ↓A	1 + 2 A A 1 + 3 A a **or** 3 + 4 a a 2 + 4 A a

Total:
8A a : 2A A : 2a a = 4A a : 1A A : 1a a

Figure 13.33 The genetics of tetraploidy. (a) Tetraploids arise from a failure of chromosomes to separate into two daughter cells during mitosis in a diploid. **(b)** In successful tetraploids, the pairing of chromosomes as bivalents generates genetically balanced gametes. **(c)** Gametes produced in an *A A a a* tetraploid heterozygous for two alleles of a centromere-linked gene, with orderly pairing of bivalents. The four chromosomes can pair to form two bivalents in three possible ways. For each pairing scheme, the chromosomes in the two pairs can assort in two different orientations. If all possibilities are equally likely, the expected genotype frequency in a population of gametes will be 1 *A A* : 4 *A a* : 1 *a a*.

(a)

(b)

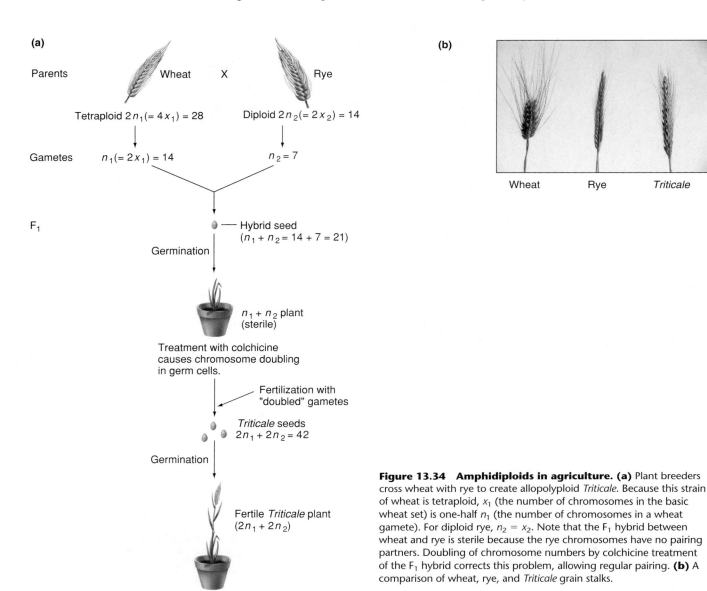

Figure 13.34 Amphidiploids in agriculture. (a) Plant breeders cross wheat with rye to create allopolyploid *Triticale.* Because this strain of wheat is tetraploid, x_1 (the number of chromosomes in the basic wheat set) is one-half n_1 (the number of chromosomes in a wheat gamete). For diploid rye, $n_2 = x_2$. Note that the F_1 hybrid between wheat and rye is sterile because the rye chromosomes have no pairing partners. Doubling of chromosome numbers by colchicine treatment of the F_1 hybrid corrects this problem, allowing regular pairing. **(b)** A comparison of wheat, rye, and *Triticale* grain stalks.

A Glimpse of the Future: Emergent Technologies in the Analysis of Chromosomal Rearrangements and Changes in Chromosome Number

Two main problems occur when searching for chromosomal rearrangements and changes in chromosome number by karyotype analysis. First, it is a tedious procedure that depends on highly trained technicians to identify chromosomal alterations under the microscope. Because of the subjective nature of the analysis, mistakes can reduce the accuracy of results. Second, even in the hands of the best technicians, there is a limit to the viewing res-

olution. Even under optimal circumstances, it is not possible to detect deletions or duplications of less than 10 Mb in human karyotypes. Human populations no doubt have many chromosomes with as yet undetected smaller deletions or duplications.

To overcome the limitations of karyotype analysis, researchers have recently developed a microarray-based hybridization protocol that can scan the genome for deletions, duplications, and aneuploidy with much greater resolution, very high accuracy, and much greater throughput and without the need for a subjective determination of the result. The technique is called *Comparative Genomic Hybridization (CGH).*

The protocol works as follows. First, a series of 20,000 BAC clones with DNA inserts averaging 150 kb that collec-

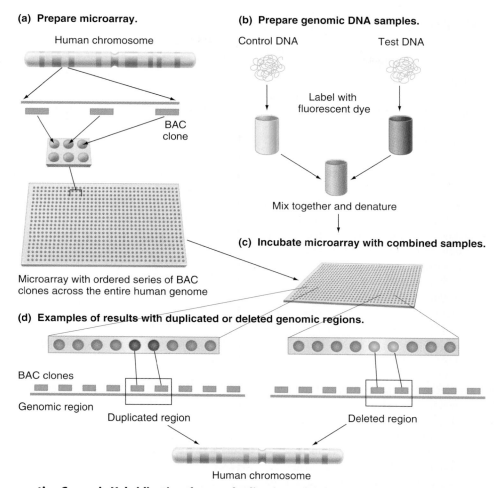

Figure 13.35   Comparative Genomic Hybridization detects duplications, deletions, and aneuploidy. (a) BAC clones representing the human genome are spotted in order onto a microarray. **(b)** The genomic sample to be tested is labeled with one color dye (here, *red*), and the control genome sample is labeled with a second color dye (*yellow*). **(c)** The two samples are mixed together, denatured, and then incubated on the microarray. **(d)** Automated analysis of each spot on the microarray detects the ratio of the two dyed probes that hybridize. *Orange* indicates a 1:1 ratio; other colors indicate deletion (0.5:1 ratio; *yellow*) or duplication (1.5:1 ratio; *red*) of BAC clone sequences in the test sample.

tively represent the entire human genome are spotted onto a microarray. These BAC clones were characterized in the course of the Human Genome Project. Next, genomic DNA from a control sample with a normal genome content is labeled with a yellow fluorescent dye, while the genomic DNA from the test sample is labeled with a red fluorescent dye. The two genomic DNA samples are mixed together in equal amounts, denatured, and applied to the microarray as a probe. After hybridization is complete and unhybridized material is washed away, the fluorescence emission from each microarray dot is analyzed automatically by a machine designed for this task.

If the genomic region probed with a particular BAC clone is present in a diploid two copies in the test sample, then the ratio of

red to yellow dyes on that dot will be 1:1. However, if a particular genomic region is duplicated or deleted from one homolog in the test sample, the ratio of red to yellow will be 1.5 : 1 or 0.5 : 1, respectively. An example of this analysis is shown in Fig. 13.35.

CGH provides a powerful clinical tool to detect any type of aneuploidy or any deletion or duplication of 50 kb or more anywhere in the genome. Clinicians can use it in conjunction with amniocentesis or preimplantation genetic analysis. They can also use CGH to screen tissue biopsies for cancerous cells that have deleted or duplicated regions containing oncogenes or tumor suppressor genes. The technique thus holds great promise for the detection of new genes that contribute to the genesis of cancer.

CONNECTIONS

The detrimental consequences of most changes in chromosome organization and number cause considerable distress in humans (Table 13.2). Approximately 4 of every 1000 individuals has an abnormal phenotype associated with aberrant chromosome organization or number. Most of these abnormalities result from either aneuploidy for the X chromosome or trisomy 21. By comparison, about 10 people per 1000 suffer from an inherited disease caused by a single-gene mutation.

The incidence of chromosomal abnormalities among humans would be much larger were it not for the fact that many fetuses with abnormal karyotypes abort spontaneously early in pregnancy. Fully 15% to 20% of recognized pregnancies end with detectable spontaneous abortions; and half of the spontaneously aborted fetuses show chromosomal abnormalities, particularly trisomy, sex chromosome monosomy, and triploidy. These figures almost certainly underestimate the rate of spontaneous abortion caused by abnormal chromosomal variations, since fetuses carrying aberrations for larger chromosomes, such as monosomy 2 or trisomy 5, may abort so early that the pregnancy goes unrecognized.

But despite all the negative effects of chromosomal rearrangements and changes in chromosome number, a few departures from normal genome organization survive to become instruments of evolution by natural selection.

As we see in the next chapter, chromosomal rearrangements occur in bacteria as well as in eukaryotic organisms. In bacteria, transposable elements catalyze many of the changes in chromosomal organization. Remarkably, the reshuffling of genes between different DNA molecules in the same cell catalyzes the transfer of genetic information from one bacterial cell to another.

TABLE 13.2   Aneuploidy in the Human Population

Chromosomes	Syndrome	Frequency at Birth
Autosomes		
Trisomic 21	Down	1/700
Trisomic 13	Patau	1/5,000
Trisomic 18	Edwards	1/10,000
Sex chromosomes, females		
XO, monosomic	Turner	1/5,000
XXX, trisomic		
XXXX, tetrasomic		1/700
XXXXX, pentasomic		
Sex chromosomes, males		
XYY, trisomic	Normal	1/10,000
XXYY, tetrasomic		
XXXY, tetrasomic		
XXXXY, pentasomic	Klinefelter	1/500
XXXXXY, hexasomic		

About 0.4% of all babies born have a detectable chromosomal abnormality that generates a detrimental phenotype.

ESSENTIAL CONCEPTS

1. *Rearrangements* reorganize the DNA sequences within genomes. Like point mutations, rearrangements are subject to natural selection and thus serve as instruments of evolution.

   a. *Deletions* remove DNA from a chromosome. Homozygosity for a large deletion is usually lethal, but even heterozygosity for a large deletion can create a deleterious *genetic imbalance*. Deletions may uncover recessive mutations on the homologous chromosome and are thus useful for gene mapping at the cytological and molecular levels.

   b. *Duplications* add DNA to a chromosome. The additional copies of genes contained in a duplication are a major source of new genetic functions. Duplications sometimes produce changes in phenotype that allow their detection. Homozygosity or heterozygosity for duplications causes departures from normal gene dosage that are often harmful to the organism. *Unequal crossing-over* between duplicated regions expands or contracts the copy number of genes and may lead to multigene families.

   c. *Inversions* alter the order, but not the number, of genes on a chromosome. They may produce novel phenotypes by modifying the activity of genes near the rearrangement breakpoints or by disrupting genes at the breakpoints. Inversion heterozygotes act as *crossover suppressors* because progeny formed from recombinant gametes are genetically imbalanced.

   d. In *reciprocal translocations*, parts of two chromosomes trade places without the loss or gain of chromosomal material. By relocating pieces of DNA, translocations may modify the function of genes at or near the translocation breakpoints. Heterozygosity for translocations in the germ line results in *semisterility* and *pseudolinkage*.

e. *Transposable elements (TEs)* are short, mobile segments of DNA that reshape genomes by generating mutations, causing chromosomal rearrangements, and relocating genes.

2. Changes in chromosome number also reshape genomes.

a. *Aneuploidy,* the loss or gain of one or more chromosomes, creates a genetic imbalance. Aneuploidy can result from mistakes in meiosis, which produce aneuploid gametes, or from mistakes in mitosis, which generate aneuploid clones of cells. Autosomal aneuploidy is usually lethal to the organism. Aneuploidy for sex chromosomes is better tolerated because of dosage compensation mechanisms; nevertheless, it can have significant consequences.

b. *Euploid* organisms contain complete sets of chromosomes. Organisms with three or more sets of chromosomes are *polyploids. Autopolyploids* derive all their chromosome sets from the same species; *allopolyploids* are hybrids in which the chromosome sets come from two or more distinct, though related, species.

c. Organisms containing odd numbers of complete chromosome sets are sterile because the chromosomes cannot pair properly during the first meiotic division.

d. Polyploids with even numbers of chromosome sets can be fertile if proper chromosome segregation occurs sufficiently frequently. *Amphidiploids,* which are allopolyploids produced by chromosome doubling of two genomes derived from different diploid parental species, are often fertile and are sometimes useful in agriculture.

S O C I A L   A N D   E T H I C A L   I S S U E S

1. Until recently, female athletes were gender tested to confirm they were female using the Barr body test. In 1985, the Spanish hurdler Maria Patinez failed the Barr body test. She was encouraged by her coaches to fake an injury, quietly withdraw from the race, and retire from competition. Maria, indignant that someone suggested she was not a woman, spoke out against the testing policy. What are the problems scientifically with the Barr body test for gender identification? What do you think is a reasonable, accurate way to ascertain the sex of an individual? How should sex be defined? by karyotype? by examination of external genitalia? by internal gonadal structure?

2. Gary is a livestock farmer and has 2000 head of cattle. An outbreak of bacterial infection would be devastating to his livelihood, so he uses feed that contains antibiotics. Recently he heard a report indicating that the use of antibiotics in livestock food was contributing to the increase in antibiotic resistance in bacteria that infect humans. (The antibiotic resistance genes are contained on transposable elements that can be transferred from one bacterium to another.) Being a conscientious citizen, Gary is now considering eliminating the use of antibiotic feed. Will discontinuing the use of this feed make enough of a difference to justify the risk he takes? What should he do?

What are the stakes for him and for the public in this issue?

3. Jason and Sylvia received some unsettling news from the amniocentesis of their baby. Karyotype analysis showed a partial trisomy in which one chromosome 8 homolog carried a duplication of a particular region. Genetic imbalances that can be observed cytologically are not well tolerated in humans and usually lead to medical conditions or developmental abnormalities. However, this specific partial trisomy had not been previously observed nor associated with any medical conditions or developmental abnormalities. The parents decided to continue the pregnancy. Their doctor alerted other medical colleagues who now would like to follow the progress of this child after birth. In this way, more knowledge can be gained for future similar situations. Jason and Sylvia would prefer not to have their child subjected to intense medical scrutiny and are upset that their doctor gave information to other colleagues without their permission. Should the doctor have obtained permission before sharing the details with colleagues? Do Jason and Sylvia and their unborn child have a responsibility to provide information on the effects of this partial trisomy so others in the future will know what to expect if the duplication occurs?

S O L V E D   P R O B L E M S

I. Male *Drosophila* from a true-breeding wild-type stock were irradiated with X rays and then mated with females from a true-breeding stock carrying the following recessive mutations on the X chromosome: yellow body (*y*), crossveinless wings (*cv*), cut wings (*ct*), singed bristles (*sn*), and miniature wings (*m*). These markers are known to map in the order:

$$y - cv - ct - sn - m$$

Most of the female progeny of this cross were pheno-typically wild type, but one female exhibited *ct* and *sn* phenotypes. When this exceptional *ct sn* female was mated with a male from the true-breeding wild-type stock, there were twice as many females as males among the progeny.

a. What is the nature of the X-ray-induced mutation present in the exceptional female?

b. Draw the X chromosomes present in the exceptional *ct sn* female as they would appear during pairing in meiosis.

c. What phenotypic classes would you expect to see among the progeny produced by mating the exceptional *ct sn* female with a normal male from a true-breeding wild-type stock? List males and females separately.

Answer

To answer this problem, you need to think first about the effects of different types of chromosomal mutations in order to deduce the nature of the mutation. Then you can evaluate the consequences of the mutation on inheritance.

a. Two observations indicate that *X rays induced a deletion mutation.* The fact that two recessive mutations are phenotypically expressed in the exceptional female suggests that a deletion was present on one of her X chromosomes that uncovered the two mutant alleles (*ct* and *sn*) on the other X chromosome. Second, the finding that there were twice as many females as males among the progeny of the exceptional female is also consistent with a deletion mutation. Males who inherit the deletion-bearing X chromosome from their exceptional mother will be inviable (because other essential genes are located in the region that is now deleted), but sons who inherit a nondeleted X chromosome will survive. On the other hand, all of the exceptional female's daughters will be viable: Even if they inherit a deleted X chromosome from their mother, they also receive a normal X chromosome from their father. As a result, there are half as many male progeny as females from the cross of the exceptional female with a wild-type male.

b. During pairing, *the DNA in the normal (nondeleted) X chromosome will loop out* because there is no homolo-gous region in the deletion chromosome. In the simplified drawing of meiosis I that follows, each line represents both chromatids comprising each homolog.

c. *All daughters of the exceptional female will be wild type* since the father contributes wild-type copies of all the genes. Each of the surviving sons must inherit a nondeleted X chromosome from the exceptional female. Some of these X chromosomes are produced from meioses in which no recombination occurred, but other X chromosomes are the products of recombin-ation. *Males can have any of the genotypes listed here and therefore the corresponding phenotypes.* All contain the *ct sn* combination because no recombin-ation between homologs is possible in this deleted region. Some of these genotypes require multiple crossovers during meiosis in the mother and will thus be relatively rare.

y	cv	ct	sn	m
+	+	ct	sn	+
+	cv	ct	sn	m
y	+	ct	sn	+
y	cv	ct	sn	+
+	+	ct	sn	m
+	cv	ct	sn	+
y	+	ct	sn	m

II. One of the X chromosomes in a particular *Drosophila* female had a normal order of genes but carried recessive alleles of the genes for yellow body color (*y*), vermilion eye color (*v*), and forked bristles (*f*) as well as the dominant X-linked Bar eye mutation (*B*). Her other X chromosome carried the wild-type alleles of all four genes, but the region including y^+, v^+, and f^+ (but not B^+) was inverted with respect to the normal order of genes. This female was crossed to a wild-type male in the cross diagrammed here.

The cross produced the following male offspring:

y	v	f	B	48
y^+	v^+	f^+	B^+	45
y	v	f	B^+	11
y^+	v^+	f^+	B	8
y	v^+	f	B	1
y^+	v	f^+	B^+	1

a. Why are there no male offspring with the allele combinations $y\ v\ f^+$, $y^+\ v^+\ f$, $y\ v^+\ f^+$, or $y^+\ v\ f$ (regardless of the allele of the Bar eye gene)?

b. What kinds of crossovers produced the $y\ v\ f\ B^+$ and $y^+\ v^+\ f^+\ B$ offspring? Can you determine any genetic distances from these classes of progeny?

c. What kinds of crossovers produced the $y^+\ v\ f^+\ B^+$ and $y\ v^+\ f\ B$ offspring?

Answer

To answer this question, you need to be able to draw and interpret pairing in inversion heterozygotes. Note that this inversion is paracentric.

a. During meiosis in an inversion heterozygote, a loop of the inverted region is formed when the homologous genes align. In the following simplified drawing, each line represents both chromatids comprising each homolog.

If a single crossover occurs within the inversion loop, a dicentric and an acentric chromosome are formed. Cells containing these types of chromosome are not viable. The resulting allele combinations from such single crossovers are not recovered. The four phenotypic classes of missing male offspring would be formed by single crossovers between the y and v or between the v and f genes in the female inversion heterozygote and therefore are not recovered.

b. *The* $y\ v\ f\ B^+$ *and* $y^+\ v^+\ f^+\ B$ *offspring are the result of single crossover events outside of the inversion loop, between the end of the inversion (just to the right of* f *on the preceding diagram) and the* B *gene.* This region is approximately 16.7 m.u. in length (19 recombinants out of 114 total progeny).

c. *The* $y^+\ v\ f^+\ B^+$ *and* $y\ v^+\ f\ B$ *offspring would result from two crossover events within the inversion loop, one between the* y *and* v *genes and the other between the* v *and* f *genes.* You should note that these could be either two-strand or three-strand double crossovers, but they could not be four-strand double crossovers.

III. In maize trisomics, $n + 1$ pollen is not viable. If a dominant allele at the B locus produces purple color instead of the recessive phenotype bronze, and a $B\ b\ b$ trisomic plant is pollinated by a $B\ B\ b$ plant, what proportion of the progeny produced will be trisomic and have a bronze phenotype?

Answer

To solve this problem, think about what is needed to produce trisomic bronze progeny: three b chromosomes in the zygote. The female parent would have to contribute two b alleles, since the $n + 1$ pollen from the male is not viable. What kinds of gametes could be generated by the trisomic $B\ b\ b$ purple female parent, and in what proportion? To track all the possibilities, rewrite this genotype as $B\ b_1\ b_2$, even though b_1 and b_2 have identical effects on phenotype. In the trisomic female, there are three possible ways the chromosomes carrying these alleles could pair as bivalents during the first meiotic division so that they would segregate to opposite poles: B with b_1, B with b_2, and b_1 with b_2. In all three cases, the remaining chromosome could move to either pole. To tabulate the possibilities:

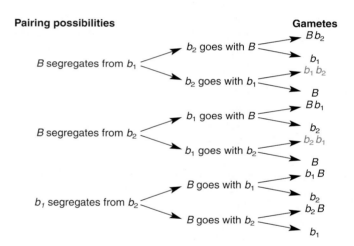

Of the 12 gamete classes produced by these different possible segregations, only the 2 classes written in red contain the two b alleles needed to generate the bronze ($b\ b\ b$) trisomic zygotes. There is thus a $2/12 = 1/6$ chance of obtaining such gametes.

Although segregation in the $B\ B\ b$ male parent is equally complicated, remember that males cannot produce viable $n + 1$ pollen. The only surviving gametes would thus be B and b, in a ratio ($2/3\ B$ and $1/3\ b$) that must reflect their relative prevalence in the male parent genome. *The probability of obtaining trisomic bronze progeny from this cross is therefore the product of the individual probabilities of the appropriate* b b *gametes from the female parent (1/6) and* b *pollen from the male parent (1/3): 1/6 × 1/3 = 1/18.*

PROBLEMS

13-1 For each of the terms in the left column, choose the best matching phrase in the right column.

a.	reciprocal translocation	1. lacking one or more chromosomes or having one or more extra chromosomes
b.	gynandromorph	2. movement of short DNA elements
c.	pericentric	3. having more than two complete sets of chromosomes
d.	paracentric	4. exact exchange of parts of two nonhomologous chromosomes
e.	euploids	5. excluding the centromere
f.	polyploidy	6. including the centromere
g.	transposition	7. having complete sets of chromosomes
h.	aneuploids	8. mosaic combination of male and female tissue

13-2 A diploid strain of yeast was made by mating a haploid strain with a genotype w^-, x^-, y^-, and z^- with a haploid strain of opposite mating type that is wild type for these four genes. The diploid strain was phenotypically wild type. Four different X-ray-induced mutants with the following phenotypes were produced from this diploid yeast strain. Assume there is a single mutation in each strain.

Strain 1	w^-	x^+	y^-	z^+
Strain 2	w^+	x^-	y^-	z^-
Strain 3	w^-	x^+	y^-	z^-
Strain 4	w^-	x^+	y^+	z^+

When these mutant diploid strains of yeast go through meiosis, each ascus is found to contain only two viable haploid spores.

a. What kind of mutations were induced by X rays to make the listed diploid strains?

b. Why did two spores in each ascus die?

c. Are any of the genes *w, x, y,* or *z* located on the same chromosome?

d. Give the order of the genes that are found on the same chromosome.

13-3 In flies that are heterozygous for either a deletion or a duplication, there will be a looped-out region in a preparation of polytene chromosomes. How could you distinguish between a deletion or a duplication using polytene chromosome analysis?

13-4 A series of chromosomal mutations in *Drosophila* were used to map the *javelin* gene, which affects bristle shape, and *henna,* which affects eye pigmentation. Both the *javelin* and *henna* mutations are recessive. A diagram of region 65 of the *Drosophila* polytene chromosomes is shown here.

The chromosomal breakpoints for six chromosome rearrangements are indicated in the following table. (For example, deletion A has one breakpoint between bands A2 and A3 and the other between bands D2 and D3.)

		Breakpoints in region 65	
Deletion	A	A2–3;	D2–3
	B	C2–3;	E4–F1
	C	D2–3;	F4–5
	D	D4–E1;	F3–4

		Breakpoints	
Inversions	A	Band 65A6	Band 82A1
	B	Band 65B4	Band 98A3

Flies with a chromosome containing one of these six rearrangements (deletions or inversions) were mated to flies homozygous for both *javelin* and *henna.* The phenotypes of the heterozygous progeny (that is, *rearrangement / javelin, henna*) are shown here.

		Phenotypes of F_1 flies
Deletion	A	javelin, henna
	B	henna
	C	wild type
	D	wild type
Inversions	A	javelin
	B	wild type

Using this data, what can you conclude about the cytogenetic location for the *javelin* and *henna* genes?

13-5 Four strains of *Drosophila* were constructed in which one autosome contained recessive mutant alleles of the four genes *rolled eyes, thick legs, straw bristles,* and *apterous wings,* and the homologous autosome contained one of four different deletions (deletions 1–4). The phenotypes of the flies were as follows:

Deletion	Phenotype
1	rolled eyes, straw bristles
2	apterous wings, rolled eyes
3	thick legs, straw bristles
4	apterous wings

Whole-genome DNA was prepared from the flies. The DNA was digested to completion with the restriction enzyme *Bam*HI, run on an agarose gel, and transferred to nitrocellulose filters. The filters were then probed with a 20 kb cloned piece of wild-type genomic DNA obtained by partially digesting the plasmid clone with *Bam*HI (so the ends of the probe were *Bam*HI ends, but the piece was not digested into all the possible *Bam*HI fragments). The results of this whole-genome Southern blot are shown here. Dark bands indicate fragments present twice in the diploid genome; light bands indicate fragments present once in the genome.

6.3	6.3	6.3	6.3
5.6	5.6	5.6	5.6
5.1			
			4.9
4.2	4.2	4.2	4.2
3.8			
	3.3		
3.0	3.0	3.0	3.0
		1.7	
0.9	0.9	0.9	0.9

Deletion   1   2   3   4

a. Make a map of the *Bam*HI restriction sites in this 20 kb part of the wild-type *Drosophila* genome, indicating distances in kilobases between adjacent *Bam*HI sites. (Hint: The genomic DNA fragments in wild type are 6.3, 5.6, 4.2, 3.0, and 0.9 kb long.)

b. On your map, indicate the locations of the genes.

13-6 The partially recessive, X-linked z^1 mutation of the *Drosophila* gene *zeste* (*z*) can produce a yellow (zeste) eye color only in flies that have two or more copies of the wild-type *white* (*w*) gene. Using this property, tandem duplications of the w^+ gene called w^{+R} were identified. Males with the genotype $z^1\ w^{+R}$ / Y thus have zeste eyes. These males were crossed to females with the genotype $y\ z^1\ w^{+R}\ spl\ /\ y^+\ z^1\ w^{+R}\ spl^+$. (These four genes are closely linked on the X chromosome, in the order given in the genotype, with the centromere to the right of all these genes: *y* = yellow bodies; y^+ = tan bodies; *spl* = split bristles; spl^+ = normal bristles.) Out of 81,540 male progeny of these females, the following exceptions were found:

Class A	2430 yellow bodies, zeste eyes, wild-type bristles
Class B	2394 tan bodies, zeste eyes, split bristles
Class C	23 yellow bodies, wild-type eyes, wild-type bristles
Class D	22 tan bodies, wild-type eyes, split bristles

a. What were the phenotypes of the remainder of the 81,540 males from the first cross?

b. What events gave rise to progeny of classes A and B?

c. What events gave rise to progeny of classes C and D?

d. On the basis of these experiments, what is the genetic distance between *y* and *spl*?

13-7 Two yeast strains were mated and sporulated (allowed to carry out meiosis). One strain was a haploid with normal chromosomes and the linked genetic markers *ura3* (requires uracil for growth) and *arg9* (requires arginine for growth) surrounding their centromere. The other strain was wild type for the two markers (*URA3* and *ARG9*) but had an inversion in this region of the chromosome (pink) as shown here:

During meiosis, several different kinds of crossover events could occur. For each of the following events, give the genotype and phenotype of the resulting four haploid spores. Assume that any chromosomal deficiencies are lethal in haploid yeast. Do not consider crossovers between sister chromatids.

a. a single crossover outside the inverted region

b. a single crossover between *URA3* and the centromere

c. a double crossover involving the same two chromatids each time, where one crossover occurs between *URA3* and the centromere and the other occurs between *ARG9* and the centromere

13-8 Suppose a haploid yeast strain carrying two recessive linked markers *his4* and *leu2* were crossed with a strain that was wild type for *HIS4* and *LEU2* but had an inversion of this region of the chromosome as shown here in blue.

Several different kinds of crossover events could occur during meiosis in the resulting diploid. For each of the following events state the genotype and phenotype of the resulting four haploid spores. Do not consider crossover events between chromatids attached to the same centromere.

a. a single crossover between the markers *HIS4* and *LEU2*

b. a double crossover involving the same chromatids each time, where both crossovers occur between the markers *HIS4* and *LEU2*

c. a single crossover between the centromere and the beginning of the inverted region

13-9 In the mating between two haploid yeast strains depicted in the preceding problem 13–8, describe a scenario that would result in a tetratype ascus in which all four spores are viable.

13-10 Genes *a* and *b* are 21 m.u. apart when mapped in highly inbred strain 1 of corn, and 21 m.u. apart when mapped in highly inbred strain 2. But when the distance is mapped by testcrossing the F₁ progeny of a cross between strains 1 and 2, the two genes are only 1.5 m.u. apart. What arrangement of genes *a* and *b* and any potential rearrangement breakpoints could explain these results?

13-11 In the following group of figures, the pink lines indicate an area of a chromosome that is inverted relative to the normal (black line) order of genes. The diploid chromosome constitution of individuals 1–4 is shown. Match the individuals with the appropriate statement(s) that follow. More than one diagram may correspond to the following statements, and a diagram may be a correct answer for more than one question.

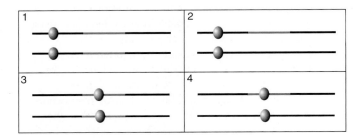

a. An inversion loop would form during meiosis I and in polytene chromosomes.
b. A single crossover involving the inverted region on one chromosome and the homologous region of the other chromosome would yield genetically imbalanced gametes.
c. A single crossover involving the inverted region on one chromosome and the homologous region of the other chromosome would yield an acentric fragment.
d. A single crossover involving the inverted region yields four viable gametes.

13-12 Three strains of *Drosophila* (Bravo, X-ray, and Zorro) are obtained that are homozygous for three variant forms of a particular chromosome. When examined in salivary gland polytene chromosome spreads, all chromosomes have the same number of bands in all three strains. When genetic mapping is performed in the Bravo strain, the following map is obtained (distances in map units).

Bravo and X-ray flies are now mated to form Bravo / X-ray F₁ progeny, and Bravo flies are also mated with Zorro flies to form Bravo / Zorro F₁ progeny. In subsequent crosses, the following genetic distances were found to separate the various genes in the hybrids:

	Bravo / X-ray	Bravo / Zorro
a–b	5.2	5.2
b–c	6.8	0.7
c–d	0.2	<0.1
d–e	<0.1	<0.1
e–f	<0.1	<0.1
f–g	0.65	0.7
g–h	3.2	3.2

a. Make a map showing the relative order of genes *a* through *h* in the X-ray and Zorro strains. Do not show distances between genes.
b. In the original X-ray homozygotes, would the physical distance between genes *c* and *d* be greater than, less than, or approximately equal to the physical distance between these same genes in the original Bravo homozygotes?
c. In the original X-ray homozygotes, would the physical distance between genes *d* and *e* be greater than, less than, or approximately equal to the physical distance between these same genes in the original Bravo homozygotes?

13-13 Semisterility in corn, as seen by unfilled ears with gaps due to abortion of approximately half the ovules, is an indication that the strain is a translocation heterozygote. The chromosomes involved in the translocation can be identified by crossing the translocation heterozygote to a strain homozygous recessive for a gene on the chromosome being tested. The ratio of phenotypic classes produced from crossing semisterile F₁ progeny back to a homozygous recessive plant indicates whether the gene is on one of the chromosomes involved in the translocation. For example, a semisterile strain could be crossed to a strain homozygous for the *yg* mutation on chromosome 9. (The mutant has yellow-green leaves instead of the wild-type green leaves.) The semisterile F₁ progeny would then be backcrossed to the homozygous *yg* mutant.
a. What types of progeny (fertile or semisterile, green or yellow-green) would you predict from the backcross of the F₁ to the homozygous *yg* mutant if the gene was not on one of the two chromosomes involved in the translocation?
b. What types of progeny (fertile or semisterile, green or yellow-green) would you predict from the backcross of the F₁ to the homozygous mutant if the

yg gene is on one of the two chromosomes involved in the translocation?

c. If the *yg* gene is located on one of the chromosomes involved in the translocation, a few fertile, green progeny and a few semisterile, yellow-green progeny are produced. How could these relatively rare progeny classes arise? What genetic distance could you determine from the frequency of these rare progeny?

13-14 During ascus formation in *Neurospora,* any ascospore with a chromosomal deletion dies and appears white in color. How many ascospores of the eight spores in the ascus would be white if the octad came from a cross of a wild-type strain with a strain of the opposite mating type carrying:

    a. a paracentric inversion, and no crossovers occurred between normal and inverted chromosomes?

    b. a pericentric inversion, and a single crossover occurred in the inversion loop?

    c. a paracentric inversion, and a single crossover occurred outside the inversion loop?

    d. a reciprocal translocation, and an adjacent-1 segregation occurred with no crossovers between translocated chromosomes?

    e. a reciprocal translocation, and alternate segregation occurred with no crossovers between translocated chromosomes?

    f. a reciprocal translocation, and alternate segregation occurred with one crossover between translocated chromosomes (but not between the translocation breakpoint and the centromere of any chromosome)?

13-15 In the following figure, black and pink lines represent nonhomologous chromosomes. Which of the figures matches the description? More than one diagram may correspond to the following statements, and a diagram may be a correct answer for more than one question.

    a. gametes produced by a translocation heterozygote

    b. gametes that could not be produced by a translocation heterozygote

    c. genetically balanced gametes produced by a translocation heterozygote

    d. genetically imbalanced gametes that can be produced (at any frequency) by a translocation heterozygote

13-16 In *Drosophila,* the gene for cinnabar eye color is on chromosome 2 and the gene for scarlet eye color is on chromosome 3. A fly homozygous for both recessive *cinnabar* and *scarlet* alleles (*cn / cn ; st / st*) is white-eyed.

    a. If male flies (containing chromosomes with the normal gene order) heterozygous for *cn* and *st* alleles are crossed to white-eyed females homozygous for the *cn* and *st* alleles, what are the expected phenotypes and their frequencies in the progeny?

    b. One unusual male heterozygous for *cn* and *st* alleles, when crossed to a white-eyed female, produced only wild-type and white-eyed progeny. Explain the likely chromosomal constitution of this male.

    c. When the wild-type F_1 females from the cross with the unusual male were backcrossed to normal *cn / cn ; st / st* males, the following results were obtained:

wild type	45%
cinnabar	5%
scarlet	5%
white	45%

Diagram a genetic event at metaphase I that could produce the rare cinnabar or scarlet flies among the progeny of the wild-type F_1 females.

13-17 Human chromosome 1 is a large, metacentric chromosome. A map of a cloned region from near the telomere of chromosome 1 is shown here. Three probe DNAs (A, B, and C) from this region were used for *in situ* hybridization to human mitotic metaphase chromosome squashes made with cells obtained from individuals with various genotypes. The breakpoints of chromosomal rearrangements in this region are indicated on the map. The black bars for Deletions (*Del*) 1 and 2 represent DNA that is deleted. The breakpoints of Inversions (*Inv*) 1 and 2 not shown in the figure are near but not at the centromere. For each of the following genotypes, draw chromosome 1 as it would appear after *in situ* hybridization. An example is shown on the next page for hybridization of probe A to the two copies of chromosome 1 in wild type (*+/ +*).

Genotype: Wild type
Probe: A

Hybridization →

a. genotype: *Del1/Del2;* probe: B
b. genotype: *Del1/Del2;* probe: C
c. genotype: *Del1/+;* probe: A
d. genotype: *Inv1/+;* probe: A
e. genotype: *Inv2/+;* probe: B
f. genotype: *Inv2/Inv2;* probe: C

13-18 A proposed biological method for insect control involves the release of insects that could interfere with the fertility of the normal resident insects. One approach is to introduce sterile males to compete with the resident fertile males for matings. A disadvantage of this strategy is that the irradiated sterile males are not very robust and can have problems competing with the fertile males. An alternate approach that is being tried is to release laboratory-reared insects that are homozygous for several translocations. Explain how this strategy will work. Be sure to mention which insects will be sterile.

13-19 A *Drosophila* male is heterozygous for a translocation between an autosome originally bearing the dominant mutation *Lyra* (shortened wings) and the Y chromosome; the other copy of the same autosome is *Lyra⁺*. This male is now mated with a true-breeding, wild-type female. What kinds of progeny would be obtained, and in what proportions?

13-20 The picture at the beginning of this chapter on p. 441 shows a polytene chromosome preparation from a fruit fly heterozygous for a chromosome carrying a very large composite transposon and a wild-type homolog. Suppose you had a probe made from wild-type DNA sequences that span the site into which the transposon is inserted. Diagram the pattern of *in situ* hybridization you would expect on the polytene chromosome preparation shown on p. 441.

13-21 a. Among the selfed progeny of a semisterile corn plant that is heterozygous for a reciprocal translocation, what ratio do you expect for progeny plants with normal fertility versus those showing semisterility? In this problem, ignore the rare gametes produced by adjacent-2 segregation.
b. Among the selfed progeny of a particular semisterile corn plant heterozygous for a reciprocal translocation, the ratio of fertile to semisterile plants was

1:4. How can you explain this deviation from your answer to part (a)?

13-22 Explain how transposable elements can cause movement of genes that are not part of the transposable element.

13-23 In the 1950s, Barbara McClintock found a transposable element in corn she called *Ds* (*Dissociator*). When inserted at a particular location, this element could often cause chromosomal breaks at that site, but these breaks occurred only in the presence of another unlinked genetic element she called *Ac* (*Activator*). She found further that in the presence of *Ac, Ds* could jump to other chromosomal locations. At some of these locations (and in the presence of *Ac*) *Ds* would now cause chromosomal breakage at the new position; at other positions, it appeared that *Ds* could cause new mutations that were unstable as shown by their patchy, variegated expression in kernels. Interestingly, the position of the *Ac* element seemed to be very different in various strains of corn. Explain these results in terms of our present-day understanding of transposons.

13-24 Gerasimova and colleagues in the former Soviet Union characterized a mutation in the *Drosophila* cut wing (*ct*) gene called *ct^{MR2}*, which is associated with the insertion of a transposable element called *gypsy*. This allele is very unstable: approximately 1 in 100 of the progeny of flies bearing *ct^{MR2}* show new *ct* variants. Some of these are *ct⁺* revertants, while others appear to be more severe alleles of *ct* with stronger effects on wing shape. When the *ct⁺* revertants themselves are mated, some of the *ct⁺* alleles appear to be stable (no new *ct* mutants appear) while others are highly unstable (many new mutations appear). What might explain the generation of stable and unstable *ct⁺* revertants as well as the stronger *ct* mutant alleles?

13-25 In sequencing a region of the human genome, you have come across a segment of about 200 A residues. You suspect that the sequence preceding the A residues may have been moved here by a transposition event mediated by reverse transcriptase. If the adjacent sequence is in fact a retroposon, you might expect to find other copies in the genome. How could you determine if other copies of this DNA exist?

13-26 The *Eco*RI restriction map of the region in which a coat color gene in mice is located is presented on the next page. The left-most *Eco*RI site is arbitrarily labeled 0 and the other distances in kilobases are given relative to this coordinate. Genomic DNA was prepared from one wild-type mouse and 10 mice homozygous for various mutant alleles. This genomic DNA is digested with *Eco*RI, fractionated on agarose gels, and then transferred to nitrocellulose filters. The filters were probed with the radioactive DNA fragment indicated by the purple bar, extending from coordinate 2.6 kb to coordinate 14.5. The resultant autoradiogram is shown schematically.

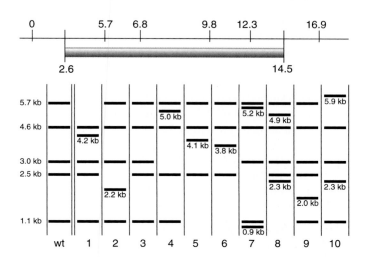

Assume that each of the mutations 1–10 is caused by one and only one of the events on the following list. Which event corresponds to which mutation?

a. a point mutation exactly at coordinate 6.8
b. a point mutation exactly at coordinate 6.9
c. a deletion between coordinates 10.1 and 10.4
d. a deletion between coordinates 6.7 and 7.0
e. insertion of a transposable element at coordinate 6.2
f. an inversion with breakpoints at coordinates 2.2 and 9.9
g. a reciprocal translocation with another chromosome with a breakpoint at coordinate 10.1
h. a reciprocal translocation with another chromosome with a breakpoint at coordinate 2.4
i. a tandem duplication of sequences between coordinates 7.2 and 9.2
j. a tandem duplication of sequences between coordinates 11.3 and 13.3

13-27 The number of chromosomes in the somatic cells of several oat varieties (*Avena* species) are: sand oats (*Avena strigosa*)—14; slender wild oats (*Avena barata*)—28; and cultivated oats (*Avena sativa*)—42.
a. What is the basic chromosome number (*x*) in *Avena?*
b. What is the ploidy for each of the different species?
c. What is the number of chromosomes in the gametes produced by each of these oat varieties?
d. What is the *n* number of chromosomes in each species?

13-28 Common red clover, *Trifolium pratense,* is a diploid with 14 chromosomes per somatic cell. What would be the somatic chromosome number of:
a. a trisomic variant of this species?
b. a monosomic variant of this species?

c. a triploid variant of this species?
d. an autotetraploid variant?

13-29 Fred and Mary have a child with Down syndrome. A probe derived from chromosome 21 was used to identify RFLPs in Fred, Mary, and the child (darker bands indicate signals of twice the intensity). Explain what kind of nondisjunction events must have occurred to produce the child if the child's RFLP pattern looked like that in lanes A, B, C, or D of the following figure:

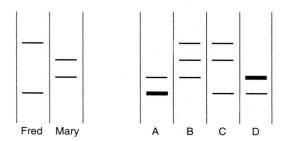

13-30 *Uniparental disomy* is a rare phenomenon in which only one of the parents of a child with a recessive disorder is a carrier for that trait; the other parent is homozygous normal. By analyzing DNA polymorphisms, it is clear that the child received both mutant alleles from the carrier parent but did not receive any copy of the gene from the other parent.
a. Diagram at least two ways in which uniparental disomy could arise. (Hint: These mechanisms all require more than one error in cell division, explaining why uniparental disomy is so rare.) Is there any way to distinguish between these mechanisms to explain any particular case of uniparental disomy?
b. How might the phenomenon of uniparental disomy explain rare cases in which girls are affected with rare X-linked recessive disorders but have unaffected fathers, or other cases in which an X-linked recessive disorder is transmitted from father to son?
c. If you were a human geneticist and believed one of your patients had a disease syndrome caused by uniparental disomy, how could you establish that the cause was not instead mitotic recombination early in the patient's development from a zygote?

13-31 Human geneticists interested in the effects of abnormalities in chromosome number often karyotype tissue obtained from spontaneous abortions. About 35% of these samples show autosomal trisomies, but only about 3% of the samples display autosomal monosomies. Based on the kinds of errors that can give rise to aneuploidy, would you expect that the frequencies of autosomal trisomy and autosomal monosomy should be more equal? Why or why not? If you think the frequencies should be more equal, how can you explain the large excess of trisomies as opposed to monosomies?

13-32 Among adults with Turner syndrome, it has been found that a very high proportion are genetic mosaics. These are of two types: In some individuals, the majority of cells are 45, XO but a minority of cells are 46, XX. In other Turner individuals, the majority of cells are 45, XO but a minority of cells are 46, XY. Explain how these somatic mosaics could arise.

13-33 The *Drosophila* chromosome 4 is extremely small; there is virtually no recombination between genes on this chromosome. You have available three differently-marked chromosome 4s: one has a recessive allele of the gene *eyeless* (*ey*), causing very small eyes; one has a recessive allele of the *cubitus interruptus* (*ci*) gene, which causes disruptions in the veins on the wings; and the third carries the recessive alleles of both genes. *Drosophila* adults can survive with two or three, but not with one or four, copies of chromosome 4.

   a. How could you use these three chromosomes to find *Drosophila* mutants with defective meioses causing an elevated rate of nondisjunction?

   b. Would your technique allow you to discriminate nondisjunction occurring during the first meiotic division from nondisjunction occurring during the second meiotic division?

   c. What progeny would you expect if a fly recognizably formed from a gamete produced by nondisjunction were test-crossed to a fly homozygous for a chromosome 4 carrying both *ey* and *ci?*

   d. Geneticists have isolated so-called *compound 4th chromosomes* in which two entire chromosome 4s are attached to the same centromere. How can such chromosomes be used to identify mutations causing increased meiotic nondisjunction? Are there any advantages relative to the method you described in part (a)?

13-34 In *Neurospora, his2* mutants require the amino acid histidine for growth, and *lys4* mutants require the amino acid lysine. The two genes are on the same arm

of the same chromosome, in the order centromere - *his2* - *lys4*. A *his2* mutant is mated with a *lys4* mutant. Draw all of the possible ordered asci that could result from meioses in which the following events occurred, accounting for the nutritional requirements for each ascospore. Ascospores without any copy of a chromosome will abort and die, turning white in the process.

   a. a single crossover between the centromere and *his2*

   b. a single crossover between *his2* and *lys4*

   c. nondisjunction during the first meiotic division

   d. nondisjunction during the second meiotic division

   e. a single crossover between the centromere and *his2,* followed by nondisjunction during the first meiotic division

   f. a single crossover between *his2* and *lys4,* followed by nondisjunction during the first meiotic division

13-35 You have haploid tobacco cells in culture and have made transgenic cells that are resistant to herbicide. What would you do to obtain a diploid cell line that could be used to generate a new fertile herbicide-resistant plant?

13-36 An allotetraploid species has a genome composed of two ancestral genomes, A and B, each of which have a basic chromosome number (x) of seven. In this species the two copies of each chromosome of each ancestral genome pair only with each other during meiosis. Resistance to a pathogen that attacks the foliage of the plant is controlled by a dominant allele at the F locus. The recessive alleles F^a and F^b confer sensitivity to the pathogen, but the dominant resistance alleles present in the two genomes have slightly different effects. Plants with at least one F^A allele are resistant to races 1 and 2 of the pathogen regardless of the genotype in the B genome, and plants with at least one F^B allele are resistant to races 1 and 3 of the pathogen regardless of the genotype in the A genome. What proportion of the self progeny of an $F^A F^a F^B F^b$ plant will be resistant to all three races of the pathogen?

13-37 Using karyotype analysis, how could you distinguish between autopolyploids and allopolyploids?

14

The Prokaryotic Chromosome: Genetic Analysis in Bacteria

Some species of bacteria can live in environments as hostile as hot springs (such as this beautiful pool in Yellowstone National Park, Wyoming). Comparative genome analyses of bacteria that live in unusual environments will increase our understanding of the adaptations that allow survival in different niches.

Gonorrhea, a sexually transmitted infection of the urogenital tract in men and women, is on the rise in many parts of the world. Caused by the bacterium *Neisseria gonorrhoeae,* gonorrhea is rarely fatal, but in men it can spread from an initial site, usually the urethra, to the prostate gland and the epididymis (two structures that play a role in sperm production), diminishing sperm count; and in women, it can move from the cervix to the uterine lining and fallopian tubes, leading to sterility. A common symptom is painful urination, although many infected people have no symptoms at all. Because infants passing through a gonorrhea-infected birth canal can contract severe eye infections, hospitals routinely treat the eyes of newborns with a few drops of silver nitrate solution or penicillin. If left untreated, gonorrhea can spread through the bloodstream to the joints, skin, bones, tendons, and heart. Until the late 1970s, a few shots of penicillin were a surefire cure for gonorrhea, but by 1995, more than 20% of *N. gonorrhoeae* bacteria isolated from patients worldwide was resistant to penicillin.

Geneticists now know that the agent of this alarming increase in antibiotic resistance was the transfer of DNA from one bacterium to another. According to epidemiologists, penicillin-resistant *N. gonorrhoeae* bacteria first appeared in Asia in the 1980s, in a patient receiving penicillin treatment for gonorrhea who was also fighting an infection caused by another species of bacteria—*Haemophilus influenzae.* Some of the patient's *H. influenzae* bacteria apparently carried a plasmid, a small, circular molecule of double-stranded DNA (see Chapter 9), that contained a gene encoding penicillinase, an enzyme that destroys penicillin. When the doubly infected patient mounted a specific immune response to *H. influenzae* that degraded these cells, the broken bacteria released their plasmids. Some of the freed circles of DNA entered *N. gonorrhoeae* cells, transforming them to penicillin-resistant bacteria. The transformed gonorrhea bacteria then multiplied, and successive exposures to penicillin selected for the resistant bacteria. As a result, the patient transmitted penicillin-resistant *N. gonorrhoeae* to subsequent sexual partners. Thus, while penicillin treatment does not create the genes for resistance, it fast-forwards the spread of those genes. After continuous, long-term exposure to the drug, only cells resistant to penicillin survive. Today in the United

States, many *N. gonorrhoeae* are simultaneously resistant to penicillin and two other antibiotics—spectinomycin and tetracycline.

In this chapter, we survey the remarkable diversity of bacteria and discuss the structure of bacterial genomes. We also examine the mechanisms by which bacteria transfer genes between cells of the same species, between cells of distantly related species, and between bacterial cells and bacterial viruses. Geneticists can use their knowledge of bacterial genetics in general and of various forms of bacterial gene transfer in particular to identify, map, and characterize the genes that contribute to biological processes. Such genetic analysis has made it possible to dissect and reconstruct, for example, the biochemical pathways by which bacteria move toward food and away from toxins.

As we describe how bacteria replicate their DNA, reproduce, and transfer genes between cells, we present:

- A general overview of bacteria: their numbers; their range of sizes, habitats, and metabolic activity; and how researchers grow them for study.

- The bacterial genome: the structure, organization, transcription, replication, and evolution of large, circular bacterial chromosomes; and the structure and function of smaller circles of DNA known as plasmids and episomes.

- Gene transfer in bacteria: the mechanisms of transformation, conjugation, and transduction.

- A comprehensive example: using the tools of bacterial genetics to dissect bacterial chemotaxis.

- The impact of genomics on bacterial genetics.

A General Overview of Prokaryotes

Bacteria such as *Neisseria gonorrhoeae* and *Haemophilus influenzae* constitute one of the three major evolutionary lineages of living organisms (Fig. 14.1). The two other lineages are the **eukaryotes** (organisms whose cells have nuclei encased in a membrane) and the **archaea.** Bacteria and archaea are both classified as **prokaryotes:** organisms lacking a nuclear membrane, whose chromosomes are therefore not enclosed in a nucleus. Archaea and bacteria are so similar in appearance that it is often difficult to distinguish them in the light microscope. Most of the few known groups of archaea live under extreme conditions in unusual ecological niches such as very salty surroundings, anaerobic environments that do not contain oxygen, or environments with extremely high temperatures.

In 1996 researchers determined the complete DNA sequence of the genome of one organism in the archaea lineage: *Methanococcus jannaschii,* a methane-producing anaerobe. The results provided evidence that bacteria, archaea, and eukaryotes are of equal taxonomic rank. Some genes in archaea resemble those in bacteria, while others are similar to genes in eukaryotes. Remarkably, more than 50% of *M. jannaschii* genes are completely different from any bacterial or eukaryotic genes. On the basis of these comparisons, many evolutionary biologists proposed that the earliest organisms, presumably single-celled prokaryotes, existed roughly 3.5 billion years ago. A billion years later, bacteria and archaea diverged from their common ancestor, and somewhat later eukaryotes split off from the archaea evolutionary line (Fig. 14.1). The complete sequences of several additional archaea species, obtained since 1996, have shown that most of the genes for replication, transcription, and translation are similar to eukaryotic genes, while the genes involved in some of the metabolic pathways that allow organisms to survive in unusual habitats are similar to bacterial genes. Archaea's mixed genetic heritage has lead to the hypothesis that the lateral transfer of genes between species was a common occurrence, especially early in the evolution of the three lineages. For example, by some estimates, up to 18% of the *E.coli* genome consists of laterally transferred genes.

Despite the evolutionary distance separating bacteria from archaea, living representatives of the two lineages show marked similarities in morphology, genome structure, and the mechanisms of gene transfer. Thus, although our discussion in the remainder of this chapter focuses on bacteria (sometimes referred to as "eubacteria"), it is possible to apply to archaea many of the lessons learned from the other prokaryotes.

Figure 14.1   A family tree of living organisms. There are many diverse present-day forms of bacteria and archaea. The three major evolutionary lineages of living organisms are bacteria and archaea (both prokaryotes) and eukaryotes. DNA sequencing results suggest that all living forms descended from a common prokaryotic ancestor that lived 3.5 billion years ago.

The Immense Diversity of Bacteria

Bacteria are crucial to the maintenance of earth's environment. Various species release oxygen into the atmosphere; recycle carbon, nitrogen, and other elements; and digest human and other animal wastes as well as pesticides and other pollutants, which would otherwise eventually poison the air, soil, and water. Bacteria also cause hundreds of animal and plant diseases, but even so, harmful species are a small fraction of all bacteria. Many species, in fact, produce vitamins and other materials essential to the health and survival of humans and other organisms. Two key features of bacteria are their astounding ability to proliferate and their enormous diversity.

Bacteria Outnumber All Other Organisms on Earth

A human adult carries at least 100 g (roughly a quarter pound) of live bacteria, mainly in the intestines. An estimated 10^{14} bacteria make up those 100 g, a number many thousands of times as great as the number of people on the planet.

Researchers Have Identified 10,000 Species of Bacteria Exhibiting an Amazing Range of Sizes and Lifestyles

The smallest bacteria are about 200 nanometers (nm = billionths of a meter) in diameter. The largest are 500 micrometers (μm = millionths of a meter) in length, which makes them 10 billion times larger in volume and mass than the smallest bacterial cells. These large bacteria are visible without the aid of the microscope. Some bacteria live independently on land, others float freely in aquatic environments, and still others live as parasites or symbionts inside other life-forms (Table 14.1).

Bacteria Display a Remarkable Metabolic Diversity That Enables Them to Live Almost Everywhere

Some soil bacteria obtain the energy to fuel their metabolism from the chemical ammonia, while other, photosynthesizing bacteria obtain their energy from sunlight. Because of their metabolic diversity, bacteria play essential roles in many natural processes, such as the decomposition of materials essential for nutrient cycling. The balance of microorganisms is a key to the success of these ecological processes, which help maintain the environment. In the cycling of nitrogen, for example, decomposing bacteria break down plant and animal matter rich in nitrogen and produce ammonia. Nitrifying bacteria then use this ammonia as a source of energy and release nitrate (NO_3), which some plants can use as is; denitrifying bacteria convert the nitrate not used directly by plants to atmospheric nitrogen (N_2); and nitrogen-fixing bacteria, such as *Rhizobium,* that live in the roots of peas and other leguminous plants convert N_2 to ammonium, which their host plants can use.

Recently, geneticists and molecular biologists have used microbes to isolate unusual enzymes that carry out natural and industrial processes. They then clone and manipulate the genes encoding these enzymes. One use of unusual microorganisms and their enzymes is the development of bioremediation bacteria that can, for example, break down the hydrocarbons found in oil. In 1989 when the tanker *Exxon Valdez* spilled millions of barrels of oil along the Alaskan coastline, cleanup crews used oil-digesting bacteria to help revive the environment.

Despite Their Diversity, Bacterial Cells Have a Number of Features in Common

Although bacteria come in a variety of shapes and sizes adapted to a range of habitats, all lack a defined nuclear membrane as well as membrane-bounded organelles, such as the mitochondria and chloroplasts found in eukaryotic cells. Bacterial chromosomes fold to form a dense **nucleoid body** that appears to exclude ribosomes, which function in the surrounding cytoplasm. A membrane encloses bacterial cells, and a membranous structure known as the *mesosome,* which apparently serves as a source of new membranes during cell division, projects from the membrane into the cell's interior. In most species of bacteria, the membrane is supported by a cell wall composed of carbohydrate and peptide polymers. Some bacteria have, in addition to the cell wall, a thick, mucuslike coating called a *capsule* that helps the cell resist attack by the immune system. Many bacteria have flagella that propel them toward food or light.

The Power of Bacterial Genetics Is the Potential for Studying Rare Events

Researchers grow bacteria in liquid media or on media solidified by agar (Fig. 14.2). In a liquid medium, the cells of commonly studied species, such as *E. coli,* grow to a concentration of 10^9 cells per milliliter within a day. In agar-solidified medium, a single bacterium will multiply to a visible colony containing 10^7 or 10^8 cells in less than one day. The ability to grow large numbers of cells is one advantage that has made bacteria, especially *E. coli,* so attractive for genetic studies.

Genetic studies of bacteria require techniques to count these large numbers of cells and to isolate individual cells of interest. Researchers can use a solid medium to calculate the number of cells in a liquid culture. They begin with sequential dilutions (illustrated in Fig. 7.15a) of cells in the liquid medium. They then spread a small sample of the diluted solutions on agar-medium plates and count the number of colonies that form. Although it is still difficult to work with a single bacterial cell, except in very specialized studies, the cells constituting a single colony contain the genetically identical descendants of the one bacterial cell that founded the colony.

Focus on E. coli

The most studied and best understood species of bacteria is *Escherichia coli,* a common inhabitant of the intestines of warm-blooded animals (Fig. 14.3). *E. coli* cells can grow in the complete absence of oxygen—the condition found in the intestines—or in air. The *E. coli* strains studied in the laboratory are not pathogenic, but other strains of the species can cause a variety of intestinal diseases, most of them mild, a few life-threatening.

TABLE 14.1   Characteristics and Significance of Various Groups of Bacteria

Domain and Evolutionary Group	Groups Based on Structure and Physiology	Characteristics and Significance
Bacteria		
Purple bacteria and relatives	Purple photosynthetic bacteria	Some can form a purple layer in anaerobic zone of lakes; mitochondria arose from ancient purple bacteria similar to *Rhodospirillum*
	Pseudomonads	Versatile bacteria in the soil; useful in cleaning up industrial pollutants
	Enteric bacteria	Inhabit intestinal tracts, aiding digestion (*Escherichia coli*) or causing disease (*Salmonella*)
	Myxobacteria	Gliding bacteria whose colonies make complex multicellular spore-forming stalks
	Nitrogen fixers	Free in soil or associated with plants (*Rhizobium*); fix nitrogen in biological form
	Rickettsias	Live only inside other cells; cause typhus fever and Rocky Mountain spotted fever
Cyanobacteria	Cyanobacteria	Have photosynthetic system like a plant and are nitrogen fixing; most are self-sufficient organisms; earliest organism to release oxygen; precursors of chloroplasts
Gram-positive bacteria	Endospore formers	Heat-resistant spores can cause botulism or tetanus (*Clostridium* and *Bacillus*)
	Lactic acid bacteria	*Lactobacillus* make yogurt; *Streptococcus* causes strep throat
	Gram-positive cocci	*Staphylococcus* causes toxic shock syndrome and other infections
	Actinomycetes	Cause leprosy and tuberculosis; produce antibiotics
	Mycoplasmas	Smallest of all cells, no cell wall; one cause of pneumonia
Chlamydiae	Chlamydia	Leading cause of blindness in humans, cause of most common sexually transmitted disease, cells live inside other cells, can't make ATP
Bacterioides and relatives	A branch of the gram-negative rods	Major anaerobic cells in the large intestine
Green sulfur bacteria	A branch of the phototrophic bacteria	Live on the warm fringes of sulfur hot springs
Spirochetes	Spirochetes	Corkscrew shape; cause Lyme disease and syphilis
Deinococci and relatives	A branch of gram-positive cocci	Highly resistant to radiation; found near atomic reactors
Green nonsulfur bacteria	A branch of phototrophic bacteria	Grow in mats at the fringes of hot springs
Thermotogas	Anaerobic fermenters	Live in oceanic vents at highest temperature of any true bacteria

(a)

(b)

Figure 14.2 Bacteria in the laboratory. Bacteria grow as **(a)** colonies on solid nutrient-agar plates or **(b)** as suspension in a liquid medium.

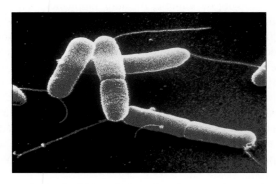

Figure 14.3 *Escherichia coli.* Scanning electron micrograph of *E. coli* (14,000 ×).

Because *E. coli* encodes all the enzymes it needs for amino-acid and nucleotide biosynthesis, it is a *prototrophic* organism that can grow in *minimal media,* which contain glucose as the only source of carbon and energy, and inorganic salts as the source of the other elements that compose bacterial cells. In a minimal medium, *E. coli* cells divide every hour, redoubling their numbers 24 times a day. In a richer, more complex medium containing sugars and amino acids, *E. coli* cells divide every 20 minutes to produce 72 generations per day. Two days of logarithmic growth at this rate, if unchecked by any limiting factor, would generate a mass of bacteria equal to the mass of the earth.

The rapidity of bacterial multiplication makes it possible to grow an enormous number of cells in a relatively short time and, as a result, to examine very rare genetic events. For example, wild-type *E. coli* cells are normally sensitive to the antibiotic streptomycin. By spreading a billion wild-type bacteria on an agar-medium plate containing streptomycin, you could isolate a few extremely rare mutants that are resistant to streptomycin and arose by chance among the 10^9 cells. It is not as easy to find and examine such rare events with nonmicrobial organisms; in multicellular animals, it is almost impossible.

The Bacterial Genome

The study of bacteria played a critical role in the development of the field of genetics. From the 1940s to the 1970s (considered the era of classical bacterial genetics), virtually everything researchers learned about gene organization, gene expression, and gene regulation came from analyses of bacteria and the *phages* (bacterial viruses) that infect them. The advent of recombinant DNA technology in the 1970s and 1980s depended on an understanding of genes and chromosomes in bacteria.

One Circular Chromosome

The essential component of a typical bacterial genome is the **bacterial chromosome:** a single circular molecule of double-helical DNA. This single closed circle of double-stranded DNA is 4–5 Mb long in commonly studied species. The circular chromosome of *E. coli,* if broken at one point and laid out in a line, would form a DNA molecule 2.4 nm wide and 1.6 mm long, almost a thousand times longer than the *E. coli* cell in which it is found (Fig. 14.4). Inside the cell, the long, circular DNA molecule condenses by supercoiling and looping into a densely packed nucleoid body. During the bacterial cell cycle, each bacterium replicates its circular chromosome and then divides by binary fission into two identical daughter cells, each with its own chromosome, generating two organisms from one. While the majority of bacteria contain a single circular chromosome, there are exceptions. Genomic analyses have shown that some bacteria, such as *Vibrio cholera* (the cause of the disease cholera) carry two essential circular chromosomes. Other bacteria contain linear DNA molecules.

Finding Mutations in Bacterial Genes

Because, with a few exceptions, the bacterial genome consists of one circular chromosome, it is a haploid genome, and the relation between gene mutation and phenotypic variation is relatively straightforward; that is, in the absence of a second, wild-type allele for each gene, all mutations express their phenotype.

Bacteria are so small that the only practical way to examine them is in the colonies of cells they form on a petri plate. Within this constraint, it is still possible to identify many different kinds of mutations. These include:

1. Mutations affecting *colony morphology,* that is, whether a colony is large or small, shiny or dull, round or irregular.

2. Mutations conferring *resistance* to bactericidal agents such as antibiotics or bacteriophages.

3. Mutations that create *auxotrophs* unable to grow and reproduce on minimal medium; auxotrophic mutations are in genes encoding enzymes required to synthesize relatively complex compounds such as amino acids or nucleotide components from simpler materials in the environment.

4. Mutations affecting the *ability of cells to break down and use complicated chemicals* in the environment; for example, the *lacZ* gene in *E. coli* encodes the enzyme

(a)

(b)

Diagrammatic representations
of bacterial chromosome

**Figure 14.4 The bacterial chromosome: A circle of
double-stranded DNA. (a)** An electron micrograph of an *E. coli*
cell that has been lysed, allowing its chromosome to escape.
(b) Chromosomal DNA is shown either as a double helix or as a single
ring in this chapter.

β-galactosidase needed to break down the sugar lactose
into glucose and galactose. Wild-type cells can grow if
lactose, rather than glucose, is the sole source of carbon in
the medium, but *lacZ⁻* mutants cannot.

5. Mutations in *essential* genes whose protein products are
required under all conditions of growth; because a null
mutation in an essential gene would prevent a colony
from growing in any environment, bacteriologists must
work with conditional lethal mutations such as

temperature-sensitive (*ts*) mutations that allow growth at
one temperature but not at another.

Bacteriologists use different techniques to isolate rare mutations of the various classes described previously. With mutations conferring resistance to a particular agent, they can do a
straightforward **selection,** that is, establish conditions in which
only the desired mutant will grow. For example, if wild-type
bacteria are streaked on a petri plate containing the antibiotic
streptomycin, the only colonies to appear will be streptomycin
resistant (Str^r). It is also possible to select for prototrophic revertants of strains carrying auxotrophic mutations by simply
plating cells on minimal medium agar, which does not contain
the compounds auxotrophs require for growth.

Because the key characteristic of most of the other types of
mutants just described is their inability to grow under particular
conditions, it is not possible to select for them directly. Instead,
researchers must identify these mutations by a **genetic screen:**
an examination of each colony in a population for its phenotype.
They can, for example, use a toothpick to transfer cells from a
colony growing on minimal medium supplemented with methionine to a petri plate containing minimal medium without
methionine. Failure of those cells to grow on the unsupple-
mented medium would indicate that the corresponding colony
on the original plate is auxotrophic for methionine.

Spontaneous mutations in specific bacterial genes occur
very rarely, in 1 in 10^6 to 1 in 10^8 cells, depending on the gene.
It would be virtually impossible to identify such rare mutations
if you had to check the phenotype of a million to a hundred million colonies through the individual transfer of each one with a
toothpick. A number of techniques simplify the process. We describe four.

1. *Replica plating* allows the simultaneous transfer of
thousands of colonies from one plate to another (see
Fig. 7.5).

2. *Treatments with mutagens* increase the frequency with
which a mutation in a gene appears in the population
(see Fig. 7.10).

3. *Enrichment* increases the proportion of mutant cells in a
population by exposing the population to agents that kill
wild-type cells. Penicillin is one agent of enrichment; it
acts by disrupting the formation of the cell wall in
growing cells and thus kills cells that can grow but not
cells that cannot grow. Figure 14.5 shows how researchers
use this property of penicillin to enrich the proportion of
auxotrophic cells in a mixed population of auxotrophic
and prototrophic cells.

4. *Testing for visible mutant phenotypes* on a petri plate. In
an important example, *E. coli* producing functional
β-galactosidase (the product of the *lacZ* gene) cleaves the
colorless artificial compound X-Gal, producing a blue
product. Thus *lacZ⁺* colonies turn blue on medium
containing X-Gal, while *lacZ⁻* colonies remain white (the
usual color of the colonies; review Fig. 9.8 on p. 289).

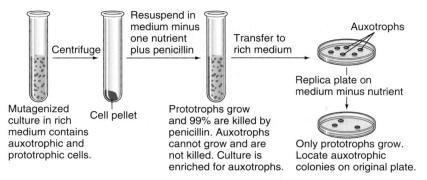

Figure 14.5   Penicillin enrichment for auxotrophic mutants. Penicillin selectively kills growing cells that are making new cell walls, but not bacteria whose growth is arrested. Auxotrophs will not be killed by the penicillin. After enrichment, cells must be screened by replica plating to identify auxotrophs, because penicillin does not kill 100% of the prototrophs.

Researchers designate the genes of bacteria by three lowercase, italicized letters that signify something about the function of the gene. For example, genes in which mutations result in the inability to synthesize the amino acid leucine are *leu* genes. In *E. coli,* there are four *leu* genes—*leuA, leuB, leuC,* and *leuD*—that correspond to the three enzymes (one constructed from two different polypeptides) needed for the synthesis of leucine from other compounds. A mutation in any one of the *leu* genes changes a bacterium into an auxotroph for leucine, that is, into a cell unable to synthesize leucine. Such a cell can grow only in media supplemented with leucine. Mutations in genes required for the breakdown of a sugar (for example, the *lacZ* gene) produce cells unable to grow in medium containing only that sugar (lactose) as a source of carbon. Other types of mutations give rise to antibiotic resistance; str^r is a mutation producing streptomycin resistance. To designate the alleles of genes present in wild-type bacteria, researchers use a superscript "+": leu^+, str^+, $lacZ^+$. To designate mutant alleles they use a superscript "−", as in $leuA^-$ and $lacZ^-$, or a superscript description, as in str^r.

The phenotype of a bacterium that is wild type or mutant for a particular gene is indicated by the three letters that designate the gene, written, however, with an initial capital letter, no italics, and a superscript of minus, plus, or a one-letter abbreviation: Leu^- (requires leucine for growth); Lac^+ (grows on lactose); Str^r (is resistant to streptomycin). A Leu^- *E. coli* strain cannot multiply unless it grows in a medium containing leucine; a Lac^+ strain can grow if lactose replaces the usual glucose in the medium; a Str^r strain can grow in the presence of streptomycin.

species. Some of the sequences identified as genes, however, encode proteins with functions that have not yet been assigned. Such putative genes are known as **open reading frames,** or **ORFs;** they consist of long stretches of codons in the same reading frame uninterrupted by stop codons.

Close to 90% of *E. coli* DNA encodes proteins; on average, every kilobase (kb) of the chromosome contains one gene. This contrasts sharply with the human genome, in which less than 5% of the DNA encodes proteins and there is roughly one gene every 30 kb. One reason for this discrepancy is that *E. coli* genes have no introns. In addition, there is very little repeated DNA in bacteria, and intergenic regions tend to be very small.

The complete sequence of the *E. coli* genome identified 4288 genes, and surprisingly, the function of 40% of the genes remains a mystery at this time. Given the small genome size and the tools developed over the years for genetic analysis in *E. coli,* however, researchers can easily mutate the genes and examine the resulting cells for phenotypic effects. So far, they have grouped genes whose function is known or has been deduced on the basis of sequence into broad functional classes. The 427 genes that are postulated to have a transport function make up the largest class. Of the 182 genes involved in translation, 6 were identified by the sequence analysis. Other classes include the genes for amino-acid biosynthesis and the genes for DNA replication and recombination.

Another interesting feature of the *E. coli* genome is the existence in eight different locations of remnants of bacteriophage genomes. The presence of these sequences suggests an evolutionary history of bacteria that includes invasion by viruses on several occasions.

Structure and Organization of the *E. coli* Chromosome

The circular *E. coli* chromosome consists of about 4.6 million base pairs. In 1997, molecular geneticists completed sequencing the genome of the *E. coli* strain known as K12. From previous genetic work, they knew many of the genes within this sequence; and they could identify others because the polypeptides they encode have amino-acid sequences similar to those of sequenced proteins found in other bacteria or even eukaryotic

Insertion Sequences Dot the *E. coli* Chromosome

DNA sequence analysis of bacterial genomes also revealed the position of several small transposable elements called **insertion sequences (IS's).** These elements, which dot the chromosomes of many types of bacteria, are transposons that do not contain selectable markers (such as genes conferring antibiotic resistance). Researchers have identified several distinct elements ranging in length from 700–5000 bp; they named the elements

IS1, IS2, IS3, and so forth, with the numbers designating the order of discovery. Like the ends of transposons in eukaryotic cells (see Chapter 13), the ends of IS elements are inverted repeats of each other (Fig. 14.6a); and each IS includes a gene encoding a transposase that initiates transposition by recognizing these mirror-image ends. Because insertion sequences can move to other sites on the bacterial chromosome when they transpose, their distribution differs in different strains of the same bacterial species. For example, one strain of *E. coli* may have 15 insertion sequences of five different kinds, while a second strain isolated from a different population may have 25 insertion sequences, lack one of the types found in the first strain, and have a different distribution of IS's around the chromosome. Some bacterial species, such as *Bacillus subtilis,* carry no insertion sequences.

Insertion sequences were first identified in the 1970s as elements that caused inactivation of genes required for galactose metabolism (Gal⁻ mutants). When an IS transposes and lands within the coding region of a gene, it disrupts the coding region and inactivates the gene (Fig. 14.6b). We now know that many of the spontaneous mutations isolated in *E. coli* are the result of IS transposition into a gene. Researchers have exploited this ability to cause mutation by using a more complex type of transposable element in bacteria: a Tn element. In addition to carrying a gene for transposase, Tn elements contain genes conferring resistance to antibiotics or toxic metals such as mercury. One Tn element known as Tn10 consists of two IS10 elements flanking a gene encoding resistance to tetracycline (Fig. 14.6c). After the introduction of Tn10 into a cell, its transposition into, for example, the *lacZ* gene, produces a *lacZ⁻* mutant that is phenotypically both Lac⁻ and Tet^r (resistant to tetracycline). Because of these effects, the Tn element in the gene is an easily scored genetic marker for mapping experiments (as described later). In addition, because researchers know the sequence of Tn10, they can make a primer corresponding to the end of the Tn10 element to determine the base sequence of the adjacent DNA, and thereby identify the gene that was mutated by the Tn10 insertion. This is only one of the many ingenious uses of transposons that researchers have devised.

Transcription in *E. coli*

During the transcription of the *E. coli* chromosome, different strands of DNA serve as the template for different genes. The transcription machinery moves clockwise on one DNA strand for a few kilobases, transcribing several contiguous and often related genes. In an adjacent region, RNA polymerase transcribes the other strand of genomic DNA, moving counterclockwise for a few kilobases. The only rule seems to be that the most highly transcribed genes tend to be oriented in the direction in which the replication fork moves.

DNA Replication in *E. coli*

DNA replication begins at a site called *oriC,* for origin of chromosomal replication, and proceeds simultaneously in both directions (clockwise and counterclockwise) from that point to a termination region called *terC* (Fig. 14.7a). Genes that must undergo a particularly high rate of expression, such as the genes for ribosomal RNA and ribosomal proteins, are often concentrated near the origin of replication. The reason is that during rapid growth, new rounds of replication are initiated before the previous one is completed. This means that there are several replication forks progressing along the chromosome from the single origin of replication at any one time (Fig. 14.7b). As a result, genes near the origin are present in four to eight copies, while genes near the terminator are present in only one copy. The increase in copy number of genes near the origin allows the cell to produce more of the corresponding proteins.

Strain and Species Differences in Bacterial Chromosomes

Chromosomes of *E. coli* strains retrieved from different sources have minor differences in the sequences of individual genes and somewhat greater differences in the regions between genes. Most *E. coli* chromosomes carry a total of one to two dozen copies of four or five different insertion sequences. As we have seen, the IS positions and numbers vary from strain to strain.

Studies comparing the chromosome of a typical *E. coli* strain with the chromosome of a typical strain of *S. typhimurium,* a related species that causes food poisoning in humans and more severe infections in some other mammals,

Figure 14.6 Transposable elements in bacteria. (a) An IS element showing the inverted repeats at each end. **(b)** Insertion of an IS into a gene. Here, insertion of an IS inactivates the *lacZ* gene because the IS contains a transcription termination signal. **(c)** The composite transposon Tn10, in which two slightly different IS10s (IS10L and IS10R) flank 7 kb of DNA including a gene for tetracycline resistance. Because it is flanked by IS10 inverted repeats, Tn10 can be mobilized by the IS10 transposase.

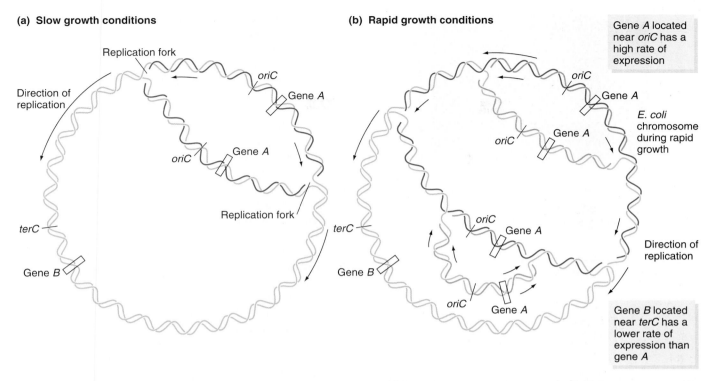

(a) Slow growth conditions

Replication fork

Direction of replication

oriC

Gene *A*

Gene *A*

oriC

Replication fork

terC

Gene *B*

(b) Rapid growth conditions

oriC

Gene *A*

oriC

Gene *A*

terC

Gene *B*

oriC

Gene *A*

oriC

Gene *A*

Direction of replication

Gene *A* located near *oriC* has a high rate of expression

E. coli chromosome during rapid growth

Gene *B* located near *terC* has a lower rate of expression than gene *A*

Figure 14.7 Replication of the *E. coli* chromosome. Replication begins at the origin (*oriC*) and proceeds to the terminator (*terC*). The *arrow* indicates the direction in which the replication fork moves. **(a)** Under conditions of slow growth, the origin of replication "fires" once in each cell division cycle. **(b)** Under conditions of rapid growth, replication initiates more times than the cell divides. As a result, there will be more copies of genes near the origin than genes near the terminator.

reveal more sequence differences between related genes in the two species than between the same genes in two strains of *E. coli.* The insertion sequences found in *S. typhimurium* are also different from those found in *E. coli.* Larger chromosomal differences that distinguish the two species include several inversions of extensive chromosomal regions, the deletion from the *S. typhimurium* chromosome of whole groups of genes that are present in the *E. coli* chromosome; and *S. typhimurium*'s inclusion of several genes that *E. coli* lacks, notably genes involved in pathogenesis. Several factors help maintain the boundaries between even closely related bacterial species. For example, differences in DNA sequence make recombination between two species less efficient than recombination within a species. Moreover, unlike different strains of the same species, different species of bacteria encode restriction enzymes that can destroy the DNA of other species (see the Fast Forward box in Chapter 6 and the Genetics and Society box in Chapter 9 for more about restriction enzymes).

Plasmids: Smaller Circles of DNA That Do Not Carry Essential Genes

Bacteria carry their essential genes—those necessary for growth and reproduction—in their large circular chromosome. In addition, some bacteria carry genes not needed for growth and reproduction under normal conditions in smaller circles of double-stranded DNA known as **plasmids** (Fig. 14.8a). Plasmids

come in a range of sizes. The smallest are 1000 bp long; the largest are several megabases (Mb) in length. Bacteria usually harbor no more than one extremely large plasmid, but they can house several or even hundreds of copies of smaller DNA circles.

Although plasmids carry genes not normally needed by their bacterial hosts for growth and reproduction, these same genes may benefit the host cell under certain conditions. For example, the plasmids in many bacterial species carry genes that protect their hosts against toxic metals such as mercury. The plasmids of various soil-inhabiting *Pseudomonas* species encode proteins that allow the bacteria to metabolize chemicals such as toluene, naphthalene, or petroleum products. Since the 1980s, natural and genetically engineered plasmids of this type have become part of the tool kit for cleaning up oil spills and other contaminated sites. Plasmids thus help expand the capabilities of bacteria in nature, and they also provide a rich source of unusual and useful proteins for commercial purposes.

Many of the genes that contribute to pathogenicity (that is, the capacity for causing disease in plants or animals) reside in plasmids. For example, the toxins produced by *Shigella dysenteriae,* the causative agent of dysentery, are encoded by plasmids. Genes encoding resistance to antibiotics are also often located on plasmids. The plasmid-determined resistance to multiple drugs was first discovered in *Shigella* in the 1970s. Multiple antibiotic resistance is often due to composite IS/Tn elements on a plasmid (Fig. 14.8b). As we see later, plasmids can be transferred from

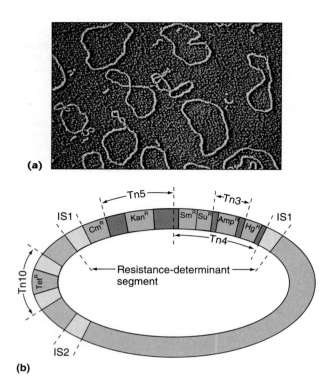

Figure 14.8 Plasmids are small circles of double-stranded DNA. (a) Electron micrograph showing circular plasmid DNA molecules. **(b)** Some plasmids contain multiple antibiotic resistance genes (shown in yellow: *Cm^R* for chloramphenicol, *Kan^R* for kanamycin, *Sm^R* for streptomycin, *Su^R* for sulfonamide, *Amp^R* for ampicillin, *Hg^R* for mercury, and *Tet^R* for tetracycline). Transposons (IS and Tn elements, shown in *pale peach*) facilitate the movement of the antibiotic resistance genes onto the plasmid. Note that many antibiotic resistance genes are located between two IS1 elements, allowing them to transpose subsequently as a unit.

one bacterium to another, sometimes even across species. Plasmids thus have terrifying implications for medicine. If transferred to new strains of pathogenic bacteria, the new hosts acquire resistance to many antibiotics in a single step. We encountered an example of this potential in the opening story on gonorrhea.

One important group of plasmids allows the bacterial cells that carry them to make contact with another bacterium and transfer genes—both plasmid and bacterial—to the second cell. We describe this cell-to-cell mating, known as conjugation, in the next section.

Gene Transfer in Bacteria

Gene transfer from one individual to another plays an important role in the evolution of new variants in nature. Many cases of such horizontal, or lateral, gene transfer have come to light through recent molecular and DNA sequencing analyses. The phrase **lateral gene transfer** means that the traits involved are not transferred by inheritance from one generation to the next; rather, they are introduced from unrelated individuals or from different species. Comparative genomic analysis of many dif-

ferent genes in various bacterial species has shown similarities of genes in species thought to be only distantly related. The simplest explanation is that significant transfer of DNA between bacteria has occurred throughout evolution. A close examination of the known mechanisms of DNA transfer will help illuminate this phenomenon. In addition, you will see that researchers can use the various methods of gene transfer to map genes and to construct bacterial strains with which to test the function and regulation of specific genes.

Bacteria can transfer genes from one strain to another by three different mechanisms: transformation, conjugation, and transduction (Fig. 14.9). In all three mechanisms, one cell—the **donor**—provides the genetic material for transfer, while a second cell—the **recipient**—receives the material. In **transformation,** DNA from a donor is added to the bacterial growth medium and is then taken up from that medium by the recipient. In **conjugation,** the donor carries a special type of plasmid that allows it to come in contact with the recipient and transfer DNA directly. In **transduction,** the donor DNA is packaged within the protein coat of a bacteriophage and transferred to the recipient when the phage particle infects it. The recipients of a gene transfer are known as **transformants, exconjugants,** or **transductants,** depending on the mechanism of DNA transfer that created them.

All bacterial gene transfer is asymmetrical in two ways. First, transfer goes in only one direction, from donor to recipient. Second, most recipients receive 3% or less of a donor's DNA; only some exconjugants contain a greater percentage of donor material. Thus, the amount of donor DNA entering the recipient is small relative to the size of the recipient's chromosome, and the recipient retains most of its own DNA. We now examine each type of gene transfer in detail.

Transformation: Fragments of Donor DNA Enter the Recipient and Alter Its Genotype

A few species of bacteria spontaneously take up DNA fragments from their surroundings in a process known as **natural transformation.** By contrast, the large majority of bacterial species can take up DNA from the surrounding medium only after laboratory procedures make their cell walls and membranes permeable to DNA in a process known as **artificial transformation.**

In Natural Transformation, the Recipient Cell Has the Enzymatic Machinery for DNA Import

Researchers have studied several species of bacteria that undergo natural transformation, including *S. pneumoniae,* the pathogen in which transformation was discovered by Frederick Griffith (see Chapter 6) and that causes pneumonia in humans; *B. subtilis,* a harmless soil bacterium; *H. influenzae,* a pathogen causing various diseases in humans; and *N. gonorrhoeae,* the microbial agent of gonorrhea.

In one study of natural transformation, investigators isolated *B. subtilis* bacteria with two mutations—*trpC2* and *hisB2*—that made them Trp$^-$, His$^-$ double auxotrophs. These double auxotrophs served as the recipients in the study; wild-type cells (Trp$^+$, His$^+$) were the donors. The experimenters ex-

Transformation

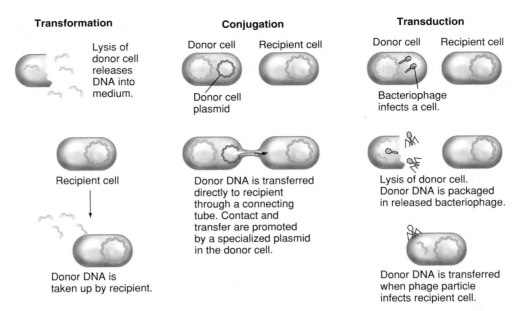

Conjugation

Transduction

Figure 14.9   Three mechanisms of gene transfer in bacteria: An overview. In this figure, and throughout this chapter, the donor's chromosome is *blue,* and the recipient's chromosome is *orange.* In transformation, fragments of donor DNA released into the medium enters the recipient cell. In conjugation, a specialized plasmid (*shown in red*) in the donor cell promotes contact with the recipient and initiates the transfer of DNA. In transduction, DNA from the donor cell is packaged into bacteriophage particles that can infect a recipient cell, transferring the donor DNA into the recipient.

tracted and purified donor DNA and grew the *trpC hisB* recipients in a suitable medium until the cells became **competent,** that is, able to take up DNA from the medium. Different bacterial species require different regimens to achieve competence. For *B. subtilis* (Fig. 14.10a), competence occurs only in nearly starving cells at very specific times in the growth of the culture. Investigators can starve the cells by growing them in a glucose-salts medium containing a limited amount of tryptophan and histidine. As growth of the culture slows toward the end of what is known as the stage of logarithmic growth, a fraction of the bacteria—1% to 5% in *B. subtilis*—become competent and will take up DNA added to the medium at this time.

The mechanism of DNA uptake in *B. subtilis* is more complicated than was originally anticipated (Fig. 14.10b). The donor DNA fed to the competent cells usually consists of linear double-stranded molecules with an average length of 20 kb. The competent cells have enzyme complexes at their surfaces that take up one strand of a DNA fragment and simultaneously break down the other strand to its constituent nucleotides. The wild-type DNA strand admitted to the doubly auxotrophic cell quickly finds a homologous region of the bacterial chromosome and recombines with it, replacing the almost identical portion of one of the two strands of the recipient DNA (which is subsequently degraded). After the chromosome replicates and the cell divides, one of the two daughter cells becomes a transformant, carrying wild-type *trpC⁺* and/or *hisB⁺* genes from the donor.

To observe and count Trp⁺ transformants, researchers decant the liquid containing newly transformed recipient cells onto petri plates containing a simple glucose-salts solid medium with histidine. Recipient cells that did not take up donor DNA

are unable to grow on this medium because it lacks tryptophan, but the Trp⁺ transformants can grow and be counted. To select for His⁺ transformants, researchers pour the transformation mixture on glucose-salts solid medium containing tryptophan, instead of histidine. In this study, the numbers of Trp⁺ and His⁺ transformants were equal. In conditions where *B. subtilis* bacteria become highly competent, 10^9 cells will produce approximately 10^5 Trp⁺ transformants and 10^5 His⁺ transformants.

To discover if any of the Trp⁺ transformants were also His⁺, the researchers used sterile toothpicks to transfer colonies of Trp⁺ transformants to a glucose-salts solid medium containing neither tryptophan nor histidine. Forty of every 100 Trp⁺ transferred colonies grew on this minimal medium, indicating that they were also His⁺. Similarly, tests of the His⁺ transformants showed that roughly 40% are also Trp⁺. Thus, in 40% of the analyzed colonies, the *trpC⁺* and *hisB⁺* genes had been cotransformed. **Cotransformation** is the simultaneous transformation of two or more genes.

Since during transformation, donor DNA replaces only a small percentage of the recipient's chromosome, it might seem surprising that the two *B. subtilis* genes are cotransformed with such high frequency. The explanation is that the *trpC* and *hisB* genes lie very close together on the chromosome and are thus genetically linked. The entire *B. subtilis* chromosome is approximately 4700 kb long. Only genes in the same chromosomal vicinity can be cotransformed; the closer together the genes lie, the more frequently they will be cotransformed. Thus, although the donor chromosome is fragmented into small pieces of about 20 kb during its extraction for the transformation process, the wild-type *trpC⁺* and *hisB⁺* alleles are so close that they often

(a) Donor and recipient genomes

Wild-type donor cell $trpC^-$ / $hisB^-$ double auxotrophs
                          Recipient cell

(b) Mechanism of natural transformation

Donor DNA binds to recipient cell at receptor site.

One donor strand is degraded. Admitted donor strand pairs with homologous region of bacterial chromosome. Replaced strand is degraded.

Donor strand is integrated into bacterial chromosome.

After cell replication, one cell is identical to original recipient; the other carries the mutant gene.

Transformed cell

Figure 14.10 Natural transformation in B. subtilis. (a) A wild-type donor and a $hisB^-$ $trpC^-$ double auxotroph recipient. Selection for His$^+$ and/or Trp$^+$ phenotypes identifies transformants. **(b)** Mechanism of natural transformation in *B. subtilis*. One strand of a fragment of donor DNA enters the recipient, while the other strand is degraded. The entering strand recombines with the recipient chromosome, producing a transformant when the recipient cell divides.

appear together in the same donor DNA molecule. Sequence analysis shows that the *trpC* and *hisB* genes are only about 7 kb apart. By contrast, genes sufficiently far apart that they cannot appear together on a fragment of donor DNA will almost never be cotransformed, because transformation is so inefficient that recipient cells usually take up only a single DNA fragment.

Transformation usually incorporates a single strand of a linear donor DNA fragment into the bacterial chromosome of the recipient through recombination. However, if the donor DNA includes plasmids, recipient cells may take up an entire plasmid and acquire the characteristics conferred by the plasmid genes. Bacteriologists suspect that penicillin-resistant *N. gonorrhoeae,* described in the introduction to this chapter, originated through transformation by plasmids. The donors of the plasmids were *H. influenzae* cells disrupted by the immune defenses of a doubly infected patient. The plasmids carried the gene for penicillinase; and the recipient *N. gonorrhoeae* bacteria, transformed by the plasmids, acquired resistance to penicillin.

In Artificial Transformation, Damage to Recipient Cell Walls Allows Donor DNA to Enter the Cells

Researchers have devised many methods to transform bacteria that do not undergo natural transformation. The existence of these techniques was critical for the development of the gene-cloning technology described in Chapter 9. All the methods include treatments that damage the cell walls and membranes of recipient bacteria so that donor DNA can diffuse into the cells. With *E. coli,* the most common treatment consists in suspending the cells in a high concentration of calcium at cold temperature. Under these conditions, the cells become permeable to single- and double-stranded DNA. Because *E. coli* cells have efficient DNA exonucleases that degrade free ends, circular DNA is much more efficient than linear DNA at transforming *E. coli.*

Another technique of artificial transformation is *electroporation,* in which researchers mix a suspension of recipient bacteria with donor DNA and then subject the mixture to a very brief high-voltage shock. The shock most likely causes holes to form in the cell membranes. With the proper shocking conditions, recipient cells take up the donor DNA very efficiently. Transformation by electroporation works with most bacteria.

Conjugation: Donor Cells Carrying Specialized Plasmids Establish Contact with and Transfer DNA to Recipients

In the late 1940s, Joshua Lederberg and Edward Tatum analyzed two *E. coli* strains that were each multiple auxotrophs and made the striking discovery that genes seemed to transfer from one type of *E. coli* cell to the other (Fig. 14.11a). Neither strain grew on a minimal glucose-salts medium. Strain A required supplementation with methionine and biotin; strain B required supplementation with threonine, leucine, and thiamine (vitamin B$_1$). Lederberg and Tatum grew the two strains together on supplemented medium. When they then transferred a mixture of the two strains to minimal medium, about 1 in every 10^7 transferred cells proliferated to a visible colony. What were these colonies and how they did they arise?

(a) Demonstration of gene transfer

(b) Conjugation requires cell–to–cell contact

Figure 14.11   Conjugation: A type of gene transfer requiring cell-to-cell contact. (a) Neither of two multiple auxotrophic strains analyzed by Lederberg and Tatum formed colonies on minimal medium. When cells of the two strains were mixed, gene transfer produced some prototrophic cells that formed colonies on minimal medium. **(b)** The U-tube experiment demonstrating that direct contact between cells is needed for conjugation. Here, the two strains were separated by a filter that allowed the passage of all molecules in the medium, including DNA fragments, but prevented contact between the cells. The filter abolished gene transfer between the two strains.

To answer these questions, they performed the necessary control for this experiment. If they plated either strain alone on a minimal medium, nothing grew. Because they observed colonies only when they plated a mixture of the two strains on a simple medium, Lederberg and Tatum concluded that some sort of genetic exchange—rather than a highly unlikely set of simultaneous reversions—had taken place. *E. coli,* however, do not undergo natural transformation. Further experiments showed that instead, direct cell-to-cell contact was necessary for the exchange. In one experiment, Bernard Davis showed that the genetic exchange did not occur when cells of one strain were mixed with a filtrate containing DNA of the other strain, or when a porous filter allowing the passage of liquid medium but not bacteria cells separated strain A from strain B in a U-shaped tube (Fig. 14.11b). More than a decade of further experiments confirmed that Lederberg and Tatum had observed what became known as bacterial conjugation: a one-way DNA transfer from donor to recipient initiated by **conjugative plasmids** in donor strains. Many different plasmids can initiate conjugation because they carry genes that allow them to transfer themselves (and sometimes some of the donor's chromosome) to the recipient.

The F Plasmid and Conjugation

Figure 14.12 illustrates the type of bacterial conjugation initiated by the first conjugative plasmid to be discovered—the F plasmid of *E. coli.* Cells carrying an F plasmid are called F⁺ cells; cells without the plasmid are F⁻. The F plasmid carries many genes required for the transfer of DNA, including genes for formation of an appendage, known as the pilus, by which a donor cell contacts a recipient cell, and a gene encoding an endonuclease that nicks the F plasmid's DNA at a specific site (the origin of transfer). Once a donor has contacted a recipient cell (lacking the F plasmid) via the pilus, retraction of the pilus pulls the donor

and recipient close together. The F plasmid DNA is then nicked and a single strand moves across a bridge between the two cells. Movement of the F DNA into the recipient cell is accompanied by synthesis in the donor of another copy of the DNA strand that is leaving. When the donor DNA enters the recipient cell, it reforms a circle and the recipient synthesizes the complementary DNA strand. In this F⁺ × F⁻ mating, the recipient becomes F⁺ and the donor remains F⁺. By initiating and carrying out conjugation, the F plasmid acts in bacterial populations the way an agent of sexually transmitted disease acts in human populations. When introduced via a few donor bacteria into a large culture of cells that do not carry the plasmid, the F plasmid soon spreads throughout the entire culture and all the cells become F⁺.

The F Plasmid Occasionally Integrates into the *E. coli* Chromosome Where It Promotes the Conjugational Transfer of Chromosomal Genes

The F plasmid contains three different IS transposable elements: one copy of IS2, two copies of IS3, and one copy of the particularly long IS1000. These IS sequences on the F plasmid are identical to copies of the same IS elements found at various positions along the bacterial chromosome. In roughly 1 of every 10^5 (100,000) F⁺ cells, homologous recombination (that is, a crossover) between an IS on the plasmid and the same IS on the chromosome integrates the entire F plasmid into the *E. coli* chromosome (Fig. 14.13a). This insertion by recombination enlarges the 4700 kb chromosome by 100 kb of DNA. Cells whose chromosomes carry an integrated plasmid are called **Hfr** bacteria, because, as we will see, they produce a <u>h</u>igh <u>f</u>requency of <u>re</u>combinants for chromosomal genes in mating experiments with F⁻ strains. Since the recombination event that results in the F plasmid's insertion into the bacterial chromosome can occur

The F plasmid and conjugation.

a. The F plasmid contains genes for synthesizing connections between donor and recipient cells. The F plasmid is a 100-kb-long circle of double-stranded DNA. Host cells that carry it generally have one copy of the plasmid. By analogy with sexual reproduction, researchers think of F⁺ cells as *male bacteria* because the cells can transfer genes to other bacteria. About 35% of F plasmid DNA consists of genes that control the transfer of the plasmids. Most of these genes encode polypeptides involved in the construction of a structure called the F **pilus** (plural, **pili**): a stiff thin strand of protein that protrudes from the bacterial cell. Other regions of the plasmid carry IS's and genes for proteins involved in DNA replication.

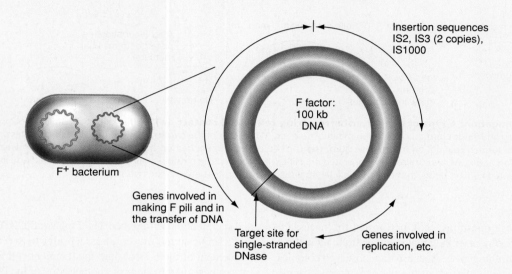

F⁺ bacterium

Genes involved in making F pili and in the transfer of DNA

Target site for single-stranded DNase

Insertion sequences IS2, IS3 (2 copies), IS1000

F factor: 100 kb DNA

Genes involved in replication, etc.

b. The process of conjugation.
1. The **pilus.** An average pilus is 1 μm in length, which is almost as long as the average *E. coli* cell. The distal tip of the pilus consists of a protein that binds specifically to the cell walls of F⁻ *E. coli* not carrying the F factor.
2. **Attachment to F⁻ cells (female bacteria).** Because they lack F factors, F⁻ cells cannot make pili. The pilus of an F⁺ cell, on contact with an F⁻ cell, retracts into the F⁺ cell, drawing the F⁻ cell closer. A narrow passageway forms through the now adjacent F⁺ and F⁻ cell membranes.
3. **Gene transfer: A single strand of DNA travels from the male to the female cell.** Completion of the cell-to-cell corridor signals an endonuclease encoded by the F plasmid to cut one strand of the F plasmid DNA at a specific site (*the origin of transfer*). The F⁺ cell extrudes the cut strand through the passageway into the F⁻ cell at a rate of about 47,000 bases per minute. As it receives the single

---

between any of the IS elements on the F plasmid and any of the corresponding IS elements in the bacterial chromosome, geneticists can isolate 20–30 different strains of Hfr cells (Fig. 14.13b). Various Hfr strains are distinguished by the location and orientation (clockwise or counterclockwise) of the integrated F plasmid with respect to the bacterial chromosome. These different Hfr strains have different names; one, for example, is called HfrH after its discoverer William Hayes. Plasmids like the F plasmid that can integrate into the host chromosome are sometimes called **episomes**.

During bacterial reproduction, the integrated plasmid of an Hfr cell replicates with the rest of the bacterial chromosome. As a result, the chromosomes in daughter cells produced by cell division contain an intact F plasmid at exactly the same location that the plasmid originally integrated into the chromosome of the parental cell. All progeny of an Hfr cell are thus identical, with the F plasmid inserted into the same chromosomal location and in the same orientation.

The integrated F plasmid still has the capacity to initiate DNA transfer via conjugation, but now that it is part of a bacterial chromosome, it can promote the transfer of some or all of that chromosome as well (Fig. 14.14). The transfer of DNA from an Hfr cell mated to an F⁻ cell starts with a single-strand nick in the middle of the integrated F plasmid at the origin of transfer. Beginning at this site, half of the F plasmid moves into the recipient, followed by donor chromosomal DNA carrying bacterial genes. However, since in conjugation, the cell-to-cell connection is labile and easily disrupted, the cells often separate before transfer of the whole *E. coli* chromosome has been completed. As a result, the exconjugant is a partially diploid bacterium that carries half of the F plasmid attached to a variable amount of DNA from the Hfr cell's chromosome, as well as the strand of F plasmid DNA, the F$^-$ cell synthesizes a complementary strand so that after 2–3 minutes of contact, the formerly F$^-$ cell contains a double-stranded F plasmid and is now an F$^+$ cell.

4. **Meanwhile, back in the original F$^+$ cell, newly synthesized DNA replaces the single strand transferred**

to the previously F$^-$ cell. When the two bacteria separate at the completion of DNA transfer and synthesis, they are both F$^+$.

To summarize: Conjugation passes the F$^+$ plasmid and therefore maleness—the ability to transfer DNA to other cells—from F$^+$ to F$^-$ bacteria.

1. F pilus extends and binds to F$^-$ cell wall.

F$^+$ F$^-$

F pilus

2. On contact with F$^-$ cell, pilus retracts. Cells are drawn together. Protein corridor forms between cells.

F$^+$ F$^-$

3. DNase cuts plasmid DNA at target site. Single strand moves across corridor. F$^-$ cell synthesizes complementary strand. F$^+$ cell replaces transferred strand.

F$^+$ F$^-$

4. Cells separate at completion of transfer and synthesis. Both are now F$^+$ and can conjugate with other F$^-$ cells.

F$^+$ F$^+$

F$^-$ cell's entire genome. Because the transfer is usually interrupted in midstream, the second half of the F plasmid is not transferred and the recipient thus remains F$^-$.

The single-stranded DNA fragment transferred to the recipient cell is copied into a double-stranded linear fragment by DNA polymerases in the new host cell. But because bacterial cells cannot properly replicate and transmit to subsequent generations linear fragments of DNA, DNA from the donor will be lost unless it is incorporated into the recipient's chromosome. Incorporation can occur through a double recombination event (catalyzed by bacterial recombination enzymes) between homologous regions on the linear DNA fragment from the donor and the circular chromosome of the recipient.

Only about 1 in 10,000 Hfr cells transfers an entire strand of its chromosome to a recipient F cell. And only those rare F$^-$ cells that receive the entire Hfr chromosome, including the sec-

ond half of the F plasmid, have the full 100 kb of F plasmid DNA that make them F$^+$. Therefore, most of the progeny produced by an Hfr × F$^-$ conjugation remain F$^-$. This contrasts sharply with the result of conjugation initiated by an F$^+$ cell, where almost all the exconjugants become F$^+$. Thus, Hfr cells are poor donors of the entire F plasmid but efficient donors of chromosomal genes. The situation is reversed for F$^+$ donors, which can transfer bacterial genes only in those rare cells in which the F plasmid integrates to form an Hfr.

Analyses of Hfr × F$^-$ Matings Help Map Genes
Because genes in an Hfr chromosome are transferred into the recipient in a consistent order, researchers realized that they could use Hfr crosses to map genes. To gain useful information from a conjugational mating between an Hfr "male" and an F$^-$ "female," they analyze strains that differ from each other in the alleles of

(a) Creation of Hfr chromosome.

(b) Many different Hfr strains can form.

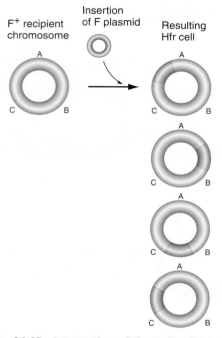

Figure 14.13 Integration of the F plasmid into the bacterial chromosome forms an Hfr bacterium. In this figure, the *filled bar* represents both strands of DNA. **(a)** Recombination between an IS on the F plasmid and the same kind of IS on the bacterial chromosome creates an Hfr chromosome. **(b)** Recombination can occur between any IS on the F plasmid and any corresponding IS on the bacterial chromosome to create many different Hfr strains.

several genes. In early studies of the *E. coli* genome, for example, Elie Wollman and François Jacob used the following genotypes:

HfrH	F⁻
str^s (sensitive to streptomycin)	*str^r* (resistant to streptomycin)
thr^+ (able to synthesize the amino acid threonine)	*thr^-* (threonine auxotroph)
azi^r (resistant to sodium azide)	*azi^s* (sensitive to sodium azide)
ton^r (resistant to bacteriophage T1)	*ton^s* (sensitive to phage T1)
lac^+ (able to grow with lactose as sole source of carbon)	*lac^-* (unable to grow on lactose)
gal^+ (able to grow with galactose as sole source of carbon)	*gal^-* (unable to grow on galactose)

To isolate and analyze the exconjugants in which transfer and recombination had occurred, the researchers had to be able to select for only those particular cells. In one experiment using the donor-recipient combination just outlined, Wollman and Jacob began by mixing the two strains in a rich nonselective liquid medium that allowed the growth of both strains; they also established conditions that favored the rapid formation of cell-to-cell connections. Next, at 1-minute intervals, they violently stirred samples of the mating mixture in a kitchen blender, reasoning that the extreme agitation would separate pairs of cells before all genes from the donor had passed to the recipient, thereby interrupting the mating (Fig. 14.15a). (The experiment derives its name of *interrupted-mating experiment* from this part of the procedure.) They then poured samples of the terminated matings onto petri plates containing streptomycin, which killed the original donor cells. The plates also lacked threonine, to select against F⁻ cells that had not mated. Finally, after the exconjugants grew on these streptomycin plates lacking threonine, they used replica plating to copy each exconjugant colony onto media that would test for the four other markers present in the donor and recipient strains. In this way, they could determine which donor genes had transferred into the F⁻ recipients before the interruption of mating.

Figure 14.15b shows the frequency of recombinant colonies containing various alleles from the donor strain with each increase in the length of conjugation before violent agitation. After mating has proceeded for 8 minutes, a small fraction of the recombinants are *azi^r*, but not one carries the other donor alleles. At about 10 minutes, some of the recombinants also have the donor's *ton^r* allele. By 15 minutes, the first *lac^+* alleles appear in the recombinant colonies; and by 17 minutes, *gal^+* arrives. The percentages of exconjugant colonies containing a particular gene from the Hfr donor increase with time until they reach a plateau characteristic of the gene: 90% for *azi^r*, 85% for *ton^r*, 40% for *lac^+*, and 20% for *gal^+*.

The presence of an integrated F plasmid in Hfr strains explains the two attributes of each gene revealed by the interrupted mating experiment: the time at which transfer of the gene is first seen, and the plateau percentages of exconjugants that carry the donor gene. The time of gene transfer depends on the fact that all HfrH donor cells carry the F plasmid DNA at the same site on their chromosomes, and they all start to transfer their DNA into the F cells at about the same time and in the same orientation. Thus, the time a gene first enters the recipient cell reflects the distance of that gene from the origin of transfer located within the integrated F plasmid. The first few HfrH donors to establish a mating connection with an F⁻ partner inject their *azi^r* allele through the intercellular connection by 8 minutes; more donors make the connection and transfer *azi^r* thereafter. The first *ton^r* alleles pass into F⁻ cells at about 10 minutes, the first *lac^+* alleles at about 15 minutes, and the first *gal^+* alleles at about 17 minutes after the mixing together of Hfr and F⁻ cells. This interpretation not only predicts the order of genes on the *E. coli* chromosome, it also makes it possible to map the distances between the genes. The units of distance are defined as *minutes of chromosome transfer*, and each minute translates to approximately 47,000 bases transferred (Fig. 14.15c).

Figure 14.14 Gene transfer in a mating between Hfr donors and F⁻ recipients. In an Hfr × F⁻ mating, single-stranded DNA is transferred into the recipient, starting with the origin of transfer on the integrated F plasmid. Within the recipient cell, this single-stranded DNA is copied into double-stranded DNA. If mating is interrupted, the recipient cell will contain a double-stranded linear fragment of DNA plus its own chromosome. Genes from the donor are retained in the exconjugant only if they recombine into the recipient's chromosome.

The plateaus for the percentage of colonies carrying each transferred gene derive from the fragility of cell-to-cell connections. Transfer between Hfr donors and F⁻ recipients can spontaneously abort if the connected cells separate or if the transferring chromosome breaks. As a result, not all exconjugants receive even early arriving genes such as *azi*^{r}. In addition, the more time required to transfer a marker, the greater the chance that mating pairs will separate before the transfer occurs.

As a result, the percentage of exconjugants receiving later-arriving genes becomes successively smaller.

Comparisons of Interrupted Mating Studies Using Different Hfr Strains Confirm That the Bacterial Chromosome Is a Circle

The HfrH strain transfers genes in the order *azi-ton-lac-gal*. A different Hfr strain (strain 1 in Fig. 14.16a) transfers these genes

Figure 14.15 Mapping genes by interrupted-mating experiments. (a) Hfr and F⁻ cells were mixed to initiate mating. Samples were agitated at 1-minute intervals in a kitchen blender to disrupt gene transfer. Cells were plated onto a medium that contained streptomycin (to kill the Hfr donor cells) and that lacked threonine (to prevent growth of F⁻ cells that had not mated). The phenotypes of the exconjugants for other markers was established by replica plating. **(b)** Results of the Wollman-Jacob interrupted-mating experiment. **(c)** Gene order established from the data in part (b), with positions determined by the time a donor gene first enters the recipient.

in the opposite order: *gal-lac-ton-azi*. The maps obtained using these and other Hfr strains as donors can be related on the basis of the fact that in various Hfr strains, the F plasmid integrates into the bacterial chromosome at different locations and in different orientations (Fig. 14.16b). The location of the origin of transfer within each type of F plasmid determines where transfer begins and thus the start of that particular map. The orientation of the plasmid with respect to the bacterial chromosome determines whether transfer occurs clockwise or counterclockwise.

Comparisons of maps drawn from many Hfr × F⁻ crosses provided the first clues that the bacterial chromosome is circular (Fig. 14.16b). The maps from different Hfr matings are all permutations of the same order of genes that would occur if the genes were located on a circular chromosome, but with transfer starting at a different position and moving in either the clockwise or counterclockwise direction. Many physical studies examining the DNA of the bacterial genome later confirmed the circularity of the bacterial chromosome.

Researchers now use well-studied collections of Hfr strains with different sites and orientations of F plasmid insertion to determine the approximate position on the *E. coli* genetic map of any gene they can follow on the basis of its phenotype. They designate the gene's position by the minute at which its transfer begins. The circumference of the entire circular map is 100 minutes, the time required for transfer of the complete *E. coli* chromosome. The top of the circle, where 0 minutes

meets 100 minutes, is arbitrarily set at the *thr* gene, the site of F plasmid insertion in HfrH. About 1400 of the estimated 4000–5000 *E. coli* genes have so far been located in this way. A selection of these mapped genes appears in Fig. 14.16c.

Recombination Analyses of Hfr Crosses Improve Mapping Accuracy

Interrupted mating experiments cannot distinguish the relative positions of genes within about 2 minutes of each other and thus give only a crude idea of gene location. However, analyses of conjugation experiments can provide better resolution if they take into account what happens within an F⁻ exconjugant after it has received genes from the Hfr donor. Because bacteria cannot completely replicate and transmit the linear DNA from the Hfr donor remaining inside the F⁻ cell at the interruption of mating, the only way the progeny of the recipient F⁻ cell can inherit donor genes is if those genes recombine into the F⁻ cell's chromosome. Note that a single recombination event, or any odd number of recombination events, between the circular bacterial chromosome and the linear DNA fragment would be lethal: By making the chromosome linear, it would prevent multiplication of the cell. As a result, conjugation can produce recombinant cells only if an even number (two, four, six, and so forth) of crossovers occurs.

To map genes on the basis of the frequencies of recombinant classes emerging from a conjugation experiment, you must con-

(a) Gene transfer in different Hfrs

Hfr strain	Order of transfer ⟶
H	thr **azi ton** lac pur **gal** his gly thi
1	thr thi gly his **gal** pur lac **ton azi**
2	lac pur **gal** his gly thi thr **azi ton**
3	**gal** pur lac **ton azi** thr thi gly his

(b) Data interpretation

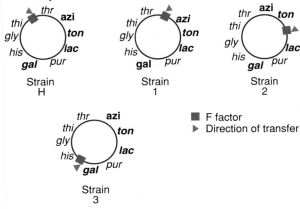

■ F factor
▶ Direction of transfer

(c) The circular *E.coli* chromosome

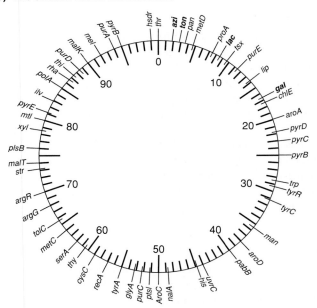

Figure 14.16 Order of gene transfers in different Hfr strains. (a) Comparison of the order of gene transfers in crosses between F⁻ recipients and various Hfr strains. The genes from Fig. 14.15 are indicated in *bold*; genes mapped later by other investigators are in *normal font*. **(b)** Interpretation of the data shown in part (a). In different Hfr strains, the F plasmid integrates in different locations and in different orientations into the circular donor chromosome. **(c)** A partial genetic map of the *E. coli* chromosome.

sider only exconjugants that received from the donor all of the genes to be mapped. Otherwise, the resulting progeny would reflect not only recombination events but also the likelihood that a particular gene from the Hfr donor was transferred to recipient cells. To ensure that all the recipient cells you are examining received all pertinent genes from the donor, you select exconjugants that contain the last of the genes transferred from the Hfr donor, as these cells must also have received the earlier arriving genes.

With this understanding of how to select suitable exconjugants and what type of recombination occurs inside those F⁻ cells, it is possible to map closely linked genes by a type of three-point cross. Recall that in three-point crosses in eukaryotes, you can infer that the gene in the middle is the gene that recombined with respect to the other two in the double-crossover class. You can do the same in prokaryotes, although here, the equivalent of the double-crossover class requires four crossovers. In the example shown in Fig. 14.17, the order of the closely linked genes *met* and *arg* is determined by a three-point cross in which you select for a nearby marker known to be transferred later—in this case, *mal⁺*. After selecting for Mal⁺ and then testing for the Arg and Met phenotypes, you tally the numbers for each type of exconjugant (Fig 14.17b). The least frequently occurring class was produced by quadruple crossovers. By setting up all possible orders of the genes (Fig 14.17c) you can see which order leads to the genotype corresponding to the smallest number of exconjugants. Comparisons of recombination frequencies between genes in a particular bacterial cross provide a measure of relative map distances.

Bacterial Geneticists Can Use a Special Type of F Plasmid Known As F′ for Genetic Complementation Studies

We saw earlier in this chapter that insertion of the F plasmid into the bacterial chromosome occurs in about 1 F⁺ cell per 100,000 to produce an Hfr cell. In approximately that same proportion of Hfr cells, an excision event causes the Hfr cell to revert to an F⁺ cell; and in a small fraction of these excision events, an error in recombination generates a plasmid containing most of the F plasmid genes plus a small region of the bacterial chromosome that had been adjacent to the integrated F plasmid. The newly formed plasmid carrying most of the genes of the F plasmid plus some bacterial DNA is known as an **F′ plasmid** (Fig. 14.18a). F′ plasmids replicate as discrete circles of DNA inside *E. coli* cells. They are transferred to recipient (F⁻) cells in the same way that F plasmids are transferred. The difference is that a few chromosomal genes will always be transferred as part of the F′ plasmid. This ability to transfer chromosomal genes on a molecule that can replicate independently of the bacterial chromosome makes the F′ plasmid a perfect vehicle for creating partial diploids for complementation analysis.

Recall from Chapter 7 that complementation tests depend on cells that are diploid for the genes under analysis. Although bacteria are haploid, F′ plasmids carrying bacterial genes can create specific regions of partial diploidy. In fact, it is possible to find F′ plasmids carrying *E. coli* genes that lie close to any of

(a) Hfr mapping: *mal⁺* transfers last

(b) Transfer of genes in *mal⁺* exconjugants

Cross between
arg⁺, met⁺, mal⁺, Hfr
and *arg⁻, met⁻, mal⁻*
F⁻ recipient

Exconjugant phenotype	Number
met⁺ arg⁺ mal⁺	375
met⁻ arg⁻ mal⁺	149
met⁻ arg⁺ mal⁺	32
met⁺ arg⁻ mal⁺	2

(c) Types of crossover events

Crossover event	Phenotype of colony
	met⁺arg⁺mal⁺
	met⁻arg⁻mal⁺
	met⁻arg⁺mal⁺
	met⁺arg⁻mal⁺

Figure 14.17 Mapping genes using a three-point cross.
(a) The order of the closely linked genes *met* and *arg* can be determined by a three-point cross, selecting for a nearby marker known to be transferred later (in this case, *mal⁺*). **(b)** The percentage of *mal⁺* exconjugants having all other combinations of the *arg* and *met* alleles are counted. **(c)** Crossover events that produce the various types of exconjugants in particular proportions. The *met⁺ arg⁻ mal⁺* exconjugant requires quadruple crossovers, so it is the least frequently observed class of exconjugants.

the chromosomal positions where F plasmids integrate. For example, some F′ plasmids carry the *trp* genes that control the biosynthesis of tryptophan. An F′ plasmid carrying these genes is called an F′ *trp* plasmid.

To create partial diploids in bacterial cells using F′ plasmids, researchers must transfer the F′ plasmid into a strain whose chromosome is not deleted for the genes carried by the

F′ plasmid (the chromosomes of the cells in which the aberrant plasmid excision occurred are now deleted for those genes). Because the F′ plasmids can still promote conjugation, experimenters easily accomplish this by mating F′-carrying cells with F⁻ cells (Fig. 14.18b). The exconjugants from these matings are partial diploids containing two copies of certain bacterial genes—one on the F′ plasmid, the other on the bacterial chromosome. Partial diploids in which the two gene copies are identical are known as **merozygotes;** partial diploids carrying different alleles of the same gene are **heterogenotes.**

As an example of complementation studies using F′ plasmids, consider an analysis of mutations affecting tryptophan biosynthesis (Fig. 14.18c). All these mutations map very close to each other. You first construct a partial diploid by introducing an F′ plasmid carrying the entire *trp* region with a particular *trp⁻* mutation into a bacterial strain that carries a different *trp⁻* mutation in the chromosome. Growth of a particular partial diploid on minimal medium without tryptophan would indicate that the mutations complement each other and are thus in different genes. If the cells do not grow, the mutations must be in the same gene. Complementation studies using F′ *trp* heterogenotes have shown that there are five different *trp* genes: *A, B, C, D,* and *E,* each one corresponding to one of the five polypeptides required for the biosynthesis of tryptophan.

Summary of Conjugation in *E. coli*
The F plasmid that initiates conjugation in *E. coli* functions in one of three ways. In F⁺ cells, it remains a small, independent, self-replicating circle of DNA that during conjugation transfers only its own genes. In Hfr cells, it integrates via homologous recombination into the bacterial chromosome so that during conjugation, it transfers some or all of the chromosomal DNA. In F′ cells, the F plasmid carries a small amount of chromosomal DNA, and during conjugation, only the chromosomal DNA on the plasmid is transferred, along with the plasmid itself. Table 14.2 compares the conjugational cycles of the three F-plasmid incarnations.

Transduction: Gene Transfer via Bacteriophages
The bacteriophages, or phages, that infect, multiply in, and kill various species of bacteria are widely distributed in nature. Most bacteria are susceptible to one or more such viruses. During infection, a virus particle may incorporate a piece of the bacterial chromosome and introduce this piece of bacterial DNA into other host cells during subsequent rounds of infection. The process by which viral particles transfer bacterial DNA from one host cell to another is known as transduction.

Bacteriophage Particles Are Produced by the Lytic Cycle
When a bacteriophage injects its DNA into a bacterial cell, the phage DNA takes over the cell's protein synthesis and DNA replication machinery, forcing it to express the phage genes, produce phage proteins, and replicate the phage DNA (see Fig. 7.17 on p. 210). The newly produced phage proteins and DNA

(a) F' plasmid formation

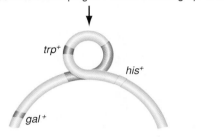

A rare recombination event between regions of limited sequence homology permits out-looping of F factor including *trp⁺* locus.

Separation of F creates F' *trp⁺* plasmid and a chromosome deleted for the *trp* genes.

(b) F' plasmid transfer

(c) Complementation testing using F' plasmids

Figure 14.18 F' plasmids. (a) Formation of an F' plasmid by improper excision of an integrated F plasmid from an Hfr chromosome. **(b)** Transfer of an F' *trp* plasmid into a *trp⁺* F⁻ recipient creates a merozygote with two copies of *trp⁺*. **(c)** Using F' plasmids for complementation analysis. Different *trp⁻* mutations are induced on the F' *trp* plasmid (*trp⁻ x*), and in the F⁻ chromosome (*trp⁻ y*). Matings between these strains create partial diploids (heterogenotes) with both mutations; the phenotype of the partial diploid establishes whether the two mutations complement each other or not.

assemble into phage particles, after which the infected cell bursts, or lyses, releasing 100–200 new viral particles ready to infect other cells. The cycle resulting in cell lysis and release of progeny phage is called the **lytic cycle** of phage multiplication. The population of phage particles released from the host bacteria at the end of the lytic cycle is known as a **lysate.**

Generalized Transduction

Many kinds of bacteriophages encode enzymes that destroy the chromosomes of the host cells. Digestion of the bacterial chromosome by these enzymes sometimes generates fragments of bacterial DNA about the same length as the phage genome, and these phage-length bacterial DNA fragments occasionally get incorporated into phage particles in place of the phage DNA (Fig. 14.19). After lysis of the host cell, the phage particles can attach to and inject the DNA they carry into other bacterial cells,

thereby transferring genes from the first bacterial strain (the donor) to a second strain (the recipient). Recombination between the injected DNA and the chromosome of the new host completes the transfer. This process, which can result in the transfer of any bacterial gene between related strains of bacteria, is known as **generalized transduction.**

Mapping Genes by Generalized Transduction

As with cotransformation, two genes close together on the bacterial chromosome may be cotransduced. The frequency with which two bacterial genes are cotransduced depends directly on the distance between them: the closer two genes are, the more likely they are to appear on the same short DNA fragment and be packaged into the same transducing phage. Two genes that are farther apart than the amount of DNA that can be packaged into a single phage particle can never be cotransduced. For

TABLE 14.2   Comparing the Conjugational Cycles of F⁺, Hfr, and F′ Cells

	In an F⁺ Cell	In an Hfr Cell	In an F′ Cell
Donor and Recipient Before Conjugation	F⁺ / F⁻ (F plasmid)	Hfr / F⁻	F′ / F⁻ (F′ plasmid)
Events of Conjugation			
Donor and Recipient After Conjugation	F⁺ / F⁺	Hfr / F⁻ partial diploid (transient)	F′ / F′ merozygote
Comments	Efficient transfer of F factor; poor transfer of bacterial genes.	Transfer of entire F factor only inefficiently. Efficient transfer of bacterial genes but requires an even number of crossovers in recipient.	Efficient transfer of F′ element, which includes a few bacterial genes, generating stable partial diploids. Poor transfer of other chromosomal genes.

bacteriophage P1, a phage often used for generalized transduction experiments with *E. coli,* the maximum separation allowing cotransduction is about 90 kb of DNA, which corresponds to about 2% (or 2 minutes) of the bacterial chromosome. You will recall that researchers cannot use interrupted conjugation to define the relative location of genes injected less than 2 minutes apart. Thus, the two methods of genetic mapping can be used in concert to cover the entire spectrum of distances. In mapping a new *E. coli* mutation, investigators first find the approximate position of the mutation on the *E. coli* chromosome map by mating the mutant strain to a small collection of different Hfr strains. They then use P1 phage transduction to map the new mutation relative to other known genes in the region.

Consider, for example, the three genes—*thyA, lysA,* and *cysC*—that all map by interrupted mating experiments to the region between 53 and 56 minutes of the *E. coli* chromosome. Where do they lie in relation to one another? You can find out by using a P1 generalized transducing lysate from a wild-type strain to infect a *thyA⁻, lysA⁻, cysC⁻* strain and then selecting the transductants for either Thy⁺ or Lys⁺ phenotypes. After replica plating, you test each type of selected transductant for alleles of the two nonselected genes. As the phenotypic data in Fig. 14.20a indicate, *thyA* and *lysA* are close to each other but far from *cysC; lysA* and *cysC* are so far apart that they never appear in the same transducing phage particle; *thyA* and *cysC* are only rarely cotransduced. Thus, the order of the three genes must be *lysA, thyA, cysC* (Fig. 14.20b).

Temperate Phages Can Integrate into the Bacterial Chromosome

The types of bacteriophages we have discussed so far are **virulent:** After infecting a host, they always enter the lytic cycle, multiplying rapidly and killing the cell. Other types of bacteriophages are **temperate:** Although they can enter the lytic cycle, they can also enter an alternative **lysogenic cycle,** during which their DNA integrates into the host genome and multiplies along with it, doing little harm to the host (Fig. 14.21). The integrated copy of the temperate bacteriophage is called a **prophage** (Fig. 14.22). The choice of lifestyle—lytic or lysogenic—occurs when a temperate phage injects its DNA into a bacterial cell and depends on many factors, including environmental conditions. Normally when temperate phages inject their DNA into host cells, some of the cells undergo a lytic cycle, while others undergo a lysogenic cycle.

Under certain conditions, it is possible to induce an integrated viral genome to excise from the chromosome, undergo replication, and form new viruses. In a small percentage of excision events, some of the bacterial genes adjacent to the site where the bacteriophage integrated may be cut out along with the viral genome and be packaged as part of that genome. Viruses produced by the faulty excision of a lysogenic virus from the bacterial genome are called **specialized transducing phage** (Fig. 14.22c). During the production of such phage, bacterial genes may become passengers along with the viral DNA. When the specialized transducing phage then infects other cells, these

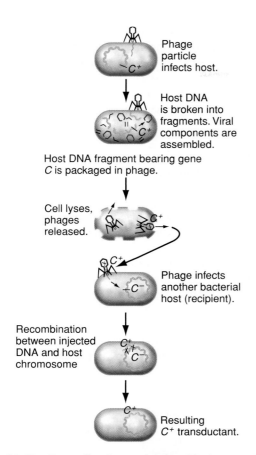

Host DNA fragment bearing gene C is packaged in phage.

Figure 14.19 Generalized transduction. The incorporation of random fragments of bacterial DNA from a donor into bacteriophage particles yields generalized transducing phages. When these phage particles infect a recipient, donor DNA is injected into the recipient's cell. Recombination of donor DNA fragments with the recipient cell chromosome yields transductants.

(a) Donor: *thyA*⁺ *lysA*⁺ *cysC*⁺

↓ make P1 lysate; infect recipient

Recipient: *thyA*⁻ *lysA*⁻ *cysC*⁻

Selected marker	Unselected marker
thy⁺	47% *lys*⁺; 2% *cys*⁺
lys⁺	50% *thy*⁺; 0% *cys*⁺

(b)   *lysA*   *thyA*   *cysC*

Figure 14.20 Mapping genes by cotransduction frequencies. (a) A P1 lysate of a *thyA*⁺ *lysA*⁺ *cysC*⁺ donor is used to infect a *thyA*⁻ *lysA*⁻ *cysC*⁻ recipient. Either *Thy*⁺ or *Lys*⁺ cells are selected and then tested for the unselected markers. **(b)** Genetic map based on the data in part (a). The *thyA* and *cysC* genes were cotransduced at a low frequency, so they must be closer together than *lysA* and *cysC*, which were never cotransduced.

transducing phages can transfer just those genes near the site where the bacteriophage inserted into the bacterial genome.

Bacterial Genetic Analysis Today

As geneticists learn more about genome structure and the mechanisms of gene transfer, including the transposition of DNA sequences, transduction, and conjugation, they are able to devise ever more clever ways of carrying out genetic analysis. Here we describe how an *E. coli* geneticist might approach the genetic dissection of a phenomenon. Many of the principles we present are applicable to other bacteria in which similar gene transfer mechanisms exist.

The Isolation of Mutants Is the Geneticist's Top Priority

Transposons have played the largest role in simplifying genetic analysis because of their role in creating mutations. The insertion of a transposon into a gene, resulting in the gene's inactivation, is the basis of many mutant screens. Transposons are useful as mutagenic agents because they contain genes for easily selectable antibiotic resistance. To carry out transposon mutagenesis, geneticists introduce a transposon into a cell as part of a DNA molecule that is not able to replicate on its own inside the cell. The mechanism of gene transfer can be transformation, transduction, or conjugation. For the transposon to be passed on during cell division, it must transpose from the incoming DNA molecule to the bacterial chromosome. By growing cells on a medium containing antibiotics, it is possible to select for those cells in which transposition has occurred. One can then screen the resulting population of cells, which contain transposons at different locations around the chromosome, for the mutant phenotype.

Transposon insertion usually inactivates the gene receiving the insertion, thereby creating a knockout or a null mutation. Such mutations can be useful, but if a gene is essential for a bacterium's survival, it will not be possible to isolate the knockout

few bacterial genes may be transferred into the infected cells. The phage-mediated transfer of a few bacterial genes is known as **specialized transduction.** Temperate phages are thought to be a significant vehicle for the lateral transfer of genes from one bacterial strain to another or even from one species to another.

Comparing Generalized and Specialized Transduction

Phage particles that act as agents of generalized transduction differ in critical ways from particles that carry out specialized transduction.

1. Generalized transducing phages pick up donor bacterial DNA during the lytic cycle, at the point when DNA is packaged inside a phage protein coat; specialized transducing phages pick up the donor bacterial DNA during the transition from the lysogenic to the lytic cycle.

2. Generalized transducing phages can transfer any bacterial gene or set of genes contained in the right size DNA fragment into the bacterial chromosome; specialized

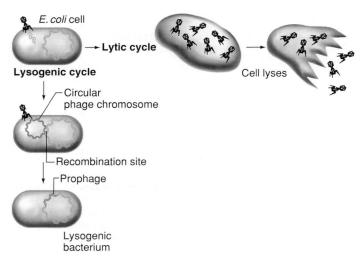

Figure 14.21 Temperate phages can choose between lytic and lysogenic modes of reproduction. Cells infected with temperate bacteriophages (whose chromosomes are shown in *green*) enter either the lytic or lysogenic cycles. In the lytic cycle, phages reproduce by forming new bacteriophage particles that lyse the host cell and can infect new hosts. In the lysogenic cycle, the phage chromosome becomes a prophage incorporated into the host chromosome.

mutation. For genes encoding essential proteins, conditional mutants (for example, temperature-sensitive mutants) are isolated. Even for nonessential genes, conditional mutations may be the most informative, since cells grown under permissive conditions can be shifted to nonpermissive conditions and then observed for changes in phenotype.

Mapping

Researchers can now map mutations originally identified by transposon insertion through DNA sequence analysis. By creating a primer from a transposon DNA sequence, they can use PCR to amplify a bit of the transposon DNA, and DNA adjacent to the transposon, and then sequence the amplified hybrid DNA segment. They then match the sequence with the genome sequence database to determine which gene was interrupted by the transposon. This technique not only indicates map position but also identifies the gene involved.

The standard Hfr mapping procedures described earlier are still used for mapping a mutation that is not caused by transposon insertion. To facilitate such Hfr mapping, researchers have created a set of Hfr strains with different origins of transfer around the chromosome and a Tn10 transposon in a location in each strain that ensures transfer of the transposon-based tetracycline resis-

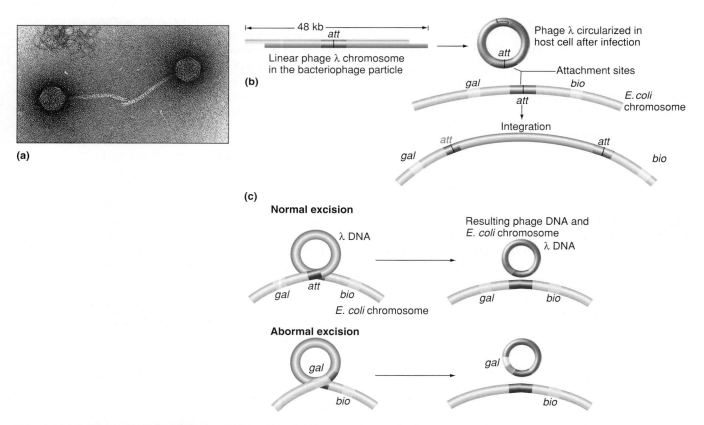

Figure 14.22 Bacteriophage lambda and lysogeny. (a) Electron micrograph of a temperate phage, bacteriophage lambda (λ). **(b)** Integration of the phage DNA initiates the lysogenic cycle. Recombination between *att* sites on the phage and bacterial chromosomes allows integration of the prophage. **(c)** Errors in prophage excision produce specialized transducing phages. Normal excision produces circles containing only lambda DNA. If excision is inaccurate, adjacent bacterial genes are included in the circles that form and in the resulting bacteriophages. Illegitimate recombination between the prophage and bacterial chromosome forms a circle that lacks some phage genes but has acquired the adjacent *gal* genes.

tance marker early in a mating. Each Hfr strain is mated for a short period of time to an F cell containing the mutation under analysis, and the resulting exconjugants are screened for antibiotic resistance. For one of the Hfr strains, the mating will result in loss of the mutant phenotype in most of the Tet^r exconjugants. The map location of the mutation is therefore near the origin of transfer (the site at which the F plasmid integrated into the bacterial chromosome) in this particular Hfr strain. The use of 10–12 Hfrs can localize the mutation to a region of the chromosome. The use of P1 transductions can then refine the map position of the gene. This quick method for mapping a mutation eliminates the need for a series of interrupted matings and is thus much less labor intensive.

The Reverse Genetics Approach to the Study of Gene Function

The sequence analysis of bacterial genomes has led to the identification of genes whose functions are not yet known. One approach to determining the function of such genes is to make a knockout mutation in the chromosomal gene, using recombinant DNA techniques and the homologous recombination machinery of the cell (Fig. 14.23). The gene to be mutated is first cloned in a plasmid. Next, a gene for antibiotic resistance is in-

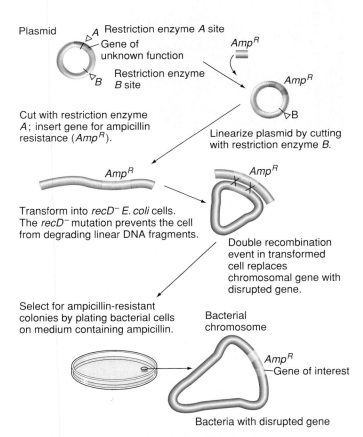

Plasmid
A Restriction enzyme A site
Gene of unknown function
Amp^R
B Restriction enzyme B site

Cut with restriction enzyme A; insert gene for ampicillin resistance (Amp^R).

Amp^R

Linearize plasmid by cutting with restriction enzyme B.

Amp^R

Amp^R

Transform into $recD^-$ E. coli cells. The $recD^-$ mutation prevents the cell from degrading linear DNA fragments.

Amp^R

Double recombination event in transformed cell replaces chromosomal gene with disrupted gene.

Select for ampicillin-resistant colonies by plating bacterial cells on medium containing ampicillin.

Bacterial chromosome

Amp^R
Gene of interest

Bacteria with disrupted gene

Figure 14.23 One way to characterize the roles of uncharted bacterial genes. First, an ampicillin resistance gene (Amp^R) is inserted into the gene of interest, located on a plasmid. The linearized plasmid is then transformed into a wild-type E. coli strain, where it recombines to leave a disrupted gene.

serted into the middle of the cloned gene through use of a restriction enzyme that cuts only within the gene. The resulting circular molecule is then cut with a restriction enzyme that recognizes only a vector sequence; the cut creates a linear fragment of DNA that cannot replicate on its own in the cell and has to integrate into the chromosome to be passed on to further generations. This linear DNA fragment is used to transform a population of cells, after which selection for antibiotic resistance identifies those cells in which recombination between the mutated gene on the linear DNA fragment and the wild-type gene on the bacterial chromosome resulted in replacement of the wild-type gene with the mutated copy. The phenotype of the cell containing the knockout mutation provides clues about the function of the gene.

Comprehensive Example: Genetic Dissection Helps Explain How Bacteria Move

Bacteria can move up gradients of chemical attractants, such as food particles, and down gradients of chemical repellents, such as toxins. This sophisticated behavior of seeking out or avoiding chemicals diffusing through their growth medium is known as **bacterial chemotaxis.** Researchers have applied the powerful procedures for mutating, mapping, and cloning bacterial genes to identify the genes and proteins that function in bacterial chemotaxis and to discover how the proteins work together to accomplish their task.

How Bacteria Move to Achieve Chemotaxis

With specially constructed microscopes that make it possible to record the motion of individual bacteria, scientists have observed that cells swim rapidly in one direction for about 1 second and then abruptly reorient and swim in another direction. The 1-second, rapid, one-way movement is called a "straight run"; the abrupt reorienting movement is a "tumble" (Fig. 14.24). The change in direction is random; thus bacteria move through space in a **random walk.** With the addition of an attractant or repellent, the behavior of individual bacteria changes. They still move in a random walk, but the time spent in a straight run is longer immediately after the addition of an attractant and shorter immediately after the addition of a repellent. This reactive behavior is known as a **biased random walk:** Each bacterium still executes a random walk with frequent changes in direction, but its movement over time is biased toward or away from a chemical gradient. The biased behavior of individual bacteria determines the net movement of a population toward an attractant and away from a repellent.

If the concentration of attractant or repellent does not change, within a short time, the bacteria resume the unbiased random walk they were doing before the addition of either type of chemical. The ability to stop responding when the stimulus is present in an unchanging concentration allows a bacterium to "sense" when it is moving up or down a concentration gradient.

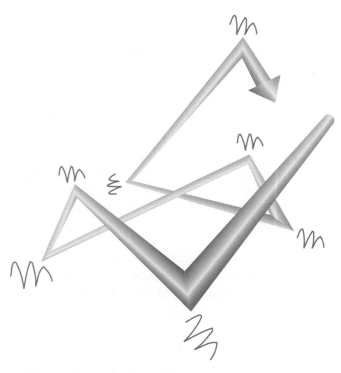

Figure 14.24   How bacteria move. The lack of a consistent direction is seen in the run and tumble style of movement. *Squiggles* indicate tumbles.

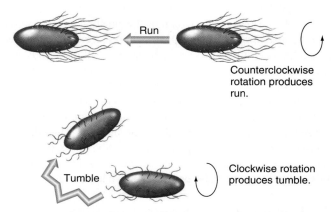

Counterclockwise rotation produces run.

Clockwise rotation produces tumble.

Run

Tumble

Figure 14.25   Counterclockwise rotations of bacterial flagella produce runs; clockwise rotations cause tumbles. During a run, each flagellum turns counterclockwise, allowing all the flagella to bundle together and rotate counterclockwise as a unit. When flagella turn clockwise, they cannot bundle and thus cannot go in a straight line, which produces a tumble.

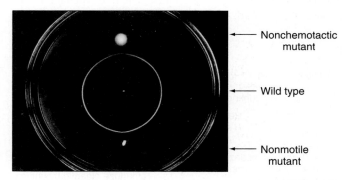

Nonchemotactic mutant

Wild type

Nonmotile mutant

Figure 14.26   Isolating bacterial mutants that cannot move toward food. The wild-type bacteria move away from the center of the plate toward more food as nutrients are used up, but the nonchemotactic and nonmotile mutants are unable to sense or move toward food, respectively.

This aspect of chemotaxis is known as **adaptation.** It is analogous to the way you adjust to the bright light of day when you emerge from a dark movie theater—after a while in the new environment, your eyes adapt and you no longer blink.

Rotating Flagella Are the Mechanical Basis of the Random Walk

Bacteria have many long, rotating flagella distributed over their surface. When the flagella all rotate counterclockwise, they bundle together, propelling the cell in a forward run. When they rotate clockwise, the helical structure of each flagellum prevents bundling, and the cell undergoes a chaotic motion producing a tumble (Fig. 14.25).

You can observe the direction of rotation of the flagella on individual bacteria by attaching the cells to the surface of a petri plate with an antibody against the flagella and then looking down through a microscope at the rotating cells. As with runs and tumbles, the cells spend about 1 second rotating counterclockwise and then switch and rotate clockwise for a brief period. Immediately after the addition of an attractant, cells spend longer periods rotating counterclockwise; immediately after the addition of a repellent, they spend shorter periods rotating counterclockwise. From these observations, it is possible to conclude that gradients of attractants and repellents influence the direction of flagellar rotation.

Many Bacterial Mutants Cannot Carry out Chemotaxis

Bacteria inoculated into the center of a nutrient plate that is wet enough for them to swim in will multiply and use up the nutrients in the immediate vicinity. As they do, the population of cells swarms out in concentric circles, with the leading edge moving toward the unused nutrient supply. This behavior is the basis of a quick test for mutants that cannot move toward food; it also provides a way to isolate such mutants (Fig. 14.26). After the motile cells have swarmed, it is possible to lift from the center of the plate the cells that did not move. Given the known details of how bacteria move in chemotaxis, one would anticipate a variety of mutants incapable of chemotaxis. Some might be defective in making flagella; some in turning the flagella; others in coordinating the behavior of flagella; and still others in detecting gradients of chemicals. Investigators have isolated all of these types of mutants. Genetic and biochemical analyses of the mutants' properties have produced a detailed understanding of how bacteria accomplish the sophisticated task of moving toward and away from chemicals.

Flagellum Mutants

More than 20 *fla* genes are required to generate the complex structure of a flagellum (Fig. 14.27). Most of each flagellum's extracellular portion is composed of a single protein, but electron microscopy of the intracellular structure reveals a series of rings. Mutations in the *fla* genes usually prevent production of functional flagella.

Motor Mutants

The *mot* genes are required to turn the flagellum. Mutant *mot* cells have normal-looking flagella, but they are paralyzed.

Signal Transduction Mutants

Mutants in the *che* (*chemotaxis*) genes have flagella that rotate, but only in one direction—either clockwise or counterclockwise. The *che* genes encode components of a **signal transduction system,** which relays messages from the cell surface, where nutrients are detected, to the motor, where the messages control the frequency with which the direction of rotation changes. The signal transduction system must decide whether the concentration of a chemical is increasing or decreasing, relay the appropriate information to the motor, and then undergo adaptation to bring the system back to its unstimulated state. The proteins encoded by the *che* genes send signals by phosphorylating (that is, by adding a phospate group onto) one an-

other. Phosphorylation is the signal that tells the next protein in the cascade whether to be active or inactive.

Receptor Mutants

All the mutants described so far have a defective response to all attractants and repellents. But some mutants respond in a defective way to a particular attractant or repellent, while responding normally to others. A simple "capillary test" reveals why. If a capillary pipette containing a nutrient is placed in a solution of bacteria, the cells swim into the tube (Fig. 14.28). Researchers leave the capillary tube in the solution for a standard amount of time and then remove it and determine the number of cells it contains by diluting and plating the bacteria for colony count. With this assay, they can discover the response of bacteria to specific nutrients and determine whether particular bacteria can respond to one nutrient in the capillary in the presence of a high concentration of another nutrient (both in the bacterial solution and in the capillary). Their observations confirm that bacteria normally respond to many different attractants and repellents independently. This suggests that the bacterial systems for detecting these chemicals are composed, in part, of different molecules.

Other studies have shown that mutants defective in specific responses have normal flagella that rotate normally and that switch from clockwise to counterclockwise with the expected frequency for an unstimulated cell. These mutants do not, however, respond when a particular stimulus is added to their

Figure 14.27 More than 20 genes are needed to generate a bacterial flagellum. Some proteins anchor the flagellum in the cell membrane; others implement the switch in rotational direction; still others serve as structural protein for the part of the flagellum that is outside the cell.

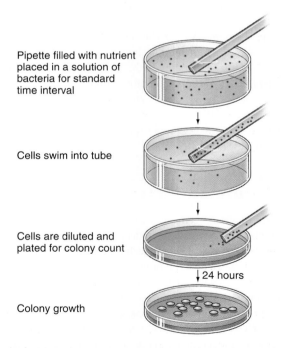

Figure 14.28 The capillary test for chemotaxis. The basis of this test is to see whether bacteria swim into a capillary tube containing attractants and/or repellents. The number of bacteria retained in the tube after a standard amount of time allows for quantitative measurements of chemotactic behavior.

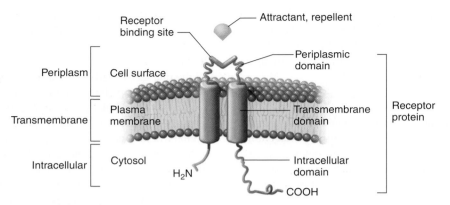

Figure 14.29   Bacteria have cell surface receptor proteins that recognize particular attractants or repellents. The binding of an attractant or repellent molecule to the membrane domain of a receptor specific for the compound causes changes in the conformation of the receptor's intracellular domain. The intracellular domain then interacts with the general components of the signaling pathway.

growth medium, because the membrane proteins that act as receptors for the attractant or repellent in question are defective. These receptor proteins have an extracellular domain that interacts with one or a few attractants or repellents, a transmembrane domain, and an intracellular domain that interacts with the Che proteins to stimulate signal transduction (Fig. 14.29).

Not surprisingly, the cell surface receptors that bind particular chemicals are the molecules that are modified during adaptation to bring the cell back to an unstimulated state. The protein product of the *cheR* gene reversibly methylates positions on the intracellular domains of the receptors; the protein product of the *cheB* gene reversibly demethylates those same sites. The methylation status of the intracellular domain, in turn, determines the receptors' ability to respond to the binding of attractant or repellent by the extracellular receptor domain. The CheW protein phosphorylates CheB, CheY, and CheA, thereby activating them. Phosphorylated CheY sends a signal to the flagella for clockwise rotation. CheZ dephosphorylates CheY to restore the initial state (Fig. 14.30).

The Future of Bacterial Genetics

E. coli was the first organism in which the detailed genetic analysis of cellular processes was practical on a large scale. Studies of *E. coli* using the classical genetic techniques described in this chapter helped establish the basic principles of cell physiology as well as protocols for genetic analysis. With the advent of recombinant DNA technology in the 1970s, *E. coli* served as the host into which genes were cloned and as a result, became once again the organism of choice for genetic experimentation. *E. coli* mutants (such as Lac⁻ strains and recombination deficient mutants) that had been characterized in classical genetic experiments were put to use in protocols for selecting and maintaining collections of clones.

The recent genomic phase of exploration has produced a wealth of data on bacteria as well as other organisms. Researchers now know the complete genomic sequence of several pathogenic and nonpathogenic bacteria as well as the DNA

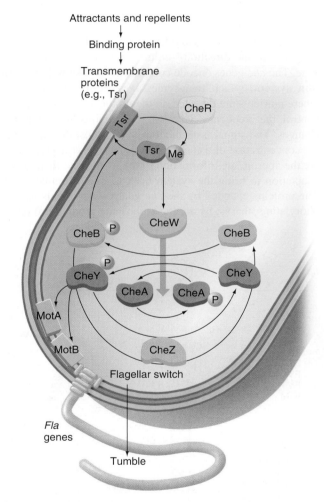

Figure 14.30   The genetic and molecular basis of bacterial chemotaxis. The *fla* genes are necessary for the structure of the flagella; *mot* genes are involved in the flagellar motor and determine the direction of rotation; *tsr* genes encode the transmembrane receptors; *che* genes process the signals from receptors and communicate with the Mot proteins to cause the appropriate response. The methylation (Me) of receptors prevents them from sustaining continual stimulation.

sequences of bacteria with unusual metabolic activities. These complete genomic sequences not only provide interesting new information; they also suggest new avenues of inquiry and exploration.

Comparative Genome Analysis Can Help Dissect the Genetic Basis of Bacterial Behavior

Comparative genome analysis is the examination and comparison of different species' genomes. Researchers can use the comparative genome approach to study many different bacterial functions, including survival at high temperatures, the invasion of specific tissues, growth, and pathogenesis. Genes shared among thermophilic bacteria but not with nonthermophilic bacteria could be candidates for further study as genes necessary for survival at high temperatures. Pathogens that invade similar regions of the human body might carry a set of genes required for the invasion of those tissues; to discover what those genes are, one could perform a subtractive analysis of the genomes of several different pathogens that invade mucosal tissues. Such an analysis could lead to the identification of unique genes that bacteria express to evade host defenses and successfully establish themselves in human mucosal cells.

Recent studies have suggested that several bacteria streamline their genomes and retain only genes essential for growth and pathogenesis. In these studies, comparative genome analysis of isolates of the same bacterium, *Mycobacterium tuberculosis,* showed deletions of regions of the genome that must not be essential for growth and pathogenesis. Comparative analyses of obligate parasitic bacteria and their nonparasitic relatives might shed a different light on necessary genes.

Significant genomic findings in pathogenic bacteria include the observation that pathogenic determinants are often clustered in "pathogenicity islands." With such an arrangement, the transfer of a "package" of genes from one species to another can create a pathogenic strain. In the cholera-producing *Vibrio cholerae,* for example, a single region of one chromosome is bounded by inverted repeats, and the DNA of the region encodes a recombinase enzyme. This DNA organization and content suggest that the entire chromosomal region may have been transferred into *V. cholerae* intact from another source. In another example, DNA array analysis of the genomes of several isolates of *Staphylococcus aureus,* including antibiotic-resistant strains and strains that cause toxic shock syndrome, showed that several distinct strains carry added regions of DNA from 4–50 kb in size. Antibiotic resistance genes were present in one of these regions, and most of the added regions encoded genes homologous to transposases. These findings suggest that the added regions arrived in the genomes of the *S. aureus* strains through lateral gene transfer.

The Fruits of Genomic Analysis May Help Protect Human Health

One hope is that genomic knowledge of pathogenic bacteria will lead to the identification of vaccine candidates, an ever more pressing concern in this era when the efficacy of antibiotics is di-minishing because bacterial resistance to such drugs is on the rise. Another hope in the fight against pathogens is that genomic analysis will aid in the discovery of new drug targets. Identification of the genes and gene products essential for growth in a pathogenic species, for example, could allow rational drug design wherein pharmacologists synthesize compounds that target only those proteins not found in the host species.

Genomic technology and information may also aid epidemiologists by providing new avenues for the unambiguous identification of specific bacterial strains, which would make it possible to trace the history of an infection. If, for example, outbreaks of a disease occur in New York City, Los Angeles, and Seattle in the same year, comparative analysis of the genomes of the disease-causing microbes could show how closely related the strains are and therefore how likely the disease is to spread from one location to another. An understanding of the origins and pathogenesis of new strains could also shed light on how the organisms evolve and what mechanisms propel their dispersal.

In the future, human needs will undoubtedly influence the choice of which organisms to sequence. The bacteria chosen for study may be major pathogens or organisms with unusual and perhaps commercially useful gene products.

Genomic Information Has Forced Scientists to Reevaluate Their Notions of Bacterial Evolution

Geneticists had thought that bacteria started out with a set of genes, which evolved slowly through point mutation, some deletion, and some duplication within the species. The latest data, however, show that in some cases, a gene or set of genes in one bacterial species is very similar to genes in another bacterial species. This suggests that the bacterial genome has picked up DNA from several different sources during the course of evolution. Data from the genomes of other organisms, including humans, point to the same phenomenon. Because of these observations, biologists now recognize that lateral gene transfer between species is a significant evolutionary factor. The data also suggest that sudden changes in an organism's metabolic capability may be mediated by the acquisition of sets of genes.

Challenges for the Postgenomic Era

Bacterial geneticists in the postgenomic era will seek to use the new genomic data together with classical and recombinant DNA methodology to discover the function of genes and address new questions that arise from analysis of the genome data. The availablility of genomic information has already made it possible to approach the question, What is the minimal set of genes required for a cell to function? The smallest known genome of a free-living cell is that of the bacterium *Mycoplasma genitalium,* a parasite of the human genital and respiratory tracts. With a genome size of 580 kb, it has been established that the cell has 484 coding regions. Using transposon insertion mutagenesis, investigators showed that about 100 of these coding regions are nonessential. Another approach to identifying essential genes comes from an examination of several distinct pathogenic isolates of *Mycobacterium tuberculosis*

Genetics and Society

How Bacteria Cause Disease

Bacteria that cause disease in other organisms are known as *pathogens*. Their relative ability to initiate illness is their *virulence*. Mild pathogens cause relatively mild symptoms; virulent pathogens inflict serious damage that may threaten life. To be successful, a pathogen must outwit the varied defenses of a host organism, multiply inside the host, and be transmitted to other hosts. How do bacterial pathogens accomplish all this in humans?

The Steps of Bacterial Pathogenesis

Entry into the host. The human skin is a protective barrier that excludes most bacteria, but there are several ways to breach its defenses. Some species of bacteria gain entry to the body via wounds that damage the skin; *Clostridium tetani,* the agent of tetanus, is one such species. Other bacteria pass through openings established by insect bites; *Yersinia pestis,* the agent of bubonic plague, for example, travels from rodent hosts to humans via flea bites. A third avenue of access to the human host is transmission to mucosal surfaces. *Mycobacterium tuberculosis,* the agent of tuberculosis, infects the respiratory tract via aerosols produced by sneezing or coughing; *Neisseria gonorrhoeae,* the agent of gonorrhea, infects the urogenital tract via sexual transmission; *Vibrio cholerae,* the agent of cholera, infects the digestive tract via ingestion. *Once inside the host, some bacteria adhere to the surface of eukaryotic cells in the region of entry.* Attachment depends on interactions between protein structures on the bacterial cell surface—similar to the pili that help initiate conjugation—and specific receptors, usually composed of carbohydrate, on the surface of the host's eukaryotic cells (Fig. A). Adherence is essential to pathogenesis. One illustration of this fact is the observation that the few pathogenic strains of *E. coli* are able to synthesize attachment proteins, while the large majority of non-pathogenic *E. coli* strains make no attachment proteins and are unable to adhere to cells of the intestinal mucosa. After adhesion, pathogenic bacteria begin to replicate.

Some bacteria penetrate to deeper layers of host cells. *Helicobacter pylori,* the bacteria associated with many cases of ulcers, twists through the layer of mucus protecting the stomach and adheres to cells on the surface of the stomach wall. After penetrating the stomach's mucosal lining, *H. pylori* secretes urease, an enzyme that generates ammonia, which in turn, neutralizes the stomach acid in the area. Other penetrating bacteria forge a path into tissues deep within the body by secreting enzymes such as collagenase, which break down the molecules that help hold tissues together. These penetrating bacteria then adhere to cells and begin to replicate.

Some bacteria invade host cells. Multiplication inside those cells leads to cell lysis (Fig. B). *Shigella dysenteriae,* the agent of bacterial dysentery, invades the epithelial cells of the intestines. *Legionella pneumophila* (the agent of Legionnaire's disease) and *M. tuberculo-*

Figure A   Adherence. Bacteria can adhere to the surface of eukaryotic cells. Shown here are spirochetes (*dark purple*) attached to bowel cells (*pink*).

Figure B   Invasion. Some bacteria, such as *Mycobacterium tuberculosis* (*green cells*), can invade and multiply within host cells.

sis live and replicate within the large white blood cells known as *phagocytes,* which normally destroy bacteria.

Bacterial replication can damage the host in several ways. First, the presence of a large number of bacteria at one site can block the normal flow of materials. In the respiratory system, for example, a mass of bacteria can block the flow of oxygen through the smaller air passages; in the heart, large numbers of bacteria can clog a valve, hindering the flow of blood. Second, bacterial multiplication inside invaded cells destroys those cells, causing inflammation and disarray, which undermine the function of essential tissues. Finally, some bacteria produce toxins that act on-site or travel to other parts of the body. Tetanus bacteria, for example, produce a neurotoxin that interferes with the transmission of nerve impulses and causes muscle spasms in the neck and jaw. Dysentery bacteria produce a toxin that poisons intestinal cells and causes diarrhea.

In summary, adherence to the surface of host cells near the point of entry, penetration and adherence to tissues deep within the body, and invasion of host cells allow bacteria to replicate, using nutrients provided by the host. Sufficient replication produces large colonies that cause disease.

The Human Host Puts up a Fight Based on Natural and Synthetic Defenses

Several natural components of the human immune system as well as human-made drugs act to deter and cure bacterial infections.

The skin and mucosal membranes. We have already seen that bacteria must breach the skin or mucosal linings of the mouth, nose, or digestive and urogenital tracts to gain entry into the host. The skin is a tough, nearly impervious barrier. The mucosal membranes are very acidic, providing an environment inhospitable to most bacteria. Mucus also contains sterilizing factors such as the enzyme lysozyme, which destroys bacterial cell walls.

Antibodies. The immune system responds to bacterial infection with the production of immense numbers of antibodies that bind specifically to various proteins on the invading bacteria. Some antibodies block the surface attachment molecules the bacteria use to adhere to eukaryotic cells; others may agglutinate the toxin molecules produced by the bacteria, rendering them nontoxic. The binding of antibodies to bacteria also activates the complement cascade—a series of reactions among a set of 11 proteins that ultimately form a complex on the bacteria and punch holes in their cell membranes.

Phagocytes and other cells of the immune system. Phagocytes are relatively large white blood cells that engulf and destroy bacteria. Highly differentiated phagocytes known as macrophages become activated by other cells of the immune system to pursue and dispose of specific invaders.

Drugs. In the last hundred years, pharmaceutical research has produced a large number of bactericidal drugs, including the sulfa drugs, penicillin, and other antibiotics. If administered soon enough after infection, these drugs can kill susceptible bacteria before they replicate sufficiently to cause serious symptoms.

Bacteria Fight Back via Evasion of Immune Responses and the Development of Drug Resistance

The success of a pathogen depends on effective host entry strategies, efficient multiplication within the host, and the possibility of transmission to a new host. By these measures, *N. gonorrhoeae* is one of the most successful bacterial pathogens of humans in the world today. It has evolved impressive stratagems for bypassing the skin barrier and overcoming human immune defenses; it has also become resistant to several antibiotics.

For *N. gonorrhoeae,* transmission via sexual intercourse circumvents the skin barrier, while the unprotected sexual practices of many human cultures ensure entry into the mucosal linings of the urogenital tract of a large number of hosts worldwide. Once inside the host, the bacteria secrete an enzyme that destroys immunoglobulin A, the type of antibody that is active in the bodily secretions it encounters in the urogenital tract. The bacteria then buy time for replication by varying their cell surface molecules, the very molecules the human immune system recognizes and reacts against. Because an antibody response against one molecule has little effect against a different molecule, the variation tactic helps *N. gonorrhoeae* evade the immune system. The bacteria also avoid immune destruction through the special construction of their cell membrane. Although components of the complement cascade can still bind to the bacterial surface, the spatial configuration of the bound complement is aberrant and, as a result, the proteins are unable to punch holes in the bacterial membrane. The gonorrheal bacterium escapes destruction by phagocytes by invading and replicating inside other types of host cells. Finally, as we have seen, many populations of *N. gonorrhoeae* have developed resistance to three widely used antibiotics—penicillin, spectinomycin, and tetracycline—through genes acquired from other bacterial species. Thus, for the moment, *N. gonorrhoeae* seems to be winning the battle between the bacteria and their human hosts. New drugs, however, could temporarily tip the balance in the other direction.

Knowledge of Bacterial Pathogenesis Has Led to an Understanding of the Genetic Underpinnings of Virulence

Virulence, the relative ability of a pathogen to cause disease, depends on a bacterium's production of proteins that promote adherence to host cells, penetration of host tissues, the invasion of host cells, the production of bacterial toxins, and the avoidance of host defenses. Bacterial geneticists have begun to determine which genes contribute to the pathogenicity of particular bacterial species by using recombinant DNA techniques to transfer genes from virulent (relatively pathogenic) strains into nonvirulent strains. If the nonvirulent strains become somewhat pathogenic, the likely conclusion is that the transferred genes conferred virulence. A significant insight from this work is that the genes conferring virulence are often present as clusters on plasmids or transposons (as we saw previously). Their tandem arrangement on small pieces of DNA that move from strain to strain underlies the one-step conversion of nonpathogenic bacteria to virulent pathogens. The medical implications of this fact merit close attention and emphasize the need to learn more about gene transfer.

(another species with a relatively small genome). DNA array analysis revealed deletions in several, but not all, of the strains. One conclusion is that the deleted regions must contain genes that are not essential for growth.

As knowledge of essential genes and of genes that carry out specialized functions increases, geneticists will try to under-stand how a bacterial cell integrates various gene products into its physiology. This broader knowledge of bacterial physiology will, in turn, put within reach a better understanding of the complex interactive world of microbial ecology.

C O N N E C T I O N S

The study of bacterial genetics underscores the unity of genetic phenomena in all types of living organisms. Double-stranded DNA serves as the genetic material in bacteria as it does in eukaryotes. The general mechanisms of gene expression, DNA replication, and recombination are also similar in prokaryotes and eukaryotes. The study of bacterial genetics shows, however, that within the unity of basic biological phenomena, there is a remarkable diversity of detail. Unlike most eukaryotic cells, bacteria do not produce gametes through meiosis or new generations of cells by the union of gametes. Nevertheless, bacteria can exchange genes between different strains through transformation, conjugation, and transduction. Together, these three modes of gene transfer increase the potential for the evolution of prokaryotic genetic material.

Knowledge of bacterial genetics raises the possibility of combatting bacteria harmful to humans, animals, and crops and using beneficial bacteria for positive purposes, such as degrading oil slicks and protecting crops from freezing. Knowledge of bacterial genetics also increases our understanding of the organelles of eukaryotic cells. Biologists believe that mitochondria, the cell organelles that produce energy for metabolic processes, and chloroplasts, the photosynthetic organelles of plant cells, are descendants of bacteria that fused with the earliest nucleated cells. Mitochondria are similar in size and shape to today's aerobic bacteria and have their own DNA, which replicates independently of the cell's nuclear genetic material. Chloroplasts are similar in shape and size to certain cyanobacteria and have self-replicating DNA carrying bacteria-like genes. Based in part on these observations, the **endosymbiont theory** proposes that chloroplasts and mitochondria originated when free-living bacteria were engulfed by primitive nucleated cells. Host and guest formed cellular communities in which each member adapted to the group arrangement and derived benefit from it. Chapter 15 examines the structure and function of the chromosomes of mitochondria and chloroplasts.

E S S E N T I A L   C O N C E P T S

1. Bacteria are prokaryotic cells with no membrane-enclosed nucleus or other cell organelles. The bacterial genome consists of a single circular chromosome in which the genes are tightly packed, with about one gene per kilobase pair.

2. During transcription in *E. coli,* different strands of DNA serve as the template for different genes, with the most frequently transcribed genes oriented in the direction of replication fork movement. Replication begins at the origin of chromosomal replication (*oriC*) and proceeds bidirectionally around the circular genome to the termination region (*terC*).

3. In addition to their chromosome, most bacteria carry plasmids: small circles of double-stranded DNA. Plasmids may include genes that benefit the bacterial host under certain conditions. One important group of plasmids promotes conjugative gene transfer between two bacteria.

4. Bacterial genomes contain IS and Tn elements, transposons that can move between sites on any DNA molecule in the cell.

5. *Transformation* is a form of gene transfer in which donor DNA that is floating free in the growth medium enters a recipient cell. Some bacteria have cellular machinery that supports efficient natural transformation. Species that do not undergo natural transformation can be induced to take up DNA by treatments that disrupt their cell walls in a process known as artificial transformation.

6. *Conjugation* is a second form of gene transfer. It depends on direct cell-to-cell contact between a donor carrying a conjugative plasmid (the F plasmid is one example) and a recipient lacking such a plasmid. In crosses between F$^+$ and F$^-$ strains, only the plasmid is transferred. In crosses between Hfr and F$^-$ strains or between F′ and F$^-$ strains, the plasmid also transfers chromosomal genes.

7. *Transduction* is a third form of gene transfer in bacteria. It depends on the packaging of bacterial donor DNA in the protein coat of a bacteriophage. In generalized transduction, phages can package any part of the donor genome. Specialized transduction is a property of lysogenic bacteriophages, which can package only host genes adjacent to the integrated prophage genome.

8. It is possible to map bacterial genes using any of the three forms of gene transfer. Interrupted-mating experiments can map the approximate positions of genes. Measures of frequencies of cotransformation or cotransduction provide better resolution: The closer two genes are, the more likely they are to appear on the same short DNA fragment. Sequences of complete bacterial genomes are becoming increasingly important as mapping resources.

SOCIAL AND ETHICAL ISSUES

1. Bacteria have now been engineered that will break down components of oil, potentially providing a biological way to clean up an oil spill. Many environmentalists have been upset by and are opposed to this action. Why would they be disturbed by this action? How widespread do you think effects of seeding a spill site with oil-eating bacteria might be? Who would have to approve this type of action in cleaning an oil spill? What characteristics might you want to engineer into these bacteria to make them safer for the environment?

2. Sarah has been using Southern blot analysis to determine whether genes homologous to specific ones found in *E. coli* are present in a rare thermophilic (heat-loving) bacterium. Since she had very little of the bacterial DNA, she was able to make only one filter, and she had to reprobe that filter several times. After many months of work, she is getting her answer on what genes are present. The final blot that she does gives a large smudge in an important lane, but she feels she can still tell where there are bands and where there are not bands. The blot is not of high enough quality for reproduction in a publication. Sarah knows that reviewers and editors would probably reject it, but she thinks she could refer to this piece of evidence in her paper without showing a photo of the blot (listed as data not shown in the paper). It would take several months to again grow enough of the bacteria to repeat the experiment, and she knows there are competitors with similar experiments to report. Should she go ahead and submit the paper for publication even though most people would say the data are messy?

3. *E. coli* strain O157 is a unique strain of this bacterial species that has caused serious outbreaks of human illness, in some cases leading to death of small children. Yet *E. coli* is part of the normal flora of the gut of all individuals, and researchers work on harmless laboratory strains all the time. In a college microbiology course in a large university, one student got upset at the prospect of working with *E. coli* in the lab. Even with an explanation of the difference between the toxic and nontoxic bacteria, he was not convinced that his safety was ensured. He says he will not do the lab work himself but will do analysis of the lab data obtained by his partner. The laboratory is a required part of the course and accounts for 25% of the final grade. What should and/or could the professor do with this student? Should the student have to forfeit the laboratory points in the course? How far should the professor or university go to accommodate his needs?

SOLVED PROBLEMS

I. You have cloned the gene encoding the major protein in the flagella of a new bacterial strain. In screening for mutant bacteria that have a defective flagellar protein, you found mutants at an exceptionally high frequency (1 in 10^3 bacterial cells). You suspect these may have been caused by insertion of a transposable element into the gene. How could you determine if this had occurred?

Answer

One way to determine if the high-frequency mutants result from insertion into the gene is to perform a Southern hybridization. *The cloned gene would be used as a probe to hybridize with DNA from the wild-type and the mutant cells.* If the mutant arises from insertion of a transposable element, the size of fragments containing the interrupted gene will be different from fragments containing the normal gene.

II. You have an F^- strain of *E. coli* that is resistant to streptomycin but requires the following amino acids for growth on minimal medium: arginine, cysteine, methionine, phenylalanine, and proline. You do a series of interrupted-mating experiments with two different Hfr strains (Hfr1 and Hfr2), both of which are wild type for all of the amino-acid markers but are Str^s.

a. The following data are the times in minutes at which you first observed recombinant progeny containing the indicated amino-acid marker. Draw a chromosome map showing all the genes, the position of the origin of each Hfr, the direction of transfer of each Hfr, and the distances between each marker. Use a directional arrowhead to indicate each origin of transfer.

	Hfr1 × F⁻	Hfr2 × F⁻
arginine	5	40
cysteine	65	70
methionine	75	60
phenylalanine	35	10
proline	10	35

b. You do a new mating experiment with the same strains; you interrupt the mating after 60 minutes and plate on a medium that selects for progeny that are recombinant for the phenylalanine gene (*phe*⁺). What medium would you use to select for these progeny?

Answer

a. The order of transfer of the five genes is indicated by the time at which cells showing the marker first appear. The origin of these two Hfr strains is different; that is, the location of the inserted F factor is different and the direction of transfer is different in the two Hfr strains.

b. When you repeat the experiment selecting for progeny recombinant for phenylalanine, you plate *the cells on medium that contains arginine, cysteine, methionine, proline, and streptomycin.* The streptomycin selects against the Hfr strain, and only phenylalanine is left out of the media to select for F⁻ cells that have received genes from the donor.

III. Using bacteriophage P22 you performed a three-factor cross in *Salmonella typhimurium.* The cross was between an Arg⁻ Leu⁻ His⁻ recipient bacterium and bacteriophage P22, which was grown on an Arg⁺ Leu⁺ and His⁺ strain. You selected for 1000 Arg⁺ transductants and tested them on several selective media by replica plating. You obtained the following results:

Arg⁺ Leu⁻ His⁻	585
Arg⁺ Leu⁻ His⁺	300
Arg⁺ Leu⁺ His⁺	114
Arg⁺ Leu⁺ His⁻	1

a. What is the order of the three markers?

b. What are the cotransduction frequencies?

Answer

a. The order can be determined by looking at the relative frequencies of each phenotypic class. Arg⁺ Leu⁻ His⁻ is the largest class, with only the *arg* gene transferred. The next largest class is Arg⁺ and His⁺. Therefore *arg* and *his* are closer to each other than *arg* and *leu. The order of the genes is* arg-his-leu.

b. Cotransduction frequency is the percentage of cells that received two markers. For *arg* and *his,* this includes the Arg⁺ Leu⁻ His⁺ cells (300) and Arg⁺ Leu⁺ His⁺ cells (114). The cotransduction frequency of *arg* and *his* is 414/1000 = 41.4%. The cotransduction frequency of *arg* and *leu* is 114 + 1 or 115/1000 = 11.5%.

P R O B L E M S

14-1 Choose the phrase from the right column that best fits the term in the left column.

a. transformation 1. requires supplements in medium for growth

b. conjugation 2. transfer of DNA between bacteria via virus particles

c. transduction 3. small circular DNA molecule that can integrate into the chromosome

d. lytic cycle 4. transfer of naked DNA

e. lysogeny 5. transfer of DNA requiring direct physical contact

f. episome 6. integration of phage DNA into the chromosome

g. auxotroph 7. infection by phages in which lysis of cells releases new virus particles

14-2 A liquid culture of *E. coli* at a concentration of 2×10^8 cells/ml was diluted serially, as shown in the following diagram, and 0.1 ml of cells from the last two test tubes were spread on agar plates containing rich media. How many colonies do you expect to grow on each of the two plates?

0.1 ml   0.1 ml   1.0 ml   1.0 ml

9.9 ml   9.9 ml   9.0 ml   9.0 ml

Starting
culture
(2 x 10^8 cells/ml)

0.1 ml   0.1 ml
Spread onto rich media

14-3 Pick out the media onto which you would spread cells from a Lac$^-$ Met$^-$ *E. coli* culture to
   a. select for Lac$^+$ cells
   b. screen for Lac$^+$ cells
   c. select for Met$^+$ cells
      i. minimal media + glucose + methionine
      ii. minimal media + glucose (no methionine)
      iii. rich media + X-Gal
      iv. minimal media + lactose + methionine

14-4 Now that the sequence of the entire *E. coli* genome (about 5 Mb) is known, you can determine exactly where a cloned fragment of DNA came from in the genome by sequencing a few bases and matching that data with genomic information.
   a. How many nucleotides of sequence information would you need to determine exactly where a fragment is from?
   b. If you had purified a protein from *E. coli* cells, roughly how many amino acids of that protein would you need to know to establish which gene encoded the protein?

14-5 The numbers of IS1 elements in different laboratory strains of *E. coli* vary. There are no recognition sites for the enzyme *Eco*RI in IS1. How could you determine the number of IS1 elements in the two strains *E. coli B* and in *E. coli K?*

14-6 There is usually one copy of the F plasmid per cell in an *E. coli* strain. You suspect you have isolated a cell in which a mutation increases the copy number of F to three to four per cell. (The copy number is determined by hybridization experiments.) How could you distinguish between the possibility that the copy number change was due to a mutation in the F plasmid versus a mutation in a chromosomal gene?

14-7 Genes encoding toxins are often located on plasmids. There has been a recent outbreak in which a bacterium

that is usually nonpathogenic is producing a toxin. Plasmid DNA can be isolated from this newly pathogenic bacterial strain and separated from the chromosomal DNA. To determine if the plasmid DNA contains a gene encoding the toxin, you could determine the sequence of the entire plasmid and search for a sequence that looks like other toxin genes previously identified. There is an easier way to determine whether the plasmid DNA carries the gene(s) for the toxin that does not involve DNA sequence analysis. Describe an experiment using this easier method.

14-8 DNA sequencing of the entire *H. influenzae* genome was completed in 1995. When DNA from the nonpathogenic strain *H. influenzae Rd* was compared to that of the pathogenic *b* strain, eight genes of the fimbrial gene cluster (located between the *purE* and *pepN* genes) involved in adhesion of bacteria to host cells were completely missing from the nonpathogenic strain. What effect would this have on cotransformation of *purE* and *pepN* genes using DNA isolated from the nonpathogenic versus the pathogenic strain?

14-9 In an Hfr × F$^-$ cross, the *pyrE* gene enters the recipient in 5 minutes, but at this time point there are no exconjugants that are Met$^+$, Xyl$^+$, Tyr$^+$, Arg$^+$, His$^+$, or Mal$^+$. The mating is now allowed to proceed for 30 minutes and Pyr$^+$ exconjugants are selected. Of the Pyr$^+$ cells, 32% are Met$^+$, 94% are Xyl$^+$, 7% are Tyr$^+$, 59% are Arg$^+$, 0% are His$^+$, 71% are Mal$^+$. What can you conclude about the order of the genes?

14-10 You have strains with the following genotypes:

Hfr *cys$^-$ str^s*

F$^-$ *trp$^-$ man$^-$ his$^-$ tyr$^-$ thr$^-$ str^r*

These mutations have the following phenotype:

Cys$^-$	cannot synthesize cysteine
Str^s	sensitive to streptomycin
Trp$^-$	cannot synthesize tryptophan
Man$^-$	cannot utilize the sugar mannose
His$^-$	cannot synthesize histidine
Tyr$^-$	cannot synthesize tyrosine
Thr$^-$	cannot synthesize threonine
Str^r	resistant to streptomycin

   a. What is the simplest medium you can use to grow the Hfr strain? the F$^-$ strain?
   b. You mixed the two strains and did two experiments. In experiment 1, you took samples and then diluted and plated on minimal medium supplemented with

histidine, tyrosine, threonine, and streptomycin and glucose as a carbon source. In experiment 2, you did essentially the same experiment but before you diluted you subjected the mating mixture to treatment in the food blender. In experiment 2, you never got any colonies until after 8 minutes. In experiment 1 you got some colonies at an earlier time. Why was there a difference in the results of these two experiments?

c. What medium would you use to select the exconjugants from experiment 2 that are Man$^+$?

d. The following table presents some of the data from experiment 2 in part *b*.

Marker	Time when first recombinants were found in minutes
Man$^+$	15.5
His$^+$	24
Thr$^+$	16
Tyr$^+$	16

Make the best map of the Hfr origin and genes.

e. To resolve ambiguities, you did the following crosses:

Cross 3   Hfr *man$^+$ thr$^-$ tyr$^+$* × F$^-$ *man$^-$ thr$^+$ tyr$^-$*

Cross 4   Hfr *man$^+$ thr$^+$ tyr$^-$* × F$^-$ *man$^-$ thr$^-$ tyr$^+$*

You note how many wild-type recombinants are produced and find many more in cross 3 than in cross 4. Show the order of these three genes.

14-11 You have recovered three independent mutations in one of the arginine biosynthesis genes. You now wish to map the site of the three mutations (*arg101, arg102,* and *arg103*) in the *arg* gene via the following Hfr × F$^-$ crosses in which Lac$^+$ cells were selected and then screened for Arg phenotype.

Cross	Hfr (*lac$^+$*)	F$^-$ (*lac$^-$*)	% of Lac$^+$ that are Arg$^+$
A	*arg101$^-$, arg102$^+$*	*arg101$^+$, arg102$^-$*	0.50
B	*arg101$^+$, arg102$^-$*	*arg101$^-$, arg102$^+$*	0.02
C	*arg101$^+$, arg103$^-$*	*arg101$^-$, arg103$^+$*	0.50
D	*arg101$^-$ arg103$^+$*	*arg101$^+$, arg103$^-$*	0.04
E	*arg102$^+$, arg103$^-$*	*arg102$^-$, arg103$^+$*	0.50
F	*arg102$^-$, arg103$^+$*	*arg102$^+$, arg103$^-$*	0.06

Draw a genetic map of the three *arg* mutations. Show the order of the three sites relative to the *lac* gene.

14-12 In a cross between an Hfr that has the genotype *ilv$^+$ bgl$^+$ mtl$^+$* and an F$^-$ that is *ilv$^-$ bgl$^-$ mtl$^-$*, *ilv* is known to be transferred later than *bgl* and *mtl*. To determine the order of *bgl* and *mtl* with respect to *ilv*, Ilv$^+$ exconjugants were selected and these colonies were screened for Bgl and Mtl phenotypes. Based on the following data, what is the order of the three genes?

	Number of exconjugants
Ilv$^+$ Mtl$^+$ Bgl$^+$	220
Ilv$^+$ Mtl$^-$ Bgl$^-$	60
Ilv$^+$ Mtl$^+$ Bgl$^-$	0
Ilv$^+$ Mtl$^-$ Bgl$^+$	18

14-13 An F′ element containing the maltose (*mal$^+$*) gene is discovered in *E. coli.*

a. Starting with an F$^+$ strain of *E. coli*, diagram the chromosomal events that must have occurred to produce this element.

b. The F′ *mal$^+$* element was introduced into a F$^-$ *mal$^-$* strain. Most of the resulting cells transferred the element to other F$^-$ cells. However, occasionally cells appeared that transferred the entire *E. coli* chromosome, beginning with the maltose gene. Draw the chromosomal events that must have occurred to produce such a strain.

14-14 Starting with an F$^+$ strain that was prototrophic (that is, had no auxotrophic mutations) and Str^s, several independent Hfr strains were isolated. These Hfr strains were mated to an F$^-$ strain that was Str^r Arg$^-$ Cys$^-$ His$^-$ Ilv$^-$ Lys$^-$ Met$^-$ Nic$^-$ Pab$^-$ Pyr$^-$ Trp$^-$. Interrupted mating experiments showed that the Hfr strains transferred the wild-type alleles in the following order as a function of time. The time of entry for the markers within parentheses could not be distinguished from one another.

Hfr strain	Order of transfer →
HfrA	*pab ilv met arg nic (trp pyr cys) his lys*
HfrB	*(trp pyr cys) nic arg met ilv pab lys his*
HfrC	*his lys pab ilv met arg nic (trp pyr cys)*
HfrD	*arg met ilv pab lys his (trp pyr cys) nic*
HfrE	*his (trp pyr cys) nic arg met ilv pab lys*

a. From these data derive a map of the relative position of these markers. Indicate with labeled arrows the position and orientation of the integrated F plasmid for each Hfr strain.

b. To determine the relative order of the *trp, pyr,* and *cys* markers and the distances between them, HfrB was mated with the F$^-$ strain long enough to allow transfer of the *nic* marker, after which Trp$^+$ recombinants were selected. The unselected markers *pyr* and *cys* were then scored in the Trp$^+$ recombinants, yielding the following results:

Number of recombinants	Trp	Pyr	Cys
790	+	+	+
145	+	+	−
60	+	−	+
5	+	−	−

Draw a map of the *trp, pyr,* and *cys* markers relative to each other. (Note that you cannot determine the order relative to the *nic* gene using this data.) Express map distances between adjacent genes as the frequency of crossing-over between them.

c. Detailed study of the genes in the *trp, pyr, cys* region would require an F′ plasmid carrying the three genes. Which of the preceding Hfr strain(s) would you use to isolate such an F′ plasmid?

14-15 In *E. coli,* four streptomycin-sensitive, prototrophic Hfr strains were mated with a streptomycin-resistant F⁻ strain auxotrophic for a number of nutritional requirements. Matings were interrupted at various intervals, and cells were plated on streptomycin-containing minimal medium supplemented with particular nutrients to test for gene transfer. The results are plotted in the following graphs.

a. Draw the best map for the *E. coli* genome consistent with the data. Include distances between the genes, and label arrowheads to show the origin of transfer for the four Hfr strains.

b. Suppose you wanted to generate an F′ (N^+) plasmid for use in complementation experiments. Describe how you would obtain such a plasmid, indicating which of the four Hfr strains in the graphs you would use and briefly listing the steps involved in isolating the F′ (N^+) plasmid. (Hint: Think about how the time of transfer of gene *N* could differ in an Hfr and an F′(N^+) strain.)

14-16 You can carry out matings between an Hfr and F⁻ strain by mixing the two cell types in a small patch on a plate and then replica plating to selective media. This methodology was used to screen hundreds of different cells for a recombination-deficient *recA⁻* mutant. Why is this an assay for RecA function? Would you be screening for a *recA* mutation in the F⁻ or Hfr strain using this protocol?

14-17 In two isolates (one is resistant to ampicillin and the other is sensitive to ampicillin) of a new bacterium, you found that genes encoding ampicillin resistance are being transferred into the sensitive strain. To determine if the gene transfer is transduction or transformation, you treat the mixed culture of cells with DNase. Why would this distinguish between these two modes of gene transfer? Describe the results predicted if the gene transfer is transformation versus transduction.

14-18 Suppose you have two Hfr strains of *E. coli* (HfrA and HfrB), derived from a fully prototrophic streptomycin-sensitive (wild-type) F⁺ strain. In separate experiments you allow these two Hfr strains to conjugate with an F⁻ recipient strain (Rcp) that is streptomycin resistant and auxotrophic for glycine (*gly⁻*), lysine (*lys⁻*), nicotinic acid (*nic⁻*), phenylalanine (*phe⁻*), tyrosine (*tyr⁻*), and uracil (*ura⁻*). By using an interrupted mating protocol you determined the earliest time after mating at which each of the markers can be detected in the streptomycin-resistant recipient strain, as shown here.

Earliest appearance of marker from Hfr

	gly⁺	*lys⁺*	*nic⁺*	*phe⁺*	*tyr⁺*	*ura⁺*
HfrA × Rcp	3	*	8	3	3	3
HfrB × Rcp	8	3	13	8	8	8

(The * indicates that no Lys⁺ cells were recovered in the 60 minutes of the experiment.)

a. Draw the best map you can from these data, showing the relative locations of the markers and the origins of transfer in strains HfrA and HfrB. Show distances where possible.

b. To resolve ambiguities in the preceding map, you studied cotransduction of the markers by the generalized transducing phage P1. You grew phage P1 on strain HfrB and then used the lysate to infect strain Rcp. You selected 1000 Phe⁺ clones and tested them for the presence of unselected markers, with the following results:

Number of transductants			Phenotype			
	Gly	Lys	Nic	Phe	Tyr	Ura
600	−	−	−	+	−	−
300	−	−	−	+	−	+
100	−	−	−	+	+	+

Draw the order of the genes as best you can based on the preceding cotransduction data.

c. Suppose you wanted to use generalized transduction to map the *gly* gene relative to at least some of the other markers. How would you modify the cotransduction experiment just described to increase your chances of success? Describe the composition of the medium you would use.

14-19 Generalized and specialized transduction both involve bacteriophages. What are the differences between these two types of transduction?

14-20 You have isolated a new temperature-sensitive mutation that affects the ability of *E. coli* to insert proteins into the membrane. At the restrictive temperature, the cells are unable to grow and thus do not form colonies.

a. Using a set of Hfr strains with origins every 5′ around the chromosome and a Tn10 insertion as a marker that is transferred in the first 5′, describe how you would rapidly narrow down the location of the new mutation. How many matings would you have to do?

b. If the Hfr set was not available, you could locate the mutation by transforming with a set of plasmids (average size: 20 kb). About how many plasmids would you have to transform into the mutant to just cover the genome? (Assume that your plasmids each overlap by 1 kb.)

14-21 a. Using the following pieces of technical information in the order given, explain how you would be able to identify the genes encoding proteins in *E. coli* cells that could bind directly to β-galactosidase: (1) β-galactosidase protein binds very tightly to a resin called APTG-agarose; (2) the 20 amino acids found in proteins vary widely in molecular weight; (3) the enzyme trypsin can cleave proteins into smaller peptides that are in the range of 3–40 amino acids long; the enzyme cleaves in a very predictable and reproducible way (after lysine and arginine amino acids); (4) modern techniques of mass spectrometry can measure the molecular weight of peptides to an accuracy of 0.01%; these machines measure the molecular weights of a large number of peptides in a complex mixture at the same time; (5) the entire *E. coli* genome has been sequenced.

b. Generalize the technique you described in part *a* to identify the genes encoding proteins that bind to any other particular protein in *E. coli*. (Hint: Use the fact that β-galactosidase binds to the APTG-agarose in your scheme.)

REFERENCES

Boucher, Y., and F. W. Doolittle. (2000). The role of lateral gene transfer in the evolution of isoprenoid biosynthesis pathways. *Molecular Microbiology* 37:703–16.

Kolsto, A-B. (1999). Time for a fresh look at the bacterial chromosome. *Trends in Micro.* 7:223–26.

15

The Chromosomes of Organelles Outside the Nucleus Exhibit Non-Mendelian Patterns of Inheritance

False color electron micrograph of an isolated chloroplast in a leaf cell of timothy grass (*Phleum pratense*) (11,000×).

Just nine years after the rediscovery of Mendel's laws, plant geneticists reported a perplexing phenomenon that challenged one of Mendel's basic assumptions. In a 1909 paper, they described the results of reciprocal crosses analyzing the transmission of green versus variegated leaves in flowering plants known as four-o'clocks (Fig. 15.1). Fertilization of eggs from a plant with variegated leaves by pollen from a green-leafed plant produced uniformly variegated offspring. Surprisingly, the reciprocal cross—in which the leaves of the mother plant were green and those of the father variegated—did not lead to the same outcome; instead all of the progeny from this cross displayed green foliage. From these results, it appeared that offspring inherit their form of the variegation trait from the mother only. This type of transmission, known as maternal inheritance, challenged Mendel's assumption that maternal and paternal gametes contribute equally to inheritance. Geneticists thus said that the trait in question exhibited **non-Mendelian inheritance.**

Another example of a non-Mendelian trait emerged 40 years later. In 1949 French researchers published studies on the size of yeast colonies in laboratory strains of the single-celled organism *Saccharomyces cerevisiae.* Mitotically dividing cultures of these cells, when grown on plates containing the fermentable sugar glucose as the source of carbon, produced colonies of two distinctly different sizes. Ninety-five percent of the colonies were large (in French, *grande*); the remaining 5% were small (*petite*). Cells from *grande* colonies, when separated and grown on fresh plates containing glucose, yielded some *petite* colonies, but cells from *petite* colonies never generated *grande* colonies. From these observations, the researchers deduced that the founder cells of *petite* colonies arose from frequent—1 in 20 cells—mutations in cells of the *grande* colonies.

The French researchers pursued their study of the genetic basis of this difference in colony size by analyzing various matings using haploid cells from *grande* and *petite* colonies. As described in Chapter 4, the diploid cells formed by mating haploid cells of opposite mating types may, under stressful conditions (not enough nutrients, for example), enter meiosis. When the French researchers mated *grande* cells of one mating type with *grande* cells

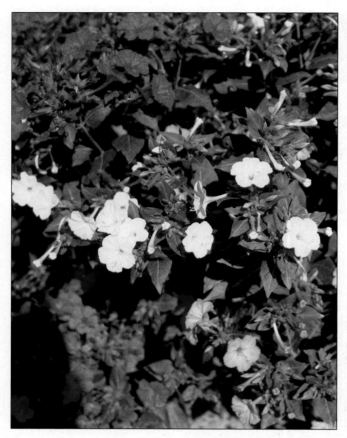

Figure 15.1 Non-Mendelian inheritance in four-o'clocks. The first example of non-Mendelian inheritance uncovered by geneticists was seen in the flowering plants known as four-o-clocks.

mined by genes that do not reside in the nucleus but instead lie in the genomes of nonnuclear organelles. Mutations resulting in leaf variegation occur in chloroplast DNA (cpDNA), in genes that encode proteins active in photosynthesis. Mutations that diminish yeast colony size occur in mitochondrial DNA (mtDNA), in regions of the genome that influence the efficiency of a cell's energy use. While chloroplasts are the sites of photosynthesis in plant cells, mitochondria are the sites of cellular respiration—the aerobic conversion of energy stored in nutrient molecules to energy stored in the high-energy bonds of ATP—in all eukaryotic cells. The *petite* mutants form smaller colonies because they are unable to carry out cellular respiration and must obtain the energy they need for survival from the less efficient, anaerobic energy conversion pathway of fermentation.

In this chapter, we see that non-Mendelian inheritance results from the fact that the genomes of chloroplasts and mitochondria are transmitted from generation to generation separately from the nuclear genome and in a very different fashion.

Three themes surface during our detailed discussion of the genes and genomes of mitochondria and chloroplasts. First, unlike the rules governing the transmission of nuclear genes, the formal rules for the transmission of organelle genomes can vary from organism to organism. Most diploid organisms, for example, inherit non-Mendelian traits from their mother, but in some species, the traits are transmitted by the father, and in other species, by both parents. Second, the maintenance of organelles requires cooperation between two genomes: that of the organelle itself and that in the nucleus of the cell in which the organelle functions. Some of the molecules in an organelle are encoded by organelle DNA and synthesized in the organelle; other molecules active in the organelle are encoded by nuclear genes and imported into the organelle. Finally, the genomes and biochemical processes of organelles are more similar in many ways to those of bacteria than to those in other parts of the eukaryotic cell. These observations formed the basis of the now accepted endosymbiont theory of the origin of extranuclear organelles. The theory proposes that organelles are the evolutionary remnants of bacteria that became symbionts in the ancient precursors of the earliest eukaryotes.

As we examine the structure, function, and transmission of the DNA carried by mitochondria and chloroplasts, we present:

■ The structure and function of mitochondrial and chloroplast genomes, including a description of the size, shape, replication, and expression of those genomes and a brief discussion of the evolutionary origin of mitochondria and chloroplasts.

■ How genetic studies of transmission revealed and explained non-Mendelian patterns of inheritance, including descriptions of uniparental and biparental inheritance and mitotic segregation.

■ A comprehensive example of how mutations in mitochondrial DNA affect human health.

of the opposite mating type, the resulting diploids were *grande;* and when these *grande* diploids sporulated via meiosis, each one yielded four *grande* spores (that is, spores that after germination produced *grande* colonies) and zero *petite* spores. A cross of two cells from *petite* colonies produced only *petite* diploids, which, however, could not sporulate because of their deficiency in respiration. Matings of *grande* with *petite* generated only *grande* diploids, and each sporulation of those diploids yielded four *grande* spores and zero *petites.* This 4:0 ratio consistently replaced the 2:2 ratio predicted by Mendelian genetics. From these observations, the researchers concluded that a genetic factor necessary for respiratory growth is present in *grande* cells but absent from *petite* cells. They named the factor "rho" (symbolized by "ρ"); and they designated grande cells "ρ^+" and petite cells "ρ^-." They also noted that because of the non-Mendelian inheritance pattern, the rho factor did not segregate at meiosis.

Is there a connection between the maternal inheritance of leaf variegation in four-o'clocks and the unusual 4:0 inheritance pattern of *grande* and *petite* colony sizes in yeast? The answer is yes. Decades of experiments have shown that both traits are deter-

The Structure and Function of Mitochondrial and Chloroplast Genomes

Although geneticists pieced together the molecular foundation of non-Mendelian inheritance—that is, the structure and function of organelle genomes—beginning only in the 1960s, we present a summary of that molecular context up front to provide a clear framework for understanding the earlier genetic evidence. We begin with a brief description of the structure and function of mitochondria and chloroplasts, and then we describe the genomes that encode the RNA and protein machinery the organelles use to sustain their activities.

Mitochondria and Chloroplasts Are Organelles of Energy Conversion That Carry Their Own DNA

Every time a bird takes flight, a person speaks, a worm turns, or a flower unfurls, the organism's cells use energy that a plant captured from sunlight and stored in nutrient molecules. The chloroplasts of plants and other photosynthesizers capture solar energy and store it in the chemical bonds of carbohydrates. The mitochondria found in most eukaryotic cells release energy from those nutrients and convert it to ATP.

Mitochondria Are Sites of the Krebs Cycle and an Electron Transport Chain That Carries out the Oxidative Phosphorylation of ADP to ATP

Each eukaryotic cell houses many mitochondria, with the exact number depending on the energy requirements of the cell as well as the chance distribution of mitochondria during cell division. In humans, nerve, muscle, and liver cells carry more than a thousand mitochondria.

Figure 15.2 reveals the organelle's detailed structure. Separating the mitochondrion from the cell's cytoplasm is a smooth *outer membrane* perforated by protein-lined pores; molecules up to the size of small proteins can readily pass through these pores to an *intermembrane space*—a gap between the outer membrane, and a second, *inner membrane.* The inner membrane has many infoldings called *cristae* and is studded with enzymes of the electron transport chain and ATP synthesis. It is unusually rich in proteins (it is 75% protein by weight) and is impermeable to most substances, which can traverse it only by active transport. Tightly sequestered inside the inner membrane is a core compartment known as the *matrix,* which contains many enzyme systems, including one that carries out the tricarboxylic acid (TCA) cycle, commonly referred to as the Krebs cycle after its discoverer. The matrix also carries one or more circles of mitochondrial DNA.

Biologists call mitochondria the powerhouses of the cell because these organelles produce most of the cell's usable energy in the form of ATP molecules. There are two stages by which mitochondria help convert food into energy. In the first stage, mitochondria employ the Krebs cycle to metabolize pyruvate and fatty acids (the breakdown products of carbohydrates

Figure 15.2 Anatomy of a mitochondrion. The organization and structure of a single mitochondrion is shown. Core regions that are entirely enclosed within an inner membrane are known as the matrix (*shown in blue*). The matrix contains the mitochondrial DNA and enzymes of the Krebs cycle. Individual inner-membrane foldings are called cristae. A single crista is magnified to show how enzymes of the electron transport chain carry out oxidative phosphorylation between the matrix and the intermembrane space.

and fats, respectively) and produce the high-energy electron carriers NADH and $FADH_2$. In the second stage, four multisubunit enzyme complexes (referred to as I, II, III, and IV), which are embedded in the inner mitochondrial membrane, harness the energy in the high-energy electrons carried by the NADH and $FADH_2$. The process by which they harness this energy is **oxidative phosphorylation:** a set of reactions requiring oxygen that creates portable packets of energy in the form of ATP. In carrying out oxidative phosphorylation, molecular complexes I, II, III, and IV form a chain that transports the high-energy electrons from NADH and $FADH_2$ to their final electron acceptor, oxygen. A fifth protein complex (V) attached to the inner mitochondrial membrane uses the energy released by the electron transport chain to produce ATP.

To summarize: Starting with O_2 and the breakdown products of fats and carbohydrates, mitochondria produce packets of energy in the form of ATP.

Chloroplasts Are the Sites of Photosynthesis: The Capture, Conversion, and Storage of Solar Energy in the Bonds of Carbohydrates

In corn, one of the many crop plants adept at carrying out photosynthesis, each leaf cell contains 40–50 chloroplasts, and each square millimeter of leaf surface carries more than 500,000 of

Chloroplast

Figure 15.3 Anatomy of a chloroplast. Like a mitochondrion, a chloroplast also has an outer membrane and an inner membrane, but the inner membrane is not folded. The space found within the inner membrane—containing the chloroplast DNA and photosynthetic enzymes—is called a stroma. The stroma contains thylakoid membranes surrounding the thylakoid lumens. The magnified portion of a thylakoid membrane indicates the presence of photosynthetic enzymes.

the organelles. Figure 15.3 illustrates the structure of a chloroplast. Like mitochondria, chloroplasts have an *outer membrane,* an *intermembrane space,* and an *inner membrane* enclosing a core compartment. In chloroplasts, this core compartment is called the *stroma.* It contains the chloroplast's DNA and various enzyme systems, including the photosynthetic enzymes that fix carbon dioxide into carbohydrates. Within the stroma lies a third set of membranes, the *thylakoid membranes,* which surround a space called the *thylakoid lumen.* In places, portions of the thylakoid membranes fuse into stacks of disklike sacs known as *grana* (singular, *granum*). Embedded in the thylakoid membranes are chlorophyll and other light-absorbing proteins as well as proteins of the photosynthetic electron transport system. The light-absorbing proteins capture light energy from the sun and transfer this trapped solar energy to a photosynthetic electron transport chain.

Photosynthesis takes place in two parts: a light-trapping phase and a sugar-building phase. During the light-trapping phase, solar energy becomes trapped as it boosts electrons in chlorophyll or another pigment molecule to higher energy lev-

els. The energized electrons are then conveyed to an electron transport system that uses the energy to convert water to oxygen and H^+. Photosynthetic electron transport forms NADPH and drives the synthesis of ATP via an ATP synthase similar to the one in mitochondria. During the sugar-building phase of photosynthesis, enzymes of the Calvin cycle, located in the chloroplast stroma, use the ATP and NADPH generated by the conversion of solar to chemical energy to fix atmospheric carbon dioxide into carbohydrates. The energy stored in the bonds of these nutrient molecules fuels the activities of both the plants that make the carbohydrates and the animals that ingest them.

Mitochondria and Chloroplasts Carry Their Own DNA

Cells stained with DNA-specific dyes, when viewed under the light microscope, reveal DNA molecules in the matrix of mitochondria and the stroma of chloroplasts, as well as in the nucleus. Using methods for purifying mitochondria and chloroplasts, researchers have extracted DNA directly from these organelles and shown by analyses of base composition and buoyant density that an organism's organelle DNA differs from its nuclear DNA.

With this background in mind, we examine the structure and function of mitochondrial and chloroplast genomes. As we do, we see that although both organelles replicate and express all the genes in their own DNA, their genomes encode only some of the proteins they require for their activities.

The Genomes of Mitochondria

Mitochondrial DNA lies within the matrix of the organelle, where it appears in highly condensed structures called nucleoids. Each haploid yeast cell, for example, has about 20 mtDNA molecules in a variable number of mitochondria and nucleoids. In cells with just one to a few mitochondria, each organelle contains a number of nucleoids. By contrast, in cells grown on lactate medium, which carry up to 20 mitochondria, each organelle contains only one to a few nucleoids. Thus, the mtDNA of most cells does not reside in a single location.

Variations in the number of mitochondria, nucleoids, and mtDNA molecules are regulated by complicated means that researchers do not yet understand. Mitochondria can fuse with each other as well as divide. In general, however, mitochondria double in size and then divide in half in each cell generation. The replication of mtDNA molecules, as well as the division of the mitochondria, can occur throughout the cell cycle independent of the replication of genomic nuclear DNA (which occurs only during S phase) and of the cell division at the end of mitosis. Interestingly, which mtDNA molecules undergo replication seems to be determined at random; as a result, some molecules replicate many times in each cell cycle, while others do not replicate at all. This is one cause of the mitotic segregation of mitochondrial genomes discussed later in the chapter.

The size and gene content of mitochondrial DNA vary from organism to organism. The mtDNAs in the malaria parasite

TABLE 15.1   **Mitochondrial DNA Sizes**

Organism	Size (kb)
Plasmodium	6
Yeast	75
Drosophila	18
Pea	110
Human	16.5

Figure 15.4   Kinetoplast DNA network. In certain protozoan parasites, there is a single mitochondrion, or kinetoplast, that contains a large interlocking network of DNA molecules present in mini- and maxicircles.

Plasmodium falciparum are only 6 kb in length; those in the free-living nematode *Ascaris suum* are 14.3 kb; those in the muskmelon *Cucumis melo* are a giant 2400 kb long. These mtDNA size differences do not necessarily reflect comparable differences in gene content. Although the large mtDNAs of higher plants do contain more genes than the smaller mtDNAs of other organisms, the 75 kb mtDNA of baker's yeast encodes fewer proteins of the respiratory chain than does the 16.5 kb mtDNA of humans. Tables 15.1 and 15.2 summarize the size and gene content of mtDNAs from organisms representative of the plant, animal, and fungus kingdoms.

Like size and gene content, the shape of mtDNAs varies. Biochemical analyses and mapping studies have shown that the mtDNAs of most species are circular; but the mtDNAs of the ciliated protozoans *Tetrahymena* and *Paramecium*, the alga *Chlamydomonas*, and the yeast *Hansenula* are linear. The difficulty in isolating unbroken mtDNA molecules from some organisms makes it hard to tell the shape of their mtDNA *in vivo*.

To complete this snapshot of the mitochondrial genome's variation in size, gene content, and shape, it is necessary to mention the unusually organized mtDNAs of protozoan parasites of the genera *Trypanosoma*, *Leishmania*, and *Crithidia*. These single-celled eukaryotic organisms carry a single mitochondrion known as a kinetoplast. Within this structure, the mtDNA exists in one place (contrary to the mtDNA of most other cells) as a large network of 10–25,000 "minicircles" 0.5–2.5 kb in length interlocked with 50–100 maxicircles 21–31 kb long (Fig. 15.4). The maxicircles contain most of the genes usually found on mtDNA, while the minicircles play a role in RNA editing, as described later.

A comparison of mitochondrial genome organization and function in humans, yeast, and the liverwort (a species of moss) illustrates some of the details of mtDNA diversity. Molecular biologists have sequenced the mtDNAs of all three organisms.

Human mtDNA Carries Closely Packed Genes

The 16.5 kb human mitochondrial genome, which accounts for 0.3% of a cell's DNA, is a circular DNA molecule (like most mtDNAs) that carries 37 genes. Thirteen of these genes encode polypeptide subunits of the protein complexes that make up the oxidative phosphorylation apparatus embedded in the mitochondrial inner membrane. These same 13 genes are present in the mtDNAs of all metazoans analyzed to date. In addition to the 13 polypeptide-encoding genes, there are 22 tRNA genes. The remaining 2 genes are for the large and small rRNAs found in mitochondrial ribosomes (Fig. 15.5a).

A significant feature of the human mitochondrial genome is the compactness of its gene arrangement. As Fig. 15.5a shows, adjacent genes either abut each other or slightly overlap. With virtually no nucleotides between them and no introns within them, the genes are packaged very tightly. The reason for this compact arrangement is not yet known.

The Larger Yeast Mitochondrial Genome Contains Spacers and Introns

The mitochondrial genome of the yeast *S. cerevisiae* is more than four times longer than human and other animal mtDNA (Fig. 15.5b). Two DNA elements account for the larger size of

TABLE 15.2   **Comparison of Some Functions Encoded in mtDNA**

Organism	Oxidative Phosphorylation Genes	tRNAs	Genome Size (kb)
Yeast	7	25	75
Marchantia (liverwort)	14	29	186.0
Human	13	22	16.5

(a) Human mtDNA (16.5 kb)

(b) Yeast mtDNA (75 kb)

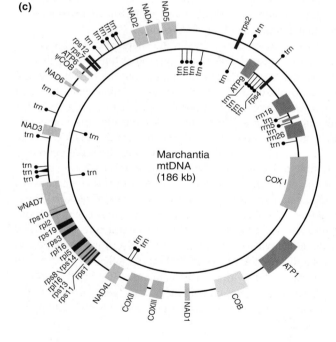

(c) Marchantia mtDNA (186 kb)

Figure 15.5 Mitochondrial genomes of three species.
Comparison of mitochondrial genomes of humans, baker's yeast, and liverwort. Human mtDNA is 16.5 kb long, while that of baker's yeast is 75 kb long, and liverwort is 121 kb long. Important differences illustrated in the diagrams are the presence of long intergenic sequences in yeast mtDNA and their absence from human mtDNA and differences in the numbers of genes in the mt genomes from these three organisms. The color scheme allows you to quickly see differences in numbers of genes for various functions in these three species: *green* for genes coding for cytochrome oxidase proteins; *red* for genes coding for ATPase subunit proteins; *yellow* for genes coding for NADH complex proteins; *tan* for genes coding for cytochrome complex proteins; *purple* for genes coding for ribosomal proteins or ribosomal RNAs. Each tRNA gene is indicated by a *black ball and stick*. Genes shown on the outer and inner circles are transcribed in opposite directions.

the yeast mitochondrial genome: long intergenic sequences and introns. Long A–T-rich sequences called "spacers" separate the genes in yeast mtDNA and account for more than half of the additional DNA. Most of the spacer DNA is transcribed, and some of it is retained in the mRNAs as long untranslated 5′ and 3′ extensions. Introns, the second DNA-lengthening element, form about 25% of the yeast mitochondrial genome and account for most of the remaining size difference between yeast and human mtDNA.

The 186 kb mtDNA of the Liverwort *Marchantia polymorpha* Carries Many More Genes Than Its Animal and Fungal Counterparts

The mtDNA of *M. polymorpha* was the first plant mtDNA to be entirely sequenced. Although it is one of the smallest plant mitochondrial genomes, it is far larger and has many more genes than nonplant mtDNAs (Fig. 15.5c). The genes it carries encode 12 of the 13 electron-transport-chain proteins found in human mtDNA (lacking only the gene for subunit 8 of ATP synthase), as well as subunit 9 of ATP synthase (which is absent from humans) and subunit A of ATP synthase (which is in the nuclear genomes of all animals and fungi). In addition, the moss mitochondrial genome contains 16 genes for ribosomal proteins and 29 genes encoding proteins of unknown function.

Thus, although mitochondria in different eukaryotic organisms play similar roles in the conversion of food to energy, evolution has produced mtDNAs with an astonishing diversity in the content and organization of their genes. As we see next, mitochondrial evolution has also led to some remarkable variations on the basic mechanisms of gene expression.

Mitochondrial Transcripts Undergo RNA Editing, a Rare Variation on the Basic Theme of Gene Expression

Researchers discovered the unexpected phenomenon of RNA editing in the mitochondria of trypanosomes. As already noted, these protozoan parasites have a single, large mitochondrion—the kinetoplast—which contains much more DNA than the mitochondria of other organisms; this kDNA exists as a series of interlocking maxi- and minicircles. DNA sequencing shows that the minicircles carry no protein-encoding genes. The detection of transcripts from maxicircle DNA, however, confirms that these larger circles do carry and express genes.

Surprisingly, the sequencing of maxicircle DNA reveals only short, recognizable gene fragments, instead of whole mitochondrial genes. Furthermore, the sequencing of RNA molecules in the kinetoplast revealed both RNAs that looked like the strange fragments of kinetoplast genes and related RNAs that could encode recognizable mitochondrial proteins. From these observations, investigators concluded that kDNA encodes a precursor (the strange fragment observed) for each mRNA. After transcription, the cellular machinery turns these precursors into functional mRNAs through the insertion or deletion of nucleotides.

The process that converts pre-mRNAs to mature mRNAs is **RNA editing.** It is essential for the expression of these mitochondrial genes because without RNA editing, the pre-mRNAs do not encode polypeptides. Some pre-mRNAs lack a first codon suitable for translation initiation; others lack a stop codon for the termination of transcription. RNA editing creates both types of sites, as well as many new codons within the genes.

In addition to the mitochondria of trypanosomes, the mitochondria of some plants and fungi carry out RNA editing. The extent of RNA editing varies from mRNA to mRNA and from organism to organism. In trypanosomes, the RNA editing machinery adds or deletes uracils. In plants, the editing adds or deletes cytosines. At present, researchers understand the general mechanism of uracil editing, but not that of cytosine editing. As Fig. 15.6 shows, uracil editing occurs in stages in which enzymes use an RNA template as a guide for correcting the pre-mRNA. The guide RNAs are encoded by short stretches of kDNA on both maxi- and minicircles, and a structure known as an "editosome" is the workbench where the RNA editing takes place.

Figure 15.6   RNA editing in trypanosomes. Example of a portion of a pre-mRNA sequence is shown at the *top.* This pre-mRNA forms a double-stranded hybrid with a guide RNA through both standard Watson-Crick A–U and G–C base pairing, as well as atypical G–U base pairing. Unpaired G and A bases within the guide RNA initiate the insertion of U's within the pre-mRNA sequence, bringing about the final edited mRNA.

Translation in Mitochondria Shows That the Genetic Code Is Not Universal

As mtDNA carrying its own rRNA and tRNA genes would suggest, mitochondria have their own distinct translational apparatus. Mitochondrial translation is quite unlike the cytoplasmic translation of mRNAs transcribed from nuclear genes in eukaryotes. Many aspects of the mitochondrial translational system resemble details of translation in prokaryotes. For example, as in bacteria, *N*-formyl methionine and tRNA[fMet] initiate translation in mitochondria. Moreover, inhibitors of bacterial translation, such as chloramphenicol and erythromycin, which have no effect on eukaryotic cytoplasmic protein synthesis, are potent inhibitors of mitochondrial protein synthesis.

We saw in Chapter 8 that the genetic code is almost, but not quite, universal. The mtDNA sequences of tRNAs and protein-encoding genes in several species cannot explain the sequences of the resulting proteins in terms of the "universal" code. For example, in human mtDNA, the codon UGA specifies tryptophan rather than stop (as in the standard genetic code); AGG and AGA specify stop instead of arginine; and AUA specifies methionine rather than isoleucine (Table 15.3). No single mitochondrial genetic code functions in all organisms, and the mitochondria of higher plants use the universal code. Moreover, while an f-Met-tRNA usually initiates translation in mitochondria by reading AUG or AUA, other triplets, which do not specify methionine, often mark the site of initiation. The genetic codes of mitochondria probably diverged from the universal code by a series of mutations occurring some time after the organelles became established components of eukaryotic cells.

As we see next, chloroplast DNA, although similar in many ways to mtDNA, has some remarkable features of its own.

The Genomes of Chloroplasts

Chloroplasts occur in plants and algae. The genomes they carry are much more uniform in size than are the genomes of mitochondria. Although cpDNAs range in size from 120–217 kb, most are between 120 and 160 kb long (Table 15.4). Chloroplast DNA contains many more genes than mtDNA. Like the genes of bacteria and human mtDNA, these genes are closely packed,

TABLE 15.3 Variations in the Genetic Code of Mitochondria

Characteristic	Universal Code	mtDNA Code
Number of tRNAs	32	22
UGG	Trp	Trp
UGA	Stop	Trp
AGG	Arg	Stop
AGA	Arg	Stop
AUG	Met	Met
AUA	Ile	Met

Altered genetic code. The mtDNA genetic code is simplified such that a modified U in the tRNA "wobble" position can read all four codons in a codon family (that is, UUU,UUC, UUA, UUG). An unmodified U can read both purines and G can read both pyrimidines. Tryptophan tRNA has a U in the wobble position so that it will read both the traditional UGG codon and the associated UGA stop codon as tryptophan. Similarly, the methionine codon reads both AUG and the associated AUA as methionine. Finally, one of the two arginine tRNAs has been deleted such that two of the six arginine codons (AGG and AGA) now function as stop codons.

TABLE 15.4 Chloroplast DNA Sizes

Organism	Size (kb)
Chlamydomonas reinhardtii	196
Marchantia (liverwort)	121
Nicotiana tabacum (tobacco)	156
Oryza sativa (rice)	135

with relatively few nucleotides between adjacent coding sequences. Like the genes of yeast mtDNA, they contain introns. The shape of most cpDNAs is a circle.

A close look at the chloroplast genome of the liverwort *M. polymorpha,* the first cpDNA to be sequenced completely, illustrates details of cpDNA structure and function (Fig. 15.7). Liverwort cpDNA contains 92 protein-encoding genes and 36 genes for tRNAs and rRNAs, which is more genes than are carried by the mitochondrial genomes of much larger plants. The rRNA genes of the chloroplast encode four ribosomal RNAs—16S, 23S, 4.5S, and 5S. These same four rRNAs are found in eubacteria. Another similarity with the bacterial genome is that some functionally related cpDNA genes are organized in clusters that resemble bacterial operons.

The 92 polypeptides encoded by *M. polymorpha* cpDNA account for only a small fraction of the proteins active in chloroplasts. However, cpDNA encodes proteins for a wider range of processes than does mtDNA. The cpDNA-encoded proteins include many of the molecules that carry out photosynthetic electron transport and other aspects of photosynthesis, as well as RNA polymerase, translation factors, ribosomal proteins, and other molecules active in chloroplast gene expression. The RNA polymerase of chloroplasts is similar to the multisubunit bacterial RNA polymerases. Inhibitors of bacterial translation, such as chloramphenicol and streptomycin, inhibit translation in chloroplasts, as they do in mitochondria.

The function of some chloroplast genes is not yet known. Because most of these unidentified reading frames (URFs) are well conserved in many cpDNAs and have counterparts in bacteria, it is likely that they are key elements in chloroplast gene expression and organelle function.

Mitochondria and Chloroplast Functions Require Cooperation Between the Organelle and Nuclear Genomes

The maintenance and assembly of functional mitochondria and chloroplasts depend on gene products from both the organelles themselves and from the nuclear genome (Fig. 15.8). In most organisms, for example, cytochrome *c* oxidase, the terminal protein of the mitochondrial electron transport chain, is composed of seven subunits, three of which are encoded by mitochondrial genes, whose mRNAs are translated on mitochondrial ribosomes. The remaining four are encoded by nuclear genes whose messages are translated on ribosomes in the cytoplasm. In all organisms, nuclear genes encode the majority of the proteins active in mitochondria and chloroplasts. For example, although mitochondrial genomes carry the rRNA genes, nuclear genomes carry the genes for most (in yeast and plants) or all (in animals) of the proteins in the mitochondrial ribosome.

Because mitochondria and chloroplasts do not carry all the genes for the proteins (and in some organisms, the tRNAs) they need to function and reproduce, they are **semiautonomous,** requiring the constant provision of proteins (and tRNAs) encoded by nuclear genes.

Origin and Evolution of Organelle Genomes: The Molecular Evidence

Mitochondria are remarkably similar in size and shape to today's aerobic bacteria, while chloroplasts in many ways resemble living photosynthetic bacteria. In addition, the genomes of both organelles bear many striking resemblances to the genomes of bacteria. Because of these morphological and molecular likenesses, it is probable that mitochondria and chloroplasts started out as free-living bacteria that merged with the ancestors of modern eukaryotic cells to form a cellular community in which host and guest benefited from the group arrangement.

The "Endosymbiont Theory"

In the 1970s, Lynn Margulis was one of the first biologists to propose that mitochondria and chloroplasts originated more than a billion years ago when ancient precursors of eukaryotic cells engulfed and established a symbiotic relationship with

Figure 15.7 Chloroplast genome of the liverwort _M. polymorpha._ The relative locations and symbols of some of the 128 genes are indicated. Genes are color-coded according to general function.

some bacteria. The primitive cells carrying a mitochondrion-like or chloroplast-like bacterial cell would have gained an edge in the fierce competition for energy production and eventually evolved into complex eukaryotes. So much evidence now supports this hypothesis that it is accepted as the **endosymbiont theory.** The molecular evidence for the theory includes the following facts: (1) Both mitochondria and chloroplasts have their own DNA, which replicates independently of the nuclear genome. (2) Like the DNA of bacteria, mtDNA and cpDNA are not organized into nucleosomes by histones. (3) Mitochondrial genomes, as we have seen, use _N_-formyl methionine and tRNA^fMet in translation. And (4) inhibitors of bacterial translation, such as chloramphenicol and erythromycin, block mitochondrial translation but have no effect on eukaryotic cytoplasmic protein synthesis. Comparisons of DNA sequences from organelle and bacterial rRNA genes and comparisons of sequences from organelles of different organisms suggest that mitochondrial genomes derive from a common ancestor of present-day gram-negative nonsulfur purple bacteria, while chloroplast genomes derive from cyanobacteria (formerly referred to as blue-green algae).

It is likely that the precursors of eukaryotes and the ancient bacteria that evolved into organelles interacted many times and in many different ways en route to the present-day symbiotic relations. The first interactions may have been transient; indeed, many present-day eukaryotes containing mitochondria and chloroplasts also have bacteria living inside them. Once a stable symbiotic state formed in a given lineage, however, if the engulfed cell lost some genes critical for life on the outside, it would have also lost the ability to function and compete on its own. The loss of genes could have occurred through transfer to the nucleus or deletion of redundant sequences. We have already seen that some of the genes required for oxidative phosphorylation and photosynthesis reside in the nuclear genome; they may have been transferred there from the organelle genome. The deletion of redundant sequences might have occurred when organelle genes encoding a biosynthetic pathway critical for autonomous growth were discarded or lost from the organelle genome because the nuclear genome of the host encoded its own version of the pathway. In the evolution of the eukaryote's symbiotic community, the engulfed cell may have not only lost genes but also contributed coding sequences to the host

Number and genomic location of oxidative phosphorylation genes

	Number of polypeptides					
	Electron transport chain				ATP synthase	
Genomic location	I	II	III	IV	V	Total
Mitochondrion	7	0	1	3	2	13
Nucleus	≥33	4	10	10	10	≥67
Total	≥40	4	11	13	12	≥80

Figure 15.8 Mitochondria and chloroplasts depend on gene products from the nucleus. Although some organelles in some species have many more genes than others, all are dependent on RNA and protein products encoded by nucleus genes. The location of oxidative phosphorylation (OXPHOS) genes is shown.

cell; these sequences could have encoded enzymes that added to or replaced proteins encoded by the genes of the nuclear chromosomes.

Gene Transfer Occurs Through an RNA Intermediate or by Movement of Pieces of DNA

Researchers have some understanding of the mechanisms by which genes transfer between an organelle and the nucleus. In many plants, the mitochondrial genome encodes the *COXII* gene of the mitochondrial electron transport chain; in other plants, the nuclear DNA encodes that same gene; and in several plant species where the nuclear *COXII* gene is functional, the mtDNA

still contains a recognizable, but nonfunctional, copy of the gene (that is, a *COXII* pseudogene). Remarkably, the mtDNA gene contains an intron, while the nuclear gene does not. Geneticists have interpreted this finding to mean that the *COXII* gene transferred from mtDNA to nDNA via an RNA intermediate; the RNA would have lacked the intron, and when the mRNA was copied into DNA and integrated into a chromosome in the nucleus, the resulting nuclear gene also had no intron.

Good evidence also exists for the transfer of many genes at the DNA level. The fact that some plant mtDNAs carry large fragments of cpDNA shows that pieces of cpDNA can move from one organelle to another. Similarly, nonfunctional, intact or partial copies of organelle genes litter the nuclear genomes of eukaryotes. DNA sequencing reveals strong similarities between the organelle and nuclear DNAs, which means that the nuclear copies are relatively young. This, in turn, suggests that the organelle-to-nucleus transfer of DNA is still going on.

In this evolutionary perspective, the properties of mitochondrial and chloroplast genomes that vary among the organelles of present-day species are probably relatively new. These recently established features include long stretches of cpDNA incorporated in the mtDNAs of many plants, as well as many of the introns in organelle genomes. Some of these introns may have originated in the earliest bacterial symbionts; or they may have been incorporated into the organelle genomes after horizontal transfers between organelle DNAs long after the organelles were established.

Mitochondrial DNA Has a High Rate of Mutation

In the 1980s, surveys of DNA sequence variations among individuals of a given species and between closely related species showed that the mtDNA of vertebrates evolves almost 10 times more rapidly than the nuclear DNA of those same organisms. The higher rate of DNA mutation in mitochondria probably reflects more errors in replication and less efficient repair mechanisms.

Because of mtDNA's high mutation rate, the variation among mitochondrial genomes provides a valuable tool for studying the evolutionary relationships of organisms whose nuclear DNAs are very similar. Conversely, mtDNA variation, because it accumulates so rapidly, is of little value in evaluating the relationships of distant evolutionary relations, but here sequence variation data for nuclear genomes are useful (see Chapter 20). Sequence analyses of mtDNA have shown that the maternal lineage of all present-day humans, no matter what ethnic group they belong to, traces back to a few women who lived in Africa some 200,000 years ago. The Fast Forward box "Mitochondrial DNA Sequences Shed Light on Human Evolution" describes the evidence for this hypothesis of a "mitochondrial Eve." It also explains how the early representatives of modern humans may have interacted with archaic humans and other hominid species in the genus *Homo* during the evolution of modern human populations.

Cloning and PCR of mtDNA, which enable the amplification of small biological samples for analysis, have helped identify the closest mammalian relatives of the extinct *Equus quagga,* a horselike African animal that died out in 1883 (see Chapter 6). Analyses of quagga mtDNA from preserved museum samples show that the quagga's closest living relatives are African plains zebras, and that it is less closely related to the mountain zebra and the horse.

Genetic Studies of Organelle Genomes Clarify Key Elements of Non-Mendelian Inheritance

Mutations in organelle genes produce readily detectable whole-organism phenotypes because the altered proteins and RNAs they encode disrupt the production of cellular energy. The mtDNA mutations that cause *petite* colonies in yeast, for example, disrupt the oxidative-phosphorylation system; the cpDNA mutations that cause variegation in four-o'clocks incapacitate proteins essential for photosynthesis. Although most mutations in the genes for these energy-producing systems are lethal, in both plants and animals, some organelle gene mutations yield detectable, nonlethal phenotypes that researchers can study genetically. Data from such genetic studies show that modes of organelle gene transmission vary among organisms. We describe the main modes of transmission from one generation to the next, which determine what is inherited. We also examine how parental contributions to the zygote and mitotic segregation influence what is available for inheritance by determining which organelles are passed on to progeny cells during cell divisions.

In Most Species, Progeny Inherit Organelle DNA from Only One—Usually the Maternal—Parent

Genetic and biochemical analyses of thousands of crosses have shown that most species transmit mtDNA and cpDNA via one parent, usually the mother; that is, they use a uniparental, mainly maternal, mode of transmission. Researchers have analyzed **uniparental inheritance** in several ways.

Maternal Inheritance of Preexisting Differences in Wild-Type mtDNAs

In a classic experiment documenting maternal inheritance in vertebrates, investigators purified mtDNA from frog eggs, which contain a large number of mitochondria, and used hybridization tests to distinguish the mtDNA of one frog species, *Xenopus laevis,* from the mtDNA of the closely related *Xenopus borealis*. In these tests, probes from *X. laevis* hybridized more efficiently with *X. laevis* mtDNA than with *X. borealis* DNA and vice versa. Because interspecific crosses between the two species yield viable progeny, the analysis of F_1 mtDNA was one way to trace the inheritance of that DNA in frogs. Figure 15.9 diagrams the reciprocal crosses and mtDNA typing that formed the basis of the study. The first-generation progeny of both crosses carried mtDNA like that of the maternal parent. Although the analysis might have missed small contributions from the paternal genome, these *Xenopus* crosses confirmed the predominantly maternal inheritance of mtDNA in these species.

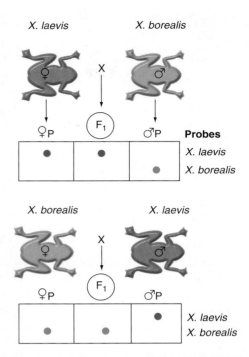

Figure 15.9 Maternal inheritance of *Xenopus* mtDNA. *X. laevis* and *X. borealis* mtDNA can be distinguished by strong hybridization only to probes made from the same species. Reciprocal crosses between two species produce F_1 hybrids. Each F_1 hybrid retains mtDNA only from its mother.

They also showed that it is possible to follow preexisting differences in functionally wild-type mtDNAs in a cross. Since the 1980s, analysis of crosses using DNA polymorphisms confirmed maternal mtDNA inheritance among horses, donkeys, and many other vertebrates.

Maternal Inheritance of Specific Genes in cpDNA

Interspecific crosses tracing biochemically detectable, species-specific differences in several chloroplast proteins provided further evidence of maternal inheritance. In the mtDNA inheritance studies just cited, the identity of the organelle gene containing the markers was not known and did not matter. By contrast, in cpDNA studies researchers identified specific organelle genes through the analysis of proteins. They began by isolating from tobacco plants (*Nicotiana* species) proteins in which interspecies differences could be distinguished by gel electrophoresis. To determine each protein's mode of inheritance, they evaluated the allele expressed in the progeny of a controlled cross, carefully noting the maternal (ovum) and paternal (pollen) contributions. In one set of experiments, ribulose bisphosphate carboxylase (Rubisco, for short), the first enzyme of photosynthetic carbon fixation in plants and the most abundant protein in tobacco leaves, was the protein under observation. Rubisco has a 55 kDa large subunit (called LSU) and a 12 kDa small subunit (called SSU). The researchers purified Rubisco from many species of tobacco plants, digested the purified proteins with trypsin, and analyzed the digests for usable differences. When they followed the inheritance of these differences, they found that LSUs

Fast Forward

Mitochondrial DNA Sequences Shed Light on Human Evolution

The mitochondrial DNA of all humans alive today traces back through maternal lineages to the mtDNA of a human population living in Africa some 200,000 years ago. Such is the startling conclusion of two papers published in 1987 and 1991 by Allan C. Wilson and colleagues. The carrier of this ancestral mtDNA, dubbed "mitochondrial Eve," probably lived in a population of 10,000–50,000 people. (In this context, a *population* is a group of interbreeding individuals of the same species who inhabit the same space.)

How mtDNA Variations Suggest the Region Where Modern Humans Emerged

In their studies supporting an African origin for modern humans, Wilson and coworkers first looked at restriction fragment length polymorphisms (RFLPs) in the mtDNA of 143 subjects, including Americans of African, Asian, European, and Middle Eastern origin as well as Aboriginal women in New Guinea and Australia. Four years later, the group followed up their original 1987 study with sequence analyses of a rapidly evolving, highly polymorphic noncoding segment of mtDNA from 189 individuals, including 121 native Africans from various parts of the continent, Papua New Guineans, Europeans, Asians, African-Americans, and a native Australian. In both studies, the researchers found greater sequence differences among Africans, particularly sub-Saharan Africans, than among Asians or Europeans. Because mutations accumulate over time, they concluded that the African population has had the longest time to evolve variation and thus modern humans originated in Africa.

Statistical Calculations of the Number and Rate of Mutations Suggest When Modern Humans Appeared

Having proposed that modern humans first appeared in Africa, the researchers calculated the probable date of their origin by extrapolating the unknown from the known. They had observed the greatest human mtDNA variation in a sub-Saharan African population; about 2.8% of the base pairs in the mitochondrial genome varied among the individuals they studied from that population. They also knew that chimpanzees and humans diverged approximately 5 million years ago and that human mtDNA differs from that of chimpanzees in about 15% of the genome. Adjusting these data to account for multiple substitutions at the same base pair, they estimated that the mtDNA of humans and chimpanzees has been diverging at an average rate of about 13.8% per million years. To determine approximately how long ago the human population containing mitochondrial Eve lived, they divided the percentage of maximal human variation by the rate of chimp-human divergence:

$$2.8/13.8 = 0.20 \text{ million} = 200,000 \text{ years ago}$$

Although there has been some controversy over the statistical methods and assumptions that formed the basis of this analysis, most geneticists now agree with the conclusion that the women carrying our ancestral mtDNA lived roughly 200,000 years ago in sub-Saharan Africa. Recent studies of parts of the Y chromosome and other pieces of nuclear DNA support this conclusion. In one 1997 study, evolutionary geneticists examined two nuclear DNA segments on chromosome 12 in the genomes of 1600 individuals from 42 populations around the world. In populations outside of Africa, one combination of the two elements showed up almost exclusively; by contrast, in sub-Saharan Africa, almost all the possible variants appeared. Recently, analyses of Y chromosome polymorphisms uncovered two ancient DNA markers shared by nonhuman primates and a small group of sub-Saharan African men. These markers sustained mutations about 100,000–200,000 years ago. Today all men outside of Africa as well as most African men carry the mutated markers.

Comparisons of Sequence Variations Point to How *Homo sapiens* Evolved

In addition to providing evidence for where and when modern humans first appeared, mtDNA may help evolutionary biologists determine what happened to the human populations that migrated out of Africa to other areas of the globe. At present, there are two models. One, the replacement theory, posits that as early modern humans (*Homo sapiens*) migrated to Europe, Asia, and elsewhere, they eventually replaced the archaic hominid groups living in those areas. In this view, although early modern *H. sapiens* spread to parts of the world already populated by archaic species in the genus *Homo* (such as *H. erectus*), the early moderns did not intermingle with these other hominid groups. Rather, they outcompeted and eventually replaced them such that anatomically modern *H. sapiens* became the only extant human (*Homo*) species (Fig. A). An opposing view, the regional continuity theory, proposes that humans evolved continuously in many parts of the world, with early modern humans interbreeding with archaic peoples in Europe and Asia. These archaic peoples would have included the Neanderthals, who coexisted with modern humans in the Middle East and parts of Europe and Asia for close to 70,000 years.

The first Neanderthal skeleton to be dug up was unearthed in Germany's Neander Valley (in German, *Neander Tal*) in 1856; its estimated age was 30,000 years. In 1997, a group of German and American biologists used state-of-the-art tools of DNA analysis to amplify and examine a minute amount of mtDNA from that skeleton. Even though mitochondrial DNA has a much greater chance of surviving in a long-dead specimen because of its abundance (500–1000 copies per cell rather than the usual 2 for nuclear DNA),

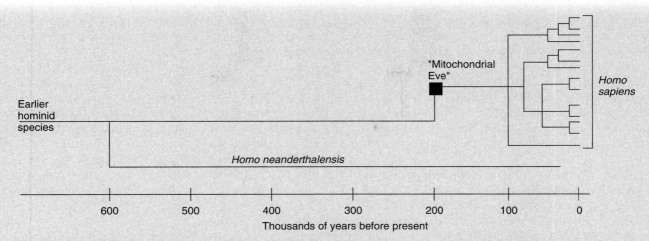

Figure A   Evolutionary trees of *Homo sapiens* and *Homo neanderthalensis* based on analysis of mitochondrial DNA. Comparisons of mitochondrial DNA sequence in existing humans led to the hypothesis of a mitochondrial Eve existing about 200,000 years ago. Analysis of mtDNA from Neanderthal skeletons indicates that *Homo neanderthalensis* evolved as a separate line much earlier.

it was a technical triumph to recover enough untainted mtDNA—379 bp in all—for PCR amplification and subsequent analysis. mtDNA from a second Neanderthal skeleton, this one about 29,000 years old and unearthed in the Northern Caucasus mountains, was examined in 2000. When scientists compared the Neanderthal sequences with more than 900 distinct mtDNA sequences from living humans, they found, on average, three times as many differences between Neanderthal and modern human DNAs as between mtDNAs from random pairs of modern humans or mtDNAs from the single random pair of Neanderthals. The examined mtDNA pairs from modern humans, for example, differed at an average of 8 positions, while the human–Neanderthal-from-Germany pairs differed at 25.6 positions. Furthermore, the types of nucleotide substitution and their locations were different in the two groups. According to the researchers, these data "make it highly unlikely that Neanderthals contributed to the human mtDNA pool." Using the same type of calculations that the Wilson group used, including the time when humans and chimps diverged and the rate at which the particular segments of mtDNA under analysis accumulate mutations, the German and American researchers estimated that the mtDNA sequence ancestral to that of contemporary humans and Neanderthals began to diverge 550,000–690,000 years ago, compared with only 120,000–150,000 years ago for divergence of the ancestral sequence of all modern humans. In other words, the last common ancestor of Neanderthals and modern humans is roughly four times older than the common ancestor of all modern humans.

These calculations, although again based on assumptions that are open to question, nevertheless suggest that humans and Neanderthals diverged before the first known Neanderthal appeared some 300,000 years ago and long before the first modern humans evolved around 200,000 years ago. If further evidence and analyses support these inferences, the logical conclusion would be that there was no mixing of humans and Neanderthals, at least not in the mitochondrial genes. This, in turn, would support the replacement theory of modern human evolution.

Opponents of the replacement theory maintain that the close similarity of modern human mtDNAs from all parts of the world and its difference from that of Neanderthals could have arisen in other ways—for example, from a population bottleneck, such as a flood or drought, that left few survivors carrying Neanderthal genes, or from the disappearance by random genetic loss or negative selection of Neanderthal genes that may have been part of the early modern human gene pool. (Chapter 20 explains the details of these possibilities.)

The information needed to put this controversy to rest is not yet in hand. One step in that direction would be the unearthing and analysis of mtDNA from a North African Neanderthal, who would represent an early stage in the migration out of Africa. Another step would be the analysis of mtDNA from the already discovered 30,000-year-old Cro-Magnon, judged to be a representative of the first undisputed early modern humans. Evidence gathered in such studies could narrow the possibilities. If it supports the replacement theory, the icon of human evolution would be a bush with many twigs, of which all but modern *H. sapiens* have dried up. If it doesn't support the theory, the image would be much more complex.

manifested patterns of maternal inheritance, while SSUs showed biparental inheritance. From these results, they hypothesized that a chloroplast gene encodes the LSU polypeptide, while a nuclear gene encodes the SSU.

These studies of the two Rubisco subunits reveal that organelle and nuclear genomes cooperate in specifying even a relatively simple enzyme with only two different subunits. Both genomes contribute essential information for most photosynthetic activities, including those whose elements are much more complex.

The inheritance studies just described followed preexisting differences in functionally wild-type organelle genomes. To verify and understand the details of uniparental inheritance, organelle geneticists followed the inheritance of mutations affecting phenotype at both the biochemical and the organismal levels.

A Mutation in a Human Mitochondrial Gene Generates a Maternally Inherited Neurodegenerative Disease

Leber's hereditary optic neuropathy, or LHON, is a disease in which flaws in the mitochondria's electron transport chain lead to optic nerve degeneration and blindness (Fig. 15.10). Family pedigrees show that LHON passes only from mother to offspring. In the late 1980s and early 1990s, a series of molecular studies showed that a G-to-A substitution at nucleotide 11,778 in the human mitochondrial genome is a main cause of the condition. The substitution alters an arginine-specifying codon in the NADH dehydrogenase subunit 4 gene to a histidine codon. The resulting protein product diminishes the efficiency of electron flow down the respiratory transport chain, reducing the cell's production of ATP sufficiently to cause a gradual decline in cell function and ultimately cell death. Since optic nerve cells have a relatively high requirement for energy, the genetic defect affects vision before it affects other physiological systems.

In large families, not all offspring show signs of the disease, and not all siblings manifesting the condition have symptoms of the same severity. The random allotment to daughter cells of a large number of mitochondria during mitosis (described later) helps explain these observations.

Cells Can Contain One Type or a Mixture of Organelle Genomes

A diploid cell contains dozens to thousands of organelle DNAs. It is therefore not possible to use the terms "homozygous" and "heterozygous" to describe a cell's complement of mtDNA or

cpDNA. Instead, geneticists use the terms "heteroplasmic" and "homoplasmic" to describe the genomic makeup of a cell's organelles. **Heteroplasmic** cells contain a mixture of organelle genomes. **Homoplasmic** cells carry only one type of organelle DNA. Except for the rare appearance of a new mutation, the mitotic progeny of homoplasmic cells carry a single type of organelle DNA. By contrast, the mitotic progeny of heteroplasmic cells may be heteroplasmic, homoplasmic wild type, or homoplasmic mutant. In most people affected by LHON, for example, the optic nerve cells are homoplasmic for the disease mutation; but in some LHON patients, these optic nerve cells are heteroplasmic. Homoplasmy causes earlier appearance of the disease as well as more severe symptoms.

Mitotic Segregation Produces an Uneven Distribution of Organelle Genes in Heteroplasmic Cells

The random partitioning of organelles during cell divisions is the basis of the mitotic segregation of organelle genomes. This pattern of random segregation is in sharp contrast to the mitotic transmission pattern of nuclear genes: In diploid cells that are heterozygous for a nuclear gene, the alleles of the gene almost never segregate during mitotic divisions; the rare exception in which a heterozygous diploid produces homozygous wild-type and homozygous mutant progeny is the result of mitotic crossing-over.

The mitotic segregation of organelle genomes has distinct phenotypic consequences. In a woman whose cells are heteroplasmic for the LHON mutation, some ova may carry a few mitochondria with the LHON mutation and a large number of mitochondria with the wild-type gene for subunit 4 of NADH dehydrogenase; other ova may carry mainly mitochondria with LHON mutations; still others may carry only wild-type organelles. The precise combination depends on the random partitioning of mitochondria during the mitotic divisions that gave rise to the germ line. After fertilization, as a result of the mitotic divisions of embryonic development, the random segregation of mutation-carrying mitochondria from heteroplasmic cells can produce tissues with completely normal ATP production and tissues of low energy production. If cells homoplasmic for low energy production happen to end up in the optic nerve, LHON will result.

In plants where cpDNA mutations leading to a defect in photosynthesis would be lethal, heteroplasmy for chloroplast genomes is prevalent. In fact, mitotic segregation of the chloroplasts of heteroplasmic cells explains the transmission of variegation in four-o'clocks, described at the beginning of this chapter (see Fig. 15.1). Most female gametes from a variegated plant are heteroplasmic for mutant and wild-type cpDNAs. Zygotes resulting from fertilization with pollen from a wild-type green plant will develop into variegated progeny. Segregation of wild-type and mutant chloroplasts during F_1 plant development may, however, generate some female gametes with only mutant cpDNA. Fertilization of these homoplasmic mutant gametes with wild-type pollen produces zygotes with only mutant cpDNA; the seedlings that develop from these zygotes cannot carry out photosynthesis and eventually die.

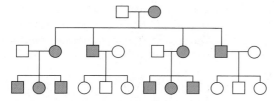

Figure 15.10   Hypothetical example of LHON pedigree.
Characteristic pedigree of mitochondrial disease. All offspring of diseased mothers can express the disease phenotype, while none of the offspring of diseased fathers express the disease phenotype.

Experiments with Mutants of cpDNA in *Chlamydomonas reinhardtii* Reveal Uniparental Inheritance of Chloroplasts

Chlamydomonas, a unicellular alga, is easy to handle, grows abundantly in defined media, and has a tractable sexual reproduction cycle. It can live in different environmental conditions, functioning photosynthetically in the light on media where atmospheric CO_2 is the sole source of carbon, or without photosynthesis in the dark on media containing acetate as a source of carbon. These features make it a useful experimental organism. In the laboratory, *Chlamydomonas* usually grows as a haploid cell, and each haploid strain comes in one of two mating types: mt^+ or mt^- (the minus in this case does not denote a mutant). Researchers can induce cultures of mitotically dividing haploid cells (Fig. 15.11a) to develop into haploid gametes by a special physiological regime (Fig. 15.11b). When they then mix gametes of opposite mating types, an elaborate mating process results in the fusion of gametes of opposite types to form diploid zygotes. On suitable medium, these zygotes undergo meiosis and form unordered tetrads. For simplicity, we describe all experiments with this organism as leading to tetrads with four spores, even though in reality sometimes a mitotic division after meiosis generates eight spores.

Pioneering research on the genetics of chloroplasts in *Chlamydomonas* led to the isolation of mutants with various defects in photosynthetic growth. The first such mutant analyzed genetically was resistant to the antibiotic streptomycin, that is, it carried the *sm-r* allele, which encodes an inhibitor of prokaryotic (and also chloroplast) protein synthesis. These mutants could grow photosynthetically on medium containing streptomycin, while the drug-sensitive *sm-s* parental strain could not. The mutant strain also grew well in the absence of the drug. Researchers could thus carry out crosses in media lacking the drug and then grow the progeny on drug-containing media to determine their phenotype.

Consider a cross between mt^+ *sm-r* gametes and mt^- *sm-s* gametes in which researchers scored the resulting tetrads for the transmission of drug sensitivity (Fig. 15.11c). The investigators included a known nuclear marker affecting colony color (*yellow 1,* with alleles *y1$^+$* and *y1$^-$*) in the parental strains and monitored the marker in the progeny. In this experiment, alleles of the mating-type gene and the yellow-marker gene segregated 2:2 in the spores, but the ratio of *sm-r* to *sm-s* was 4:0. This lack of segregation was a striking departure from Mendelian principles. Subsequent reciprocal crosses showed that the *sm-r* allele did not itself determine the nonsegregating outcome. In a reciprocal cross of mt^+ *y1$^+$ sm-s* gametes with mt^- *y1$^-$ sm-r* gametes, the tetrads were 4 *sm-s* : 0 *sm-r*. Because *Chlamydomonas* lacks distinct male and female gametes, it is an *isogamous* species in which the "sex" cells are similar in size and shape but come in two mating types. In speaking of isogamous species, geneticists say that a 4:0 ratio of spore genotypes reflects *uniparental,* rather than maternal, inheritance. Researchers have isolated many other *Chlamydomonas* mutants resistant to various antibiotics

or deficient in photosynthesis, and all show the uniparental mode of transmission.

Many Mechanisms Contribute to Uniparental Inheritance of Mitochondrial and Chloroplast Genes

Differences in gamete size help explain maternal inheritance in some species. In most higher eukaryotes, the male gamete is much smaller than the female gamete. As a result, the zygote receives a very large number of maternal organelles and, at most, a very small number of paternal organelles. In some organisms, cells degrade the organelles or the organelle DNA of male gametes. In some plants, the early divisions of the zygote distribute most or all of the paternal organelle genomes to cells that are not destined to become part of the embryo. In some animals, details of fertilization prevent a paternal cell from contributing its organelles to the zygote. With tunicates, for example, events of fertilization allow only the sperm nucleus to enter the egg, physically excluding the paternal mitochondria. In some organisms where the complete gametes fuse, the zygote destroys the paternal organelles after fertilization.

These explanations do not apply to *Chlamydomonas,* in which the two gametes are of equal size and fuse completely. In this organism, the single chloroplast of one gamete fuses with the chloroplast of the other gamete shortly after cell fusion. However, the zygote, able to distinguish between cpDNAs from mt^+ and mt^- cells, degrades most of the cpDNAs from the mt^- parent. Interestingly, *Chlamydomonas* mtDNA is also inherited from one parent, but here, the mitochondrial genome from the mt^- cell type remains.

Summary of Uniparental Inheritance

The transmission of organelles and their genomes via one parent, mainly the mother, in most plants and animals results in two significant departures from Mendelian patterns of inheritance. First, organelle genes show no meiotic segregation: all F_1 individuals carry the same organelle genotype—that of the transmitting parent. Second, when the parent transmitting the organelle genome is heteroplasmic and therefore passes along organelles of more than one genotype, mitotic segregation of those genotypes occurs in the offspring. This segregation of genotypes during mitosis is a consequence of the random partitioning of organelles during cell division.

The Genetics and Society box "Mitochondrial DNA Tests Replace HLA Typing As Evidence of Kinship in Argentine Courts" describes how a human rights organization in Argentina eventually used mtDNA sequences as the legal basis for reuniting kidnapped children with their biological families. The maternal inheritance of mitochondria makes it possible to compare and match the DNA of a grandmother and a grandchild.

Some Organisms Exhibit Biparental Inheritance

Although uniparental inheritance is the norm among most metazoans and plants, single-celled yeast and some plants inherit

(a)

Figure 15.11 **Genes in the cpDNA of *C. reinhardtii* do not segregate at meiosis. (a)** A nuclear gene (*yl*) segregates 2:2 in the gametes but a chloroplast-encoded gene (*sm*) segregates 4:0. **(b)** When haploid cultures of gametes of opposite mating type are mixed, cells aggregate and form pairs, leading to another round of life cycles. **(c)** A cross of *C. reinhardtii* gametes is illustrated. Vegetative cells of this organism are usually haploid. Mating of those cells is controlled by the mating-type locus that exists in two alleles (*mt*⁺ and *mt*⁻) indicated by a nucleus colored *blue* or *yellow*, respectively. Each haploid cell has one large chloroplast (*light* or *dark green* for different alleles of cpDNA) and several mitochondria (*light* or *dark red* for different alleles of mtDNA). Mating occurs by pairwise fusion of gametes of opposite mating type. Shortly after cell fusion, the nuclei and chloroplasts fuse so that each zygote has one diploid nucleus and one large chloroplast. The zygote matures, during which time chloroplast markers from the *mt*⁻ parent and mitochondrial markers from the *mt*⁺ parent are inactivated. Mature zygotes then undergo meiosis, forming four haploid meiotic products, each of which can germinate. If light or ammonium ions are present, the haploid cells will grow vegetatively. If ammonium ions are absent, the cells will undergo gametogenesis to prepare for another round of mating.

their organelle genomes from both parents in a **biparental** fashion. The earliest report of biparental inheritance is a 1909 description of reciprocal crosses between green and variegated geraniums (*Pelargonium zonale*). Unlike what happens in four-o'clocks (as described earlier), both reciprocal crosses yielded green, white, and variegated seedlings in varying proportions (Fig. 15.12). Thus, variegated leaves in geraniums are a chimeric condition that results from the chloroplast traits inherited from both parents. Many other plants as well as yeast similarly inherit their organelle genomes from both parents. Geneticists have devised experimental systems for analyzing biparental organelle inheritance in several organisms. We now look at some of the studies in yeast.

A Genetic System for Studying Mitochondrial Genes in Yeast

We have seen that translation in mitochondria resembles translation in bacteria, and that inhibitors of bacterial translation, such as chloramphenicol and erythromycin, are potent inhibitors of mitochondrial protein synthesis. In the early 1960s, researchers discovered that chloramphenicol inhibits the growth of wild-type yeast on media containing a nonfermentable source of carbon (either glycerol or ethanol); but the same or higher levels of the drug does not inhibit yeast growth on medium containing glucose, a fermentable carbon source. They concluded that the inhibitor acts on the then recently discovered mitochondrial translation system and began to develop a system for studying the inheritance of yeast mtDNA. The initial step was to isolate mutants that could grow on medium containing glycerol in the presence of a drug that inhibits mitochondrial protein synthesis. The first useful mutants were resistant to chloramphenicol or erythromycin (C^r or E^r); they derived from wild-type cells that were sensitive to one drug or the other (C^s or E^s).

Like *Chlamydomonas,* yeast is isogamous. Each haploid strain has one of two mating types: MATa or MATα. The mixing of cells of opposite mating type results in the fusion of dissimilar haploid cells to form diploid zygotes (Fig. 15.13a). Unlike *Chlamydomonas,* yeast cells can mate without first differentiating into gametes, and yeast crosses generate cultures of diploid cells because the zygote produced by mating does not usually enter meiosis; rather, it continues to multiply by mitosis until some sort of stress induces its diploid descendants to enter the meiotic cycle.

In Yeast, mtDNA-Encoded Traits Show a Biparental Mode of Inheritance and Mitotic Segregation

To analyze the inheritance pattern of mtDNA in yeast, researchers mixed C^r and C^s parental cells, allowed them to grow for many generations, and isolated diploids, taking advantage of nuclear auxotrophic markers in the parental strains for the selection of diploids (Fig. 15.13b). They then scored cells from the diploid progeny for the C^r or C^s phenotype by replica plating the cells to petri plates containing glycerol medium with or without an inhibitory level of chloramphenicol.

In most progeny resulting from the crosses, about half of the diploid cells were C^s and half were C^r. This result shows that both parents transmitted organelles to the progeny, that is, inheritance was biparental. It also shows that the organelles from parental cells of different mating types segregated during the mitotic divisions of vegetative growth so that after several generations colonies were either C^s or C^r, but not both. Importantly, all of the progeny scored, whether C^s or C^r, had the same nuclear-determined phenotype (that is, they were prototrophic for

(a)

(b)

Figure 15.12 Biparental inheritance of variegation in geraniums. (a) Examples of green and variegated *P. zonale* plants. **(b)** Reciprocal crosses between green and variegated geraniums yield the same classes of offspring, indicating that the gene is inherited from both parents.

Figure 15.13   Mitochondrial transmission from the yeast *S. cerevisiae*. **(a)** General life cycle of *S. cerevisiae*. **(b)** One-factor crosses in *S. cerevisiae* show biparental inheritance of mtDNA.

two nutrients). This meant that the diploids carried nuclear genes from both parents in the cross, and those nuclear genes did not segregate during this period of vegetative growth.

Researchers Can Now Use Recombinant DNA Techniques to Study the Genetics of Organelles

In the early days of recombinant DNA technology, organelle researchers were frustrated by an inability to transfer cloned genes and mutated DNA fragments into organelle genomes. Researchers could deliver plasmid DNAs to yeast cell nuclei very efficiently, but they could not get it to enter the mtDNA of yeast mitochondria or the cpDNA of *Chlamydomonas* chloroplasts. Because recombination occurs readily in both organelle genomes, the problem seemed to be in getting the foreign DNA into the organelles themselves.

In the late 1980s, development of the *gene gun* (Fig. 15.14) and a gene delivery method known as "biolistic transformation" solved the problem. The basic idea is to coat small (1 μm) metal particles with DNA and then shoot these DNA-carrying "bullets" at cells. First attempted in plant cells, biolistic transformation occurs when a particle lands within a cell, such as a plant protoplast, without killing it and releases the DNA from the metal; rarely, the released DNA enters the nucleus. If the DNA shot into the cell contains a strong selective marker, plant geneticists can isolate the rare transformed plant cells in which the released DNA has entered the nucleus.

Figure 15.14   Genetic transformation of organelle genomes. Foreign DNA can be delivered into an organelle with the use of a gene gun that deposits tiny biolistic pellets having micrometer diameters.

Shortly after the use of the gene gun to transfer genes into plants, two groups of researchers, one working with yeast mitochondria, the other with *Chlamydomonas* chloroplasts, successfully transformed organelle DNA. In one study, researchers constructed a special, mitochondrion-translatable form of a gene encoding green fluorescent protein (GFP); the gene product fluoresces brightly when stimulated by light of the right wavelength. The codons of the specially constructed gene were based on the mitochondrial genetic code, instead of on the "universal" genetic code. To this specially constructed gene, the experimenters fused that part of the mitochondrial *COX3* gene extending from the promoter through sequences encoding the first three amino acids of the COX3 protein. When they then bombarded mitochondria with this plasmid, they obtained transformants that expressed the *GFP* gene via the mitochondrial gene expression system. Fluorescence of the mitochondria showed that the transformed cells retained the GFP protein in their mitochondria.

Current researchers can use yeast with this GFP construct to select for mutations that affect mitochondrial gene expression. In fact, the transformation of organelle genomes is now a reliable way to manipulate and study organelle genes in yeast and *Chlamydomonas*. Recently, researchers have achieved chloroplast transformation in higher plants and think they may some day be able to use such transformation in engineering crop plants resistant to certain herbicides.

Summary of the Genetic Principles of Non-Mendelian Inheritance

Three features distinguish the non-Mendelian traits encoded by organelle genomes from the Mendelian traits encoded by nuclear genomes.

1. In the inheritance of organelle genomes from one generation to the next, there is a 4:0 segregation of parental alleles, instead of the 2:2 pattern seen for the alleles of nuclear genes.

2. In most organisms, transmission of organelle-encoded traits is uniparental, mainly maternal, although in a few organisms, notably yeast, transmission is biparental.

3. With both uniparental and biparental inheritance, when the parents transmit organelles of more than one genotype, mitotic segregation of those genotypes occurs in the offspring. This segregation of genotypes during mitosis is a consequence of the random partitioning of organelles during cell division.

Comprehensive Example: How Mutations in mtDNA Affect Human Health

Some debilitating diseases of the human nervous system pass from mother to daughters and sons, from affected daughters to granddaughters and grandsons, and so on down through the maternal line. Unexpectedly, the symptoms of these diseases vary enormously among family members, even among very close relatives. Pedigrees tracing general transmission patterns, in combination with molecular studies, help clarify what is happening.

Individuals with Certain Rare Diseases of the Nervous System Are Heteroplasmic for Wild-Type and Mutant mtDNAs

People with a rare inherited condition known as myoclonic epilepsy and ragged red fiber disease (MERRF) have a range of symptoms: uncontrolled jerking (the myoclonic epilepsy part of the condition), muscle weakness, deafness, heart problems, kidney problems, and progressive dementia. Affected individuals often have an unusual "ragged" staining pattern in regions of their skeletal muscles, which explains the ragged red fiber part of the condition's name (Fig. 15.15a). As the pedigree in Fig. 15.15b shows, family members inherit MERRF from their mothers; in the pedigree, none of the offspring of the affected male sibling (II-4) has symptoms of the disease. The family history also reveals individual variations in the number and severity of

(a)

(b)

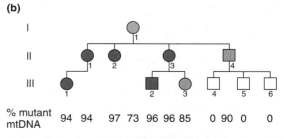

Figure 15.15 Maternal inheritance of the mitochondrial disease MERRF. (a) Transmission electron micrograph of muscle. Mitochondria from patients expressing MERRF. Mutant mitochondria are highly abnormal, showing paracrystalline arrays and crista degeneration. **(b)** Pedigree of family showing inheritance of MERRF. Pedigree shows typical pattern of maternal transmission observed with mitochondrial mutations. The percent of mutant mtDNA in the cells of individuals varies and corresponds with the severity of the condition (indicated by different color coding).

Mitochondrial DNA Tests Replace HLA Typing As Evidence of Kinship in Argentine Courts

Between 1976 and 1983 the military dictatorship of Argentina kidnapped, incarcerated, and killed more than 10,000 university students, teachers, social workers, union members, and others who did not support the regime. Many very young children disappeared along with the young adults, and close to 120 babies were born to women in detention centers. In 1977, the grandmothers of some of these infants and toddlers began to hold vigils in the main square of Buenos Aires to bear witness to and inform others about the disappearance of their children and grandchildren. They soon formed a human rights group—the "Grandmothers of the Plaza de Mayo."

The grandmothers' goal was to locate the more than 200 grandchildren they suspected were still alive and reunite them with their biological families. To this end, they gathered information from eyewitnesses, such as midwives and former jailers, and set up a network to monitor the papers of children entering kindergarten. Questionable birth certificates could reveal a possible missing grandchild. They also publicized their work inside Argentina and contacted organizations outside the country, including the United Nations Human Rights Commission and the American Association for the Advancement of Science (AAAS).

What the grandmothers asked of AAAS was help with genetic analyses that would stand up in court. By the time a democracy had replaced the military regime and the grandmothers could argue their legal cases before a relatively impartial court, children ab-

ducted at age 2 or 3 or born in 1976 were 7–10 years old. Although the grandmothers had compiled an enormous amount of circumstantial evidence, the Argentine courts did not accept such evidence as proof of a young person's identity and biological relatedness. The courts did acknowledge, however, that although the size and other external features of the children had changed, their genes—relating them unequivocally to their biological families—had not. The grandmothers, who had educated themselves about the potential of genetic tests, sought help with the details of obtaining and analyzing such tests. Starting in 1983, the courts agreed to accept their test results as proof of kinship.

In 1983, the best way to confirm or exclude the relatedness of two or more individuals (for instance, a child and her grandparents) was to compare the gene-encoded proteins called human lymphocyte antigens (HLAs). People carry a unique set of HLA markers on their white blood cells, or lymphocytes, and these markers are diverse enough to form a kind of molecular fingerprint. HLA analyses can be carried out even if a child's parents are no longer alive, because for each HLA marker, a child inherits one allele from the maternal grandparents and one from the paternal grandparents. Statistical analyses can establish the probability that a child shares genes with a set of grandparents because he or she is their grandchild compared to the probability that the child shares the genes by chance.

symptoms. From these two features of transmission, clinical researchers suspected that MERRF results from mutations in the mitochondrial genome.

Molecular analyses confirm this hypothesis. The mtDNA from patients affected by MERRF carries a mutation in the gene for tRNA^{Lys} (review Fig. 15.5) or one of the other mitochondrial tRNAs. These tRNA mutations disrupt the synthesis of proteins in multiplexes I and IV of the mitochondrial electron transport chain, thereby decreasing the production of ATP.

The Proportion of Mutant mtDNAs and the Tissue in Which They Reside Influence Phenotype

In a second large family in which clinicians looked at the mitochondria of muscle cells, an individual carrying 73% mutant and 27% normal mtDNAs showed no symptoms of MERRF; a relative with 85% mutant mtDNA showed no external signs of the disease but lab tests revealed some muscle tissue abnormalities; and two family members with 98% mutant mitochondria

showed serious symptoms of MERRF. This suggests that a relatively small percentage of normal mitochondria can have a strong protective effect. The random partitioning of mitochondria during the mitotic cell divisions of development, both *in utero* and after birth, explains the variation in the percentage of mutant mitochondria in individuals of the second, third, and fourth generations of family pedigrees.

It is likely that many tissues in individuals affected by MERRF are heteroplasmic for the tRNA mutations causing the disease. Within each person, the ratio of mutant to wild-type mtDNA varies considerably from tissue to tissue, and because each tissue has its own energy requirements, even the same ratio can affect different tissues differently (Fig. 15.16). Muscle and nerve cells have the highest energy needs of all types of cells and are therefore the most dependent on oxidative phosphorylation. This explains why mitochondrial mutations that by chance segregate to these tissues generate the defining features of MERRF.

The AAAS put the grandmothers in touch with Mary Claire King, then at the University of California. In the 1980s, King taught Argentine medical workers to analyze the diverse HLA markers on white blood cells. The grandmothers then obtained the HLA types of as many living members as possible of the missing children's families and stored that information in an HLA bank. When a child whom they believed to be one of the missing turned up, they analyzed his or her HLA type and tried to find a match among their data. Depending on the number of different alleles carried and the rarity of the variations, the probability that a tested child belonged to the family claiming him or her on the basis of eyewitness accounts of a birth or abduction varied from 75% to 99%.

By the late 1980s, the combination of circumstantial and genetic evidence had helped reunite 49 children with their biological families. In one case, a grandmother who had been kidnapped September 5, 1976, along with her pregnant daughter and then released from prison contacted the Grandmothers of the Plaza de Mayo for help. Ten and a half years later, in April 1987, she was reunited with her granddaughter. The child had been born on November 5, 1976, and registered in the name of a detention center police officer. Her place of birth remains unknown. But, her grandmother recounted in an April 1987 interview, "It was very easy to prove she was my granddaughter because of the blood tests. They analyzed my blood, the blood of my husband and the other grandparents and they [the results] proved more than 99% positive."

As time passed and the "easier" cases had been settled, the limitations of the HLA approach became apparent. By the mid-1980s, for example, there were too few living relatives in some families to establish a reliable match through HLA typing. But the advent of new tools such as PCR and DNA sequencing made it possible to look at DNA directly.

King and two colleagues—C. Orrego and A. C. Wilson—used the new techniques to develop an mtDNA test based on the PCR amplification and direct sequencing of a highly variable noncoding region of the mitochondrial genome. Mitochondrial DNA has several characteristics that make it more powerful than HLA typing for confirming or excluding family relatedness: maternal inheritance, a lack of recombination, a highly variable noncoding region of 331 bp, and a large number of mtDNAs per cell. The maternal inheritance and lack of recombination mean that as long as a single maternal relative is available for matching, the approach can resolve cases of disputed relatedness. The extremely polymorphic noncoding region makes it possible to identify grandchildren through a direct match with the mtDNA of only one person—their maternal grandmother, or mother's sister or brother—rather than through statistical calculations assessing data from four people; the probability that any two unrelated individuals would be identical for all 331 bp by chance is almost nil. Finally, the large number of mtDNAs per cell increases the chances of isolating mtDNA from and identifying the bodies of presumed parents that are found after many years.

To validate their approach for matching maternal grandmothers to their grandchildren, King and colleagues amplified sequences from three children and their three maternal grandmothers without knowing who was related to whom. The mtDNA test unambiguously matched the children with their grandmothers. Thus, after 1989, the grandmothers included mtDNA data in their archives.

Today, the grandchildren—the children of "the disappeared"—have reached adulthood and attained legal independence. Although most of their grandmothers have died, the grandchildren may still discover their biological identity and what happened to their families through the HLA and mtDNA data the grandmothers left behind.

Mitochondrial Inheritance in Identical Twins

In some pairs of identical twins, one twin manifests symptoms of neurodegenerative disease, while the other twin does not. This is because even though their nuclear genomes are identical, their mitochondrial genomes are not.

Consider the following. A fertilized human egg cell carries up to 2000 mtDNAs, 99.9% of which come from the mother. If 10% of those mtDNAs carry a MERRF mutation, at the first mitotic division, the mutant mtDNAs could by chance be partitioned in any one of a number of ways: The cell that will develop into one twin may get all the mutant mitochondria (this would be a very rare occurrence); the mtDNAs with mutations may be equally divided between the two twin-producing cells; the mutated mtDNAs may be unevenly distributed to the two twin-generating cells. Even with an even distribution at the first cell division, during the subsequent divisions of development, the mutant mitochondria in one twin may end up in cells that eventually die during normal development, while the mutant mitochondria in the other twin may end up in cells that differentiate into heart and optic nerve tissue.

Mitochondrial Mutations and Aging

We have seen that like the nuclear DNA mutations that cause cancer, some mutations in mtDNA are inherited through the germ line, while others arise sporadically in somatic cells as a result of random events, such as radiation or chemical insults. We have also seen that the rate of somatic mutations is much higher in mitochondrial DNA than in nuclear DNA. This is in part because the oxidative phosphorylation system in mitochondria generates a relatively high concentration of DNA-damaging free radicals in a membrane-confined space. In one study, mtDNA accumulated 16 times more oxidative damage than nDNA.

Some researchers focusing on the genetics of aging think that the accumulation of mutations (both base substitutions and deletions) in mtDNA over a lifetime, and the progressive

Individual mtDNA genotypes		Tissues Affected				
					Skeletal Muscle	
		Brain	Heart	Type I	Type II	Skin
I	20% mutant mtDNAs	+	−	−	−	−
II	40% mutant mtDNAs	+	+/−	−	−	−
III	60% mutant mtDNAs	+	+	+	−	−
IV	80% mutant mtDNAs	+	+	+	+/−	+/−

Figure 15.16 Disease phenotypes reflect the ratio of mutant-to-wild-type mtDNAs and the reliance of cell type on mitochondrial function. The proportion of mutant mitochondria determines the severity of the MERRF phenotype and the tissues that are affected. Tissues with higher energy requirements (for example, brain) are least tolerant of mutant mitochondria. Tissues with low energy requirements (for example, skin) are only affected when the proportion of wild-type mitochondria is greatly reduced.

enrichment of deleted mtDNAs through their tendency to replicate more rapidly than longer, nondeleted mtDNAs, results in an age-related decline in oxidative phosphorylation. This decline, in turn, accounts for some of the symptoms of aging, such as decreases in heart and brain function.

Proponents of this hypothesis suggest that individuals born with deleterious mtDNA mutations start life with a diminished capacity for ATP production, and as a result, several of their tissues may cross the threshold from function to nonfunction early or in the middle of life; by comparison, people born with a normal mitochondrial genome start life with a high capacity for ATP production and may die before a large number of tissues dip below the required energy threshold.

Evidence in support of an association between mtDNA mutations and aging comes from a variety of studies. In one study, researchers looked at 140 hearts obtained from autopsies and found significant decreases in cytochrome *c* oxidase, a respiratory enzyme largely encoded by mtDNA. In another study, researchers analyzed a 7.4 kb and a 5 kb deletion in heart and brain mtDNAs in people of different ages. The percentage of hearts that had the 7.4 kb deletion increased with age, and the number of 5 kb deletions increased in normal heart tissue after age 40. Moreover, the 5 kb deletion was absent from the brain tissue of children but present in the brain tissue of adults. Finally, although biomedical researchers had known for decades that the brain cells of people showing symptoms of Alzheimer's disease (AD) have an abnormally low energy metabolism, they recently discovered that 20% to 35% of the mitochondria in the brain cells of most AD patients carry mutations in two of their three cytochrome *c* oxidase genes; the inefficiency of the resulting mutant cytochrome *c* oxidase enzyme impairs the brain's energy metabolism. To confirm an association between this enzymatic abnormality and AD, the researchers transferred mitochondria from AD patients into normal cultured cells from which they had removed the native mitochondria, and they found that the engineered cells had defective energy production.

These data suggest that if it were possible to assess all forms of mtDNA damage, it might turn out that a significant proportion of mtDNAs are defective in elderly people. On the basis of this hypothesis, clinicians have proposed that the restoration of enzymes encoded by wild-type mtDNA but absent from mitochondria with deleted and otherwise mutated genomes might ease some of the symptoms of aging. Further research will be necessary to discover whether mitochondrial damage makes a significant contribution to the aging process.

CONNECTIONS

Mitochondria, the sites of oxidative phosphorylation, and chloroplasts, the sites of photosynthesis, exhibit non-Mendelian patterns of inheritance that are remarkably similar across a wide range of species.

Various regulatory mechanisms help correlate the day-to-day operations of a cell's organelles with the needs and circumstances of different tissues. One unusual example of the fine-tuning of these controls is seen in the voodoo lily (*Sauromatum guttatum*), a native of Southeast Asia that belongs to the same plant family as the skunk cabbage. The day the flower opens, a burst of mitochondrial activity in some of its cells increases the temperature inside the bloom by 22°C (72°F); the heat so generated causes odorous molecules to evaporate and attract insect pollinators, increasing the plant's chances for reproduction. Although researchers currently know very little about the regulation of mitochondrial genes, they believe that, given the endosymbiont theory of organelle evolution, the regulation of genes in prokaryotes may serve as a model for what happens in mitochondria.

In Chapter 16 we examine the molecular mechanisms that regulate gene expression in prokaryotes. Then, in Chapter 17, we explore the regulation of nuclear genes in eukaryotes. In both these chapters, we see that gene regulation controls the metabolic activities of all cells as well as the differentiation of cells in multicellular organisms.

ESSENTIAL CONCEPTS

1. Mitochondria and chloroplasts are semiautonomous organelles of energy conversion. They carry their own double-stranded DNA in circular or linear chromosomes whose size and gene content vary from species to species.

2. Translation in the mitochondria of many species depends on an alternative genetic code.

3. Chloroplast genomes are relatively uniform in size and carry many more genes than mitochondrial genomes. In some chloroplasts, the genes are organized in clusters that resemble the operons of bacteria.

4. According to the *endosymbiont theory,* mitochondria and chloroplasts evolved from bacteria engulfed by the precursors of eukaryotic cells. The genomes of these organelles have probably lost more than two-thirds of their original bacterial genes in the course of evolution.

5. In most species, organelle genomes show *uniparental inheritance,* mainly through the maternal line. Cells containing a mixture of organelle genomes are *heteroplasmic.* Cells carrying only one type of organelle DNA are *homoplasmic.* The genomes of heteroplasmic cells are not evenly partitioned at mitosis.

6. The organelles of yeast and several other species exhibit biparental inheritance. The genomes of organelles inherited in a biparental fashion show mitotic segregation.

SOCIAL AND ETHICAL ISSUES

1. A researcher has promising results that could lead to a treatment for a neurodegenerative mitochondrial disease. When she applies for a grant from a federal agency, the reviewers respond that the research proposed is more applied than basic in nature and therefore should be funded by a commercial pharmaceutical company. When the researcher approaches drug companies with her proposal, she finds they are not interested in developing a drug that would have a rather limited market. Should the government create mechanisms to cover the expenses of development and clinical trials for drugs of relatively low use that do not have a potential for producing a profit but that will nonetheless greatly benefit affected people?

2. During repressive political regimes in many countries, children often are separated from their parents and/or become orphaned. Human rights groups have been in the forefront of assisting the identification and reunion of relatives using genetic markers such as mitochondrial DNA. Should the government of the country in which the atrocities were committed be obligated to pay some or all of the costs of these searches, even if the current regime is far removed in political ideology and action from the repressive regime?

3. It has been suggested that a databank consisting of DNA fingerprints of previously convicted felons be established. Both nuclear and mitochondrial fingerprints would be included to accommodate situations in which the nuclear DNA was too degraded to be used. Who should have access to such a database? Law enforcers? Psychologists? Sociologists? Is this an invasion of privacy or a protection for the public?

SOLVED PROBLEMS

I. Differential hybridization of a probe to mitochondrial DNA from two *Xenopus* species was the methodology employed to demonstrate maternal inheritance in vertebrates. However, as noted in this chapter, this hybridization technique was not sensitive enough to detect small amounts of paternal DNA. What technique that is more sensitive to small amounts of DNA could be used today? How could you use this technique to determine if paternal mitochondrial DNA was present in the progeny of the interspecies cross?

Answer

Polymerase chain reaction (PCR) is a sensitive technique that detects very small amounts of DNA. Oligonucleotide primers that are specific for each of the mitochondrial DNAs in each of the two different species could be used to determine if paternal DNA is present in the offspring from the interspecies cross.

II. A newly isolated *petite* (respiration deficient) mutant of yeast, when mated to a wild-type strain and sporulated, produced two *grande* and two *petite* spores. What does this meiotic segregation pattern tell you about the *petite* mutation?

Answer

The 2:2 segregation of mutant and wild-type phenotypes from meiosis indicates that the *mutation is in a nuclear gene.* The finding that a *petite* phenotype is caused by a nuclear mutation is not surprising, since functioning of the mitochondria requires both nuclear and mitochondrial genes. If the mutation were in mitochondrial DNA, a 4:0 segregation pattern would be seen instead of the 2:2 pattern.

III. a. Does the following pedigree suggest mitochondrial inheritance? Why or why not?

   b. Is there another mode of inheritance that is consistent with this data?

Answer

a. The data presented in this pedigree *are consistent with mitochondrial inheritance since the trait is transmitted by females; the affected males in this family did not transmit the trait; and all of the females' progeny have the trait.*

b. This inheritance pattern is *also consistent with transmission of an autosomal dominant trait.* According to this hypothesis, individuals I-1 and II-2 passed on the dominant allele to all children, but II-4 did not pass on the dominant allele to either child.

PROBLEMS

15-1 Choose the phrase from the right column that best fits the term in the left column.

a. cristae	1. results in production of ATP
b. grana	2. yeast cells that can respire
c. *petite*	3. results in production of sugar
d. *grande*	4. cell containing organelles that are genetically identical
e. oxidative phosphorylation	5. contains multiple linked mini- and maxicircles of DNA
f. Calvin cycle	6. folded inner membrane in mitochondria
g. kinetoplast	7. yeast cells that cannot respire
h. heteroplasmic	8. stacked thylakoid membranes in chloroplasts
i. homoplasmic	9. cell containing two genetically different organelles

15-2 Does each of the following processes occur in the chloroplast, mitochondria, both, or neither?
  a. electron transport
  b. pyruvate and fatty acid breakdown
  c. sugar synthesis
  d. O_2 generation
  e. O_2 consumption

15-3 Is each of these statements true of chloroplast or mitochondrial genomes, both, or neither?
  a. contain tRNA genes
  b. exist as condensed structures called nucleoids
  c. all genes necessary for function of the organelle are present
  d. vary in size from organism to organism

15-4 Some genes required for chloroplast function are encoded in the nuclear genome; others are encoded in the chloroplast genome. Nuclear and chloroplast DNA have different buoyant densities and can therefore be separated from each other by centrifugation based on these differences. There is a small amount of cross-contamination in the separation of nuclear and chloroplast DNAs using this technique. You have just found that a probe for a photosynthetic gene that is present in the chloroplast genome of plants hybridizes to nuclear DNA of a red alga. What experimental control(s) would you do to be more confident that your result means that the gene is encoded in the nucleus of this organism?

15-5 "Reverse translation" is a term given to the process of deducing the DNA sequence that could encode a particular protein. If you had the amino-acid sequence Trp His Ile Met,
  a. What mammalian nuclear DNA sequence could have encoded these amino acids? (Include all possible variations.)
  b. What mammalian mitochondrial DNA sequence could have encoded these amino acids? (Include all possible variations.)

15-6 a. Results from hybridization using a probe for the small subunit gene of the Rubisco protein and a probe for the large subunit gene of the Rubisco protein to chloroplast and nuclear DNAs from a green, a red, and a brown alga are shown here. What conclusions would you reach about the location of the small and large subunit genes in each of the three types of algae?

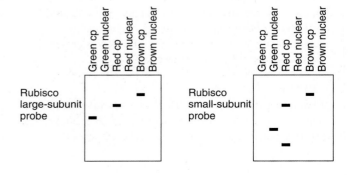

b. When RNA was extracted from the same three algal species and hybridized with a large subunit Rubisco probe and also with a small subunit probe, the following results were obtained. What conclusion would you reach about large and small subunit gene transcription in red and golden brown algae? Is this consistent with your answer in part *a?*

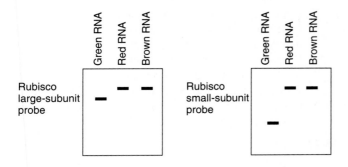

15-7 Which of the characteristics of chloroplasts and/or mitochondria listed here make them seem more similar to bacterial cells than to eukaryotic cells?
   a. Translation is sensitive to chloramphenicol and erythromycin.
   b. Alternate codons are used in mitochondria genes.
   c. Chloroplasts and mitochondria have inner and outer membranes.
   d. Introns are present in organelle genes.
   e. DNA in organelles is not arranged in nucleosomes.

15-8 A chloroplast translation elongation factor gene (*tufA*) hybridized to a fragment of nuclear DNA in the green alga *Coleochaete*. How could you determine if the entire coding sequence of the gene was intact in the nuclear fragment that hybridized with the probe?

15-9 A form of male sterility in corn is inherited maternally. Marcus Rhoades first described this cytoplasmic male sterility by crossing female gametes from a male sterile plant with pollen from a male fertile plant. The resulting progeny plant was male sterile. Female gametes from this male sterile progeny were backcrossed again with pollen from the same male fertile plant of the first cross. The process was repeated many times. What was the purpose of the many backcrosses? (Hint: Think about what is happening to the nuclear genome with repeated backcrosses.)

15-10 Which of the two methods listed would you choose to determine if organelles in an organism are heteroplasmic or homoplasmic and why?
   a. hybridize probes to cells immobilized on a slide
   b. PCR amplify DNA isolated from a population of cells

15-11 Studies distinguishing between uniparental and biparental inheritance of organelles employed a variety of detection methods. Match the system studied with the method used.

a. *Xenopus laevis* and *X. borealis* mitochondrial DNA	1. protein analysis
b. *Nicotiana* large and small subunits of Rubisco	2. pedigrees
c. LHON phenotype in humans	3. differential hybridization

15-12 Describe two ways in which the contribution of mitochondrial genomes from male parents is prevented in different species.

15-13 If a human trait is determined by a factor in the cytoplasm, would an offspring more resemble its mother or its father? Why?

15-14 Why are very severe mitochondrial or chloroplast mutations usually found in heteroplasmic cells instead of homoplasmic cells?

15-15 You have isolated a new erythromycin-resistant mutant of *Chlamydomonas* in an mt^+ strain.
   a. If the mutation was in a chloroplast gene, what predictions would you make about genotype and phenotype of the diploid after mating the mutant with a $mt^-\ ery^s$ strain?
   b. What would be the genotype and phenotype of the diploid if the mutation was in a nuclear gene?
   c. What additional step would enable you to determine if the mutation was in chloroplast or nuclear DNA?

15-16 You have isolated a new C^r mutant in yeast, starting with a MATa strain. How could you test the new C^r mutant to determine if the mutation is in nuclear DNA or mitochondrial DNA?

15-17 An example of a cloning vector used for biolistic transformation of chloroplasts is shown in the following diagram. The vector DNA can be prepared in large quantities in *Escherichia coli*. Once "shot" into a chloroplast, the vector DNA integrates into the genome. Match the component of the vector with its function.

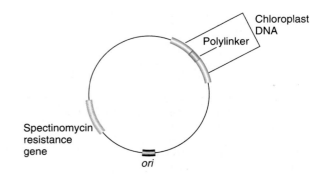

a. spectinomycin resistance gene

b. chloroplast DNA

c. polylinker (multiple restriction sites)

d. ori

1. homologous DNA that mediates integration

2. gene used to select chloroplast transformants

3. sequence for replication in *E. coli*

4. site at which DNA can be inserted

15-18 What characteristics in a human pedigree suggest a mitochondrial location for a mutation affecting the trait?

15-19 The first person in the family represented by the pedigree shown here who exhibited symptoms of the mitochondrial disease MERFF was II-2.

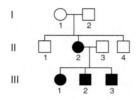

a. What are two possible explanations of why the mother I-1 was unaffected but daughter II-2 was affected?

b. How could you differentiate between the two possible explanations?

15-20 Which of the pedigrees shown could be consistent with each of the following mutation scenarios?

a. Mutation occurred in a somatic cell in an embryo.

b. Mutation occurred during development in I-1 in a precursor to all germ cells.

c. Mutation occurred during development in an egg cell.

15-21 Kearns-Sayre is a disease in which mitochrondrial DNA carries deletions of up to 7.6 kb of the mitochrondrial genome. Although Kearns-Sayre is due to a mitochondrial DNA defect, it does not show maternal inheritance but arises as a new mutation in an individual. The severity of symptoms range from mild to severe and affected people can have defects in some tissues but not in others. How can you explain the variation in tissues affected and severity of symptoms? (Assume that the size of the deletion does not contribute to phenotypic differences.)

15-22 Deletions of various sizes in the mitochondrial genome have been found to increase with age in humans. Which technique would you choose to use to analyze human mtDNA samples for deletions—PCR or gel electrophoresis? Why?

REFERENCES

Estevez, A. M., and Stimpson, L. (1999). Uridine insertion/deletion RNA editing in trypanosome mitochondria—a review. *Gene 240*:247–60.

Gray, M. W., Burger, G., and Lang, B. F. (1999). Mitochondrial Evolution. *Science 283*:1476–81.

Sager, R. (1972). *Cytoplasmic Genes and Organelles.* New York: Academic Press.

Wallace, D. C. (1999). Mitochondrial Diseases in Man and Mouse. *Science 283*:1482–88.

CHAPTER

16

Gene Regulation in Prokaryotes

Lac repressor protein (*violet*) binds to specific sites in the DNA to turn off expression
of the *lac* operon in *E. coli*. Lac repressor is a tetramer with two subunits binding
to each of two operator sites 93 bases apart, causing a loop (*blue* and *green*) to form in the
DNA. This model also shows where the CRP protein (*dark blue*) binds to *lac* DNA.

Among the many types of bacteria that thrive in sewage water is the species *Vibrio cholerae,* the cause of the life-threatening diarrheal disease cholera. The last worldwide cholera pandemic occurred in 1961. Today, cholera is nearly absent from areas of the globe where secure sanitation systems are in place, but epidemics of the disease still devastate human populations in regions where sewage treatment and water purification programs are inadequate or nonexistent.

When a person drinks water contaminated by *V. cholerae,* the bacteria enter the digestive tract. Soon after, the bacteria encounter the "perilous" environment of the stomach, whose acidity kills the majority of them. The bacteria respond to this hostile environment by curtailing production of several proteins they will use later but do not need for passage through the stomach. Only a large initial *V. cholerae* population ensures that at least a small group of cells will survive to exit the stomach and enter the small intestine.

Upon their arrival in the small intestine, these survivors come face to face with a thick mucus that coats their ultimate target—

the intestinal epithelial cells. To penetrate this protective mucous layer, the *V. cholerae* cells navigate by chemotaxis (see the comprehensive example in Chapter 14) and by using flagella (Fig. 16.1). They also make and secrete proteases that ease their passage by degrading the protein component of mucus.

When the bacteria at last reach their destination, they attach to the epithelial cells of the small intestine, stop fabricating flagellin (the chief protein component of flagella), and begin production of several virulence proteins, including a potent toxin that is the actual agent of cholera. Mutant bacteria that produce no toxin do not generate symptoms of disease.

The toxin secreted by cholera bacteria causes chloride ions (Cl^-) to leak from the intestinal cells. To reestablish the osmotic balance, these same cells secrete water. Symptoms of this ionic disruption and fluid flow are watery diarrhea and severe dehydration, which can lead to death within a few hours. The most effective life-saving therapy is oral rehydration: administration of an electrolyte solution consisting of glucose, table salt (NaCl), sodium bicarbonate ($NaHCO_3$), and potassium chloride (KCl) dissolved in purified

551

Figure 16.1 *V. cholerae* **bacteria.** *V. cholerae* invade cells in the intestine.

water. Once toxin production is in full swing, antibiotics without oral rehydration are of little benefit.

The story of *V. cholerae* infection illustrates two key aspects of the life of a unicellular prokaryote: direct contact with the external environment and the ability to respond to changes in that environment by changes in gene expression. *V. cholerae* bacteria that have entered a human host respond to rapid changes in external conditions in part by diminishing or abolishing the production of proteins not required for survival (thereby conserving energy and nutrients) and in part by initiating or increasing synthesis of proteins required in new environments (such as proteases when they contact mucus and toxin and other virulence proteins in the vicinity of intestinal epithelial cells). Such coordinated control of gene expression in a bacterial cell is an example of **prokaryotic gene regulation,** the subject of this chapter.

Prokaryotes regulate gene expression by activating, increasing, diminishing, or preventing the transcription and translation of specific genes or groups of genes. Most metabolic pathways in bacteria require several gene products, and bacterial cells have evolved coordinate controls that enable them to turn the genes encoding those products on and off in concert. Bacteria attune their controls to gene function. With the catabolic pathways that break down complex substances to usable units—for example, lactose to glucose and galactose—the appearance of the substrate (lactose) induces, or turns on, coordinate expression of genes whose protein products will accomplish the substrate's breakdown. With anabolic (biosynthetic) pathways that build needed

compounds—for example, the amino acid tryptophan—the control mechanisms coordinately repress expression of the pathway's multiple genes when the product of the pathway (tryptophan) is present. This means that bacterial cells do not waste energy making unneeded proteins, such as the enzymes in a catabolic pathway whose substrate is absent or the enzymes in an anabolic pathway whose end product is already present in the environment. In fact, many aspects of prokaryotic gene regulation enable bacterial cells to conserve energy by distinguishing housekeeping proteins that are synthesized continuously, irrespective of environmental conditions, from proteins required only in situations signaled by specific cues.

One overarching theme emerges from our discussion. In unicellular organisms like bacteria, the regulatory mechanisms that turn genes on and off in response to environmental conditions enable the organisms to adapt and survive in a constantly changing world.

As we examine the regulation of gene expression in bacteria we describe:

- An overview of prokaryotic gene regulation: There are many steps in gene expression, and regulation can occur at any one of them.

- The regulation of gene transcription: Genetic and molecular studies, particularly of the genes for lactose utilization, show that most regulation affects the initiation of RNA transcripts; "negative regulation" blocks transcription, while "positive regulation" increases it. DNA-binding proteins able to inhibit or enhance the activity of RNA polymerase at the promoter are the main agents of both positive and negative regulation.

- The attenuation of gene expression: Controls regulating the genes of the tryptophan pathway include mechanisms that fine-tune gene expression by prematurely terminating transcription.

- Global regulatory mechanisms: *Escherichia coli*'s response to heat shock is an example of the bacterial ability to coordinate the expression of different sets of genes dispersed around the chromosome. Microarray analysis is an important new tool for detecting changes in gene expression in response to environmental changes, such as a rise in temperature.

- A comprehensive example of how *Vibrio cholerae* regulate their virulence genes.

An Overview of Prokaryotic Gene Regulation

We saw in Chapter 8 that *gene expression* is the production of proteins according to instructions encoded in DNA. During gene expression, the information in DNA is transcribed into RNA, and the RNA message is translated into a string of amino acids.

RNA Polymerase Is the Key Enzyme for Transcription

To begin the process of gene expression in prokaryotes, RNA polymerase transcribes a gene's DNA into RNA. RNA polymerase participates in all three phases of transcription: initiation, elongation, and termination. Initiation requires a special subunit of RNA polymerase—the sigma (σ) subunit—in addition to the

Figure 16.2 Role of RNA polymerase in transcription. The core RNA polymerase enzyme plus sigma factor bind to a promoter sequence to initiate transcription. Once productive copying of the DNA into RNA begins, RNA polymerase moves along the DNA to elongate the transcript, leaving sigma factor behind. Transcription is terminated when rho factor recognizes a sequence on the mRNA and pulls the message away from the enzyme or a stem loop (Rho-independent signal) forms in the mRNA, causing release of the enzyme and message.

two alpha (α), one beta (β), and one β' subunits that make up the core enzyme (Fig. 16.2). When bound to the core enzyme, the σ subunit recognizes and binds specific DNA sequences at the promoter; in its free form, σ does not bind DNA because the σ DNA-binding site is obscured by its own C-terminal tail. The full RNA polymerase—core enzyme plus σ—when bound to the promoter, functions as a complex that both initiates transcription by unwinding the DNA and begins polymerization of bases complementary to the DNA template strand.

The switch from initiation to elongation requires the movement of RNA polymerase away from the promoter and the release of σ. Interaction between the short nascent RNA chain and a site on the polymerase surface probably triggers this release. During elongation, the core RNA polymerase catalyzes the polymerization of ribonucleotides onto the 3' end of the growing chain, eventually forming a complete mRNA. Elongation continues until the RNA polymerase encounters a signal in the RNA sequence that triggers termination (Fig. 16.2). Two types of termination signals are found in prokaryotes: Rho dependent and rho independent. In Rho-dependent termination, a protein factor called Rho recognizes a sequence in the newly transcribed mRNA and terminates transcription by binding to the RNA and pulling it away from the RNA polymerase enzyme. In Rho-independent termination, a sequence of about 20 bases in the RNA, with a run of 6 or more Us at the end, forms a sec-

ondary structure, known as a *stem loop,* that serves as a signal for the release of RNA polymerase from the completed RNA.

Translation in Prokaryotes Starts Before Transcription Ends

Because there is no membrane enclosing the bacterial chromosome, translation of the RNA message into a polypeptide can begin while mRNA is still being transcribed. Ribosomes bind to special initiation sites at the 5' end of the reading RNA frame while transcription of downstream regions of the RNA is still in progress. Signals for the initiation and termination of translation are distinct from signals for the initiation and termination of transcription. Because prokaryotic mRNAs are often polycistronic, that is, contain the information of several genes, ribosomes can initiate translation at several positions along a single mRNA molecule. See Fig. 8.24 for a review of how ribosomes, tRNAs, and translation factors mediate the initiation, elongation, and termination phases of mRNA translation to produce a polypeptide that grows from its N terminus to its C terminus, according to instructions embodied in the sequence of mRNA codons.

The Regulation of Gene Expression Can Occur at Many Steps

Many levels of control determine the amount of a particular polypeptide in a bacterial cell at any one time. Some controls affect an aspect of transcription: the binding of RNA polymerase to the promoter, the shift from transcriptional initiation to elongation, or the release of the mRNA at the termination of transcription. Other controls are posttranscriptional and determine the stability of the mRNA after its synthesis, the efficiency with which ribosomes recognize the various translational initiation sites along the mRNA, or the stability of the polypeptide product.

As we see next, the critical step in the regulation of most bacterial genes is the binding of RNA polymerase to DNA at the promoter. The other potential points of control, while sometimes important in the expression of certain genes, serve more often to fine-tune the amount of protein produced.

The Regulation of Gene Transcription

Researchers delineated the principles of gene regulation in prokaryotes through studies of various metabolic pathways in *Escherichia coli*. In this section, we focus our attention on regulation of the lactose utilization genes in *E. coli* because genetic and molecular experimentation in this system established a fundamental principle of gene regulation: The binding of regulatory proteins to DNA targets controls transcription. The DNA binding of these regulatory proteins either inhibits or enhances the effectiveness of RNA polymerase in initiating transcription. In our discussion, we consider the inhibition of RNA polymerase activity as "negative regulation" and the enhancement of RNA polymerase activity as "positive regulation."

The Utilization of Lactose by *E. coli*: A Model System for Studying Gene Regulation

Proliferating *E. coli* can use any one of several sugars as a source of carbon and energy. If given a choice, however, they prefer glucose. *E. coli* grown in medium containing both glucose and lactose, for example, will deplete the glucose before gearing up to utilize lactose.

Lactose is a complex sugar composed of two monosaccharides: glucose and galactose. A membrane protein, lac permease, transports lactose in the medium into the *E. coli* cell. There, the enzyme β-galactosidase splits the lactose into galactose and glucose (Fig. 16.3).

The Presence of Lactose Induces Expression of the Genes Required for Lactose Utilization

The two proteins, lac permease and β-galactosidase, both required for lactose utilization are present at very low levels in cells grown without lactose. The addition of lactose to the bacterial medium induces a 1000-fold increase in the production of these proteins. The process by which a specific molecule stimulates synthesis of a given protein is known as **induction.** The molecule responsible for stimulating production of the protein is called the **inducer.** In the regulatory system under considera-

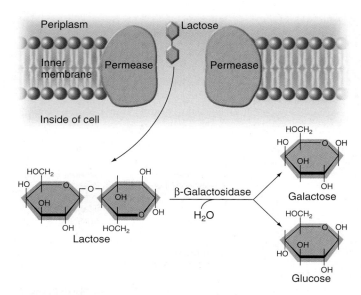

Figure 16.3 Lactose utilization in an *E. coli* cell. Lactose passes through the membranes of the cell via an opening formed by the lactose permease protein. Inside the cell, β-galactosidase splits lactose into galactose and glucose.

tion, lactose modified to a derivative known as allolactose is the inducer of the genes for lactose utilization.

How lactose in the medium induces the simultaneous expression of the proteins required for its utilization was the subject of a major research effort in the 1950s and 1960s—a period some refer to as the golden era of bacterial genetics.

Lactose Utilization in *E. coli*: A Wise Choice for the Study of Gene Regulation

Like Mendel's decision to study seven easily identifiable characteristics in peas, the decision to focus on the induction of genes for lactose utilization in *E. coli* was well considered. The possibility of culturing large numbers of the bacteria made it easy to isolate rare mutants. Once isolated, the mutations responsible for the altered phenotypes could be located by Hfr mapping techniques developed in the 1950s (see Fig. 14.15). Another advantage was that the lactose utilization genes are not essential for survival because the bacteria can grow using glucose as a carbon source. In addition, there is a striking 1000-fold difference between protein levels in induced and uninduced cells. This makes it easy to see the difference between the mutant and wild-type states, and it also allows the identification of mutants that have partial—not just all-or-none—effects.

The ability to measure levels of expression was critical for many of these experiments. To this end, chemists synthesized compounds other than lactose, such as *o*-nitrophenyl-galactoside (ONPG), that could be split by β-galactosidase into products that were easy to assay. One product of ONPG splitting has a yellow color, whose intensity is proportional to the amount of product made, and thus reflects the level of activity of the β-galactosidase enzyme. A spectrophotometer can easily measure the amount of cleaved yellow product in a sample. Other substrates of the β-galactosidase enzyme that produce a color change upon cleavage include X-Gal, whose cleavage produces a blue substance; as we have seen, X-Gal is often used to indicate whether a piece of DNA has been cloned into plasmid vectors containing parts of the β-galactosidase gene (see Fig. 9.8).

We now look at some of the experiments that suggested how the presence of lactose induces expression of the genes required for its own utilization. The techniques designed for these experiments and the ideas that emerged from the results were extremely influential because of their broad application to the study of gene regulation in other systems.

Experiments Analyzing the Behavior of Lactose-Utilization Mutants Reveal the Coordinate Repression and Induction of Three Genes

Jacques Monod (Fig 16.4), a man of diverse interests, was a catalyst for research on the regulation of lactose utilization. A political activist and a chief of French Resistance operations during World War II, he was also a fine musician and esteemed writer on the philosophy of science. Monod led a research effort centered at the Pasteur Institute in Paris, where scientists from around the world came to study enzyme induction. Results from many genetic studies led Monod and his close collaborator François Jacob to propose a model of gene regulation known as the *operon theory*, which suggested that a single signal can simultaneously regulate the expression of several genes that are clustered together on a chromosome and involved in the same process. They reasoned that because these genes form a cluster, they can be transcribed together into a single mRNA, and thus anything that regulates the transcription of this mRNA will affect all the genes in the cluster. Clusters of genes regulated in this way are called *operons*. We first describe key experiments that influenced Jacob and Monod's thinking, as well as data that supported components of their theory, and then summarize the theory itself.

Figure 16.4   Jacques Monod. A key scientist in discovering principles of gene regulation, Jacques Monod was also a talented musician, philosopher, and political activist.

Complementation Analysis of Dozens of Mutants Identifies a Set of Lactose-Utilization Genes

Monod and his collaborators isolated many different Lac⁻ mutants, that is, bacterial cells unable to utilize lactose. Then, using complementation analysis, they showed that the cells' inability to break down lactose resulted from mutations in two genes: *lacZ*, which encodes β-galactosidase, and *lacY*, which encodes lac permease. They also discovered a third *lac* gene, *lacA*, which encodes a transacetylase enzyme that adds an acetyl (CH_3CO) group to lactose and other β-galactoside sugars. Genetic mapping showed that the three genes appear on the bacterial chromosome in a tightly linked cluster, in the order *lacZ-lacY-lacA* (Fig. 16.5). Because the *lacA* gene product is not required for the breakdown of lactose, most studies of lactose utilization have focused on *lacZ* and *lacY*.

Experimental Evidence for a Repressor Protein

Mutations in another gene named *lacI* produce **constitutive** mutants that synthesize β-galactosidase and lac permease even in the absence of lactose. Constitutive mutants synthesize certain enzymes irrespective of environmental conditions. The existence of these constitutive mutants suggested that *lacI* encodes a negative regulator, or **repressor.** Cells would need such a repressor to prevent expression of *lacY* and *lacZ* in the absence of inducer. In constitutive mutants, however, a mutation in the *lacI* gene generates a defect in the repressor protein that prevents it from carrying out this negative regulatory function.

The historic PaJaMo experiment—named after Arthur Pardee (a third collaborator), Jacob, and Monod—provided further evidence that *lacI* indeed encodes this hypothetical negative regulator of the *lac* genes. Hfr matings in which the chromosomal DNA of a donor cell is transferred into a recipient cell served as the basis of the PaJaMo study. The researchers transferred the *lacI⁺* and *lacZ⁺* alleles into a cytoplasm devoid of LacI and LacZ proteins by crossing a *lacI⁺ lacZ⁺* Hfr strain with a *lacI⁻ lacZ⁻ F⁻* strain, in a medium containing no lactose (Fig. 16.6). Shortly after the transfer of the *lacI⁺ lacZ⁺* genes, the researchers detected synthesis of β-galactosidase. Within about an hour, this synthesis stopped.

Pardee, Jacob, and Monod interpreted these results as follows. When the donor DNA is first transferred to the recipient, there is no repressor (LacI protein) in the cytoplasm because the recipient cell's chromosome is *lacI⁻*. In the absence of repressor, the *lacY* and *lacZ* genes are expressed. Over time, however, the host cell begins to make the LacI repressor protein from the *lacI⁺* gene introduced by the Hfr cross.

	lacZ	lacY	lacA	

Figure 16.5   Lactose utilization genes in *E. coli.* Three genes, *lacZ, lacY,* and *lacA,* are involved in lactose metabolism in *E. coli.*

Figure 16.6   The PaJaMo experiment. When DNA carrying *lacI⁺ lacZ⁺* genes was introduced into a *lacI⁻ lacZ⁻* cell, β-galactosidase was synthesized from the introduced *lacZ⁺* gene initially, but as repressor (made from the introduced *lacI⁺* copy of the gene) accumulates, the synthesis of β-galactosidase stops (and only residual β-galactosidase is seen). If inducer is added (*dotted line*), the synthesis of β-galactosidase resumes.

On the basis of the described experiments, Monod and company proposed that the repressor protein prevents further transcription of *lacY* and *lacZ* by binding to a hypothetical site: a DNA sequence near the promoter of the lactose-utilization genes. They suggested that the binding of repressor to this site, termed the **operator site,** blocks the promoter and occurs only when lactose is not present in the medium.

How the Inducer Releases Repression to Trigger Enzyme Synthesis

In the final step of the PaJaMo experiment, the researchers added the lactose inducer to the culture medium. With this addition, the synthesis of β-galactosidase resumed.

Their interpretation of this result was that the inducer binds to the repressor. This binding changes the shape of the repressor so that it can no longer bind to DNA. When the inducer is removed from the environment, the repressor, free of inducer, reverts to its DNA-bindable shape. Proteins that undergo reversible changes in conformation when bound to another molecule are called **allosteric proteins.** The binding of inducer to repressor causes an allosteric effect that abolishes the repressor's ability to bind the operator. In this sequence of events, the inducer is an effector molecule that releases repression without itself binding to the DNA.

In summary, repressor bound to inducer cannot bind to DNA; repressor not bound to inducer can bind to the DNA of the operator. The binding of repressor to the operator keeps RNA polymerase from recognizing the promoter.

The Repressor Has Distinct Binding Domains—One for the Operator and the Other for the Inducer

If the repressor protein interacts with both the operator and the inducer, what outcome would you predict for mutations that disrupt one of these interactions without affecting the other? Biochemical studies showed that the constitutive *lacI⁻* mutations we discussed earlier produce defects in the repressor's ability to bind DNA (Fig. 16.7a). A different type of *lacI* mutant cannot undergo induction. Researchers designate the noninducible mutations as *lacIˢ* or superrepressor (Fig. 16.7b). The *lacIˢ* mutants, although they cannot bind inducer, can still bind to DNA and repress transcription of the operon. The mapping of large numbers of these two types of mutations by DNA sequencing has shown that *lacI⁻* missense mutations, which generate proteins incapable of binding DNA, are clustered in the codons that determine the amino (N) terminus of the repressor, while the *lacIˢ* mutations, which generate proteins incapable of binding inducer, cause amino-acid alterations throughout much of the rest of the repressor. Subsequent structural analyses of the repressor protein confirmed what these mutational analyses suggest: that the repressor protein has at least two separate domains, one that binds to DNA, another that binds inducer (Fig. 16.7c).

To summarize, the repressor has separate regions, or domains, correlating with two functions defined by mutations. Defects in either domain as well as the presence or absence of the inducer can affect the function of the repressor.

Changes in the Operator DNA to Which the Repressor Binds Can Affect Repressor Activity

While mutations in the DNA-binding domain of the repressor can erase repressor activity, mutations that alter the specific nucleotide sequence of the operator recognized by the repressor can have the same effect (Fig. 16.8). When mutations change the nucleotide sequence of the operator, the repressor is unable to recognize and bind to the site; the resulting phenotype is the constitutive synthesis of the lactose-utilization proteins. Researchers have isolated constitutive mutants whose genetic defects map to the *lac* operator site, which is adjacent to the *lacZ* and *lacY* genes. They call the constitutive operator DNA alterations *lacOᶜ* mutations.

How can one distinguish the constitutive operator (*lacOᶜ*) mutants from the previously described constitutive *lacI⁻* mutants, considering that both prevent repression?

Proteins Act in *Trans,* but DNA Sites Act Only in *Cis*

Elements that act **in *trans*** can diffuse through the cytoplasm and act at target DNA sites on any DNA molecule in the cell. Elements that act **in *cis*** can only influence the expression of adjacent genes on the same DNA molecule. Studies of partial diploids constructed using F′ plasmids (see Fig. 14.18) helped distinguish mutations in the operator site (*lacOᶜ*), which act in *cis,* from mutations in *lacI,* which encodes a protein that acts in *trans.*

(a) Repression

lacI⁺ normal repressor — Can bind to operator. Enzymes are not produced.

lacI⁻ mutant repressor — Cannot bind to operator. Enzymes produced constitutively.

(b) Induction

lacI⁺

Inducer

*lacI*ˢ superrepressor mutant

Inducer can bind to repressor and this complex cannot bind to operator. Enzymes are produced.

Inducer cannot bind to repressor. Repressor binds to operon, even when inducer is present. Enzymes are not produced.

(c)

DNA-binding domain

N terminus

Inducer-binding domain

C terminus

Figure 16.7 Mutational and structural analysis of repressor.
(a) In the *lacI⁻* mutants, the repressor cannot bind to the operator site and therefore cannot repress the operon. **(b)** Superrepressor mutants *lacI*ˢ bind to operator but cannot bind inducer, so the repressor cannot be removed from the operator and genes are continually repressed. **(c)** X-ray crystallographic data enables the construction of a model of repressor structure that shows a region to which operator DNA binds and another region to which inducer binds.

Recall from Chapter 14 that F′ factors are plasmids that carry a few chromosomal bacterial genes. When F′ *lac* plasmids are present in a bacterium, the cell has two copies of the lactose-utilization genes—one on the plasmid and one on the bacterial chromosome. Using F′ *lac* plasmids, Monod's group could create bacterial strains with diverse combinations of regulatory

Mutant operator (*oᶜ*)

Nucleotide sequence is changed

Repressor cannot recognize and bind to operator. *lac* enzymes are synthesized constitutively.

Figure 16.8 Operator mutants. The repressor cannot recognize the altered DNA sequence in the *lacOᶜ* mutant site, so it cannot bind and repress the operon. The result: Lactose genes are constitutively expressed.

(*lacOᶜ* and *lacI*) and structural-gene (*lacZ* and *lacY*) mutations. As described later, the phenotype of these partially diploid cells allowed Monod and his collaborators to determine whether particular constitutive mutations were in the genes that produce diffusible, *trans*-acting proteins or at *cis*-acting DNA sites that affect only genes on the same molecule.

In one experiment, Monod and colleagues used a *lacI⁻ Z⁺* bacterial strain that was constitutive for β-galactosidase production because it could not synthesize repressor (Fig. 16.9a). The introduction of an F′ *lacI⁺ Z⁻* plasmid into this strain created a partial diploid that was phenotypically wild type with respect to β-galactosidase expression: both repressible in the absence of lactose and inducible in its presence. Its capacity for repression and subsequent induction indicated that *lacI⁺* is dominant to *lacI⁻*, and that the LacI protein produced from the *lacI⁺* gene on the plasmid can bind to the operator on the bacterial chromosome. Thus, the product of the *lacI* gene is a *trans*-acting protein able to diffuse inside the cell and bind to any operator site it encounters, regardless of the chromosomal location of the operator.

In a second experiment, introduction of a *lacI*ˢ plasmid into a *lacI⁺* strain of bacteria that was both repressible and inducible created bacteria that were still repressible but were no longer inducible (Fig. 16.9b). This was because the mutant LacIˢ repressor, while still able to bind to the operator, could no longer bind inducer. The noninducible repressor was dominant to the wild-type repressor because after a while, the mutant repressor, unable to bind inducer, occupied all the operator sites and blocked all *lac* gene transcription in the cell.

In a third set of experiments, the researchers used *lacI⁺ lacOᶜ lacZ⁺* bacteria that were constitutive for lactose utilization because the wild-type repressor they produced could not bind to the altered operator (Fig. 16.9c). Introduction of an F′ *lacI⁺ lacO⁺ lacZ⁻* plasmid did not change this state of affairs—the cells remained constitutive for β-galactosidase production. The explanation is that the *lacO⁺* operator on the plasmid had no effect on the *lacZ⁺* gene on the chromosome DNA because the operator DNA acts only in *cis*. Since it was able to influence gene expression only of the *lacZ⁻* gene on its own DNA molecule, the wild-type operator on the plasmid could not override

(a) F' *lacI⁺ o⁺ Z⁻* plasmid in *lacI⁻ o⁺ Z⁺* bacteria

lacI⁺ gene encodes a diffusible element that acts in *trans*.

Inducible synthesis

(b) F' *lacIˢ Z⁻* plasmid in *lacI⁺ Z⁺* bacteria

Noninducible—all *o⁺* sites eventually occupied by superrepressor.

(c) F' *lacI⁺ o⁺ lacZ⁻* plasmid in *lacI⁺ oᶜ lacZ⁺* bacteria

Constitutive—presence of *o⁺* in plasmid has no effect on expression of *lacZ⁺* gene in bacterial chromosome.

Figure 16.9   *Trans*-acting proteins and *cis*-acting sites.
(a) LacI⁺ protein acts in *trans*. Repressor protein, made from the *lacI⁺* gene on the plasmid, can diffuse in the cytoplasm and bind to the operator on the chromosome as well as to the operator on the plasmid. **(b)** LacIˢ protein acts in *trans*. The superrepressor encoded by *lacIˢ* on the plasmid diffuses and binds to operators on both the plasmid and chromosome to repress the *lac* operon. **(c)** *lacOᶜ* acts in *cis*. The *lacOᶜ* constitutive mutation affects only the operon of which it is a part. In this cell, only the chromosomal copy will be expressed constitutively. The copy on the plasmid will be inducible.

the mutant chromosomal operator to allow repression of genes on the bacterial chromosome.

A general rule derived from these experiments is that if a gene encodes a diffusible element—usually a protein—that can bind to target sites on any DNA molecule in the cell, whichever allele of the gene is dominant will override any other allele of that gene in the cell (and therefore act in *trans*). If a mutation is *cis*-acting, it affects only the expression of adjacent genes on the same DNA molecule; it does this by altering a DNA site, such as a protein-binding site, rather than by altering a protein-encoding gene.

The *lac* Genes Are Transcribed Together

Many of the experiments that led to an understanding of the *lac* genes focused on the expression of *lacZ* because the level of β-galactosidase is easy to measure. But Monod and coworkers also showed that the repression and induction of *lacY* and *lacA* occur in tandem with the repression and induction of *lacZ*.

Observation of the coordinate expression of the genes for lactose utilization led to the proposal that the three genes are transcribed as part of the same polycistronic mRNA. Although researchers in the 1960s hypothesized that RNA was the intermediate between DNA and protein, they had not yet demonstrated the existence of such an intermediary for any gene. In the 1970s, however, biochemical studies showed that RNA polymerase initiates transcription of the tightly linked *lac* gene cluster from a single promoter. During transcription, the polymerase produces a single polycistronic mRNA containing the *lac* gene information in the order 5'-*lacZ-lacY-lacA*-3'. As a result, mutations in the promoter (which must be located just upstream of *lacZ*) affect the transcription of all three genes.

The Operon Theory

Based on the genetic evidence for three lactose-utilization genes that function as an ensemble, as well as the genetic evidence for a repressor that binds both an operator site in the DNA and an inducer, Monod and Jacob proposed the **operon theory** of gene regulation. Figure 16.10 presents the players in the theory and how they interact to achieve the coordinate regulation of the genes for lactose utilization. As shown, the structural genes (*lacZ, lacY,* and *lacA*) encoding proteins needed for lactose utilization together with two regulatory elements—the promoter (*P*) and the operator (*O*)—make up the **lac operon**: a single DNA unit enabling the simultaneous regulation of the three structural genes in response to environmental changes. Molecules that interact with the operon include the repressor, which binds to the operon's operator, and the inducer, which when present, binds to the repressor and prevents it from binding to the operator.

The 1961 Paper Describing the Operon Theory Was a Landmark

In proposing the operon theory, Jacob and Monod took enzyme induction, which most of their contemporaries considered a biochemical problem, and used genetic analysis to develop a mol-

Feature Figure 16.10

The lactose operon in *E. coli.*

(a)

(b) Repression

a. The players.   The coordination of various elements enables bacteria to utilize lactose in an energy-efficient way. These elements include:

1. A closely linked cluster of three genes—*lacZ*, *lacY*, and *lacA*—that encode the enzymes active in splitting lactose into glucose and galactose.
2. A promoter site, from which RNA polymerase initiates transcription of a polycistronic mRNA. The promoter acts in *cis*, affecting the expression of only downstream structural *lac* genes on the same DNA molecule.
3. A *cis*-acting DNA operator site lying very near the *lac* operon promoter.
   The three structural genes together with the promoter and the operator constitute the *lac operon*: a regulatory unit that allows the *lacZ*, *lacY*, and *lacA* genes to respond to changes in the environment in the same way.
4. A *trans*-acting repressor that can bind to the operator; the repressor is encoded by the *lacI* gene, which is separate from the operon.
5. An inducer that prevents the repressor's binding to the operator. Although early experimenters thought lactose was the inducer, we now know that the inducer is actually allolactose, a molecule derived from and thus related to lactose.

(c) Induction

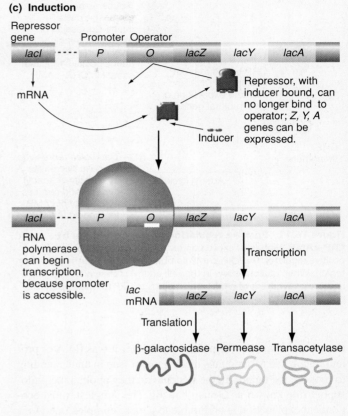

How the players interact to regulate the lactose-utilization genes.

b. Repression
   In the absence of lactose, the repressor binds to the DNA of the operator, and this binding prevents transcription. The repressor thus serves as a negative regulatory element.

c. Induction
1. When lactose is present, allolactose, an inducer derived from the sugar, binds to the repressor. This binding changes the shape of the repressor, making it unable to bind to the operator.
2. With the release of the repressor from the operator, RNA polymerase gains access to the *lac* operon promoter and initiates transcription of the three lactose-utilization genes into a single polycistronic mRNA.

---

ecular model explaining how environmental changes could provoke changes in gene activity. Their theory is remarkable because the authors were working with a very abstract sense of the molecules in the bacterial cell: The Watson-Crick model of DNA structure was only eight years old, mRNA had only recently been identified, and the details of transcription had not

yet been described. In fact, some of the research described earlier in the chapter that confirmed the operon theory followed the 1961 Jacob and Monod publication by years and decades. For example, although Monod was a biochemist with a special interest in allostery and its effects, the repressor itself was a purely conceptual construct; at the time of publication, it had not yet

Figure 16.11 Positive regulation of the *lac* operon by CRP–cAMP. High-level expression of the *lac* operon requires that a positive regulator, the CRP–cAMP complex, be bound to the promoter region. When glucose is low in the cell, adenyl cyclase activity is high and causes synthesis of cAMP, which binds to CRP to make up the positive regulator.

Within the figure:
- *I* P O *lacZ* *lacY* *lacA* DNA
- Inducer is bound to repressor; doesn't inhibit expression of *lac* genes, but high-level expression also needs positive regulator complex, CRP–cAMP.
- Positive regulation: CRP bound to cAMP can bind to DNA and promote expression of *lac* structural genes.
- CRP monomers
- cAMP
- Adenyl cyclase
- ATP
- Catabolite repression: Glucose decreases availability of adenyl cyclase and reduces availability of cAMP for binding to CRP.

been isolated and it was unknown whether it was RNA or protein. Jacob and Monod thus made a major leap in understanding to propose the theory. In their endeavor, they applied the philosophic rule known as Occam's razor: The simplest, most economical explanation is preferable to a more complex one. As is often true in nature, there is a beauty, or elegance, in the simplicity of the solution.

We now know that a key concept of the theory—that proteins bind to DNA to regulate gene expression—holds true for the positive as well as the negative regulation of the *lac* operon. It also applies to many prokaryotic genes outside the *lac* operon and to eukaryotic genes as well. Our next section examines positive regulation of the *lac* operon.

A Positive Control Increases Transcription of *lacZ*, *lacY*, and *lacA*

In focusing on the repression and subsequent induction of the genes for lactose utilization, the Jacob and Monod model did not address a key question: Why do *E. coli* grown in a medium containing both glucose and lactose not initiate *lac* gene transcription? If the lactose is present, why doesn't it act as an inducer? Answers to these questions emerged from molecular studies carried out long after publication of the theory. These studies showed that transcriptional initiation at the *lac* operon is a complex event. In addition to the release of repression, initiation depends on a positive regulator protein that assists RNA polymerase in the start-up of transcription. Without this assist, the polymerase does not open up the double helix very efficiently. As we see next, the presence of glucose indirectly blocks the function of this positive regulator.

The CRP Protein in a Complex with cAMP Is a Positive Regulator of Catabolic Operons

Inside bacterial cells, the small nucleotide known as cAMP (cyclic adenosine monophosphate) binds to a protein called cAMP receptor protein, or CRP (also known as CAP for catabolite activator protein). The binding of cAMP to CRP enables CRP to bind to DNA in the regulatory region of the *lac* operon, and this DNA binding of CRP increases the ability of RNA polymerase to transcribe the *lac* genes (Fig. 16.11). Thus, CRP functions as a positive regulator that enhances the transcriptional activity of RNA polymerase at the *lac* promoter, while cAMP is an effector whose binding to CRP enables CRP to bind to DNA near the promoter and carry out its regulatory function.

Glucose indirectly controls the amount of cAMP in the cell by decreasing the availability of adenyl cyclase, the enzyme that converts ATP into cAMP. Thus, when glucose is present, the level of cAMP remains low; when glucose is absent, cAMP synthesis increases. As a result, when glucose is present in the culture medium, there is little cAMP available to bind to CRP and therefore little induction of the *lac* operon, even if lactose is present in the culture medium. The overall effect of glucose in preventing *lac* gene transcription is known as *catabolite repression*, because the presence of a preferred catabolite (glucose) represses transcription of the operon.

In addition to functioning as a positive regulator of the *lac* operon, the CRP–cAMP complex increases transcription in several other catabolic gene systems, including the *gal* operon (whose protein products help break down the sugar galactose) and the *ara* operon (contributing to the breakdown of the sugar arabinose). As you would expect, these other catabolic operons are also sensitive to the presence of glucose, exhibiting a low level of expression when glucose is present and cAMP is in short supply. Mutations in the gene encoding CRP that alter the DNA-binding domain of the protein reduce transcription of the *lac* and other catabolic operons. The binding of the CRP–cAMP complex is an example of a global regulatory strategy in response to limited glucose in the environment.

Some Positive Regulators Increase Transcription of Genes in Only One Pathway

AraC, for example, is a positive regulatory protein specific for all the arabinose genes responsible for the breakdown of the sugar arabinose. Three arabinose structural genes, *araB, araA,* and *araD,* appearing on the chromosome in that order, constitute an operon (*araBAD*) that is regulated as a single transcription unit. Like the genes for lactose utilization, the arabinose genes are induced when their substrate (arabinose) is present. Evidence that the AraC protein is a positive regulator of the *araBAD* operon came from studies in which *araC⁻* mutants did not express high levels of these three arabinose genes in either the presence or absence of arabinose (Fig. 16.12). (We describe later the mechanisms by which AraC acts.) The mutations were recessive, loss-of-function mutations. When the loss of function of a regulatory protein results in little or no expression of the regulated genes, the protein must be a positive regulator. (By

(a) araC⁺

araO

Ara C

Ara C

RNA polymerase

araI₁ araI₂ Promoter araB araA araD

Transcription of
araBAD

(b) araC⁻

Ara C⁻

Ara C⁻

Promoter araB araA araD

No transcription
of araBAD

Figure 16.12 Loss-of-function mutations show that AraC is a positive regulator. Expression of the arabinose genes in *E. coli* requires the AraC protein to be bound next to the promoter. In an *araC⁻* mutant, the defective protein cannot bind and RNA polymerase will not transcribe the genes.

contrast, loss of function of a negative regulator causes constitutive production of the operon's gene products.)

Summary of How DNA-Binding Proteins Control the Initiation of Transcription at the Lactose and Other Operons

In bacteria, the initiation of transcription by RNA polymerase is under the control of regulatory genes whose products bind to specific DNA sequences in the vicinity of the promoter. The binding of negative regulatory proteins prevents the initiation of transcription; the binding of positive regulators assists the initiation of transcription. Regulation of the *lac* operon depends on at least two proteins: the repressor (a negative regulator) and CRP (a positive regulator). Maximum induction of the *lac* operon occurs in media containing lactose but lacking glucose. Under these conditions, the repressor binds inducer and becomes unable to bind to the operator, while CRP complexed with cAMP binds to a site near the promoter to assist RNA polymerase in the initiation of transcription.

Operons that function in the breakdown of other sugars are also under the control of negative and positive regulators. Transcription of the arabinose operon, for example, which is induced in the presence of arabinose, receives a boost from two positive regulators: the CRP–cAMP complex and AraC.

Thus, proteins that bind to DNA affect RNA polymerase's ability to transcribe a gene. The activity of multiple regulators that respond to different cues increases the range of gene regulation.

Molecular Studies Help Fill in the Details of Control Mechanisms

In 1966, Walter Gilbert and coworkers purified the *lac* repressor protein and determined that it is a tetramer of four *lacI*-encoded subunits, with each subunit containing an inducer-binding do-

main as well as a domain that recognizes and binds to DNA. (Note that we use the term "domains" for the functional parts of proteins but the term "sites" for the DNA sequences with which a protein's DNA-binding domain interacts.) Gilbert and colleagues then used radioactively labeled repressor protein and a bacterial virus DNA that contained the *lac* operon to show that the repressor binds to operator DNA. When they combined the labeled protein and viral DNA and centrifuged the mixture in a glycerol gradient, the radioactive protein cosedimented with the DNA (Fig. 16.13). If the viral DNA contained a *lac* operon that had an *lacOᶜ* mutation, the protein did not cosediment with the DNA, because it could not bind to it. Subsequent sequence analysis of the isolated DNA revealed that the *lac* operator is about 26 bp in length, and it includes the first nucleotides used as a template for the mRNA.

With the development of cloning, DNA sequencing, and techniques for analyzing protein-DNA interactions in the 1970s, researchers increased their ability to isolate specific macromolecules, determine the structure of each molecule, and analyze the interactions between molecules. This new experimental potential revolutionized the study of genetics even though researchers still used mutants isolated in earlier genetic analyses as the bases of many of their studies. The new generation of gene-regulation analyses confirmed the basic principles of the operon theory and elucidated compelling details of their components. Such mutants continue to be valuable research tools today.

(a)

*

Radioactively labeled
repressor protein

+

lacO⁺

Bacterial virus DNA

Mix and sediment
in glycerol gradient

* — DNA + radioactivity
(repressor protein)

(b)

*

Radioactively labeled
repressor protein

+

lacOᶜ — Mutated *lac* operon

Bacterial virus DNA

Mix and sediment
in glycerol gradient

— DNA (no radioactivity present)

Figure 16.13 The *lac* repressor binds to operator DNA. A radioactive tag is attached to the *lac* repressor protein so it can be followed in the experiment. **(a)** When repressor protein from *lacI⁺* cells was purified and mixed with DNA containing the *lac* operator (on bacterial virus DNA), the protein cosedimented with the DNA. **(b)** When wild-type repressor was mixed with DNA containing a mutant operator site, no radioactivity sedimented with the DNA.

Many DNA-Binding Proteins Contain a Helix-Turn-Helix Motif

It is now possible to predict a protein's secondary structures—such as α helices and β sheets—by comparing the amino-acid sequence of a newly isolated protein with sequences of proteins whose secondary structures have already been determined by X-ray crystallography. Several of the polypeptides that make up repressor proteins, including the subunits of the *lac* repressor, have the identifiable feature of two α-helical regions separated by a turn in the protein structure. This helix-turn-helix (HTH) motif in the protein fits well into the major groove of the DNA. One of the α helices in an HTH carries amino acids that recognize and interact with a specific DNA sequence of nucleotides; thus each HTH has a specificity for DNA binding based on its sequence of amino acids (Fig. 6.14a).

Mark Ptashne and colleagues used cloned DNA to construct made-to-order mutations in the gene encoding the repressor of a bacterial virus known as 434. (This bacterial virus is similar to the temperate bacteriophage λ, which infects *E. coli*, and bacteriophage P22, which infects *Salmonella enterica*.) The 434 repressor binds to DNA of the 434 viral DNA that has integrated into the bacterial genome and prevents transcription and production of viral particles. After predicting that a region of the α helix of the 434 repressor recognizes the DNA of its specific operator site, Ptashne and coworkers altered the DNA sequence of the gene region encoding this α helix so that it now encoded most of the amino acids in the corresponding α helix of the repressor for another bacterial virus known as P22 (Fig. 16.14b). The resulting hybrid 434-P22 repressor protein, encoded by the altered gene, contained a P22 α helix that recognized the P22 operator *in vivo*. Binding of the hybrid repressor to the P22 prophage DNA operator region that had integrated into the bacterial host genome shut down transcription of most P22 viral proteins and prevented subsequent infection by the P22 virus. This experiment showed that specific amino acids in this α he-

lix determine the binding specificity of the repressor protein. The P22-like α helix in the hybrid protein is sufficient to convert the binding specificity of the 434 repressor to that of the P22 repressor.

The HTH motif is found in hundreds of DNA-binding proteins. Surprisingly, more than 20 different DNA-binding proteins in bacteria are very similar to the LacI repressor, not only in the HTH DNA-binding domain but throughout much of the protein. This group of repressors is known as the LacI repressor family of proteins. Their structural similarity suggests that they evolved from a common ancestral gene whose duplication and divergence produced a family of transcriptional repressor proteins with similar overall structures but unique recognition regions. The uniqueness of their DNA-recognition regions means that they interact with different operators to regulate different groups of genes.

Most Regulatory Proteins Are Oligomeric and Contain More Than One Binding Domain

In the 1970s, geneticists studying gene regulation developed new *in vitro* techniques to determine where regulatory proteins bind to the DNA. Purified proteins that bind to fragments of DNA protect the region of the double helix to which they bind from digestion by enzymes such as DNase I that break the phosphodiester bonds between nucleotides. If a sample of DNA, labeled at one end of one strand and bound by a purified protein, is partially digested with DNase I, the enzyme will cleave phosphodiester bonds in at least some DNA molecules in the sample, except for those phosphodiester bonds that are in regions protected by the binding protein. After the DNase I and binding proteins are removed and the DNA is denatured into single strands, gel electrophoresis of the DNA followed by autoradiography of the gel will reveal bands at positions corresponding to the cleavage at each phosphodiester bond, except in the region where bound protein protected the DNA. Portions of the

Figure 16.14 DNA recognition sequences in a helix-turn-helix section of a protein. (a) A protein motif that has the shape of a helix-turn-helix (*helices shown here inside a cylindrical shape*) fits into the major groove of the DNA helix. Specific amino acids within the helical region of the protein recognize a particular base sequence in the DNA. **(b)** The amino acids inside the recognition helix for phages 434 and P22 and for the hybrid 434-P22 repressor. The amino acids shown in *red* in the hybrid repressor helix section are ones that were modified to be like those of the P22 repressor.

gel without bands are thus "footprints," indicating the nucleotides of the DNA fragment that were protected by the DNA-binding protein (Fig. 16.15a).

When analyzed in this way, many of the DNA sequences to which a negative or positive regulator protein binds exhibit rotational symmetry; that is, their DNA strands have a similarity of sequence when read in the 5′-to-3′ direction (although these sequences are usually not perfect palindromes). An example of such symmetry is in the *lac* operon's CAP-binding site whose sequence is

5′ GTGAGTTAGCTCAC 3′

3′ CACTCAATCGAGTG 5′

Most regulatory proteins that bind to DNA exist as oligomers composed of two to four polypeptide subunits. The regulatory proteins, which are present in very low numbers in the cell, gain an advantage from this multimeric form: Since each polypeptide subunit of an oligomer has a DNA-binding domain,

an assembled oligomer has multiple DNA-binding domains. If the sites to which an oligomeric protein can bind are clustered in a gene's regulatory region, many contacts can be established between the protein and the regulatory region. By increasing the stability of protein-DNA interactions, these multiple binding domains collectively produce the strength of binding necessary to maintain repression or activate transcription.

The CRP protein binds to its DNA site (for example, the site in the *lac* operon just shown) as a dimer at a sequence with rotational symmetry, with one monomer of CRP binding to each side of the sequence (Fig. 16.15b). Thus, the DNA site that is referred to as the CRP-binding site actually consists of two recognition sequences, each able to bind one subunit of the CRP dimer.

The *lac* repressor exists as a tetramer with each of its four subunits containing a DNA-binding HTH motif. This tetramer binds to two operators located far apart on the DNA, with each operator containing two recognition sequences. The binding of the tetrameric repressor to the two operators causes a loop of

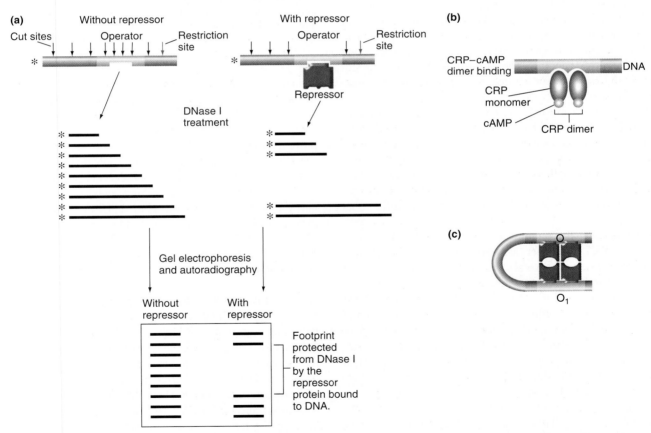

Figure 16.15 **Proteins binding to DNA.** **(a)** DNase footprint establishes the region to which a protein binds. A partial digestion with DNase I produces a series of fragments. If a protein is bound to DNA, DNase cannot digest at sites covered by the protein. Gel electrophoresis of digested products shows which products were not generated and indicates where the protein binds. **(b)** CRP–cAMP binds as a dimer to a regulatory region. **(c)** The Lac repressor is a tetrameric protein. For simplicity we previously showed a single repressor object binding to one operator site, but in reality there are two identical LacI subunits that bind to each operator site. Two of the subunits bind to the sequence in one operator site (O_1), and the other two subunits bind to a second operator (either O_2 or O_3).

DNA to form between the two operator sites (Fig. 16.15c); formation of the loop, in turn, facilitates the two-position binding. There are actually three operator sites in the *lac* operon to which the repressor can bind: O_1 (the site originally identified by *lacO^c* mutations), O_2, and O_3. Site O_1 has the strongest binding affinity for the repressor, and two subunits of the tetramer always bind at this site. The other two subunits bind at either O_2 or O_3. The distance between operator sites—multiples of 10 bases—allows repressor binding to the same side of the helix and thus formation of the loop. Mutations in *either* O_2 *or* O_3 have very little effect on repression. By contrast, mutations in *both* O_2 *and* O_3 make repression 50 times less effective. The conclusion is that for maximal repression, all four of the repressor's subunits must bind DNA simultaneously. Binding at four recognition sequences (in two operator sites) increases the stability of the protein-DNA interactions. In fact, the DNA binding of the *lac* repressor is so efficient that only 10 repressor tetramers per cell are sufficient to maintain repression.

The Looping of DNA Is a Common Feature of Regulatory Systems

We have seen that the DNA binding of the *lac* repressor tetramer to two distant sites on the DNA results in formation of a loop. This ability to form a loop shows that DNA is not a static, stiff molecule. It has a flexibility that enables it to bend, and this potential for bending enables all four DNA-binding domains of the *lac* repressor tetramer to work in tandem for maximal repression.

Looping first came to light in work on AraC, the regulatory protein that helps control the arabinose operon described previously. AraC functions as a dimer. As we have seen, in the presence of the inducer arabinose, AraC is a positive regulator that helps initiate transcription of the *araBAD* operon. Unexpectedly, in the absence of arabinose, AraC acts as a repressor. In this capacity, the AraC dimer binds to two sites—*araO* and *araI*—that are 194 nucleotide pairs apart (Fig. 16.16a). In one set of experiments analyzing the binding to two sites with concomitant looping of DNA, researchers altered the distance between *araO* and *araI* by inserting several base pairs. The introduction of 11 or 31 bp—alterations that are close to integral changes in the number of turns of the double helix (a full turn of the helix = 10.5 bp)—had little effect on repression. The introduction of 5, 15, or 24 bp, however, noticeably reduced repression. These results suggest that the orientation of the binding sites is a significant variable; only when two sites have an orientation that puts them on the same side of the helix can a dimer bind simultaneously to both and cause formation of a DNA loop.

What enables the AraC dimer to function as both an activator and a repressor? The answer most likely involves allostery and AraC's different binding affinities for recognition sequences at *araI* and *araO*. The *araI* site contains two recognition sequences (*araI_1* and *araI_2*) to which AraC can bind; the *araO* site contains one recognition sequence. The AraC dimer is an allosteric protein whose structure changes with the binding of inducer (arabinose). One hypothesis of how this allosteric

(a) No arabinose present

AraC dimer binding at both *araO* and *araI* sites. No *araBAD* genes are transcribed.

(b) Arabinose present

Figure 16.16 How AraC acts as both a repressor and an activator. The AraC protein can bind to sites *araI_1*, *araI_2*, and *araO*. **(a)** When no arabinose is present, the binding of *araC* to *araO* and the *araI_1* sites causes looping of the DNA and prevents RNA polymerase from transcribing the genes. **(b)** When arabinose (inducer) is present, it binds to *araI_1* and *araI_2* but not to *araO*. RNA polymerase interacts with *araC* at the *araI* sites and transcribes the genes.

change alters function is that in the absence of arabinose, the size and shape of the protein unbound to inducer allow the AraC dimer to bind to two recognition sequences—*araO* and one of the two sites within *araI*—at the same time; this double binding prevents AraC from binding in a way that would enable it to assist RNA polymerase in the initiation of transcription. When arabinose is present, binding to the inducer changes the shape of the AraC dimer. In this inducer-bound conformation, the regulatory molecule does not bind to *araI* and *araO* at the same time; instead, the inducer–dimer complex binds exclusively to recognition sequences in *araI* (Fig. 16.16b). When bound to DNA at only this site, AraC's positive regulatory domain is free to interact with RNA polymerase and increase transcription.

How Regulatory Proteins Interact with RNA Polymerase

Many negative regulators, such as the *lac* repressor, prevent initiation by physically blocking the DNA-binding site of RNA polymerase. For example, the O_1 operator site to which the *lac* repressor tightly binds consists of 27 nucleotides centered 11 bp downstream from the transcriptional start site. The operator thus includes part of the region where RNA polymerase has to bind to initiate transcription (Fig. 16.17a). When repressor is bound to the operator, its presence on the DNA prevents the binding of RNA polymerase.

Positive regulators, by contrast, usually establish a physical contact with RNA polymerase that enhances the enzyme's ability to initiate transcription (Fig. 16.17b). For several positive regulators, researchers have identified points of contact between the regulator and the α, β, or σ subunits of RNA polymerase.

Figure 16.17 How regulatory proteins interact with RNA polymerase. (a) The *lac* repressor bound to the operator prevents RNA polymerase from binding. The binding sites for RNA polymerase and repressor (determined by DNase digestion experiments) show that there is overlap between the two sites. **(b)** The CRP–cAMP complex contacts RNA polymerase directly to help in transcription initiation.

Although RNA polymerase will bind to a promoter in the absence of a positive regulator, it is less likely to unwind DNA and begin polymerization than when it receives assistance from a positive regulator.

Using the *lacZ* Gene As a Reporter of Gene Expression

The extensive molecular knowledge of *lacZ* has enabled its use as a "reporter" gene in the study of a large variety of regulatory regions in both prokaryotes and eukaryotes. A **reporter gene** is a protein-encoding gene whose expression in the cell is quantifiable by sensitive and reliable techniques of protein detection. Fusion of the coding region of the reporter gene to *cis*-acting regulatory regions (including promoters and operators) of other genes creates a DNA construct that enables researchers to assess the activity of the regulatory elements by monitoring the amount of reporter gene product appearing in the cell. For example, with the fusion of gene *X*'s regulatory region to the *lacZ* gene, one can assess the activity of the regulatory elements of gene *X* by monitoring the level of β-galactosidase expression. With this construct, conditions that induce expression of gene *X* will generate β-galactosidase (Fig. 16.18a). The use of reporter genes makes it possible to identify the DNA sites necessary for regulation as well as the genes and signals involved in that regulation. For example, you could mutate gene *X*'s control region *in vitro,* and then transform the construct back into bacterial cells and look for mutations that disrupt a particular aspect of control (as measured by levels of *lacZ*); this protocol would identify *cis*-acting sites important for the regulation. Or, you could mutate the bacteria themselves, and then introduce a reporter construct into the mutated cells and look for changes in level of *lacZ* expression as measured by blue or white colony color; this protocol would identify *trans*-acting genes.

Reporter constructs not only provide the basis for analyzing the regulation of one specific gene; they also make it possible to identify many genes regulated by the same stimulus. To this end, researchers can use transposition to insert the *lacZ* gene without its regulatory region at various sites around the bacterial chromosome. For example, they can engineer the bacteriophage Mu to contain *lacZ* inside its genome and then infect a population of *E. coli* cells with the specially engineered phage. The phage will insert at sites around the bacterial chromosome at random, generating a collection of *E. coli* cells, some of which contain *lacZ* fused to genes (and their regulatory regions). On exposure of this collection to a stimulus such as UV light, the cells containing *lacZ* fused to the regulatory regions of genes induced by UV light will produce β-galactosidase (Fig. 16.18b). Researchers identified a set of genes activated by exposure to DNA-damaging agents by this method.

In addition to using *lacZ* as a reporter gene, geneticists studying gene regulation can use their extensive knowledge of the *lac* operon regulatory region to construct recombinant molecules carrying genes whose expression can be controlled. For example, by fusing the *lac* operon control DNA to a human gene expressed in *E. coli,* they could cause overproduction of the human protein in response to an induction cue. The ability to control expression of a foreign gene is important because it provides a way to ensure that protein production is not turned on until cells have proliferated to a high density. The culture will thus contain many cells making the desired foreign protein; and even if the protein has deleterious effects on the growth of *E. coli,* the culture can still grow to high density before addition of the inducer. The production in *E. coli* of human proteins, such as human growth hormone, human insulin, and other pharmacologically useful proteins, is based on this strategy (Fig. 16.18c).

(a) Fusion used to perform genetic studies of the regulatory region of gene *X*.

cis-acting gene *X* regulatory region

lacZ (Reporter gene)

Transcription ——————→

lacZ expression is under control of gene *X* regulatory region.

Translation

β-galactosidase

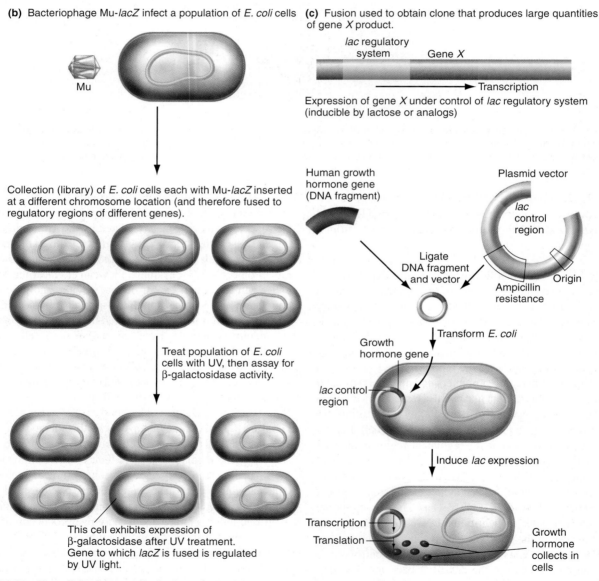

(b) Bacteriophage Mu-*lacZ* infect a population of *E. coli* cells

Mu

Collection (library) of *E. coli* cells each with Mu-*lacZ* inserted at a different chromosome location (and therefore fused to regulatory regions of different genes).

Treat population of *E. coli* cells with UV, then assay for β-galactosidase activity.

This cell exhibits expression of β-galactosidase after UV treatment. Gene to which *lacZ* is fused is regulated by UV light.

(c) Fusion used to obtain clone that produces large quantities of gene *X* product.

lac regulatory system

Gene *X*

Transcription ——————→

Expression of gene *X* under control of *lac* regulatory system (inducible by lactose or analogs)

Human growth hormone gene (DNA fragment)

Plasmid vector

lac control region

Ligate DNA fragment and vector

Ampicillin resistance

Origin

Transform *E. coli*

Growth hormone gene

lac control region

Induce *lac* expression

Transcription

Translation

Growth hormone collects in cells

Figure 16.18   Use of the *lac* gene fusions to study gene regulation. (a) The *lacZ* structural gene can be fused to a regulatory region of gene *X*. Expression of β-galactosidase will be dependent on signals in the regulatory region to which *lacZ* is fused. **(b)** Creating a collection of *lacZ* insertions in the chromosome. Bacteriophage Mu containing the *lacZ* gene without its promoter integrates randomly around the chromosome. If the Mu-*lacZ* integrates within a gene in the orientation of transcription, *lacZ* expression will be controlled by that gene's regulatory region. The library of clones created can be screened to identify insertions in genes regulated by a common signal. **(c)** Use of fusions to overproduce a gene product. The *lac* regulatory region can be fused to gene *X* to control expression of genes. The gene encoding human growth hormone is cloned next to the *lac* control region and transformed into *E. coli*. Conditions that induce *lac* expression will cause expression of growth hormone that can be purified from the cells.

The Attenuation of Gene Expression: Fine-Tuning of the *trp* Operon Through the Termination of Transcription

In bacteria, the multiple genes of both catabolic and anabolic pathways are clustered together and coregulated in operons. We have seen that regulators of the catabolic *lac* and *ara* operons respond to the presence of lactose or arabinose, respectively, by inducing gene expression. By contrast, regulators of anabolic operons respond to the presence of the pathway's end product by shutting down the genes of the operon that encode proteins needed to manufacture the end product.

There are many anabolic bacterial operons involved in the synthesis of amino acids. A well-studied example is the *E. coli* tryptophan (*trp*) operon, a group of five structural genes—*trpE, trpD, trpC, trpB,* and *trpA*—required for construction of the amino acid tryptophan. Maximal expression of the *trp* genes occurs when tryptophan is absent from the growth medium.

The Presence of Tryptophan Activates a Repressor of the *trp* Operon

The *trp* operon is regulated by a protein repressor that is the product of the *trpR* gene. In contrast to the *lac* operon, where lactose functions as an inducer that prevents the repressor from binding to the operator, tryptophan functions as a **corepressor:** an effector molecule whose binding to the actual TrpR repressor protein allows the negative regulator to bind to DNA and inhibit transcription of the genes in the operon. The binding of tryptophan to the TrpR repressor causes an allosteric alteration in the repressor's shape, and only with this alteration can the TrpR protein bind to the operator site (Fig. 16.19a). Mutations in the *trpR* gene that change either the protein's tryptophan-binding domain or its DNA-binding domain destroy the TrpR repressor's ability to bind DNA, and they result in the expression of the *trp* genes even when tryptophan is present in the growth medium. The TrpR-mediated repression of the *trp* operon is critical, but it is only one of the regulatory components controlling expression of the *trp* genes in *E. coli*.

The Termination of Transcription Fine-Tunes Regulation of the *trp* Operon

One would expect *trpR⁻* mutants to show constitutive expression of their *trp* genes. With or without tryptophan in the medium, if there is no repressor to bind at the operator, RNA polymerase would have uninterrupted access to the *trp* promoter. Surprisingly, studies show that the *trp* genes of *trpR⁻* mutants are not completely de-repressed (that is, turned on) when tryptophan is present in the growth medium. As Table 16.1 shows, the removal of tryptophan from a medium in which *trpR⁻* mutants are growing causes expression of the *trp* genes to increase threefold. What control mechanism is responsible for this repressor-independent change in *trp* operon expression?

(a) **Tryptophan present**

(b) **Tryptophan not present**

Figure 16.19 Tryptophan acts as a corepressor. (a) When tryptophan is available, it binds to the *trp* repressor causing the molecule to change shape so that the repressor can bind to the operator of the *trp* operon and repress transcription. **(b)** When tryptophan is not available, the repressor cannot bind to the operator and the tryptophan biosynthetic genes are expressed.

TABLE 16.1	**Expression of *trp* Operon in *trpR⁺* and *trpR⁻* Strains**	
	With Tryptophan* (%)	**Without Tryptophan*** (%)
trpR⁺	8	100
trpR⁻	33	100

*In the growth medium.

Two Alternative Transcripts Lead to Different Transcriptional Outcomes

In a series of elegant experiments analyzing transcription of the *trp* operon, Charles Yanofsky and coworkers found that initiation at the *trp* promoter can produce two alternative transcripts (Fig. 16.20a). Sometimes initiation at the promoter leads to transcription of a truncated mRNA about 140 bases long from a short DNA region immediately preceding the first *trp* structural gene (*trpE*); this pregene DNA region is called a **leader sequence** and the RNA transcribed from it is the RNA leader. At other times, transcription continues beyond the end of the leader sequence to produce a full operon-length transcript. In analyzing

(a)

Alternate stem-loop structures

Figure 16.20 **Attenuation in the tryptophan operon of** ***E. coli.*** **(a)** Stem loops form by complementary base pairing in the *trp* leader RNA. Two different secondary structures are possible in the mRNA from the *trp* operon. Stem loops using complementary base pairs between regions 1 and 2 will enable the formation also of the stem-loop 3–4, which is a termination signal for RNA polymerase. Base pairing between regions 2 and 3 leads to a stem loop that prevents formation of stem-loop 3–4. In the early portion of the transcript there are two codons for tryptophan. **(b)** When tryptophan is present, the ribosome follows quickly along the transcript, preventing stem-loop 2–3 from forming. Stem-loop 3–4 can form and transcription is terminated. **(c)** If tryptophan is absent, the ribosome stalls at the *trp* codons, allowing formation of stem-loop 2–3, preventing stem-loop 3–4 formation. Transcription continues.

why some mRNAs terminate before they can transcribe the structural *trp* genes, while others do not, the researchers discovered **attenuation:** control of gene expression by premature termination of transcription. Whether or not transcription terminates prematurely depends on how the translation machinery reads the secondary structure of the RNA leader.

That RNA leader can fold into two different stable conformations, each one based on the complementarity of bases in the same molecule of RNA. The first structure contains two stem-loop structures with regions 1 and 2 associated by base pairing and regions 3 and 4 similarly associated by base pairing. When the transcriptional machinery "sees" the 3–4 stem-loop configuration, which has seven Us at the end, it stops transcription, producing a short, "attenuated" RNA. The alternative RNA structure forms by base pairing between regions 2 and 3. In this conformation, the leader RNA does not display the 3–4 stem-loop termination signal, and as a result, the transcription machinery reads right through it to produce a full-length transcript that includes the structural-gene sequences.

The Simultaneous Translation of a Small Peptide in the Leader Determines Which Outcome Occurs

The early translation of a short portion of the RNA leader (while transcription of the rest of the leader is still taking place) determines which of the two alternative RNA structures forms. That key portion of the RNA leader includes a short open reading frame containing 14 codons, 2 of which are *trp* codons (Fig. 16.20b). When tryptophan is present, the ribosome moves quickly past the *trp* codons in the RNA leader and proceeds to the end of the leader's codons, allowing formation of the stem-loop 3–4 structure and preventing formation of the alternative RNA structure. As we have seen, this 3–4 RNA structure causes the termination of transcription and hence the attenuation (that is, lessening) of *trp* gene expression. In the absence of tryptophan, the ribosome stalls at the two *trp* codons in the RNA leader because of the lack of charged tRNA^{Trp} in the cell. The 2–3 stem-loop structure depicted in Fig. 16.20c is then able to form, and its formation prevents formation of the 3–4 stem-loop RNA structure recognized by the transcriptional terminator. As a result, transcription proceeds through the leader into the structural genes.

How do we know these secondary RNA structures exist *in vivo* and that the translation of the leader RNA plays a significant role in their formation? Several experiments support this model of attenuation. First, deletion of the complete leader sequence results in the loss of control by attenuation; in *trpR*$^-$ mutants that also do not contain the leader sequence, there is no difference in *trp* expression with or without tryptophan in the medium. This double mutant makes the *trp* biosynthetic enzymes constitutively at maximal levels. Second, mutations that weaken the stems of the RNA stem-loop secondary structures alter regulation, but they can be compensated for by a second-site mutation that restores base pairing. Third, mutations that change the RNA terminator structure increase the readthrough of transcription and thus enhance gene expression. Finally, mutations of the translation-initiating AUG codon that prevent translation of the leader sequence produce an increase in the amount of short transcript, as predicted from the model.

Why has such a complex system evolved in the regulation of the *trp* operon and other biosynthetic pathways? While the TrpR repressor shuts off transcription in the presence of tryptophan and allows it in the amino acid's absence, the attenuation mechanism provides a way to fine-tune this off/on switch. It allows the cell to sense the level of tryptophan by "reading" the level of charged tRNA^{Trp} and to adjust the level of *trp* mRNA accordingly. In *E. coli,* systems for regulation by attenuation similar to that observed for tryptophan exist for several other amino-acid biosynthetic operons, including histidine, phenylalanine, threonine, and leucine.

The attenuation mechanism is unique to prokaryotes because only in cells without a membrane-enclosed nucleus can the expression machinery couple transcription and translation. The opportunity for some aspect of the translational apparatus to directly affect the outcome of transcription does not exist in eukaryotes.

Global Regulatory Mechanisms Coordinate the Expression of Many Sets of Genes

Dramatic shifts in environmental conditions can trigger the expression of sets of genes or operons dispersed around the chromosome. The absence of glucose, we have already seen, increases the expression of several catabolic operons that are at least partially controlled by a common factor, the CRP–cAMP complex. Another example of such global regulation is *E. coli*'s response to heat shock, which results from exposure to extremely high temperatures (up to 45°C).

An Alternative Sigma (σ) Factor Mediates *E. coli's* Global Response to Heat Shock

At high temperatures, most proteins denature or aggregate, or both. In *E. coli,* exposure to high temperature induces the expression of several proteins that alleviate heat-shock-related damage. The induced proteins include those that recognize and degrade aberrant proteins as well as so-called *chaperone proteins,* which assist in the refolding of other proteins and also prevent their aggregation.

E. coli's induction of the proteins that combat heat shock is a highly conserved stress response. Organisms as different as bacteria, flies, and plants induce similar proteins, notably the chaperones, in response to high temperatures.

Studies of Mutants Unable to Generate a Heat-Shock Response Helped Reveal a Mechanism of Global Regulation

Conditional lethal *E. coli* mutants in which high temperatures do not induce transcription of the heat-shock genes provided critical evidence for the global regulatory mechanism. These conditional lethal mutants have a defect in the *rpoH* gene that encodes an alternate RNA polymerase sigma factor known as σ^{32}. The normal housekeeping sigma factor, σ^{70}, is active in the cell under normal physiological conditions. By contrast, the alternative σ^{32} can function at high temperatures; it also recognizes different promoter sequences than those recognized by σ^{70}. Genes induced by heat shock contain nucleotide sequences in their promoters that are recognized by σ^{32} (Fig. 16.21a). σ^{32} mediates the heat-shock response by binding to the core RNA polymerase and thereby allowing the polymerase to initiate transcription of the genes encoding the heat-shock proteins. The RNA polymerase σ^{32} holoenzyme is relatively resistant to heat inactivation compared to the heat-sensitive σ^{70}-dependent RNA polymerase. As a result, when temperatures rise, genes with a σ^{32} promoter undergo transcription, while genes with a σ^{70} promoter do not.

Levels of the σ^{32} protein and the σ^{32} RNA polymerase holoenzyme increase immediately after heat shock. Several factors cause this increase in σ^{32} activity, including:

■ An increase in the transcription of the *rpoH* gene.

■ An increase in the translation of σ^{32} mRNA stemming from greater stability of the *rpoH* mRNA.

■ An increase in the stability and activity of the σ^{32} protein. Chaperones DnaJ/K bind to and inhibit σ^{32} under normal physiological conditions. When the temperature rises, these proteins bind to the large number of cellular proteins that become denatured, leaving σ^{32} free to associate with RNA polymerase.

(a)

σ^{70} recognizes this promoter sequence

σ^{32} recognizes this promoter sequence

(b) At high temperatures σ^{24} recognizes different promoter sequence on *rpoH* gene and σ^{32} is transcribed

Figure 16.21 Alternate sigma factors. (a) Base sequences recognized by σ^{32} and σ^{70}. (The *N* indicates that any base can be found at this position.) **(b)** At high temperature, the *rpoH* gene (encoding σ^{32}) is transcribed. The σ^{32} interacts with RNA polymerase and transcribes the heat-shock genes.

Genetics and Society

Nitrogen Fixation Depends on Many Levels of Gene Regulation

Nitrogen, an essential component of amino acids, chlorophylls, and nucleic acids, is a growth-limiting plant nutrient—the more nitrogen available, the faster most plants grow. However, although gaseous nitrogen (N_2) makes up 78% of earth's atmosphere, plants cannot use nitrogen in this form. They can use only nitrogen that has been *fixed,* that is, converted to ammonia (NH_3) or another nitrogen-containing compound.

Several types of bacteria are agents of nitrogen fixation. These bacteria may be free-living cells (such as those in the genus *Klebsiella*) or plant symbionts (such as those in the genus *Rhizobium*). Of the many symbiotic rhizobial species, each one is able to form a working relationship with only one or a few plants, mainly legumes like peas, beans, and alfalfa. For hundreds of centuries, without understanding the microbiological and molecular bases of nitrogen fixation, farmers made use of the nitrogen-fixing abilities of rhizobial bacteria via the rotation of crops, observing that if they planted alfalfa one year, the next year's wheat crop, for example, would be much more abundant. Here we describe a few aspects of the multilayered regulatory mechanisms controlling the nitrogen-fixation activities of *Rhizobium meliloti*, a small, heterotrophic, rhizobial bacterium that lives in the soil and acquires a source of energy and amino acids through the induction of root nodules in alfalfa.

Bacteria-Plant Interactions Lead to Nitrogen Fixation
In the symbiotic relationship that develops between *R. meliloti* and alfalfa, bacterial genes produce the enzymatic machinery for nitrogen fixation while the plants provide a low-oxygen environment that allows the nitrogen-fixation enzymes to function. Achievement of the symbiotic state depends on a series of communications between plant and bacteria that lead to dramatic changes in both the anatomy of the plant and the structure of the bacteria.

Alfalfa's secretion of flavenoids triggers the events leading to nitrogen fixation. *R. meliloti* responds to this environmental signal by expressing *nodulation* (*nod*) genes whose protein products are enzymes active in the synthesis of liposaccharides known as Nod factors. Different Nod factors determine the specificity of interaction between different bacterial species and the plants with which they establish symbiosis. Release of the Nod factors from the rhizobial cells elicits a curling of root hairs and cell division in the meristem, which lead to the formation of root nodules in the alfalfa plant (Fig. A). The bacteria now navigate by chemotaxis to the alfalfa host nodules and penetrate the host's root cortex with the help of a gelatinous filament secreted by the plant itself. Once inside the plant, the *R. meliloti* enter root cells, where they divide and differentiate into *bacteroids,* cells that produce nitrogenase, an enzyme complex that catalyzes the conversion of N_2 to NH_3. The host plant monitors the concentration of oxygen in the area of the nodule where the bacteroids thrive to ensure that it is much lower than in the surrounding plant cells or soil. A nearly anaerobic environment in the nodule is crucial to the survival and function of nitrogenase, which is extremely sensitive to oxygen and becomes irreversibly inactivated in air. The alfalfa plant uses the nitrogen fixed by *R. meliloti* as its source of nitrogen and, in return, provides the bacteria with photosynthetic products and amino acids.

The Genetic Components and Mechanisms That Mediate Nitrogen Fixation in Rhizobial Bacteria
Three sets of genes, a special promoter, and a specific activator. The steps of nodule formation and nitrogen fixation just outlined require the coordinated expression of at least three types of *R. meliloti* genes: *nod* genes, which elicit the early steps of nodule formation; *fix* genes, which contribute to the development and metabolism of bacteroids and are essential to nitrogen fixation (a mutation in a *fix*

- The inactivity of σ^{70} at high temperatures. Because of this inactivity, σ^{70} does not compete with σ^{32} in forming the RNA polymerase holoenzyme.

Given that high temperatures render σ^{70} inactive, what enables the transcription of σ^{32} during heat shock? The *rpoH* gene, which encodes σ^{32}, has a promoter sequence that is recognized by σ^{70} and used for transcription at lower temperatures; however, at high temperatures, another sigma factor, σ^{24}, recognizes a differ-

ent promoter sequence at *rpoH* and transcribes the *rpoH* gene from that promoter (Fig. 16.21b). Although σ^{24} is always present in the cell, its own transcription (mediated by the σ^{24} holoenzyme) increases with heat shock and the appearance of denatured proteins.

In summary, *E. coli*'s heat-shock response depends on the regulation of transcription by alternative sigma factors. These alternative factors, which recognize different promoter sequences, complex with the core RNA polymerase as the temperature rises, leading to the transcription of heat-shock proteins.

Figure A **_R. meliloti's release_ of Nod factors induces the formation of root nodules in alfalfa.** Nod⁻ mutants cannot form nodules.

Figure B **How environmental signals influence the expression of _nif_ genes in _R. meliloti._** Low oxygen activates the _nif_ genes via RNA polymerase associated with a σ^{54} factor. A regulatory cascade promotes expression of the _nifA_ gene only under appropriate conditions. Pluses and minuses indicate the turning on or off of genes.

gene destroys the ability to fix nitrogen); and _nif_ genes, which encode the polypeptide subunits of the nitrogenase complex. The _nif_ genes carry a special type of promoter that RNA polymerase recognizes only when the polymerase is associated with the σ^{54} factor. Initiation of transcription at the σ^{54}-dependent promoters of the _nif_ genes depends on NifA, an activator protein responsive (via various intermediaries) to specific environmental signals (Fig. B).

Cascades of regulatory events make nitrogen fixation as energy efficient as possible. The process of nitrogen fixation has high energy demands, and the _R. meliloti_ bacteroids can only proceed using energy and materials supplied by the host plant. Under conditions of low oxygen concentration and ample nitrogen, the FixL protein in the membrane senses the O_2 concentration and activates FixJ, which then turns on expression of other _fix_ genes as well as _nifA_. The NifA protein product interacts with RNA polymerase associated with σ^{54} to activate the _nif_ genes encoding ni-

trogenase. If the oxygen level rises too high, the FixK gene product turns off _nifA_ expression. No NifA means no synthesis of the _nif_-gene-directed nitrogenase complex; this lack of nitrogenase production saves energy because the enzyme complex would be incapacitated by the excess oxygen. The transcriptional activator NifA is thus turned on and off in response to environmental cues reflecting the concentration of oxygen (Fig. B).

This brief description of the process of nitrogen fixation by _R. meliloti_ gives some idea of the complex layering of gene regulation that ensures nitrogen fixation moves forward under favorable conditions but comes to a halt under conditions of too much oxygen. Other symbiotic, nitrogen-fixing bacteria also have intricate controls over the expression of their _fix_ and _nif_ genes. The coordinate expression of these bacterial genes contributes to the bacteria's ability to respond to environmental signals and become nitrogen-generating symbionts in their host plants.

The Induction of Alternative σ Factors That Recognize Different Promoter Sequences Serves As a Global Regulatory Mechanism in Many Bacteria

By coordinating the expression of different sets of bacterial genes in response to specific cues from the environment, this global regulatory mechanism contributes to the control of such complex processes as sporulation, the synthesis of flagella, and nitrogen fixation (see the Genetics and Society box "Nitrogen

Fixation Depends on Many Levels of Gene Regulation"). For example, the bacterium _Bacillus subtilis,_ under the adverse conditions of nutrient deprivation (such as nitrogen or carbon starvation), uses a cascade of sigma factors, induced in a temporal order, to turn on successive sets of genes needed to form spores. With the proper expression of these genes, the bacterial cell becomes a metabolically inert spore able to withstand heat, aridity, extreme cold, toxic chemicals, and radiation.

Microarrays Provide a New Tool for Studying Genes Regulated As Part of a Global Response

Microarrays (review Fig. 1.13, p. 10) are an important new tool for microbial geneticists studying cellular responses to changing environmental conditions. These environmental variations include such phenomena as the rising temperatures that produce heat shock, changes in osmotic pressure, and the presence or absence of glucose. The cellular responses to these conditions often involve a global change in gene expression that is measurable by microarray analysis of mRNA isolated from cultures of cells grown under different conditions.

For example, to study changes in gene expression when lactose is substituted for glucose in the extrinsic environment (growth medium), scientists grow one culture of *E. coli* in medium containing glucose as a carbon source and another culture in medium containing lactose as a carbon source. They then isolate RNA from each culture and synthesize labeled DNA complementary to the collection of RNAs from the two different cultures. By comparing the hybridization of the two sets of cDNAs to microarrays containing oligonucleotide spots for each *E. coli* gene, they can see which genes are turned on and which are turned off in cells grown in lactose versus glucose. In the lactose-bred cells, one would expect to see an increase in the mRNA of the lactose operon genes and therefore an increase in hybridization to the *lacZ, Y*, and *A* gene spots on the microarray.

Many genes that are not specifically involved in lactose utilization may also show altered expression in a lactose medium. Recall that when glucose is not present, cAMP levels rise and the CRP–cAMP complex binds and increases expression of other genes, including several catabolic genes. Thus the experiment just outlined is not specific enough to reveal changes in gene expression related only to a change from glucose to lactose in the medium. Rather, its results, which encompass changes that occur whenever glucose is not present, are more general.

To circumvent the experimental problem arising from the fact that environmental changes often produce a general physiological reaction (the slowing of growth rate being another example), investigators use bacterial strains with mutations in specific genes: the regulatory genes that serve as the main on/off switch for numerous genes in the pathway they are analyzing. Instead of treating cells to two different growth conditions and measuring RNA levels under the different conditions, they use microarrays to compare RNA levels in a wild-type culture to RNA levels in a culture of cells containing a mutation in the key regulatory gene. They can grow the two cultures under the same environmental conditions because the mutation itself simulates a different environmental condition. For example, a *lacI⁻* cell behaves as if lactose is in the medium, even if the mutant is grown in a glucose medium.

Microbiologists have successfully used the combination of specific mutants and microarrays to identify genes that *E. coli* expresses specifically in response to nitrogen limitation. Under the best of circumstances, *E. coli* uses ammonia as its source of nitrogen. A lack of ammonia in the external environment, however, activates a master control gene called *ntrC*. The NtrC protein, in turn, activates many genes whose expression enables *E. coli* to use sources of nitrogen other than ammonia.

Researchers identified many of the genes activated by nitrogen limitation through molecular analyses. To carry out a more comprehensive analysis, however, they compared RNA levels of all *E.coli* genes in a cell containing a null mutation in *ntrC* to RNA levels in a mutant strain that produced a greater than normal amount of the NtrC protein. The resulting microarray data confirmed a finding from earlier studies that the *glnA* gene, which encodes glutamine synthetase, is regulated by NtrC. A second, very striking finding from the microarray analysis was that about 2% of the *E. coli* genome is under NtrC control.

Arranging the spots on microarrays in the order in which the genes occur in the genome makes it easier to identify coregulated adjacent genes that might form an operon. This type of microarray analysis has revealed several additional genes that are regulated by NtrC. The additional genes are involved in the scavenging of proteins and nitrogen through the transport of nitrogen-containing compounds from the cell wall into the cell and the breakdown of amino acids. NtrC's regulation of these genes has been confirmed by other types of analysis. Microarrays have thus provided an avenue for uncovering changes in gene expression that investigators can confirm and expand on through other methodologies.

A Comprehensive Example: The Regulation of Virulence Genes in *V. cholerae*

The principles and mechanisms of gene regulation in *E. coli* apply to gene regulation in other prokaryotics as well, including the bacteria *V. cholerae,* which we described at the beginning of this chapter. As we saw, these bacterial agents of cholera are able to sense changes in their environment and transmit signals about those changes to regulators that initiate, enhance, diminish, or repress the expression of various genes. Of particular interest to epidemiologists and medical practitioners seeking to prevent or treat the symptoms of cholera are the genes bestowing virulence.

Three Regulatory Proteins—ToxR, ToxS, and ToxT—Turn on the Genes for Virulence

To understand the regulation of the genes for virulence, researchers first cloned the two genes that encode the polypeptide subunits of cholera toxin: *ctxA* and *ctxB*, which are transcribed and regulated together as an operon. They next made a *ctxA-lacZ* reporter gene fusion construct that could detect changes in regulation of the operon by measuring β-galactosidase expression. They then cut *V. cholerae* genomic DNA into pieces and cloned these into a vector that would replicate in *E. coli*. With the construction of these tools, they were able to perform experiments in *E. coli* cells, which are more amenable than *V. cholerae* to some types of genetic manipulation.

To isolate a gene that regulates expression of the *ctx* operon, they transformed *E. coli* cells already containing the *ctx-lacZ* construct with the clones containing *V. cholerae* DNA. A clone that contains a gene encoding a positive cholera regulator should turn on expression of the *lacZ* fusion hybrid in *E. coli*. Clones that turned on expression contained the regulatory *toxR* gene, which encodes a membrane protein (ToxR) with an N-terminal end in the cytoplasm and a C-terminal end in the periplasm (the space between the inner and outer membranes of the bacterium). This is a surprising location for a regulatory molecule, since most gene regulators must diffuse through the cytoplasm until they bind to DNA. However, in bacteria, unlike in eukaryotes, the chromosome, which is not enclosed in a nuclear membrane, is free to make contact with the inner cell membrane. In *V. cholerae*, *toxR⁻* mutants do not induce virulence, and the *toxR⁻* mutation is recessive as you would predict of a positive regulator.

The *V. cholerae*-DNA-containing clone that was isolated based on its ability to activate the *ctx-lacZ* fusion construct contained a second gene, designated *toxS*. Transcriptional activation of the *ctx-lacZ* fusion in *E. coli* required the ToxS protein. The postulated function of ToxS, which is active in the periplasm, is to help ToxR monomers form dimers. When researchers engineered a *toxR* gene to express a product that spontaneously formed dimers, the resulting ToxR protein no longer required ToxS to carry out its function; it could activate transcription in *E. coli* on its own. The formation of dimers may be important for the tight binding of ToxR to DNA, as is true for most other regulatory proteins in bacteria.

To determine what genes ToxR regulates other than those in the *ctx* operon, researchers fused the *toxR* gene to a constitutive promoter, and they introduced this fusion construct into a collection of *V. cholerae* strains in which copies of the *lacZ* gene had randomly inserted around the chromosome. Those colonies expressing β-galactosidase (as shown by the blue color resulting from the splitting of the X-Gal substrate) contained *lacZ* genes adjacent to a promoter region regulated by *toxR*. In *V. cholerae*, these *lacZ* fusion genes must have been regulated by ToxR (at least indirectly) because all bacteria in the study contained the *toxR* gene fused to a constitutive promoter and thus were constitutive synthesizers of ToxR. However, when transferred into *E. coli*, this collection of genes was not regulated by ToxR. Something required to make these genes respond to ToxR was present in the *V. cholerae* genome but was missing in *E. coli*. The lack of direct regulation by ToxR in *E. coli* triggered a search that culminated in the identification of an intermediate regulatory gene named *toxT*. The ToxT protein is a transcriptional activator that carries out its function by binding to the promoters of many genes, including *ctx* and the other virulence genes. While either ToxR/S or ToxT can activate the *ctx* genes that produce toxin, ToxT alone activates the additional

virulence genes, which encode pili and other proteins that enable the bacteria to colonize the small intestine. Transcription of the *toxT* gene is regulated by ToxR/S.

A Model of Virulence Regulation Leaves Unanswered Questions

On the basis of these studies, researchers proposed the following model (Fig. 16.22): The membrane location of the ToxR protein enables it to respond to environmental cues that signal *V. cholerae*'s arrival in the small intestine; in response to these cues, the cytoplasmic domain of ToxR binds to the *ctx* promoter. In addition, ToxR binds to the promoter of *toxT*, activating the transcription of *toxT*, which then activates the expression of *ctx* and the other virulence genes. Experiments in which investigators monitored gene expression in the animal (mouse) disease model for *V. cholerae* confirmed the requirement of ToxR as a positive regulator of *ctxA* expression during pathogenesis in the animal. In these animal studies, when *V. cholerae* strains containing the *ctx* reporter gene fusion were injected into mice, expression of *ctxA* occurred in the *toxR⁺* strain but not in the *toxR⁻* mutant. Interestingly, in the mouse model, the specifics of some of the other regulatory pathways for the *V. cholerae* pathogenesis genes did not coincide with the results obtained from the isolated bacterial cultures. From these studies, we can conclude that studying gene expression in pathogenic strains in culture, where it is easier to manipulate genes and measure their expression, provides a valuable first analysis. Once the potential regulators and pathways have been identified, experiments to detect

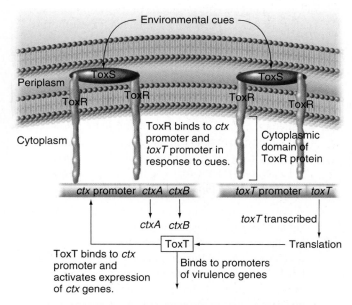

Figure 16.22 Model for how *V. cholerae* regulates genes for virulence. Environmental cues are sensed by ToxR, a transmembrane protein. ToxR is held in its active dimeric form by ToxS. In the cytoplasm, ToxR interacts with promoter of the *ctxA*, *ctxB*, *toxT* gene. ToxT in turn regulates the expression of many other virulence genes.

gene expression during pathogenesis in model animal systems provide critical tests of the models.

Several intriguing questions remain about the regulatory system that controls the expression of the virulence genes in *V. cholerae.* Why is there a cascade (ToxR, ToxT) of regulatory factors? Since the promoter regions in the two genes recognized by ToxR do not have a common sequence of rotational symmetry (as has been found in other regulatory systems), what DNA sequence in the promoters does ToxR recognize? What is the

signal that makes the cholera bacteria stop swimming and start to colonize (that is, adhere to the cells of) the small intestine? What molecular events differentiate swimming and colonization? How does the ToxR regulatory protein, located in the membrane, find binding sites on the chromosome?

Answers to these questions will help scientists complete the picture of how *V. cholerae* generate disease. With a better understanding of pathogenesis, they will be able to devise more effective treatments for cholera as well as measures to prevent it.

C O N N E C T I O N S

Regulation in prokaryotes depends on the binding of regulatory proteins to specific DNA segments in the vicinity of a gene or group of genes. The existence of these regulatory elements adds another notch to the concept of the gene. Most geneticists would say that a gene consists of the nucleotides that specify amino acids in the gene's protein product or the ribonucleotides in the gene's RNA product, as well as the regulatory elements that influence the gene's transcription.

Some of the ways in which bacteria regulate their genes are available to eukaryotes as well. For example, both types of or-

ganisms can use diffusible regulatory proteins to start or stop transcription. By contrast, eukaryotes cannot regulate transcription by the attenuation mechanism described for the *trp* operon because their nuclear membrane prevents access to the growing transcript by the translational machinery. However, eukaryotic cells, with their larger genomes, have evolved many other mechanisms of gene regulation that go beyond those found in prokaryotic systems. In Chapter 17 we examine the special regulatory needs of eukaryotes and some of the solutions they have evolved.

E S S E N T I A L   C O N C E P T S

1. Most mechanisms of gene regulation in prokaryotes block or enhance the initiation of transcription. Later steps in gene expression are potential targets for fine-tuning the amount of gene products that accumulate in cells.

2. In the *lac operon* model proposed by Jacob and Monod, the binding of a *repressor* protein (encoded by the *lacI* gene) to the DNA *operator* prevents transcription of the structural genes *lacZ, lacY,* and *lacA* in the absence of the inducer lactose. When lactose is present, its binding to the repressor induces expression of the structural genes by causing the repressor to change its shape and lose its ability to bind to the operator.

3. A critical, general principle emerges from the *lac* operon studies: Regulatory genes usually encode *trans*-acting regulatory proteins that interact with *cis*-acting regulatory DNA elements located near the promoter (such as the *operator*). Negative regulatory proteins prevent or diminish the rate of transcription, while positive regulatory proteins enhance transcription.

4. Many types of coordinate gene regulation result from the clustering of genes into *operons* that are transcribed into a single polycistronic mRNA from a single promoter.

5. The binding of *repressor* proteins to *operators* can be influenced by either inducers (as for the *lac* repressor) or corepressors (as for the *trp* operon).

6. *Catabolite repression* regulates certain catabolite operons by preventing the CRP protein, a positive regulator, from binding to the operons' promoter region in the presence of high concentrations of glucose.

7. Many regulatory proteins, both positive and negative, contain a helix-turn-helix motif, function as oligomers that bind to more than one DNA site, and interact with RNA polymerase to prevent or assist its function.

8. *Attenuation,* a form of fine-tuning for operons involved in the biosynthesis of amino acids, is based on the amount of premature termination of mRNA transcription. The termination, in turn, is determined by the intracellular concentration of tRNAs charged with the amino acid produced by the enzyme products of the structural genes in the operon.

9. Cells can express different sets of genes at different times or under different conditions by using alternate sigma factors or by producing novel RNA polymerases that recognize different classes of promoters.

S O C I A L   A N D   E T H I C A L   I S S U E S

1. The bacterium *Bacillus thuringiensis* (Bt) produces a toxin capable of killing many insects that attack crop plants. For more than 20 years, Bt has served as a biological pesticide that is sprayed on crops. The constant spraying, however, has led to the development of resistant insect strains in the field. A cloned version of Bt toxin consists of the toxin gene fused to a regulatory region that allows expression of the toxin only when crops carrying the gene are under stress—as in an insect infestation. Should farmers be forced to plant the transgenic crops rather than spray their fields so that this biological pesticide will remain effective? Are there ways to enforce this policy in the United States or worldwide?

2. Many scientists who are employed as professors at universities have become involved in biotechnology ventures outside the university. Companies often set up their laboratories near universities to take advantage of the intellectual resources there. Industries that develop near universities employ nonscientists in the community as well as scientists and are generally looked upon as an asset to the community. The president of a major state university is encouraging faculty members to become active in the development of associated industries, viewing their mission as academics to include community related activities. But legislators and the public, whose tax dollars fund the university, believe that the professors should concentrate on teaching. Should faculty be encouraged to participate in outside ventures because it is a service to the community and potentially to humankind through development of new biotechnology, or should they be held more strictly to a solely academic, teaching mission?

3. In studying virulence in bacteria, a group of researchers stumbled on ways of modifying a particular bacterial species in a way that increased the virulence. While their finding helped them understand the mechanism of virulence better, the group of researchers were split on whether to publish their findings or not. Some fear that this new information could be misused in developing biological warfare. Do scientists have a responsibility to withhold information if it could be used in a harmful way?

S O L V E D   P R O B L E M S

I. In the galactose operon in *E. coli,* a repressor, encoded by the *galR* gene, binds to an operator site, *O,* to regulate expression of three structural genes, *galE, galT,* and *galK.* Expression is induced by the presence of galactose in the media. For each of the merodiploids listed, would the cell show constitutive, inducible, or no expression of each of the structural genes? (Assume that *galR⁻* is a loss of function mutation.)

a. *galR⁻ galO⁺ galE⁺ galT⁺ galK⁺*

b. *galR⁺ galOᶜ galE⁺ galT⁺ galK⁺*

c. *galR⁻ galO⁺ galE⁺ galT⁺ galK⁻/galR⁺ galO⁺ galE⁻ galT⁺ galK⁺*

d. *galR⁻ galOᶜ galE⁺ galT⁺ galK⁻/galR⁺ galO⁺ galE⁻ galT⁺ galK⁺*

Answer

This problem requires an understanding of how regulatory sites and proteins that bind to regulatory sites behave. To predict expression in merodiploid strains, look at each copy of the operon individually, and then assess what effect alleles present in the other copy of the operon could have on the expression. After doing that for each copy of the operon, combine the results.

a. The *galR* gene encodes a repressor, so the lack of a GalR gene product would lead to constitutive expression of the *galE, T,* and *K.*

b. The *galOᶜ* mutation is an operator site mutation. By analogy with the *lac* operon, the designation *galOᶜ* indicates that repressor cannot bind and there is constitutive expression of *galE, T,* and *K.*

c. The first copy of the operon listed has a *galR⁻* mutation. Alone, this would lead to constitutive synthesis *galE* and *galT.* (*galK* is mutant, so there will not be constitutive expression of this gene.) The other copy is wild type for the *galR* gene, so it produces a repressor that can act in *trans* on both copies of the operon, overriding the effect of the *galR⁻* mutation. Overall, there will be *inducible expression of the three* gal *genes.*

d. The first copy of the operon contains an *galOᶜ* mutation, leading to constitutive synthesis of *galE* and *galT.* The other copy has a wild-type operator so it is inducible, but neither operator has effects on the other copy of the operon. The net result is constitutive *galE* and *galT* and inducible *galK* expression.

II. The *araI* site is required for induction of *araBAD. I⁻* mutants do not express *araBAD.* In an *I⁻* mutant, a second mutation arose that resulted in constitutive arabinose synthesis. A Southern blot using a probe from the regulatory region and early part of the *araB* gene showed a very different set of restriction fragments than were seen in the starting strain. Based on the altered restriction pattern and constitutive expression, propose a hypothesis about the nature of the second mutation.

Answer

To answer this question, you need to consider how changes in restriction patterns could arise, what effects they could have, and what is necessary to get expression. The fact that the experiment began with a strain that lacked the inducing site, *I*, and that there is constitutive synthesis mean that the normal regulation is lacking. Constitutive synthesis could result from a *deletion that fused the araBAD genes to another promoter* (one that is on under the growth conditions used). A deletion would lead to a different pattern of restriction fragments that could be observed in the Southern hybridization analysis.

III. Bacteriophage λ, after infecting *E. coli,* can take one of two routes. It will either produce many progeny that are released by lysis of the cell (lytic growth) or the phage DNA will integrate into the chromosome because transcription from the major phage promoters of the phage will have been shut down. The repressor protein cI, encoded by phage λ, binds to two operator regions to shut down expression and therefore no phages are produced. Mutations in the *cI* gene that destroy the binding ability of the repressor lead to the lytic type of life cycle

exclusively; that is, all cells infected by the phage will burst and release progeny phages. Another type of mutation gives the same phenotype-lytic growth only. Such mutations, called λ*vir,* arise at a much lower frequency than the *cI* mutations (about 1 in 10^{12} compared with 1 in 10^6 for *cI* mutants). What do you think these mutations are, and why are they less frequent than *cI* mutations?

Answer

This problem requires an understanding of the types of regulatory mutations that can affect negative regulation. The lack of negative regulation (by *cI*) in the life cycle leads to the lytic cycle of growth only. Such mutations could be either in the gene encoding the negative regulator or the site to which the repressor binds. You were told that the *cI* mutations are defects in the gene encoding the repressor. The λ*vir* mutations could be mutations in the site to which the repressor binds, but since the repressor has to bind to two sites, *there must be two mutations in a* λvir *mutant. Therefore these would arise less frequently* (at a frequency predicted for two independent mutational events combined: 1 in 10^6 × 1 in 10^6, or 1 in 10^{12}).

PROBLEMS

16-1 For each of the terms in the left column, choose the best matching phrase in the right column.

a. induction
b. repressor
c. operator
d. allostery
e. operon
f. catabolite repression
g. reporter gene
h. attenuation

1. glucose prevents expression of catabolic operons
2. protein undergoes a reversible conformational change
3. often fused to regulatory regions of genes whose expression is being monitored
4. stimulation of protein synthesis by a specific molecule
5. site to which repressor binds
6. gene regulation involving premature termination of transcription
7. group of genes transcribed into one mRNA
8. negative regulator

16-2 All mutations that abolish function of the Rho (termination) protein in *E. coli* are conditional mutations. What does this tell you about the *rho* gene?

16-3 Bacteriophage λ, after infecting a cell, can integrate into the chromosome of the cell if the repressor protein, cI, binds to and shuts down phage transcription immediately. (A strain containing a bacteriophage integrated in the chromosome is called a lysogen.) The alternative fate is the production of many more viruses and lysis of the cell. In an Hfr cross, a donor strain that

is a lysogen was crossed with a lysogenic F^- recipient cell and no phages were produced. However, when the Hfr lysogen donor strain transferred its DNA to a nonlysogenic F^- recipient cell, the recipient cell burst, releasing a new generation of phages. Why did infection of a nonlysogenic cell result in phage growth and release but infection of a lysogenic recipient did not?

16-4 The promoter of an operon is the site to which RNA polymerase binds to begin transcription. Some base changes in the promoter result in a mutant site to which RNA polymerase cannot bind. Would you expect mutations in the promoter that prevent binding of RNA polymerase to act in *trans* on another copy of the operon on a plasmid in the cell or only in *cis* on the copy immediately adjacent to the mutated site?

16-5 For each of the *E. coli* merodiploid strains containing the *lac* operon alleles listed, indicate whether the strain is inducible, constitutive, or unable to express β-galactosidase and permease.
a. $I^+ O^+ Z^- Y^+ / I^+ O^c Z^+ Y^+$
b. $I^+ O^+ Z^+ Y^+ / I^- O^c Z^+ Y^-$
c. $I^+ O^+ Z^- Y^+ / I^- O^+ Z^+ Y^-$
d. $I^- P^- O^+ Z^+ Y^- / I^+ P^+ O^c Z^- Y^+$
e. $I^s O^+ Z^+ Y^+ / I^- O^+ Z^+ Y^-$

16-6 Mutants were isolated in which the constitutive phenotype of a missense *lacI* mutation was suppressed.

That is, the operon was now inducible. These mapped to the operon but were not in the *lacI* gene. What could these mutations be?

16-7 For each of the following mutant *E. coli* strains, plot a 30-minute time course of concentration of β-galactosidase, permease, and acetylase enzymes grown under the following conditions. For the first 10 minutes, no lactose is present; at 10′ lactose becomes the sole carbon source. Plot concentration on the y-axis, time on the x-axis. (Don't worry about the exact units for each protein on the y-axis.)

a. $I^- P^+ O^+ Z^+ Y^+ A^+ / I^+ P^+ O^+ Z^- Y^+ A^+$

b. $I^- P^+ O^c Z^+ Y^+ A^- / I^+ P^+ O^+ Z^- Y^+ A^+$

c. $I^s P^+ O^+ Z^+ Y^+ A^+ / I^- P^+ O^+ Z^- Y^+ A^+$

d. $I^- P^- O^+ Z^+ Y^+ A^+ / I^- P^+ O^c Z^+ Y^- A^+$

e. $I^- P^+ O^+ Z^- Y^+ A^+ / I^- P^- O^c Z^+ Y^- A^+$

16-8 You have isolated two different mutants (*reg1* and *reg2*) causing constitutive expression of the *emu* operon (*emu1 emu2*). One mutant contains a defect in a DNA-binding site, and the other has a defect in the gene encoding a protein that binds to the site (loss of function).

a. Is the DNA-binding protein a positive or negative regulator of gene expression?

b. To determine which mutant has a defect in the site and which one has a mutation in the binding protein, you decide to do a merodiploid analysis. Assuming you can assay levels of Emu1 and Emu2 proteins, what results do you predict for the two strains following (*i* and *ii*) if *reg2* encodes the regulatory protein and *reg1* is the regulatory site?

c. What results do you predict for the two strains (*i* and *ii*) if *reg1* encodes the regulatory protein and *reg2* is the regulatory site?

   i. $F'\ reg1^-\ reg2^+\ emu1^-\ emu2^+ / reg1^+\ reg2^+$ $emu1^+\ emu2^-$

   ii. $F'\ reg1^+\ reg2^-\ emu1^-\ emu2^+ / reg1^+\ reg2^+$ $emu1^+\ emu2^-$

16-9 You have isolated a protein that binds to DNA in the region upstream of the promoter sequence of the *sys* gene. If this protein is a positive regulator, which would be true?

a. Loss-of-function mutations in the gene encoding the DNA-binding protein would cause constitutive expression.

b. Loss-of-function mutations in the gene encoding the DNA-binding protein would result in no expression.

16-10 Seven strains of *E. coli* (mutants 1–6) that had one of the following mutations affecting the *lac* operon were isolated.

a. deletion of *lacY*

b. *lacO^c* mutation

c. missense mutation in *lacZ*

d. inversion of the *lac* operon (but not an inversion of the *lacI* gene)

e. superrepressor mutation

f. inversion of *lacZ, Y,* and *A* but not *lacI, P, O*

The entire *lac* operon (including the *lacI* gene and its promoter) from each of the seven *E. coli* strains was cloned into a plasmid vector containing an ampicillin resistance gene. Each recombinant plasmid was transformed into each of the seven strains to create merodiploids, and each was plated on ampicillin media in which lactose is the only carbon source. (Ampicillin was included to ensure maintenance of the plasmid.) Growth of the transformants is scored here (a + sign = growth, a − sign = no growth). Synthesis of β-galactosidase and permease are required for growth on this medium. Results of this merodiploid analysis are shown here. Which mutant bacterial strain contained each of the alterations listed previously? (Mutation 1 was found to be a deletion of the *lacY* gene.)

	1	2	3	4	5	6
1	−	+	−	+	−	+
2	+	−	−	+	−	+
3	−	−	−	+	−	+
4	+	+	+	+	−	+
5	−	−	−	−	−	+
6	+	+	+	+	+	+

16-11 Given the following data, explain which strains and growth conditions are important for reaching the following conclusions.

a. Arabinose induces coordinate expression of the *araBAD* genes (encoding kinase, isomerase, and epimerase).

b. The *araC* gene encodes a positive regulator of *araBAD* expression.

Genotype	Arabinose in medium	Kinase	Isomerase	Epimerase
1. $C^+ B^+ A^+ D^+$	no	−	−	−
2. $C^+ B^+ A^+ D^+$	yes	+	+	+
3. $C^- B^+ A^+ D^+$	no	−	−	−
4. $C^- B^+ A^+ D^+$	yes	−	−	−

16-12 For each of the growth conditions listed, what proteins would be bound to *lac* operon DNA? (Do not include RNA polymerase.)

a. glucose

b. glucose + lactose

c. lactose

16-13 Seven *E. coli* mutants were isolated. The activity of the enzyme β-galactosidase produced by cells containing each mutation alone or in combination with other mutations was measured when the cells were grown in medium supplemented with different carbon sources.

	Glycerol	Lactose	Lactose + Glucose
Wild type	0	1000	10
Mutant 1	0	10	10
Mutant 2	0	10	10
Mutant 3	0	0	0
Mutant 4	0	0	0
Mutant 5	1000	1000	10
Mutant 6	1000	1000	10
Mutant 7	0	1000	10
F′ *lac* from mutant 1/ mutant 3	0	1000	10
F′ *lac* from mutant 2/ mutant 3	0	10	10
Mutants 3 + 7	0	1000	10
Mutants 4 + 7	0	0	0
Mutants 5 + 7	0	1000	10
Mutants 6 + 7	1000	1000	10

Assume that each of the seven mutations is one and only one of the genetic lesions in the following list. Identify the type of alteration each mutation represents.
a. superrepressor
b. operator deletion
c. nonsense (amber) suppressor tRNA gene (assume that the suppressor tRNA is 100% efficient in suppressing amber mutations)
d. defective CRP–cAMP binding site
e. nonsense (amber) mutation in the β-galactosidase gene
f. nonsense (amber) mutation in the repressor gene
g. defective *crp* gene (encoding the CRP protein)

16-14 For each of the *E. coli* strains that follow, indicate the effect of the genotype on expression of the *trpE* and *trpC* genes in the presence and absence of tryptophan. (In the wild type [$R^+ P^+ O^+ att^+ trpE^+ trpC$], *trpC* and *trpE* are fully repressed in the presence of tryptophan and are fully induced in the absence of tryptophan.)

R = repressor gene; R^n product cannot bind tryptophan; R^- product cannot bind operator
O = operator; O^- cannot bind repressor
att = attenuator; att^- is a deletion of the attenuator
P = promoter; P^- is a deletion of the *trp* operon promoter
$trpE^-$ and $trpC^-$ are null (loss of function) mutations

a. $R^+ P^- O^+ att^+ trpE^+ trpC^+$
b. $R^- P^+ O^+ att^+ trpE^+ trpC^+$
c. $R^n P^+ O^+ att^+ trpE^+ trpC^+$
d. $R^- P^+ O^+ att^- trpE^+ trpC^+$
e. $R^+P^+O^-att^+trpE^+trpC^-/R^-P^+O^+att^+trpE^-trpC^+$
f. $R^+P^-O^+att^+trpE^+trpC^-/R^-P^+O^+att^+trpE^-trpC^+$
g. $R^+ P^+O^-att^-trpE^+trpC^-/R^-P^+O^-att^+trpE^-trpC^+$

16-15 Clones of three adjacent genes involved in arginine biosynthesis have been isolated from a bacterium. If these three genes together make up an operon, what result do you expect when you use the DNA from each of these genes as probes in a Northern analysis? What result do you expect if the three genes do not make up an operon?

16-16 A molecular geneticist is investigating an operon by measuring the amount of expression of the four structural genes (*A, B, C,* and *D*) produced in wild-type and mutant bacterial cells after the addition of compound Z to a minimal medium. An additional protein (E) is of very small size (less than 20 amino acids) and cannot be measured by the same analytical system employed for the other proteins. Several of the mutations are nonsense mutations that have an effect on the genes transcribed after them in the operon. In addition to stopping translation of the gene in which the mutations lie, these nonsense mutations (called polar mutations) cause termination of transcription of the next genes in the operon. This means that genes downstream of the mutation are not expressed. (For example, in the *lac* operon, some *lacZ* nonsense mutations can result in no expression of *lacY* and *lacA*.) The investigator has also obtained mutations in two other sites, *F* and *G,* closely linked to *A–D*. The graphs shown are all semilogarithmic. The percentage of maximal possible expression for a particular protein is plotted on the y-axis, while the x-axis coordinate is time. Compound Z is added at the point specified by the arrow.

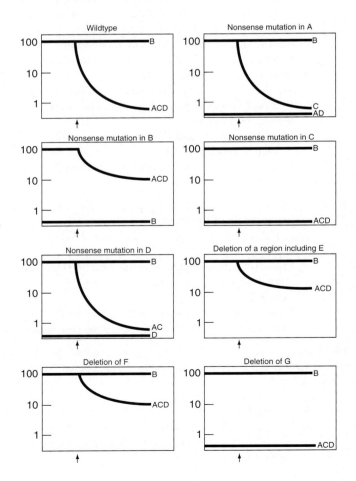

a. Is this operon likely to be involved in a pathway of biosynthesis or a pathway of degradation? Is the operon inducible or repressible?

b. Construct a map of this operon. Indicate the relative positions of genes *A, B, C, D,* and *E* as well as the sites *F* and *G*. List possible functions for all the genes and sites. (Possible functions for the genes and sites *A–G:* promoter, operator, enzyme structural gene, CRP-binding site, *crp* gene, attenuator region. This list is not necessarily all-inclusive.)

16-17 Following is a sequence of the leader region of the *his* operon mRNA in *Salmonella typhimurium.* What bases in this sequence could cause a ribosome to pause when histidine is limiting (that is, there is very little of it) in the medium?

5′ AUGACACGCGUUCAAUUUAAACACCACCAUCAUCACCAUCAUCCUGACUAGUCUUUCAGGC 3′

16-18 Maltose utilization in *E. coli* requires the proteins encoded by genes in three different operons: *malE malF malG, malK lamB,* and *malP malQ* operons. MalT is a positive regulator protein that regulates expression from each of these operons and the operon is catabolite sensitive. Bacteriophage λ uses one of the maltose transport proteins, LamB, that is found in the outer membrane. Synthesis of the λ receptor is induced by maltose in the media. *E. coli* cells that are resistant to infection by phage λ have been isolated. List the types of mutations in the maltose regulon that λ-resistant mutants could contain.

16-19 In an effort to determine the location of an operator site for a negatively regulated gene, you have made a series of deletions within the regulatory region. The extent of each deletion is shown by the line underneath the sequence and the resulting expression from the operon (*i* = inducible; *c* = constitutive; − = no expression).

```
. . . GGATCTTAGCCGGCTAACATGATAAATATAA...
. . . CCTAGAATCGGCCGATTGTACTATTTATATT...
1  i  ─────────
2  −  ─────────────────────────────
3  c  ──────────────────────────
4  −  ──────────────────────────────
5  c  ────────────
```

a. What can you conclude from this data about the location of the operator site?

b. Why do you think deletions 2 and 4 show no expression?

16-20 The following data is from a DNase footprinting experiment in which either RNA polymerase or a repressor protein was added to a labeled DNA fragment, and then the complex was digested with DNase. DNA sequencing reactions were also performed on the same DNA so the bases that were

protected by proteins binding could be identified. (Notice that DNase does not cut after each base in the DNA fragment.)

a. What is the sequence of the DNA in this fragment?

b. Mark on the sequence the region where the repressor binds.

c. Mark on the sequence the region where RNA polymerase binds.

16-21 The original constitutive operator mutations in the *lac* operon were all base changes in O_1. Why do you think mutations in O_2 or O_3 were not isolated in these screens?

16-22 Cells containing mutations in the *crp* gene (encoding the positive regulator CRP) are Lac^-, Mal^-, Gal^-, etc. To find suppressors of the *crp* mutation, cells were screened to find those that were both Lac^+ and Mal^+.

a. What types of suppressors would you expect to get using this screen compared with a screen for Lac^+ only?

b. All suppressors isolated were mutant in the α-subunit gene of RNA polymerase. What hypothesis could you propose based on this analysis?

16-23 An operon fusion consists of a regulatory region cloned next to the coding region of the genes of an operon. A gene fusion consists of a regulatory region of a gene such as *lacZ* and the DNA encoding the first amino acids of the β-galactosidase protein cloned next to the coding region of another gene. What additional feature do you have to consider to create a functional gene fusion that is not necessary for an operon fusion?

16-24 Many genes, whose expression is turned on by DNA damage, have been isolated. Loss of function mutations in the *lexA* gene leads to expression of many of these genes, even when there has been no DNA damage. Would you hypothesize that LexA protein is a positive or negative regulator? Why?

REFERENCES

DiRita, V. J., Parsot, C., Jander, G., and McKalanos, J. J. (1991). Regulatory cascade controls virulence in *Vibrio cholerae. Proc. Natl. Acad. Sci.* USA 88:5403–7.

Jacob, F., and Monod, J. (1961). Genetic regulatory mechanisms in the synthesis of proteins. *J. Mol Biol.* 3:318–56.

Lewis et al. (1996). Crystal Structure of the Lactose Operon Repressor and Its Complexes with DNA and Inducer. *Science* 271:1247–54.

Zimmer et al. (2000). Nitrogen regulatory protein C-controlled genes on *Escherichia coli:* Scavenging as a defense against nitrogen limitation. *Proc. Natl. Acad. Sci.* USA 97:14674–79.

Gene Regulation in Eukaryotes

IRF-1

NF-κB Dimerization Domain

NF-κB Specificity

The family of interferon regulatory factor (IRF) transcription factors is important in the regulation of interferons.

When a *Drosophila* male courts a *Drosophila* female, he sings a species-specific song and dances an ancient dance. If successful, his instinctive behaviors culminate in mating. The male senses a female's presence by visual and tactile cues as well as by the pheromones she produces. After orienting himself at a precise angle with respect to his prospective mate, he taps his partner's abdomen with his foreleg and then performs his song by stretching out his wings and vibrating them at a set frequency; when the song is over, he begins to follow the female. If she is unreceptive (perhaps because she has recently mated with another male), she will run away, but if she is receptive, she will let him overtake her. When he does, he licks her genitals with his proboscis, curls his abdomen, mounts the female, and copulates with her for about 20 minutes.

Various mutations in a gene called *fruitless* produce behavioral changes that prevent the male from mating properly. Some mutant alleles alter the song to an unfamiliar air. Others diminish the male's ability to distinguish females from males; male flies with this mutation court each other. Still others reduce the male's ability to

court either sex. Finally, some mutations create lethal null alleles that cause male flies to die just before they emerge from the pupal case, showing that the *fruitless* gene has other functions in addition to its effects on courtship and mating.

Cloned in 1996, *fruitless* is a large gene—roughly 150 kb in length—encoding a product that can regulate transcription at multiple promoters. Alternative RNA splicing of the gene's several different transcripts can produce many variations of the fruitless protein. All of these gene products contain zinc-finger peptide motifs that facilitate binding to DNA, suggesting a role for the protein in the regulation of transcription. Some forms of the fruitless protein are sex-specific, appearing only in males or only in females. By contrast, some fruitless proteins are expressed at low levels in many kinds of cells in both sexes.

The male version of the fruitless protein, although very similar to the female protein in amino-acid sequence, has an extra 101 amino acids at its N terminus, and this addition almost certainly determines the observed differences in male and female behavior. Remarkably, the male-specific fruitless mRNA is synthesized in only a few hundred of the tens of thousands of neurons that make up the male *Drosophila*'s nervous system. Most of these *fruitless*-expressing cells are located near motor neurons that control either wing movements (and thus possibly the song) or abdominal movements (and thus possibly the abdominal curling that immediately precedes mating). Work on the *Drosophila fruitless* gene has provided strong evidence that differences in gene expression can directly influence complex behaviors.

In this chapter, we see that **eukaryotic gene regulation**—the control of gene expression in the cells of eukaryotes—depends on an array of interacting regulatory elements that turn genes on and off in the right places at the right times. Some of these elements are specific DNA sequences in the vicinity of the gene to be regulated; others are DNA-binding proteins encoded by other genes; still others are proteins that bind to and modify the activity of DNA-binding proteins.

During the embryonic development of multicellular eukaryotic organisms like *Drosophila* or humans, gene regulation controls not only the elaboration of sex-related characteristics and behaviors but also the differentiation of tissues and organs, as well as the precise positioning of these tissues and organs within the multicellular animal. For example, because of the tightly controlled differential regulation of genes in different types of cells, red blood cell precursors selectively turn on the genes for making hemoglobin, while pancreatic cells turn on the genes for making insulin, even though both types of cells, like nearly all somatic cells, carry the same number of chromosomes with the same amount of DNA and the same complete set of genes.

In exploring the intricacies of eukaryotic gene regulation, it is helpful to bear in mind key similarities and differences between eukaryotes and prokaryotes (Table 17.1). In both types of cells, tran-

scription is regulated through the attachment of DNA-binding proteins to specific DNA sequences that are adjacent to or near the transcription unit itself; and in both eukaryotes and prokaryotes, the same polypeptide motifs appear in many DNA-binding proteins. Indeed many, though not all, of the basic principles of gene regulation in prokaryotes apply to eukaryotes, but additional levels of complexity tailor regulation in eukaryotes to their requirements. First, because eukaryotes have up to 700 times more DNA and 10 times more genes than prokaryotes, they require additional biochemical mechanisms to achieve specificity; the mechanisms they have evolved include the higher-affinity binding of proteins to DNA. Second, because eukaryotic DNA is packaged in chromatin, the transcriptional machinery must be able to control chromatin compaction and decompaction. The control of chromatin structure over defined distances along the chromosomes provides eukaryotic cells with a level of gene regulation that is unavailable to prokaryotic cells. Third, eukaryotes accomplish a type of regulation, through the differential splicing of exons, that is also unavailable to prokaryotes. Moreover, although the separation of transcription and translation into different cellular compartments in eukaryotes precludes certain forms of prokaryotic regulation, such as attenuation, it opens up the possibility of regulating gene function at different points in the expression process, from the initiation of transcription, through mRNA processing before transport from the nucleus, to protein synthesis and modification in the cytoplasm. In addition to these basic differences between eukaryotes and prokaryotes, multicellular eukaryotes must be able to use gene regulation to control cellular differentiation and the complex interactions of various types of differentiated cells within tissues and organs.

One multifaceted overarching theme emerges from our discussion of eukaryotic gene regulation. Like complex behaviors, most

TABLE 17.1 **Key Regulatory Differences Between Eukaryotes and Prokaryotes**

Characteristic	Prokaryote	Eukaryote
Control of transcription through specific DNA-binding proteins	Yes	Yes
Reutilization of same DNA-binding motifs by different DNA-binding proteins	Yes	Yes
Activator proteins	Yes	Yes
Repressor proteins	Yes	Yes
Specificity of binding to DNA by regulatory protein	Specific	Highly specific
Affinity of binding	Strong	Very strong
Role played by chromatin structure	No	Yes
Coordinate control achieved with operons	Yes	Rare
Differential splicing	No	Yes
Attenuation	Yes	No
mRNA processing	No	Yes
Differential polyadenylation	No	Yes
Differential transport of RNA from nucleus to cytoplasm	No	Yes

biological functions arise from the regulated interactions of large networks of genes. And because each gene in a network has multiple potential points of regulation, the possibilities for regulatory refinement are enormous. Thus, eukaryotic gene regulation is a modular system in which different combinations of elements interact with each other at specified times and places, as well as in response to changes in the cellular environment, to allow the synthesis of precisely modulated amounts of gene products.

As we examine the components and mechanisms of eukaryotic gene regulation, we describe:

■ The use of genetics to study gene regulation, focusing on mutations that identify *cis*-acting control elements and *trans*-acting proteins.

■ Gene regulation at the initiation of transcription, including a discussion of how three different RNA polymerases

recognize three different classes of promoters; how *trans*-acting proteins help control class II promoters, causing the activation or repression of gene expression; how chromatin structure affects expression; how signal transduction systems work; and how DNA methylation, an epigenetic process, can regulate gene expression through silencing and genomic imprinting.

■ Regulation after transcription, including a discussion of the control of RNA splicing, RNA stability, mRNA editing, and translation, as well as the control of protein activity by posttranslational modification.

■ A comprehensive example of gene regulation during sex determination in *Drosophila*.

The Use of Genetics to Study Gene Regulation

Genetic studies of various mutations have helped demystify the complexities of gene regulation. The techniques for analyzing these mutations in eukaryotes are similar to those researchers use in examining gene control in bacterial cells. The studies have established that for most, but not all, eukaryotic genes, the critical decisions controlling the amount of gene product synthesized are made during the initiation of transcription, when RNA polymerase starts to make a primary transcript, or RNA copy, of a gene's template strand (review Fig. 8.11 on p. 248).

The Analysis of Regulatory Components Focuses on Mutations That Affect a Gene's Function But Do Not Affect the Amino Acids in the Gene's Product

Geneticists can study regulation by choosing a gene whose control they want to study (the "target gene") and then looking for mutations that affect the expression of that gene. To find such mutations, they observe changes in a phenotype that is extremely sensitive to the amount of gene product synthesized.

Regulatory Mutations That Map at or Near the Target Gene Help Define *cis*-Acting DNA Sequences That Influence Transcription

cis-**acting** elements at a gene are DNA sequences that serve as attachment sites for the DNA-binding proteins that regulate the initiation of transcription. The *promoter* is a *cis*-acting element that is very close to a gene's initiation site; its binding of RNA polymerase allows a basal level of transcription (Fig. 17.1a). *Enhancers* are another class of *cis*-acting elements; their binding of proteins augments or represses basal levels of transcription. Enhancers are defined by their ability to retain original function even when moved far from the gene whose transcription they influence or when reversed in orientation.

(a) *cis*-acting elements

(b) *trans*-acting gene products interact with *cis*-acting elements

trans-acting genetic elements

Figure 17.1 How *cis*-acting and *trans*-acting elements influence transcription. (a) *cis*-acting regulatory elements are regions of DNA sequence that lie nearby to the gene they control. Promoter elements typically lie directly adjacent to the gene that they control. Enhancers that regulate expression can sometimes lie thousands of base pairs away from a gene. **(b)** *trans*-acting genetic elements encode protein products called transcription factors that interact with *cis*-acting elements, either directly through DNA binding or indirectly through protein-protein interactions. A stable multimeric transcription factor–enhancer complex is called an enhancesome.

Reporter constructs are one tool for studying gene regulation. They are synthesized in the laboratory to contain a gene's postulated regulatory region but not its coding region. Investigators can systematically identify promoters and enhancers by altering reporter constructs through *in vitro* mutagenesis across a presumed regulatory region and then reintroducing the reporter constructs into the genome by transformation. In assembling a reporter construct for this purpose, they replace the coding region of the gene whose regulation they are studying with the coding region of an easily identifiable product (the "reporter"), such as β-galactosidase or green fluorescent protein (GFP). (Recall that β-galactosidase turns blue in the presence of a substrate known as X-Gal [review Fig. 9.8]; similarly, GFP luminesces green when exposed to light of a particular wavelength.) Reporter constructs are particularly valuable for looking at mutations that affect gene expression rather than the amino-acid composition of the gene's polypeptide product. Mutations that alter the amount of reporter synthesized help define the elements necessary for a gene's regulation.

Regulatory Mutations That Map Far from a Target Gene Help Define *trans*-Acting Genetic Elements

***trans*-acting** elements are genes located elsewhere in the genome, that is, somewhere other than at the target gene; they encode proteins (or, in rare cases, RNAs) that interact directly or indirectly with the target gene's *cis*-acting elements to activate or repress expression of the target gene (Fig. 17.1b). *trans*-acting genes operate through their gene products, known as *trans*-acting factors. The *trans*-acting proteins that influence transcription are known generically as **transcription factors.**

Researchers can identify *trans*-acting elements in many ways. Mutations that alter phenotypes dependent on either the level of target gene expression or the level of expression of a reporter construct previously introduced into the genome, and that map far from the target gene or reporter construct, are likely to reside in *trans*-acting elements. Biochemical procedures are useful for isolating proteins that bind *in vitro* to *cis*-acting DNA sequences. Once researchers identify a *trans*-acting element, they can clone it for further study.

We now examine how *cis*-acting and *trans*-acting regulatory elements interact to initiate, increase, diminish, or prevent gene expression.

Most Gene Regulation Occurs at the Initiation of Transcription

During transcription in eukaryotes, RNA polymerase makes a single-stranded RNA molecule—known as the primary transcript—that is complementary in base sequence to a gene's DNA template strand (see Chapter 8, Fig. 8.18). Many *cis*-acting DNA elements and *trans*-acting regulatory proteins help initiate or prevent this first step of gene expression.

In Eukaryotes, Three RNA Polymerases Transcribe Different Sets of Genes

One of the first indications that events at the initiation of transcription usually determine gene regulation was the finding that different classes of genes in eukaryotic cells are transcribed by different RNA polymerases. The three types of RNA polymerases have some subunits in common, but they contain other subunits that are distinct to each type. Each of the three RNA polymerases transcribes a nonoverlapping class of genes (Fig. 17.2). RNA polymerase I (pol I) transcribes so-called class I genes, which encode the rRNAs. RNA polymerase II (pol II) transcribes class II genes, by far the largest class; they encode all proteins. RNA polymerase III (pol III) transcribes class III genes, which encode the tRNAs as well as other small RNA molecules.

RNA Polymerase I Is Responsible for the Transcription of rRNA

In most, if not all, eukaryotic genomes, the genes for rRNA are arranged in tandem repeats at sites on one or more chromosomes. As described in Chapter 12, these sites loop out from their respective chromosomes and come together in the nucleolus. Pol I transcribes rRNA from these clusters of class I genes in the nucleolus (review Fig. 12.16). Each tandem repeat has a promoter region just upstream of the transcription unit (Fig. 17.2a). The transcription of rRNA by pol I produces a large rRNA primary transcript. Subsequent RNA processing in the nucleolus breaks that transcript into a 28S rRNA and a 5.8S rRNA that become incorporated into the large ribosomal subunit and an 18S rRNA that becomes incorporated into the small subunit (Fig. 17.2a). These rRNAs associate with ribosomal proteins during assembly of the ribosome.

RNA Polymerase III Transcribes the tRNAs and Other Small RNAs

In addition to transcribing the tRNAs, which bring amino acids to the ribosome in response to codons in the mRNA, pol III transcribes a 5S rRNA that is yet another component of the large ribosomal subunit. It also transcribes some of the small nuclear RNAs (snRNAs) found in spliceosomes (Fig. 17.2b). As with class I genes, the promoter elements in some class III genes are just upstream of the gene's transcribed sequence. Surprisingly, however, studies of the transcriptional effects of small deletions in tRNA genes showed that the pol III promoter for these genes consists of sequences within the DNA transcribed to produce the tRNA. These sequences thus serve two functions: They are promoter elements recognized by specific transcription factors; and, after the initiation of transcription, they serve as part of the tRNA gene that is transcribed into the primary transcript.

RNA Polymerase II Transcribes All Protein-Encoding Genes

Most of the primary transcripts produced by RNA polymerase II (pol II) undergo further processing to generate mRNAs (Fig. 17.2c). During mRNA formation, introns are spliced out. In addition, ribonuclease cleaves pol II–transcribed primary transcripts to form a new 3′ end, to which the enzyme poly-A polymerase adds a poly-A tail; and the chemical modification of the 5′ end of the transcript produces a "5′GTP cap," which protects the molecule from degradation.

(a) Tandem repeats of rRNA genes are transcribed by RNA polymerase I

(b) Small RNA genes are transcribed by RNA pol III

(c) Protein-encoding genes are transcribed by RNA pol II

Figure 17.2 The three RNA polymerases of eukaryotic cells have different functions and recognize different promoters. **(a)** RNA polymerase I recognizes promoters associated with ribosomal RNA (rRNA) genes, which always exist in tandem arrays. The primary rRNA transcripts are processed into smaller RNA molecules that form components of the ribosome. **(b)** RNA polymerase III recognizes promoters associated with a limited number of genes that encode small RNA molecules involved in the translation process. **(c)** RNA polymerase II recognizes promoters associated with all of the diverse protein-encoding genes in the genome.

Measurements of mRNA and protein levels in eukaryotic cells have revealed that, usually, the more mRNA of a gene that accumulates in the cell, the greater the production of that gene's protein product. Further biochemical experiments showed that the control of mRNA production almost always reflects the rate at which transcription is initiated, rather than the rate of elongation or termination of transcription, or the stability of the mRNA.

cis-Acting Regulatory Regions Recognized by Pol II Consist of a Promoter and One or More Enhancers

Although each of the regulatory regions of the thousands of class II genes in a eukaryotic genome is unique, they all contain two kinds of essential DNA sequences (Fig. 17.2c). The **promoter** is always very close to the gene's protein-coding region. It includes an initiation site, where transcription begins, and a "TATA" box, consisting of roughly seven nucleotides of the sequence T–A–T–A–(A or T)–A–(A or T). **Enhancers** are regulatory sites that can be quite distant—up to tens of thousands of nucleotides away—from the promoter. In yeast, enhancer elements are often called *upstream activation sites,* or *UAS*s.

The enhancer regions of some class II genes are very large, containing multiple elements that make it possible to fine-tune regulation of the gene. This is particularly true of those genes in multicellular eukaryotes that must be expressed in many different tissues. The *string* gene in *Drosophila* is an example. The gene encodes a protein product that activates the fourteenth mitosis of embryonic development. This fourteenth mitosis begins just after membranes simultaneously form around the roughly 6000 nuclei of the giant syncytium that resulted from the first 13 mitoses (review the Fast Forward box in Chapter 4). What is interesting about the fourteenth mitosis is that cells in different areas, or domains, of the embryo enter it at different times in an intricate but reproducible temporal pattern (Fig. 17.3a). Thus, although the cells of each domain sustain the fourteenth nuclear division together, the time at which that mitosis takes place is different for different domains. Remarkably, the cells within each domain simultaneously express the *string* gene just before they enter mitosis; in fact, expression of the *string* gene induces their entry into the mitotic cycle. In *string* mutants, all embryonic cells arrest in the G2 stage of cycle 14 and never undergo mitosis. A roughly 35 kb region upstream of the *Drosophila string* gene contains binding sites for many transcription factors known to regulate formation of the *Drosophila* body pattern (Fig. 17.3b). The complex interaction of these factors ensures that the *string* gene is turned on in the cells of each embryonic domain at the correct time. Interestingly, the approximately 25 embryonic mitotic domains correspond, in part, to specific tissues or structures in the *Drosophila* larva.

trans-Acting Proteins Control Transcription from Class II Promoters

The binding of *trans*-acting proteins (transcription factors) to a gene's promoter and enhancer (or enhancers) controls the

(a)

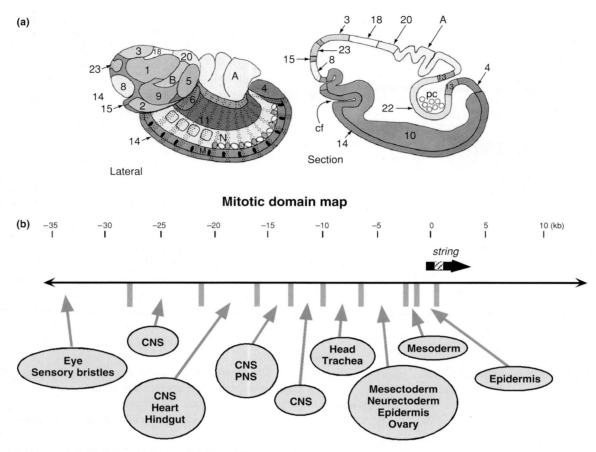

Mitotic domain map

Figure 17.3   The large enhancer region of the *Drosophila string* gene helps create discrete mitotic domains during embryogenesis. (a) *Colors* indicate individual mitotic domains during the fourteenth cell cycle of the fruit fly embryo. The cells within each domain divide synchronously, but different domains initiate their divisions at different times. Both lateral and cross-sectional views of the embryo are shown. **(b)** A variety of enhancer regions ensure that *string* is turned on at the right time in each mitotic domain and tissue type (CNS: central nervous system; PNS: peripheral nervous system).

initiation of transcription from each class II gene. Different types of proteins bind to each of the *cis*-acting regulatory regions: *Basal factors* bind to the promoter; *activators* and *repressors* bind to the enhancers.

Basal Factors Are Required to Maintain a Basal Level of Transcription

Basal factors assist the binding of RNA polymerase II to the promoter and the initiation of low levels of transcription called basal transcription (from which the basal factors get their name). The key component of the basal factor complex that forms on most promoters is the TATA box–binding protein, or **TBP** (so named because it binds to the TATA box described previously). The TBP is essential to the initiation of transcription from all class II genes that have a TATA box in their promoter. TBP associates with several other basal factors called TBP-associated factors, or **TAFs** (Fig. 17.4). The complex of basal factors bind to the proximal promoter in an ordered pathway of assembly. Once the complex has formed, basal transcription is initiated. Researchers have determined the structure of the TBP-TAF complex on the DNA at the TATA box (Fig. 17.4). The

most striking feature of the structure is a sharp bend in the DNA at the TATA box, induced by TBP.

The primary sequence and three-dimensional structure of the basal factors are highly conserved in all eukaryotes, from yeast to humans, and this sequence conservation has facilitated identification of the factors. For example, researchers isolated yeast TBP—the first basal factor to be purified—through its ability to substitute for mammalian TBP *in vitro*.

Activators Bind to Enhancer Sequences and Can Increase Transcription 100-Fold Above the Basal Level

Although similar sets of basal factors bind to all the promoters of the tens of thousands of genes in the eukaryotic genome, a cell can express different genes in widely varying amounts of mRNA. This enormous range of gene regulation occurs through the binding of different transcription factors to different enhancer elements associated with different genes. When the transcription factors bind to an enhancer element, they can interact directly or indirectly with basal factors at the promoter to cause an increase in transcriptional initiation (Fig. 17.5a). Because of this transcription-increasing activity, the factors are called

Figure 17.4 Basal factors bind to promoters of all protein-encoding genes. (a) Schematic representation of the binding of the TATA box–binding protein (TBP) to the promoter DNA, the binding of two TBP-associated factors (TAFs) to TBP, and the binding of RNA polymerase (pol II) to these basal factors. **(b)** Computer-generated image of the actual molecular structure of basal factors complexed to the promoter.

Figure 17.5 Transcriptional activators bind to specific enhancers at specific times to increase transcriptional levels. (a) In the presence of basal factors alone bound to the promoter, low levels of transcription occur. The binding of activator proteins to an enhancer element leads to an increase in transcription beyond the basal level. **(b)** Examples of peptide motifs that appear in the DNA-binding domains of activator proteins and their interaction with enhancer elements. A zinc-finger domain extends into the groove of the double helix. A helix-loop-helix protein wraps around the DNA molecule. **(c)** Some activator domains are themselves activated into a DNA-binding conformation through allosteric changes caused by the binding of a steroid hormone molecule to another domain within the activator protein.

activators. Researchers have already identified hundreds of eukaryotic activators, and it is likely that each eukaryotic genome encodes several thousand of them.

To carry out their function, activator proteins (the transcription factors) must bind to enhancer DNA in a sequence-specific way; and after binding, they must be able to interact with other proteins to activate transcription. Two structural domains within the activator protein—the DNA-binding domain and the transcription-activator domain—mediate these two biochemical functions.

A small number of peptide motifs appear over and over again in the DNA-binding domains of many different transcription factors (Fig. 17.5b). The best characterized of these motifs are the zinc-finger, the helix-loop-helix, and the helix-turn-helix conformations. The general function of each of these motifs is to promote binding to the DNA double helix. Subtle differences in amino-acid sequence among activators can specify high-affinity binding to different DNA sequences associated with different enhancer elements. Extensive repetition of coding regions for the same peptide motif increases the genomic potential for evolving new activator proteins with new DNA-binding specificity. The duplication of an existing gene, followed by a slight divergence that by chance increases the binding affinity for a different DNA sequence, could generate a new protein that acts at a different target gene.

Many of the transcription-activation domains associated with different activators also have features in common. One subset is relatively acidic and rich in the amino acid asparagine; another is rich in prolines. Other transcription-activation domains are unique to the protein in which they occur.

Some activators have a third domain that is responsive to specific signals from the environment. An example of activators with this type of domain are the steroid hormone receptors (Fig. 17.5c). Each receptor has a domain that is unique for a particular steroid. The binding of this steroid causes an allosteric change that greatly increases the affinity of the DNA-binding domain of the protein for its target enhancer sequence. Once bound, the hormone-receptor complex activates transcription of its target genes. In the absence of hormone, DNA binding does not occur, and target genes remain unactivated, that is, transcribed only at basal rates. There are no universal features of signal response domains.

Polypeptides and other molecules that play a role in transcriptional activation without binding directly to DNA are called **coactivators.** The hormone component of a DNA-bound hormone-receptor activation complex is one example of a coactivator. The term **enhancesome** is used to describe a multimeric complex of proteins and other small molecules associated with an enhancer element (see Fig. 17.1b); the multimeric complex of proteins can include activators, coactivators, and other types of transcription factors known as *repressors* and *corepressors* (which we define shortly).

Molecular biologists have identified and characterized the different domains associated with different activators through the use of recombinant DNA constructs. These constructs encode proteins with a well-defined domain from a well-characterized

Figure 17.6 Localization of activator domains within activator proteins can be achieved with recombinant DNA constructs. The figure shows a strategy to determine which portion of an activator protein contains the actual domain that interacts with basal proteins to bring about activation. In this simplified example, the gene encoding the activator protein is divided into three parts, A, B, and C, and each part is fused to a portion of the bacterial *lexA* repressor gene that encodes a well-defined DNA-binding domain, and to a yeast cell promoter. Each three-part fusion construct is placed into different yeast cells, where the production of a fusion protein takes place. The yeast cell also contains a reporter gene with a downstream enhancer element that will bind to all three fusion proteins through their lexA-encoded binding domain. The reporter gene can only be transcribed if an activator domain is also present in the fusion construct. In this example, the results of the reporter gene readout demonstrate the presence of a complete activation domain in part B of the activator protein, but not in parts A or C.

transcription factor fused to a series of polypeptide regions from the activator under analysis (Fig. 17.6). For example, the bacterial repressor lexA, which contains only a DNA-binding domain and no activator domain, binds to the same well-defined DNA sequence whether it is in its native prokaryotic cell or in an engineered eukaryotic cell. One can fuse different segments of an activator protein gene to *lexA* and produce this fusion protein in an engineered yeast cell in which the lexA-binding site has been placed upstream of a special reporter gene. Activation (or lack of activation) of the reporter gene can then signal that the segment from the activator under analysis does (or does not) have a complete activation domain.

Most Eukaryotic Activators Must Form Dimers to Function

Molecular analyses have shown that many eukaryotic transcription factors are **homomers** (that is, multimeric proteins composed of identical subunits) or **heteromers** (multimeric proteins composed of nonidentical subunits; review Fig. 7.24). Among

Figure 17.7 Most activator proteins function in the cell as dimers. (a) Homodimers contain two identical polypeptides, while heterodimers contain two different polypeptides. **(b)** The polypeptides within a dimer bind to each other through a dimerization domain. A common peptide motif present within dimerization domains is the leucine zipper.

the best-characterized transcription factors of this type is Jun, which can form dimers (multimers composed of two subunits) with either itself or with another protein called Fos (Fig. 17.7a). The Jun-Jun dimers are *homodimers;* the Jun-Fos dimers are *heterodimers.*

Dimerization occurs through yet another transcription factor domain, the **dimerization domain,** which is specialized for specific polypeptide-to-polypeptide interactions. As with other transcription factor domains, certain motifs recur in dimerization domains. One of the most common is the leucine zipper motif (Fig. 17.7b), an amino-acid sequence that twirls into an α helix with leucine residues protruding at regular intervals. The motif received its name from the propensity of one leucine zipper motif to interlock like a zipper with a leucine zipper motif on another polypeptide. The ability of two leucine zippers to interlock depends on the specific amino acids that lie between the leucines.

The Jun and Fos polypeptides both contain leucine zippers in their dimerization domains. A Jun leucine zipper can interact with another Jun leucine zipper or with a Fos leucine zipper. But the Fos leucine zipper *cannot* interact with its own kind to form a homodimer. Neither Jun nor Fos alone can bind DNA, so neither can act as a transcription factor as a monomer. Thus the Jun-Fos transcription factor system can produce only two types of transcription factors: Jun-Jun proteins or Jun-Fos proteins. Both bind to the same enhancer elements, but with different affinities.

The ability to form heterodimers greatly increases the number of potential transcription factors a cell can assemble from a set number of gene products. In theory, 100 polypeptides could combine in different ways to form 5000 different transcription factors; with 500 polypeptides, the number jumps to 125,000.

Repressors Diminish Transcriptional Activity

Some transcription factors suppress the activation of transcription caused by activator proteins. Any transcription factor that has this effect is considered a **repressor.** Different repressors act in different ways. Some compete with activator proteins for binding to the same enhancer (Fig. 17.8a). When a repressor binds to an enhancer, it blocks the activator's access to the same sequence. The Myc-Max system described in the following section provides an example of this type of activator-repressor competition.

(a) Competition for binding between repressor and activator proteins

(b) Quenching

Type I: Repressor binds to and blocks the DNA-binding region of an activator.

Type II: Repressor binds to and blocks the activation domain of an activator.

Figure 17.8 Repressor proteins reduce transcriptional levels through competition or quenching. (a) Some repressor proteins act by *competing* for the same enhancer elements as activator proteins. But repressor proteins have no activation domain, so when they bind to enhancers, no activation of transcription can occur. **(b)** A second class of repressors act by binding directly to the activator proteins themselves to *quench* activation in one of two ways. Type I quenching is achieved when the repressor prevents the activator from reaching the enhancer. Type II quenching is achieved when the activator can bind to the enhancer but the repressor prevents the activation domain from binding to basal proteins.

Some repressors operate without binding DNA at all. Instead, in a mechanism called *quenching,* they bind directly to a specific activator (Fig. 17.8b). In one type of quenching, a repressor binds to and blocks the DNA-binding region of an activator, thereby preventing the activator from attaching to its enhancer. In another type of quenching, a repressor binds to and blocks the activation domain of an activator. These blocked activators still bind to their enhancers, but once bound, they are unable to carry out activation. Quenching polypeptides that operate in this manner are **corepressors.** Just like coactivators, corepressors associate indirectly with enhancers through their interaction with DNA-binding proteins. An example of this type of repression is seen in the yeast GAL system, which we discuss later.

The repression resulting from both activator-repressor competition and quenching reduces activation, but it has no effect on basal transcription. As in prokaryotes, however, some eukaryotic repressors act directly on the promoter to eliminate almost all transcriptional activity. They can do this by binding to DNA sequences very close to the promoter and thereby blocking RNA polymerase's access to the promoter. Or they can bind to DNA sequences farther from the promoter, and then reach over and contact the basal factor complex at the promoter, causing the DNA between the enhancer and promoter to loop out and allow contact between the repressor and the basal factor complex. This second mechanism also denies RNA polymerase access to the promoter and reduces transcription below the basal level.

Whether a transcription factor acts as an activator or a repressor, or has no effect at all, depends not only on the cell type in which it is expressed but also on the gene it is regulating. This is one reason why *cis*-acting elements that bind either activators or repressors are all referred to as enhancers, even though some may actually repress transcription when associated with the appropriate protein. As mentioned earlier, all enhancers are defined by their ability to function when placed at variable distances from a promoter and when reversed in orientation.

The specificity of transcription factors can be altered by other molecules in the cell. One example of this phenomenon is observed with the yeast α2 repressor, which helps determine the mating type of a cell. Yeast cells can be either haploid or diploid, and haploid cells come in two mating types: α and a. In α cells, the α2 repressor binds to enhancers that control the activity of a set of a-determining genes, whose expression would make the cell type a. The binding of the α2 repressor to these a-determining genes is one step in the generation of α cells (Fig. 17.9a). In diploid yeast cells, however, the same α2 repressor plays a different role. In such cells, expression of the polypeptide known as a1 occurs; the binding of a1 to the α2 repressor alters the repressor's DNA-binding specificity such that α2 now binds to enhancers associated with a set of haploid-specific genes, repressing the expression of those genes (Fig. 17.9b). To summarize, in diploid cells, the α2 repressor maintains the diploid state by repressing haploid-specific genes.

The Myc-Max System Is a Regulatory Mechanism for Switching Between the Activation and Repression of Transcription

The ratio of heterodimer partners within a cell can have profound regulatory consequences. The Myc-Max transcription factor system, which evolved in multicellular eukaryotes, illustrates this point (Fig. 17.10). Through the identification of mutations affecting *myc* gene expression in one class of lymphocytes (found in a form of cancer called Burkitt's lymphoma), researchers

(a) Haploid α cell

(b) Diploid α/a cell

Figure 17.9 The same transcription factors can play different roles in different cells. (a) In haploid α yeast cells, the α2 factor acts to silence the set of "a" genes. **(b)** In α/a diploid yeast cells, the α2 factor dimerizes with the a1 factor and acts to silence the set of haploid-specific genes.

showed that *myc* plays a critical role in the regulation of cell proliferation. However, although the genetic data suggested that the Myc protein is a transcription factor, biochemists could find no evidence for this function *in vitro*. Their experiments revealed that even though the Myc polypeptide contains both a helix-loop-helix (HLH) motif and a leucine zipper, it cannot bind to DNA or form homodimers. The apparent contradiction between genetic and biochemical results stymied the scientists who first associated mutations in the *myc* gene with Burkitt's lymphoma and other forms of cancer.

The discovery of the *max* gene product helped resolve this dilemma. Like Myc, Max contains a HLH motif and a leucine zipper (Fig. 17.10a). Moreover, both Myc and Max contain another, more recently defined DNA-binding motif called a "basic motif" (because it contains mostly basic amino acids). Unlike Myc, however, the Max polypeptide can form homodimers. When one mixes Max with Myc, heterodimers of the two polypeptides form.

In the Myc-Max system, the HLH and leucine zipper are both components of the dimerization domain (Fig. 17.10a). The basic motif, by comparison, defines the DNA-binding domain, which functions only in dimeric proteins, never in monomers. This restriction of binding ability is a common feature of all transcription factor systems based on dimerization.

The Myc polypeptide contains an activation domain, but when the molecule is on its own, it cannot bind DNA and thus cannot serve as an activator. The Max polypeptide, on the other hand, can form homodimers and can bind DNA when present without Myc; but Max has no activation domain, so it cannot function as an activator even when it does bind to DNA. It is only when Myc and Max come together in a heterodimer that

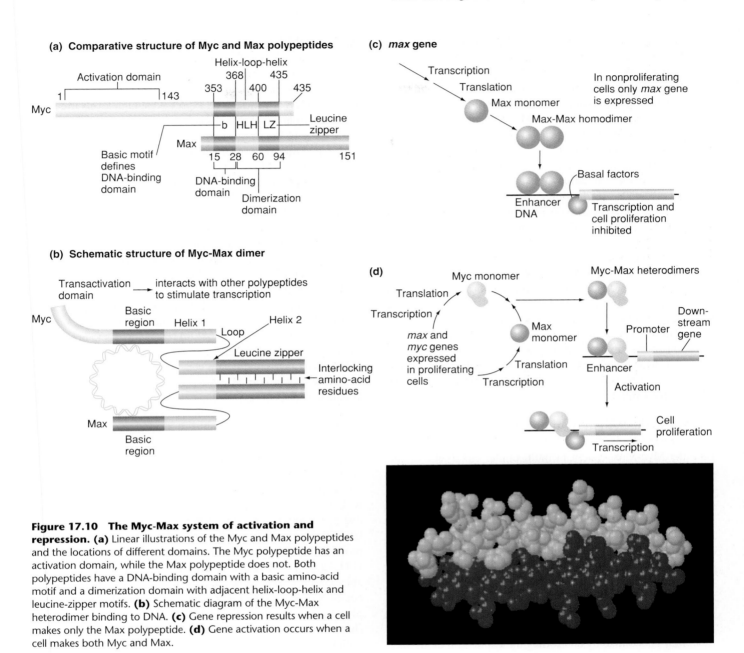

(a) Comparative structure of Myc and Max polypeptides

(c) *max* gene

(b) Schematic structure of Myc-Max dimer

(d)

Figure 17.10 The Myc-Max system of activation and repression. (a) Linear illustrations of the Myc and Max polypeptides and the locations of different domains. The Myc polypeptide has an activation domain, while the Max polypeptide does not. Both polypeptides have a DNA-binding domain with a basic amino-acid motif and a dimerization domain with adjacent helix-loop-helix and leucine-zipper motifs. **(b)** Schematic diagram of the Myc-Max heterodimer binding to DNA. **(c)** Gene repression results when a cell makes only the Max polypeptide. **(d)** Gene activation occurs when a cell makes both Myc and Max.

both DNA binding and Myc-directed activation become possible, and a transcriptional activator is born (Fig. 17.10b).

Myc-Max heterodimers and Max-Max homodimers both bind to the same enhancer sequences associated with multiple genes that contribute to cell proliferation. The binding of a heterodimer results in transcriptional activation, whereas the binding of a homodimer results in transcriptional repression.

One final characteristic of this system is that Myc polypeptides have a much higher affinity for Max polypeptides than Max has for its own polypeptides. Thus, when Myc and Max are in solution together, the predominant dimer is a heterodimer. With this extensive background, we are ready to see how the cell uses the Myc-Max system to respond rapidly to signals that tell it to proliferate or stop proliferating.

The *max* gene is expressed in all cells at all times, but since its protein product does not carry an activation domain, Max-Max homodimers, when bound to enhancer DNA, inhibit transcription and therefore inhibit cell proliferation (Fig. 17.10c). By contrast, the *myc* gene is not universally expressed; rather the Myc polypeptide is normally synthesized in cells undergoing proliferation but not in cells at rest.

As soon as a cell expresses its *myc* gene, all its Max-Max homodimers convert to Myc-Max heterodimers that bind to the enhancers previously bound by Max-Max homodimers. Since the heterodimers include the Myc activation domain, the binding of Myc-Max complexes induces the expression of genes required for cell proliferation. Although researchers have not yet characterized all the genes activated by the Myc-Max dimer, they know that the genes guide the cell through its mitotic cycle. Thus, each cell in which *myc* is active divides to produce two daughter cells.

Interestingly, when cells reach a state of terminal differentiation, they stop expressing *myc,* and since the Myc protein is unstable and easily degraded, any Myc remaining from earlier *myc* expression quickly disappears. The loss of Myc allows the remaining Max polypeptides to form homodimers, whose binding to the enhancer once again represses the expression of genes required for cell proliferation. When this happens, terminally differentiated cells stop dividing.

In short, the Myc-Max system provides a rapid genetic switch for regulating cell division during the cell proliferation and terminal differentiation phases of development. The system functions by keeping the *max* gene turned on in all cells, carefully regulating the expression of *myc,* allowing homodimerization of Max but not Myc, allowing Myc-Max heterodimerization to outcompete Max-Max homodimerization, and allowing the rapid degradation of Myc. As a result, the expression of *myc* quickly turns on genes required for cellular proliferation and the cessation of *myc* expression just as quickly turns those same genes off. Mutations in the *myc* regulatory region can lead to constitutive expression of this critical gene, causing abnormal cellular proliferation and cancer.

The Yeast GAL System Is Another Complex Regulatory Mechanism

Like the Myc-Max system in multicellular eukaryotic organisms, the GAL system in unicellular yeast has been well characterized genetically and biochemically. The biochemical function of the GAL system is to modulate the expression of three linked genes—*GAL1, GAL7,* and *GAL10*—that encode the enzymes galactokinase, transferase, and epimerase (Fig. 17.11a). A cell must produce all three enzymes to use galactose as a source of energy. When galactose is present, the cell activates the genes encoding these enzymes. When galactose is absent, repression of the three genes allows the cell to avoid the unnecessary expenditure of energy and materials.

Since yeast cells always need the three enzymes at the same time, they economize by controlling the expression of all three structural genes from a single enhancer. Mutations in this enhancer can be detected by their effect on the regulation of all three gene products rather than just one.

Researchers identified the *GAL4* locus as a *trans*-acting genetic regulator of *GAL1, GAL7,* and *GAL10* through mutations that prevent galactose from activating expression of the three structural genes. Biochemical studies showed that the *GAL4* product has a DNA-binding domain containing multiple zinc-finger motifs and an acidic activation domain; it binds to an enhancer several hundreds of base pairs upstream of the *GAL1, GAL7, GAL10* genes and activates their transcription. Mutations in the *GAL4* coding region prevent its product from functioning as an activator.

Researchers identified another *trans*-acting genetic element—*GAL80*—through mutations that caused continuous expression of the three GAL enzymes, even in the absence of galactose. This genetic result suggested that the product of the *GAL80* gene is a repressor of *GAL1, GAL7,* and *GAL10.* When this repressor is absent, constitutive expression can occur. (Recall that a *constitutive mutation* is one that causes a specified gene, or group of genes, to be expressed at high levels irrespective of environmental changes that can affect transcription in the wild-type cell. *Constitutive expression* refers to an active state of gene expression that is unaffected by environmental conditions.)

On the basis of further genetic and biochemical studies, the researchers proposed that the GAL80 product binds to the activation domain of *GAL4* (Fig. 17.11b). GAL80 would thus be a *corepressor* which does not bind directly to DNA; instead it quenches the activating ability of the GAL4 product by sequestering its activation domain. To test this model, they constructed double mutants in which they had knocked out both *GAL4* and *GAL80.* In these *GAL4⁻* / *GAL80⁻* cells, no induction of GAL1, GAL7, and GAL10 occurred. This means that, in genetic terms, the GAL4 phenotype is epistatic to the GAL80 phenotype. Such a result is consistent with the biochemical model because to function as a repressor, the GAL80 product needs to operate on the GAL4 product. In the absence of *GAL4,* mutations in *GAL80* will have no effect on gene regulation.

How does a normal yeast cell use this genetic regulatory system to control the expression of the *GAL1, GAL7,* and *GAL10* structural genes in response to the presence or absence of galactose? The answer is by controlling the physical placement of the GAL80 gene product. When galactose is present in the cell, it attaches to both the GAL1 product (a very small amount of which is produced through basal transcription), as well as to another constitutively expressed polypeptide called GAL3. Association with galactose causes an allosteric change in GAL1 and GAL3 that gives both polypeptides a high-affinity binding site for the GAL80 product. When either polypeptide binds to the GAL80 product, this binding somehow dislodges GAL80 from the GAL4 activation domain. With its activation domain now available, GAL4 is able to activate the *GAL1, GAL7, GAL10* gene complex (Fig. 17.11c). Later, when the cell has depleted its supply of galactose, the GAL1 and GAL3 products again change shape, reverting to the conformation in which they do not have a high affinity for GAL80 and falling off. GAL80, when no longer bound to GAL1 or GAL3, moves back into position to block the GAL4 activation domain. The binding of GAL80 to the activation domain of GAL4 once again reduces transcription of the three *GAL* structural genes to basal levels.

A Locus Control Region Is a *cis*-Acting Regulatory Sequence That Operates Sequentially on a Cluster of Related Genes

Promoters and enhancers are not the only kinds of *cis*-acting regulatory regions active in eukaryotic genomes. Clues that another kind of *cis*-acting element exists came from experiments in which the promoter and adjacent enhancer region of the β-globin gene were not sufficient to mediate its full transcriptional activation.

As we saw in Chapter 9 (introduction and comprehensive example), the β-globin gene is part of a cluster of related genes that function at different times during mammalian development

Figure 17.11   The GAL1, GAL7, GAL10 gene system is a model of complex transcriptional regulation. (a) The three genes have independent promoters but are all controlled by the same enhancer element. **(b)** The activator protein GAL4 binds to this enhancer, but its activity is quenched by the repressor protein GAL80. **(c)** Galactose molecules in the cell bind to GAL1 and GAL3 polypeptides, which undergo allosteric changes that allow them to bind to GAL80, causing GAL80 to move into a position that is no longer quenching the activator GAL4. This allows activation of the GAL1, GAL7, and GAL10 genes, whose products bring about the enzymatic digestion of galactose. Once galactose is depleted, GAL1 and GAL3 can no longer bind to GAL80, and it moves back into position to quench GAL4.

to produce polypeptides that become incorporated into hemoglobin molecules with different oxygen-binding kinetics (Fig. 17.12a). The β-globin gene was among the first mammalian genes to be cloned, characterized, and used in the production of transgenic mice. Results from the earliest transgenic studies were disappointing. Even when researchers introduced into transgenic mice the entire region of noncoding genomic DNA 5′ of the β-globin gene (which separates that gene from the flanking gene in the cluster), the level of β-globin transgene transcription in the mice was 10 times less than that observed with the endogenous gene.

Subsequent experiments showing that other transgenes with just the 5′-flanking sequences (containing the promoter and all of the enhancer elements) were fully expressed suggested that the β-globin gene is different. For full transcription, it requires a *cis*-acting element that was missing from the DNA containing both the gene and the flanking sequence that separates it from its nearest neighbor (Fig. 17.12b).

(a) Gene complex

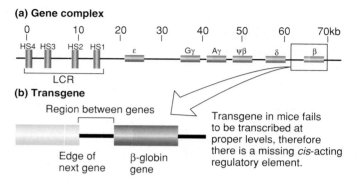

(b) Transgene

Region between genes

Edge of next gene

β-globin gene

Transgene in mice fails to be transcribed at proper levels, therefore there is a missing *cis*-acting regulatory element.

(c) Mechanism of transcriptional activation

[figure showing gene loop with activator proteins, RNA polymerase, basal factors, transcription, β globin, δ, Gγ, ε, Aγ, δ, β]

Activator proteins

Basal factors

RNA polymerase

Transcription

β globin

δ

Figure 17.12 A locus control region (LCR) operates over a cluster of genes to activate transcription. **(a)** The human β-globin gene cluster contains five functional genes that can all undergo *cis*-regulation by a distant LCR. The LCR associated with this particular gene cluster contains four separate *cis*-acting DNA elements designated HS1 through HS4. **(b)** A transgene composed of an isolated β-globin gene together with the entire 5′ flanking region between it and the δ gene fails to undergo proper expression when it is inserted into the mouse genome. **(c)** The LCR at the 5′ end of the β-globin gene cluster is required for transcriptional activation of each of the genes in the cluster in hemoglobin-producing cells at different points in development. The four DNA elements of the LCR (HS1 through HS4) could effect activation of the distant δ-globin gene through DNA looping, which brings each of these elements and their associated transcription factors into contact with the β-globin promoter and its associated basal factors and RNA polymerase.

Surprisingly, investigators found that the missing element, located 50 kb upstream of the β-globin gene, is separated from the β-globin gene by four other genes in the cluster. This regulatory element is called the **locus control region (LCR)** because it exerts a higher level of control, influencing the transcription of all genes in the β-globin cluster. Like an enhancer, the LCR functions by binding to transcription factors with activation domains. Unlike most enhancers, however, the LCR–transcription factor complex interacts sequentially during development with other transcription factors at *cis*-regulatory regions that are directly adjacent to each gene within the β-globin gene cluster (Fig. 17.12c). An assembled LCR–transcription factor complex—like an assembled enhancer-transcription factor complex—is known as an *enhancesome*. If any component of the enhancesome is missing, the program does not work, and no activation of any genes in the cluster takes place. Cells other than the precursors of red blood cells do not express the genes for β-globin-LCR-specific enhancesome components; as a result, there is no functional enhancesome and no expression of any of the β-globin genes in these cells. The LCR for the β-globin cluster thus pro-

vides a single mechanism for ensuring that none of the genes in the cluster is expressed in cells outside the red blood cell lineage.

Complex Regulatory Regions Enable an Organism to Fine-Tune Gene Expression

In complex multicellular organisms, gene regulation is not just a matter of turning genes on and off. It also entails fine-tuning the precise level of transcription—higher or lower in different cells, higher or lower in cells of the same tissue at different stages of development. It also includes mechanisms that allow each cell to modify its program of gene activity in response to constantly changing signals from its neighbors. Organisms accomplish the orchestration of transcription from each of tens of thousands of genes through *cis*-acting regulatory regions that are often far more complex than those we have so far described.

A regulatory region may contain a dozen or more enhancer elements, each with the ability to bind different activators and repressors, with varying affinities. At any moment, there may be dozens of transcription factors in the cell whose affinities for DNA or other polypeptides are being modulated by binding to hormones or other molecules. Different sets of these transcription factors compete for different enhancers within the regulatory region. And different sets of coactivators and corepressors compete with each other for binding to different activators or repressors. The biochemical integration of all this information yields a precise level of transcriptional activation or repression.

As we saw earlier, a complex of regulatory factors together with the enhancer to which they are bound is an enhancesome. Slight changes in a cell's environment can dispatch signal molecules that cause changes in the balance of transcription factors, or in their relative affinities for DNA or for each other. These changes, in turn, lead to the assembly of an altered enhancesome, which recalibrates gene activity.

Molecular biologists and biochemists are not yet able to unravel the details of such complex regulatory networks, but their knowledge of them increases every day.

Chromatin Structure Plays a Role in Eukaryotic Gene Regulation

We know from Chapter 12 that the DNA of eukaryotic genomes does not float freely in the nucleus but is normally packaged into chromatin (review Figs. 12.4 and 12.5 and Table 12.2). The basic repeating structural unit of chromatin is the nucleosome, which consists of a ball of histone proteins (two each of H2A, H2B, H3, and H4) around which is wrapped approximately 200 bp of DNA. Histone H1, which binds to short stretches of DNA that lie between nucleosomes, helps maintain chromatin structure. The binding of nonhistone proteins is the basis of higher-order structures that further compact DNA.

The Normal Structure of Chromatin Provides a Brake on Runaway Basal Transcription

In vitro experiments show that basal factors and RNA polymerase readily bind to promoters on naked DNA and initiate high levels of transcription in the absence of activator proteins (Fig. 17.13a). One significant function of chromatin, then, is the

(a) Naked promoter binds RNA polymerase and basal factors.

(b) Chromatin reduces binding to basal factors and RNA pol II to very low levels.

(c) Chromatin remodeling can expose promoter.

(d) DNase hypersensitive sites are at 5' ends of genes.

(e) The SWI-SNF multisubunit complex is an example of a remodeling complex.

The SWI-SNF multisubunit destabilizes chromatin structure and gives transcription machinery access to the promoter.

Figure 17.13 Chromatin structure plays a critical role in eukaryotic gene regulation. (a) DNA molecules containing a promoter and an associated gene can be purified away from chromatin proteins *in vitro.* The addition of basal factors and RNA polymerase to this purified DNA induces high levels of transcription. **(b)** In its natural state within the eukaryotic nucleus, DNA is present within chromatin. The underlying structure of chromatin is based on repeating nucleosomes that each contain two loops of DNA wrapped around a histone core. Promoter regions are generally sequestered within the nucleosome and only rarely bind to basal factors and RNA polymerase. Thus, the chromatin structure maintains basal transcription at very low levels. **(c)** Chromatin remodeling can expose the promoter region. Remodeling proteins can cause specific nucleosomes to unravel in specific cells at specific times during differentiation or development. Exposed promoter regions more readily bind basal factors. **(d)** Exposed regions of DNA sequence can be recognized by their hypersensitivity to DNase digestion. **(e)** The SWI-SNF protein complex is an example of a well-characterized remodeling apparatus that functions within yeast cells to expose promoter regions to basal factors, RNA polymerase, and transcriptional activation.

reduction of basal transcription from all genes to a very low level. In contrast to transcriptional modulation in prokaryotes, which requires active repression through the binding of repressors to *cis*-acting elements, in eukaryotes, the normal structure of chromatin is sufficient in and of itself to maintain transcriptional activity at the minimal, basal level (Fig. 17.13b). Thus, eukaryotic repressors play a different role than their prokaryotic counterparts. The main role of eukaryotic repressors is to modulate the activation of transcription caused by transcriptional activators. They do this through competition for DNA-binding sites, as described earlier for the Myc-Max system, or through quenching, as described for the yeast GAL system.

The Remodeling of Chromatin Mediates the Activation of Transcription

An initial component of gene activation in eukaryotes is the remodeling of chromatin structure in the promoter region. Specialized proteins that unravel promoter DNA sequences away from the histone core are the agents of this remodeling. The freed DNA becomes much more accessible to transcription factors that then control transcription (Fig. 17.13c).

Chromosomal regions from which the nucleosomes have been eliminated are experimentally recognizable through their hypersensitivity to the enzyme DNase (as described in Chapter 12). When one scans a chromosome with the enzyme for the presence of DNase hypersensitive (DH) sites, the sites show up at the 5′ ends of genes that are either undergoing transcription or are being prepared for transcription in a later step of cellular differentiation (Fig. 17.13d). For example, DH sites appear at the 5′ end of the β-globin gene in human cells that are precursors to those in which the gene will be activated, but not in cells from other differentiative pathways.

Studies of chromatin structure show that nucleosomes sit atop the promoters of most inactive genes. Protein complexes that remove these promoter-blocking nucleosomes, or reposition them in relation to the gene, help put a gene in an activatable state. For example, the SWI and SNF proteins in yeast form a multisubunit complex that disrupts chromatin structure by removing or repositioning nucleosomes. The resulting chromatin decompaction gives basal factors much greater access to promoter regions, and consequently, transcription rapidly accelerates (Fig. 17.13e). Human cells contain related multisubunit protein complexes that also influence nucleosome position or structure, suggesting that this particular nucleosome-disrupting machinery has been conserved throughout evolution. The SWI-SNF protein complex represents just one of the many that help remodel chromatin at specific chromosomal locations in specific cells at particular points of development.

Hypercondensation over Chromatin Domains Causes Transcriptional Silencing

The heterochromatic regions of eukaryotic chromosomes, including parts of centromeres and telomeres and the whole of Barr bodies, are highly condensed. As a result, the genes contained in heterochromatin are completely silent without even basal transcription (review Figs. 12.14 through 12.16).

The **transcriptional silencing** of heterochromatin makes it impossible to activate genes within a heterochromatic region, no matter what transcription factors are active in the cell, as long as the silencing structure is in place. By contrast, as we have seen, substantial expression of genes in unsilenced regions depends on the outcome of competition between activators and repressors. The appropriate combination of signals can always activate these genes.

In mammals, the chromosomal regions that are silenced by heterochromatin vary with the cell type, but they are often associated with DNA methylation (Fig. 17.14a). Methylation—the addition of a methyl (CH_3) group—is specific for CG dinucleotides in the DNA and occurs at the fifth carbon of the cytosine base (see Fig. 6.7 on p. 162). It is possible to determine the state of methylation of a DNA region by using two restriction enzymes that both cleave at a sequence containing a CG dinucleotide but have different sensitivities to the methylation of the DNA substrate. For example, *Hpa*II and *Msp*I both cleave at CCGG, but *Hpa*II will not cleave if the middle C of this site is methylated, while *Msp*I will (Fig. 17.14b). Thus, by digesting genomic DNA with *Hpa*II and *Msp*I and using a specific DNA probe on a Southern blot, you can determine whether a given CCGG sequence is methylated. Although methylation is associated with silencing, we know it cannot be the only mechanism because some organisms that show silencing, such as yeast, do not contain methylated DNA.

Deeper insight into the mechanism of silencing comes from studies of mutations that give rise to sterility in yeast. Recall that yeast cells come in two mating types—α and a—and the α2 gene product represses certain a-determining genes. The chromosomal locus of the α2 gene is known as MAT (for <u>mat</u>ing <u>type</u>). In normally mating yeast cells, there are two additional copies of the MAT locus called HML and HMR; located near the telomeres on each arm of chromosome III, these loci are transcriptionally silent. Mutations that reduce or destroy silencing at these loci cause sterility because they allow the simultaneous expression of α and a information, and the resulting cells, which behave as diploids, do not mate. Analysis of these mutations identified the family of *SIR* genes. The SIR polypeptide products associate to form a *trans*-acting complex that mediates silencing by acting at *cis*-acting sites near HML and HMR. Null mutations that eliminate the activity of any *SIR* gene or mutations that delete a *cis*-acting site abolish silencing. The SIR complex binds to other polypeptides and these larger complexes interact with histones H3 and H4 (Fig. 17.14c). These interactions with the histones establish a silenced chromosomal domain that remains hidden from the activators and repressors of transcription.

Genomic Imprinting Results from Chromosomal Events That Selectively Silence the Expression of Genes Inherited from One Parent or the Other

A major tenet of Mendelian genetics is that the parental origin of an allele—whether it comes from the mother or the father—does not affect its function in the F_1 generation. For the vast majority of genes in plants and animals, this principle still holds

Figure 17.14 Heterochromatin formation can lead to transcriptional silencing. (a) Normal chromatin in standard nucleosome conformation can be converted into tightly packed heterochromatin with the addition of methyl groups to a series of cytosine bases within a local DNA region. **(b)** A determination of the methylation status of a DNA region can be made using a pair of restriction enzymes that both recognize the same base sequence, with one being able to digest methylated DNA while the other can't. In this example, the restriction site is CCGG, and the enzymes are *Msp*I, which can digest both methylated and unmethylated sites, and *Hpa*II, which cannot digest methylated sites. If a methylated site is present between two unmethylated sites, *Hpa*II digestion will leave a larger fragment than *Msp*I. After electrophoresis and Southern blot analysis with a probe that hybridizes to a sequence on one side of the methylated site, there will be a clearly observable difference in band size. **(c)** The SIR complex of polypeptides can bind to basal factors associated with the promoters of the *HML* and *HMR* genes. This binding, in turn, causes the SIR complex to interact with the histones H3 and H4 present in downstream nucleosomes associated with the gene itself. The result is complete transcriptional silencing.

true today. Surprisingly, however, experiments and pedigree analyses have uncovered convincing evidence of exceptions to this general rule for some genes in mammals.

In the 1980s, when embryologists learned to manipulate the haploid nuclei, or pronuclei, of eggs and sperm, they created *in vitro* fertilized eggs carrying one maternally derived and one paternally derived pronucleus; or two maternal pronuclei; or two paternal pronuclei. When placed in a mouse uterus, embryos emerging from the fusion of one maternal and one paternal pronucleus developed into viable, fertile adult mice, like embryos that arise in a similar way in nature. But embryos formed from two maternal or two paternal haploid nuclei did not develop normally. Since all of the fertilized eggs providing the pronuclei were produced by matings between inbred animals of the same strain, the only possible genetic differences between the embryos lay in the number and type of sex chromosomes. Yet, even XX embryos with two maternal or two paternal pronuclei failed to thrive. The experimenters concluded that something other than the DNA itself differentiated maternal and paternal pronuclei.

An understanding of the mechanism behind these failures came from studies of the transmission of a deletion in the chromosome 7 insulin-like growth factor gene (*Igf2*) in mice. Mice inheriting the deletion from the paternal side were small, while mice inheriting the same deletion from the maternal side were normal size. The simplest explanation of these results is based on a model in which the *Igf2* gene copy inherited from the mother is normally silenced (Fig. 17.15a). Thus a deletion inherited from the mother produces no phenotypic effect because the maternal allele is not expressed anyway. If the deletion comes from the father, however, it produces a phenotypic effect because the animal is now unable to make any IGF2 products.

With the advent of molecular tools able to distinguish the transcription of a gene on a maternally derived chromosome from transcription of the same gene on the paternally derived homolog, geneticists observed that expression of a small number of genes—scattered around the genome, but often found in clusters—depends on whether the copy of the gene comes from the mother or the father. The phenomenon in which a gene's expression depends on the parent that transmits it is known as **genomic imprinting.** In most cases of genomic imprinting, the copy of a gene inherited from one parent is transcriptionally inactive in all or most of the tissues in which the copy from the other parent is active. The term "imprinting" signifies that whatever silences the maternal or paternal copy of an imprinted gene is not encoded in DNA sequence; rather it exercises its effect through some **epigenetic** (that is, outside-the-genes, but heritable) alteration of the DNA or chromatin during gametogenesis. This epigenetic imprint remains throughout the life of the mammal, but it is erased and regenerated during each passage of the gene through the germ line into the next generation (Fig. 17.15b). Some genes receive an imprint in the maternal germ line; others receive it in the paternal germ line. For each gene, imprinting occurs in either the maternal or paternal line, never in both.

Although the biochemical mechanism of genomic imprinting is not yet completely understood, one important component

is the methylation of cytosines in CG dinucleotides within the imprinted region (Fig. 17.15c). The methylated Cs silence the gene or genes in the region by preventing RNA polymerase and other transcription factors from gaining access to the DNA. A particular pattern of methylation can be transmitted during DNA replication, with the presence of a methyl group on one strand of a newly synthesized double helix signaling methylase enzymes to add a methyl group to the other strand.

While this simple model accounts for many instances of imprinting, some cases are more complex. One example involves *Igf2* and a closely linked gene called *H19*. Insight into the mechanism of *Igf2* imprinting came from the surprising finding that *H19*, found just 70 kb downstream of *Igf2*, is also imprinted, but in the opposite way. With *H19*, it is the copy inherited from the father that is silenced and the copy inherited from the mother that is active in normal mice.

A model of how imprinting works at both *H19* and *Igf2* is based on detailed biochemical and genetic studies of a 100 kb region encompassing both genes. Researchers identified an enhancer region downstream of the *H19* gene that can interact with promoters for both genes (Fig. 17.15d). In the region between the two genes lies another type of transcriptional regulation element called an **insulator.** When an insulator becomes functional, it stops communication between enhancers on one side of it with promoters on the other side. Insulators exist throughout the genome, where they play an important role in limiting the chromatin region over which an enhancer can operate. Without insulators, enhancers could wreak havoc in a cell by turning on genes at DNA distances of hundreds of kilobases.

In the *Igf2-H19* region, the insulator DNA becomes functional by binding a protein called CTCF. The binding normally occurs on the maternal chromosome. As a result, the enhancer element on the maternal chromosome can only interact with the promoter of *H19;* this interaction, of course, turns on the *H19* gene. In such a situation, the *Igf2* gene remains unexpressed. On the paternal chromosome, by comparison, both the insulator and the *H19* promoter are methylated. Since methylation of the insulator prevents the binding of CTFC, the insulator is not functional; and without a functional insulator, the enhancer downstream of *H19* can reach over a great distance to activate transcription from the *Igf2* promoter. In addition, methylation of the *H19* promoter suppresses transcription of the paternal *H19* gene. Imprinting of the paternal chromosome by methylation thus turns on transcription of *Igf2* and prevents transcription of *H19*.

Although biochemical analyses suggested the preceding model, results of a genetic experiment greatly strengthened the evidence for the model. In the experiment, researchers knocked out the gene for methylase—the enzyme that adds methyl groups to DNA—in mouse embryos. The experimental embryos lacking the methylase gene died before birth. The reason is as follows: Since methylation of the paternally inherited *H19* gene and insulator region could not occur, the CTCF protein could bind to the insulator on both maternal and paternal chromosomes; this binding induced transcription of the *H19* gene from both chromosomes while preventing transcription of both copies of *Igf2*.

Genetics and Society

A Promising Medical Tool: Synthetic Oligonucleotides That Selectively Reduce the Expression of Specific Gene Products

Oligonucleotides, often referred to as *oligos,* are short segments of DNA chemically synthesized to precise specifications. Most oligos are 20–50 nucleotides in length. Hybridization of an oligo to mRNA via complementary base pairing creates a DNA-mRNA duplex that can inhibit gene product formation by blocking translation (Fig. A). The DNA-mRNA duplex can also serve as a substrate for the cellular enzyme ribonuclease H, which destroys the integrity of the mRNA. In this duplex, the mRNA is often referred to as the "sense strand" because it contains information specifying the order of amino acids in a polypeptide; and the complementary sequence in the oligo is considered an "antisense sequence." The use of oligos to diminish the level of noxious gene products is known as "antisense therapy."

Clinical trials using antisense oligos as antiviral and anticancer agents have shown that antisense therapies may be able to help control both types of disease.

Oligos As Treatments for Serious Viral Infections
A key to the therapeutic use of oligos is their specificity. In theory, a well-designed antisense oligo complementary to one of a virus's

mRNAs would not bind to RNAs transcribed from the host cell genome. By binding to the viral RNA as it enters the host cell, such an antisense drug might help protect humans against infection by RNA viruses like the HIV retrovirus. Formation of the DNA-RNA duplex would block the reverse transcription on which the virus depends for creation of the proviral genome. If antisense oligos are not applied until later in the viral cycle, after the viral genome has integrated into the host genome and become activated for transcription, they could operate instead to keep the mRNAs from leaving the nucleus for translation in the cytoplasm. Still later, antisense oligos to mature mRNAs could block translation in the cytoplasm. Preliminary clinical trials of antisense drugs against the AIDS virus are in the process of evaluating an oligo against the RNA product of the HIV *gag* gene, whose expression is required for viral replication.

Oligos As Cancer Therapies
The specificity and resultant selectivity of oligos also makes them good candidates for cancer therapy. In addition to their selectivity (which could enable them to generate fewer side effects than

Figure A   Antisense oligonucleotides are designed to hybridize around the start codon (AUG) on the targeted mRNA. Hybridization between mRNA and antisense forms a double helix that blocks association of ribosomes with the mRNA and thus prevents translation.

Before the late 1980s, clinical geneticists were used to seeing sex-linked differences in inherited phenotypes related to the sex of the affected individual. With imprinting, however, it is the sex of the parent carrying a mutant allele that counts, and not the sex of the individual inheriting the mutation. After the discovery of imprinting in mice, medical geneticists reanalyzed human pedigrees and determined retroactively that what appeared to be instances of incomplete penetrance were actually manifestations of imprinting (Fig. 17.15e). An inactivating mutation in a maternally imprinted gene could pass unnoticed from mother to daughter for many generations (since the maternally derived gene copy is inactive in any case because of imprint-

ing). If, however, the mutation passed from mother to son, the son would have a normal phenotype (because he received an active wild-type allele from his father), but the son's children, both boys and girls, would each have a 50% chance of receiving a mutant paternal allele and of expressing the mutant phenotype resulting from the absence of any gene activity.

Evidence for imprinting as a contributing factor now exists for a variety of human developmental disorders, including the related pair of syndromes known as Prader-Willi syndrome and Angelman syndrome. Children with Prader-Willi syndrome have small hands and feet, underdeveloped gonads and genitalia, a short stature, and mental retardation; they are also obese chemotherapy), they are orders of magnitude smaller than the vesicles used to deliver some anticancer toxins to specific cellular addresses; as a result, they can penetrate capillary walls and the molecular meshes of intercellular space to reach tumor cells that otherwise may be inaccessible.

In theory, antisense oligos could fight cancer at several levels. They could suppress the expression of oncogenes—dominant mutant alleles that contribute to a cell's progression to malignancy (see Chapter 18). They could induce the redifferentiation of tumor cells by turning off expression of some of the transcription factors that allow proliferation, thus pushing the cell to terminal differentiation. They could suppress expression of protein factors that induce the growth of blood vessels into solid tumors (without these new vessels, the tumors cannot grow larger than 1 mm in diameter). And they could reverse the immunosuppression characteristic of many cancers by blocking production of the molecules that tumor cells secrete to shield themselves from destruction by the host's immune system.

Trials are underway in several laboratories to assess whether antisense oligos can prevent the progression of various cancers. In one study, clinicians are evaluating an antisense drug to see if it can eliminate cancer cells that have been removed from the bone marrow of leukemia patients. The target mRNA is from the *c-myb* gene, which promotes the proliferation of blood cells. Researchers think that abnormal regulation of the gene may contribute to several human leukemias.

Hurdles Remain in the Development of Oligo Drugs

While the preliminary laboratory tests on model animals and on small numbers of human patients have been very promising, antisense oligo developers are still grappling with crucial molecular and clinical considerations. In the molecular arena, some oligos work much better than others. Investigators trying to understand the basis for these differences are learning that oligos complementary to some points on an mRNA (usually those closer to the start-of-translation site) are more effective. The researchers are also trying to determine the most appropriate length for an antisense

oligo—short enough to retain its therapeutic properties, but long enough to avoid chance complementarity to points on untargeted mRNAs. In one study, for example, an oligo consisting of nine nucleotides whose target was a mutated *ras* gene mRNA bound to several hundred other completely complementary sites in a variety of mRNAs in the cell. Such an oligo would not be a useful therapeutic agent.

Once their molecular quirks are worked out, oligos must pass clinical muster, meeting the two basic criteria of safety and effectiveness. Preliminary trials have given them high marks in both areas, but some critical concerns remain. A few trials have shown inflammation of the injection site and toxicity of the liver, kidney, and blood in some animals at doses only slightly above therapeutic levels. For example, in one trial, monkeys suffered adverse cardiovascular effects after rapid infusion of 20–26-nucleotide oligos of varying sequences (the higher the dose, the worse the effects). Moreover, researchers do not yet know whether oligo breakdown products can produce adverse effects.

Thus, the clinically determined goals in the development of therapeutic oligos are identification of a safety threshold below which few side effects occur, and the production of oligos that are strong enough and specific enough to have a therapeutic effect at doses far below this threshold. A corollary is the identification of nonproblematic methods of delivery that make the drugs available to the organism at safe doses.

In summary, molecular and clinical studies suggest that antisense oligos able to attenuate the expression of specific gene products hold promise as short-term therapies for life-threatening conditions such as AIDS and cancer. If, with minor adjustments, they pass further clinical tests, they could become notable additions to the array of antiviral and anticancer drugs already available. Doctors would most likely prescribe them in combination with one or more of these other drugs in an effort to halt the progression of serious disease. However, since oligos as currently synthesized are costly to produce and seem to have some cumulative side effects, they are probably not yet suitable for the treatment of less serious diseases or chronic conditions that require continuous, long-term therapy.

and are compulsive overeaters. Children affected by Angelman syndrome have red cheeks, a large jaw, a large mouth with a prominent tongue, and a happy disposition accompanied by excessive laughing; they also show severe mental and motor retardation. Both syndromes are often associated with small deletions in the q11–13 region of chromosome 15. When the deletions are inherited from the father, the child gets Prader-Willi syndrome; when the deletions come from the mother, the child has Angelman syndrome. The explanation is that at least two genes in the region of these deletions are differently imprinted. One gene is maternally imprinted: Children receiving a

deleted chromosome from their father and a wild-type (nondeleted) chromosome with an imprinted (inactivated) copy of this gene from their mother will have Prader-Willi syndrome. The other gene is paternally imprinted: Children receiving a deleted chromosome from their mother and an imprinted (inactivated) gene from their father will have Angelman syndrome.

Among vertebrate animals, only mammals imprint their genomes. Thus imprinting is a rather late evolutionary development. By 1995, researchers had evidence of imprinting in about 15 genes in the mouse and human genomes. They estimate that about 1% of the mammalian genome—roughly 500 genes—

Feature Figure 17.15

Genomic imprinting.

(a) Deletion of *Igf2* causes mutant phenotype only when transmitted by father.

M – maternal chromosome
P – paternal chromosome
Deletion inherited from mother
Normal levels of *Igf2* expression.
Normal size

Deletion inherited from father.
No active *Igf2*.
Small size

The phenotypic effect of an *Igf2* deletion is determined by the parent transmitting the mutant locus. This parent-of-origin effect can be demonstrated in the two-generation cross illustrated here.

(b) The resetting of genomic imprints during meiosis.

Follow the transmission of a pair of homologous chromosomes (homolog 1 from the mother and homolog 2 from the father) from gametes through fertilization and the development of female and male progeny, to meiosis and the creation of a new set of gametes. Maternally imprinted genes are shown in red, paternally imprinted genes in black. The cellular machinery erases the old imprints and establishes new ones in germs cells during meiosis. Note that in the second generation, one of the chromosomes in both egg (homolog 2) and sperm (homolog 1) will be differently imprinted than the way it was in the first generation.

(c) Methylation of complementary strands of DNA.

Methylase recognizes methyl group on one strand of the helix and then adds a methyl group to the other strand.

An epigenetic state of DNA methylation can be maintained across cell generations. This is accomplished by the activity of DNA methylases that recognize methyl groups on one strand of a double helix and respond by methylating the opposite strand.

(d) Methylation of paternally inherited *H19* promoter

Mouse embryo lacking methylase

Reciprocal parent-of-origin expression occurs with the *Igf2–H19* gene pair. Only the unmethylated insulator between the two genes can bind to the protein CTCF. On the maternal chromosome, the enhancer only has access to the *H19* promoter. On the paternal chromosome, methylation occurs only at the insulator and the *H19* promoter (indicated with darkened circles). This serves the double purpose of blocking transcription of *H19* and allowing access of the enhancer to *Igf2*. In mouse embryos lacking methylase, the paternal chromosome behaves biochemically like the maternal chromosome.

(e) Genomic imprinting and human disease

Paternal imprinting

Maternal imprinting

In each pedigree, affected individuals (represented by filled-in, yellow circles and squares) are heterozygotes for a deletion removing a gene that has either a paternal or a maternal imprint. In these pedigrees, a dotted symbol indicates individuals carrying a deleted chromosome but not displaying the mutant phenotype.

will turn out to be imprinted. Most genes imprinted in one mammalian species are also imprinted in other species of mammals, but there are exceptions.

The biological role of genomic imprinting is still under debate. Some scientists believe it is one of the many specialized tools that cells use for gene regulation. Other scientists believe that genomic imprinting is not fundamentally a mechanism of gene regulation but, instead, an evolutionary response to competition between the sexes. In support of this evolutionary view, they point out that genomic imprinting is not universal, since as mentioned, it is found only among mammals.

From an evolutionary standpoint, genomic imprinting seems to place individuals at a disadvantage for survival. Individuals with two potentially active copies of a gene have one copy to spare to random mutations. Most null mutations, for example, are recessive because with only one copy mutated, the remaining functional copy of the gene provides enough gene product for viability. By comparison, with imprinted genes, the single functional gene copy makes an individual susceptible to all mutations.

Why then does imprinting exist? Does it confer some advantage on a population of organisms that counterbalances the obvious evolutionary disadvantages? One proposed answer is that imprinting represents the endgame in a struggle between the sexes. The argument is as follows: In most mammalian species, a female in estrus can mate with multiple males, generating multiple embryos fathered by different males. If a male could cause the embryos he sires to underexpress genes that normally retard embryonic growth, those embryos would use more maternal resources and grow more rapidly *in utero* than the sibling embryos; at birth, the larger, more robust offspring would have a fitness advantage. But the overgrowth of embryos and fetuses, while potentially beneficial to some offspring, would be severely draining to the mother. She would compensate by down-regulating her copies of genes that normally enhance growth. This down-regulation would give all the embryos she is carrying an equal chance of survival and prevent a multiple pregnancy from sapping all her strength. At the evolutionary endpoint of this male-versus-female tug-of-war, imprinting will have silenced growth-enhancing genes in all females and growth-retarding genes in all males.

Known as the Haig hypothesis for David Haig, its originator, this line of reasoning explains why imprinting is found only in mammals: It is only in mammals that embryos can compete for maternal resources *in utero*. Interestingly, many of the imprinted genes uncovered to date are involved in embryonic growth, as predicted by the hypothesis. However, other imprinted genes are not. Only many more hours of experiment and analysis will confirm or reject the Haig hypothesis.

Irrespective of the reason for its original emergence, genomic imprinting now plays a role in gene regulation, for a tiny number of genes in a small number of eukaryotic species.

As we saw earlier, genomic imprinting is an example of an epigenetic phenomenon: a variant condition that does not involve a change in DNA sequence but is nevertheless inherited from one somatic cell generation to the next during the growth of the organism. Other examples of epigenetic phenomena described in earlier chapters include X-chromosome inactivation and position-effect variegation. Methylation seems to play an important role in maintaining all three of these epigenetic conditions, which arise from gene or chromosomal inactivation events that occur early in development or in each gamete before fertilization.

Regulation After Transcription Influences RNA Production, Protein Synthesis, and Protein Stability

Gene regulation can take place at any point in the process of gene expression. While most regulatory mechanisms influence the initiation of transcription, some affect posttranscriptional events such as RNA splicing and stability, mRNA editing, and protein synthesis and stability.

RNA Splicing Helps Regulate Gene Expression

One example of how RNA splicing contributes to the regulation of gene expression is the role it plays in maintaining sexual identity throughout the life of *Drosophila*. As we see in the comprehensive example at the end of this chapter, transcription factors in very early female (XX) embryos activate the expression of a key gene called *Sxl* through a promoter called the early promoter (P_e). The cellular machinery splices the resulting transcript to create an mRNA that is translated into the Sxl protein, which is essential to the female-specific developmental program. The *Sxl* gene is not transcribed in early male (XY) embryos, so these embryos do not make the Sxl protein (Fig. 17.16a).

Later in development, the transcription factors activating the *Sxl* early promoter in females disappear; yet, to develop as females, these animals still need the Sxl protein. How can they still make the Sxl protein they need? The answer is that later in embryogenesis, the *Sxl* gene in both males and females is transcribed from another promoter—the late promoter (P_L) (Fig. 17.16b). In males, splicing of the primary Sxl transcript generates an RNA including an exon (exon 3) that contains a stop codon in the reading frame of the protein. As a result, this RNA in males is not productive, that is, it does not generate any Sxl protein. In females, however, the Sxl protein previously produced by transcription from the early promoter influences the splicing of the primary transcript initiated at the late promoter. When the earlier-made Sxl protein binds to the later-transcribed RNA, this binding alters the splicing pathway such that exon 3 is no longer part of the final mRNA. Without exon 3, the mRNA is productive; that is, it can be translated to make more Sxl protein. Thus, a small amount of Sxl protein synthesized very early in development establishes a positive feedback loop that ensures more fabrication of Sxl protein later in development.

RNA Stability Provides a Mechanism for Controlling the Amount of Gene Product Synthesized

We saw in Chapter 8 and again in this chapter that most polypeptide-encoding RNAs transcribed in eukaryotes (by

(a) Early embryo

(b) Sxl protein regulates the splicing of its mRNA by blocking male acceptor site

Figure 17.16 Differential RNA splicing can regulate gene expression at the polypeptide level. (a) In the early female—but not the male—*Drosophila* embryo, transcriptional activators initiate transcription from the P_e promoter of *Sxl* to produce an mRNA that encodes the Sxl protein. **(b)** Later in development, transcriptional activators that bind the P_L promoter are produced in both male and female animals. When the Sxl protein is present, as it is in females, it causes the splicing apparatus to skip over this exon and splice exon 2 directly to exon 4. The resulting RNA molecule has an intact coding sequence and can be translated into more Sxl protein. This results in a feedback loop that maintains the presence of Sxl protein in females but not males.

pol II) are modified at their 3′ ends by the addition of a poly-A tail (not encoded in the DNA) that can vary in length from fewer than 20 adenines to more than 200. Cellular enzymes in the cytoplasm slowly shorten the poly-A tails of all mRNAs after they leave the nucleus. Once the tail has disappeared, the core mRNA molecule is quickly degraded. This process provides the cell with a mechanism for fine-tuning the half-life of mRNAs produced from different genes. mRNAs that receive a longer poly-A tail last longer in the cytoplasm than mRNAs fitted with a shorter tail.

It might seem more efficient for a cell to keep mRNAs around for a long time to avoid using more energy in the continuous production of new transcripts. This is most likely true for the synthesis of many gene products in most cell types.

During development and in fully differentiated cells, however, there are many instances in which a cell needs to implement a quick change in its genetic program by eliminating a gene product that it made and used earlier. Repression of transcriptional activation is part of the solution. But it is not enough if preexisting mRNAs and proteins are stable.

An extreme example of mRNA instability is observed in the class of histone transcripts, which are among the few protein-encoding RNAs that receive no poly-A tail at all. A cell requires high levels of histones during its S phase when it is adding these proteins to newly synthesized DNA. But as soon as S phase ends, the cell must rapidly eliminate all remaining histone messages. It achieves this goal by simply repressing transcription. Within minutes, all histone mRNAs, which are already without poly-A tails, are rapidly degraded.

In another more specialized example of regulation through changes in RNA stability, mammalian cells maintain the optimal concentration of iron for their activities by modifying the stability of mRNAs encoding a membrane-bound transferrin receptor that transports iron into the cell. The number of transferrin-receptor molecules that a cell makes depends on how much receptor mRNA is available to ribosomes for translation and that, in turn, depends in part on how long a transferrin receptor mRNA lasts in the cell.

Figure 17.17 illustrates how cells regulate the stability of transferrin-receptor mRNA. The binding of extracellular iron (Fe) to an extracellular protein called transferrin renders the iron soluble. The iron-transferrin complex gains entry into a cell via the transferrin receptor. The mRNA encoding the transferrin receptor contains sequences in the 3′ untranslated region known

Figure 17.17 How the concentration of iron in a cell regulates transferrin receptor levels.

as iron response elements (IREs), and these IREs form stem-loop structures targeted by a cytoplasmic protein known as IRE-binding protein (IRE-BP). In the absence of iron, IRE-BP molecules bind to the IRE and stabilize the transferrin-receptor mRNA by protecting it from degradation. This makes biological sense because when little iron is available in the extracellular environment, a cell needs more transferrin receptors to transport as much as possible of the small amount of extracellular iron bound to transferrin. A high concentration of iron in the cytoplasm inactivates IRE-BP, preventing its binding to IRE, and thereby causing the degradation of the transferrin-receptor mRNA. This also makes biological sense, because when iron is abundant, it can gain access to the cell via other, lower-affinity receptors.

mRNA Editing Can Affect the Biological Properties of a Gene's Product

Cells in the central nervous system of mammals carry AMPA membrane channel receptors that mediate excitatory postsynaptic currents. Comparisons of genomic DNA encoding the AMPA receptor with some cDNA copies of the mRNA show that an AGA (Arg) codon in the genomic DNA has become a GGA (Gly) codon in the mRNA (Fig. 17.18). This type of base substitution, occurring after transcription and RNA splicing, is known as **RNA editing.** Researchers do not yet understand how the cellular machinery accomplishes such editing. They do know, however, that the biological properties of a Gly-containing membrane channel receptor differ from those of an Arg-containing channel, and that the extent of mRNA editing changes during brain development.

Figure 17.18 mRNA editing can regulate the function of the protein product that ensues. The AMPA receptor gene is transcribed into a sequence that contains an arginine codon at a particular point in the open reading frame. RNA editing replaces the first adenosine residue in this codon with a guanidine residue, and in so doing, converts this Arg codon into a Gly codon. The codon change has a significant effect on the activity of the protein product.

The observation of multiple RNAs for the AMPA receptor raises the possibility that RNA editing may fine-tune the regulation of neuronal signaling by allowing cells to synthesize—from the same DNA blueprint—proteins with different properties, at different times in development and different stages of neuronal activity.

Noncoding Sequences in mRNA Can Help Modulate Translation

Many mRNAs carry untranslated sequences that help modulate their translation. The human body stores iron for future use in an iron-storage protein complex called ferritin. When there is no iron to store, the IRE-BP molecule just described binds to an IRE in the 5′ untranslated region of the ferritin mRNA, and this binding represses ferritin translation. When iron is abundant, the inactivated IRE-BP does not bind to the ferritin IRE, and translation proceeds. This on/off switch makes biological sense because the cell needs ferritin only when iron is present in the cytoplasm. Note that the interaction of IRE-BP with other IREs can influence mRNA stability as well as translation.

Protein Modifications After Translation Provide a Final Level of Control over Gene Function

The activity of a gene is reflected in the activity of its protein product, and various posttranslational modifications affect protein function. Many of these modifications occur extremely rapidly compared to the time it takes to activate gene transcription and accumulate sufficient protein product for a particular process, or to deactivate transcription and await the slow disappearance of a protein product. Thus, cells often rely on posttranslational modification in situations that require a rapid response to a stimulus.

Ubiquitination Targets Proteins for Degradation

Cells have many enzyme systems that destroy proteins. In one, ubiquitin—a small, highly conserved protein—functions as a marker. The covalent attachment of chains of ubiquitin to other proteins marks the ubiquitinized proteins for degradation by a large multienzyme complex known as the *proteosome.*

During the cell cycle, for example, the proper separation of sister chromatids at the beginning of anaphase depends on ubiquitination and its aftermath. As Fig. 17.19a shows, the separation of sister chromatids requires the elimination of specialized proteins that, like a glue, keep the chromosomes together. Elimination of these adhesive proteins requires the action of at least two large multienzyme complexes—the proteosome and the anaphase-promoting complex (APC). The APC brings together the gluelike proteins that must undergo ubiquitination with enzymes that catalyze the attachment of chains of ubiquitin to these targets. Inhibitors of sister chromatid separation are among the various targets of the APC. Ubiquitination of these inhibitors at the onset of anaphase marks them for degradation by proteosomes, which, in turn, allows the sister chromatids to separate. Researchers do not yet know exactly how the inhibitors carry out their function. One hypothesis is that some act

Figure 17.19 Protein modifications can affect the final level of gene function. (a) The covalent addition of ubiquitin peptide chains to the gluelike proteins that hold together sister chromatids during the metaphase stage of mitosis marks them for digestion by proteosomes. **(b)** Covalent phosphorylation of the β-adrenergic receptor has no effect on its binding to epinephrine, but it blocks its downstream function of modulating heart rate.

as part of a proteinaceous glue that attaches sister chromatids at their centromeres; degradation of this glue as a result of APC and proteosome activity allows the sister chromatids to separate. Whatever the means, ubiquitination swiftly lifts the inhibition on sister chromatid separation, allowing the cell cycle to proceed.

Phosphorylation and Dephosphorylation Are Other Regulators of Protein Activity

In a process known as *sensitization,* many tissues exposed to hormones for a long time lose their ability to respond to the hormone. An example is the exposure of heart muscle to the stress hormone epinephrine. The binding of epinephrine to proteins called β-adrenergic receptors, which are located in the plasma membrane of heart muscle cells, normally increases the rate at which the heart contracts. But after several hours of continuous exposure to epinephrine, the heart muscle cells no longer respond in this way. Their sensitization is due to phosphorylation of the β-adrenergic receptors. The phosphorylation (addition of a phosphate) does not affect a receptor's ability to bind epi-

nephrine, but it does prevent the receptor from transmitting the hormone signal into the heart muscle cells (see Chapter 18 for a detailed discussion of signal transduction). The phosphorylation itself depends in large part on the activity of kinase (phosphate-adding) enzymes that phosphorylate the β-adrenergic receptor only when the receptor is bound to epinephrine (Fig. 17.19b). With the removal of epinephrine from the heart tissue, the kinases no longer act on the receptors, and phosphatase enzymes remove any phosphates already on them. The removal of phosphate from the β-adrenergic receptors eventually restores the heart muscle's ability to respond to new doses of epinephrine.

Sex Determination in *Drosophila*: A Comprehensive Example of Gene Regulation

Male and female *Drosophila* exhibit many sex-specific differences in morphology, biochemistry, behavior, and function of the germ line (Fig. 17.20). By examining the phenotypes of flies

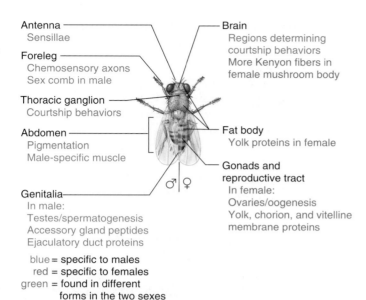

Antenna
Sensillae

Foreleg
Chemosensory axons
Sex comb in male

Thoracic ganglion
Courtship behaviors

Abdomen
Pigmentation
Male-specific muscle

Genitalia
In male:
Testes/spermatogenesis
Accessory gland peptides
Ejaculatory duct proteins

Brain
Regions determining
courtship behaviors
More Kenyon fibers in
female mushroom body

Fat body
Yolk proteins in female

Gonads and
reproductive tract
In female:
Ovaries/oogenesis
Yolk, chorion, and vitelline
membrane proteins

blue = specific to males
red = specific to females
green = found in different
forms in the two sexes

Figure 17.20 Sex-specific traits in *Drosophila*. Objects or traits shown in *blue* are specific to males. Objects or traits shown in *red* are specific to females. Objects or traits shown in *green* are found in different forms in the two sexes.

TABLE 17.2	How Chromosomal Constitution Affects Phenotype in *Drosophila*	
Sex Chromosomes	**X:A**	**Sex Phenotype**
Autosomal Diploids		
XO	0.5	Male (sterile)
XY	0.5	Male
XX	1.0	Female
XXY	1.0	Female
Autosomal Triploids		
XXX	1.0	Female
XYY	0.33	Male
XXY	0.66	Intersex

TABLE 17.3	*Drosophila* Mutations That Affect the Two Sexes Differently	
Mutation	**Phenotype of XY**	**Phenotype of XX**
Sx^{fl*}	Male	Dead
Sxl^{ML**}	Dead	Female
transformer (*tra*)	Male	Male (sterile)
doublesex (*dsx*)	Intersex	Intersex
fruitless	Male with aberrant courtship behavior	Female

*Sx^{fl} is a recessive mutation of *Sex lethal*.
**Sxl^{ML} is a dominant mutation of *Sex lethal*.

with different chromosomal constitutions, researchers confirmed that the ratio of X:A chromosomes helps determine sex, fertility, and viability (Table 17.2). They then carried out genetic experiments that showed that the X:A ratio influences sex through three independent pathways: One determines whether the flies look and act like males or females; another determines whether germ cells develop as eggs or sperm; and a third produces dosage compensation through doubling the rate of transcription of X-linked genes in males. (Note that this strategy of dosage compensation is just the opposite of that seen in mammals, where the inactivation of one X chromosome in females equalizes the expression of X-linked genes with that in males.)

To simplify our discussion of sex determination in *Drosophila*, we focus on the first-mentioned pathway: the determination of somatic sexual characteristics. An understanding of this pathway emerged from analyses of mutations affecting particular sexual characteristics in one sex or the other. For example, as we saw at the beginning of the chapter, XY flies carrying mutations in the *fruitless* gene exhibit aberrant male courtship behavior, while XX flies with the same *fruitless* mutations appear to behave as normal females. Table 17.3 shows that mutations in other genes also affect the two sexes differently. Clarification of how these mutations influence somatic sex determination came from a combination of genetic experiments (studying, for example, whether one mutation in a double mutant is epistatic to the other) and molecular biology experiments (in which investigators cloned mutant and normal gene products for analysis). Through such experiments, *Drosophila* geneticists dissected various stages of sex determination to delineate the following complex regulatory network.

The X:A Ratio Regulates Expression of the *Sex Lethal* (*Sxl*) Gene

Recall from Chapter 4 that it is the ratio of X chromosomes to autosomes (A) that determines sex in *Drosophila*. Since in normal diploids, there are two copies of each autosome, the X:A ratio is $2/2 = 1.0$ in a normal XX female and $1/2 = 0.5$ in a normal XY male. In short, when the X:A ratio is 1.0, females develop; when the ratio is 0.5, males develop.

Key factors of sex determination are helix-loop-helix proteins encoded by genes on the X chromosome. Sisterless-A (Sis-A) and sisterless-B (Sis-B) are two such proteins. Referred to as *numerator elements*, these two proteins monitor the X:A ratio through the formation of homodimers containing two of the same kind of subunit or heterodimers containing two different subunits. The homodimers consists of two numerator elements, while the heterodimers are composed of one numerator element and one denominator element. *Denominator elements* are helix-loop-helix proteins that are encoded by genes on autosomes.

Figure 17.21 The X:A ratio determines the expression of *Sxl*, the first gene in the sex-determination pathway. Numerator elements are produced by the X chromosome at a slightly higher level than denominator elements are produced by autosomes. Denominator elements cannot form homodimers but they can bind tightly to, and sequester, numerator elements. When the X:A ratio is 1 (in females), there are too many numerator elements to be occupied by denominators, and those not sequestered can form homodimers, which act as activators of the *Sxl* gene. When the X:A ratio is 1/2 (in males), there are fewer numerators than denominators, and all the numerators become sequestered. Thus males are unable to produce numerator homodimers, and *Sxl* remains turned off.

Since the number of X chromosomes determines the ratio of numerator homodimers to numerator/denominator heterodimers, the homodimers of numerator elements provide a measure of the X:A ratio (Fig. 17.21).

The observation that in flies with a greater number of numerator homodimers, transcription of the *Sxl* gene occurs early in development suggests the following hypothesis for operation of the X:A ratio.

Numerator Subunit Homodimers May Function As Transcription Factors That Turn On *Sxl*

In this hypothesis, the association of denominator subunits with numerator subunits sequesters the numerator elements in inactive heterodimers that cannot activate transcription. Females produce enough numerator subunits, however, that some remain unbound by denominator elements. Homodimers formed from these free numerator elements act as transcriptional activators of *Sxl* at the P_e promoter early in development. Males, by contrast, carry only half as many X-encoded numerator subunits; thus, the abundant denominator proteins tie up all the numerator elements, and as a result, there are no free numerator elements in males to turn on the P_e promoter of the *Sxl* gene.

Although this model is likely to be an oversimplification, it suggests how different X:A ratios might activate and repress transcription of the *Sxl* gene.

The Sxl Protein Expressed in Early Development in Females Regulates Its Own Later Embryonic Expression Through RNA Splicing

We have seen that the Sxl protein produced early in the development of female embryos participates later in an autoregulatory feedback loop (review Fig. 17.16). In this self-regulating system, the Sxl protein catalyzes the synthesis of more of itself through RNA splicing of the P_L-initiated transcript, which results in a productive mRNA. By contrast, in males where there is no transcription of *Sxl* early in development, activation of the P_L promoter later in embryonic development results in an unproductive *Sxl* transcript containing a stop codon near the beginning of the message. Since no Sxl protein is present to splice out the problem stop codon, this unproductive transcript is not translated to protein; and males thus have no Sxl protein at any point in development.

The Effects of *Sxl* Mutations

Recessive *Sxl* mutations that produce nonfunctional gene products have no effect in XY males, but they are lethal in XX females (see Table 17.3). This is because males, which do not normally express the *Sxl* gene, do not miss its functional product, but females, which depend on the Sxl protein for sex determination, do. The absence of the Sxl protein in females allows the expression of certain male-specific dosage-compensation genes that increase transcription of genes on the X chromosome; and the hypertranscription of these X-linked genes on two X chromosomes in mutant females proves lethal.

By comparison, rare dominant *Sxl* mutations that allow production of Sxl protein even in XY embryos are without effect in females, which normally produce the product, but lethal to males, which normally do not. In these mutants, the *Sxl* gene product indirectly represses transcription of genes that males need to express for dosage compensation. Without the products of these male-specific dosage-compensation genes, males cannot hypertranscribe X-linked genes and thus do not have enough X-linked gene products to survive.

Summary: The *Sxl* Gene Acts As an On/Off Switch That Is Sensitive to the X:A Ratio

Different proportions of numerator and denominator proteins reflecting the X:A ratio determine whether or not the *Sxl* P_e promoter is active in the early embryo. Later, when other transcription factors activate the upstream P_L promoter in both males and females, alternative RNA splicing after transcription generates more Sxl protein in females and none in males. Female development requires Sxl protein; male development requires its absence.

Sxl Triggers a Cascade of Splicing

In addition to splicing its own transcript, the Sxl protein influences the splicing of RNAs transcribed from other genes. Among these is the *transformer* (*tra*) gene. In the presence of the Sxl protein, the *tra* primary transcript undergoes productive splicing that produces an mRNA translatable to a functional protein. In the absence of Sxl protein, the splicing of the *tra* transcript results in a nonfunctional protein (Fig. 17.22).

The cascade continues. The functional Tra protein synthesized only in females, along with another protein encoded by the *tra2* gene (which is transcribed in both males and females), influences the splicing of the *doublesex* (*dsx*) gene's primary transcript. This splicing pathway results in the production of a female-specific Dsx protein called Dsx-F. In males, where there is no Tra protein, the splicing of the *dsx* primary transcript produces the related, but different, Dsx-M protein (Fig. 17.22b).

The N-terminal parts of the Dsx-F and Dsx-M proteins are the same, but the C-terminal parts of the proteins are different.

Effects of Mutations in *tra*

Like mutations in *Sxl*, mutations in *tra* affect the two sexes differently. Since males do not normally express *tra*, null mutations of the gene have no effect on them. In females, however, the Tra protein helps implement female-specific patterns of RNA splicing. In the absence of Tra protein, *dsx* primary transcripts can only be spliced to make Dsx-M. XX *Tra2* flies expressing Dsx-M are converted into flies having the phenotype of a male. The resulting XX males exhibit proper male sexual behavior, but without the Y chromosome, they are sterile because certain Y-linked genes are prerequisites of male fertility.

The Dsx-F and Dsx-M Proteins Are Transcription Factors That Determine Somatic Sexual Characteristics

Although both Dsx-F and Dsx-M function as transcription factors, they have opposite effects. In conjunction with the protein encoded by the *intersex* (*ix*) gene, Dsx-F functions mainly as a repressor that prevents the transcription of genes whose expression would generate the somatic sexual characteristics of males. Dsx-M, which works independently of the intersex protein, accomplishes the opposite: the activation of genes for the somatic sexual characteristics of males and the repression of genes that determine female somatic sexual characteristics. Interestingly, the two Dsx proteins can bind to the same enhancer elements, but their binding produces opposite outcomes (Fig. 17.23). For example, both bind to an enhancer upstream of the promoter for the *YP1* gene, which encodes a yolk protein; females make this protein in their fat body organs and then transfer it to developing eggs. The binding of Dsx-F stimulates transcription of the

(a) *tra* splicing
Results of *tra* splicing when Sxl protein is present (♀)

Functional Tra protein

Full length open reading frame

Results of *tra* splicing in absence of Sxl protein (♂)

No functional Tra protein

Stop codon
Truncated open reading frame

(b) *dsx* splicing
Results of splicing when *tra* is present (♀)

Dsx–F

Results of splicing when *tra* is absent (♂)

Dsx–M

Figure 17.22 The expression or nonexpression of *Sxl* leads to downstream differences in expression of products that also play a role in the sex-determination pathway. (a) The presence of *Sxl* alters the splicing of *tra* mRNA. Female transcripts produce functional Tra protein, while male transcripts have a truncated open reading frame and are unable to produce Tra. **(b)** Tra protein in turn plays a role in altering the splicing pattern of the *dsx* mRNA. A different Dsx product results in males (Dsx-M) than in females (Dsx-F).

Figure 17.23 Alternative forms of the *dxs* gene product both bind to the same *YP1* enhancer, but they have opposite effects on the expression of the *YP1* gene. Dsx-F acts as a transcriptional activator, while Dsx-M acts as a transcriptional repressor.

YP1 gene in females; the binding of Dsx-M to the same enhancer region inactivates (in conjunction with other transcription factors) transcription of *YP1* in males.

The Effect of *dsx* Mutations

Mutations in *dsx* affect both sexes because in both males and females, the production of Dsx proteins represses certain genes specific to development of the opposite sex. Null mutations in *dsx* that make it impossible to produce either functional Dsx-F or Dsx-M result in intersexes that cannot repress either certain male-specific or certain female-specific genes.

The Tra and Tra-2 Proteins Also Help Regulate Expression of the *Fruitless* Gene

We saw at the beginning of this chapter that the courting song and dance of male *Drosophila* are among the sexual behaviors under the control of the *fruitless* (*fru*) gene. As it turns out, the *fru* primary transcript is another regulatory target of the Tra and Tra2 splicing factors (Fig. 17.24). In females, whose cells make both Tra and Tra2 proteins, splicing of the *fru* transcript produces an mRNA that encodes a protein we refer to as Fru-F. In males, whose cells carry no Tra protein, alternative splicing of the *fru* transcript generates a related Fru-M protein with 101 additional amino acids at its N terminus. As we mentioned at the beginning of this chapter, these additions almost certainly determine some of the observed differences between male and female behavior. Since both Fru-F and Fru-M have the zinc-finger motifs characteristic of transcription factors, they probably activate and repress genes whose sex-specific products help generate courting behaviors.

The sex-specific products of *fru,* we have seen, appear in only a few cells in the nervous system, and the location of these neurons is significant. Some are in regions known to help regulate the courtship song; others are in areas that process chemosensory information from the antennae (perhaps in neurons that receive pheromone signals); still others are in regions that control abdominal movements (suggesting how *fruitless* may influence the male's curling of the abdomen during mating). To understand precisely how changes in gene expression in these few cells control sexual behaviors, *Drosophila* researchers are now trying to discover which genes are the targets of transcriptional regulation by the fruitless protein.

Summary: A Complex Network of Molecular Interactions Regulates the Determination of Somatic Sexual Characteristics in *Drosophila*

Several kinds of gene regulation make the embryonic "decisions" that determine the sex of the fly. Controls of both transcriptional initiation and RNA splicing play critical roles in this biological process. First, X-encoded numerator and autosome-encoded denominator elements allow the early embryo to measure the X:A ratio. Presumably, numerator homodimers, present in sufficient quantity only in females, activate *Sxl* transcription in early embryos. Next, the Sxl protein acts as a splicing factor that perpetuates its own synthesis in females. There it initiates a cascade of splicing factors, each of which determines the splicing patterns of the primary transcript of the next gene in the pathway. These splicing cascades eventually produce female-specific versions of at least two transcription factors: Dsx-F and Fru-F. In males, default splicing of these genes produces the Dsx-M and Fru-M transcription factors. The female- and male-specific versions of the Dsx and Fru proteins help turn on and off genes whose products elaborate the sexual appearance and behavior of male and female *Drosophila*.

Genes other than those just described also contribute to sex-specific behaviors. For example, a gene called *dissatisfaction* affects courtship behavior in males and females. Mutant females resist copulating with males; mutant males are maladroit in copulation because they don't properly curl their abdomen. Genetic experiments suggest that the *dissatisfaction* gene is part of the same sex-determination pathway as *fru*, but its position in the pathway is not yet known.

Figure 17.24 The primary *fru* RNA transcript is made in both sexes. Splicing occurs unhindered in males to produce an mRNA, which is translated into the Fru-M protein product. But tra protein (present only in females) causes alternative splicing of the *fru* transcript to produce an alternative mRNA, which encodes an alternative protein product Fru-F.

C O N N E C T I O N S

Multiple controls regulate gene activity and function from imprinting, to chromatin remodeling, to the initiation of transcription, to the processing of RNA transcripts, to the chemical modification of final gene products. At the outset, the regulation of transcription occurs through the interaction of *cis*-acting DNA regions and a variety of transcription factors.

Accurate regulation of gene function is crucial for proper control of the cell cycle. Indeed, a critical network of *cis*-acting control regions, *trans*-acting factors, and protein modifications promotes cell growth, DNA replication, and cell division in response to certain environmental signals and delays these events in response to other signals, such as DNA damage. In Chapter 18 we describe the regulatory network controlling the cell cycle in normal cells and explain how mutations that disrupt one or more aspects of that network can result in cancer.

ESSENTIAL CONCEPTS

1. The activity of most genes in eukaryotic cells is regulated primarily at the point of transcriptional initiation. Analyses of mutations that affect a gene's function without changing the sequence of its product provided insight into this level of regulation. Through these mutations, researchers defined *cis*-acting DNA regulatory elements and *trans*-acting *transcription factors.*

2. The eukaryotic genome contains three classes of genes, each associated with a different type of promoter and transcribed by a different RNA polymerase. Class I genes encode the rRNAs. Class III genes encode the tRNAs as well as other small RNA molecules. Class II genes, by far the largest class, encode all proteins.

3. Two types of *cis*-acting regulatory regions—*promoters* and *enhancers*—are associated with class II genes. All class II promoters are located at the 5′ end of the gene they influence. The enhancers have a more variable location in relation to the genes they control.

4. The binding of nonspecific *basal factors* to promoters is a prerequisite for the transcription of a gene. Basal factors alone allow a low, nonspecific basal level of gene transcription.

5. The association of transcription factors with enhancer elements can modulate levels of transcriptional initiation. Activation is mediated by transcription factors called *activators,* which bind to enhancers, and *coactivators,* which bind to activators. Activators and coactivators can interact with basal factors at the promoter to increase transcription levels above the basal level. *Repressors* can compete with activators for enhancer binding or quench the ability of activators to carry out their function. *Corepressors* operate by binding to DNA-bound transcription factors. Many transcription factors can function as either activators or repressors, depending on the situation. Activators and repressors that bind directly to DNA typically form homodimers and/or heterodimers. Formation of these dimers is a prerequisite for them to function as transcription factors.

6. Although promoters and enhancers are the most common and best characterized types of *cis*-acting regulatory elements, there are other elements that do not fit into either of these classes. One such element is the *locus control region* associated with the β-globin gene cluster. Another is the *insulator,* which restricts the genomic region over which an enhancer can operate.

7. The regulation of transcription is a complex process dependent on the interactions of multiple enhancer elements with many competing transcription factors, and responsive to signals originating within and outside the cell. The integration of signals and regulatory components determines the precise level of transcription of a particular gene in a particular cell at a particular time.

8. The structure of chromatin around a gene contributes to the regulation of transcription. Normal chromatin structure prevents runaway transcription from genes that are not under activation. The unraveling of the DNA in chromatin is an initial step in activation. Hypercompaction of chromatin domains causes *transcriptional silencing* by blocking access to the promoter and enhancers of a gene and thereby preventing its activation even in the presence of activator proteins.

9. *Genomic imprinting* is an example of epigenetic control over gene expression. Imprinting operates on the copy of a gene received from one parent but not the other. *DNA methylation* plays a role in the maintenance of imprinting from one mammalian somatic cell generation to the next.

10. Although the regulation of most genes depends primarily on controls over transcription, in some cases, further regulation down the path to protein production also plays a role. Modulation of gene function can occur through changes in RNA splicing or RNA stability, mRNA editing, changes in the efficiency of translation, and chemical modification of the gene product.

S O C I A L   A N D   E T H I C A L   I S S U E S

1. Since the cloning of the human growth hormone gene (*HGH*), the hormone has become available as a drug and is prescribed for children with growth hormone defects. Fred and Susannah's 12-year-old son Sam is below average in height, and his parents want the family doctor to prescribe growth hormone. They feel that Sam is suffering socially because of his small size and that his height will have a lasting psychological impact on him. The doctor believes that medicine with a lifelong effect is only appropriate when actual disease is present, and he does not consider Sam to have a disease. Who should set the limits for the use of HGH to treat potential growth problems in children—a governmental agency, the child's physician, or the parents? If a governmental agency or an individual physician regulates the use of HGH, how should they decide on a cutoff height at which to begin treatment, and how should they determine a final height at which treatment should end?

2. Statins are a new class of drugs that lower the level of a harmful form of cholesterol called LDL, which plays a significant role in atherosclerosis, a form of heart disease. The drug acts by inhibiting the liver enzyme HMG-CoA reductase, which converts HMG-CoA to mevalonate, an early step in the biosynthetic pathway for cholesterol. Unfortunately, in some people, some forms of the drug can cause grossly elevated levels of the muscle enzyme creatine kinase, leading to muscle failure and even, rarely,

death. Sean has borderline high levels of LDL and wants to begin taking statins. His doctor thinks that the side effects of the drug are not well enough known, and she doesn't want to prescribe the drug to Sean. Should Sean listen to his doctor, or should he try to find another doctor who is more willing to prescribe the drug?

3. One difficulty with the use of transgenic technology in animals is the inability to direct where DNA integrates and how many copies of the DNA integrate. Yet germ cell treatments are proposed for improving livestock. Animal rights groups are protesting that there are too many unknowns involved in these experiments. Should the same criteria for experimentation be applied to animals as is applied to humans?

4. Experimentation has shown that fetal tissue can be used to treat conditions such as the neurological disorder of Parkinson disease. However, because the fetal tissue is at a relatively early stage in development, it could contain many proteins that will affect the expression of genes in the tissue into which it is transplanted. If this is an experimental technology, should the individuals taking part in the trials be obliged to undergo testing that monitors changes in their physiology or should they be able to choose which follow-up studies they want to participate in?

S O L V E D   P R O B L E M S

I. You are studying expression of a gene whose protein product is made after UV irradiation. You cloned the gene and made antibody to the protein.

   a. If expression is regulated by turning on transcription after UV exposure, what results would you predict from hybridizing a DNA probe to RNA isolated from cells before and after UV irradiation (Northern analysis) and from incubating the antibody to proteins isolated from cells before and after UV treatment (Western analysis)?

   b. If expression is regulated by preventing translation, what results would you predict from doing similar Northern and Western analyses?

Answer

To answer this question, you need to consider the consequences of transcriptional and translational regulation on expression and think through what happens experimentally in Northern and Western analyses.

   a. If a gene is transcriptionally regulated, the mRNA will not be present in cells that were not exposed to

UV. *There will be no hybridizing band in the Northern analysis of mRNA from unexposed cells. The mRNA will be present in cells that have been treated with UV and there will be a hybridizing band. Similarly, the protein will only be found in cells that were exposed to UV and the antibody will bind to its protein target only in the protein preparation from exposed cells.*

   b. If expression is regulated at the translation step, mRNA will be present in the cells whether they have been exposed to UV or not. *Hybridizing bands will be found in both RNA samples. The protein will be present only in those cells that were exposed to UV so signal will be seen only in the exposed preparation.*

II. The retinoic acid receptor (RAR) is a transcription factor that is similar to steroid hormone receptors. The substance (ligand) that binds to this receptor is retinoic acid. One of the genes whose transcription is activated by retinoic acid binding to the receptor is *myoD*. The following diagram shows a schematic of the RAR protein into which two different 12-base oligonucleotides were inserted at the sites indicated. For constructs a–e, oligonucleotide 1

(TTAATTAATTAA) was inserted. For constructs f–m, oligonucleotide 2 (CCGGCCGGCCGG) was inserted into the RAR gene. Each mutant protein was tested for its ability to bind retinoic acid, bind to DNA, and activate transcription of *myoD* gene. Results are presented.

Mutant	Retinoic acid binding	DNA binding	Transcriptional activation
a	−	−	−
b	−	−	−
c	−	−	−
d	−	+	+
e	+	+	+
f	+	+	+
g	+	+	−
h	+	+	−
i	+	−	−
j	+	−	−
k	−	+	+
l	−	+	+
m	+	+	+

a. What is the effect of inserting oligonucleotide 1 anywhere in the protein?

b. What is the effect of inserting oligonucleotide 2 anywhere in the protein?

c. Indicate the three protein domains on a copy of the preceding drawing.

Answer

This question involves the concepts of domains within proteins and use of the genetic code to understand effects of oligonucleotide insertions.

a. Oligonucleotide 1 contains a stop codon in any of its three reading frames. This means it will *cause termination of translation of the protein wherever it is inserted.*

b. Oligonucleotide 2 does not contain any stop codons and so will *just add amino acids to the protein.* Because there are 12 bases in the oligonucleotide, it will not change the reading frame of the protein. *Insertion of the oligonucleotide can disrupt the function of a site in which it inserts.*

c. Looking at the data overall, notice that all mutants that are defective in DNA binding are also defective in transcriptional activation, as would be expected for a transcription factor that binds to DNA. The mutants

that will be informative about the transcriptional activation domain are those that do not have a DNA-binding defect. Inserts a, b, and c using oligonucleotide 1, which truncates the protein at the site of insertion, are defective in all three activities. The protein must be made at least as far as point d before DNA binding or transcription activation are seen. These two activities must lie before d. Truncation at d is negative for retinoic acid binding but the truncation at e does bind to retinoic acid. The retinoic acid-binding activity must lie before e. Using the oligonucleotide 2 set of insertions, transcriptional activation was disrupted by insertions at sites g and h, indicating that this region is part of the transcriptional domain; i and j insertions disrupted the DNA binding; and k and l insertions disrupted the retinoic acid binding. The minimal endpoints of domains as determined from these data are summarized in the following schematic.

III. A cDNA clone that you isolated using pituitary gland mRNA from mice was used as a probe against a blot containing RNAs from embryonic heart (EH), adult heart (AH), embryonic pituitary (EP), adult pituitary (AP), and testis (T). The results of the hybridization are shown here.

a. What would you conclude about this gene based on the result with AH RNA?

b. How would you explain the result with testis RNA?

Answer

This problem requires an understanding of RNA and transcription.

a. No RNA from adult heart (AH) hybridized with the probe, indicating that the *gene is not transcribed in this tissue.*

b. *A different-sized RNA is seen in the testis sample.* This could be due to *alternate splicing of the transcript or a different start site* in testis compared to other tissues.

PROBLEMS

17-1 For each of the terms in the left column, choose the best matching phrase in the right column.

a. basal factors	1. influences transcription from multiple genes over a large region of DNA
b. transcriptional silencing	2. marks a protein for degradation
c. locus control region	3. pattern of expression is dependent on which parent transmitted the allele
d. activators	4. multimers of nonidentical subunits
e. imprinting	5. heterochromatin
f. coactivators	6. multimers of identical subunits
g. homomers	7. bind to enhancers
h. heteromers	8. bind to promoters
i. ubiquitination	9. bind to activators

17-2 Does each of the following types of gene regulation occur in eukaryotes only? in prokaryotes only? in both prokaryotes and eukaryotes?
a. differential splicing
b. positive regulation
c. chromatin compaction
d. attenuation
e. negative regulation

17-3 What events occur during processing of a primary transcript?

17-4 Which fusion gene would you use for which purpose? (The slash indicates the fusion, and the parts of each type of fusion are given in the order in which they would occur.)

Types of fusions:
a. DNA-binding domain of *lexA* gene/random mouse genome fragments
b. random mouse sequences/*lacZ* gene
c. mouse metallothionein promoter/a mouse gene

   Uses:
   i. try to identify genes turned on in specific cell types
   ii. turn on expression of gene by including Zn in the diet
   iii. search for transcription-activation domains

17-5 Which eukaryotic RNA polymerase (RNA pol I, pol II, or pol III) transcribes which genes?
a. tRNAs
b. mRNAs
c. rRNAs

17-6 You have isolated a gene expressed in differentiated neurons in mice. You then fused the upstream DNA and beginning of the gene to *lacZ* (reporter gene) so that you could monitor expression. Different fragments (shown as *dark lines* in the following figure) were cloned next to the *lacZ* gene that lacked a promoter. The clones were introduced into neurons in tissue culture to monitor expression. From the results that follow, which region contains the promoter and which contains an enhancer?

17-7 Is each of the motifs listed associated with DNA binding, transcription activation, or dimer formation?
a. zinc finger
b. helix-loop-helix
c. leucine zipper
d. acidic region
e. helix-turn-helix

17-8 From Northern analysis, you find that the *ADAG* gene is expressed only in the brain. You examine expression in glial and neuronal cell lines (two types of cells in the brain) and find that only glial cells make *ADAG* mRNA. No one has characterized the *cis-* or *trans-*acting elements required for glial specific expression, so you decide to do so. You make a set of deletions in the regulatory region and fuse these to the *lacZ* gene so you can easily monitor the expression after introducing the clones into tissue culture cells. Deletions beginning at a site upstream of the gene and extending to base -85 (with the transcription start site considered -1 and bases prior to the start having negative designations) still retain full activity, but a deletion to -75 leaves only 1% of the original activity. A DNase-I footprinting experiment reveals that a 19 bp sequence is protected. You also obtain evidence that the factor binds as a dimer although each monomer binds to a similar sequence. The sequence between -90 and -60 is shown here.

-90 -60
TAGGCTTCTGCGGACCGCAAACTGTGCTGGC

a. What is the most likely region that was protected from DNase-I treatment in the footprint experiment?
b. What is a likely consensus sequence for the monomer binding site?

17-9 A single enhancer site regulates expression of three adjacent genes *GAL1, GAL7,* and *GAL10,* but the genes are not cotranscribed as one mRNA. How could you show experimentally that each gene is transcribed separately?

17-10 In yeast, the GAL4 protein binds to DNA to activate transcription of GAL7, or GAL10. GAL80 represses expression by binding to GAL4 protein and preventing it from binding to DNA. In which genes should you be able to isolate galactose constitutive mutations, and what characteristics of the protein would the mutation disrupt?

17-11 *MyoD* is a transcriptional activator that turns on the expression of several muscle-specific genes in human cells. The *Id* gene product inhibits *MyoD* action. How could you determine if *Id* acts by quenching *MyoD* (as *GAL80* does to *GAL4* in yeast) or by blocking access to the enhancer? What differences would you expect to see experimentally?

17-12 How could you make a library of genes expressed during sporulation in yeast?

17-13 What experimental evidence indicates that chromatin structure acts to reduce basal levels of transcription?

17-14 You isolated nuclei from liver cells, treated them with increasing amounts of DNase-I, stopped the reactions, isolated the DNA, treated it with restriction enzyme *Eag*I, electrophoresed the DNA, transferred it to a blot, and hybridized it with a probe from the gene you are studying. With no DNase-I treatment, there was a 20 kb *Eag*I fragment that hybridized with your probe. With trace amounts of DNase-I, two bands of 16 and 4 kb were present. The same DNase-I treatment of DNA from muscle cells produced only a 20 kb fragment. What does this result tell you about the region of DNA?

17-15 Which of the following would be suggested by a DNase hypersensitive site?
a. No transcription occurs in this region of the chromosome.
b. The chromatin is in a more open state than a region without the hypersensitive site.
c. Transcription terminates at this site.

17-16 Match the gene expression phenomenon with molecular components that modulate each.
a. transcriptional silencing   1. insulator
b. imprinting   2. heterochromatin
c. restricting the range of   3. TBF, TAFs
   enhancer activity
d. basal transcription   4. methylation

17-17 A boy expresses a mutant phenotype because he has received a mutation in a paternally imprinted gene. From which parent did the boy inherit the mutant allele?

17-18 Prader-Willi syndrome is caused by a mutation in a maternally imprinted gene. Answer the following questions as true or false, assuming that the trait is 100% penetrant.
a. Half of the sons of affected males will show the syndrome.
b. Half of the daughters of affected males will show the syndrome.
c. Half of the sons of affected females will show the syndrome.
d. Half of the daughters of affected females will show the syndrome.

17-19 Assume that the disease illustrated with this pedigree is due to expression of a rare allele of an autosomal gene that is a paternally imprinted. What would you predict is the genotype of individuals (a) I-1, (b) II-1, and (c) III-2?

17-20 The *IGF-2R* gene is autosomal and maternally imprinted. Copies of the gene received from the mother are not expressed, but copies received from the father are expressed. You have found two alleles of this gene that encode two different forms of the IGF-2R protein distinguishable by gel electrophoresis. One allele encodes a 60K blood protein; the other allele encodes a 50K blood protein. In an analysis of blood proteins from a couple named Bill and Joan, you find only the 60K protein in Joan's blood and only the 50K protein in Bill's blood. You then look at their children, Jill and Bill Jr., Jill is producing only the 50K protein, while Bill Jr. is producing only the 60K protein.
a. With these data alone, what can you say about the *IGF-2R* genotype of Bill Sr. and Joan?
b. Bill Jr. and a woman named Sara have two children, Pat and Tim. Pat produces only the 60K protein and Tim produces only the 50K protein. With the accumulated data, what can you now say about the genotypes of Joan and Bill Sr.?

17-21 Follow the expression of a paternally imprinted gene through three generations. Indicate whether the copy of the gene from the male in generation I is expressed in the germ cells and somatic cells of the individuals listed.

a. generation I male (I-2) germ cells
b. generation II daughter (II-2): somatic cells
c. generation II daughter (II-2): germ cells
d. generation II son (II-3): somatic cells
e. generation II son (II-3): germ cells
f. generation III grandson (III-1): somatic cells
g. generation III grandson (III-1): germ cells

17-22 You are studying muscle cells and have found a protein that is only made in this tissue. The data here are from analysis of RNA (Northern blot) using a DNA probe from the gene and analysis of protein (Western blot) using antibody directed against the protein. Is this gene transcriptionally or translationally regulated?

RNA analysis			Protein analysis		
Muscle cell RNA	Nerve cell RNA	Skin cell RNA	Muscle cell protein	Nerve cell protein	Skin cell protein

17-23 a. Assume that two transcription factors are required to get expression of the blue pigmentation genes in pansies. (Without the pigment, the flowers are white.) What phenotypic ratios would you expect from crossing strains heterozygous for each of the genes encoding these transcription factors?

b. Now assume that either transcription factor is sufficient to get blue color. What phenotypic ratios would you expect from crossing strains heterozygous for each of the genes encoding these transcription factors?

17-24 You isolated a cDNA from skin cells, and when you hybridized that cDNA as a probe with a blot containing mRNAs from skin cells and nerve cells you saw a 1.2 and a 1.3 kb fragment, respectively. How could you explain the different-sized cDNAs?

17-25 The *hunchback* gene, one of the genes necessary for setting up the dorsal-ventral axis of the *Drosophila* embryo, is translationally regulated. The position of the coding region within the transcript is known, and there is additional sequence beyond the coding region at the 5′ and 3′ ends of the mRNA. How could you determine if the sequences at the 5′ or 3′ or both ends are necessary for proper regulation of translation?

17-26 From Northern and Western hybridization, you know that the mRNA and protein produced by a tissue-specific gene are present in brain, liver, and fat cells, but you are detecting an enzymatic activity associated with this protein only in fat cells. Give an explanation for this phenomenon.

REFERENCES

Buratowski, S., Hahn, S., Sharp, P. A., and Guarente, L. (1988). Function of a yeast TATA element-binding protein in a mammalian transcription system. *Nature* 334(6177):37–42.

Choi, O-R., and Engel, J. D. (1988). Developmental regulation of β-globin gene switching. *Cell 55*:17–26.

Cote, J., Quinn, J., Workman, J. L., and Peterson, C. L. (1994). Stimulation of GAL4 derivative binding to nucleosomal DNA by the yeast SWI/SNF complex. *Science* 265(5168):53–60.

Eden, S., and Cedar, H. (1995). Action at a distance. *Nature 375*:16–17.

Jones, N. (1990). Transcriptional regulation by dimerization: Two sides to an incestuous relationship. *Cell 61*: 9–11.

Murre, C., McCaw, P. S., and Baltimore, D. (1989). A new DNA binding and dimerization motif in immunoglobulin enhancer binding, daughterless, MyoD, and myc proteins. *Cell* 56(5):777–83.

Tremblay, K. D., Saam, J. R., Ingram, R. S., and Tilghman, S. M. (1995). A paternal-specific methylation imprint marks the alleles in the mouse *H19* gene. *Nature Genet. 9*: 407–13.

Vinson, C. R., Sigler, P. B., and McKnight, S. L. (1989). Scissors-grip model for DNA recognition by a family of leucine zipper proteins. *Science 246*: 911–16.

Viville, S., and Surani, M. A. (1995). Towards unraveling the *Igf2/H19* imprinted domain. *BioEssays 17*: 835–38.

18

Cell-Cycle Regulation and the Genetics of Cancer

Killer cells surround a large cancer cell. Note the extended cellular processes of the cancer cell.

The roughly 300 different types of cells that compose the brain, heart, lungs, liver, immune system, and other tissues of a healthy human adult proliferate only when necessary—to replace cells lost through normal attrition or injury and, in the case of the immune system, to attack infectious agents. Many cells, such as those that constitute the inner skin, blood, and intestinal lining, multiply daily to compensate for the senescence and death of their counterparts. Other cells, such as those in the liver, rarely divide under normal circumstances but proliferate rapidly when toxins destroy some among them. Still others, such as specific cells of the immune system, reproduce in response to particular bacteria and viruses. Finally, some cells, such as the terminally differentiated cells of nerves and surface skin, never divide.

Most of us give little thought to this life-sustaining cycle of cell death and renewal, because despite the continuous comings and goings of our body's cells, we generally remain the same shape and size. But the appearance of an abnormal growth or a set of symptoms diagnosed as cancer startles us into the realization that we take for granted the intricate checks and balances that control cell division and behavior. These controls enable all cells of the body to function as part of a tightly organized, cooperative society. Cancer results when some cells divide out of control and eventually acquire the ability to spread beyond their prescribed boundaries. Cancer rarely occurs in children, but it is a common disease of older adults, in whom it accounts for about one-fourth of all deaths. Although there are many types of cancer, they all result from excessive and inaccurate cellular proliferation (Fig. 18.1). Thus, the body's ability to regulate cell division is a foundation of health.

Scientists who study cancer seek information about the normal controls that govern cell proliferation as a basis for understanding what goes wrong in cancer. But they often work in reverse, using the abnormalities of cancer to retrieve information about the normal operation of the cell cycle. Observations of the differences in behavior between normal and cancerous cells, obtained from diseased individuals or from cultured cell lines, have made it clear that cancer cells can divide when normal cells cannot, and cancer cells can migrate from one tissue to another

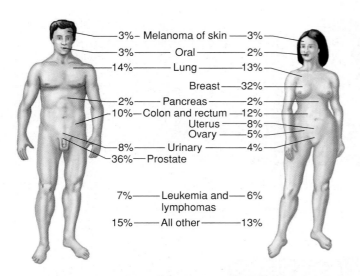

Figure 18.1 The relative percentages of new cancers in the United States that occur at different sites in the bodies of men and women.

chinery of cell division, molecules that integrate the repair of DNA damage with progression through the cell cycle, and molecules that relay messages about whether conditions are right for cell division. We then describe how mutations in these genes and gene products lead to cancer.

Two unifying themes emerge from our discussion. First, cancer is ultimately a disease of the genes: The multiple phenotypes collectively referred to as cancer all result from mutations in genes that regulate a cell's passage through the cycle of growth and division. Chemicals in the environment that raise the rate of gene mutation increase the probability of cancer incidence. Second, cancer differs in two ways from cystic fibrosis, Huntington disease, and other genetic conditions caused by the inheritance of one or two copies of a single defective gene: (1) Most mutations that lead to cancer occur in the somatic cells of one tissue, and (2) multiple mutations in an array of genes must accumulate over time in the clonal descendants of a single cell before the cancer phenotype appears. By contrast, the mutations that cause cystic fibrosis and Huntington disease are transmitted through the germ line; thus the mutant alleles of one particular gene appearing in all cells of all somatic tissues cause the disease in an affected individual.

Our presentation of the correlation between cell-cycle regulation and cancer covers:

■ The normal control of cell division, including a description of the normal cell cycle and the molecular signals, machinery, and checkpoints that regulate passage through the cell cycle.

■ How cancer arises from malfunctions in controls over cell division, including a description of cancer phenotypes, an analysis of the clonal nature of tumors, and an examination of the mutations in protooncogenes and tumor-suppressor genes that underlie all cancers.

■ A comprehensive example describing the progression of mutations leading from low-grade brain tumors to aggressive brain cancer.

whereas normal cells cannot. Unfortunately, it is hard to study cellular behavior over time *in vivo* and *in vitro*. To study cells *in vivo*, it is necessary to see and manipulate them; but this requires cutting open an animal and carrying out extended observations, an approach that is rarely possible in humans and not easy even in model organisms like mice. It is also difficult to compare most types of tumor cells with normal laboratory-grown cells because most normal cell types do not grow in culture, and even those that do have limited life spans. Genetic studies on the genomes of various eukaryotic organisms, from yeast to humans, have helped overcome these experimental impasses by identifying genes in which mutations lead to cancer; such studies continue to provide insights that are unavailable from other types of analyses.

In this chapter we examine the genes and gene products that control cell proliferation, including molecules that control the ma-

■ ■ ■

The Normal Control of Cell Division

A variety of genes and proteins control the events of the cell cycle. These genes and proteins allow progression to the next stage of the cycle when all is well, but they cause the cellular machinery to brake when damage to the genome or to the machinery itself requires repair. We now describe the molecules that control cell division.

Cyclin-Dependent Kinases Collaborate with Cyclins to Ensure the Proper Timing and Sequence of Cell-Cycle Events

Cell division, we saw in Chapter 4, requires the duplication of chromosomes and other cellular components as well as the precise partitioning of the duplicated elements to two daughter cells. During this complicated process, the cell coordinates the

function of hundreds of different proteins. To see how the cell orchestrates the events of cell division, we first review the stages of the cell cycle and then look at some of the proteins that control progression through that cycle.

The Cell Cycle Has Four Phases: G_1, S, G_2, and M

G_1 is the period, or gap, between the end of a mitosis and the DNA synthesis that precedes the next mitosis (Fig. 18.2). During G_1, the cell grows in size, imports materials to the nucleus, and prepares in other ways for DNA replication. S is the period of DNA synthesis, or replication. It requires the timely, sequential activation of proteins that function at the replication fork, as well as the timely inactivation of these same proteins. G_2 is the gap between DNA synthesis and mitosis. During G_2, the cell prepares for division. M, the phase of mitosis, includes the breakdown of the nuclear membrane, the condensation of the

G_1: Cell grows in size, prepares for DNA replication

Centrosome
Cell membrane
DNA
Nuclear membrane

S: DNA replication

Daughter cells

G_2: Cell prepares for division

Spindle pole
Condensed chromosome
Mitotic spindle

M: Mitosis
1. Breakdown of nuclear membrane
2. Condensation of chromosomes
3. Attachment of chromosomes to mitotic spindles

Figure 18.2 The cell cycle is the series of events that transpire between one cell division and the next. After division, a cell begins in the G_1 phase, progresses into S phase, where the chromosomes replicate, to the G_2 phase, and to the M phase, where replicated sister chromosomes segregate to daughter cells. In M phase, the nuclear membrane breaks down, centrosomes form the poles of the spindle, and microtubules construct a scaffold on which chromosomes migrate.

chromosomes, their attachment to the mitotic spindle, and the segregation of chromosomes to the two poles; at the completion of mitosis, the cell divides. During M, the cell must coordinate the activities of a variety of proteins: those that cause chromosome condensation (see Chapters 4 and 12), tubulins that polymerize to form the mitotic spindle on which the chromosomes move, motor proteins in the centromeres that power chromosome movement, proteins that dissolve and re-form the nuclear membrane at the beginning and end of mitosis, and others.

Experiments with Yeast Helped Identify Genes That Control Cell Division

The budding yeast *Saccharomyces cerevisiae* and the fission yeast *Schizosaccharomyces pombe* have been instrumental in identifying the genes that control cell division. Several properties of both yeast species make them particularly useful for the genetic dissection of cellular processes. Both can grow as haploid or diploid organisms. As a result, it is possible to identify recessive mutations in the haploid cells, which carry no second, wild-type allele to obscure the mutant phenotype; and it is possible to construct diploid cells containing two mutations, allow them to proliferate, and test the resulting cell populations to de-

termine the number of complementation groups defined by the mutations. The budding yeast *S. cerevisiae* has yet another property that facilitates cell-cycle analysis. At the beginning of the cell cycle, toward the end of G_1, a new daughter cell arises as a bud on the surface of the mother cell. As the mother cell progresses through the division cycle, the bud grows in size; it is small during S phase and large during mitosis. Bud size thus serves as a marker of progress through the cell cycle. One can order cells in an asynchronously cycling population according to position in the cell cycle by observing the relative sizes of their buds. A normal population of growing yeast cells contains nonbudding cells as well as cells with buds of all sizes.

Mutations that interfere with the cell cycle are lethal; and because cell proliferation depends on successive repeats of the cell cycle, a mutant unable to complete the cell cycle cannot grow into a population of cells. Researchers have obtained cell-cycle-defective mutants by isolating cells with temperature-sensitive mutations (see the Fast Forward box in Chapter 7). In these mutants, a protein needed for cell division functions normally at a low *permissive temperature* but loses function at a higher *restrictive temperature*. At the permissive temperature, the mutants grow nearly normally, producing a population of cells for study. A shift to the restrictive temperature causes the temperature-sensitive gene in the mutant population to become nonfunctional; researchers can then study the consequences of its loss.

To isolate temperature-sensitive mutations, investigators expose haploid cells to a mutagen and then plate them at the permissive temperature, allowing them to form colonies (Fig. 18.3). After the colonies grow up, each from a single mutant cell, the experimenters use replica plating to imprint them on two plates. They incubate one plate at the permissive temperature and the other plate at the restrictive temperature. Cells that have sustained a temperature-sensitive mutation grow at the permissive temperature but not at the restrictive temperature.

With this protocol, researchers have isolated thousands of temperature-sensitive mutations. These mutations could occur in any gene required for cell reproduction. Genes of particular interest for understanding the cell cycle are those whose protein product functions at only one stage in the cell cycle. It was possible to identify such mutants in *S. cerevisiae* by observing in the light microscope the shape and behavior of cells shifted from the permissive to the restrictive temperatures (Fig. 18.4). A population of cells growing at the permissive temperature includes unbudded cells as well as cells with the full range of bud sizes. After a cell-cycle mutant has grown at the restrictive temperature for about two cycles, however, the cells have a uniform appearance. In the mutant population shown in Fig. 18.4, for example, all cells have a single large bud. Moreover, the nuclei (not visible in the figure) are uniformly located at a position between the mother cell and the daughter cell, as if beginning to divide. This uniformity identifies a particular cell-cycle mutant. Other cell-cycle mutants would arrest with different but also uniform morphologies, for example, with all unbudded cells. Thus, mutants that acquire a uniform bud-related morphology at the restrictive temperature are each defective at one stage of the cell cycle.

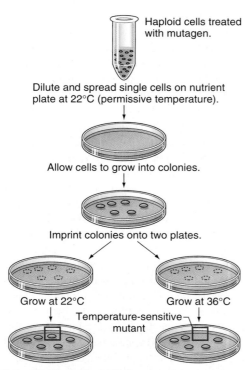

Figure 18.3   The isolation of temperature-sensitive mutants of yeast. Mutations are induced in a culture of haploid cells by treating them with a chemical mutagen. The treated cells are distributed onto solid medium. Each cell proliferates to a colony (clone) of cells, passing on the mutation. Replicas of the colonies are imprinted onto solid medium. One is grown at the permissive temperature (22°C), one at the restrictive temperature (36°C). Colonies that grow on the former, but not the latter, carry a temperature-sensitive mutation.

(a)

(b)

Figure 18.4   A cell-cycle mutant of yeast. Cells of a temperature-sensitive mutant growing at the permissive temperature **(a)** display buds of all sizes. After incubation at the restrictive temperature **(b)**, the same cells have arrested—all with a large bud. Cells that are early in the cell cycle at the time of the temperature shift (small buds in [a]) arrest in the first cell cycle and cells that are later in the cell cycle (large buds in [a]) finish the first cell cycle and arrest in the second.

Further examination of cells transferred from permissive to restrictive temperatures illustrates another property of cell-cycle mutants—a requirement for the normal gene product at a particular stage of the cell cycle. Note that some of the cells in Fig. 18.4 formed one cell with a large bud at the restrictive temperature, while others formed two cells, each with a large bud. Note also that the former all had smaller buds than the latter at the time of the shift to the restrictive temperature. This observation indicates that cells early in the cell cycle at the moment of temperature shift arrested division in the first cell cycle, while those later in the cell cycle at the time of temperature shift finished the first cycle and became arrested only in the second cell cycle. The point at which a cell acquires the ability to complete a cell cycle is the moment at which the temperature-sensitive protein has fulfilled its function in that cycle.

By analyzing the morphology of buds on cells shifted from permissive to restrictive temperatures, yeast geneticists have identified about 70 cell-cycle genes (Table 18.1). Epistasis experiments comparing the phenotype of a double mutant to that of each of the two single mutants have enabled them to order some of these mutations with respect to one another (Fig. 18.5). One of the most important mutations, in the *CDC28* gene, identified the first step in the cell cycle, a step occurring prior to that

catalyzed by the *CDC7* gene: After the *CDC28* step, the cell is committed to finishing the cycle it has begun before pursuing an alternative fate. Alternative fates include the arrest of cell division in response to nutrient starvation or fusion with a cell of opposite mating type to form a diploid cell (Fig. 18.6). The decision to pursue a particular fate occurs in the G_1 phase of the cell cycle at a step called "start." If cells have completed this *CDC28*-determined step, they cannot consider the alternatives until they have finished the cell cycle and arrived at the next G_1. The *CDC28* gene thus controls the commitment to proceed through the cell cycle.

TABLE 18.1 Some of the Cell-Cycle Genes in Which Mutations Contribute to Cancer

Genes	Gene Products and Their Function
CDKs	Enzymes known as cyclin-dependent protein kinases that control the activity of other proteins by phosphorylating them
CDC28	A CDK discovered in the yeast *Saccharomyces cerevisiae* that controls several steps in the *S. cerevisiae* cell cycle
CDC2	A CDK discovered in the yeast *Schizosaccharomyces pombe* that controls several steps in the *S. pombe* cell cycle; also the designation for a particular CDK in mammalian cells
CDK4	A CDK of mammalian cells important for the G_1-to-S transition
CDK2	A CDK of mammalian cells important for the G_1-to-S transition
cyclins	Proteins that are necessary for and influence the activity of CDKs
cyclinD	A cyclin of mammalian cells important for the G_1-to-S transition
cyclinE	A cyclin of mammalian cells important for the G_1-to-S transition
cyclinA	A cyclin of mammalian cells important for S phase
cyclinB	A cyclin of mammalian cells important for the G_2-to-M transition
E2F	A transcription factor of mammalian cells important for the G_1-to-S transition
RB	A mammalian protein that inhibits E2F
p21	A protein of mammalian cells that inhibits CDK activity
p16	A protein of mammalian cells that inhibits CDK activity
p53	A transcription factor of mammalian cells that activates transcription of DNA repair genes as well as transcription of *p21*
RAD9	A protein that inhibits the G_2-to-M transition of *S. cerevisiae* in response to DNA damage
E6	A protein of the HPV virus that inhibits p53
E7	A protein of the HPV virus that inhibits Rb

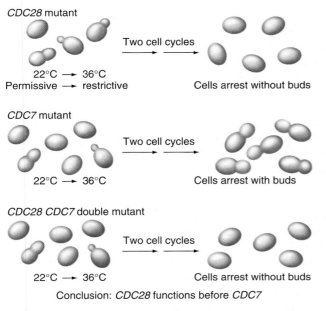

Figure 18.5 A double mutant reveals which gene is needed first. A temperature-sensitive *cdc28* mutant arrests as unbudded cells and a *cdc7* mutant as budded cells after a shift from permissive to restrictive temperature. The double mutant arrests with the phenotype of the *cdc28* mutant, revealing that this step occurs before the *cdc7* step.

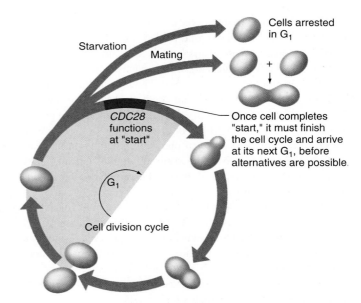

Figure 18.6 Yeast cells become committed to the cell cycle in G_1. A haploid yeast cell prior to "start" in the G_1 phase of the cell cycle is capable of arresting growth due to starvation, mating with a cell of opposite mating type, or dividing. Once the cell has passed "start" it is committed to completing the cell cycle.

The significance of the *CDC28* gene became apparent when geneticists identified related genes in other organisms. They found, for example, that the *CDC2* gene in fission yeast controls a step of commitment in that cell. They also learned that in extracts of *Xenopus laevis* (African clawed frog) embryos, the activity of a protein known as MPF (for maturation-promoting factor) controls the rapid early divisions. Sequences of the cloned budding and fission yeast genes revealed that they encode *protein kinases:* enzymes that add phosphate groups to their substrates. The *Xenopus* MPF also turned out to be a protein kinase. Moreover, genetic swapping experiments showed that the budding yeast *CDC28* gene and the fission yeast *CDC2* gene can replace one another in either organism, demonstrating that they encode proteins that carry out the same activity; the same is true of the *Xenopus* MPF-encoding gene.

Thus in three different organisms, genes that seem to be the central controlling element of the cell cycle encode functionally homologous protein kinases. Further work has shown that these kinases are **cyclin-dependent kinases (CDKs),** that is, they require another protein known as a **cyclin** for their activity.

Cyclin-Dependent Kinases and Cyclins Set the Times of Cell-Cycle Events

A protein kinase is an enzyme that adds a phosphate group to a target protein molecule; the addition either activates or inactivates the target protein. The CDKs are a family of kinases that regulate the transition from G_1 to S and from G_2 to M through phosphorylations that activate or inactivate target proteins. As mentioned and as their name implies, cyclin-dependent kinases function only after associating with another type of protein known as a cyclin. The cyclin portion of a CDK-cyclin complex specifies which set of proteins a particular CDK will phosphorylate; the CDK portion of the complex then performs the phosphorylation (Fig. 18.7a). One CDK-cyclin, for example, activates target proteins required for DNA replication at the onset of the S phase, whereas another CDK-cyclin activates target proteins necessary for chromosome condensation and segregation at the beginning of the M phase. The cyclins that guide the CDK phosphorylations appear on cue at each phase of the cell cycle, associate with the appropriate CDKs, point out the proper protein targets, and then disappear to make way for the succeeding set of cyclins. The cycle of precisely timed cyclin appearances and disappearances is the result of two mechanisms: gene regulation that turns on and off the synthesis of particular cyclins and regulated protein degradation that removes the cyclins.

The cyclin-dependent kinases are ideal enzymes for the coordination of complex events. Many copies of a particular CDK can simultaneously activate many different proteins throughout the cell, such as the proteins necessary for chromosome condensation, spindle assembly, and nuclear membrane breakdown. Binding with the appropriate cyclin guides the CDK to these targets as the cell prepares for mitosis.

The target of an enzyme is its **substrate.** Although there is little direct information about most of the cellular proteins that serve as CDK substrates, scientists who study the cell cycle have confirmed one: the **nuclear lamins,** a group of proteins that un-

Figure 18.7   The cyclin-dependent kinases (CDK) control the cell cycle by phosphorylating other proteins. (a) A CDK combines with a cyclin and acquires the capacity to phosphorylate other proteins. Phosphorylation of a protein can either inactivate or activate it. **(b)** CDK phosphorylation of the nuclear structure proteins, lamins, is responsible for the dissolution of the nuclear membrane at mitosis.

derlie the inner surface of the nuclear membrane (Fig. 18.7b). The nuclear lamins probably provide structural support for the nucleus and possibly provide sites for the assembly of proteins that function in DNA replication, transcription, RNA transport, and chromosome structure. During most of the cell cycle, the lamins form an insoluble structural matrix. At mitosis, however, the lamins become soluble, and this solubility allows dissolution of the nuclear membrane into vesicles. Lamin solubility requires phosphorylation; mutant lamins that resist phosphorylation do not become soluble at mitosis. Thus, one critical mitotic event— dissolution of the nuclear membrane—is most likely triggered by CDK phosphorylation of nuclear lamins. It is highly likely that CDK phosphorylation of particular chromatin proteins activates other events of mitosis, such as chromosome condensation.

Genetic studies of yeast provided much of the evidence that CDK-cyclin complexes are key controlling agents in all eukaryotic cell cycles. In one series of studies, geneticists used yeast mutants that carry defective CDKs or cyclins to find the corresponding human genes (Fig. 18.8) and to show that the human CDKs and cyclins can function in yeast in place of the native proteins.

How Molecular Interactions Control the G_1-to-S Transition

Cell-cycle investigators have identified many of the molecular events controlling the transition from G_1 to S in human cells by

Figure 18.8 Mutant yeast permit the cloning of a human CDK gene. A culture of yeast cells containing a temperature-sensitive mutation in the yeast CDK gene were transformed with a library composed of human cDNA cloned into a yeast centromere-containing vector. The transformed yeast cells were spread on solid medium at the restrictive temperature. Only the rare transformants with a functional copy of the human CDK gene were able to grow.

Figure 18.9 CDKs mediate the transition from the G_1 to the S phase of the cell cycle. In human cells CDK4 complexed to cyclinD and CDK2 complexed to cyclinE phosphorylate the Rb protein, causing it to dissociate from, and activate, the E2F transcription factor. E2F stimulates transcription of many genes needed for DNA replication. At the transition into S phase, cyclinD is destroyed, cyclinA is synthesized, and the CDK2–cyclinA complex activates DNA replication.

analogy with similar events in the cell cycle of yeast. From their analyses, they have pieced together the following scenario.

The first CDK-cyclin complexes to appear during G_1 in humans are CDK4-cyclinD and CDK2-cyclinE (Fig. 18.9). These complexes initiate the transition to S by a programmed succession of specific phosphorylations, among which are phosphorylations of the protein product of the retinoblastoma (*RB*) gene. Unphosphorylated Rb protein inhibits a transcription factor, E2F. Phosphorylated Rb no longer inhibits E2F. Rb phosphorylation thus indirectly activates DNA synthesis by releasing the brakes on E2F and thereby allowing it to activate the transcription of genes necessary for DNA synthesis.

As cells initiate DNA synthesis, cyclinD and cyclinE are degraded, and a new cyclin—cyclinA—is synthesized. CyclinA serves as a kinase guide during S. As the cyclinA complexes with the appropriate CDK, the cell rapidly degrades the cyclinD it is replacing. The importance of cyclin degradation is evident from yeast cells in which mutations in genes controlling cyclin degradation prevent the destruction of G_1 cyclins; in these mutants, cells remain in G_1, unable to initiate DNA replication.

How Molecular Interactions Control the G_2-to-M Transition

Human cells appear to make the transition from G_2 to mitosis much as the well-studied cells of the yeast *S. pombe* accomplish the same transition. In the yeast, a CDK known as CDC2 (the second C replaces the K for historical reasons) forms a complex with cyclinB. Both the CDC2 kinase and cyclinB are present throughout G_2, but phosphorylation of a specific tyrosine

residue on the cyclin-dependent kinase (by another protein kinase) keeps it inactive. When the time comes to initiate mitosis, a phosphatase enzyme removes the phosphate group from the CDC2 tyrosine, this removal activates the CDK, and the cell enters mitosis (Fig. 18.10).

Summary of Controls Intrinsic to the Cell Cycle

The activation and inactivation of CDK-cyclin complexes trigger transitions of the cell cycle. A variety of controls set the timing of these events. For example, phosphorylation and dephosphorylation at specific sites inactivate and activate CDKs. The transcriptional regulation of cyclin synthesis and the rapid degradation of cyclins at the completion of their specified tasks make these proteins appear and disappear at appropriate times.

Cell-Cycle Checkpoints Integrate Repair of Chromosomal Damage with Events of the Cell Cycle

Damage to a cell's genome, whether caused by environmental agents or random errors of the cellular machinery as it attempts to replicate and segregate the chromosomes, can cause serious problems for the cell. Damage to the cell-cycle machinery can also cause problems. It is therefore not surprising that elaborate mechanisms have evolved to arrest the cell cycle while repair takes place. These additional controls are called **checkpoints** because they check the integrity of the genome and cell-cycle machinery before allowing the cell to continue to the next phase of the cell cycle.

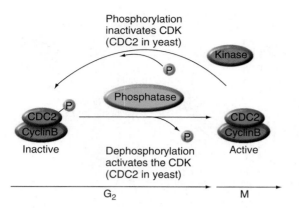

Figure 18.10 CDK activity in yeast is controlled by phosphorylation and dephosphorylation. The CDC2 protein complexed with cyclinB is inactivated prior to mitosis by phosphorylation of a specific kinase and then activated at the onset of mitosis through dephosphorylation by a specific phosphatase.

The G_1-to-S Checkpoint

When radiation or chemical mutagens damage DNA during G_1, DNA replication is postponed. This postponement allows time for DNA repair before the cell proceeds to DNA synthesis. Replication of the unrepaired DNA could exacerbate the damage; for example, replication over a single-strand nick or gap would produce a double-strand break. In mammals, cells exposed to ionizing radiation or UV light during G_1 delay entry into S phase by activating the p53 pathway (Fig. 18.11a). p53 is a transcription factor that induces expression of DNA repair genes as well as expression of the **CDK inhibitor** known as p21. Like other CDK inhibitors, p21 binds to CDK–cyclin complexes and inhibits their activity; specifically, p21 prevents entry into S by inhibiting the activity of CDK4-cyclinD complexes.

Mutations in *p53* disrupt the G_1-to-S checkpoint. One sign of this disruption is a propensity for **gene amplification:** an in-

Figure 18.11 Cellular responses to DNA damage. (a) DNA damage activates the p53 transcription factor, which in turn induces expression of the *p21* gene. The p21 protein inhibits CDK activity, producing an arrest of the cell cycle in the G_1 phase. **(b)** Tumor cells exhibit amplified regions of DNA, unlike normal cells, that can appear as homogeneously staining regions (HSR) within a chromosome or as double minutes, small pieces of extrachromosomal DNA. **(c)** When the *p53* gene is mutated in cancers, *p21* expression is not induced, cell-cycle progress is not arrested, and cells replicate damaged DNA, producing DNA double-strand breaks from single-strand nicks or gaps. **(d)** DNA damage in normal cells often leads to cell death or apoptosis, in which cellular DNA is degraded and the cell is engulfed and digested by neighboring cells.

crease from the normal two copies to hundreds of copies of a gene. This amplification is visible under the microscope, appearing as an enlarged area within a chromosome known as a homogeneously staining region (HSR) or as small chromosome-like bodies (called *minutes*) that lack centromeres and telomeres (Fig. 18.11b). Normal human cells do not generate gene amplification in culture, but *p53* mutants exhibit high rates of such amplification; *p53* mutants also exhibit many types of chromosome rearrangements. The explanation is as follows. Cells carrying mutations in the *p53* gene most likely have a defective G_1-to-S checkpoint that allows the replication of single-strand nicks. This replication produces double-strand breaks, which, in turn, lead to chromosome rearrangements. Some of the rearrangements generate gene amplification (Fig. 18.11c).

Wild-type cells able to produce functional p53 not only arrest in G_1 in the presence of DNA damage; if the damage is great enough, they also commit suicide in a process known as **programmed cell death (PCD),** or **apoptosis.** During apoptosis, the cellular DNA is degraded and the nucleus condenses. The cell may then be devoured by neighboring cells or by phagocytes (Fig. 18.11d). Programmed cell death and the proteins that regulate it—including those that are part of the p53 pathway—appear in multicellular animals from roundworms to humans. It makes sense for multicellular organisms to have a mechanism for eliminating cells that have sustained chromosomal damage. The survival and reproduction of such cells could generate cancer cells.

The G_2-to-M Checkpoint

Damage to DNA during G_2 delays mitosis, allowing time for repair before chromosome segregation (Fig. 18.12a). Researchers have isolated yeast mutants that, after sustaining radiation-induced DNA damage, do not arrest in G_2 and used these mutants to identify many genes that contribute to this checkpoint. One of these genes is *RAD9*. Whereas wild-type yeast cells can pause to repair as many as 100 double-strand breaks before entering mitosis, *RAD9* mutants fail to arrest in G_2 and die as a result of any double-strand breaks that were not repaired before mitosis.

A Checkpoint in M

During mitosis, one checkpoint oversees formation of the mitotic spindle and proper engagement of all pairs of sister chromatids (Fig. 18.12b). Observations of living cells reveal that as chromosomes condense and attach to the spindle, sometimes a single chromosome fails to attach at the expected time. When this happens, the cell does not initiate sister chromatid separation or anaphase chromosome movement until the lagging chromosome attaches to the spindle. Other studies show that in yeast cells exposed to an inhibitor that prevents assembly of a functional spindle, sister chromatids remain firmly attached. These observations suggest the presence of a checkpoint that prevents chromosome segregation until all chromosomes are properly attached to the spindle. Mutations that eliminate the surveillance of chromosome behavior during mitosis have helped researchers identify several genes in yeast responsible for this checkpoint.

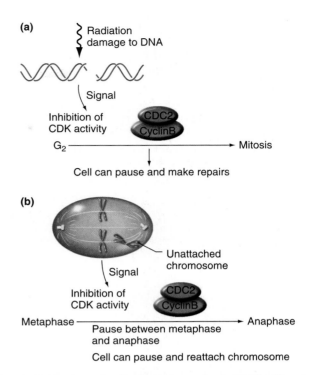

Figure 18.12 Two checkpoints act at the G_2-to-M cell-cycle transition. (a) DNA damage, particularly double-strand breaks, induce a signal that inhibits CDK activity, preventing entry into mitosis. **(b)** Spindle damage resulting from the failure of a chromosome to attach to the mitotic spindle generates a signal that inhibits CDK activity and thereby prevents the metaphase-to-anaphase transition.

Checkpoints Ensure Genomic Stability

Checkpoints are not essential for cell division. In fact, experiments in mice and other animals demonstrate that mutant cells with one or more defective checkpoints are viable and divide at a normal rate. These mutant cells, however, are much more vulnerable to DNA damage than normal cells.

Knowledge of how checkpoints work hand in hand with repair processes to ensure the fidelity of DNA replication and chromosome segregation clarifies how checkpoints help prevent transmission of three types of genomic instability (described in Chapter 13): chromosome aberrations; aneuploidy (the loss or gain of one or more chromosomes); and changes in ploidy, for example, from $2n$ to $4n$ (Fig. 18.13). Single-strand nicks resulting from oxidative or other types of DNA damage are probably fairly common. A cell normally repairs such nicks to DNA in G_1 before it enters S phase. If the checkpoint that makes time for and coordinates this repair fails to function, however, the copying of single-strand breaks during DNA replication would produce double-strand breaks that could lead to chromosome rearrangements. Chromosome loss or gain can occur if a chromosome fails to attach to the spindle or if sister chromatids fail to disjoin. Normally, the G_2-to-M checkpoint recognizes such failures and prevents mitosis until the cell has fixed the problem. Cells without a functional checkpoint will produce daughter cells carrying one too few or one too many chromosomes. Changes in ploidy can occur if a cell begins

(a)

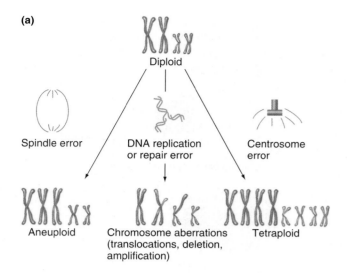

(b)

Figure 18.13   Three classes of error lead to aneuploidy in tumor cells. (a) Spindle errors can mis-segregate chromosomes resulting in whole chromosome aneuploidy; DNA replication and/or repair damage can lead to chromosome aberrations; centrosome errors can result in changes in cell ploidy. **(b)** The diagnosis of aneuploidy in tumors is improved by "chromosome painting" techniques that use a variety of fluorescent dyes attached to chromosomal DNA sequences. An appropriate choice of dyes and probes can cause each normal chromosome to appear relatively homogeneous with a unique color (*top*), while cancer cell chromosomes reveal many rearrangements and whole chromosome changes (*bottom*).

S phase before completing mitosis or if a cell fails to replicate or to properly segregate its microtubule-organizing centers, or centrosomes. Checkpoints also recognize these errors, ensuring integration of the centrosome cycle with DNA replication and the formation and function of the mitotic spindle.

A Cascade of External and Internal Molecules Tells Cells Whether or Not to Initiate Division

How do cells know when to divide? To function according to the needs of the body as a whole, cells depend on signals sent from one tissue to another. These signals tell them whether to divide or metabolize (that is, make the product[s] they are programmed to make) or die. There are two basic types of signals: extracellular and cell-bound. *Extracellular signals* in the form of steroids, peptides, and proteins act over long or short distances and are collectively known as *hormones* (Fig. 18.14a). The thyroid-stimulating hormone (TSH) produced by the brain's pituitary gland, for example, travels through the bloodstream to the thyroid gland, where it stimulates cells to produce another

hormone, thyroxine, which increases metabolic rate. *Cell-bound signals,* such as the histocompatibility proteins that, like fingerprints, distinguish an individual's cells from all foreign cells and molecules, require direct contact between cells for transmission (Fig. 18.14b). The macrophages, helper T cells, and antibody-producing B cells of the immune system communicate via cell-bound signals about the presence of viral particles, bacteria, and toxins.

The Molecular Components of Each Signaling System: Growth Factors, Receptors, Intracellular Transducers, and Transcription Factors

Extracellular hormones and cell-bound signals that stimulate or inhibit cell proliferation are known as **growth factors.** Most growth factors deliver their message to specific **receptors** embedded in the membrane of the receiving cell (Fig. 18.15a). The receptors are proteins that have three parts: a signal-binding site outside the cell, a transmembrane segment that passes through the semipermeable cell membrane, and an intracellular domain that relays the signal (that is, the binding of growth factor) to

(a)

Pituitary gland

Secretory cell

TSH molecules produced by pituitary travel through bloodstream to thyroid

TSH molecule

Receptor site on thyroid cell surfaces

Thyroid

Thyroxine

Thyroid cells produce thyroxine, which increases metabolic rate

(b)

Virus particle

These are T cell receptors for histo-compatibility protein and for virus protein

Host cell

T cell

Antigen

Infected cell is destroyed by T cell

T cell binds to histo-compatibility-antigen complex

Figure 18.14 Extracellular signals can diffuse from one cell to another or be delivered by cell-to-cell contact. (a) The pituitary gland produces thyroid-stimulating hormone (TSH) that diffuses through the circulation to the thyroid gland, which produces another hormone, thyroxine, that acts on many cells throughout the body. **(b)** A killer T cell recognizes its target cell by direct cell-to-cell contact.

proteins inside the cell's cytoplasm. These cytoplasmic proteins, which relay signals inside the cell, are known as **signal transducers.** The final link in the relay system is usually a **transcription factor** that activates the expression of specific genes in the nucleus, either to promote or to inhibit cell proliferation (Fig. 18.15b and c).

How the Molecules Interact to Relay a Signal
Binding of a growth factor to its specific receptor elicits a cascade of biochemical reactions inside the cell, often involving a

large number of molecules. Each molecule in the cascade transmits the receptor's binding-of-messenger signal by activating or inhibiting another molecule. The activation and inhibition of intracellular targets after growth-factor binding is called **signal transduction.**

One example of a signal transduction system includes the product of the *RAS* gene (Fig. 18.15d). The RAS protein is a molecular switch that exists in two forms: an inactive form in which it is bound to guanosine diphosphate (RAS-GDP) and an active form in which it is bound to guanosine triphosphate (RAS-GTP). (GTP is also the nucleotide used to insert a G residue into an RNA molecule.) Once a growth factor activates a receptor, the receptor flicks the RAS switch to active by exchanging GDP for GTP. Next RAS-GTP activates a series of three protein kinases, and this trio, known as a MAP kinase cascade, activates a transcription factor.

The proteins in a signal transduction system are like the neurons in a nerve fiber: Each one serves as a link in a message-relay chain. In deciding whether or not to divide, the cell, like the brain, combines messages from many signal transduction systems and adjusts its behavior in response to the integrated information.

As we see in the next section, cells that have lost the ability to reproduce their genomes faithfully, because of defects in the DNA repair systems described in Chapter 7 or defects in checkpoints, are prone to mutations that cause cancer.

Cancer Arises When Controls over Cell Division No Longer Function Properly

An understanding of the molecular basis of cell-cycle regulation sheds light on the life-threatening proliferative disease of cancer. The many genes contributing to the normal control of cell proliferation are all subject to mutations. Such mutations may produce inappropriate signals about the need for cell division, malfunctions in the CDK-cyclin complexes that control cell-cycle transitions, or checkpoint breakdowns that lead to genomic instability.

The Cancer Phenotype Results from the Accumulation of Multiple Mutations in the Clonal Progeny of a Cell
Cancer biologists now believe that most cancers result from the accumulation of many mutations during the proliferation of somatic cells. When enough mutations accumulate in genes controlling proliferation and other processes within a single clone of cells, that clone overgrows the normal cells that surround it, disseminates through the bloodstream to other parts of the body, and forms a life-threatening tumor, or cancer. (In this chapter, we use the term "tumor" to designate cancerous tissue and the term "growth" to designate a benign mass.)

Epidemiological data, clinical studies, and experimental analyses of a range of cell types in a variety of species provide

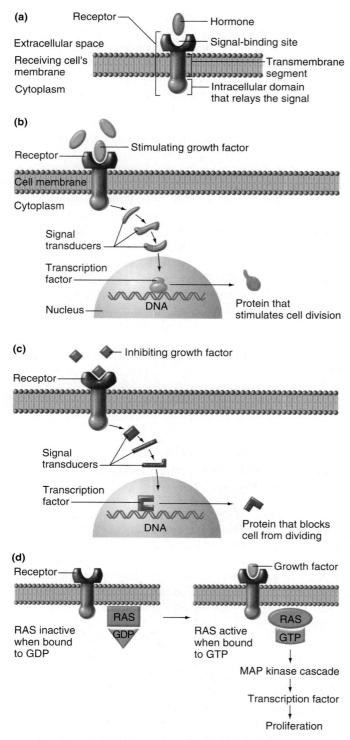

Figure 18.15 Many hormones transmit signals into cells through receptors that span the cellular membrane.
(a) Hormones bind to a cell surface receptor that specifically recognizes their structure. The extracellular surface of the receptor transmits a signal to the intracellular domain of the receptor, which in turn interacts with other signaling molecules in the cell either to stimulate growth **(b)** or to inhibit growth **(c).** The RAS protein is an intracellular signaling molecule that is induced to exchange a bound GDP (inactive) for a bound GTP (active) when a growth factor binds to the cellular receptor with which RAS interacts **(d).**

evidence for this gene-based view of cancer. Here we present some of that evidence as well as an interpretation of the evidence that explains how a cell progresses from normal to cancerous.

The General Cancer Phenotype Includes Many Types of Cellular Abnormalities

Theodor Boveri, one of the architects of the chromosome theory of inheritance, observed as early as 1914 that cells excised from malignant tumors have abnormal chromosomes. By the 1970s, when new staining techniques and improved equipment made it possible to distinguish each of the 23 different chromosome pairs in the human genome by their specific banding patterns, investigators noted that many different chromosomal abnormalities appear in tumor cells. Using tools developed in the 1980s, geneticists confirmed that most tumor cells exhibit karyotypic instability.

Figure 18.16 shows the main characteristics that distinguish tumor cells from normal cells. The cancer phenotype includes uncontrolled cell growth, genomic and karyotypic instability, the potential for immortality, and the ability to invade and disrupt local and distant tissues. Although no one cancer cell necessarily manifests all the phenotypic changes illustrated in Fig. 18.16, each cancer cell displays a number of them.

More Than One Mutation Must Fuel the Progression from Normal to Cancerous

The large catalog of phenotypic changes seen in tumor cells suggests that many mutations—at least one for each phenotypic change—are necessary to convert a normal cell into a cancerous cell. In other words, a normal cell must change many aspects of its phenotype to become a life-threatening cancer cell, and this set of changes probably requires the alteration of many genes. Clinical analyses of human cells from colon tumors show that these cancer cells contain mutations in at least 5–10 genes. Since the intricate web of controls that organize and maintain the society of cells constituting our body consists of many gene-encoded proteins, the number of genes in which mutations can fuel the progression to cancer is probably quite large—at least 100 and possibly several hundred.

To study mutations associated with cancer, researchers initially identify and isolate a mutation of interest by linkage analysis of markers, traditional genetic mapping to a chromosome, and positional cloning (all techniques described in Chapter 11). It is possible to test in mice whether a mutation in a single gene associated with cancer is sufficient to induce a tumor. If the mutation acts in a dominant fashion, researchers insert a copy of the mutant allele into the mouse genome of a fertilized egg; if the mutation is recessive, they delete one copy of the homologous gene from the early embryonic mouse genome and then breed animals homozygous for the deletion. In gene transfer experiments where a dominant cancer-causing mutation was inserted into a mouse genome under the control of a breast-cell-specific promoter, the transgenic mice produced a few breast tumors (Fig. 18.17a). Doubly transgenic mice made by breeding these

Feature Figure 18.16

Phenotypic changes that distinguish tumor cells from normal cells.

(a) MOST NORMAL CELLS   MANY CANCER CELLS

1.   **Autocrine stimulation**

Absent   Present

2.   **Contact inhibition**

Present   Absent

3.   **Cell death**

Irradiation   Irradiation

Cell
death

Present   Absent

4.   **Gap junctions**

Present   Absent

a. Changes that produce uncontrolled cell growth

1. *Autocrine stimulation.* Most cells "decide" whether or not to divide only after receiving signals from neighboring cells, either positive signals that stimulate division or negative signals that prevent proliferation. Many tumor cells, by contrast, make their own stimulatory signals, in a process known as **autocrine stimulation,** or are insensitive to negative signals.

2. *Loss of contact inhibition.* Normal cells stop dividing when they come in contact with one another, as evidenced by the fact that the few normal cell types that grow in culture form sheets one cell thick. Tumor cells, which have lost the property of contact inhibition, climb all over each other to produce piles that are many cells thick. This change in behavior contributes to the disordered array of cells seen in tumors, which is a significant departure from the highly ordered patterns seen in normal tissues.

3. *Loss of cell death.* Normal cells die when starved of growth factors or when exposed to agents such as toxins or X rays that damage them. Their programmed cell death is activated by the expression of certain genes in the cell. Because it keeps cells from proliferating when they should not and eliminates badly damaged cells, apoptosis is probably a safeguard against the early stages of cancer. Most cancer cells are much more resistant than normal cells to programmed cell death.

4. *Loss of gap junctions.* Normal cells connect to their neighbors by small pores, or gap junctions, in their membranes. The gap junctions permit the transfer of small molecules that may be important in controlling cell growth. Most tumor cells have lost these channels of communication.

(continued)

Phenotypic changes that distinguish tumor cells from normal cells. (continued)

b. Changes that produce genomic and karyotypic instability

1. *Defects in the DNA replication machinery.* Cancer arises most often in cells that have lost the ability to reproduce their genomes faithfully. We saw in Chapters 6 and 7 that cells have elaborate systems for repairing DNA damage; these systems include the enzymatic machinery for mismatch repair and the repair of damage caused by radiation or ultraviolet light. Work on yeast and bacteria has shown that mutant organisms defective in DNA repair have enormously increased rates of mutation. And these increased mutation rates often lead to cancer. An example is the hereditary syndrome of xeroderma pigmentosum, which predisposes to skin cancer.

2. *Increased rate of chromosomal aberrations.* Tumor-cell karyotypes often carry gross rearrangements, including broken chromosomes, with some of the pieces rejoined to other chromosomes; multiple copies of individual chromosomes, rather than the normal two; and deletions of large chromosomal segments and of whole chromosomes. The changes can be revealed by chromosome painting. Studies have confirmed that the fidelity of chromosome reproduction is greatly diminished in tumor cells. Normal fibroblast cells, for example, have an undetectable rate of gene amplification (an increase in the number of copies of a gene), whereas tumor cells have amplification rates as high as 1 in 100 cells. Some of the genes that control cell proliferation are inappropriately activated by such amplification.

Although research has confirmed that tumor cells have an increased rate of chromosomal aberration, probably only a small fraction of these chromosomal rearrangements lead to cancer; for example, tumors from solid tissues typically carry many chromosomal rearrangements, but most of these aberrations do not recur in all tumors. A few rearrangements, however, regularly appear in specific tumor types. Examples include the translocation between chromosomes 8 and 14 found in patients with certain kinds of lymphoma and the translocation between chromosomes 9 and 22 found in certain types of leukemias (see Fig. 13.17).

(b.1)

(b.2)

c. Changes that produce a potential for immortality

1. *Loss of limitations on the number of cell divisions.* Most normal cells (except for the rare stem cells) die spontaneously after a specifiable number of cell divisions. The senescence and natural death of normal cells is evident both in culture and in the body. Tumor cells, by contrast, can divide indefinitely.

2. *Ability to grow in culture.* Cells derived from tumor cells usually grow readily in culture, making cancerous cell lines available for study. Normal cells do not grow well in culture. This difference between abnormal and normal is greatest in cells grown as isolated clones in soft agar, most likely because the normal cells must undergo more genetic changes to become competent to divide under the artificial conditions in culture than tumor cells.

3. *Restoration of telomerase activity* (not shown). Most normal human somatic cells do not express the enzyme telomerase, and this lack of telomerase expression prevents them from replicating the repeated sequences in the telomeres at the ends of their chromosomes (see Chapter 12). As a result, after a certain number of cell divisions, the telomeres shorten to the point where they contribute to cell senescence and death. Tumor cells have the ability to express telomerase, a feature that most likely contributes to their immortality.

(c.1)

Most normal cells · Cumulative cell number · Number of passages

Many cancer cells · Cumulative cell number · Number of passages

Immortality

(c.2)

Growth in soft agar

No cell growth

Cell growth

d. Changes that enable a tumor to disrupt local tissue and invade distant tissues

1. *The ability to metastasize.* Normal cells stay within rigidly defined boundaries. Tumor cells, by comparison, often acquire the capacity to invade surrounding tissues and eventually to travel through the bloodstream to colonize distant tissues. Metastasis— the invasion of other tissues—is a complicated behavior requiring many genetic changes.

2. *Angiogenesis.* Once the adult human body has developed, new blood vessels do not normally form except to heal a wound. Tumor cells, however, secrete substances that cause blood vessels to grow toward them. The new vessels serve as supply lines through which the tumor can tap new sources of nutrients; they also serve as escape routes through which tumor cells can metastasize.

3. *Evasion of immune surveillance* (not shown). The human immune system may recognize cancer cells as foreign and attack them, thereby helping to eliminate tumors before they become large enough for clinical detection. As evidence, cancer patients often have antibodies and/or killer T cells directed against their cancer cells. Successful tumor cells, however, somehow develop the ability to evade detection by the immune system.

(d.1) Metastasis

Tumor cells

Basement membrane

(d.2) Angiogenesis

Blood vessel

(a)

(b)

Figure 18.17 The percent of mice still alive as a function of age. (a) The activated *myc* oncogene produces tumors more slowly than the *ras* oncogene. Mice containing both oncogenes develop tumors even faster than mice with *ras*. **(b)** Homozygous *p53⁺* mice rarely get life-threatening tumors, while those heterozygous for a *p53⁻* mutation develop tumors late in life. Mice homozygous for the *p53⁻* mutation develop tumors early in life.

transgenic mice carrying one mutated gene with transgenic mice carrying a different mutated gene implicated in cancer generated more tumors earlier. Even in these mice, however, only a small percentage of the transformed cells proliferated abnormally to form a tumor. These results support the idea that it takes several mutations in different genes to produce a cancer.

Studies of recessive mutations in the *p53* gene point to the same conclusion. Mice with both copies of the *p53* gene deleted from their genome develop relatively normally. The *p53* mutant mice, however, have shortened life spans and get a variety of tumors more frequently than wild-type mice (Fig. 18.17b). This experiment shows that the *p53* gene is not essential for development or for normal cell function, but it does play a role in preventing tumor formation. Consequently, deleting both copies of the wild-type gene from all of a mouse's cells does not convert every cell to a tumor cell, but it does increase the probability that at least one cell will become cancerous. The conclusion is that

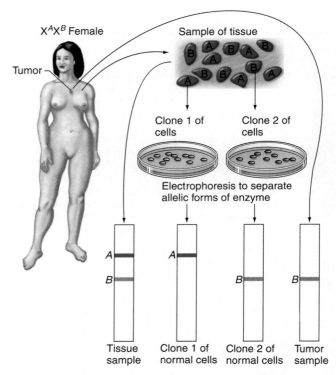

Figure 18.18 Polymorphic enzymes encoded by the X chromosome reveal the clonal origin of tumors. Each individual cell in a female expresses only one form of a polymorphic X-linked gene because of X-chromosome inactivation. A patch of tissue will usually contain both types of cells. If single cells are grown into a clone they will exhibit one or the other enzyme form. A tumor also exhibits only one form, demonstrating that it arose from a single cell. The two forms are distinguished by electrophoresis.

mutations in *p53* are just one of the many genetic changes that may occur in a cell to produce cancer.

Cancers Are Clonal Descendants of One Cell

Karyotypes of cells from women heterozygous for X-linked alleles provide evidence that cancer originates in a single somatic cell (Fig. 18.18). Although the random inactivation of one of the two X chromosomes in each cell of a female means that individual cells express only one of the two X-linked alleles for any gene, in small samples of normal somatic tissues, one usually finds both alleles expressed. This is because most somatic tissues are constructed from many clones of cells. In contrast to normal tissue, tumors from females invariably express only one allele of an X-linked gene (review discussion of X inactivation in Chapter 12). This finding suggests that the cells of each tumor are the clonal descendants of a single somatic cell that sustained a rare mutation.

Most Cancers Result from Exposure to Environmental Mutagens

Several epidemiological surveys support the hypothesis that most cancers arise by chance in somatic cells during their division and differentiation from fertilized egg to adult. The mutations that produce these cancers are not inherited through the germ line in a dominant or recessive pattern; rather they arise sporadically in a population as a result of chemicals or viruses in the environment. The evidence is as follows. First, the degree of concordance for cancers of the same type among first-degree relatives, such as sisters and brothers or even identical twins, is low for most forms of cancer in the population as a whole (we discuss specific exceptions later). If one sibling or twin gets a cancer, the other usually does not. Second, although rates for the incidence of specific cancers vary worldwide (Table 18.2), when populations migrate from one place to another, their profile of cancer incidence becomes more like that of the people indigenous to the new location. The change in cancer profile often takes decades, suggesting that the environment acts over a long period of time to induce the cancer. Third, epidemiological studies have established that numerous environmental agents increase the likelihood of cancer, and many of these agents are mutagens. These mutagens include cancer-causing viruses, some of which carry mutant forms of normal genes that control the cell cycle, as well as cigarette smoke. People who smoke for many years have a higher risk of lung cancer than people who do not smoke; and their risk increases with the number of cigarettes and the length of time they smoke.

Cancer Develops over Time

The data on lung cancer shows that decades elapse between the time a population begins smoking and the time that lung cancer begins increasing. In the United States, cancer incidence in men rose dramatically after 1940, roughly two decades after men began frequent smoking; women did not begin frequent smoking until several decades after men, and lung cancer incidence in women did not begin its dramatic increase until after 1960 (Fig. 18.19a).

Epidemiological data also show that the incidence of cancer rises with age. The prevalence of cancer in older people supports the idea that cancer develops over time as well as the idea that the accumulation of many mutations in the clonal descendants of a somatic cell fuels the progression from normal to cancerous. If you assume that the rate of accumulation of cancer-causing mutations is constant over a lifetime, the slope of a curve plotting cancer incidence against age is a measure of the number of mutations required for cancer (Fig. 18.19b). Interestingly, the data for many types of tumors generate a similar curve in which the evolution of cancer requires 6–10 mutations. Thus, the correlation between cancer incidence and aging, as well as the time lag between exposure to carcinogens and the appearance of tumors, suggests that the mutations that produce cancer accumulate over time.

TABLE 18.2 The Incidence of Some Common Cancers Varies Between Countries

Site of Origin of Cancer	High-Incidence Population		Low-Incidence Population	
	Location	Incidence*	Location	Incidence*
Lung	USA (New Orleans, blacks)	110	India (Madras)	5.8
Breast	Hawaii (Hawaiians)	94	Israel (non-Jews)	14.0
Prostate	USA (Atlanta, blacks)	91	China (Tianjin)	1.3
Cervix	Brazil (Recife)	83	Israel (non-Jews)	3.0
Stomach	Japan (Nagasaki)	82	Kuwait (Kuwaitis)	3.7
Liver	China (Shanghai)	34	Canada (Nova Scotia)	0.7
Colon	USA (Connecticut, whites)	34	India (Madras)	1.8
Melanoma	Australia (Queensland)	31	Japan (Osaka)	0.2
Nasopharynx	Hong Kong	30	UK (southwestern)	0.3
Esophagus	France (Calvados)	30	Romania (urban Cluj)	1.1
Bladder	Switzerland (Basel)	28	India (Nagpur)	1.7
Ovary	New Zealand (Polynesian Islanders)	26	Kuwait (Kuwaitis)	3.3
Pancreas	USA (Los Angeles, Koreans)	16	India (Poona)	1.5
Lip	Canada (Newfoundland)	15	Japan (Osaka)	0.1

*Incidence indicates number of new cases per year per 100,000 population, adjusted for a standardized population age distribution (so as to eliminate effects due merely to differences of population age distribution). Figures for cancers of breast, cervix, and ovary are for women; other figures are for men. Adapted from V. T. DeVita, S. Hellman, and S. A. Rosenberg (eds.), *Cancer: Principles and Practice of Oncology*, 4th ed. Philadelphia: Lippincott, 1993; based on data from C. Muir et al., *Cancer Incidence in Five Continents*, Vol. 5. Lyon: International Agency for Research on Cancer, 1987.

(a) Lung cancer death rates, United States, 1930-91

Rates are per 100,000 and are age-adjusted to the 1970 U.S. census population.

(b)

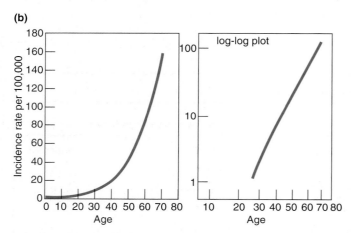

Figure 18.19 Lung cancer death rates and incidence of cancer with age. (a) Lung cancer death rates in the U.S. during the twentieth century began increasing rapidly for men in the 1940s and for women in the 1960s. This reflects the fact that smoking became prevalent among men about 20 years before it did among women. **(b)** The incidence of most cancers show a dramatic increase with age, a result thought to reflect the accumulation of mutations in somatic cells.

Although Most Cancers Are Sporadic, Some Cancers Run in Families

In some families, a specific type of cancer recurs in many members, indicating the inheritance of a predisposition through the germ line. We saw in Chapters 3, 5, and 13 (pp. 62, 142, and 447) that retinoblastoma is an example of this type of cancer. Half the individuals in families affected by retinoblastoma inherit a mutation in the *RB* gene from one parent. Since all their somatic cells carry one defective copy of the gene, a mutation

Figure 18.20 Individuals who inherit one copy of the RB⁻ allele are prone to cancer of the retina. During the proliferation of retinal cells the *RB⁺* allele is lost or mutated and cancers grow out of the *RB⁻ / RB⁻* clone of cells.

in the single remaining wild-type copy of the *RB* gene in the cells that proliferate to produce the retina predisposes these cells to develop retinal cancer (Fig. 18.20). People who do not inherit a mutation in the *RB* gene need to experience a mutation in both copies of the gene in the same cell to develop cancer; this type of double hit is very rare. Interestingly, for nearly all common types of cancer that occur sporadically in a population, rare families can be found that exhibit an inherited predisposition to that cancer.

Interpretation of the Evidence: Cancer Develops When Cells in a Single Lineage Accumulate Enough Mutations to Overgrow Normal Cells

The epidemiological, clinical, and experimental evidence demonstrates that cancer cells have many phenotypic abnormalities. Underlying these phenotypic abnormalities are mutations. More than one mutation in the clonal descendants of a single cell is necessary to produce cancer. These mutations occur primarily in somatic cells and accumulate over time. Most mutations result from exposure to environmental mutagens and appear sporadically in a population, but the inheritance of certain mutations predisposes some families to cancers associated with those mutations.

A multistep model of cancer formation accounts for the evidence and provides a coherent interpretation of the disparate observations. According to this model, mutations accumulate over time in the cells of a clone, and this accumulation fuels the progression from normal to cancerous (Fig. 18.21).

The Mutations That Lead to Cancer Create Dominant Oncogenic Alleles or Recessive Tumor-Suppressor Alleles

Research has not only revealed that cancer results from multiple genetic changes in the clonal descendants of one cell, it has also established that the mutations found in tumors are of two general types: those that improperly activate genes (for example, the genes responsible for promoting cell proliferation under clearly circumscribed conditions) and those that improperly inactivate genes (for example, the genes responsible for preventing excessive cell proliferation).

The mutant alleles that lead to cancer are referred to as **cancer genes;** but the term "genes" is a misnomer. *All cancer genes are, in fact, mutant alleles of normal genes.* When present in all or a subset of cells within an organism, these mutant alleles pre-

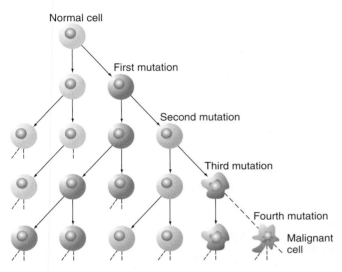

Figure 18.21 Cancer is thought to arise by successive mutations in a clone of proliferating cells.

dispose the individual to develop cancer over a lifetime. Mutant alleles that act dominantly are known as **oncogenes;** in a diploid cell, only one mutant oncogenic allele alters the cell phenotype (Fig. 18.22a). Mutant alleles that are recessive are known as **mutant tumor-suppressor genes;** in a diploid cell, both copies of a tumor-suppressor gene must be mutant to make the cell abnormal (Fig. 18.22b).

Mutations That Create Oncogenes Often Increase Cell Proliferation

Two approaches to identifying oncogenes are the study of tumor-causing viruses and the study of tumor DNA itself.

Tumor viruses are useful tools for studying cancer-causing genes because they carry very few genes and because they infect and change cultured cells to tumor cells, which makes it possible to study them *in vitro.* A large number of the viruses that generate tumors in animals are retroviruses whose RNA genome, upon infecting a cell, is copied to cDNA, which then integrates into the host chromosome (review the Genetics and Society box in Chapter 8). Later, during excision from the host chromosome, the virus can pick up copies of host genes. These normal genes change to abnormally activated oncogenes either through mutations that occur during viral propagation or through their placement near powerful promoters and enhancers in the viral genome (Fig. 18.23a). The oncogenes carried by tumor-producing viruses are thus mutated versions of normal host cell genes. The wild-type, or normal, genes that become oncogenes on mutation are known as **protooncogenes.** When a

Figure 18.22 Cancer-producing mutations occur in two forms. (a) Dominant mutations generate oncogenes that exhibit abnormal activity or produce an excessive amount of protein. **(b)** Recessive mutations produce altered tumor suppressors that usually generate little or no phenotype when they are heterozygous with a wild-type allele, but they affect cell proliferation when a second mutation inactivates the wild-type allele.

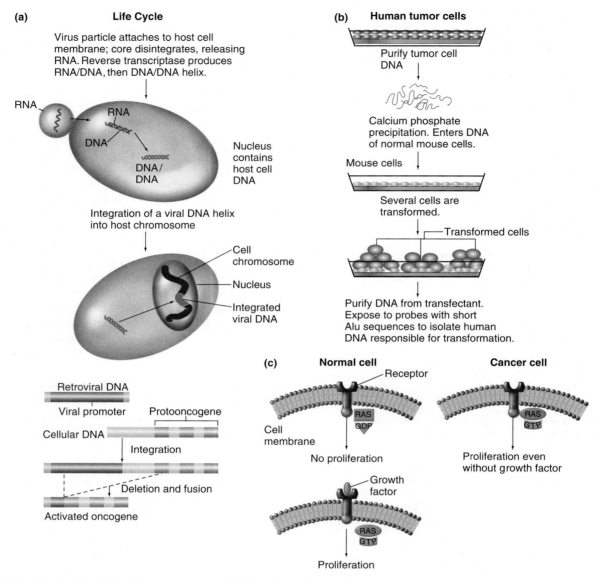

Figure 18.23 Two methods to isolate oncogenes. (a) Retroviruses that cause cancer carry a mutant or overexpressed copy of a cellular growth-promoting gene. If the genome of a retrovirus integrates into the host chromosome near a protooncogene, the cellular gene may be packaged with the viral genome when the virus leaves the cell. **(b)** DNA isolated from some human cancers is able to transform mouse cells into cancer cells. These cells are found to contain a human oncogene. **(c)** The *RAS* oncogene, a mutant form of the *RAS* protooncogene, produces a protein that becomes locked into the GTP-activated form.

virus carrying one or more oncogenes infects a cell, the onco-genes cause abnormal proliferation that can lead to the accumu-lation of more mutations and eventually to cancer. The analysis of tumor-causing retroviruses led to the discovery of oncogenes in a variety of species (Table 18.3).

Some DNA viruses also carry oncogenes. An example is the human papilloma virus (HPV). HPV infection of a woman's cervical cells is probably the first step in the development of cer-vical cancer. The papilloma virus carries at least two oncogenes capable of transforming appropriate recipient cells in culture: *E6* and *E7*. The E6 and E7 proteins bind to and inactivate the normal products of the *p53* and *RB* genes. Only those HPV sub-

types whose E6 and E7 proteins bind p53 and RB proteins are associated with cervical cancer in women. Progression of HPV-initiated cells to cancer requires additional mutations in genes not yet identified.

Scientists also identify oncogenes by isolating DNA from tumor cells and exposing noncancerous cells in culture to this tumor DNA. Some tumor DNA transforms cultured cells into cells capable of producing tumors (Fig. 18.23b). If one uses hu-man tumor DNA in the transformation of mouse cells, for ex-ample, it is possible to identify the DNA responsible for the transformation by reisolating the human DNA from the trans-formed mouse cells with probes for the short interspersed ele-

TABLE 18.3 Retroviruses and Their Associated Oncogenes*

Virus	Species	Tumor	Oncogene
Rous sarcoma	Chicken	Sarcoma	*src*
Harvey murine sarcoma	Rat	Sarcoma and erthyroleukemia	*H-ras*
Kristen murine sarcoma	Rat	Sarcoma and erthyroleukemia	*K-ras*
Moloney murine sarcoma	Mouse	Sarcoma	*mos*
FBJ murine osteosarcoma	Mouse	Chondrosarcoma	*fos*
Simian sarcoma	Monkey	Sarcoma	*sis*
Feline sarcoma	Cat	Sarcoma	*sis*
Avian sarcoma	Chicken	Fibrosarcoma	*jun*
Avian myelocytomatosis	Chicken	Carcinoma, sarcoma, and myleocytoma	*myc*
Ableson leukemia	Mouse	B cell lymphoma	*abl*

*Retroviruses identified as causative agents of tumors in animals contain oncogenes that were derived from a cellular gene. Adapted from Lewin, *Genetics,* 1e, Oxford University Press, Inc. by permission.

ments known as Alu sequences that appear only in the human genome (see Chapter 13, p. 461). The oncogenes identified in this way, like those discovered in studies of tumor viruses, are oncogenic alleles of normal cellular genes that have mutated to abnormally active forms. Sometimes the two approaches have identified the same oncogene, for example, *RAS*. The oncogenic forms of the *RAS* gene generate proteins that are always (or constitutively) in the GTP-activated form; hence, whether or not growth factor is present, a cell carrying a *RAS* oncogene receives signals to divide (Fig. 18.23c). Like the mutated *RAS*, many oncogenes continuously turn on one or more of a cell's many signal transduction systems. They do this by encoding receptors, signal transmitters, and transcription factors that are active with or without growth factor (Table 18.4).

Excessive Proliferation Enhances the Potential for Mutation

Like the oncogenic *RAS* gene, many of the oncogenes so far identified affect cell-signaling pathways that tell a cell whether or not to proliferate. The importance of these genes in generating cancer is not just that they cause cells to proliferate, because an increase in proliferation alone, without other changes, generates benign growths that are not life threatening and can be removed by surgery. Rather, increased proliferation provides a large clone of cells within which further mutations can occur, and these further mutations may eventually lead to malignancy. The more cells that exist in a clone, the more likely that rare mutations will occur in the clone; and these additional mutations appear in a clone that already has the potential for rapidly propagating them.

Cancer researchers believe that oncogenes make up only a small percentage of the cancer genes identified to date; yet analyses of tumor viruses and of *in vitro* transformations by tumor DNA can identify only this subset. This is because both ap-

proaches depend on the addition of cancer-causing DNA to the wild-type genome of a normal cell. Under such circumstances, cancer can arise only if the cancer-causing allele is dominant to the normal counterpart in the host cell.

Mutations That Create Mutant Tumor-Suppressor Alleles Release a Brake on Cell Division and Often Decrease the Accuracy of Cell Reproduction

Mutant tumor-suppressor genes are recessive alleles of genes whose normal alleles help put cell division on hold, whether in terminally differentiated cells or in cells with DNA damage. Targets for tumor-suppressor mutations include *RB, p53, p16,* and *p21*. One wild-type copy of these genes apparently produces enough protein to regulate cell division; the loss of both wild-type copies releases a brake on proliferation (see Fig. 18.22b). Researchers have identified dozens of tumor-suppressor genes through the genomic analysis of families with an inherited predisposition to specific types of cancer or through the analysis of specific chromosomal regions that are reproducibly deleted in certain tumor types.

Retinoblastoma provides an example of this identification process. A cancer of the color-perceiving cone cells in the retina, retinoblastoma is one of several cancers inherited in a dominant fashion in human families (Fig. 18.24a). Roughly half the children of a parent with retinoblastoma get the disease. Retinoblastoma tumors are easy to diagnose and remove before they become invasive. As we saw in Chapter 13, karyotypes of normal, noncancerous tissues from many people suffering from retinoblastoma reveal heterozygosity for deletions in the long arm of chromosome 13; that is, the patients carry one normal and one partially deleted copy of 13q. Karyotypes of the cancerous retinal cells from some of these same patients show homozygosity for the same chromosome 13 deletions that are heterozygous in the noncancerous cells (Fig. 18.24b). Although

TABLE 18.4 **Oncogenes Are Members of Signal Transduction Systems***

Name of Oncogene	Tumor Associations	Mechanism of Activation	Properties of Gene Product
hst	Stomach carcinoma	Rearrangement	Growth factor
erb-B	Mammary carcinoma, glioblastoma	Amplification	Growth factor receptor
trk	Papillary thyroid carcinomas	Rearrangement	Growth factor receptor
Ha-ras	Bladder carcinoma	Point mutation	GDP/GTP binding signaling protein
raf	Stomach carcinoma	Rearrangement	Cytoplasmic serine/threonine kinase
myc	Lymphomas, carcinomas	Amplification, chromosomal translocation	Nuclear transcription factor

*The roles of several of the oncogene products that are members of the signal transduction pathway and the ways in which they get activated in human cells are shown.

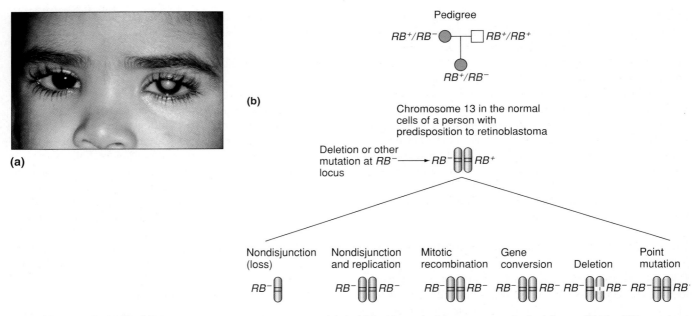

Figure 18.24 **The retinoblastoma tumor-suppressor gene. (a)** A child with a retinoblastoma tumor in the left eye. **(b)** The RB^- gene is inherited through the germ line as an autosomal recessive mutation. Subsequent changes to the RB^+ allele during somatic divisions generates a clone of cells homozygous or hemizygous for the RB^- allele.

the deletions vary in size and position from patient to patient, they all remove band 13q14.

These observations indicate that band 13q14 includes a gene whose removal contributes to the development of retinoblastoma. *RB* is the symbol for this gene. The heterozygous cells in a patient's normal tissues carry one copy of the gene's wild-type allele (RB^+), and this one copy prevents the cells from becoming cancerous. Tumor cells homozygous for the deletion, however, do not carry any copies of RB^+, and without it, they begin to divide out of control.

Geneticists used their understanding of retinoblastoma inheritance to find the *RB* gene. They cloned DNA carrying the gene by looking for DNA sequences in band 13q14 that were lost in all of the deletions associated with the hereditary condition. They then identified the gene by characterizing a very small deletion that affected only one transcriptional unit—the *RB* gene itself. Analysis of the gene's function showed that it encodes a protein involved (along with many other proteins) in regulating the cell cycle. *RB* thus fits our definition of a tumor-suppressor gene; the protein it determines helps prevent cells from becoming cancerous. Cancer can arise when cells heterozygous for an *RB* deletion lose the remaining copy of the gene.

This picture of the genetics of retinoblastoma raises a perplexing question: How can the retinoblastoma trait be inherited in a dominant fashion if a deletion of the *RB* gene is recessive to the wild-type *RB*$^+$ allele? The answer depends on the specific phenotype under consideration: susceptibility to cancer versus the progression from a normal to a cancerous cell. At the level of the organism, *RB* deletions are dominant because there is a strong likelihood that in at least one of the hundreds of thousands of retinal cells heterozygous for the deletion, a subsequent genetic event, such as mitotic nondisjunction, mitotic recombination, or a new mutation, will disable the single remaining *RB*$^+$ allele, resulting in a mutant cell with no functional tumor-suppressor gene. This one cell lacking any *RB*$^+$ gene will then multiply out of control, eventually generating a clone of cancerous cells. Although the aberrant events that create mutant cells with no functional *RB* gene happen very rarely, there are enough cells in the retina that most patients who inherit one 13q14-deleted chromosome 13 develop multiple retinoblastoma tumors in both eyes. At the level of a single cell, however, *RB* deletions are recessive to *RB*$^+$ because a cell with only one copy of *RB*$^+$ produces enough functional protein to remain normal. Hence mutations in tumor-suppressor genes appear dominant when the phenotype under consideration is susceptibility of individuals to cancer, but they act in a recessive manner when the phenotype under examination is the progression from normal to cancerous of individual cells in a clone.

Geneticists first recognized the recessive *RB* mutation that leads to retinoblastoma through the genomic analysis of families inheriting a predisposition to the cancer. More recently, they noted that both copies of the *p16* gene on chromosome 9 are deleted in roughly 75% of all melanomas (a malignant skin cancer) and in approximately 85% of all gliomas (the most common form of brain cancer). The *p16* gene, we have seen, encodes a protein that binds to and inactivates CDK4. Finally, observations of deletions of both copies of a specific region of chromosome 18 in all colon cancers led to identification of the *DCC* (deleted in colorectal cancer) gene.

Many tumor-suppressor mutations occur in genes that control the cell cycle and with it, the accuracy of genomic replication. It is important to distinguish those mutations that determine how the cell cycle is completed from mutations in genes that control proliferation. Alterations in genes that control proliferation result in an enlarged clone of cells, but aside from their increase in number, these cells—if they sustain no further mutations—are normal and thus form a benign growth. By contrast, mutations in genes that control the cell cycle can alter the accuracy with which a cell reproduces its genome. The resulting mutant cells can produce offspring with many more mutations than occur in normal cells, and this increase in the frequency of mutation vastly increases the probability that the cascade of mutations necessary to produce the phenotypic changes of tumor cells will occur.

Because cancer arises in cells that have lost the ability to reproduce their genomes faithfully, it seems reasonable to conclude that a cell's primary safeguard against cancer lies in maintaining the integrity of its genome. Cells have extensive, elaborate systems for repairing damage to their DNA. Studies of yeast and bacteria have shown that mutant organisms defective in DNA repair often have enormously increased rates of mutation. One important system for repairing DNA damage is the mismatch repair system, described in Chapter 7. This system recognizes mismatches that result from errors in DNA replication and removes the mismatched bases, allowing substitution of the correct bases. Proper functioning of the mismatch repair system increases the accuracy of DNA replication 100- to 1000-fold over the accuracy obtained by DNA polymerase alone. Mutants with a defective system have mutation rates several orders of magnitude greater than wild-type cells. In the 1990s, cancer researchers discovered that some people with a hereditary predisposition to colon cancer are heterozygous for a mutation that inactivates a gene required for the normal functioning of the mismatch repair system; the cancers that develop in these individuals consist of cells that have lost the single remaining wild-type allele. This mismatch repair gene thus behaves like a classical tumor-suppressor gene. Presumably, the greatly increased mutation rate in a homozygous cell that has lost both wild-type alleles makes it easier for progeny cells to accumulate the large number of mutations necessary to produce a cancer cell. Why these cancers develop mainly in the colon rather than in other tissues is not clear. Table 18.5 describes several other tumor-suppressor genes that affect the accuracy of cell division.

Summary: The Accumulation of Oncogenic and Tumor-Suppressor Mutations Produces Cancer Cells with Grossly Altered Genomes

Cancer-causing mutations disrupt the normal controls that create a balance between activation and inhibition of cell division. Dominant mutations that change protooncogenes to oncogenes may overactivate expression of proteins that promote proliferation. Recessive mutations in tumor-suppressor genes may release the brakes that keep cells from proliferating. Both types of mutations may tip the balance toward excessive and inaccurate cell proliferation.

Mutations that disable one part of a cell's elaborate DNA repair system increase its mutation rate and thus its likelihood of becoming cancerous. Although no single mutation converts a normal cell to a cancer cell, if a cell has a mutation in one gene that predisposes to cancer, that cell has a higher than normal probability of becoming cancerous because it is already one step along the way. The early mutations in a cell's progression from normal to cancerous may lead to increased proliferation and affect the accuracy of cellular reproduction, allowing the accumulation of several mutations. Other subsequent or simultaneous mutations may enable the abnormally and inaccurately proliferating cells of a single lineage to avoid programmed cell death, evade the immune system, increase formation of blood vessels supplying the abnormal clone, alter the proteins that control tissue architecture, and invade nearby

TABLE 18.5 Mutant Alleles of These Tumor-Suppressor Genes Decrease the Accuracy of Cell Reproduction*

Gene	Normal Function of Gene (if known), or Disease Syndrome Resulting from Mutation	Function of Normal Protein Product
p53	Controls G_1-to-S checkpoint	Transcription factor
RB	Controls G_1-to-S transition	Inhibits a transcription factor
p21	Controls G_1-to-S transition	Inhibits CDK
ATM	Controls G_1-to-S phase, and G_2-to-M checkpoint	DNA-dependent protein kinase
BS	Recombinational repair of DNA damage	DNA/RNA ligase
XP	Excision of DNA damage	Several enzymes
hMSH2, hmLH1	Correction of base-pair matches	Several enzymes
FA	Fanconi anemia	Unknown
BRCA1	Repair of DNA breaks	Unknown
BRCA2	Repair of DNA breaks	Unknown

*Many tumor-suppressor genes have been associated with a specific function in the cell cycle necessary for accuracy of cell division.

or distant tissues (that is, metastasize). Environmental factors such as radiation and mutagenic chemicals cause most of the mutations that result in cancer, but rare inherited defects can contribute the first step (see the Genetics and Society box "Genetic Testing Has Some Use in Predicting and Treating Cancer").

Comprehensive Example: The Genetics of Brain Cancer

Abnormal proliferation of cells at a critical site in the brain can threaten life even if cells in the resulting tumor do not acquire the ability to metastasize. Like tumorigenesis in other tissues, the development of a brain tumor depends on mutations that change protooncogenes to oncogenes or inactivate tumor-suppressor genes.

Cellular and Clinical Background

The most malignant form of brain cancer—glioblastoma multiforme (GBM)—is an aggressive cancer of the glial cells. The term "multiforme" signifies that GBM is a heterogeneous condition in which the same phenotype can result from mutations in different subsets of genes.

GBM Is the Highest Grade of Glial Cell Tumor

Glial cells, the most numerous cells in the human brain, come in three varieties: *astrocytes,* which provide support for the neurons; *oligodendrocytes,* which produce the myelin sheaths that vastly increase the speed of nerve impulse conduction; and *ependymal cells,* which line the brain cavities known as ventricles and help regulate production of cerebrospinal fluid

(Fig. 18.25). Any of these glial cells can mutate to form a glial-cell tumor, or glioma. Gliomas containing only mutated astrocytes are called *astrocytomas,* tumors consisting of abnormal oligodendrocytes are *oligodendrogliomas,* and gliomas carrying both mutant astrocytes and mutant oligodendrocytes are *oligoastrocytomas.* Tumors composed of abnormal ependymal cells are extremely rare and will not be part of the following discussion.

Cancer specialists recognize different grades of gliomas—low grade (designated by a Roman numeral II), anaplastic (III), and GBM (IV)—with each higher grade representing an increased likelihood of aggressiveness of growth. Unfortunately, low-grade gliomas have a tendency to undergo malignant transformations. Most eventually develop into aggressive, fast-growing GBMs.

A person's prognosis depends on his or her age and the grade and type of tumor. For example, a 25-year-old with a low-grade oligodendroglioma can survive for as long as 10 years, whereas a 70-year-old with a GBM probably has less than a year to live.

A Large Number of Genes in Various Combinations Produce GBMs

A close look at the genetic basis of glioblastoma multiforme tumors clarifies how a cancer phenotype can arise from different combinations of changes in several genetic pathways (Fig. 18.26).

Some GBMs Evolve from Low-Grade Astrocytomas

Many early astrocytomas contain at least three types of mutations: those that inactivate the *p53* tumor-suppressor gene on chromosome 17; those that oncogenically activate expression of a growth factor known as PDGF (for platelet-derived growth

Genetics and Society

Genetic Testing Has Some Use in Predicting and Treating Cancer

Genetic tests for mutations in protooncogenes and tumor-suppressor genes can reveal whether a person has a higher probability of getting cancer at some point in his or her lifetime than a person without the mutations. But of those with an increased risk, some will and some will not get cancer. This is because, unlike many single-gene disorders where inheriting one dominant or two recessive mutant alleles means you will get some form of the condition, cancer is a complex disease resulting from mutations in many genes in the same cell. While a person who acquires one of these mutations through inheritance is pushed one step along the road to cancer, the other mutations must occur in the same cell by chance; and nongenetic factors such as exposure to radiation influence if, where, and when the mutations occur. Given this situation, what good is it to learn from a genetic test that you have an increased probability of getting cancer sometime in your life?

Predictive testing is useful if the means of medical surveillance make it possible to detect the cancer to which a mutation predisposes at an early stage. Thus testing for a genetic predisposition to skin, breast, or colon cancers would be useful because careful examination of the skin, breast, or colon with existing technology (microscopes, mammograms, or colonoscopies) can often detect cancers of these tissues in their earliest stages. A person whose genetic test shows a predisposition to colon cancer, for example, could undergo a colonoscopy each year, and if one of these colon exams disclosed a small cancer, medical practitioners could remove it by surgery or other means. Predictive genetic testing is not useful at this time for pancreatic cancer because there is as yet no way to detect tumors of the pancreas when they are small; by the time doctors identify this cancer, it has almost always reached the aggressive stage and metastasized to other tissues.

Once a cancer has been diagnosed, genetic testing of tumor cells may be useful for making a prognosis and determining a course of therapy. For example, if a breast tumor is very small (less than 1 mm in diameter) and has not spread to the lymph nodes, surgical removal and follow-up radiation usually give the patient a good chance of overcoming the cancer. But if the tumor, however small, carries mutations in the *p53* gene, the prognosis is poor. Breast tumors with absent or mutated p53 proteins tend to resist treatment with radiation and many anticancer drugs.

Genetic tests for oncogenic and mutated tumor-suppressor alleles thus have specific predictive and therapeutic uses. Understanding the limitations of such tests is important. The predictive tests can indicate only increased risk. If knowledge of such risks makes it possible to take preventive measures, the tests make sense. Tests to help determine the course of therapy are possible for only a few diseases in which specific mutations have been linked to specific prognoses.

factor) as well as its receptor; and those that delete an as yet unidentified tumor-suppressor gene on chromosome 22. We know that *p53* encodes a transcription factor active in cell-cycle arrest, response to DNA damage, and apoptosis. Most astrocytoma cells with no active *p53* gene have a mutation in one copy of the gene and have lost the other copy. These same low-grade astrocytoma cells also overexpress both the PDGF and its receptor, suggesting that the *p53* mutation may produce tumors only in the presence of PDGF overexpression. Finally, low-grade astrocytoma cells with no or inactive *p53* and an overactive PDGF system have lost both copies of part of chromosome 22. Initial linkage studies suggest that the deleted region is near a telomere.

Astrocytomas that progress from low grade to anaplastic have three additional deletions—from 13q, 9p, and 19q. The deletion from the long arm of chromosome 13 carries the *RB* tumor-suppressor gene, which, as we have seen, encodes a protein that contributes to cell-cycle arrest; this deletion thus re-

moves a brake on abnormal proliferation. The region deleted from the short arm of chromosome 9 carries genes for two CDK inhibitors: p15 and p16; deletion of these inhibitors releases a brake on the activity of some of the cyclin-dependent kinases that help regulate the cell cycle. The loss of part of the long arm of chromosome 19 is a feature of all gliomas that does not appear in any other type of cancer. Linkage analysis has mapped the crucial tumor-suppressor region to 19q13.3. (Bear in mind throughout this discussion that if a cell loses a region containing a tumor-suppressor gene from one homologous chromosome, the remaining wild-type copy of the same gene must sustain a mutation to produce a cancer cell.)

Progression from aplastic astrocytoma to glioblastoma multiforme involves the loss of genes at particular points on both the long and short arms of chromosome 10. Deletion mapping shows that a primary tumor-suppressor gene resides at 10q25, while a secondary suppressor is located at an as yet unidentified point on 10p.

Figure 18.25 Mutations in three types of glial cells cause brain cancer. (a) Astrocyte brain cells from the cortex of a mammalian brain. **(b)** Oligodendrocyte cell (*orange structure in the middle*) attaching to an axon nerve cell at the start of myelination. **(c)** Ependymal cell. The astrocyte cells in (a) mutate to form astrocytomas (low-grade) shown in **(d)**. The oligodendroglioma shown in **(e)** is the result of mutated oligodendrocyte cells (b). The tumors resulting from mutated ependymal cells are extremely rare and are not shown here.

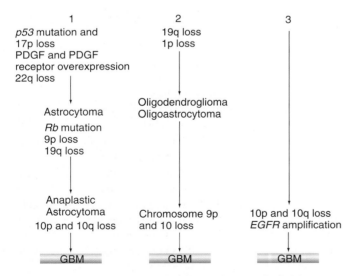

Figure 18.26 Three routes for the evolution of glioblastoma multiforme have been identified. Different combinations of mutations along this genetic pathway give rise to a cancer phenotype.

Thus, a high-grade GBM tumor evolves from a normal astrocyte through three levels of malignancy by the accumulation of at least 10 mutations: 2 generate oncogenes that overactivate the platelet-derived growth factor system and 8 inactivate tumor-suppressor genes on chromosomes 9, 10, 13, 17, 19, and 22, releasing a number of controls on cell division (Fig. 18.27).

Clinicians have compared mutations from early astrocytomas with mutations found in full-blown GBMs that appeared months or years later at the same site as the low-grade tumor. They found that the initial, low-grade tumors carried fewer mutations than the later, more malignant GBMs; and the GBMs carried all the mutations found in the lower-grade cancers and then some. These data demonstrate that an accumulation of mutations in the clonal descendants of a single astrocyte cause the cell-cycle deregulation that produces this malignant cancer.

Some GBMs Arise from Oligodendrogliomas or Oligoastrocytomas

Such GBMs carry a different combination of mutations from those found in the GBMs that develop from astrocytomas. Low-grade gliomas containing oligodendrocytes have lost part of the long arm of chromosome 19 (a deletion also seen in anaplastic astrocytomas). They have also lost a region on the short arm of

Low-grade astrocytoma (II)

RB mutation and 13q loss. Loss of 9p and 19q

Anaplastic astrocytoma (III)

10p and 10q loss

GBM (IV)

Figure 18.27 Pathway from grade II astrocytoma to malignant GBM. These photos illustrate the development of a tumor from a low-grade (II) astrocytoma to a high-grade (IV) GBM. Deletions from the tumor-suppressor regions of chromosomes 9, 13, and 19 and *RB* mutation mark the transition from grade II to grade III. Deletions from the tumor-suppressor regions of chromosome 10 mark the step from grade III to IV. Note that necrosis (cell and tissue death) has begun in the GBM photo.

chromosome 1 (see Fig. 18.26). The loss of one copy of both 9p and chromosome 10 converts these low-grade tumors to aggressive GBMs; the deletions are thought to carry at least three tumor-suppressor genes whose remaining, wild-type copies have no doubt sustained a mutation.

Some Rapidly Arising GBMs Have No Apparent Precursors

These glioblastoma multiforme tumors have an oncogenic amplification of the *epidermal-growth-factor-receptor* (*EGFR*) gene, and they have also lost regions from both 10p and 10q (review Fig. 18.26). The tumors with these genetic alterations arise *de novo* or so rapidly that no precursors are detectable. Amplification of the *EGFR* gene almost never occurs in the astrocytoma-derived GBM tumors carrying *p53* mutations and 17p deletions. Moreover, GBMs with the *EGFR* gene amplification occur in significantly older adults than GBMs with mutant or deleted *p53* and chromosome 17 genes.

Summary

The GBM phenotype develops via different combinations of mutations in different genetic pathways: from lower-grade astrocytomas via *p53* and *RB* gene inactivations, from oligodendroglial tumors via deletions of chromosomes 1 and 19, or *de novo* via *EGFR* gene activation. This neat separation of muta-

tional pathways is, however, an oversimplification. Not every GBM shows all the genetic changes described, and some GBMs derived from one type of cell exhibit mutations usually associated with another cell type. For example, rare astrocytoma-derived GBMs show a loss of chromosome 10 and amplification of the *EGFR* gene. The different combinations of mutations that subvert a cell's normal progression through the cell cycle can accumulate in the clonal descendants of a single cell either slowly over decades or rapidly in the course of a few months.

Knowledge of the Genetics of Brain Cancer May Lead to More Selective Therapies

Although cancer researchers have identified and isolated only a few of the genes that play a role in malignant gliomas, their insights have begun to point the way to more customized therapies for this type of brain cancer. Identification of the precise mutations in the cells of a patient's tumor can pinpoint the stage of disease, thus clarifying the benefits and disadvantages of certain therapeutic choices. An understanding of the genetic basis of cancer might also lead to the development of drugs that slow tumor growth by counteracting the effects of specific mutations in target cancer cells. Such drugs would make it possible to zero in on a small subset of tumor cells and thereby produce fewer side effects than current chemotherapies, which indiscriminately kill normal and cancerous cells.

C O N N E C T I O N S

The existence of numerous controls in each of several cell-cycle pathways suggests that evolution has erected many barriers in multicellular animals to the uncontrolled reproduction of "selfish" cells. At the same time, the hundreds of genes contributing to normal cell-cycle regulation provide hundreds of targets for cancer-producing mutations.

Variations on the theme of cell-cycle regulation play a key role in the development of eukaryotic organisms. During the development of multicellular organisms, cells must not only control their cell cycles, they must also adopt different fates and differentiate into different tissues. In *Drosophila,* for example, after fertilization, nuclear division occurs without cell division for the first 13 cycles; during these cycles, the nuclei go through many rapid S and M phases without any intervening G_1 or G_2 (Fig. 18.28). In cycles 10–13, the synthesis and degradation of cyclinB regulates mitosis. Sometime during cycles 14–16, a G_2 phase appears, and distinct patches of cells with different-length cycles become evident within the embryo. The differences in cycle time between the different cell types is the result of variable G_2 phases. Late in G_2, *CDC25* activates cyclin-dependent kinases to control the timing of mitosis. Many tissues stop dividing at cycle 16, but a few continue. In the still-dividing cells, a G_1 phase appears. Some of these cells will arrest in G_1 during larval growth, only to start dividing again in response to signals relayed during metamorphosis.

In Chapter 19, we present the basic principles of development and describe how biologists have used genetic analysis in various model organisms to examine development at the cellular and molecular levels.

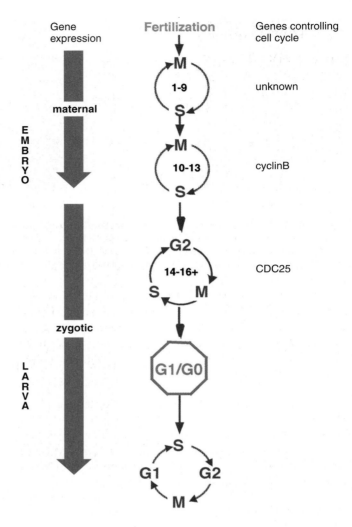

Figure 18.28 Regulation of the cell-cycle changes during *Drosophila* development. Each step of development has built-in regulators that act as barriers to uncontrolled reproduction of "selfish" cells. Some of these regulators, such as cyclinB and *CDC25,* are known; others are not.

E S S E N T I A L   C O N C E P T S

1. Several genetic pathways help control cell division.
   a. The inhibition or activation of CDKs inhibits or activates G_1-to-S and G_2-to-M transitions.
   b. The measured synthesis and degradation of different cyclins guides CDKs to the appropriate targets at the appropriate times.
   c. Checkpoints that integrate repair of chromosomal damage with events of the cell cycle minimize the replication of DNA damage.
   d. The genes and proteins of various signal transduction systems relay signals about whether or not to enter the cell cycle.

2. Cancer is a genetic disease resulting from the growth of a clone of mutant cells.

3. A cell requires many mutations to become cancerous. Exposure to environmental mutagens probably generates most of these mutations.

4. Many mutations that lead to cancer jeopardize cell-cycle regulation.
   a. Mutations in growth factors, receptors, and other elements of signal transduction pathways can release cells from the signals normally required for proliferation.
   b. Mutations in CDKs and the proteins that control them may also lead to inappropriate proliferation or genomic instability.
   c. Mutations in DNA repair and checkpoint controls lead to genomic instability and, often, to loss of the surveillance system that kills aberrant cells by apoptosis.
   d. Mutations that lead to genomic instability permit the rapid evolution of abnormal tumor cells.

S O C I A L   A N D   E T H I C A L   I S S U E S

1. Taxol is a chemical produced in the bark of Pacific yew trees. In experimental trials, taxol reduced ovarian tumors in up to 50% of terminally ill patients. When this effect was first discovered, ovarian cancer patients were extremely anxious to get the drug. Unfortunately, the Pacific yew tree is a slow-growing tree found in old growth forests of the Pacific Northwestern United States. For a series of treatments, each patient would require the bark of approximately three trees. Because of the scarcity of trees and the difficulty in obtaining taxol from the bark, the expense was high. In 1993, the cost of production by the National Cancer Institute was $1200 for the 2 grams needed for each patient. When a drug is scarce, who should get the drug? The clinical trials involved patients for whom no other treatment had worked. Is this the best use of limited resources of a potentially powerful anticancer drug? Who should decide who gets the drug? On what basis should they make their decision?

2. In 1976, cancer patient John Moore had his spleen removed at a university teaching hospital as a treatment for the hairy cell leukemia from which he suffered. Some of his spleen cells were cultured at that time. During subsequent treatment other cell types were removed from the same patient and preserved in culture. Researchers eventually used some of these cells to establish a very useful T-cell line. The cell line produced the lymphokine GM-CSF, used in treating cancers and AIDS. A patent, granted to the university and the doctors for the cell line, was sold to two biotechnology companies. Moore filed a claim against the doctors, companies, and the university stating that he was entitled to some of the profits and that there had been a lack of informed consent. He had signed a consent form but was never informed of the potential for commercial exploitation. Several questions arise from the landmark legal case, *Moore v. Regents of California.* To what extent must a doctor disclose interests in a commercial venture to a patient? Does a person own the rights to cells that have been removed from his or her body?

3. Suitable animal models for the effects of exposure to radiation on humans are not available. Only tests on humans can make it possible to evaluate the dangers of chromosome damage and cancer resulting from various types of exposure. Under these circumstances, is human experimentation acceptable? Who should participate in human trials? Volunteers who are well apprised of the risks? Prisoners on death row? Are there valid alternatives?

S O L V E D   P R O B L E M S

I. The addition of growth factors to tissue culture cells stimulates cell division. A number of candidate drugs can be tested for their ability to stop this stimulation of cell division. What do you think the target of these drugs could be?

Answer

This question concerns the regulation of cell division. Growth factors are made by one cell and bind to receptors of another cell to stimulate the cell division cycle. *A drug that binds to receptors* would block access and prevent growth factors from binding. Alternatively, *the drug could bind to the growth factor,* thereby preventing its interaction with the receptor. (These are the most obvious targets. If you are familiar with the signal transduction pathway inside the cell, you might also propose that proteins in this pathway could be targets for drug development.)

II. The *p53* gene has been cloned and you are using it to analyze DNA in patients in which *p53* defects are involved in the development of their tumors. DNA samples were obtained from normal and tumor tissue of three different cancer patients, digested with *Bam*HI, electrophoresed on an agarose gel, and transferred to filter paper that was probed with a labeled *p53* fragment. Each of the patients inherited a *p53* mutation. Both wild-type alleles produce the same three fragments. Thin bands indicate half the DNA content of thick bands. DNA from an individual who did not inherit a *p53* mutation is shown in the lane labeled wild type. Assuming the model of *p53* acting as a tumor-suppressor gene is correct and that *p53* defects are involved in each of these cancer patients, how would you describe the genetic makeup of the *p53* gene in the normal and tumor tissue of each of the three patients?

Answer

This question requires knowledge of tumor-suppressor genes. The wild-type *p53* region (as seen in the "wild type" individual) has three hybridizing bands. Because *p53* is a tumor suppressor, it is recessive at the cellular level and both copies must be defective in the tumor cells. No observable changes are apparent in patient 1, so this patient must have inherited a small mutation in *p53* and in the tumor cells the second copy would also contain a small mutation, thereby inactivating both copies of *p53* in the tumor. In patient 2, a small mutation must have been inherited. In the tumor, the whole region containing *p53* was deleted (thereby removing the second copy of the gene), as seen by the loss of restriction fragments. In patient 3, a deletion mutation (smaller than that in the tumor cells of individual 2), as seen in the altered restriction pattern, was inherited and in the tumor, the second copy of the gene was similarly deleted (probably by gene conversion since the pattern is the same for both copies).

III. The CDC28 protein of budding yeast *S. cerevisiae* and CDC2 protein of fission yeast *S. pombe* are protein kinases required at the "start" of the cell cycle. Both genes were identified by mutational analysis

(temperature-sensitive mutations in each gene cause arrest in early G_1) and both genes have been cloned. How could you determine if one could substitute for the other functionally? (Be sure to mention sources of DNA and genotypes involved.)

Answer

The *CDC28* gene of *S. cerevisiae* could be cloned into a vector and transformed into a temperature-sensitive *cdc2* mutant of *S. pombe*. If *CDC28* has the same role (function) as *CDC2*, the transformed cell will now grow and divide at nonpermissive temperatures. Conversely, a clone of the *CDC2* gene of *S. pombe* could be cloned into a vector able to transform a temperature-sensitive *cdc28* mutant of *S. cerevisiae*. If *CDC2* of *S. pombe* can substitute for *CDC28* of *S. cerevisiae*, the transformed cells would grow and divide at nonpermissive temperatures.

PROBLEMS

18-1 For each of the terms in the left column, choose the best matching phrase in the right column.

a. growth factor	1. mutations in these genes are dominant for cancer formation
b. tumor-suppressor genes	2. programmed cell death
c. cyclin-dependent protein kinases	3. series of steps by which a message is transmitted
d. apoptosis	4. proteins that are active cyclically during the cell cycle
e. oncogenes	5. control progress in the cell cycle in response to DNA damage
f. receptor	6. mutations in these genes are recessive at the cellular level for cancer formation
g. signal transduction	7. signals a cell to leave G_0 and enter G_1
h. checkpoints	8. cell-cycle enzymes that phosphorylate proteins
i. cyclins	9. protein that binds a hormone

18-2 During which phase(s) of the cell cycle would the following enzymes or proteins be most active?
a. tubulins in the spindle fibers
b. centromere motor
c. DNA polymerase
d. CDC28 of *S. cerevisiae* or CDC2 of *S. pombe*

18-3 Conditional mutations are useful for genetic analysis of essential processes. For example, temperature-sensitive cell-cycle mutations in yeast do not divide at 37°C (nonpermissive temperature) but will divide at 30°C. An alternative type of conditional mutation is a cold-sensitive mutation in which the nonpermissive temperature is low (23°C). List the steps you would go through to isolate cold-sensitive cell-cycle mutants of yeast.

18-4 Many temperature-sensitive yeast mutants that showed defects in the cell cycle were isolated in the 1970s. The mutants that arrested at the unbudded stage were mated with each other to do a complementation analysis. A + sign on the chart indicates that the resulting diploids grew at the high (nonpermissive) temperature. How many complementation groups (how many genes) are represented by these mutants?

	1	2	3	4	5	6	7	8	9
1	−	+	+	−	−	+	+	+	+
2	+	−	+	+	+	+	+	−	+
3	+	+	−	+	+	−	−	+	−
4	−	+	+	−	−	+	+	+	+
5	−	+	+	−	−	+	+	+	+
6	+	+	−	+	+	−	−	+	−
7	+	+	−	+	+	−	−	+	−
8	+	−	+	+	+	+	+	−	+
9	+	+	−	+	+	−	−	+	−

18-5 In 1951 a woman named Henrietta Lacks died of cervical cancer. Just before she died a piece of her tumor was taken and put into culture in a laboratory in an attempt to induce the cells to grow *in vitro*. The attempt succeeded, and the resulting cell line (known as HeLa cells) is still used today in laboratories around the world for studies of various aspects of cell biology. In the cell cycle of typical HeLa cells, G_1 lasts about 11 hours, S lasts about 8 hours, G_2 lasts 4 hours, and mitosis (M) takes about 1 hour.
a. Cultured cells do not typically grow synchronously. That is, the individual cells in a culture are randomly distributed throughout the cell cycle. If you looked through the microscope at a sample of

HeLa cells, in approximately what proportion of them would you expect the chromosomes to be visible? (The cells do not split apart completely after cytokinesis, and each joined double cell should be counted as one.)

b. Approximately what proportion would be in interphase?

18–6 The activity of key cell-cycle regulatory proteins is cyclical, appearing only when needed. What are three ways by which a cell can achieve this cyclical nature of protein activity?

18–7 True or false?

a. CDKs phosphorylate proteins in the absence of cyclins.

b. Degradation of cyclins is required for the cell cycle to proceed.

c. CDKs are involved in checking for aberrant cell-cycle events.

18–8 Checkpoints occur at several different times during the cell cycle to check that the DNA content of the cell has not been damaged or altered. Match the defect in a checkpoint with the consequences of that defective checkpoint.

Defective checkpoint	Consequences of defect in checkpoint
a. G_1 to S	1. aneuploidy
b. G_2 to M	2. single-strand nicks get replicated and amplification occurs
c. M	3. unrepaired double-strand breaks result in broken chromosomes

18–9 Molecules outside and inside the cell regulate the cell cycle, making it start or stop.

a. What is an example of an external molecule?

b. What is an example of a molecule inside the cell that is involved in regulation?

18–10 a. Would you expect a cell to continue or stop dividing at a nonpermissive high temperature if you had isolated a temperature-sensitive *RAS* mutant that remained fixed in the GTP-bound form at nonpermissive temperature?

b. What would you expect if you had a high-temperature-sensitive mutant in which *RAS* stayed in the GDP-bound form at high temperature?

18–11 Put the following steps in the correct ordered sequence.

a. kinase cascade

b. activation of a transcription factor

c. hormone binds receptor

d. expression of target genes in the nucleus

e. RAS molecular switch

18–12 Mouse tissue culture cells infected with the SV40 virus lose normal growth control and become transformed. If transformed cells are transferred into mice, they grow into tumors. The SV40 protein responsible for this transformation is the T antigen. T antigen has been found to associate with the cellular protein p53. If the *p53* gene fused to a high-level expression promoter is transfected into tissue culture cells, the cells are no longer transformed by infection by SV40.

a. Propose a hypothesis to explain how the high expression of *p53* saves the cells from transformation by T antigen.

b. You have decided to examine the functional domains of the p53 protein by mutagenizing the cDNA, fusing it to the high-level promoter, and transfecting into cells. Results are shown here. How would you explain the effects of mutations 1 and 2 on p53 function?

c. What is the effect of mutation 3 on p53 function?

	Morphology	
p53 construct	*Noninfected cells*	*SV40-infected cells*
None	Normal	Transformed
Wild type	Normal	Normal
Mutation 1	Normal	Transformed
Mutation 2	Normal	Transformed
Mutation 3	Normal	Normal

18–13 What are four characteristics of the cancer phenotype?

18–14 Amplification of DNA sequences in *p53* mutants can be visualized using electron microscopy.

a. Using a different technique, how could you detect amplification of a *specific sequence?*

b. How could you detect gross rearrangements (>10 Mb) of chromosomal DNA?

18–15 Some germ-line mutations predispose to cancer yet often environmental factors (chemicals, exposure to radiation) are considered major risks for developing cancer. Are these conflicting views of the cause of cancer or can they be reconciled?

18–16 The incidence of colon cancer in the United States is 30 times higher than it is in India. Differences in diet and/or genetic differences between the two populations may contribute to these statistics. How would you assess the role of each of these factors?

18–17 Put the following steps in the order appropriate for the positional cloning of *BRCA1*, a gene involved in predisposition to breast cancer.

a. Locate transcripts corresponding to the DNA.

b. Use the physical map to get clones.

c. Determine the tissues in which the transcripts are present.

d. Look for homologous DNA in other organisms.

e. Determine linkage to RFLPs and other molecular markers.

f. Sequence the DNA from affected individuals.

18–18 Because mutations occur in the development of cancer, researchers suspected that defects in DNA repair machinery might lead to a predisposition to cancer. Place the following steps in appropriate order for following a candidate gene approach to determine if defects in mismatch repair genes lead to cancer.

a. Use molecular markers near the homologous gene to determine if the candidate gene is linked to a predisposition phenotype.

b. Isolate a human homolog of a yeast mismatch repair gene.

c. Compare the DNA sequence of the mismatch repair gene of affected and unaffected persons in a family with predispositions.

d. Determine the map location of the human homolog of the yeast mismatch gene.

18–19 Which of the following events is unlikely to be associated with cancer?

a. mutations of a cellular protooncogene in a normal diploid cell

b. chromosomal translocations with breakpoints near a cellular protooncogene

c. deletion of a cellular protooncogene

d. mitotic nondisjunction in a cell carrying a deletion of a tumor-suppressor gene

e. incorporation of a cellular oncogene into a retrovirus chromosome

18–20 You have decided to study genetic factors associated with colon cancer. An extended family from Morocco in which the disease presents itself in a large percentage of family members at a very early age has come to your attention. (The pedigree is shown.) In this family, individuals either get colon cancer before the age of 16, or they don't get it at all.

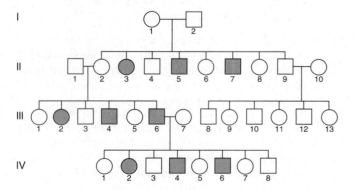

a. Based on the information you have been given, what evidence, if any, suggests an inherited contribution to the development of this disease?

b. You decide to take a medical history of all of the 36 people indicated in the pedigree and discover that a very large percentage drink a special coffee on a daily basis, while the others do not. The only ones who don't drink coffee are individuals numbered I-1, II-2, II-4, II-9, III-7, III-13, IV-1, and IV-3. Could the drinking of this special coffee possibly play a role in colon cancer? Explain your answer.

18–21 To further understand the basis for colon cancer, you find a family from the United States in which two members also get the disease before the age of 16. If there were a dominant inherited mutation segregating in this family, which of the individual(s) would you predict had the mutation in their colon cells but did not develop the disease?

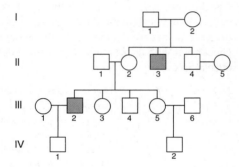

18–22 You suspect that a very specific point mutation in the *p53* gene is responsible for the majority of *p53* mutations found associated with tumors. Which combination of these techniques would you be most likely to use in developing a *simple assay* for predictive testing?

a. polymerase chain reaction with oligos flanking the mutation

b. restriction enzyme digestion followed by Southern blot

c. RNA isolation followed by Northern blot

d. hybridization with allele-specific oligonucleotides

CHAPTER

19

Using Genetics to Study Development

A young patient with polydactyly, a genetically caused developmental defect of humans that produces extra toes and/or fingers.

The union of a human sperm and egg (Fig. 19.1a) initiates the amazing process of development in which a single cell—the fertilized egg—divides by mitosis into trillions of genetically identical cells. These cells differentiate from each other during embryonic development to form hundreds of different cell types. Cells of various types assemble into wondrously complex yet carefully structured systems of organs, including two eyes, a heart, two lungs, and an intricate nervous system. Within a period of three months, the human embryo develops into a fetus whose form anticipates that of the baby who will be born six months later (Fig. 19.1b). At birth, the baby is already capable of crying, breathing, and eating; and the infant's development does not stop there. New cells form and differentiate throughout a person's growth, maturation, and even senescence.

The development of a child from a fertilized egg is so miraculous that it is difficult to look at the process from a reductionist point of view, seeking to explain it on the basis of physical, chemical, and biological principles. Yet the more biologists study development at the microscopic level, the more clearly the miracle

emerges as a precisely patterned set of cellular behaviors controlled by the genes. Cells divide, change shape, specialize, and interact in highly reproducible ways.

Biologists now accept that genes direct the cellular behaviors underlying development, but as recently as the 1940s, this idea was controversial. Many embryologists could not understand how cells with identical chromosome sets, and thus the same genes, could form so many different types of cells if genes were the major determinants of development. As we now know, the answer to this riddle is very simple: Not all genes are "turned on" in all tissues. Cells regulate the expression of their genes so that each gene's protein product appears only when and where it is needed. Two central challenges for scientists studying development are to identify which genes are critical for the development of particular cell types or organs, and to figure out how these genes work together to ensure that each is expressed at the right time, in the right place, and in the right amount.

Biologists who use genetics to study how the fertilized egg of a multicellular organism becomes an adult are called **developmental**

(a) Fertilization

(b) A human fetus three months after fertilzation

Figure 19.1 Human development. Fertilization of an egg by a sperm **(a)** creates a single cell (the zygote), which undergoes many rounds of division and cell differentiation to produce a recognizable fetus **(b)** by the end of the first trimester of pregnancy.

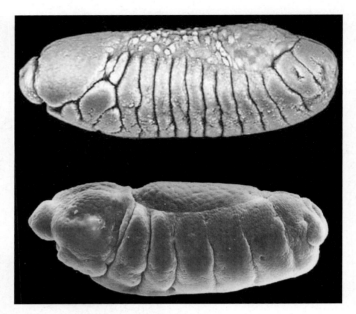

Figure 19.2 Mutations in specific *Drosophila* genes can affect early development. Wild-type embryo (*top*); embryo homozygous for a mutation in a gene called *ftz* (*bottom*). The mutant embryo has fewer body segments than normal.

geneticists. Like other geneticists, they analyze mutations; in this case, mutations that produce developmental abnormalities. An understanding of such mutations helps clarify how normal genes control cell growth, cell communication, and the emergence of specialized cells, tissues, and organs.

Given the great diversity of the 6 billion humans on our planet, it is not hard to find rare individuals carrying mutations that alter their development. For example, pedigree analysis shows that one form of *polydactyly* (hands or feet with more than five fingers or toes) is inherited in families as a simple Mendelian autosomal dominant trait (see the photograph at the beginning of this chapter). There are, however, significant ethical limitations on the study of developmental genetics in humans, including taboos on the deliberate production of mutants, on the experimental manipulation of affected individuals, and on forced matings between specific in-

dividuals with various abnormalities. But perhaps the most important limitation is not so obvious: Mutations that disrupt the earliest (and to some, the most interesting) stages of development almost always cause the spontaneous abortion of the affected embryo or fetus, often before the mother knows she is pregnant.

As a result, most modern developmental geneticists, even those whose primary interest is in human development, study mutations affecting the development of model organisms which are more amenable to experimentation. In *Drosophila,* for example, only a few dozen genes guide the formation of the early embryo's segmented body plan. Mutations in some of these genes eliminate specific body segments (Fig. 19.2). Once the embryo has divided into segments that will become parts of the head, thorax, and abdomen, the activation and inactivation of different sets of genes direct the development of specialized organs, such as wings and legs, in each segment. Surprisingly, many of the genes controlling segmentation and body structure in *Drosophila* have counterparts in humans, and some of the human genes can substitute for the corresponding insect genes during fruit fly development. From this observation, we can infer that evolution has conserved many of the genes and genetic pathways underlying development. Insights from model organisms can thus illuminate general principles applicable to many species, including humans.

We examine in this chapter how the single cell of the fertilized egg, or zygote, differentiates into hundreds of different cell types. It is impossible to do justice to this complicated issue in a short section of a textbook. What we offer here is an overview of some of the experimental strategies scientists use to study this question, and a synopsis of key results illustrating some of the genetically controlled mechanisms that direct development. The genetic portraits in the References at the end of this book provide more details; each portrait focuses on one model organism's contribution to our understanding of development.

In this chapter, we describe:

■ How model organisms serve as prototypes for research in developmental genetics, and how investigators extrapolate the insights from this research to all living forms.

■ The experimental strategies developmental geneticists use to dissect the development of complex multicellular organisms.

■ Key experimental results that provide a framework for understanding how genes help control cellular differentiation during development; that is, how they direct most of the changes that turn a fertilized egg into a functioning adult.

Model Organisms: Prototypes for Developmental Genetics

Throughout the twentieth century, developmental geneticists concentrated their research efforts on a small number of organisms that sampled a range of species from different phyla. The organisms that have contributed most to our understanding of development include the yeast *Saccharomyces cerevisiae;* the plant *Arabidopsis thaliana;* two invertebrate animals (the fruit fly *Drosophila melanogaster* and the nematode [or round worm] *Caenorhabditis elegans*); and one vertebrate (the house mouse *Mus musculus*). Although we focus here on these five eukaryotic organisms, some researchers have made major findings in other model systems, such as corn and the zebrafish; even prokaryotic organisms and viruses have provided paradigms for tackling certain developmental problems in eukaryotes.

Why These Model Organisms?

The five model organisms we discuss in this chapter and in the genetic portraits at the back of the book share three characteristics of value to geneticists: ease of cultivation, rapid reproduction, and small size. These properties satisfy the geneticist's need to examine large numbers of individuals, in order to isolate mutants and to study the segregation of those mutants through successive generations. Each organism also possesses idiosyncratic features specific to itself that make it valuable for particular types of genetic or developmental analyses.

Take yeast, for example. Because *S. cerevisiae* cells can grow as haploids or diploids, researchers can identify mutations in haploid cells and then combine mutations in diploid cells for complementation analysis. It is also possible to look at more cells (and thus find more mutations) in a shorter period of time with yeast than with other eukaryotic organisms. And although *S. cerevisiae* is a single-celled eukaryote, yeast cells signal each other and differentiate into two mating types using variations of processes observed during the development of multicellular eukaryotes. In another example, the roundworm *C. elegans* is transparent (Fig. 19.3) and contains an invariant number of somatic cells as an adult—959 in the female/hermaphrodite and 1031 in the male. Because of these unusual properties, researchers can discern the lineage of every cell as the fertilized egg develops into the multicellular adult.

Figure 19.3   The transparency of C. *elegans* facilitates study of the worm's development.

One of the most important reasons for the focus on the five model organisms is simple momentum. Each organism has attracted a dedicated cadre of researchers. Once initial studies have made a model look attractive, other investigators realize that further work on this organism would be synergistic and mutually supportive. As a result, there is a long tradition of cooperative study, during which researchers gather a great deal of information about each model's normal development. This information is essential to an analysis of mutants with developmental defects. Moreover, the identification of many mutants provides a basis for the isolation and study of additional mutants. Stock centers maintain these mutants and make them available to the whole community of geneticists even decades after their original isolation.

More recently, the remarkable success of genome projects has culminated in the cloning and sequencing of each model organism's genome. As we saw in Chapter 10, the completion of these genome projects makes it much easier for geneticists to identify genes whose alteration by mutation produces a phenotypic effect on the organism's development. The combination of a large group of researchers, an extensive base of knowledge, a large collection of available mutations, and computer databases containing the entire sequence of the genome is now accelerating the pace at which geneticists are gaining knowledge of the genes that influence development in these model organisms.

Alignment of histone H4 from humans and four model systems.

H. sapiens	MSGRGKGGKGLGKGGAKRHRKVLRDNIQGITKPAIRRLARRGGVKRISGLIYEETRGVLKVFLENVIRDAVTYTEHAKRKT
M. musculus	MSGRGKGGKGLGKGGAKRHRKVLRDNIQGITKPAIRRLARRGGVKRISGLIYEETRGVLKVFLENVIRDAVTYTEHAKRKT
C. elegans	MSGRGKGGKGLGKGGAKRHRKVLRDNIQGITKPAIRRLARRGGVKRISGLIYEETRGVLKVFLENVIRDAVTYCEHAKRKT
D. melanogaster	MTGRGKGGKGLGKGGAKRHRKVLRDNIQGITKPAIRRLARRGGVKRISGLIYEETRGVLKVFLENVIRDAVTYTEHAKRKT
A. thaliana	MSGRGKGGKGLGKGGAKRHRKVLRDNIQGITKPAIRRLARRGGVKRISGLIYEETRGVLK IFLENVIRDAVTYTEHARRKT
	• •

* Indicates identical amino acid
• Indicates similar amino acid

Figure 19.4 The amino-acid sequence of histone H4 is highly conserved in evolution. The histone H4 proteins in the five organisms depicted here are nearly identical; the first 12 amino acids in these proteins (*tan shading*) differ only in the second position of the *D. melanogaster* protein. Geneticists use such sequence similarities to generalize results from research on model organisms.

All Living Forms Are Related. . .

In the last 150 years, biologists have come to realize that life-forms are related on many levels. For example, the cells of all eukaryotic organisms have many features in common that are recognizable in the light or electron microscope. These include a membrane-bounded nucleus containing linear chromosomes, mitochondria containing their own chromosome, and a network of membranous structures such as the endoplasmic reticulum which facilitates the intracellular transport of molecules. Moreover, the metabolic pathways by which cells make or degrade organic molecules are virtually identical in all living organisms, and almost all cells use the same genetic code to synthesize proteins. The relatedness of organisms is even visible at the level of the amino-acid sequence of individual proteins. For example, evolution has conserved the sequence of the histone protein H4 so that the H4 proteins of widely divergent species are identical at all but a few amino acids (Fig. 19.4). Most other proteins are not as invariant as H4, but scientists can nonetheless usually trace the evolutionary descent of a protein through the amino-acid similarities of its homologs in various species.

Of particular importance to this chapter is the fact that evolution has conserved many basic strategies of development in all multicellular eukaryotes, even in organisms with body plans that look quite different. A graphic example is seen in studies of the genetic control of eye development in fruit flies, mice, and humans. *Drosophila* homozygous for mutations of the *eyeless* (*ey*) gene have either no eyes at all or, at best, very small eyes (Fig. 19.5a). Mutations in the *Pax-6* gene in mice (Fig. 19.5b) and the *Aniridia* gene in humans also reduce or totally abolish eye formation. When researchers cloned the *ey, Pax-6,* and *Aniridia* genes, they found that the amino-acid sequences of all three encoded proteins were closely related. This is surprising because the eyes of vertebrates and insects are so dissimilar: Insect eyes are composed of many ommatidia, while the vertebrate eye is a single camera-like organ (see Fig. 1.9 on p. 7 in Chapter 1). Biologists had thus long assumed that the two types of eyes evolved independently. However, the homology of *ey, Pax-6,* and *Aniridia* suggests instead that the eyes of insects and vertebrates evolved from a single prototypical light-sensing organ whose development required a gene ancestral to *ey* and its

(a)

(b)

Figure 19.5 The *eyeless/Pax-6* gene is critical for eye development. (a) Hypomorphic or null mutations of the *eyeless* gene reduce the size of eyes or completely abolish them in adult flies. **(b)** Hypomorphic or null mutations in the homologous mouse *Pax-6* gene also reduce or abolish eye development. *Top:* wild-type mouse fetus. *Bottom: Pax-6* mutant fetus.

mouse and human homologs. Providing an even more emphatic demonstration of the conservation of this genetic program through more than 300 million years of evolution, the wild-type mouse and human genes can function in flies to stimulate the formation of eyes. We will describe this experiment in more detail later in this chapter.

. . .Yet All Species Are Unique

Although the conservation of developmental pathways makes it tempting to conclude that humans are simply large fruit flies, this is obviously not true. Evolution is not only conservative, but it is also innovative. Organisms sometimes use disparate strategies to accomplish the same developmental goal. One example is the difference between the two-cell embryos that form in *C. elegans* and humans upon completion of the first mitotic division in the zygote. If one of the two cells is removed or destroyed in a *C. elegans* embryo at this stage, a complete nematode cannot develop. Because each of the two cells has already received a different set of molecular instructions to guide development, the descendents of one of the cells can only differentiate into certain cell types and the descendent of the other cell into other types. The situation is very different in humans: If the two embryonic cells are separated from each other, two complete individuals (identical twins) will develop. In fact, as we saw on pp. 371–373 in Chapter 11, removal of a cell from a 6–10-cell human embryo has no effect on the development of the remainder of the embryo. There is thus an intrinsic difference in the way worm and human embryos develop at these early stages. Each cell in a *C. elegans* embryo has already been assigned a specific fate, whereas the cells in a human embryo can adjust their development to make up for missing cells.

A divergence of developmental principles can also occur at the molecular level. For example, the genetic and molecular strategies for determining sexual differentiation in *D. melanogaster* are quite different from those in humans. As we described in the comprehensive example on pp. 605–609 in Chapter 17, the major guiding principle in fly sex determination is a cascade of splicing factors, in which the splicing of one gene's primary transcript governs the splicing of the primary transcript for the next gene in the cascade. By contrast, in humans, the initial stages of sex determination depend on the expression of testis-determining factor (TDF), encoded by the *SRY* gene on the Y chromosome (see p. 186 in Chapter 6 and pp. 395–396 in Chapter 11). The TDF protein does not act as a splicing factor, as one might expect from the *Drosophila* model. Instead, it functions as a transcription factor that activates or represses the transcription of other genes. Moreover, many of the secondary sexual characteristics in humans such as the external genitalia (penis or vagina) are determined by hormones secreted from the gonads (testes or ovaries). Flies do not have corresponding sexual hormones.

To summarize, genetic studies of development in model organisms often provide important lessons that can be generalized to all eukaryotes. The differences observed in the general strategies directing cell fates and in the molecular underpinnings of developmental processes highlight the distinctive ways in which evolution has molded divergent gene actions to fulfill similar purposes.

Genetics Simplifies the Study of Development

Because proteins are the basic elements of cellular function, biologists can try to understand development by defining the roles played by individual proteins. To do this, they eliminate all copies of a single protein species from a cell or organism and determine the consequences. From these consequences, they can often infer the function of the normal protein in normal development.

Genetics makes this experimental strategy possible. All an investigator has to do is isolate a mutant cell or organism with a specific, inactivated gene. As described later, such mutants are usually found in the course of a genetic screen to look for animals whose development is aberrant in interesting ways. A mutant with an altered gene will lack the wild-type protein encoded by that gene. Careful analysis of the mutant phenotype can pinpoint what the protein does in development. As we will see, this basic strategy, while not the only way researchers can harness genetics to study development, is almost always the first and most important step.

The Genetic Dissection of Development Depends on a Comprehensive Set of Mutants

Once a gene affecting a developmental process has been identified by a mutation, geneticists try to isolate many additional mutations in that gene. If each mutation affects the function of the corresponding protein in a different way, studies of the phenotypes associated with the various mutations may shed light on the diverse roles the protein plays in the organism.

Loss-of-Function Mutations

Most mutations are loss-of-function mutations: They disrupt gene function by altering the amino acid sequence (and thus the three-dimensional structure) of the protein product, or by interfering with any step of gene expression (transcription, translation, or RNA processing). As a result, such mutations give rise to proteins with diminished (or no) biochemical activity; or they decrease (or stop) production of an otherwise normal protein. We describe four kinds of loss-of-function mutations.

Null Mutations.   The best way to draw legitimate conclusions about the importance of a protein in development is to study an organism that completely lacks the function provided by that protein. In an analogy, if you tried to ride a bicycle with no chain, you might conclude that the chain is required to move the wheels. If, however, you tried to ride a bicycle whose chain had a missing link (the equivalent of a partially defective protein) such that the bicycle would move but respond only erratically to your peddling, you might conclude that the chain is not critical to wheel movement and instead affords the cyclist some control over wheel movement. In Chapter 7 we saw that mutations that remove all function act as *null alleles*, and that such null alleles are usually (but not always) recessive to wild-type alleles.

Hypomorphic Mutations. Although the use of null mutations allows investigators to infer the most straightforward explanations for the function of the wild-type protein in development, there are situations in which it is actually more desirable to have a partial loss-of-function (hypomorphic) mutant allele. This is because many molecules function at multiple times in development. For example, the *wingless* (*wg*) gene in *Drosophila* is needed both for the formation of a proper embryo early in development and for the formation of an adult wing much later in development. An animal homozygous for a null allele of *wg* will die during embryogenesis. This allele thus points to the importance of *wg* for early development. But because the animal dies before wings form, you could not infer from this homozygote that the gene also functions to generate wing structures. On the other hand, flies homozygous for a certain hypomorphic allele of the *wg* gene survive to adulthood, but they have no wings. Observing the effects of this allele alone, you would conclude that the gene is involved in wing formation, but you could not infer its role in early development. This example illustrates the importance of obtaining as many different mutant alleles as possible of a gene whose function you wish to study.

Conditional Mutations. Another way to study genes with effects on diverse developmental processes is to isolate *conditional mutations* that cause a loss of function only under special circumstances. The most commonly studied type of conditional mutation, the temperature-sensitive mutation, produces a protein that is functional at a lower, *permissive* temperature but defective at a higher, *restrictive* temperature. In contrast, the protein product of the wild-type allele functions at both temperatures. It is best if the conditional mutation produces completely nonfunctional proteins at the restrictive temperature, but it is sometimes difficult to determine whether the resulting protein is completely nonfunctional or partly functional.

Temperature-sensitive mutations have one main experimental advantage. They make it possible to raise an animal at the permissive temperature (which allows the early stages of development to proceed normally), and then to increase the temperature to assess the importance of the gene product at later developmental stages. This temperature-shift procedure can help define the time at which the protein product is needed for any developmental process in which it participates. Figure 19.6 shows a temperature-shift analysis of a mutant strain of *C. elegans* carrying a temperature-sensitive lethal allele of the *zyg-9* gene, which helps determine the basic polarity of the early embryo. This temperature-shift study established that the ZYG-9 protein is required only in a very narrow window of about 15 minutes during the period between fertilization and the completion of the first mitotic division. If the protein is inactivated prior to the end of the first mitosis at any time outside the 15-minute window, development is normal.

Dominant-Negative Mutations. Most loss-of-function mutations have recessive effects because heterozygotes have about 50% of the gene function of wild-type homozygotes, and this level of gene function is sufficient for a normal phenotype. There are two exceptions to this rule. First, as discussed on pp. 223–224 in Chapter 7, for a small number of developmentally important genes, one wild-type copy is insufficient for normal development. The mutant allele will thus be dominant to wild type; this type of dominance is called **haploinsufficiency.** A second situation in which a loss-of-function mutation can have a dominant effect occurs with so-called **dominant-negative mutants.** Here, the inactive protein encoded by a mutant allele "poisons" or otherwise counteracts the function of the protein encoded by the wild-type allele. Figure 7.27b on p. 223 illustrates one of several ways in which this can occur: In a multimeric protein, the presence of one abnormal subunit might block the function of the protein even if the protein's other subunits are wild type.

Dominant-negative mutations can be particularly valuable when researchers suspect that a gene has an impact on development, but they have not yet found a loss-of-function mutation in the gene to test their hypotheses. In such a situation, it is sometimes possible to engineer a dominant-negative mutant transgene *in vitro*, and then introduce this transgene back into a

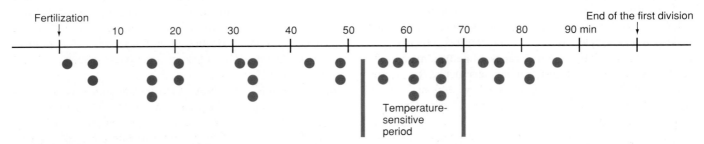

Figure 19.6 Time-of-function analysis. *C. elegans* embryos from mothers homozygous for a temperature-sensitive allele of the *zyg-9* gene develop properly if they are subjected to a 15-minute pulse of high temperature starting at any of the times indicated by the *green circles*. They develop incorrectly and subsequently die only if the high temperature begins at one of the times indicated by the *red circles*. (Each circle represents an experiment with one embryo.) These data show that the ZYG-9 protein is needed only during 15 of the first 100 minutes of development.

wild-type organism. In an interesting example of this technology, one research group made a dominant-negative fibroblast growth factor receptor (FGFR). These receptors are normally found on the surface membranes of many cell types (Fig. 19.7a). One part of the receptor molecule faces the outside of the cell, where it can bind to a molecule called fibroblast growth factor (FGF). FGF is a **ligand:** a molecule involved in cell-to-cell communication that is produced by the cell sending the signal; its binding to a receptor in the membrane of the cell receiving the signal alters the behavior of that cell. The binding of FGF to the extracellular part of FGFR causes several changes in the receptor, including the dimerization of two FGFR subunits and the activation of a kinase in each subunit that adds a phosphate group to the other subunit. This reciprocal phosphorylation in turn initiates a complicated intracellular signaling mechanism that changes the receiving cell's developmental fate.

To make a dominant-negative FGFR mutant, the investigators synthesized a gene that gives rise to an abnormal form of the receptor. This mutant receptor cannot localize to the cell membrane and is instead secreted out of the cell. The researchers reasoned that the secreted form of FGFR would bind to FGF and thereby prevent the ligand from reaching the normal membrane-bound FGFR (Fig. 19.7b). When they injected their engineered transgene into early mouse embryos, they observed a number of defects, including problems in limb development (Fig. 19.7c). These results demonstrated that FGF signaling contributes to the developmental pathway leading to normal limbs.

RNA Interference: Genetics Without Mutations

As we will see, the genetic screens performed to find mutations in developmentally important genes require a considerable investment of effort and are often subject to unanticipated difficulties. As a result, geneticists have to date identified mutations in only a subset of the genes that play a role in development. But within the last few years, researchers have been able to employ a new strategy to deplete the protein products of specific genes from developing organisms. This strategy, called **RNA interference (RNAi),** is based on the following discovery. When cells ingest or are injected with double-stranded RNA (dsRNA) corresponding to the sequence of a gene's mRNA, the intracellular presence of the dsRNA triggers the degradation of the corresponding mRNA into short fragments. In the absence of intact mRNA, the cell cannot synthesize the protein. The details of RNAi are just being worked out, but it appears that many kinds

Figure 19.7 Engineering a dominant-negative mutation in a mouse fibroblast growth factor receptor (FGFR) gene. (a) A dimer of fibroblast growth factor (FGF) binds two FGFR molecules in the cell membrane, causing them to dimerize. As a result, the protein kinase domains of the two FGFRs phosphorylate (add a phosphate group to) each other; they also phosphorylate and dimerize two STAT molecules, which initiates a signal necessary for development. **(b)** A truncated mutant soluble form of FGFR can bind to FGF, preventing it from binding to normal FGFR in the cell membrane. **(c)** Phenotypic effects. *Top:* wild-type mouse limb. *Bottom:* limb from a mouse engineered to contain a dominant-negative version of the *FGFR* gene. Note the poor development of the digits in the mutant mouse (*insets*).

of cells have enzymes that degrade long dsRNAs into shorter dsRNAs 21 nucleotides in length. These shorter dsRNAs then serve as templates for the degradation of homologous mRNAs into similar 21 bp fragments (see pp. 798–799 of the *C. elegans* portrait in Reference C).

To employ this RNAi strategy, researchers first synthesize a dsRNA and then deliver it into the cells of a developing organism. They usually carry out the dsRNA synthesis *in vitro*. For example, they clone a cDNA corresponding to a gene's mRNA into a vector such that the cDNA is located between strong promoters (Fig. 19.8a). They next use purified DNA from the recombinant clone as a template for transcribing the cDNA. The addition of RNA polymerase and the four nucleotide triphosphates (ATP, CTP, GTP, and UTP) initiates transcription, which then proceeds in both directions and produces RNAs for both strands of the cDNA. These complementary RNA strands can anneal together to form dsRNA. There are several methods for getting the dsRNA into developing animals, including injection of dsRNA into the body cavity or soaking the animal in a dsRNA-containing solution. Investigators working with *C. elegans* can simply feed larvae with *E. coli* cells that contain a plasmid like the one shown in Fig. 19.8a. RNA polymerase within the *E. coli* cells containing such a plasmid will synthesize the desired dsRNA, which is then taken up by *C. elegans* larval cells as the bacteria are digested in the worm's gut.

Figure 19.8b shows an RNAi experiment in which the dsRNA corresponded to the mRNA for a *C. elegans* gene called *par-1*. The result of this dsRNA treatment was an abnormal vulva (the structure through which fertilized eggs are released) that protruded outside of the animal. Though worm researchers already knew that *par-1* functions very early in development to help establish the anterior-posterior axis of the animal, the results shown in Fig. 19.8b showed that the gene also functions later in development in the patterning of the vulva.

RNA interference is an extremely useful technique for creating a phenocopy that mimics a loss-of-function mutation. Recall from Chapter 3, however, that a phenocopy is not a true, heritable mutation; thus RNAi-based protein depletion is not passed from generation to generation. Another slight drawback of the RNAi method is that results may vary from experiment to experiment, because they depend on the relative level of dsRNA uptake. Consequently, even if the results of an RNAi experiment provide clues to a developmentally interesting gene's function, it is usually desirable to obtain and study a classical, heritable mutation in the gene as well.

Gain-of-Function Mutations

Mutations that produce too much protein, or proteins with a new function not present in the wild-type protein, are **gain-of-function mutations.** The alleles resulting from gain-of-function mutations are often distinguishable from recessive null alleles by the fact that they are frequently dominant to the wild-type allele. It is hard to understand unambiguously the role of a protein in development from a gain-of-function allele. This is because the mutation, rather than taking something away, adds something unusual to the organism, which might behave in an

(a) Synthesis of dsRNA

(b) A result of *par-1* dsRNA treatment

Figure 19.8 RNA interference (RNAi): A new tool for studying development. (a) How to make double-stranded RNA (dsRNA). A cDNA is cloned between two promoters, allowing transcription from both cDNA template strands. Complementary RNA transcripts will anneal with each other to make dsRNA. **(b)** Abnormal structure of the vulva in *C. elegans* treated with dsRNA for the *par-1* gene. *Top:* wild-type vulva. *Bottom:* protruding vulva in an animal treated with *par-1* dsRNA.

unpredictable way. Nevertheless, gain-of-function mutants can help identify developmentally important genes and clarify the roles they play in development. We now look at two kinds of gain-of-function mutations.

Mutations Causing Excessive Gene Function. Such mutations are rare because they result only from highly specific changes in a gene, in contrast with loss-of-function mutations, which can disrupt gene function in many ways. Nonetheless, there are several ways in which mutations can lead to increased gene activity. One mechanism involves changes to promoters that make the promoters more accessible to transcription factors and RNA polymerase. Examples of other scenarios include various dominant mutations in *FGFR3,* one of the four genes in mice and humans that encode related yet distinct fibroblast growth factor receptors (review Fig. 19.7a on p. 655). Some of these mutations increase the affinity of the FGFR3 receptor protein for its ligand FGF, inappropriately turning on the developmental signal when the concentration of the FGF ligand would normally be too low to accomplish this. Other *FGFR3* mutations allow the developmental signal to be turned on in the absence of FGF. These mutations cause the constitutive activation of the phosphate group-adding kinase domain of FGFR3; they accomplish this by altering a part of the protein that normally blocks the kinase function in the absence of the ligand.

Interestingly, a single amino-acid substitution in the FGFR3 protein is sufficient to cause a gain-of-function dominant phenotype through the constitutive activation of the kinase domain. This substitution causes achondroplasia, the most common form of short-limb dwarfism in humans. Researchers have recently engineered mice with exactly the same amino-acid substitution in their homologous *FGFR3* gene. Remarkably, this mutant gene produces what appears to be the same dominant dwarf phenotype seen in human achondroplasia (Fig. 19.9). You might wonder how a gain-of-function mutation in a growth factor receptor gene results in dwarfism, since this is not what you would expect at first glance. The answer is that the FGFR3 protein is made in cartilage precusor cells called chondrocytes. Activated FGFR3 signals the chondrocytes to stop dividing and begin differentiating first into cartilage and then into bone. A gain-of-function FGFR mutation thus causes the chondrocytes to stop proliferating sooner than normal, which leads to fewer bone precursor cells and thus less bone.

Mutations Causing Ectopic Gene Expression. Suppose you suspect that a particular protein plays an important role in initiating the development of some structure like the legs or eyes. If that were true, it might be possible that expression of this protein in tissues in which it is not normally made could lead to the development of legs or eyes in unusual locations in the animal. The expression of a gene at an abnormal place or time is called **ectopic gene expression.** Rarely, spontaneous mutations cause ectopic expression of genes important to development. An interesting example of such a mutation occurred in *Drosophila* when a chromosomal inversion moved the *Antennapedia* gene (normally transcribed in tissues destined to

Figure 19.9 Achondroplastic dwarfism in the mouse. (a) The dwarf mouse at the *right* is heterozygous for an *FGFR3* allele with the same amino-acid change as that causing achondroplasia in humans. A control littermate is at the *left.* **(b)** Skeletal abnormalities in the dwarf mouse at the *top* include a shorter face, overgrowth of the incisor teeth (*arrowhead*), and improper connection of the head to the spine (*arrow*) as compared to a control littermate at the *bottom.*

become legs) next to a specific kind of enhancer, which turned the gene on in tissues normally destined to develop into antennae. Animals carrying this *Antennapedia* mutation have legs growing out of their heads in place of antennae (review Fig. 7.27d on p. 223). The phenotype of this ectopic mutant shows that the wild-type protein encoded by *Antennapedia* plays a critical role in leg development.

Instead of relying on rare and unpredictable mutations that might cause ectopic gene expression, researchers can now use recombinant DNA technology to make such mutations in a systematic way. They can change the promoter of a cloned gene by adding enhancers or other elements that might cause it to be transcribed at inappropriate places or times, and then they introduce this altered gene back into the organism's genome by transformation. One research group placed the *eyeless* gene of *Drosophila* (recall Fig. 19.5a on p. 652) under the control of a promoter for a "heat-shock" gene whose transcription in any tissue is turned on by higher than normal temperatures. Flies bearing this recombinant gene that were grown at high temperature made the *eyeless*-encoded protein throughout their bodies. These animals had eye tissue growing at many different locations, even on their wings and antennae (Fig. 19.10). This result demonstrates that the Eyeless protein is a master developmental switch that can activate a cellular program causing eye development. Ectopic eyes also arise when the mouse *Pax-6* or the human *Aniridia* gene is expressed in *Drosophila* under the control of the same heat-shock gene promoter. This means that both the amino-acid sequence and the actual function of this master switch have been conserved throughout animal evolution.

Mutant Screens Help Identify the Components of Development

To understand a developmental event, biologists seek to discover as many as possible of the genetically encoded components that participate in the event.

Saturation Mutagenesis and Complementation Studies Are the First Steps

Saturation mutagenesis is the production of as many different mutants with related phenotypes as possible in an attempt to identify all the genes relevant to a process. Many mutations found in a genetic screen will of course identify the same gene, so it is necessary to group the mutations into genes by complementation analysis. When bacterial geneticists looked at the motility of *Escherichia coli* cells through saturation mutagenesis and complementation analysis, they identified 20 genes that contribute to the construction of the bacterial flagellum (see the comprehensive example on pp. 511–514 of Chapter 14). Although this example applies to bacterial motility rather than the development of a multicellular eukaryote, the same basic principles hold for any genetic screen in which researchers look at many individuals to find mutations of interest. In saturation

Ectopic red eye tissue

Ectopic red eye tissue

Figure 19.10 Ectopic expression of the *eyeless* gene produces ectopic eye tissue. This fly carries a synthetic *eyeless* gene that is turned on in inappropriate tissues. As a result, eye tissue grows in unexpected places, such as at the end of the antennae and on the thorax above the wings.

mutagenesis, geneticists try to isolate so many mutations that every complementation group contains several. Once they've done this, it is unlikely that they have missed any genes purely by chance. Thus the number of complementation groups provides a good estimate of the complexity of the biological process.

Because it is rare to find spontaneous mutations that affect development, geneticists usually screen for developmental mutations in populations treated with mutagens. In the simplest screens, they look for mutations causing visible developmental deformities in adults. But to identify genes with earlier roles in development, they look through collections of lethal mutants to find those in which the embryonic developmental abnormality is so severe that the individuals cannot live to adulthood. Many of the mutants described in this chapter were found through saturation mutagenesis in screens for visible phenotypes or phenotypes that cause embryonic lethality.

Saturation mutagenesis does not necessarily identify all the genes in a developmental pathway, because some functions can be performed by either of two *redundant genes* (Fig. 19.11a). In such situations, a single mutation in either gene is without phenotype. Only when both genes sustain mutations in the same individual does the mutant phenotype manifest itself. Another reason for missing important genes in a mutant hunt is that some proteins perform more than one function, and a mutation affecting one function may mask other functions (Fig. 19.11b). For example, as we have seen, the protein encoded by the *wingless* gene in *Drosophila* plays a role both early in embryonic development and later during the development of the adult. If you were looking for mutations that affected the structure of the adult wing, you might miss an important gene like *wingless* if you could only find mutations in that gene that arrest development during formation of the embryo.

Are there ways of finding genes missed for reasons of protein redundancy or the masking of phenotypes? The answer is yes; geneticists can use several techniques to reveal additional genes involved in a developmental process.

Overexpression or Misexpression Screens

We have already seen that the excessive expression of a gene, or its ectopic expression at unexpected places or times, can be detrimental and produce a recognizable phenotype. Overexpression phenotypes are usually dominant, and the redundancy of protein function does not prevent their appearance. As a result, a technique that causes overexpression or ectopic expression can serve as the basis for a systematic genetic screen for novel genes of consequence to development. One such methodology recently developed in *Drosophila* works as follows. A promoter region that can be induced to activate transcription to high levels under certain conditions is cloned within a transposable element (TE). The TE is then reintroduced into the fly genome as a transgene. When the TE moves from place to place within the genome, it brings the promoter region with it. If the TE lands next to a gene of developmental importance, it causes overexpression or ec-

(a) Redundant gene function

Proteins perform same function.

Gene 1 mutant

No functional protein

Functional protein

Gene 2 mutant

Functional protein

No functional protein

Cells still perform the function, even with a mutation in one of the genes.

(b) Multiple roles in development

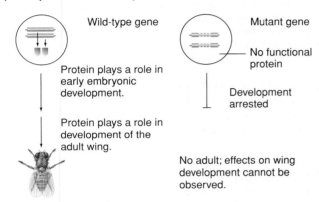

Wild-type gene

Protein plays a role in early embryonic development.

Protein plays a role in development of the adult wing.

Mutant gene

No functional protein

Development arrested

No adult; effects on wing development cannot be observed.

Figure 19.11 Why saturation mutagenesis may miss some genes. (a) If two genes encode proteins that perform the same function, a mutation in only one of these genes may not cause a mutant phenotype. **(b)** If a gene plays important roles at different stages of development, a mutation in the gene may arrest development in the early stages, obscuring the gene's function in later stages.

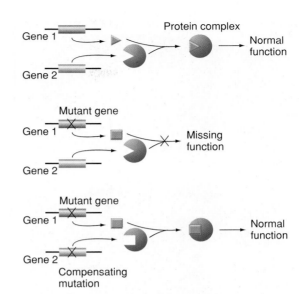

Gene 1

Gene 2

Protein complex

Normal function

Mutant gene

Gene 1

Gene 2

Missing function

Mutant gene

Gene 1

Gene 2

Normal function

Compensating mutation

Figure 19.12 Suppressor mutations. A mutation in one gene can suppress a mutation in a second gene. This may happen if the two gene products form a protein complex and the two mutations counterbalance each other such that the two mutant polypeptides form three-dimensional structures that can associate with each other. This figure assumes the individual is homozygous for mutant alleles of both genes, although only one copy of each gene is shown.

topic expression of that gene, which may produce an aberrant phenotype, such as roughened eyes or irregular wings.

Screens for Suppressor Mutations

If you begin with a mutation that causes a mutant phenotype, you can often isolate additional mutations that correct the aberration. The correction may result from a reversion at the original mutant locus or from a compensating mutation in another gene active in the same process, which suppresses the effects of the first mutation. A mutation in one gene that compensates for or counteracts the effects of a mutation in another gene is known as a **suppressor mutation.** An individual whose genome carries both the original mutation and a suppressor mutation will have a phenotype more similar to wild type than an individual with only the original mutation. Figure 19.12 suggests one of several ways in which suppression may occur. In the figure, the two genes encode subunits of the same dimeric protein. Amino-acid

substitutions in either subunit alone can prevent association with the other wild-type subunit, but two mutant subunits can dimerize together.

Other types of suppressor mutations can help identify genes whose products might not associate directly but that nevertheless operate in the same developmental pathway. Figure 19.13a–c compares three types of *Drosophila* wings: from a wild-type fly, from an animal carrying a mutant *Notch* gene, and from a double mutant of *Notch* and a suppressor of *Notch* found in a genetic screen. The phenotype of the double mutant wing is closer to wild type than that of the *Notch* mutant alone, indicating the suppression. Subsequent work has shown that *Notch* and its suppressor have antagonistic functions in wing development.

Screens for Enhancer Mutations

There are several ways in which developmental geneticists can use mutations in one gene to find a mutation in a second gene that worsens the phenotype. The double mutant thus displays a more aberrant phenotype than that caused by the original mutation alone. A mutation that intensifies the effects of a first mutation is called an **enhancer mutation.** In spite of their opposite effects on mutant phenotypes, enhancer mutations have the same practical value as suppressor mutations: Both can identify genes that participate in the same genetic pathway as the original mutant gene. To screen for enhancer mutations, you would start with a mutation causing a visible defect such as abnormal wing development in *Drosophila.* You would then look for enhancer mutations in other genes that worsen the phenotype (Fig. 19.13d).

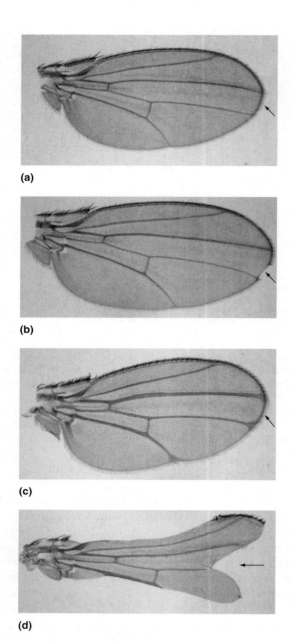

(a)

(b)

(c)

(d)

Figure 19.13 Suppressor and enhancer mutations modify the phenotype caused by mutations in another gene. (a) A wild-type *Drosophila* wing. **(b)** A wing from a fly hemizygous for a particular mutation of the X-linked *Notch* gene. Notice the notch in the margin of the wing (*arrow*). **(c)** A wing from a fly hemizygous for the same *Notch* mutation whose genome also carries a suppressor mutation in another gene. The wing margin defect is corrected (*arrow*). **(d)** A wing from a fly hemizygous for the same *Notch* mutation and also heterozygous for an autosomal dominant enhancer mutation. The notch in the wing is much deeper (*arrow*), and other wing defects are also visible.

If a double mutant carrying the original mutation plus an enhancer mutation dies, even though individuals with either mutation alone survive, the effect of the double mutation is called **synthetic lethality.** Figure 19.14 shows one of the many possible molecular explanations for synthetic lethality. The

Figure 19.14 Synthetic lethality: Mutations in two genes together cause lethality, but a mutation in either gene alone does not. In this example, the two genes are redundant and encode identical or nearly identical proteins. This figure assumes individuals are homozygous for wild-type or mutant alleles of each gene.

model depicted suggests that a screen for enhancer mutations or for synthetic lethality is of particular value when two genes have redundant, overlapping functions such that a severe phenotype results only when both genes are mutant.

Screens for Nonallelic Noncomplementation

Most mutations are recessive because an organism carrying one mutant and one wild-type allele has 50% of the normal gene product, and this is usually enough to generate a wild-type phenotype. In some cases, however, an individual simultaneously heterozygous for mutations in two genes may exhibit an abnormal developmental phenotype that does not appear when the individual is heterozygous for either mutation alone. This phenomenon is known as **nonallelic noncomplementation** (sometimes called *second-site noncomplementation*). It usually implies that the two polypeptides normally cooperate in some way. Decreasing to 50% the amount of each of two polypeptides that cooperate, for example, in building a multimeric protein may reduce to 25% of normal the amount of multimer that is made (Fig. 19.15). In a genetic screen for nonallelic noncomplementation, you would start with an animal heterozygous for a mutation in one gene, and then look for a second mutation in a different gene that fails to complement the first and thereby produces an aberrant phenotype.

Analyzing How Genes Work Together in Developmental Pathways

Once you have isolated a comprehensive set of mutations and identified as many as possible of the genes that function in the biological process of interest, the next step is to establish the order of the functions performed by these genes. The ultimate aim of such studies is to discern a *developmental pathway:* a detailed description of how the products of these many genes interact and cooperate with each other to produce a particular outcome in development.

Figure 19.15 **Nonallelic noncomplementation: When mutations in two different genes fail to complement each other.** One way in which nonallelic noncomplementation can occur. Heterozygosity for mutations in two genes can sufficiently lower the concentration of a multimeric protein so as to produce a mutant phenotype. In this figure, both homologous chromosomes are shown.

Figure 19.16 **The homeodomain: The DNA-binding domain found in many transcription factors that regulate development.** The amino-acid backbone of a homeodomain (*yellow*) interacts with specific sequences in a DNA double helix (*red* and *blue*).

Characterizing Genes Important for Development

Before looking at complicated pathways, however, it is crucial to learn as much as possible about each of the genes identified in the screen. Answers to the following five questions can help fill in significant details.

What Is the Developmental Phenotype Caused by Mutations in the Gene? There are many different ways to look at phenotypes in order to understand which tissues a mutation affects and how it affects them. It is particularly important to define the earliest stage of development disrupted by a mutation, as this may provide clues about the order in which the gene functions in a developmental pathway.

What Is the Nature of the Protein Encoded by the Mutant Gene? With the completion of genome projects for key model organisms, it is now often possible to identify the mutant gene within a few months of finding the mutation. Once you know the nucleotide sequence of a gene, you automatically know the amino-acid sequence of the protein it encodes. You can then use computer programs to search the amino-acid sequence for motifs that offer clues to the protein's function. For example, computer programs can often predict whether a protein resembles known membrane-bound receptors, or whether a protein acts as a kinase that phosphorylates other proteins. One motif seen in many proteins with developmental significance is the **homeodomain** (Fig. 19.16). It is found in the proteins encoded by the *eyeless/Pax6* and *Antennapedia* genes discussed earlier in this chapter. The homeodomain binds to specific DNA sequences, so its presence suggests that a protein might be a transcription factor.

When and in What Tissues Is the Gene Expressed? One way to answer this question is to perform an RNA *in situ* hybridization experiment. To do this, you label cDNA sequences corresponding to the gene's mRNA, and then use the labeled cDNA as a probe for the mRNA on preparations of thinly sectioned tissues. Where the probe is retained, the signal indicates cells containing the gene's mRNA (Fig. 19.17). Defining the tissues in which the gene is expressed can help formulate hypotheses concerning the gene's role in development. For example, if a mutation in the gene affects the development of a tissue other than that in which the gene is transcribed, you might hypothesize that the gene encodes a signaling molecule like a hormone. Such molecules of cellular communication are made in one tissue but influence the fate of cells in other tissues that contain receptors for the hormone.

Where Is the Protein Product of the Gene Within the Animal and Within the Cell? It is often technically easier to find and evaluate the tissues in which a gene is expressed by following the gene's protein product than by using RNA *in situ* hybridization to look for the gene's mRNA. In addition, an mRNA may be found in a tissue that doesn't contain the protein. This would point to the existence of regulatory controls that prevent translation of the mRNA. Finally, the intracellular localization of a protein often provides clues to its function. For example, concentration of the protein in the nucleus would be consistent with a role as a transcription factor.

Methods to follow a protein usually involve the generation of antibodies against parts of the protein. One way to do this is

Figure 19.17 *In situ* **hybridization locates cells expressing a gene of interest.** This example shows that mRNA for the *Pax-6* gene (*yellow signals*) accumulates in the eye of a human fetus in the seventh week of gestation. Hybridization is specific to the developing neural retina (nr) and the developing eye lens above it.

to use recombinant DNA techniques to construct a fused gene (Fig. 19.18a). In this construct, part of a cDNA for the gene of interest is cloned downstream of, and in the same reading frame as, part of a gene encoding a protein that can be made at high levels in bacteria. If you transform a plasmid containing this fused gene into *E. coli,* the bacterial cells will make large amounts of a **fusion protein** whose N-terminal amino acids are from the bacterial protein and whose C-terminal amino acids are from the eukaryotic developmental protein. If you inject this fusion protein into rabbits or other animals, they will synthesize antibodies against it. And once you label these antibodies with a fluorescent tag, the tagged antibodies will react with the corresponding protein of developmental interest in preparations of tissues and cells (Fig. 19.18b).

A new way to follow a protein is to use recombinant DNA technology to construct a gene encoding a tagged protein that will itself fluoresce. The idea, illustrated in Fig. 19.18c, is to synthesize an open reading frame that encodes not only the entire protein of interest but also (at the protein's N or C terminus) the amino acids composing a naturally fluorescent protein from jellyfish called green fluorescent protein (GFP). When this recombinant gene is reintroduced into the genome by transformation, the organism will make the GFP fusion protein in the same places and times it makes normal untagged protein. Investigators can keep track of the fusion protein by following GFP fluorescence (Fig. 19.18d). A major advantage of this approach is that researchers can use it to follow a GFP-tagged protein in living cells or animals, which is generally not possible with tagged antibodies for technical reasons. With the GFP fusion protein, researchers can even make movies that reveal subtle changes in the location of the protein over time.

What Cells or Tissues Are Affected by the Loss of Gene Function? The answer to this question might seem obvious: Only cells that make a particular protein would show the phenotypic effects of mutations that prevent synthesis of that protein. But this "obvious" answer is misleading because cells often communicate with each other to influence developmental decisions. In one simple example, if a gland synthesizes a hormone that circulates through the blood, and the gland can no longer make the hormone, the phenotypic effects might not show up in the gland itself but rather in target cells elsewhere in the body that contain receptors for the hormone.

To address a variety of issues involving communication between cells, developmental geneticists construct **genetic mosaics:** organisms in which some cells (like those in the gland just described) have one genotype while other cells (such as those in the hormone's target tissues) have a different genotype. Researchers can use several techniques to make such genetic mosaics. The technique chosen often depends on the species. *Drosophila* geneticists usually employ mitotic recombination (see pp. 820–823 of the fly portrait in Reference D), those working with *C. elegans* use methods based on the loss of small extra chromosomes during mitosis (see pp. 800–801 of the worm portrait in Reference C), and investigators studying mice mix embryonic cells from mutant and wild-type strains to make *chimeric mice* with two different cell types (see pp. 849–851 of the mouse portrait in Reference E). **Chimeras** are genetic mosaics in which the cells of different genotype originate from two different individuals.

In the genetic mosaics most useful for studying development, some cells have a mutant genotype, while other cells have a wild-type genotype. Most mosaics are constructed with markers, different from the mutant developmental gene, that allow investigators to differentiate between the two types of tissues. Figure 19.19a shows mosaic seedlings of the plant *Arabidopsis* in which blue tissue contains both a marker gene resulting in blue color and a wild-type gene called *AGAMOUS⁺,* while white tissue lacks the marker gene and is simultaneously mutant for *AGAMOUS.* Figure 19.19b diagrams how researchers used such marked mosaics to show that cells from a particular layer of undifferentiated cells (called L2) in the apical meristem send a signal needed for the proper differentiation of cells in a different layer (L1). This signal depends on the presence of a wild-type *AGAMOUS⁺* allele in L2 cells. In other words, even *AGAMOUS⁺* genotypically wild-type L1 cells develop abnormally if the adjacent L2 cells are mutant for this gene.

Try to Determine How Well-Defined Genes Work Together in a Developmental Pathway

There is no one correct way to analyze how several genes act together to orchestrate a developmental process, but it is generally easiest to focus first on the interaction of the genes in pairs: How does one gene influence the other and *vice versa?* We mention here three of the most common approaches to answering this question. You will see other examples later in the chapter.

(a) Fusion protein gene in *E. coli*

(b) A tissue stained with fluorescent antibodies

(c) Tagging a protein with GFP

(d) A mouse with a GFP-tagged transgene

Figure 19.18   Using antibodies and GFP tagging to follow the localization of proteins. (a) This synthetic gene encodes a fusion protein that will be made at high levels when transformed into *E. coli* cells. Animals injected with purified fusion protein will make antibodies against the protein of interest. **(b)** A *Drosophila* "imaginal disk" is stained with antibodies against several proteins. Each antibody is tagged with a dye that fluoresces in a particular color. **(c)** Making a GFP-tagged protein. The recombinant gene encodes a fusion protein that contains GFP at its C terminus. **(d)** This mouse contains a GFP-labeled transgene expressed thoughout the body; the entire mouse becomes fluorescent when illuminated with UV light as at the *bottom*. The same mouse is shown in normal light at the *top*.

Order-of-Function Analysis.   In the study of a developmental process, you may find that some mutants arrest development at early stages and some at later stages (Fig. 19.20). The differences suggest the order in which the genes function.

Double Mutant Analysis.   If two mutations define successive steps in a process, the double mutant will often arrest with the phenotype characteristic of the earliest block in the process; that is, the earlier-acting mutation is epistatic to the other. Figure 19.21a diagrams an analysis of mutations in yeast that disrupt the secretion of molecules from the cell. Such secretion is important for many developmental events. The phenotype of

the double mutant in this example makes sense. We would expect that a molecule needs to be loaded into secretory vesicles before the fusion of these vesicles with the cell membrane allows the molecule to be secreted from a cell.

It would be incorrect, however, to conclude that an epistatic gene always governs an earlier step than the gene whose mutant phenotype is masked. In many pathways, the opposite is true. For example, in *C. elegans* the pathway leading to formation of the vulva includes three genes: *let-60, lin-45,* and *mek-2* (Fig. 19.21b). The LET-60 protein becomes activated in cells that receive an extracellular signal. Active LET-60 then activates the LIN-45 protein, which subsequently activates the

(a) Mosaic seedlings of *Arabidopsis*

(b) Using mosaics to study cell signaling

Figure 19.19 Mosaic analysis. (a) In these mosaic seedlings, *blue* tissue contains both a marker gene and the *AGAMOUS⁺* gene, while *white* tissue contains neither (it is *AGAMOUS⁻*). **(b)** A signal from blue *AGAMOUS⁺* L2 cells is needed for the proper differentiation of nearby L1 cells (symbolized by a change from a *circle* to a *square*). If L2 cells lack the *AGAMOUS⁺* gene (*white*), nearby L1 cells do not differentiate properly, even if they are themselves *AGAMOUS⁺*.

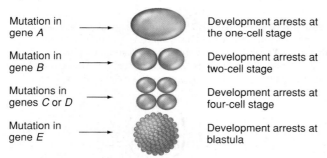

Suggests that gene *A* acts first, then gene *B*, then genes *C* and *D*, then gene *E*.

Figure 19.20 When do mutations arrest development? The stages at which various mutants stop developing may suggest the order in which the corresponding genes act.

MEK-2 protein; active MEK-2 protein leads through several steps to vulva formation. (Reference C, the genetic portrait of *C. elegans,* describes this pathway in more detail.) Gain-of-function *let-60* alleles cause the over-activation of LIN-45 protein, which in turn produces too much active MEK-2 protein, so

(a) Epistasis in the secretion pathway

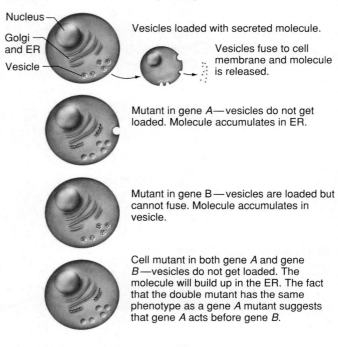

(b) Epistasis in the pathway for vulva formation

Figure 19.21 Double mutant analysis. The phenotype of individuals simultaneously mutant for two genes can suggest hypotheses for the functional relationship of the genes in a pathway. **(a)** In this simple example, mutations in gene *A* are epistatic to those in gene *B,* suggesting that gene *A* controls an earlier step of the process. **(b)** In this signal transduction pathway, a mutation in the gene controlling a later step is epistatic to a mutation in a gene whose product acts earlier.

the eventual phenotype is the formation of too many vulvas. Loss-of-function *mek-2* alleles have the opposite phenotype of no vulvas. Double mutant animals (with both the gain-of-function *let-60* allele and the loss-of-function *mek-2* allele) have no vulvas, so the *mek-2* mutation is epistatic even though *let-60* encodes a protein that acts earlier in the process. This result makes sense because if there is no MEK-2 protein, the relative activity of LET-60 cannot affect vulva formation.

As you can see, it is dangerous to interpret the results of epistasis tests taken in isolation. You first need to know something about the nature of the mutations involved: Are they loss-of-function or gain-of-function alleles? You also need to know something about the general nature of the pathway: Does blockage of each step cause a different aberrant outcome (as is the case with secretion in yeast), or does the whole pathway have a single output (like the formation of a vulva in the *C. elegans* example)? It also helps to know something about the potential biochemical activities of the gene products in the pathway: Are they kinases that phosphorylate other proteins (as do the LIN-45 and MEK-2 proteins), or are they transcription factors (as are other proteins that operate later in the vulva formation pathway), or do they have other recognizable roles? True understanding of the pathway can only be achieved by intergrating the results of many kinds of analysis.

Epistasis, where the double mutant resembles either one of the single mutants, is only one of several possibilities for the phenotype of a double mutant. We have already seen that developmental mutations in one gene can be counteracted by mutations in a second, suppressor gene. If both mutations are due to loss of function, the suppression suggests that the two genes encode proteins with antithetical activities: One might for example be an activator of transcription, whereas the other might repress transcription of the same genes. Yet another potential outcome is that mutations in one gene worsen, that is, enhance the phenotype caused by alleles of the other gene. In this case, if the mutations both cause loss of function, the two gene products might for example be transcription factors that activate the same suite of target genes. However, just as in the case of epistasis, one must exercise caution in interpreting double mutant phenotypes suggestive of suppression or enhancement; it is important to gather as much information as possible before reaching conclusions.

Analyzing How One Gene Affects the Expression of Another Gene. Once you have defined the tissue distribution and intracellular location of one gene's mRNA or protein, you can ask how mutations in a different gene affect this distribution or localization. For example, consider the *wingless* and *vestigial* genes of *Drosophila*. Certain mutations in either gene result in the loss or reduction in size of the adult wings. The protein products of these genes are expressed only in small, overlapping subsets of the cells within tissues called "imaginal discs" that will eventually develop into wings (Fig. 19.22a and b). Flies mutant for *wingless* not only fail to make the Wingless protein, but they also fail to produce the Vestigial protein in many of the cells where it would normally be found (Fig. 19.22c). This suggests that the expression of the *vestigial* gene in those cells is dependent upon *wingless* gene function in adjacent cells.

In summary, many different kinds of analysis can help understand the function of developmentally important genes as well as the ways in which they interact to allow the proper development of structures in the body. A full description of development from a genetic perspective will ultimately require a chart describing all the genes necessary for development and the phenotypes caused by many kinds of mutant alleles of these genes. This genetic overview will provide a blueprint for further studies that seek to explain development in terms of biochemical pathways involving the products of these genes.

Figure 19.22 A mutation in one gene can affect the expression of another gene. (a) Diagram of a *Drosophila* wing imaginal disc. The cells in the *blue-shaded* area will become cells in the adult wing blade. **(b)** A wild-type wing disc stained for Wingless (Wg, *green*) and Vestigial (Vg, *red*) proteins. A thin band of cells expressing both Wg and Vg is *yellow*. **(c)** A *wingless* mutant wing disc stained as in (b). Not only is there no Wg protein, but Vg protein is made only in a narrow band about two cells wide, not in the broader region about 12 cells wide where it is normally made (as in part b).

How Genes Help Control Development: A Mechanistic Framework

Most animals begin development by building a ball of cells, or blastula, that folds in on itself during gastrulation. In the early stages of embryogenesis, three different layers of cells appear: the endoderm, mesoderm, and ectoderm. Later, descendents of these cells give rise to sets of more differentiated cells that form gut, muscles, blood cells, cartilage, bone, nerves, and so forth.

Given the similarities of embryonic development among animals, developmental biologists seek to answer the following questions: What mechanisms regulate cell proliferation? What mechanisms ensure that cells differentiate by taking on different fates? What mechanisms control cell migration? How do cells influence one another? How do cells cooperate to form higher-level structures like organ systems? How do cells die? Insights gained from the study of various model organisms will help provide answers.

A full description of progress in understanding all these topics would require much more than a few pages of text. In the remainder of this chapter, we instead focus exclusively on the genetic basis of differentiation, looking at how genes help control development and why different cell types express different genes. Our viewpoint is that cells in multicellular organisms continually receive a variety of information concerning their history, their location within the organism, and the identity of their closest cellular neighbors. Each cell integrates this information to influence the transcription of a large suite of target genes in its genome; it is the expression of these target genes that leads the cell to a particular path of differentiation.

The following discussion concentrates on development in animals. However, although there are many differences in the way plants and animals develop, organisms in both kingdoms use the same basic strategies to control development through their genes.

Development Requires Progressive Changes in Gene Expression

The enormous diversity of cells within the body of a multicellular organism results in a remarkable variety of cell shapes and functions. Even a single tissue, such as the blood, harbors many different kinds of cells (Fig. 19.23). Differentiation into so many types requires, in part, that cells undergo a successive restriction in developmental potential that affects both themselves and their descendents. For example, the two daughter cells of a human zygote can each generate descendents able to fulfill any fate in the adult. But later in development, cells must "decide" whether they and their descendents will adopt one kind of fate (say that of neurons) or a different kind of fate (say that of epidermal cells). Once a developmental decision is made, a cell and its descendents embark on a particular pathway of differentiation that excludes them from an alternative fate.

How do cells that contain the same genes make such varied developmental decisions and become so different? As we saw in the introduction to this chapter, the reason is simple: Different cell types express different, characteristic subsets of genes. It is easy to understand this point when comparing cells whose function depends on the production of a large amounts of a particular gene product. Red blood cells produce copious amount of hemoglobin, the cone and rod cells in our eyes synthesize vast numbers of photoreceptor molecules, and certain pancreatic cells produce insulin and secrete it into the bloodstream. But the biochemical differences between cell types are not restricted to the expression of a single key gene. Instead, the differentiation of these various cells requires changes in the expression of many genes. Figure 19.24 illustrates how complex these developmental patterns of expression can be. It shows that in *Drosophila*, for example, the many different proteins necessary for generating the structure of an adult wing are expressed in very precise, partially overlapping subsets of cells in the larval imaginal discs that give rise to the wing.

Most Developmental Decisions Ultimately Affect Transcription

To decide which proteins a particular cell type will or will not produce, it makes the most sense for those cells to regulate the

Figure 19.23 Different types of blood cells. The *red* cells are erythrocytes, the oxygen-carrying red blood cells. The cells colored in *green* are macrophages that ingest and destroy invading microbes. The *yellow* cells are T lymphocytes involved in the immune system. The single *blue* cell is a monocyte, an immature cell that can develop into a macrophage.

Figure 19.24 Development requires precise control of the expression of many genes. Each imaginal disk was stained with a fluorescently-tagged antibody against a different specific protein important for patterning of the wing. Each protein is expressed in a unique set of cells in these imaginal disks.

first step in gene expression: the initiation of transcription. Indeed, most of the processes that influence cellular fates culminate in decisions to turn on or off the transcription of "target" genes (such as those for hemoglobin) whose expression is important to that cell type (the red blood cell precursor).

We make this generalization based on two observations. First, RNA *in situ* hybridization experiments such as the one shown in Fig. 19.17 on p. 662 demonstrate that the mRNA for many developmentally important genes appears only in certain cells at certain times in development. Second, many of the genes that play key roles in developmental decision-making encode proteins that function as transcription factors. The *Eyeless/Pax-6* gene whose expression helps initiate eye development in flies, mice, and humans (Fig. 19.5 on p. 652), and the *Antennapedia* gene whose ectopic expression in *Drosophila* can cause cells normally destined to become an antenna to develop instead into a leg (Fig. 7.27d on p. 223), both encode transcription factors. Experiments have demonstrated that these two proteins, as well as many other developmentally significant regulatory factors, actually bind to DNA elements in the vicinity of promoters such as enhancers, altering the transcription of target genes. Because the same DNA elements occur near the promoters of many genes, each transcription factor can simultaneously regulate a suite of target genes.

The transcriptional regulation of development is often very complex. Many different transcription factors may bind to DNA in the vicinity of a target gene's promoter. The cell needs to integrate information about the effects of all the various transcription factors to determine whether or not transcription of the target gene will proceed. For example, the *Drosophila* gene called *even skipped* (*eve*) helps generate the proper number of body segments in the animal; *eve* mutant embryos resemble those at the bottom of Fig. 19.2 on p. 650. Correct body segmentation depends upon the expression of *eve* in seven narrow stripes in early *Drosophila* embryos (Fig. 19.25). These stripes form only in regions of the embryo where the concentration of proteins that activate *eve* transcription are sufficiently high and the concentration of proteins that repress *eve* transcription are sufficiently low. One of the factors that activates *eve* transcription in some of these stripes is a protein called Hunchback (Hb), whose distribution in the embryo is also shown in Fig. 19.25.

As this example of early *Drosophila* development illustrates, the genes encoding transcription factors function in *developmental hierarchies* that successively provide more refined information for the specification of cell fates. In the *eve*-containing hierarchy, the transcription factors such as Hb that regulate *eve* are expressed in specific regions covering roughly one-quarter to one-third of the embryo—a much broader area than the narrowly restricted location of each *eve* stripe (compare the red and green signals in Fig. 19.25). On a lower level of the hierarchy, the protein product of *eve*—itself a homeodomain-containing transcription factor—regulates the transcription of target genes that are expressed in 14 even narrower stripes. Each of these 14 stripes corresponds to a segment of the animal, such as a particular segment of the thorax or abdomen. The genes expressed in the 14 stripes determine particular patterns of cells that

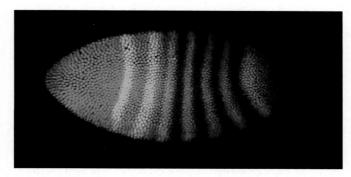

Figure 19.25 Part of a developmental hierarchy in *Drosophila*. This early embryo is stained with fluorescent antibodies to highlight the Even-skipped (Eve; *green*) and Hunchback (Hb; *red*) proteins. Nuclei containing both proteins are *yellow*. To express Eve in the leftmost two bands (out of seven), nuclei in these positions must first express Hb, a transcription factor that binds to enhancers near the *eve* gene's promoter.

are repeated in each segment. Hierarchies of transcription factors allow an organism to provide its cells with progressively more detailed information on their relative positions, and this increasingly specific information guides the cells to very specific fates.

Some Developmental Processes Require Other Types of Gene Regulation

Although regulation of transcriptional initiation is the most general strategy by which cells control gene expression during development, it is by no means the only one. The progression from gene to protein involves many subsequent steps, each of which is amenable to regulation. A gene's primary transcript has to be spliced into a mature mRNA. This mRNA must be translocated from the nucleus into the cytoplasm, and then it must be translated into a protein. The relative stability of an mRNA or protein can affect its concentration in the cell. And finally, once made, a protein can be altered after translation in ways that affect its activity. Researchers have learned of various molecular mechanisms underlying development that exploit each of these steps of gene regulation.

Nevertheless, many developmental pathways involving the control of these later steps in gene expression still culminate in the control of target gene expression by transcription factors. We presented an example of this principle on pp. 605–609 of Chapter 17 in our discussion of sex determination in *Drosophila*. Recall that several of the genes contributing to the process of sex determination encode a cascade of splicing factors. Females make a splicing factor called Sxl that triggers the splicing of the *tra* gene's primary transcript so as to make a functional Tra protein. The Tra protein is itself a splicing factor needed to splice the primary transcript of a third gene (*dsx*) into an mRNA for a female-specific form of the Dsx protein. The splicing cascade does not operate in males, where the *dsx* gene's primary transcript is instead alternatively spliced into an mRNA for a male-specific form of the Dsx protein. Importantly, the male- and female-specific forms of Dsx are themselves transcription factors that turn on and off the genes responsible for many characteristics of the two sexes. Thus, although this pathway for sex determination makes extensive use of the regulation of alternative splicing, the ultimate goal of the pathway is to make male- or female-specific transcription factors.

In *C. elegans,* one example of the posttranscriptional control of gene expression involves effects on translation rather than on splicing. After *C. elegans* embryos hatch into larvae, certain cells must divide and their descendents must differentiate for the animal to become an adult. For these crucial developmental events to occur, the level of a protein called LIN-14 must decrease during the larval stages. This decrease in LIN-14 depends on expression of the *lin-4* gene, whose product is a small, 22-nucleotide RNA. The sequence of this small RNA is complementary to the 3'-untranslated region (3'-UTR) of the *lin-14* mRNA. The binding of this tiny *lin-4* RNA to *lin-14* mRNA prevents translation of the LIN-14 protein. The amounts of the small *lin-4* RNA increase during larval life because the expression of *lin-4* is triggered when larvae feed. Thus, the more the larvae feed, the higher the amount of the *lin-4* RNA and the lower the amount of LIN-14 protein. Interestingly, the *lin-4* RNA does not affect the amount of *lin-14* mRNA, but instead it blocks the translation of the *lin-14* mRNA by some unknown mechanism (see the *C. elegans* portrait in Reference C for more details on the *lin-4/lin-14* system). Just as in the example of *Drosophila* sex determination, the ultimate result of the *lin-4/lin-14* system is to control the transcription of a battery of target genes that, in turn, regulate events during larval development. The exact role of the LIN-14 protein in this regulation is not yet understood.

It's the Chicken *and* the Egg: Both Maternal and Zygotic Genes Impact Early Development

The earliest stages of development require not only the regulation of the expression of genes in the developing individual's genome but also the regulation of gene expression in the mother's genome. Before fertilization, the egg in most organisms already contains many of the mRNAs and proteins needed for the earliest stages of development. The egg must load up on these molecules because transcription of the zygotic genome (that is, of the diploid genome containing an equal number of maternal and paternal chromosomes) does not begin immediately after fertilization. In *Drosophila,* for example, transcription of zygotic genes does not begin in earnest until the embryo contains roughly 6000 cells and has completed some of the earliest steps determining cell fates.

The delayed transcription of the zygotic genome means that the mRNAs and proteins the egg receives from the mother during oogenesis are critical to the earliest stages of development. As a result, developmental geneticists studying embryogenesis must analyze not only the genes of the embryo but also the genes of the mother. Two kinds of genetic screens for mutations that affect early development enable them to do this. In one, they look for zygotic mutations that disrupt the development of embryos whose genomes contain the mutations. In the second type of screen, they seek to identify **maternal-effect mutations:** mutations in the mother's genome that disrupt the development of her progeny. Females carrying maternal-effect

mutations in genes essential for embryogenesis are sterile because their embryos cannot develop. For example, female flies homozygous for loss-of-function mutations in a gene called *bicoid* (see the following section for details) produce embryos that never hatch into larvae because they lack head structures.

Development Exploits Asymmetries

For cells to differentiate into different types, they must either be exposed to different signals from their environment or they must be intrinsically biochemically distinct. Nature has used both strategies for the differentiation of cell types during the development of multicellular organisms.

The Developmental Origins of Asymmetry

In some species, the egg is inherently asymmetric, providing a way for cells in the early embryo to receive information about their relative position. The *Drosophila* egg cell, for example, is part of a more complicated structure called an *egg chamber* (Fig. 19.26a). (You should note that we use the term "egg cell" somewhat incorrectly for simplicity's sake: it is more properly called the oocyte.) Certain cells known as *nurse cells* act as factories that synthesize large amounts of mRNAs and proteins; the nurse cells then deposit these molecules into the egg. The egg has an anterior-to-posterior (that is, head-to-tail) sense of direction in large part because it is connected to the nurse cells only at its anterior end.

Figure 19.26b shows one way in which this asymmetry affects differentiation. The mRNA of the *bicoid* gene is transcribed in the nuclei of the nurse cells and then transported into the egg. The *bicoid* mRNA, in association with certain proteins that bind to its 3'-UTR, appears to become ensnared by microtubules. These microtubules act as tracks along which the mRNA and its associated proteins are transported to the cortex (the cytoplasm just beneath the cell membrane) at the anterior end of the egg cell. At fertilization, translation of the *bicoid* mRNA occurs at the anterior end of the egg, where the mRNA is located. Because the Bicoid protein (as opposed to its mRNA) is not tethered at the anterior end, the protein diffuses away from its source. Early in embryogenesis, then, the embryo has a gradient of Bicoid protein, high at the anterior end and tailing off toward the posterior end (see Fig. D.20b on p. 829 for a picture of this gradient). The Bicoid gradient is critical for *Drosophila* development because the Bicoid protein acts as a **morphogen:** a molecule whose concentration determines the developmental fate of a cell. Cells in regions containing high levels of Bicoid develop into head structures, whereas cells in regions of low Bicoid concentration form abdominal structures. This scheme explains why the embryos produced by *bicoid* mutant mothers lack head structures.

In other species, the first asymmetries important for development occur after fertilization. For example, in *C. elegans,* the site at which the sperm enters the egg to effect fertilization defines the anterior end of the embryo. Before fertilization, the egg has no polarity; sperm entry initiates rearrangements of the cytoplasm that establish the anterior-to-posterior axis of the embryo. The asymmetries affecting early mammalian development emerge even later, after four rounds of mitosis have produced a 16-cell embryo.

(a) Diagram of a *Drosophila* egg chamber

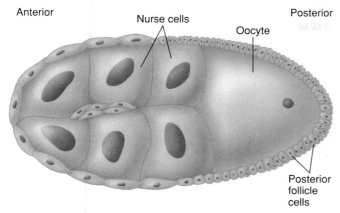

(b) RNA *in situ* hybridization results

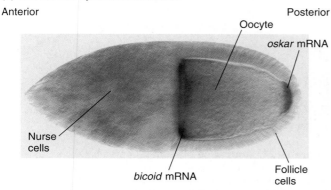

Figure 19.26 ***Bicoid* mRNA localizes to the anterior cortex of the oocyte. (a)** Large nurse cells at the anterior of the egg chamber synthesize mRNA and proteins and transport them to the oocyte. The nurse cells and the oocyte are surrounded by a layer of *follicle cells*. **(b)** One of the mRNAs transported into the oocyte from the nurse cells is *bicoid* mRNA (*black*), which localizes to the anterior cortex of the oocyte. Interestingly, the mRNA for a different gene called *oskar* localizes to the posterior cortex (*red*). Thus, the egg already has a defined polarity.

In Some Mitoses, the Daughter Cells Are Not Created Equal

Surprisingly, certain special mitotic divisions of a single cell can generate two daughter cells that have different molecular components. These differences may cause the daughter cells to develop along different pathways.

Stem cell divisions, which give rise to most of the differentiated cells in the body, provide particularly important examples of the asymmetric distribution of cellular components during cell division. **Stem cells** are relatively undifferentiated cells that have the ability to divide indefinitely yet also give rise to more differentiated cells that eventually cease dividing. Stem cells can play these two seemingly opposite roles simultaneously because of the unusual nature of stem cell divisions. Stem cells divide asymmetrically, creating two very different daughter cells. One of the daughter cells is itself a stem cell, while the other daughter cell and its descendents can differentiate (Fig. 19.27a). In *Drosophila*,

(a) Asymmetric neuroblast stem cell divisions

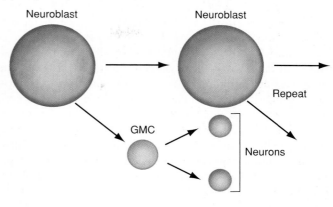

(b) Asymmetric distribution of Prospero protein

Figure 19.27 Asymmetric cell division. (a) A neuroblast divides into a large and a small cell. The small ganglion mother cell (GMC) subsequently divides symmetrically, producing two neurons. The larger cell retains the neuroblast identity. **(b)** The Prospero protein (*red*) is found only at the cortex underneath one spindle pole at early anaphase (*top*). Later in anaphase (*bottom*), Prospero protein goes exclusively to the presumptive small GMC cell. (*Blue* = chromosomes; *green* = mitotic spindle.)

the neural stem cells called *neuroblasts* can repeatedly divide to form a daughter neuroblast and another daughter called a *ganglion mother cell (GMC).* GMCs subsequently divide, giving rise to the *neurons,* which transmit electrical signals.

Figure 19.27b shows that a key aspect of the neuroblast stem cell divisions is the asymmetric distribution of certain molecules during mitosis. During mitosis, the Prospero protein becomes localized to one region of the neuroblast's cortex. When the cell divides, Prospero goes exclusively to the GMC daughter cell. Prospero is one of several proteins that behaves in this fashion. Other proteins, distributed on the other side of the mitotic spindle in the dividing neuroblast, do not go to the GMC cell and instead stay in the emerging neuroblast daughter cell. Although the mechanisms that ensure these asymmetric protein distributions are complex and not yet completely understood, they clearly contribute to the developmental differences between the daughter cells. Prospero, for example, is a homeobox-containing transcription factor whose action affects the fate of the GMC cells into which it is distributed. *Drosophila* mutant for the *prospero* gene thus have a highly aberrant nervous system.

Cell-to-Cell Communication Is Essential for Proper Development

Construction of a large, complicated multicellular organism depends on more than broad, asymmetric cues such as the morphogen gradients laid down by the mother and the asymmetric distribution of molecules during cell division. Cells must "talk" to each other to obtain enough information about their relative positions in the organism. The information obtained from cell-to-cell communication enables cells to refine the decisions that guide their subsequent development.

Signals Come from Near and Far

Cells can communicate with each other either by direct contact or by diffusible factors (usually proteins) released from one cell and received by a second cell (review Fig. 19.7a on p. 655). As we have seen, cell-to-cell communication usually takes place at the surface of the second cell when a ligand made by the signaling cell binds to a receptor embedded in the membrane of the receiving cell. One type of cell-to-cell communication, called *juxtacrine signaling,* takes the form of direct contact. In such signaling, the ligand is a cell surface molecule anchored in the membrane and extending outside of the signaling cell (Fig. 19.28).

Other cellular interactions are mediated by *paracrine factors:* ligands secreted by the signaling cell. Some secreted ligands diffuse over small distances and thus affect only nearby cells. The concentration of the ligand, and thus the strength of the signal, is greater the closer the receiving cell is to the ligand-producing cell. The Wingless (Wg) protein visualized in Fig. 19.22 on p. 665 is an example of a paracrine factor. In wild-type imaginal disks, Wg is made in a thin band of cells (stained yellow in Fig. 19.22b) while Vestigial (Vg) protein is made in a broad region about six cells to either side of this band. In the

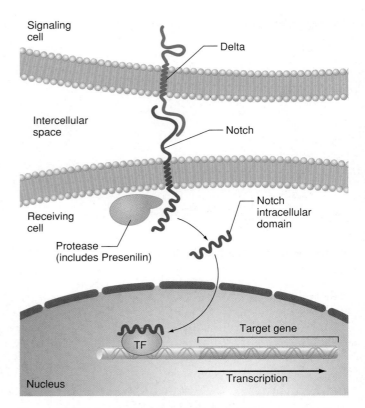

Figure 19.28 Juxtacrine signaling. Contact between a signaling cell expressing the Delta ligand on its surface and an adjacent receiving cell expressing the Notch receptor on its surface initiates changes affecting the fate of the receiving cell. These changes include the cleavage of Notch, whose intracellular domain can then enter the nucleus to help a latent transcription factor (TF) activate the transcription of target genes.

absence of Wg (Fig. 19.22c), Vg is made just in a few cells, instead of its usual broad region. The reason is that Wg normally diffuses from the thin band of cells that produce it to signal nearby cells. Because of this diffusion, cells throughout a broad region are close enough to receive the signal, and will become proper wing cells that make Vg. In the absence of Wg signaling, only a few cells can still make Vg because of signaling by factors other than Wg.

Yet other ligands, called *hormones,* or *endocrine factors,* circulate throughout the body in the blood and can affect tissues far removed from the gland that produces them.

Signal Transduction Pathways Move Signals from a Cell's Surface to Its Interior

The binding of ligand to receptor occurs at the cell surface, yet as we have seen, the developmental signal conveyed by that binding ultimately activates or represses batteries of genes in the nucleus of the receiving cell. What happens in the receiving cell to convert the binding of ligand into an action by a nuclear transcription factor? The answer is a series of biochemical steps known as a **signal transduction pathway.** Central to the development of multicellular organisms, such pathways have been highly conserved during evolution.

Because there are many kinds of cell-to-cell communication involving different ligands, there are many types of signal transduction pathways. Here we discuss only two of these: the highly conserved juxtacrine signaling pathway initiated by the binding of the ligand Delta to the receptor Notch (Fig. 19.28), and the pathway turned on by the binding of the paracrine fibroblast growth factor (FGF) to its receptors in the FGFR family (review Fig. 19.7a on p. 655).

When the Delta ligand on one cell binds to the Notch receptor on an adjacent cell, the Notch protein undergoes a conformational change. This change activates a proteolytic activity in the membrane of the responding cell that cleaves the Notch protein, separating its intracellular domain from the remainder of the protein (which passes through the cell membrane to the exterior of the cell). The cleavage of Notch is critical to the signal transduction pathway, because the cleaved intracellular domain of Notch can now enter the nucleus of the cell. Once in the nucleus, it associates with and activates a latent transcription factor. The activated factor initiates transcription of cell-fate-specific target genes.

Interestingly, the complex protease enzyme responsible for the cleavage of Notch includes a subunit called Presenilin. Dominant mutations in the human Presenilin gene cause familial Alzheimer's disease by altering the cleavage of a different protein than Notch, called the amyloid precursor protein.

The FGF paracrine signal transduction pathway is similar to the Notch system in that the binding of ligand to receptor causes translocation of a protein into the nucleus; and the protein's presence in the nucleus helps activate transcription. However, as Fig. 19.7a on p. 655 shows, the protein transported into the nucleus is not a part of the receptor itself. Instead, as we saw previously, the binding of the FGF ligand to the fibroblast growth factor receptor (FGFR) activates the receptor's kinase activity. This kinase activity transfers phosphate groups not only to other FGFR molecules but also to so-called STAT proteins. Phosphorylated STAT subunits can dimerize with each other, and this dimerization enables their translocation into the nucleus, where they directly activate transcription of target genes specific for FGF-stimulated cells.

A Signal Transduction Pathway Can Have Multiple Effects.
Notch signaling plays many roles in *Drosophila* development. During embryogenesis, it helps cells decide whether to become neurons or epidermal cells. During oogenesis, it is required for the proper differentiation of the follicle cells that surround the egg chamber (see Fig. 19.26). *Notch* mutations also affect the shape of the developing wing (see Fig. 19.13). Like the Notch pathway, many other signal transduction pathways govern diverse developmental processes. How can a single signal transduction pathway have so many disparate developmental functions if the signal itself originates with the simple binding of ligand to receptor?

There are many answers to this question. One is that the effects of a signal transduction pathway can differ depending on the local concentrations of ligand and receptor. A second answer is that individual ligands often bind to several related but different receptor molecules, which may be present in varying amounts on recipient cell surfaces. An example of related receptors are the four FGFR receptors in mice and humans. The binding of a ligand to different receptors may turn on diverse signal transduction pathways with varying effects on gene expression. Third, recipient cells can react differently to the binding of the same receptor to the same ligand. For example, some transcription factors, when transported into the nucleus by a signal transduction pathway, activate transcription of certain genes in one cell but repress transcription of the same genes in another cell. This disparity could reflect the fact that various cell types contain different sets of nuclear proteins capable of interacting with the transcription factor. Finally, it appears that there is a considerable amount of "cross-talk" between different signal transduction pathways. Cells can simultaneously receive signals from several ligands, with each signal activating its own signal transduction pathway. A gene turned on in one pathway might encode a component of another pathway. With so many possible interconnections, there is an almost limitless number of ways in which signal transduction pathways can interact to influence cell fates.

Genes Explain Much, But Not Everything, About Development

Throughout this chapter, we have considered cells in developing multicellular organisms as complex computers. These cellular computers integrate a variety of inputs: the cell's history, its location within the organism, signals from neighboring cells, and signals from more distant cells. The outputs of the cellular computer are alterations to the transcription of a large suite of target genes, which determine the developmental fate of the cell. The central processors that convert the inputs into the outputs are located near the promoters of the target genes, where assessment of the combinatorial effects of many transcription factors determines the time and rate of target gene transcription.

This reductionist point of view has been remarkably successful in building our understanding of development. We now have lists of many genes that play important roles in development, and we are beginning to fathom how each of these genes work and how they interact with each other. Particularly remarkable in the recent past has been our growing appreciation for the way in which evolution has conserved critical genes and pathways, while at the same time creating new twists that underlie the enormous complexity and diversity of life-forms on earth.

Although genes clearly set the ground rules for an individual's development, the same set of genes does not inevitably lead to precisely the same result. Many events in development reflect the strong influence of environmental factors or chance. For example, the name of the first kitten cloned from an adult cell (Fig. 19.29) is "cc" for "carbon copy," but this is somewhat

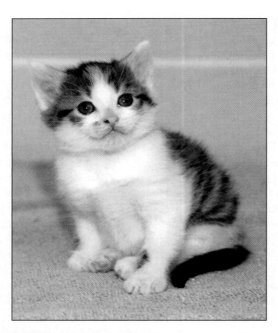

Figure 19.29 "cc," the cloned kitten.

of a misnomer. Though the kitten has exactly the same alleles of all genes as the cat that donated the adult cell, the coats of the two animals are dissimilar due to different prenatal environments. Chance occurrences often influence expression of the genome as well as cellular behaviors. In mammalian females, for example, the decision of which X chromosome is inactivated in which cell is determined by stochastic events. Similarly, as will be apparent from the problem set at the end of the chapter, the choice of which cells adopt neural or epidermal fates depends on small chance fluctuations in the concentrations of certain ligands and receptors. Finally, the incredibly complex connections between neurons in the developing brain are highly plastic and can be influenced by the environment, particularly through learning.

C O N N E C T I O N S

This chapter has presented ample evidence for the conservation of genes that play important roles in development. The *eyeless/Pax-6/Aniridia* gene, for example, acts as a master switch to initiate the development of eyes in many types of organisms. Yet the eyes of various species show tremendous differences, from the compound eyes of *Drosophila* to the single camera-like organs of humans.

The themes of conservation and change have been central to our understanding of evolution since Darwin. Evolution creates and then preserves genetic solutions to problems organisms encounter in their development, biochemistry, physiology, and

behavior; but evolution also tinkers with these solutions to produce novel outcomes.

In the next two chapters, we shift our focus from the analysis of gene activity in individuals to an analysis of gene transmission in whole populations and an examination, at the molecular level, of how genes and genomes evolve over time. Chapter 20 describes why an understanding of evolution requires an understanding of gene transmission in populations. Chapter 21 then builds on ideas presented throughout the book to reconstruct the molecular strategies by which genes and genomes have evolved throughout the roughly 4 billion years of life on this earth.

E S S E N T I A L   C O N C E P T S

1. Developmental geneticists use model organisms as the basis for studying how a fertilized egg becomes a multicellular adult. Each model has special attributes that give it value as an experimental system. The accumulation of knowledge, mutants, and techniques from long years of study enhances the utility of model organisms.

2. The evolutionary relatedness of all organisms, revealed by the conservation of genes, structures, and functions often makes it possible to extrapolate from model organisms to all living forms.

3. Genetics simplifies the study of development. A key to the genetic dissection of development is the isolation of a comprehensive set of mutations affecting a particular process. Multiple alleles of any one gene make it possible to analyze the many roles a protein might play in development. Saturation mutagenesis and screens for suppressors or enhancers of a phenotype can help identify many of the genes participating in a developmental pathway.

4. To understand a gene's role in development, researchers analyze mutant phenotypes. They also determine where

the gene is expressed through RNA *in situ* hybridization experiments or by following the protein product with fluorescently tagged antibodies or with GFP labels. The construction of *genetic mosaics* can help determine which cells need to express the gene so that the organism can develop normally. Careful examination of individuals mutant for two or more genes can help reveal how proteins cooperate in developmental pathways.

5. Cellular differentiation requires progressive changes in gene expression. These changes usually result from decisions concerning the transcription of batteries of genes. Some developmental decisions, however, involve posttranscriptional gene regulation. The earliest stages of development require control of gene expression in both the maternal and zygotic genomes.

6. Differentiation requires either that cells have intrinsic differences at the biochemical level, or that they are exposed to different information in their environment. Asymmetries in early embryonic development, or the asymmetric distribution of molecules during cell division, can generate intrinsic differences. Cell-to-cell communication, effected by the binding of ligands to receptors and mediated by signal transduction pathways, supplies cells with complex information about their relative position in the organism.

PROBLEMS

This chapter's problem set encompasses a single comprehensive example. The example concerns some of the cellular decisions required for the development of the mechanosensory bristles on the *Drosophila* adult body (Fig. A). These bristles help the fly detect foreign materials on their bodies; contact of the bristles with such materials initiates an intricate grooming response in which the fly uses its legs to scrape off the foreign matter. As shown in Fig. A, the bristles are typically composed of four cells: (1) a *neuron,* which generates electrical impulses that are transmitted to the central nervous system; (2) a *trichogen* cell that creates the visible shaft of the bristle; (3) a *tormogen* cell that produces the socket surrounding the base of the shaft; and (4) a *thecogen* cell that makes a sheath that wraps around and protects the neuron. All four of these cells are the mitotic descendants of a single cell called a *sensory organ precursor (SOP)*. In this comprehensive example, we consider a single question: How does the developing animal form the right number of bristles in the right places on the body? Because the complete answer to this question is very complex, we focus even more narrowly on just one step in the process.

The bristles form within an epithelium—a two-dimensional sheet of cells. Most of these cells are destined to become epidermal cells that form the "skin" of the animal, although some (the presumptive SOP cells) become committed to a "neural" fate that allows their descendents to differentiate into the bristle-forming cells. We start this description at a stage when groups of cells in the epithelium called *proneural clusters* have already formed (Fig. B). All the cells in each proneural cluster have the potential to become SOP cells, because they make certain transcription factors encoded by the so-called *achaete-scute* genes. (Expression of the *achaete-scute* genes is a prerequisite for neural fates, including that of the SOP.) The key to the following discussion is that only one of the cells in each proneural cluster will actually become an SOP, while the others will switch to an epidermal fate similar to that of the cells outside the proneural clusters within the epithelium.

The presumptive SOP cell appears to be selected from among the cells of a proneural cluster by random events that are not well understood. Once the decision is made, the future SOP cell continues to express the *achaete-scute* transcription factors, and this allows it to retain a neural fate. The single SOP then communicates with the surrounding cells in the proneural cluster through a form of juxtacrine signaling. The end product of

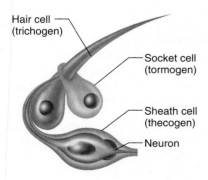

Figure A   A mechanosensory bristle in *Drosophila*.

Hair cell (trichogen)

Socket cell (tormogen)

Sheath cell (thecogen)

Neuron

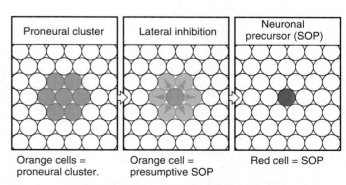

Proneural cluster	Lateral inhibition	Neuronal precursor (SOP)
Orange cells = proneural cluster.	Orange cell = presumptive SOP	Red cell = SOP

Figure B   Lateral inhibition: Each proneural cluster gives rise to only one SOP cell and thus one bristle. Colored cells have the potential to assume a neural fate. The white cells in each diagram are committed to an epidermal, rather than a neural, fate.

this signaling is to turn off the *achaete-scute* genes in the surrounding cells, which prevents them from becoming SOPs and instead causes them to adopt epidermal fates. The ability of the presumptive SOP cell to prevent adjacent cells from becoming SOPs is called *lateral inhibition* and is essential for ensuring that each proneural cluster makes only a single mechanosensory bristle.

The juxtacrine signaling underlying lateral inhibition is mediated by the Delta (ligand)/Notch (receptor) pathway shown in Fig. 19.28 on p. 670. To answer the problems below, you need to know that in the *Drosophila* lateral inhibition pathway, the transcription factor depicted at the bottom of Fig. 19.28 is encoded by the *Supressor of Hairless (Su[H])* gene, and that the pathway regulates several related target genes clustered in one part of the genome called the *Enhancer of split (E[spl]) complex*. In these problems, assume that the Notch and Delta proteins are found on the surfaces of all cells within a proneural cluster. You should also assume that each SOP cell will eventually give rise to a bristle on the adult body, even though this is an oversimplification because the lateral inhibition pathway also influences the subsequent divisions that generate the four bristle cells from their SOP ancestor.

19-1 Suppose you were able to isolate an epithelial layer that could give rise to SOP cells. Using the information just presented, how could you identify the locations in this epithelium of proneural clusters? How could you identify cells in the epithelium that were the presumptive SOP cells?

19-2 Based on the description provided above, do you think that for lateral inhibition, it is more important to express the *Delta* gene in the presumptive SOP cell or in the surrounding cells that will be inhibited and adopt epidermal fates? In which cells is it most important to express the *Notch* gene?

19-3 Knowing that the Delta and Notch proteins are involved in lateral inhibition, can you suggest a simple hypothesis that predicts which of the cells in a proneural cluster will become the presumptive SOP cell? (Another way to think of this question is to ask why the presumptive SOP cell isn't itself subject to lateral inhibition.)

19-4 Suppose that an epithelium contains a clone of cells that is homozygous for a null allele of *Notch*.
a. Describe the phenotypes of the epithelium, both within the mutant clone and outside of it. Describe in particular what would happen to the cells in a proneural cluster if the mutant clone bisected the cluster (that is, some of the cells in the cluster made no Notch protein while other cells in the same cluster made some Notch protein).
b. Developmental geneticists often distinguish between *autonomous traits,* in which only genotypically mutant cells show the mutant phenotype, and *nonautonomous traits,* in which

mutant cells can cause other cells (even genotypically wild-type cells) to exhibit a mutant phenotype. Is the *Notch* phenotype an autonomous or a nonautonomous trait?
c. According to these criteria, is the phenotype of *Arabidopsis* caused by null mutations in the *AGAMOUS* gene (see Fig. 19.19b on p. 664) autonomous or nonautonomous?

19-5 Repeat parts (a) and (b) of problem 19-4, now supposing that the mutant clone is homozygous for a null allele of *Delta*.

19-6 The proteins encoded by genes of the *Enhancer of split (E[spl]) complex* are also transcription factors. It has been proposed that one target of these E(spl) proteins are the *achaete-scute* genes. If this is true, would you expect that the E(spl) proteins act as activators or repressors of *achaete-scute* gene transcription? Why?

19-7 There is another protein, not shown in Fig. 19.28, that also plays a role in lateral inhibition. This protein is encoded by the *Hairless (H)* gene. In clones homozygous for null mutations in *H,* none of the cells in the epithelium become SOP cells. It has also been shown that the H protein binds to the Su(H) protein described previously.
a. Given this information, formulate a hypothesis for the action of the H protein.
b. Do you predict that the mutant *H* phenotype is autonomous or nonautonomous? (See problem 19-4b for definitions of these terms.)

19-8 Flies heterozygous for null mutants in the *H* gene described in the previous problem (19-7) have fewer sensory bristles than wild type. As indicated by its name, null mutations in the *Su(H)* gene are dominant suppressors of the mutant phenotype seen in flies heterozygous for *H* mutations. Thus, flies simultaneously heterozygous for null mutations in *H* and *Su(H)* have a nearly wild-type number of bristles. However, one group of investigators found an abnormal allele of *Su(H)* that actually enhances the effect of heterozygosity for *H*. Flies simultaneously heterozygous for this unusual allele of *Su(H)* and for a null mutation in *H* have even fewer bristles than *H* heterozygotes alone. The investigators found that the protein encoded by the unusual *Su(H)* allele is unable to activate the transcription of the *E(spl)* genes but retains the ability to bind to H protein. How can you explain why this allele of *Su(H)* actually acts as a dominant enhancer of *H?*

19-9 As seen in the previous problem (19-8), animals that are simultaneously heterozygous for null mutations in *H* and in *Su(H)* have a bristle number closer to wild type than that of heterozygotes for *H* alone.

a. Interestingly, animals heterozygous for null mutations in *Su(H)* alone have a wild-type number of bristles. How can you explain this?

b. Suppose now that a clone was generated in a region of an epithelium that is simultaneously homozygous for null mutations in both *H* and *Su(H)*. What would you expect the phenotype of such a clone to be?

19-10 Animals heterozygous for null mutants in *Notch* have several aberrant phenotypes, including an increase in the number of bristles, that in total are much less severe than the phenotype of mutant homozygotes. (In fact, *Notch* mutant homozygotes die early in embryogenesis because the Notch signaling pathway plays important roles in many developmental events beyond lateral inhibition.) Why do you think animals heterozygous for null mutants in *Notch* show any aberrant phenotype?

19-11 In which of the following genes might null mutants be expected to act as dominant enhancers of the extra bristle phenotype seen in heterozygous *Notch* mutants? In which genes would null mutations act as dominant suppressors? Why?

a. *E(spl)*
b. *H*
c. *Su(H)*
d. *Delta*

19-12 *Drosophila* geneticists often call *Notch* a "neurogenic" gene and the *achaete-scute* genes "proneural." Both of these terms imply that neuronal differentiation is favored, but some commentators have called this nomenclature perverse.

a. What aspects of this nomenclature are so confusing? Why do you think *Drosophila* geneticists originally used the terms "neurogenic" and "proneural"?

b. Which of the following genes would you classify as neurogenic, which as proneural, and which as neither? (i) *E(spl)*; (ii) *H*; (iii) *Su(H)*; (iv) *Delta*.

19-13 Developmental geneticists working with *Drosophila* have developed methods to make clones in which the germ-line cells that produce the eggs are homozygous for null mutations in either *Delta* or *Notch*. (These methods are described in detail in Reference D.) So-called *holonull* embryos are homozygous mutant embryos produced by such homozygous mutant germ-line clones. *Zygotic null* embryos are homozygous mutant embryos produced from heterozygous mutant germ-line cells. It has been observed that: (1) *Notch* holonull embryos have more severe mutant phenotypes than *Notch* zygotic null embryos; (2) *Delta* holonull embryos and *Delta* zygotic null embryos have mutant phenotypes of about equal severity; and (3) *Notch* holonull embryos have mutant phenotypes more severe than those of *Delta* holonull embryos.

a. Why did these developmental geneticists need to make clones of homozygous mutant germ-line cells in order to produce holonull embryos?

b. What conclusions can you draw from these three observations?

19-14 Heterozygosity for a particular mutation in *Notch* called N^{414} causes a reduction in the number of bristles, opposite to what is seen in the case of heterozygosity for a null mutation in *Notch*. Presumably, this reflects an increase in the amount of Notch signaling that prevents any of the cells in a proneural cluster from becoming an SOP. There are two potential models to explain this gain-of-function: (1) The mutation results in an increase in the affinity of the Notch receptor for the Delta ligand, or (2) the mutation causes a constitutive activation of Notch signaling independent of binding to the ligand.

a. It was found that the number of bristles in N^{414} / + ; $Delta^{null}$ / + double heterozygotes is the same as that in N^{414} / + single heterozygotes. Which of the two models described is more likely given this finding?

b. Do you expect flies of the genotype N^{414} / $Dp(N^+)$ (where $Dp[N^+]$ is a duplication of the wild-type N^+ gene) to have more or fewer bristles than flies of genotype N^{414} / +?

19-15 Loss-of-function mutations in the *hairy* gene (not previously discussed) cause bristles to appear in locations on the wings and legs where bristles are not usually found. Scientists have established that the Hairy protein binds to DNA sequences in the vicinity of the *achaete-scute* genes. With this information, suggest a hypothesis for the function of the wild-type Hairy protein.

19-16 Devise a genetic screen for *Drosophila* unable to groom themselves properly after their bodies come into contact with foreign matter.

The Genetic Analysis of Populations and How They Evolve

The enormous range of genetic diversity within our own species is easy to see.

Tuberculosis (TB) is an ancient and persistent human disease. Bone deformities typical of those produced by the infection are found in Egyptian mummies dating to 2000–4000 B.C.; and as recently as the mid–nineteenth century, TB was the leading cause of death in Europe and the urban United States. The microbe that causes TB is the bacterium *Mycobacterium tuberculosis*. In humans, populations of *M. tuberculosis* most often infect the lungs and lymph nodes, but sometimes they colonize the bones and skin of a patient (Fig. 20.1a). Severe infection by the tuberculosis bacterium damages the lungs, resulting in weakness, a persistent cough, bloody sputum, and ultimately death. *M. tuberculosis* bacteria can spread from person to person through the air when an infected individual exhales bacteria from his or her lungs during coughing.

Beginning in the late nineteenth century, improved sanitation and the quarantine of TB patients in sanitaria led to a steady decline in the death rate from TB in Europe and the United States (Fig. 20.1b). The introduction of antibiotics during the 1940s and 1950s further reduced TB mortality in those areas, and in the 1960s, when the death rate from TB had decreased to less than 2

deaths per 100,000, many people believed that the disease had been eradicated as a threat to public health, at least in the United States, where the drugs were available. But 25 years later, the incidence of TB began to rise in urban areas around the globe. By 1994, in the Harlem area of New York City, there were 700 cases per 100,000 people. By 2000, TB accounted for more deaths worldwide than any other identifiable infectious disease, claiming close to 3 million lives annually; and epidemiologists estimated that *M. tuberculosis* had infected an astonishing 2 billion people, roughly one-third of the global population. Three factors contributed to this rapid increase in TB incidence: the emergence of AIDS, which among other effects, weakens the human immune system; protein deficiencies among the malnourished in many areas of the globe, which also weaken the immune system; and the widespread occurrence of *M. tuberculosis* strains that are resistant to one or more antibiotics.

Despite the cultural, technological, and medical advances of the twentieth century, infectious microorganisms such as *M. tuberculosis* are still among the leading causes of death in many human

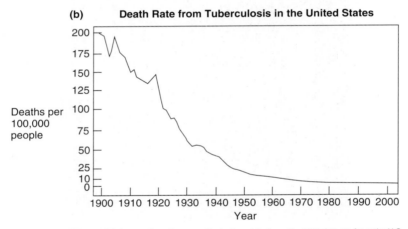

Figure 20.1 **Tuberculosis in human populations. (a)** Photograph of *M. tuberculosis* bacteria colonizing the lungs and bones. **(b)** Death rate from tuberculosis in the United States from 1900–1990.

populations. And in a related arena, populations of plant pests (such as the mites that prey on strawberry plants and almond trees) destroy a substantial fraction of human food supplies.

How do new diseases emerge in human populations? Why do diseases persist in all living organisms? What causes diseases and pests long under control to resurge in frequency and intensity?

One way to answer these questions is to examine genetic variation and its expression as phenotypes within populations of organisms. The scientific discipline that studies what happens in whole populations at the genetic level is known as **population genetics.** It encompasses the evolutionary ideas of Darwin, the laws of Mendel, and the insights of molecular biology. To population geneticists, a **population** is a group of interbreeding individuals of the same species that inhabit the same space at the same time. An example would be all the white-tailed deer on Angel Island in San Francisco Bay in 1990 or all the rock cod at the mouth of the bay. The sum total of all alleles carried in all members of a population is that population's **gene pool.** In nature, the genetic

makeup of a population changes over time as new alleles arise by mutation or are introduced by immigration, and as rare, preexisting alleles disappear when all individuals carrying them leave the population or die. Changes in the frequency of alleles within a population are the basis of **microevolution:** alterations of a population's gene pool.

In this chapter, we explore the nature of genotypic and phenotypic variation within populations and the role of this variation in evolution. We know from Chapter 2 that variation exists in all populations. To begin our examination of this variation, we analyze the incidence of genetic diseases, such as cystic fibrosis and retinoblastoma, that are determined by a single gene. Variations in the nature and number of wild-type and disease alleles determine the frequency of disease and how this frequency may change. Through our relatively simple analysis of single-gene disorders, we develop a framework for understanding how the frequency of a disease-causing allele determines the frequency of diseased individuals in a population.

We next examine how genetic variation within populations of pathogens and pests arises and how human efforts to control these destructive organisms force changes in allele frequencies that determine the rate at which the pathogenic and pest populations evolve.

Finally, we consider variation in multifactorial traits, that is, in traits determined by two or more genes and their interaction with the environment (see Chapter 3). In fact, most aspects of disease susceptibility are multifactorial. To analyze the inheritance of multifactorial traits, population geneticists use quantitative tools that enable them to assess the relative contributions of genetic and environmental factors to individual differences in phenotype.

One general theme emerges from our discussion: Population geneticists rely on mathematical models in predicting a population's potential for evolution because most of the scientifically useful questions they ask are statistical. These questions include: What is the frequency of genetic disease in a population? What fraction of the phenotypic variation in a trait is the result of genetic variation? Given certain quantifiable variables, how rapidly can a dis-ease gain a foothold in a population and how long is it likely to persist? Simple mathematical models not only clarify the questions, they also serve as tools for analyzing data obtained from the field, and for making predictions about future populations.

As we present ways to examine the gene pools of stable and evolving populations over time, we describe:

- The Hardy-Weinberg law: a model for understanding allele, genotype, and phenotype frequencies for single-gene traits in a genetically stable population.

- Calculations beyond Hardy-Weinberg: measuring how selection and mutations cause changes in the allele frequencies of a gene pool over time, including a comprehensive example of the evolution of drug and pesticide resistance.

- How to use the tools of quantitative genetics to analyze the inheritance of multifactorial traits.

The Hardy-Weinberg Law: A Model for Understanding Allele, Genotype, and Phenotype Frequencies for a Single-Gene Trait in a Genetically Stable Population

Suppose you were to look at a human population of 16 in which 6 people suffer from the recessive disease of cystic fibrosis caused by the *r* variant of the wild-type *R* allele. To predict how the number of diseased individuals in the population will change over time, you need to determine the frequencies of each genotype (homozygous *RR*, heterozygous *Rr*, and homozygous *rr*), each phenotype (normal and diseased), and each allele (*R* and *r*). Population geneticists define **phenotype frequency** as the proportion of individuals in a population that are of a particular phenotype. For our hypothetical population, the phenotype frequencies are 6/16 = 3/8 diseased (the number of homozygous *r r* individuals affected by the recessive condition) and 10/16 = 5/8 normal (the remaining fraction with either *RR* or *Rr* genotypes).

Genotype frequency is the proportion of total individuals in a population that are of a particular genotype. To determine genotype frequencies, you simply count the number of individuals of each genotype and divide by the total number of individuals in the population. For recessive traits like cystic fibrosis, it is not possible to distinguish between wild-type and heterozygous genotypes: Both give rise to nondiseased individuals. Thus, the only way to determine genotype frequencies directly at loci like the cystic fibrosis locus is to use molecular probes that distinguish between different alleles. For our hypothetical population, molecular analyses showed that 8 individuals (of 16) are of geno-type *RR*, 2 are of genotype *Rr*, and 6 are *rr*. This means that the *RR* genotype frequency is 8/16 = 4/8 = 1/2; the *Rr* genotype frequency is 2/16 = 1/8; and the *rr* genotype frequency is 6/16 = 3/8. Note that these three frequencies (4/8 + 1/8 + 3/8) sum to 1, the totality of genotypes in the whole population.

Finally, the definition of **allele frequency** is the proportion of all copies of a gene in a population that are of a given allele type. Because each individual in a population has two copies of each gene, the total number of gene copies is two times the number of individuals in the population. Thus, for our hypothetical population of 16 people, there would be 32 copies of the cystic fibrosis gene. Of course, both homozygotes and heterozygotes contribute to the frequency of an allele. But homozygotes contribute twice to the frequency of a particular allele, while heterozygotes contribute only once. To find the frequencies of *R* and *r*, you first use the number of people with each genotype to compute the number of *R* and *r* alleles.

$$8 \; RR \rightarrow 16 \text{ copies of } R$$
$$2 \; Rr \rightarrow 2 \text{ copies of } R$$
$$6 \; rr \rightarrow 0 \text{ copies of } R$$

Together, 16 + 2 + 0 = 18 copies of the *R* allele.

$$8 \; RR \rightarrow 0 \text{ copies of } r$$
$$2 \; Rr \rightarrow 2 \text{ copies of } r$$
$$6 \; rr \rightarrow 12 \text{ copies of } r$$

Together, 0 + 2 + 12 = 14 copies of the *r* allele. Next, you add the 18 *R* alleles to the 14 *r* alleles to find the total number of gene copies.

18 + 14 = 32 copies of the gene, which is twice the number of people in the population

Finally, you divide the number of each allele by the total number of gene copies to find the proportion, or frequency, for each allele.

For the R allele, it is $18/32 = 9/16 = 0.56$

For the r allele, it is $14/32 = 7/16 = 0.44$

Note that here again, the frequencies sum to a 1, representing all the alleles in the gene pool.

We have spent time defining and deriving the genotype, phenotype, and allele frequencies of a hypothetical population because population geneticists use these frequencies as the basis for calculating the genetic and phenotypic variations of a gene in a population and how they may change over time.

The Hardy-Weinberg Law Correlates Allele and Genotype Frequencies Through a Binomial Equation

Many inherited diseases result from disease-causing alleles that are recessive to wild-type alleles. For scientists seeking to predict the potential incidence of such recessive conditions, an important question is, How common are the heterozygous carriers of the disease-causing allele in a population? At the start, the scientists know only the phenotypic frequencies of healthy and diseased individuals. Can they use this information to predict the frequency of heterozygous carriers?

The key to answering this question lies in establishing a quantitative relationship between phenotype, genotype, and allele frequencies within and between generations. The **Hardy-Weinberg law,** named for the two men—G. H. Hardy and W. Weinberg—who independently developed it in 1908, clarifies the relationships between genotype and allele frequencies within a generation and from one generation to the next. To derive a general law describing the inheritance of alleles in a population, it is necessary to make certain simplifying assumptions about the nature of the population, the individuals within it, and the genes these individuals carry. Derivation of the Hardy-Weinberg law depends on five such assumptions:

1. The population includes a very large, virtually infinite number of individuals.

2. The individuals mate at random in the sense that each individual's genotype at the locus of interest does not influence his or her choice of a mate.

3. No new mutations appear in the gene pool.

4. There is no migration into or out of the population.

5. There are no genotype-dependent differences in the ability to survive to reproductive age and transmit genes to the next generation.

Populations that satisfy all five of these assumptions are said to be at Hardy-Weinberg equilibrium.

The assumptions behind the Hardy-Weinberg equilibrium enable the mathematical derivation of an equation for calculating genotype (and thence phenotype) frequencies, even though no actual population is at Hardy-Weinberg equilibrium. All populations are finite; mating is not always at random; mutations occur constantly; migration into and out of a population is common; and many genotypes of interest, such as those that cause lethal childhood diseases, affect the ability to survive and reproduce. Nevertheless, even when many of the assumptions of the Hardy-Weinberg equilibrium do not apply, the equation derived on the basis of these assumptions is remarkably robust at providing estimates of genotype and phenotype frequencies in real populations. Indeed, the equation has always been the most powerful mathematical tool available to population geneticists.

For a population of sexually reproducing diploid organisms, there are two steps in translating the genotype frequencies of one generation into the genotype frequencies of the next generation (Fig. 20.2).

First, if the likelihood that an individual will grow into an adult does not depend on the genotype (that is, if there is no difference in fitness among individuals), then the allele frequencies in the adults should be the same as in their gametes. For example, if p is the frequency of allele R, and q is the frequency of allele r in the adults, p and q will also be the frequencies of the two alleles in the gametes produced by the whole population of those adults.

Next, you can use the allele frequencies in the gametes to calculate genotype frequencies in the zygotes of the next generation—the Punnett square, which provides a systematic means of considering all possible combinations of uniting gametes, is the tool of choice (Fig. 20.3). For example, if the probability that an egg and a sperm unite does not depend on the genotype of the egg or sperm (that is, if fertilization is random among genotypes) and if the population of gametes is very large, then the following pattern emerges. Recall that RR zygotes result from fertilization of R-carrying eggs by R-carrying sperm. If p is the frequency of R gametes (eggs and sperm), then, applying the product rule, the frequency of RR zygotes is p (frequency of R eggs) $\times$ p (frequency of R sperm) $= p^2$. Similarly, rr zygotes result from fertilization of r-carrying eggs by r-carrying sperm. If q is the frequency of r gametes (eggs and

Figure 20.2 Computing allele frequencies from genotype frequencies.

sperm), the frequency of *rr* zygotes will be *q* (the frequency of *r* eggs) × *q* (the frequency of *r* sperm) = q^2. Finally, *Rr* zygotes result either from fertilization of *R* eggs by *r* sperm, with a frequency of *p* × *q* = *pq*, or from the fertilization of *r* eggs by *R* sperm, occurring at a frequency of *q* × *p* = *pq*. The total frequency of *Rr* zygotes is thus *pq* + *pq* = 2*pq*.

To summarize: The genotype frequencies of zygotes arising in a large population of sexually reproducing diploid organisms are p^2 for *RR*, 2*pq* for *Rr*, and q^2 for *rr* (see Fig. 20.3). These genotype frequencies are known as the Hardy-Weinberg proportions; they exist in populations that satisfy the Hardy-Weinberg assumptions of a large number of individuals, mating at random, with no new mutations, no migration, and no genotype-dependent differences in fitness. Since these genotype frequencies represent the totality of genotypes in the population, they must sum to 1. Thus the binomial equation representing the Hardy-Weinberg proportions is

$$p^2 + 2pq + q^2 = 1 \qquad (20.1)$$

Because we have assumed no differences in fitness, the genotype frequencies of the zygotes will be the genotype frequencies of the adult generation that develops from those zygotes.

This equation thus enables us to use information on genotype and allele frequencies to predict the genotype frequencies of the next generation. Suppose, for example, that in a population of 100,000 people carrying the recessive allele *a* for albinism, there are 100 *aa* albinos and 1800 *Aa* heterozygous carriers. To find what the frequency of heterozygous carriers will be in the next generation, you compute the allele frequencies in the parent population.

98,100 *AA* individuals; 1800 *Aa* individuals,
and 100 *aa* individuals → 196,200 *A* alleles +
1800 *A* alleles; 1800 *a* alleles + 200 *a* alleles

Sperm

		R		r	
Allele		R		r	
Frequency		p		q	
Allele R		RR		Rr	
Frequency p		p^2		pq	
Allele r		Rr		rr	
Frequency q		pq		q^2	

Eggs (left margin label)

Figure 20.3   Derivation of the Hardy-Weinberg equation.
After mating, the genotype frequencies of the next generation will be p^2 for *RR*, 2*pq* for *Rr*, q^2 for *rr*.

Out of 200,000 total alleles the frequency of the *A* allele is

$$198,000/200,000 = 99/100 = 0.99; \text{ thus } p = 0.99$$

and the frequency of the *a* allele is

$$2,000/200,000 = 1/100 = 0.01; \text{ thus } q = 0.01$$

The Hardy-Weinberg equation for the albino gene in this population is

$$p^2 + 2pq + q^2 = 1$$
$$(0.99)^2 + 2(0.99 \times 0.01) + (0.01)^2 = 1$$
$$0.9801 + 0.0198 + 0.0001 = 1$$

It thus predicts that in the next generation of 100,000 individuals, there will be

$$100,000 \times 0.9801 = 98,010 \ A \ A \text{ individuals}$$
$$100,000 \times 0.0198 = 1980 \ A \ a \text{ individuals}$$
$$100,000 \times 0.0001 = 10 \ a \ a \text{ individuals}$$

This example shows that in one generation, the genotype frequencies have changed slightly. A natural question is, Have the allele frequencies also changed? Recall that the initial frequencies of the *R* and *r* alleles are *p* and *q*, respectively, and that *p* + *q* = 1. You can use the rules for computing allele frequencies from genotype frequencies (see Figs. 20.2 and 20.3) to compute the allele frequencies of the next generation. From the Hardy-Weinberg equation, you know that p^2 of the individuals are *RR*, whose alleles are all *R*, and 2*pq* of the individuals are *Rr*, 1/2 of whose alleles are *R*. Similarly, q^2 of the individuals are *rr*, whose alleles are all *r*, and 2*pq* of the individuals are *Rr*, 1/2 of whose alleles are *r*. If *p* + *q* = 1, then *q* = 1 − *p*, and the frequency of the *R* allele in the next-generation population is

$$p^2 + 1/2[2p(1 - p)] = p^2 + p(1 - p)$$
$$= p^2 + p - p^2 = p \qquad (20.2)$$

Similarly, *p* = 1 − *q*, and the frequency of the *r* allele in the next-generation population is

$$q^2 + 1/2[2q(1 - q)] = q^2 + q(1 - q)$$
$$= q^2 + q - q^2 = q \qquad (20.3)$$

Using these equations to calculate the allele frequencies of *A* and *a* in the second generation of 100,000 individuals, some of whom are albinos, we find

for *p* 0.9801 + 0.99 − 0.9801 = 0.99
= the frequency of the *A* allele

for *q* 0.0001 + 0.01 − 0.0001 = 0.01
= the frequency of the *a* allele

These frequencies are the same as those in the previous generation. Thus, even though the genotype frequencies have changed slightly from the first generation to the next, the allele frequencies have not. Note that this is true of both the dominant and the recessive alleles.

Consider, for example, the dominant human condition of *brachydactyly,* which causes people to have short, stubby fingers. If a group of *BB* persons homozygous for the allele causing short, stubby digits were all to conceive children with *bb* mates homozygous for the allele determining normal finger length, all the progeny will be *Bb* heterozygotes showing the brachydactyly phenotype. But even though the phenotype and genotype frequencies have changed drastically, the allele frequencies have not. In both generations, one half of the alleles are *B* and one half are *b*. If no new alleles appear in the population by mutation or immigration and the heterozygotes of the second generation mate with each other at random (with respect to the presence or absence of brachydactyly, the third generation produced by *Bb* × *Bb* matings will contain the 3:1 ratio of brachydactyly to normal phenotypes, determined by the Mendelian genotype ratios of 1*BB* : 2*Bb* : 1*bb*. Within this third generation, the allele frequency is 4/8*B* and 4/8*b*, the same as in the first generation in which *BB* mated with *bb*.

The Hardy-Weinberg law generates two significant conclusions.

Allele Frequencies Do Not Change from Generation to Generation in a Population at Hardy-Weinberg Equilibrium

This is one of the most important insights of the Hardy-Weinberg law: As long as a population is free of all influences that introduce or remove alleles, that is, as long as it is at Hardy-Weinberg equilibrium, its allele frequencies remain unchanged. The Hardy-Weinberg law thus provides a model for looking at genotype, allele, and phenotype frequencies in a genetically stable population.

A Hardy-Weinberg Population Achieves the Genotype Frequencies of p^2, $2pq$, and q^2 in Just One Generation and Maintains These Frequencies in Subsequent Generations

This second significant implication of the Hardy-Weinberg law, which applies regardless of the genotype frequencies in the initial generation, enables us to answer the question posed at the beginning of this discussion: If you know only the phenotypic frequencies of healthy and diseased individuals in a population, can you use this information to predict the frequency of heterozygous carriers? For an autosomal recessive condition, all affected individuals are homozygotes for the recessive disease-causing allele. If you assume that the genotype and allele frequencies for the condition follow the Hardy-Weinberg law and that the frequencies of the normal and disease-causing alleles are *p* and *q*, respectively, then the genotype frequency of affected homozygotes is simply q^2, and the square root of this genotype frequency, *q*, is the allele frequency. Because the frequency of heterozygotes according to Hardy-Weinberg is $2pq$ and $p = 1 - q$, you can compute the frequency of heterozygous carriers of the disease.

Consider the inherited condition of phenylketonuria (PKU), which is caused by an autosomal recessive mutation that eliminates activity of the enzyme that converts the amino acid

phenylalanine to the amino acid tyrosine. As we saw in Chapter 3, without the enzyme, phenylalanine builds up, and this buildup interferes with normal brain development. Medical studies show that about 1 in 3600 Caucasians in the United States have PKU. Because PKU is an autosomal recessive condition, all the affected individuals are homozygotes for the recessive allele. Thus the genotypic frequency of PKU is 1/3600. Since according to the Hardy-Weinberg equation, the genotypic frequency of homozygotes for the recessive allele is q^2, you can say that $q^2 = 1/3600$, and then compute the frequency of the recessive PKU allele in the whole population by finding the square root of this number.

$$q = \sqrt{q^2} = \sqrt{1/3600} = 1/60 = 0.0167$$

Next, because $p = 1 - q$, you can compute *p*.

$$p = 1 - 1/60 = 59/60 = 0.9833$$

Knowing the frequencies of both *p* and *q*, we can compute the frequency of heterozygotes according to Hardy-Weinberg.

$$2pq = 2 \times (0.0167) \times (0.9833) = 0.0328$$

Thus, if one assumes that the U.S. Caucasian population can be approximated by the conditions that define a Hardy-Weinberg equilibrium, it becomes possible to determine simply from observations of phenotypes that within this population, the frequency of heterozygous carriers is 0.0328 or roughly 1 in 30 people.

Beyond Hardy-Weinberg: Measuring How Mutation and Selection Cause Changes in Allele Frequencies

One definition of **evolution** is changes in allele frequency in a population over multiple generations. Evolutionary scientists refer to this level of evolution as *microevolution* in comparison with **macroevolution,** which occurs over geologic time. Macroevolution refers to the process by which new species emerge from existing species, while microevolution refers to genetic changes that occur within a species. Many evolutionary biologists believe that macroevolution is simply a consequence of extended periods of microevolution. Population geneticists can use the Hardy-Weinberg law to look at microevolution. Because the law models populations in which genetic changes do not occur, it provides a theoretical standard against which to measure changes that do occur. In natural populations, conditions almost always deviate from the Hardy-Weinberg assumptions: Some populations are not very large, individuals do not always mate at random, new mutations do arise, there is migration into and out of the population, and different genotypes do generate differences in survival rates. We now examine how simple modifications of the Hardy-Weinberg law make it possible to analyze the effects of these deviations, particularly those resulting from genotype-dependent differences in fitness, and new mutations.

Natural Selection Acts on Differences in Fitness to Alter Allele Frequencies

Contrary to the Hardy-Weinberg assumption that the probability of survival and reproduction does not vary with the genotype, for many phenotypic traits, including inherited diseases, genotype does influence survival and the ability to reproduce. Thus in real populations, as opposed to model populations constructed according to Hardy-Weinberg assumptions, not all individuals survive to adulthood, and there is always some probability that an individual will not live long enough to reproduce. As a result, the genotype frequencies of real populations change as their individual members mature from zygotes to adults.

To population geneticists, an individual's relative ability to survive and transmit its genes to the next generation is its **fitness.** But although fitness is an attribute associated with each genotype, it cannot be measured within the individuals of a population because each animal with a particular genotype survives and reproduces in an unpredictable way. However, by considering all the individuals of a particular genotype together as a group, it becomes possible to measure the relative fitness for that genotype. Thus, for population geneticists, fitness is a statistical measurement only. Nevertheless, differences in fitness can have a profound effect on the allele frequencies of a population.

Fitness has two basic components: viability and reproductive success. The fitness of individuals possessing variations that help them survive and reproduce in a changing environment is relatively high; the fitness of individuals without those adaptive variations is relatively low. In nature, the process that progressively eliminates individuals whose fitness is low and chooses individuals of high fitness to survive and become the parents of the next generation is known as **natural selection.** The mechanisms of selection act independently of any individual; often, interactions between genetically determined phenotypes and environmental conditions are the agents of natural selection. For example, in a hypothetical population of giraffes browsing on the leaves of an ancient savanna, suppose that some of the animals had long necks and some short necks, and that each of these phenotypes resulted from variations in the genes contributing to neck length. If during a long drought, there were fewer low-hanging leaves than usual on the flora of the savanna, those long-necked individuals able to reach the higher morsels would have been able to harvest more food and survive to produce more offspring than individuals with shorter necks. Natural conditions would thus have selected the better adapted, long-necked giraffes—those with the higher fitness—to survive and become the parents of the next generation (Fig. 20.4). Similarly, in the laboratory, scientists can establish experimental conditions, such as the absence of a nutrient or the presence of an antibiotic, that become the agents of **artificial selection.** In both natural and artificial selection, environmental and other external conditions interact with preexisting variations to "select" the parents of the next generation.

Field studies show that natural selection occurs for many phenotypic traits in a variety of natural populations. Since in human populations, genetic diseases frequently affect a person's

Figure 20.4   Giraffes on the savanna. A visible example of one outcome of natural selection for increased fitness.

Gametes	R		r
Allele frequencies	p		q
Zygotes' genotype frequencies	$G_{RR} = p^2$	$G_{Rr} = 2pq$	$G_{rr} = q^2$
Relative fitness: basis of selection	w_{RR}	w_{Rr}	w_{rr}
Adults' genotype frequencies	$\dfrac{p^2 w_{RR}}{\overline{w}}$	$\dfrac{2pq w_{Rr}}{\overline{w}}$	$\dfrac{q^2 w_{rr}}{\overline{w}}$
Gametes' (next generation) allele frequencies	$p' = \dfrac{p^2 w_{RR} + pq w_{Rr}}{\overline{w}}$		$q' = \dfrac{q^2 w_{rr} + pq w_{Rr}}{\overline{w}}$

$$\overline{w} = p^2 w_{RR} + 2pq w_{Rr} + q^2 w_{rr}$$

Figure 20.5   Changes in allele and genotype frequencies when selection acts on genotype-dependent differences in fitness.

survival and reproduction, the potential for natural selection in people affected by such diseases is considerable. How does selection alter our conclusions from Hardy-Weinberg?

We begin to answer this question with an analysis of the R gene in a population of zygotes in Hardy-Weinberg equilibrium. In this population, the genotype frequencies RR, Rr, and rr are p^2, $2pq$, and q^2 respectively. Now suppose that the viability, that is, the probability of surviving from zygote to adult, depends on genotype, while the second component of fitness—success at productive mating—is independent of genotype. If we define the relative fitness of the three genotypes as w_{RR}, w_{Rr}, and w_{rr}, respectively, the relative frequencies of the three genotypes at adulthood is $p^2 w_{RR}$, $2pq w_{Rr}$, and $q^2 w_{rr}$ (Fig. 20.5). These frequencies do not sum to 1 because they are relative frequencies for the three genotypes. To make the frequencies sum to 1, even with the inclusion of modifying terms that change the value of each genotypic frequency, it is necessary to "recalibrate" the Hardy-Weinberg equation.

This recalibration is accomplished by calculating the value of a numerical factor that each Hardy-Weinberg term can be divided into so that the terms all add up to 1. The calculation consists simply in setting the sum of the terms in the modified equation to a new variable, designated $\overline{w}$.

$$p^2 w_{RR} + 2pq w_{Rr} + q^2 w_{rr} = \overline{w} \qquad (20.4a)$$

Since $\overline{w}$ represents the sum of the individual fitness values, each multiplied by their relative occurrence in the population, it in fact represents the average fitness of the population.

In populations that satisfy the conditions of the original Hardy-Weinberg equilibrium, when the fitness for each genotype is 1, the value of $\overline{w}$ is also 1. However, when fitness varies from one genotype to another, $\overline{w}$ can also vary but in a way that can be calculated as long as the value of each of the variables in the equation is known. The modified Hardy-Weinberg equation is thus recalibrated by dividing each term by $\overline{w}$ such that the new equation becomes

$$\frac{p^2 w_{RR}}{\overline{w}} + \frac{2pq w_{Rr}}{\overline{w}} + \frac{q^2 w_{rr}}{\overline{w}} = 1 \qquad (20.4b)$$

Each term in this recalibrated equation represents the actual frequency that each genotype will assume in the generation following the one used for the original calculation. Table 20.1 summarizes this process of calculation.

As an example, let us use the variables p' and q' to represent the frequencies of the R and r alleles in this next generation. Among the gametes produced by the original population, the frequency of allele r will be the result of contributions of r alleles from both Rr and rr adults relative to the number of individuals in the entire adult population. If q' represents the frequency of the r allele in the next generation adults, then

$$q' = \frac{q^2 w_{rr} + \frac{1}{2}(2pq w_{Rr})}{\overline{w}} = \frac{q(q w_{rr} + p w_{Rr})}{\overline{w}} \qquad (20.5)$$

Keep in mind:

1) Δq is defined as $q' - q$
2) A useful identity in moving from equation 20.5 to equation 20.6 is $1 - 2q = 1 - q - q = (p - q)$

Thus, in one generation of selection, the allele frequency of r has changed from q to q'. It is often useful to know the change in allele frequency over one generation of selection. We can represent this change as

$$\Delta q = \frac{pq[q(w_{rr} - w_{Rr}) + p(w_{Rr} - w_{RR})]}{\overline{w}} \qquad (20.6)$$

The derivation of equation 20.6 from 20.5 is given in the solutions manual for this book. As these equations show, selection can cause the frequency of an allele to change from one generation to the next. Equation (20.6) shows that the change in allele frequency resulting from one generation of selection depends on both the allele frequencies and the relative fitnesses (in this case, the viability component of fitness) of the three genotypes. Note that if the fitnesses of all genotypes are the same, as in populations at Hardy-Weinberg equilibrium, then the change in q (Δq) = 0; in other words, if there are no genotype-related differences in fitness, there is no possibility of selection and if there is no possibility of selection there can be no evolution (defined as changes in allele frequency).

As an example, let us use Eq. (20.6) to look at how a recessive genetic condition, such as thalassemia, influences the allele frequencies of a population. If the disease, which results from an rr genotype for the R gene, decreases fitness by decreasing the probability of surviving to adulthood, then the fitness of RR and Rr individuals is the same, while the fitness of rr individuals is reduced. Because only the relative values of the fitnesses are important, it is useful to set the values of $w_{RR} = 1$, $w_{Rr} = 1$, and $w_{rr} = 1 - s$, where s is the *selection coefficient* against the rr genotype. This selection coefficient can vary from 0 (no selection against rr) to 1 (rr is lethal and no rr individuals survive to adulthood). For this example, we can rewrite Eq. (20.6) as

$$\Delta q = \frac{pq[q(1 - s - 1) + q(1 - 1)]}{\overline{w}} = \frac{-spq^2}{\overline{w}} \qquad (20.7)$$

Equation (20.7) has three interesting features. First, unless there is no selection and $s = 0$, Δq is always negative and the frequency of the r allele decreases with time.

Second, the rate at which q decreases over time depends on the allele frequencies; in particular, because Δq varies with q^2,

TABLE 20.1 **Calculating Natural Selection's Effect on Genotype Frequency**

Genotypes	RR	Rr	rr	
Initial Genotypic Frequency	p^2	$2pq$	q^2	
Fitness	w_{RR}	w_{Rr}	w_{rr}	
Frequency After Selection	$p^2 w_{RR}$	$2pq w_{Rr}$	$q^2 w_{rr}$	$\overline{w} = p^2 w_{RR} + 2pq w_{Rr} + q^2 w_{rr}$
Relative Genotypic Frequency After Selection	$\dfrac{p^2 w_{RR}}{\overline{w}}$	$\dfrac{2pq w_{Rr}}{\overline{w}}$	$\dfrac{q^2 w_{rr}}{\overline{w}}$	

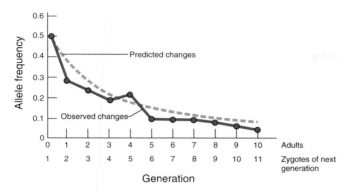

Figure 20.6 Decrease in the frequency of a recessive allele over time. The *dotted line* represents the mathematical prediction. The *blue line* represents the actual data obtained with an autosomal recessive allele.

the rate at which q decreases diminishes as q becomes smaller. (Recall that because q is always less than 1, $q^2 < q$.) To understand the effect of this correlation between the allele frequency and the rate at which q decreases over time, consider the special case of a lethal recessive disease for which $s = 1$. The dotted line in Fig. 20.6 shows the decrease in allele frequency predicted by Eq. (20.7), starting from an initial allele frequency of 0.5. The decrease in allele frequency is rapid at first, and then slows. After 10 generations, the predicted frequency of the recessive disease allele is nearly 10%, even though the homozygous recessive genotype is lethal. The solid line in Fig. 20.6 plots actual data for the decrease in frequency of an autosomal lethal allele in an experimental population of *Drosophila melanogaster.* In Fig. 20.6, the predicted and observed changes in allele frequency match quite closely.

Why is selection unable to reduce the frequency of a recessive lethal allele to zero? The answer goes back to our consideration of the frequency of heterozygous carriers of a recessive disease allele. When q is small, individuals homozygous for the disease allele are very rare because most copies of the r allele occur in $R\,r$ heterozygotes who do not experience negative selection. By contrast, a lethal dominant allele will disappear from a population in a single generation of selection.

The third feature of Eq. (20.7) is that it predicts that the allele frequency q will continue to decline, albeit more and more slowly over time as q moves closer and closer to a value of zero. However, as we noted earlier, the reality is that many recessive genetic diseases persist in human populations at low but stable frequencies. What maintains these diseases despite continuing selection against them? One answer is that sometimes heterozygotes have a higher fitness than either homozygote, a situation referred to as **heterozygous advantage.**

It Is Possible to Calculate the Effect of Heterozygous Advantage on Allele Frequencies

We have seen that sickle-cell anemia, which includes episodes of severe pain, serious anemia, and a probability of early death, is a recessive condition resulting from two copies of the sickle-cell allele at the β-globin locus. The disease allele has not disappeared from several African populations, where it seems to have existed for a very long time. One clue to the maintenance of the sickle-cell allele in human populations lies in the observation that heterozygotes for the normal and sickle-cell alleles are resistant to malaria. This resistance is due, in part, to the fact that red blood cells infected by the malaria parasite, if they also contain a sickle-cell allele, break open, destroying the parasite as well as the red blood cell itself; by contrast, in cells with two normal hemoglobin alleles, the malaria parasite thrives. To set up a model of heterozygous advantage, let B_1 represent the normal β-globin allele and B_2 stand for the abnormal recessive sickle-cell allele; and for simplicity, assume that $B_1 \, B_2$ heterozygotes have the maximum relative fitness of 1, while the selection coefficient (representing the selective disadvantage) for $B_1 \, B_1$ homozygotes is $1 - s_1$ and the selection coefficient for $B_2 \, B_2$ homozygotes is $1 - s_2$. We can then represent the changes in allele frequency resulting from selection as

$$\Delta q = \frac{pq(s_1 p - s_2 q)}{\overline{w}} \qquad (20.8)$$

To maintain both alleles in the population using this equation, Δq must be 0 for some value of q between 0 and 1. The q value at which $\Delta q = 0$ is known as the **equilibrium frequency** of allele B_2. The value of q when $\Delta q = 0$ and both alleles are present occurs when the term inside the parentheses of Eq. (20.8) is 0, that is, when

$$s_1 p - s_2 q = 0$$

Substituting $1 - q$ for p ($p = 1 - q$) and solving this equation for q reveals that the equilibrium frequency of B_2 (represented by q_e) is reached when

$$q_e = \frac{s_1}{s_1 + s_2} \qquad (20.9)$$

Note that to find the equilibrium frequency, that is, the value of q at which $\Delta q = 0$ such that both alleles B_1 and B_2 persist in the population, you need know only the selection coefficients for the two homozygotes.

To understand the relationship between q, the change in q, and the equilibrium frequency q_e, you can formulate Δq using q_e.

$$\Delta q = \frac{-\,pq(s_1 + s_2)(q - q_e)}{\overline{w}} \qquad (20.10)$$

From this formulation, you can see that when q is greater than q_e, Δq is negative. Under these circumstances, q, or the frequency of allele B_2, will decrease toward the equilibrium frequency. By contrast, when q is less than q_e, Δq is positive and the frequency of B_2 will increase toward the equilibrium. Thus, the equilibrium frequency stabilizes, because a change away from it is always followed by a change toward it.

Now, if you assume that the African populations in which sickle-cell anemia is prevalent are currently at equilibrium relative to their alleles at the β-globin locus, you can use the observed frequency of the sickle-cell allele in these populations to calculate the relative values of the selection coefficients. Field studies show that the actual value of q_e lies between 0.15 and 0.2 for an average value of 0.17. If you plug this number into Eq. (20.9), you get

$$0.17 = \frac{s_1}{(s_1 + s_2)}$$

This equation makes it possible to express either selection coefficient in terms of the other. For example, $s_1 = 0.2s_2$.

If you assume that $s_2 = 1$ (that is, those with sickle-cell trait never reproduce), as was most probably true before medical advances enabled the survival of children expressing the sickle-cell trait, you will find that $s_1 = 0.2$, which in turn means that the rel-

ative fitness of the wild-type genotype is 0.8. Recall, however, that we set the fitness of the heterozygote at 1.0. By dividing 1.0 by 0.8, you get 1.25, which represents the relative advantage in fitness that heterozygotes for the sickle-cell allele have over people who do not carry this allele in African populations exposed to malaria. The use of simple statistical methods to calculate this heterozygous advantage demonstrates how medical geneticists can use the tools of population genetics (Fig. 20.7).

Heterozygous advantage is one phenomenon that maintains recessive disease alleles in human populations. We now look at other phenomena that can counteract the effects of selection and maintain recessive alleles within a population.

Figure 20.7   Frequency of the sickle-cell allele across Africa where malaria is prevalent.

Evolutionary Equilibrium: A Balance Between Mutation to a New Allele and Selection Against That Allele

Mutations, we have seen, are the ultimate source of genetic variation in a population. They arise from DNA damage due to environmental agents and from errors during DNA replication and the transmission of genetic information during cell division. Various studies show that in most animals, the mutation rate is roughly 10^{-4} to 10^{-6} mutations per gene per generation. Since we know that the human genome carries about 40,000–60,000 genes, if we take the average of 50,000, we can infer that 0.05–5 new mutations arise somewhere in the genes of each human gamete per generation.

Many mutations that alter the amino-acid sequence of a gene product have deleterious consequences on survival and reproduction, that is, on fitness. Indeed, many genetic diseases result from mutations that give rise to a nonfunctional gene product. Individuals homozygous for such disease alleles produce no functional copies of an enzyme or other polypeptide required for normal development or metabolism.

How Selection Against a Disease Interacts with Mutation to the Disease Allele to Influence the Frequency of the Disease Allele in a Population

To address this question, let us again consider the example of a recessive genetic disease that occurs at low frequency in a population. Suppose that mutations from allele R (the dominant, normal allele) to allele r (the recessive, disease allele) occur at a rate of μ per generation. As a first approximation, we can ignore the very low level of reverse mutations from r to R because the frequency of r is so small. With this background, we can calculate Δq using Eq. (20.7) as the basis, but accounting for the fact that a fraction μ of the R alleles has mutated to r in each generation. In this equation, we multiply μ by $(1 - q)$ because only R alleles can mutate to r, and the frequency of R alleles is $(1 - q)$.

$$\Delta q = \frac{- spq^2 + \mu(1 - q)}{\overline{w}} \qquad (20.11)$$

Note that the effects of selection (s) and mutation (μ) on allele frequency are in opposite directions. Allele frequencies will thus evolve until $\Delta q = 0$. The allele frequency, labeled $\hat{q}$, at which evolution no longer continues, is called the **evolutionary equilibrium.** By rearranging Eq. (20.11), you can see that this equilibrium occurs when

$$\hat{q}/\sqrt{(1 - \hat{q})} = \sqrt{(\mu/sp)} \qquad (20.12)$$

In this equation, $\hat{q}$ represents the evolutionary equilibrium.

This simple equation shows how a balance between selection and mutation can maintain recessive genetic diseases in a population. Note that the equilibrium frequency of the disease allele can be quite high. For example, consider a lethal recessive disease (for which $s = 1$) in which all affected individuals die before reproducing and the mutation rate $= 1 \times 10^{-6}$. Substitution of these values into Eq. (20.12) reveals that the equilibrium frequency of the disease allele is 1×10^{-3}, or 1 in 1000.

1. Let $\mu = 10^{-6}$ and $s = 1$.

2. Plug into Eq. (20.12).

3. Assume that $\hat{q}$ is small, which enables you to approximate $p = 1$ and $1 - \hat{q} = 1$.

4. Following these steps, you can conclude that $\hat{q} = 10^{-3}$.

Since the frequency of heterozygotes among zygotes is $2pq$, as predicted by the Hardy-Weinberg law, the frequency of heterozygous carriers of this deadly recessive condition will be approximately 2 per 1000.

For the many genetic diseases that have less severe effects on fitness and therefore sustain less negative selection, the allele frequency at evolutionary equilibrium will be even higher. For example, an inherited recessive disease that decreases the likelihood of surviving to reproductive age by only 10% ($s = 0.1$) would result in an equilibrium frequency of heterozygous carriers of about 63 per 1000 individuals.

In short, this simple mathematical model suggests that, like heterozygous advantage (that is, selection for the heterozygotes), mutation is quite effective in maintaining recessive alleles in a population. As a result, at evolutionary equilibrium, substantial numbers of deleterious recessive alleles can remain in the heterozygous carriers of a population. However, as we see later, the mathematical model does not accurately predict what actually happens in human and other populations, because a critical assumption—namely, infinite size—does not reflect reality. Thus, the human population, with its millions of generations of evolution, is still nowhere near equilibrium.

The Time of Onset of a Disease Can Also Influence the Frequency of Disease Alleles

The effects of many inherited human diseases become manifest only during middle age or later in life; the effects of others may appear earlier than middle age but not cause death until much later. From an evolutionary viewpoint, the important issue is how a disease reduces survival to reproductive age, for it is its impact on reproduction that determines a disease's potential effect on the strength of selection. Diseases that cause death after the completion of reproduction will sustain little or no negative selection; and with lower selection against them, their equilibrium frequency will be higher. One significant consequence of the relationship between the time of onset of disease and the strength of selection is that genetic diseases whose phenotypic effects appear late in life will experience little selection and may increase in frequency simply as a result of mutation. Aging—the increased rate of mortality with age that occurs after the completion of reproduction—may be the natural consequence of mutations whose effects appear late in life and accumulate in a population unopposed by natural selection.

Genetic Drift Has an Unpredictable Effect on the Evolutionary Equilibrium

Equation (20.12) models the balance between selection and mutation in a Hardy-Weinberg world where populations are very, very large—virtually infinite. But in addition to clarifying how mutations help maintain recessive diseases in a population, Eq. (20.12) leads to the untenable conclusion that in the absence of selection ($s = 0$), $\hat{q}$ will become infinite; that is, new alleles with no effect on fitness will invariably spread through the entire population. This is not what actually happens. In nature, where no population is infinite in size, new alleles with a neutral effect on fitness arise continuously, but most of these alleles drift to extinction in the course of several generations. Loss of an allele occurs when the individuals carrying that allele, by chance, fail to reproduce. In contrast, very rarely, a new recessive allele may by chance increase to high frequency, but even when it does, it will be replaced later by another new allele.

Population geneticists use the expression **genetic drift** to refer to unpredictable, chance fluctuations in allele frequency that have a neutral effect on fitness (Fig. 20.8). Genetic drift occurs in populations of any size, but the smaller the population, the more clearly visible are the effects of drift. To understand why this is so, consider the perpetuation of allele r, which occurs in 10% of the individuals in a population of sunflowers. If a chance occurrence such as a severe summer hailstorm strikes a population of 1,000,000 plants and half die, it is likely that among the 500,000 survivors, roughly 50,000 (10%) will still carry the r allele. If the same storm hits a field of just 10 plants, only 1 of which carries the r allele, there is a 50% chance that the single plant bearing the r allele will not be among the survivors. If it isn't, the total loss of this one allele from the population will reduce the allele frequency in an unpredictable way to 0. If, however, the lone individual carrying r does by chance survive, the frequency of allele r will increase unpredictably to 20%.

One example of genetic drift in human populations is the **founder effect,** which occurs when a few individuals separate from a larger population and establish a new one that is isolated from the original population. The small number of founders in the new population carry only a fraction of the alleles from the original population, and this discrepancy may change the allele frequencies. The 14,000 people living in the Amish community in eastern Pennsylvania are the descendants of about 200 individuals who emigrated from Germany in the early eighteenth century. Among these Amish, there is a much higher incidence

of manic-depressive illness than among the larger original European population, most likely because several founders carried alleles producing this disease.

Summary of the Balance Between Mutation and Selection

New alleles (some of which are disease alleles) arise in a population by mutation. When a new allele has an effect on fitness that differs from the effect of the wild-type allele, selection will drive the frequency of the new allele toward an equilibrium with the wild-type allele. The equilibrium value will be determined by the relative selection coefficients of individuals heterozygous and homozygous for the new allele compared to individuals who are wild type. In contrast, when an allele has a neutral effect on fitness, its frequency will depend on unpredictable genetic drift. Most neutral alleles become extinct, but occasionally one "drifts up" to become the new wild-type allele in the population.

Comprehensive Example: How Human Activity Affects the Evolution of Human Pathogens and Crop Pests

Infectious diseases have been a major killer throughout human history, and as the AIDS epidemic illustrates, previously unknown diseases continue to emerge. In the twentieth century, discovery of a variety of vaccines, antibiotics, and other drugs made it possible to combat infectious diseases such as smallpox (vaccines) and tuberculosis (antibiotics) with great success. In the last 25 years, however, many formerly surefire drugs have lost their effectiveness because populations of pathogens have evolved resistance to them. Similarly, populations of agricultural pests have evolved resistance to the pesticides used to control or eradicate them.

At the beginning of this chapter we posed three questions: How do new diseases emerge in human populations? Why do diseases persist in all living organisms? What causes diseases and pests long under control to resurge in frequency and intensity? We have answered the first two. New diseases emerge in human populations as a consequence of new mutations. Diseases persist because changes in allele frequency tend toward an evolutionary equilibrium in which mutation balances selection. To answer the third question, we turn to an examination of how pathogens and pests interact with their hosts.

The Evolution of Drug Resistance in Pathogens

We have seen that many of the bacterial agents of tuberculosis are resistant to several antibiotics. We now know that a major factor contributing to the evolution of multidrug-resistant TB strains is the failure of patients to complete the lengthy drug regimens required for a cure. The two most widely used drugs for TB, isoniazid and rifampicin, require ingestion for six months and have side effects that include nausea and loss of appetite. However, the symptoms of TB can begin to disappear after only two to four weeks of treatment.

Imagine a TB patient with a persistent cough, shortness of breath, and general weakness. This individual harbors a large,

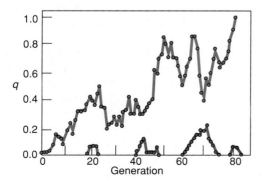

Figure 20.8   The effect of genetic drift. New alleles with no effect on fitness are usually lost rapidly from a population (*red and green lines*). Rarely, a new allele will increase in frequency (*blue line*).

actively growing and dividing population of TB bacteria in his lungs. At first, these bacteria are susceptible to antibiotics, but occasional mutations conferring partial resistance appear at random (Fig. 20.9). The patient's physician prescribes a six-month course of treatment with the antibiotic isoniazid. After a few weeks of treatment, the bacterial population in the patient's lungs has decreased considerably, and the patient's symptoms have abated, although the negative side effects of the drug continue to cause discomfort. However, the composition of the bacterial population has now changed so that the remaining bacteria are likely to include a high proportion of mutant bacteria possessing partial resistance to the antibiotic. If the patient continues his course of treatment, the persistent dose of antibiotic will eventually kill all of the bacteria, even those with partial resistance, eliminating the infection. By contrast, if the patient stops

treatment prematurely, the remaining (partially resistant) bacteria will proliferate and within three to four weeks reestablish a large population. Subsequent treatment of the same individual upon relapse or of a new patient to whom the partially resistant bacteria have spread would permit a second cycle of selection. New mutations could then convert the partially resistant bacteria to fully resistant microbes.

It is easy to see how repeated cycles of antibiotic treatment with multiple drugs, coupled with premature cessation of treatment, can promote the evolution of fully resistant bacterial populations and even bacterial strains resistant to more than one drug within a single patient. Initially individual mutant cells that express genes conferring partial resistance increase in frequency; subsequent mutations in some of the mutant bacteria will increase resistance, and incomplete drug dosages will select for the resistant strains. Surveys of resistance illustrate the effects of this process on the frequency of resistance. A 1982–1986 U.S. Centers for Disease Control (CDC) survey showed that 9% of TB strains isolated from patients who had not previously received treatment were resistant to antibiotics, compared with 25% of strains isolated from previously treated, relapsed patients. More recently, a 1991 survey in New York City showed that 23% of strains isolated from previously untreated patients were resistant to one or more antibiotic drugs and 7% were resistant to both isoniazid and rifampicin; moreover, 44% of isolates from relapsed patients were resistant to one or more drugs and 30% were resistant to both isoniazid and rifampicin. These data support the idea that patient noncompliance with drug treatments is a major factor in the evolution of antibiotic resistance.

Several factors contribute to the rapid evolution of resistance in bacterial pathogens. The short generation times—often only a few hours—and rapid rate of reproduction under optimal conditions allow evolution to proceed quickly relative to a human life span. The large population densities typical of bacteria, which may exceed $10^9/cm^3$, ensure that rare resistance-conferring mutations will appear in the population. The strong selection imposed by antibiotics increases the rate of evolution in each generation, unless the bacterial population is entirely eliminated.

The large variety of ways by which bacteria can acquire genes also contributes to the rapid evolution of resistance. Many genes for resistance are found on plasmids, and, as we saw in Chapters 9 and 14, the capacity of plasmids to replicate and be transmitted among bacteria allows the amplified expression of resistance genes in bacterial populations. Plasmids also provide a means for the genetic exchange of resistance genes among bacterial populations and species through transformation, conjugation, and transduction. Laboratory studies have demonstrated the ready transfer of plasmid-borne resistance genes to new bacterial species.

The Evolution of Pesticide Resistance

Like infectious bacteria, many agricultural pests spawn large populations because of their short generation times and rapid rates of reproduction. These large, rapidly reproducing populations

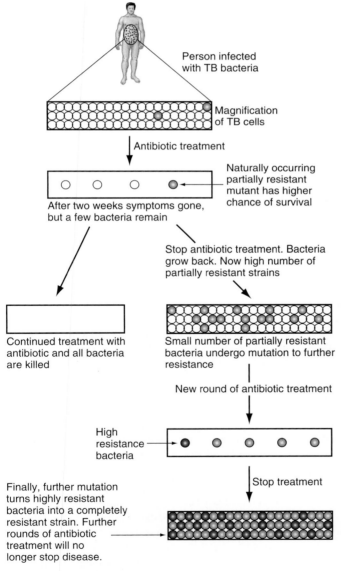

Figure 20.9 The evolution of resistance in TB bacteria.

Person infected with TB bacteria

Magnification of TB cells

Antibiotic treatment

Naturally occurring partially resistant mutant has higher chance of survival

After two weeks symptoms gone, but a few bacteria remain

Stop antibiotic treatment. Bacteria grow back. Now high number of partially resistant strains

Continued treatment with antibiotic and all bacteria are killed

Small number of partially resistant bacteria undergo mutation to further resistance

New round of antibiotic treatment

High resistance bacteria

Finally, further mutation turns highly resistant bacteria into a completely resistant strain. Further rounds of antibiotic treatment will no longer stop disease.

Stop treatment

evolve resistance to the chemical pesticides used to control them via selection for resistance-conferring mutations. Our understanding of this familiar pattern of selection and rapid evolution is most complete for certain insecticides, that is, pesticides that target insects.

The large-scale, commercial use of DDT and other synthetic organic insecticides, begun in the 1940s, was initially highly successful at reducing crop destruction by agricultural pests, such as the boll weevil, and medical pests, such as the mosquitoes that transmit malaria and yellow fever. Within a few years, however, resistance to these insecticides was detectable in the targeted insect populations. Since the 1950s, resistance to every known insecticide has evolved within 10 years of its commercial introduction. By 1984 there were reports of more than 450 resistant species of insects and mites (Fig. 20.10). Because different populations within a species can become resistant independently of other populations, the number of times insecticide resistance has evolved probably exceeds 1000.

Genetic studies show that insecticide resistance often results from changes in a single gene, and that several significant mechanisms of resistance are similar to those seen in infectious bacteria. DDT, for example, is a nerve toxin in insects. House flies and some mosquitoes develop resistance to DDT from dominant mutations in a single, enzyme-encoding allele. The mutant enzyme detoxifies DDT, rendering it harmless to the insect. As we saw earlier, even at low frequencies, dominant alleles can experience strong selection, because of heterozygous advantage. Consider, for example, dominant mutation R (for insecticide resistance), which occurs initially at low frequency in a population. Soon after the mutation appears, most of the R alleles are in $S R$ heterozygotes (in which S is the wild-type susceptibility allele). With the application of insecticide, strong selection favoring $S R$ heterozygotes rapidly increases the frequency of the resistance allele in the population. A field study of the use of DDT in Bangkok, Thailand, to control *Aedes aegypti* mosquitoes, the carriers of yellow fever, illustrates the rapid evolution of resistance. Spraying of the insecticide began in 1964 (Fig. 20.11). Within a year, DDT-resistant genotypes emerged and rapidly increased in frequency. By mid-1967, the frequency of resistant $R R$ homozygotes was nearly 100%.

The Biological Costs of Resistance

Since DDT no longer controlled mosquito populations in the region, the insecticide program was stopped. The response of the mosquito population to the cessation of spraying was intriguing: The frequency of the R allele decreased rapidly, and by 1969, $R R$ genotypes had virtually disappeared. The precipitous decline of the R allele suggests that in the absence of DDT, the $R R$ genotype produces a lower fitness than the $S S$ genotype. In other words, the homozygous resistance genotype imposes a **fitness cost** on individuals such that in the absence of insecticide, resistance is subject to a negative selection that decreases the frequency of R in the population.

To understand the biological basis of fitness costs, consider how rats evolve resistance to warfarin, a pesticide introduced in the 1940s and 1950s to control small mammals, among other

(a)

(b)

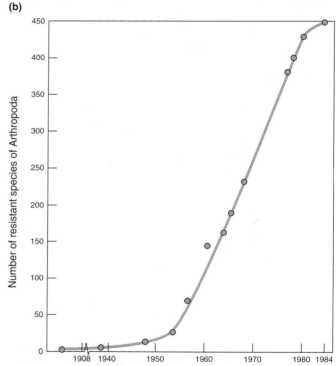

Figure 20.10   Increase of insecticide resistance from 1908–1984. (a) Insecticide resistance evolved with the aerial spraying of DDT, which began in the 1940s. Within 10 years, resistance to an insecticide becomes widespread as the insects evolve defenses to the insecticides just as they would evolve defenses against infectious bacteria. **(b)** The evolution of resistance among Arthropoda.

pests. Warfarin interferes with blood clotting by blocking the recycling of vitamin K (a cofactor in the clotting cascade). When a rat ingests warfarin, the inability to form a clot leads to a fatal loss of blood following any internal or external injury. In Europe in the 1960s, the extensive use of warfarin for rat control fueled the evolution of a single-gene resistance allele in many rat populations. The frequency of resistance, however, did not increase to 100%; instead, in most populations, it leveled off at 30% to 60%. Apparently, some mechanism was maintaining both the R (resistance) and the S (susceptibility) alleles in the presence of warfarin.

(a)

(b)

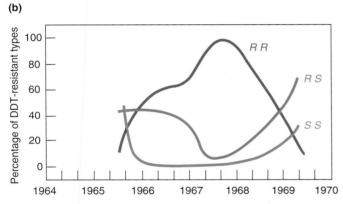

Figure 20.11 How genotype frequencies among populations of *A. aegypti* mosquito larvae change in response to insecticide. (a) Mosquitoes and larvae. **(b)** Changing proportions of resistance genotypes of *A. aegypti* (larvae) under selection with DDT (1964–1967), and after selection was relaxed (1968), in a suburb of Bangkok, Thailand.

Further study showed that in the presence of warfarin, the relative fitnesses were 0.37 for the *S S* genotype, 1.0 for the *S R* genotype, and 0.68 for the *R R* genotype. These figures suggest that heterozygous advantage, the higher fitness of heterozygotes, was maintaining both alleles in the populations. Further investigation revealed that two factors caused the observed differences in fitness. Both the *S R* and the *R R* genotypes were relatively resistant to the effects of warfarin, and thus they provided higher fitnesses than the susceptible *S S* genotypes. However, *R R* homozygotes suffered from a vitamin K deficiency because of the less efficient vitamin K recycling during blood clotting, and this deficiency reduced the rate of survival when the diet did not contain a large amount of vitamin K. In the absence of warfarin, therefore, *R R* homozygotes had a lower fitness than *S S* homozygotes because of a nutritional deficiency. The biological costs of fitness, which are widespread and occur by various mechanisms, are very likely a major reason why resistance alleles occurred in very low or undetectable frequencies before the routine use of pesticides.

Ecological Considerations

Selection for resistance in the presence of pesticides is not the only factor influencing the rate of pesticide resistance evolution among pest populations. The ecological context also plays a significant role. One illustration of this phenomenon comes from studies of insect predators and parasites, and insect pests. Most predators and parasites on agricultural insect pests are also insects. It would be logical to expect that applications of insecticides to agricultural fields would expose pests and predators alike, fostering the evolution of resistance in both classes of insects. Observations, however, do not support this logic. Of the more than 450 species of insects, mites, and ticks known to be resistant to insecticides, less than 10% are predators or parasites. What causes this pattern of resistance and what are its consequences?

Simple ecological and evolutionary models provide an explanation. Consider a pest species and a predator species that coexist in an agricultural field. In the absence of insecticide, the population densities of pest and predator change over time, but both persist (Fig. 20.12a). Suppose that in each species, a single

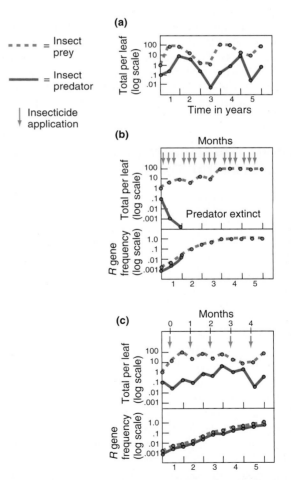

Figure 20.12 Changes in population density (insects per leaf) of prey and pest over time, with and without insecticide spraying. Fluctuations **(a)** in the absence of insecticide, **(b)** with biweekly applications of insecticide, and **(c)** with monthly applications of insecticide.

allele *R* at one locus determines resistance to a particular insecticide, and that initially the frequency of the *S* susceptibility allele is nearly 100% in both populations. The grower then applies heavy doses of insecticide to the field every two weeks (Fig. 20.12b). Because most individuals in both the pest and the predator populations are *S S*, high mortality from the insecticide decreases the numbers of pests and predators. For the few surviving pests, conditions are now ideal: Plant food is abundant, while competitors as well as predators are few. These pest survivors grow and reproduce at a rapid rate. By contrast, the predators remaining after exposure to pesticide at first face low densities of prey, which retards their growth and reproduction. This initial impairment may be so great that the predator population dies out within a few generations, before it has had a chance to evolve pesticide resistance. Now, the pest population, which has no predators to contend with, goes on to evolve a high frequency of resistance as a result of strong selection promoted by the insecticide. In these circumstances, the application of large doses of insecticide yields a worst-case scenario of a resistant pest population and a suppressed or extinct predator population. In sum, the different ecological consequences of insecticide application for prey and predators have created differences in the rate and likelihood that each population will evolve resistance. As Fig. 20.12c shows, changes in the intensity and timing of insecticide application may diminish these differences by allowing predators more time to recover before the next exposure. These models suggest that both genetics and ecology have an impact on the use of predators to control pests.

Analyzing the Quantitative Variation of Multifactorial Traits

Unlike cystic fibrosis, albinism, PKU, and thalassemia, the majority of inherited human disorders are influenced by more than one gene. For example, at least 12 separate regions of the human genome are associated with the development of insulin-dependent diabetes. Indeed, in addition to the many inherited abnormalities, most aspects of normal size, shape, physiology, and behavior in humans and other organisms are complex traits determined by multiple genes that interact with each other as well as with the environment.

Complex traits of this type are called **multifactorial** because more than one factor, either genetic or environmental, contributes to the phenotype. We saw in Chapter 3 that many multifactorial traits are continuous, or **quantitative,** that is, they vary over a continuous range of measurement—such as the length of a tobacco flower in millimeters, the amount of milk produced by a cow per day in liters, or the height of a person in meters. Figure 3.21 on p. 66 shows how quantitative variation can arise when the alleles of two or more genes affect the same trait. From a review of the Chapter 3 figure, you can see that two alleles—one dominant and one recessive—at one locus produce two discrete (discontinuous) phenotypes (*A A* and *A a* versus *a a*); two alleles at two genes generate five discrete phenotypes; two alleles at six genes can produce thirteen phenotypes that appear to vary over a continuous range. As the number of phenotypic categories for a trait increases, the trait's variation appears more and more continuous.

Quantitative traits controlled solely by the alleles of two or more genes are **polygenic traits** that are, by definition, multifactorial. But not all multifactorial traits are polygenic (some may be determined by interactions of the alleles of one gene with each other and with the environment). Moreover, not all quantitative traits are polygenic (continuous variation may result from environmental effects on the expression of one gene), and not all polygenic traits are quantitative (some traits determined by the interaction of several genes may have discrete, discontinuous phenotypes).

With these subtle distinctions in mind, we now examine how population geneticists study quantitative traits. The continuous variation of such traits depends on the number of genes that generate the trait as well as the genetic and environmental factors that affect the penetrance and expressivity of these genes (see Chapters 3 and 10). One of the goals of quantitative analysis is to discover how much of the variation in a particular trait is the result of genotypic differences among individuals in a population and how much arises from differences in the environment.

Genes Versus the Environment

To sort out the genetic and environmental determinants of phenotypic variation in a population, consider a series of experiments on a population of dandelions, a common weedy plant in lawns and other disturbed areas throughout North America (Fig. 20.13a). Dandelions have a long tubular stem and a large, yellow composite flowering structure composed of many small individual flowers; each of these flowers can produce a single, tufted, diploid seed, dispersible by the wind. Most dandelion seeds arise from mitotic, rather than meiotic, divisions such that all the seeds from a single plant are genetically identical. Your goal is to compare the influence of genes and environment on the length of the stem at flowering.

To distinguish environmental from genetic effects on phenotypic variation, you need to quantify one variable, say the environment, while controlling for the other one; that is, while holding the genetic contribution steady. You could begin by planting half of the genetically identical seeds from a single plant on a grassy hillside and allowing them to grow undisturbed until they flower. You then measure the length of the stem of each flowering plant and determine the mean and variance of the distribution of values for this trait in this dandelion population. As Fig. 20.13b shows, you find the **mean** by summing the values of all stem lengths and dividing by the number of stems. You then find the **variance** by expressing the stem lengths as plus or minus deviations from the mean, squaring those deviations, and again dividing by the number of stems. Because all members of this population are genetically identical, any observed variation in stem length among individuals should be mainly a consequence of environmental variations, such as different amounts of water and sunlight at different locations on the hillside. When represented as a variance from the mean, these

(a)

(b)
Genetically identical plants grown on hillside

Mean ($\bar{x}$)

Number of plants

$\sqrt{V_E}$

Stem length (x)

Finding the mean:

Let x_i = the stem length of the plant i in a population of N plants. The mean of stem length, $\bar{x}$, for the population is defined as:

$$\bar{x} = \frac{\sum\limits_{i=1}^{N} x_i}{N}$$

Finding the variance:

The variance V_E of stem length for the population is defined as:

$$V_E = \frac{\sum\limits_{i=1}^{N} (x_i - \bar{x})^2}{N}$$

(c)
Genetically identical seeds

Mean ($\bar{x}$)
Grown in greenhouse
Grown on hillside

$\sqrt{V_E}$

$\sqrt{V_E}$

Number of plants

Stem length

V_E is smaller among greenhouse-grown plants.

(d)
Plants grown in greenhouse

Mean ($\bar{x}$)

Plants grown from genetically identical seeds

Plants grown from genetically diverse seeds

$\sqrt{V_E}$

$\sqrt{V_G}$

Number of plants

Stem length

V_G due to genetic variance

(e)
Plants grown on hillside

Mean ($\bar{x}$)

Genetically identical plants

Genetically diverse plants

$\sqrt{V_E}$

$\sqrt{V_G}$

Number of plants

Stem length

Phenotypic variance ($V_E + V_G$)

Figure 20.13 Studies of dandelions can help sort out the effects of genes versus the environment. (a) The familiar dandelion (*Taxaracum* sp.) is a useful model when studying population variations. **(b)** Finding the mean and variance of stem length. **(c)** Variance of genetically identical plants grown in a greenhouse and grown on a hillside. **(d)** Genetic variance of plants grown in a greenhouse. **(e)** Phenotype variance of plants grown on a hillside.

observed environmentally determined differences in stem length are called the **environmental variance,** or V_E.

To refine your estimate of environmental variance, you plant the second half of the genetically identical seeds from the single test plant in a controlled greenhouse in which growth conditions are everywhere the same (Fig. 20.13c). Because en-

vironmental conditions are very similar for all these genetically identical plants, the amount of environmental variance (V_E) among greenhouse plants is much smaller than among hillside plants. In theory, in a perfectly controlled greenhouse, growth conditions would be the same for all plants, the V_E would be zero, and all plants would have identical stem lengths (within measurement error). In reality, there is no such thing as perfect control, and the greenhouse V_E will have some value greater than zero. Nonetheless, the difference between the V_E of the dandelions grown on the hillside and the V_E of dandelions grown in the greenhouse is a measure of the impact of the more diversified hillside environment on the phenotypic variation of stem length.

Even though the greenhouse V_E will have some value greater than zero, for the sake of simplicity in the following discussion, we assume that it is small.

To examine the impact of genetic differences on stem length, you take seeds from many different dandelion plants growing in many different locations, and you plant them in a controlled greenhouse (Fig. 20.13d). Because you are raising genetically diverse plants in a relatively uniform environment, observed variation in stem length (beyond that found in the genetically identical population) is the result of genetic differences promoting **genetic variance, or V_G**.

Now, to determine the total impact on phenotype of genes and environment, you take the seeds of many different plants from many different locations and grow them on the same hillside (Fig. 20.13e). For the population of dandelions that grow up from these seeds, the **total phenotype variance (V_P)** in stem length will be the sum of the genetic variance (V_G) and the environmental variance (V_E). The environmental variance is determined directly from the phenotypic variance found in the initial population of genetically identical plants grown on the hillside. It thus becomes possible for you to calculate the genetic variance in the second, mixed population as the difference between the phenotypic variance found in this genetically mixed population and the phenotypic variance found in the genetically identical population. For natural populations of dandelions, both genetic variation among individuals and variation in the environmental conditions experienced by each plant contribute to the total phenotypic variation.

Heritability Is the Proportion of Total Phenotype Variance Attributable to Genetic Variance

With the ability to determine the relative contributions to phenotypic variation of genes and environment, geneticists have developed a mathematical definition of the **heritability (h^2)** of a trait: it is the proportion of total phenotypic variance ascribable to genetic variance.

$$h^2 = \frac{V_G}{V_G + V_E} = \frac{V_G}{V_P} \qquad (20.13)$$

Because the amounts of genetic, environmental, and phenotypic variation may differ among traits, among populations, and among different environments, the heritability of a trait is always defined for a specific population and specific set of environmental conditions. If you know any two of the three variables of total phenotype variance, genetic variance, and environmental variance, you can find the remaining unknown variance.

How to Measure Heritability

In analyzing the contributions of genes and environment to dandelion stem length, you measured the phenotypic variation among genetically identical individuals in a range of specified environments and compared it to the phenotypic variation among all individuals in the population. Of course, most organisms are not as easy to clone as dandelions. The key to generalizing from the dandelion example, however, is to recognize that genetic clones are simply a special case of the broader notion of genetically related individuals, or genetic relatives. Various kinds of genetic relatives share certain alleles because they have one or more common ancestors. To quantify this idea, we can define the **genetic relatedness** of two individuals as the average fraction of common alleles at all gene loci that the individuals share because they inherited them from a common ancestor.

To determine the genetic relatedness between two siblings, for example, all you have to do is calculate the probability they received the same allele at any locus from the same parent. If you assume that one sibling received allele *A1* from an *A1 / A2* heterozygous parent, the probability that the second sibling received the same allele is 0.5. Because this simple calculation holds for every locus transmitted by both parents, the total genetic relatedness of two siblings is 0.5.

If genetic similarity contributes to phenotypic similarity for some trait, it is logical to expect that a pair of close genetic relatives will be more phenotypically similar than a pair of individuals chosen at random from the population at large. Thus, by comparing the phenotypic variation among genetic relatives with the phenotypic variation of the entire population over some range of environments, it is possible to estimate the heritability of a trait.

The finches observed by Darwin in the Galápagos Islands (often referred to as "Darwin's finches") provide an example of a population for which geneticists have measured the heritability of a trait under natural conditions in the field. Scientists studied the medium ground finch, *Geospiza fortis,* on the island of Daphne Major by banding many of the individual birds in the population (Fig. 20.14a). They then measured the depth of the bill for the mother, father, and offspring in each nest on the island and calculated how the bill depth of the offspring correlated with the average bill depth of the mother and father (called the *midparent value*). The results, depicted in Fig. 20.14b, show a clear correlation between parents and offspring; parents with deeper bills had offspring with deeper bills, while parents with smaller bill depth had offspring with smaller bill depth. In the figure, the heritability of bill depth, as represented by the slope of the line correlating midparent bill depth to offspring bill depth, is 0.82. This means that roughly 82% of the variation in bill depth in this population of Darwin's finches is attributable to genetic variation among individuals in the population; the other 18% results from variation in the environment. If the environment had no influence at all on the trait, the slope of the line representing the heritability of bill depth, that is, correlating bill depth in parents with bill depth in offspring, would be 1.0 (Fig. 20.14c).

(a)

(b) Correlation between parents and offspring

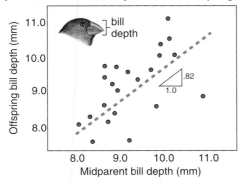

(c) If the heritability were 1.0

(d) If the heritability were 0.0

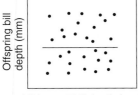

Figure 20.14 Measuring the heritability of bill depth in populations of Darwin's finches. (a) *G. fortis* with bands placed by scientists. **(b)** Results of the field study with a heritability of 0.82. **(c)** Results if heritability were 1.0. **(d)** Results if heritability were 0.0.

Now consider a population in which the bill depth for parents and their offspring is, on average, no more or less similar than the bill depths for any pair of individuals chosen from the population at random. In such a population, there is no correlation between the bill depth trait in parents and in offspring, and

a plot of midparent and offspring bill depths produces a circular "cloud" of points with a slope of zero (Fig. 20.14d). This zero slope indicates that the heritability of this trait in this population is zero.

From these examples, you can conclude that phenotypic similarity among genetically related individuals may provide evidence for the heritability of a trait. However, conversion of the phenotypic similarity among genetic relatives to a measure of heritability depends on a crucial assumption: that there is a random distribution of genetic relatives with respect to environmental conditions experienced by the population. In the finch example, we assumed that parents and their offspring do not experience environments that are any more similar than the environments of unrelated individuals. In nature, however, there may be reasons why genetic relatives violate this assumption by inhabiting similar environments. With finches, for example, all offspring produced by a mother and father during a breeding season normally hatch and grow in a single nest where they receive food from their parents. Because bill depth affects a finch's capacity to forage for food, the amount of feeding in a nest correlates with parental bill depth, for reasons quite distinct from genetic similarities. One way to circumvent the confounding variable of environmental similarity is to remove eggs from the nest of one pair of parents and randomly place them in nests built by other parents in the population; this random relocation of eggs is called **cross-fostering.** In heritability studies of animals that receive parental care, cross-fostering helps randomize environmental conditions. Controlling for both environmental conditions and breeding crosses is a fundamental part of the experimental design of heritability studies carried out on wild and domesticated organisms.

Measuring the Heritability of Polygenic Traits in Humans

Mating does not occur at random with respect to phenotypes in human populations, yet researchers cannot apply techniques for controlling environmental conditions and breeding crosses to studies of such populations. Nonetheless, in most human societies, family members share similar family and cultural environments. Thus phenotypic similarity between genetic relatives may result either from genetic similarities or similar environments (or both). How can you distinguish the effects of genetic similarity from the effects of a shared environment?

One way is to study monozygotic, "identical" twins given up for adoption shortly after birth and raised in different families. In such a pair of identical twins, any phenotypic similarity should be the result of genetic similarity. At first glance, then, the study of adopted identical twins eliminates the confounding effects of a similar family environment. Further scrutiny shows that this is usually not true. Many pairs of twins are adopted by different genetic relatives; the adoptions often occur in the same geographic region (usually in the same state and even the same city); and families wishing to adopt must satisfy many criteria, including job and financial stability and a certain family size. As a result, the two families adopting a pair of twins are likely to be much more similar than a pair of families chosen at random, and this similarity can reduce the phenotypic differences between the twins.

A better approach is to compare the phenotypic differences between different sets of genetic relatives, particularly different types of twins (Fig. 20.15a). For example, monozygotic (MZ) twins, which are the result of a split in the zygote after fertilization, are genetically identical because they come from a single sperm and a single egg; they share all alleles at all loci and thus have a genetic relatedness of 1.0. By contrast, dizygotic (DZ) twins, which are the result of sperm from a single father fertilizing two different maternal eggs, are like any pair of siblings born at separate times; they have a genetic relatedness of 0.5. Comparing the phenotypic differences between a pair of MZ twins with the phenotypic differences between a pair of DZ twins can help distinguish between the effects of genes and family environment.

Consider a trait in which the differences in phenotype among individuals in the population arise entirely from differences in the environment experienced by each individual, that is, a trait for which the heritability is 0.0 (Fig. 20.15b). For this trait, you would expect the phenotypic differences among many pairs of MZ twins to be as great as the differences among many pairs of DZ twins. The fact that the MZ twins are more closely related genetically has no effect.

Now consider a trait for which differences in phenotypes among individuals in a population arise entirely from genetic differences, that is, a trait for which the heritability is 1.0 (Fig. 20.15c). Because DZ twins have a genetic relatedness of only 0.5 while MZ twins have a genetic relatedness of 1.0, you would expect the phenotypic differences among many pairs of DZ twins to be twice as great as the differences among many pairs of MZ twins. Thus, by comparing measures of differences among MZ and DZ twins, it is possible to assess the relative effects of genetic relatedness and family environment on phenotypic variation. Recent studies comparing MZ and DZ twins to measure the heritability of some aspects of personality and cognition have shown that there is a genetic component to memory, extroversion, and verbal reasoning, among other behavioral traits. It is important to remember, however, that although heritability is a useful measure of the relative contributions of genetics and environment, as with any measurement, it must be properly interpreted. The heritability of a trait applies only for a specific set of environments and specific population; the heritability of the same trait may be different in a different environment or in a different population.

The Fact That a Trait Has a Large Heritability Does Not Mean That the Trait Is Genetically Determined and Therefore Unaffected by the Environment

Changes in the environment can, in fact, cause large changes in a phenotypic trait, even if that trait has a high heritability. For example, the trait of adult height in humans has a heritability greater than 0.8 in many populations; but European adult men in the twentieth century are much taller, on average, than European adult men were in the Middle Ages. This increase in average

(a) **Monozygotic (MZ) twins**
Single ovulated egg fertilized by one sperm

Dizygotic (DZ) twins
Two ovulated eggs fertilized by different sperm

Embryo splits into two

Monozygotic twins
100% genetic identity

Dizygotic twins
Increase in genetic identity relative to unrelated individuals
25% increase in genotypic identity
50% increase in allele sharing

(b) **Examples of results expected with traits of 0.0 heritability**

Very common trait

Moderately common trait

Rare trait

MZ
DZ

(No difference in correlation levels between MZ and DZ classes)

(c) **Examples of results expected with traits of 1.0 heritability**

Very common trait

Common trait

Rare trait

MZ
DZ

(Differences observed in correlation levels of genotypic traits between MZ and DZ classes)

Figure 20.15 Comparing phenotypic differences between sets of MZ and DZ twins to show the heritability of a trait.
(a) The difference between MZ and DZ twins. **(b)** Bar graphs of a trait with 0.0 heritability (MZ versus DZ twins). **(c)** Bar graphs of a trait with 1.0 heritability (MZ versus DZ twins).

height is largely the result of vast improvements in diet. Similar changes in height associated with dietary changes have also occurred in other human populations. Because heritability is defined as the ratio of genetic variation to total phenotypic variation, the heritability of a trait decreases as the proportion of environmental variation increases. Thus, changes in the environment can alter the heritability of a trait. Unfortunately, many reports of supposed genetic determinism in the popular and scientific literature fail to take this fact into account.

The Heritability of a Trait Determines Its Potential for Evolution

We saw earlier how the selection of preexisting mutations generates evolutionary change. Since the heritability of a multifactorial trait is a measure of the genetic component of its variation, heritability quantifies the potential for selection and thus the potential for evolution from one generation to the next. A trait with high heritability has a large potential for evolution via selection, because the amount of evolution (change in allele frequency) generated by selection depends on the degree of heritability. To grasp the relationship between heritability, selection, and evolution, consider the number of bristles on the abdomens of fruit flies in a laboratory population of *D. melanogaster.* This fruit fly population exhibits some phenotypic variation in the trait of bristle number. If the trait has a high heritability in the population, the offspring of this original population will closely resemble their parents in bristle number (Fig. 20.16). If, however, you select as parents of the next generation only those flies among the top 15% in bristle number, the average bristle number among these breeders of the next generation will be greater than the average bristle number in the population as a whole. This artificial selection in conjunction with the high heritability of the trait will produce an F_1 generation in which the average bristle number is greater than the average bristle number in the previous generation. In other words, the artificial selection imposed by the experimenter will induce an evolutionary change whose magnitude is related to the heritability of the trait. If the heritability of bristle number were zero, there would be no evolutionary change.

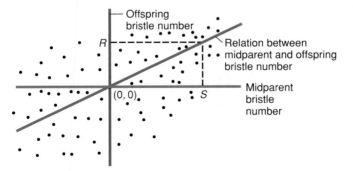

Figure 20.16 Relationship between midparent number of abdominal bristles and bristle number in offspring for a hypothetical laboratory population of *Drosophila*.

A Mathematical Model of the Relation Between Heritability and Evolution

If you let S represent the average trait value (in this case, the average bristle number) of breeding individuals in the parental population, measured as the difference between the value of this trait for *parents* and the value of the trait in the entire parental population of both breeding and nonbreeding individuals, then S is a measure of the strength of selection on the trait; as such, it is called the **selection differential.** Now let R represent the average trait value in the offspring of these breeding parents, this time measured as the difference between the trait's value for *offspring* and its value in the entire parental population of breeding and nonbreeding individuals. Used in this way R signifies the **response to selection,** that is, the amount of evolution, or change in mean trait value, resulting from selection. The heritability of the trait (arbitrarily designated as h^2), as seen in the slope of the line relating parental to offspring trait values, determines the relationship between S and R.

$$R = h^2 S \qquad (20.14)$$

In other words, the strength of selection (represented by S) and the heritability of a trait (h^2) directly determine the trait's amount or rate of evolution in each generation. This is the primary reason population and evolutionary geneticists consider the ability to measure heritability so important.

Variations in Polygenic Traits Arise at a Rapid Rate Because a Change at Any One of Many Loci Can Cause a Change in Phenotype

Geneticists have long used bristle number in *Drosophila* as a model for understanding the variation, selection, and evolution of quantitative traits. Early laboratory studies of selection acting on bristle number showed that the trait has substantial heritability in *Drosophila,* and it evolves rapidly in response to selection for either high or low bristle number (Fig. 20.17). Two results from these early studies were particularly striking. First, selection can rapidly lead to phenotypes not seen in the original population. After 35 generations of artificial selection for high or low bristle number, there was no overlap in the unselected and the selected populations. Some of the change in bristle number probably arose from reassortment and changes in frequency of existing alleles at multiple genes affecting the trait, without the appearance of new alleles. However, traits such as bristle number continue to evolve in response to selection for many generations. This observation suggests that new mutation is an additional source of variation in the population.

Experimenters have examined the contribution of mutation to genetic variation in bristle number (and by extension, other quantitative traits) through studies of highly inbred lines of *Drosophila* that at first had low or no genetic variation in bristle number. With these inbred lines, selection could occur only in the presence of new mutations affecting bristle number. Quantitative analyses revealed a significant selection-driven evolution of bristle number, which means that new mutations affecting bristle number arise in a population at a substantial rate

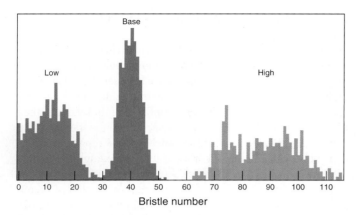

Figure 20.17 Evolution of abdominal bristle number in response to artificial selection in *Drosophila*. Bristle number distributions in different populations under selection for low or high values, compared to distributions in a population not subjected to selection (base).

(Fig. 20.18). These results highlight a key characteristic of polygenic traits: If many genes contribute to a trait, new variation in the trait may arise rapidly even if the mutation rate per gene is low, because a change at any one of many loci can cause a phenotypic difference.

Populations Eventually Reach a Selective Plateau

The bristle-number experiments with *Drosophila* showed that after many generations, populations eventually reach a selective

plateau at which, even with continued selection, the average bristle number does not change for many more generations. The existence of such evolutionary plateaus suggests that selection can, for a time, eliminate all genetic variation in a trait and that the potential for new mutations allowing further extremes in phenotypes has been exhausted.

How Many Loci Determine a Quantitative, Polygenic Trait?

In theory, if a quantitative trait were determined by thousands or millions of loci, selection would never exhaust the trait's genetic variation. The observation of evolutionary plateaus thus begs the question of how many loci determine a trait. While it is hard to come up with an exact answer, recent studies on bristle number provide some tantalizing clues. According to genetic and molecular characterization of QTLs (quantitative trait loci, see Chapter 10) affecting bristle number in *Drosophila*, 10–20 loci contribute the majority of genetic variation in bristle number, and substantial variation is attributable to alleles with large phenotypic effects at a small number of loci. It is thus likely that many polygenic traits are determined by tens, rather than hundreds or thousands, of loci, and that alleles of large effect at relatively few loci have considerable influence on such quantitative traits. If this turns out to be a general pattern, it will greatly simplify genetic analysis of normal and disease-causing variation in human and other populations.

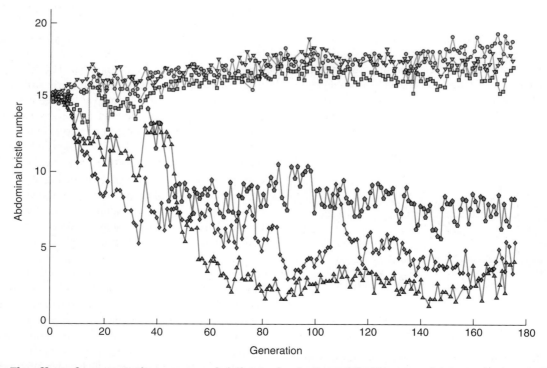

Figure 20.18 The effect of new mutations on mean bristle number in *Drosophila*. The average bristle number in a population under artificial selection for a reduced number over many generations is indicated with the *diamond, pentagon,* and *up-triangle* data points. Populations not under selection are indicated with the *down-triangle, circle,* and *square* data points.

CONNECTIONS

We have seen that populations at Hardy-Weinberg equilibrium have unchanging allele frequencies, and in one generation they achieve genotype frequencies of p^2, $2pq$, and q^2, which are subsequently maintained. In nature, where populations are rarely at Hardy-Weinberg equilibrium, however, natural selection acts on differences in fitness to alter allele frequencies. New mutations, genetic drift, and heterozygous advantage can also alter the allele frequencies of a population. For quantitative traits influenced by the alleles of two or more genes as well as by the in-

teraction of those alleles with the environment, the heritability of a trait, that is, the proportion of its phenotypic variation attributable to genetic variation, determines its potential for evolution by mutation, selection, and genetic drift.

In Chapter 21, we look at evolution from the point of view of the molecular mechanisms that propel it. In that chapter, we examine the various ways in which changes at the genomic level continually reshuffle the genetic deck to create the ever-changing abundance of life-forms that inhabit the earth.

ESSENTIAL CONCEPTS

1. The *Hardy-Weinberg law* uses the binomial equation of $p^2 + 2pq + q^2 = 1$ to correlate allele, genotype, and phenotype frequencies in a very large ideal population. In this ideal population, individuals mate at random, no new mutations appear, no individuals enter or leave, and there are no genotype-dependent differences in fitness. In the binomial equation, p represents the frequency of one allele and q represents the frequency of the other, and p^2, $2pq$, and q^2 represent the frequencies of the two homozygous and one heterozygous genotypes.

2. A population satisfying the Hardy-Weinberg assumptions is said to be at *Hardy-Weinberg equilibrium.* In such a population, allele frequencies remain constant from one generation to the next, and the genotype frequencies of p^2, $2pq$, and q^2 appear in one generation, after which they are maintained.

3. *Evolution* consists of changes in allele frequency over time. *Selection* acting on genotype-dependent differences in fitness can drive evolution. Selection does not entirely eliminate deleterious recessive alleles from a population. One reason for this is *heterozygous advantage.*

4. The existence of an *evolutionary equilibrium* is another reason deleterious recessive alleles persist in populations. The evolutionary equilibrium is a balance between mutation to a new allele and selection against that allele. Late onset of a disease undermines selection against the disease allele, while *genetic drift* has an unpredictable effect on the evolutionary equilibrium.

5. For quantitative traits, the *environmental variance* is a measure of the influence of environment on phenotypic

variation. Similarly, *genetic variance* measures the contribution of genes to phenotypic variation. Total *phenotype variance* is the sum of genetic variance and environmental variance.

6. Measures of environmental, genetic, and total phenotype variance make it possible to define the *heritability* of a trait as the proportion of total phenotype variance attributable to genetic variance. With traits for which the number and identity of contributing genes remain unknown and there is no way to obtain genetic clones, it is possible to correlate phenotypic variation with the genetic relatedness of individuals—that is, the average fraction of common alleles at all gene loci that the individuals share because they inherited them from a common ancestor—to measure the heritability of a trait.

7. To ascertain the heritability of a human trait, population geneticists often turn to studies of twins. The most useful approach is to compare the phenotypic differences between pairs of monozygotic and dizygotic twins. Environmental changes can always influence heritability.

8. The heritability of a trait quantifies its potential for sustaining selection and thus its potential for evolution from one generation to the next.

9. Variations in polygenic traits arise at a rapid rate through selection because changes at many loci contribute to changes in phenotype. Nevertheless, it is possible that many polygenic traits are determined by tens, rather than thousands, of loci.

SOCIAL AND ETHICAL ISSUES

1. A U.S. company developed a new strain of cotton in which a gene conferring resistance to a major cotton pest was introduced. The company that developed the strain is testing it in Bolivia. Company representatives chose this

site because it is an area in which native cotton grows well. However, because native cotton grows here, there is concern that the genes from the genetically engineered plants may be transferred to the wild strains. Should there be international laws or regulations established to ensure that testing is done in an environmentally responsible way? What responsibility does the company have to the country in which the testing is carried out? What type of remuneration should there be for the testing? Who is responsible if the spread of resistance genes to native populations becomes a problem long after the testing is done?

2. Zoos often house endangered species and breed them to increase the population size and reintroduce individuals into the wild. A problem with some endangered species is that, in the wild as well as in the zoos, there is very little genetic diversity remaining. Should it be a policy to try to save a species by mating individuals from different geographic locales in the hope of generating more genetic diversity and thereby increasing the possibility of

survival? Or should the focus instead be to breed existing animals to try to increase their numbers even though the inbreeding may lead to the expression of additional recessive diseases? Are these activities that you would want your city zoo to spend money on? What is the human responsibility in preserving other species?

3. Periodically in the last 30 years there have been efforts to determine the heritability of IQ. Results on the heritability of IQ in one particular population are not applicable to other populations, but unfortunately such results have been applied sometimes to serve a political agenda. Is it ethical to even do studies on the heritability of IQ or other politically or socially charged topics when the data are easily misinterpreted or misused? Is there a way to prevent these studies from being done or to prevent their inappropriate application?

S O L V E D   P R O B L E M S

I. A population called the "founder generation," consisting of 2000 *A A* individuals, 2000 *A a* individuals, and 6000 *a a* individuals is established on a remote island. Mating within this population occurs at random, the three genotypes are selectively neutral, and mutations occur at a negligible rate.

a. What are the frequencies of alleles *A* and *a* in the founder generation?
b. Is the founder generation at Hardy-Weinberg equilibrium?
c. What is the frequency of the *A* allele in the second generation (that is, the generation subsequent to the founder generation)?
d. What are the frequencies for the *A A*, *A a*, and *a a* genotypes in the second generation?
e. Is the second generation at Hardy-Weinberg equilibrium?
f. What are the frequencies for the *A A*, *A a*, and *a a* genotypes in the third generation?

Answer

This question requires calculation of allele and genotype frequencies and an understanding of the Hardy-Weinberg equilibrium principle.

a. To calculate allele frequencies, count the total alleles represented in individuals with each genotype and divide by the total number of alleles.

Number of individuals	Number of A alleles	Number of a alleles
2000 *A A*	4000	0
2000 *A a*	2000	2000
6000 *a a*	0	12,000
Total	6000	14,000

The frequency of the A allele (p) = 6000/20,000 = 0.3.
The frequency of the a allele (q) = 14,000/20,000 = 0.7.

b. If a population is in Hardy-Weinberg equilibrium, the genotype frequencies are p^2, $2pq$, and q^2. We calculated in part *a* that $p = 0.3$ and $q = 0.7$ in this population. Therefore,

$$p^2 = 0.09$$

$$2pq = 2\ (0.3)(0.7) = 0.42$$

$$q^2 = 0.49$$

For a population of 10,000 individuals, the number of individuals with each genotype, if the population were at equilibrium and the allele frequencies were $p = 0.3$ and $p = 0.7$, would be: *A A*, 900; *A a*, 4200; and *a a*, 4900. *The founder population described therefore is not in equilibrium.*

c. Given the conditions of random mating, selectively neutral alleles, and no new mutations, allele frequencies do not change from one generation to the next; p *= 0.3 and* q *= 0.7.*

d. *The genotype frequencies for the second generation would be those calculated for part* b *because in one generation the population will go to equilibrium.* A A = p² = 0.09; A a = 2 pq = 0.42; *and* a a = q² = 0.49.

e. *Yes,* in one generation a population not at equilibrium will go to equilibrium if mating is random and there is no selection or significant mutation.

f. *The genotype frequencies will be the same in the third generation as in the second generation.*

II. Two alleles have been found at the X-linked phosphoglucomutase gene in *Drosophila persimilis* populations in California. The frequency of the Pgm^A allele is 0.25, while the frequency of the Pgm^B allele is 0.75. Assuming the population is at Hardy-Weinberg equilibrium, what are the expected genotype frequencies in males and females?

Answer

This problem requires application of the concept of allele and genotype frequencies to X-linked genes. For X-linked genes, males (XY) have only one copy of the X chromosome so the genotype frequency is equal to the allele frequency. Therefore, $p = 0.25$ and $q = 0.75$. *The frequency of male flies with genotype $X^{PgmA}Y$ is 0.25; the frequency of males with genotype $X^{PgmB}Y$ is 0.75.* Three genotypes exist for females: $X^{PgmA} X^{PgmA}$, $X^{PgmA} X^{PgmB}$, $X^{PgmB} X^{PgmB}$, and X^{PgmB} corresponding to p^2, $2pq$, and q^2. *The frequencies of female flies with these three genotypes are $(0.25)^2$, $2(0.25)(0.75)$, and $(0.75)^2$ or 0.0625, 0.375, and 0.5625.*

III. Two hypothetical lizard populations found on opposite sides of a mountain in the Arizonan desert have two alleles (A^F, A^S) of a single gene A with the following three genotype frequencies:

	$A^F A^F$	$A^F A^S$	$A^S A^S$
Population 1	38	44	18
Population 2	0	80	20

a. What is the allele frequency of A^F in the two populations?
b. Do either of the two populations appear to be at Hardy-Weinberg equilibrium?
c. A huge flood opened a canyon in the mountain range separating populations 1 and 2. They were then able to migrate such that the two populations, which were of equal size, mixed completely and mated at random. What are the frequencies of the three genotypes in the next generation of the single new population of lizards? ($A^F A^F$, $A^F A^S$, and $A^S A^S$).

Answer

This question requires calculation of allele frequencies and genotype frequencies in existing and in newly created populations.

a. The frequency of allele A is calculated in the following way:

38 $A^F A^F$	76 A^F alleles
44 $A^F A^S$	44 A^F alleles
	120 A^F alleles

120 A^F alleles/200 total alleles $= 0.6$.

b. For population 1, the allele frequencies are $p = 0.6$ and $q = 0.4$. Genotype frequencies when the population is in equilibrium are

$$p^2 = (0.6)^2 = 0.36$$

$$2pq = 2(0.6)(0.4) = 0.48$$

$$q^2 = (0.4)^2 = 0.16$$

For population 1, which consists of 100 individuals, the equilibrium would be 36 $A^F A^F$, 48 $A^F A^S$, and 16 $A^S A^S$ lizards. *Population 1 does seem to be at equilibrium.* (Sampling error and small population size could lead to slight variations from the expected frequencies.) For population 2, the allele frequency (p) is based on the number of A^F alleles from the 80 $A^F A^S$ individuals. The total number of alleles $= 200$, so the frequency of A^F alleles is 80/200 or 0.4. The genotype frequencies for a population at equilibrium would be

$$p^2 = (0.4)^2 = 0.16$$

$$2pq = 2(0.4)(0.6) = 0.48$$

$$q^2 = (0.6)^2 = 0.36$$

Population 2 does not seem to be at equilibrium.

c. The combination of the two populations of lizards results in one population with the following allele frequencies:

A^F alleles:		
$A^F A^F$	38 × 2	76
$A^F A^S$	44	44
$A^F A^S$	80	80
Total:		200

A^S alleles:		
$A^F A^S$	44	44
$A^F A^S$	80	80
$A^S A^S$	18 × 2	36
$A^S A^S$	20 × 2	40
Total:		200

The allele frequencies are 200/400 or 0.5 for both p and q. *The genotype frequencies in the next generation will therefore be*

$$p^2 = (0.5)^2 = 0.25$$

$$2pq = 2(0.5)(0.5) = 0.5$$

$$q^2 = (0.5)^2 = 0.25$$

P R O B L E M S

20-1 Choose the best matching phrase in the right column for each of the terms in the left column.

a. fitness	1. the genotype with the highest fitness is the heterozygote
b. gene pool	2. chance fluctuations in allele frequency
c. fitness cost	3. ability to survive and reproduce
d. allele frequency	4. proportion of total phenotypic variance representation attributed to genetic variance
e. genotype frequency representation	5. collection of alleles carried by all members of a population
f. heterozygote advantage	6. p^2 and q^2
g. equilibrium frequency	7. p and q
h. genetic drift	8. the advantage of a particular genotype in one situation is a disadvantage in another situation
i. heritability	9. frequency of an allele at which $\Delta q = 0$

20-2 In a certain population of frogs, 120 are green, 60 are brownish green, and 20 are brown. The allele for brown is denoted G^B, while that for green is G^G, and these two alleles show incomplete dominance relative to each other.
  a. What are the genotype frequencies in the population?
  b. What are the allele frequencies of G^B and G^G in this population?
  c. What are the expected frequencies of the genotypes if the population is at Hardy-Weinberg equilibrium?

20-3 A large, random mating population is started with the following proportion of individuals for the indicated blood types:

0.5 MM
0.2 MN
0.3 NN

This blood type gene is autosomal and the *M* and *N* alleles are codominant.
  a. What will be the allele and genotype frequencies after one generation under the conditions assumed for the Hardy-Weinberg equilibrium?
  b. What factors cause a departure from a Hardy-Weinberg equilibrium?

20-4 A dominant mutation in *Drosophila* called *Delta* causes changes in wing morphology in *Delta / +* heterozygotes. Homozygosity for this mutation (*Delta / Delta*) is lethal. In a population of 150 flies, it was determined that 60 had normal wings and 90 had abnormal wings.
  a. What are the allele frequencies in this population?
  b. Given that there is random mating, no migration, and no mutation, and ignoring the effects of genetic

drift, what are the expected numbers of the different genotypes in the next generation if 160 viable offspring of the population in part *a* are counted?

20-5 Which of the following populations are at Hardy-Weinberg equilibrium?

Population	AA	Aa	aa
a	0.25	0.50	0.25
b	0.10	0.74	0.16
c	0.64	0.27	0.09
d	0.46	0.50	0.04
e	0.81	0.18	0.01

20-6 A gene called Q has two alleles, Q^F and Q^G, that encode alternative forms of a red blood cell protein that allows blood group typing. A different, independently segregating gene called R has two alleles, R^C and R^D, permitting a different kind of blood group typing. A random, representative population of football fans was examined, and on the basis of their blood typing, the following distribution of genotypes was inferred (all genotypes were equally distributed between males and females):

$Q^F Q^F R^C R^C$	202
$Q^F Q^G R^C R^C$	101
$Q^G Q^G R^C R^C$	101
$Q^F Q^F R^C R^D$	372
$Q^F Q^G R^C R^D$	186
$Q^G Q^G R^C R^D$	186
$Q^F Q^F R^D R^D$	166
$Q^F Q^G R^D R^D$	83
$Q^G Q^G R^D R^D$	83

This sample contains 1480 fans.
  a. Is the population at Hardy-Weinberg equilibrium with respect to either or both of the Q and R genes?
  b. After one generation of random mating within this group, what fraction of the *next* generation of football fans will be $Q^F Q^F$ (independent of their R genotype)?
  c. After one generation of random mating, what fraction of the *next* generation of football fans will be $R^C R^C$ (independent of their Q genotype)?
  d. What is the chance that the first child of a $Q^F Q^G R^C R^D$ female and a $Q^F Q^F R^C R^D$ male will be a $Q^F Q^G R^D R^D$ male?

20-7 A population with an allele frequency (p) of 0.5 and a genotype frequency (p^2) of 0.25 is in equilibrium. How can you explain the fact that a population with an allele frequency (*p*) of 0.1 and a genotype frequency (p^2) of 0.01 is also in equilibrium?

20-8 When an allele is dominant, why does it not always increase to produce the phenotype proportion of 3/4 : 1/4 (3:1) dominant to recessive individuals in a population?

20-9 It is the year 1998, and the men and women sailors (in equal numbers) on the American ship the *Medischol Bounty* have mutinied in the South Pacific and settled on the island of Bali Hai, where they have come into contact with the local Polynesian population. Of the 400 sailors that come aboard the island, 324 have MM blood type, 4 have the NN blood type, and 72 have the MN blood type. Already on the island are 600 Polynesians between the ages of 19 and 23. In the Polynesian population, the allele frequency of the *M* allele is 0.06, and the allele frequency of the *N* allele is 0.94. No other people come aboard the island over the next 10 years.
   a. What is the allele frequency of the *N* allele in the sailor population that mutinied?
   b. It is the year 2008, and 1000 children have been born on the island of Bali Hai. If the mixed population of 1000 young people on the island in 1998 mated randomly and the different blood group phenotypes had no effect on viability, how many of the 1000 children would you expect to have MN blood type?
   c. In fact, 50 children have MM blood type, 850 have MN blood type, and 100 have NN blood type. What is the observed allele frequency of the *N* allele among the children?

20-10 A new university on a Caribbean island has recruited its 700 faculty members directly after they received their Ph.D.s from colleges in France and Kenya. Five hundred came from France and 200 came from Kenya, with equal numbers of men and women in both cases. Upon arrival, you notice that 90 of the French and 75 of the Kenyans express a peculiar trait of rolling their eyes up into their sockets when asked a stupid question. Upon studying this trait, you discover that it is always due to the expression of allele 1 at a single gene called *Ugh*. This unusual phenotypic expression of allele 1 is dominant to the non-eye-rolling expression of the only other known allele (2) at the *Ugh* locus. Field trips taken to both Kenya and France indicate that the two alleles at the *Ugh* locus are at Hardy-Weinberg equilibrium in both of these *separate populations*. You also determine that the trait is only expressed with 50% penetrance under all circumstances. All of the faculty members arrived on the island single, but after teaching for a few years, they all married, and in all cases, they married other faculty members. If the faculty members marry each

other in a completely random manner to produce 1000 children in total, how many of these children do you expect will express the *Ugh* phenotype?

20-11 You have identified a new human immune-system gene that encodes a protein you refer to as antibody X. In the general *adult* population, you find only two codominant alleles of this gene. Allele 1 produces ≈100 ± 50 times as much of the same product as allele 2. Your semiquantitative survey of the general population for antibody X production shows that 50% have high levels of production (spread over a broad range) and 50% have a *very low* production. You then perform a detailed analysis of the population and discover that high levels of antibody production provide a five-year-old child with complete resistance to a newly evolved disease that attacks *only children of this age group*. Ten percent of all five-year-old children are attacked by the virus that causes this disease and 80% of those with low levels of antibody X die, whereas none of those with high levels of antibody X die. Assume that these infection rates remain constant as far as you can see into the future.
   a. What is the frequency of allele 1 in the current adult population?
   b. In the babies born in the following year, what will the frequency of allele 1 be?
   c. In individuals who were born in the following year (referred to in part *b*), what will the allele 1 frequency be after these babies grow up?

20-12 A mouse mutation with incomplete dominance (*t* = *tailless*) causes short tails in heterozygotes (t^+ / t). The same mutation acts as a recessive lethal that causes homozygotes (*t / t*) to die in utero. In a population consisting of 150 mice, 60 are t^+ / t^+ and 90 are heterozygotes.
   a. What are the allele frequencies in this population?
   b. Given that there is random mating among mice, no migration, and no mutation, and ignoring the effects of random genetic drift, what are the expected numbers of the different genotypes in this next generation if 200 offspring are born?
   c. Two populations (called Dom 1 and Dom 2) of mice come into contact and interbreed randomly. These populations initially are composed of the following numbers of wildtype (t^+ / t^+) homozygotes and tailless (t^+ / t) heterozygotes.

	Dom 1	Dom 2
Wild type	16	48
Tailless	48	36

   What are the frequencies of the two genotypes in the next generation?

20-13 The phenotypic frequency of color-blind males in the Caucasian population is .08. If the population is at equilibrium, what would be the frequency of color-blind females?

20-14 For the autosomal recessive condition cystic fibrosis, affected individuals in the African-American population are found at a frequency of 1 in 17,000. Assuming Hardy-Weinberg equilibrium conditions and no advantage or disadvantage for heterozygotes, what is the frequency with which carriers are present in the African-American population?

20-15 In *Drosophila*, the vestigial wings recessive allele, *vg*, causes the wings to be rudimentary. A geneticist crossed some true-breeding wild-type males to some vestigial virgin females. The male and female F_1 flies were wild type. He then allowed the F_1 flies to mate and found that 1/4 of the male and female F_2 flies had vestigial wings. He dumped the vestigial F_2 flies into a morgue and allowed the wild-type F_2 flies to mate and produce an F_3 generation.
  a. What will be the frequencies of wild-type and vestigial flies in the F_3?
  b. Assuming the geneticist repeated the selection against the vestigial F_3 flies (that is, he dumped them in a morgue and allowed the wild-type F_3 flies to mate at random), what will be the frequency of the wild-type allele and mutant alleles in the F_4 generation?
  c. Now the geneticist lets all of the F_4 flies mate at random (that is, both wild-type and vestigial flies mate). What will be the frequencies of wild-type and vestigial F_5 flies?

20-16 Would you expect to see a greater Δq from one generation to the next in a population with an allele frequency (q) of 0.2 or in a population with an allele frequency of 0.02? Assume relative fitness is the same in both populations and that the equilibrium frequency for q is 0.01.

20-17 Why is the elimination of a fully recessive deleterious allele by natural selection difficult?

20-18 An allele of the *G6PD* gene acts in a recessive manner to cause sensitivity to fava beans, resulting in a hemolytic reaction (lysis of red blood cells) after ingestion of the beans. The same allele also confers dominant resistance to malaria. The heterozygote has an advantage in a region where malaria is prevalent. Will the equilibrium frequency (q_e) be the same for an African and a North American country? What factors affect q_e?

20-19 Human geneticists have found the Finnish population to be very useful for studies of a variety of conditions. The Finnish population is small; Finns have extensive church records documenting lineages, and few people have migrated into the population. The frequency of

some recessive disorders is higher in the Finnish population than elsewhere in the world; and diseases such as PKU and cystic fibrosis that are common elsewhere do not occur in the Finnish population.
  a. How would a population geneticist explain these variations in disease occurrence?
  b. The Finnish population is also a source of information for the study of quantitative traits. The genetic basis of schizophrenia is one question that can be explored in this population. What advantage(s) and disadvantage(s) can you imagine for studying complex traits based on the Finnish population structure?

20-20 How can each of the following be used in determining the role of genetic and/or environmental factors in phenotypic variation in different organisms?
  a. genetic clones
  b. human monozygotic versus dizygotic twins
  c. cross-fostering

20-21 Which of the following statements would be true of a human trait that has high heritability in a population of one country?
  a. The phenotypic difference within monozygotic twin pairs would be about the same as the phenotypic differences among members of dizygotic twin pairs.
  b. There is very little phenotypic variation between monozygotic twins but high variability between dizygotic twins.
  c. The trait would have the same heritability in a population of another country.

20-22 Two alleles at one locus produce 3 distinct phenotypes. Two alleles of two genes lead to 5 distinct phenotypes; two alleles of six genes lead to 13 distinct phenotypes.
  a. Derive a formula to express this relationship. (Let *n* equal the number of genes.)
  b. Each of the most extreme phenotypes for a trait determined by two alleles at one locus are found in a proportion of 1/4 in the F_2 generation. If there are two alleles of two genes that determine a trait, each extreme phenotype will be present in the F_2 as 1/16 of the population. In common wheat (*Triticum aestivum*), kernel color varies from red to white and the genes controlling the color act additively, that is, alleles for each gene are semidominant and each gene contributes equally to the color. A true-breeding red variety is crossed to a true-breeding white variety, and 1/256 of the F_2 have red kernels and 1/256 have white kernels. How many genes control kernel color in this cross?

20-23 In a certain plant, leaf size is determined by four independently assorting genes acting additively. Thus, alleles *A, B, C,* and *D* each adds 4 cm to leaf length and alleles A', B', C', and D' each adds 2 cm to leaf length. Therefore, an A/A, B/B, C/C, D/D plant has leaves 32 cm long and an A'/A', B'/B', C'/C', D'/D' plant has leaves 16 cm long.

a. If true-breeding plants with leaves 32 cm long are crossed to true-breeding plants with leaves 16 cm long, the F_1 will have leaves 24 cm long, that is, A/A', B/B', C/C', D/D'. In the F_2, what proportion of the plants will have leaves that are 32 cm long?

b. Now assume that in a randomly mating population the following gene frequencies occur:

frequency of A = 0.9
frequency of A' = 0.1
frequency of B = 0.9
frequency of B' = 0.1
frequency of C = 0.1
frequency of C' = 0.9
frequency of D = 0.5
frequency of D' = 0.5

In the F_2, what proportion of the plants will have leaves that are 32 cm long?

21

Evolution at the Molecular Level

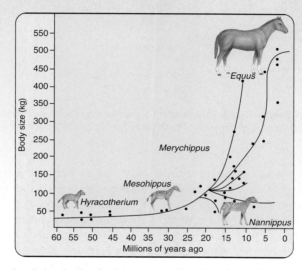

Lines show the broad outline of evolutionary relationships in the body size of horses. Most changes involved increase body size, but a few actually reflect a decrease in body size.

From December 1831 to October 1836, Charles Darwin circled the globe as naturalist for the HMS *Beagle* (Fig. 21.1). He was 22 years of age when he set sail. With indefatigable energy and insatiable curiosity, he dredged the oceans of the world for samples of the myriad organisms they concealed. He scoured the pampas of Uruguay for fossils and unusual contemporary species of lizards, birds, and mammals; scouted the no-man's-land of Tierra del Fuego for signs of unexpected life-forms; and climbed the highest peaks of the Andes, where he found deposits of seashells thousands of feet above sea level. Darwin also collected sundry specimens of the various species inhabiting different islands of the Galápagos. Whenever possible, he preserved and shipped his discoveries back to England. Some 23 years later, in 1859, he published *On the Origin of Species by Means of Natural Selection,* a comprehensive distillation of his thinking on what he had observed. The first printing of 1250 copies sold out in one day.

In *The Origin of Species,* Darwin uses an extensive comparative analysis of thousands of specimens and fossils as the basis for proposing that "the similar framework of bones in the hand of a man, wing of a bat, fin of the porpoise, and leg of the horse—the same number of vertebrae forming the neck of the giraffe and of the elephant—and innumerable other such facts, at once explain themselves on the theory of descent with slow and slight successive modifications." This observation leads him to the stunning conclusion that "all organic beings which have ever lived on this earth may be descended from some one primordial form."

While Darwin was not the first to suggest that species could undergo evolution, he was the first to suggest a mechanism by which evolution could occur. He based his theory of evolution on three principles—each obvious in and of itself—whose combination had revolutionary implications. First, within any species, there is variation among the individuals of a population in the expression of numerous traits. Second, variant forms of traits can be passed down through inheritance from one generation to the next. Third, some variant traits give the individuals that express them a greater chance of surviving and reproducing. (This is the so-called survival of the fittest principle.)

Figure 21.1 **The voyage of the HMS *Beagle*.** Charles Darwin spent 5 years circling the globe collecting specimens. The material acquired on this grand tour provided the basis for his comparative analysis in *The Origin of Species,* published some 23 years after he returned to England.

Darwin recognized that an advantageous trait that at first appears in only one or a few individuals could allow those individuals and their descendants to outcompete individuals that do not express the trait. This process of "natural selection," that is, of natural conditions selecting for an advantageous trait, would result in the transmission of the trait to a greater proportion of the population at each successive generation. Ultimately, many generations of natural selection would produce a population whose members all expressed a particular advantageous trait. Meanwhile, variant forms of other traits would arise in different individuals in each generation, and natural selection would, for each trait, determine which variants survive over time. Darwin's understanding of the process of continuous evolution by natural selection became the cornerstone of his theory of evolution. In *The Origin of Species,* he nevertheless cautions that "the natural selection of numerous successive, slight, favorable variations . . . has been the most important, but not the exclusive, means of modification."

Despite the enormous range of his revolutionary insight, Darwin was at a loss to explain the source of the visible variation on which natural selection acts. He knew inheritance plays a role: "Everyone must have heard of cases of albinism, prickly skin, hairy bodies, etc., appearing in several members of the same family. If strange and rare deviations of structure are really inherited, less strange and commoner deviations may be freely admitted to be inheritable." But he did not understand how "The laws governing inheritance are for the most part unknown." Mendel published "Experiments on Plant Hybrids" in 1866 (just seven years after Darwin's *The Origin of Species* appeared), but as we saw in Chapter 2, although Darwin received a copy of Mendel's paper, he most likely never read it.

Today, over 140 years after the publication of *The Origin of Species,* biologists accept Darwin's theory as a foundation of modern biology. Moreover, thanks to Mendel (and many other geneticists), they now understand the basic principles of heredity. And thanks to Watson and Crick (and many other molecular biologists), they know that ultimately, evolution is a process that begins at the molecular level, inside the double helix of DNA.

An example of short-term competitive evolution at the molecular level is seen in the battle between cells of the human immune system and the AIDS virus. Both evolve by the diversification of progenitor populations followed by the selective amplification of some divergent forms. When an invading virus activates the immune system, a virus-specific immune response ensues. As we saw in the Fast Forward box on p. 445 of Chapter 13, specific human immune responses depend on the cellular diversity of the immune system, which includes a trillion (10^{12}) circulating white blood cells, or lymphocytes, divided into two types—T cells and B cells. Populations of lymphocytes have the genetic capacity to synthesize a large and varied group of cell surface immune receptors collectively capable of recognizing foreign (that is, nonself) macromolecular structures on virtually any virus, bacterium, or other pathogenic microorganism (Fig. 21.2). Gene rearrangements in the cells of the developing immune system enable this enormous diversification of receptors from a relatively small repertoire of genes.

However, although populations of lymphocytes carry a great diversity of immune receptors, each individual lymphocyte synthesizes only one type of immune receptor. This attribute of one type of immune receptor per cell is the key to the specificity of an immune response. When the multiple copies of that one type of

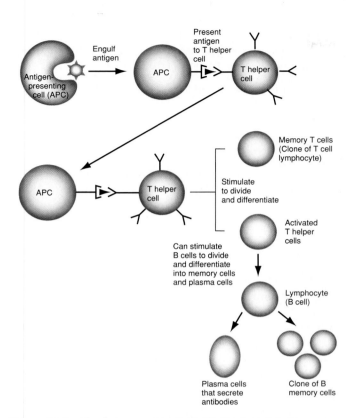

Figure 21.2 The immune response. Diverse receptors on lymphocytes recognize different molecular structures from invading pathogens. T cells employ T-cell receptors. B cells employ antibodies. Differentiated B cells, or plasma cells, secrete antibodies that destroy or neutralize antigen by a variety of mechanisms. Expanded numbers of memory T cells and B cells form the basis of a memory response that can rapidly respond to an antigen encountered a second time.

receptor in a cell's membrane encounter complementary foreign structures (called *antigenic determinants*), interaction between the immune receptors and the antigenic determinants triggers the lymphocyte to divide and differentiate.

The rapid proliferation of just a few selected lymphocytes to a few clones of genetically identical cells is the first step of an immune response. Among the cells in each expanded clone are *memory cells,* which may live as long as 40 years, and *effector cells,* which actually carry out (or effect) the immune reactions that help dispose of the microorganisms. These effector cells include T effector cells that bind to the antigenic determinants of a bacterium or virus and B effector cells that secrete antibodies, which in turn bind to antigenic determinants. As the immune response progresses, some of the B effector cells mutate their membrane receptors (generating further diversification). The altered receptors with the closest fit to the antigenic determinants bind the determinants more tightly, and this high-affinity binding drives the amplification of the lymphocytes that carry the altered receptors (selective amplification). The differentiated effector cells become involved in a multitude of effector mechanisms that ultimately de-

stroy the pathogenic organism carrying the antigenic determinants targeted by the immune response. Thus, the immune system's generation of specific immune responses is a marvelous example of molecular evolution: the diversification of lymphocytes into many variants and the selection by antigenic determinants of just one or a few of these variants for amplification over a period of weeks.

The AIDS virus (HIV) is a worthy adversary for the human immune system because it is able to diversify and amplify via selection far faster than the immune response itself. HIV, we have seen, is a retrovirus with an RNA genome (review Fig. A in the Genetics and Society box on p. 252 of Chapter 8). The virus makes its own reverse transcriptase available to the cells it infects (which include the T lymphocytes), forcing the cellular machinery first to copy the viral RNA genome into a DNA copy and then to generate a complementary DNA strand to form a proviral double helix. The infected cells integrate the double-stranded viral DNA into their own genomes (review Fig. B in the Genetics and Society box on p. 252 of Chapter 8). We have seen that the HIV reverse transcriptase has a very high mutation rate of roughly 1 in 5000 nucleotides incorporated into the viral cDNA. Since the HIV genome is about 10 kb in length, each replicated virus carries an average of two mutations (the basis of rapid diversification).

When HIV replicates explosively on infection, it generates billions of variant viruses before the host's immune response has a chance to take hold. Many of the variant viral particles are beyond the reach of the immune response. This is because the antigenic determinants targeted by the initial response have mutated beyond recognition by the time the originally activated T cells and B cells have proliferated (selection by lack of immune-cell recognition). Although some of the new antigenic determinants may be recognized by other cells among the diverse population of lymphocytes, subsequent viral mutations may alter these viral targets in good time. Eventually, the speed of viral evolution outstrips the ability of the immune response to keep pace not only because the viral genome mutates at a high frequency but also because a viral generation is so short.

A new therapeutic approach to AIDS, which tries to tip the evolutionary balance in favor of the human immune system, has been remarkably effective in prolonging the initial symptom-free phase of the disease. Called triple drug therapy, it entails the simultaneous delivery of three different anti-HIV drugs; two block the function of the HIV reverse transcriptase through different mechanisms, and the third blocks the functioning of a viral protease critical to HIV reproduction. By reducing the rate of viral replication, the triple drug therapy significantly decreases the virus's ability to develop variants that are simultaneously resistant to the human immune response and all three drugs. As a result, the HIV population in patients diminishes dramatically with the initial treatment. The triple drug therapy thus shifts the ability to diversify and amplify by selection in favor of the human immune response.

In this chapter, we examine in detail the basic components of evolution at the molecular level: **diversification** into many variants, followed by **selection** of one or a few variants for **amplification** in a population over many generations. The same process is at work in nearly all organisms, from tiny, subcellular viruses to large, multicellular plants and animals. Unlike the evolution of the human immune response and the AIDS virus, however, most molecular evolution in higher, more complex organisms occurs over millions of years.

As we describe how diversification and selection for amplification operate at the molecular level, we discuss:

■ The origin of life on earth, including a possible scenario for descent from a single ancestor; the idea that RNA may have been the original replicator; and a probable timetable for the evolution of complex life-forms from a sea of molecules.

■ The evolution of genomes, including a review of the various types and effects of mutations; a discussion of how an

increase in genome size correlates with the evolution of complexity; a description of how larger genomes evolve through duplication and divergence; and an overview of molecular archaeology based on an understanding of gene duplication, diversification, and selection.

■ The organization of genomes, including a description of genes, gene families, and gene superfamilies; repetitive nonfunctional DNA families; simple sequence repeats that dot the genome at random; and repeat sequences in centromeres and telomeres.

■ The immunoglobulin gene superfamily: a comprehensive example of molecular evolution.

■ ■ ■

The Origin of Life on Earth

To many biologists, the similarity of all living things on earth is more striking than the differences among them. Not only are all organisms composed of cells, but these cells work in essentially the same way, using the same complex molecules and the same type of genetic material read according to virtually the same genetic code. As we have seen, the flow of information in most living organisms from DNA via RNA to protein follows a well-defined pathway. DNA contains a digital code with a four-letter alphabet encoding the instructions for the construction and development of organisms. The units of information in the DNA code are genes that can be differentially expressed as RNA. Some RNA molecules (for example, mRNA) contain codes for proteins; other RNA molecules can fold into three-dimensional molecular machines and function as enzymes. The proteins synthesized from the information carried in RNA are three-dimensional objects that catalyze the chemistry of life and give organisms their shape and form. In complex organisms, the proteins, together with other macromolecules, assemble to form organelles, cells, tissues, and organ systems.

Descent from a Single Ancestor: Speculations on How the First Cell Arose

No fundamental law of biochemistry says that all living cells have to be constructed in the way that they are; indeed, an imaginative biochemist could think of an almost infinite number of ways to build a functioning cell that, like the organic life-forms around us, is based on the laws of chemistry. The observation that the cells of all plants, animals, fungi, and microorganisms analyzed so far are extremely similar in subcellular organelles, biochemistry, and genetic processes suggests that the abundant variety of life-forms alive today descend from the same original cell that happened to begin life with the particular genetic code scientists now consider universal (even with the minor differences seen in some organelles and microorganisms).

The First Step on the Pathway to Life Must Have Been a Replicator Molecule That Could Copy Itself
While the molecular evidence for the evolution of viral genomes and immune responses is accessible and available for examination, no evidence remains of the earliest steps of molecular evolution, during which self-replicating macromolecules became enclosed in membranes and ultimately formed cells, the basic units of life as we know it today. Scientists have therefore inferred and imagined these earliest steps, building their inferences and imaginings on a foundation of macro- and microbiological observations and analyses.

A key step in molecular evolution was the emergence of a molecule that could replicate itself. Although one can only speculate on the nature of the original replicator, it must have been simple enough to form spontaneously after a half billion years of atoms bumping into each other on the surface of the primordial earth. To give rise to the living cells we see today, it would have had to fulfill three requirements: (1) encode information by the variation of letters in strings of a simple digital alphabet; (2) fold in three dimensions to create molecules capable of self-replication and ultimately other functions; (3) expand the population of successful molecules through selective self-replication. What was the original self-replicating molecule and where did it come into existence?

The Original Replicator May Have Been RNA
In the 1980s, Thomas Cech and colleagues discovered that RNA, in addition to its ability to carry genetic information like DNA, can catalyze chemical reactions. Before this discovery, molecular biologists had assumed that only proteins have the chemical flexibility to fold into enzymes. RNA molecules that can act as enzymes to catalyze specific chemical reactions are called **ribozymes.** Although the substrates of most naturally occurring ribozymes are other RNA molecules, their potential range of enzymatic activity may be much broader.

The discovery of ribozymes gave molecular biologists an ideal candidate for the original replicator: a hypothetical RNA molecule that on its own could have (1) encoded information as linear strings in a four-letter alphabet; (2) folded into three-dimensional molecular machines able to execute critical functions of life (such as those currently carried out by polymerases, nucleases, and ligases); and (3) reproduced itself. The hypothetical primordial world in which this RNA became the first replicator has been termed the **RNA world.** The fact that all living organisms share the four letters of the RNA alphabet (A, U, G, and C) is testimony to the idea that successful, self-replicating RNA strings probably had a single origin, which, in turn, could have provided the common evolutionary origin of the information in all living organisms.

The earliest RNA strings most likely had informational motifs, or coding regions (for example, encoding a simple polymerase), separated by noncoding regions of "background noise." The ribozymes of contemporary organisms that can cleave, join, copy, and even modify informational strings may reflect these early RNA activities. The pattern of coding regions separated by noncoding introns could have been established by the early random assembly of RNA strings. The evolutionary challenge was how to use the information and separate it from the noninformational regions. Accordingly, there were evolutionary pressures to evolve RNA coding regions that could cut (like an endonuclease) and splice (like a ligase), as well as synthesize (like a polymerase) RNA molecules.

However, multipurpose RNA strings that both stored information and folded into molecular machines would have had several intrinsic disadvantages. First, RNA is a relatively unstable molecule, readily susceptible to chemical and enzymatic hydrolysis. Second, RNA has only four letters in its alphabet, which are chemically similar, and is thus less capable of folding into complex three-dimensional structures than other polymers with more complex alphabets. This is so because each letter in a molecular alphabet represents a particular shape and charge; proteins with their 20 chemically divergent amino-acid letters can fold into more complex structures than RNA with only four. Accordingly, there were evolutionary pressures to evolve more stable storage molecules (DNAs) and more flexible molecular machines (proteins).

No Record Exists of the Intermediates Between the RNA World and the Organized Complexity of the Cell

This lack of evidence makes it difficult to provide solid answers to questions about evolutionary events at the precellular stage. If RNA was indeed the original replicator, was it available for the first cells to use as an informational molecule, with the DNA and protein alphabets evolving subsequently? How did membrane compartments that concentrate the molecules of life and facilitate their sorting arise? What metabolic pathways gave some of the early cells an evolutionary edge? Many biologists speculate that life began around the volcanic thermal vents of the ancient seas more than 3.5 billion years ago. In this high-temperature environment, informational subunits could have emerged and then polymerized into strings of informational molecules. Each of these informational intermediates was a proto-life-form that reproduced itself into a large number of slightly different copies (diversification) until by chance one gained a better mode of survival and outcompeted all its cousins in the next round of reproduction (selection). This two-step process of diversification and selection repeated itself over and over again until the first cell appeared. From that point on, the history of life's evolution is easier to discern.

The Evolution of Living Organisms: Inferences from the Fossil Record

It is an enormous jump from a microscopically small single-celled organism to, for example, a human being composed of 100 trillion cells. However, like Darwin, no current biologist doubts that the one evolved into the other. This confidence arises not just from what Darwin's theory says *should* happen but from the possibility of seeing the critical intermediate stages of biological complexity, all the way up the ladder from single cells to sponges, to worms, to fish, to reptiles, to mammals, to primates and humans.

The Intermediate Stages of Evolution from a Single Cell to Complex Multicellular Organisms Exist in the Fossil Record and in the Living Fossils That Surround Us

While the fossil record enables us to date the initial appearance of each intermediate stage, the living representatives of each stage, in conjunction with our ability to study them with the tools of molecular biology, give us a glimpse into how evolution occurs step by step. It is fortunate for the scientists who study evolution that so many critical intermediate forms of life kept on reproducing in their less complex state, even as their cousins went on to evolve to the next stage of complexity.

Single-Celled Organisms Without a Membrane-Bounded Nucleus Emerged First

Scientists agree that planet earth coalesced some 4.5 billion years ago. By 4.2 billion years ago, enormous oceans covered the planet, and the first informational RNA molecules may have emerged around the high-temperature volcanic ocean vents. The first living organisms, consisting of a membrane surrounding information-replicating and information-executing machinery, evolved about 3.7 billion years ago; these were the precursors of present-day cells.

Fossil cells laid down 3.5 billion years ago near North Pole, Australia (a small town in an arid, rocky region) are the earliest evidence so far uncovered of distinct cells (Fig. 21.3a). Once life in cellular form emerged, living organisms evolved into three distinct domains: archaea, bacteria, and eukarya (Fig. 21.3b). Contemporary representatives of archaea and bacteria include only single-celled organisms whose genomes carry tightly packed genes. Because their genes do not contain introns, some biologists have concluded that introns were a late evolutionary elaboration, appearing with the emergence of eukaryotes. An alternative explanation is that archaea and bacteria, requiring great efficiency in their use of energy, relinquished, in the course of evolution, the flexibility of discontinuous coding segments for

(a)

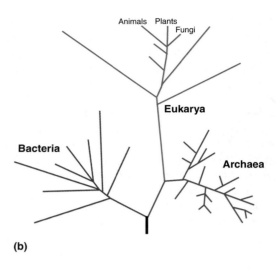

(b)

Figure 21.3 The earliest cells evolved into three kingdoms of living organisms. (a) The oldest fossilized cells. **(b)** The distinct branches represent different organisms in each kingdom. The length of the branches is proportional to the times of species divergence.

the more energy-efficient synthesis of genes without introns. (See Chapter 8, pp. 251–256 for a definition and earlier discussion of introns.)

More Complex Cells and Multicellular Organisms Appeared After More Than 2 Billion Years of Cellular Evolution

Eukarya emerged about 1.4 billion years ago with the symbiotic incorporation of certain single-celled organisms into other single-celled organisms and the complex compartmentalization of the cell's interior, including the segregation of DNA molecules into the nucleus. The incorporated single-celled organisms evolved to become intracellular organelles. The evolution of these relatively complex eukaryotes from the earliest cells thus took almost 2.3 billion years. About 1 billion years ago, the single-celled ancestors of contemporary plants and animals diverged. The first primitive multicellular organisms appeared 600–900 million years ago. Then 570 million years ago, one of the most remarkable events in the evolution of life occurred: the explosive appearance of a multitude of multicellular organisms, both plants and animals. The multicellular animals are often referred to as **metazoans.**

The Burgess Shale of Southeastern British Columbia in Canada Captured the Drama of Metazoan Evolution

The shale arose from a mud slide that trapped a wide variety of different organisms in a shallow Cambrian Sea. Amazingly, events and conditions during and after the slide conspired to

achieve the nearly perfect preservation of the three-dimensional structure of the entrapped specimens' soft body parts. Three aspects of the Burgess organisms are remarkable. First they represent myriad of very different body plans (Fig. 21.4). For example, paleobiologists have distinguished 20–30 classes of arthropods in the Burgess sea shale, a striking contrast to the three contemporary classes of arthropods. Second, this emergence of metazoan organisms occurred over a remarkably short (in evolutionary terms) period of time, perhaps just 20–50 million years. This rapid evolution is an example of **punctuated equilibrium:** the tendency of evolution to proceed through long periods of stasis (lack of change) followed by short periods of explosive change. It was as if the evolution of life represented a supersaturated salt solution where the addition of a final salt crystal catalyzed a remarkable solutionwide cooperative crystallization of many different body plans. As we will see later, this rapid change in body plans reflects an equally rapid change in the regulatory networks that control the development of organisms. Third, it seems that all the basic body plans of contemporary organisms initially established themselves in the metazoan explosion. For example, the ancestor of contemporary vertebrates depicted in Fig. 21.4c probably emerged at about the same time as the ancestor of all contemporary invertebrates.

Many Species Resulting from the Metazoan Explosion Have Disappeared

The enormous diversity of metazoan body plans that materialized about 500 million years ago has by now become tremen-

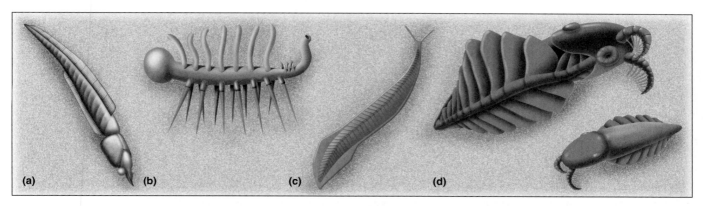

Figure 21.4 Some of the body plans found in Burgess shale organisms. Although all these life-forms are now extinct, fossils have revealed the enormous diversity of their body plans: **(a)** the *Nectocaris,* **(b)** the *Hallucigenia,* **(c)** the *Pikaia,* and **(d)** the *Anomalocaris.* The *Pikaia* is a vertebrate with a notochord, which makes it an ancestor to modern-day vertebrates.

Figure 21.5 Humans diverged from chimpanzees about 6 million years ago. Representatives of primates alive today: **(a)** orangutan, **(b)** gorilla, **(c)** chimpanzee, and **(d)** human.

dously reduced, in part through four to six abrupt extinction events that each destroyed 70% to 95% of the existing organisms. The most recent example was the global decimation 65 million years ago that led to the extinction of the dinosaurs. Many scientists believe this extinction was a consequence of a large meteorite impact in the Yucatan region of present-day Mexico that dramatically changed earth's climate by propelling enormous amounts of dust into the higher atmosphere. Scientists hypothesize that this thick cloud of dust dispersed and shrouded the globe for several years, preventing solar rays from reaching earth's surface. The lack of solar energy led to a nuclear-winter-like scenario in which the demise of all green life caused the demise of all large animals, such as dinosaurs, that depended for survival either directly on plants or indirectly on animals that ate plants.

Some smaller animals (like our mouse-sized ancestors) presumably survived this long sunless winter because their lesser size allowed them to get by on seeds alone. When the sun returned, the seeds lying dormant on the ground sprang to life

and the world again became an abundantly fertile environment. In the absence of competition from dinosaurs, mammals became the dominant large animal group, diverging into numerous species that could take advantage of all the newly unoccupied ecological niches. Some eventually evolved into our own species.

The Evolution of Humans

Humans arose from an ancestor common to most contemporary primates that existed 35 million years ago. They diverged from the ancestors of their closest primate relatives, the chimpanzees, about 6 million years ago (Fig. 21.5). While paleobiologists have not yet sorted out the immediate evolutionary ancestors of *Homo sapiens,* the recent typing of fossil DNA suggests that one previous candidate, the Neanderthal lineage, is not on the direct human evolutionary line (see the Fast Forward box on pp. 536–537 of Chapter 15). Perhaps the single most fascinating aspect of human evolution is the striking changes in the evolution of the brain since the human lineage diverged from the chimpanzee line. These changes include an increase in size, folding complexity, and nerve cell density, which together provide people with the capacity for a level of intelligence sufficient to appreciate the universe and the history of life.

Remarkably, on average, the chimpanzee and human genomes are approximately 99% similar. Moreover, as we saw in Chapter 12, the chimpanzee and human karyotypes are nearly the same. And in every comparison to date of chimpanzee and human DNA sequences, the observed differences between the two have been insignificant in terms of gene function. What these data suggest is that the evolution from a common primate ancestor to the modern human species might be accounted for by a few thousand isolated genetic changes yet to be uncovered. While these changes may have occurred in protein-coding sequences, many evolutionary biologists think it more likely that the changes occurred in regulatory sequences. Such changes would alter the regulatory sequences controlling when and how master regulatory genes produce transcription factors, and when and how ordinary structural genes or batteries of genes respond to these regulatory molecules. For example, the brains of humans and chimps are quite different from one another. The human brain is larger, is far more convoluted (folded), and contains a significantly greater density of neurons (brain cells). Hence, the regulatory networks guiding chimp and human brain development have diverged strikingly in the 6 million years since the two species split from a common ancestor. These regulatory changes are reflected in modified patterns of transcription factor binding sites in the promoter regions of genes that specify brain development. The rewiring of the regulatory networks in less than 6 million years is a very rapid change in terms of evolutionary time. The same is true of the diversity of body plans generated during the Cambrian explosion of metazoa 570 million years ago. The idea that evolution occurs primarily because of changes in regulatory networks and not structural genes is supported by the amazing ability of genes from one species to substitute for the absence of homologous genes in other species, in some cases even when the species are as different as yeast and humans. If homologous coding sequences from very different species are functionally indistinguishable, it is reasonable to speculate that species-specific differences in phenotype may arise, to a large degree, from species-specific differences in gene expression. Thus, what makes humans and chimpanzees different may not be the proteins that their genes encode but rather when, where, and at what level those proteins are expressed during development. It seems likely, for example, that the difference between chimpanzee and human consciousness can be accounted for simply by changes in the regulatory sequences of master regulatory genes and other genes encoding proteins that play a role in the development of the cerebral cortex.

The Evolution of Genomes

Although Darwin developed his theory of evolution—based on the selection of preexisting variations—without any knowledge of the molecules that make up living systems, evolution is very much a molecular process that operates on genetic information. In particular, the variation that initiates each step in the evolutionary process occurs within the genetic material itself in the form of new mutations.

DNA Alterations Form the Basis of Genomic Evolution

We have seen that mutations arise in several different ways (review Figs. 7.6–7.11). One is the replacement of individual nucleotides by other nucleotides. Replacements occurring in a coding region are silent, or *synonymous,* if because of the degeneracy of the genetic code, they have no effect on the amino acid encoded; by contrast, they are *nonsynonymous* if the change in nucleotide determines a change in amino acid or a premature termination codon, leading to a truncated gene product. Molecular biologists further distinguish between nonsynonymous changes that cause conservative amino-acid changes (for example, from one acidic amino acid to another) and those that cause nonconservative changes (for example, from a charged amino acid to a noncharged amino acid). Other gene mutations, arising from errors in replication or recombination, consist in the deletion from or insertion into genes of a DNA sequence of any length.

Changes can also occur in the order and types of transcription factor binding sites in the promoters of genes; such changes alter the patterns of gene expression. These expression-altering changes may occur very rapidly, which raises the question of whether such regulatory evolution can be explained merely by single-base mutations followed by selection.

Different mutations can be *deleterious, neutral,* or *favorable* to the organisms that inherit them. In multicellular organisms with large genomes, such as corn or humans, genes and their regulatory sequences make up only a small fraction (<3%) of the total genetic material. As a result, random mutations occur most often in DNA that plays no role in the development or function of an organism. Such mutations are presumably neutral. It might seem that synonymous mutations within coding regions would also be neutral, but there is some evidence that even changes in the codon used to produce a particular amino acid can provide a minute advantage or disadvantage to the organism based on the availability of different tRNA molecules and their associated synthetases. Thus while conservative amino-acid replacements were once considered neutral, current evidence suggests that they can have an impact on the growth and survival of an organism. Nonconservative amino-acid changes and changes such as deletions and insertions involving larger portions of a gene almost certainly have an impact on gene function.

Genetic Changes That Are Truly Neutral Are Unaffected by the Agents of Selection

Such mutations have a neutral effect on fitness and thus are not susceptible to selection. They survive or disappear from a population through *genetic drift,* which is the result of chance reproductive events. The smaller the population, the more rapidly genetic drift exerts its effect.

Mutations That Have Only Deleterious Effects Will Disappear from a Population by Negative Selection, That Is, Selection Against an Allele

However, as we saw in Chapter 20, some mutations (such as those that cause sickle-cell anemia) are deleterious to homozygotes for the mutation but advantageous to heterozygotes. Because of this *heterozygous advantage,* selection retains these mutations in a population at a low equilibrium level (review Fig. 20.7). However, even deleterious mutations for which there is no heterozygous advantage are eliminated very slowly if they are fully recessive.

Some Extremely Rare Mutations Give an Organism a Significant Advantage

Because of this advantage, individuals carrying the mutation are more likely to reproduce, and in each succeeding generation, the frequency of the mutation increases (review Fig. 20.4). This is positive selection (selection for an allele). Ultimately, the allele that began as a mutation is present in nearly every member of the population on both chromosomes. At this stage, the allele has become *fixed* in the population.

An Increase in Genome Size Generally Correlates with the Evolution of Complexity

Consider that although both bacteria and mammals evolved from a common cellular ancestor, the contemporary *Escherichia coli* genome is about 5 Mb in length, while in humans, the genome is about 3000 Mb long. How has evolution fashioned such different genomes from the same original material? The answer lies in the evolutionary potential for increasing the size of the genome through the duplication and diversification of genomic regions and, even more strikingly, through the acquisition of repetitive sequence elements that may represent more than 50% of the genome. In these processes, new DNA is born out of old. Although the increase in genome sizes from yeast to flies to vertebrates does reflect the increasing complexity of the organisms, there are occasional examples of amoeba, plants, and amphibians with considerably more genomic DNA than humans.

Genomes Grow in Size Through Repeated Duplications

Duplications can occur at random throughout the genome, and the size of the duplication unit can vary from a few nucleotides to the entire genome. When a duplication segment contains one or more genes, either the original or the duplicated copy of each gene is free to accumulate function-destroying mutations (diversify) without harm to the organism, since the other "good" copy with original function is still present. With duplications acting as such an important force in evolution, it is critical to understand the two main ways in which they arise.

Some Duplications Result from Transpositions. As we saw on p. 460 of Chapter 13, transposition, the transfer of one copy of a chromosomal sequence from one chromosomal site to another, can occur in various ways: through the direct movement of a DNA sequence from one chromosomal site to another;

through an RNA intermediate that is copied into a DNA intermediate, leaving the original DNA site intact; or through a DNA intermediate (Fig. 21.6; review Figs. 13.23 and 13.24). When the genomic DNA (rather than its RNA or DNA proxy) moves to a new site, the duplication of genetic material occurs only after the altered chromosome receiving the DNA segregates together with the unaltered homolog of the chromosome containing the original locus into an egg or sperm. When the gamete with the duplication unites with a normal gamete, the resulting zygote has three copies of the original locus (Fig. 21.7). In subsequent generations, the new transposition element may become fixed in the population.

Figure 21.6 Duplication by transposition: Three possibilities. Transposition may occur by **(a)** excising and reinserting the DNA segment; **(b)** by making an RNA copy, which is then converted to a DNA copy for insertion; or **(c)** by making a DNA copy for integration. In the second and third examples, the transposon is duplicated.

Figure 21.7 Transposition through direct movement of a DNA sequence. A DNA sequence may transpose from one chromosome to a second chromosome in a sex cell. In subsequent generations, this transposed element may become fixed.

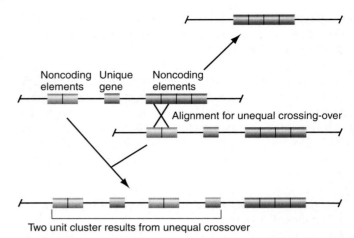

Figure 21.8 Duplications arise from unequal crossing-over. The crossover shown involved dispersed repeat elements (*blue boxes*). The *pink boxes* denote a unique gene. In the crossover event, this originally unique gene is deleted in one chromosome and duplicated in the other.

Other Duplications Result from Unequal Crossing-Over.

Normal crossing-over, or recombination, occurs between equivalent loci on the homologous chromatids present in a synaptonemal complex that forms during the pachytene stage of meiosis. Unequal crossing-over, also referred to as illegitimate recombination, occurs between nonequivalent loci (review Fig. 7.10a). Unequal crossing-over is most often initiated by related sequences located close to each other in the genome. Although the event is unequal in the exchange of nonequivalent segments of DNA, it is still mediated by the sequence similarities at the two separate loci.

So-called **nonhomologous unequal crossing-over** also occurs, although much less often than homologous crossing-over. A nonhomologous crossover may be mediated by at least

a short stretch of sequence homology, coding or noncoding, at the crossover's two sites of initiation.

An initial duplication by unequal crossing-over that produces a two-unit cluster may be either homologous or nonhomologous, but as Fig. 21.8 illustrates, once two units of related sequence are present in tandem, further rounds of homologous unequal crossing-over between nonequivalent members of the pair readily occur. Thus small clusters can easily expand to contain three, four, and many more copies of an original DNA sequence.

The result of unequal crossing-over between homologous chromosomes is always two reciprocal chromosomal products: one carrying a duplication of the region located between the two crossover sites and the other carrying a deletion that covers the exact same region (review Fig. 7.10). Unlike retrotransposition, unequal crossing-over operates on genomic regions without regard to functional boundaries. Regions duplicated by unequal crossing-over can vary from a few base pairs to hundreds of kilobases in length, and they may contain no genes, a portion of a gene, a few genes, or many genes.

Genetic Drift and Mutations Can Turn Duplications into Pseudogenes That Eventually Dissolve into Random DNA Sequences

Duplicated regions, like all other genetic novelties, must originate in the genome of a single individual. At first, the survival of a duplication in at least some animals in each subsequent generation of a population is, most often, a matter of chance. This is because the addition of one extra copy of a chromosomal region, including most genes, to the two already present in the diploid genome usually causes no significant harm to the individual animal. In the terminology of population genetics, the duplicated units are neutral with regard to genetic selection. They are thus subject to genetic drift, inherited at random by some offspring but not others. By chance again, most neutral genetic elements disappear from a population in several generations.

When a duplicated region that includes a functional gene survives for hundreds or thousands of generations, random mutations in the gene may turn it into a related gene with a different function, or into a nonfunctional *pseudogene*. Some of the mutations generating a pseudogene lead to a loss of regulatory function; others change one or more critical amino acids in the gene product; still others cause premature termination of the growing polypeptide chain, or change the translational reading frame of the gene, or alter the RNA splicing patterns. Pseudogenes, because they serve no function, are subject to mutation without selection and thus accumulate mutations at a far faster pace than the coding or regulatory regions of a functional gene. Eventually, nearly all pseudogene sequences mutate past a boundary beyond which it is no longer possible to identify the functional genes from which they derived. Thus, continuous mutation can turn a once functional sequence into an essentially random sequence of DNA.

Diversification of a Duplicated Gene Followed by Selection Can Produce a New Gene

Every so often, the accumulation of a set of random mutations in a spare copy of a gene leads to the emergence of a new functional gene that provides benefit and, consequently, selective advantage to the organism in which it resides. Because it provides a selective advantage, the new gene persists in the population. Although its function is usually related to that of the original gene, it almost always has a novel pattern of expression—in time, in space, or both—which most likely results from alterations in *cis*-regulatory sequences that occur along with codon changes. For example, a duplicated copy of the original human β-globin gene evolved into the myoglobin gene, whose protein product has a higher affinity for oxygen than the hemoglobin molecules composed in part of β-globin polypeptides. The myoglobin gene is active only in muscle cells, while the β-globin gene is expressed only in red blood cell precursors. Thus, duplication, divergence, and selection generated a new gene function from a previously functional gene.

Molecular Archaeology Based on an Understanding of Gene Diversification and Selection

Before the advent of molecular biology, researchers determined the genealogies of organisms by calculating the rates of evolution in phenotypic traits such as teeth and vertebrae. Then, in the 1950s, Linus Pauling and Emile Zuckerkandl analyzed hemoglobin and cytochrome C protein sequences from different species and noted that the rates of amino-acid substitution in each type of protein are similar for various mammalian lineages. On the basis of this observation, they postulated that for a given protein, the rate of evolution is constant across all lineages. They called this idea of a constant rate of change for each type

of molecule a **molecular clock.** Its existence enables biologists to determine from molecular data the approximate times when species diverged and then use these dates to reconstruct genealogical phylogenetic trees. Although the molecular clock hypothesis does not keep perfect time, in many instances where it has been tested, it has produced a reasonably good estimate of the time of divergence between two types of organisms. Today molecular biologists compare the nucleotide sequences of genes as well as the amino-acid sequences of proteins to determine phylogenetic divergences.

Scientists use molecular data in various ways to construct **phylogenetic trees** that illustrate the relatedness of homologous genes or proteins. A phylogenetic tree consists of **nodes** and **branches** (Fig. 21.9). The nodes represent the taxonomic units, which may be species, populations, individuals, or genes, while the branches define the relationship of these units. The branch length suggests the amount of time that has elapsed based on the number of molecular changes that have occurred. Because different genes accumulate changes at different rates, different types of genes are best suited to the construction of different types of phylogenetic trees. For example, in fibrinopeptides, the major components of wound-responsive blood clots, the exact amino-acid sequences are not critical to function. As a result, the fibrinopeptides are not under strong selective pressure, and their genes evolve quickly. In contrast, the ribosomal genes, which are highly conserved, evolve slowly. Thus, the genes encoding fibrinopeptides are useful for looking at recent evolutionary events among very closely related species, while the ribosomal genes are useful for looking at ancient evolutionary events such

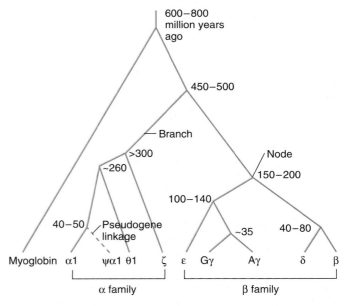

Figure 21.9 A phylogenetic tree. This phylogenetic tree diagrams the evolutionary history of human hemoglobin genes. The *broken line* denotes a pseudogene linkage. Only one of the two α-hemoglobin genes is shown because their date of divergence is so recent.

as the relationships of phyla to each other. Indeed, phylogenetic trees based on ribosomal genes helped rewrite the most fundamental domain classifications of multicellular organisms (see Fig. 21.3).

The Organization of Genomes

With this understanding of the basic mechanisms by which genomes evolve, we turn our attention to the results of genome evolution as seen in the organization of contemporary genomes. Our focus is on the various organizational features of the enormous mammalian genome, which evolved from the much simpler bacteria-like genomes through eons of duplication, diversification, and selection.

Molecular geneticists have accumulated data on genome organization by analyzing the genomic sequences of chromosomal DNAs from completed sequences of more than 60 genomes, including those of humans, mice, puffer fish, rice, *Drosophila, C. elegans* (the first animal to have its genome completely sequenced), Arabidopsis (the first plant to yield a complete genome sequence), yeast, *E. coli,* and approximately 100 other organisms (mostly microbes). The single-celled organisms exhibit genomes with densely packed genes and few, if any, introns. The mammalian genomes, with their far less densely packed genes, have several distinct features dominating their landscape: genes and families of genes; dispersed repetitive elements constituting more than a third of the genome; simple sequence repeats composed of single nucleotides or di-, tri-, tetra-, pentamers, and so forth; simple repetitive elements serving as a core for centromeres and telomeres; and unique nongene sequences. We now describe how these genomic features could have evolved.

Mechanisms Behind the Expansion from Genes to Multigene Families to Gene Superfamilies

Four levels of duplication (followed by diversification and selection) have fueled the evolution of complex genomes. At the lowest level, exons duplicate or shuffle to change the size or function of genes. At the next three increasingly complex levels, entire genes duplicate to create multigene families; multigene families duplicate to produce gene superfamilies; and the entire genome duplicates to double the number of copies of every gene and gene family (Fig. 21.10). At each of these successively higher levels of organization, the duplication of larger and larger units leads to the hierarchical generation of greater and greater amounts of new information.

Basic Gene Structure: All Genes Have Several Components

We have seen that genes consist of many different components: exons containing coding regions and 5′ and 3′ untranslated regions; introns that will be spliced out; and associated control regions (Fig. 21.11). Many of the control regions lie just 5′ to the

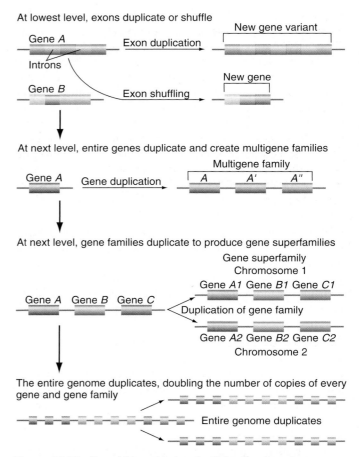

Figure 21.10 Four hierarchic levels of duplications increase genome size. Genome size increases through duplication of exons, genes, gene families, and finally of entire genomes.

Figure 21.11 The basic structure of a gene. A typical eukaryotic protein-coding gene. The *large dark boxes* represent the 5′ and 3′ untranslated mRNA sequences. The small open region represents regulatory sequences where gene-controlling transcription factors bind. A eukaryotic gene may range in size from a few hundred base pairs to more than 2 million base pairs.

transcribed region; but some, such as the locus control region of the β-globin gene family (described on p. 593 of Chapter 17), lie far outside the gene and appear to play a role in opening up the chromatin of the gene family locus so that gene expression can proceed at the appropriate time and level. Until all associated regulatory elements have been defined, the boundaries of a genetic locus remain uncertain.

The Duplication and Shuffling of Exons Can Elongate Genes and Enable Them to Encode Proteins with More Than One Functional Domain

Many proteins carry discrete compact domains, some of which perform a specific function, while others sustain molecular structure. In many genes, the discrete exons encode the structural and functional domains of a protein. Genes may elongate by the duplication of these exons to generate tandem exons that determine tandem functional domains such as those seen in antibody molecules (Fig. 21.12a). The functions of tandem domains may eventually diverge.

Entirely new genes may arise from **exon shuffling:** the exchange of exons among different genes. Exon shuffling produces mosaic proteins such as tissue plasminogen activator (TPA), a molecule with four domains of three distinct types: kringle (K), growth factor (G), and finger (F). The gene for TPA captured exons governing the synthesis of four domains from the genes for three other proteins: K from the gene for plasminogen, G from the gene for epidermal growth factor, and F from the gene for fibronecten (Fig. 21.12b). The mechanism by which exon shuffling occurs is unclear: Since each distinct domain has a different function, one protein may have two or more

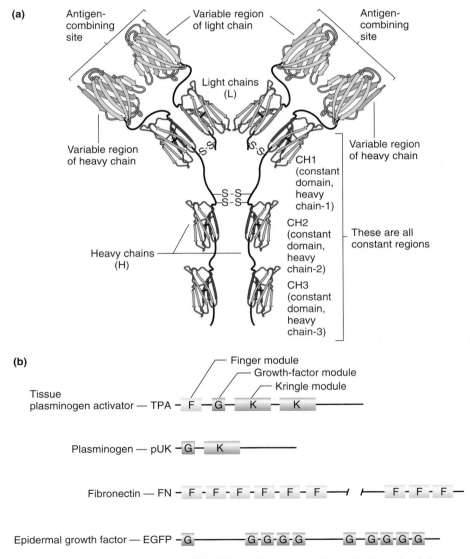

Figure 21.12 Genes can change by the duplication and shuffling of exons. (a) The antibody molecule is composed of two identical pairs of light (*inner*) and heavy (*on the outside*) chains. The variable regions of each chain, which are modified by somatic mutation, are shown in *tan;* the constant regions in *lavender*. The variable regions of a light and a heavy chain come together to form a V domain. Likewise, the pairs of the CH1, CH2, and CH3 regions come together to generate the CH1, CH2, and CH3 domains, respectively. Each domain carries out separate functions. **(b)** The tissue plasminogen activator (TPA) gene has evolved by the shuffling of exons from the genes for plasminogen, fibronectin, and epidermal growth factor.

(a)

Tandem gene family: Members of the multigene family are clustered on the same chromosome

Dispersed gene family: Members of the multigene family are on different chromosomes

(b)

Number of members in this cluster increases

Number of members in this cluster decreases

Figure 21.13 Multigene families. (a) Tandem and dispersed multigene families on segments of the indicated chromosomes. **(b)** A schematic illustration of how a crossover can expand and contract gene numbers in a multigene family.

(a)

Gene family

During synapsis gene *A* aligns with gene *C* because of homology between the two. Then a portion of gene *C* is changed to the sequence of gene *A*.

Note: Gene *A* remains unchanged, only gene *C* has a portion (dark) of the gene *A* sequence.

Converted region

Alternative resolution: unequal crossing-over

Divergence

(b)

MHC (mice)

Small set of functional class I genes

Class I pseudogenes (25 – 38) Nonfunctional pseudogenes

Information flow through gene conversion

Many alleles in population

Figure 21.14 Intergenic gene conversion. (a) In intergenic gene conversion, one gene is changed, the other is not. **(b)** How gene conversion events from pseudo-class I genes could increase the polymorphism in functional MHC class 1 genes in mice.

different functions. Exon shuffling may create new proteins with different combinations of functions.

One or More Duplications of an Entire Gene Can Create Multigene Families with Homologous Members

A **multigene family** is a set of genes descended by duplication and diversification from one ancestral gene. The members of a multigene family may be either arrayed in tandem (that is, clustered on the same chromosome) or distributed on different chromosomes (Fig. 21.13a). Unequal crossing-over can expand and contract the number of members in a multigene family cluster (Figure 21.13b).

Genetic Exchange Between Related DNA Elements by Gene Conversion Can Increase the Variation Among Members of a Multigene Family

There are many places in the genome where there appears to have been a flow of genetic information from one DNA element to other related, but nonallelic, elements located nearby or on different chromosomes. Such information flow between related DNA sequences occurs through an alternative outcome of the process responsible for unequal crossing-over. This alternative is known as **intergenic gene conversion** (Fig. 21.14a). Recall from pp. 178–185 of Chapter 6 that a recombination intermediate can resolve itself in two ways: one leads to crossing-over and the other allows gene conversion to occur without crossing-over. The same alternative outcomes can occur with unequal recombination intermediates. The gene conversion outcome of unequal

crossing-over allows the transfer of information from one gene to another. In special cases (such as the ribosomal gene family that exists in every eukaryotic species), the flow of information from such intergenic gene conversion has been so extreme that it has caused all members of a gene family to coevolve with near identity. And in at least one case—that of the class I genes of the major histocompatibility complex (MHC)—selection has acted on information flow in only one direction, causing information transfer from a series of nonfunctional pseudogenes to a small subset of just two to three functional genes (Fig. 21.14b). In this unusual case, the pseudogene family members served as a reservoir of genetic information that produced a dramatic increase in the amount of polymorphism (that is, the number of alleles) in the small number of functional gene members.

Concerted Evolution Can Make the Members of a Multigene Family Virtually Identical

A few multigene families have evolved under a special form of selective pressure that requires all family members to maintain essentially the same sequence. In these families, the high number of gene copies does not result in variations on a theme; rather it supplies a cell with a large amount of product within a short period of time. Among the gene families with identical elements is the one that produces RNA components of the cell's ribosomes, the one that produces tRNAs, and the one that produces the histones (which must rapidly generate enough protein to coat the new copy of the whole genome replicated during the S phase of every cell cycle).

Each of these gene families consists of one or more clusters of tandem repeats of identical elements. There is strong selective pressure to maintain the same sequence across all members of each family because all must produce the same product. Optimal functioning of the cell requires that the products of any individual gene be interchangeable in structure and function with the products of all other members of the same family. The problem is that the natural tendency of duplicated sequences is to drift apart over time. How does the genome counteract this natural tendency?

When researchers first compared ribosomal RNA and other gene families in this class, both between and within species, a remarkable picture emerged: Between species, there was clear evidence of genetic drift, but within species, all sequences were essentially equivalent. Thus it is not simply that some mechanism suppresses mutational changes in these gene families. Rather, there appears to be an ongoing process of **concerted evolution,** which allows changes in single genetic elements to spread across a complete set of genes in a particular family. With this knowledge, we can narrow the previously posed question about counteracting the natural tendency toward drifting apart to: How does concerted evolution occur?

Concerted evolution appears to occur through two different processes. The first is based on the expansion and contraction of gene family size through sequential rounds of unequal crossing-over between homologous sequences (Fig. 21.15a). Selection acts to maintain the absolute size of the gene family within a small range around an optimal mean. As the gene family becomes too large, the shorter of the unequal crossover products becomes selected; as the family becomes too small, the longer of the products become selected. This cyclic process causes a continuous oscillation around a size mean. However, each contraction results in the loss of divergent genes, whereas each expansion results in the indirect "replacement" of those lost genes with identical copies of other genes in the family. With unequal crossovers occurring at random positions throughout the cluster and with selection acting in favor of the least divergence among family members, this process can act to slow down dramatically the natural tendency of genetic drift between family members.

The second process responsible for concerted evolution is intergenic gene conversion between nonallelic family members. Although in each case, the direction of information transfer from one gene copy to the next is random, selection will act on this

Figure 21.15 Concerted evolution can lead to gene homogeneity. Repeat cycles of unequal crossover events **(a)** or gene conversion **(b)** cause the duplicated genes on each chromosome to become progressively more homogenized.

molecular process to ensure an increase in homogeneity among different gene family members (Fig. 21.15b). Information transfer (presumably by means of intergenic gene conversion) can also occur across gene clusters that belong to the same family but are distributed to different chromosomes.

Thus, with unequal crossing-over and intergenic gene conversion (which are actually two alternative outcomes of the same initial process) as well as selection for homogeneity, all members of a gene family can retain nearly the same DNA sequence. However, although concerted evolution maintains the close similarity of the members of multigene families within a species, divergences between the whole gene families present in different species nevertheless occur. This is because different sequences serve as the prototype for gene family homogenization in different species.

The Evolution of Gene Superfamilies

Molecular geneticists use the phrase "gene superfamily" to describe a large set of related genes that is divisible into smaller sets, or families, with the genes in each family being more closely related to each other than to other members of the larger superfamily. The multigene (or single-gene) families that compose a gene superfamily reside at different chromosomal locations. A prototypical small-size gene superfamily is the very well-studied globin genes illustrated in Fig. 21.16. The superfamily has three branches: the multigene family of β-like genes, the multigene family of α-like genes, and the single myoglobin gene. The duplications and divergences that produced the three superfamily branches occurred early in the evolution of vertebrates; as a result, the three branches of the superfamily are found in all mammals. All functional members of this superfamily play a role in oxygen transport (see Chapter 9 introduc-

tion and comprehensive example). As we have seen, the products of the α- and β-globin genes are active in red blood cells, while the product of the myoglobin gene transports oxygen in muscle tissue.

The primordial globin gene gave rise to the myoglobin and α-/β-globin precursor genes by gene duplication and transposition. The primordial α- and β-globin genes probably arose by a large-scale genome duplication (tetraploidization).

The β-like branch of this gene superfamily arose by duplication via multiple unequal crossovers. In the mouse, all the β-like genes are present in a single cluster on chromosome 7, which contains four functional genes and three pseudogenes (Fig. 21.16). The β-like chains encode similar polypeptides, each of which has been selected for optimal function at a specific stage of mouse development: one functions during early embryogenesis, one during a later stage of embryogenesis, and two function in the adult.

The α-like branch also evolved by unequal crossovers and divergences that generated a cluster of three genes on mouse chromosome 11: one functions during embryogenesis, and two function in the adult (Fig. 21.16). The two adult α-globin genes are virtually identical at the level of DNA sequence, which sug-

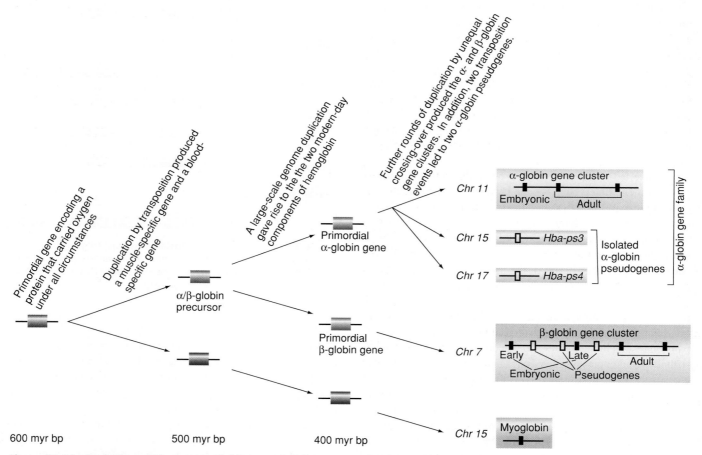

Figure 21.16 Evolution of the mouse globin superfamily. Repeated gene duplication by various mechanisms gave rise to the globin supergene family in mice, with two multigene families (α and β) and one single gene (myoglobin). The α family has both tandemly arrayed and dispersed gene members.

gests that the duplication producing them occurred very recently (on the evolutionary time scale). In addition to the primary α-like cluster, there are two nonfunctional α-like genes—pseudogenes—that have dispersed via transposition to locations on chromosomes 15 and 17. Pseudogenes existing in isolation from their parental families are called *orphons*. Interestingly, the α-globin orphon on mouse chromosome 15 (named *Hba-ps3*) has no introns and thus appears to have arisen through a retrotransposition event involving mRNA copied back to DNA; whereas the α-globin orphon on chromosome 17 (*Hba-ps4*) contains introns and may have arisen by a direct DNA-mediated transposition. The single mouse myoglobin gene on chromosome 15 has no close relatives either nearby or far away. The globin gene superfamily provides a view of the many different mechanisms that the genome can employ to evolve structural and functional complexity.

The mouse *Hox* gene superfamily (discussed in the genetic portrait of the house mouse on pp. 845–865 of Reference E) provides an alternative prototype for the evolution of a gene superfamily (Fig. 21.17). At first, there was one *Hox* gene that had evolved to produce a protein product that could bind to DNA enhancer regions and thereby regulate the expression of other genes. Unequal crossover events predating the divergence of insects and vertebrates some 600 million years ago produced a cluster of five related genes encoding DNA-binding proteins that regulated the expression of other genes encoding spatial information (that is, instructions for the spatial positioning of tissues and organs) in the developing embryo. The original *Hox* gene family then duplicated en masse and dispersed to four locations—on chromosomes in a common ancestor to all vertebrates. Because of the order of duplication events leading to the superfamily—unequal crossing-over to expand the original cluster size, followed by transposition en masse of the expanded cluster to create the superfamily—an evolutionary tree would show that a single gene family within the superfamily has actually splayed out physically across the four gene clusters, as shown in Fig. 21.17. After the en masse duplication that generated the gene superfamily, smaller duplications by unequal crossing-over added genes to some of the dispersed clusters and subtracted genes from others, thereby generating differences in gene number and type within a basic framework of homology among the different clusters. Each of the four *Hox* clusters in the mouse superfamily currently contains 9–12 homologous genes.

Repetitive "Nonfunctional" DNA Families Constitute Nearly One-Half of the Genome

Many repetitive nonfunctional DNA families consist of retroviral elements. Retroviruses, we have seen, are RNA-containing viruses that can convert their RNA genome into circular DNA molecules through the viral-associated RNA-dependent DNA polymerase known as reverse transcriptase, which becomes activated on cell infection (see Figs. A and B in the Genetics and Society box in Chapter 8). The resulting DNA can integrate itself at random into the host genome, where it becomes a provirus that retains the genetic information of the retroviral genome. Under certain conditions, the provirus can become activated to produce new viral RNA genomes and associated proteins, including reverse transcriptase, which can come together to form new virus particles that are ultimately released from the cell surface by exocytosis. By contrast, many stably integrated retroviral elements appear to be inactive.

Once integrated into a host chromosome, the provirus replicates with every round of host DNA replication, irrespective of whether the provirus itself is expressed or silent. Moreover, proviruses that integrate into the germ line, through the sperm or egg genome, segregate along with their host chromosome

Figure 21.17   Evolution of the *Hox* gene superfamily of mouse and *Drosophila*. This gene superfamily arose by a series of gene duplications. Four multigene families are present in the mouse and one in *Drosophila*.

into the progeny of the host animal and into subsequent generations of animals as well. The genomes of all species of mammals contain inactive integrated proviral elements.

It is of evolutionary interest to ask where retroviruses originated. Since retroviruses cannot propagate in the absence of cells, but cells can propagate in the absence of retroviruses, it seems likely that retroviruses evolved from nucleotide sequences that were originally present in the cell genome. The first retrovirus might have been able to free itself from the confines of the cell nucleus through an association with a small number of proteins that allowed it to coat and thus protect itself from the harsh extracellular environment. Of course the protein most critical to the propagation of the retrovirus would have been the reverse transcriptase that allows the retrovirus to reproduce. But where did this enzyme come from? Reverse transcriptase catalyzes the production of single-stranded complementary DNA molecules from an RNA template, but this enzymatic activity is not required for any normal cellular process known in mammals. So, how could an activity without any apparent benefit to the host organism arise *de novo* in a normal cell? One possible answer is that reverse transcriptase did not evolve because it benefited the organism itself; rather it evolved because it benefited selfish DNA elements within the genome that use it to propagate themselves within the confines of the genome. We now describe one such selfish element.

The LINE Family of DNA Elements Are Retroposons That Appear to Derive from a Selfish DNA Sequence Encoding Reverse Transcriptase

As we saw on p. 461 of Chapter 13, LINE is an acronym for Long INterspersed Elements. The LINE family of DNA elements is very old. Homologous families of repetitive elements exist in a wide variety of organisms, including protists and plants. Thus, LINE-related elements, or other DNA elements of a similar nature, are likely to have been the source material that gave rise to retroviruses.

Dispersion to new positions in the germ-line genome presumably begins with the transcription of LINE elements in spermatogenic or oogenic cells. The reverse-transcriptase-encoding region on the transcript is translated into an enzyme that preferentially associates with and uses the transcript it came from as a template to produce LINE cDNA sequences; however, the reverse transcriptase often stops before it has made a full-length DNA copy of the RNA transcript. The resulting incomplete cDNA molecules can nevertheless form a second strand to produce truncated double-stranded LINE elements that integrate into the genome but remain forever dormant (Fig. 21.18a).

The SINE Family Consists of a Second Type of Repetitive, Nonfunctional DNA Element That Did Not Originally Encode Reverse Transcriptase

The Alu family in the human genome is an example of a highly repetitive, widely dispersed SINE (Short INterspersed Elements) family (see p. 461 of Chapter 13 for a discussion of SINEs). Over

Figure 21.18 The creation of LINE and SINE gene families. **(a)** A complete LINE sequence can be copied into RNA. It encodes a reverse transcriptase that can make cDNA copies from the RNA. These copies may be complete or truncated. Either kind may integrate into other sites on any chromosome. **(b)** SINE elements can be transcribed and, because they form 3′ hairpin loops, can be copied into cDNAs by LINE-encoded reverse transcriptase. These cDNA copies then integrate into the genome.

500,000 Alu elements are dispersed throughout the human genome. At 300 bp in length, the Alu element is far too short to encode a reverse transcriptase. Nonetheless, like LINE elements, Alu and other SINE elements are able to disperse themselves throughout the genome by means of an RNA intermediate that undergoes reverse transcription. Clearly, SINES depend on the availability of reverse transcriptase produced elsewhere, perhaps from LINE transcripts or the proviral elements of retroviruses.

All SINE elements in the human genome, as well as in other mammalian genomes, appear to have evolved from small cellular RNA species, most often tRNAs, but also the 7S cytoplasmic RNA that is a component of the signal recognition particle (SRP) essential for protein translocation across the endoplasmic reticulum. The defining event in the evolution of a functional cellular RNA into an altered-function, self-replicating SINE element is the accumulation of nucleotide changes in the 3′ region that lead to self-complementarity with the propensity to form hairpin loops. Reverse transcriptase can recognize the open end of the hairpin loop as a primer for strand elongation. Since it is likely that the hairpin loops form only rarely among normal cellular RNAs, a cell will preferentially use its SINE transcripts as templates for the production of cDNA molecules that are somehow able to integrate into the genome at random sites (Fig. 21.18b).

Comparing LINE and SINE Families

Like LINE families, the Alu SINE family appears to be evolving by episodic amplification followed by sequence degradation. Unlike LINE families, however, SINE families present in the genomes of different organisms appear, for the most part, to have independent origins.

Some Selfish Elements May Confer a Selective Advantage

While SINE and LINE elements may have amplified themselves for selfish purposes, they have had a profound impact on whole-genome evolution. In particular, homologous SINE or LINE elements located near each other can, and will, catalyze unequal but homologous crossovers that result in the duplication of single-copy genes located between the homologous elements. Such duplications initiate the formation of multigene families. In addition, some selfish elements appear to have evolved a regulatory role—acting as enhancers or promoters—through chance insertions next to open reading frames. Thus, selfish elements may confer a selective advantage by facilitating duplication through unequal crossing-over or by becoming regulatory elements.

Genomic Stutters—the Simple Sequence Repeats Known As Microsatellites, Minisatellites, and Macrosatellites— Dot the Mammalian Genome

Through large-scale sequencing and hybridization analyses of mammalian genomes, researchers have found tandem repeats of DNA sequences with no apparent function scattered throughout the genome. The size of the repeating units ranges from two nucleotides (CACACACA . . .) to 20 kb or more, and the number of tandem repeats varies from two to several hundred.

The mechanism by which these tandem repeats originate may be different for loci having very short repeat units as compared with loci having longer repeat units. Tandem repeats of short di- or trinucleotides can originate through random changes in nonfunctional sequences. By contrast, the initial duplication of larger repeat units is likely to be a consequence of unequal crossing-over. Once two or more copies of a repeat unit exist in tandem, an increase in the number of repeat units in subsequent generations can occur through unequal crossovers or errors in replication (see Fig. 21.8). It is not yet clear whether random mechanisms alone can account for the rich variety of tandem repeat loci in mammalian genomes or whether other selective forces are at play. Either way, tandem repeat loci continue to be highly susceptible to unequal crossing-over and, as a result, tend to be highly polymorphic in overall size.

Repeat Sequences in Centromeres and Telomeres

Highly repetitive, noncoding sequences shorter than 200 bp are found in and around centromeres. For example, in the human genome, *alphoid*, a noncoding sequence 171 bp in length, is present in tandem arrays extending over a megabase in the centromeric region of each chromosome. In addition, several similar-sized repetitive sequences unrelated to alphoid are found in some centromeres (review Fig. 12.11 on p. 428). These regions are sites of interaction with the spindle fibers that segregate chromosomes during meiosis and mitosis. Selection may have acted to retain the thousands of copies of centromeric repeat elements in each centromeric region because they increase the efficiency and/or accuracy of chromosome segregation.

A second type of repeated genomic element with a special location is the hexamer TTAGGG found in the telomeres of all human chromosomes. The six-base unit is repeated in tandem arrays 5–10 kb in length at the ends of human and all other mammalian chromosomes. Selection may have conserved this repeat element because it plays an essential role in maintaining chromosomal length (review Figs. 12.9 and 12.10 on pp. 425 and 426).

Unique Nongene Sequences

The mutation and informational degradation of extra copies of duplicated genes, which produces pseudogenes, can ultimately result in unique nongene sequences. The mutation and divergence of genomic repeat elements can also lead to unique nongene sequences.

The organization of genomes in higher organisms exhibits several cardinal features. Genes, constituting less than 3% of the genome, are often organized into multigene families or even gene superfamilies. A variety of mechanisms, such as homologous but unequal crossing-over and gene conversion, permit the expansion and contraction of gene families, as well as their concerted evolution. In some organisms, such as human and mouse, approximately 50% of the genome may be repeat elements. These elements may facilitate chromosomal rearrangements or duplications and, on occasion, evolve to become genes or regulatory sequences. The telomeres and centromeres are largely composed of repeat sequence elements. Close to half of the genome may represent genes or repeat sequences that have degraded to unrecognizable unique sequences.

The Immunoglobulin Gene Superfamily: A Comprehensive Example of Molecular Evolution

More than 200 different types of genes belong to the immunoglobulin gene superfamily (Table 21.1). Most encode cell-membrane proteins that recognize either soluble molecules or molecules on the surface of other cells. All members of the immunoglobulin gene superfamily share a common feature: the

TABLE 21.1   Examples of Immunoglobulin Gene Superfamily Members

Protein	Molecular wt. (kDa)	Ig domains			Human Chromosome	Function
		C1	C2	V		
Ig heavy chain	55	3–4	—	1	14q32.33	Recognition of Ag
Ig κ light chain	24	1	—	1	2p12	Recognition of Ag
Ig λ light chain	24	1	—	1	32q11.12	Recognition of Ag
B29 (Ig-β/Ig-γ)	39	(1 Ig domain)			(Mouse)	Component of surface Ig complex
MB-1 (IgM-α)	34	(1 Ig domain)			(Mouse)	Component of surface Ig complex
TCR α chain	45–60	1	—	1	14q11.12	Recognition of peptide-MHC
TCR β chain	40–50	1	—	1	7q32	Recognition of peptide-MHC
TCR γ chain	45–60	1	—	1	7p15	Recognition of peptide-MHC
TCR δ chain	40–60	1	—	1	14q11.2	Recognition of peptide-MHC
CD3 ε chain	25–28	—	1	—	11q23	TCR signal transduction
CD3 δ chain	20	—	1	—	11q23	TCR signal transduction
CD3 ε chain	20	—	1	—	11q23	TCR signal transduction
MHC class 1 or chain	44	1	—	—	6p21.3	Peptide binding
β2 microglobulin	12	1	—	—	15q21-q22	Stabilize MHC class I α
MHC class II or chain	32–34	1	—	—	6p21.3	Peptide binding
MHC class II β chain	29–32	1	—	—	6p21.3	Peptide binding
Qa/TL heavy chain	48	1	—	—	(Mouse)	Restriction element for γ/δ TCR?
CD1	43–49	1	—	—	1q22-q23	Restriction element for γ/δ TCR?
CD2	50	—	2	—	1p13	Ligand for LFA-S (CD58)
LFA-3 (CDUB)	40–65	—	2	—	1	Ligand for CD2
ICAM-1 (CD64)	80–114	—	5	—	19	Ligand for LFA-1
CD4	55	—	1	2	12pter-q12	Binds MHC class II
CD8 α chain	34	—	—	1	2p12	Binds MHC class I
CD8 β chain	34	—	—	1	2p12	Binds MHC class I
CD7	40	—	—	1	17q25	T-cell activation
CD22	130/140	—	5	—	Unassigned	B-cell adhesion
CD28	44	—	—	1	2q33-q34	T-cell activation
CD33	67	(2 Ig domains)			19q13	Myeloid cell adhesion?
CD48 (blast-2)	41–45	—	1	1	1q21-q23	B-cell adhesion
Thy-1	17.5–18.7	—	—	1	11q23	T-cell activation and adhesion
MRC OX2	41–47	—	1	1	3	Unknown
PolyIgR	100	—	1	4	1q31-q42	Polymeric Ig receptor
FcγRI (CD6)	70	—	3	—	1q23-q24	High-affinity IgG receptor
FcγRII (CD32)	40	—	2	—	1q23-q24	Low-affinity IgG receptor

TABLE 21.1 (continued)

Protein	Molecular wt. (kDa)	Ig domains			Human Chromosome	Function
		C1	C2	V		
FcγRIII (CD16)	50–70	—	2	—	1q23-q24	Medium-affinity IgG receptor
FcγRI or chain	25	—	2	—	1q23-q24	High-affinity IgE receptor
FcαR	60	—	2	—	Unassigned	IgA receptor
IL-1 receptor (type 1)	82	(8 Ig domains)			2q12	Cytokine receptor
IL-6 receptor	80	—	1	—	Unassigned	Cytokine receptor
M-CSF receptor (*c-fma*)	165	—	4	1	5q33.2-q33.3	Colony stimulating factor (CSF-1) receptor
G-CSF receptor	150	—	1	—	ND	Colony stimulating factor (G-CSF) receptor
SLF-HGF receptor (*c-kit*)	145	—	4	1	4q31-3	Hematopoietic growth factor receptor
FDGF receptor	180	—	4	1	5q31-q32	Growth factor receptor
NCAM (CD58)	97–220	—	5	—	11q23	Adhesion of neural cells
MAG	67/72	—	5	—	(Rat)	Axon-glial interactions of myelination
P₀ myelin protein	28–30	—	—	1	(Rat)	Myelin compaction
Neuroglian (NGL)	167	—	6	—	(*Drosophila*)	Neural cell adhesion
Falciculum II	96	—	5	—	(Grasshopper)	Axonal fasciculation
Almagam	38–50	—	2	1	(*Drosophila*)	Neural cell adhesion
Contactin	180	—	6	—	(Aves)	Neurite fasciculation
F11	180	—	6	—	(Aves)	Neurite fasciculation and elongation
L1	200	—	6	—	(Mouse)	Neurite fasciculation and elongation, neuronel migration
Ng-CAM (G4)	200	—	6	—	(Aves)	Homolog of L1
NILE	200	—	6	—	(Rat)	Homolog of L1
F3	155	—	6	—	(Mouse)	Unknown
TAG-1	135	—	6	—	(Rat)	Growth and guidance of axons
Sk. muscle C-protein	140	—	6	—	(Aves)	Binds myosin H chain and F-actin
Carcinoembryonic Ag	175–200	—	2–6	1	19q13.1-q13.2	Family of molecules; adhesion?
Pregnancy-specific Ag	54–72	—	3	1	19q13.2-q13.3	Family of molecules; adhesion?
α 1B glycoprotein	83	—	5	—	Unassigned	Unknown
B, Memb. link, protein	44.5–48.8	—	—	1	Unassigned	Locks proteoglycan to haluronate chain
Twitchin	600	—	5	—	(Nematoda)	Regulation of myosin activity
CaVPT	30	—	2	—	(*Amphioxus*)	Unknown
Basigin	43–66	(1 Ig domain)			(Mouse)	Unknown
LAR	215–250	—	3	—	Unassigned	Adhesion?

Figure 21.19 Three-dimensional structures encoded by immunoglobulin homology units. Tertiary organization of V and C homology units in an immunoglobulin light chain. The β strands and their orientation are shown as *flat arrows*. Opposite faces of the β strands are *light lavender*. The disulfide bond is indicated by a *solid bar*. The V and C homology units are remarkably similar in structure. The variable loops indicated by *96, 26,* and *53* fold to form the walls of the antigen-binding site.

immunoglobulin homology unit, so-called because it was first identified in the genes for immunoglobulin (antibody) molecules (Fig. 21.19). The homology unit encodes about 100 amino acids, which fold into a characteristic three-dimensional structure termed the *immunoglobulin fold.* The unique design of the immunoglobulin fold enables it to (1) interact with other immunoglobulin folds to form dimeric molecules, (2) display an enormous diversity of recognition sites while still maintaining a highly conserved three-dimensional framework, and (3) resist proteolysis in the blood. The folds encoded by different homology units may vary by as much as 90% of their sequences while still preserving their folded framework because of a small but critical number of highly conserved structural features. The evolutionary success of the immunoglobulin homology unit is reflected in its deployment in more than 200 types of genes.

Evolution of the Immunoglobulin Homology Unit

The immunoglobulin homology unit probably emerged early in the evolution of metazoans. Scientists believe this to be true for three reasons. First, both vertebrates and invertebrates employ cell surface recognition molecules containing the immunoglobulin homology unit, but such molecules are not found in prokaryotes or single-cell eukaryotes. Since the common ancestor of present-day vertebrates and invertebrates was present when metazoan organisms first emerged, it is likely that the immunoglobulin gene superfamily also came into existence at that time. Second, the need for molecular recognition at the cell surface increased exponentially with the evolution of multicellular organisms, and it appears that the immunoglobulin homology unit evolved coincidentally, allowing metazoans to fill their need for diverse cell surface recognition units. Third, the early primordial homology unit quickly duplicated to make three dis-

tinct types of homology units: variable (V), constant (C), and homology (H) (Fig. 21.20a). The V, C, and H units are equally related to one another and are deployed in the diverse cell surface receptors used by all metazoans (Fig. 21.20b). The inescapable conclusion from these observations is that the highly diverse immunoglobulin gene superfamily evolved relatively rapidly, and its rapid creation may have been a driving force in generating the cell surface recognition required by emerging metazoan organisms.

Evolution of the Immunoglobulin Gene Superfamily

Molecular evolution, we have seen, can occur very rapidly because of the hierarchical nature of changes in structural information: Nucleotide substitutions can change exons; the duplication, divergence, and shuffling of exons can create more complex genes; the duplication of genes can create multigene families; and the duplication of multigene families (and single-gene members) can create gene superfamilies.

The immunoglobulin gene superfamily evolved by exon shuffling to create mosaic proteins (Fig. 21.20c), by exon duplication to create very long genes, by gene duplication to create multigene families, and by the duplication of whole gene families to create the gene superfamily depicted in Fig. 21.20 and Table 21.1.

The First Members of the Superfamily to Evolve Encoded Specific Antigen Receptors

These receptors included antibodies, T-cell receptors, and MHC molecules. Antibodies recognize the three-dimensional shapes of their cognate antigens. T-cell receptors recognize peptide fragments of antigen presented by MHC molecules; the T-cell receptors also react with coreceptors, including the CD4 and CD8 molecules (which are also members of the immunoglobulin gene superfamily), to orchestrate immune responses. Elaborate recognition mechanisms operating at the cell surface are the most common functions of the immunoglobulin superfamily members. Other members of the immunoglobulin gene superfamily evolved to recognize immunoglobulin molecules at the cell surface and/or to transport immunoglobulins across membrane compartments.

Surprisingly, Many Members of the Immunoglobulin Superfamily Carry out Functions Unrelated to Vertebrate Immune Recognition

Some components, such as link protein and HSP-PBPG7, help recognize the cellular matrix. Other components make up the carcinoembryonic antigen (CEA) gene family, which has 10 or more members all involved in specific cell adhesions; the recognition of specific cells by other cells is critical for the development and structural integrity of tissues and organs. Several cell-adhesion molecules guide the growth of neurons in vertebrates (CAM LI and contactin) and invertebrates (falciculum I and almagam). Two other members of the superfamily (MAG and P_0) appear to play a role in the wrapping and compaction of the myelin sheath. Interestingly, several other superfamily

(a)

(b)

Figure 21.20   Evolution of the immunoglobulin homology unit and immunoglobulin gene superfamily.
(a) Model of the evolution of different functional branches of the immunoglobulin gene superfamily showing the emergence of special functional properties. **(b)** Examples of the structures of members of the immunoglobin gene superfamily. Homology units are indicated as loops labeled *V, C,* or *H.* Loops of uncertain relationship to homology units are labeled with a question mark. Different-sized loops illustrate the relative differences in length between the conserved disulfide bond of the labeled types. Membrane-spanning peptides are shown as simple helices. Glycophospholipid linkage to the membrane is represented as an *arrow.* Intracytoplasmic regions are drawn with *wavy lines* that indicate their relative lengths. Extra- and intracellular orientations are indicated by NH_2 and COOH labels on the protein chains, respectively. Possible asparagine-linked carbohydrates are shown as *jagged lines* extending from the protein chains. Note that these sites are not necessarily conserved between alleles or across species, but they are representative of at least one known example of the labeled protein. Functionally related members are illustrated with a single structure, as indicated by the name labels above each structure. **(c)** Examples of exon shuffling using immunoglobulin homology units.

molecules, including Thy I and MRC OX2, are part of the vertebrate immune and nervous systems, raising the possibility that these molecules play a key role in communication between these two complex systems. Other members can encode growth factors or their receptors (PDGF receptor). The existence of so many immunoglobulin superfamily members that do not play a role in the immune response demonstrates how powerful the forces of duplication, divergence, and selection are at making new functional units from older units with a different function.

Immune-Cell Receptors Provide the Backbone of the Immune Response

We saw at the beginning of this chapter that the vertebrate immune response develops the capacity to recognize virtually any foreign macromolecular pattern (antigenic determinant) through the maturation of its two major functional branches. The B cell, or humoral system, manufactures antibodies that bind with and clear unprocessed antigen in viral particles and microbial macromolecules. The T cell, or cellular, immune system uses T-cell receptors and two classes of major histocompatibility complex (MHC) receptors to recognize the processed peptide products of protein antigens. There are two general types of T cells: T cytotoxic cells, which destroy foreign cells as well as intracellular infections and neoplastically transformed cancer cells, and T helper cells, which facilitate the differentiation of B-cell and T-effector-cell responses.

A Close Look at Immune-Cell Receptors

The immune receptors on both B cells and T cells are heterodimers. The antibodies on B cells consist of light and heavy chains; T-cell receptors are composed of α and β chains; class II MHC molecules carry α and β chains, while class I MHC molecules consist of a heavy chain plus a β-microglobulin chain (which is an immunoglobulin superfamily member containing only a single immunoglobulin fold) (Fig. 21.21).

The T-cell and B-cell systems employ very different strategies for the recognition of antigen (see Figs. 21.12a and 21.21). The antibodies of B cells view the native three-dimensional surface of their corresponding antigen by a direct molecular complementarity. T-cell receptors recognize the peptide fragment of their corresponding antigens as presented in a molecular groove on the surface of the MHC molecules. Class I MHC molecules on antigen-presenting cells generally present peptide fragments to T cytotoxic cells, whereas class II MHC molecules generally present peptides to T helper cells. Antibodies and T-cell receptors both employ two types of homology units: variable (V) and constant (C).

The primary structure of the antibody molecule provided the first evidence of the homologous repeat units that are the functional building blocks of the immunoglobulin gene superfamily. In three dimensions, the homology units form a sandwich of β-pleated sheets connected by loops that can be highly variable. It is in the variation of these loops that much of the immune cells' ability to recognize specific antigen resides (review Fig. 21.19).

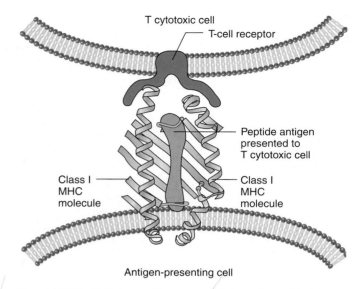

Figure 21.21 Immune-cell receptors. Schematic model of the T-cell receptor recognizing a peptide antigen presented by an MHC molecule. The T cell recognizes both the peptide fragment and portions of the MHC molecule.

Immune-Cell Receptors Define the Basic Features of an Immune Response

An immune response has three characteristics: it is highly **specific,** it is **adaptive,** and it has a **memory.** Immunologic *specificity* arises because each type of T cell or B cell expresses a single type of receptor. The binding of receptor to complementary, or cognate, antigen drives the clonal expansion of B cells and T cells carrying suitable receptors and thus provides the basis for the specificity of immune responses (see Fig. 21.2).

The fact that each lymphocyte expresses only a single receptor is interesting in two regards. It signifies that in a large multigene family, only one gene undergoes expression; and of the two copies of this gene—on the paternal and maternal chromosomes—only one copy is expressed. (This second phenomenon—the expression of only one allele of the gene—is known as *allelic exclusion.*) The DNA rearrangement process for creating the highly variable and thus highly specific antigen-recognition regions is probably responsible for the expression of just one immune receptor gene and just one allele of this gene (review the Fast Forward box on pp. 445–447 of Chapter 13).

The immune response is *adaptive* because it has a large repertoire of different lymphocytes (approximately 10^{12}) that circulate in the blood and lymph. With this constant circulation, foreign organisms receive continuous exposure to many different T-cell and B-cell receptors. This enables specific lymphocyte receptors complementary to the pathogen's antigenic determinants to recognize and interact with them. The large repertoire of lymphocytes collectively has sufficient diversity to respond to all antigens, even the many natural and synthetic ones that did not exist when the immune system emerged. Moreover, in the B cell, or antibody response, there is a process

Human
Heavy chain

κ light chain

λ light chain

Figure 21.22   Three families of antibody genes encode three types of antibody chains: One heavy and two light—λ and κ. The letters *L, V, D, J,* and *C* indicate leader, variable, diversity, joining, and constant regions of the genes, respectively.

called *affinity maturation* in which as the immune response progresses, the fit of the antibody becomes better and better. This affinity maturation is the consequence of *somatic hypermutation,* which we discuss later.

Immunologic *memory* arises when individual lymphocytes are stimulated by antigen to undergo clonal expansion and two types of cells result: effector cells that lead to the destruction of the antigen and memory cells that can remain in the organism for many decades, after which time the same antigen that caused the production of the memory cells can restimulate them to proliferate and produce more effector cells.

Gene Rearrangements and Other Recombinatorial Mechanisms Generate the Immense Diversity of Immune Receptors

Immunoglobulin (or antibody) molecules are encoded by three families of genes: one heavy-chain gene and two types of light-chain genes, kappa (κ) and lambda (λ) (Fig. 21.22). The genes for the variable regions of each heavy chain contain three distinct types of segments: variable (V), diversity (D), and joining (J). The genes for the light-chain variable regions encode V and J gene segments. During the differentiation of lymphocytes that occurs as an individual's immune system develops, the gene segments that will make up the gene to be expressed in each lymphocyte are brought together by DNA rearrangements (Fig. 21.23). Specific signals at the boundaries of the gene segments to be joined and site-specific recombination enzymes mediate these rearrangements. The heavy gene segments undergo rearrangement early in development—first joining a specific D segment to a specific J segment, then joining a particular V to the DJ combination. After the heavy gene rearrangements have occurred, one V to J rearrangement occurs in either the κ or the λ light-chain gene. Allelic exclusion ensures that each B cell expresses only one type of light and one type of heavy chain, which prevents the formation of unproductive antibodies. Each

light-chain and each heavy-chain rearrangement is ultimately selected to produce one useful antibody; if a cell made more than one light chain and more than one heavy chain, unproductive dimers could form. The assembled V gene for both heavy and light chains is joined with an assembled constant gene during RNA splicing to generate a complete mRNA for the immunoglobulin chain.

Five distinct mechanisms of diversification generate the enormous repertoire of specific antibody receptors:

1. *Germ-line diversity.* A multiplicity of V, D, and J gene segments exist for each of the gene families. These are the raw materials for the rearrangements that occur during development.

2. *Combinatorial joining.* With heavy chains, any D can join to any J and any V and join to any DJ. Thus, there are a large number of possible V, D, and J combinations.

3. *Junctional diversity.* Two mechanisms operate to create diversity at the junctions of joined gene segments. Exonucleases can trim back one or both of the gene segments to be joined; and the enzyme terminal deoxynucleotide transferase can add random nucleotides not encoded in the chromosome to the junctional region. These two mechanisms together generate enormous diversity in the joined regions, which in the resulting folded protein happen to represent a major portion of the antigen-binding site.

4. *Somatic hypermutation.* As the immune response proceeds with the proliferation of specific cells into large clones of millions of identical cells, the joined V genes undergo random *somatic hypermutation,* that is, somatic mutation at a rate that is orders of magnitude higher than the typical mutation rate. A single V gene may have 10–15 changes. These somatic hypermutations are often rather extensive. Hypermutation produces many different forms of the gene; those mutations encoding V regions with a better fit for the antigen will be selectively amplified through tighter binding to create "affinity maturation" during the immune response.

5. *Joining light and heavy chains.* Any light chain can associate with any heavy chain, so once again, there is a combinatorial amplification of diversity through the manipulation of information at the protein level. For example, if two cells happen to produce the same light chain, it is highly unlikely they will also produce the same heavy chain; and the joining of a different heavy chain with the same light chain produces two distinct immunoglobulin molecules with distinct antigen-binding sites.

Most of these mechanisms of diversification also apply to T-cell receptors. The two types of T-cell receptors, α/β and γ/δ, are encoded by three distinct gene families: α/δ, β, and γ. The T-cell receptors employ all the diversification mechanisms discussed

Figure 21.23 DNA rearrangements bring together segments of a gene for expression. Expression of rearranged heavy- and light-chain genes in Bμ cells. The genetic rearrangements occur during development of the Bμ. For simplicity, only a single-gene segment of each kind is represented. The letters *L, V, D, J,* and *C* are leader, variable, diversity, joining, and constant regions, respectively; *S* and *M* indicate the secreted and membrane-bound tails of the μ chain, which are generated by alternative patterns of RNA splicing. Thus the IgM molecule (made in Bμ cells) may be attached to the membrane (M), where it will serve as a receptor whose binding with antigen triggers B-cell differentiation, or secreted from the cell to the circulation, where it will serve as a soluble antibody molecule.

for the immunoglobulin receptors, except for somatic hypermutation. The MHC receptors do not employ any special mechanisms for generating diversity; they have achieved their great range of variation via evolutionary mutation and divergence.

Evolution of the Vertebrate Immune Response: A Possible Scenario

The evolution of the vertebrate immune response demonstrates the enormous flexibility and extendibility of the immunoglobulin gene superfamily. The first member of the superfamily may have encoded nothing more than a cellular glue to adjoin various cells of the earliest metazoan organisms. With gene duplication and divergence, various members of the superfamily became tissue-specific cell adhesives. From the DNA evidence, we know that the three types of homology units emerged in this early period of superfamily evolution, and their appearance initiated the rapid acquisition of a variety of different functions (review Figure 21.20).

The Key Event in the Evolution of the T-Cell Receptor Gene Family Was the Acquisition of the Ability to Rearrange Gene Segments

This probably arose from the insertion of a transposable element in the 3′ end of the primordial V gene segment. Once the DNA rearrangement machinery of this transposon came under precise developmental control, the first T-cell receptor gene family acquired the ability to rearrange gene segments. We know now that Rag1 and Rag2, the two enzymes that mediate the rearrangement and joining events of antibody and T-cell receptor genes, evolved from a repetitive element that had a gene capable of transposing DNA (that is, its protein product mediated the cutting and joining of double-stranded DNA). Since all of the rearranging gene families employ similar recognition sequences for their DNA rearrangements, it is safe to assume that the rearranging gene family for immune recognition emerged once—in the T cells of early vertebrates; then duplicated to make successive T-cell receptor gene families; and subsequently, duplicated to generate the antibody gene families. Presumably the primordial T-cell receptor gene family was able to make homodimers and then duplicate to make the early heterodimeric T-cell receptor gene families. Subsequent duplication of one or more of these gene families could have led at a later time to the acquisition of rearranging gene families for the evolution of antibody molecules.

Comparisons of the Sequences of T-Cell Receptor Gene Families Provide a Snapshot of Evolution

The Human Genome Project has created the resources to determine the DNA sequence of entire family branches of the immunoglobulin gene superfamily. As a result, immunogeneticists can study evolution by comparing, for example, the individual genes of the β T-cell receptor family in one species to one another and then comparing β T-cell receptor families of different species (Fig. 21.24). The availability of large-scale DNA sequence analyses that enable one to compare gene families is transforming our understanding of molecular evolution.

The Human β T-Cell Receptor Family

The β T-cell receptor family of humans spans 700,000 bp of DNA on chromosome 7. It contains 45 functional V β gene segments and 19 pseudogene segments, 2 clusters consisting of 1 D segment and 5 or 6 functional J gene segments, with each DJ region associated with a different C β gene (Fig. 21.24). The β locus also has trypsinogen genes or pseudogenes on either side of the V β cluster of genes. This gene family has several interesting features.

Approximately 36% of the locus consists of repeat units such as LINE and Alu sequences. Since the age of many of these repeat sequences can be approximated by noting the extent of their divergence from the ancestral repeat sequence (which itself can be determined because so many repeat element sequences are available), the repeat sequences have been very useful in reconstructing the complicated molecular archaeology of this family.

Within the human β T-cell receptor family, there are seven different genomic modules (duplicated chromosomal segments) ranging in size from 0.7–32 kb. One contains the two D, J, C clusters; a second the five trypsinogen genes that are intercalated between the V and D gene segments. The remaining five modules are associated with the amplification of two or more V gene segments. Indeed, more than 47% of the locus is composed of these modular units.

A portion of the 3′ end of the β T-cell receptor locus has been duplicated and translocated from human chromosome 7 to human chromosome 9. The translocated segment on chromosome 9 contains seven V gene segments and one functional trypsinogen gene. The V gene segments are not functional because the corresponding D, J, and C elements were not translocated. Thus, translocation is one mechanism for the birth of a new multigene family.

Analysis of the promoter regions of the 45 functional V β gene segments reveals that a conserved motif is present adjacent to most of the genes. Presumably this motif represents a site at which transcription factors bind to regulate the expression of these genes.

Rearrangement signals from the V, D, and J gene segments are strikingly similar to their counterparts in antibody genes, which reaffirms the idea that these rearranging gene families had a common origin.

Figure 21.24 Diagrams of the human and mouse β T-cell receptor loci. The *black bars* indicate T-cell receptor elements and the *orange bars* trypsinogen genes. The *full-length bars* represent functional gene elements; the *half bars,* pseudogenes; and the *quarter bars,* badly decayed genes. The *vertical lines* between the mouse and human gene families indicate orthologs.

Comparing Human and Mouse β T-Cell Receptor Gene Families Enables One to Ask and Tentatively Answer Questions About Evolution

One can search the sequences of the two species for conserved regions where recognizable sequence homology has been retained through 65 million years of mammalian species divergence. The conserved regions might represent coding regions, regulatory regions, or regions that are associated with various chromosomal functions, such as DNA replication or compaction.

Detailed comparisons of the V, D, and J genes in the two species can help reveal *orthologs*—equivalent homologs in the V gene families of different species—and make it possible to determine how these genes and their organization have changed in 65 million years of divergence. Preliminary comparisons have shown that very different repetitive sequences have been sprinkled through mouse and human loci (about 45% of the mouse locus consists of repetitive sequences), and the nature of the homology units that have expanded in each species is quite different. Nevertheless, identification of a number of different orthologs suggests that the general organization of these genes has been conserved in the two species.

In both mouse and human genomes, trypsinogen genes are associated with the V β genes. However, there has been an enormous expansion of trypsinogen genes in the mouse as compared to humans. Conversely, there has been an enormous expansion of V genes in the human genome as compared to that in the mouse. The net result of these two expansions is that the loci in both species are roughly the same 700,000 bp in length (Fig. 21.24). The trypsinogen genes are also associated with the T-cell receptor β genes in the chicken, a species that diverged from mammals more than 350 million years ago.

One as yet unanswered question raised by the conserved association of T-cell receptor genes and trypsinogen genes is whether genes that live together work together; that is, do the trypsinogen and T-cell receptor genes share common functional and/or regulatory strategies? This question must be answered separately for each genomic region. Its answer is crucial for understanding the regulation of the genes in a particular region.

CONNECTIONS

A retrospective bird's-eye view of key events that led to an understanding of evolution at the molecular level goes something like this. In 1859, Charles Darwin published *The Origin of Species* in which he inferred from the visible evidence of descent through modification that the diverse organisms alive today evolved from a single primordial form, in large part, by a process of natural selection. Several years later, Mendel published "Experiments on Plant Hybrids" in which he applied the laws of probability to the visible evidence of heredity, inferring the existence of hereditary units that segregate during gamete formation and assort independently of each other. In the early twentieth century, Thomas Hunt Morgan and coworkers gave Mendel's units of heredity a physical location in the cell, establishing the chromosomal basis of heredity and showing not only that genes have chromosomal addresses but also that recombination can separate otherwise linked genes. In the 1940s several people, including Oswald Avery, Martha Chase, and Alfred Hershey, showed that the molecule of heredity is DNA. Then in 1953, James Watson and Francis Crick deciphered the structure of DNA and proposed a mechanism by which the molecule replicates. By the end of the twentieth century, extensive genomic analyses had made it possible to explain how DNA mutates, duplicates, diverges, and is acted on by selection to generate the diversity of life we see around us.

ESSENTIAL CONCEPTS

1. All forms of life on earth are descendants of a single cell—a common ancestor that existed approximately 3.7 billion years ago.

2. Charles Darwin explained how biological evolution occurs through a process of *natural selection.*

3. Natural selection operates on variant forms of inherited traits. The particular variant of a trait that provides the highest degree of reproductive fitness is selected over many generations to become the predominant form in the entire population.

4. New mutations provide a continuous source of variation.

5. Mutations with no effect on fitness are considered neutral. *Neutral mutations* are not acted on by selection and are subject instead to *genetic drift.* Mutations with a deleterious effect on fitness are selected against, while extremely rare mutations with a positive effect on fitness are selected for. Selection can operate simultaneously at hundreds or thousands of variant loci within a population.

6. RNA has a unique combination of properties: It can carry genetic information as well as catalyze chemical reactions. These two properties led scientists to speculate that RNA may have predated the cell as the original independent replicator, or proto-life-form.

7. The fossil record as well as living organisms of all levels of complexity provide scientists with a detailed picture of the evolution of complex life from the first cell to human beings.

8. The evolution of organismal complexity depends on an increase in genome size, which occurs through repeated duplications. Some duplications result from transpositions, while others arise from unequal crossing-over.

9. Mutations rendering genes nonfunctional turn many duplicated genes into *pseudogenes* that over time diverge into random DNA sequences. However, rare advantageous mutations can turn a second copy of a gene into a new functional unit able to survive and spread through positive selection.

10. Sequence comparisons make it possible to construct *phylogenetic trees* illustrating the relatedness of species, populations, individuals, or molecules.

11. The mammalian genome contains genes, *multigene families, gene superfamilies,* genomewide repetitive elements, and simple sequence repeats; repetitive elements in centromeres and telomeres; and unique nongene sequences.

12. Complex genomes arose from four levels of duplication followed by diversification and selection: exon duplication to create larger, more complex genes; gene duplication to create multigene families; multigene family duplication to create gene superfamilies; and the duplication of entire genomes.

13. Genetic exchange between related DNA elements by intergenic gene conversion most often increases the variation among members of a multigene family. Sometimes, however, it can contribute to *concerted evolution,* which creates a family of nearly identical genes.

14. The enormously diverse immunoglobulin gene superfamily encodes cell receptors that carry out different recognition functions at the cell surface in different types of cells. In the cells of the vertebrate immune system, these receptors enable both the specificity and diversity of immune responses. Gene rearrangements and other recombinatorial mechanisms generate the immense diversity of immune receptors during the development of each individual.

SOCIAL AND ETHICAL ISSUES

1. Jill's husband's boss Mary always seems to be going to the doctor for some ailment and has built up quite a store of antibiotics. Whenever she gets a cold, she starts taking antibiotics. She continues them only until she begins to feel a little better. She is now starting to give her children medications in a similar way even though many of the illnesses may be caused by viruses, which do not respond to antibiotics. Jill knows from her training as a physician's assistant that this practice can lead to an increase in bacteria resistant to antibiotics, since the bacterial population evolves by selection. However, she hesitates to interfere with her husband's boss. Should she intervene? What should she do?

2. In the novel *Jurassic Park,* extinct dinosaurs were re-created from bits of DNA found in fossilized insect blood. While there were several unrealistic steps in the re-creation of the dinosaurs, it raised the issue whether there should be controls or laws that would prevent individuals from pursuing this type of re-creation. Who could or would enforce them? What are the problems associated with introducing an extinct organism into today's environment?

3. Evolution is a challenging area of study because many of the hypotheses are not directly testable; instead they require drawing inferences about what happened in the past from what is present today. When new data or new analyses seem to contradict or refute an existing major hypothesis, who decides when the new data are sufficient to invalidate the particular hypothesis?

S O L V E D   P R O B L E M S

I. The sequence of two different forms of a gene starting with the ATG is shown here. Which of the base differences in the second sequence are synonymous changes and which are nonsynonymous changes?

form 1   ATGTCTCATGGACCCCTTCGTTTG

form 2   ATGTCTCAAAGACCACATCGTCTG

Answer

The key to answering this question is understanding the difference between synonymous and nonsynonymous changes in DNA sequence.

Synonymous changes are nucleotide substitutions that do not change the amino acid specified by the DNA sequence. Nonsynonymous changes are nucleotide changes that result in a different amino acid in the protein.

Looking at the amino acids specified by the base sequence,

	Met	Ser	His	Gly	Pro	Leu	Arg	Leu
form 1	ATG	TCT	CAT	GGA	CCC	CTT	CGT	TTG

	Met	Ser	Gln	Arg	Pro	His	Arg	Leu
form 2	ATG	TCT	CAA	AGA	CCA	CAT	CGT	CTG

(The base substitutions in form 2 are underlined.) *The first, second, and fourth A substitutions are nonsynonymous changes; the third A substitution and the C substitution are synonymous changes.*

II. What difficulties would there be if you tried to derive a molecular clock rate using a noncoding sequence in some species and a coding sequence in other species?

Answer

To answer this question, you need to think about how molecular clocks are derived and the constraints on base changes in coding and noncoding sequences.

The evolution of coding sequences is restricted by the fact that the gene sequence needs to be maintained for the gene product to function. The sequence of noncoding regions generally can tolerate many base substitutions without selection acting on these sequences. *Therefore, you would expect more substitutions in the noncoding sequence. The result would be an inconsistency in your clock rate if you are using coding region in some species and noncoding DNA in other species.*

III. If these chromosomes misalign by pairing of repeated sequences (shown as solid blocks) and crossing-over occurs, what will be the products?

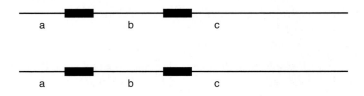

Answer

This question requires an understanding of crossing-over via homology. When you align the homologous repeated sequences out of register, one of the resulting products will have a duplication of the region between the repeats and three copies of the repeated sequence and the other product will be deleted for the DNA between the repeats and contain only one repeated sequence.

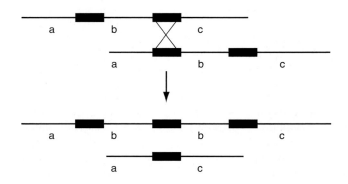

P R O B L E M S

21-1 For each of the terms in the left column, choose the best matching phrase in the right column.

a. ribozymes

b. retrotransposition

c. SINES

d. punctuated equilibrium

e. molecular clock

f. phylogenetic trees

g. exon shuffling

1. constant rate of change in amino-acid sequence

2. sudden explosive evolutionary changes

3. exchange of pieces of genes

4. RNA molecules that catalyze specific chemical reactions

5. short repeated sequences in human genomes

6. RNA intermediate in duplication of genetic material

7. representation of evolutionary relationships

21-2 What observations support the unity of life concept that all life-forms evolved from a common ancestor?

21-3   Which of the following statements is support for RNA being the first replicator molecule?
  a. RNA molecules can function as enzymes.
  b. DNA is more stable than RNA.
  c. Information can be encoded in RNA.

21-4   a. In what ways is RNA not a good information storage molecule?
  b. In what ways is RNA not as good as protein as a molecular machine?

21-5   a. A particular chemical reaction that occurs between two proteins is sensitive to RNase treatment *in vitro* (that is, it will not occur if the reaction mix is pretreated with RNase), but it is not sensitive to protease treatment. What would you propose about the nature of the enzyme that carries out the reaction?
  b. Another chemical reaction is sensitive to both RNase and protease treatment. What would you propose about the nature of the enzyme that carries out this reaction?

21-6   Humans and chimps have a 1% difference in their genomic sequence, while two humans have a 0.5% sequence difference. How could something as small as a 1% difference in DNA sequence lead to dramatic differences seen in chimps versus humans? Speculate on the types of differences the 1% variation may represent.

21–7   Rates of nonsynonymous and synonymous amino-acid substitutions for three genes that were compared in humans and mice or rats are shown here. (The rates are expressed as the average number of substitutions per base per 10^9 years, with the standard deviation given. In each case, the number shown is the rate at which the human and rat sequences have diverged from each other, by either a nonsynonymous or a synonymous substitution.)
  a. Why do the rates of nonsynonymous substitutions vary among these genes?
  b. Why are the rates of synonymous substitutions similar?

Protein	Nonsynonymous substitutions	Synonymous substitutions
Histone 3	$0.0 \pm 0.0 \times 10^9$	$6.38 \pm 1.10 \times 10^9$
Growth hormone	$1.23 \pm 0.15 \times 10^9$	$4.95 \pm 0.77 \times 10^9$
β globin	$0.8 \pm 0.13\,3 \times 10^9$	$3.05 \pm 0.56 \times 10^9$

21-8   Synonymous mutations are more prevalent than nonsynonymous mutations in most genes. The immunoglobulin (Ig) genes (encoding antibody subunits) are an exception where nonsynonymous base changes outnumber synonymous changes. Based on the function of the Ig genes, why do you think this might be true?

21-9   Mutations in the *CF* gene that cause cystic fibrosis are carried by 1 in 20 individuals of Caucasian ancestry. The disease is clearly detrimental, yet the allele is maintained at a relatively high level in the population. What does this paradox suggest about the effect of the mutation?

21-10   Unequal crossing-over between two copies of a gene can lead to duplication and deletion on the two homologs involved. How could a single gene become duplicated?

21-11   Human beings have three-color vision, while most other species of animals have two-color vision or one-color vision (that is, black and white). Three-color vision is produced by the products of a three-member, cross-hybridizing gene family that encodes light-sensitive pigments active in different ranges of the color spectrum (red, green, and blue). What is the most likely molecular explanation for the evolution of three-color vision in the ancestor to human beings?

21-12   How do transposition and unequal crossing-over differ based on the location of final copies of the duplicated sequence?

21-13   You have identified an interesting new gene that appears to be involved in human brain development. You have discovered three cross-hybridizing copies of this gene within the human genome (A, B, and C). In the mouse genome, there is only a single copy of this gene (named M) and in the frog *Xenopus*, there is also only a single copy (named X). You have sequenced the same 10,000 bp of open reading frame from each of these genes and calculated the number of base-pair differences that exist between different pairs with the following results:

Comparison	Number of base-pair differences
A versus B	300
B versus C	10
A versus C	300
A versus M	600
A versus X	3000

If you assume that a constant rate of evolution has occurred with all members of this gene family, and you assume that mice and humans evolved apart 60 million years ago:
  a. How long ago did frogs split from the line leading to mice?
  b. How many gene duplication events are observable within these data? At what time in the past did each of these duplication events occur?
  c. Mapping of the A, B, and C genes shows that A and C are very closely linked, while B assorts independently from either of these genes. With this

linkage information, what can you say about the molecular nature of the duplication events that occurred along the evolutionary line leading to human beings?

21-14 Organisms have a characteristic percentage of GC bases in their genomes. In a species of halophilic (salt-loving) bacteria containing two essential plasmids and one chromosome, the plasmids have a different GC content than the chromosome. What could you hypothesize about the origin of the plasmids?

21-15 When a phylogenetic tree was constructed using comparisons of glucose-6-phosphate isomerase amino-acid sequences from a wide variety of species including animals, plants, and bacteria, the bacterium *E. coli* was placed on a branch of the phylogenetic tree with a flower. What explanation could there be for this unusual association based on this one protein-coding sequence? (Recall that bacteria and plants are in very different locations on the evolutionary trees derived from analyses of several gene sequences or morphology and physiology.)

21-16 Phylogenetic trees of primates constructed using chromosome mutations (deletions, insertions, inversions, etc.) show the same relationships of species as those constructed using base substitutions within a gene. Which type of genetic alteration would you expect to have the greater impact on the evolution of chimp and human from a common ancestor. Why?

21-17 What is the unit that can be duplicated and modified to form
a. a new gene?
b. a multigene family?

21-18 It has been hypothesized that in the evolution of vertebrates, there were two successive doublings of the genome (tetraploidization) to produce the vertebrate genome.
a. How does the *Hox* superfamily fit with this hypothesis?
b. Among vertebrates, there is variation in the numbers of genes present within the *Hox* gene family clusters. How could this variation arise?

21-19 LINEs and SINEs are considered to be selfish DNA, yet they can, in some instances, confer a selective advantage on the organism. What are two ways in which a LINE or SINE can change the genome?

21-20 How is the size polymorphism of dinucleotide repeated sequences thought to occur?

21-21 a. What is an example of repetitive noncoding DNA for which we know no physiological function?
b. What is an example of repetitive noncoding DNA for which we know a cellular function?

21-22 Match the observation on the right with the event that it suggests occurred during evolution.

Event	Evidence in the genome
a. concerted evolution	1. genes in different individuals of the same species have 6, 8, or 10 homology units in an immunoglobulin gene family member
b. exons and introns of a gene	2. several copies of the same gene that are identical in the genome
c. unequal crossing-over	3. poly-A sequences at the end of coding regions in genomic DNA
d. retrotransposition	4. blocks of DNA sequence conserved between species separated by nonconserved blocks

Saccharomyces cerevisiae: Genetic Portrait of a Yeast

Contributing author Michael R. Culbertson *University of Wisconsin*

Yeast colonies, each containing about 10^7 cells, grow from single cells spread on solid media.

*S*accharomyces cerevisiae, commonly known as baker's yeast, has been a preferred organism for genetic research since the mid–twentieth century (Fig. A.1). The experimental value of this single-celled eukaryotic species lies in its simple life cycle, alternating haploid and diploid phases, short generation time, and easy-to-identify meiotic products. Recent technological advances enable yeast biologists to use genetic analysis to gain a deep understanding of the organization and regulation of eukaryotic cells.

Like all eukaryotic cells, yeast cells make decisions that determine their fate. They decide, for example, whether they are *a* or *α* cells, and whether to mate, enter the meiotic pathway, grow and divide by mitosis, or bud. Researchers can use genetic techniques to analyze the molecular events that influence these fate-determining decisions; and knowledge of these events, which include the activation and inactivation of transcription factors, can contribute to an understanding of how the cells of multicellular organisms differentiate.

Three main themes unify our genetic portrait of yeast. First, to identify the genes and proteins influencing development, researchers isolate and study mutants defective for a specific process. Second, master regulatory genes in yeast control the expression of other genes, whose activities determine the phenotype of a cell. Third, cascades of proteins called *signal transduction systems* (described in Chapter 18) transmit the signals responsible for shifting a cell from one phenotype to another.

Figure A.1 Yeast cells. Phase contrast micrograph of yeast cells showing small and large buds characteristic of different phases of the mitotic cell cycle.

739

In examining yeast as a model organism for understanding development, we present:

- An overview of yeast in the laboratory: current knowledge of the yeast genome, basic tools for looking at yeast, and significant details of the yeast life cycle.

- Cell differentiation: molecular mechanisms for determining cell type.

- Mating: how cell-to-cell communication through pheromones promotes the conversion of haploid cells to diploid cells.

An Overview of Yeast in the Laboratory

Yeast geneticists have gathered enough information on *S. cerevisiae* to state with confidence that single-celled eukaryotes such as yeast express genes, organize themselves, perform biological functions, and differentiate using variations of the same processes found in multicellular eukaryotes. Since cellular differentiation is a driving force behind the development of a multicellular organism from a fertilized egg, what we learn from yeast can very likely tell us a great deal about our own development, biochemistry, genetic makeup, and genetic history. This is the main reason why research on a seemingly irrelevant creature like yeast is funded by the National Institutes of Health under the umbrella of medical research. The experimental advantages of yeast include the fact that unlike the cells of most multicellular eukaryotes, yeast cells can grow as haploids or diploids. This makes it possible to identify recessive mutations in haploid cells and then combine mutations in diploid cells for complementation analysis. An equally important experimental consideration is that one can look at more cells in a shorter period of time with yeast than with any other eukaryotic organism; in fact, it is possible to obtain as many as 10^8 yeast cells in 1.0 ml of culture, which makes it feasible to study rare events (mutations).

The Nuclear Genome of Yeast

Haploid *S. cerevisiae* has 16 linear chromosomes in the nucleus, all constructed according to the same blueprint. Each chromosome contains one centromere/kinetochore that binds to one microtubule, two termini consisting of longer subtelomeric repeats followed by short telomere repeats at the very ends, and multi-

ple origins of replication spaced approximately 30–40 kb apart (of which only a small subset are used to initiate replication at any one time) (Fig. A.2a). Yeast chromosomes are packaged into nucleosomes consisting of the core histones H2A, H2B, H3, and H4.

Physical Maps and the Complete Sequence of the Genome

Pulse-field gel electrophoresis, which separates intact yeast chromosomes, produces molecular "karyotypes" of the yeast genome, such as the one shown in Fig. A.2b. Geneticists use these karyotypes to confirm the number of yeast chromosomes and to observe major alterations in chromosome structure. As Fig. A.2b indicates, chromosome I, about 235 kb in length, is the smallest yeast chromosome. Chromosome XII is the largest; its size varies between about 2060 and 3060 kb because of a variable number of tandem ribosomal RNA genes (rDNA)—usually in the range of 100–200 copies—in different strains. The total genome length is 12,000 kb plus the 2000–3000 kb of rDNA. DNA sequence data from the International Genome Project became available for 100% of the genome by April of 1996. Current information on the yeast genome can be found at http://genome-www.stanford.edu/Saccharomyces/.

Genetic Maps

The classical genetic map of yeast currently shows the locations of over 1000 markers determined by tetrad analysis. The total genetic map length, a function of the frequency of meiotic recombination, is about 4400 cM. With an average of 3 kb/cM, genetic distance is roughly proportional to physical distance (Fig. A.3). Exceptions are the rDNA repeats, in which there is

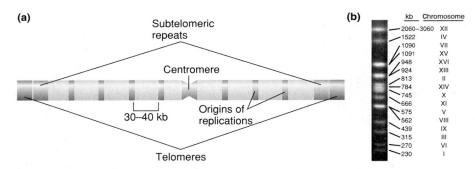

Figure A.2   Yeast chromosomes. (a) Elements of a chromosome. Each linear chromosome contains one centromere, multiple origins of replication, subtelomeric repeats, and telomeres at the ends of the linear DNA. **(b)** Molecular karyotype of yeast chromosomes. Using pulse-field gel electrophoresis, the chromosomes can be separated by size.

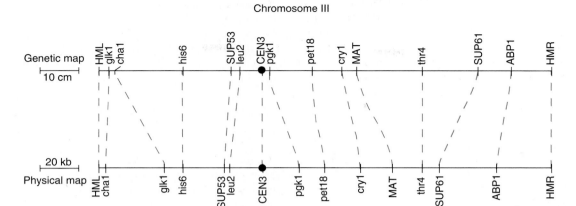

Figure A.3 Comparison of genetic and physical maps. The location of genes on chromosome III as determined by recombination distances (genetic map) and sequencing DNA (physical map) are shown. While the order of genes is the same, some distances determined by these two measurements differ.

virtually no meiotic recombination, and the regions surrounding the centromeres, where recombination is modestly suppressed. Because *S. cerevisiae* has an unusually active recombination system, the total genetic map of yeast is extraordinarily long compared to the genetic map of other fungal organisms. *Neurospora crassa,* for example, has about the same DNA content as yeast but a genetic map that is only 15% as long as yeast's. The extremely active homologous recombination system in yeast might explain why plasmids in yeast integrate primarily by homologous recombination, whereas plasmids in other organisms integrate mainly by nonhomologous recombination.

Determining the Number of Genes in Yeast

The Human Genome Project, completed in 2001, has taught researchers that estimating the number of protein-coding genes in an organism is surprisingly difficult. Before completion of the project, human geneticists estimated the existence of 100,000 human genes, about 16 times as many as were predicted for yeast. After the sequence became available, the estimate dropped to about 40,000–60,000 human genes or 6–10 times as many as in yeast. The estimate of the number of human genes will undoubtedly undergo repeated revisions in the coming years. In yeast, there is still debate over the number of protein-coding genes. To predict the number of genes that code for proteins from the number of open reading frames (ORFs) identified in the yeast genomic DNA sequence, an arbitrary cutoff of 100 amino acids was established as the minimum length of a functional open reading frame. The need for a cutoff derived from the fact that about 260,000 ORFs were identified assuming a length of 2–99 amino acids and about 114,000 ORFs assuming a length of 15–99 amino acids. Most of the short ORFs do not code for proteins (Fig. A.4). Using the arbitrary cutoff of 100 amino acids, the genome contains 7505 ORFs that potentially code for proteins. However, the cutoff is useful only as an approximation. Two hundred ten ORFs less than 100 amino acids in length now appear to code for proteins and a substantial num-

ber of ORFs greater than 100 amino acids in length may not code for proteins. Thus, since the completion of the yeast genome project in 1996, it is still not possible to say exactly how many protein-coding genes exist in yeast. The number is likely to be between 6000 and 6500 genes. Among the protein products of these genes, about 2000 have no known function, have not been identified by genetic or biochemical approaches, and share no sequence similarity to other characterized proteins. For this reason the genes encoding them are called "orphan genes." Following completion of the DNA sequence, a systematic effort was launched to disrupt the function of every yeast gene one at a time to determine how many genes are required for growth. Only 20% of yeast genes cause lethality when disrupted. Although this indicates that the majority of yeast genes are not required for growth under laboratory conditions, they presumably confer an evolutionary advantage for yeast in its natural habitat.

Among the protein-coding genes, introns occur in only 4% to 5% of all yeast genes (276 genes contain a single intron and 7 genes contain two separate introns). By comparison, most genes of higher eukaryotes are composed of large numbers of noncontiguous coding regions (exons) separated by introns, sometimes hundreds of them. The low frequency of introns in yeast may be related to a similarly low frequency of "processed" pseudogenes, which, like introns, are common in higher eukaryotes. Pseudogenes, we have seen, contain coding sequences that resemble those of real genes, but because the pseudogenes lack introns and transcriptional promoters, the coding sequences are not expressed.

Protein-coding genes are not the only types of functional genes. The yeast genome also contains numerous genes that produce noncoding RNAs, including reiterated genes coding for the ribosomal RNAs, 274 tRNA genes of which 80 contain a special class of introns, 71 small nucleolar RNAs that function in ribosomal RNA processing, 5 small nuclear RNAs that function in intron splicing, a few RNAs of unknown function, and three RNAs that serve as functional subunits of the enzymes RNase P, endoribonuclease MRP, and telomerase. The former

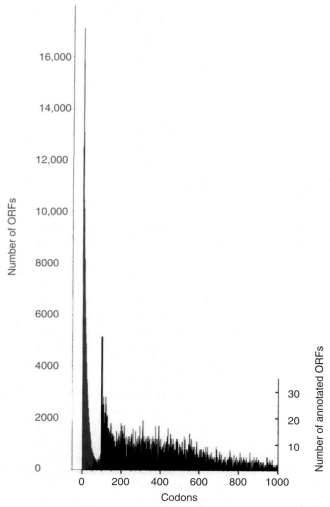

Figure A.4 Analysis of open reading frames in the yeast genome. The total number of ORFs of the indicated length encoded in the *S. cerevisiae* genome are shown in *red*. The total number of annotated ORFs in the *Saccharomyces* Genome Database (SGD) are shown in *green*. The scale for the total number of ORFs is compressed 10-fold relative to the number of annotated ORFs. The difference in magnitude is therefore underrepresented by 100-fold. The *black vertical line* indicates the 100-amino-acid cutoff chosen for annotating the genes.

are involved in RNA processing, whereas telomerase is required for replication and maintenance of telomeres at the end of each chromosome arm.

Gene Arrangement, Genome Organization, and Genome Evolution

The bulk of the yeast genome is composed of unique, nonreiterated DNA sequences. Approximately 75% of the genome is transcribed into RNAs; and most of these RNAs are nonoverlapping. On average, there is one yeast gene for every 1.5–2.0 kb of DNA. Genes related in function are often coregulated, but they are not usually clustered according to function or organized like bacterial operons. In the rare instances where functional clustering does occur, as with the *GAL* genes that encode enzymes active in the transport and degradation of galactose, the clustered genes are usually transcribed into mRNA from separate, divergent transcriptional promoters.

Although the yeast genome does contain some reiterated DNA sequences, the proportion of the genome devoted to reiterated DNA is much lower than in multicellular eukaryotes. The 100–200 copies of rDNA rank highest in abundance. Other repeated sequences of note include tRNA genes, which occur on average in five to seven dispersed copies per tRNA isoacceptor (that is, per group of different tRNAs that accept the same amino acid but possess different anticodons); the subtelomeric repeats, which are reiterated one to four times near the ends of the chromosomes; and a complex family of retrotransposons called Ty elements, of which there are about 50 dispersed complete copies per haploid genome. In addition, there are 385 incomplete dispersed copies that contain only the ends of the intact Ty elements (called long terminal repeats, or LTRs). These solo LTRs are the sequences left behind when an intact Ty element is deleted from the genome through recombination between the LTRs at either end of the intact element. Solo LTRs represent the evolutionary footprints of intact Ty elements during evolution. The genetically mobile Ty elements resemble animal cell retroviruses except for the inability to undergo a lytic infection cycle. Intact Ty elements are 5.4–6.3 kb in length and can pair through homologous sequences and recombine with each other even when copies of the elements are on different chromosomes. Such recombination between dispersed transposable elements may be a major source of reciprocal translocations, inversions, and deletions in yeast. Multicellular eukaryotes have a variety of transposable elements that may also be a significant source of these types of chromosomal aberrations.

It has become clear from yeast genome data that the genome evolved through duplications of large chromosomal segments followed by reciprocal translocations and other large-scale rearrangements. Figure A.5 shows a map of the yeast genome with 53 pairs of otherwise nonhomologous chromosomal segments identified as having been originally derived by duplication. One theory explaining the structure of the modern yeast genome is that the organism evolved from an ancestral degenerate tetraploid. This tetraploid resulted from whole-genome duplication that occurred perhaps 100 million years ago, after the divergence of *Saccharomyces* and the related yeast *Kluveromyces.* Following the whole-genome duplication, many segments of the duplicated chromosomes were deleted, mutated, and rearranged by reciprocal translocations between nonhomologous chromosomes. Eventually the minimal ploidy needed for survival was reduced to the haploid state. The genome was left, however, with the remnants of the ancient duplications and still retains many duplicated genes scattered around the genome. Yeast geneticists estimate that about 15% of all yeast proteins are encoded by duplicated, unlinked genes.

Figure A.5 Chromosomal duplications in the yeast genome. View of 53 clustered gene duplications among the 16 chromosomes of yeast. Each chromosome is represented by a *bold horizontal line*. The chromosomes are aligned according to the positions of the centromeres shown as a *vertical line*. Gene cluster duplications are represented by *colored polygons*. Ty elements and telomeres that represent multiply repeated sequences are not included in the diagram.

Examples of duplicated genes include gene pairs that code for transcription factors, translation factors, protein kinases, contractile proteins like myosin and tubulin, the cyclins, the histones, and the mating pheromones.

The genes identified by the genome project can be thought of as a blueprint from which we can generate a parts list for building yeast cells. The parts include the noncoding RNAs and all the proteins that form structures, catalyze reactions, and regulate gene expression. Knowing all the parts, however, does not by itself reveal how the parts work together to accomplish a biological process such as mating. The completion of the genome sequence has thus given birth to a new field called *functional genomics* in which it is possible to assay the expression of all genes simultaneously in a single experiment using DNA chips (see Chapter 10). The ability to generate genomewide gene expression profiles with DNA chips is transforming the way we approach genetic analyses and broadening our ability to visualize how all the parts of a cell function together. Examples of functional genomic approaches are described in the final section on cell-to-cell communication.

In addition to genomics, *proteomics*—the characterization of all proteins in terms of their structures, functional domains, interactions, and expression levels—represents another challenge for the future. The merging of information from genomics and proteomics will bring us to the next level of understanding yeast as an organism. Information on yeast proteomics can be found at the Munich Information Center for Protein Sequences (MIPS, http://mips.gsf.de/).

Basic Tools of the Yeast Geneticist

Several techniques of yeast genetics enable researchers to answer questions that are difficult, if not impossible, to resolve in other organisms. However, although many of the techniques described next were developed for the study of yeast, creative adaptations have turned them into similar tools for the study of other organisms.

Transformation: An Efficient System Relying on Homologous Recombination

Yeast cells are susceptible to efficient transformation by DNA molecules from external sources. To execute a transformation experiment, researchers construct a plasmid carrying a marker in the form of a functional yeast gene and expose a population of yeast cells to a solution of the plasmids. They then confirm the presence of the plasmid in transformed cells by ascertaining whether the marker gene has supplied a function missing in the transformation recipients, or host cells. For example, a plasmid carrying the yeast *LEU2* gene, coding for an enzyme in leucine biosynthesis, causes a host cell carrying a loss-of-function mutation in the *LEU2* gene to become Leu$^+$ in phenotype. The plasmids that transform yeast fall into two categories: those that cannot replicate on their own in yeast cells and those that can. YIp plasmids are circular plasmids that lack the ability to replicate on their own in yeast cells (Fig. A.6a). YIps produce stable transformants by integrating into the chromosomal DNA through genetic recombination. Such integration in yeast almost always occurs through homologous recombination, that is, through alignment (pairing) and genetic recombination with a similar sequence. By contrast, autonomously replicating plasmids maintain transforming genes in an extrachromosomal state. Examples are yeast episomal plasmids (YEps) and yeast centromeric plasmids (YCps), both of which replicate in the yeast cell nucleus (Fig. A.6b and c).

YEps replicate autonomously because they carry the replication origin from an endogenous yeast DNA plasmid known as the 2 μm circle. YEps are present in about 20 copies per cell. Because of this relatively high copy number, YEps are useful in experiments requiring the overproduction of a gene. YCps replicate autonomously because they carry DNA corresponding to a chromosomal origin of replication. They also carry the centromere of a yeast chromosome and, consequently, segregate like a chromosome. Because they contain a centromere, YCps have a copy number of one per cell, and the markers they transport segregate like chromosomes during meiosis. However, even though YCps behave like chromosomes, they are more prone than natural chromosomes to nondisjunction events that lead to duplications and deletions.

Moving Genes Between Yeast and Bacteria

The autonomously replicating plasmids just described are designed to function as "shuttle vectors" able to ferry genes into both bacteria and yeast for replication. To this end, they all carry an origin of replication from a bacterial plasmid that allows replication as a multicopy plasmid in *Escherichia coli*. Most also carry genes conferring on *E. coli* resistance to ampicillin or tetracycline. Such resistance makes it possible to select for *E. coli* transformants that carry the plasmids. Completing the prerequisites necessary for shuttle vectors to move genes between yeast and bacteria are design features that enable the vectors to replicate and be selected for in yeast. With vectors carrying all these features, researchers can use recombinant DNA techniques on DNA prepared in *E. coli* and then place the DNA manipulated by recombinant technology back into yeast to test its function. For example, studies of the deletion of DNA sequences around the centromere have shown that a functional yeast centromere need contain only 120 bp of DNA.

Gene Replacement, or Genetic Microsurgery, Makes It Possible to Alter Genes Without Resorting to a Formal Genetic Cross

Gene replacement is a very powerful technique that has a great advantage over classical methods of strain construction: Researchers can use it to alter at will a chromosomal gene in a living cell without changing the genome in any other way. This level of control over gene alteration makes it possible to compare the phenotypic effects of mutations in an otherwise isogenic (that is, genetically identical) background and attribute any phenotypic differences to specific alleles.

For example, researchers can replace copies of genes under analysis with alleles constructed *in vitro*. To create such alleles, either they use recombinant DNA technology to delete or insert DNA in a cloned version of a gene, or they use a variety of *in vitro* methods to induce specific mutations. They then introduce the altered gene into yeast via a transformation that replaces the wild-type allele with the modified gene. Figure A.7 illustrates gene replacement via transformation using a DNA fragment carrying a disrupted gene. The disrupted gene includes a functional *URA3* gene (used to select for transformants) inserted into

(a)   Plasmid   Properties

YIp

- Cannot replicate autonomously
- Integrates at region of homology, that is, *LEU2*
- One copy per cell

(b)

YEp

- Replicates autonomously
- Many copies per cell

(c)

YCp

- Replicates autonomously
- One copy per cell
- Segregates in meiosis like a chromosome

LEU2	Yeast gene	CEN	Yeast chromosomal centromere
2 μm	2 μm origin of replication	AMP	Bacterial antibiotic resistance gene
ARS	Yeast chromosomal origin	ORI	Bacterial origin of replication

Figure A.6   Plasmids in yeast. (a) YIp plasmids do not contain an origin of replication and therefore cannot replicate autonomously (separate from the chromosome). If the plasmid integrates into a chromosome by homology it is maintained and transmitted to subsequent generations of cells. **(b)** YEp plasmids contain an origin of replication from the 2 μm circle and therefore replicate autonomously. **(c)** YCp plasmids contain a centromere in addition to a replication origin, so they are maintained autonomously and segregate during meiosis as a linear chromosome does.

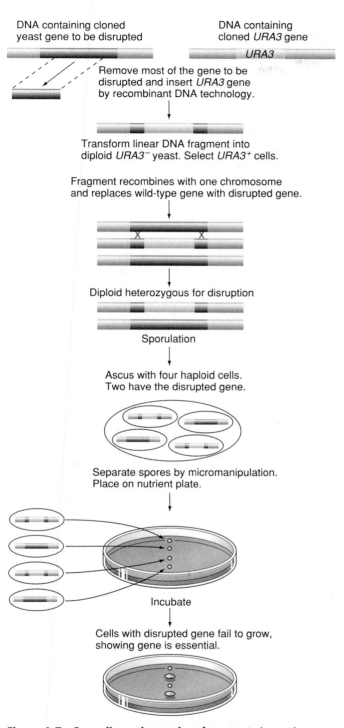

Figure A.7 Gene disruption and replacement. A gene is disrupted by first constructing a linear fragment in which the *URA3* gene (to be used as a selectable marker) replaces the middle of the gene of interest. When the fragment is transformed into a diploid yeast cell, it can recombine with the gene in one of the chromosomes by homology, replacing the good copy with a mutant. The heterozygous diploid, when sporulated, produces two haploid spores containing the disrupted copy and two spores with the normal copy of the gene. If the gene is needed for growth, the spores with the disrupted gene copy will die.

the gene to be disrupted. Chromosomal integration of the disrupted gene occurs through the equivalent of a double crossover.

Since some genes are essential for growth and viability, researchers generally use disruption alleles to transform wild-type diploids. Such transformation produces $-/+$ heterozygotes containing one disruption allele and one wild-type allele. If the disruption allele confers recessive loss of viability, the diploid will be viable. In every tetrad generated by sporulation of the diploid, however, the two haploid spores that carry the disruption mutation will not survive, whereas the two spores that carry the wild-type allele will divide to form colonies (Fig. A.7).

The Yeast Life Cycle

Yeast cells alternate between haploid (1*n*) and diploid (2*n*) phases in which new daughter cells arise mitotically as buds that grow in size and eventually split from the mother cell (Fig. A.8). Haploid cells occur in two mating types, *a* and α. Both mating types can reproduce mitotically as stable haploid cells. Or they can engage in sexual reproduction, in which cells of opposite mating types communicate with one another by proteins known as pheromones. The pheromones induce the dissimilar cells to

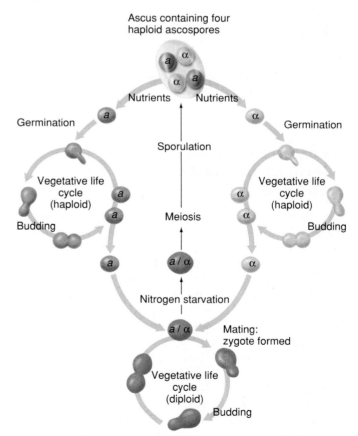

Figure A.8 Life cycle of yeast. Yeast can grow vegetatively as either haploid or diploid cells. The transition from haploid to diploid occurs via mating, and the transition from diploid to haploid occurs via meiosis during sporulation.

undergo cell fusion followed by nuclear fusion. The zygote produced in this way has a single diploid nucleus and buds to produce diploid progeny. Diploid yeast cells can propagate as stable diploid cells by mitotic division. Starvation, however, induces them to undergo meiosis and sporulation. The ability to undergo meiosis allows the yeast cells to "reshuffle" their genes when conditions are poor, perhaps enabling them to find a combination more suitable for survival in the environment at hand. Meiosis reduces the diploid nucleus to four haploid nuclei, which become encapsulated in four haploid spores. On the addition of nutrients, the haploid cells bud and can divide mitotically.

Yeast populations must experience an alternation of feast and famine to complete the life cycle. Nutrient depletion induces meoisis and sporulation, while the subsequent availability of nutrients promotes spore germination and gamete production. The association of haploid and diploid phases with periodic fluctuations in environmental conditions is common among eukaryotic microorganisms.

Sex in yeast is determined by the mating type locus (designated as MAT) on chromosome III. As already mentioned, there are two mating types: *a* and α. Mating ability segregates $2a : 2\alpha$ in tetrads derived from MAT*a* / MATα heterozygous diploids, indicating that the *a* and α mating types are specified by alleles of a single locus (MAT). MAT*a* or MATα cells mate efficiently with cells of the opposite sex. Heterozygous MAT*a* / MATα diploids are sterile, but it is possible to derive MAT*a* / MAT*a* or MATα / MATα diploid cells. These diploids will mate with cells of the opposite mating type, either haploids or diploids. The ability to mate is thus determined by the genetic configuration at the MAT locus and is not related to ploidy per se.

Cell Differentiation in Yeast: Mechanisms for Determining Cell Type

Like the specialized liver, muscle, and nerve cells of animals, the *a*, α, and *a* / α cells of yeast are differentiated cells that express different sets of genes. The three cell types in yeast differ in how they respond to signals from the environment and in their developmental potential. Three different genetic programs establish the differentiation pathways leading to each cell type. Because the establishment of cell identity is basic to the differentiation and development of multicellular organisms, our understanding of the process in yeast serves as a model for development in multicellular eukaryotes.

The MAT Locus Controls Expression of the Genes That Determine Cell Type

A cell with MAT*a* information at the MAT locus has an *a* mating type. By contrast, a cell carrying MATα information at the MAT locus has an α mating type. What at the MAT locus determines these different mating types?

Clues to the role of the MAT locus first emerged from mutant analyses. When sterile mutants of α cells were isolated, one class of mutants had defects at the MAT locus. Surprisingly, the

MATα mutations fell into two complementation groups, indicating that instead of a single gene at the MATα locus, there are two genes, α*1* and α*2*. Mutations in the α*1* and α*2* complementation groups produce different molecular effects, which inform us of the roles of the genes. Strains carrying α*1* mutations fail to synthesize α pheromone, a protein normally made by α cells. Strains carrying α*2* mutations synthesize α pheromone but degrade it before it can be secreted; normally, only *a* cells express the phenotype of α-pheromone degradation.

What hypothesis could explain the roles of α*1* and α*2* in determining cell type? According to the proposal illustrated in Fig. A.9a, MATα is a complex locus consisting of two genes, α*1* and α*2*. The α*1* gene encodes an activator of the α-specific function required to elicit the α cell type. In the absence of a functional α*1* product, α cells are sterile, in part because they do not produce α-specific proteins such as α pheromone. The α*2* gene encodes a repressor of *a*-specific functions. Cells lacking a functional α*2* gene product express both α-specific and *a*-specific genes, and the simultaneous expression of both sets of genes is incompatible with mating. $\alpha1^-\alpha2^-$ double mutants mate like *a* cells because they fail to turn off *a*-specific functions and fail to turn on α-specific functions. In effect, a simple network depending on two regulatory proteins—one an activator, the other a repressor—accounts for the *a* and α cell types. In this network, α*1* and α*2* act as master regulatory genes whose products control the expression of a large number of genes required for the appropriate behavior of *a* or α cells.

What is the role and makeup of the MAT locus in an *a* cell? Mutant analyses of the MAT*a* locus indicate that it has a slightly different configuration from that of the MATα locus (Fig. A.9b). No sterility-producing mutations have been isolated that map at the MAT*a* locus. Haploid strains carrying MAT*a* mutations mate normally and fall into a single complementation group known as *a1*. The lack of any detectable haploid phenotype for MAT*a* mutations suggests that the *a1* protein performs no function necessary for mating or differentiation into *a* cells. It is the lack of α*1* and α*2* functions that makes *a* cells what they are. *a* cells do not express α-specific functions because they lack the α*1* activator, and they constitutively express *a*-specific functions because they lack the α*2* repressor. In short, the *a* cell type arises through a default pathway requiring no action on the part of the MAT-encoded regulatory proteins.

What happens in a MAT*a* / MATα heterozygous diploid? According to the model (Fig. A.9c), the *a*1 and α*2* regulators act in concert to perform a series of negative functions. The *a*1-α*2* coregulators, acting together, repress expression of α*1*; as a result, the α-*specific* genes (or α*sg* genes) are not activated. The *a*1-α*2* coregulators also repress expression of *haploid-specific* genes (*hsg* genes) that normally function in both *a* cells and α cells but not in *a* / α diploids, such as *HO* (see later). The coregulators further repress expression of the *rme1* gene. The presence of the Rme1 protein inhibits the meiotic pathway; its absence, therefore, confers competence for meiosis. Finally, the α*2* regulator alone represses *a-specific* genes (*asg* genes). The net result is that *a* / α cells are sterile but competent to undergo meiosis.

(a) MAT locus of α cell

Repressor ⊣ a̶s̶g̶ (*a-specific* genes)

αsg (*α-specific* genes)

hsg (*haploid-specific* genes)

Activator

RME ⊣ Meiosis blocked

(b) MAT locus of a cell

asg

α̶s̶g̶ No α1 activator.
αsg not activated

hsg

RME ⊣ Meiosis blocked

(c) MAT loci of a / α cell

Repressor

α2 ⊣ a̶s̶g̶

α̶s̶g̶ No α1 activator.

a1, α2 corepressors

⊣ h̶s̶g̶

RME ⟶ Meiosis

Figure A.9 MATa and MATα loci. Two different versions of the MAT locus, MATa and MATα, share some common DNA sequences (shown in *pink* and *green*) but also contain unique sequences known as Ya and Yα. The presence of Ya, Yα, or Ya + Yα in a cell has different consequences on expression of the *a-specific, α-specific, haploid-specific,* and *meiosis (RME)* genes because activators and repressors are encoded by the different versions of the MAT locus. *Arrows* indicate the location of transcripts for the α1, α2, and a1 genes at the MAT locus.

MATα   Yα   α2   α1

1427 bp   747 bp   327 bp

MATa   Ya   a1

1427 bp   642 bp   327 bp

Figure A.10 Structure of the MAT locus. The MAT locus consists of sequences that are found in both MATa and MATα versions with the Ya or Yα block of DNA between them. The Y DNA contains promoters and early sequences for genes α1 and α2 in the Yα version, and a1, in the Ya version. (Part of each of these genes is encoded in the common sequences that surround Ya or Yα.)

Verification of the α1-α2 hypothesis, which was proposed on the basis of mutant analysis, has come from various molecular studies, including the cloning and sequencing of the MATa and MATα alleles (Fig. A.10). These alleles have blocks of similar sequences at each end totaling about 2500 bp, but they differ by the insertion of a unique 747 bp DNA sequence, known as Yα, in MATα and a unique 642 bp DNA sequence, known as Ya, in MATa. The unique segments contain the promoters for transcription and parts of the coding regions of the α1, α2, and a1 genes. Compared to other loci, the MAT locus is unique in that its two forms—MATa and MATα—carry dramatically different genetic information.

Molecular analyses of RNA in mutant and wild-type cells have shown that the α1, α2, and a1 gene products act at the level of transcription. MATα1⁻ mutants (lacking a functional α1 gene) do not transcribe the α-specific genes known as *STE3* (which encode an *a* pheromone receptor) and *MFα1* and *MFα2* (which encode α pheromone); normal α cells do express these genes. Similarly, MATα2⁻ cells (lacking a functional α2 gene) transcribe and produce functional products of the *a*-specific genes that encode an α pheromone receptor and *a* pheromone; normal α cells repress expression of these genes. MATa /

MATα cells normally repress transcription of haploid-specific genes; however, loss of *a1* or *α2* function in these cells results in the transcription of the haploid-specific genes, including genes that influence the pheromone response, cell fusion, and meiosis.

Studies of the binding to DNA of isolated α1 and α2 proteins have shown that the transcriptional activation of α-specific genes in an *a* cell requires the α1 gene product together with a protein known as Mcm1. In this same *a* cell, the DNA binding of the α2 gene product together with the Mcm1 protein represses the transcription of *a*-specific genes. By contrast, the transcriptional activation of *a*-specific genes depends only on Mcm1, requiring no product encoded by the MAT locus (see Fig. A.9b). The Mcm1 protein thus functions as a transcription factor in both *a* and α cells but actually has opposite effects in the two haploid cell types. This single transcription factor becomes an activator when combined with α1 protein and a repressor when combined with α2 protein. Furthermore, in *a* / α cells, a combination of the *a1* and α2 proteins forms a new transcriptional repressor whose binding to the promoters of haploid-specific genes represses their transcription with no requirement for the Mcm1 protein. Thus the three cell types, in which several cell-type-specific genes are expressed, arise from the activities of only three specific proteins that bind directly to the promoters of cell-type-specific genes to enhance or repress their transcription.

Mating Type Switching

Yeast strains differ from each other according to the stability of their mating type. Most laboratory strains have stable haploid mating types such that the progeny of successive mitotic divisions always have the same mating type as their parents. These strains with stable haploid mating types are known as **heterothallic** strains. By contrast, strains that can switch mating types are known as **homothallic** strains (Fig. A.11a). In homothallic strains, switching occurs at a much higher frequency than one would expect if the switch were due to a point mutation. The homothallic phenotype arises from a dominant, functional

Figure A.11   Mating type switching in homothallic strains.
(a) Homothallic strains are able to switch mating types. After production of an *a* / α diploid through mating, the diploid can undergo meiosis during sporulation, producing two *a* and two α spores. Progeny of these spores can switch to the opposite mating types and mate back with a cell of opposite mating type. **(b)** Cassette model of mating type switching. Three copies of mating type information exist on chromosome III: HML, MAT, and HMR. Only the information at the MAT locus is expressed. HML and HMR contain the stored, silent information that can be copied into the MAT locus.

HO allele at the *HO* locus on chromosome IV; by comparison, heterothallism results from a recessive loss-of-function allele (*ho*) at this same locus.

The Roles of HML, HMR, and HO in Mating Type Switching

The key to understanding the mechanism by which homothallic (HO) strains switch mating type came from the discovery of two additional copies of the MAT locus called HML and HMR. Located near the telomeres on each arm of chromosome III, the HML and HMR loci are transcriptionally silent.

In wild-type strains, HML contains α information and HMR contains *a* information. The mechanism of mating type switching involves a transposition of mating type information from the transcriptionally silent HML or HMR locus into the transcrip-

tionally active MAT locus. Switching occurs only in the *a* and α cell types. *a* / α cells do not switch even if they carry HO because the a1 and α2 regulators repress transcription of the *HO* gene.

A simple way to understand the mechanism of mating type switching in homothallic strains is to imagine an analogy with a cassette tape recorder. A cell can retrieve a cassette of mating type information located in an HML or HMR storage locus and insert it into the MAT playback locus. The cell retrieves information from the storage locus by a series of steps. The product of the *HO* gene, a site-specific endonuclease, initiates the process. The *HO*-encoded enzyme makes a double-stranded cut in chromosome III at a specific 18 bp recognition sequence just to the right of a segment known as the Y segment in MAT. Next, the double-stranded break is repaired through a recombination-like process in which homologous pairing occurs between sequences from one of the storage loci, such as HML, and identical sequences surrounding MAT (Fig. A.11b). In this example, HML DNA then serves as a template to resynthesize DNA at MAT. The result is the replacement of MAT DNA with DNA from the storage locus. Overall the process is conservative, with the duplication of storage-locus DNA balanced by an equal loss of DNA from MAT. The information in the storage locus remains unchanged.

The storage loci are silent in two respects. They are not transcribed, and even though they contain the 18 bp HO-protein-recognition sequence, they are not susceptible to double-stranded cutting by the HO enzyme. Both these properties result from the presence of "silencer" sequences at the boundaries of the storage loci, which are not present at the boundaries of MAT. In a process that may resemble the silencing of genes in higher eukaryotes, the silencer sequences in yeast contribute in a complex way to the configuration of the chromatin structure in the vicinity of each storage locus. RNA polymerases, transcription factors, and the HO endonuclease cannot gain access to these regions. As a result, the HO endonuclease cannot make double-stranded cuts at the storage loci. The silencing effect is general: Genes from other chromosomes that are normally active become transcriptionally silent when inserted into a storage locus.

Are there gene products that function to keep these storage loci silent? Mutational analyses show that the answer is yes. Loss-of-function mutations isolated in four *SIR* (*silent information regulator*) genes cancel repression and result in the transcriptional activation of the α*1,* α*2,* or *a1* genes located at HML or HMR. In haploid cells, transcription of the storage loci results in a phenotype resembling that of an *a* / α diploid because both MAT*a* and MATα alleles are expressed. Moreover, in *sir⁻* strains, the HO enzyme cleaves the chromosome at both MAT and the storage loci. Depending on which site is cut in a given cell, HO-mediated transposition can occur from MAT to one of the storage loci or in the other direction. This observation demonstrates that in normal *SIR⁺* strains, the silencing of the storage loci ensures that MAT serves as recipient and the storage loci as donors in the switching process. As a consequence, in wild-type cells, mating type information flows in only one direction—from the storage loci to MAT.

The Role of HO in Mating Type Determination and Stability Serves As a Paradigm for Development

Pedigrees of cells undergoing HO-mediated mating type switching have revealed a few rules governing mitotic cell lineages during development (Fig. A.12). In a typical pedigree analysis, investigators tease apart the mother and daughter cells in each cell division by delicate micromanipulation under a microscope so that they can examine each one individually. The first generation arises when a spore germinates to produce the spore (S) and the first daughter (D1) (Fig. A.12a). In the second generation, the spore gives birth to the second daughter (D2), and the first daughter gives birth to its first daughter (D1–1).

Two general rules for HO-mediated switching have emerged from observations of many generations. First, cells always switch in pairs; if a mother cell switches, its immediate daughter will have the same switched mating type (Fig. A.12b). Second, cells must gain competence to switch before actually switching; and they gain competence through the experience of undergoing at least one cell division. Thus, an "inexperienced" spore never switches at the first cell division. In addition, a daughter cell will not switch when it produces its first daughter cell. Once a cell has gained experience through cell division, however, it becomes competent to switch in subsequent generations (Fig. A.12c).

What aspect of cell division triggers competence to switch? The answer, although not yet entirely clear, relates to the transcriptional activation of the *HO* gene. Occurring only during S phase, the transcription of *HO* is controlled in part by a tran-

scription factor that regulates oscillations in the expression of other cell-cycle genes. The transcription factor regulates expression such that HO transcripts are made only in experienced mother cells and never in inexperienced daughters. Thus the inexperienced cells undergo a change of state in the process of cell division. There is an analogy here with the stem cells that give rise to differentiated cells in multicellular eukaryotes. The inexperienced daughter cells are like stem cells; they do not themselves express HO, but once they undergo cell division, they differentiate by acquiring the capacity to produce the HO protein.

In many cell lineages that arise during the development of an organism, programmed cell death, or apoptosis, destroys some cells before the organism achieves adulthood. Studies of the nematode *Caenorhabditis elegans* have provided the most elegant demonstration of the role of cell death in development. But HO pedigrees in yeast provide clues to the genetic programming of cell death. The pedigree in Fig. A.12d illustrates how this happens. This pedigree is identical to that shown in Fig. A.12a, except that the initial spore carries complete deletions of the HML and HMR storage loci. Since there are no donor sequences available for mating type switching, the HO-mediated, double-stranded break at MAT cannot be repaired and becomes a lethal event. The resulting "pedigree of death" follows the rules described earlier; for example, experienced mothers who could switch now have a double-stranded break and die. Some multicellular organisms use genetic mechanisms like the introduction of lethal breaks in DNA to eliminate cells that outlive their usefulness during development.

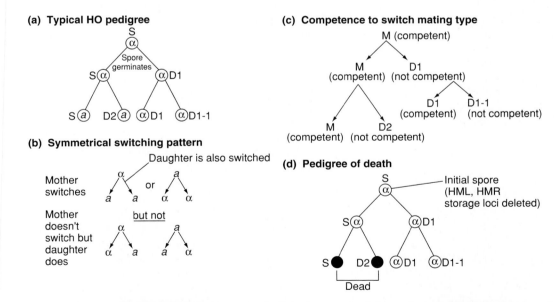

Figure A.12   Pedigrees of mating type switching. (a) Pedigree of mating types in divisions of a homothallic (HO) spore cell. Yeast cells reproduce by budding. The inexperienced α spore cell (one that has not yet divided) produces a daughter cell (D1) in addition to the mother (S) and both are still α. With the next division of the spore cell, producing S and D2, mating type can change. **(b)** When a mother cell divides to produce a mother and a daughter cell, switching can occur and will be seen in both resulting cells. When a daughter cell is produced, it always has the same mating type as the mother cell. **(c)** The ability of cells to switch is dependent on the cell having undergone at least one cell division. Therefore, a newly arising daughter cell is not competent to switch, whereas the mother cell from which it arose is competent to switch. The competency to switch is therefore asymmetric. **(d)** In a strain containing a deletion of the storage cassettes (HML and HMR), a cell that tries to switch mating type will end up with a broken chromosome and therefore die. Again the symmetry of switching pattern but asymmetry of competence to switch is seen in this pedigree of death.

Mating: Cell-to-Cell Communication Through Pheromones Triggers the Conversion of Haploid *a* and α Cells to *a* / α Diploids

Mating is the result of a genetic program that allows *a* cells and α cells to become *a* / α cells. During mating *a* and α cells communicate with one another through **pheromones:** hormones that play a role in eliciting sexual behavior. *a* cells secrete *a* pheromone and α cells secrete α pheromone. Each haploid cell type carries receptors for the pheromone of the opposite cell type; *a* cells carry receptors for α pheromone, and α cells carry receptors for *a* pheromone (Fig. A.13). When bound to pheromone, the receptors activate a signal transduction system whose cascade of protein interactions ultimately promotes transcription of a specific set of genes. The yeast pheromones provide an example of how small molecules produced in one cell type and secreted into the extracellular environment can influence the behavior of a neighboring cell. Through the isolation of mutants, it has been possible to dissect genetically the processes by which cells produce and respond to pheromones.

Researchers discovered many of the genes responsible for the pheromone response system by the isolation of sterile mutants. Some mutants are sterile because they cannot signal to a mating partner; others are sterile because they cannot respond to a signal. A simple test enables one to distinguish between the two. Both *a* and α mating pheromones elicit an arrest of cell division in the receptive cell type. If one places a patch of cells of one mating type on a patch of cells of the opposite mating type, a zone of growth inhibition appears where the pheromone from one cell type has caused the arrest of division of the other cell type. Sterile mutants that produce no zone of growth inhibition either fail to secrete mating pheromone or they secrete mating pheromone but are unresponsive to pheromone from the opposite cell type (Fig. A.14).

Cells Respond to the Binding of Pheromone Through a Signal Transduction System

By analyzing mutants that are sterile but still synthesize pheromone, yeast researchers identified many components of the signal transduction systems through which cells respond to the binding of pheromone at their surface. Plasma membrane receptors at the cell surface mediate the pheromone response. *a* cells produce α-pheromone receptors, which are products of the *a*-specific *STE2* gene. Similarly, α cells produce *a*-pheromone receptors, which are products of the α-specific *STE3* gene. The receptors contain seven noncontiguous, hydrophobic domains long enough to span the lipid bilayer of the plasma membrane; as a result, each protein snakes back and forth through the plasma membrane, exposing some parts of the entire polypeptide to the cell's exterior and others to the cell's interior (Fig. A.15). Although the receptors are cell-type specific, the remainder of the pheromone response pathway, depicted in Fig. A.15, is controlled by the *haploid-specific* genes, or *hsg*, which are expressed in both *a* and α cells (review Fig. A.9).

a cells and α cells communicate with each other for the purpose of mating through extracellular segments of the receptors that interact directly with the appropriate mating pheromone. Receptor segments exposed to the inside of the cell are coupled to a membrane-bound G protein composed of α, β, and γ subunits, encoded respectively by the *SCG1, STE4,* and *STE18* genes. The mating pheromones, receptors, and G proteins act in concert as a molecular switch, with the binding of pheromone

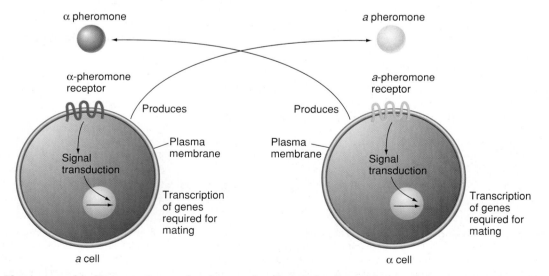

Figure A.13 Pheromones bind to receptors and activate a signal transduction pathway. Chemical signals (pheromones) released by *a* and α cells bind to receptors on the cell surface of cells of the opposite mating type. Once bound to the receptor, a signal is sent to the nucleus via a signal transduction pathway and transcription of genes required for mating is turned on.

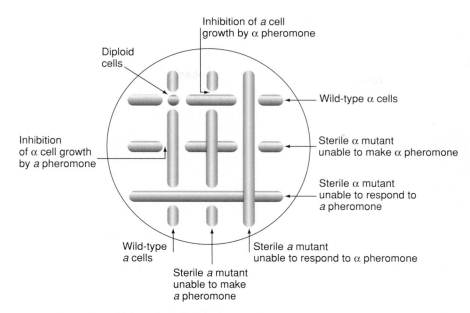

Figure A.14 Mutants defective in producing or responding to mating signals. Strains can be cross-streaked on plates to test for defects in sending or receiving pheromone messages. Pheromones from a strain of one mating type bind to a strain of the opposite mating type and this binding leads to the arrest of cell division. If a cell cannot produce or respond to one of the pheromones, there will be no inhibition where the streaks cross.

causing the conversion of Gα-GDP to Gα-GTP and this conversion, in turn, causing the dissociation of the Gβ/Gγ subunits from the heterotrimeric G-protein complex. The dissociated but still membrane-bound Gβ/Gγ dimeric complex initiates a signal that propagates through the series of proteins in the signal transduction pathway. The pathway ultimately leads to the transcriptional activation of genes whose products are required for G_1 cell-cycle arrest, cell fusion, and nuclear fusion.

As Fig. A.15 shows, Gβ/Gγ transmits the signal initiated by pheromone binding through a series of three protein kinases, known collectively as a MAP kinase cascade. In the pheromone response pathway, the trio of kinases first phosphorylate each other and then promote phosphorylation of the *STE12* gene product, a transcription factor that is active only when phosphorylated. Since transcription occurs in the nucleus, the phosphorylation of Ste12 shows that the signal transduction pathway has transduced the pheromone-induced signal from the plasma membrane to the nucleus. Phosphorylated Ste12 activates the transcription of several genes whose products contribute to G_1 cell-cycle arrest (for example, *FAR1*), cell fusion (for example, *FUS1*) and nuclear fusion (for example, *KAR5*). All these processes are necessary for mating. Components of the type of signal transduction pathway dissected in the yeast mating response by genetic and molecular analyses are remarkably conserved in all eukaryotes.

The protein kinase cascades are so complicated that it has only been possible to unravel how they work using new genomewide assays for gene expression made possible by the completion of the genomic sequence. In one approach, researchers used DNA chips (see Chapter 10) to assay simultaneously the abundance of cDNA's corresponding to virtually all cellular mRNAs in cells before and after treatment with pheromone (Fig. A.16a). The chips that were used are called spotted arrays; they contain oligonucleotides corresponding to all the annotated ORFs known to code for yeast proteins (about 6000 of them). In a representative experiment, cDNA pools from two sets of *a* cells—one treated with α pheromone, the other untreated—were differentially labeled with the fluorescent dyes Cy3 (green) or Cy5 (red) and then mixed and probed by competitive hybridization to the genomic arrays. Changes in the abundance of an individual mRNA produces a spot that fluoresces red or green, depending on whether the corresponding mRNA increases or decreases in abundance in response to α-pheromone treatment. A genomic view of the data can be obtained by plotting the intensity of the signal for each mRNA versus the expression ratio (that is, expression in α-pheromone treated cells versus expression in untreated cells). The data showed that more than 200 genes increase expression and nearly as many decrease expression as a result of α-pheromone treatment. Genes required for all stages of the mating response were represented among those whose expression increased. α-pheromone treatment repressed the expression of genes required for cell-cycle progression, DNA replication, budding, and mitosis. Similar results were found when α cells were treated with *a* pheromone.

Another way to study the response of cells to pheromone treatment is to assay expression from strains that carry specific defects in genes known to be involved in the response. Figure A.16b shows a genomic expression profile generated in the

Figure A.15 Signal transduction cascade in the mating response. Binding of pheromone to a receptor causes the conversion of GDP to GTP, which then activates the MAP kinase cascade, leading to activation of the STE12 transcriptional activator. Activated STE12 turns on genes involved in G_1 arrest of cells, cell fusion, and nuclear fusion. Mutations in genes in any of these pathways cause defects in mating.

same way as the one described previously except that the response of two different *a* strains were compared, a wild-type strain and one carrying a deletion of the gene coding for the Ste12 transcription factor. The experiment shows that the entire transcriptional response to pheromone treatment depends on

Ste12 because the changes in expression of all of the genes that respond to α pheromone are abolished by loss of Ste12 function. Although Ste12 activates transcription, negative regulators may be involved in the circuit downstream of Ste12 action, which could explain why pheromone treatment activates some genes but represses others.

A third genomic approach consists in the identification of Ste12-binding sites across the genome. Transcription factors like Ste12 activate transcription by binding to regulatory sequences upstream of the 5′ end of regulated transcripts. To map the genomic locations for Ste12 binding, cells were treated to chemically cross-link DNA to proteins and then broken open. Enrichment for Ste12-DNA complexes was accomplished by immunoprecipitation using an antibody that recognizes and binds to Ste12. The DNA in the enriched sample was then amplified by PCR and labeled with the fluorescent dye Cy5 (red). A sample that was not enriched by immunoprecipitation was amplified and labeled with Cy3 (green). The DNA samples were hybridized to DNA arrays containing oligonucleotides corresponding to over 6000 intergenic DNA sequences (the spaces between genes). When the Cy5 and Cy3 signals were merged, spots corresponding to sequences that bound to Ste12 showed red fluorescence, whereas those that do not bind Ste12 were yellow or green (Fig. A.16c). Twenty-nine genes were identified whose transcription is controlled through the binding of Ste12 to the corresponding 5′ regulatory regions. The list includes a number of genes shown in Fig. A.14, including *FUS3* (MAP kinase), *FAR1* (G_1 cell-cycle arrest), *FUS1* (cell fusion), and *KAR5* (nuclear fusion).

Two findings from these genomic studies indicate that additional circuitry exists in the pheromone response pathway downstream of Ste12. First, Ste12 is an activator of transcription and all signaling for mating appears to flow through Ste12, yet pheromone treatment causes the repression of many genes involved in cell-cycle progression, DNA replication, budding, and mitosis. Secondly, the number of genes directly regulated by Ste12 is much smaller than the number of genes whose expression is altered by pheromone treatment. These findings suggest the existence of negative regulators that function downstream of Ste12 in the pheromone response circuitry.

As this discussion shows, genomic analysis using DNA arrays will bring us closer to a complete molecular picture of the sequence of developmental events that lead to mating following exposure to pheromone.

(a) Genomic profile of yeast grown in the presence of α pheromone.

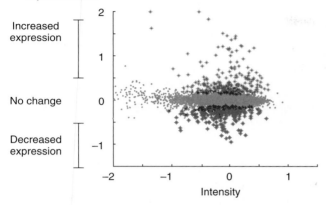

(b) Genomic profile of a *Ste12⁻* mutant.

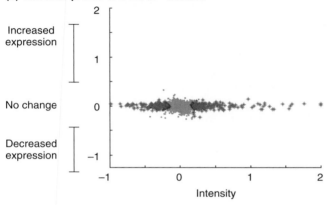

(c) Mapping of Ste12 DNA binding sits.

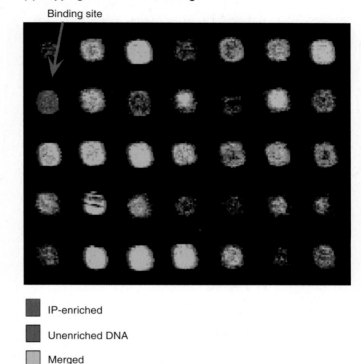

Figure A.16   DNA array analysis of the pheromone response. **(a)** Fluorescently labeled cDNAs derived from wild-type cells that were untreated or treated with α pheromone were hybridized to spotted DNA arrays representing over 6000 ORFs. Data are displayed as intensity versus red:green ratio for each ORF. *Grey dots* represent genes that show no change in the level of gene expression. *Red* or *green dots* represent genes that were induced or repressed, respectively. **(b)** cDNAs derived from a wild-type strain and a *ste12⁻* mutant strain were hybridized to the arrays. *Grey dots* represent genes whose expression was unchanged. *Green dots* represent mRNAs whose expression was shown to change in (a). **(c)** The figure shows a close-up of a scanned image of a DNA chip containing DNA fragments representing over 6000 intergenic regions of the yeast genome. The *arrow* points to a spot where the red intensity is overrepresented, identifying a region bound *in vivo* by the Ste12 protein.

CONNECTIONS

Because it is relatively easy to find and analyze mutant yeast cells, as well as to replace yeast genes with engineered variants, yeast has become the model organism for studying eukaryotic cell biology. Three different genetic programs, each governing a particular pattern of gene expression, trigger the differentiation pathways that generate *a*, α, and *a* / α cell types in yeast. Mating between *a* and α cells is directed by another genetic program. Research has shown that the mechanisms for cell communication, the control of gene expression, and the differentiation of cell types are based on the same principles in yeast and multicellular, higher eukaryotes.

What we learn from yeast will very likely tell us a great deal about ourselves. Mounting evidence suggests that similar proteins in yeast and humans carry out most cellular processes. It also indicates that human genes closely related in sequence and function to the regulatory genes of yeast play similar regulatory roles in human cells. Geneticists have already sequenced the entire yeast genome and thereby produced a comprehensive catalog of more than 6000 gene functions required to make a eukaryotic cell.

In the next genetic portrait (Reference B), we move from the differentiation of single-celled eukaryotes to the development of multicellular eukaryotic organisms as we examine the simple, multicellular plant known as *Arabidopsis*.

E S S E N T I A L   C O N C E P T S

1. Yeast geneticists use transformation studies, the ability to move genes between yeast and bacteria, and gene replacement to answer questions about eukaryotic cell functions that are difficult to answer with other organisms.

2. The nuclear genome of yeast contains 16 linear chromosomes carrying about 6000–6500 genes; very few of these genes have introns. The yeast genome also contains mobile Ty elements, whose capacity for recombination may be a source of reciprocal translocations, inversions, and deletions.

3. During the yeast life cycle, cells alternate between haploid and diploid phases. Haploid cells occur in *a* and α mating types, which communicate with each other via pheromones; the fusion of cells of opposite mating types creates diploid zygotes.

4. Genes at the MAT locus act at the level of transcription to control the expression of other genes that determine cell type. MATα consists of two genes: α*1* and α*2*. α*1* encodes an activator of α-specific genes encoding proteins that elicit the α-type cell. α*2* encodes a repressor of *a*-specific functions. A simple network of the α1 activator and the α2 repressor accounts for both α and *a* cell types. By contrast, the *a*1 protein expressed in *a*-type cells performs no function necessary for mating or differentiation into *a* cells. The interaction between *a*1 and α2 coregulators makes MAT*a* / MATα heterozygous diploids sterile.

5. Heterothallic yeast strains are unable to switch mating type, while homothallic yeast strains can switch from *a* to α and back again. Mating type switching in homothallic strains is controlled by the activity of the MAT locus, the *HO* gene, the alleles at HML and HMR, and the four *SIR* genes. Two general rules govern HO-mediated switching: (1) Cells always switch in pairs, and (2) cells must gain competence to switch by undergoing at least one cell division.

6. *a* and α cells communicate through pheromones determined by *a*-specific and α-specific genes. *a* cells secrete *a* pheromone, while α cells secrete α pheromone. Each cell type carries receptors for the opposite type pheromone. The binding of pheromone from the opposite cell type activates a signal transduction system that results in the transcription of genes whose products help arrest the cell cycle and contribute to nuclear and cellular fusion.

S O L V E D   P R O B L E M S

I. You want to study metabolism of purines in *S. cerevisiae* and begin your analysis by isolating auxotrophic mutants that cannot synthesize adenine. To achieve this, you mutagenize haploid *a* HO cells and screen for cells that require adenine for growth.
   a. Will you know if your mutations are dominant or recessive? If not, how could you find out?
   b. If you wanted to know if any of the mutations had an effect on sporulation, how could you test this? What strain would you need to construct?

Answer

This problem requires an understanding of the yeast life cycle, ploidy and the expression of dominant and recessive alleles, and the segregation of alleles during meiosis.

   a. Because the newly isolated mutations are present in haploid strains you will not know immediately if they are dominant or recessive. *To determine if a mutation is dominant or recessive, mate the mutant haploid strain with a wild-type haploid α strain.* If the diploid cannot grow without adenine in the medium, the mutation is dominant. If the diploid cells have the wild-type phenotype, the adenine mutation is recessive.

   b. *If the mutation is dominant, you can determine its effects on sporulation using the diploid strain constructed for part* a. *If the mutation is recessive, you would have to construct a homozygous recessive diploid.* You could do this by switching the mating type of the original strain to obtain an α haploid cell and mating this with the original *a* haploid cell.

II. You have cloned a gene encoding the histone subunit H2A.
   a. Now you want to do a gene replacement with a nonfunctional copy of the gene and observe the phenotype of the knockout mutation. Which of the following steps will you do and in what order?
      i. Transform haploid *a* cells.
      ii. Transform haploid α cells.
      iii. Select Ura$^+$ cells.
      iv. Transform diploid *a* / *a* cells.
      v. Transform diploid *a* / α cells.
      vi. Sporulate the transformed cells.
      vii. Mate the transformed cells with cells of the opposite mating type.
      viii. Add the *URA3* gene to the plasmid vector.
      ix. Clone the *URA3* gene into the *H2A* gene on your plasmid.

b. You have obtained haploid cells that have the *H2A* gene knockout but there is no phenotype. The cells live and appear to grow normally. Offer two possible explanations for this result.

c. How could you determine if there is more than one copy of the *H2A* gene in the yeast genome?

d. You located a second copy of the *H2A* gene, determined the sequence, and found that the two copies had only a few base differences. Transcription of the first copy of *H2A* that you isolated is transcriptionally regulated during the cell cycle. How could you determine if the other copy is regulated in the same way?

e. How could you determine if there are sites in the DNA 5′ to the *H2A* coding sequence that are responsible for the cell-cycle transcriptional regulation?

f. *HIR2* and *HIR3* are two genes that were identified as regulators of transcription of the histone genes based on the properties of mutations in these genes. Null mutations in either gene lead to constitutive expression of *H2A-H2B*. Do *HIR2* and *HIR3* encode positive or negative regulators?

g. Histone gene transcription is also sensitive to the number of copies of the histone genes in the cell. Introduction of a high copy number plasmid containing *H2A* and *H2B* into the cell will repress transcription of *H2A* and *H2B*. How could you determine if the copy number control of transcription is dependent on *HIR2* and *HIR3*?

Answer

This problem involves molecular analysis of gene structure and expression and the use of mutants to determine the role of a gene.

a. Because histones have such an essential role in packaging DNA in the cell, you might predict that a knockout of function would be a lethal event. Therefore, the gene replacement should be done in a diploid cell, which can then be sporulated to

determine the phenotype of a knockout in a haploid cell. To do a gene replacement with a nonfunctional gene, the gene should be disrupted by a selectable marker like *URA3*. *The steps you would take and the order in which you would take them are as follows:* Clone *URA3* into the *H2A* gene on your plasmid (ix). Transform diploid *a* / α cells (v). Select Ura$^+$ cells (iii). Sporulate the transformed cells (vi).

b. *The H2A protein may not be an essential protein for cell viability* or *there may be another copy of the* H2A *gene in yeast.*

c. To determine if there is another copy of the *H2A* gene, *the cloned* H2A *gene could be used as a probe in a Southern hybridization using total yeast genomic DNA.*

d. The few base differences between the copies of the *H2A* gene could be used to distinguish transcription of the two genes. *Oligonucleotides that contain the specific base differences between the two genes could be used as probes in two separate hybridizations to mRNAs that had been separated by electrophoresis and transferred to filter paper (Northern hybridization).*

e. *If the DNA upstream to (5′ to) the* H2A *gene was cloned next to a reporter gene such as* lacZ, *this construct could be used to analyze the role of the DNA in cell-cycle regulation. Different portions of the upstream region could be deleted, the resulting clones transformed into yeast, and expression (as measured by production of β-galactosidase) analyzed at different times in the cell cycle.*

f. Because the lack of functioning protein in the null mutations led to constitutive expression, *HIR2 and* HIR3 *must encode negative regulators.*

g. To determine if *HIR2* and *HIR3* are also involved in copy number regulated transcriptional control, the *high copy number histone gene plasmids should be transformed into* hir2 *or* hir3 *mutants. If the level of expression is repressed to the same extent as in* HIR$^+$ strains, then the *HIR2* and *HIR3* gene products are not required for copy number sensitive transcription.

PROBLEMS

A-1 Choose the matching phrase in the right column for each of the terms in the left column.

a.	pseudogene	1.	silent copy of mating type information
b.	shuttle vector	2.	expressed copy of mating type information
c.	homothallic	3.	remnant of a gene
d.	pheromone	4.	plasmid that replicates in both *E. coli* and yeast
e.	MAT	5.	able to switch mating type
f.	HML	6.	unable to switch mating type
g.	heterothallic	7.	chemical signal

A-2 The yeast genome is approximately 12,000 kb in length and contains about 6000 open reading frames encoding

an average of 500 amino acids. Using these numbers, calculate the proportion of the yeast genome that is protein-coding DNA.

A-3 Chromosome XII in yeast is about 2060–3060 kb in length depending on the number of the rDNA repeats (100–200 copies of a 9 kb repeating unit). If you grow up 10 individual yeast cells as clones in liquid culture, how could you analyze each clone to see if there is variation in the size of the rDNA repeat in the clones?

A-4 Expression of genes involved in the same process (for example, amino-acid biosynthesis) is often coregulated

in yeast. For the following questions, assume that you have cloned the genes involved in histidine metabolism.
  a. How could you show if the *HIS* genes are transcriptionally coregulated?
  b. If there is transcriptional coregulation, what might you expect to see in the sequence in the regulatory region of these genes and why?

A-5 Several copies of subtelomeric repeats may be present at the ends of chromosomes, resulting in more than 20 kb of repeat DNA.
  a. How could you show which chromosomes contain repeated DNA sequences near the ends of each chromosome?
  b. If you had a genomic library of plasmid clones with an average insert size of 15 kb and were trying to make a physical map, how would you know which chromosome your cloned DNA came from?

A-6 In screening a population of diploid yeast for a defect in chromosome segregation, you found that the chromosome segregation defect segregated $2^+{:}2^-$ in the spores of the ascus.
  a. Why is this evidence that the mutant phenotype is due to an alteration in one gene?
  b. Is the mutation dominant or recessive?

A-7 If a tetraploid $a\,/\,a\,/\,\alpha\,/\,\alpha$ yeast strain was sporulated, what mating type phenotypes would you predict in the asci?

A-8 Which of the following statements apply to
  a. YIp vectors
  b. YEp vectors
  c. YCp vectors
      i. contains a yeast origin of replication
      ii. high copy number
      iii. one copy per cell
      iv. contains a selectable marker
      v. integrates into chromosome via homologous recombination
      vi. exists autonomously

A-9 a. Describe the model for the roles of the Mcm1 protein in determining expression in *a* cells, α cells, and $a\,/\,\alpha$ diploids.
  b. What would be the mating phenotype of an $\alpha\ mcm1^-$? $a\ mcm1^-$? $a\,/\,\alpha\ mcm1^-$? Would they be sterile or fertile?

A-10 Seven haploid yeast strains auxotrophic for the compound sneezan were isolated. These strains were mated with each other, and the ability of the resultant $a\,/\,\alpha$ diploids to grow on the intermediate compounds dopan, sleepan, and happan are scored here.

Growth on sneezan

	1	2	3	4	5	6	7
1	+	+	+	+	+	+	+
2		+	+	+	+	+	+
3			+	+	+	+	+
4				+	+	+	+
5					+	+	+
6						+	+
7							+

Growth on dopan

	1	2	3	4	5	6	7
1	+	+	+	+	+	+	+
2		−	+	+	+	−	+
3			+	+	+	+	+
4				+	+	+	+
5					+	+	+
6						−	+
7							+

Growth on sleepan

	1	2	3	4	5	6	7
1	−	+	+	+	+	+	+
2		−	+	+	+	−	+
3			−	+	+	+	−
4				−	−	+	+
5					−	+	+
6						−	+
7							−

Growth on happan

	1	2	3	4	5	6	7
1	−	+	+	+	+	+	+
2		−	+	+	+	−	+
3			+	+	+	+	+
4				+	+	+	+
5					+	+	+
6						−	+
7							+

a. Are these results derived from a complementation or from a recombination experiment?

b. The $a\,/\,\alpha$ diploids from the experiment in part *a* are now induced to sporulate. Out of 400 spores examined from each $a\,/\,\alpha$ diploid, the numbers that were able to grow on minimal medium are scored here.

	1	2	3	4	5	6	7
1	0	100	5	100	100	100	5
2		0	100	3	3	0	100
3			0	100	100	100	0
4				0	0	3	100
5					0	3	100
6						0	100
7							0

How can you explain the two classes of results here (100 spores or ≤ 5)?

c. How many complementation groups involved in the biosynthesis of sneezan are represented in this collection of seven mutations? Which genes are in which complementation groups?

d. The biochemical pathway involved in the biosynthesis of sneezan is linear. One of the steps of the pathway is catalyzed by a heterodimeric enzyme, with each subunit encoded by a different gene. What is the biochemical pathway? Identify the genes that encode the enzyme involved in each step. Indicate which enzyme is dimeric.

A-11 Six genes, each conferring the ability to use the sugar sucrose have been mapped to different loci in different *Saccharomyces* strains.

a. You have cloned the *SUC2* gene from a *S. cerevisiae* strain. How can you tell if the *SUC* loci in other strains contain DNA homologous to that at the *SUC2* locus?

b. Other Suc^+ strains have DNA homologous to the *SUC2* gene but, in some cases, there is not enough information for full gene function. What would you call these partial genes?

c. Starting with an $SUC2^+$ strain, you have identified cells incapable of fermenting sucrose. Some of the mutants are defective in the gene at the *SUC2* locus ($suc2^-$). Other Suc^- strains are pleiotropic. These genes, called *snf* (*sucrose nonfermenting*) genes, affect expression of galactose and maltose as well as sucrose. The common feature of the affected genes is that all are glucose repressible (not expressed when glucose is present). Expression of the *SUC2* gene was examined in the wild type and the mutant under repressing and nonrepressed conditions. How would you describe the transcription pattern?

Wild type		snf1	
−Glucose	+Glucose	−Glucose	+Glucose

A-12 To study DNA repair mechanisms, researchers isolated yeast mutants that were sensitive to various types of radiation. For example, they isolated mutants that were more sensitive to damage by UV light (and therefore died at a lower dose of UV radiation). Ten haploid mutants were isolated.

a. How would you do the complementation tests for these mutants in yeast?

b. Based on the following complementation chart, how many genetic loci are represented by these mutants?

	1	2	3	4	5	6
1	−	−	+	−	+	+
2		−	+	−	+	+
3			−	+	−	+
4				−	+	+
5					−	+
6						−

c. Double mutants were constructed containing some of the UV-sensitive mutations. If the UV sensitivity of the double mutant is greater than that of the individual mutants, the mutations are in different repair pathways. How many different repair pathways do these mutations represent? What mutations are in each pathway?

Single mutant	Sensitivity to UV
1	+
3	+
6	+
Double mutant	
1,3	++
1,6	+
3,6	++

d. Assume you have cloned the genes and now want to ask whether genes identified in these mutant searches are expressed constitutively in cells or only in response to damage. How would you design your experiment?

e. To identify additional genes that might be responding to DNA damage, you have decided to construct a library of yeast DNA fused to the *lacZ* gene. How would you use this library to search for more genes that are induced by DNA-damaging agents such as radiation?

REFERENCES

Mewes, H. W. et al (1997). The Yeast Genome Directory. *Nature* 387 (supplement).

Arabidopsis thaliana:
Genetic Portrait of a Model Plant

Contributing authors Patrick H. Masson *University of Wisconsin*
Anthony Bleeker *University of Wisconsin*

The flower of *Arabidopsis thaliana.*

A. thaliana, a tiny weed whose common name is "mouse ear cress," grows low to the ground and produces clusters of small white flowers in meadows and laboratories around the globe (Fig. B.1). Like more than 150,000 other species of flowering plants, including roses, daisies, tomatoes, peas, beans, and maple trees, *Arabidopsis* is a dicotyledonous angiosperm—**dicotyledonous** because the mature embryo carries two leaves, and **angiosperm** because its seeds are enclosed in an ovary within a flower. *Arabidopsis* shares three basic characteristics with other plants: It is an *autotroph,* that is, a nutritionally self-sufficient organism that produces its own food via photosynthesis; it is *nonmotile,* that is, rooted to one spot; and it produces new organ systems continuously throughout its life cycle (which consists of alternations between a diploid [2*n*] sporophyte generation and a haploid [*n*] gametophyte generation).

The impetus for studying a small flowering weed with no commercial potential is straightforward: Plants are the chief source of food for life on our planet, and an understanding of how they grow and reproduce may make it possible to grow more nutri-

tious, more copious crops in a wide range of environments. Classified in the same family as mustard and cabbage plants, *A. thaliana* has several features that make it an excellent experimental model. Its general strategies of growth, development, flowering, and seed production are the same as those of other higher plants, including many crop plants. However, it has a shorter generation time than most other plants, requiring only six weeks for seeds to germinate and develop into mature plants that produce more seeds. Moreover, like the peas that Mendel studied, it reproduces mainly by self-fertilization, yet cross-pollination—through the removal of stamen from one plant's flowers and the application of pollen from another plant's flowers onto the stigma of the stamen-deprived flowers—is possible. Whether self- or cross-pollinated, wild-type *Arabidopsis* plants produce a very large number of seeds—from 10,000–40,000 per plant—and these seeds have a high rate of germination. Such rapid, abundant reproduction makes it possible for geneticists to screen large populations of seedlings for specific phenotypes. Finally, *A. thaliana* grows well in the laboratory, requiring relatively little light (illumination from cool, white

Figure B.1 *A. thaliana* **in the palm of a hand.**

(a) **(b)**

Figure B.2 The Pink Perfection camellia as a horticultural mutant. (a) Wild-type camellia. **(b)** Pink Perfection flower.

fluorescent bulbs is sufficient) and temperatures in the range of 22–26°C. Because the plant is small, researchers can grow hundreds of thousands of them a year in a modest-sized laboratory with no special growth facilities.

The goal of geneticists studying *A. thaliana* is to understand the physiology, biochemistry, growth, and development of a plant at the molecular level. With this knowledge, they may someday be able to produce plants containing more of particular parts prized for their nutritional content or floral display. Ornamental plant breeders have already accomplished this goal with some species. For example, the flowers of wild camellias contain stamens (male reproductive organs) and carpels (female reproductive organs) that enable them to produce seeds; but the flowers of a cultivated variety known as Pink Perfection lack both male and female reproductive organs and carry many extra petals in their place (Fig. B.2).

In our examination of *A. thaliana* as a model plant, we present:

- The structure and organization of the genome.

- The plant's anatomy and life cycle.

- Techniques of mutational analysis, including chemical and radiation procedures, insertional mutagenesis, and the analysis of mutations to identify gene function.

- Genetic analysis applied to various aspects of development, including embryogenesis, hormonal control systems, and responses to environmental signals.

- The genetic analysis of flowering: a comprehensive example.

Genome Structure and Organization

The genome of a model organism for genetic studies should be relatively small and share the genomic characteristics of the organisms it represents. *A. thaliana* fills this bill to perfection. Its genome, composed of 125 Mb, is one of the smallest genomes known in the plant kingdom, notably smaller than the genomes of most other angiosperms. By contrast, the genome of the angiosperm maize (*Zea mays*) is 45 times larger. The *Arabidopsis* genome is about as large as the genome of *Caenorhabditis elegans* (see Reference C) and is 60% the size of the *Drosophila melanogaster* genome (see Reference D). The nuclear DNA of *Arabidopsis* is carried by five pairs of small chromosomes with well-defined banding patterns. Researchers completed sequencing the whole genome of *Arabidopsis* in the year 2000.

Comparing Genetic and Physical Maps

Geneticists study an array of *A. thaliana* **ecotypes:** plant varieties analogous to animal strains. The members of each ecotype share a common origin, shape, and other phenotypes. The uniformity of shape and characteristics among related individuals is a direct consequence of the plant's propensity to reproduce by self-pollination. The most commonly used *Arabidopsis* ecotypes—Columbia (Col) and Landsberg *erecta* (Ler)—are

derived from the same heterogeneous population of seedlings, originally named Landsberg. Columbia arose by self-pollination from a plant isolated from the Landsberg population because of its ability for prolific reproduction. Landsberg *erecta*, by contrast, arose from a plant of the same population that, after mutagenesis by gamma irradiation, was selected for its small size and erect stature. A number of other ecotypes with well-defined phenotypes are natural isolates that come from and are adapted to specific ecosystems or regions around the world.

Arabidopsis ecotypes differ from each other not only in phenotypes but also in a significant number of DNA polymorphisms that are identifiable by RFLP, PCR, or SNP analysis (as described in Chapter 11). Different ecotypes have different alleles at many different DNA marker loci. Researchers can use allelic variation at DNA markers to follow the segregation of chromosomal regions carrying the markers in the progeny of self-pollinated plants derived from a cross between different ecotypes. They can also analyze the linkage between DNA markers as well as between DNA markers and morphological loci (defined by two or more alleles conferring alternative phenotypes on the different ecotypes). With the resulting data, researchers can create DNA marker and phenotypic maps, and ultimately they can integrate the two types of maps for all five chromosomes (Fig. B.3).

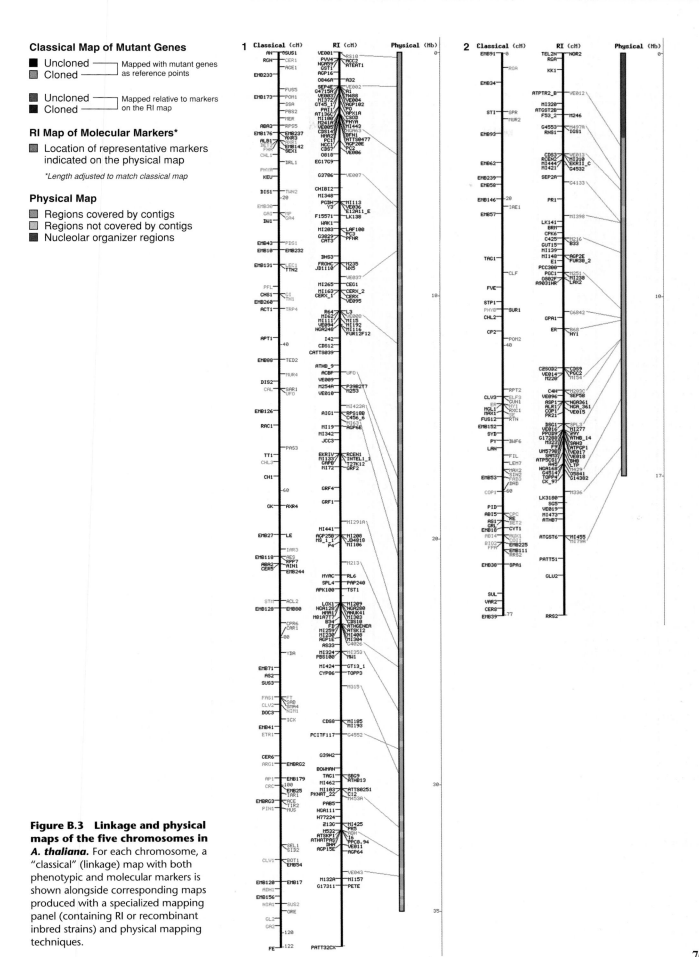

Figure B.3 Linkage and physical maps of the five chromosomes in *A. thaliana*. For each chromosome, a "classical" (linkage) map with both phenotypic and molecular markers is shown alongside corresponding maps produced with a specialized mapping panel (containing RI or recombinant inbred strains) and physical mapping techniques.

Classical Map of Mutant Genes

■ Uncloned ⎱ Mapped with mutant genes
■ Cloned ⎰ as reference points

■ Uncloned ⎱ Mapped relative to markers
■ Cloned ⎰ on the RI map

RI Map of Molecular Markers*

▨ Location of representative markers indicated on the physical map

Length adjusted to match classical map

Physical Map

▨ Regions covered by contigs
▨ Regions not covered by contigs
▨ Nucleolar organizer regions

761

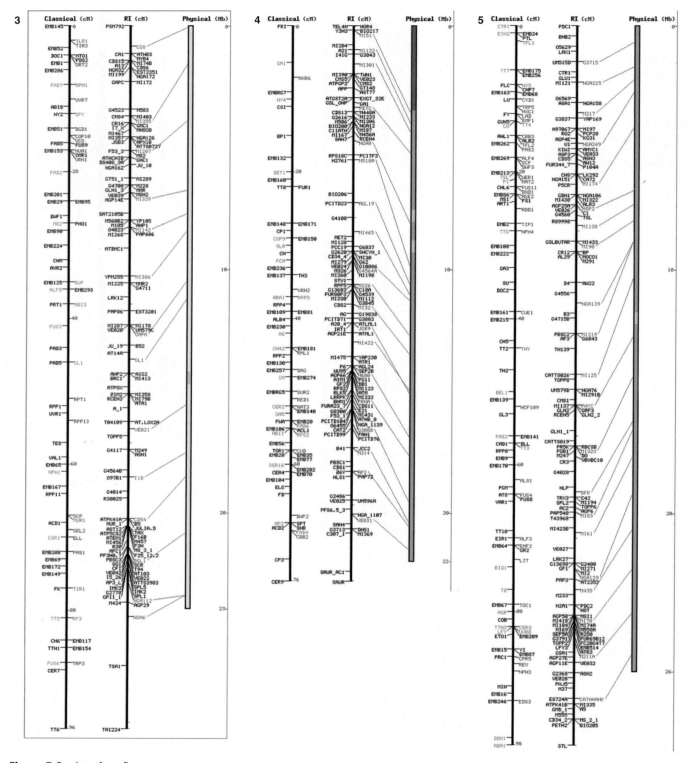

Figure B.3 *(continued)*

Little Repetitive DNA and a Tight Arrangement of Genes

The small size of the *A. thaliana* genome stems from the fact that it contains much less repetitive DNA than most other angiosperms. In fact, the *Arabidopsis* genome contains 25,498 protein-coding genes that make up close to 40% of the total genome and can be grouped into functional categories by sequence analysis (Fig. B.4). This means there is one gene for every 4 kb of genome, on average—an unusually tight arrangement. In most plants, the vast majority of the genome consists of repetitive DNA.

Sixty-five percent of the proteins in *Arabidopsis* are encoded by more than one gene; and analysis of large duplications in the genome suggests that *Arabidopsis* derived from an ancient tetraploid. Taken together, these observations suggest a large extent of functional redundancy in *Arabidopsis*.

Like other species, the *A. thaliana* genome contains multiple copies of transposable elements, as well as several pieces of DNA that appear to have originated in the mitochondrial and chloroplast genomes. As in animals, the centromeres are heterochromatic regions where recombination is strongly supressed.

Comparisons of the genetic and physical maps of the *A. thaliana* genome reveal that genetic and physical distances vary widely along the plant's chromosomes. Within each chromosome, the frequency of recombination varies greatly such that a centimorgan represents anywhere from 30–1600 kb (or 1.6 Mb) (see Fig. B.3).

Summary

A. thaliana has a small genome carried by five chromosomes. Each chromosome contains a limited amount of repetitive DNA outside the centromeric and telomeric regions. Its genes, though similar in structure to the genes of other plants, are more tightly packed and contain only small introns. As a result, a much higher percentage of the genome is associated with coding and genetic function than is true of most plants. This higher signal-to-noise ratio is one of the features that make *A. thaliana* an excellent model for genetic analysis.

Anatomy and Life Cycle

To understand a plant's growth and development at the molecular level, geneticists use mutational analyses to identify genes that influence the plant's anatomy and life cycle. A brief overview of the anatomy and life cycle of *A. thaliana* prepares the way for examining the techniques of genetic analysis in this model plant.

Anatomy: The Basic Body Plan

Like most higher plants, the adult *A. thaliana* is composed of only a few relatively simple organs and tissues (Fig. B.5). The aboveground shoot connected to the belowground root system represents the plant's main axis. Both root and shoot contain several cell types organized into the outer epidermal layer; the middle ground, or cortical, layer; and the inner vascular cylinder (which contains the water-conducting xylem vessels and the food-conducting phloem elements). Attached to the shoot are mature leaves also composed of three layers: an epidermal layer, containing both hair cells, or trichomes, and small pores, or stomata (singular, *stoma*) that regulate gas exchange; a mesophyll layer responsible for photosynthesis; and vascular tissues that conduct water and nutrients between the leaf and the stem.

The Apical Meristem Promotes Continuous Growth

Cells at the growing points of both shoot and root are organized into a tissue called the **apical meristem,** a group of undifferentiated cells that divide continuously. The root apical meristem, which consists of a central zone of meristem cells and a peripheral zone of differentiated cells, produces the cylindrical root through cell division, cell expansion, and cell differentiation. The shoot apical meristem, which consists of a central zone where cell divisions also maintain the meristem and a peripheral zone where cell divisions give rise to mature organs such as leaves and lateral shoots, generates the entire aboveground plant.

Vegetative Versus Reproductive Development

During **vegetative development**—the period of growth before flowering—the shoot apical meristem initiates a spiral of leaves in precise sequence. When the plant switches to **reproductive development,** the shoot apical meristem becomes an inflorescence meristem, which, in turn, produces a series of lateral floral meristems. Each floral meristem produces a floral primordium that gives rise to a flower. The **inflorescence** is the flower or group of flowers at the tip of a branch. In *Arabidopsis,* the flowers are arranged in a spiral around the inflorescence stalk, as depicted in Fig. B.5.

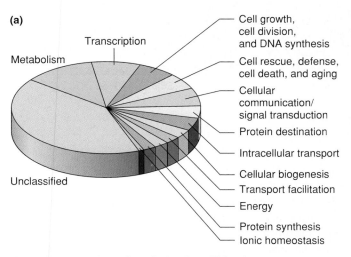

(a)

Metabolism

Transcription

Cell growth, cell division, and DNA synthesis

Cell rescue, defense, cell death, and aging

Cellular communication/ signal transduction

Protein destination

Intracellular transport

Cellular biogenesis

Transport facilitation

Energy

Protein synthesis

Ionic homeostasis

Unclassified

Figure B.4 Functional analysis of *Arabidopsis* genes.
Proportion of predicted *Arabidopsis* genes in different categories.

Figure B.5 **The anatomy of *A. thaliana*.** Components of the plant are labeled.

The *Arabidopsis* Flower Is the Most Complex Set of Organs in the Plant

The flower consists of a modified stem with four concentric regions, or **whorls,** of modified leaves, as shown in Fig. B.5. The first whorl (whorl 1) consists of four green leaflike **sepals;** whorl 2 is composed of four white **petals** that are leaflike in shape but contain no photosynthetic cells; whorl 3 is made of six **stamens** (four long and two short) bearing the male gametes in the form of pollen; and whorl 4 is a cylinder composed of two fused **carpels** housing the female gametes in the form of ovules. The fused carpels are part of a cylinder known as the **pistil** that consists of pollen-receptive **stigma** at the top and a short neck, or **style,** leading to the **ovary,** which houses roughly 50 gamete-bearing ovules. Because the flower contains both male and female reproductive structures, it is considered a *perfect flower.* After fertilization, the plant sheds the organs of the outer three whorls, and the pistil develops into the seed-bearing fruit.

Life Cycle: From Fertilization to Flowering to Senescence

As we have seen, *Arabidopsis,* like a hermaphrodite animal, carries male and female gametes; it is thus capable of self-fertilization, although cross-fertilization by artificial means is easy to accomplish. Fertilization is the first step in a life cycle that also includes embryonic development, seed germination and vegetative growth, reproductive development, and senescence.

Arabidopsis Undergoes Double Fertilization— A Process Unique to Higher Plants

In *Arabidopsis,* each mature pollen grain contains two coupled sperm cells, while each ovule includes six mononucleate cells, each with one nucleus and a central cell containing two nuclei, all encased within a common cell wall (Fig. B.6a). The landing of a pollen grain on the receptive tissue of the stigma initiates fertilization by triggering germination of the pollen grain (Fig. B.6b). The emerging pollen tube migrates through the short neck of the pistil, also known as the *transmitting tract,* to an ovule in the ovary. There, one $1n$ sperm nucleus from the pollen fuses with a $1n$ egg nucleus from the ovule to form the $2n$ nucleus of the zygote; the second $1n$ sperm nucleus fuses with the two $1n$ nuclei within the ovule to form a $3n$ endosperm nucleus (Fig. B.6c). After fertilization, the zygote divides mitoti-

cally to form the embryo, and the endosperm nucleus divides mitotically to form endosperm tissue that, like an animal yolk sac, will nourish the developing embryo. Meanwhile, the remaining $1n$ cells of the ovule degenerate, and the outer layer of the ovule sac hardens to form a seed coat.

The Embryo Develops Within the Protective Maternal Seed Coat

Embryogenesis begins with a single fertilized egg cell, or zygote, which elongates. A sequence of well-defined cell divisions then leads the embryo through a series of stages (Fig. B.7). The first zygotic division gives rise to two cells, one small and one larger. The small cell constitutes the embryo proper, while the larger cell will divide only a few more times to produce the **suspensor,** a structure analogous to the umbilical cord in mammals (whose divisions we will not describe here). The embryo divides following a plane at right angles to the first plane of division to create a vertical wall and two side-by-side cells. The next division, which is at 90° from the previous one, generates four cells that can be seen end-on from the top. The third embryonic division is horizontal and makes an eight-cell, or **octant stage,** embryo with four cells above and four below. Additional divisions parallel to the surface of the embryo result in the formation of a discrete outer cell layer that will give rise to the

Figure B.6   Double fertilization in *Arabidopsis*. (a) Micrographs of pollen tube growth (*left*) and of a pollen grain landing on a stigma. **(b)** Diagram of a pollen grain and an embryo sac. **(c)** Fertilization and postfertilization events.

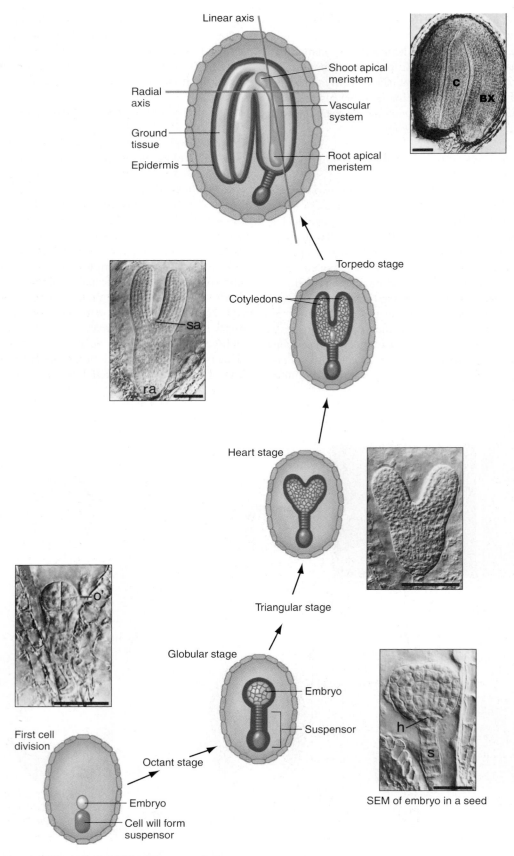

Figure B.7   Stages of embryonic development in *A. thaliana*.

epidermis of the mature embryo, and an innermost group of cells that acquire an elongated shape, the first evidence of differentiation toward the vasculature found in the mature embryo. The embryo has now reached the **globular stage.** The globular embryo rapidly loses its globular shape and acquires a more triangular form that defines the **transient** or **triangular stage** embryo, at which point the embryo becomes self-sufficient for growth, having used up most of its maternally deposited reserves. Next, the embryo moves through the **heart stage** to the **torpedo stage** as two protuberances expand and differentiate into two well-defined, discrete cotyledons.

By the end of this progression from octant to torpedo stage, cellular differentiation has established all the major tissues required for the future vegetative development of the plant along two main axes: the linear axis that makes up the shoot/root continuum and a radial axis moving from the center out to the edge of that continuum (Fig. B.7). The shoot/root axis includes two cotyledons, which store nutrients in the mature seed and will differentiate into the first rudimentary leaves after germination. The axis also includes a *hypocotyl,* which is a short section of embryonic stem between the two cotyledons and the embryonic root, and the root and shoot meristems. The radial axis carries the central vascular system, a middle layer of ground tissues, and the outer epidermis.

As the embryo reaches maturity, growth and development cease and desiccation (water loss) occurs. When the loss of water becomes extreme enough, the embryo enters a dormant state in which virtually all metabolic activity ceases. An embryo in the dormant state can withstand adverse environmental conditions such as winter in Madison, Wisconsin, or Buffalo, New York.

Favorable Environmental Conditions Trigger Seed Germination and Vegetative Growth

Under appropriate conditions, the *Arabidopsis* embryo absorbs water and rehydrates within the seed, which enables growth and development to resume. The enlarging embryo bursts out of the seed coat and begins to grow along its vertical axis. At this point, environmental cues influence the form the seedling will take. Both the root and the shoot can sense gravity; the root responds by bending toward it while the shoot bends away from it. Light provides another developmental cue. If germination occurs in the dark, for example, under the surface of the soil, the hypocotyl, or embryonic stem, elongates rapidly; but development of the leaflike cotyledons and activity of the shoot apical meristem are suppressed. The elongating hypocotyl is hooked near its apex, an adaptation that enables the seedling to push its way up through the soil while protecting the delicate meristem. When the elongating hypocotyl senses light, elongation ceases, the apical hook opens, and the cotyledons expand and turn green as chloroplasts begin to develop and establish the potential for self-nourishment through photosynthesis (Fig. B.8a). This light-regulated developmental program is called **photomorphogenesis.**

With the establishment of the self-nourishing autotrophic lifestyle, the shoot apical meristem produces a compact spiral, or rosette, of leaves. Early flowering varieties of *Arabidopsis* may only produce 6–10 leaves before the shoot apical meristem converts to reproductive development. Later-flowering varieties may produce a rosette containing dozens of leaves before flowering begins (Figure B.8b).

Reproductive Development Begins When the Leaf-Producing Apical Meristem Switches to a Flower-Producing Meristem

Under the right conditions, the shoot meristem enlarges and switches from producing only cells that differentiate into leaves to producing floral meristems on its flanks. At this point, the shoot apical meristem becomes an **inflorescence meristem.** The floral meristems arise in a spiral around the inflorescence meristem and, in turn, give rise to flower primordia that differentiate into the four whorls of organs that make up a flower. Initiated one at a time, the flowers develop sequentially. Thus a view from the apex of the inflorescence meristem reveals the progression of floral developmental stages spiraling away from the shoot apex. Elongation of the main stem between each flower ultimately produces the inflorescence branch containing about 35 flowers (Fig. B.9).

Both environmental cues and endogenous developmental signals (resulting from the genetic program) control the timing of the switch from vegetative to reproductive development in *Arabidopsis.* Day length, or **photoperiod,** is a particularly important regulator of flowering. Laboratory strains, such as Columbia and Landsberg *erecta,* when grown under constant light or a long, 16-hour day, will flower after producing less than 10 rosette leaves. When grown in an 8-hour photoperiod, these same strains will produce over 30 leaves before flowering. Another significant environmental signal is temperature. If exposed to extended periods of low temperature during early vegetative development, normally late-flowering varieties of *Arabidopsis* will flower early. This phenomenon of exposing plants to the cold early in vegetative development is known as **vernalization.**

Since neither photoperiod stimulation nor vernalization is strictly necessary for flowering, researchers have inferred that an endogenous developmental clock is active in this process. However, this endogenous program remains poorly characterized.

Senescence: The Vegetative Plant Ages and Dies

Because plants can produce new organ systems continuously, they do not necessarily display the finite life spans characteristic of animal species. However, some plants, including *Arabidopsis,* do senesce and die after a single reproductive effort. In *Arabidopsis,* somatic tissues such as leaves and stems senesce a few weeks after they are initiated at the meristem. In leaves, senescence is a genetically programmed process in which yellowing occurs as salvage systems convert cellular constituents into mobile nutrients for transport out of the aging leaf (Fig. B.9). The limited longevity of individual leaves does not limit the life

(a)

(b)

Figure B.8 Emerging shoot in *A. thaliana*. (a) The embryonic stem reaches for the surface of the soil (*right seedling*), where the apical hook will open, allowing the greening of the cotyledons (*left seedling*). **(b)** In early-flowering varieties, the shoot apical meristem may produce only a few leaves before developing flowers (*left*). Late-flowering varieties may produce dozens of leaves before flowering (*right*).

span of the plant in the vegetative phase because the shoot apical meristem continuously produces new leaves. Rather, it is the conversion of the shoot apical meristem to reproductive development that delimits the production of new photosynthetic organs. In addition, the inflorescence meristem ceases proliferative activity after a set number of seeds have been produced. The parent plant fails to survive the massive reproductive effort that results in thousands of progeny-carrying seeds.

Figure B.9 Reproductive development and senescence in *A. thaliana*. Young flowering plants next to older plants undergoing senescence.

Techniques of Mutational Analysis

Researchers have used various techniques to identify mutations affecting many aspects of *A. thaliana* anatomy and development. Some of the mutations characterized occur spontaneously in natural populations, but as described in Chapter 7, these spontaneous mutations appear at a low rate. Geneticists use mutagens to increase their frequency. The means of inducing mutations in *Arabidopsis* include exposure to mutagenic chemicals, exposure to radiation, and insertional mutagenesis.

Mutagenesis by Chemical and Irradiation Procedures Produces Different Ratios of Various Mutations

A mature seed carries a dormant embryo in which one to three cells from the apical meristem are destined to form the germ line (gametes) of the mature plant. If a mutation is to be transmitted to the progeny, it has to affect one of the two alleles carried by any of these cells. The phenotype derived from a recessive mutation induced by seed mutagenesis in *A. thaliana* will segregate in different ratios depending on how many gametophyte (gamete-producing) progenitor cells are present in the embryo at the time of mutagenesis. With a parental plant grown from a mutagenized seed in which only one gametophyte progenitor cell was present during mutagenesis, the phenotype associated with the recessive mutation will segregate into 1/4 of the progeny resulting from self-pollination. If two gametophyte progenitor cells were present in the parental seed during mutagenesis, the mutant phenotype will segregate into 1/8 of the selfed progeny. And if three progenitor cells were present, 1/12 of the progeny resulting from self-pollination will show the phenotype.

As in microorganisms, fungi, and animals, mutations resulting from mutagenesis vary with the nature of the mutagen. For instance, chemical mutagenesis using ethyl methane sulfonate (EMS) usually generates simple G-to-A transitions in the DNA, whereas radiation often generates deletions.

Arabidopsis Researchers Use Two Types of Insertional Mutagenesis: Transformation by T-DNA and Transposon Tagging

Strategies for insertional mutagenesis depend on the ability of certain classes of DNA molecules to integrate into the *Arabidopsis* genome. If an integration disrupts a gene, it represents a mutation.

In the T-DNA Approach, the Bacterium *Agrobacterium tumefaciens,* Which Causes Crown Gall Disease in Plants, Is the Agent of Transformation

A. tumefaciens can transfer a piece of plasmid DNA known as the T-DNA into the genome of wounded plant cells (Fig. B.10a).

Experimenters can modify the T-DNA to eliminate its oncogenes and insert transformational marker genes that make the plant resistant to antibiotics (such as kanamycin and hygromycin) or herbicides (such as chlorsulfuron). In one protocol using modified T-DNA, *A. thaliana* is transformed by soaking the shoots of mature three-week-old seedlings or germinating seedlings in an *Agrobacterium* culture to which a strong vacuum or a weak detergent has been applied. The vacuum or detergent enable the bacteria to infiltrate the shoots. As the bacterium-infiltrated plants grow and eventually self-pollinate, a few gametes or zygotes are transformed by *Agrobacterium,* and these produce a transformed seed that is resistant to antibiotics or herbicides. This resistance provides a means of selecting for transformed plants that have inserted the transgenic DNA into their genomes. Because the *A. tumefaciens* T-DNA integrates more or less randomly into the plant genome, it has the potential to serve as a mutagenic agent for any gene.

Researchers have used insertional mutagenesis to create collections of independent transgenic *A. thaliana* lines, which can be screened for specific morphological, developmental, or biochemical phenotypes. They have also modified the T-DNA to carry a *GUS* reporter gene with no promoter of its own. *GUS* encodes the enzyme β-glucuronidase, which converts a colorless soluble compound (X-Glu) into a blue insoluble product. Cells expressing *GUS* stain blue. Because the *GUS* gene in the modified T-DNA has no promoter, its expression depends on the insertion of the T-DNA into a plant-coding sequence; when that happens, the regulatory region of the disrupted gene controls the expression of *GUS*. The expression of *GUS* thus reports the expression pattern of the disrupted gene (Fig. B.10b). The use of the *GUS* reporter allows not only the mutation of genes but also the determination of the disrupted genes' expression pattern (see the discussion of "enhancer-trapping" in *Drosophila* in Reference D).

Figure B.10   The natural process of *A. tumefaciens* transformation can be exploited for use as a method of gene transfer into *A. thaliana*. (a) Natural integration of T-DNA from *Agrobacterium tumefaciens* into *A. thaliana*. **(b)** Blue-staining cells expressing *GUS* gene.

Transposon-Tagging Strategies Rely on the Transposable Elements from Corn (*Zea mays*) and *Arabidopsis*

Transposable elements, originally identified and characterized in corn, can transpose to a new position in the *A. thaliana* genome when introduced by the *Agrobacterium*-mediated DNA transformation just described. For example, corn transposable elements of the families known as *Ac/Ds* and *Spm/dSpm* transpose when introduced into *Arabidopsis*. Similarly, transposable elements endogenous to *Arabidopsis* transpose in the plant's own genome.

Insertional Mutagenesis Not Only Allows the Generation of Mutations but Also Facilitates Their Molecular Characterization

One advantage of insertional mutagenesis over chemical and radiation procedures is that insertional mutagenesis produces mutated alleles carrying a specific DNA insert (the T-DNA or the transposable element). A copy of the specific T-DNA or transposable element that became the insert can therefore serve as a probe to identify and clone the gene the insert disrupts.

From Gene to Phenotype: Analyzing Mutations to Identify Gene Function

The genome sequencing project has identified many genes with specific sequences, but the sequence alone is often not sufficient to attribute a specific function to a particular gene. One way to solve this problem is to mutate the gene under study and analyze the phenotype of plants homozygous for the mutation. Since knocking out genes by homologous recombination is not very efficient in plants, *Arabidopsis* researchers use strategies based on insertional mutagenesis and gene silencing to obtain mutations in genes of interest. We now describe these strategies.

Using PCR Protocols to Find Plants Carrying a Mutated Gene of Interest

Collections of insertional mutant lines contain many independent inserts, but it is now possible to identify in these collections particular lines carrying a DNA insertion (T-DNA or transposable element) in any gene of interest whose sequence is not known. The first step is to transform a large number of plant genomes with *A. tumefaciens* (Fig. B.11a) and pool the purified DNA from 100 different plant genomes. The T-DNA integrates at a random location in each genome, and from 1000 genomes there would be 10 pools. Pooling reduces the amount of work required to screen a very large number of genomes for a rare insertion in the gene of interest. The next step is to attempt a PCR amplification in which one oligonucleotide hybridizes with a sequence in the T-DNA or transposon used to generate the collection, while a second oligonucleotide hybridizes with a sequence

Figure B.11   A PCR screening protocol enables rapid screening of a large collection of transformed plant genomes for a rare one containing disruption of the gene of interest. **(a)** Transformation of 1000 plant genomes with T-DNA. **(b)** Each genome has T-DNA integrated at a different location. **(c)** DNA is purified from each genome. One hundred samples are combined, or pooled. With 1000 genomes, there are 10 pools. **(d)** PCR amplification is attempted on each pool with one primer from the gene of interest and one primer from the T-DNA. Only pools that contain a genome with T-DNA inserted inside the gene yield a positive result. **(e)** Each individual within a positive pool is separately subjected to PCR to identify the one plant of interest. This one plant is bred for genetic analysis of the gene of interest.

in the gene of interest. If the pool of individuals under study contains just one plant with an insertion in the gene of interest, the PCR protocol will amplify one DNA fragment carrying part of the disrupted gene and part of the DNA insertion. With the finding of such a band in one pool of lines, the next step is to repeat the strategy on smaller and smaller subsets to identify the one plant that carries the disrupted allele of the gene under study. With the plant in hand, it becomes possible to produce homozygotes for the mutation by allowing the heterozygous plant to self-pollinate. Study of these homozygous mutants can shed light on gene function.

Silencing Specific Genes Through the Expression of Antisense RNAs, or RNAs Carrying Inverted Duplications of Gene Fragments

To silence the expression of specific plant genes or groups of genes, researchers can activate a mechanism that evolved to protect plant and animal cells against invading viruses or overly active transposons. In this mechanism, double-stranded RNAs generated in or introduced into the cells are targeted by a ribonuclease, named *Dicer*, that cleaves them into 21 bp fragments. The resulting small pieces of dsRNA then recognize homologous molecules of mRNA expressed within the cell and promote their degradation (Fig. B.12). Eventually, most of the homologous mRNAs in the cell are degraded, regardless of whether or not the genes encoding them remain actively transcribed. This *dicer*-based system thus silences any gene or genes encoding mRNA containing regions homologous to the 21 bp dsRNAs.

To knock out expression of a specific gene (*endogene*) within a plant, geneticists induce the plant to express a recombinant gene (*transgene*) that carries an inverted duplication of the endogene, or of a fragment thereof. Transgene expression produces an RNA molecule that folds on itself, forming a double-stranded RNA by base pairing between the two inverted repeats (Fig. B.12). The *dicer* ribonuclease then cleaves this dsRNA into 21 bp fragments, and these fragments, in turn, recognize homologous mRNAs within the cell and promote their degradation. Experimenters obtain a similar—though less efficient—result by making plants express an antisense fragment of the gene to be silenced. RNA produced by this transgene can base-pair with the mRNA encoded by the endogene to form double-stranded structures that initiate the silencing process.

The silencing strategy, heavily used in plants and animals (such as *Caenorhabditis elegans*), offers two key advantages. It is not restricted by the types of mutations available in current collections; and it allows the isolation of transgenic lines in which the endogene under investigation is not completely silenced. This partial silencing is quite similar to the effect of hypomorphic mutations, which only partially alter the function of a particular gene product (as described on p. 654 of Chapter 19). Hypomorphs or partially silenced lines can bypass the early developmental arrests associated with the total silencing of a gene required for early development, and thereby they permit a functional characterization of the gene of interest at later phases of development.

Figure B.12 A cellular defense mechanism against invading viruses can be induced and targeted against mRNAs encoded by cellular genes. The double-stranded RNA produced by replication of an invading RNA virus (*right*) or by folding of a transgenic mRNA containing inverted repeats (*left*) is recognized by the *Dicer* ribonuclease (*green* and *yellow circles*) and cleaved into 21 bp fragments. Homologous mRNAs (*pink line*) encoded by either the invading virus, the transgene, or an endogene are recognized by the *Dicer*-carrying RNA fragments and cleaved into smaller, noncoding fragments. The transgenic DNA is represented by a *light-blue* double helix; inverted repeats are indicated in *red*, and the direction of transgene transcription is shown by *yellow arrows*. RNA polymerase is represented by *yellow circles*.

Another advantage of the silencing strategy over more classical knockout methods is that the same transgene can simultaneously knock out several endogenes. Indeed, if several homologous genes encode similar proteins, it is possible to construct a transgene that expresses an inverted duplication of the DNA fragment conserved between these genes. When introduced into plants, such a transgene will silence all genes that encode mRNAs sharing partial nucleotide homology with it. This eliminates the need to isolate numerous mutations in each one of these genes, and to develop a complex scheme of crosses aimed at bringing together in the same plant multiple mutations in members of the gene family under investigation. This strategy makes it possible to determine the phenotype associated with the disappearance of a specific protein function that is encoded by multiple, functionally redundant genes.

Insertional mutagenesis and gene-silencing strategies for moving from gene to phenotype have found wide use, now that the *Arabidopsis* genome has been completely sequenced. Indeed, geneticists currently not only identify the genes involved in specific functions, they also use their knowledge of gene sequences and mutation-induced phenotypes to define what specific genes do.

The Genetic Analysis of Development in *Arabidopsis*

Although geneticists only partially understand the mechanisms of development in plants and animals, they do know that instructions for the formation of a body plan and the elaboration of specific functions are ultimately encoded in the DNA. Mutations can thus change the position, structure, or function of an organ or alter an individual's orderly passage through the life cycle. We now look at the use of genetic analysis in the study of embryogenesis, hormone control systems, and responses to environmental signals.

The Genetic Analysis of Embryogenesis
Mutations that kill the embryo or arrest its development have provided a basis for understanding embryonic development in *A. thaliana.* Screenings for mutations resulting in developmental arrest at a specific stage of embryogenesis have led to the identification of more than 500 embryo-lethal and embryo-defective mutations. A significant portion of the embryo-lethal mutants are arrested at the transient stage. Many of them carry mutations in housekeeping genes.

Researchers Have Found Very Few Maternal-Effect Mutations in *Arabidopsis*
Such mutations occur in maternal genes whose protein products are deposited in the egg during oogenesis. Maternal-effect mutations disrupt embryogenesis in all progeny carrying a particular mutant genotype.

The small number of maternal-effect mutations constitutes a major difference between plants and certain kinds of animals, such as *Drosophila,* that develop from large eggs. (In *Drosophila,* numerous maternal-effect mutations alter the early stages of embryogenesis, as described in Reference D.) A major reason for the near absence of maternal effects in *Arabidopsis* is that the plant's embryonic development does not depend on a large cytoplasm filled with maternal products (as happens in *Drosophila*). Instead, it depends from a very early stage on a cytoplasm newly produced from the zygote's own genome (as in mammals).

Also unlike what happens in *Drosophila,* in *Arabidopsis,* very few zygotic mutations (that is, mutations in the zygote's own genes) affect embryogenesis before the globular stage. At least part of this difference between *Arabidopsis* and *Drosophila* may derive from the fact that in plants many zygotic genes are expressed very early in embryogenesis (as they are in mammals),

and these zygotic genes may encode protein products that are functionally redundant with those of maternal origin. In addition, many plant genes are members of multigene families in which each member encodes a protein whose function duplicates that of some of the other members. Hence, as in mammals, the scarcity of maternal-effect and early zygotic mutations in plants may simply reflect a large amount of functional redundancy between maternally deposited and zygotically derived products.

Screenings for Mutations That Arrest Development at a Specific Stage of Embryogenesis Enabled the Identification of Important Regulatory Processes in Plant Development
A type of mutation known as *leafy cotyledon* (*lec*) results in embryos with cotyledons resembling mature leaves rather than wild-type cotyledons (Fig. B.13a). As described earlier, cotyledons are embryonic structures that first appear late in the globular stage; leaves, by contrast, are products of the apical meristem that appear only after germination. Unlike cotyledons, leaves have trichomes—small, hairlike projections from the epidermis—and a leaf-specific pattern of vascularization. In addition to causing the conversion of cotyledons into leaves, *lec* mutations produce seeds unable to germinate because they cannot withstand the desiccation that accompanies seed maturation. These observations suggest that the wild-type *LEC* genes encode molecules that help regulate the process of seed maturation in *A. thaliana.* One of these *LEC* genes, *LEC2,* was cloned and shown to encode a transcription factor that plays a fundamental role in the differentiation of cells into an embryonic state. The *twin* mutation revealed an example of negative regulation during embryogenesis. In embryos carrying that mutation, some cells in the suspensor modify their developmental fate, resulting in the formation of a second embryo within the seed (Fig. B.13b). From this observation, *Arabidopsis* geneticists inferred that the *TWIN* gene probably encodes a molecule that represses the embryo development program in suspensor cells. Additional studies also demonstrated that *TWIN* also functions later in development to influence organ formation at the embryonic shoot apex. Hence, *TWIN* appears to convey local positional information in the apical and basal regions of the embryo.

The Genetic Analysis of Hormonal Control Systems
The control of plant development requires complex communications within cells, between neighboring cells, and over long distances between different organ systems. **Plant hormones** are the main internal chemical signals that influence plant development. One group of hormones, the *auxins* (from the Greek *auxein,* to increase or augment), promotes shoot growth through cell elongation and lateral root formation and helps prevent senescence. The auxins also inhibit growth in lateral buds in favor of growth in the apical meristem (apical dominance), and they inhibit cell elongation in roots. A second set of hormones, the *cy-*

(a)

(b)

Figure B.13 Examples of two mutations that arrest development. (a) Micrograph of *lec* mutant (*right*) versus a plant carrying *LEC* wild type (*left*). The cotyledons in a *lec* mutant are abnormally rounded. **(b)** Micrograph of *twin* (*right*) versus *TWIN* embryos (*left*). *a* = axis; *c* = cotyledons.

tokinins, promotes growth by stimulating cell division, including cytokinesis (hence their name). The cytokinins also play a role in the prevention of senescence but, unlike the auxins, promote the growth of lateral buds. Together, the auxins and cytokinins help regulate a range of processes including meristem induction, apical dominance, and root-shoot communication. A third type of hormone, *abscisic acid,* controls aspects of seed development and dormancy; it also mediates responses to water stress (too much or too little water). Yet another group of plant hormones, the *gibberellins,* regulates seed germination and stem elongation. The *brassinosteroids* enhance cell growth and expansion, and they suppress leaf organogenesis in dark-grown seedlings. *Ethylene* promotes maturation of the seedling and influences the timing of leaf senescence. Finally, *salicylic acid, jasmonic acid,* and *methyl-jasmonate* are associated with disease resistance. Jasmonic acid also inhibits seed and pollen germination, inhibits seedling growth, and induces fruit ripening and abscission of flowers.

Mutations That Disrupt Hormone Activity Can Clarify a Hormone's Biological Significance and the Biosynthetic Pathways Involved in Its Production

Researchers originally identified the *gal-1* mutation, an X-ray-induced deletion, through a dwarf phenotype that could be rescued by gibberellin application (Fig. B.14). Later, analysis of the cloned gene indicated that it encodes an enzyme that operates early in the gibberellin biosynthetic pathway. A second mutation, the *gai* mutation, produces a phenotype similar to that produced by *gal-1,* but *gai* mutants are not rescued by gibberellin application. The inference from this observation is that *gai* most likely affects a plant's response to gibberellin rather than the hormone's biosynthesis. Researchers are currently using similar mutational analyses to identify the structural and regulatory genes contributing to the biosynthesis of other plant hormones.

Mutations That Render *Arabidopsis* Insensitive to a Hormone Help Reveal the Mechanisms by Which Plants Perceive and Transduce Hormone Signals

In the most advanced studies of this type, investigators look at ethylene signaling, using ethylene's effects on early seedling

Wild type   *gal-1*

Figure B.14 The *gal-1* mutation. The wild-type plant on the *left* compared to a dwarf plant with the *gal-1* mutation (*right*).

development as the basis for mutant screens. The exposure to ethylene of seedlings grown in the dark produces an inhibition of shoot and root elongation and an accentuation of the apical hook. Mutant seedlings that are insensitive to ethylene grow tall in the presence of ethylene.

Several loci that affect responsiveness to ethylene have been identified in *Arabidopsis* (Fig. B.15). For example, dominant mutations at the *ETR1* locus cause insensitivity to ethylene,

as do mutations at *EIN2* and *EIN3*. By contrast, mutations at the *CTR1* locus result in the constitutive activation of ethylene responses. All these mutations affect a wide range of ethylene responses in various tissues, suggesting that the genes influence the primary steps of ethylene signal processing. Analysis of various combinations of double mutants reveals that *CTR1* is epistatic to *ETR1*, while *EIN2* and *EIN3* are epistatic to *CTR1* (Fig. B.16).

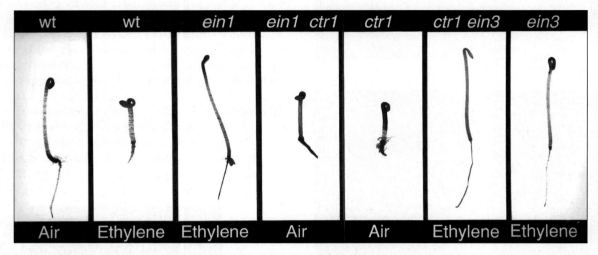

Figure B.15   The effects of ethylene exposure on wild-type and mutant plants. Ethylene inhibits shoot and root elongation in wild-type seedlings. Single- and double-mutant analyses allow the ordering of genes in the signal pathway. The Ein3 protein acts before *Ctr1*, which acts before *Ein1*.

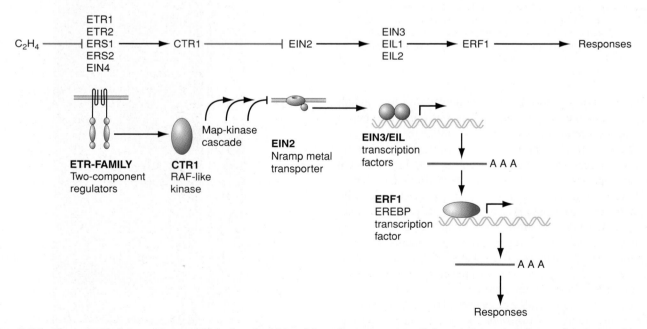

Figure B.16   The genetic interactions and biochemical identities of components of the ethylene signal transduction pathway. Progressing from *left* to *right,* ethylene is thought to regulate negatively a family of membrane-associated receptors. The histidine-kinase transmitter domains of members of the receptor family interact with the regulatory domain of the Raf-like kinase *CTR1*. The receptor/*CTR1* complex negatively regulates a membrane protein (*EIN2*) related to a superfamily of metal transporters. The cytoplasmic C-terminal domain of *EIN2* positively signals downstream to the *EIN3* family of transcription factors. *ERF1* is rapidly induced in response to ethylene and is capable of activating a subset of ethylene responses when expressed.

The genes involved in ethylene signaling have been identified by gene tagging and map-based cloning strategies. The *ETR1* gene is a member of a small family of *Arabidopsis* genes that are related to the genes for receptor protein kinases typical of bacterial signaling systems. Expression of the *ETR1* gene in yeast confers ethylene-binding activity on the cells, demonstrating the *ETR1* protein serves as the ethylene receptor. *CTR1*, by contrast, is related to eukaryotic genes for protein kinases that initiate MAP kinase cascades in yeast and mammals (see Chapter 18), while *EIN2* encodes a membrane protein related to metal transporters and *EIN3* encodes a member of a small family of plant-specific transcription factors. The genetic analysis of this pathway has identified components of a complex signaling pathway that leads from signal perception at the membrane to a transcriptional cascade in the nucleus.

Arabidopsis also generates mutants that are insensitive to the other plant hormones. The isolation of genes determining the various mutant phenotypes suggests that genes for protein phosphatases, farnesyl transferases, transcription factors, and the ubiquitin protein degradation pathway play a role in plant hormone signaling, while genes for membrane-based permeases and transporters contribute to hormone transport. The fact that many of the genes identified are homologous to genes in yeast and mammals is evidence that the basic information-processing systems arose very early in the evolution of eukaryotes. On the other hand, receptors for ethylene, cytokinins, and red light (the phytochromes) are related to bacterial sensory receptors. These systems were likely inherited from the bacterial endosymbiont ancestor of the chloroplast.

The Genetic Analysis of Photomorphogenesis—The Regulation of Growth and Development in Response to Lighting Cues

During growth and development, plants continuously alter their responses to changing environmental signals. One of the most dramatic series of developmental changes takes place when the seedling grows up through the soil and encounters light. Collectively known as *photomorphogenesis,* these changes include the opening of the apical hook and the generation of chloroplasts. Two classes of photoreceptors—phytochromes and cryptochromes—mediate the changes. Receptors in the phytochrome family sense light primarily in the red and far-red wavelengths. Receptors in the cryptochrome family perceive blue wavelengths of light.

Mutational Analyses Have Helped Characterize the Photoreceptor Molecules by Which Plants Receive Light Signals

White light inhibits the elongation of the hypocotyl, or seedling stem, in *Arabidopsis*. A class of photomorphogenic mutants referred to as *hy* (for long hypocotyl) show reduced sensitivity to white light (Fig. B.17a). Linkage map locations for two of the *hy* mutations, *hy3* and *hy8,* coincide with the locations of *phyA* and *phyB* members of the phytochrome gene family. The *hy3* mutant has an altered response to red wavelengths of light,

while the *hy8* mutant has an altered response to far-red wavelengths of light, indicating that these two genes have different roles in mediating photomorphogenesis. Two other mutations, *hy1* and *hy2,* block the biosynthesis of the tetrapyrolle prosthetic group associated with all phytochromes.

The *hy4* mutant in *Arabidopsis* shows a decreased sensitivity to blue light, suggesting that the normal *HY4* gene product plays a role in the reception of blue light. Cloned using a T-DNA-tagged allele and then sequenced, the *HY4* locus exhibits homology with a bacterial enzyme that uses absorbed light energy to repair thymidine dimers in DNA. This conservation of a light-absorbing function is compelling evidence that the *HY4* gene product is one of two blue-light receptors active in photomorphogenic responses.

Mutations in the *HY5* locus also block photomorphogenesis. Molecular cloning of the *HY5* gene revealed that it encodes a transcription factor that appears to control expression of a number of genes contributing to morphogenesis.

Mutational Studies Have Also Clarified How Plants Process Light Signals

While *Arabidopsis* researchers have analyzed *hy* mutants for information on light perception, they have studied other sets of mutants known as *cop* (for constitutive photomorphogenesis) and *det* (for de-etiolated, that is, released from development without chlorophyll) to understand how plants process light signals into the multitude of changes that characterize morphogenesis. Mutant *cop* and *det* seedlings grown in the dark behave as though they have received the light signal—their apical hook opens, the cotyledons expand, light-regulated genes are derepressed, and chloroplasts develop, all in the absence of light (Fig. B.17b). The *det* and *cop* mutations are recessive. From characterizing the phenotypes generated by these recessive mutations, researchers have inferred that the normal *COP* and *DET* gene products somehow suppress all or some photomorphogenic responses until a light signal is received.

The *COP* genes affect all morphogenic responses. Several *COP* loci have been cloned, and biochemical analyses of the gene products reveal that some COP polypeptides associate to form a large protein complex (the signalosome) in the nucleus. Interestingly, in the dark, the *COP1* gene product localizes in the nucleus, where it directly interacts with the HY5 and related transcription factors, targeting them for degradation. In this capacity, *COP1* functions as a negative regulator, mediating the degradation of transcription factors that are required for photomorphogenesis. *COP1* also directly interacts with the cryptochrome blue-light photoreceptors, raising the possibility that a short signal transduction pathway, through which light promotes a change in cryptochrome conformation, helps inactivate the cryptochrome-associated *COP1* protein. With this inactivation, *HY5* and related transcription factors accumulate in the nucleus, where they turn on the expression of downstream genes whose protein products promote photomorphogenesis.

The *DET* genes affect only some photomorphogenic responses. The *det2* mutant, for example, shows derepression of some light-regulated genes, but it does not develop chloroplasts

(a)

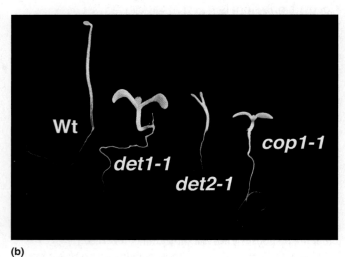

(b)

Figure B.17   Mutations that affect the response to light.
(a) Comparison of the response of a wild-type seedling to light, versus seedlings with the *hy* mutation in one of five different "long hypocotyl" loci. **(b)** Comparison of the response of wild-type and mutant seedlings to dark. The three different mutant seedlings shown behave as if they have been exposed to light when they have not.

in the dark. This mutant develops into a dwarf adult, indicating that the wild-type gene influences more than morphogenesis. The cloning of *DET2* led to the discovery that the gene is related to mammalian genes that operate in steroid biosynthetic pathways. Later, researchers found that the application of brassinosteroids, a class of steroidlike compounds found in plants, can rescue *det2* mutants. While botanists had long suspected that brassinosteroids contribute to the regulation of plant growth and development, genetic analysis clarified that they play a role in the regulation of gene expression by light.

The Genetic Analysis of Flowering: A Comprehensive Example

One of the major life events in the development of an angiosperm is the switch to reproductive growth that culminates in flowering. During the switch, reprogramming of the apical meristem from a leaf-producing to a flower-producing tissue gives rise to an astonishing series of events: The meristem no longer generates mature leaves in a specific spiral pattern. Instead, it becomes an inflorescence meristem (IM) that produces smaller (cauline) leaves, an elongated stem, and many side shoots. The meristems of the side shoots develop into floral meristems (FMs) that produce a single flower of specific design. The IM is *indeterminate* because it produces FMs indefinitely. By contrast, the FMs are *determinate* because they normally produce a fixed number of floral organs in a precise pattern of concentric rings, and in the process, they lose the potential for meristematic activity as cells of the meristem become incorporated into the maturing pistil (Fig. B.18).

Arabidopsis accomplishes the reprogramming of the meristem through regulation of a complex network of intricately overlapping genetic pathways in response to environmental cues. Some of these determine the body plan of the flower; others specify the identity of the floral meristem; still others control the timing of floral meristem formation.

How Genes Determine the Body Plan of a Flower

The floral primordium is an outgrowth of undifferentiated cells near the tip of an inflorescence meristem. As cells in the primordium divide, differentiate, and elongate according to genetic instructions, a flower emerges. A wild-type *Arabidopsis* flower, we have seen, carries four green sepals in the outside whorl, four white petals just inside the sepals, six stamens in the next-to-center whorl, and a cylinder of two fused carpels in the center (review Fig. B.5). Cascades of genetic signals alert cells to their position within the developing flower and thereby direct the proper growth and differentiation of the organs that make up a flower. Mutants can have an abnormal arrangement of organs.

Analyses of Homeotic Mutations Reveal That in *Arabidopsis,* Three Types of Single-Gene Products Influence Floral Pattern Formation

A **homeotic gene** is one that plays a role in determining a tissue's identity during development. A **homeotic mutation** causes cells to misinterpret their position in the blueprint and become normal organs in inappropriate positions. Homeotic mutations thus alter the overall body plan. In one study, researchers looked at mutants with an abnormal order and selection of floral organs. They obtained some of the mutants for their study from laboratory populations, and they used the mutagen ethyl methane sulfonate to generate more. Phenotypic screens of a large number of mutants showed they fell into three classes (Fig. B.19a). Class A displayed carpels instead of sepals in the first whorl, and stamens

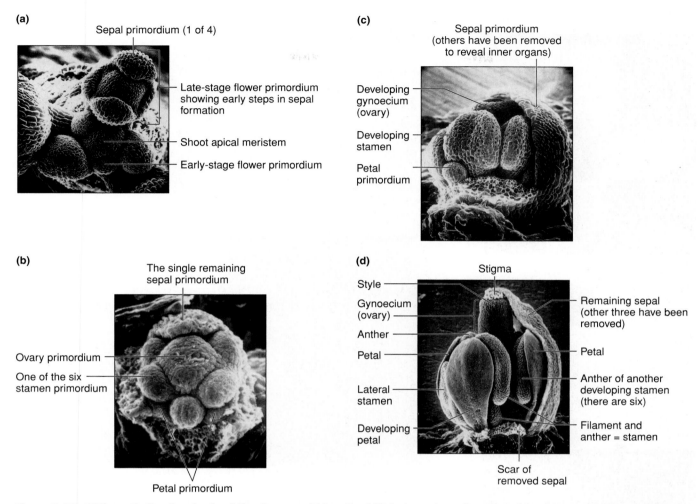

(a)

Sepal primordium (1 of 4)

Late-stage flower primordium showing early steps in sepal formation

Shoot apical meristem

Early-stage flower primordium

(b)

The single remaining sepal primordium

Ovary primordium

One of the six stamen primordium

Petal primordium

(c)

Sepal primordium (others have been removed to reveal inner organs)

Developing gynoecium (ovary)

Developing stamen

Petal primordium

(d)

Stigma

Style

Gynoecium (ovary)

Anther

Petal

Lateral stamen

Developing petal

Remaining sepal (other three have been removed)

Petal

Anther of another developing stamen (there are six)

Filament and anther = stamen

Scar of removed sepal

Figure B.18 Differentiation in whorls of floral organs. This series of SEMs shows the earliest stage of development of the *A. thaliana* flower. **(a)** Sepals are just beginning to form. Three of the four developing sepals have been removed to **(b)** reveal the other organ primordia. Petal and stamen formation are noticeable in **(c).** The structure of the stamen and anthers become more defined in **(d).** Again, three outer sepals have been removed to show the developing flower parts; mature flower carrying four sepals, six stamen, and one gynoecium.

instead of petals in the second whorl, for an overall radial pattern of carpels, stamens, stamens, carpels. In controlled crosses, the ratio of normal to abnormal phenotypes indicated that one gene was responsible for this abnormal radial pattern. Researchers named the gene carrying the mutations that caused this aberrant pattern *APETALA2* (*AP2*) for the lack of petals arising as a result of the mutation. Class B mutants carried sepals in the first and second whorls, and carpels in the third and fourth whorls. Mutations in either *APETALA3* (*AP3*) or *PISTILLATA* (*PI*) caused this deviation from the normal pattern (in botanical terminology, pistillate flowers lack stamen). Class C mutants had an abnormal radial pattern of sepals, petals, petals, sepals. Mutations in the *AGAMOUS* (*AG*) gene, named for the lack of gamete-forming tissues resulting from the mutations, determined these mutant phenotypes. The wild-type *AGAMOUS* gene also stops flower development after establishment of the fourth whorl.

The Radial Pattern Model Suggests How Three Classes of Genes Could Determine the Identity of Floral Organs
For the sake of simplicity, consider that class A mutants are missing class *A* gene activity, B mutants are missing *B* gene activity, and C mutants are missing *C* gene activity. The radial pattern model makes three sets of assumptions about these three classes of genes (Fig. B.19b):

1. *A* genes are active early in the emergence of the first and second whorls; *B* genes are active early in the definition of the second and third whorls; and the *C* gene is active early in the differentiation of the third and fourth whorls.

2. The gene products of these three types of genes, alone and in combination, determine the identity of the floral organs as they develop from the flower primordium. The *A* product determines sepals, *A* and *B* together determine

(a)

Class A, B, and C mutants

Wild type

Class A mutants

Lack of activity of *APETALA2 (AP2)*

Class B mutants

Lack of activity of *APETALA3 (AP3)*
or *PISTILLATA (PI)*

Class C mutants

Lack of activity of *AGAMOUS (AG)*

Schematic key:

Sepal

Petal

Carpel

Stamen

(b)

Inference about the activity of Class A, B, and C mutants

A gene product ⟶ sepals

A+B gene product ⟶ petals

B+C gene product ⟶ stamens

C gene product ⟶ carpels

If A is active, C is repressed   result: sepals

If C is active, A is repressed   result: carpels

If a mutation incapacitates A, C is abnormally active   result: carpels

If a mutation incapacitates C, A is abnormally active   result: sepals

(c)

(d)   **(e)**

Figure B.19   Homeotic mutants and flower formation in *Arabidopsis*. (a) Class A, B, and C mutants. **(b)** Diagram showing how three types of genes determine the identity of the floral organs. **(c)** *ap2 / ap3, ap2 / ag,* and *ap3 / ag* double mutants. **(d)** *ap2 / ap3 / ag* triple mutant. **(e)** Hybridization probe visualizing *AP3* activity in the flower primordium.

petals, the combination of *B* plus *C* determines stamens, and the *C* product alone determines carpels.

3. The activities of *A* and *C* are mutually exclusive; that is, in areas where *A* is active, *C* is repressed, and in areas where *C* is active, *A* is repressed. Moreover, if a mutation incapacitates *A, C* is abnormally active and vice versa.

The model provides an elegant explanation of the mutational evidence. To confirm its predictions, researchers carried out further genetic investigations. In one line of experiments, they bred double and triple mutants and predicted what they would look like. For example, they bred plants with mutations in both *AP2* (the *A* gene) and *AP3* (a *B* gene) and predicted that with only normal *C* activity in the flower primordium, the emerging flower would consist of all carpels. The resulting flowers confirmed the prediction. Similarly, they eliminated both *B* and *C* activity by breeding plants with mutations in the *AP3* and *AG* genes, and as predicted, the resulting mutants bore flowers consisting of only sepals (Fig. B.19c). In triple mutants lacking *A, B,* and *C* gene activity, the flower organs differentiated as leaves (Fig. B.19d).

The cloning and sequencing of the homeotic genes for floral identity have further validated the radial pattern model. All of the genes produce DNA-binding transcription factors. *In situ* hybridizations using DNA fragments complementary to A-, B-, and C-type mRNAs have monitored the activity of the normal genes and shown that the *B* and *C* genes (*AP3, PI,* and *AG*) are expressed specifically in the whorls predicted by the model (Fig. B.19e). The *A* gene, *AP2,* however, is expressed in all four whorls of the flower although its product affects organ identity only in the first and second whorls. In *ap2* mutants, *AG* RNA appears in all four whorls of the flower, as predicted by the model. In these *ap2* mutants, however, only the first and second whorls have an altered phenotype. The explanation is as follows: *AP1* (one of the *A*-function genes) is on everywhere in the early flower primordium as is *AP2. AG* turns on in whorls 3 and 4, repressing *AP1* and thus excluding *A* function from whorls 3 and 4. In *ap2* mutants, *AG* comes on everywhere and represses all *AP1* activity, so there is *C* function in all whorls, but no *A* function. Thus in *ap2* mutants, you expect to see only whorl 1 and whorl 2 phenotypes.

Studies of the radial pattern model in snapdragons have shown that what happens in *Arabidopsis* also happens in other higher plants.

The data discussed here indicate that the *A, B,* and *C* genes are necessary for radial patterning in flowers. However, it appears that they are not sufficient to determine flower organ identity. For instance, the expression of *C* genes in leaves does not convert them into carpels. Recent genetic analysis of flower patterning in *Arabidopsis* has demonstrated that three functionally redundant genes (*SEPALLATA1, 2,* and *3,* abbreviated *sep1, 2,* and *3*) are essential for the activity of the *B*- and *C*-type genes. Indeed, triple *sep1 sep2 sep3* mutants form flowers that carry an indeterminate number of whorls. Interestingly, single *sep1, sep2,* or *sep3* mutants, or double mutants carrying different combinations of these genes, have wild-type flowers, indicating

that the three genes are functionally redundant. Molecular analyses of these genes have provided an explanation for these striking observations. Such analyses show that *sep1, 2,* and *3* encode putative transcription factors that are highly similar to each other, as well as to *AG.* These three proteins can bind directly to type B and C proteins, forming heteropolymers that are likely to regulate the expression of different sets of target genes.

In short, the A, B, C model explains one part of a complicated genetic process—the interpretation of cellular position within the flower primordium—but leaves unanswered the question of what establishes the positions. As we see next, the differentiation of floral organs, in addition to the appropriate expression of the *A, B, C,* and *sep* genes, requires the activity of many other interacting genes at the proper time and place in an immense hierarchy of genetic control.

Earlier-Acting Genes Specify the Identity of the Floral Meristem

The floral meristem produces the flower primordium that differentiates into four whorls of organs, two of which contain the gametes. But what tells cells in the side shoots of an inflorescence meristem to become a floral meristem? To answer this question, researchers analyzed loss-of-function single and double mutants (Fig. B.20a and b) and gain-of-function transgenic plants. They constructed the transgenic plants by first fusing a nonspecific promoter called 35S with the DNA of the cloned gene under study and then inserting the promoter/gene fusion product into otherwise wild-type *Arabidopsis* plants. Table B.1 describes six of the floral meristem development genes uncovered by these loss- and gain-of-function strategies. By comparing the effects of each gene's activity with that of the others, the researchers were able to suggest how they interact.

Model of the Progressive Specification of the Floral Fate

The first sign of floral development at the molecular level is expression of the *LEAFY* (*LFY*) gene. Activation of *LFY* marks the position of a future floral meristem and takes place before a visible FM appears. With inadequate *LFY* function, the meristem remains an inflorescence meristem. Next comes the expression *APETALA1* (*AP1*) and *CAULIFLOWER* (*CAL*), whose activities overlap in newly formed meristems. The orderly expression of *LFY, AP1,* and *CAL* promotes emergence of an unbranched flowering structure with four whorls of organs. Loss-of-function mutations in any of these genes prevents the shoot from flowering. Interestingly, mutations in the *AP2* gene enhance the effects of *lfy* and *ap1* mutations.

Expression of *AG* occurs at the center of the floral meristem where its product represses *AP1* expression and further commits the meristem to the determinate production of flowers (Fig. B.20c). A lack of *AG* activity can cause the developing FM to revert to an IM. With the expression of *LFY, AP1,* and *AG,* cells in the lateral, floral meristems become increasingly committed to reproductive development.

Wild-type *Arabidopsis* plants produce only lateral flowers because the *TFL1* gene prevents the expression of FM-producing genes in the center of the IM. As a result the indeterminate in-

(a)

(b)

(c)

Figure B.20 Identifying genes that influence development of the floral meristem. (a) Single mutants *lfy-6* (*left*), *ap1–1* (*middle*), and *tfl-2* (*right*). **(b)** Double mutant *ap1–1 / cal-1*. **(c)** Hybridization probes visualize the activity of *AG* RNA in the center of the flower primordium.

florescence meristem continues to grow and produce new side shoots even as older lateral shoots are activating the genetic program for flowering.

Some Genes Control the Timing of FM Formation and Flowering

Laboratory-grown, wild-type *Arabidopsis* plants adjust their flowering time in response to the amount of light they receive. Those that receive 16 hours of light a day begin to flower three weeks after germination, while those that receive only 8–10 hours of light do not initiate flowering until at least six weeks after germination. Moreover, the shorter days increase the duration of all phases of development, causing the production of more leaves in the rosette and more flowers on the inflorescence. Exposure of hydrated seeds (which have taken up water and are ready to germinate) or young plants to low temperatures for three to six weeks after germination accelerates the flowering of wild-type plants.

As Table B.2 shows, two types of mutations influence flowering time: those that delay it, and those that precipitate it. Researchers identified many of the mutations that disrupt normal flowering by screening laboratory populations for their flowering responses. They also found that some mutations identified on the basis of other phenotypes affect flowering. For example, *phyB* mutants, which have an altered response to far-red wavelengths of light, and *cop1* mutants, which grow in the dark as if they have received a light signal, both flower early.

To understand the function of genes that affect flowering time, researchers have analyzed single mutants, double mutants, and transgenic plants containing extra copies of a particular gene. Among their findings is the observation that *cop1, phyB* double mutants show the Cop1 phenotype of short hypocotyls (rather than the phyB phenotype of expanded hypocotyls). This suggests that *COP1* acts downstream of *PHYB*. Since both mutations accelerate flowering, the normal gene products may be part of a pathway required to repress flowering and several

TABLE B.1   **Six of the *Arabidopsis* Genes That Control Meristem Identity**

Arabidopsis Gene	Loss of Function (Mutant Phenotypes in Meristem Identity)	Gain of Function (Phenotypes of Transgenic Plants)
LEAFY (LFY)	Conversion of flower to shoot or shoot-like structures; floral reversion	Conversion of lateral shoots to flower; terminal flower; early flowering
APETALA1 (AP1)	Conversion of flower to shoot; branched flowers	Conversion of lateral shoots to flower; terminal flower; early flowering
CAULIFLOWER (CAL)	Enhances ap1 phenotypes	Not tested
APETALA2 (AP2)	Enhances ap1 and lfy phenotypes	Not tested
AGAMOUS (AG)	Indeterminate flowers; axillary buds; floral reversion	Conversion of lateral shoots to flower; terminal flower; early flowering; suppression of late flowering mutants
TERMINAL FLOWER (TFL1)	Terminal flower; early flowering	Not tested

Reference: TIG January 1998 Vol. 14 No. 1, "To Be, Or Not To Be, A Flower—Control of Floral Meristem Identity" by H. Ma.

TABLE B.2   **Some of the *Arabidopsis* Mutations That Affect Flowering Time**

Full Name	Abbreviated Name	Probable Function of Gene Product
Mutations causing late flowering		
constans	co	Transcription factor
luminidependens	ld	Transcription factor
phytochrome A	phyA	Light receptor
gibberellic acid 1	ga1	Enzyme required for gibberellic acid biosynthesis
gibberellic acid insensitive	gai	Unknown
gigantea, frigido, fha, fy, fpa, fve, fca, fe, fwa, fd, ft	gi, fri, fha, fy, fpa, fve, fca, fe, fwa, fd, ft	Unknown
Mutations causing early flowering		
phytochrome B	phyB	Light receptor
constitutive photomorphogenic 1	cop1	Transcriptional repressor
early flowering 3	elf3	Unknown
early short day 4	esd4	Unknown
embryonic flower 1	emf1	Unknown
embryonic flower 2	emf2	Unknown
terminal flower	tfl	Unknown

Reference: TIG, October 1995 Vol. 11 No. 10 p. 395 "Genetic and environmental control of flowering time in *Arabidopsis*" by George Coupland.

overlapping pathways most likely lead to flowering. Another finding is that environmental influences on flowering time affect the interactions among floral meristem identity genes. The *lfy* mutants, for example, show a stronger phenotype under short days than under long days; and there is more *COP1* RNA in young seedlings grown under long days than under short days, suggesting that the *COP1* gene helps promote flowering.

Hundreds of Mutational Analyses Have Generated a Preliminary Model to Guide Future Research on Flowering

Since there are probably hundreds of genes that contribute to the control of floral development, the genetic analysis of flowering is a work in progress. Nevertheless, certain observations enable us to build a preliminary genetic model of the process. It appears that at least four pathways promote flowering in *Arabidopsis:* a pathway regulated by day length (photoperiod-dependent pathway), an autonomous pathway, a vernalization-dependent (cold-responsive) pathway, and a gibberellin-dependent pathway (Fig. B.21).

Photoperiod-Dependent Pathway. Genetic analysis of late-flowering mutants revealed that mutations in one group of genes (*PHYTOCHROME A, CRYPTOCHROMES A* and *B, CONSTANS [CO], CCA1, ELF3, FD, FKF1, FE, FHA, FT, FWA, GIGANTEA [GI], LHY, TOC1, ZTL*) delay flowering in periods of long, but not short, days. Hence these late-flowering plants do not respond to photoperiod (in both long and short days, they behave as if they are in short days); and the corresponding genes appear to promote flowering in response to long days. *CONSTANS* is a key regulator in the photoperiodic pathway. Researchers have observed that its overexpression in plants under the control of the strong 35S promoter is sufficient to promote early flowering; thus plants with *CO* overexpression are insensitive to day length. The *CO* gene encodes a transcription factor that most likely regulates expression of downstream genes.

Figure B.21 Flowering time is determined by at least four different pathways: (1) the autonomous pathway, **(2)** the vernalization-dependent (cold-responsive) pathway, **(3)** the photoperiod-dependent pathway, and **(4)** the gibberellin-dependent pathway. *Green arrows* indicate positive regulation. *Red lines* ending with a T represent negative regulation.

Autonomous Pathway. Mutations in a second group of genes (*FCA, FLOWERING LOCUS D [FLD], FPA, FVE, FY, LUMINIDEPENDENS [LD]*) result in late flowering in both long and short days. This suggests that these genes promote flowering in an autonomous fashion, independent of photoperiod. Interestingly, autonomous-pathway mutants respond to vernalization (they flower early when exposed to long cold treatments), which implies that a third pathway is responsible for flower promotion by vernalization.

Vernalization-Dependent Pathway. To look at the vernalization-dependent pathway, scientists crossed early-flowering plants with plants belonging to late-flowering winter-annual ecotypes (which flower early if exposed to vernalization), and they showed that the late-flowering phenotype was due to dominant mutations at two loci: *FRIGIDA (FRI)* and *FLOWERING LOCUS C (FLC)*. When they set up to clone these loci, they found that most of the early-flowering ecotypes carry loss-of-function mutations. They cloned *FRI* and *FLC* by chromosome walking and showed that *FRI* encodes a novel protein of unknown function while *FLC* encodes a putative transcription factor in shoot and root tip tissues.

Analysis of *FLC* expression in plants subjected to vernalization, as well as in plants carrying mutations in *FRI* or in some of the genes belonging to the autonomous pathway, increased researchers' understanding of the mode of action of vernalization pathway genes. *FLC* expression appears to decrease to undetectable levels in *fri* mutants and autonomous-pathway mutants, suggesting that *FRI* helps promote *FLC* expression. Furthermore, vernalization suppresses expression of *FLC,* and this *FLC* suppression is maintained over the entire life cycle of the plant. It is not, however, transmitted to the next generation. Thus, *FLC* expression correlates with an early-flowering phenotype, suggesting that *FLC* is a key negative regulator of the transition between the vegetative and reproductive phases in *Arabidopsis* (Fig. B.21). To test this hypothesis, researchers created plants that overexpress *FLC* under the control of the strong 35S promoter. These plants flowered late and could not be induced to flower early by mutations in *FRI* and in the autonomous pathway, nor by vernalization. Conversely, loss-of-function mutations at *FLC* resulted in an early-flowering phenotype, as discussed previously.

It is interesting that, upon vernalization, *FLC* is silenced not only during cold treatment but also during the entire life cycle of the plant. This means that the plant maintains a memory of the cold treatment, but, as we saw, that memory is not transmitted to the next generation. To understand this long-term memorization of vernalization, scientists isolated mutations that eliminate a plant's responsiveness to vernalization. One such mutation, named *vrn2*, gave rise to a plant that still responded to vernalization by decreasing *FLC* expression but that did not maintain *FLC* suppression after vernalization ended. Instead, the gene became active again, and the plant flowered late despite having been subjected to vernalization. When these researchers cloned the *VRN2* gene by chromosome walking, they found that it encodes a regulatory protein with homology to animal proteins of the Polycomb-complex. Polycomb-complex proteins suppress the expression of multiple developmental genes by modifying the chromatin structure in their periphery.

Gibberellin-Dependent Pathway.   Gibberellins also promote flowering, through a fourth pathway. Mutations that affect gibberellin biosynthesis or response often result in later-flowering phenotypes (Fig. B.21).

Genetic analyses of flowering in *Arabidopsis* have uncovered a number of mechanisms that help control the transition from vegetative to reproductive development in plants. Some of the key regulators identified to date also appear to contribute to the control of flowering in other plants, including crops. For instance, *FRI* and *FLC* homologs seem to confer a winter-annual habit (in which cold treatment gives rise to an early-flowering phenotype) or a biennial habit (in which cold treatment is required for flowering) on cultivars of many different types of domesticated plants. Further research will lead to a better understanding of flowering time in plants; it will also aid in the design of strategies for engineering different flowering habits in crops and horticultural species.

To Flesh out the Multiple Pathway Model, It Will Be Necessary to Identify More of the Genes That Influence Flowering

There are several reasons why some of these genes have not yet been identified: The genome is not yet saturated with mutations; some genes have redundant functions; and mutations in regulatory genes controlling earlier stages of floral development do not necessarily cause unique floral phenotypes. To find these hard-to-identify genes that influence flowering, researchers can look for new mutants, try to isolate suppressors and enhancers of existing mutations, and devise protocols that test for the late function of early-acting genes (see pp. 663–665 of Chapter 19).

In addition to identifying more of the genes active in the pathways that promote flowering, researchers will continue to analyze the mechanisms by which these genes work. To this end they will seek to demonstrate the activation of transcription by encoded proteins and to discover the target genes of the transcription factors. Because the food consumed by people around the globe consists largely of grains, fruits, and other flower products, an understanding of flowering at the molecular level may increase the chances of feeding an ever-expanding human population.

C O N N E C T I O N S

Plants are fundamentally different from animals at the level of the whole organism. In contrast to animals, plants continue to produce new organ systems throughout their life cycles, are autotrophic, and do not move about. Despite these differences, the cellular machinery and genetic mechanisms of plants and animals are surprisingly similar. In fact, many of the genes involved in basic metabolic functions and information processing show a striking degree of homology between plants and animals. For example, the *AGAMOUS* gene specifies a DNA-binding protein that most likely regulates downstream genes. A human relative of the *AGAMOUS* gene makes a DNA-binding protein, known as serum response factor, that regulates the *c-fos* oncogene.

The remaining genetic portraits (References C, D, and E) look at genetic mechanisms that propel development in three model animals: the nematode *C. elegans,* the fruit fly *Drosophila,* and the house mouse *Mus musculus.*

E S S E N T I A L   C O N C E P T S

1. *A. thaliana* makes an excellent model organism for genetic analysis because of its capacity for prolific reproduction and the relatively small size of its genome. The genome is carried in five chromosomes that contain little repetitive DNA and a tight arrangement of genes with a few small introns.

2. *Arabidopsis* switches from vegetative to reproductive development when the leaf-producing apical meristem becomes a flower-producing meristem. After flowering, it undergoes double fertilization to produce an embryo that develops within a protective maternal seed coat. In response to proper conditions, including favorable light and temperature, the seed germinates and vegetative

growth begins again. The vegetative plant eventually ages and dies.

3. Researchers use several techniques of genetic analysis in *Arabidopsis* to generate mutations and characterize them at the molecular level. Mutagenesis by treatment with chemicals or X rays produces a variety of mutations for study. Insertional mutagenesis by *A. tumefaciens* and transposons produces mutated alleles carrying a specific DNA insert that can then serve as a probe for identifying and cloning the gene of interest. Researchers can also start with a cloned and sequenced gene, pool DNA-transformed genomes, and use PCR screening to identify single plants with

mutations in the gene of interest. The ability to generate mutations makes it possible to move from identification of a gene to the phenotype it produces. It is also possible to determine gene function with the use of novel RNA-based silencing techniques, which do not depend on the recovery of mutations in the gene of interest.

4. The genetic analysis of mutations that affect embryogenesis has revealed that in one respect, *Arabidopsis* is more similar to mammals than to arthropods such as flies. Like mammals and unlike flies, plants have very few maternal-effects mutations and few zygotic mutations before the globular stage. Mutational analyses have also helped identify some of the regulatory processes involved in plant development. Future research will try to pin down the molecular mechanisms that control embryogenesis.

5. The genetic analysis of hormonal control systems has clarified the biological significance of hormones and the biosynthetic pathways that produce them. It has also shown how *Arabidopsis* perceives and transduces hormone signals. Some of the genes involved in hormone production and hormone responses are homologous to those found in yeast and mammals.

6. The genetic analysis of growth and development in response to light conditions, known as photomorphogenesis, has clarified how plants receive and process light signals. Some of the genes that contribute to photomorphogenesis are related to mammalian genes that operate in steroid biosynthetic pathways.

7. From the genetic analysis of flowering, researchers have proposed that three types of genes determine the identity of floral organs. These genes operate in the middle of a regulatory cascade in which earlier-acting genes specify the identity of the floral meristem, while other genes control the timing of both floral meristem formation and flowering.

S O L V E D   P R O B L E M S

I. The St ecotype of *A. thaliana* switches from vegetative to reproductive (flower) development late in the season compared to ecotype Limburg-5, which flowers early. How could you determine if flowering time is based on alleles of a single or multiple genes? Describe the results you would expect for the two different cases.

Answer

This problem requires an understanding of single- and multigene inheritance. The two different ecotypes could be crossed and the hybrid selfed to produce an F_2 generation. *If the phenotype is due to a single gene, you would expect discrete phenotypes in the F_2 generation—* either two or three phenotypes, depending on whether there was complete or incomplete dominance. *If the trait is determined by several genes, there would be a spectrum of phenotypes in the F_2 generation* ranging from early to late flowering.

II. *HY5* is a gene that encodes a positive regulator of gene expression in response to light in *Arabidopsis.*

a. What do you predict would happen to the expression of light-regulated genes in a loss-of-function *hy5* mutant?

b. To examine the action of the Hy5 protein at the molecular level, the expression of two promoter-GUS reporter gene fusions was monitored. One of the fusions contained a light response element (G box) and the other lacked the light response element.

Results are shown here. (A + indicates an increase in expression.)

Promoter elements	Light-activated GUS expression			
	$HY5^+$		$hy5^-$	
	No light	+ Light	No light	+ Light
G box	−	+	−	−
No G box	−	−	−	−

What would you conclude about the action of Hy5?

c. If you wanted to determine if *HY5* regulates expression in nonphotosynthetic as well as photosynthetic tissues of the plant, how would you design the experiment?

Answer

This question requires an understanding of gene regulation.

a. Loss of function of a positive regulator would result in *no expression of genes in response to light.*

b. Both the *G box in the promoter and the* HY5 *gene product are needed to get increased gene expression in response to light.* (It is reasonable to hypothesize that *HY5* interacts directly or indirectly with the G box sequence.)

c. There are several ways in which you could examine the tissue specificity of light-regulated, *HY5*-dependent transcription. In all cases, the wild type

($HY5^+$) and $HY5^-$ mutant should be compared with and without the light stimulation. *RNA can be isolated from photosynthetic and nonphotosynthetic tissues in the two strains, with and without light, and hybridized using a probe from a light-regulated, HY5-dependent gene; or DNA from a* *HY5-dependent gene can be hybridized to fixed tissues in the different plants* (in situ *hybridization) or a reporter gene linked to a promoter of an HY5-regulated gene could be used to monitor expression in tissues.*

PROBLEMS

B-1 For each of the terms in the left column, choose the best matching phrase in the right column.

a. apical meristem	1. light-regulated developmental program
b. photomorphogenesis	2. variety of plant
c. whorls	3. growing point of a plant containing undifferentiated cells
d. inflorescence	4. concentric portions of a flower
e. ecotype	5. group of flowers at the tip of a branch

B-2 Why are the following considered to be advantages for the geneticist or molecular biologist using *Arabidopsis* as a model plant system compared to corn or peas?
a. generation time of six weeks
b. 10,000–40,000 seeds per plant
c. genome size similar to *C. elegans*

B-3 *Arabidopsis* has a small genome yet carries out the same functions as plants such as tobacco and pea, which have genomes 50–100 times larger. How can *Arabidopsis* accomplish similar physiology with so little DNA?

B-4 Different ecotypes produce different numbers of seeds (seed sets).
a. If you wanted to determine whether this trait was due to a single gene or multiple genes, how would you test this?
b. If the trait was due to multiple genes, what characteristic of the different ecotypes would you use to locate genes involved?

B-5 What is the ploidy of each of the following tissues, what nuclei combined to form them, and how are they related genetically to the plant on which they are found?
a. endosperm
b. zygote
c. embryo sac cells

B-6 After germinating mutagenized seed, there may be some mutant phenotypes in that first generation. However, these are not usually the plants that a geneticist will choose to study. Why not?

B-7 Match the type of mutagenic agent with the best use (assume there is only one match for each agent).

a. promoterless GUS-containing vector	1. generate a collection of plants with different loss-of-function mutations
b. EMS	2. study expression of genes
c. kan^r Ti	3. generate temperature-sensitive mutants

B-8 a. Order the following steps for a reverse genetics approach to exploring the function of a gene:
   i. PCR amplify using primers from the gene and from the T-DNA sequence.
   ii. Use insertion to make a plant a homozygote.
   iii. Obtain a library of mutant lines containing T-DNA insertions.
   iv. Find a potential gene based on DNA sequence information.
b. An alternative to insertion mutagenesis with T-DNA is an RNA-based silencing protocol for determining gene function. Explain how the phenotype associated with a null mutation at a particular gene can be ascertained even when no mutations in that gene are actually available to the researcher.

B-9 A sterile mutant was identified in which there was a random instead of directed growth of pollen tubes toward the ovary. The mutant was discovered after mutagenesis of seeds and self-fertilization of the progeny plants. It was not known therefore whether the sterility defect was a male or female defect. How could you determine that?

B-10 How could you find out if an *APETALA* homolog exists in petunia, snapdragons, or potatoes?

B-11 The gravitropic response in plants includes upward curve of the hypocotyl and flowering stalk and downward growth of the root.
a. It has been postulated that this response is mediated by auxin. What is a genetic experiment to test this hypothesis?
b. Auxin-regulated mRNAs have been identified in *Arabidopsis*. How would you determine the distribution of these auxin-regulated mRNAs in the part of the plant responding to gravity?

B-12 Ethylene affects plant growth and development in many ways including inhibition of stem and root elongation, stimulation of seed germination, and ripening of fruits. Ethylene biosynthesis occurs in the plant at several different developmental times. Mutants that are defective in ethylene stimulation might be defective in the biosynthesis of ethylene by the plant or in the signaling pathway in cells that receives and processes the signal.

a. The *ctr1⁻* mutant has a defect in the signaling pathway resulting in a constitutive ethylene response. If you treat this mutant with an inhibitor of ethylene biosynthesis, what effect do you expect?

b. Ethylene overproduction mutants *eto1⁻*, *eto2⁻*, and *eto3⁻* are sensitive to inhibitors of ethylene biosynthesis. Would you conclude that the *eto* genes are biosynthetic or signaling mutants?

c. No loss-of-function mutations have been isolated in the *ETR1* gene. This could indicate that the gene is required for growth or that there is redundancy of this gene. How could you look for redundancy of this gene in the genome?

d. A series of bases in the promoter region of several ethylene responsive genes have been identified that are essential for proper DNA binding to specific transcription factors in the ethylene-responsive element-binding protein (EREBPs) family. The *APETALA2* gene product has a polypeptide domain that is homologous to a polypeptide domain found in the EREBPs. Why would these proteins show evidence of homology in one of their domains?

e. Genetic engineering of *ETR1* and/or *CTR1* genes may prevent or modify the ethylene response and could be very helpful in regulating the ripening of fruits and vegetables. If you need to control expression of the genes, what would you want to be part of the DNA construct that you injected into the plant?

Caenorhabditis elegans:
Genetic Portrait of a Simple
Multicellular Animal

Contributing author William B. Wood *University of Colorado*

An adult *C. elegans* hermaphrodite surrounded by larvae of various stages.

The nematode *Caenorhabditis elegans,* one of the simplest multicellular organisms, lives in soils worldwide and feeds on soil bacteria. Adults are about 1 mm in length and contain an invariant number of somatic cells (Fig. C.1). The mature "female," which is actually a hermaphrodite able to produce both eggs and sperm, has precisely 959 somatic cells that arose from progenitor cells by a reproducible pattern of cell division. The mature male, which produces sperm and has genitalia that enable it to mate with the hermaphrodite, includes precisely 1031 somatic cells that also arose by a reproducible pattern of cell division. *C. elegans* has a short life cycle and an enormous reproductive capacity, progressing in just three days from the fertilized egg of one generation to between 250 and 1000 fertilized eggs of the next generation. It is transparent at all stages, so that investigators can use the light microscope to track development at the cellular level throughout the life cycle. Its small size and small cell number, precisely reproducible and viewable cellular composition, short life cycle, and capacity for prolific reproduction make *C. elegans* an ideal subject for the genetic analysis of development. The fact that the

genome for *C. elegans* was sequenced in 1998 makes it an even more appealing organism to study.

Although *C. elegans* and most other free-living species of nematodes are generally beneficial, they are related to nematodes that parasitize animals and plants, causing human disease and agricultural damage. Knowledge gained from the study of *C. elegans* will help combat these problems.

Three unifying themes surface in our discussion of *C. elegans.* First, the invariance of cell number and fates forms the basis of many experimental protocols used to study nematode development. Second, the invariant specification of cellular divisions and fates depends on a varied palette of developmental strategies. These include the segregation of particular molecules to particular daughter cells at division, inductive signals sent from one cell to influence the development of an adjacent cell, signal transduction pathways within each cell that respond to the arrival of an inductive signal, and a genetically determined program that causes the death of specific cells. Third, genetic studies on the development of *C. elegans* reveal the simultaneous conservation and innovation

Figure C.1 Photomicrograph of adult *C. elegans* hermaphrodite (*top*) and male (*bottom*) caught *in flagrante delicto*. The animals are oriented in opposite directions during their mating: the symbol denoting the male is placed near the head, while that for the hermaphrodite is near the tail. Note that the body is transparent, enabling visualization of internal structures. In this photo, it is easy to see the eggs in the hermaphrodite.

of evolution. Because the nematode exhibits many features of development, physiology, and behavior found in other complex animals such as *Drosophila* and humans, studies of *C. elegans* can help elucidate developmental pathways and genes conserved throughout animal evolution. But because other features of *C. elegans* development, such as the invariant spatial and temporal pattern of cell positions, divisions, and fates, are quite different from those found in more complex animals, studies of *C. elegans* provide a comparative counterpoint that deepens our understanding of the full range of genetic controls over development in multicellular eukaryotes.

Our genetic portrait of *C. elegans* presents:

■ An overview of *C. elegans* as an experimental organism, including descriptions of its genome, its life cycle and anatomy, the precise patterns of its cell lineages throughout development, and techniques of genetic and molecular analysis that biologists use to study its development.

■ The genetic dissection of several developmental processes in *C. elegans,* including the specification of early embryonic blastomeres, the role of programmed cell death, and the timing of decision making during larval development.

■ A comprehensive example on the use of genetics to probe a signaling pathway that helps control development of the hermaphrodite vulva.

■ ■ ■

An Overview of C. *elegans* As an Experimental Organism

As one of the model organisms selected for genomic analysis under the Human Genome Project, *C. elegans* is under intensive investigation.

The Nuclear Genome of *C. elegans*

The completely sequenced *C. elegans* genome is small for a multicellular animal, only 97 Mb in size. While this is about 10 times the size of the yeast genome, it is only two-thirds the size of the *Drosophila* genome. The *C. elegans* genome is packaged into six chromosomes, designated I, II, III, IV, V, and X. These chromosomes are small and of approximately the same size, which makes it difficult to obtain much genetic information from cytogenetic and karyotypic analyses. Figure C.2 shows simplified genetic maps of all six chromosomes.

Note that the maps lack centromeres, an important feature found on all other eukaryotic chromosomes described in this book. The *C. elegans* maps lack this feature because the chromosomes do not have a defined centromere. Instead, like the chromosomes of some insects and ciliates, *C. elegans* chromosomes are **holocentric;** that is, they have a diffuse centromere. The distribution of centromeric activity along the length of the chromosome allows many points of spindle attachment during both meiosis and mitosis. This unusual centromere structure does not in general affect genetic analysis because the end result

of chromosome segregation and recombination in *C. elegans* is similar to that of chromosomes in organisms with defined centromeres. As we see later, however, the behavior of holocentric chromosomes provides the basis of a technique important in the construction of *C. elegans* genetic mosaics.

Aligning the Physical and Genetic Maps

Researchers have constructed a detailed physical map of the *C. elegans* genome showing the order of cosmid and YAC clones covering each of the six chromosomes. The physical map has been useful not only for genomic sequencing but also for the positional cloning of genes important in development.

Ongoing mapping and molecular identification of genes and DNA markers are providing an increasingly accurate alignment of the genetic and physical maps. Interestingly, the relationship between recombination frequency and physical distance varies considerably in different parts of the *C. elegans* genome. In central regions of some chromosomes, for example, one map unit (or 1 cM) as determined by recombination is the equivalent of 1500 kb of DNA, while in the more distal regions of these same chromosomes, one map unit equals 50 kb of DNA. The reasons for these discrepancies are not clear. One possibility is that the holocentric organization of the chromosome somehow impedes the resolution of recombination events occurring near the center of the centromere, so that fewer of the resulting recombinant gametes survive. Another possibility is that recombination-promoting sequences are found more fre-

Figure C.2 Skeleton genetic maps of the six *C. elegans* chromosomes. Only a small subset of the more than 1600 currently mapped genes is shown. Distances on these maps represent recombination frequencies in centimorgans. The zero position on each chromosome was chosen arbitrarily, as these chromosomes do not have defined centromeres. Each of the five autosomes has a central cluster (*darker blue*) in which the gene density appears to be unusually high, reflecting a lower rate of recombination per kilobase of DNA.

quently near the tips than near the centers of the chromosomes. Whatever the cause for the depression of recombination frequencies near the center of the chromosomes, the genetic map is clearly not an accurate reflection of the physical distribution of *C. elegans* genes. One symptom of this anomaly is that genetic maps based on linkage analysis show a marked clustering of genes near the centers of the autosomes (Fig. C.2), although the genomic sequence has shown that the average gene density is no greater in these regions.

Genome Sequence and Organization

The sequencing of the *C. elegans* genome, completed in 1998, provided researchers with the first complete DNA sequence of a multicellular organism. The results established a genome size of 97 Mb, and computer analysis predicted that there were about

19,000 genes in the genome, making the average gene density about 1 gene per 5 kb of DNA. Compared with the gene densities seen in larger animals, *C. elegans* genes are closely packed, in part because its introns are smaller on average than those of most other animals.

Annotation of the complete genome sequence, that is, the confirmation and cataloging of all the predicted genes, will take many years. Already, however, the completed sequence has enabled researchers to make some generalizations about the *C. elegans* **proteome:** the complete set of proteins encoded by the 19,000 known and predicted *C. elegans* genes. The majority of these proteins match homologous proteins in current databases derived from sequencing the genomes of other organisms. Based on the known functions of these homologs, most of the *C. elegans* proteins can be assigned to a functional class such as transcription factor, protein kinase, membrane-bound receptor, and so on.

About 20% of the predicted *C. elegans* proteins carry out "core" biological functions that are common to all living cells. The enzymes of intermediary metabolism; the machinery for DNA, RNA, and protein synthesis; and components of the cytoskeleton are all in this category. These proteins have functional homologs, in about the same proportions, in yeast as well as in other metazoans. Interestingly, genes for the core-function proteins are found preferentially in the central regions of the chromosomes, where recombination frequencies are lower.

The remainder of the *C. elegans* proteins whose functions can be surmised are involved in processes required only in multicellular organisms; such processes include specialized signal transduction pathways and programmed cell death. These multicellular-based proteins have homologs in other animals, but not in yeast. Almost all the signaling pathways found in other animals are represented in *C. elegans*.

In addition to the 19,000 known and predicted genes, there are several kinds of repetitive DNA sequences dispersed throughout the *C. elegans* genome. The best characterized of these sequences are seven kinds of transposable elements, named Tc1 through Tc7. Different strains of *C. elegans* carry different numbers of these elements. In most strains, the elements are stable, but in some strains, they actively transpose, allowing transposon mutagenesis.

Gene Expression

Two aspects of *C. elegans* gene expression are unique among multicellular animals. First, more than 70% of mRNAs have had one of two splice-leader sequences trans-spliced onto the 5′ end of the message (Fig. C.3). Pre-mRNAs do not contain these splice-leader sequences, but during processing in the nucleus, they are trans-spliced to acceptor sites near the 5′ ends of the corresponding primary transcripts. The function of splice leaders is not known, but it is striking that these 22-nucleotide sequences have been completely conserved among all nematode genera examined, suggesting functional significance. The second aspect of *C. elegans* gene expression unique to mulitcellular animals is that about 25% of adjacent genes are transcribed as operons to produce a polycistronic primary transcript (Fig. C.3). During the subsequent processing of this transcript, trans-splicing with splice-leader sequences produces constituent single-gene mRNAs. Before the discovery of these unique aspects of *C. elegans* gene expression, polycistronic transcripts had been reported only in prokaryotic cells and trans-splicing had been reported only in trypanosomes.

Life Cycle, Development, and Anatomy

We saw in the introduction to this chapter that *C. elegans* females are actually hermaphrodites that produce both oocytes and sperm. In the absence of males, they can reproduce by self-fertilization. Mating with males, however, produces cross-fertilized progeny. In the wild, mating promotes mixing of the gene pool; in the laboratory, it makes experimental genetics possible. Since a typical hermaphrodite produces 200–300 sperm and a larger number of oocytes, it can generate 200–300 progeny by self-fertilization and over 1000 progeny when fertilized by a male. One external feature that distinguishes the sexes is the tail, which is simple and tapered in the hermaphrodite but fan-shaped and specialized for mating in the male (see Fig. C.1).

Gamete Production and Fertilization

The *C. elegans* hermaphrodite manufactures both oocytes and sperm in its bilobed gonad (Fig. C.4a). The far (distal) end of each lobe contains germ-line nuclei, not separated by cell membranes, which divide mitotically in a common core of cytoplasm (forming a syncytium). The nuclei move toward the other (prox-

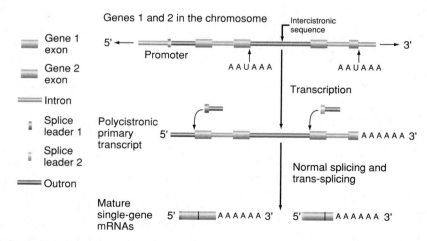

Figure C.3   Polycistronic transcription and trans-splicing in *C. elegans*. A process of trans-splicing and cleavage of multigenic transcripts results in the addition of splice leaders (encoded elsewhere in the genome) to the 5′ ends and poly-A tracts to the 3′ ends of what will become the individual mature mRNAs for each gene. The splice leader SL1 trans-splices to the most 5′ gene of the operon, while SL2 trans-splices to the internal genes. The sequence found at the 5′ end of the primary transcript that is removed by trans-splicing is technically not an intron (because it is not flanked on both sides by exons); it is sometimes called an *outron*.

(a)

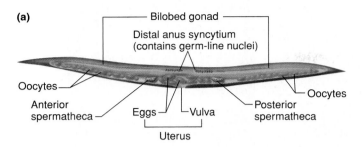

Bilobed gonad

Distal anus syncytium (contains germ-line nuclei)

Oocytes

Anterior spermatheca

Eggs — Vulva

Oocytes

Posterior spermatheca

Uterus

(b)

Figure C.4 Reproduction in *C. elegans*. (a) The hermaphrodite reproductive system. During the life of a hermaphrodite, the first 40 or so germ cells that enter meiosis in each arm of the gonad develop into approximately 150 sperm that are deposited into the spermatheca. Thereafter, the sexual fate switches so that all subsequent germ cells differentiate as oocytes. **(b)** Scanning electron micrograph of *C. elegans* sperm. These amoeboid cells do not swim, but instead they crawl along surfaces by extending and retracting *pseudopods*.

imal) end of the lobe, which terminates in the uterus. As they move toward the bend between the distal and proximal arms of the lobe, the nuclei enter into meiosis. At the bend, individual nuclei are pinched off as cells, first to form sperm during a brief period just before adulthood, and then to form oocytes once the hermaphrodite has become an adult. The oocytes move through the oviducts toward the central uterus.

At the end of each oviduct lies a spermatheca, the storage site for sperm (from either the hermaphrodite or the male). These sperm are unusual in that they are amoeboid rather than flagellated like the cells in most animals (Fig. C.4b). Each spermatheca of a young, unmated adult hermaphrodite contains about 150 sperm. Fertilization occurs as oocytes pass through the spermatheca into the uterus, where the fertilized eggs begin to develop. Eggs exit the uterus through the vulva, which forms a characteristic protrusion on the ventral side of the hermaphrodite.

To effect cross-fertilization, the male lies next to the hermaphrodite and slides its fan-shaped tail along the hermaphrodite's body surface until it contacts the vulva. The male then inserts specialized structures at the base of its tail into the hermaphrodite's vulva (see Fig. C.1). Sperm move from the male's tail through the vulva into the hermaphrodite's uterus. From the

uterus, the sperm migrate to the spermathecae. There the male sperm somehow gain an advantage over the resident hermaphrodite sperm so that for several hours after mating, the hermaphrodite produces almost exclusively outcross progeny.

Development from Zygote to Adult

At room temperature, *C. elegans* progresses from a fertilized egg to the fertilized eggs of the next generation in about three days. Embryonic development begins in the hermaphrodite's uterus with secretion of a tough chitinous shell around the fertilized egg and continues after the hermaphrodite lays her eggs about 2 hours later. Figure C.5a shows several stages of embryonic development. The large cells produced by early embryonic cleavages are called **blastomeres.** Many of the early cleavages are unequal, producing daughter blastomeres of different sizes. Until the 28-cell stage, all blastomeres contact the surface of the embryo. At the 28-cell stage, two gut precursor cells begin to move to the embryo's interior, initiating gastrulation. Continued cell proliferation and cell movements during gastrulation produce the three embryonic germ layers (endoderm, mesoderm, and ectoderm) that give rise to the basic body plan of the larval and adult stages. Later in embryogenesis, cell divisions cease while the various tissues differentiate and become further organized. Contractile proteins arranged into filaments around the circumference of the animal literally squeeze the embryo into its final wormlike shape. Roughly 14 hours after fertilization, the first-stage juvenile, known as the L1 larva, hatches from the egg shell. It is 250 mm long and consists of 558 cells.

Over the next 50 hours, development proceeds through three additional larval stages—L2, L3, and L4—separated from each other by molts (Fig. C.5b). During each molt, *C. elegans* synthesizes a new cuticle under the old cuticle, and then it sheds the old. The structures of these cuticles differ from one stage to the next. In parasitic nematodes, the cuticles of different stages help the animal adapt to existence in a particular plant or animal host.

During larval development, most of the cells present at the hatching of the L1 larva do not undergo further cell divisions, although they may increase in size as the animal grows to adulthood. A few blast cells, however, do divide further during larval development to produce additional neurons, muscles, and the structures involved in mating and reproduction. Some of these blast cells are referred to by one-letter names, for example, the laterally situated P cells that give rise to ventral cord neurons and the vulva. Sperm production begins in both hermaphrodites and males during the L4 stage. After the final L4-to-adult molt, the hermaphrodite germ line switches to oocyte production only. At this molt in the male, the development of specialized mating structures in the tail reaches completion, and the animal becomes competent to mate.

In the laboratory, investigators usually grow *C. elegans* on agar plates covered with a lawn of *Escherichia coli* bacteria, which serve as food. They transfer individual animals from one plate to another with a flattened platinum wire. Under laboratory conditions, adult nematodes continue to reproduce for three to four days. However, they live up to two more weeks.

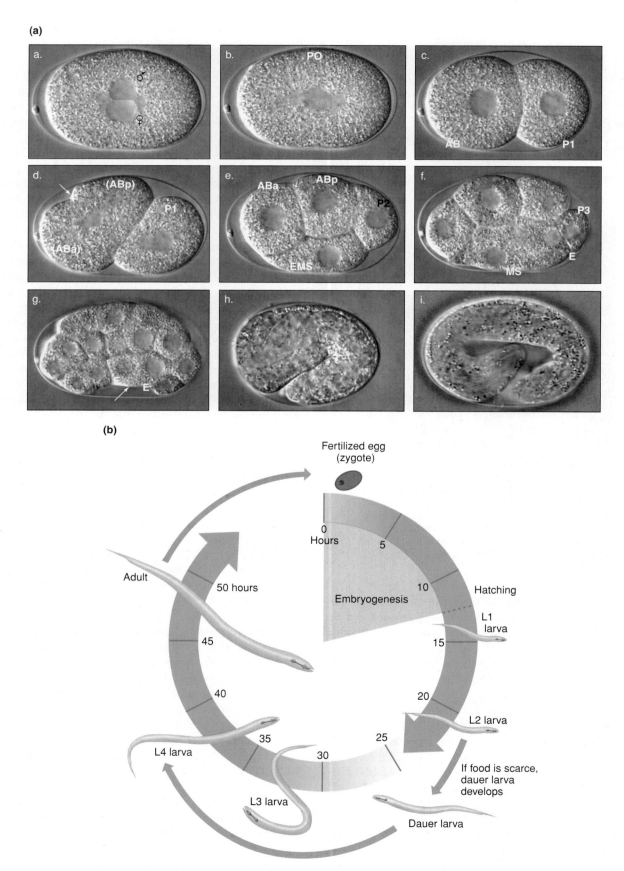

Figure C.5 Stages of _C. elegans_ development. (a) Embryonic development of _C. elegans._ Several of the earliest cell cleavages in the embryo are unequal; for example, the first mitotic division produces a larger AB daughter cell and a smaller P₁ daughter cell (_panel c_). Gastrulation begins at the 28-cell stage as the two cells marked with a small _E_ move into the embryo's interior (_panel g_). Later on, the embryo adopts a wormlike shape prior to hatching (_panel i_). **(b)** The _C. elegans_ life cycle. Embryogenesis, the time period illustrated in part (a), is highlighted in _yellow._ Under unfavorable environmental conditions, L2 larvae can molt into dormant, long-lived dauer larvae.

The short life span makes *C. elegans* an excellent model system for the study of aging and its genetic control. One curious aspect of the life cycle adds interest to the question of how life span is determined. If food is scarce during early larval development, the L2 larva can molt to an alternative L3 form known as the **dauer larva** (German for long-lasting larva; Fig. C.5b). Dauer larvae do not feed and have a specialized cuticle that resists desiccation. They move around rapidly at first and then become more dormant, although they are still able to respond to stimuli. Dauer larvae that do not experience excessive dehydration can survive for at least six months. Whenever food becomes available within that time, they molt to L4 larvae and resume normal development. Apparently the *C. elegans* aging clock can stop and then restart in response to environmental signals.

Activation of the alternative dauer-larva pathway in response to starvation allows *C. elegans* to survive when conditions for reproduction are unfavorable and resume development later when conditions for reproduction improve. Interestingly, initiation of dauer-larva formation is controlled by proteins homologous to components of the insulin-signaling pathway in mammals, which also controls responses to changes in the abundance of available food.

Anatomy of the Adult

At 1 mm in length, *C. elegans* adults are just visible to the naked eye. As we learned earlier, hermaphrodites have 959 somatic cells, while males have 1031. The *C. elegans* body plan, typical of nematodes in general, is in essence two concentric tubes (Fig. C.6a). The outer tube is made up of a tough cuticle covering the *hypodermis* (the "skin" of the worm), under which are neurons and muscles. The inner tube consists of the intestine, which runs from the anterior *pharynx* to the posterior *anus*. The space between the two tubes is called the *pseudocoelom;* in adults, it contains the gonad.

Lineage Diagrams Show the Life History of Every Somatic Cell

Because *C. elegans* is transparent at all stages of its life cycle, it is possible to use light microscopy to observe and analyze development at the cellular level, following living animals from fertilization to adulthood. A microscope equipped with differential interference contrast (Nomarski) optics provides a shallow depth of field with clear imaging of any plane in the specimen and little disturbance from out-of-focus information. With such a microscope, nematode investigators were able to observe and record the positions, movements, and divisions of every cell throughout the life cycle (review Fig. C.5a). In this way, they learned that both embryonic and larval development proceed by an invariant temporal sequence of cell divisions and spatial patterns of cell positions. From their observations, they succeeded in piecing together the entire **cell lineage** of *C. elegans,* establishing the ancestry of every cell in the adult animal all the way back to the fertilized zygote. Among animals of the same sex, the cell lineage is identical for every individual with one exception: The germ-line cell divisions that occur during larval development and into adulthood are variant, because dif-

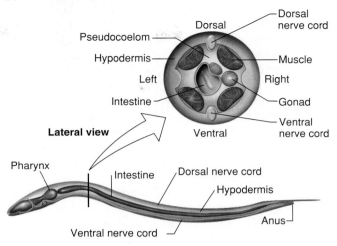

(a) Cross section

Dorsal
Dorsal nerve cord
Pseudocoelom
Hypodermis
Left
Intestine
Muscle
Right
Gonad
Ventral nerve cord
Ventral

Lateral view

Pharynx
Intestine
Dorsal nerve cord
Hypodermis
Anus
Ventral nerve cord

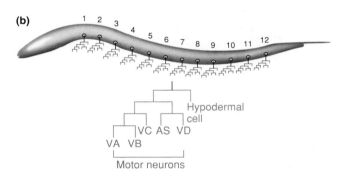

(b)

1 2 3 4 5 6 7 8 9 10 11 12

Hypodermal cell
VC AS VD
VA VB
Motor neurons

Figure C.6 The body plan of *C. elegans*. (a) The basic features of *C. elegans* adult anatomy in cross-section and lateral view. **(b)** Development along the anterior-posterior axis includes repeating sublineages. In the L1 larva, 12 precursor cells in the ventral nerve cord (the P cells) each initiate a similar series of divisions that will create a group of cells related by ancestry (a sublineage). Within each of these sublineages, one of the cells usually becomes a hypodermal cell (*green lines*), while the remaining cells become motor neurons (*purple lines*). The ultimate fate of corresponding cells in the 12 sublineages is not as invariant as suggested by this diagram. For example, one or more of the motor neurons undergoes programmed cell death in some sublineages, while in certain others the hypodermal cell undergoes additional rounds of cell division later in development (*not shown*).

ferent animals of either sex produce slightly different numbers of gametes. In the somatic tissues, the invariant lineage differs somewhat between hermaphrodites and males, reflecting their anatomical differences.

To study particular developmental processes in detail, *C. elegans* researchers focus on a small part of the cell lineage, such as the one diagrammed in Fig. C.6b. Although *C. elegans* is not segmented, its body plan is partially *metameric*, that is, composed of repeating patterns of cell groups with similar functions. For example, along the ventral nerve cord, a set of 12 precursor cells located along the body of the L1 larva undergo similar patterns of division to generate similar groups of cells known as *sublineages.* The cells produced by each of these 12 sublineages generally include five types of motor neurons and a hypodermal cell. There are, however, variations in cell fate specific to particular sublineages that can be likened to the differences in segment identity that arise during the embryogenesis of truly segmented animals such as *Drosophila* (described in Reference D).

The diagram in Fig. C.7a shows the complete lineage for all of the 959 somatic cells in the hermaphrodite. Such diagrams provide an unusual view of the developmental process. The vertical axis represents time, each horizontal line represents a cell division, and the end of each branch represents a cell in the adult. Because the history and fate of each of these cells are reproducible from one individual to another, every somatic cell present at every stage of development can be given its own unique name, using a nomenclature we describe next.

Another important part of the *C. elegans* cell lineage is diagrammed in Fig. C.7b. This higher-resolution view depicts the early divisions after fertilization that give rise to the six so-called *founder cells,* the progenitors of the major embryonic lineages. These cells arise during the first 100 minutes of cleavage, between fertilization and the 28-cell stage when gastrulation begins (corresponding to the stages seen earlier in panels a–g of Fig. C.5a). Daughters of the founder cells (not shown in Fig. C.7b but seen in Fig. C.5a) and their descendants are given lineage names according to their relative positions after division; for example, the anterior and posterior daughters of the AB founder cell are named ABa and ABp, respectively.

Three of the founder cells—E (gut), D (muscle), and P_4 (germ line)—give rise to only single tissues, while the other three—AB, MS, and C—contribute to more than one of the three embryonic germ layers and thus to more than one tissue. Conversely, several tissues include cells from more than one founder cell lineage. For example, in different parts of the body, the body-wall muscle derives from progeny of the MS, C, D, and AB founder cells.

The cell divisions that produce the founder cells are asynchronous (Fig. C.7b). For example, though EMS and P_2 are both the daughters of P_1, EMS divides a few minutes before P_2. In contrast, the later cell divisions within a founder lineage tend to be synchronous. The timing of cell division in *C. elegans* is under genetic control, and considerable progress has been made in identifying the genes responsible for this kind of developmental regulation.

Laser Ablation: An Experimental Tool for Investigating Development and Cell Lineage

Because *C. elegans* is transparent researchers can deploy a laser beam to kill, or *ablate,* specific cells and then analyze the consequences to determine the functions of the ablated cells in development, physiology, or behavior. Used in conjunction with an understanding of the genetics and molecular biology of the animal, and based on the invariance of cell lineages, this laser technology is a powerful tool. In ablation experiments, a beam of laser light is directed into the microscope through the optical system so that it is focused on the same point and focal plane

that the viewer sees (there is no danger to the viewer). When the laser is fired, the beam can fatally damage a single cell without harming the cells around it. If appropriate for the experiment, an investigator can individually kill as many as 20 or 30 cells in a single animal. Much of the knowledge of development and neurobiology presented in this chapter depends on laser ablation technology.

The central finding of cell ablation experiments in which researchers destroy a single cell is that every cell in the early embryo is essential for normal development, and the organism cannot replace destroyed cells. Thus the developmental potential of these cells is already restricted to a limited set of fates at the time each cell is born. This suggests that each cell has received a specific set of molecular instructions that differ from those received by its neighbors, instructions that autonomously define its subsequent development. However, this is not the whole story. Ablation of a particular cell can also affect the subsequent fates of adjacent cells in the embryo, indicating that early development depends not only on specific internal cell-autonomous signals but also on inductive signaling, in which one cell transmits instructions to others in its neighborhood. Examples of both these mechanisms for specifying cell fates during *C. elegans* development have been studied using genetic analysis and are now understood at the molecular level, as we see later in this chapter.

It is interesting to note that the results of cell ablation experiments in *C. elegans* are very different from the results obtained in embryos of mammals, including humans. For example, on pp. 371–372 of Chapter 11, where we describe how to determine the genotype of an early human embryo by removing and examining a blastomere from the 6–10-cell stage, we mention that the remaining five to nine blastomeres are still able to develop into an adult. Identical twins, formed when early blastomeres separate from each other, are another case in point. Cells in the early human embryo are still **totipotent:** They retain the complete developmental potential of the fertilized egg. This property is responsible for the enormous potential utility of cell lines derived from such early embryos, **embryonic stem cells,** which theoretically can be made to differentiate into any human tissue type in the laboratory. Only later in development do the cells in mammalian embryos become restricted to specific fates.

The Use of Genetic Analysis and Recombinant DNA Technology to Study Development

C. elegans is a diploid organism, except for the sex chromosome in males. Males have only a single X chromosome and are described as XO; hermaphrodites are XX. Males arise spontaneously as the result of a low frequency of X-chromosome nondisjunction during hermaphrodite gametogenesis, which produces 0.1% to 0.2% XO embryos. Once males are present in a population, mating generates more of them in the following manner. When male sperm fertilize a hermaphrodite, half the progeny result from sperm that carry no X chromosome; these progeny are XO males.

(a)

Figure C.7   Cell lineage diagrams. (a) The complete somatic cell lineage for *C. elegans* hermaphrodites. Development begins at the top with the zygote and progresses to adulthood at the bottom of the diagram. Each horizontal line represents a cell division. The timing and pattern of these divisions are invariant from animal to animal. Colors highlight the lineages established by the six founder cells. **(b)** Lineage diagram of the early cleavages in the *C. elegans* embryo that give rise to the six founder cells, which are indicated by different colors. Cells labeled in *black*—P_0 (the original zygote), P_1, P_2, P_3, and EMS—are each precursors of two or more founder cells. Below each founder cell is a list of the kinds and numbers of cells in its lineage at the time of hatching, color-coded as in part (a).

Mutant Isolation

Chemical mutagens such as ethylmethane sulfonate (EMS) induce a high forward mutation rate in *C. elegans,* and mutants are simple to isolate. Hermaphrodite genetics takes some getting used to, because a mated hermaphrodite can produce both self-fertilization and outcross progeny, but it has several advantages. It is easy to recover recessive alleles following mutagenesis because a hermaphrodite heterozygous for a new mutation will produce 1/4 homozygous mutant progeny by self-fertilization. As a further convenience, because hermaphrodites can reproduce without having to mate, it is possible to recover and propagate mutations causing very severe defects, for example, complete paralysis of the body-wall muscles. Any mutant hermaphrodite that can feed well enough to survive and has a functional reproductive system can produce progeny.

By mutagenesis, screening, and genetic analysis, researchers have identified and placed on the genetic map more than 1600 genes defined by mutation. Easily recognized mutant phenotypes often used as markers in mapping experiments include animals with abnormal body shape, such as Dumpy (Dpy) or Long (Lon), and animals that move abnormally, designated Unc for "Uncoordinated." By convention, *C. elegans* phenotypes are capitalized and not italicized, while genes and alleles are in lowercase, italic type. An example is *dpy-10(e128)*, which designates the *e128* mutant allele of the *dpy-10* gene; this allele results in the Dpy phenotype. Protein gene products are indicated by all uppercase letters; for example, the protein encoded by the *dpy-10* gene is designated DPY-10.

DNA Transformation

Many aspects of *C. elegans* genetic and molecular analysis depend on DNA transformation, the ability to create transgenic animals by introducing cloned DNA sequences into the germ line. To transform *C. elegans,* experimenters generally microinject DNA through the cuticle into the syncytial distal gonad of the hermaphrodite (Fig. C.8a). The injected DNA transgenes recombine with each other to form large extrachromosomal arrays that are incorporated into oocytes and retained in most cells during embryogenesis. It is possible to demonstrate gene expression from such transgenes by the phenotypic rescue of a mutant phenotype, if the corresponding wild-type gene is present in the injected DNA. It is also possible to demonstrate expression visually if a suitable reporter gene is present in the injected DNA. One reporter gene is the *E. coli* β-galactosidase gene *lacZ,* whose expression can be detected by X-Gal staining of fixed preparations (Fig. C.8b). In *C. elegans,* the most widely used reporter gene is the jellyfish **green fluorescent protein (GFP)** gene, which has the advantage of being visible in living animals. An animal expressing a GFP reporter is shown in Fig. C.8c. Researchers use reporter genes to study the regulation of gene expression in *C. elegans* as they do in other organisms (see p. 584 in Chapter 17).

If an embryo from an injected hermaphrodite retains the transgenic array in its germ line, it can give rise to a *transmitting line* in which the array passes from one generation to the next. However, at a low frequency that seems to depend on the individual array, it will occasionally be lost at cell division. Loss can occur during meiosis, resulting in progeny that lack the array. Loss can also occur at a mitosis during development to produce a clone of somatic cells that do not carry the array. This property can be useful for mosaic analysis of gene function.

Irradiation of transmitting lines with X or γ rays can induce integration of a portion of the array at random into one of the chromosomes. Such integrated arrays are inherited in a completely stable fashion like any other chromosomal marker. However, integration of an injected transgene by homologous

(a)

Injection of cloned DNA into hermaphrodite gonad

Syncytial distal gonad

(b)

(c)

Figure C.8   DNA transformation in *C. elegans.* (a) Researchers inject DNA into the distal syncytial gonad of hermaphrodites, where it can be taken up by any of the germ cell nuclei in the common syncytial cytoplasm. Irradiation promotes the integration of transgenes into random positions in the genome, stabilizing their transmission. **(b)** Reporter constructs show the expression of transgenes. Here, the *lin-3* gene (including its regulatory region and part of its coding sequence) was fused in frame to the gene for β-galactosidase, which can be detected by X-Gal staining (blue color). Expression of this reporter gene in transformed animals is specific to the anchor cell. **(c)** A commonly used reporter gene is green fluorescent protein (GFP), which can serve as a tag in living animals.

recombination at the site of the identical *endogenous gene* occurs only at extremely low frequencies. (An endogenous gene is a copy of a gene at its normal place in the genome.)

RNA-Mediated Interference (RNAi)

Another powerful method for analyzing gene function became available recently, when researchers discovered that injection of double-stranded copies of a specific *C. elegans* mRNA (dsRNA) into adult hermaphrodites could cause silencing of the corresponding gene in the injected animal and its progeny. This surprising phenomenon has been called **RNA-mediated interference (RNAi).** Further work has revealed some of the underlying mechanism. An enzyme complex, named "dicer," cleaves the incoming dsRNA into small segments of about 22 bp, called *trigger RNAs* or *small interfering RNAs (siRNAs),* which appear to be the active agents conferring specificity. These siRNAs are transported from one cell to another, apparently through specialized membrane channels, so that the siRNAs can spread from the initial site of entry to all cells in the body except possibly neurons. Because of this transport, the initial entry site of the dsRNA is not important, and researchers have even found strong specific inhibitory effects on gene expression when a dsRNA is fed to the worms. Feeding can be conveniently accomplished using bacteria that express the dsRNA as the worms' food. A second group of proteins promotes unwinding of the siRNAs and complexing of their antisense strands with complementary sequences of the corresponding endogenous mRNA. This sense-antisense complex can meet either of two fates. In one, destruction of the mRNA accomplishes gene silencing. In the second, the bound antisense strand acts as a primer for reverse transcription to form a new dsRNA, which is then cleaved by the dicer complex to produce a new generation of siRNAs, still specific for the same gene (Fig. C.9). This second alternative leads to amplification of the siRNA population. One consequence of amplification is that an RNAi effect can persist through two or even three generations after initial exposure to a dsRNA.

Although RNAi was originally discovered and analyzed in *C. elegans,* it is related to phenomena described earlier in plants and fungi and appears to be widespread among biological systems. Most important, researchers have found they can use it in several other model organisms, including *Drosophila,* mouse, and even cultured mammalian cells, as a powerful new tool for specific gene silencing.

Cloning of *C. elegans* Genes

Genes defined by transposon insertion mutations can be cloned directly by transposon tagging, as in other organisms. However, to clone a gene defined only genetically by point mutations, the gene must be isolated by **positional cloning,** that is, cloning based only on the gene's position in the genome (see pp. 389–396 of Chapter 11). Positional cloning begins by precise genetic mapping relative to known markers and extrapolation to an approximate position on the physical map. Next comes fine mapping using not only phenotypic markers but also DNA markers such as polymorphisms resulting from transposon insertion or **single-nucleotide polymorphisms** (see Chapter 11, p. 376), of which thousands have now been identified in *C. elegans.* DNA markers have a significant advantage: Because their precise physical locations are known from the genomic sequence, positioning a new mutation between two DNA markers precisely defines the physical interval in which the new gene must lie.

Once the smallest possible physical interval has been established, there are several approaches by which the gene can be

Figure C.9 Mechanism of RNA interference (RNAi). RNA interference can be mediated by RNA that enters the cell from outside (*shown here*) or it can be generated by viruses, transposons, or endogenous genes within the cell. Once siRNAs (small interfering RNAs) are produced, they can exert their effects within the same cell and/or be transported between cells.

identified. In the classical approach, still often the method of choice, researchers inject into mutant hermaphrodites each of the cosmids corresponding to the defined interval. This allows them to identify first the cosmid and then smaller subclones of the cosmid that complement ("rescue") the mutant phenotype. From the annotated genomic sequence, researchers can then look up the predicted gene present in the smallest rescuing clone.

Availability of the complete genomic sequence makes possible a sometimes simpler *candidate gene* approach. If the physical interval is small, containing only a few candidate genes, researchers can determine the sequences of these genes in DNA from the mutant, and they can identify which of the genes contains the mutation by comparison to the known wild-type sequence in the database. A variation of this approach is made possible by RNAi. If the gene being identified is defined by a loss-of-function mutation, then candidate genes can be tested by silencing them one at a time using RNAi and determining which RNAi-silenced candidate gene produces the mutant phenotype.

Defining the Biological Function of a Gene Identified Only by Its DNA Sequence

The isolation of a mutant on the basis of phenotype and subsequent identification of the gene producing the aberrant developmental phenotype through positional cloning have driven the field of developmental genetics until recently. Because this methodology is an extension of the original approach taken by geneticists since Mendel, it is sometimes called **forward genetics.** One begins with a mutant phenotype, which defines the normal functions of the mutant gene, and the task is to isolate the gene itself for molecular analysis. With the advent of complete genome sequencing, an opposite approach called **reverse genetics** is becoming more important. Here, one begins with a gene of possible developmental interest that has been predicted from the genomic DNA sequence, and perhaps assigned to a functional class based on its protein product's homology with other known proteins. The task is then to determine its function in the animal. To assign functions to the thousands of genes identified by genome sequencing projects, efficient methods for determining the function of genes identified through sequence analysis have become a high priority for all model organisms.

The obvious way to determine whether a gene identified from genomic sequence data is essential for normal development or physiological function in *C. elegans* is to somehow create mutations of the gene in the animal and then analyze the resulting phenotype. But while yeast and mouse geneticists have long been able to obtain knockout mutations by homologous recombination of suitably modified transgenes with their corresponding endogenes, researchers have not yet devised a reliable method for targeted homologous recombination in *C. elegans.* Nevertheless, because of the worm's small genome size, suitability for large genetic screens, and ability to survive freezing, it is possible to obtain mutations in a predicted *C. elegans* gene through PCR methods (described on pp. 297–302 of Chapter 9). In this approach, suitable primers are designed to detect novel amplification products from the gene of interest after random mutagenesis, often using a transposon, which generates mutations upon insertion into a gene, or mutagens such as the DNA cross-linker trimethylpsoralen, which causes a high frequency of small deletions. Figure C.10 shows the basis of these techniques.

Figure C.10 Identification of mutations caused by insertion of a transposable element. Researchers first construct many independent worm strains, starting with worms in which the Tc1 transposon is highly mobile. They isolate genomic DNA from a small sample of each line and freeze the remaining worms (*C. elegans* can survive indefinitely in the frozen state). Using one primer specific to Tc1 and a second primer specific to their gene of interest (*YFG* for "your favorite gene"), a PCR product will be formed only if Tc1 has inserted into the gene. The worms in lines identified in this way can be thawed and grown for study.

An alternative approach to determining function has become available with the development of RNAi. Researchers can make dsRNA *in vitro* from a cDNA corresponding to the gene of interest, inject it into or feed it to wild-type worms, and observe phenotypes among the progeny. For many genes, RNAi results in the null phenotype, as well as in less severe hypomorphic phenotypes as the RNAi is taking effect or wearing off. Nevertheless, since some genes are not efficiently silenced by RNAi, it is preferable to check RNAi results by obtaining and characterizing a deletion predicted to be a null mutation.

The availability of procedures for defining the biological function of genes known only by their DNA sequence substantially increases the usefulness of *C. elegans* as an experimental system. For example, researchers can mutate *C. elegans* genes that are homologous to human genes involved in genetic diseases and analyze the nematode genes genetically and phenotypically. It may even be possible to develop an animal model in *C. elegans* for some human genetic diseases far more quickly and at much less expense than in a mammalian species such as the mouse.

Genetic Mosaic Analysis

Genetically constructed animals composed of both wild-type and mutant cells are called **genetic mosaics.** The analysis of such mosaics is one way to determine the *focus of action* for a gene of interest, that is, the cells in which the gene must be active to allow the animal to develop and function normally. For example, if a mutant *Unc* gene causes paralysis as the result of a muscle defect, then mosaics whose muscle cells are all wild type will not have the uncoordinated Unc phenotype even if the rest of the cells in the animal are mutant. By contrast, if a mutant gene causes paralysis as the result of a neuronal defect, only mosaics in which the neurons that control the body-wall muscle cells have the mutant gene will be uncoordinated (Unc), and the genotype of the muscle cells themselves will be irrelevant. This result would demonstrate that the focus of action for this gene is in the neurons.

To construct genetic mosaics in *C. elegans,* it is possible to use chromosomal fragments generated by X or γ irradiation. Because *C. elegans* chromosomes are holocentric, even small chromosomal fragments have sufficient centromeric activity to segregate fairly normally during meiosis and mitosis. These fragments can thus be maintained as extrachromosomal, *free duplications* in a genetic stock. Even so, cells lose them at relatively high frequencies, ranging from one loss per 200 cell divisions for small fragments to much lower frequencies for larger fragments.

To carry out a mosaic analysis, a researcher constructs a *C. elegans* strain that is chromosomally homozygous for a recessive mutation in a gene of interest and introduces into that strain a free duplication including the wild-type allele of the same gene. If the free duplication is lost during mitosis at some point during somatic development, a clone of mutant cells will arise. The inclusion of appropriate marker mutations in the experimental strain makes it possible to identify in adult mosaic animals those cells that have lost the free duplication. And since the entire cell lineage is known, it is usually possible to establish the precise division at which the fragment was lost and thereby ascertain the genotype of every cell in the animal. Phenotypic analysis of a large number of mosaic animals can then reveal which cells must express the wild-type gene for an animal to develop and function normally.

Figure C.11 diagrams a genetic mosaic experiment using a free duplication. In the developing somatic gonad, the cells Z1.ppp and Z4.aaa make cell-fate decisions requiring the function of the *lin-12* gene. In half of all wild-type animals, Z1.ppp

(a)
Lin-12⁺ Mrk⁺ zygote carrying a free duplication

becomes the anchor cell (AC) and Z4.aaa becomes a ventral uterine precursor cell (VU). The other half of the time, Z1.ppp becomes a VU and Z4.aaa becomes the AC. In *lin-12* loss-of-function (null) mutants, both cell types become ACs. In *lin-12* gain-of-function (constitutively active) mutants, both cell types become VUs. Using mosaic analysis of *lin-12,* researchers showed that Z1.ppp and Z4.aaa cells must express the *lin-12* gene product to become a VU. If only one of these two cells lacked *lin-12* function, that cell always became the AC, and the other became a VU cell. If neither cell could manufacture LIN-12 protein, then both became AC. This is consistent with the anticipated role of the *lin-12* gene product, which is similar in amino-acid sequence to the membrane-bound Notch growth-factor receptor in *Drosophila*. The LIN-12 protein presumably receives the signal that induces Z1.ppp or Z4.aaa to become a VU cell.

This procedure depends on obtaining a free duplication that includes the gene of interest, which is sometimes difficult. If the gene of interest has been cloned, instead of using a free duplication, it is possible to use a transgenic array that includes a wild-type copy of the gene as well as a GFP reporter made with a gene that is expressed in most cells throughout development. Mutant animals carrying such an array should be phenotypically normal, because the array rescues the mutant defect, and almost all their cells will fluoresce. Mosaic animals arising in the population that have lost the array in some of their cells can be analyzed under the microscope, and losses in particular cells can be correlated with appearance of the mutant phenotype to determine which cells must express the wild-type gene for development to be normal.

C. elegans Research in the Postgenomic Era

The complete genomic sequence of *C. elegans* has given researchers an overall picture of the genes and provided the data with which to explore not only individual gene functions but also the entire genome as a functional system. This pursuit, sometimes called **functional genomics,** includes the identification of *all* the genes that are expressed and *all* the genes that are required during various stages of development or in response to a physiological stimulus.

Global analysis of gene expression at specific times in development is possible using DNA chips that contain ordered microarrays of DNA sequences representing all of the 19,000 predicted genes in the genome, each at a known location on the chip. If investigators take populations of mRNAs from, for example, two different stages of development, label them with fluorescent tags, and hybridize them to the chip, they can analyze the fluorescence intensities at each location to determine which transcripts are present as well as their relative abundance for each developmental stage (see Figs. 1.13 and 10.28). By making many such comparisons, the investigators can create a temporal expression profile for the entire genome at each stage of development. Analysis of spatial expression is more difficult, but projects are also underway to determine *in situ* hybridization patterns at various stages of development using probes for every predicted gene.

Figure C.11 Mosaic analysis in *C. elegans*. (a) Mosaic analysis is set up so that a free duplication carries a wild-type allele of a gene of interest (*lin-12⁺* in this example) in a mutant (*lin-12⁻*) worm. Mitotic loss of the duplication creates a clone of cells that lack *lin-12* function. An additional gene (*mrk⁺* on the duplication, *mrk⁻* in the background) provides a phenotypic marker that indicates which cells have lost the *lin-12⁺* free duplication. By examining many mosaic animals, researchers can determine which cells must have the wild-type allele for the animal to develop normally. **(b)** Results of mosaic analysis showing that a Z1.ppp or Z4.aaa cell must have *lin-12⁺* activity to assume a VU fate. The 50/50 distribution of fates when both cells are *lin-12⁺* occurs as a consequence of cross-talk between these cells that randomly turns up expression of *lin-12* in one of the cells (which will become VU) and down in the other (the presumptive AC).

The Genetic Dissection of Developmental Processes in *C. elegans*

The techniques of functional genomics described previously hold great promise for the future, but most current knowledge of *C. elegans* developmental and physiological mechanisms has come from the application of forward genetics (isolation of mutants by phenotype followed by identification of the responsible genes), often supplemented with laser ablation experiments. Through mutagenesis and a variety of clever screening strategies, geneticists have identified genes that control a range of developmental and physiological processes in *C. elegans*. To understand how the protein products of these genes work together at the molecular level, these scientists have carried out more refined genetic analyses (for example, epistasis tests and mosaic analysis), and they have cloned the genes to allow molecular investigations into cellular functions of the corresponding encoded proteins.

In addition to confirming that many developmental processes in *C. elegans* have been conserved in more complex animals, these studies have revealed that genes controlling aspects of development that seem to be unique in the nematode often have counterparts in other animals. This finding demonstrates that closely related proteins can play substantially different roles in different organisms.

We now describe genetic studies of several aspects of development in *C. elegans*. These examples illustrate the power of genetic analysis in the dissection of development.

Specification of Early Embryonic Blastomeres

The lineage diagrams in Fig. C.7 show that successive cleavages of embryonic cells produce blastomeres that give rise to different sets of tissues and structures during embryogenesis. Genetic screens for mutations that cause abnormal early cleavage patterns have identified many genes that control early cell fates. Analyses of the functions of these genes suggest that the determination of these fates depends on two factors: the segregation of cell-autonomous determinants, which act independently in single cells; and the transmission of inductive signals between cells.

Early Cleavages Depend on Maternally Supplied Components

As in most animals, the transcription of all but a few genes in the embryonic nuclei of *C. elegans* begins only after several rounds of cell division have occurred. Therefore, the fates of the earliest blastomeres must be guided by mRNAs or proteins that were present in the egg before fertilization. These molecules—the *maternal components* that the mother supplied to the egg during oogenesis—are the products of *maternal-effect genes*. Mutations in maternal-effect genes (known as **maternal-effect mutations**) disrupt the early events of embryogenesis in *C. elegans*.

A genetic screen devised to identify the maternal-effect genes transcribed during oogenesis is based on the fact that fer-

tilization and early embryonic development occur in the hermaphrodite's uterus. If, for example, a hermaphrodite is unable to lay her fertilized eggs because of mutations that block development of the vulva, the eggs will grow into larvae within the uterus. These larvae will eventually eat the mother up from the inside and cause her to burst open. However, if the mother is homozygous for a mutation in a gene encoding a maternal component needed for early development (a maternal-effect lethal mutation), the embryos in her uterus will be unable to grow into larvae, and she will survive. Investigators can then look in the microscope at the defective embryos to determine how the mutation has disrupted their development (Fig. C.12).

Cell-Autonomous Determinants Must Be Distributed to the Cells in Which They Act

The unfertilized *C. elegans* oocyte has no predetermined anterior or posterior end. Instead, the polarity of the embryo depends on the site of sperm entry, which becomes the posterior end. The

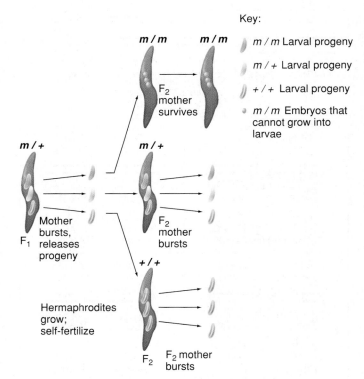

Figure C.12 Identification of maternal-effect lethal mutations in *C. elegans*. Mutagenized parental animals (*not shown*) produce F₁ hermaphrodites heterozygous for a maternal-effect lethal mutation (*m*). One-quarter of the self-fertilization progeny of such an *m / +* F₁ mother will be *m / m* F₂ hermaphrodites. Embryos made within these *m / m* mothers cannot hatch into larvae when they lack a maternally supplied component needed for embryogenesis. The hermaphrodites also have a mutation that prevents formation of the vulva, and therefore they will be eaten from the inside if their self-fertilized progeny hatch into larvae. In a screen for maternal-effect lethal mutations, researchers look for cultures derived from individual F₁ hermaphrodites in which some F₂ hermaphrodites (those that are *m / m*) survive because their progeny arrest development before hatching. The mutation can be maintained in *m / +* heterozygotes from the same population.

sperm entry point somehow directs flowing movements of the cytoplasm within the fertilized egg, and these cytoplasmic movements cause some molecules and molecular complexes to localize in an asymmetric pattern. An example of this phenomenon are the so-called P granules: ribonucleoprotein particles that are necessary for the development of the germ line. Before fertilization, P granules are uniformly distributed in the egg. After sperm entry, the granules become localized to the posterior part of the one-cell zygote (Fig. C.13a). After the first cleavage, they are found in the posterior P_1 daughter cell (from which the germ line descends), but not in the anterior AB daughter cell. Another example of the polarity of the first division is the fact that the P_1 and AB cells are not the same size (see panel c in Fig. C.5). How does

an embryo that starts its life without polarity develop anterior and posterior ends by the first cleavage?

To answer this question, researchers looked through collections of hermaphrodites for maternal-effect lethal mutants that produce two-cell embryos with no evidence of polarity. The investigators identified mutations in six *par* (for "partitioning") genes. Mothers bearing these mutations produce embryos that have P granules in both the P_1 and AB cells and/or P_1 and AB cells that are the same size. Because this lack of polarity disrupts the patterning of the embryo, the embryos from mutant mothers eventually arrest their growth as amorphous masses of partially differentiated cells.

How the proteins encoded by the *par* genes respond to the cue given by entry of the sperm is not yet known, but molecular analysis of these genes is providing clues. Many of the PAR proteins become localized to the embryonic cortex (the part of the cytoplasm just under the cell membrane) at one or the other end of the embryo. For example, PAR-2 accumulates in the cortex at the posterior end, while PAR-3 is restricted to the cortex at the anterior end (Fig. C.13b). Localization of PAR proteins occurs after fertilization but before the P granules move to the posterior part of the zygote, as would be expected if the PAR proteins regulate the positioning of the P granules. Further experiments show that interactions between PAR proteins are required for the establishment of polarity. For example, in eggs produced by a *par-2* mutant mother, the PAR-3 protein is distributed uniformly around the cortex; the same is true of the PAR-2 protein in eggs from a *par-3* mutant hermaphrodite. Thus, the PAR-2 protein excludes PAR-3 from its half of the embryo's cortex, and vice versa. Current research is focusing on the possibility that the PAR proteins influence microfilaments in the cortex of the one-cell embryo that are responsible for cytoplasmic flows.

Interestingly, even animals such as *Drosophila* and humans, whose early embryonic development contrasts sharply with that of *C. elegans*, have genes that are closely related to and probably homologous with many of the *par* genes. The *Drosophila* and human PAR-like proteins no doubt play some role in development, but not exactly the same role as that performed by the products of the *C. elegans par* genes.

(a)

(b)

Figure C.13 The PAR proteins help direct embryonic polarity. (a) In wild-type zygotes, P granules (*green*) are initially uniformly distributed (*top*), but then they become localized to the posterior part of the zygote (*bottom*). **(b)** Shortly after fertilization, PAR-2 protein is found in the posterior cortex and PAR-3 protein in the anterior cortex.

Inductive Signals Control Cell Fates in Early Embryogenesis

The PAR proteins exert their effects only within the individual cells that contain them. Because of this cellular localization, researchers consider them *cell autonomous.* Early *C. elegans* development also depends on other asymmetrically partitioned factors that control signals sent via cell contact between adjacent blastomeres. Because the patterns of early cleavage and later gastrulation cell movements are invariant, blastomere contacts and cell-to-cell interactions are reproducible from embryo to embryo.

An example of signaling via cell contact is the process by which different fates are established for the ABa and ABp daughter cells produced by cleavage of the AB founder cell. The two daughter cells are developmentally equivalent when they arise at the second cleavage. They become different as a result of interactions between ABp and the P_2 cell. Normally P_2 contacts ABp but not ABa (Fig. C.14a). If P_2 is removed by laser ablation, ABp adopts an ABa pattern of differentiation. Alternatively, when the embryo is manipulated so that ABa contacts P_2, the ABa lineage develops in an ABp-like fashion. These observations suggest that the P_2 cell provides an inductive signal receivable by AB cells that contact it.

To identify molecules involved in this signaling, researchers searched collections of maternal-effect lethal mutations for mutations that disrupt the fate of the ABp lineage. They found that the receptor for the inductive signal is encoded by the *glp-1* gene. Like *lin-12*, *glp-1* is closely related to the *Drosophila Notch* gene. *Notch* encodes a membrane-bound receptor for a membrane-bound **ligand** (signaling molecule) encoded by the *Delta* gene, which is expressed in adjacent cells. *glp-1* mRNA is present in all cells at the four-cell stage, but asymmetrically partitioned factors repress its translation in P_2 and EMS, so that it is expressed only in AB cells and their descendants. The signaling ligand for ABp determination is encoded by the *apx-1* gene. Asymmetrically segregated factors allow its expression only in P_2 cells, where the APX-1 protein localizes to the membrane junction between P_2 and ABp and contacts the GLP-1 protein (Fig. C.14b). *apx-1* is a homolog of *Drosophila Delta.* Thus, very similar ligands and receptors mediate signaling through cell contact in both *C. elegans* and *Drosophila.*

Programmed Cell Death

About one-sixth of the cells that arise during *C. elegans* embryogenesis are eliminated by a process of cellular suicide called *programmed cell death* or **apoptosis.** During apoptosis,

the affected cell shrinks, degrades its DNA, and dies, after which it is engulfed and digested by a neighboring cell. In *C. elegans,* the pattern of apoptosis, like most other aspects of development, is invariant: The identity of the cells that die and the timing of their deaths is the same from animal to animal. The function of programmed cell death seems to be the elimination of unneeded cells that could be detrimental to the organism.

Mutations that cause the failure of all programmed cell deaths during development have identified the *ced* (for <u>c</u>ell <u>d</u>eath) genes that control apoptosis. For example, loss-of-function mutations in the *ced-3* and *ced-4* genes and gain-of-function mutations in the *ced-9* gene cause the complete failure of programmed cell death. Surprisingly, although animals bearing such mutations have 12% more somatic cells than normal, they show only very subtle phenotypic defects in their growth rate and fertility. By contrast, loss-of-function *ced-9* mutations cause the death of many cells that would normally have lived, and this extensive cell death is lethal to the embryo.

Epistasis tests and phenotypic analysis of *ced* mutations led researchers to formulate the developmental pathway shown in Fig. C.15. Cells make a decision to die based on the action of genes upstream from the *ced* genes that control cell fates and function differently in different types of cells. These upstream genes activate or repress *ced-9.* Since most if not all cells require the function of the *ced-9* gene to prevent activation of the cell-death process, repression of *ced-9* allows activation of the events leading to apoptosis. Next, the *ced-3* and *ced-4* genes function to carry out the cell-death program. CED-3 is a member of the caspase family of proteases, with homologs in both *Drosophila* and humans that are also essential for apoptosis. The CED-4 protein is required for activation of CED-3. Targets of the CED-3 protease, whose cleavage must trigger the subsequent events of apoptosis, are still being identified. Other genes, such as *ced-1* and *ced-2,* act in the last stages of the cell-death process. These genes do not directly contribute to cell death itself. Instead, they are required for the engulfment and digestion of dead cells by neighboring cells. Mutations in *ced-1* and *ced-2* cause a "persistent corpse" phenotype in which cells die as usual but remain in the worm for several days.

Programmed cell death in *C. elegans* appears very similar to apoptosis in mammals. In both types of organisms, the process helps control the size of cell populations by eliminating excess cells. As we saw in Chapter 18 (pp. 625–626), the investigation of genes that control apoptosis is an important aspect of current cancer research. Although tumor formation often involves the release of normal controls on cell proliferation, it can also result from the failure of apoptosis. Sequence similarity between the membrane-associated protein product of the *C. elegans ced-9* gene and the mutant protein product of the *bcl-2* oncogene, which helps cause tumors by blocking apoptosis in humans, underscores the close relation between the control of programmed cell death in *C. elegans* and mammals. In fact, although nematodes and humans are separated by almost a billion years of evolution, their apoptosis-controlling genes are still

Key: ▬▬▬ GLP-1 protein
═══ APX-1 protein

(a) ABa ABp P_2 EMS

(b) ABa ABp P_2 EMS

Figure C.14 Intercellular signaling determines the fates of the ABa and ABp cells. (a) Contact between P_2 and ABp enables P_2 to send a signal to ABp that is required for the proper fate of the ABp lineage. **(b)** Molecular basis of the intercellular signal. The P_2 cell expresses APX-1 protein on its surface, while ABa and ABp both have GLP-1 protein, the receptor for the APX-1 ligand, on their surfaces. Binding of APX-1 to GLP-1 in the membrane junction between P_2 and ABp initiates the proper developmental program within ABp. It is not clear why the APX-1 ligand localizes only to the surface of P_2 in contact with ABp; perhaps it is stabilized there by interaction with GLP-1.

Figure C.15 Control of programmed cell death in *C. elegans*. The *blue-shaded area* describes four steps in the cell-death program and some of the genes involved in each of these steps. In the genetic pathway, → indicates positive regulation (the protein encoded by the gene to the left activates the gene to the right or its encoded protein), while ⊣ indicates negative regulation (the protein encoded by the gene on the left represses the gene to the right or inhibits its encoded protein). Inputs from several pathways can activate or deactivate a sensor encoded by the *egl-1* gene, which in its active state represses *ced-9*. All these genes act within the cell that is fated to die except for *ced-1* and *ced-2*, which function in the engulfing cell. *Green-* and *orange-shaded areas* detail gene activity and subsequent events within cells that are fated to live or undergo apoptosis.

functionally interchangeable: The normal human allele of the *bcl-2* gene can restore the normal process of cell death when injected into a *C. elegans ced-9* mutant.

Control of Timing During Larval Development

Throughout development, control over the timing of a cellular event is as important as control over where it occurs. Researchers have used *heterochronic mutations,* that is, mutations resulting in the inappropriate timing of cell divisions and cell-fate decisions during larval development, to identify *C. elegans* genes involved in temporal controls.

The genes identified by heterochronic mutations often behave as *switch genes* in which loss-of-function and gain-of-function mutations have opposite temporal effects. For example, loss-of-function mutations in the *lin-14* gene cause precocious development: Cell lineages that normally emerge late in development appear too early. Conversely, a gain-of-function mutation in the *lin-14* gene causes retarded development, in which lineage changes that normally appear only in early larval stages continue to appear at later larval stages (Fig. C.16a). Mutations in *lin-14* not only change the timing of certain developmental programs, but they can also alter the cell cycle. Cells that normally divide at a specific time in larval development will divide earlier than normal in *lin-14* null mutants and later than normal in *lin-14* gain-of-function mutants (Fig. C.16b). In addition, heterochronic mutations can affect the formation of dauer larvae in starving animals. Instead of occurring normally at the L2 molt, the dauer larvae can form at the L1 or L3 molts. Investigators have identified mutations in several other genes that produce effects either similar or opposite to those associated with *lin-14*. Interactions between these genes control the timing mechanisms that specify many aspects of larval development.

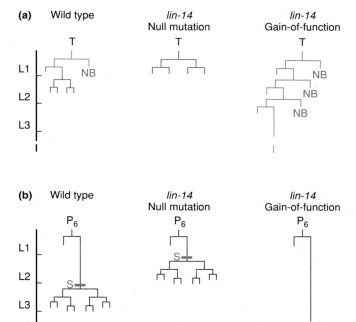

Figure C.16 Effects of mutations in the heterochronic gene *lin-14*. (a) Transformations of cell fates in the T lineage. In wild type, the T cell undergoes a pattern of division specific for the L1 larval stage (*red; NB* signifies a neuroblast), and one of the granddaughters of T undergoes an L2-specific cleavage pattern (*blue*). In *lin-14* null mutants, the T cell precociously initiates the L2 pattern in the L1 stage. In *lin-14* gain-of-function mutants, the lineage inappropriately reiterates the L1 pattern at later larval stages. **(b)** Effects on the timing of cell divisions in the hypodermal P6 lineage. Loss of *lin-14* function causes one of the daughters of P6 to start cell division too early (*S* indicates the DNA synthesis phase of this division). Gain of *lin-14* function can delay this division indefinitely.

The LIN-14 protein, which shows no similarity to other proteins of known function, behaves as a central clock component whose concentration descends from a high in L1 animals to a low in L4 animals. During larval development, cells somehow use the relative abundance of LIN-14 to "tell time." The level of *lin-14* transcript, however, remains constant during larval development, which suggests that control of LIN-14 protein concentration is at the level of translation or protein stability. The pathway defined by the heterochronic genes appears to act ultimately through regulation of the *lin-14* gene.

Epistasis tests have shown that one of the pathway's temporal control genes, *lin-4,* negatively regulates expression of *lin-14* by a novel regulatory mechanism. The *lin-4* gene does not encode a protein. Instead, the small *lin-4* gene transcript is processed to produce a still smaller 22-nucleotide RNA whose sequence is complementary to sequences in the 3′ untranslated region of the *lin-14* mRNA. The unprocessed *lin-4* transcript, which includes self-complementary sequences, is predicted to form a double-stranded hairpin structure. This feature, and the size similarity between the *lin-4* RNA and the siRNAs generated by RNAi that we described earlier, suggests a hypothesis that appears to be correct. Protein components of the RNAi machinery appear to promote the complexing of the mature *lin-4* RNA with the 3′ untranslated region of *lin-14* mRNA, rendering this mRNA untranslatable (rather than degrading it). This inhibition, coupled to a constant rate of LIN-14 protein degradation, results in decreasing levels of LIN-14 during development.

Small regulatory RNAs like *lin-4* RNA may play a much larger role in development than originally suspected, and not only in nematodes. A second such RNA, derived from the transcript of the genetically identified *let-7* gene, plays a role similar to that of *lin-4* RNA in controlling the expression of additional *C. elegans* timing proteins that mediate the molt from L4 to adult. Recently, researchers have identified more than 50 other noncoding RNAs of about 22 nucleotides in length that could also play regulatory roles in *C. elegans.* The genes for these RNAs, identified by comparison of their sequences with the genomic DNA sequence, are predicted to encode somewhat larger, self-complementary RNAs that must be processed to produce the 22-nucleotide products. Moreover, database searches have yielded the exciting result that many of these sequences are highly conserved in the *Drosophila* and human genomes, suggesting broad biological significance. It is not yet known whether these unexpected small RNAs play roles in regulation of gene expression or in other processes. However, their existence may help to explain the wide prevalence of the machinery that mediates RNAi and its normal function in cells.

Using Genetics to Probe the Development of the Hermaphrodite Vulva: A Comprehensive Example

A major contribution of developmental genetic studies in *C. elegans* has been the demonstration that a signaling pathway for cell-fate determination includes a gene similar to the *ras* onco-

gene in humans, which is responsible for a substantial fraction of malignant tumors arising from connective tissue. The *C. elegans* signal transduction pathway operates during larval development in the formation of the hermaphrodite vulva, the passageway through which sperm from the male enter and fertilized eggs exit the gonad.

Mutant Screens Identify Genes Involved in Vulva Formation

Overlying the developing hypodermis just below the somatic gonad on the ventral side of the L3 larva, there are six cells capable of giving rise to the vulva. These cells are numbered P3p through P8p and are known as vulval precursor cells, or VPCs. Only three of the VPCs (P5p, P6p, and P7p) normally form the vulva in response to an inductive signal from the anchor cell (AC) in the gonad, as shown in Fig. C.17a. Researchers inferred the existence of this signal from experiments showing that even though the AC does not become part of the vulva, elimination of this cell by laser ablation prevents vulva formation. The remaining three VPCs (P3p, P4p, and P8p) normally produce only additional hypodermal cells, but they can adopt vulval fates under certain experimental conditions.

Worm geneticists have devised several powerful genetic screens to identify the genes involved in the AC induction of P5p, P6p, and P7p to produce vulval cell lineages. The screens all take advantage of the fact that a hermaphrodite can survive and propagate without a functional vulva.

In one procedure, investigators search for Vulvaless mutants. Although these hermaphrodites cannot lay their eggs, they can self-fertilize them normally. The resulting zygotes, which develop and hatch within the uterus, grow by devouring the mother's organs to produce a "bag of worms" that is easy to spot in a mutant hunt (Fig. C.17b). Loss-of-function mutations that prevent formation of the vulva define genes required for vulval development.

In another type of genetic screen, researchers look for mutants with a Multivulva phenotype. These animals form additional, ectopic vulvalike structures from descendants of the P3p, P4p, and P8p cells, which do not normally participate in the process. Loss-of-function mutations that cause ectopic vulva formation define genes whose products act as negative regulators of vulval development.

From these screens, researchers identified various genes that participate in the inductive signaling pathway that specifies the fates of vulval precursor cells. Interestingly, many of these genes are essential for viability. The discovery of their role in vulval determination was possible only because the screens were so powerful that investigators could recognize animals with rare hypomorphic (partial loss-of-function) or hypermorphic (gain-of-function) mutations.

Organizing the Genes Contributing to Vulva Formation into a Signaling Pathway

By combining epistasis tests, mosaic analysis, and phenotypic characterization, researchers constructed a genetic pathway that clarifies the relationship between many of the genes involved in

(a)

(b)

(c) Extracellular

Intracellular

Figure C.17   Cell signaling and signal transduction cascades in vulval development. (a) The anchor cell (AC) of the gonad induces vulval fates in the vulval precursor cells (VPCs) P5p, P6p, and P7p adjacent to the underlying hypodermis. The signaling molecule produced by the AC is the LIN-3 protein; the receptor for this ligand on the VPCs is the LET-23 protein. **(b)** A "bag of worms" in which a Vulvaless mother is consumed by her progeny. **(c)** The signal transduction pathway activated by binding of the LIN-3 ligand to the LET-23 receptor. Earliest events are to the *left,* subsequent events to the *right.* The signal-bound receptor activates the adaptor proteins SEM-5 and LET-341 (a homolog of Drosophila Sos), which in turn help catalyze the conversion of LET-60 into an active GTP-bound form. Activated LET-60 initiates a cascade of protein kinases; the phosphorylated form of the last of these kinases (MPK-1) regulates the activity of several transcription factors in the nucleus including LIN-1 and LIN-31. These transcription factors in turn regulate the expression of the ultimate targets of the pathway: those genes whose products are required to make vulval structures.

vulva formation (Fig. C.17c). Subsequent cloning of the genes elucidated many of the pathway's protein components and led to an understanding of the pathway at the molecular level.

The Binding of a Signal Molecule to a Cell Surface Receptor Initiates Events Leading to Vulva Formation

In this first step of the signaling pathway, the anchor cell emits a signal that must be received by the P5p, P6p, and P7p vulva precursor cells (Fig. C.17). This AC signal is a small molecule encoded by the *lin-3* gene. The LIN-3 protein is similar in structure to mammalian epidermal growth factor (EGF). The receptor for the LIN-3 ligand on the VPC cells is encoded by the *let-23* gene, and the LET-23 protein it encodes is similar to mammalian EGF receptors. As expected for genes whose products are needed for vulval specification, partial loss-of-function mutations in both *lin-3* and *let-23* confer Vulvaless phenotypes. However, overproduction of the LIN-3 protein causes a Multivulva phenotype. As required by the pathway model, LIN-3 (the ligand) is expressed in the AC cell, while LET-23 (the receptor) is synthesized in the VPCs.

A Signal Transduction System Conveys the Signal Received at the Cell Surface to the Cell's Nucleus

After the LIN-3 signal binds to the LET-23 receptor, subsequent steps in the vulva-induction pathway occur within the P5p, P6p, and P7p cells (Fig. C.17c). These intracellular steps together constitute a signal transduction pathway that converts the signal received at the cell surface into changes in the transcription of particular genes in the cell nucleus. It is these changes in gene activity that cause the cells to adopt vulval fates.

A key player in the signal transduction system for vulva induction is the *let-60* gene, which encodes a *C. elegans* RAS protein. Gain-of-function mutations in *let-60* produce alterations in the LET-60 protein identical to those that convert mammalian *ras* products to their oncogenic form. The resulting mutant LET-60 becomes active even without a signal, and the ensuing inappropriate induction of vulval cell fates in P3p, P4p, and P8p leads to a Multivulva phenotype. By contrast, partial loss-of-function mutations in *let-60* cause a Vulvaless phenotype.

Like other RAS proteins, LET-60 is a guanine nucleotide-binding protein whose activity is indirectly controlled by an EGF receptor. Binding of the ligand signal activates the receptor, which in turn activates certain adaptor proteins. These adaptor proteins help transform the inactive GDP-bound form of LET-60 into an active GTP-bound form.

Also genetically identified in *C. elegans* are the downstream genes through which *ras* activity eventually influences events in the nucleus. These genes include *lin-45*, which encodes a protein kinase similar to one encoded by the mammalian protooncogene *raf*, as well as *mek-2* and *mpk-1*, which encode a MAP-kinase kinase and a MAP kinase, respectively. These proteins form a protein kinase cascade in which each member catalyzes the phosphorylation of the next member. Although the details are still unclear, epistasis tests and biochemical evidence suggest strongly that this cascade modifies the nuclear transcription factors encoded by *lin-1, lin-25,* and *lin-31* to control their activities.

This Type of Signaling Pathway Has Been Conserved in Evolution

Signaling pathways including a ligand, its receptor, and a signal transduction system activated by the binding of ligand to receptor are a feature of intercellular signaling in many organisms. In *Drosophila,* for example, a system similar to the one we have described here operates in the development of the eyes; in humans, defects in such signaling systems contribute to cancer formation.

Geneticists are thus very interested in extending the genetic dissection of vulval development to identify additional genes that influence this ubiquitous signaling pathway. They have re-

cently begun two new types of searches. First, they are looking for mutations that suppress the Multivulva phenotype caused by *let-60* gain-of-function mutations. Second, they are screening for mutations that enhance the Vulvaless phenotype caused by *let-60* partial loss-of-function mutations. Through these mutant screens, they have already identified the kinase-encoding *ksr-1* gene, a component of the signal transduction system that eluded detection by other means (such as biochemical analysis). The results of these investigations may be of value in understanding and eventually controlling the production of tumors caused by the *ras* and *raf* oncogenes in humans.

C O N N E C T I O N S

The special features of *C. elegans,* including its small, invariant number of cells and transparency at all stages of the short life cycle, make it an excellent model for the genetic analysis of development in multicellular animals. With knowledge of the progenitors and position of every cell in the embryo, larvae, and adult, researchers have been able to identify many of the genes that specify cell fates during development. The completion of the genomic DNA sequence and the development of high-throughput methods for determining gene-expression patterns and functions should make it possible in the foreseeable future to understand the roles of most of the 19,000 predicted *C. elegans* genes in development and physiology.

The next genetic portrait is of another multicellular animal: *Drosophila melanogaster.* In *Drosophila,* the developmental fates of individual cells are determined by a much more flexible mechanism than the invariant cell lineage seen in *C. elegans.* Through the analysis of these two model systems for animal development, biologists can compare the ways in which evolution has generated various structures and functions. Sometimes the strategies different organisms use to perform the same basic task are similar, but as the fundamental dichotomy in the mechanisms of cell determination suggests, sometimes they are very different.

E S S E N T I A L   C O N C E P T S

1. *C. elegans* is an ideal subject for genetic analysis because of its transparency, invariant cell lineage, short life cycle, and capacity for prolific reproduction.

2. The small genome of 97 Mb has been completely sequenced. It is packaged into six small chromosomes. The 19,000 genes contain small introns. *Trans-splicing* and the production of *polycistronic transcripts* are two unusual aspects of *C. elegans* gene expression.

3. RNA-mediated interference (RNAi), discovered in *C. elegans,* also appears to function in many other organisms. It provides researchers with a powerful method for silencing genes predicted from genomic sequencing, in order to determine their functions.

4. The life cycle from the fertilized egg of one generation to the fertilized eggs of the next takes about three days. Mating results in cross-fertilization. Hermaphrodites can self-fertilize.

5. Because the cell lineage of *C. elegans* is almost completely invariant, developmental biologists have been able to construct complete *cell lineage diagrams*

that show the routes by which all the somatic cells in the adult arise from the fertilized zygote. Six *founder cells* give rise to the major embryonic lineages. The study of mutants with abnormal cell lineage patterns is a major focus of *C. elegans* research.

6. Determination of blastomere fate depends on inductive signaling between cells as well as on the segregation of cell-autonomous factors acting only in the cells that contain them.

7. Mutations that cause the failure of all programmed cell deaths during development helped researchers identify the genes that control *apoptosis*. Although nematodes and humans are separated by almost a billion years of evolution, some of their apoptosis-controlling genes are functionally interchangeable.

8. The interactions of various *heterochronic "switch" genes* set the timing of different stages of larval development. Novel RNAs involved in control of this process may be representative of a widespread class of small regulatory RNA molecules.

9. Induction of the hermaphrodite vulva depends on a *signal transduction pathway* that includes a gene similar to the *ras* oncogene in humans. Identification of the genes active in this pathway in *C. elegans* may help researchers understand tumor formation in humans.

10. Despite some unusual features, *C. elegans* shares many of its regulatory pathways and developmental mechanisms with other multicellular animals.

SOLVED PROBLEMS

I. You need to construct a strain of *C. elegans* that is homozygous for the *dpy-10* and *unc-54* mutations. The *dpy-10⁻* mutation is a defect in the collagen gene that results in a short, stubby-shaped worm (Dumpy). The *unc-54⁻* mutation, a defect in a myosin heavy-chain gene, affects muscle control so severely that the worm is paralyzed (Uncoordinated). *unc-54* and *dpy-10* are unlinked, autosomal genes and both mutations are recessive. Stock hermaphrodite worms that are homozygous for *dpy-10⁻* and another stock that is homozygous for *unc-54⁻* are available for your use. The first step is to cross wild-type males to homozygous *dpy-10⁻* hermaphrodites.

   a. What genotypes and phenotypes (including sex phenotype) do you predict in the progeny?

   b. The wild-type male progeny from the cross-fertilization are crossed to virgin hermaphrodites that are homozygous for the *unc-54⁻* mutation. (To isolate virgins, L4 larval-stage worms are transferred individually to petri plates, ensuring that mating has not occurred yet with any available males in the population.) What are the genotypes and phenotypes that result from this cross?

   c. What worms will you use from the second cross to get your double recessive *dpy-10⁻ unc-54⁻* strain? Will you be able to identify worms with the specific genotype you need for the next step?

   d. What is the phenotype of worms with the desired double recessive genotype? What proportion of the worms emerging from the last step of the crossing scheme will have the double recessive genotype?

Answer

This problem requires an understanding of basic Mendelian ratios and of *C. elegans* matings.

   a. The cross can be represented:

$$\frac{dpy\text{-}10^+}{dpy\text{-}10^+}\text{ males} \times \frac{dpy\text{-}10^-}{dpy\text{-}10^-}\text{ hermaphrodites}$$

Either the hermaphrodite will self-fertilize or there will be cross-fertilization between males and hermaphrodites. *When the hermaphrodite self-fertilizes, all progeny will be Dumpy hermaphrodites with the homozygous* dpy-10⁻ / dpy-10⁻ *genotype.*

(Males [XO] will also arise from self-fertilization due to nondisjunction of sex chromosomes during formation of gametes, but the frequency is low enough that we will generally ignore this group of males. These male worms would also have a Dumpy phenotype.) In a mating between the male and hermaphrodite, the male can produce only *dpy-10⁺* gametes. Half of these gametes contain an X chromosome and the other half contain no sex chromosome. *The outcross with males results in heterozygous* dpy-10⁺ / dpy-10⁻ *hermaphrodites and males that are normal in phenotype and heterozygous at the dpy locus.*

   b. The second cross is:

$$\frac{dpy\text{-}10^+}{dpy\text{-}10^-}\ \frac{unc\text{-}54^+}{unc\text{-}54^+}\text{ males}$$

$$\times \frac{dpy\text{-}10^+}{dpy\text{-}10^+}\ \frac{unc\text{-}54^-}{unc\text{-}54^-}\text{ hermaphrodites}$$

The hermaphrodite will produce, by self-fertilization, hermaphrodites that are normal shaped (nondumpy) but paralyzed and have the genotype dpy-10⁺ / dpy-10⁺ unc-54⁻ / unc-54⁻. *From the outcross to males, the male can contribute either the* dpy-10⁺ *or* dpy-10⁻ *allele with the* unc-54⁺ *allele; whereas the hermaphrodite can produce only* dpy-10⁺ unc-54⁻ *alleles. Half of the sperm from males have an X chromosome and half do not have a sex chromosome. Results of cross-fertilization:*

XX *dpy-10⁺ / dpy-10⁺ unc-54⁺ / unc-54⁻*   wild-type hermaphrodites
XO *dpy-10⁺ / dpy-10⁺ unc-54⁺ / unc-54⁻*   wild-type males
XX *dpy-10⁺ / dpy-10⁻ unc-54⁺ / unc-54⁻*   wild-type hermaphrodites
XO *dpy-10⁺ / dpy-10⁻ unc-54⁺ / unc-54⁻*   wild-type males

   c. *To obtain a doubly homozygous recessive, a double heterozygote must be selfed. Notice that half of the wild-type hermaphrodites will have the genotype desired for this cross, but you cannot distinguish between the two genotypes. Pick several wild-type hermaphrodites and allow them to self-fertilize.*

   d. *The phenotype of the desired doubly recessive worm is short, stubby (Dpy), and paralyzed (Unc). The proportion of worms with a double recessive phenotype from selfing heterozygotes is 1/4 for each gene. But only half of the worms picked were heterozygous. (1/4)(1/4)(1/2) = 1/32 worms are expected to have the double homozygous recessive genotype.*

II. In *C. elegans,* the cells at the tip of the gonad go through several mitotic divisions before switching to meiotic divisions and differentiating into germ cells. As a result of the mitotic and meiotic divisions, up to 1500 germ cells are produced. When the somatic cell at the very tip of the developing gonad (distal tip cell) is killed by laser ablation, the few germ cell precursors surrounding it will immediately go into meiosis instead of mitosis.

a. What effect would this ablation have on the number of germ cells produced?

b. The mutation *glp-1⁻* mimics the laser ablation results. In the *glp-1⁻* mutant, cells immediately go into meiosis, without mitotic divisions, and only a few germ cells are produced. The hypothesis is that the distal tip cell produces a molecule (inducer) that binds to a receptor to control mitotic and meiotic divisions. In what cells would you expect to see *glp-1* expression if it encodes a receptor?

c. The GLP-1 protein functions in other processes in addition to germ cell development. For example, GLP-1 is required in the earliest cell divisions in the embryo to establish the fate of cells that develop into the pharynx. In addition, some mutations in the *glp-1* gene affect only a subset of the *glp-1* activities. Speculate on how the GLP-1 can act as a receptor in processes that have a very different outcome.

Answer

The question involves thinking through cell-to-cell interactions and their consequences.

a. Without mitotic divisions, only a few cells are available to go through meiosis and produce gametes. *Many fewer gametes are produced.*

b. If *glp-1* encodes a receptor, it should be expressed in cells that receive and interpret the signal. Those cells are the *germ cell precursors.*

c. GLP-1 protein, because of its role in various cell-to-cell interactions, could be considered a general responding protein. This general protein could be *interacting with different cell-type-specific proteins to give an appropriate, specific response.* Alternatively, or perhaps in addition, the GLP-1 receptor *may have different regions that respond to different inducing signals* and send specific messages into the cell based on which region of the receptor has been contacted.

P R O B L E M S

C-1 Choose the best matching phrase in the right column for each of the terms in the left column.

a. ablation	1. having centromere activity along the length of the chromosome
b. apoptosis	2. cell death caused by outside forces such as lasers
c. heterochronic	3. extrachromosomal fragments of DNA
d. polycistronic	4. change in the time at which an event occurs
e. trans-splicing	5. programmed cell death caused by signals inside the cell
f. holocentric	6. sections of different RNAs are joined together
g. free duplication	7. containing several genes

C-2 To construct a library for which there is a 99% probability that any one genomic sequence will be represented, the amount of DNA represented in the independent clones should be approximately five times the genome size. Assuming an average cosmid size of 50 kb and a genome size of 100 Mb, how many cosmids would you need to have in your library to ensure that each sequence in *C. elegans* was represented at least once?

C-3 What is (are) the advantage(s) in using a hermaphrodite when screening for mutants?

C-4 Vulval development in the *C. elegans* hermaphrodite has provided an excellent model system for studying the molecular basis of cell interaction and signaling during development. Genetic analysis and cell ablation studies have been critical in identifying genes and determining the role of their gene products in vulval development. For which of the following purpose(s) would you want to

a. combine different vulval mutations into one strain?

b. introduce RNAi specific for a gene expressed in vulval precursor cells?

c. isolate suppressors of a vulval mutant phenotype?

d. ablate a cell?

   i. to identify mutations that bypass a normal mechanism

   ii. to identify genes that work in the same pathway

   iii. to determine the function of a cell during development

   iv. to determine if mutations are in the same gene

   v. to determine if a gene is needed for proper vulval development

C-5 You isolated a cDNA for an actin gene. Which of the following results would suggest that trans-splicing occurs during processing?

a. the 5′ end of the cDNA hybridized with many DNA clones from the genome

b. the 5′ end of the cDNA hybridized with many RNAs in the worm

c. the 3′ end of the cDNA hybridized with many DNA clones from the genome

d. the 3′ end of the cDNA hybridized with many RNAs in the worm

e. the 5′ end of the cDNA hybridized with one DNA clone from the genome

f. the 5′ end of the cDNA hybridized with one RNA in the worm

g. the 3′ end of the cDNA hybridized with one DNA clone from the genome

h. the 3′ end of the cDNA hybridized with one RNA in the worm

C-6 Laser ablation of cells can be done during embryonic development, during larval stages, or in the adult. To determine if the anchor cell is involved in establishing the identity of surrounding vulval precursor cells, describe (in general) the time at which you would ablate the anchor cell.

C-7 The concentration of LIN-14 protein, high during the L1 larval stages and decreasing in later stages, regulates the timing of cell divisions in the T-cell lineage in *C. elegans*. According to the hypothesis presented in this chapter, *lin-4* RNA regulates the level of LIN-14 protein by binding to the *lin-14* mRNA and preventing translation. If you could introduce extra copies of *lin-4* into the cells and more *lin-4* RNA was produced, what effect do you think the extra copies of *lin-4* would have on translation of LIN-14? What effect could this have on cell lineages during development?

C-8 The *C. elegans* strain Bergerac has a mutation rate significantly higher than that seen in other strains. For example, spontaneous mutations in the *unc-22* gene arise at a frequency 100 times higher than the frequency of mutation in the Bristol strain. You think the mutations may be caused by insertion of transposons. How could you test whether the transposition of the Tc1 element caused some or all of the mutations in the *unc-22* gene? (Assume that the *unc-22* gene and Tc1 DNA have been cloned.)

C-9 How would you construct a hermaphrodite with the genotype *dpy-10⁻ / dpy-10⁻ daf-1⁻ / daf-1⁻*? You are given stocks of wild-type males, *dpy-10⁻ / dpy-10⁻* hermaphrodites and *daf-1⁻ / daf-1⁻* hermaphrodites. (*dpy-10* and *daf-1* are unlinked autosomal genes.)

a. What is the first cross you would do?

b. Which worms would you use in the second cross?

c. What is the final cross you do to get the doubly homozygous worms? What proportion of the hermaphrodites do you expect to have the doubly homozygous genotype?

C-10 The *unc-54* (uncoordinated) and *daf-8* (dauer constitutive) genes are 18 m.u. apart on chromosome I of *C. elegans*. You want to construct a worm strain homozygous for mutations in both genes. Stocks of wild-type males and *unc-54⁻ / unc-54⁻* hermaphrodites

and *daf-8⁻ / daf-8⁻* hermaphrodites are available for your use. The first cross is wild-type males X *unc-54⁻ / unc-54⁻* hermaphrodites.

a. What progeny of this first cross will you use for the second cross?

b. What progeny arise from your second cross?

c. What worms will you use for the final cross to get the desired double homozygote?

d. What event has to happen to get a double homozygote?

e. What proportion of the progeny will have the desired genotype?

C-11 Muscles in the worm can be classified in four major groups: body muscles involved in locomotion, enteric muscles involved in defecation, pharynx muscles involved in feeding, and egg-laying muscles in the hermaphrodite. Uncoordinated (Unc) worms have been isolated that are defective in neuromuscular control, resulting in a variety of phenotypes from slow-moving to completely paralyzed.

a. *unc-103⁻* is a mutation that causes defects in all muscle groups except pharyngeal muscles and has the phenotype of complete paralysis. The approximate location of the gene was determined by genetic mapping. One gene already identified in the region encodes a potassium (K^+) channel protein. Because ion channels and membrane depolarization are involved in muscle response, it was hypothesized that the *unc-103* mutation might be in the channel gene. Taking a candidate gene approach, how would you proceed to determine if the *unc-103* mutation was in fact a mutation in the K^+ channel gene?

b. Some of the *unc-103* alleles were constructed by insertion of the Tc1 transposon. How could you use this type of mutation and the physical map information on the location of the K^+ channel gene to determine if *unc-103* is in fact the K^+ channel gene?

c. Depolarization of the membrane (increase in the positive charge inside the cell) is necessary for muscle contraction. Opening of a K^+ channel is involved in the repolarization event, allowing K^+ to flow out of the cell and reestablish the negative charge on the inside of the cell. *unc-103* gain-of-function alleles lead to an inappropriately open K^+ channel so potassium continues to flow out of the cell. Do you think the gain-of-function *unc-103* allele is dominant or recessive and why?

d. Another gene identified by mutations that cause defects in locomotion, defecation, and egg-laying is the *egl-2* gene. This gene also encodes a K^+ channel. When a fusion was made between the 5′ end of the *egl-2* gene and a reporter gene, the expression of the fusion gene could be traced in specific cells. It was found to be expressed in muscles and nerves required for egg-laying and defecation but also in a subset of the chemosensory neurons—the AWC but not the AWA cells. Each of these cell types is involved in sensing different chemoattractants. The AWC neurons sense volatile substances such as isoamyl alcohol, and AWA neurons sense soluble substances such as diacetyl alcohol. How could you determine if the *egl-2* mutation disrupts chemosensory function in these cells?

C-12 Guidance nerve cell axons growing toward their ultimate targets are models for cellular movement during development. Because the location of all neurons in the *C. elegans* adult and the ancestry of all cells is known, much can be learned about the process of axonal growth using mutants defective in the final function of the nerve cells. Molecules involved in axon guidance have been identified by isolation of *unc* (Uncoordinated) mutants that are defective in locomotion. Proteins called netrins that act as guidance molecules along the path of growth have been identified in flies, worms, and humans. The UNC-6 gene product in *C. elegans* is a netrin, and the UNC-5 protein is a receptor protein on some nerve cells that interacts with UNC-6 protein in the basal lamina (extracellular matrix) and receives guidance for growth.

a. Mouse Netrin-1 has similar properties to *C. elegans* UNC-6. How could you determine if mouse Netrin-1 could substitute for UNC-6 in *C. elegans?*

b. Other cells migrate during development in *C. elegans.* How would you know if these migrations require *unc-5* or *unc-6* function?

c. UNC-5 protein is present on the surface of the growing neuron and receives guidance from the UNC-6 protein for axon growth. Starting with an *unc-5* defective mutation (but not one completely lacking the protein), how could you isolate genes encoding proteins that interact with (and perhaps relay messages from) the UNC-5 receptor to the inside of the cell?

d. The appropriate growth of axons is hypothesized to involve a series of attractions and repulsions that may depend on the type of receptor expressed at a particular time by a nerve cell. If the time of expression of receptor/sensing genes such as *unc-5* is altered, there may be an effect on guidance. Describe a general approach you could take to express *unc-5* constitutively in the cells of *C. elegans.*

REFERENCES

Sijen, T., Fleenor, J., Simmer, F., Thijssen, K. L., Parrish, S., Timmons, L., Plasterk, R. H. A., and Fire, A. (2001). On the Role of RNA Amplification in dsRNA-Triggered Gene Silencing. *Cell 107* (4):465–76.

Drosophila melanogaster:
Genetic Portrait of the Fruit Fly

Contributing author Ian Duncan *Washington University*

Adult fruit flies expressing β-galactosidase reporter genes.

A wild-type *Drosophila melanogaster,* or fruit fly, has multifaceted brick red eyes, a tan thorax studded with arched black bristles, a striped abdomen, and a pair of translucent wings (Fig. D.1). The fly's rapid life cycle, low chromosome number, small genome size, and giant salivary gland chromosomes (review Fig. 12.15 on p. 433) are among the experimental advantages that made *D. melanogaster* the central organism in the study of transmission genetics in the first half of the twentieth century. These same features also made *Drosophila* a major model organism in genetic studies of animal development and behavior in the last quarter of the century.

Modern *Drosophila* genetics originated with Thomas Hunt Morgan's discovery of the sex-linked *white* eye mutation in 1910. In the next few decades, *Drosophila* studies by Morgan and his students at Columbia University in New York City laid the experimental foundation for the field of genetics (Fig. D.2). Calvin Bridges, for his Ph.D. with Morgan, proved the chromosome theory of inheritance by showing that aberrations in the inheritance of the sex-linked *white* eye gene correlate perfectly with the nondisjunction of the X chromosome (see Chapter 4). Alfred H. Sturtevant, as an undergraduate in Morgan's lab, generated the first chromosome map by measuring recombination frequencies among the known sex-linked genes (see Chapter 5). Herman J. Muller demonstrated the mutagenicity of X rays (see Chapter 7). These and other members of Morgan's "fly room" went on to establish many other basic features of transmission genetics, including the structure and genetic consequences of each major type of chromosome rearrangement (described in Chapter 13).

From the 1940s through the 1960s, bacteria and their viruses replaced *Drosophila* as the dominant organisms in genetic studies. The simplicity of these microbes and the ease with which they can be grown in vast numbers made them ideally suited for genetic and biochemical work on the basic structure and function of genes. During this time, *Drosophila* research went into almost total eclipse. Then, in the early 1970s, when many geneticists began to investigate how genes control the development and behavior of higher organisms, interest in *Drosophila* increased dramatically as researchers realized the great potential of combining the sophisticated genetic

Figure D.1 A colorized scanning electron micrograph of a *Drosophila* adult's head and wings.

inally developed for *Drosophila,* researchers have made stunning progress in understanding how genes control the development and behavior of higher organisms. The impact of this work has been immense, and the accelerating pace of *Drosophila* genetics suggests that the fruit fly will remain a key model organism for the foreseeable future.

A major theme that emerges from our discussion of the genetic analysis of *Drosophila* is that many of the developmental genes and mechanisms operating in the fly have been conserved in a wide variety of animals. For example, most of the developmentally important genes first found in *Drosophila* have turned out to play similar roles in other animals, including humans. In fact, researchers identified most of the genes known to influence the development of humans as homologs of genes already discovered in *Drosophila.*

As we examine *D. melanogaster*'s use as a model organism, we present:

■ The structure and organization of the genome.

■ The life cycle.

■ Techniques of genetic analysis, including the construction and use of balancer chromosomes, the uses of P-element transposons, the production of mosaics, ectopic gene expression, targeted gene knockouts, and the *Drosophila* Genome Project.

■ The genetic analysis of body plan development in *Drosophila,* a comprehensive example.

techniques of *Drosophila* with the newly developed methods of recombinant DNA technology. The successful marriage of these approaches depended in part on the fortunate fact that a handful of *Drosophila* geneticists had continued to work and maintain stock collections through the "dark" years.

In this chapter, we see that by using modern molecular methods in conjunction with the rich legacy of genetic techniques orig-

Figure D.2 The fountainhead of *Drosophila* genetics. Thomas Hunt Morgan and his students in the fly room at Columbia University, at a party in 1919 celebrating the return of Alfred Sturtevant from World War I military service. Individuals whose work is discussed in this book include Morgan (*back row, far left*), Sturtevant (*front row, third from the right*), Calvin Bridges (*back row, third from the right*), and Herman J. Muller (*back row, second from the left*). The "honored guest" between Muller and Bridges has clearly seen better days.

■■■

Structure and Organization of the *Drosophila* Genome

The *Drosophila* genome is about 5% the size of the human genome and is packaged in far fewer chromosomes. It contains roughly 13,600 genes, about a third of the number found in the human genome. This means that the density of genes on *Drosophila* chromosomes is higher than that of genes in the human genome. Even so, about one-third of the fly genome consists of repetitive sequences that do not encode proteins or that act as transposable elements.

The Chromosomes of *Drosophila*

The haploid genome of *D. melanogaster* contains four chromosomes, designated numerically as 1–4. Chromosome 1 is the X chromosome; chromosomes 2–4 are autosomes. Sex determination, as we saw in Chapter 4, is of the XY type, with females being XX and males XY. Unlike the situation in mammals, however, the Y plays no role in sex determination; sex is instead determined solely by the ratio of the number of X chromosomes to the number of copies of each autosome (the X:A ratio). Thus, XXY zygotes, which have an X:A ratio of 1 (two copies of the X : two copies of each autosome), develop as normal females, while XO zygotes, which have an X:A ratio of 0.5, develop as morphologically normal, but sterile, males. (Review Chapters 4 and 17 for more details about sex determination in *Drosophila*.) The Y chromosome, which by convention is not given a number, is required only to confer male fertility. The low chromosome number is one of *Drosophila*'s key advantages for genetic studies, as it simplifies most genetic manipulations.

Figure D.3 shows the structure of *Drosophila* chromosomes prior to S phase DNA replication. The X chromosome is large and acrocentric; chromosomes 2 and 3 are large and metacentric, and chromosome 4 is a tiny acrocentric "dot" chromosome that is only about 2% the size of the major autosomes. Each chromosome contains a large block of heterochromatin near its centromere. Although a few genes are present in these heterochromatic blocks, for the most part, these regions contain highly repetitive DNA sequences and are transcriptionally silent. Centromeric heterochromatin comprises about a quarter of the length of the X chromosome and each of the large autosomes, as well as the majority of the small chromosome 4. The large submetacentric Y chromosome is almost entirely heterochromatic, and it carries only a few (probably seven or eight) genes, which function in the male germ line. Nucleolus organizer regions are located within the short arm of the Y and the centromeric heterochromatin of the X. Each nucleolus organizer region carries 200–250 copies of the 18S and 28S ribosomal RNA genes in tandem array.

The Giant Polytene Chromosomes of the Larval Salivary Gland Are Key Tools for *Drosophila* Genetics

As we saw in Chapter 12, chromosomes in the fruit fly's salivary glands replicate 10–11 times without cell division. The 1024 or 2048 sister chromatids generated during these rounds of

Figure D.3   The *Drosophila* chromosome complement. Major blocks of all chromosomes, particularly in regions surrounding the centromeres, are heterochromatic. Key: *dark blue,* heterochromatin; *light blue,* euchromatin; Y^L and Y^S, the long and short arms of the Y chromosome; and 2L, 2R, 3L, and 3R, the left (L) and right (R) arms of chromosomes 2 and 3. See the Fast Forward box in Chapter 4 (p. 88) for a picture of these same chromosomes in mitotic cells.

replication stay intimately associated with one another in perfect lateral register, producing a many-stranded polytene chromosome. These chromosomes are truly gigantic: Polytene chromosomes are about as wide as normal metaphase chromosomes are long.

Regions of local condensation along a chromosome become amplified laterally in polytene chromosomes and are visible as distinct crossbands (separated by less condensed interbands) when viewed through a light microscope. A striking feature seen in cells carrying polytene chromosomes is that homologous chromosomes are tightly synapsed with each other (review Fig. 12.15 on p. 433). This is unusual because in most organisms, extensive synapsis of homologs occurs only in prophase I of meiosis. In contrast to the euchromatin, regions of centromeric heterochromatin are not replicated during polytenization, and they coalesce into a common region called the *chromocenter* (also shown in Fig. 12.15). As a result, in squash preparations of salivary gland nuclei, five major arms (the left and right arms of chromosomes 2 and 3, and the X chromosome) radiate out from the chromocenter. The much smaller chromosome 4 also associates with the chromocenter, but it is often hard to see. The Y chromosome is completely heterochromatic and therefore is not visible.

The importance of polytene chromosomes in *Drosophila* genetics cannot be overemphasized. As we saw in Chapters 12 and 13, they have provided *Drosophila* geneticists with a ready-made detailed physical map of the fly genome. Since the finer visible bands contain only a few kilobases of DNA per chromatid, rearrangement breakpoints can often be mapped with very high precision simply by microscopic examination. Cloned

DNA sequences can also be mapped very rapidly and precisely by *in situ* hybridization to polytene chromosomes. This technique was of particular utility when the *Drosophila* Genome Project determined the sequence of the fruit fly genome. Researchers could determine the location and extent of the large genomic clones (in YAC, BAC, or P1 vectors), whose sequences they determined in the genome project, by using the clones as probes for *in situ* hybridization to polytene chromosomes.

The *Drosophila* Genome

The haploid genome contains about 170,000 kb of DNA. Roughly 21% of this DNA consists of highly repetitive "satellite" DNA that is concentrated in the centromeric heterochromatin and the Y chromosome. Another 3% consists of repeated genes that encode rRNA, 5S RNA, and the histone proteins. Approximately 9% is composed of around 50 families of transposable elements, which are roughly 2–9 kb in length; the number of copies of each type of element varies from strain to strain but is usually somewhere between 10 and 100. In addition to residing at scattered sites within the euchromatin, many transposable elements are found in the centromeric heterochromatin, while some are concentrated in telomeres. An unusual feature of *Drosophila* chromosomes is that the telomeres do not consist of simple repeats, such as the TTGGGG or TTAGGG repeats found at the ends of chromosomes in other well-characterized organisms. Instead, it appears that the rapid jumping into the telomeric region of certain kinds of transposable elements maintains the telomeres. These insertions into the telomeric region of the transposable elements counter the chromosomal shortening that inevitably occurs during the replication of the ends of linear DNA molecules.

The remaining 67% of the genome consists of unique DNA sequences in the euchromatic chromosome arms. We describe the nature and organization of the approximately 13,600 genes present within these unique DNA sequences later in the chapter, in our discussion of the *Drosophila* Genome Project.

Life Cycle

Fruit flies can be raised on a wide variety of media, but the standard formulation is a mixture of corn meal, brewer's yeast, sugar, agar, and a small amount of mold inhibitor. Flies bred on such a medium at 25°C complete their life cycle from fertilization to emergence of the adult fly in 10 days (Fig. D.4).

Drosophila adults have a very high reproductive capacity. A female can produce 3000 progeny in her lifetime; a single male can sire well over 10,000 offspring. The female has special sperm storage organs (the spermatheca and seminal receptacle) that allow her to produce several hundred progeny after a single mating. Each *Drosophila* egg is about half a millimeter long and is well supplied with yolk that supports embryonic development. Fertilization normally occurs as the egg is being laid.

The embryo completes its development in just under 24 hours, and some of the early events happen so rapidly they can

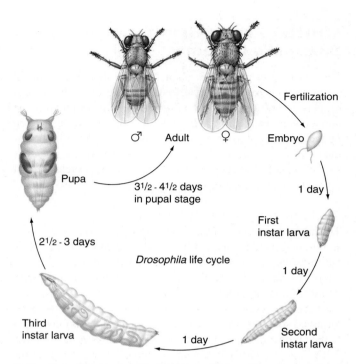

Figure D.4 The *Drosophila* life cycle. The transition from an embryo to a first instar larva is called *hatching.* The transitions between larval instars are *molts.* The process that converts a third instar larva to a pupa is *pupariation.* Emergence of the adult from the pupal case is called *eclosion.*

be watched in real time. For example, in the early stages of embryonic development, the zygotic nuclei divide every 10 minutes, and the morphogenic movements of gastrulation that produce the endoderm, mesoderm, and ectoderm tissue layers take only 20 minutes. Embryogenesis ends with the hatching of a wormlike first instar (or first-stage) larva that is specialized for feeding and grows dramatically. To allow for an increase in body size, the larva molts 24 and 48 hours after hatching to produce second and third instar larvae.

About three days after the second molt, the third instar larvae complete their growth, crawl out of the food, and pupate. Once inside the protective pupal case, the larvae undergo **metamorphosis:** a dramatic reorganization of the fly body plan. This reorganization takes about four days and consists of the disintegration of most larval tissues and their replacement through the proliferation and differentiation of cells that produce adult structures. Most structures specific to adults, such as the wings, legs, eyes, and genitalia, are generated from **imaginal discs:** flattened epithelial sacs that develop from small groups of cells set aside in the early embryo (Fig. D.5). Imaginal discs undergo extensive growth and development during the larval and pupal stages. Other adult-forming cells, clustered in nests within each abdominal segment, produce the adult abdominal epidermis. At the end of metamorphosis, when the adult emerges from the pupal case, its wings expand and the entire exoskeleton hardens and becomes pigmented.

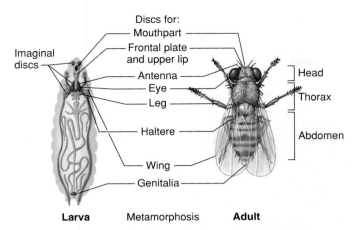

Figure D.5 **Many adult structures develop from imaginal discs in larvae and pupae.** Lines connect particular imaginal discs in the larva with the structures they generate in the adult.

Imaginal discs

Discs for:
Mouthpart
Frontal plate and upper lip
Antenna
Eye
Leg
Haltere
Wing
Genitalia

Head
Thorax
Abdomen

Larva   Metamorphosis   **Adult**

Techniques of Genetic Analysis

The practice of genetics in *Drosophila* has a unique character that in part reflects the tools created by *Drosophila* researchers over many years and in part reflects certain distinctive features of fruit fly inheritance. For a review of the symbols used by *Drosophila* geneticists, see the Chapter 4 discussion of eye color inheritance on pp. 99–101. Of particular importance to this chapter, the names or abbreviations of genes are always italicized, while the names or abbreviations of proteins are not italicized and begin with a capital letter.

In *Drosophila,* Crossing-Over Occurs Only in Females

Although the reason for the restriction to females of crossing-over is not clear, in the few other known species in which recombination is restricted to one sex, it is always the *heterogametic sex* (that is, the gender characterized by dissimilar sex chromosomes) that lacks crossing-over. This suggests that the prevention of crossing-over between different sex chromosomes has been of selective importance.

The Absence of Crossing-Over in the Male Has Considerable Technical Significance

The absence of crossing-over in the male makes it possible to maintain linkage relationships simply by ensuring that a chromosome is inherited strictly through the male parent. This may seem trivial, but it greatly simplifies many genetic manipulations and allows a variety of selective genetic screens.

In Females, Only a Moderate Amount of Crossing-Over Occurs

Typically, only one or two crossovers take place in each major chromosome arm per meiosis, and essentially no crossing-over occurs in the centromeric heterochromatin or anywhere in the fourth chromosome. As in other organisms, however, crossing-over is required for the accurate disjunction of the major chromosomes during meiosis I in females. In the absence of chiasmata, bivalents fall apart at the end of prophase I, causing nondisjunction and chromosome loss. A special mechanism, currently under active study, ensures the segregation of the fourth chromosomes and of the occasional large chromosome bivalents in which crossing-over fails to occur. How the segregation of homologs is directed in males is not well understood, but it appears to depend on special sites along the chromosomes that mediate the association of homologs.

Balancer Chromosomes Help Preserve Linkage

Drosophila geneticists have used X rays as a mutagen to create many special chromosome types useful for genetic manipulations. By far the most important of these is the **balancer chromosome,** which carries multiple, overlapping inversions (Fig. D.6a). Most balancers also contain a dominant marker that enables researchers to follow the balancer through crosses, and a recessive lethal mutation that prevents the survival of homozygotes. The utility of balancer chromosomes is that, like all inversion-containing chromosomes, they prevent the recovery of crossovers between normal sequence chromosomes and themselves. As a result, both balancer chromosomes and their homologs are inherited through crosses as intact units, and no recombinant chromosomes are passed on to the next generation.

Balancer Chromosomes Are Useful in Stock Keeping

The term "balancer" derives from the extensive use of these chromosomes in stock keeping. Experimenters establish a typical *Drosophila* stock by crossing males and females that carry a desired mutation heterozygous with a balancer chromosome homolog. If the mutation is a recessive lethal, as is often the case, and the balancer chromosome carries a different recessive lethal, the only surviving progeny will, like their parents, be heterozygous for the chromosome bearing the desired mutation and the balancer (Fig. D.6b). The only requirement for maintaining the mutation is to transfer progeny to fresh medium periodically. Stocks of this type are said to be "balanced lethals," hence the name "balancer."

Balancer Chromosomes Are Also Useful in Genetic Manipulations and Mutant Screens

As an example, Fig. D.6c shows a cross to determine by complementation analysis whether two independent recessive lethal mutations (*lethal₁* [*l₁*] and *lethal₂* [*l₂*]) on the second chromosome are allelic. The cross involves a second chromosome balancer, designated SM1, which carries the dominant wing mutation *Curly* (*Cy*). Males of genotype l_1/SM1 are crossed to l_2/SM1 females. If the mutations are allelic, only progeny with the Curly wing phenotype will emerge, whereas if the mutations complement each other, non-Curly (l_1/l_2) progeny will also appear.

(a) SM1 balancer chromosome

Key:

[ ] Breakpoints of pericentric inversion

( ) Breakpoints of paracentric inversions

(b) A balanced lethal stock

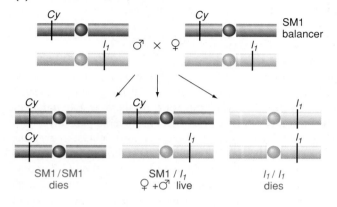

SM1/SM1 dies

SM1 / l_1 ♀ +♂ live

l_1 / l_1 dies

(c) Complementation analysis

♂ SM1 / l_1 × SM1 / l_2 ♀

SM1 / SM1 Dies

SM1 / l_1 SM1 / l_2 — Viable, *Cy*-winged

l_1 / l_2
Dies if l_1 and l_2 are allelic
Viable non-*Cy*-winged if l_1 and l_2 complement

Figure D.6 Balancer chromosomes in *Drosophila*. (a) SM1, a chromosome 2 balancer, contains overlapping inversions to suppress recombination, and it also carries the dominant marker *Cy* (Curly wings). *Cy* is also a recessive lethal, so that SM1/SM1 homozygotes do not survive. **(b)** When flies heterozygous for a lethal mutation (l_1) and a balancer chromosome are intercrossed, the only adult survivors are l_1/Balancer heterozygotes identical to their parents. **(c)** Are two independently isolated lethal mutations (l_1 and l_2) allelic? The diagrammed cross will produce l_1/l_2 progeny only if the mutations are in different genes.

P-Element Transposons Are Critical Tools in *Drosophila* Molecular Genetics

The discovery of P elements arose from research on a phenomenon called **hybrid dysgenesis**. P elements (Fig. D.7a) are the causative agent of hybrid dysgenesis, which occurs when males from *Drosophila* strains carrying P elements (referred to as P strain males) are crossed with females that lack P elements (known as M strain females). The P elements become highly mobile in the germ line of the F_1 hybrids produced from such crosses (Fig. D.7b and c). Chromosome breakage resulting from this high transposon mobility causes reduced fertility in the hybrids; the few progeny of the F_1 flies that do survive carry many new mutations induced by the insertion of P elements at new

(a) P-element structure

Inverted repeats — *Drosophila* chromosomal DNA

5' — Exon — 3' Primary transcript
Intron
P element

(b) Hybrid dysgenesis

Nondysgenic crosses

P♂ × P♀ → Normal progeny

M♂ × M♀ → Normal progeny

M♂ × P♀ → Normal progeny

Dysgenic crosses

P♂ × M♀ → Semisterile progeny × P or M

↓

Second-generation progeny with many mutations and chromosome rearrangements

(c) P-element transposition in dysgenic crosses

P♂ × M♀

Embryo

Germ-line pole cells

Germ-line cell

P elements transpose

Grows into

Semisterile fly × P or M mate

Progeny with many mutations and chromosome rearrangements due to P-element transpositions

Figure D.7 P elements and hybrid dysgenesis. (a) Transposition requires the action of the P-element transposase enzyme on the inverted repeats at P-element ends. **(b)** Hybrid dysgenesis occurs only when P males mate with M females. The progeny of such a mating appear wild type but are semisterile. **(c)** P elements on sperm brought into embryos derived from M females transpose at high rates, but only in germ-line cells.

sites in the genome. Hybrid dysgenesis does not occur in any other combination of M or P males and females, such as in crosses of P males and P females (Fig. D.7b).

Molecular details of P-element biology help clarify the hybrid dysgenesis phenomenon. The P-element primary transcript encodes a transposase enzyme that acts on the 31 bp inverted repeats at P-element ends to catalyze transposition. Hybrid dysgenesis obviously does not occur in M × M matings without P elements. The reason hybrid dysgenesis does not take place in crosses involving P females is that the eggs produced by P females contain high amounts of a repressor protein that prevents transcription of the transposase gene. This repressor is actually translated from an alternatively spliced P-element mRNA. In contrast, eggs from M strain mothers lack this cytoplasmic repressor of P-element transposition. As a result, P elements in sperm from P fathers will transpose rapidly when they fertilize eggs from M mothers, explaining the hybrid dysgenesis seen in crosses of P strain males to M strain females (Fig. D.7b and c). In the embryos from dysgenic crosses, transposition occurs only in germ-line cells because a splicing event needed to make transposase mRNA does not take place in somatic cells. The adult progeny of a dysgenic mating thus appear normal, but they are semisterile because their gametes carry many chromosomal aberrations and mutations caused by P-element movement.

The high mobility of P elements during hybrid dysgenesis makes it possible to exploit these elements for two purposes: as transformation vectors and as genetic tags.

Transformation: The Introduction of Cloned DNA into Fruit Flies

Drosophila geneticists realized from the following experiment that P elements can serve as vectors to introduce cloned DNA into the organism. Researchers first constructed a recombinant plasmid that contained a fragment of *Drosophila* genomic DNA including an intact P element. When they injected this plasmid into early syncytial embryos from M strain mothers, the P element began to transpose and insert itself into the chromosomes of host germ-line cells at a high frequency. This artificial situation mimicked the phenomenon of hybrid dysgenesis. Additional experiments demonstrated that this transposition is catalyzed by the P-element-encoded transposase and requires the presence of the inverted terminal repeats normally found at P-element ends.

For germ-line transformation, *Drosophila* geneticists use vectors that contain the P-element ends but not the P-element gene that encodes the P transposase. They clone novel DNA transgene sequences between the P-element ends and inject the resulting DNA construct into eggs from M strain mothers. A helper P element is coinjected to provide a source of the P-element transposase; this helper cannot itself integrate into a chromosome because it lacks both of the P-element ends. Figure D.8 illustrates one protocol for using P elements to introduce cloned genes into fruit flies. Note that pCaSpeR, one of the most frequently used transformation vectors, contains a *white*$^+$ marker gene that makes it easy to trace the presence of the vector in the genome by the red eye color it confers.

P-Element Tagging Has Enabled the Cloning of Many Genes of Developmental Interest

The basic strategy is to isolate a mutant allele of the desired gene that arose from insertion of a P element, and then to recover sequences from the gene by isolating DNA flanking the P element. P insertion alleles are usually generated and the flanking DNA recovered using two special strains of flies. One carries a chromosomal copy of a defective P element that provides a large amount of transposase, but it is itself not mobile because it lacks

(a) Transformation of *Drosophila*

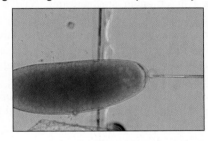

(b) Injecting a transgene into a *Drosophila* embryo

Figure D.8 P-element-mediated transformation of *Drosophila*. (a) A transgene (Gene) is cloned into a transformation vector (pCaSpeR) containing the *white*$^+$ (*w*$^+$) marker gene between the P-element inverted ends. Researchers coinject this construct, along with a helper P element that encodes transposase but has no P-element ends, into M-type *w*$^-$ fertilized embryos. Transposition of the *w*$^+$ gene and the transgene occurs in some germ-line cells. If adults with these germ cells are mated with *w*$^-$ flies, some progeny will have red (*w*$^+$) eyes and an integrated transgene. **(b)** Injection of DNA into the posterior end of a *Drosophila* embryo.

P-element ends. The other strain carries a specially engineered P element that contains both a marker such as the *w*$^+$ eye color gene that makes it possible to keep track of this P element, as well as bacterial plasmid sequences with an antibiotic resistance gene, whose purpose will become apparent in a moment. When the two strains are crossed, transposase in the germ-line cells of the F$_1$ flies catalyzes transposition of the specially marked P element. Researchers can identify movements of the marked P element to new sites on other chromosomes by following the *w*$^+$ marker in the progeny of the F$_1$ flies.

(a) An enhancer trap construct

(b) Detecting tissue-specific enhancers

Figure D.9 **Plasmid rescue of a transposon-tagged gene.** An engineered P element contains bacterial plasmid DNA including an ampicillin resistance gene (*Amp^R*). To identify genes mutated by P-element insertion, researchers digest genomic DNA from mutant flies with an enzyme like *Eco*RI that cuts on one side of the bacterial plasmid but nowhere else in the P element. When the *Eco*RI-cut genomic DNA is circularized by DNA ligase and transformed into *Amp^S* bacteria, all ampicillin-resistant colonies should harbor a plasmid that includes *Drosophila* DNA adjacent to the P-element insertion.

When a P insertion in an interesting gene is found, researchers can rapidly clone DNA from the gene by a technique called *plasmid rescue,* illustrated in Fig. D.9. In this technique, they simply obtain genomic DNA from the mutant insertion strain, cut the DNA with a restriction enzyme, circularize the restriction fragments with DNA ligase, and use the circularized DNA to transform *Escherichia coli.* The only bacterial colonies able to grow on antibiotic-containing medium will contain the bacterial plasmid sequences plus the gene-containing *Drosophila* DNA flanking the P-element insertion.

Enhancer Trapping: Using P Elements to Identify Genes by Their Expression Patterns

Drosophila developmental geneticists have constructed P elements in which the *E. coli lacZ* gene is placed downstream of the P-element promoter (Fig. D.10a). Remarkably, when an element is mobilized, over 65% of the new insertions isolated show patterned expression of the *lacZ* reporter during development, with the *lacZ* gene on in some tissues but off in others. The patterned expression results from the fact that the P-element promoter by itself is very weak. The promoter can only activate *lacZ* transcription when it comes under the control of nearby regulatory sequences (enhancers) of genes near its new insertion site. (See Chapter 17 for a definition and discussion of enhancers.) Many such insertions into known genes have been recovered, and in most of these cases, the expression pattern of β-galactosidase (the protein product of *lacZ*) accurately reflects the normal expression of the gene into which the element has been inserted (Fig. D.10b). The identification of P-element insertion lines with particular β-galactosidase patterns is known

Figure D.10 **Enhancer trapping. (a)** This "enhancer trap" contains a *lacZ* reporter gene encoding β-galactosidase downstream of a P element's promoter. Expression of β-galactosidase in transformed animals reflects enhancers near the P-element insertion site. **(b)** Late embryos from two lines of flies with the P-element "enhancer trap" inserted into different genomic locations were stained for β-galactosidase (*blue*). Expression of the reporter is seen in the salivary glands (*top*) or in the Malpighian tubules (*bottom*).

as **enhancer trapping.** It is a powerful way to detect genes that are turned on in specific tissues. Since the most frequently used enhancer trap elements carry *E. coli* plasmid sequences, genes showing β-galactosidase expression patterns of particular interest can be rapidly isolated by plasmid rescue.

A particularly good example of the utility of P-element tagging and enhancer trapping is seen in the work of Thomas Kornberg and his colleagues, who have studied the function of a gene called *engrailed* (*en*), which plays a key role in establishing and patterning body segments. As described in a later section of this chapter, *en* encodes a transcription factor that is expressed in a stripe in the posterior portion of each body segment. To identify target genes regulated by *en,* Kornberg's group screened for enhancer trap insertions that are expressed in the same or a complementary striped pattern. In this way, they identified several key targets of *en.* Among them is a crucially important gene called *hedgehog* (*hh*), which, as explained later, encodes a secreted protein responsible for patterning not only body segments but many other structures as well.

The Production of Genetic Mosaics

Genetic mosaics are individuals composed of cells of more than one genotype. Studies of mosaics allow developmental geneticists to answer many types of questions about the developmental fate of various cells and the functions of particular genes.

The most important method for generating mosaics in flies is mitotic recombination, previously introduced on pp. 140–143 of Chapter 5. Although crossing-over is normally restricted in *Drosophila* to the prophase of meiosis I in females, after irradiation of embryos or larvae, crossing-over between homologs occurs quite frequently in somatic cells. It is thought that X-ray-induced chromosome breakage initiates mitotic recombination. If mitotic recombination occurs in the G_2 portion of the cell cycle, and is followed by the appropriate segregation of chromatids in the subsequent mitotic division, the resulting daughter cells are each homozygous for all loci distal to the crossover (that is, for all genes farther away from the centromere than the site of recombination). The mitotic descendants of these cells then produce genetically distinct patches in the emerging adults (review Fig. 5.26 on p. 141).

We discuss here three of the many important applications of genetic mosaics in the analysis of *Drosophila* development.

Testing the Potential of a Cell and Its Mitotic Descendants to Contribute to Particular Structures

A cell produced by mitotic recombination and its clonal descendants may have a phenotype that marks them as different from their neighbors and thus enables them to be tracked through the organism's development. If the fate of the initial cell produced by mitotic recombination is already restricted to a tissue or region, all of the cells in the clone will be found only in that one part of the body. In contrast, if the initial cell produced by mitotic recombination has a less restricted developmental potential, the cells of the clone will end up in two or more different tissues or regions of the body.

To test the developmental potential of cells, geneticists have devised a method to give clones a growth advantage over neighbouring cells by employing dominant mutations known as *Minutes*. There are about 50 *Minute* genes scattered throughout the genome; those that have been studied in detail turn out to encode ribosomal proteins. Heterozygotes for mutations in these genes show developmental delay and reduced bristle size, presumably because the *Minute* mutations cause a reduced rate of protein synthesis. Investigators use mitotic recombination to generate *Minute⁺* / *Minute⁺* cells in a *Minute⁺* / *Minute⁻* background (Fig. D.11). These wild-type (+/+) cells have a strong growth advantage over their neighbors and produce very large clones. Remarkably, these large *Minute⁺* clones can fill either the anterior half of the wing or the posterior half, but they never cross a boundary line separating the two. These two parts of the wing thus constitute two distinct regions known as **lineage compartments.** This experiment indicates that the *Minute⁺* / *Minute⁺* cells formed at the time of mitotic recombination already have a restricted developmental potential: Their descendants can populate either the posterior wing compartment or the anterior wing compartment, but not both. By inducing mitotic recombination later in embryogenesis, researchers observed that the developmental potential of clones of cells becomes increasingly restricted to smaller and smaller subcompartments (for example, the anterior ventral or anterior dorsal regions of the wing) as development proceeds.

Using Mitotic Recombination to Determine When and Where a Gene Must be Expressed to Generate a Normal Phenotype

Ommatidia are the facets of the fly's compound eye (Fig. D.12a). *Drosophila* geneticists have used mosaic analysis to study the

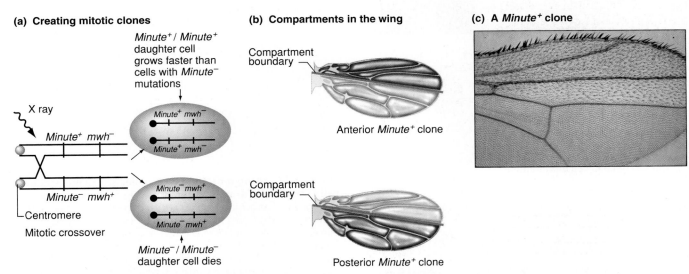

(a) Creating mitotic clones

Minute⁺ / *Minute⁺* daughter cell grows faster than cells with *Minute⁻* mutations

X ray

Minute⁺ mwh⁻

Minute⁻ mwh⁺

Centromere

Mitotic crossover

Minute⁺ mwh⁻
Minute⁺ mwh⁻

Minute⁻ mwh⁺
Minute⁻ mwh⁺

Minute⁻ / *Minute⁻* daughter cell dies

(b) Compartments in the wing

Compartment boundary

Anterior *Minute⁺* clone

Compartment boundary

Posterior *Minute⁺* clone

(c) A *Minute⁺* clone

Figure D.11 Mosaic analysis demonstrates the existence of developmental compartments. (a) X-ray-induced mitotic crossovers in cells with the genotype at the *left* can generate daughter cells homozygous for *Minute⁺* and *mwh⁻*. These *Minute⁺* cells and their mitotic descendants will outgrow all other cells, and they are marked by *mwh⁻*, which causes each wing cell to produce several tiny hairs instead of the normal one. The homozygous *Minute⁻* daughter cells die. **(b)** Large *Minute⁺* clones (*dark gray*) can fill the anterior or the posterior half of the wing, but they will not cross the compartment boundary. **(c)** Photograph of an anterior clone; the wing cells in the clone are large (because they are *Minute⁺*) and have several small hairs (as they are homozygous for *mwh⁻*).

normal development of the ommatidia. During the development of the eye, undifferentiated cells are recruited in a series of steps to become the eight light-sensing cells or photoreceptors (designated R1–R8) of each eye facet. The last photoreceptor cell to be recruited in each group is R7 (Fig. D.12b). To understand how R7 is specified, geneticists have isolated mutations that cause ommatidia to develop without an R7 cell. One gene defined by such mutations is *sevenless* (*sev*); another is *bride of sevenless* (*boss*). In what cells must these two genes be expressed if R7 is to become part of the eye facet?

To answer this question for the *sevenless* gene, biologists generated mosaics with marked cells that were mutant for both *white* and *sevenless* in an eye where other cells carried the dominant wild-type alleles for the two genes. Some facets contained a mix of red and white photoreceptor cells (Fig. D.12c). But in every facet with an R7 cell, the R7 cell was red (that is, it had both the w^+ and sev^+ genes), even if all the other photoreceptor cells in the

same facet were homozygous for w^- and sev^-. There were no facets in which the R7 cell was white. These results suggest that the requirement for the sev^+ gene product is **cell autonomous:** To develop correctly, a presumptive R7 cell must make the Sev$^+$ protein, and it does not matter whether the protein is made in any other photoreceptor cell in the same facet (Fig. D.12c). The Sev$^+$ protein therefore affects only the cell in which it is made.

By contrast, when researchers performed the same type of analysis for *boss*, the results were very different. An R7 cell develops properly only if an R8 cell makes the Boss$^+$ protein; this is true even when the R7 cell itself is *boss$^-$*. If the R8 cell is *boss$^-$*, however, the facet does not have an R7 cell. These results show that the Boss$^+$ protein is not cell autonomous. Instead, it is produced in one cell (R8), but affects the differentiation of a different cell (R7).

If a gene is cell autonomous, only cells with a mutant genotype will express the mutant phenotype, which indicates that the

(a) The compound eye of *Drosophila*

(b) Recruitment of cells into an ommatidium

Undetermined cells

(c) The *sev* gene is cell autonomous

(d) How Sev$^+$ and Boss$^+$ specify R7 cells

Figure D.12 Using mosaics to determine if a gene product is cell autonomous or nonautonomous. (a) Scanning electron micrograph of an adult *Drosophila* eye, showing the individual facets (ommatidia). **(b)** Eight photoreceptor cells are sequentially recruited into each ommatidium. *Blue* indicates the commitment of cells to a photoreceptor fate. The sev^+ and $boss^+$ gene products specify photoreceptor 7 (R7). **(c)** If the presumptive R7 cell is sev^+, it develops into an R7 photoreceptor regardless of whether any or all of the other cells in the ommatidium are sev^+ or sev^-. If the presumptive R7 cell is sev^-, it never develops into an R7 cell and instead develops into a cone cell (*not shown*). **(d)** The Boss$^+$ protein is a ligand expressed on the surface of the R8 cell. Binding of the Boss$^+$ ligand to the Sev$^+$ receptor on the surface of the R7 precursor cell activates a signal cascade that regulates the expression of genes determining an R7 fate.

gene product acts within the cell in which it is made. With a *nonautonomous gene,* the phenotype of a cell does not depend strictly on its own genotype; rather, its phenotype depends on the genotype of neighboring cells. Nonautonomy indicates that the gene product involved can influence cells other than those in which it is made (for example, by diffusing to other cells). The cloning of the *sev* and *boss* genes made it possible to understand in molecular terms the different behaviors of the two genes in the mosaic analysis. The *sev* gene encodes a receptor protein that is embedded in the membrane of the developing R7 cell, while the *boss* gene encodes a protein located on the surface of the R8 cell. The binding of the Boss$^+$ protein on R8 (the ligand) to the Sev$^+$ receptor on the presumptive R7 cell initiates a cascade of gene activity (a signal transduction pathway; see Chapters 18 and 19) in the potential R7 cell that triggers its differentiation into a mature R7 cell (Fig. D.12d).

Using Mosaic Analysis to Determine Whether Genes Function at Multiple Times in Development

Suppose you are studying a gene you know contributes to early embryogenesis; animals homozygous for mutations in this gene die before they hatch into larvae. You want to know whether or not this gene plays a role in the production of structures that appear much later in development, for example, the adult eyes. You cannot answer this question by simply looking at the homozygous mutant, because it dies before the eyes form. The use of mosaics circumvents this problem. Mitotic recombination is used to produce clones of cells homozygous for the mutation in adults that are otherwise heterozygous for the mutation and hence, viable. Assuming the gene studied is autonomous in mosaics, the role of the gene in eye development can then be inferred from the phenotype of the homozygous mutant clones within the eyes.

Until recently, mitotic recombination was almost always induced by X-irradiation. However, researchers have developed a more efficient means to promote mitotic recombination by using FLP, a site-specific recombination enzyme from yeast (Fig. D.13). The normal function of FLP is to catalyze reciprocal crossing-over at specific recombination targets (*FRT*s) contained within inverted repeats of a yeast DNA plasmid called the 2 micron circle. Investigators cloned the *FLP* gene downstream of, and therefore under the control of, a heat-shock promoter whose

expression is turned on at elevated temperatures, and they introduced this construct into *Drosophila* by P-element-mediated transformation. To insert the *FRT* target into *Drosophila* chromosomes, they cloned *FRT* into a P element and also introduced this second construct into the fly genome by transformation. In this way, they have obtained *FRT* insertions close to the centromere on each chromosome arm of *Drosophila.* If *FRT*s are present at the same location on both homologs, FLP recombinase (whose synthesis is induced by heat shock) catalyzes recombination between them. The resulting frequency of clones is extremely high, with several clones on each fly. This technology is so efficient that mutant screens can now be performed in which new mutations are identified by their effects in homozygous patches.

Drosophila geneticists often use mitotic recombination to produce mosaics in which the female germ line is homozygous for a mutation. They do this to remove the maternally supplied products of a gene from embryos. In *Drosophila,* many genes are expressed in the female germ line and their products are deposited into the developing egg. For most such genes, the maternally supplied products are sufficient to sustain development into the larval stages even if the zygote itself completely lacks the gene. To determine the effect on development of a complete lack of a gene product, it is thus usually necessary to generate homozygous mutant zygotes from homozygous mutant mothers. However, this strategy cannot be employed for lethal mutations, because homozygous mutant mothers cannot be obtained.

To circumvent this problem, geneticists generate homozygous mutant clones in the germ cells of heterozygous mothers. Since germ cells give rise to the eggs, such clones often produce eggs completely lacking the gene product of interest. As Fig. D.14 shows, geneticists can use a dominant mutation called *ovo*^{D} to ensure that the only eggs produced by heterozygous mothers are made from homozygous mutant germ-line cells. Researchers can increase the frequency of mutant clones by using the just-described FLP/*FRT* recombination system (not shown). For many genes, homozygous mutant zygotes developing from eggs made by homozygous mutant germ-line cells show a much more severe phenotype than homozygous mutant zygotes produced by mothers having heterozygous germ lines. Such a result indicates the importance of the maternally supplied protein to early development.

Figure D.13 Efficient induction of mitotic recombination in *Drosophila*. When a fly of the genotype shown is subjected to elevated temperature, FLP recombinase is produced. This enzyme catalyzes recombination between *FRT* sites in mitotic cells. Note that the *FRT* sites must be closer to the centromere than the gene for which one wishes to obtain homozygous mutant clones. In this and subsequent figures in this chapter, certain features of modified P elements such as the *w*$^+$ gene (needed so investigators can identify transgene-containing flies) are not shown for simplicity.

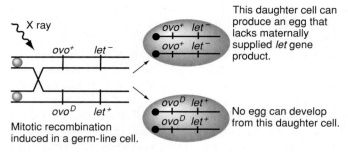

This daughter cell can produce an egg that lacks maternally supplied *let* gene product.

No egg can develop from this daughter cell.

Figure D.14 Using mosaics made by mitotic recombination to determine the importance of a maternally supplied gene product for embryonic development. A female of the genotype on the *left* cannot produce any eggs because of the dominant *ovo^D* mutation, unless mitotic recombination creates an *ovo$^+$/ovo$^+$* germ-line cell. Such a cell will also be homozygous for the lethal mutation (*let$^-$*) in question. If maternally supplied *let$^+$* gene product is needed for early embryogenesis, fertilized eggs cannot develop into adults.

Ectopic Expression

Although the effects of eliminating function (as in a null mutant) are of paramount importance in assessing the function of a gene, it is often helpful to determine the effects of **ectopic expression;** that is, expression outside the cells or tissue where the gene is normally expressed. One of the most remarkable examples of the use of technology to achieve ectopic expression was described in Chapter 19. As explained in that discussion, when investigators placed a *Drosophila* cDNA sequence encoding the Eyeless (Ey) protein under the control of a heat-shock promoter and introduced this artificial gene into the genome by P-element-mediated transformation, they could turn on expression of the Ey protein in all cells by subjecting the flies to elevated temperatures. Figure 19.10 on p. 658 illustrates the dramatic results: the development of ectopic eye tissue virtually anywhere on the fly's body—such as on the antennae, legs, or wings! The experiment showed that the Ey protein is a master regulator able to activate a program of eye development. This conclusion fits with the fact that the Ey protein appears to be a transcription factor that can turn on a number of target genes.

Although placing a gene under heat-shock control was the first method used to generate ectopic expression, a much improved technique emerged in the 1990s. In the newer method, investigators replace the *E. coli lacZ* reporter normally used in enhancer trap vectors (see Fig. D.10 on p. 820) with a yeast gene that encodes the GAL4 transcriptional activator protein (Fig. D.15). Although GAL4 has no effect on its own in *Drosophila*, it can function in flies to activate transcription from genes that carry a yeast DNA sequence called the *GAL4 Upstream Activator Sequence (UAS$_G$)*. To drive ectopic expression of a gene under study, researchers first clone the gene's cDNA downstream of a *UAS$_G$* and a promoter, and they introduce this synthetic gene into the germ line by P-element-mediated transformation. *UAS$_G$*-cDNA transgenes are not transcribed in the absence of the GAL4 protein and so have no effect on the flies that carry them. However, when UAS-cDNA flies are crossed to a *GAL4* enhancer trap line, the cDNA is expressed in the F$_1$ progeny, specifically in those cells making GAL4 protein. The pattern of *GAL4* expression is determined by enhancers ("drivers") near the site where the *GAL4*-containing P element inserted into the genome. Because a large number of *GAL4* enhancer trap lines expressed in many different patterns are now available, the *GAL4* method for ectopic expression is a highly versatile tool for studies of *Drosophila* development.

The *Drosophila* Genome Project

Drosophila, like other model organisms, is the subject of a very active genome project. The sequencing of the euchromatic portion of the genome is now essentially complete.

What the *Drosophila* Genome Sequence Has Taught Us

About 13,600 known or predicted genes are present. This gene number compares with 6000–6500 in yeast, 18,425 in the nematode, and 40,000–60,000 in humans. The average gene density in *Drosophila* is about one gene for every 9 kb. Surprisingly, the *Drosophila* gene set appears more similar to that of mammals than to that of nematodes: About half of fly proteins show homology with mammalian proteins, whereas only about a third of fly proteins appear to be homologous to nematode proteins.

Figure D.15 Driving ectopic expression in particular cell types. This technique requires flies containing two constructs. The first is an enhancer trap that expresses GAL4 protein in tissues dictated by nearby enhancers (drivers). The second construct contains a cDNA of interest under control of *UAS$_G$*. The protein of interest will be made ectopically, specifically in those tissues producing GAL4.

(a) Starting construct (anywhere in genome)

Figure D.16   Targeted gene knockouts by homologous recombination. Heat shock induces synthesis of FLP recombinase and I-SceI endonuclease from appropriate transgenes (review Fig. D.15). FLP excises an engineered gene of interest as a circle; I-SceI cuts the circle at the I-site to produce a recombinogenic linear DNA fragment. Recombination replaces the endogenous wild-type gene with the defective engineered gene. This figure shows only one of several recombination events that could lead to a gene knockout.

Thus only about a third of the genes in the fly genome are essential, a far smaller proportion than had previously been thought. This number is particularly surprising when you consider that most of the genes are present only once in the genome. If so many genes are present in only a single copy and yet are not essential for viability, what are they doing?

The availability of the *Drosophila* genome sequence is having a significant impact on the practice of *Drosophila* genetics. First, investigators have located many polymorphisms in the genome that have enormously expedited the mapping of point mutations to the affected gene. Second, using DNA chips, fly geneticists can examine global changes in gene transcription caused by mutations in key regulatory genes or by environmental changes, as discussed in Chapter 10. This approach promises to revolutionize our understanding of gene regulatory networks. Finally, investigators can "mine" the *Drosophila* genome sequence using the methods of bioinformatics. For the first time, they can identify all the members of gene families and study their individual and overlapping functions. In a different type of data mining, researchers are identifying important *cis*-regulatory regions by searching the genome sequence for small regions near the 5′ ends of genes that contain clustered binding sites for known transcription factors.

A New Direction for the *Drosophila* Genome Project: Obtaining Mutants in All Fly Genes

The goals of the *Drosophila* Genome Project go well beyond genome sequencing and seek to exploit the many genetic advantages of the fly to obtain functional as well as structural information about genes. One approach has been a large-scale effort to isolate P-element-insertion alleles of as many genes as possible. Studies of the phenotypes associated with these mutations have provided important information about the biological functions of the disrupted genes. The P elements used carry bacterial plasmid sequences that allow the easy cloning of flanking sequences by plasmid rescue. Many of the P elements employed in the genome project are enhancer traps that enable the rapid determination of the expression pattern of the disrupted gene (see Figs. D.9 and D.10 on p. 820). So far, investigators have isolated P-element disruptions for about 25% of the estimated 4000 or so essential *Drosophila* genes. Mobilization of existing insertions with P-element transposase should make possible the rapid recovery of insertions in many of the remaining genes.

For many years the most significant drawback to *Drosophila* as an experimental system was the lack of any method for generating targeted gene knockouts; that is, techniques to block the function of particular genes of interest. In the past few years, two such methods have been developed. In the first, knockouts are produced by homologous recombination. The procedure, which is quite complex, is as follows. First, a fragment of DNA from the gene of interest is engineered in several ways (Fig. D.16a). (1) Researchers insert into the fragment an 18 bp sequence called an *I-site* recognized by the site-specific endonuclease enzyme I-SceI from yeast. (2) The gene of interest is altered by *in vitro* mutagenesis to destroy its function; in the example shown on Fig. D.16, a nonsense mutation is introduced near the beginning

Moreover, 61% of human disease genes have apparent homologs in the fly. This conservation of fly and human genes suggests that flies have a very bright future as a model for understanding the functioning of human genes. An additional advantage of *Drosophila* for genetic experiments is that most fly genes are present in just one copy in the genome. It is thus more likely that a mutation in *Drosophila* will cause an aberrant phenotype than a mutation in mammals, whose genomes often contain several related genes that have overlapping or redundant functions.

The genome sequence has provided other, major surprises. One is that about 30% of the genes in *Drosophila* appear unrelated to genes in other organisms or to other genes within *Drosophila*. The functions of most of these fast-evolving "orphan genes" remain to be determined. Yet another surprise is the total gene number. Earlier genetic studies had indicated that roughly 4000 genes are essential for viability in *Drosophila*.

of the reading frame. (3) The fragment is further modified by flanking it with *FRT* sites (which are recognized by the FLP recombinase of yeast as described previously). A construct containing this triply modified gene is then inserted at random into the *Drosophila* genome by P-element-mediated transformation.

The next parts of the procedure are designed to generate a linear fragment containing the altered gene within the germ line of flies containing the construct (Fig. D.16b–d). The rationale is that linear fragments of DNA are known to stimulate recombination. Linearization requires that flies carrying the insert are crossed to a strain carrying two P-element transgenes: one with the coding sequence for FLP recombinase, the other with the coding sequence for I-SceI. Both of these P-element transgenes are under the control of heat-shock promoters. When progeny carrying all three transgenes are heat shocked, the FLP recombinase that is produced catalyzes recombination between the FRT sites flanking the introduced gene fragment. As a result, this fragment excises from the chromosome as a circle. I-SceI (whose synthesis is also stimulated by heat shock) then opens this circle by cutting at the introduced cut site. The linearized piece ultimately recombines with the wild-type endogenous gene from which the fragment is derived, causing a targeted knockout (Fig. D.16e–f).

The gene targeting method just described is quite laborious, so it is fortunate that a more convenient and flexible way to knock out gene function has been developed that depends upon double-stranded RNA interference (RNAi). As described in Chapter 19, double-stranded RNAs (dsRNAs) can reduce gene expression in diverse organisms by targeting homologous mRNAs for degradation. Injection of dsRNAs into embryos effectively disrupts homologous gene expression in the embryo, but injection of dsRNAs have at best only modest effects in larvae or adults. To knock out gene expression in later developmental stages, investigators construct transgenes that encode inverted repeat or "snap-back" RNAs (Fig. D.17). When these transgenes are transcribed in the animal, the snap-back RNAs form double-stranded regions that can almost completely suppress expression of the endogenous gene.

The development of gene targeting techniques for *Drosophila* is very exciting, as these methods should facilitate the systematic inactivation of all *Drosophila* genes. By combining these technologies with those for driving ectopic expression, we should have in the not-too-distant future a nearly complete picture of basic *Drosophila* gene functions. Even so, there will still be much to do, since experience to date suggests that many genes will have no obvious effect when inactivated or ectopically expressed.

The Genetic Analysis of Body Plan Development in *Drosophila*: A Comprehensive Example

The immensely powerful combination of genetic and molecular techniques available in *Drosophila* has revolutionized our understanding of development. And because most of the develop-

Figure D.17 RNA interference (RNAi) in *Drosophila*. Investigators clone a cDNA of interest in forward and reverse orientations, separated by a 10–100 bp linker of random DNA. When these sequences are transcribed, the resulting RNA can "snap back" by complementary base pairing to form double-stranded RNA (dsRNA) that can inhibit gene expression. Here, transcription is controlled by UAS_G so that dsRNA is made only in GAL4-producing tissues (see Fig. D.15 on p. 824).

mental regulators identified in *Drosophila* are highly conserved in other animals, the *Drosophila* model is having a profound influence on other fields, including medicine. In fact many developmentally important genes in humans were first identified as homologs of genes found in *Drosophila*. A vivid illustration of the impact of *Drosophila* is the work on genes controlling the fly's body plan.

Contributions from many laboratories have revealed how the body of *Drosophila* becomes specialized along both its anterior-posterior (AP) and dorsal-ventral (DV) axes. Together, these studies constitute one of the most successful genetic dissections ever performed. Here we consider only development of the AP body axis, that is, how the body differentiates along the line running from the head to the tail of the animal.

Very early in development, the action of a large group of genes known as the **segmentation genes** subdivides the body into an array of essentially identical body segments. Expression of a second set of genes called **homeotic genes** then assigns a unique identity to each body segment. In studies of the genes that specify and regulate development of the AP body axis, researchers have tried to answer two basic questions. First, how does the developing animal establish the proper number of body segments? And second, how does each body segment "know" what kinds of structures it should form and what role it should play in the animal's biology?

Early Development of the Basic Body Plan

To understand how the segmentation and homeotic genes function, it is helpful to consider some of the basic events that take place in the first few hours of *Drosophila* development (Fig. D.18a and b). The egg is fertilized in the uterus as it is being laid, and the meiotic divisions of the oocyte nucleus, which had previously arrested in the metaphase of meiosis I, resume at this time. After fusion of the haploid male and female pronuclei, the diploid zygotic nucleus of the embryo undergoes 13 rounds of nuclear division at an extraordinarily rapid rate, with the average time of mitotic cycles 2 through 9 being only 8.5 minutes. Nuclear division in early *Drosophila* embryos, unlike most mitoses, is not accompanied by cell division, so that the early embryo becomes a multinucleate syncytium. During the first eight division cycles, the multiple nuclei are centrally located in the egg; during the ninth division, most of the nuclei migrate out to the cortex—just under the surface of the embryo—to produce the **syncytial blastoderm.** During the tenth division, nuclei at the posterior pole of the egg are enclosed in membranes that invaginate from the egg cell membrane to form the first embryonic cells; these "pole cells" are the primordial germ cells. At the end of the thirteenth division cycle, about 6000 nuclei are present at the egg cortex.

During the interphase of the fourteenth cycle, membranes in the egg's cortex grow inward between these nuclei, creating an epithelial layer called the **cellular blastoderm** that is one cell deep (Figs. D.18 and D.19a). The embryo completes formation of the cellular blastoderm about 3 hours after fertilization. At the cellular blastoderm stage, no regional differences in cell shape or size are apparent (with the exception of the pole cells at the posterior end). Experiments in which blastoderm cells have been transplanted from one location to another, however, show that despite this morphological uniformity, the segmental identity of the cells has already been determined. Consistent with this finding, molecular studies reveal that most segmentation and homeotic genes function during or even before the cellular blastoderm stage.

Immediately after cellularization, **gastrulation** and establishment of the embryonic germ layers begin. The *mesoderm* forms by invagination of a band of midventral cells that extends most of the length of the embryo. This infolding (the ventral furrow; Fig. D.19b) produces an internal tube whose cells soon divide and migrate to produce a mesodermal layer. The *endoderm* forms by distinct invaginations anterior and posterior to the ventral furrow; one of these invaginations is the cephalic furrow seen in Fig. D.19b. The cells of the endodermal infoldings migrate over the yolk to produce the gut. Finally, the nervous system arises from neuroblasts that segregate from bilateral zones of the ventral *ectoderm.* The first visible signs of segmentation are periodic bulges in the mesoderm, which appear about 40 minutes after gastrulation begins. Within a few hours of gastrulation, the embryo is divided into clear-cut body segments that will become the three head segments, three thoracic segments, and eight major abdominal segments of the larva that hatches from the eggshell (Fig. D.19c). Even though the animal eventually undergoes metamorphosis to become an adult fly, the same basic body plan is conserved in the adult stage (Fig. D.19d).

The genes responsible for the formation of segments fall into four classes and function in a regulatory hierarchy that progressively subdivides the embryo into successively smaller units. In the order of their expression, these classes are (1) maternal genes, (2) gap genes, (3) pair-rule genes, and (4) segment-polarity genes.

(a) The first three hours after fertilization

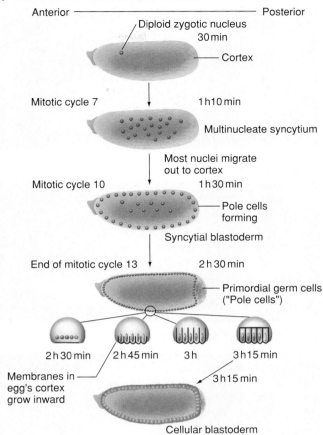

(b) Early embryonic stages in cross section

Figure D.18   Early *Drosophila* development: From fertilization to formation of the cellular blastoderm. (a) The original zygotic nucleus undergoes 13 very rapid mitotic divisions in a single syncytium. A few nuclei at the posterior end of the embryo become the germ-line *pole cells.* At the *syncytial blastoderm* stage, the egg surface is covered by a monolayer of nuclei (except for the pole cells at the posterior end). At the end of the thirteenth division cycle, cell membranes enclose the nuclei at the cortex into separate cells to produce a *cellular blastoderm.* **(b)** Photomicrographs of early embryonic stages stained with a fluorescent dye for DNA. (For another view of the syncytial blastoderm, see Fig. 4.10 on p. 86.)

(a) Cellular blastoderm

(b) Gastrulation

(c) Segmentation

(d) Segment identity is preserved throughout development.

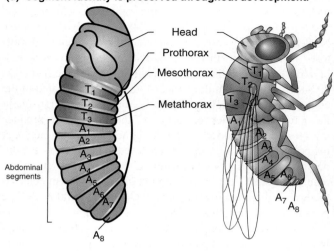

Figure D.19 *Drosophila* **development after formation of the cellular blastoderm. (a)** Scanning electron micrograph of a cellular blastoderm; individual cells are visible at the periphery of the embryo, and the pole cells at the posterior end can be distinguished (*arrow*). **(b)** A ventral view of some of the furrows that form during gastrulation, roughly 4 hours after fertilization: vf, ventral furrow; cf, cephalic furrow. **(c)** By 10 hours after fertilization, it is clear that the embryo is subdivided into segments. Ma, Mx, Lb are the three head segments (mandibular, maxillary, and labial, respectively). CL, PC, O, and D refer to nonsegmented regions of the head. The three thoracic segments (labeled T_1, T_2, and T_3) are the prothorax, the mesothorax, and the metathorax, respectively, while the abdominal segments are labeled A_1–A_8. **(d)** The identity of embryonic segments (*left*) is preserved through the larval stages and is also retained through metamorphosis into the adult (*right*). For simplicity, head segments are not distinguished here.

Specification of Segment Number: Maternal Genes Interact to Produce Morphogen Gradients

There is very little transcription of genes in the embryonic nuclei between fertilization and the end of the 13 rapid syncytial divisions that immediately follow fertilization. Because of this near (but by no means total) lack of transcription, developmental biologists suspected that formation of the basic body plan initially requires **maternally supplied components** deposited by the mother into the egg during oogenesis. How could they identify the genes encoding these maternally supplied components? Christiane Nüsslein-Volhard and Eric Wieschaus realized that the embryonic phenotype determined by such genes does not depend on the embryo's own genotype; rather it is determined by the genotype of the mother. By asking whether mothers homozygous for candidate mutations would produce defective embryos, they devised genetic screens to identify

recessive mutations in maternal genes that influence embryonic development; these recessive mutations are often called **maternal-effect mutations.**

To carry out their screens, Nüsslein-Volhard and Wieschaus established individual balanced stocks for thousands of mutagen-treated chromosomes, and then they examined the phenotypes of embryos obtained from homozygous mutant mothers. They focused their attention on stocks in which homozygous mutant females were sterile, because they anticipated that the absence of maternally supplied components needed for the earliest stages of development would result in embryos so defective that they could never grow into adults. Through these large-scale screens, Nüsslein-Volhard and Wieschaus identified a large number of maternal genes that are required for the normal patterning of the body. For their work, they shared the Nobel Prize for physiology or medicine with Edward B. Lewis—whose work we describe later. We focus here on two groups of the genes they found. One group is required for normal patterning of the embryo's anterior; the other group is required mainly for normal posterior patterning. The genes in these two groups are the first genes activated in the process that determines segment number.

The finding that separate groups of maternal genes control anterior and posterior patterning in the embryo is consistent with the conclusions of classical embryological experiments. Studies in which polar cytoplasm from the embryo's ends was transplanted, or in which preblastoderm embryos were separated into two halves by constriction of the embryo with a fine thread, suggested that the insect body axis is patterned during cleavage by the interaction of two signaling centers located at the anterior and posterior poles of the egg. In a specific model, Klaus Sander proposed that each pole of the egg produces a different substance, and that these substances form opposing gradients by diffusion. He suggested that the concentrations of these substances then determine the types of structures produced at each position along the body axis. Molecular characterization of the maternal genes of the anterior and posterior groups indicates that the Sander model for body axis patterning

is correct. Substances that define different cell fates in a concentration-dependent manner are known as **morphogens.**

In *Drosophila,* the *bicoid* (*bcd*) Gene Encodes the Anterior Morphogen

Embryos from mothers homozygous for null alleles of *bcd* lack all head and thoracic structures. The protein product of *bcd* is a DNA-binding transcription factor whose transcript is localized near the anterior pole of the egg cytoplasm (Fig. D.20a). Translation of the *bcd* transcripts takes place after fertilization. The newly made Bcd protein diffuses from its source at the pole to produce a high-to-low, anterior-to-posterior concentration gradient that extends over the anterior two-thirds of the embryo by the ninth division cycle (Fig. D.20b). This gradient determines most aspects of head and thorax development.

One of the first lines of evidence that the Bcd protein functions as a morphogen came from experiments in which the maternal dosage of the *bcd* gene varied (Fig. D.20c). Mothers that carried only one dose of the *bcd* gene, instead of the normal diploid dose, incorporated about half the normal amount of *bcd*

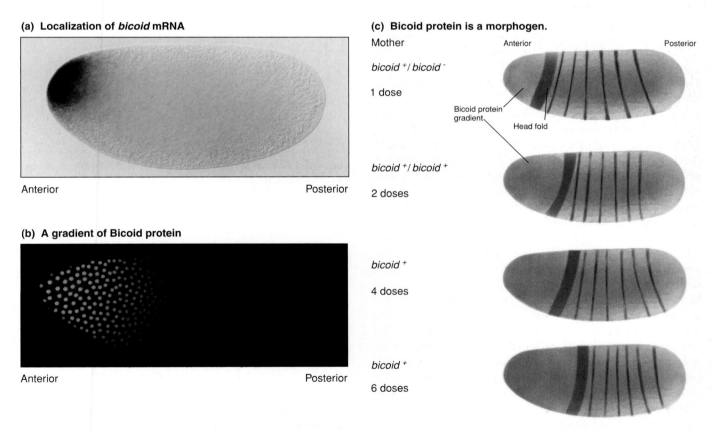

Figure D.20   The Bicoid protein is the anterior morphogen. (a) The *bicoid* (*bcd*) mRNA (visualized by *in situ* hybridization in *purple*) concentrates at the anterior tip of the embryo. **(b)** The Bicoid (Bcd) protein (seen by *green* antibody staining) is distributed in a gradient: high at the anterior end and trailing off toward the posterior end. It can be seen that the Bcd protein (a transcription factor) accumulates in the nuclei of this syncytial blastoderm embryo. **(c)** The greater the maternal dosage of *bcd+*, the higher the concentration of Bcd protein in the embryo, and the more of the embryo that is devoted to anterior structures. Head structures will develop anterior of the head fold invagination; thoracic and abdominal structures posterior to it.

RNA into their eggs. As a result, translation yielded less Bcd protein, and the Bcd gradient was shallower and shifted to the anterior. In these Bcd-deficient embryos, the thoracic segments developed from more anterior regions than normal, and less of the body was devoted to the head. The opposite effect occurred in mothers carrying extra doses of the *bcd* gene. These and other observations suggested that the level of Bcd protein is a key to the determination of head and thoracic fates in the embryo. Three other genes work with *bcd* in the anterior group of maternal genes; the function of the protein products of these three genes is to localize *bcd* transcripts to the egg's anterior pole.

The Bcd protein itself works in two ways: as a transcription factor that helps control the transcription of genes farther down the regulatory pathway, and as a translational repressor. The target of its repressor activity is the transcript of the *caudal (cad)* gene, which also encodes a DNA-binding transcription factor. The *cad* transcripts are uniformly distributed in the egg before fertilization, but because of translational repression by the Bcd protein, translation of these transcripts produces a gradient of Cad protein that is complementary to the Bcd gradient. That is, there is a high concentration of Cad protein at the posterior end of the embryo and lower concentrations toward the anterior (Fig. D.21). The Cad protein plays an important role in activating genes expressed later in the segmentation pathway to generate posterior structures.

The *nanos* (*nos*) Gene Encodes the Primary Posterior Morphogen

The *nos* RNA is localized to the posterior egg cytoplasm by proteins encoded by other posterior group maternal genes. Like *bcd* RNAs, *nos* transcripts are translated during the cleavage stages. After translation, diffusion produces a posterior-to-anterior Nos protein concentration gradient. The Nos protein, unlike the Bcd protein, is not a transcription factor; rather, the Nos protein functions only as a translational repressor. Its major target is the maternally supplied transcript of the *hunchback* (*hb*) gene, which is deposited in the egg during oogenesis and is uniformly distributed before fertilization. For development to occur properly, the Hb protein (which is another transcription factor) must be present in a gradient with high concentrations at the embryo's anterior and low concentrations at the posterior. The Nos protein, which represses the translation of *hb* maternal mRNA and is present in a posterior-to-anterior concentration gradient, helps construct the anterior-to-posterior Hb gradient by lowering the concentration of the Hb protein toward the embryo's posterior pole (Fig. D.21). The embryo also has a second mechanism for establishing the Hb protein gradient that functions somewhat later: It only transcribes the *hb* gene from zygotic nuclei in the anterior region (see following).

Figure D.21 summarizes how the coordinated activity of the maternal genes establishes polarities in the embryo that eventually result in proper embryonic segmentation. In addition to distributing the Bcd and Cad transcription factors in gradients within the embryo, the maternal genes ensure that the maternally supplied *hb* RNA is not translated in the posterior part of the embryo.

Figure D.21   Distribution of the mRNA and protein products of maternal-effect genes within the early embryo. *Top:* In the oocyte prior to fertilization, *bicoid* (*bcd*) mRNA is concentrated near the anterior tip and *nanos* (*nos*) mRNA at the posterior tip, while maternally supplied *hunchback* (*hb*) and *caudal* (*cad*) mRNAs are uniformly distributed. *Bottom:* In early cleavage stage embryos, the Bicoid (Bcd) and Hunchback (Hb) proteins are found in concentration gradients high at the anterior and lower toward the posterior (A to P), while the Nanos (Nos) and Caudal (Cad) proteins are distributed in opposite P-to-A gradients.

Specification of Segment Number Through the Activation of Zygotic Genes in Successively More Sharply Defined Regions

The maternally determined Bcd, Hb, and Cad protein gradients control the spatial expression of zygotic segmentation genes. Unlike the products of maternal genes, whose mRNAs are placed in the egg during oogenesis, the products of zygotic genes are transcribed and translated from DNA in the nuclei of embryonic cells descended from the original zygotic nucleus. The expression of zygotic segmentation genes begins in the syncytial blastoderm stage, a few division cycles before cellularization (roughly cycle 10).

Most of the zygotic segmentation genes were identified in a second mutant screen also carried out in the late 1970s by Christiane Nüsslein-Volhard and Eric Wieschaus. In this screen, the two *Drosophila* geneticists placed individual ethyl methane sulfonate (EMS)-mutagenized chromosomes into balanced stocks and then examined homozygous mutant embryos from these stocks for defects in the segmentation pattern of the embryo. Such homozygous mutant embryos were so aberrant that they were unable to grow into adults; thus the mutations causing these defects would be classified as recessive lethals. After screening several thousand such stocks for each of the *Drosophila* chromosomes, they identified three classes of zygotic segmentation genes: gap genes (9 different genes); pair-rule genes (8 genes); and segment-polarity genes (about 17 genes). These three classes of zygotic genes fit into a hierarchy of gene expression.

Gap Genes Are Expressed in Broad Zones Along the Anterior-Posterior Axis

The gap genes are the first zygotic segmentation genes to be transcribed. Embryos homozygous for mutations in the gap genes show a gap in the segmentation pattern caused by an absence of particular segments that correspond to the position at which each gene is transcribed (Fig. D.22). How do the maternal transcription factor gradients ensure that the various gap genes are expressed in their broad zones at the proper position in the embryo? Part of the answer is that the binding sites in the promoter regions of the gap genes have different affinities for the maternal transcription factors. For example, some gap genes are activated by the Bcd protein (the anterior morphogen). Gap genes such as *hb* with low-affinity Bcd protein-binding sites will be activated only in the most anterior regions, where the concentration of Bcd is at its highest; by contrast, genes with high-affinity sites will be activated farther toward the posterior pole.

Another part of the answer is that the gap genes themselves encode transcription factors that can influence the expression of other gap genes. The *Krüppel* (*Kr*) gap gene, for example, appears to be turned off by high amounts of Hb protein at the anterior end of its band of expression; activated within its expression band by Bcd protein in conjunction with lower levels of Hb protein; and turned off at the posterior end of its expression zone by the products of the *knirps* (*kni*) gap gene (Fig. D.22c). (Note that the *hb* gene is usually classified as a gap gene, despite the maternal supply of some *hb* RNA, because the protein translated from the transcripts of zygotic nuclei actually plays the more important role.)

Pair-Rule Genes Subdivide the Body into Units That Are Two Segments in Length

After the gap genes have divided the body axis into rough, generalized regions, activation of the pair-rule genes generates more sharply defined sections. These genes encode transcription factors that are expressed in seven stripes in preblastoderm and blastoderm embryos (Fig. D.23a). The stripes have a two-segment periodicity; that is, there is one stripe for every two segments. Mutations in pair-rule genes cause the deletion of

similar pattern elements from every alternate segment. For example, larvae mutant for *fushi tarazu* ("few segments" in Japanese) lack parts of abdominal segments A1, A3, A5, and A7 (see Fig. 19.2 on p. 650). Mutations in *even-skipped* cause loss of even-numbered abdominal segments.

There are two classes of pair-rule genes: primary and secondary. The striped expression pattern of the three primary pair-rule genes depends on the transcription factors encoded by the maternal genes and the zygotic gap genes. Specific elements within the upstream regulatory region of each pair-rule gene drive the expression of that pair-rule gene within a particular stripe. For example, as Fig. D.23b and c shows, the DNA region responsible for driving the expression of *even-skipped* (*eve*) in the second stripe contains multiple binding sites for the Bcd protein and the proteins encoded by the gap genes *Krüppel, giant* (*gt*), and *hb*. The transcription of *eve* in this stripe of the embryo is activated by Bcd and Hb, while it is repressed by Gt and Kr; only in the stripe 2 region are Gt and Kr levels low enough and Bcd and Hb levels high enough to allow activation of the element driving *eve* expression. In contrast with the primary pair-rule genes, the five pair-rule genes of the secondary class are controlled by interactions with transcription factors encoded by other pair-rule genes.

Segment Polarity Genes Occupy the Lowest Level of the Segmentation Hierarchy

Many segment polarity genes are expressed in stripes that are repeated with a single segment periodicity, that is, there is one stripe per segment (Fig. D.24a on p. 834). Mutations in segment polarity genes cause deletion of part of each segment and its replacement by a mirror image of a different part of the next segment. For example, in *engrailed* mutants, the posterior compartment of each segment is replaced by a reversed duplication of the anterior compartment of the subsequent segment. The segment polarity genes thus function to determine certain patterns that are repeated in each segment.

The regulatory system that directs the expression of segment polarity genes in a single stripe per segment is quite complex. In general, the transcription factors encoded by pair-rule genes initiate the pattern by directly regulating certain segment polarity genes. Interactions between various cell polarity genes then maintain this periodicity later in development. Significantly, activation of segment polarity genes occurs after cellularization of the embryo is complete. Hence, the diffusion of transcription factors within the syncytium, of central importance in segmentation up to this point, ceases to play a role. Instead, intrasegmental patterning is determined mostly by the diffusion of proteins secreted between cells.

Two of the segment polarity genes, *hedgehog* (*hh*) and *wingless* (*wg*), encode secreted proteins. These proteins, together with the transcription factor encoded by the *engrailed* (*en*) segment polarity gene, are responsible for many aspects of segmental patterning (Fig. D.24b). A key component of this control is that a one-cell-wide stripe of cells secreting the Wg protein is adjacent to a stripe of cells expressing the En protein and secreting the Hh protein. The interface of these two types of

(a) Zones of gap gene expression

(b) Phenotypes caused by gap gene mutations

(c) Gap genes: a summary

Figure D.22 Gap genes. (a) Zones of expression of four gap genes (*hunchback* [*hb*], *Krüppel* [*Kr*], *knirps* [*kni*], and *giant* [*gt*]) in late syncytial blastoderm embryos, as visualized with fluorescently labeled antibodies. **(b)** Defects in segmentation caused by mutations in selected gap genes, as seen in late embryos. Only the remaining thoracic and abdominal segments are labeled; the head segments at the anterior end are highly compressed and not labeled. **(c)** Mutation of a particular gap gene results in the loss of segments corresponding to the zone of expression of that gap gene in the embryo.

(a) Distribution of pair-rule gene products

Anterior   Posterior

■ Even-skipped (Eve)
■ Fushi tarazu (Ftz)

(b) Proteins regulating *eve* transcription

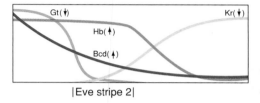

Gt(↓)   Kr(↓)

Hb(↑)

Bcd(↑)

|Eve stripe 2|

(c) Upstream regulatory region of *eve*

-1500 base pairs   -800

Kr   Gt   Gt   Kr   Gt   Kr   Kr   Kr

Bcd   Bcd   Bcd   Bcd   Hb   Bcd   Hb   Hb

Figure D.23   Pair-rule genes. (a) Zones of expression of the proteins encoded by the pair-rule genes *fushi tarazu* (*ftz*) and *even-skipped* (*eve*) at the beginning of the cellular blastoderm stage. Each gene is expressed in seven stripes. Eve stripe 2 is the second green stripe from the *left*. **(b)** The formation of Eve stripe 2 requires activation of *eve* transcription by the Bcd and Hb proteins and repression at its left and right ends by Gt and Kr proteins respectively. **(c)** The 700 bp upstream regulatory region of the *eve* gene that directs the Eve second stripe contains multiple binding sites for the four proteins shown in part (b).

cells is a self-reinforcing, reciprocal loop. The Wg protein secreted by the more anterior of the two adjacent stripes of cells is required for the continued expression of *hh* and *en* in the adjacent posterior stripe. The Hh protein secreted by the more posterior stripe of cells maintains expression of *wg* in the anterior stripe. Gradients of Wg and Hh proteins made from these adjacent stripes of cells control many aspects of patterning in the remainder of the segment. This is because the products of both *wg* and *hh* appear to function as morphogens; that is, responding cells appear to adopt different fates depending on the concentration of Wg or Hh protein to which they are exposed.

Other segment polarity genes encode proteins involved in signal transduction pathways initiated by the binding of Wg and Hh proteins to receptors on cell surfaces. Recall from Chapters 18 and 19 that signal transduction pathways enable a signal received from a receptor on the cell's surface to be converted through a series of intermediate steps to a final intracellular regulatory response, usually the activation or repression of particular target genes. The signal transduction pathways initiated by

the Wg and Hh proteins determine the ability of cells in portions of each segment to differentiate into the particular cell types characteristic of that segment portion.

Homologs of the segment polarity genes are key players in many important patterning events in vertebrates. For example, the chicken *sonic hedgehog* gene (related to the fly *hh*) is critical for the initiation of the left-right asymmetry in the early chicken embryo as well as for the processes that determine the number and polarity of digits produced by the limb buds. The mammalian homolog of this gene has the same conserved functions.

Summary of Segment Number Specification

For each class of segmentation gene, the pattern of expression of its members is controlled either by classes of genes higher in the hierarchy or by members of the same class, never by genes of a lower class (Fig. D.25). In this regulatory cascade, the maternal genes control the gap and pair-rule genes, the gap genes control themselves and the pair-rule genes, and the pair-rule genes control themselves and the segment polarity genes.

The expression of genes in successively lower parts of the hierarchy is restricted to more sharply defined embryonic regions. During the early embryonic cycles of nuclear divisions, the spatial restriction of gene expression is a response to maternally established transcription factor gradients. In the syncytial blastoderm embryo, refinement of the spatial restriction of gene expression depends on the hierarchical action of the transcription factors encoded by the zygotic genes. Once cellularization is complete, intercellular communication mediated by secreted proteins encoded by certain segment polarity genes becomes the major mode of pattern formation. Although the cellular blastoderm, when viewed from the outside, looks like a uniform layer of cells (as seen in Fig. D.19a on p. 828), the coordinated action of the segmentation genes has already divided the embryo into segment primordia. A few hours after gastrulation, these primordia become distinguishable morphologically as clear-cut segments (as in Fig. D.19c).

Each Segment Establishes Its Own Unique Identity Through the Activation of Homeotic Genes

After the segmentation genes have subdivided the body into a precise number of segments, the homeotic genes help assign a unique identity to each segment. They do this by functioning as master regulators that control the transcription of batteries of genes responsible for the development of segment-specific structures. The homeotic genes themselves are regulated by the gap, pair-rule, and segment polarity genes so that at the cellular blastoderm stage, or shortly thereafter, each homeotic gene becomes expressed within a specific subset of body segments. Most homeotic genes then remain active through the rest of development, functioning continuously to direct proper segmental specialization.

Mutations in homeotic genes, referred to as **homeotic mutations,** cause particular segments, or parts of them, to develop as if they were located elsewhere in the body. Because some of the mutant homeotic phenotypes are quite spectacular,

(a) **Distribution of Engrailed protein**

(b) **Segment polarity genes establish compartment borders.**

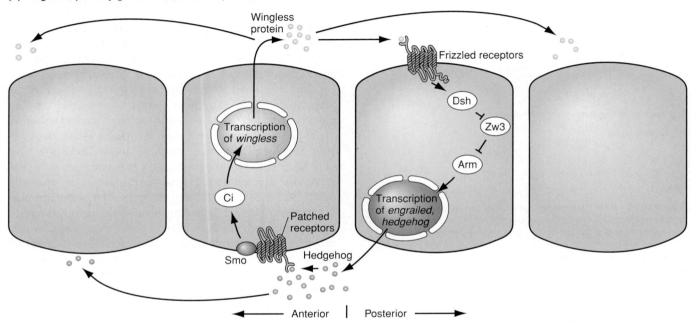

Figure D.24 **Segment polarity genes.** **(a)** Wild-type embryos express the segment polarity gene *engrailed* in 14 stripes. **(b)** The border between a segment's posterior and anterior compartments is governed by the *engrailed* (*en*), *wingless* (*wg*), and *hedgehog* (*hh*) segment polarity genes. Regulation by pair-rule genes ensures that cells in posterior compartments express *en*. The En protein activates the transcription of the *hh* gene, which encodes a secreted protein ligand. Binding of this Hh protein to the Patched receptor in the adjacent anterior cell initiates a signal transduction pathway (through the Smo and Ci proteins) leading to the transcription of the *wg* gene. Wg is also a secreted protein that binds to a different receptor in the posterior cell, which is encoded by *frizzled*. Binding of the Wg protein to this receptor initiates a different signal transduction pathway (including the Dsh, Zw3, and Arm proteins) that stimulates the transcription of *en* and of *hh*. The result is a reciprocal loop stabilizing the alternate fates of adjacent cells at the border.

researchers noticed them very early in *Drosophila* research. In 1915, for example, Calvin Bridges found a mutant he called *bithorax* (*bx*). In homozygotes for this mutation, the anterior portion of the third thoracic segment (T3) develops like the anterior second thoracic segment (T2); in other words, this mutation transforms part of T3 into the corresponding part of T2, as illustrated in Fig. D.26a. This mutant phenotype is very dramatic, as T3 normally produces only small club-shaped balancer organs called *halteres,* whereas T2 produces the wings. Another homeotic mutation is *postbithorax* (*pbx*), which affects only posterior T3, causing its transformation into posterior T2.

(Note that in this context, *Drosophila* geneticists use the term "transformation" to mean a change of body form.) In the *bx pbx* double mutant, all of T3 develops as T2 to produce the now famous four-winged fly (Fig. D.26b).

In the last half of the twentieth century, researchers isolated many other homeotic mutations, most of which map within one of two gene clusters. Mutations affecting segments in the abdomen and posterior thorax lie within a cluster known as the **bithorax complex (BX-C);** mutations affecting segments in the head and anterior thorax lie within the **Antennapedia complex (ANT-C)** (Fig. D.27).

(a) The segmentation hierarchy

(b) Mutations in segmentation genes cause segment loss.

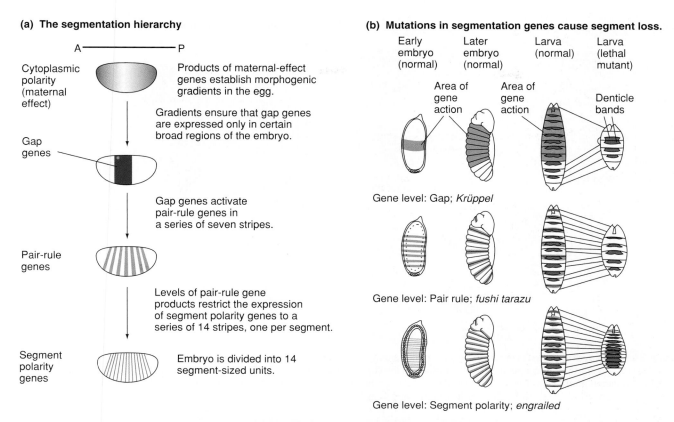

Figure D.25   A summary of the genetic hierarchy leading to segmentation in *Drosophila*. (a) Genes in successively lower parts of the hierarchy are expressed in narrower bands within the embryos. **(b)** Mutations in segmentation genes cause the loss of segments that correspond to regions where the gene is expressed (*shown in yellow*). The denticle bands (*dark brown*) are features that help researchers identify the segments remaining in the mutant animals.

The Bithorax Complex

Edward B. Lewis shared the 1995 Nobel Prize for physiology or medicine with Christiane Nüsslein-Volhard and Eric Wieschaus for his extensive genetic studies of the BX-C. In his work, Lewis isolated BX-C mutations that, like *bx* and *pbx*, affected the posterior thorax; he also found novel BX-C mutations that caused anteriorly directed transformations of each of the 10 abdominal segments. Lewis named mutations affecting abdominal segments *infra-abdominal* (*iab*) mutations, and he numbered these according to the primary segment they affect. Thus, *iab-2* mutations cause transformations of A2 toward A1, *iab-3* mutations cause transformations of A3 toward A2, and so forth.

Researchers initiated molecular studies of the BX-C in the early 1980s, and in 15 years they not only extensively characterized all of the genes and mutations in the BX-C at the molecular level, but they also completed the sequencing of the entire 315 kb region. Figure D.28 summarizes the structure of the complex. A remarkable feature of the BX-C is that mutations map in the same order on the chromosome as the anterior-posterior order of the segments each mutation affects. Thus, *bx* mutations, which affect anterior T3, lie near the left end of the complex, while *pbx* mutations, which affect posterior T3, lie

immediately to their right. In turn, *iab-2*, which affects A2, is to the right of *pbx* but to the left of the A3-determining *iab-3*.

Because the *bx*, *pbx*, and *iab* elements are independently mutable, Lewis thought that each was a separate gene. However, the molecular characterization of the region revealed that the BX-C actually contains only three protein-coding genes: *Ultrabithorax* (*Ubx*), which controls the identity of T3; *abdominal-A* (*abd-A*), which controls the identities of A1–A5; and *Abdominal-B* (*Abd-B*), which controls the identities of A5–A8 (Fig. D.28). The expression patterns of these genes are consistent with their roles. *Ubx* is expressed in segments T3–A8 (but most strongly in T3); *abd-A* is expressed in A1–A8 (most strongly in A1–A4); and *Abd-B* in A5–A8. The *bx*, *pbx*, and *iab* mutations studied by Lewis affect large *cis*-regulatory regions that control the intricate spatial and temporal expression of these genes within specific segments. The average size of each of these regulatory regions is about 15 kb. Researchers have not yet completely worked out the mechanisms by which single BX-C genes control multiple segment identities. However, it is likely that segment identity is determined by the level or timing of BX-C gene expression, by the production of segment-specific BX-C protein isoforms through alternate

(a) Effects of *bx* or *pbx* mutations

(b) A fly with both *bx* and *pbx* mutations

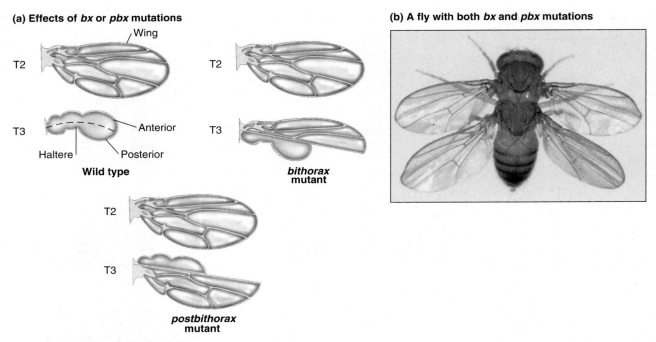

Figure D.26   Homeotic transformations. (a) In animals homozygous for the mutation *bithorax* (*bx*), the anterior compartment of T3 (the third thoracic segment that makes the haltere) is transformed into the anterior compartment of T2 (the second thoracic segment that makes the wing). The mutation *postbithorax* (*pbx*) transforms the posterior compartment of T3 into the posterior compartment of T2. **(b)** In a *bx pbx* double mutant, T3 is changed entirely into T2. The result is a four-winged fly.

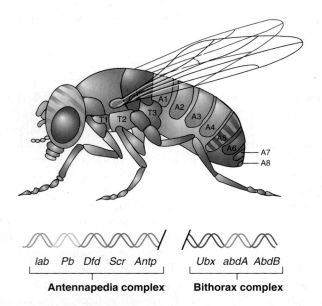

Figure D.27   Homeotic selector genes. Two clusters of genes on chromosome 3—the Antennapedia complex and the bithorax complex—are responsible for determining most aspects of segment identity. Interestingly, the order of genes in these complexes is the same as the order of the segments each gene controls.

splicing, or by the differential expression of cofactors that influence BX-C protein function.

The Antennapedia Complex and the Homeobox

Genetic studies in the early 1980s showed that a second homeotic gene cluster, the Antennapedia complex (ANT-C), specifies the identities of segments in the head and anterior thorax of *Drosophila*. The five homeotic genes of the ANT-C are *labial* (*lab*), which is expressed in the intercalary region; *proboscipedia* (*Pb*), expressed in the maxillary and labial segments; *Deformed* (*Dfd*), expressed in the mandibular and maxillary segments; *Sex combs reduced* (*Scr*), expressed in the labial and T1 segments; and *Antennapedia* (*Antp*), expressed mainly in T2, although it is also active at lower levels in all three thoracic and most abdominal segments. (Figure D.19c on p. 828 shows these head and thoracic segments, while Fig. D.27 illustrates the order of the homeotic genes in the ANT-C.) As with the BX-C, the order of genes in the ANT-C is the same (with the exception of *pb*) as the order of segments each controls. In addition to homeotic genes, the ANT-C contains several other important genes, including *zerknullt* (*zen*), which specifies dorsal embryonic structures; *fushi tarazu* (*ftz*), a segmentation gene of the pair-rule class; and *bicoid* (*bcd*), which, as we have seen, encodes the maternally supplied anterior determinant.

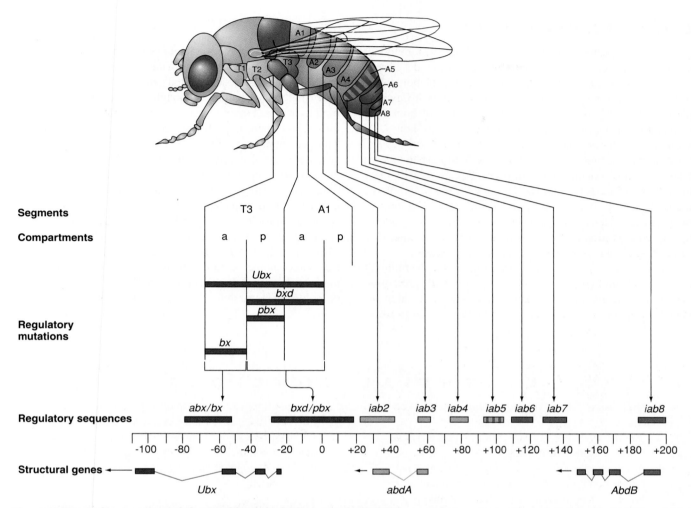

Figure D.28   A close-up view of the 300 kb of the bithorax complex. There are only three homeotic genes in this complex: *Ubx, abd-A,* and *Abd-B.* Many homeotic mutations such as *bx* and *pbx* affect regulatory regions that influence the transcription of one of these three genes in particular segments. For example, *bx* mutations (*at the left end of the complex*) prevent the transcription of *Ubx* in the anterior compartment of the third thoracic segment, while *iab-8* mutations (*far right*) affect the transcription of *Abd-B* in segment A8. Note that the order of these regulatory regions corresponds to the anterior-to-posterior order of segments in the animal. *iab-5* mutations disrupt a regulatory sequence needed for the expression of both *abd-A* and *Abd-B* in segment A5.

In the course of characterizing the ANT-C at the molecular level, researchers detected a region of homology between *Antp* and *ftz*. They subsequently found closely related sequences in all of the homeotic genes of the ANT-C and the BX-C, as well as in *zen* and *bcd* and many genes important for development located outside the homeotic gene complexes. In each gene, the region of homology, called the **homeobox,** is 180 bp in length and is located within the protein-coding sequence of the gene. The 60 amino acids encoded by the homeobox constitute a DNA-binding domain called the **homeodomain** that is structurally related to the helix-turn-helix motif of many bacterial regulatory proteins (review Fig. 19.16 on p. 661). We now know that the homeotic proteins of the ANT-C and BX-C are transcription factors in which the homeobox-encoded homeodomain is responsible for the sequence-specific binding of the proteins to their target genes. Surprisingly, however, DNA-binding studies have shown that most of these homeodomains have very similar binding specificities. The homeodomains of the Antp and Ubx proteins, for example, bind essentially the same DNA sequences. Since different homeotic proteins are thought to regulate specific target genes, this lack of DNA-binding specificity seems paradoxical. Much current research is directed toward understanding how the homeotic proteins target specific genes that dictate different segment identities.

CONNECTIONS

The discovery of the homeobox was one of the most important advances in developmental biology in the last two decades of the twentieth century. Some scientists compare the homeobox to the Rosetta stone (which led to the deciphering of Egyptian hieroglyphics) because it has enabled the isolation by homology of many other developmentally important genes in both *Drosophila* and other organisms. In the late 1980s and early 1990s, for example, the biological community was astonished to learn that the mouse contains clustered homeobox genes with clear homology to the ANT-C and BX-C genes of *Drosophila.* In the mouse, the homeobox genes are called *Hox* genes. The finding of *Hox* genes in mice is yet another demonstration of the evolutionary conservation of gene function. Reference E presents a detailed discussion of homeobox genes in the mouse.

Some geneticists have voiced the opinion that the future of *Drosophila* is limited because the four- to eightfold higher gene number in mice, humans, and other mammals means that mammals have numerous genes with no homologs in *Drosophila.*

However, more recent studies indicate that this difference in gene number results mainly from the fact that mice and humans often have several closely related genes that all show homology to a single gene in flies. Although many of the multiple homologs in mice have undergone some functional divergence, in most cases, there is probably strong redundancy of function. This is certainly true of the mouse *Hox* genes. Since redundant genes are very difficult to detect by mutation, it is likely that *Drosophila*'s low gene number (and hence low level of genetic redundancy) will prove to be a major experimental advantage, rather than a disadvantage. The approach of identifying genes by mutation in *Drosophila,* followed by isolation and study of homologs in mice and humans, is likely to remain a central research paradigm. Thus, the evolutionary conservation of gene function combined with the many experimental advantages of the fruit fly ensure that *Drosophila* will remain at the forefront of biological research for the foreseeable future.

ESSENTIAL CONCEPTS

1. Several characteristics of *Drosophila* are useful for genetic analysis.
   a. The life cycle is relatively short, and each fly can produce thousands of progeny.
   b. A moderate amount of crossing-over occurs in females and no crossing-over occurs in males.
   c. *Balancer chromosomes* help preserve linkage.
   d. *P-element transposons* are tools of molecular manipulation useful for transformation, gene tagging, and enhancer trapping.
   e. *Genetic mosaics* resulting from mitotic recombination help track the roles of individual genes in the development of specific body structures.
   f. *Ectopic expression* can help pin down the function of a gene by making it possible to examine the effects of overexpressing the gene product or expressing the gene product in the wrong tissues.

2. The *Drosophila* Genome Project has already sequenced the euchromatic portion of the genome. Analysis of this sequence suggests that there are approximately 13,600 genes in the fly genome. A major remaining goal of the *Drosophila* Genome Project is to obtain information about the function of each of these genes. Large collections of P-element-induced mutations, as well as new techniques for targeted gene knockouts by homologous recombination and RNA intereference, should make the inactivation of each of these 13,600 genes possible in the near future.

3. The development of the anterior-posterior (AP) axis in the fruit fly depends on the coordinated action of the *segmentation genes,* which divide the body into three head, three thoracic, and eight abdominal segments; and the *homeotic genes,* which assign a unique identity to each segment.

4. The four classes of segmentation genes are expressed in the order maternal, gap, pair-rule, and segment polarity. The *maternal genes* produce gradients of *morphogens.* The *gap genes,* the first zygotic segmentation genes, are expressed in broad zones along the anterior-posterior axis. The *pair-rule genes* subdivide those broad areas into units that span two of the ultimate body segments. The *segment polarity genes* subdivide the two-segment units into individual segments. In the hierarchy of segmentation gene expression, each gene class is controlled by classes of genes higher in the hierarchy or by members of the same class.

5. The homeotic genes are master regulators of other genes that control the development of segment-specific structures. The *homeobox* in each homeotic gene encodes a *homeodomain* that allows the gene products to bind to specific target genes and control their expression. The homeotic genes are, in turn, regulated by the earlier acting gap, pair-rule, and segment polarity genes. Homeotic mutations cause particular segments, or parts of them, to develop as if they were located

elsewhere in the body. Most homeotic mutations map to the bithorax and the Antennapedia complexes.

6. The finding of homeobox genes in other organisms such as the mouse has made it possible to identify develop-mentally important genes shared by all animals, demonstrating the evolutionary conservation of gene function.

S O L V E D   P R O B L E M S

I. *Drosophila* geneticists try to minimize the work needed to maintain the many lines of flies they investigate. Ideally, they establish genetic stocks in which the flies in a culture bottle mate with each other so they produce the same kinds of progeny generation after generation. Because females homozygous for a maternal-effect lethal (*mel*) mutation are sterile (their offspring die during embryogenesis), such mutations can be propagated only when females are heterozygous. A particular *mel* mutation on the second chromosome (an autosome) is kept as a heterozygote with a balancer chromosome carrying the *Cy* mutation, which is a recessive lethal but also produces a dominant Curly wing phenotype.

a. Suppose you start with equal numbers of male and female flies, both of genotype *mel/Cy*. What kinds of progeny would be produced in the next generation, and in what proportion? Would the culture bottles containing these progeny constitute a genetic stock?

b. What kind of flies would you select from the culture bottle to study the phenotypic effect of the maternal-effect mutation in detail?

Answer

This problem involves an understanding of balancer chromosomes and maternal-effect lethal mutations.

a. First determine all the ways gametes from these flies could combine and what the phenotype (including sex) would be. From a mating of *mel/Cy* females with *mel/Cy* males, the following kinds of zygotes would be formed: 1/8 *mel/mel* females (which survive but are sterile), 1/4 *mel/Cy* females (which survive and are fertile), 1/8 *Cy/Cy* females (which die before emerging as adults), 1/8 *mel/mel* males (which survive and are fertile because the mutation affects female but not male fertility), 1/4 *mel/Cy* males (which survive and are fertile), and 1/8 *Cy/Cy* males (which die before emerging as adults).

   Of the adults emerging from these zygotes, you would find: 1/6 mel/mel *females (sterile with wild-type wings), 1/3* mel/Cy *females (fertile with Curly wings), 1/6* mel/mel *males (fertile with wild-type wings), 1/3* mel/Cy *females (fertile with Curly wings). If all the flies in this culture bottle were allowed to mate at random, the same types of adult flies would emerge in the subsequent generation (you should work this out for yourself, remembering that*

only the heterozygous females are fertile). As a result, *this bottle constitutes a genetic stock: the same flies will be present in every generation.*

b. *You want* mel/mel *females, which can be recognized as the non-Curly (wild-type) females.* These females could be mated to any type of fertile male and the developing progeny examined to study the maternal-effect lethal phenotype.

II. A particular gene in *Drosophila* called *no eclosion* (*nec*) encodes an enzyme needed for the fly to escape from the pupal case. The wild-type allele (*nec+*) is dominant to a null mutant allele (*nec−*). Animals homozygous for *nec−* die because they cannot get out of the pupal case. The *nec−* mutation (on an otherwise normal chromosome) is maintained in a stock over the autosomal balancer chromosome TM6C, which carries the *nec+* allele and two dominant markers: *Stubble,* which affects bristle shape in adults, and *Tubby,* which causes larvae and pupae to be short and squat. *Stubble* also behaves as a recessive lethal mutation, causing death during embryonic development (that is, before the larva hatches). Several deletions (A–D) in the vicinity of the *nec* gene were generated by X rays. Each of these deletions is maintained in a stock over the same TM6C balancer chromosome. Crosses are made between the *nec−* / TM6C stock and each of the deletion / TM6C stocks. Crosses with deletions A and D yield only Stubble adult flies, while crosses with deletions B and C yield both wild-type (non-Stubble) as well as Stubble flies.

a. Which deletions remove the *nec+* gene?

b. If 1000 zygotes were generated from the crosses with deletions A and D that yielded only Stubble flies, how many of these zygotes should eventually produce adult flies?

c. Of the 1000 zygotes produced in part (b), how many would be expected to develop into animals that die in the pupal case? Would these pupal cases be short and squat (Tubby) or long and thin (wild type)?

d. From the preceding crosses with deletions B and C that yielded both wild-type (non-Stubble) as well as Stubble flies, what percentage of the adult flies that emerge should be non-Stubble?

e. *Stubble* and *Tubby* are located 10 m.u. apart. In a cross between *nec+* / TM6C females and males true-breeding wild type for all genes under consideration, how many of the 1000 adult progeny should have emerged from long, thin pupal cases and had Stubble bristles?

Answer

This problem involves an understanding of balancer chromosomes, the *Drosophila* life cycle, and complementation analysis.

a. In the cross between nec^- / TM6C and the deletion / TM6C strains, the following zygotes would be produced in equal proportions, each 25% of the total. For simplicity, these are labeled as classes 1–4.

Class 1	Class 2	Class 3	Class 4
nec^- / deletion	nec^- / TM6C	deletion / TM6C	TM6C / TM6C

Regardless of the relationship between nec^+ and a particular deletion, classes 2 and 3 should always survive, with animals developing as short, squat larvae and pupae (Tubby) that emerge as Stubble adults. In addition, class 4 should always die during embryonic development. Class 1 will form wild-type larvae, pupae, and adults if nec^- and the deletion in question complement each other. If nec^- and the deletion in question fail to complement each other, class 1 will not emerge from the pupal case; the pupal case itself will be long and thin.

Deletions A and D remove the nec *gene.* Because there are no wild-type adults in the matings with deletions A and D, we know that class 1 dies (these animals have no functional nec^+ product). No complementation occurred, so the deletions include the *nec* gene. With deletions B and C, there was complementation (class 1 survived) so the *nec* gene is not within the deleted areas.

b. We are looking here at crosses in which class 1 dies. In these crosses, zygotes of classes 1 and 4 would not develop into adults, but those of classes 2 and 3 would. Thus, *500 zygotes (1/2 of the total) would produce adult flies.*

c. Looking at the same kinds of crosses as in part (b), only class 1 would die in the pupal case. Thus, *1/4 of the zygotes (250 animals) would die as pupae.* These animals do not have the balancer chromosome, so *their pupal cases would be long and thin.*

d. Here, we are looking at crosses in which class 1 survives and produces wild-type animals. In these crosses, classes 1, 2, and 3 would produce adults. Of these, only class 1 should be wild type (non-Stubble), while classes 2 and 3 would have the *Stubble*-bearing TM6C balancer. Thus, of the emerged adults, *1/3 should be non-Stubble.*

e. *None.* Because of crossover suppression between the balancer chromosome and its homolog, viable recombinants that separate the *Tubby* and *Stubble* dominant markers on the balancer chromosome cannot be obtained.

P R O B L E M S

D-1 For each of the terms in the left column, choose the best matching phrase in the right column.

a. morphogens

b. chromocenter

c. balancer chromosome

d. enhancer trapping

e. ectopic expression

f. metamorphosis

1. contains multiple overlapping inversions

2. reorganization of the body plan

3. making gene products in tissues where they are not normally made

4. define cell fate in a concentration-dependent manner

5. heterochromatin in polytene chromosomes

6. locating regulatory regions based on patterns of expression of a reporter gene

D-2 A balancer chromosome that is often used in *Drosophila* genetics is called TM3. This balancer chromosome is a variant of the third chromosome (an autosome), containing multiple inversions and the dominant mutation *Stubble*, which also has a recessive lethal phenotype. Ten different third chromosomes carrying various recessive lethal point mutations were isolated and kept in stock as TM3 heterozygotes (that is, lethal mutation / TM3). These stocks were then crossed to each other, and the number of wild-type progeny per 150 total adult progeny emerging from each of these crosses is recorded here.

Mutation	1	2	3	4	5	6	7	8	9	10
1	0									
2	47	0								
3	51	49	0							
4	45	54	50	0						
5	53	44	51	48	0					
6	62	47	39	59	39	0				
7	55	50	43	0	48	52	0			
8	53	44	42	0	60	54	0	0		
9	56	42	38	57	0	50	51	48	0	
10	0	58	59	38	46	49	51	49	55	0

a. Does this experiment measure complementation or recombination between recessive lethal mutations?

b. In the cross between mutant 1 and mutant 2, there are 103 adult progeny that are not wild type. What is their phenotype?

c. In the cross between mutant 1 and mutant 2, there are an additional 50 zygotes that never became adults (and therefore were not counted). What is the genotype of these zygotes?

d. How many genes are represented in this collection of 10 mutations? Which mutations are in the same genes?

e. Five different deletions in the third chromosome were isolated and balanced by TM3. These deletion stocks were crossed to the TM3-balanced point mutation stocks described previously. The emergence of wild-type progeny is indicated in the following chart as +; the absence of wild-type progeny is indicated by −. Does this second experiment measure complementation or recombination between the recessive lethal mutations and the deletions?

				Point mutation number						
Deletion	1	2	3	4	5	6	7	8	9	10
A	+	−	+	−	+	−	−	−	+	+
B	−	−	−	−	+	−	−	−	+	−
C	−	+	−	+	−	+	+	+	−	−
D	+	+	−	−	+	−	−	−	+	+
E	−	+	−	−	+	+	−	−	+	−

f. Using this new information, show the best possible map for the 10 point mutations and 5 deletions in the region.

D-3 For a particular P-element mutagenesis experiment, you want to obtain new insertions of a P element marked with the w^+ gene into any of the autosomes. Starting with a stock of flies containing a w^+ marked P element on the X chromosome, design an experiment that would enable you to obtain flies containing P-element insertions in autosomes.

D-4 Which of the following animals would show mutant phenotypes in somatic cells as a result of a *Drosophila* cross involving hybrid dysgenesis?
a. progeny of a P male × M female cross
b. progeny of a M male × P female cross
c. progeny of the progeny of a P male × M female cross
d. progeny of the progeny of a M male × P female cross
e. progeny of the progeny of a P male × P female cross

D-5 Which of the following is not a property of the *hunchback* gene in *Drosophila*?
a. The *hunchback* mRNA is uniformly distributed in the egg by the mother.
b. Transcription of *hunchback* is enhanced by Bicoid (the anterior morphogen).
c. Translation of the *hunchback* mRNA is inhibited by Nanos (the posterior morphogen).
d. The Hunchback protein eventually is distributed in a gradient (anterior high; posterior low).
e. Hunchback protein directs the distribution of *bicoid* mRNA.

D-6 The *hunchback* gene contains a promoter region, the structural region (the amino-acid coding sequence), and a 3′ untranslated region (DNA that will be transcribed into sequences appearing at the 3′ end of the mRNA that are not translated into amino acids).
a. What important sequences required to control *hunchback* gene expression are found in the promoter region of *hunchback*?
b. What sequence elements that encode specific protein domains are found in the structural region of *hunchback*?

c. There is another important kind of sequence that turns out to be located in the part of the gene transcribed as the 3′ UTR (untranslated region) of the *hunchback* mRNA. What might this sequence do?

D-7 How do the segment polarity genes differ in their mode of action from the gap and pair-rule genes?

D-8 One important demonstration that Bicoid is an anterior determinant came from injection experiments analogous to those done by early embryologists. Injection experiments involve introduction of components such as cytoplasm from an egg or mRNA that is synthesized *in vitro* into the egg by direct injection. Describe injection experiments that would demonstrate that Bicoid is the anterior determinant.

D-9 In flies developing from eggs laid by a *nanos⁻* mother, development of the abdomen is inhibited. Flies developing from eggs that have no maternally supplied *hunchback* mRNA are normal. Flies developing from eggs laid by a *nanos⁻* mother that also have no maternally supplied *hunchback* mRNA are normal. If there is too much Hunchback protein in the posterior of the egg, abdominal development is prevented.
a. What do these findings say about the function of the Nanos protein and of the *hunchback* maternally supplied mRNA?
b. What do these findings say about the efficiency of evolution?

D-10 Mutant embryos lacking the gap gene *knirps* (*kni*) are stained at the syncytial blastoderm stage to examine the distributions of the Hunchback and Krüppel proteins. The results of the *knirps⁻* and wild-type embryos stained for the Hunchback and Krüppel proteins is shown schematically here.

Hunchback protein   **Krüppel protein**

Wild type   Wild type

kni mutant   *kni* mutant

a. Based on these results, what can you conclude about the relationships among these three genes?

b. Would the pattern of Hunchback protein in embryos from a *nanos⁻* mutant mother differ from that shown? If yes, describe the difference and explain why. If not, explain why not.

D-11 In the analysis of eye development, mosaic *Drosophila* were constructed in which some cells were w^+ sev^+ and others were w^- sev^-. Some eye facets (made up of eight cells) contained both white and red cells. When the results from many different mosaic facets were tallied, all possible combinations of red and white cells were seen.

a. What does this indicate about the origin of cells recruited for any one facet?

b. What does this indicate about the expression of the *white* gene?

D-12 Nüsslein-Volhard and Wieschaus used 10 different mutant stocks of the form mutant / CyO where CyO is a *Drosophila* second chromosome balancer. Seven of these mutations were point mutations (mutations 1–7) and three of the mutations were cytologically detectable deletions (Del A, B, C), each removing a large number of genes. They crossed these stocks to each other and noted whether they obtained progeny without the *Cy* dominant marker on the CyO balancer chromosomes, and if so, whether such progeny females were fertile when crossed with wild-type males. The results of this analysis are tabulated here.

Key   F: non-*Cy* progeny are obtained, and the females are fertile
      S: non-*Cy* progeny are obtained, but the females are sterile
      N: no non-*Cy* progeny are obtained

	Del A	Del B	Del C	1	2	3	4	5	6	7
Del A	N	N	N	N	F	S	S	F	F	F
Del B		N	F	F	N	F	S	F	F	N
Del C			N	N	F	F	F	S	S	F
1				N	F	F	F	F	F	F
2					N	F	F	F	F	N
3						S	F	F	F	F
4							S	F	F	F
5								S	S	F
6									S	F
7										N

a. Give the best genetic map of this region of the *Drosophila* second chromosome, showing the extent of the deletions with bars and positions of point mutations with an "X."

b. How many different genes are indicated by this data set? Which point mutations are in which genes?

c. How many genes needed for female fertility are indicated by this data set? Which point mutations are in which of these genes? How could you distinguish genes needed for oogenesis from maternal-effect lethal genes that make otherwise normal eggs that cannot develop into adults after fertilization?

d. How many zygotic lethal genes are indicated by this data set? Which point mutations are in which of these genes?

e. In one sentence, explain why you could not determine the map distance between any two of the preceding point mutations that both lie within the same gene, for any of the genes revealed by this data set, regardless of how many flies you analyzed or whatever experiment you used to try to determine this distance.

D-13 A particular variant of a P element was constructed containing a selectable marker (the *rosy⁺* gene, where *rosy⁻* homozygotes have rose-colored eyes) and the *lacZ* gene, whose transcription is controlled by the weak P-element promoter. When the P element inserts into a gene, the gene is inactivated, producing a mutant phenotype. The insertion of these P elements was used to produce and isolate a series of female sterile mutants and study cell activities during oogenesis. Egg development occurs in an egg chamber containing a single oocyte and 15 nurse cells all derived from a germ cell precursor. The chamber is surrounded by somatic follicle cells. Follicle cells and oocyte and nurse cells interact in establishment of polarity in the egg chamber (review Fig. 19.26 on p. 669).

a. What other information could you get about a developmental gene if the P element carrying the *lacZ* gene inserted into it?

b. How would you maintain stocks containing recessive female sterile mutations? (What would be the genotype of the stock flies and how could you study the mutants?)

c. One of the sterile mutants isolated contained occasional mispositioned oocytes within the egg chamber and some defective egg chambers. The defective gene, called *homeless,* encodes a protein that shows similarity to RNA dependent ATPases and helicases, leading to the hypothesis that the protein may interact with, and help in the positioning of, particular RNAs. *bicoid* mRNA is a major anterior determinant for which localization is important. What experiment could you do to determine if *bicoid* mRNA is appropriately localized within the developing egg chambers in *homeless⁻* mutants? (Be sure to mention your control.)

d. What experiment could you perform to see whether *homeless* gene function is required either in the germ-line-derived cells (the nurse cells and/or oocyte) or is instead required in the somatic follicle cells? This experiment is made possible by the fact that various stocks are available in which the dominant *ovo^D* mutation, whose expression in germ-line-derived cells blocks ovarian development, has been placed on different chromosome arms. Describe how you would set up this experiment and the expected results.

e. Another P-element-induced mutation isolated in the same screen was a semilethal female sterile. By *in situ* hybridization, it mapped to a chromosomal region where the *ras1* gene had been mapped. Complementation analysis established that this mutation is an allele of *ras1*. Other alleles of *ras1* affect viability, eggshell development, and eye morphology. Moderate alleles affect eggshell appearance and viability but not eye morphology. Different alleles also show different levels of Ras1 activity. What does this information suggest about Ras1 protein requirements in different signaling pathways?

D-14 A protein called Bruno was found to bind to the 3′ UTR of several developmental mRNAs, including *oskar* mRNA (necessary for proper formation of the posterior part of the embryo). When the *bruno* gene was cloned, the predicted protein sequence contained an RNA-binding domain consistent with the biochemical analysis of binding. The *bruno* gene mapped to a chromosomal region where the *arret* gene had previously been located. The *arret* gene, which was identified genetically in a screen for sterile mutants, is required for fertility in both sexes. How could you determine if *bruno* and *arret* are in fact the same gene that had been identified in different ways? What particular phenotypes would you examine?

Mus musculus: Genetic Portrait of the House Mouse

A member of the 129 strain of inbred mice commonly used in targeted mutagenesis studies.

The common house mouse, *Mus musculus,* has played a prominent role in the study of genetics ever since Carl Correns, Hugo De Vries, and Erich Von Tschermak independently rediscovered Mendel's laws at the beginning of the twentieth century. Because these three scientists, as well as Mendel himself, performed their research entirely on plants, many in the scientific community questioned whether Mendel's laws could explain the basis for inheritance in animals, especially humans. The reason for this skepticism is easy to see. People, for example, differ in the expression of many commonly inherited traits—such as skin color, eye color, curliness of hair, and height—that show no evidence of transmission according to Mendel's laws. We now know that these traits result from the interaction of many genes with multiple alleles that each segregate according to Mendel's first law even though the traits themselves do not. At the beginning of the twentieth century, however, a demonstration of the applicability of Mendel's laws to animal inheritance required the analysis of simple traits controlled by single genes.

M. musculus has many features that enhance its value as a model organism for genetic analysis, and foremost among these is the availability of hundreds of single-gene mutations. These mutations arose during the mouse's long history of domestication as a pet. Over the centuries, dealers in what became known as the "fancy mouse" trade selected and bred mice with numerous coat colors and other visible mutations, first in China and Japan, later in Europe (Fig. E.1a). In contrast to the variation that occurs naturally in wild populations, new traits that appear suddenly in captive-bred mice are almost always the result of single-gene mutations. Early animal geneticists made note of this fact and used fancy mice to demonstrate that Mendel's laws apply to mammals and, by extrapolation, to humans.

In addition to providing a ready source of single-gene mutations, the house mouse has several other features that make it the mammal of choice for genetic analysis. Mice have a very short generation time of just eight to nine weeks. They are small enough so that thousands can live in relatively small rooms. They have large litters of eight or more pups. They breed readily in captivity. Fathers do not harm their young. And after centuries of artificial selection, domesticated mice are docile and easy to handle (Fig. E.1b).

(a)

(b)

Figure E.1   The mouse is a model system for human biology.
(a) Examples of visible phenotypes caused by single-gene mutations.
(b) Mother mouse with her pups.

(a)

(b)

Figure E.2   Hirschsprung disease: A human developmental disease that causes deformities of the colon (a) and **(b)** a mouse model of the disease with a classical mutation (piebald coat).

But why study a mammal at all when animals like fruit flies and nematodes are even smaller and more amenable to genetic analysis? The answer is that a major goal of current biological research is the understanding of human beings. And although many features of human biology, especially at the cellular and molecular levels, are common to a broad spectrum of life-forms, the most advanced organism-level human characteristics appear in a limited subset of animals. In fact, many aspects of human development and disease are common only to placenta-bearing mammals such as the mouse. Thus, the mouse provides a powerful model system for investigating the genetic basis of simple and complex human traits, especially those related to development and disease (Fig. E.2).

Two general themes emerge from our presentation of *M. musculus.* First, because of the many similarities between mouse and human genomes, researchers can use *homology analysis* to identify and locate the same genes in both species. In this context, **homologs** are genes or regulatory DNA sequences that are similar in different species because of descent from a common ancestral sequence. Second, genetic analysis in mice exemplifies the combined use of molecular (that is, recombinant DNA) and clas-

sical breeding techniques to identify and understand the function of complex genetic systems.

Our genetic portrait of the house mouse describes:

- An overview of *M. musculus* in the laboratory, including a look at the mouse genome, the mouse life cycle, and two powerful transgenic protocols: the addition of specific genes to the mouse genome by nuclear injection and the removal of specific genes from the mouse genome by targeted mutagenesis.

- The uses of transgenic technology in determining the function of gene products, characterizing regulatory regions, establishing links between mutant phenotypes and particular transcription units, and creating a mouse model for a human disease.

- The *Hox* genes: a comprehensive example.

An Overview of *Mus musculus* in the Laboratory

The Mouse Genome

The most important feature of the mouse genome for contemporary geneticists is its close resemblance to the human genome (Table E.1). The haploid genomes of humans and mice (and other placental mammals as well) contain approximately 3 billion base pairs of DNA. Nearly every human gene has a homolog in the mouse genome. This does not mean that the two genomes are equivalent in content. Nearly all differences, however, appear to result from species-specific additions to gene families that already existed in the common ancestor of mice and humans.

The human genome is distributed among 22 autosomes and 2 sex chromosomes, while the mouse genome is contained within 19 autosomes and 2 sex chromosomes. As we saw at the beginning of Chapter 13, examinations of mouse and human karyotypes under the microscope reveal no evidence of chromosome banding similarities between the two species. With the mapping of thousands of homologous genes in both species, however, a remarkable pattern has emerged. Genes that are closely linked in one species are usually closely linked in the other. When two or more loci are found to be linked in one species, they are said to be **syntenic** (meaning "on the same thread," or chromosome). When the same set of loci are also found to be linked in a second species, they are said to exist in a state of **conserved synteny.** A comparison of genetic maps that comprise the whole mouse genome with genetic maps that comprise the whole human genome shows that regions of conserved synteny extend across nearly the complete lengths of both. The average size of each conserved syntenic region is roughly 17.6 Mb. The implication of this finding is that during the 75 million years that mice and humans have been evolving apart from a common ancestor, their genomes have broken apart and rearranged some 170 times (17.6 Mb × 170 = about 3000 Mb = the size of the mammalian genome). Conversely, if the proper genome-scale scissors and glue were available, one could break the mouse genome into about 170 pieces and reassemble those pieces—like a puzzle—in the form of the human genome.

In addition to its powerful evolutionary implications, conserved synteny is a useful tool for practicing geneticists. Once a researcher has mapped a locus in one species, he or she can look at a homology map and immediately identify its likely map position in the other species. Of course, for genes that have already been cloned, it is possible to use DNA-DNA hybridization, or computer analysis of a whole-genome sequence, to pick out gene homologs from the other species. But for loci characterized only by their phenotypic expression, conserved synteny enables geneticists to move back and forth between the analysis of a trait in humans and the analysis of a model for that trait in mice.

The discovery of a locus that predisposes female mice to excessive consumption of 10% ethanol (the concentration of alcohol found in most wines) provides an example of the use of conserved synteny for locating human homologs of mouse genes. Mouse geneticists used DNA markers (as described in Chapter 11) to map the *Alcohol-preference-2* (*Alcp2*) locus to the middle of mouse chromosome 11. Even though they have not yet identified and cloned the gene, scrutiny of a conserved synteny homology map shows that the most likely location for the human homolog of *Alcp2* is on the short arm of human chromosome 17, close to the centromere (Fig. E.3). With this information about the likely location of an alcohol-preference locus, human geneticists can now look for linkage between DNA markers on human chromosome 17 and a phenotypic expression of a predisposition to alcohol abuse.

The Mammalian Life Cycle

The mouse's life cycle is similar to that of humans and all other placental mammals, although the timing of events is unique to each species (see Table E.1). For example, for mice, the average life span is just 2 years; for humans it averages 78 years. Gestation—the period from fertilization to birth—lasts only 21 days in mice but 8.9 months in humans. After birth, mice reach puberty in just 5–6 weeks, 100 times sooner than humans,

---

TABLE E.1   **Comparison of Mice and Humans**

Trait	Mice	Humans
Average weight	30 g	77,000 g (170 lb)
Average length	10 cm (without tail)	175 cm
Genome size	~3,000,000,000 bp	~3,000,000,000 bp
Haploid gene number	~50,000	~50,000
Number of chromosomes	19 autosomes + X and Y	22 autosomes + X and Y
Gestation period	3 weeks	Average, 38 weeks (8.9 months)
Age at puberty	5–6 weeks	Average, 624–728 weeks (12–14 years)
Estrus cycle	4 days	Average, 28 days
Life span	2 years	Average, 78 years

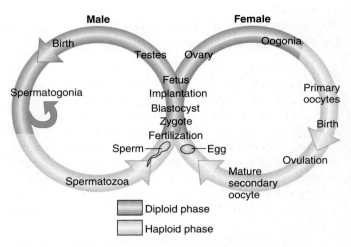

Figure E.3   Conserved synteny: An example with mouse chromosome 11. The mouse chromosome, on the *left,* shows positions of loci with mapped homologs in the human genome. The map locations of human homologs are on the *right.*

Figure E.4   The mammalian life cycle. The life cycles of males and females are shown separately in the traditional format. Notice that primary germ cells are formed in the female before birth but in the male after birth. Notice also that male germ cells are self-renewing, while in females, no new germ cells are formed after birth.

who mature for roughly 12 years before they attain puberty and the ability to conceive children of their own (although in most societies, they wait 5–10 years longer). It is thus possible to go from the birth of one mouse to the birth of its offspring in just 8 weeks, whereas with humans, completion of the same cycle would take 13–25 years.

Even with significant differences in timing, however, the details of each stage of development, both before and after birth, are remarkably similar in all mammalian species. As a result, the overview presented here applies equally to mice and humans.

The mammalian life cycle, like that of all sexually reproducing species, can be visualized as a continuous circle (Fig. E.4), with any point along the circumference marking the start. For ease of presentation, we begin with the haploid phase in both males and females.

Male Germ Cell Development

As we saw in Chapter 4, once a male mammal has reached puberty, he continuously produces a large number of haploid germ cells for the rest of his life. The mature haploid cell is a *sperm cell,* or *spermatozoa,* and the process by which it arises is *spermatogenesis* (review Fig. 4.21). Mature spermatozoa released

into the lumen at the center of the seminiferous tubule join millions of other sperm cells, which together pass through countless passageways to reach the epididymis; there they mature further and then continue on to the vas deferens, where they await ejaculation during copulation.

Female Germ Cell Development

Unlike males, females are born with all the haploid cells they will ever have (~50,000 in the mouse; 1 million in women). When mature, these haploid cells are known as *eggs,* or *oocytes,* and the process by which they arise is *oogenesis* (review Fig. 4.18). As we saw in Chapter 4, oogenesis begins inside the newly formed ovaries of the developing fetus. Long before birth, primordial germ cells differentiate into oogonia and enter meiosis, but they stop at the diplotene stage of the first meiotic prophase. These primary oocytes remain arrested in suspended animation—for weeks in mice and many years in humans—until after puberty.

From this time on, the female progresses through an *estrus cycle* that lasts about 4 days in mice and about 28 days in humans. During each cycle, *primary oocytes* (8–10 in mice, usually only 1 in women) are stimulated to complete the first meiotic division and extrude the first polar body at the end of this division; the resulting *secondary oocyte* begins the second meiotic division but stops at metaphase and is released from the ovary in a process known as *ovulation.* Following ovulation, the secondary oocyte passes into an *oviduct* (called a *fallopian tube* in humans), where for a brief time, known as *estrus,* it remains alive and receptive to fertilization.

In nature, most mammals die while they still have the ability to reproduce. Many human females, however, live long enough to pass through *menopause,* during which they stop cycling through estrus, no longer ovulate, and thus lose the ability to reproduce.

Fertilization

Just before and during the estrus phase of the estrus cycle, female mammals of nonhuman species release species-specific chemical signals, or *pheromones.* In a behavioral response to these pheromones, a male will copulate with a female and ejaculate semen containing millions of sperm into her reproductive tract. The sperm swim from the vagina into the uterus and thence up the oviducts. Only 100 or fewer sperm survive this journey to the waiting eggs.

Fertilization is a multistep process illustrated in Fig. E.5. First, surviving sperm bind to the zona pellucida—the thick solid shell composed of glycoproteins that surrounds the egg proper. The act of binding induces each sperm to release special proteases that enable it to "burn" its way through the zona pellucida into the space that surrounds the egg membrane. Although multiple sperm can make it into this space, usually only one fuses with the egg. This fusion causes rapid electrochemical changes in the egg membrane that prevent the entry of additional sperm and activate the newly fertilized egg to enter the pathway of animal development.

After fusion, the fertilized egg, or zygote, contains two haploid *pronuclei.* The two pronuclei never merge; instead, replication occurs within both of the pronuclei. The one-cell embryo carries two replicated pronuclei right up to the moment of the first mitosis, at which time the membranes of the two pronuclei break down, and the two sets of chromosomes, one from the paternal pronucleus, the other from the maternal pronucleus, align along the midplane of the fertilized egg and thence segregate chromatids into the two daughter cells.

For the purposes of analysis, scientists divide mouse development into two distinct stages of unequal length, separated by the moment of embryonic implantation into the uterus: a preimplantation stage that lasts 4–5 days in mice, and a postimplantation stage that lasts about 16.5 days in mice. During the preimplantation phase, the embryo is a free-floating object within the female's body. It is easy to remove this naturally free-floating preimplantation embryo from the animal, culture it in a petri plate, and expose it to genetic manipulation before placing it back in the reproductive tract of an adult female for development to a newborn animal. After implantation, however, such manipulation is no longer possible because the embryo, if removed from the adult's body, cannot be returned. The accessibility of the preimplantation embryo provides the basis for many of the genetic manipulations researchers use to study mammalian development.

For Most of the Preimplantation Stage, the Embryonic Cells Remain Undifferentiated

The preimplantation stage starts with the zygote at conception (Fig. E.5). Development proceeds slowly in the beginning, with the first 22 hours devoted to the expansion of the highly compacted sperm head into a paternal pronucleus that matches the size of the egg's maternal pronucleus. After the paternal pronucleus has completed its expansion and replicated its chromosomes and the maternal pronucleus has replicated its chromosomes, the embryo undergoes the first of four equal divisions, or cleavages, that increase the number of cells from 1 to 16 over 60 hours.

The period of these four equal divisions is called the **cleavage stage.** During this stage, all the cells in the developing embryo are equivalent and **totipotent,** that is, they have not yet differentiated and each one retains the ability, or *potency,* to produce *every* type of cell found in the developing embryo and adult animal. This is very different from the developmental patterns

Preimplantation development and Implantation

Figure E.5   Early development of mammals from fertilization to implantation.

found in most nonmammalian animal species, including *Caenorhabditis elegans,* where totipotency disappears as early as the two-cell stage. Because of the mouse cells' totipotency, cleavage-stage embryos can be divided into smaller groups of cells that each have the potential to develop into a normal individual. Identical human twins, or more rarely, identical triplets or quadruplets, are examples of the outcome of this process. (Twinning is impossible in *C. elegans* or *Drosophila melanogaster.*) In the laboratory, scientists have obtained completely normal mice from individual cells that they dissected out of the four-cell-stage mouse embryo and placed back into the female reproductive tract (Fig. E.6a). This experimental feat demonstrates the theoretical possibility of obtaining four identical clones from a single embryo of any mammalian species.

Another more bizarre consequence of the equivalency of cleavage-stage cells is the formation of **chimeras,** which are the opposite of clones (Fig. E.6b). The term "chimera" comes from the Greek word for a mythological beast that is part lion, part goat, and part serpent. Geneticists use the term to designate an embryo or animal composed of cells from two or more different origins. The Polish embryologist Andrezej Tarkowski reported the first mouse chimeras in 1961. To construct them, he removed the zona pellucida from two cleavage-stage mouse embryos, obtaining denuded cell masses that are naturally sticky; he then pushed the sticky denuded embryos up against each other. Denuded embryos pressed together in this way form a single chimeric cell mass that is capable of undergoing normal development within the female reproductive tract. If the two embryos of a chimera come from different females mated to different males, the resulting individual is *tetraparental,* that is, has four parents. It is also possible to produce *hexaparental* animals derived from a combination of three embryos. Every organ and tissue in the adult—including the germ line—can contain cells derived from all three original embryos. As we see later, the production of chimeric mice has been an essential component of the targeted mutagenesis technology that has revolutionized the use of the mouse as a model organism for studying human diseases.

A comparison of the early developmental program of placental mammals with that of other animals, including *C. elegans* and *D. melanogaster,* shows how different these programs can be. In nematodes, embryonic cells are highly restricted in their developmental potential, or fate, beginning at the two-cell stage; and in fruit flies, polarization of the egg before fertilization generates distinct cytoplasmic regions dedicated to supporting different developmental programs within the nuclei that end up in these locations. Consequently, half a nematode embryo or half a fly embryo can never give rise to a whole animal.

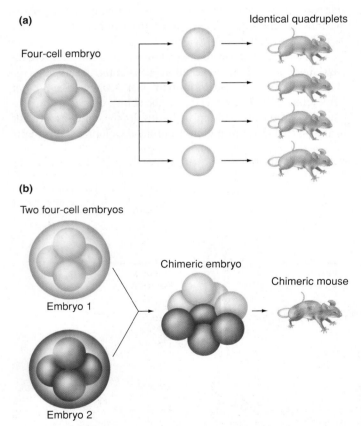

Figure E.6 Early mammalian embryos are highly malleable in their development. There is no requirement for a one-to-one correspondence between embryo and adult. **(a)** Creating *identical* quadruplets from a single fertilized *egg.* **(b)** Creating a single chimeric animal from the fusion of two embryos.

Events Restricting the Developmental Potency of Individual Cells Occur Near the End of the Preimplantation Stage

The first differentiation events of mouse embryogenesis occur in the 16-cell embryo (see Fig. E.5). The cells on the outside of the embryo turn into a trophectoderm layer that will eventually take part in the formation of the placenta. Soon thereafter, the cells on the inside of the embryo compact into a small clump called the *inner cell mass,* or *ICM,* that remains attached to one spot along the inside of the hollow trophectoderm sphere. The entire fetus and animal are derived entirely from the cells of the ICM. As Fig. E.5 shows, compaction of the ICM causes the appearance of a fluid-filled space that is devoid of cellular material and surrounded by trophectoderm. This space is called the *blastocoel cavity.* The embryo is now called a **blastocyst.** Two more rounds of cell division occur during the blastocyst stage, producing the 64-cell embryo that implants.

Throughout normal preimplantation development, the embryo remains protected within the inert zona pellucida. As a result, there is no difference in size between the 1-cell zygote and the 64-cell embryo. To accomplish implantation, the embryo must first "hatch" from the zona pellucida so that it can make direct membrane-to-membrane contact with cells in the uterine wall. Embryonic and fetal development within a uterus inside the body of a female is a characteristic unique to all mammals except the primitive egg-laying platypus.

After Implantation, the Placenta Develops, the Embryo Grows, and the Tissues and Organs Emerge

Implantation initiates development of the placenta, a mix of embryonic and maternal tissues that mediates the flow of nutrients entering the embryo from the maternal blood supply and the flow of waste products exiting the embryo to the maternal circulation. The placenta maintains this intimate connection between mother and embryo, and later between mother and fetus, until the time of birth. Development of the placenta enables a period of rapid embryonic growth. Cells from the ICM differentiate into the three germ layers of endoderm, ectoderm, and mesoderm during a stage known as gastrulation. The foundation of the spinal cord is put into place, and the development of the various adult tissues and organs begins. With the appearance of organs, the embryo becomes a fetus, which continues to grow rapidly. Birth occurs at about 21 days after conception. Newborn animals remain dependent on their mothers during a suckling period that lasts 18–25 days. By five to six weeks after birth, mice reach adulthood and are ready to begin the next reproductive cycle.

Two Powerful Transgenic Techniques for Analyzing the Mouse Genome

Geneticists have capitalized on certain features of the mouse genome and the mouse life cycle to develop protocols that make it possible to add and remove specific genes from embryonic or germ cells.

The Addition of Genes to the Mouse Genome by Nuclear Injection

The 1981 development of a method for inserting foreign DNA into the germ line of mice thrust a primarily observational discipline into the realm of genetic engineering with all its implications. Yet the incredibly powerful transgenic technology is based on a very simple process. Recall that a *transgene* is any piece of foreign DNA that researchers have inserted into the genome of a complex organism, such as a mouse or a pea plant, through experimental manipulation of early stage embryos or germ cells; any individual carrying a transgene is known as a *transgenic* animal or plant.

To create a transgenic mouse carrying a foreign DNA sequence integrated into one of its chromosomes, a researcher simply injects foreign DNA into a pronucleus of a fertilized egg and then places the injected one-cell embryo back into a female oviduct, where it can continue its development. Roughly 25% to 50% of the time, depending on the skill of the investigator, the injected DNA will integrate at random into a chromosomal location. Integration can occur while the embryo is still in the one-cell stage, in which case the transgene will appear in every cell of the adult body. Or integration may occur somewhat later, after the embryo has completed one or two cell divisions; in this case, the mouse will be a mosaic of cells, some with the transgene and some without it. The relatively high rate of integration appears to be a consequence of naturally occurring DNA repair enzymes present in all eukaryotic cells. During evolution, these

Several embryos recovered from sacrificed female

Embryos transferred to a depression slide containing culture medium

Culture medium
Oil

As embryo is held in place, DNA is injected into pronucleus.

Holding pipette

Pronucleus
DNA to be injected
Injection pipette

Several injected embryos are placed into oviduct of receptive female.

Figure E.7 How transgenic mice are created. About 12 hours after conception, the female mouse is sacrificed and the embryos recovered. They are transferred to a depression on a specialized microscope slide containing a drop of culture medium under oil to prevent evaporation. The slide is placed on the stage of an inverted microscope (as the name implies, the objective lens is beneath the stage rather than above it, as in a typical microscope). This arrangement gives the researcher space to manipulate the embryos from above. Suction holds the embryo in place on the blunt end of a special "holding pipette." A second type of pipette with a very narrow bore (the injection pipette) is used to inject transgenic DNA through the plasma membrane and into the pronucleus, where the foreign DNA is released. The altered embryos are then placed into the oviduct of a physiologically receptive female.

enzymes acquired the ability to seek out and ligate together open-ended DNA molecules, which can result from mutagenesis. Figure E.7 illustrates the details of the transgenic procedure.

Up to 50% of the mice born from injected embryos have the foreign DNA stably integrated into their genomes. They will thus transmit this DNA to their offspring. There are no limits to the type of DNA that can be incorporated. It can come from any natural source—animal, plant, or microbial—or directly from a DNA synthesizer. It is very common for investigators to construct DNA molecules (called "DNA constructs") composed of genetic elements from different sources. For example, a DNA construct might have a coding region that is a composite of human and *Escherichia coli* sequences flanked by an upstream regulatory region that is a composite of mouse and synthetic sequences.

Although embryonic nuclear injection is a powerful transgenic tool, it has two significant limitations. First, it can only add—not subtract—genetic material. Second, experimenters cannot target the insertion of foreign DNA to specific genomic locations. Consequently, transgenic mice produced by embryonic nuclear injection are useful only for the analysis of dominant phenotypes. By 1989, geneticists had developed a way to circumvent these limitations.

Targeted Mutagenesis, a Second Transgenic Technology, Makes It Possible to Remove, or Knock out, Specific Sequences from the Mouse Genome

In addition to knocking out the function of a sequence altogether, targeted mutagenesis can produce an allele with an altered function. The emergence of targeted mutagenesis, which is technically more demanding and more complex than the nuclear injection technology just described, depended on two advances in cell culture techniques that occurred during the 1980s.

The first advance was the establishment of *in vitro* conditions that enable researchers to place mouse embryos at the blastocyst stage into culture such that the embryonic cells from the ICM continue to divide without differentiating. Cultured cells that behave in this way are called **embryonic stem cells,** or **ES cells** for short. ES cells appear to be similar in their state of differentiation to cells from the ICM. It is possible to grow cultures containing many millions of ES cells from a single embryo and then recover a handful of cells from this culture for injection back into the blastocoel cavity of a normal embryo. Once inside the cavity, the ES cells can become incorporated into the ICM, and they can contribute to all of the tissues in the mouse that develops from the embryo. Most importantly for geneticists, the ES cells even contribute to the germ lines of these chimeric mice so that reproducing adults can transmit mutated genes present in the ES cells to future generations.

The second critical advance that provided a foundation for targeted mutagenesis was development of a protocol for homologous recombination in ES cells. The transformation of mammalian cells is known as **transfection.** During the transfection of mouse cells with mouse-derived DNA, the foreign mouse DNA almost always integrates at random into a chromosome at a site other than its point of origin. Occasionally, however, the added DNA will "find" and replace its homolog by homologous recombination. The frequency of homologous recombination events as a fraction of the total number of integrations is on the order of 10^{-3}–10^{-5}.

If researchers transfect mouse ES cells with unaltered, cloned fragments of mouse DNA, homologous recombination events do not cause genomic changes. But with the recombinant DNA technology described in Chapter 9, investigators can modify cloned genes so that they no longer function; cloned genes modified in this way are known as **knockout constructs.** When homologous recombination occurs with a knockout construct, the nonfunctional knockout allele replaces the endogenous wild-type allele (Fig. E.8). To construct mice in which homologous recombination has knocked out specific genes, researchers

developed protocols for identifying and recovering the very rare ES cells in which homologous recombination occurs.

How Biologists Use Transgenic Tools to Study Mice and Create a Mouse Model for Human Disease

Both add-on and knockout transgenic technologies have tremendous value in genetic research. The protocols enable researchers to determine the function of gene products, characterize genetic regulatory regions, establish links between mutant phenotypes and particular transcriptional units as an aid to verifying the identification of a cloned gene, and create mouse models of human genetic diseases.

Using Transgenic Technology to Determine Gene Function

In Chapter 11 we saw how geneticists used transgenic technology to demonstrate that the cloned *SRY* (for sex-determining region on the Y chromosome) gene could confer maleness on an animal without a Y chromosome. Incorporation of the *SRY* gene's coding region and regulatory sequences into a mouse embryo with two X chromosomes produced an animal with male genitalia and testes. This result demonstrated that the product of a single gene on the Y chromosome is all that is needed to switch the developmental pathway of the fetus from female to male (Fig. E. 9)

Biologists have used this same transgenic technology to examine the functions of many other genes. By combining a mouse gene of interest with regulatory regions from other mouse genes, they can cause transgenic mice to express the natural transgene product in an unnatural manner: at a higher than normal level, in an alternative tissue, or at an alternative developmental stage. They can then use the aberrant level, time, or place of expression to elucidate the normal function of the wild-type gene.

Experiments analyzing the *myc* gene demonstrate the power of transgenic technology for uncovering the function of genes in ectopic expression (that is, expression at an abnormal place, time, or level). Investigators originally discovered the *myc* gene in the genome of the myelocytomatosis chicken retrovirus; exposure to this virus caused cultured chicken cells to become tumorigenic. Hybridization studies demonstrated the existence of a *myc* gene homolog in the genomes of mice and other vertebrates, including humans (where it resides on the long arm of chromosome 8) but not in the genomes of nonvertebrate model organisms, such as *Drosophila*. Moreover, in human cancer cells obtained from patients with Burkitts lymphoma (a cancer of the immune system's B cells), the *myc* gene often appears close to one of the breakpoints of a reciprocal translocation characteristic of these cancer cells between the long arms of chromosomes 8 and 14; in this translocated position, the gene is usually expressed at a higher than normal level. Noncancerous animal cells have a very low level of *myc* gene expression. These findings

(a) Construction of a knockout allele in ES cells

Early blastocyst
(× 10,000)

Culture into millions
of embryonic-like
(ES) cells.

Clone containing gene of interest

5'   3'

Build knockout construct by
adding in selectable marker.

5'   3'

Marker disrupts transcription
unit.

← Add cloned DNA to
culture of cells.

Homologous recombination

Added DNA construct

5'   3'

ES cell chromosome with wild-type allele

ES cell chromosome with knockout allele

Finding the cell with the knockout allele.
Subject culture to drug that kills all cells that do not contain
selectable marker.

Survivor cells have knockout allele (1% or less).
Begin new culture with survivor cells.

(b)

Figure E.8 Knocking out a mouse gene in ES cells. (a) An early mouse blastocyst can be grown in culture under conditions that allow the cells of the ICM (inner cell mass) to remain undifferentiated as ES cells. A DNA clone, containing the gene of interest, can be modified in the laboratory into a disrupted allele with a selectable marker. This "DNA construct" is added to the ES cell culture, where homologous recombination will occur. **(b)** A chimeric mouse composed of cells derived from a normal embryo (albino) and ones derived from the mutated ES cell (dark agouti).

Two one-cell female mouse embryos (with two X chromosomes)

Pronuclei

No injection

Inject *SRY* DNA

19 pairs autosomes,
two X chromosomes

FEMALE

19 pairs autosomes,
two X chromosomes,
and *SRY* transgene

MALE

Figure E.9 The transgenic mouse protocol proves that *SRY* is the "testis-determining" locus responsible for the production of maleness during embryogenesis. DNA fragments containing the mouse *SRY* gene and its regulatory sequence were injected into a series of embryos that were allowed to develop into live animals. Normal XX animals without a transgene develop as females. But the presence of the *SRY* transgene in an XX embryo induces development as a male.

constitute circumstantial evidence that abnormally high levels of *myc* expression might help transform a cell to a cancerous state. (The biochemical mechanism by which the *myc* gene product functions is described in Chapter 17.)

The results of experiments using cells grown in culture support this hypothesis. In almost every case studied, however, cultured mammalian cells display programs of gene expression that do not correspond with those of the cells they are supposed to model. Thus, the results of cell culture studies do not necessarily reflect how cells *in vivo* (that is, within the body) behave in response to a change in gene activity. In addition to differences in gene activity, there are differences in chromatin structure and patterns of DNA methylation between cells growing *in vitro* and *in vivo*. These discrepancies are not surprising since cells in the body, unlike those in culture, exist in a complex environment that includes constant exposure to molecular signals released by other body cells. The living organism also has a pervasive immune system that is impossible to imitate *in vitro*. It is therefore possible that a phenotype observed in responses to the abnormal expression of a gene in cultured cells might be a consequence not of one gene's abnormal expression but of interactions with other genes that are expressed differently in cultured cells than in cells *in vivo*. For these reasons, it is not possible to

(a)

(1) The *myc* locus found in the mouse genome.

(2) Hybrid DNA construct containing the *myc* coding region regulated by an inducible promoter

(b)

Figure E.10 Transgenic expression of the *myc* gene provides information on the gene's role in tumor formation.
(a) Construction of a transgene containing the *myc* gene under the control of an inducible promoter: (1) structure of the endogenous *myc* gene and (2) transgene construct with the MTV (dexamethasone-inducible) promoter attached to a portion of the *myc* gene that contains the coding region. **(b)** Northern blot showing induction of transgene expression in a range of adult tissues.

rely on cell culture results for an explanation of the true function of a gene; rather it is necessary to examine the effects of aberrant gene expression *in vivo.*

To learn whether increased expression of the *myc* gene affected tumor formation in various tissues of the mouse, researchers used transgenic technology (Fig. E.10). In one experiment, they attached the immunoglobulin gene promoter to the *myc* coding sequence to produce a transgenic mouse line that expressed *myc* at high levels in the precursors to immunoglobulin-producing B cells. In another experiment, they attached tissue-specific promoters, including one for mammary gland expression, to the *myc* coding region. And in yet another experiment, they attached to the *myc* coding region a promoter that is recognized in all tissues but only after the embryo's or animal's exposure to dexamethasone, a glucocorticoid hormone.

The results of all these experiments showed that overexpression of the *myc* gene does not have any effect on normal developmental processes. Even when the gene was expressed at high levels in many developing tissues, normal animals were born. Moreover, even in the adult, most cells that overexpress the *myc* gene never display an aberrant phenotype. However, the rate of tumor formation increased significantly in most, but not all, types of tissue. The conclusion was that aberrant expression of the *myc* gene alone does not cause cells to become cancerous; but it can operate with other somatic mutational events to transform a cell to the cancerous state. These observations support the hypothesis that the cancer phenotype results from the accu-

mulation of multiple mutations in the clonal progeny of one cell (see Chapter 18).

Geneticists still do not completely understand the normal function of the wild-type *myc* gene. But biochemical and genetic studies indicate that the gene product is a component of a transcription factor that helps control the growth of cell populations in almost every tissue of the body (as described in Chapter 17).

Using Transgenic Technology to Characterize Regulatory Regions

To understand embryonic development, it is essential to learn not only how the proteins critical to this complex process function but also how the genes that encode these proteins are regulated. The pattern of gene expression in a normally differentiating cell is finely tuned, with each gene expressed at exactly the right level and each gene turned on and off at exactly the right times. Geneticists know that cis-*acting regulatory regions* (that is, regulatory regions on the same chromosome) that are closely linked to each gene control the timing and intensity of each gene's expression. Transgenic technology can help determine the location and sequence of such regions.

Researchers can characterize the *cis*-acting regulatory regions of single-cell organisms, such as bacteria and yeast, by random mutagenesis experiments. In these experiments, they select from cultures of millions of mutagenized cells those cells with single base changes or small deletions in genetic regions that influence the *cis*-regulation of a gene of interest. They can then use the regulatory mutations to identify the base pairs necessary for proper gene function. Because of the large number of organisms required, it is not possible to use a similar approach to study gene regulation in mice. Instead, mouse geneticists must turn to transgenics.

Once they have obtained a genomic clone of a mouse gene of interest, researchers can subclone flanking sequences that are likely to contain its regulatory region. In most cases studied to date, the regulatory regions have been confined to the 5–10 kb of DNA just upstream (to the 5′ side) of the coding sequence; thus, this is the first region an investigator examines at the start of a project. Although it might seem counterintuitive at first, the researcher has no use for the actual coding region of a gene when the aim is to study the gene's regulatory region. In fact, it is best to replace the true coding region with that of a "reporter gene" whose protein product is easy to detect. The most commonly used reporter gene is the β-*galactosidase* (*lacZ*) gene from *E. coli* bacteria (Fig. E.11). As we saw in Chapter 9, the β-galactosidase enzyme encoded by this gene will convert a special substrate known as X-Gal into a colored product that can be visualized under the microscope.

To characterize the regulatory region associated with a gene, you first create a series of different transgene constructs, each formed by splicing together different fragments or mutated forms of the putative regulatory region with the β-galactosidase reporter gene. You next establish a series of transgenic lines by injecting the transgene constructs into different mouse embryos, and then, after setting up timed matings within each of these

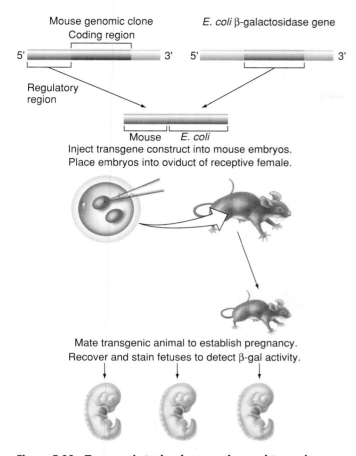

Mouse genomic clone
Coding region

E. coli β-galactosidase gene

Regulatory region

Mouse E. coli

Inject transgene construct into mouse embryos.
Place embryos into oviduct of receptive female.

Mate transgenic animal to establish pregnancy.
Recover and stain fetuses to detect β-gal activity.

Figure E.11 Transgenic technology can be used to analyze *cis*-acting regulatory regions. A DNA construct containing the mouse regulatory region of interest is attached to the *E. coli* reporter gene. The function of the regulatory region can be ascertained by observing β-gal expression in transgene fetuses. In the example shown here, expression is observed in the developing forelimb of three independently derived transgenic mice containing the complete regulatory region of the mouse *Tbx5* gene, which is known to play a specific role in forelimb development of all vertebrates. © John Schiementi, The Jackson Laboratory. Reproduced from "Promoter Mapping of the Mouse TcP-10bt Gene I Transgenic Mice Identifies Essential Male Germ Cell Regulatory Sequences," Ewulonu et al., *Molecular Reproduction and Development* 43:290–297, © 1996. Reproduced by permission of Wiley-Liss, Inc., a subsidiary of John Wiley & Sons, Inc.

lines, you recover embryos at the developmental stage or stages during which the mouse gene normally undergoes expression. By examining the distribution of the colored β-galactosidase product at each embryonic stage and comparing it to the distribution of the natural gene product, it is possible to map the extent of the *cis*-acting regulatory region. The distribution of the natural gene product can be tracked with specific antibodies labeled for examination by immunohistochemistry. If the antibody label is a different color than the β-galactosidase label, the natural gene product and the β-galactosidase enzyme can be simultaneously observed and distinguished under the microscope. Further studies of the effects of individual base-pair substitutions or small deletions on details of the expression pat-

tern can lead to a highly sophisticated understanding of the various *cis*-acting regulatory elements that together may determine the complex patterns of spatial and temporal expression of genes that play a role in development.

An example of the use of transgenic mice to study gene regulation is the analysis of a gene called *T complex protein-10bt* (*Tcp10b^t*). Only differentiating male germ cells in the process of maturing into spermatozoa express *Tcp10b^t*. Cells in the seminiferous tubules, called sertoli cells, mediate sperm differentiation. When investigators disrupt the testes and place sperm cells in culture, they cannot simulate natural conditions completely; as a result, germ cell differentiation continues *in vitro* for only a brief time. But differentiation from stem cell to mature spermatozoa takes six weeks in mice.

Sperm differentiation is of interest to cell biologists because the mechanisms that regulate it may differ significantly from those that control the differentiation of other cells. This is, in part, because the transformations of differentiating sperm cells are much more dramatic than those experienced by other types of cells. Not only do the spermatogenic cells change in shape and size from large round stem cells to tiny, sleek spermatozoa almost without cytoplasm, they also drastically change their genetic program: Stem cells have a normal program of gene expression as well as chromosomes with a normal chromatin structure; in contrast, the chromosomes of differentiated sperm cells have a unique chromatin structure with no histones attached, and they exhibit no gene activity. To understand these differences, mouse geneticists have tried to characterize the regulatory regions associated with gene activity in spermatogenic cells.

To analyze the regulatory region associated with the *Tcp10b^t* gene, geneticists first made a series of transgene constructs carrying different lengths of DNA from the 5' flanking region of the *Tcp10b^t* gene, ligated to the *lacZ* coding sequence. The result was six DNA constructs containing from 0.6–1.6 kb of DNA from the putative *Tcp10b^t* regulatory region. The researchers injected copies of each DNA construct into multiple mouse embryos, obtaining at least four independent transgenic lines of mice for each construct. (It is important to use multiple transgenic lines in an experiment of this type to verify that sequences in the transgene construct itself, rather than sequences that coincidentally flank the transgene insertion site, are responsible for a particular phenotype.)

Figure E.12 depicts how it was possible to map the regulatory region associated with *Tcp10b^t* by simply testing for the presence of *lacZ* transcripts in Northern blots of testicular RNA obtained from each transgenic line of mice. As the figure shows, there was no detectable transcription of the *lacZ* gene in testes from transgenic mice that carried 0.75 kb or less of flanking sequence 5' to the *Tcp10b^t* gene; but with 0.97 kb or more of the flanking sequence, high levels of transcription occurred in the testes, but in no other tissue, of all mouse lines.

These observations located a critical testes-specific, *cis*-regulatory sequence within a 227 bp region between 746 and 973 bases upstream of the *Tcp10b^t* gene. With this information, it became possible to design additional experiments to examine

(a)

		β-galactosidase
1	1.6 kb	
2	1.3 kb	
3	1.16 kb	
4	0.97 kb	
5	0.75 kb	
6	0.6 kb	

R   Bg   H Bsp Bam Sfa X Bam   H   NheI Bam

(b)

0.75 0.97 1.16 1.3 1.6
T   T   T   T   T

Figure E.12 An example of the use of transgenic technology to map the *cis*-acting regulatory region associated with the *Tcp10bᵗ* gene. (a) Different hybrid DNA constructs were made with varying lengths of the 5′ flanking sequence adjacent to the *Tcp10bt* gene fused to the *E. coli* β-galactosidase gene (which acts as a reporter). **(b)** Testicular RNA was obtained from transgenic mice containing the various lengths of 5′ flanking region (shown in kilobases). With 0.75 kb of flanking region, no transcription of the reporter gene was observed, but with all larger flanking regions, transcription did occur. *Source E.12b:* © John Schiementi, The Jackson Laboratory. Reproduced from "Promoter Mapping of the Mouse TcP-10bt Gene I Transgenic Mice Identifies Essential Male Germ Cell Regulatory Sequences," Ewulonu et al., *Molecular Reproduction and Development* 43: 290-297, © 1996. Reproduced by permission of Wiley-Liss, Inc., a subsidiary of John Wiley & Sons, Inc.

Transgene construct contains pme75 gene with its regulatory sequences

Inject into wild-type mouse embryos

Create transgenic line

X

Transgenic animal with Tg (pme75) Normal phenotype

T/+ genotype Short-tail phenotype

Normal tail

Offspring: *T*/+ genotype Tg (pme75) Normal phenotype

Figure E.13 Transgenic technology can be used to identify the locus responsible for a mutant phenotype. A dominant deletion mutation at the *T* locus causes a short tail. A transgenic animal containing the *pme75* transgene is mated with a mutant animal to create animals containing both the deletion and the transgene. A normal phenotype demonstrates that the deletion of the *pme75* gene is responsible for the short-tail phenotype.

the regulatory region in more detail and identify transacting proteins that bind to this region. Through these additional experiments, researchers identified specific testicular proteins that activate the *Tcp10bᵗ* gene.

Using Transgenic Technology to Link Mutant Phenotypes to Specific Transcription Units

A third significant use of transgenic tools is in establishing whether a cloned gene corresponds to a locus previously defined by a mutant phenotype. Consider, for example, the mouse *Brachyury* locus, symbolized by a *T*. This locus is defined genetically by a dominant, X-ray-induced mutation that causes heterozygous *T* / + animals to develop short tails. Through high-resolution linkage analysis and positional cloning (described in Chapter 11), researchers showed that the *T* locus mutation is associated with a 200 kb deletion on chromosome 17. They also identified a transcription unit called pme75 that is located in this same 200 kb deletion; normally, pme75 is expressed in the embryonic tissue that develops into the tail. While highly suggestive, these data do not prove that the absence of the *pme75* gene causes the short-tail phenotype. It is very likely that the 200 kb deletion also removes other genes, and on the basis of the genetic data alone, you cannot rule out the possibility that the absence of one of these undiscovered genes produces the mutant phenotype.

You can resolve this impasse with the transgenic protocol illustrated in Fig. E.13. The first step is to make a transgene construct containing the complete pme75 coding region together with its regulatory sequences. You next inject the construct into wild-type mouse embryos to create a transgenic line. By breeding this transgenic line to animals with the *T* locus mutation, you can obtain offspring that have the 200 kb deletion over the *T* locus as well as a functional *pme75* gene on a different chromosome (at the locus where the transgene construct integrated at random). These animals sport a tail of normal length. By demonstrating that the transgene can correct the mutant phenotype, this result proves that the *T* locus is the equivalent of the *pme75* gene.

Using Targeted Mutagenesis to Create a Mouse Model for Human Disease

Researchers can use targeted mutagenesis, in combination with other protocols, to create a mouse model for human diseases that result from a loss of gene function. As an example, Fig. E.14 illustrates the step-by-step creation of a mouse model for cystic fibrosis. With the identification and cloning of the human cystic fibrosis gene (designated *CFTR*), the first step toward a mouse model was the cloning of the mouse homolog. Development of the mouse model also required fabrication of a *CFTR* knockout construct, derivation of an ES cell culture from a mouse blastocyst, transfection of the ES cells with the *CFTR* knockout construct, selection of cells in which homologous recombination had replaced the wild-type *CFTR* gene with the mutant knockout allele, and finally, the production and analysis of chimeric mice and their offspring. Animals homozygous for the *CFTR* knockout allele display a mutant phenotype that is very similar to that expressed by humans suffering from cystic fibrosis. Thus, in developing drugs to alleviate CF symptoms in humans, pharmaceutical researchers can first test new products in mice to determine their efficacy.

The *Hox* Genes: A Comprehensive Example

A particularly striking example of how mouse biologists used both nuclear injection and targeted mutagenesis technologies to decipher gene function comes from an analysis of the mouse *Hox* gene family. The *Hox* genes are **homeotic selector genes,** that is, genes that control the development of body segments. Members of the *Hox* family are distributed among four unlinked gene clusters that each contain 9–11 genes (Fig. E.15).

Researchers discovered the mouse *Hox* gene family through cross-hybridization studies using the homeotic selector genes of *D. melanogaster* as probes. In fact, homeotic selector genes were first identified on the basis of mutations that produced flies with four wings instead of two, flies whose mouthparts developed incorrectly as legs, and flies with legs instead of antenna growing out of their heads. The proteins encoded by these genes turned out to be transcription factors that act as on/off switches, instructing segments of the fly to develop into one type of tissue or another. William Bateson, the same man who coined the term "genetics," chose the designation of "homeotic selector" from the Greek word *homoios,* which describes a type of variation in which "something has been changed into something else." Homeotic genes appear to control the development of each *Drosophila* body segment (as discussed on pp. 813–843 of the *Drosophila* portrait, Reference D). The bizarre phenotypes just described result when expression of a particular homeotic gene does not occur at the proper time and place. Lack of appropriate expression flicks the binary switch, transforming the recipient body segment into a different type of tissue.

Drosophila homeotic genes were first cloned in the early 1980s. By the end of that decade, it had become clear from cross-hybridization and cloning studies that homologs of these genes are likely to exist in every species of multicellular animal, from *C. elegans* to *Homo sapiens.*

In segmented animals such as flies, homeotic genes are active in the discrete segments that define the body plan, where they determine the proper differentiation of tissues. But what do they do in mice and humans, organisms that do not have obvious body segments? To answer this question, researchers had to overcome a serious drawback: the lack of known mutations at any of the *Hox* loci. This problem was not unique to understanding the functions of mammalian *Hox* genes. Rather, since the discovery of homeotic gene homologs in the mouse genome, it has become routine for developmental geneticists to use cross-hybridization protocols to look for mouse homologs of every *Drosophila* gene found to have a role in development. This strategy has led to the discovery of dozens of new mouse genes, most of which were not associated with any known mutant phenotypes.

How Scientists Determine the Function of a Gene in the Absence of Previously Characterized Mutations

Analyses of Expression Patterns in Developing Embryos Can Provide a Clue to the Time and Location of Gene Action

A clone of a gene can be a tool for analyzing the gene's expression. One way to convert a clone into this type of tool is to label it, denature it, and use the resulting DNA strands as probes in *in situ* hybridization. To study development, investigators can perform *in situ* hybridization on the RNA present in fixed tissue sections obtained from embryos at different stages of development. When developmental geneticists examined the expression patterns of the *Hox* genes, they discovered that each one is transcribed along a portion of the developing embryonic axis that extends from the same most posterior point to a specific anterior boundary (Fig. E.16). Analysis of the pooled data on *Hox* gene expression showed that the extent of spatial expression corresponds with the position of each gene in its cluster. Genes at the 5′ end of a cluster (for example, *D13*) have the least extensive expression patterns, while genes at the 3′ end of the cluster (for example, *A1* and *B1*) have the most extensive expression patterns. These data suggest that different *Hox* genes might be involved in controlling the development of different sections of the embryonic axis. Expression data alone, however, cannot provide conclusive evidence of function.

Ultimately, Only Genetic Tests Can Determine Gene Function

To understand what role a particular *Hox* gene plays in development, a scientist must be able to examine embryos that do not express that gene at all, or express it outside its normal time or place. Examining the changes in development that arise as a

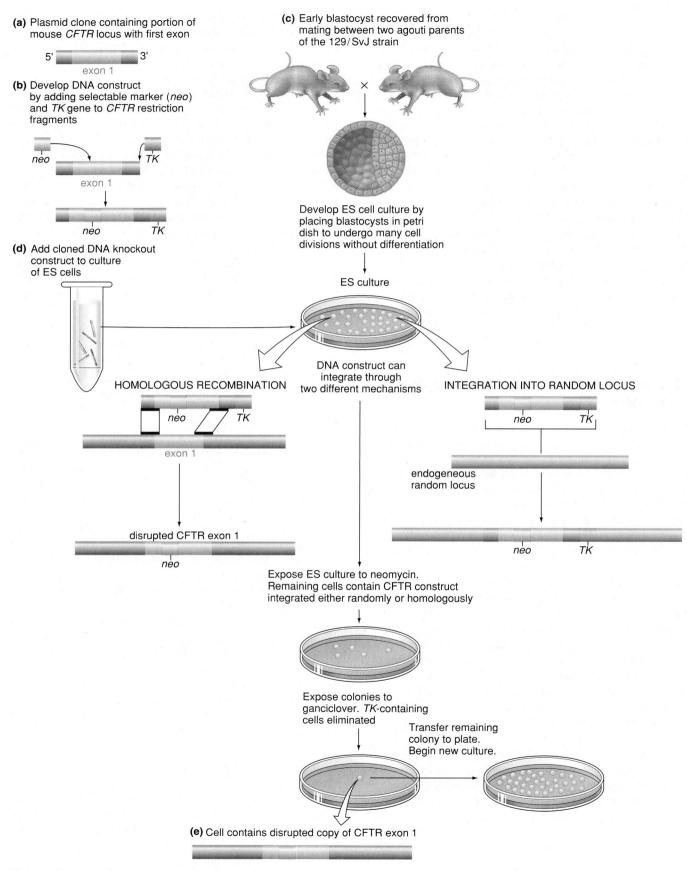

(a) Plasmid clone containing portion of mouse *CFTR* locus with first exon

(b) Develop DNA construct by adding selectable marker (*neo*) and *TK* gene to *CFTR* restriction fragments

(c) Early blastocyst recovered from mating between two agouti parents of the 129/SvJ strain

Develop ES cell culture by placing blastocysts in petri dish to undergo many cell divisions without differentiation

ES culture

(d) Add cloned DNA knockout construct to culture of ES cells

DNA construct can integrate through two different mechanisms

HOMOLOGOUS RECOMBINATION

INTEGRATION INTO RANDOM LOCUS

endogeneous random locus

disrupted CFTR exon 1

Expose ES culture to neomycin. Remaining cells contain CFTR construct integrated either randomly or homologously

Expose colonies to ganciclover. *TK*-containing cells eliminated

Transfer remaining colony to plate. Begin new culture.

(e) Cell contains disrupted copy of CFTR exon 1

Figure E.14 Creating a mouse model for cystic fibrosis.

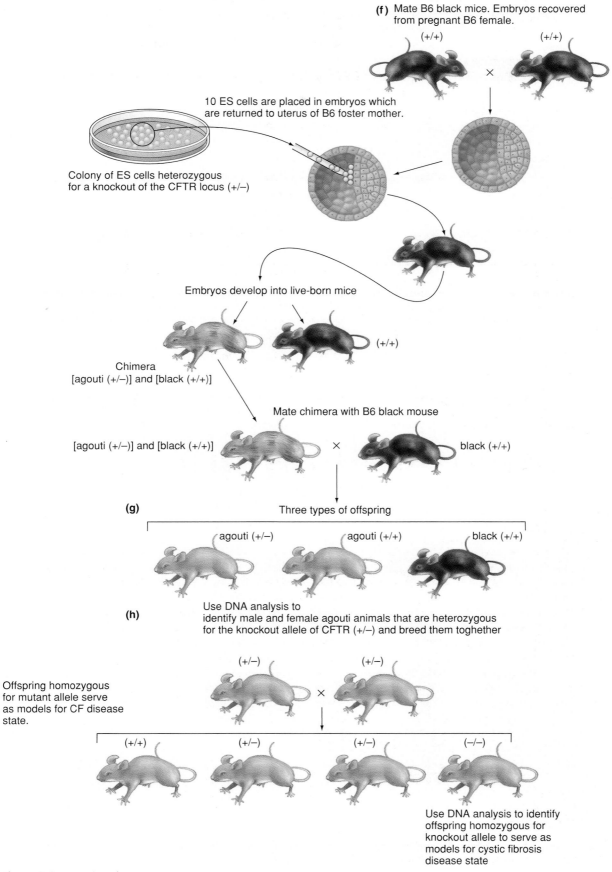

(f) Mate B6 black mice. Embryos recovered from pregnant B6 female.

(+/+)   (+/+)

10 ES cells are placed in embryos which are returned to uterus of B6 foster mother.

Colony of ES cells heterozygous for a knockout of the CFTR locus (+/−)

Embryos develop into live-born mice

(+/+)

Chimera [agouti (+/−)] and [black (+/+)]

Mate chimera with B6 black mouse

[agouti (+/−)] and [black (+/+)]   ×   black (+/+)

(g)   Three types of offspring

agouti (+/−)   agouti (+/+)   black (+/+)

(h)   Use DNA analysis to identify male and female agouti animals that are heterozygous for the knockout allele of CFTR (+/−) and breed them toghether

(+/−)   (+/−)

Offspring homozygous for mutant allele serve as models for CF disease state.

(+/+)   (+/−)   (+/−)   (−/−)

Use DNA analysis to identify offspring homozygous for knockout allele to serve as models for cystic fibrosis disease state

Figure E.14   continued

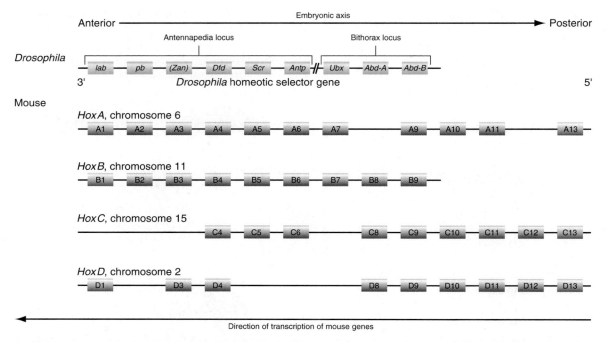

Figure E.15 The mouse *Hox* gene superfamily contains multiple homologs of each member of the *Drosophila* homeotic selector gene family. Genes within the four mouse *Hox* clusters are lined up according to their homology with each other and specific *Drosophila* genes. Not all clusters have homologs of each *Drosophila* gene. From gene 9 and higher, there are multiple homologs of the *Drosophila Abd-B* gene in *Hox* clusters *A, C,* and *D*.

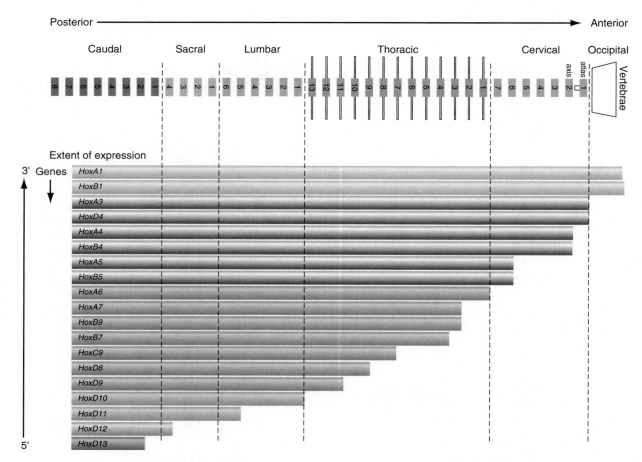

Figure E.16 Spatial extent of expression of some representative *Hox* genes along the developing spine. The top of the mature spine is shown to the *right,* bottom *to the left,* with vertebrae numbered and grouped by name.

result of these genetic changes makes it possible to decipher the normal role of the gene and, in the case of the *Hox* family, test general hypotheses concerning the functional interactions of different *Hox* genes. We now present two examples of this approach.

Validating the Hypothesis That Expression of the 5′ Gene in a *Hox* Cluster Is Epistatic to Expression of the More 3′ Genes

The expression data show that some *Hox* genes are expressed across many segments of the embryo, even as those segments develop differently from each other. To account for this difference in the simplest way, researchers proposed that the only gene that counts in any particular embryonic segment is the most 5′ gene in a *Hox* cluster. In other words, expression of the most 5′ gene is epistatic to expression of the other, more 3′ *Hox* genes. By examining Figs. E.15 and E.16, you can see that this hypothesis could explain how each spinal segment develops in a different manner.

A Transgene Test Confirmed the Prediction of a Homeotic Transformation

As one test of this hypothesis, investigators made a transgene construct with a *HoxA1* regulatory region attached to a *HoxD4* coding sequence (Fig. E.17a). According to the hypothesis, *HoxD4* normally controls the development of the C1 and C2 vertebrae in the cervical region of the spinal column, while *HoxA1* normally controls the development of the occipital bone at the base of the skull. In a transgenic fetus, however, the presence of the transgene construct causes expression of the *HoxD4* gene in the occipital region along with *HoxA1;* and as predicted by the hypothesis, a homeotic transformation converts the occipital bone into cervical vertebrae (Fig. E.17b and c).

Knockout Studies Confirmed Predictions of Aberrant Phenotypes

The hypothesis of 5′ epistasis also leads to the prediction that aberrant phenotypes will arise when different *Hox* genes are knocked out by homologous recombination. With each knockout, one would expect a particular embryonic segment to become transformed to a more anterior-like structure. The data obtained from knocking out various *Hox* genes support the hypothesis of 5′ epistasis. For example, a knockout of *HoxB4* produces a partial homeotic transformation of the second cervical vertebra from axis to atlas (see Fig. E.16).

Transgenic Studies Lead to an Understanding of the Developmental Role of *Hox* and Other Homeotic Genes

With the accumulation of transgene and knockout data for many of the mouse *Hox* genes, a general understanding of the developmental role played by this gene family, not only in mice but in other animals as well, has emerged. What the *Hox* genes and their homologs in other species apparently do is establish signals identifying the position of each region along the embryo's anterior-posterior axis. The genes do not determine the differentiation of

(a)

(b)

(c)

Figure E.17   Transgenic technology provides support for the mode of action of the *Hox* gene family. (a) A transgenic construct is produced to missexpress *HoxD4* in a more anterior region where *HoxA1* is normally expressed (refer to Fig. E.16 for the normal extents of expression of each of these genes). In panel **(b),** the complete skeleton of a wild-type animal is shown on the *left,* and that of an animal expressing the transgene construct is shown on the *right.* A blowup of the cervical regions from both skeletons is shown in panel **(c),** again the wild type is on the *left* and the transgenic construct is on the *right.* In transgenic newborns, what would have been occipital bone (region E in wild-type animal) has been transformed into ectopic arches (E1) that look like cervical vertebrae.

any particular cell type or tissue. Rather, they provide positional information that other genes act on to promote the differentiation of particular tissues. The positional information as well as the genes that act on it vary from species to species.

It is likely that the emergence of the *Hox*-like gene family in our most recent single-cell ancestor set the stage for the evolution of developmental complexity, with the consequent ap-pearance of metazoan organisms sometime between 1 billion and 600 million years before the present time.

Thus, the analysis of the *Hox* gene family is an example of how detailed genetic studies in one model species can provide a general understanding of gene function across large segments of the animal kingdom as well as clues to the conserved mechanisms by which complex developmental processes are carried out.

CONNECTIONS

Geneticists can now produce transgenic animals by combining the classic tools of mutagenesis with an understanding of molecular biology and embryology. Transgenic technology has become so sophisticated that, in theory, it is possible to make any genetic change imaginable to the mouse genome and deter-mine its effect on the individual that emerges.

With the ability to produce mice carrying add-ons and knockouts of *Hox* and other genes that play a role in develop-ment, mouse geneticists have begun to dissect the process by which the mouse embryo develops from the one-cell zygote, through gastrulation, and into the organ-building stage. Remarkably, the development of all mammals is similar, espe-cially in these early stages. For example, all the genes impor-tant to mouse development are conserved in the human genome. Thus, much of what we learn about the genetic basis for mouse development will apply to normal and abnormal hu-man development.

ESSENTIAL CONCEPTS

1. The availability of hundreds of single-gene mutations and a short life cycle contribute to *M. musculus*'s value as a model organism.

2. The mouse genome closely resembles the human genome in size, gene content, and syntenic loci. Researchers can thus use *homology* and *conserved synteny* analysis to identify, locate, and determine the function of genes in both species.

3. The mouse life cycle is representative of the mammalian life cycle, although the timing of events is unique to each species. The *totipotency* of preimplantation cells in mammals makes it possible to create *chimeras*.

4. Researchers can use the *transgenic technology* of adding genes to the mouse genome by nuclear injection to determine gene function, characterize regulatory regions, and correlate mutant phenotypes with specific transcription units.

5. *Targeted mutagenesis* is the basis for creating a mouse model of human diseases caused by a loss of gene function.

6. Studies using transgenic technology have revealed that the *Hox* genes generate signals that identify the position of each region along a mouse embryo's anterior-posterior axis. Normal development of the embryo depends on this information.

7. Knowledge of how the *Hox* genes function in mice has elucidated the general role played by homeotic selector genes in the evolution of all metazoan organisms, including humans.

SOLVED PROBLEMS

I. Gain-of-function mutations can produce a novel phenotype and act in a dominant fashion. In loss-of-function mutations no functional gene product is made; most, but not all, loss-of-function mutations are recessive.

   a. Which of these types of mutations would you study using add-on transgenic technology?

   b. Which type of mutation would you have to study with the implementation of homologous recombination in ES cells? Why?

Answer

This problem requires an understanding of add-on transgenic and homologous recombination techniques and their applications.

   a. Dominant alleles caused by gain-of-function mutations are expressed phenotypically when a single copy of the dominant allele is present. *The effect of a dominant gain-of-function mutation can be characterized in transgenic mice* where a transgene containing the dominant allele has been inserted into the genome and the original gene copy is still present.

b. *Loss-of-function mutations would have to be studied using homologous recombination,* where a normal copy of the gene is replaced with a defective version. To observe a phenotype of a recessive loss-of-function mutation, there should not be any normal version of the gene present. After replacing one gene copy with a mutant version, heterozygous mice would be mated to obtain a mouse homozygous for the recessive allele and to observe the resulting phenotype.

II. You have isolated muscle cell RNA and made a muscle cDNA library. You want to study the regulatory region of one of the genes you identified as being expressed in muscle cells.

a. You need to isolate a genomic clone of this gene. Why is the cDNA clone not sufficient for your studies?

b. Outline the steps to isolate the genomic clone of this gene.

c. You isolated a genomic clone containing the entire coding region and 1.2 kb upstream of the coding region. A map of the genomic clone is shown here. To make a fusion construct of the presumed regulatory region attached to the reporter gene *lacZ,* you cloned the 1.3 kb *Bam*HI fragment into a vector containing the *lacZ* gene. The fusion construct, when injected into cultured muscle cells, expressed β-galactosidase protein. What experiment would you do next to determine if the fusion construct is developmentally regulated?

d. Using an antibody against the β-galactosidase portion of the fusion protein you analyzed the tissues of a transgenic mouse and found that the protein encoded by this gene is also present in bone cells. How could you determine if the same transcript was expressed in both cell types?

e. You decided to analyze the promoter region of the gene to identify sequences responsible for the expression in the two different cell types. The results obtained with constructs containing different fragments from the 5′ region flanking the gene to the *lacZ* coding region are shown here. What region(s) are important for expression in muscle? in bone?

Answer

This problem requires familiarity with several molecular techniques including analysis of cDNAs and regulatory sequences in the promoter.

a. *The cDNA clone was derived from the mRNA and therefore it will probably not contain the regulatory region usually found upstream of (5′ to) the transcription start site.*

b. The cDNA clone is a good source of a probe for finding the genomic clone in a genomic library.
   1. *Isolate a DNA fragment from the cDNA clone and label for use as a probe.*
   2. *Grow clones (bacteriophages or cosmids) from a genomic library on plates.*
   3. *Transfer clones from plates to filter paper.*
   4. *Incubate the labeled probe with the filters to find, through hybridization, a genomic clone containing the gene.*

c. *To study developmental expression of this gene, the fusion construct would be transferred into the mouse genome, using transgenic techniques. At different stages in development the tissues of the mouse can be tested for the presence of the mRNA or protein using the lacZ DNA or antibody that recognizes the β-galactosidase protein respectively.*

d. To examine the transcripts in these two cell types, you would *prepare mRNA from isolated muscle and bone cells and use the cDNA clone to probe separated RNAs that have been transferred to a filter* (Northern hybridization analysis). Transcripts may be different sizes because of different transcription start sites or different processing (splicing) events.

e. The presence or absence of bone or muscle transcripts in mice that carry different transgene constructs provides a way to identify DNA fragments containing tissue-specific regulatory sequences. There are three constructs (4, 5, and 6) that express the transgene in bone tissue. The only portion of the regulatory region that they all have in common is a *50 bp* Sau3A *fragment. The bone-specific regulatory element must be within this fragment.* For muscle expression, there are *three clones (5, 6, and 7) that express the transgene in muscle tissue and these share the adjacent 350 bp* Sau3A *fragment.* This analysis shows that expression of the same gene is subject to different mechanisms of regulation in two tissue types that use different regulatory sites in the 5′ region flanking the gene.

DNA in subclone	Expression in bone	Expression in muscle
1.	−	−
2.	−	−
3.	−	−
4.	+	−
5.	+	+
6.	+	+
7.	−	+

PROBLEMS

E-1 Choose the matching phrase in the right column for each of the terms in the left column.

a.	stem cells	1.	cells destined to become the fetus
b.	conserved synteny	2.	animals derived from two or more genetically different embryonic cells
c.	inner cell mass	3.	undifferentiated cells that serve as a source of two or more types of differentiated cells
d.	trophectoderm	4.	experimental elimination of gene function
e.	embryonic stem cells	5.	the same genes are genetically linked in two different species
f.	chimeras	6.	cells destined to become extraembryonic tissue such as the placenta
g.	knockout	7.	undifferentiated cells isolated from the blastocyst that are able to be reconstituted with normal cells in an embryo and differentiate normally into every tissue type

E-2 The mouse genome (3000 Mb) is 30 times larger than the *C. elegans* genome (~100 Mb).
a. What challenges are there for a geneticist studying an organism that has a large genome size?
b. Despite the large genome size, many geneticists choose to use the mouse as a model system? Why?

E-3 The *CFTR* gene, which is defective in humans with cystic fibrosis, encodes a membrane protein that acts as a channel for the passage of Cl^-. Although one particular mutation (a deletion called D508) is predominant in Caucasians, more than 200 different mutations in the gene have been identified that cause the disease. Many Cl^- channel genes, including *CFTR* homologs, have been identified in other organisms such as *C. elegans* and yeast. For each of the following research questions, indicate whether you would be more likely to pursue an answer using yeast or mouse and why you would choose that organism.
a. Do mutations in different regions of the gene have the same effect on Cl^- channel function?
b. Do mutations in different regions of the gene affect channel function in all organs normally involved in cystic fibrosis?
c. What portion of the protein receives a signal to open the channel?
d. Are there drugs that can cause an opening of the channel?

E-4 A mouse gene was localized to a region of chromosome 4. From the synteny map, it appears this gene would localize to chromosome 8 in humans.
a. How could you determine if the gene is in this region of human chromosome 8? (Do not use the approach of cloning the gene here.)
b. Briefly outline how you could obtain the human clone containing the homolog of this mouse gene.

E-5 Why do identical twins occur in mice (and other mammals) but not in *Drosophila?*

E-6 How are different coat color alleles used in the protocol for making a gene knockout in mice?

E-7 You have discovered a gene, called *CTF,* that is expressed for several days during development of the embryonic heart and pituitary in mice. Expression continues in the pituitary in the adult and also is seen in the germ cells of the adult testis.
a. You bred a male mouse carrying one normal *CTF* allele ($+$) and one disrupted *CTF* allele ($-$) with a female heterozygous for the same alleles and examined 30 of their offspring. (The disrupted allele contains an insertion within the gene.) Twenty-one of the offspring were $+/-$ and 9 were $+/+$. What would you conclude about the *CTF* gene in this case?
b. Suppose instead that you obtained 30 offspring and 7 of the mice that had the $(-/-)$ genotype were half the normal size. Of these, all 4 males were sterile, whereas the 3 females were fertile. What would you conclude about the *CTF* gene from these results?
c. In the preceding experiments, what technique(s) were used to determine if the offspring mice were $(+/+), (+/-),$ or $(-/-)$?

E-8 Retinoic acid, which acts on cells via a protein receptor, is thought to be important for limb development in mammals.
a. Several different strains of mice that have recessive defects in limb development are available. You have tested each of these strains for the presence of the retinoic acid receptor (RAR protein), RAR mRNA, and the RAR gene using Western (protein), Northern (RNA), and Southern (DNA) blots, respectively. Based on the following data, give a reasonable hypothesis regarding the defect in each mutant.

Strain	RAR protein	RAR mRNA	RAR gene
A	Normal*	Normal	Normal
B	Absent	Normal	Normal
C	Absent	Short	Normal
D	Absent	Short	Altered†
E	Absent	Absent	Altered†
F	Absent	Absent	Absent

*Normal means that the protein, RNA, and DNA bands were normal in size and abundance; normal does not refer to function of the protein.

†One of the bands hybridizing with the RAR cDNA clone migrated to a different position in strains D and E.

b. The limb deformity in strain A could be due to a defect in some gene other than *RAR*. Suggest a simple genetic experiment that would enable you to test this possibility.

c. If you wanted to test the developmental consequences of *RAR* gene expression in neurons, which do not normally make RAR, how would you do this?

E-9 When DNA is injected into the fertilized mouse egg, the DNA can insert at random in any of the chromosomes. Subsequent matings produce animals homozygous for the transgene insertion. Sometimes an interesting mutant phenotype is generated by the insertion event. In one case, after injection of DNA containing the mouse mammary tumor virus (MMTV) promoter fused to the *c-myc* gene, investigators identified a recessive mutation that causes limb deformity. In this mouse, the distal bones were reduced and fused together; the mutation also caused kidney malfunction.

a. The mutant phenotype could be due to insertion of the transgene in a particular region of the chromosome or a chance point mutation that arose in the mouse. How could you distinguish between these two possibilities?

b. The mutation in this example was in fact caused by insertion of the transgene. How could you use this transgene insertion as a tag for cloning?

c. The insertion mutation was mapped to chromosome 2 of mice in a region where a mutation called limb deformity (*ld*) had previously been identified. Mice carrying this mutation are available from a major mouse research laboratory. How could you tell if the *ld* mutation was in the same gene as the transgenic insertion mutation?

d. Analysis of transcripts from the *ld* gene showed that many different transcripts (formed by alternate splicing) were present in both the embryo and adult. Is this consistent with a role of the *ld* gene product in limb development in the embryo?

E-10 Several different mouse mutants have been identified that have an obese phenotype. The *Ob* gene, defective in one class of these mutants, was the first gene involved in obesity to be cloned. The *Ob* gene product is made in fat cells and is transported to the brain, where it informs the brain that the animal is satiated (full).

a. You made an antibody to the Ob protein and used this to test predictions of the hypothesis that Ob is a satiety factor. You isolated protein from mice that had been eating normally, from mice that had been starved, and from mice that had been force-fed a high-calorie diet. You did a Western analysis using your antibody as a probe against proteins from these animals. Results are shown here. Are these results

consistent with the hypothesis for the role of the Ob protein? Why or why not?

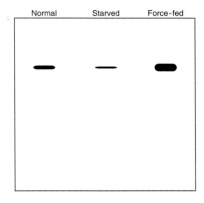

b. How could you determine if the *Ob* gene is transcriptionally regulated?

c. If the amount of mRNA was the same for all three types of mice, what type of regulation is involved?

d. You isolated RNA from several different *Ob* mutants and analyzed the RNA using Northern analysis. What conclusions could you reach about each mutation based on the results shown here?

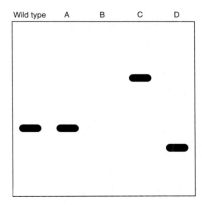

e. You have now decided to clone the receptor for the Ob protein and express the gene in mammalian cells. Which is the correct alternative for each of the following steps?

  1. (a) Isolate brain mRNA.
        (b) Isolate fat mRNA.
  2. (a) Clone cDNA into a plasmid vector next to a mouse metallothionein promoter.
        (b) Clone cDNA into a plasmid vector next to the yeast LEU2 promoter.
  3. (a) Treat cells with zinc.
        (b) Deprive cells of leucine.

f. You identified the *Ob* receptor clone using the screen outlined in part *e* and decided to make a knockout of the gene in mice. What phenotype would you predict for mice that lack the *Ob* receptor? (obese, slim, normal)

g. If you inject Ob protein into *Ob* mutant mice, what effect do you predict there will be on the phenotype?

There are inconsistencies within the various branches of genetics on some nomenclature—because it is a relatively new area of scientific investigation, the consistency present in more basic sciences has not been established. The authors debated whether they should try to impose a consistency on the entire topic area and decided against that path. As the study of genetics matures, the process itself will create a more consistent nomenclature. The following guidelines can be applied to all chapters in this book.

General Rules

- Names of genes are in italics (*lacZ, CDC28*)

- Names of proteins are in regular (Roman) type with an initial cap (LacZ, Cdc28)

- Chromosomes: sex chromosomes are represented by a capital letter in Roman type (X, Y); autosomes are designated by a cardinal number (1, 2, 21, 22)

- Names of transposons are in roman type (Tn10)

Specific Rules for Different Organisms

- **Bacteria:** lowercase italics for genes (*lac, ara*), with the addition of a capital letter to designate a specific gene in a pathway or operon (*lacZ, lacA, araB*); numbers (not superscript) for alleles (*trpC2, hisB2*); superscript "+" (plus) for wild-type alleles; superscript "−" (minus) for mutant alleles (*lacZ$^+$, lacZ$^-$*)

- **Yeast:** all caps for wild-type alleles (*CDC28*); all lowercase for mutant alleles, with a cardinal number indicating the specific mutation (*cdc28-1*)

- **Arabidopsis:** all caps for wild-type alleles (*LEAFY*, abbreviated *LFY*); all lowercase for mutant alleles (*lfy*)

- **C. elegans:** lowercase italics for genes and wild-type alleles (*dpy-10*); mutant alleles in parentheses following the gene [*dpy-10(e128)*]

- **Drosophila:** Many genes are named for the mutant phenotypes that revealed them. If the mutation causing the phenotype is dominant, the allele designation has an initial cap (*Deformed*, abbreviated *Dfd*); if the mutation that revealed the gene is recessive, the allele designation is all lowercase (*white*, abbreviated *w; wingless,* or *wg*)

- **Mice:** Genes are now named for the protein they encode and designated by an initial cap followed by any mix of letters or numbers (for example, *Tcp1* designates the gene for T complex protein 1); alleles are designated by superscripts. For multiple wild-type alleles, the superscripts may be *a, b, c,* or *1, 2, 3* (*Tcp1^a*, Tcp1^b); some wild-type alleles are indicated by a superscript plus "+" (*Kit$^+$*). For mutant alleles, the superscripts often describe the resulting phenotype (the *Kit^w* mutant allele causes white spotting).

- **Humans:** All caps for genes (*HD, CF*); wild-type alleles designated by a superscript plus, "+" (*HD$^+$*, CF$^+$); mutant alleles designated by a superscript minus "−" (*HD$^-$*, CF$^-$).

Chapter 2

1. a. 4; b. 3; c. 6; d. 7; e. 11; f. 13; g. 10; h. 2; i. 14; j. 9; k. 12; l. 8; m. 5; n. 1.

3. The dominant trait (short tail) is easier to eliminate from the population by selective breeding. You can recognize every animal that has inherited the allele, because only one dominant allele is needed to see the phenotype. Those mice that have inherited the dominant allele can be prevented from mating.

5. a. *aa Bb Cc DD Ee*; b. *a B C D E* or *a B c D E* or *a B C D e* or *a B c D e* or *a b C D E* or *a b C D e* or *a b c D E* or *a b c D e*.

7. Short hair is dominant to long hair.

9. a. Dry is recessive, sticky is dominant; b. The 3:1 and 1:1 ratios are obscured because the offspring are then combined results of different crosses.

11. a. $1/64$; b. $1/8$; c. $(3/4)^4$; d. $1-(3/4)^4$; e. $9/256$; f. $9/64$; g. $1/4$

13. The genotype can be determined by performing a testcross; that is, crossing your fly with the dominant phenotype (but unknown genotype) to a fly with the recessive (short wing) phenotype. If your fly has the homozygous dominant genotype the progeny in this case would be *Ww* and would have the dominant phenotype. If your fly had a heterozygous genotype, 1/2 of the progeny would be normal (*Ww*) and 1/2 of the progeny would be short (*ww*).

15. No. If you mate several red horses to each other and also mate several black horses to each other, the crosses that always yield offspring with the parental phenotype must have been between homozygous recessives.

17. *P* = purple, *p* = white; *S* = spiny, *s* = smooth. a. *Pp Ss × Pp Ss*; b. *PP Ss × P– ss* or *P– Ss × PP ss*; c. *Pp S– × pp SS* or *Pp SS × pp S–*; d. *pp Ss × Pp Ss*; e. *Pp ss × Pp ss*; f. *pp Ss × pp Ss*.

19. a. Rough and black are the dominant alleles (*R* = rough, *r* = smooth; *B* = black, *b* = white); b. a ratio of 1/4 rough black: 1/4 rough white: 1/4 smooth black: 1/4 smooth white.

21. a. 3/16; b. 1/16.

23. a. Recessive. Two unaffected individuals have an affected child. It was a consanguineous marriage that produced the affected child. II-1 and V-2 are affected (*aa*); all unaffected individuals except II-2, II-4, III-4, III-5, and possibly V-1 are carriers (*Aa*). b. Dominant. The trait is seen in each generation and each affected child has an affected parent; if the trait were recessive it would not be possible for III-3 to be unaffected even though both his parents are affected. All affected individuals are *Aa*, though III-4, III-5, and III-6 could be *AA*; carrier, is not applicable when the mutation is dominant; c. Recessive. Unaffected parents have an affected child. I-2 and III-4 are affected (*aa*); II-4 and II-5 are carriers (*Aa*); all others could be *AA* or *Aa*, but I-1 is almost certainly *AA* if the disease is rare.

25. a. 2/3; b. 1/9; c. 4/9.

27. Recessive; common.

29. a. $1/16 = 0.0625$; b. 0.0625.

31. In about 40% of the families, both parents were *Mm* heterozygotes. In the remaining 60% of the families, at least one parent was *MM*.

Chapter 3

1. a. 2; b. 6; c. 11; d. 8; e. 7; f. 9; g. 12; h. 3; i. 5; j. 4; k. 1; l. 10.

3. One gene, 2 alleles, incomplete dominance; red 5 $c^r c^r$, yellow 5 $c^r c^w$, white 5 $c^w c^w$.

5. Long is completely dominant to short. Flower color trait shows incomplete dominance of two alleles.

7. a. *ii* (phenotype O) or *i I^A* (phenotype A) or *i I^B* (phenotype B); b. $I^B I^B$, $I^B i$ or $I^B I^A$; c. *ii* (phenotype O).

9. a. 1/4 spotted dotted: 1/2 marbled: 1/4 spotted; b. marbled and dotted.

11. a. Coat color is determined by three alleles of a single gene arranged in a dominance series with *C* (for chinchilla) $> c^h$ (for himalaya) $> c^a$ (for albino). b. Cross 1: $c^h c^a \times c^h c^a$; Cross 2: $c^h c^a \times c^a c^a$; Cross 3: $Cc^h \times Cc^h$; Cross 4: $CC \times c^h c^h$ or $CC \times c^h c^a$; Cross 5: $Cc^a \times Cc^a$; Cross 6: $c^h c^h \times c^a c^a$; Cross 7: $Cc^a \times c^a c^a$; Cross 8: $c^a c^a \times c^a c^a$; Cross 9: $Cc^h \times c^h c^h$; Cross 10: $Cc^a \times c^h c^a$. c. 3/4 chinchilla (*CC*, *Cc^h*, and *Cc^a*) and 1/4 himalaya ($c^h c^a$).

13. 2/6 yellow: 3/6 albino: and 1/6 agouti progeny.

15. a. $A^y a Bb Cc \times Aa Bb Cc$; b. six phenotypes: albino, yellow, brown agouti, black agouti, brown, black.

17. Two genes are involved. The black mare was *AAbb* and the chestnut stallion was *aaBB*, the liver horses were *aabb* and the bay horses were *AaBb*.

19. a. There are two genes involved; homozygosity for the recessive allele of either or both genes causes yellow color.

Green parent is *AABB*; yellow parent is *aabb*. b. *AaBb, aaBb, Aabb* and *aabb* in equal proportions: 1/4 green: 3/4 yellow fruit.

21. Dominance relationships are between alleles of the same gene. Only one gene is involved. Epistasis involves two genes. The alleles at one gene affect the expression of a second gene.

23. 1/4 would appear to have O type blood, 3/8 have A, 3/8 have AB.

25.

	I-1	I-2	I-3	I-4	II-1	II-2	II-3	III-1	III-2
Phenotypes	AB	A	B	AB	O	O	AB	A	O
Genotypes	$I^A I^B$	I^A or $I^A I^A$	I^B or $I^B I^B$	$I^A I^B$	*ii*	I^B or $I^B I^B$	$I^A I^B$	$I^A i$	$I^A I^A$ or $I^A i$ $I^A I^B$ or $I^B i$
	Hh	*Hh*	*H–*	*H–*	*H–*	*hh*	*Hh*	*Hh*	*hh*

One or both of I-3 and I-4 must carry *h*.

27. a. 27/64 wild type, 37/64 mutant; b. *AA Bb Cc*.

29. a. Two genes are involved. *A–B–* and *aa B–* are WR, *A–bb* is DR, and *aa bb* is LR. b. For these true-breeding strains, WR-1 is *AA BB*; WR-2 is *aa BB*; DR is *AA bb*, and LR is *aa bb*. c. The cross was *Aa Bb* (WR) × *aa bb* (LR).

31. a. There are two genes involved. One gene has three alleles: c^R (red), c^L (lavender), and C^b (blue). c^R and c^L are both completely dominant to C^b; c^R and c^L are codominant or incompletely dominant, so c^R / c^L is bronze. There is a second gene for yellow; *Y–* allows the other colors to be visible, while *yy* produces yellow color and is epistatic to all genotypes at the other gene. b. (Parents only): Red ($c^R c^R$ *YY*) × blue ($C^b C^b$ *YY*); lavender ($c^L c^L$ *YY*) × blue ($C^b C^b$ *YY*); red ($c^R c^R$ *YY*) × lavender ($c^L c^L$ *YY*); red ($c^R c^R$ *YY*) × yellow ($C^b C^b$ *yy*); yellow ($c^L c^L$ *yy*) × blue ($C^b C^b$ *YY*). c. The F_2 will be 3 red : 6 bronze : 3 lavender : 4 yellow.

33. 44/56.

35. a. Single-gene inheritance with incomplete dominance. Heterozygotes have intermediate serum cholesterol levels; homozygotes have elevated levels. The following people must have the mutant allele but do not express it (incomplete penetrance): family 2 I-3 or I-4; family 4 I-1 or I-2. b. Other factors are involved, including environment (particularly diet) and other genes.

Chapter 4

1. a. 13; b. 7; c. 11; d. 10; e. 12; f. 8; g. 9; h. 1; i. 6; j. 15; k. 3; l. 2; m. 16; n. 4; o. 14; p. 5.

3. a. 23; b. 44 autosome s and two sex chromosomes; c. 23; d. one.

5. a. brown females and ivory-eyed males; b. females with brown eyes and males with ivory or brown eyes.

7. a. iii; b. i; c. iv; d. ii; e. v.

9. a. Mitosis, meiosis I, II; b. mitosis, meiosis I; c. mitosis; d. meiosis II and meiosis I; e. meiosis I; f. none; g. meiosis I; h. meiosis II, mitosis; i. mitosis, meiosis I.

11. 50%.

13. a. metaphase or early anaphase of meiosis I in a male (assuming X-Y sex determination in *Tenebrio molitor*); b. sister chromatids, centromeres, and telomeres (among others); c. five.

15. a. 400 spermatozoa; b. 200; c. 100; d. 100; e. 100; f. none.

17. a. Only females; b. males; c. males; d. 1/5 ZZ males and 4/5 ZW females.

19. Vestigial wings is autosomal; body color is X-linked recessive.

21. a. Nonbarred females and barred males; b. barred and nonbarred females and barred and nonbarred males.

23. a. Recessive; b. autosomal; c. *aa*; d. *Aa*; e. *Aa*; f. *Aa*; g. *Aa*; h. *Aa*.

25. a. X-linked dominant inheritance. b. Can exclude sex-linked recessive inheritance because affected females have unaffected sons. Can exclude autosomal recessive inheritance because the trait is rare and affected females have affected children with multiple husbands. Can exclude autosomal dominant inheritance because all the daughters but none of the sons of an affected male are affected. c. III-2 had four husbands and III-9 had six husbands.

27.

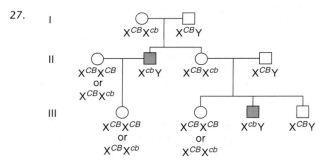

29. a. 3; b. 1 or 3.

31. Tail length is X linked with short dominant to long. XP is caused by homozygosity for recessive alleles of either or both of two autosomal genes (9:7 ratio with *A– B– XP+* and all other classes XP).

33. a. Purple is caused by homozygosity for a recessive allele of an autosomal gene (*p*), but the X-linked recessive white mutation is epistatic to *p* and to p^+. b. F_1 progeny: 1/2 white-eyed males and 1/2 wild-type (red) females; F_2 progeny: 1/4 white males, 1/4 white females, 3/16 red males, 3/16 red females, 1/16 purple males, 1/16 purple females.

35. Mate potential mutant females (with genotype X*y* / X*y*) with X*y* / $y^+ \cdot$ Y males. Nondisjunction mutants would have elevated numbers of brown-bodied female (X*y* / X*y* / $y^+ \cdot$ Y) and yellow-bodied male (X*y* / O) progeny.

Chapter 5

1. a. 8; b. 4; c. 1; d. 11; e. 2; f. 5; g. 6; h. 3; i. 10; j. 12; k. 9; l. 7.

3. a. $Oo\ Bb$; b. 9:3:3:1; c. not significant; d. between 0.5 and 0.1.

5. Both raspberry and sable are caused by recessive alleles of two X-linked genes that are separated by about 11.7 m.u. The parental generation was $r^+\ s\ /\ r^+\ s$ females and $r\ s^+\ /\ Y$ males.

7. a. Gametes: 20% $A\ b$ and $a\ B$, 30% $A\ B$ and $a\ b$. F_2 generation: 59% A–B–, 16% A–bb and $aa\ B$–, 9% $aa\ bb$.
b. Gametes: 30% $A\ b$ and $a\ B$, 20% $A\ B$ and $a\ b$. F_2 generation: 54% A–B–, 21% A–bb and $aa\ B$–, 4% $aa\ bb$.

9. 10%.

11. a. A = normal pigmentation, a = albino allele, $Hb\beta^A$ = normal globin, $Hb\beta^S$ = sickle allele

    49.5% $aHb\beta^A$, 49.5% $AHb\beta^A$, 0.5% $aHb\beta^A$, 0.5% $AHb\beta^A$;
    b. 49.5% $aHb\beta^A$, 49.5% $AHb\beta^A$, 0.5% $aHb\beta^A$, 0.5% $AHb\beta^A$;
    c. 0.0025.

13. a. 1/4 black, 1/2 albino, 1/4 brown; b. 34 m.u. apart.

15. a. 0; b. 1%; c. 5%; d.10%; e. 50%.

17. a. 39%; b. 39%; c. 0.5%; d. 8%.

19. a. 360 $a^+\ b^+\ c^+$; 360 $a\ b\ c$; 90 $a^+\ b\ c$; 90 $a\ b^+\ c^+$; 40 $a^+\ b^+\ c$; 40 $a\ b\ c^+$; 10 $a^+\ b\ c^+$; 10 $a\ b^+\ c$; b. 500 $a^+\ b^+\ c^+$; 500 $a\ b\ c$.

21. a. $sceccv\ /\ +\ +\ +$ and $b\ /\ +$; b. $sceccv\ /\ +\ +\ +$, sc–cv = 9 m.u., ec–cv = 10.5 m.u.; c. predicted DCO = 0.009, observed DCO = 0.001, interference = 0.89.

23. $\dfrac{dwp/pld^+\ rv^+\ rmp}{dwp^+/pld\ rv\ rmp^+}$ rv–rmp 10 m.u.; $pld\ /\ dwp$–rv 5 m.u.

25. a. 334;

    b.

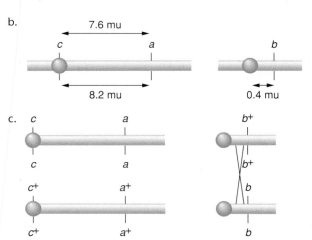

    c.

27. a. First group: met^+lys^+ and met^-lys^-; second group: met^+ lys^+, $met^+\ lys^-$, met^-lys^+, met^-lys^-. b. 5.5 m.u. c. met^-lys^+ and met^+lys^-.

29. a. 100% 4; b. 100% 2; c. 50% 0 and 50% 2; d. 40% 0, 20% 1, and 40% 2; e. 100% 0; f. 80% 0 and 20% 1; g. 2.

31. a. (Unordered tetrads): PD = 2 $lys^-\ his^+$ and 2 $lys^+\ his^-$; NPD = 2 $lys^-\ his^-$ and 2 $lys^+\ his^+$; T = 1 $lys^-\ his^+$, 1 lys^+ his^-, 1 $lys^-\ his^-$ and 1 $lys^+\ his^+$. b. 22.3 m.u. c. NCO = 222, SCO = 134; DCO = 44. d. 0.555 crossovers/meiosis. e. Two strand and three strand DCOs are missed. Map distance in map units = 1/2(T) + 3(NPD)/total asci; f. 27.8 m.u.

33. a. The sectors consist of $ade2^-/ade2^-$ cells generated by mitotic recombination. b. The sector size depends on when the mitotic recombination occurred during the growth of the colony. There should be many more small sectors because the mitotic recombinations creating them occur later in colony growth when there are many more cells.

35. a. Two mitotic crossovers occurred in succession in the same cell lineage. The first was between the sn and y genes, creating a patch of yellow tissue. The second was between the centromere and sn, creating a "clone within a clone" of yellow, singed cells. b. Yes.

Chapter 6

1. a. 6; b.11; c. 9; d. 2; e. 4; f. 8; g.10; h.12; i. 3; j.13; k. 5; l. 1; m. 7.

3. c.

5. Tube 1, nucleotides; tube 2, base pairs (without the sugar and phosphate) and sugar phosphate chains without the bases; tube 3, single strands of DNA.

7. a. 20% C; b. 30% T; c. 20% G.

9. Determine the percentage of each base.

11. 3' GGGAACCTTGATGTTTCGGCTCTAATT 5'.

13. 5' UAUACGAAUU 3'.

15. After one additional generation, 1/4 intermediate; after two additional generations, 1/8 intermediate.

17.

19. a. Relieves the stress of the overwound DNA ahead of the replication fork; b. unwinds the DNA; c. synthesizes a short RNA oligonucleotide; d. joins the sugar phosphate backbones.

21. Regardless of which strands are cut during resolution (to result in crossing-over or no crossing-over) mismatches within the heteroduplex region can be corrected to the same allele, resulting in gene conversion.

23. Would not undergo recombination.

Chapter 7

1. 1. a, b, h; 2. b, c; 3. f, j; 4. g, j; 5. a, b, h; 6. f, i, k; 7. g, k; 8. d, i; 9. e, f, i.

3. 9.5×10^{-5}; higher than normal rate.

5. If phages induce resistance, several appear in random positions on each of the replica plates. If the mutations preexist, the resistant colonies would appear at the same locations on each of the three replica plates.

7. Female A has a white-eyed mutation on the X. Female B has a recessive lethal mutation on the X. Female C is mosaic with a lethal mutation on one strand and wild-type sequence on the other strand of one X chromosome, or she on the other strand of one X chromosome, or she is heterozygenes for an incompletely penetrant lethal mutation.

9. Yes; Liver converts substance X into a mutagen.

11. If in another gene, suppressor and original mutation are separable during meiosis.

13. a. Three groups: (1,8); (2); (3,4,7); b. 3:1 ratio of lys^-:lys^+;

   c.

15. a. Deletions do not revert. Also, deletions will fail to recombine with either of two rII^- mutations that can recombine with each other to produce rII^+ phage. b. The length of the T4 chromosome predicts the number of nucleotide pairs. Recombination analysis with the mutants suggests the total map units in the T4 genome. Thus, Benzer could estimate the number of map units per nucleotide pair. c. rII^- mutations in the same nucleotide pair cannot recombine with each other to produce rII^+ phage.

17. a. 3, 6, and 7 are deletions (nonreverting);

   b.

   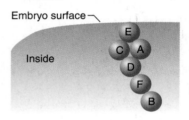

   c. Use other deletions in crosses with mutants 2 and 5.

19. a. Parental ditype: All spores Arg^-. Nonparental ditypes: 2 Arg^-: 2 Arg^+; b. Two of the PD spores grow on either ornithine, citrulline, arginosuccinate, or arginine; the other two grow with arginine only; NPD Arg^- spores grow with arginine only.

21. 45 purple : 16 green : 3 blue.

23. a. In all four crosses, there are two unlinked genes involved with complete dominance at both loci. b. (Each arrow represents a biochemical reaction catalyzed by one of the two gene products.) (Cross 1) colorless → blue → purple; (Cross 2) colorless1 → colorless2 → purple; (Cross 3) colorless1 → red and colorless2 → blue, with red + blue = purple; (Cross 4) colorless1 → purple and colorless2 → purple. c. Cross 2. d. F_2 only. (Cross 1) 2 purple: 1 blue: 1 white; (Cross 2) 1 purple: 1 white; (Cross 3) 2 purple: 1 red: 1 blue; (Cross 4) all purple.

25. a.
   $$X \xrightarrow{18} D \xrightarrow{14} B \xrightarrow{9} A \xrightarrow{10} C \xrightarrow{21} \text{thymidine}$$

   b. 9 and 10 accumulates B; 10 and 14 accumulates D.

27.

29. a. successful, immediate, prolonged; b. unsuccessful; c. successful, delayed, prolonged; d. successful, immediate, prolonged; e. unsuccessful; f. unsuccessful; g. successful, delayed, prolonged; h. unsuccessful; i. successful, immediate, short term; j. successful, immediate, prolonged.

31. One chromosome with β β/δ δ; another with β/δ only (where / signifies a protein part of which, for example, the N-terminal part, is one type of globin and the other part the other type of globin.)

33. a. β^+/β° and $\beta^+\beta^{20}$; b. β° is null, β^{20} has 20%, β^+ has 50%; c. 1/4; d. 1/4.

Chapter 8

1. a. 5; b. 10; c. 8; d. 12; e. 6; f. 2; g. 9; h. 14; i. 3; j. 13; k. 1; l. 7; m. 15; n. 11; o. 4 ; p. 16.

3. a. GU GU GU GU GU or UG UG UG UG;

   b. GU UG GU UG GU UG GU UG GU;

   c. GUG UGU GUG U etc.;

   d. GUG UGU GUG UGU GUG UGU GUG UGU GU (depends on where you start);

e. GUG UGU GU or UGU GUG UG (depends on where you start).

5. Hb^C therefore precedes Hb^S in the map of β-*globin* gene.

7. *In vitro* translation system.

9. Gene *F*: bottom strand; gene *G*: top strand.

11. a. Translation. b. Tyrosine (Tyr) is the next amino acid to be added to the C terminus of the growing polypeptide, which will be nine amino acids long when completed. c. The first amino acid at the N terminus would be f-met in a prokaryotic cell and met in a eukaryotic cell. The mRNA would have a cap at its 5′ end and a poly-A tail at its 3′ end in a eukaryotic cell but not in a prokaryotic cell. If the mRNA were sufficiently long, it might encode several proteins in a prokaryote but not in a eukaryote.

13. Three.

15. a. Mutant 1: transversion changes Arg to Pro; mutant 2: single-base-pair deletion changes Val to Trp and then stop; mutant 3: transition Thr (silent); mutant 4: single-base-pair insertion changes several amino acids then stop; mutant 5: transition changes Arg to stop; mutant 6: inversion changes identity of 6 amino acids. b. EMS: 1, 3, 5; Proflavin: 2, 4.

17. Transcription: adding appropriate ribonucleotide complementary to template; translation: codon in mRNA and anticodon in tRNA are complementary.

19. a. UGG changed to UGA, so the DNA change was G to A. b. If the second base of the Trp codon UGG changes to A, a UAG stop codon will result. If the third base of the Trp codon UGG changes to A, a UGA stop codon will result. Mutation of A to T in the first base of the Lys codon leads to UAA. If the Gly codon is GGA, mutation of the first G to T creates a UGA stop codon.

21. a. 3′ AUC 5′; b. 5′ CAG 3′, c. minimum two genes.

23. a. Missense mutations change identity of a particular amino acid inserted many times in many normal proteins but nonsense suppressors only make proteins longer b. (i) a mutation in a tRNA gene in a region other than that encoding the anticodon itself, so that the wrong aminoacyl-tRNA synthetase would sometimes recognize the tRNA and charge it with the wrong amino acid; (ii) a mutation in an aminoacyl-tRNA synthetase gene, making an enzyme that would sometimes put the wrong amino acid on a tRNA; (iii) a mutation in a gene encoding either a ribosomal protein, a ribosomal RNA or a translation factor that would make the ribosome more error-prone, inserting the wrong amino acid in the polypeptide; (iv) a mutation in a gene encoding a subunit of RNA polymerase that would sometimes cause the enzyme to transcribe the sequence incorrectly.

25. a. 5′ UAA 3′; b. 5′ UAG 3′ and 5′ UAA 3′ (due to wobble at the codon's 3′-most nucleotide); c. Gln, Lys, Glu, Ser, Leu, and Tyr.

27. a. Very severe; b. mild; c. very severe; d. mild; e. no effect; f. mild to no effect; g. severe; h. severe or mild.

29. Mutations possibly causing a detectable change in protein size: d, e, g, and i. In protein amount (assumes all mutant proteins are equally stable): e, f, j, and k. In mRNA size: i and j. In mRNA amount (assumes all mutant mRNAs with poly-A tails are equally stable): f and j.

31. Order: c e i f a k h d b j g.

Chapter 9

1. a. 10; b. 1; c. 9; d. 7; e. 6; f. 2; g. 8; h. 3; i. 5; j. 4.

3. Selectable markers are genes that allow a vector to impart protection from an antibiotic on a host cell. When cells are transformed by a vector with a selectable marker, and then exposed to the appropriate antibiotic, only cells that have the vector will survive.

5. a. all; b. 1/4; c. none; d. 1/2 chance; e. 3/4.

7. a. Five; b. divide the number of base pairs in the genome by the average insert size, then multiply by 5.

9. Where order doesn't matter, letters are bracketed: c, [j, f], a, k, i, g, b, e, h, d.

11. Shorter molecules slip through pores more easily; large molecules get caught.

13. After cloning: EcoRI: 42 and 2400 bp fragments; MboI: 705, 944, 500, and 300 bp or 905, 744, 500, and 300 bp fragments.

15. a. A; b. 10 kb; c. 10 kb.

d.

17.

19. a. 1.83 kb.

b.

Stained gel | Autoradiogram

— 0.39 kb
— 0.35 kb

— 0.26 kb

— 0.16 kb
— 0.15 kb
— 0.13 kb* Labeled fragment — 0.13 kb
— 0.11 kb
— 0.10 kb
— 0.09 kb

c.

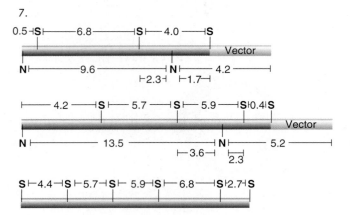

Insert | Vector

21. a. Use a probe of the same gene from another organism or if the amino-acid sequence is known, reverse translate to DNA sequence and make an oligonucleotide to use as a probe.

23. mRNA 5′
UAGAUAAGGAAUGUAAGAUAU<u>A</u>ACUGAGAUUUAAC

25. b.

Chapter 10

1. a. 9; b. 4; c. 2; d. 6; e. 3; f. 8; g. 1; h. 5; i. 7.

3. a. Cut the clone into subfragments to use as probes and look for single bands in hybridization; b. oligonucleotides used to detect STSs surrounding a microsatellite could be used to amplify DNA from a large number of individuals in the population.

5. a. Beta; b. 16 mu between alpha and delta.

7.

9. The other 2 cosmids probably contain repeated DNA also found in the YAC.

11. (1) Search for transcribed regions; (2) search for open reading frames and splice sites; (3) hybridize to look for cross-hybridizing DNA in other species; look for homoloy to STS sequences.

13. (i) 4.1 kb mRNA: the primary transcript includes all the DNA between the B, C, D, E and F bands and part or all of A and G. Bands A, D, F and G must contain exon sequences. There are a minimum of 2 introns — one that includes bands B + C and a second that includes band E. (ii) 3.4 kb mRNA was produced by a primary transcript that extended from somewhere in fragment B to somewhere in fragment E. This gene has a minimum of two exons and one intron in the region covered by fragment D and probably extending into fragments C and E. (iii) 1.8 kb transcript was produced by a primary transcript that extends from D into G. Because the mRNA is only 1.8 kb but the total fragment length in this region is 3.5, there must be at least 1.7 kb of intron in this region. This could be a 1 intron in fragment E. If not this then there must be a series of small introns because even the small DNA fragments like G have exon sequences.

15. Alternate splicing of the same primary transcript, or duplicated genes where the sequence in the middle has diverged.

17. Pseudogenes lack introns, have a poly-A tail and do not have regulatory DNA sequences upstream of the gene.

19. a. Protein motifs; b. tissue- and development-specific expression; knocking out the gene.

21. mRNAs present in all tissues could represent alternately spliced messages leading to different proteins that are not recognized by the antibody. Or, the two other messages are not translated.

23. a. Cloning a cDNA; b. clone into an expression vector.

25. (i) The region may have many genome-wide repeat sequences; (ii) the region may be duplicated at other sites in the genome; (iii) the region may have long stretches of repeated DNA (iv) centromeric and telomeric regions of the genome can not be cloned.

Chapter 11

1. a. 5; b. 3; c. 8; d. 6; e. 2; f. 7; g. 1; h. 4.

3. a. Different numbers of simple sequence repeats; b. slippage of DNA polymerase during replication; c. a different mechanism: unequal crossing-over.

5. a. The polymorphism is within the short DNA sequence that is used as a probe; b. the polymorphism is in the nucleotide adjacent to the sequence used as a primer; c. the SNP polymorphism can be kilobases away from the probe sequence in a restriction site recognized by the restriction enzyme used to digest the genomic DNA.

7. The sequences of the ASOs would be

5′ GATATTTACCCGATCCGCA and 5′ GATATTTACGCGATCCGCA.

9. Members of the disease family must be segregating two or more alleles at each DNA marker that is chosen.

11. Coworker 3 has the same DNA fingerprint as the crime sample and must be the perpetrator of the crime. The probability is essentially 100%.

13. Sperm collected from man, eggs are collected from woman. After *in vetro* fertilization, embryos are allowed to develop to the eight-cell stage. A single cell from each eight-cell embryo is removed. DNA is prepared and genotype is analyzed using PCR and *Mst* II digestion. Embryos with the desired genotype are implanted into the woman's uterus.

15. a. 10 kb, b. 7 kb. c. 07, d. 50%.

17. c, e, g, a, d, b, f.

19. Trait is autosomal dominant. Penetrance appears to be complete. There are no obvious carriers of the mutation who do not express the disease.

21. Trait is X-linked recessive. Individuals who carry the mutant allele but do not express the disease are I2, II2, and II7.

23. Identify sequences that are transcribed into RNA; use computational analysis to identify sequences that are conserved between distantly related species; use computational analysis to identify sequences that are open reading frames with appropriate codon usage and splice sites.

25. a. Choose families with many members that are afflicted with the disease at a young age; b. create a genetically engineered mouse model to confirm the role of the gene.

Chapter 12

1. a. 4; b.10; c. 5; d. 8; e. 9; f. 2; g. 3; h. 6; i. 1; j. 7.

3. Interphase: 40-fold compaction; metaphase: 10,000-fold compaction.

5. Mutate the DNA sequence so that the twelfth amino acid encoded is not lysine but another similar amino acid.

7. a. 68.75 kb per second; b. 7.3 bases per second.

9. The sequences added were specific for the species adding them (in this case, yeast)

11. a. Integrates; b. maintained separately but mitotically unstable; c. maintained separately but mitotically unstable.

13. 0.6 kb *Sau*3A.

15. More genes adjacent to boundary would be expressed.

17. a. Drosophila: Constitutive heterochromatin: centromeric DNA, Y-chromosome facultative heterochromatin: PEV with the *white* gene; b. Human: constitutive heterochromatin: centromeric DNA, Y chromosome facultative heterochromatin: inactived X (Barr body).

19. a. 1; b. 0; c.1; d. 1; e. 3; f. 0.

21. Some patches of cone cells in the eye in which the X chromosome carrying the *CB* allele was inactivated would be defective in color vision. If several patches occurred next to each other, there might be partial color vision.

Chapter 13

1. a. 4; b. 8; c. 6; d. 5; e. 7; f. 3; g. 2; h. 1.

3. In a duplication, there would be a repeated set of bands; in a deletion, bands normally found would be missing.

5.

apterous   rolled   straw   thick

7. a. 2 URA3 ARG9 spores and 2 ura3 arg9 spores;
   b. 2 spores die, 1 URA3 ARG9 and one ura3 arg9;
   c. 4 viable spores, 2 URA3 ARG9 spores and 2 ura3 arg9 spores.

9. A two-strand double crossover with both crossovers in the inversion loop. One crossover must occur between *LEU2* and *HIS4*. The other crossover must occur on the other side of either of the two genes but still within the inversion loop.

11. a. 2, 4; b. 2, 4; c. 2; d. 1, 3.

13. a. 1/4 fertile green, 1/4 fertile yellow-green, 1/4 semisterile green, 1/4 semisterile yellow-green; b. 1/2 fertile yellow-green, 1/2 semisterile green; c. from crossing-over events between the translocation chromosome and homologous region on the normal chromosome.

15. a. 1, 3, 5, 6; b. 2, 4; c. 1, 3; d. 5, 6.

17.

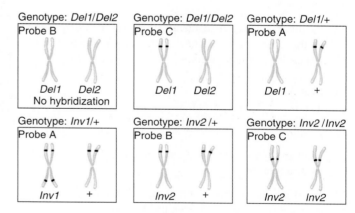

19. 1/2 *Lyra* males: 1/2 *Lyra*⁺ (wild type) females.

21. a. 1/3 fertile : 2/3 semisterile. b. Most likely, translocation homozygotes do not survive because the translocation breakpoint disrupted an essential gene.

23. *Ds* is a defective transposable element and *Ac* is a complete, autonomous copy.

25. Use a probe made of DNA from the sequence preceding the 200 A residues to hybridize to genomic DNA on Southern blots or to chromosomes by *in situ* hybridization.

27. a. 7; b. sand oats: diploid, slender wild oats: tetraploid, cultivated oats: hexaploid; c. sand oats: 7, slender wild oats: 14, cultivated wild oats: 21; d. same answer as c.

29. A: meiosis II in father; B: meiosis I in mother; C: meiosis I in father; D: meiosis II in mother.

31. You would actually expect more monosomies than trisomies, because meiotic nondisjunction would produce equal frequencies of monosomies and trisomies, but chromosome loss would produce only monosomies. The low frequency of monosomies observed is because monosomic zygotes usually arrest development so early that a pregnancy is not recognized. This may be due to a lower tolerance for imbalances involving only a single copy of a chromosome than for those involving three copies, or because recessive lethal mutations are carried on the remaining copy.

33. a. Mate putative mutants that are *ey ci⁺ / ey⁺ ci* with flies that are *ey ci / ey ci*. b. Nondisjunction during meiosis I will produce wild-type progeny; nondisjunction during meiosis II cannot be recognized. c. 2 eyeless : 2 cubitus interruptus : 1 eyeless, cubitus interruptus : 1 wild type. d. Mate putative mutants that are *ey ci⁺ / ey⁺ ci* with flies that have an unmarked attached chromosome 4. Nondisjunction during meiosis II would yield eyeless or cubitus interruptus progeny, but you could not recognize progeny resulting from nondisjunction during meiosis I. If the attached chromosome 4 carried two copies of *ey* and two copies of *ci,* you could recognize and discriminate some of the products of nondisjunction during the two meiotic divisions.

35. Treat with colchicine.

37. In autopolyploids, the banding patterns of homologs should be the same; in allopolyploids, different banding patterns will be seen for chromosomes from different species.

Chapter 14

1. a. 4; b. 5; c. 2; d. 7; e. 6; f. 3; g. 1.

3. a. iv; b. iii; c. ii.

5. Southern hybridization of *Eco*RI digested DNA using an IS1 DNA as a probe.

7. Transform the plasmid into a nontoxin-producing recipient strain, assay for toxin production.

9. *pyrE xyl mal arg met tyr his.*

11. *lac arg102 arg101 arg103.*

13. a. F plasmid must have integrated into the chromosome near the *mal* genes and then exased incorrectly.

a.

15. a.

b. Use an Hfr strain that transfers the gene late (Hfr 3) and screen for a derivative that transfers the gene early.

17. DNA being transferred by transformation is degraded; DNase has no effect on DNA being transduced.

19. Generalized transduction: transducing phage particles contain chromosomal DNA only; specialized transduction particles contain phage and chromosomal DNA.

21. a. (1) Pass extract of cells through APTG-agarose resin. (2) Digest with trypsin. (3) Use mass spectrometry to get molecular weights. (4) Compare profile of molecular weights with predicted MW from genome data. b. Make a fusion between protein coding region and lacZ and use APTG-agarose resin again.

Chapter 15

1. a. 6; b. 8; c. 7; d. 2; e. 1; f. 3; g. 5; h. 9; i. 4.

3. a. both; b. both; c. neither; d. both.

5. a.

	Trp	His	Ile	Met
mRNA	5′ UGG	CAU/C	AUU/C/A	AUG
nDNA mRNA-like	5′ TGG	CAT/C	ATT/C/A	ATG
nDNA template	3′ ACC	GTA/G	TAA/G/T	TAC

   b.

	Trp	His	Ile	Met
mRNA	5′ UGA/G	CAU/C	AUC/A	AUG/A
mtDNA mRNA-like	5′ TGA/G	CAT/C	ATC/T	ATG/A
mtDNA template	3′ ACT/C	GTA/G	TAG/A	TAC/T

7. a, c, e.

9. Demonstrates it is not a nuclear gene that is getting passed on from the fertile male.

11. a. 3; b. 1; c. 2.

13. Resemble mother because sperm does not contribute cytoplasmic material to the zygote.

15. a. Resistant to erythromycin (genotype ery^r); b. genotype ery^r / ery^s and the phenotype reflects the dominant allele; c. Sporulate the diploid. If the gene is nuclear, two spores would be ery^s and two would be ery^r. If the gene were mitochondrial, all four spores would be ery^r.

17. a. 2; b. 1; c. 4; d. 3.

19. a. Mother could have a small proportion of mutant genomes and the proportion in her daughter is higher or the mutation occurred in the germ line of I-2. b. Examine mother's somatic cells.

21. Differences in the time during development when mutation occurred or the proportion of mutant genomes in the cells in each tissue.

Chapter 16

1. a. 4; b. 8; c. 5; d. 2; e. 7; f. 1; g. 3; h. 6.

3. Nonlysogenic recipient cell did not have the cI (repressor) protein, so incoming infecting phage could go into the lytic cycle.

5.

	β-galactosidase	Permease
a.	constitutive	constitutive
b.	constitutive	inducible
c.	inducible	inducible
d.	no expression	constitutive
e.	no expression	no expression

7.

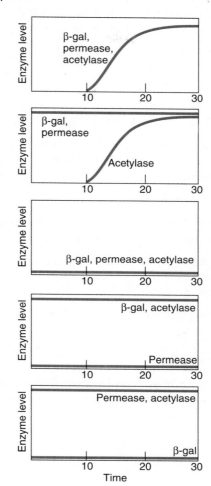

9. b.

11. a. 1 and 2; b. 2 and 4.

13. a. 4; b. 6; c. 7; d. 2; e. 3; f. 5; g. 1.

15. If the three genes make up an operon, they are cotranscribed to one mRNA and only one band should appear on a hybridization analysis using any of the three genes as a probe versus mRNA. If the genes are not part of an operon, there would be three differently sized hybridizing bands.

17. Seven His codons (CAC or CAU), in a row.

19. a. Operator begins on the left site after the endpoint of deletion 1 and before the endpoint of deletion 5. The right endpoint cannot be determined by this set. b. Removed bases within the promoter.

21. O_2 and O_3 have duplicated function, so mutations in O_2 or O_3 alone have only small effects in synthesis levels and would therefore be difficult to detect in the screens.

23. The protein-coding region of your gene must be in the same frame as the *lacZ* gene.

Chapter 17

1. a. 8; b. 5; c. 1; d. 7; e. 3; f. 9; g. 6; h. 4; i. 2.

3. Introns are spliced out, ribonuclease cleaves the primary transcript near the 3′ end and a poly-A tail is added, 5′ methl CAP is added.

5. a. pol III; b. pol II; c. pol I.

7. a. DNA binding; b. DNA binding; c. protein dimerization; d. transcription activation; e. DNA binding.

9. Using probes from each of the three genes you would expect to see different signals corresponding to three different transcripts on a Northern blot.

11. If *Id* acts by quenching it interacts with *MyoD*, whereas if it blocks access to an enhancer it binds to DNA. Experimentally, look for binding to the regulatory DNA of a gene regulated by *MyoD*.

13. RNA polymerase transcribes naked eukaryotic DNA at high levels *in vitro*; PEV and heterochromatization shut off gene expression.

15. b.

17. The mother.

19. The mutant allele is *a*. a. *Aa*; b. *AA*; c. *AA*.

21. a. No expression; b. no expression; c. expression; d. no expression; e. no expression; f. expression; g. no expression.

23. a. 9:7 ratio of blue to white flowered plants; b. 15:1 ratio of blue to white.

25. The 5′ and 3′ untranslated regions could be cloned at the 5′ or 3′ ends of a reporter gene that is transformed back into *Drosophila* early embryos to see if either of the sequences affect the translatability of the reporter protein.

Chapter 18

1. a. 7; b. 6; c. 8; d. 2; e. 1; f. 9; g. 3; h. 5; i. 4.

3. Replica plate colonies onto two sets of plates—one incubated at 30°C and the other at 23°C. Cold-sensitive mutants defective in cell-cycle genes should die at the nonpermissive temperature (23°C).

5. a. 4%; b. 96%.

7. a. F; b. T; c. F.

9. a. growth factors; b. cyclin, cyclin-dependent protein kinase, any molecule in the signal transduction pathway, receptor or molecule that transmits a second signal.

11. c; e; a; b; d.

13. Uncontrolled growth, genomic instability, potential for immortality, ability to invade and disrupt local and distant tissues.

15. The commonality is an accumulation of mutations necessary to cause cancer.

17. e; b; a or d or c; f.

Chapter 19

1. Use RNA *in situ* hybridization or antibody staining to visualize which cells express the *achaete-scute* genes, the E(spl) transcription factors, and genes expressed by cells that have acquired a neural fate.

3. Possibly the cell that by chance has the most Delta protein on its surface (and/or the least Notch) will gain an advantage over the surrounding cells in a race to become the presumptive SOP cell.

5. Outside of the clone: the epithelium should be normal. Within the clone: all of the cells in proneural clusters should become SOP cells. If the mutant clone bisects a proneural cluster: most of the wild-type and mutant cells will become epidermal cells. b. *Delta* mutants display nonautonomy.

7. a. H protein inhibits the function of Su(H) protein when the two are bound together. H and the intracellular domain of N compete with each other for binding to Su(H).
    b. *H* mutants should display autonomy.

9. Apparently, half the normal amount of Su(H) product is sufficient for lateral inhibition. b. Homozygosity for *Su(H)* null mutations should be epistatic to homozygosity for null mutations in *H*. All the cells in a proneural cluster should become presumptive SOP cells because there is no lateral inhibition.

11. Enhancers: a, c, and d. Suppressor: b.

13. a. Animals homozygous for mutations in *Notch* or *Delta* do not survive to adulthood and thus cannot make any eggs.
    b. (1) Maternally supplied Notch protein plays a role in early development that cannot be fulfilled by zygotically produced Notch protein. (2) There must be Notch ligands other than Delta.

15. Wild-type Hairy protein is likely to repress the transcription of the *achaete-scute* genes in regions of the animal that will develop into the wings and legs.

Chapter 20

1. a. 3; b. 5; c. 8; d. 7; e. 6; f. 1; g. 9; h. 2; i. 4.

3. a. *M*, 0.6; *N*, 0.4; *MM* 0.36; *MN* 0.48; *NN*, 0.16;
    b. Nonrandom mating, small population size, migration into the population, mutation, or selection.

5. a, e.

7. Each allele frequency has a different set of genotype frequencies at equilibrium.

9. a. 0.1; b. 478; c. 0.1.

11. a. 0.29; b. 0.29; c. 0.30.

13. 0.0064.

15. a. Wild type, 0.888; vestigial, 0.111; b. *Vg*, 0.75; *vg*, 0.25; c. wild type, 0.9375; vestigial, 0.0625.

17. A fully recessive allele is not expressed in a heterozygous organism, so there is no selection against the heterozygotes. Sometimes there is a heterozygous advantage, and mutation to the recessive allele will contribute to more mutant alleles.

19. a. founder effect; b. advantages: genetic homogeneity and fewer genes that may affect a polygenic trait; disadvantages: some mutations are not found in the population that are in the general population.

21. b.

23. a. 1/256 or 0.0039; b. 0.0016.

Chapter 21

1. a. 4; b. 6; c. 5; d. 2; e. 1; f. 7; g. 3.

3. a, c.

5. a. The enzyme consists of an RNA molecule; b. the enzyme has both an RNA and a protein component.

7. a. Different constraints on the functions of each of the proteins; b. rates are more constant because these base changes do not affect function of the gene product.

9. Suggests there is some benefit to the *CF* allele in the heterozygous state.

11. Duplication followed by evolutionary divergence.

13. a. 240 million years; b. two; *C* allele arose 30 million years ago; *B* allele arose 1 million years ago; c. duplication of *B*: transposition; duplication of *C*: misalignment and crossing-over.

15. This gene was introduced from a different species.

17. a. Exons; b. genes.

19. They mediate genome rearrangements or contribute regulatory elements adjacent to a gene.

21. a. SINEs or LINEs; b. centromere satellite DNA.

Reference A *Saccharomyces cerevisiae*

1. a. 3; b. 4; c. 5; d. 7; e. 2; f. 1; g. 6.

3. Isolate DNA from a culture of each of the 10 clones, use pulse-field gel electrophoresis to compare sizes of chromosome XII in the samples.

5. a. Subject to pulse-field gel electrophoresis, transfer to filter and probe with a subtelomeric repeat piece of DNA; b. use cloned piece of DNA as a probe to hybridize to a blot of chromosomes separated by gel electrophoresis.

7. *a*-maters, α-maters, and nonmaters.

9. a. α cell: α*1* + Mcm1 protein act as a positive regulator of α-specific gene transcription; α*2* + Mcm1 protein act to repress α-specific genes. *a* cell: Mcm1 protein transcribes *a*-specific genes. *a* / α cell: Mcm1 has no effect. b. α *mcm1*⁻: *a*-mater; *a mcm1*⁻: sterile; *a* / α *mcm1*⁻: sterile.

11. a. Southern hybridization using *SUC2* DNA from *Saccharomyces cerevisiae* as a probe against genomic DNA from other *Saccharomyces* species; b. pseudogene; c. larger transcript requires *SNF1* gene product, smaller one does not.

Reference B *Arabidopsis thaliana*

1. a. 3; b. 1; c. 4; d. 5; e. 2.

3. Extra DNA is repetitive DNA that is not necessary for the physiology of the plant.

5. a. 3*N* formed by fertilization of one sperm nucleus and two 1*N* nuclei in the ovule, genotype is like that of embryo (from double fertilization event); b. 2*N* formed by fertilization of one sperm nucleus with the 1*N* egg, genotype is like endosperm; c. 2*N* cells from fertilization that lead to the plant—not the same genotype as the embryo.

7. a. 2; b. 3; c. 1.

9. Do manipulated crosses in which pollen from the sterile plant is used to fertilize a wild-type plant and the reciprocal cross (pollen from fertile plant fertilizes ovule of plant with sterile genotype).

11. a. Examine the gravitropic response in an auxin mutant; b. hybridize using an auxin-regulated gene as a probe versus ground-up tissue from different parts of the plant or fixed tissue from the plant *or* make a fusion between an auxin-regulated gene and a reporter gene such as *GUS* and look at the distribution of expression of the reporter gene in the intact plant.

Reference C *Caenorhabditis elegans*

1. a. 2; b. 5; c. 4; d. 7; e. 6; f. 1; g. 3.

3. Self-fertilization makes it easier to recover recessive alleles.

5. b.

7. Lin14 protein level would be reduced further and cells influenced by Lin14 would differentiate at inappropriate times.

9. a. Cross males to the homozygous *dpy-10*⁻ / *dpy-10*⁻; b. wild-type progeny from first cross are crossed to *daf-1*⁻ / *daf-1*⁻; c. self-fertilization of the wild-type hermaphrodites from cross 2, 1/32 will have genotype desired.

11. a. Sequence of DNA from the mutant determined to look for a mutation in that particular K$^+$ channel gene or introduce a wild-type copy of the candidate gene into the mutant; b. Southern hybridization or PCR; c. dominant; d. compare the phenotypic response to isoamyl alcohol and diacetyl alcohol in the wild-type and mutant strains.

Reference D *Drosophila melanogaster*

1. a. 4; b. 5; c. 1; d. 6; e. 3; f. 2.

3. Mate males having insertion(s) of the w^+ marked *P* elements on the X chromosome with females homozygous for the w^- mutation (white eyes). Male flies in the next generation that have pigmented eyes should now have the *P* element on one of the autosomes.

5. e.

7. Segment polarity genes encode signal transduction proteins and act after cellularization has occurred. Gap and pair-rule genes encode transcription factors and act before cellularization.

9. Maternally supplied *hunchback* mRNA is completely dispensable and the function of the Nanos protein is only needed to restrict the translation of the maternally supplied *hunchback* mRNA in this posterior of the egg.

11. a. More than one progenitor cell forms any one facet. b. *White* gene is cell autonomous.

13. a. Time of expression during development and the cells in which the gene is expressed; b. heterozygous with a balancer chromosome; c. compare the pattern of *bicoid* mRNA in *homeless* versus wild-type embryos by hybridization using a *bicoid* DNA probe; d. Make females with the genotype centromere–*ovo^D*–ho$^+$/ centromere–*ovo$^+$*–ho$^-$, and induce mitotic recombination. If sterile, *homeless* is needed in the germ line. If fertile, *homeless* is needed in non-germ-line cells. e. Ras is needed for several different pathways, but the level of Ras protein required is different in the various signaling pathways.

Reference E *Mus musculus*

1. a. 3; b. 5; c. 1; d. 6; e. 7; f. 2; g. 4.

3. a. yeast; b. mouse; c. yeast; d. yeast and mouse.

5. In *Drosophila,* the zygote is a syncytium with polarity, and division of the embryo in two results in loss of important information.

7. a. *CTF* is an essential gene; b. *CTF* gene affects growth in both males and females and is necessary for fertility in males; c. different restriction patterns.

9. a. MMTV *c-myc* gene should segregate with the phenotype; b. hybridization of MMTV sequences versus a library of genomic clones from the mutant; c. in the *ld* mutant, sequence the gene into which the transgene inserted in the other mutant; d. consistent.

Note: An italicized word in a definition indicates that word is defined elsewhere in the glossary.

A

acentric fragment a chromatid fragment lacking a centromere; usually the result of a crossover in an inversion loop.

acrocentric a chromosome in which the *centromere* is close to one end.

activator a type of *transcription factor* that can bind to specific cis-*acting* enhancer elements and increase the level of transcription from a nearby gene.

adaptation the ability to stop responding when the stimulus is present.

adenine (A) a nitrogenous base, one member of the *base pair* A–T (adenine–thymine).

adjacent-1 segregation pattern one of two patterns of segregation resulting from the normal disjunction of homologs during meiosis I. Homologous centromeres disjoin so that one translocation chromosome and one normal chromosome go to each pole. Contrast with *alternate segregation pattern.*

adjacent-2 segregation pattern a pattern of segregation resulting from a nondisjunction in which homologous centromeres go to the same pole during meiosis I.

allele frequency the proportion of all copies of a gene in a population that are of a given allele type.

alleles alternative forms of a single *gene.*

allele specific oligonucleotides (ASOs) short oligonucleotides that hybridize with alleles distinguished by a single base difference.

allopolyploids polyploid hybrids in which the chromosome sets come from two or more distinct, though related, species.

allosteric proteins proteins that undergo reversible changes in conformation when bound to another molecule.

α-globin locus chromosomal region carrying all of the α-globin-like genes.

alternate segregation pattern one of two patterns of segregation resulting from the normal disjunction of homologs during meiosis I. Two translocation chromosomes go to one pole, while two normal chromosomes move to the opposite pole resulting in gametes with the correct haploid number of genes.

alternative splicing production of different mature *mRNAs* from the same *primary transcript* by joining different combinations of *exons.*

amino acids the building blocks of proteins.

aminoacyl (A) site site on a ribosome to which a *charged tRNA* first binds.

aminoacyl-tRNA synthetases enzymes that catalyze the attachment of tRNAs to their corresponding amino acids, forming *charged tRNAs.*

amniocentesis medical procedure in which a sample of amniotic fluid is taken from a pregnant woman to determine the condition of an unborn baby. A hollow needle is inserted through the woman's abdomen and uterine wall and the fluid, which contains cells shed by the embryo, is drawn off.

amphidiploids organisms produced by two diploid parental species; they contain two diploid genomes, each one derived from a different parent.

anaphase the stage of mitosis in which the connection of sister chromatids is severed, allowing the chromatids to be pulled to opposite spindle poles.

aneuploid an individual whose chromosome number is not an exact multiple of the haploid number for the species.

angiosperm plant descriptive meaning the seeds of the plant are enclosed in an ovary within the flower.

angstrom (Å) a unit of length equal to one ten-billionth of a meter.

anonymous locus a designated position on a chromosome with no known function; see *locus.*

Antennapedia complex (ANT-C) in *Drosophila;* a region containing several homeotic genes specifying the identity of segments in the head and anterior thorax.

anticodons groups of three nucleotides on *transfer RNA (tRNA)* molecules that recognize codons on the mRNA by *complementary base pairing* and *wobble.*

apical meristem a group of undifferentiated plant cells that divides continuously, located at the growing points of both shoots and roots.

apoptosis programmed cell death; a process in which DNA is degraded and the nucleus condenses; the cell may then be devoured by neighboring cells or phagocytes.

archaea one of the three major evolutionary lineages (domains) of living organisms.

artificial selection the purposeful control of mating by choice of parents for the next generation. Contrast with *natural selection.*

artificial transformation a process to transfer genes from one bacterial strain to another, using laboratory procedures to weaken cell walls and make membranes permeable to DNA. Contrast with *natural transformation.*

ascospores in some fungi, the haploid cells that result from meiosis. Also known as *haplospores.*

ascus saclike structure in some varieties of fungi that houses all four haploid products of meiosis.

autonomous elements intact transposable elements that can move from place to place in the genome by themselves. Contrast with *nonautonomous elements.*

autopolyploids a kind of polypoid that derives all of its chromosome sets from the same species.

autoradiograph an X-ray film image that records the location of radioactively labeled DNA.

autosome a chromosome not involved in sex determination. The diploid human genome consists of 46 chromosomes, 22 pairs of autosomes, and 1 pair of sex chromosomes (the X and Y chromosomes).

autotroph a plant that is nutritionally self-sufficient; it produces its own food via photosynthesis.

auxotroph a mutant microorganism that can grow on minimal medium only if it has been supplemented with one or more growth factors not required by wild-type strains.

B

bacterial chemotaxis bacterial movement up and down gradients of chemical attractants or repellents.

bacterial chromosome essential component of a bacterial genome; usually a single circular molecule of double-helical DNA.

bacteriophage See *phage.*

balancer chromosome special chromosome created for use in genetic manipulations; helps maintain recessive lethal mutations in stocks.

Barr bodies inactive X chromosomes observable at interphase as darkly stained heterochromatin masses.

basal factors a type of *transcription factor* that can associate nonspecifically with all promoters in a genome. A complex of basal factors associated with a promoter is required for the initiation of transcription by RNA polymerase.

base analogs mutagens that are so similar in chemical structure to the normal nitrogenous bases in DNA that the replication mechanism can incorporate them into DNA in place of the normal bases. This can cause base substitutions on the complementary strand synthesized in the next round of DNA replication.

base pair (bp) two nitrogenous bases on complementary DNA strands held together by hydrogen bonds. Adenine pairs with thymine and guanine pairs with cytosine.

base sequence the order of nucleotide bases in a DNA or RNA strand.

base sequence analysis a method for determining a *base sequence*.

basic chromosome number (X) the number of different chromosomes that make up a single complete set.

β-globin locus chromosomal region carrying all the β-globin-like genes.

B-form DNA the most common form of DNA in which molecular configuration spirals to the right. Compare with *Z-form DNA*.

biased random walk bacterial movement resulting from the addition of an attractant or repellant. The time spent in a straight run is longer immediately after the addition, so the movement over time is biased toward or away from the chemical gradient, though each direction change continues to be random.

biochemical pathway an orderly series of reactions that allows an organism to obtain simple molecules from the environment and convert them step by step into successively more complicated molecules.

biological systems complex interactive networks that function both within individual cells and among groups of cells within an organism.

biotechnology a set of biological techniques developed through basic research and now applied to research and product development. In particular, the use by industry of recombinant DNA, cell fusion, and new bioprocessing techniques.

bithorax complex (BX-C) in *Drosophila*; a cluster of homeotic genes that control the identity of segments in the abdomen and posterior thorax.

bivalent a pair of synapsed homologous chromosomes during prophase of meiosis I.

blastocysts describes the embryo at the 16-cell stage of development through the 64-cell stage, when the embryo implants.

blastomeres early embryonic cells.

blunt end a double-stranded DNA molecule without *sticky ends*.

branched-line diagram a method for systematically listing the expected results of multigene crosses.

branch sites special sequence of RNA nucleotides within an *intron* that helps form the "lariat" intermediate required for *RNA splicing*.

C

cancer genes genes having mutant alleles that lead to cancer.

carpels structure in a plant comprising the fourth whorl, usually two are fused together; houses the female gametes in the form of ovules. The fused carpels are part of a cylinder known as the *pistil*.

carriers *heterozygous* individuals of normal *phenotype* that have a *recessive allele* for a trait.

CDK inhibitor See *cyclin-dependent kinases*.

cDNA complementary DNA has a base sequence that is complementary to that of the mRNA template and contains no *introns*.

cDNA library a large collection of cDNA clones that are representative of the mRNAs expressed by a particular cell type, tissue, organ, or organism.

cell autonomous trait a trait for which the phenotype expressed by a cell depends solely on the genotype of that cell and not that of any other nearby cell. Compare with *nonautonomous trait*.

cell cycle repeating pattern of cell growth, replication of genetic material, and mitosis.

cellular blastoderm in *Drosophila* embryo; one-cell-deep epithelial layer resulting from cellularization of the *syncytial blastoderm*.

cellular clone an isolated colony of cells that are all descendants from a single progenitor cell.

centimorgan (cM) a unit of measure of recombination frequency. One cM is equal to a 1% chance that a marker at one genetic *locus* will be separated from a marker at a second locus due to *crossing-over* in a single generation.

centromere a specialized chromosome region at which sister chromatids are connected and to which spindle fibers attach during cell division.

centrosomes microtubule organizing centers at the poles of the *spindle apparatus*.

charged tRNA a tRNA molecule to which the corresponding amino acid has been attached by an *aminoacyl-tRNA synthetase*.

checkpoints mechanisms that prevent cells from continuing to the next phase of the cell cycle until a previous stop has been successfully completed, thus safeguarding genomic integrity.

chiasmata observable regions in which nonsister chromatids of homologous chromosomes crossover.

chimera an embryo or animal composed of cells from two or more different organisms.

chi square (χ^2) test a statistical test to determine the probability that an observed deviation from an expected outcome occurs solely by chance.

chromatin the generic term for any complex of DNA and protein found in a cell's nucleus.

chromocenter the dense heterochromatic mass formed by the fusing of the centromeres of polytene chromosomes.

chromosomal interference the phenomenon of crossovers not occurring independently. Refer to *crossing-over*.

chromosomal puff the region of a polytene chromosome that swells to form a large, diffuse structure when high rates of gene transcription cause bands to decondense.

chromosomal rearrangement change in the order of DNA sequence along one or more chromosomes.

chromosome loss a mechanism causing *aneuploidy* in which a particular chromatid or chromosome fails to become incorporated into either daughter cell during cell division.

chromosome theory of inheritance the idea that chromosomes are the carriers of genes.

chromosomes the self-replicating genetic structures of cells containing the DNA that carries in its nucleotide sequence the linear array of genes.

chromosome walking using the ends of unconnected *contigs* as probes to retrieve clones that extend into an unmapped region.

cis-acting element a DNA sequence that can bind *transcription factors* and, in so doing, control or modulate the level of transcriptional initiation from one or more nearby genes. Categories of *cis*-acting elements include promoters, enhancers, locus control regions, and boundary elements.

cistron a term sometimes used as a synonym for *complementation group* or *gene*.

cleavage stage in early embryonic development, the stage of the first four equal cell divisions.

clone a group of biological entities—cells or DNA molecules—that are genetically identical to each other.

cloning the process by which cellular clones or DNA clones are formed.

cloning vector DNA molecule into which another DNA fragment of appropriate size can be integrated without loss of the vector's capacity for replication. Vectors introduce foreign DNA into host cells, where it can be reproduced in large quantities. See *vector*.

coactivator a type of *transcription factor* or other molecule that binds to transcription factors rather than to DNA and plays a role in increasing levels of *transcription*.

codominant expression of heterozygous phenotype resulting in hybrid offspring that resemble both parents equally for a particular trait.

coefficient of coincidence the ratio between the actual frequency of double crossovers observed in an experiment and the number of double crossovers expected on the basis of independent probabilities.

coenocyte a plant cell containing multiple nuclei.

colinearity the parallel between the sequence of nucleotides in a gene and the order of amino acids in a polypeptide.

colony a mound of genetically identical cells.

competent description of state of cells able to take up DNA from the medium.

complementation process in which heterozygosity for chromosomes bearing mutant recessive alleles for two different genes produces a normal phenotype.

complementary base pairing during DNA replication, base pairing in which a complementary strand aligns opposite the exposed bases on the parent strand to create the nucleotide sequence of the new strand of DNA. Refer to *base pair, complementary sequences*.

complementary gene action genes working in tandem to produce a particular trait.

complementary sequences nucleic acid base sequences that can form a double-stranded structure by matching base pairs; the complementary sequence to 5′ GTAC 3′ is 3′ CATG 5′.

complementation group a collection of mutations that do not complement each other. Often used synonymously for gene.

complementation table a method of collating data that helps visualize the relationship among a large group of mutants. See Fig. 7.15b.

complementation test method of discovering whether two mutations are in the same or separate genes.

complete coverage in mapping of DNA, when the number of markers on a *linkage map* is sufficient so that the locus controlling any phenotype can be linked to at least one of those markers.

complete digest digesting a sample of DNA molecules with a restriction enzyme such that cleavage has occurred at every DNA site recognizable by the enzyme.

complete genomic library a hypothetical collection of DNA clones that includes one copy of every sequence in the entire genome.

complex refers to the multiple types of variation that can exist at alternative alleles, including more than one nucleotide substitution, a substitution in combination with a small deletion, a duplication, or another insertion.

complex haplotype a set of linked DNA variations along a chromosome, with the possibility of many differences in alternative alleles. See *haplotype*.

concerted evolution process that allows changes in single genetic elements to spread across a complete set of genes in a particular gene family.

conditional lethal an allele that is lethal only under certain conditions.

conjugation one of the mechanisms by which bacteria transfer genes from one strain to another; in this case, the donor carries a special type of plasmid that allows it to transfer DNA directly when it comes in contact with the recipient. The recipient is known as an exconjugant. Contrast *transformation*.

conjugative plasmids plasmids that initiate conjugation because they carry the genes that allow the donor to transfer genes to the recipient.

consanguine related by a common ancestor.

consanguineous mating mating between blood relatives sharing a recent common ancestor.

conservative substitutions mutational changes that substitute an amino acid in a protein with a different amino acid with similar chemical properties. Compare with *nonconservative substitutions*.

conserved synteny state in which the same two or more loci are found to be linked in several species. Compare *syntenic*.

constitutive expression refers to a state of gene activation that remains at a constant high level and is not subject to modulation.

constitutive heterochromatin chromosomal regions that remain condensed in heterochromatin at most times in all cells.

constitutive mutation a mutation in a *cis*-acting or *trans*-acting element that causes an associated gene to remain in a state of activation irrespective of environmental or cellular conditions that modulate gene activity in nonmutant cells.

contig a set of two or more partially overlapping cloned DNA fragments that together cover an uninterrupted stretch of the genome.

continuous trait inherited trait that exhibits many intermediate forms; determined by segregating alleles of many different genes whose interaction with each other and the environment produces the phenotype. Also called a quantitative trait. Compare with *discontinuous trait*.

core histones proteins that form the core of the nucleosome: H2A, H2B, H3, and H4.

corepressor a type of *transcription factor* or other molecule that binds to transcription factors rather than DNA and prevents transcription above basal levels.

cosmids hybrid plasmid-phage *vectors* that make use of a virus capsule to infect bacteria; constructed with plasmid-derived selectable markers and two specialized segments of phage λ DNA known as *cos* (for *co*hesive *e*nd) sites.

cotransformation simultaneous transformation of two or more genes. See *transformation*.

criss-cross inheritance inheritance pattern in which males inherit a trait from their mothers, while daughters inherit the trait from their fathers.

cross the deliberate mating of selected parents based on particular genetic traits desired in the offspring.

cross-fertilization to brush the pollen from one plant onto the female organ of another plant, thereby creating offspring.

cross-fostering random relocation of offspring to the care of other parents, typically done with animal studies to randomize the effects of environment on outcome.

crossing-over during *meiosis*, the breaking of one maternal and one paternal chromosome, resulting in the exchange of corresponding sections of DNA and the rejoining of the chromosome. This process can result in the exchange of alleles between chromosomes. Compare *recombination*.

crossover suppression result of heterozygosity for inversions, in which no viable recombinant progeny are possible.

C terminus the end of the polypeptide chain that contains a free carboxylic acid group.

cyclin-dependent kinases (CDKs) a protein kinase is an enzyme that adds a phosphate group to a target protein molecule; in this case, the kinases are dependent on proteins known as cyclins for their activity. CDKs regulate the transition from G_1 to S and from G_2 to M through phosphorylations that activate or inactivate target proteins.

cytokinesis the final stage of cell division, which begins during anaphase but is not completed until after telophase. In this stage the daughter nuclei emerging at the end of telophase are packaged into two separate daughter cells.

cytosine(C) a nitrogenous base, one member of the base pair G–C (guanine–cytosine).

D

dauer larva in nematodes, an alternate L3 form that does not feed and has a specialized cuticle that resists desiccation; dauer larvae can survive more than six months when food is scarce.

deamination the removal of an amino ($-NH_2$) group from normal DNA.

degeneracy property of the *genetic code* in which several different codons can specify the same amino acid.

degrees of freedom (df) the measure of the number of independently varying parameters in an experiment.

deletion occurs when a block of one or more nucleotide pairs is lost from a DNA molecule. Compare *insertion*.

deletion loop an unpaired bulge of the normal chromosome that corresponds to an area deleted from a paired homolog. Contrast *duplication loop*.

denaturation (denature, denatured) the disruption of hydrogen bonds within a macromolecule that normally uses hydrogen bonds to maintain its structure and function. Hydrogen bonds can be disrupted by heat, extreme conditions of pH, or exposure to chemicals like urea.

When normally soluble proteins are denatured, they unfold and expose their nonpolar amino acids, which can cause them to become insoluble. When DNA is denatured, double-stranded molecules break apart into two separate strands.

deoxyribonucleic acid (DNA) See *DNA.*

deoxyribonucleotide See *nucleotide.*

depurination DNA alteration in which the hydrolysis of a purine base, either A or G, from the deoxyribose-phosphate backbone occurs.

developmental genetics the use of genetics to study how the fertilized egg of a multicellular organism becomes an adult.

developmental hierarchy a *developmental pathway* in which the product of one gene regulates the expression of another gene.

developmental pathway a detailed description of how many genes interact biochemically to produce a particular outcome in development.

dicentric chromatid a chromatid with two centromeres.

dicotyledonous plant descriptive meaning that the mature embryo carries two leaves.

digestion the enzymatic process by which a complex biological molecule (DNA, RNA, protein, or complex carbohydrate) is broken down into smaller components.

dihybrid an individual that is heterozygous for two genes at the same time.

dimerization domain region of a polypeptide that facilitates interactions with other molecules of the same polypeptide or with other polypeptides. Certain motifs such as the leucine zipper often serve as dimerization domains.

dipeptide two amino acids connected by a peptide bond.

diploid zygotes and other cells carrying two matching sets of chromosomes are described as diploid (compare *haploid*).

discontinuous trait phenotypes that are expressed in clear-cut variations. Compare with *continuous trait.*

discovery science an approach to biology in which one seeks to identify all the elements of a biological system and place them in a database to enrich the infrastructure of biology.

discrete trait inherited trait that exhibits clearly an either/or status (that is, purple versus white flowers).

disease stratification the process of dividing a single general condition into different diseases on the basis of the underlying molecular defects.

DNA *deoxyribonucleic acid;* the molecule of heredity that encodes genetic information.

DNA clone a purified sample containing a large number of identical DNA molecules.

DNA fingerprint the multilocus pattern produced by the detection of genotype at a group of unlinked, highly polymorphic loci.

DNA markers See *marker.*

DNA probe a purified fragment of DNA labeled with a radioactive isotope or fluorescent dye and used to identify complementary sequences by means of hybridization.

DNase hypersensitive (DH) sites sites on DNA that contain few, if any, nucleosomes; these sites are susceptible to cleavage by DNase enzymes.

DNA sequence the relative order of *base pairs,* whether in a fragment of DNA, a *gene,* a *chromosome,* or an entire *genome.*

DNA topoisomerases a group of enzymes that help relax *supercoiling* of the DNA helix by nicking one or both strands to allow the strands to rotate relative to each other.

domain a discrete region of a protein with its own function. The combination of domains in a single protein determines its overall function.

domain architecture the number and order of a protein's functional domains.

dominance series dominance relations of all possible pairs of alleles are arranged in order from most dominant to most recessive.

dominant allele an allele that expresses its phenotype even when heterozygous with a recessive allele.

dominant epistasis the effects of a dominant allele at one gene hide the effects of alleles at another gene. Compare with *recessive epistasis.*

dominant negative mechanism of dominance in which some alleles of genes encode subunits of multimers that block the activity of the subunits produced by wild-type alleles.

donor in gene transfer in bacteria, the cell that provides the genetic material. See *recipient, transformation, conjugation, transduction.*

double helix the shape that two linear strands of DNA assume when bonded together.

downstream the direction traveled by RNA polymerase as it moves from the promoter to the terminator. Compare with *upstream.*

duplication events that result in the increase in the number of copies of a particular chromosomal region. See *tandem duplications, nontandem duplications.*

duplication loop a bulge in the *Dp*-bearing chromosome that has no similar region with which to pair in the unduplicated normal homologous chromosome. Contrast with *deletion loop.*

E

ecotypes plant varieties analogous to animal strains.

ectopic expression gene expression that occurs outside the cell or tissue where the gene is normally expressed.

electrophoresis a method of separating large molecules (such as DNA fragments or proteins) from a mixture of similar molecules. An electric current is passed through a medium containing the mixture and each kind of molecule travels through the medium at a different rate, depending on its size and electrical charge.

elongation phase of DNA replication, transcription, or translation that successively adds nucleotides or amino acids to a growing macromolecule. Compare *initiation.*

elongation factors proteins that aid in the *elongation* phase of *translation.*

embryonic stem cell (ES cells) cultured embryonic cells that continue to divide without differentiating.

endosymbiont theory proposes that chloroplasts and mitochondria originated when free-living bacteria were engulfed by primitive nucleated cells. Host- and guest-formed cellular communities in which each member adapted to the group arrangement and derived benefit.

enhancer mutations mutations in a *modifier gene* that worsen the phenotypic effects of a mutation in another gene.

enhancer trapping in *Drosophila,* the identification of P-element insertion lines with particular β-galactosidase expression patterns.

enhancers *cis*-acting elements that can regulate *transcription* from nearby genes. In yeast, enhancers are called upstream activation sites, or UASs. Enhancers function by acting as binding sites for transcription factors.

enhancesome a completely assembled set of transcription factors (which may include *activators, repressors, coactivators,* and *corepressors*) associated with an *enhancer* or *locus control region* in a structure that is able to modulate transcription activity.

environmental variance deviation from the mean attributed to the influence of external, noninheritable factors. Compare with *genetic variance.*

epigenetic a state of gene functionality that is not encoded within the DNA sequence but that is still inheritable from one generation to the next. It can be accomplished and maintained through a chemical modification of DNA such as methylation.

episomes plasmids, like the F plasmid, that can integrate into the host chromosomes.

epistasis a gene interaction in which the effects of an allele at one gene hide the effects of alleles at another gene.

equilibrium frequency the q value at which $\Delta q = 0$; the allele frequency required to maintain an allele in the population.

euchromatin   chromosomal region of cells that appears much lighter and less condensed when viewed under a light microscope. Contrast with *heterochromatin.*

eukaryotes   one of the three major evolutionary lineages of living organisms known as domains; organisms whose cells have a membrane-bounded nucleus. Contrast with *prokaryotes.*

eukaryotic gene regulation   the control of gene expression in the cells of *eukaryotes.*

euploid   cells containing only complete sets of chromosomes.

evolution   in the study of genetics, changes in allele frequency in a population over multiple generations; the basic components of evolution at the molecular level include diversification, selection, and amplification.

evolutionary equilibrium   the allele frequency at which evolution no longer continues; represented by p̂.

excision repair   DNA repair mechanism in which specialized proteins recognize damaged base pairing and remove the damaged section so that DNA *polymerase* can fill in the gap by complementary base pairing with the information from the undamaged strand of DNA.

exons   sequences that are found both in a gene's DNA and in the corresponding mature *messenger RNA (mRNA).* See *introns.*

exon shuffling   the exchange of exons among different genes during evolution, producing mosaic proteins with two or more distinct functions.

expression vectors   cloning vehicles which promote the expression of a polypeptide product from the gene inserts they carry.

expressivity   the degree or intensity with which a particular genotype is expressed in a phenotype.

F

facultative heterochromatin   regions of chromosomes (or even whole chromosomes) that are heterochromatic in some cells and euchromatic in other cells of the same organisms.

filial generations   the successive offspring in a controlled sequence of breeding, starting with two parents (the P generation) and selfing or intercrossing the offspring of each subsequent generation.

fine structure mapping   recombination mapping of mutations in the same gene; in some bacteriophage experiments, fine structure mapping can resolve mutations in adjacent nucleotide pairs.

first-division segregation pattern   a *tetrad* in which the arrangement of ascospores indicates that the two alleles of a gene segregated from each other in the first meiotic division.

FISH   (fluorescence *in situ* hybridization) a *physical mapping* approach that uses fluorescent tags to detect hybridization of nucleic acid probes with chromosomes.

fitness cost   negative impact of the development of a homozygous resistance genotype.

fluctuation test   the Luria-Delbrück experiment to determine the origin of bacterial resistance. Fluctuations in the numbers of resistant colonies growing in different petri plates showed that resistance is not caused by exposure to bacteriocides.

focus of action   the cells in which a gene of interest must be active to allow the animal to develop and function normally.

forward mutation   a mutation that changes a *wild-type allele* of a gene to a different *allele.*

founder cells   in nematode development, progenitors of the major embryonic lineages.

founder effect   variation of genetic drift, occurring when a few individuals separate from a larger population and establish a new one that is isolated from the original population, resulting in altered allele frequencies in the new population.

F plasmid   a *conjugative plasmid* that carries many genes required for the transfer of DNA. Cells carrying F plasmid are called F^+ cells. Cells without the plasmid are called F^- cells.

F′ plasmids   F plasmid variants that carry most F plasmid genes plus some bacterial DNA; particularly useful in genetic complementation studies.

frameshift mutations   insertions or deletions of base pairs that alter the grouping of nucleotides into codons. Refer to *reading frame.*

free duplications   small DNA fragments maintained extrachromosomally in a genetic stock.

fusion proteins   proteins encoded by parts of more than one gene.

G

gametes   specialized *haploid* cells (eggs and sperm or pollen) that carry genes between generations.

gametogenesis   the formation of *gametes.*

gastrulation   folding of the cell sheet early in embryo formation; usually occurs immediately after the blastula stage of development.

gel electrophoresis   a process used to separate DNA fragments, RNA molecules, or polypeptides according to their size. Electrophoresis is accomplished by passing an electrical current through agarose or polyacrylamide gels. The electrical current forces molecules to migrate into the gel at different rates dependent on their sizes.

gene   basic unit of biological information; specific segment of DNA in a discrete region of a chromosome that serves as a unit of function by encoding a particular RNA or protein.

gene amplification   an increase from the normal two copies to hundreds of copies of a gene; often due to mutations in *p53,* which disrupt the G_1-to-S checkpoint.

gene conversion   any deviation from the expected 2:2 segregation of parental alleles.

gene dosage   the number of times a given gene is present in the cell nucleus.

gene expression   the process by which a gene's information is converted into RNA and then (for protein-coding genes) into a polypeptide.

gene family   set of closely related genes with slightly different functions that most likely arose from a succession of gene duplication events.

gene function   generally, to govern the synthesis of a polypeptide; in Mendelian terms, a gene's specific contribution to *phenotype.*

generalized transduction   a type of *transduction* (gene transfer mediated by bacteriophages) which can result in the transfer of any bacterial gene between related strains of bacteria.

gene pool   the sum total of all alleles carried in all members of a population.

genetic code   the sequence of *nucleotides,* coded in triplets (codons) along the *mRNA* that determines the sequence of amino acids in protein synthesis.

genetic drift   unpredictable, chance fluctuations in allele frequency that have a neutral effect on fitness of a population.

genetic imbalance   situation when the genome of a cell or organism has more copies of some genes than other genes due to *chromosomal rearrangements* or *aneuploidy.*

genetic linkage   See *linkage.*

genetic markers   genes identifiable through phenotypic variants that can serve as points of reference in determining whether particular progeny are the result of recombination. Compare with *physical markers.*

genetic relatedness   the average fraction of common alleles at all gene loci that individuals share because they inherited them from a common ancestor.

genetics   the science of heredity.

genetic screen   an examination of each individual in a population for its phenotype.

genetic variance (V_G)   deviation from the mean attributable to inheritable factors. Compare with *environmental variance.*

genome   the sum total of genetic information in a particular cell or organism.

genomic equivalent   the number of clones—with inserts of a particular size—that would be required to carry a single copy of every sequence in a particular genome.

genomic imprinting the phenomenon in which a gene's expression depends on the parent that transmits it.

genomic library a collection of *DNA clones* that together carry a representative copy of every DNA sequence in the *genome* of a particular organism.

genomics the study of whole genomes.

genotype the actual alleles present in an individual.

genotype frequency proportion of total individuals in a population that are of a particular genotype.

germ cells specialized cells that incorporate into the reproductive organs, where they ultimately undergo meiosis, thereby producing haploid gametes that transmit genes to the next generation Compare *somatic cells.*

germ-line gene therapy a genetic engineering technique that modifies the DNA of *germ cells* that are passed on to progeny.

globular stage embryo stage of development in a plant in which cell divisions give rise to the first evidence of differentiation found in the mature embryo. The innermost group of cells acquire an elongated shape and a discrete outer cell layer is present.

G₁ phase stage of the cell cycle from the birth of a new cell until the onset of chromosome replication at *S phase.*

G₂ phase stage of the cell cycle from the completion of chromosome replication until the onset of cell division.

growth factors extracellular hormones and cell-bound *ligands* that stimulate or inhibit cell proliferation.

guanine (G) a nitrogenous base, one member of the base pair G–C (guanine–cytosine).

gynandromorph a rare genetic *mosaic* with some male tissue and some female tissue, usually in equal amounts.

H

hairpin loops structures formed when a single strand of DNA or RNA can fold back on itself because of *complementary base pairing* between different regions in the same molecule.

haploid a single set of chromosomes present in the egg and sperm cells of animals and in the egg and pollen cells of plants (compare with *diploid*).

haploinsufficiency a rare form of dominance in which an individual heterozygous for a wild-type allele and a *null* allele shows an abnormal phenotype because the level of gene activity is not enough to produce a normal phenotype.

haplospores See *ascospores.*

haplotype specific combination of linked alleles in a cluster of related genes. A contraction of the phrase "haploid genotype."

Hardy-Weinberg law defines the relationships between genotype and allele frequencies within a generation and from one generation to the next.

hemizygous describes the genotype for genes present in only one copy in an otherwise diploid organism, such as X-linked genes in a male.

heredity the way genes transmit biochemical, physical, and behavioral traits from parents to offspring.

heritability (h^2) the proportion of total phenotypic variance ascribable to *genetic variance.*

hermaphrodite an organism that has both male and female organs and produces both male gametes (sperm) and female gametes (eggs). The organism can have both types of organs at the same time (simultaneous hermaphrodite) or have one type early in life and the other type later in life (sequential hermaphrodite).

heterochromatin highly condensed chromosomal regions within which genes are usually transcriptionally inactive.

heterochromatic DNA genomic DNA from heterochromatic regions; this DNA is often difficult to clone.

heterochronic mutations mutations resulting in the inappropriate timing of cell division and cell-fate decisions during development.

heteroduplex region a region of double-stranded DNA in which the two strands have non-identical (though similar) sequences. Heteroduplex regions are often formed as intermediates during crossing-over.

heterogeneous trait a mutation at any one of a number of genes can give rise to the same phenotype.

heterogenotes partial diploids carrying different alleles of the same gene. Contrast *merozygotes.*

heteromers multimeric proteins composed of nonidentical subunits. Compare *homomers.*

heteroplasmic genomic makeup of a cell's organelles characterized by a mixture of organelle genomes. Contrast *homoplasmic.*

heterothallic strains of organisms with stable haploid mating types; progeny of successive mitotic divisions always have the same mating type as their parents. Contrast *homothallic.*

heterozygous a genotype in which the two copies of the gene that determine a particular trait are different *alleles.* See *hybrid.*

heterozygous advantage the situation where heterozygotes have a higher fitness than either homozygote; a possible explanation for the survival of recessive genetic diseases in a population.

heterozygous carrier unaffected parents who bear a dominant normal allele that masks the effects of an abnormal recessive one.

Hfr bacteria that produce a *high frequency* of *recombinants* for chromosomal genes in mating experiments because their chromosomes contain an integrated F plasmid.

high-density linkage map a *linkage map* that shows one gene or marker for each centimorgan of a genome.

histocompatibility antigens cell surface molecules that play a critical role in stimulating a proper immune response.

histones small DNA-binding proteins with a preponderance of the basic, positively charged amino acids lysine and arginine. Histones are the fundamental protein components of *nucleosomes.*

homeobox in *homeotic genes,* the region of homology, usually 180 bp in length, that encodes the *homeodomain.*

homeodomain a conserved DNA-binding region of *transcription factors* encoded by the *homeobox* of *homeotic genes.*

homeotic gene a gene that plays a role in determining a tissue's identity during development.

homeotic mutation mutation that causes cells to misinterpret their position in the blueprint and become normal organs in inappropriate positions; such a mutation can alter the overall body plan.

homeotic selector genes genes that control the identity of body segments.

homologous chromosomes (homologs) chromosomes that match in size, shape, and banding. A pair of chromosomes containing the same linear gene sequence, each derived from one parent.

homologs genes or regulatory DNA sequences that are similar in different species because of descent from a common ancestral sequence.

homologous genes genes in different species with enough sequence similarity to be evolutionarily related.

homomers multimeric proteins composed of identical subunits. Compare with *heteromers.*

homoplasmic genomic makeup of a cell's organelles characterized by a single type of organelle DNA. Contrast with *heteroplasmic.*

homothallic strains of organisms that can switch mating types; progeny of successive mitotic divisions may not have the same mating type as their parents. Compare with *heterothallic.*

homozygous a genotype in which the two copies of the gene that determine a particular trait are the same *allele.*

hormone a small molecule or polypeptide that is made and released by certain secretory cells in multicellular organisms. Secreted hormones diffuse or move via bodily fluids to other cell types that contain specific receptors. Hormone-receptor binding can elicit changes in gene function and differentiation in the target cells.

hot spots   sites within a gene that mutate more frequently than others, either spontaneously or after treatment with a particular *mutagen*.

Human Genome Project   initiative to determine the complete sequence of the human genome and to analyze this information.

hybrid dysgenesis   phenomenon in which high transposon mobility causes reduced fertility in hybrid progeny; the result of crossing *Drosophila* males carrying the P element with females that lack the P element.

hybridization   the propensity of complementary single strands of nucleic acid—either DNA or RNA—to form stable double helices. A powerful tool, hybridization can be used to identify sequences that are closely related to a DNA probe.

hybrids   offspring of genetically dissimilar parents; often used as synonym for *heterozygote*.

hydrogen bonds   weak electrostatic bonds that result in a partial sharing of hydrogen atoms between reacting groups.

hypermorphic mutation   produces an allele generating either more protein than the wild-type allele or the same amount of a more efficient protein. If excess protein activity alters phenotype, the hypermorphic allele is dominant. Compare with *hypomorphic mutation*.

hypomorphic mutation   produces either less of a protein or a protein with a weak but detectable function. Compare with *hypermorphic mutation*.

I

imaginal discs   flattened epithelial sacs that develop from small groups of cells set aside in the early embryo from which adult-specific structures like wings, legs, eyes, and genitalia develop; in *Drosophila*, they undergo extensive growth and development during the larval and pupal stages.

immunoglobulin homology unit   common feature of the immunoglobulin gene superfamily; encodes about 100 amino acids, which fold into a characteristic three-dimensional structure termed the immunoglobulin fold.

incomplete dominance   expression of heterozygous phenotype resulting in offspring whose phenotype is intermediate between those of the parents.

independent assortment   the random distribution of different genes during gamete formation. See *Mendel's second law*.

inflorescence   in a plant, the flower or group of flowers at the tip of a branch.

initiation   first phase of DNA replication, *transcription,* or *translation* needed to set the stage for the addition of nucleotide or amino acid building blocks during *elongation*.

initiation codon   nucleotide triplet that marks the precise spot in the nucleotide sequence of an mRNA where the code for a particular polypeptide begins. Compare *nonsense codon*.

initiation factors   a term usually applied to proteins that help promote the association of *ribosomes, mRNA,* and initiating *tRNA* during the first phase of *translation*.

insertion   the addition to a DNA molecule of one or more nucleotide pairs.

insertional mutagenesis   method of mutagenesis in which a foreign DNA sequence (viral, plasmid, transposon, or cloned fragment) is used as the mutagenic agent. Mutations result when the foreign sequence integrates into a gene. The disrupted gene can be easily identified and cloned based on its association with the foreign DNA.

insertion sequences (ISs)   small transposable elements that dot the chromosomes of many types of bacteria; they are transposons that do not contain selectable markers.

interactome   the network of protein interactions within an organism.

intercalators   class of chemical *mutagens* composed of flat, planar molecules that can sandwich themselves between successive base pairs and disrupt the machinery of replication, recombination, or repair.

intergenic gene conversion   information flow between related DNA sequences that occurs through an alternative outcome of the process responsible for unequal crossing-over.

interphase   the period in the cell cycle between divisions.

intragenic suppression   the restoration of gene function by one mutation canceling the effects of another mutation in the same gene.

introns   the DNA base sequences of a gene that are spliced out of the *primary transcript* and are therefore not found in the mature mRNA. See *exons*.

inversion   a 180-degree rotation of a segment of a chromosome relative to the remainder of the chromosome.

inversion heterozygotes   cells or organisms in which one chromosome is of a normal gene order while the homologous chromosome carries an inversion.

inversion loop   formed in the cells of an *inversion heterozygote* when the inverted region rotates to pair with the similar region in the normal homolog.

K

karyotype   the visual description of the complete set of chromosomes in one cell of an organism; usually presented as a photomicrograph with the chromosomes arranged in a standard format showing the number, size, and shape of each chromosome type.

kinetochore   a specialized structure composed of DNA and proteins that is the site at which chromosomes attach to the spindle fibers.

knockout constructs   cloned genes modified so that they no longer function.

L

***lac* operon**   a single DNA unit, composed of the *lacZ, lacY,* and *lacA* genes together with the promoter (*p*) and operator (*o*), that enables the simultaneous regulation of the three structural genes in response to environmental changes.

λ repressor protein   protein that binds to operators on phage λ DNA, preventing the transcription of genes needed for the lytic cycle; this protein makes λ lysogens immune to infection with incoming λ phage.

late-onset   a genetic condition in which symptoms are not present at birth, but manifest themselves later in life.

lateral gene transfer   the introduction and incorporation of DNA from an unrelated individual or from a different species.

law of independent assortment   See *Mendel's second law*.

law of segregation   See *Mendel's first law*.

lawn   bacteria immobilized in a nutrient agar, used as a field on which to test for the presence of viral particles.

library   see *genomic library* or *cDNA library*.

ligands   molecules such as *growth factors* or *hormones* that are produced by one cell and bind to receptors on a different cell, initiating a *signal transduction pathway*.

LINE   long *i*nterspersed *e*lements; one of the two major classes of transposable elements in mammals. Contrast with *SINE*.

lineage compartments   regions of an organism to which cells have a restricted developmental potential.

linkage   the proximity of two or more markers on a chromosome; the closer together the markers are, the lower the probability that they will be separated by recombination. Genes are linked when the frequency of parental type progeny exceeds that of recombinant progeny.

linkage group   a group of genes chained together by linkage relationships. See *linkage*.

linkage map   depicts the distances between *loci* as well as the order in which they occur on the organism.

linker DNA   a stretch of ≈40 base pairs of DNA that connect one nucleosome with the next.

locus a designated location on a chromosome. See α-*globin locus* and β-*globin locus*.

locus control region (LCR) a *cis*-acting regulatory element that operates to enhance *transcription* from individual genes within a gene complex such as the β-globin complex.

lysate population of phage particles released from the host bacteria at the end of the lytic cycle.

lysogenic bacterium bacterial cell that carries a prophage; lysogen.

lysogeny the integration of bacteriophage DNA into the host chromosome.

lytic cycle bacterial cycle resulting in cell lysis and release of progeny phage.

M

macroarray microtiter plate-based DNA array.

macroevolution in the study of genetics, the process by which new species emerge from existing species. Contrast with *microevolution*.

mapping the process of determining the locus of a gene on a particular chromosome.

mapping function mathematical equation that compensates for the inaccuracies inherent in relating recombination frequencies to physical distance.

mapping panels a set of DNA samples used in multiple linkage analysis tests.

map unit synonymous with *centimorgan*.

marker an identifiable physical location on a chromosome, whose inheritance can be monitored. Markers can be expressed regions of DNA (genes) or any segment of DNA that has variant forms which can be followed.

maternal-effect mutations mutations in genes encoding maternal components (those supplied by the egg to the mother) that disrupt the development of her progeny.

maternally supplied components molecules synthesized by the mother that are supplied to the egg and which are needed for early development of the progeny.

mean statistical average; the middle point.

meiosis the process of two consecutive cell divisions starting in the diploid progenitors of sex cells. Meiosis results in four daughter cells, each with a haploid set of chromosomes.

Mendel's first law the law of segregation states that the two alleles for each trait separate (segregate) during gamete formation and then unite at random, one from each parent, at fertilization.

Mendel's second law the law of independent assortment states that during gamete formation, different pairs of alleles (genes) segregate independently of each other.

merozygotes partial diploids in which the two gene copies are identical. Contrast with *heterogenotes*.

messenger RNA (mRNA) RNA that serves as a template for protein synthesis. See *genetic code*.

metabolism the chemical and physical reactions that convert sources of energy and matter to fuel growth and repair within a cell.

metacentric a chromosome in which the centromere is at or near the middle.

metamorphosis a dramatic reorganization of an organism's body plan; for example, transition from larval stage to adult insect stage.

metaphase a stage in mitosis or meiosis during which the chromosomes are aligned along the equatorial plane of the cell.

metaphase plate imaginary equator of the cell toward which chromosomes congress during *metaphase*.

metazoans multicellular animals, which first appeared about 0.7 billion years ago.

methylated cap formed by the action of capping enzyme and methyl transferases at the 5′ end of eukaryotic mRNA, critical for efficient translation of the mRNA into protein.

methyl-directed mismatch repair DNA repair mechanism that corrects mistakes in replication, discriminating between newly synthesized and parental DNA by the methyl groups on the parental strand.

microarray small glass-slide DNA array.

microevolution alterations of a population's *gene pool*. Contrast *macroevolution*.

microsatellite DNA element composed of 15–100 tandem repeats of one-, two-, or three-base-pair sequences.

minimal tiling path the result of the final step in the shotgun sequencing strategy, a minimally overlapping set of BAC clones.

minisatellite DNA element composed of 10–40 bp tandem repeating units of identical sequence.

missense mutations changes in the nucleotide sequence of a gene that change the identity of an amino acid in the polypeptide encoded by that gene.

mitosis the process of division that produces daughter cells that are genetically identical to each other and to the parent cell.

mitotic nondisjunction the failure of two sister chromatids to separate during mitotic anaphase generates reciprocal trisomic and monosomic daughter cells.

model organisms used in genomic analysis because they have many genetic mechanisms and cellular pathways in common with each other and with humans. These organisms lend themselves well to classical breeding

experiments and direct manipulation of the genome.

modification the phenomenon in which growth on a restricting host changes a phage so that succeeding generations grow more efficiently on that same host.

modification enzymes enzymes that add methyl groups to specific DNA sequences, preventing action of the corresponding restriction enzymes on that DNA.

modifier genes genes that produce a subtle, secondary effect on phenotype. There is no formal distinction between major and modifier genes, rather it is a continuum of degrees of influence.

molecular clock hypothesis stating that one can assume a constant rate of amino acid or nucleotide substitution as a means of determining the genealogies of organisms.

molecular cloning the process by which a single DNA fragment is purified from a complex mixture of DNA molecules, and then amplified into a large number of identical copies.

molecular marker a segment of DNA found at a specific site in a genome that has variants which can be recognized and followed. See *physical marker, genetic marker.*

monohybrid crosses crosses between parents that differ in only one trait.

monohybrids individuals having two different alleles for a single trait.

monomorphic a gene with only one wild-type allele.

monoploid describing cells, nuclei, or organisms that have a single set of unpaired chromosomes. For diploid organisms, monoploid and *haploid* are synonymous.

monosomic individual lacking one chromosome from the diploid number for the species.

morphogens substances that define different cell fates in a concentration-dependent manner.

mosaic an organism containing tissues of different genotypes.

mosaic development the embryo is a collection of self-differentiating cells that at their formation receive a specific set of molecular instructions governing their unique fates. Contrast with *regulative development*.

mRNA See *messenger RNA*.

multifactorial traits determined by two or more factors, including multiple genes interacting with each other or one or more genes interacting with the environment.

multihybrid crosses crosses between parents that differ in three or more traits.

multimeric protein a protein made from more than one polypeptide; each polypeptide in the multimeric protein is called a *subunit*.

mutagen   any physical or chemical agent that raises the frequency of mutations above the spontaneous rate.

mutations   heritable alterations in *DNA sequence.*

N

natural selection   in nature, the process that progressively eliminates individuals whose fitness is low and chooses individuals of high fitness to survive and become the parent of the next generation. Contrast with *artificial selection.*

natural transformation   a process by which a few species of bacteria transfer genes from one strain to another by spontaneously accepting DNA from their surroundings. Contrast with *artificial transformation.*

neomorphic mutations   rare mutations that produce a novel phenotype due to production of a protein with a new function or due to *ectopic expression* of the protein.

nonallelic noncomplementation   (also called second-site noncomplementation) failure of mutations in two different genes to complement each other, causing a mutant phenotype in heterozygotes. Usually indicates that the two polypeptide products cooperate with each other, for example as *subunits* of the same *multimeric protein.*

nonautonomous elements   defective transposable elements that cannot move unless the genome contains nondefective *autonomous elements* that can supply necessary functions like tranposase enzymes.

nonautonomous trait   when the phenotype expressed by a particular cell depends upon the genotypes of neighboring cells. Compare with *cell autonomous trait.*

nonconservative substitutions mutational changes that substitute an amino acid in a protein with a different amino acid with different chemical properties. Compare with *conservative substitutions.*

nondisjuction   failures in chromosome segregation during meiosis; responsible for defects such as trisomy.

nonparental ditype (NPD)   a tetrad containing four recombinant spores.

nonsense codons   the three stop codons that terminate *translation.* Compare with *initiation codon.*

nonsense mutations   mutational changes in which a codon for an amino acid is altered to a *stop codon,* resulting in the formation of a *truncated protein.*

nonsense suppressor tRNAs   tRNAs encoded by mutant tRNA genes; these tRNAs contain *anticodons* that can recognize *stop codons,* thus suppressing the effects of *nonsense mutations* by inserting an amino acid

into a polypeptide in spite of the stop codon.

nontandem duplications   two or more copies of a region that are not adjacent to each other and may lie far apart on the same chromosome or on different chromosomes. Contrast with *tandem duplications.*

Northern blot analysis   protocol for determining whether a fragment of DNA is transcribed in a particular tissue. In this protocol, RNA transcripts in the cells of a particular tissue are separated by gel electrophoresis. Compare *Southern blot.*

N terminus   the end of a polypeptide chain that contains a free amino group that is not connected to any other amino acid.

nuclear lamins   proteins that underlie the inner surface of the nuclear membrane and serve as CDK substrates.

nucleoid body   a folded bacterial chromosome.

nucleolar organizer   clusters of rRNA genes on long loops of DNA within a *nucleolus.*

nucleolus   large sphere-shaped organelle visible in the nucleus of interphase eukaryotic cells with a light microscope; formed by the *nucleolar organizer.*

nucleosome   rudimentary DNA packaging unit, composed of DNA wrapped around a *histone protein* core.

nucleotide   a subunit of DNA or RNA consisting of a nitrogenous base (adenine, guanine, thymine, or cytosine in DNA; adenine, guanine, uracil, or cytosine in RNA), a phosphate group, and a sugar (deoxyribose in DNA; ribose in RNA).

null hypothesis   a statistical hypothesis to be tested and either accepted or rejected in favor of an alternative.

null mutations   mutations that abolish the function of a protein encoded by the wild-type allele. Such mutations either prevent synthesis of the protein or promote synthesis of a protein incapable of carrying out any function.

O

octads   asci produced in *Neurospora* containing eight *ascospores* because of an extra round of mitosis that occurs after meiosis is completed.

octant stage embryo   early stage of embryo formation during cell divisions of a plant.

oligonucleotide   a short single-stranded DNA molecule (containing less than 50 bases); can be synthesized by an automated DNA synthesizer. Oligonucleotides are used as DNA probes and as primers for DNA sequencing or PCR.

oligopeptide   several amino acids linked by peptide bonds.

oncogene   a gene, one or more forms of which is associated with cancer. Many oncogenes are involved, directly or indirectly, in controlling the rate of cell growth.

oogenesis   formation of the female gametes (eggs).

oogonia   diploid germ cells in the ovary.

open reading frames (ORFs)   DNA sequences with long stretches of codons in the same reading frame uninterrupted by stop codons; suggest the presence of genes.

operator site   a short DNA sequence near a promoter that can be recognized by a repressor protein; binding of repressor to the operator blocks transcription of the gene.

operon   a unit of DNA composed of specific genes, plus a promoter and/or operator, that acts in unison to regulate the response of the structural genes to environmental changes.

ordered tetrads   *tetrads* in fungi such as *Neurospora* in which the order of *ascospores* in the *ascus* reflects the geometry of the meiotic divisions.

orthologous genes   genes, defined by their sequence similarities, in two different species that arose from the same gene in the two species' common ancestor.

ovum   haploid female germ cell (the egg).

oxidative phosphorylation   a set of reactions requiring oxygen that creates portable packets of energy in the form of ATP.

P

paracentric inversions   *inversions* that exclude the centromere. Compare with *pericentric inversions.*

paralogous genes   genes that arise by duplication within the same species, often within the same chromosome; paralogous genes often constitute a multigene family.

parental ditype (PD)   a tetrad that contains four parental class haploid cells.

parental (P) generation   pure-breeding individuals whose progeny in subsequent generations will be studied for specific traits. Refer to *filial generations.*

parthenogenesis   reproduction in which offspring are produced by an unfertilized female. Parthenogenesis is common in ants, bees, wasps, and certain species of fish and lizards.

partial digest   the result of an enzymatic digestion of a DNA sample in which only some of the available restriction sites have been cut. Compare with *complete digest.*

pedigree   an orderly diagram of a family's relevant genetic features, extending through as many generations as possible.

peptide bond   a covalent bond that joins amino acids during protein synthesis.

peptide motifs stretches of amino acids conserved in many otherwise unrelated polypeptides.

peptidyl (P) site site on a ribosome to which the initiating tRNA first binds and at which the tRNAs carrying the growing polypeptide are located during *elongation*.

peptidyl transferase the enzymatic activity of the ribosome responsible for forming peptide bonds between successive amino acids.

pericentric inversions *inversions* that include the centromere. Compare with *paracentric inversions*.

permissive condition an environmental condition that allows the survival of an individual with a *conditional lethal* allele. Contrast with *restrictive condition*.

petals the structure composing the second *whorl* of a flower, leaf-shaped but containing no photosynthetic cells, therefore not green.

phage short for *bacteriophage; a virus* for which the natural host is a bacteria cell; literally "bacteria eaters."

phage induction process of phage DNA excision from the bacterial chromosome in a *lysogen* and entry into the lytic cycle.

phage lambda (λ) a naturally occurring double-stranded DNA virus that infects *E. coli*. A common plasmid *vector* used to clone DNA from other organisms.

phenocopy a change in phenotype arising from environmental agents that mimic the effects of a mutation in a gene. Phenocopies are not heritable because they do not result from a change in a gene.

phenotype an observable characteristic.

phenotype frequency the proportion of individuals in a population that are of a particular phenotype.

pheromones molecules produced by one sex that serve as agents to elicit mating behaviors in individuals of the opposite sex.

phosphodiester bonds covalent bonds joining one *nucleotide* to another. Phosphodiester bonds between nucleotides form the backbone of DNA.

photomorphogenesis light-regulated developmental program of a plant.

photoperiod day length; period of light exposure for plants.

phylogenetic tree diagram composed of nodes, which represent the taxonomic units, and branches, which represent the relationship of these units.

physical map a map of locations of identifiable landmarks on DNA (for example, restriction enzyme cutting sites, genes). For the human genome, the lowest-resolution physical map is the banding patterns on the 24 different chromosomes; the highest resolution map is the complete nucleotide sequence of the chromosomes. See *karyotype*.

physical markers cytologically visible abnormalities that make it possible to keep track of specific chromosome parts from one generation to the next. Compare *genetic markers*.

pistil floral structure that consists of fused carpels, with pollen receptive stigma at the top, and a short neck or style, leading to the ovary.

plaque a clear area on a bacterial *lawn*, devoid of living bacterial cells, containing the genetically identical descendants of a single *bacteriophage*.

plasmids small circles of double-stranded DNA that can replicate in bacterial cells independently of the *bacterial chromosome*; commonly used as *cloning vectors*.

pleiotropy phenomenon in which a single gene determines a number of distinct and seemingly unrelated characteristics.

point mutation a mutation of one base pair.

polar body a cell produced by meiosis I or meiosis II during oogenesis that does not become the oocyte.

polarity an overall direction.

poly-A tail the 3′ end of eukaryotic mRNA consisting of 100–200 As, believed to stabilize the mRNAs and increase the efficiency of the initial steps of translation.

polycistronic mRNA an *mRNA* that contains more than one protein coding regions; often the transcriptional product of an *operon* in bacterial cells.

polygenic traits controlled by multiple genes. See *continuous trait*.

polymer a linked chain of repeating subunits that form a larger molecule; DNA is a type of polymer.

polymerase chain reaction (PCR) a fast inexpensive method of replicating a DNA sequence once the sequence has been identified; based on a reiterative loop that amplifies the products at each previous round of replication.

polymerase, DNA or RNA enzymes that catalyze the synthesis of nucleic acids on preexisting nucleic acid templates, assembling RNA from ribonucleotides or DNA from deoxyribonucleotides.

polymorphic genes that have more than one wild-type allele.

polypeptides amino-acid chains containing hundreds to thousands of *amino acids* joined by *peptide bonds*.

polyploids euploid species that carry three or more complete sets of chromosomes.

polyproteins *polypeptides* produced by *translation* that can subsequently be cleaved by protease enzymes into two or more separate proteins.

polyribosomes structures formed by the simultaneous *translation* of a single *mRNA* molecule by multiple *ribosomes*.

polytene chromosome giant chromosomes consisting of many identical chromatids lying in parallel register.

population a group of interbreeding individuals of the same species that inhabit the same space at the same time.

population genetics scientific discipline that studies what happens in whole populations at the genetic level.

positional cloning the process that enables researchers to obtain the clone of a gene without any prior knowledge of its protein product or function. It uses genetic and physical maps to locate mutations responsible for particular phenotypes.

posttranslational modifications changes such as phosphorylation that occur to a *polypeptide* after *translation* has been completed.

primary oocytes/spermatocytes germ-line cells in which meiosis I occurs.

primary structure the linear sequence of amino acids within a polypeptide.

primary transcript the single strand of RNA resulting from *transcription*.

primer short, pre-existing oligonucleotide chain to which new DNA can be added by DNA polymerase.

primer walking a common approach to directed sequencing of DNA in which primers are synthesized using the information from each previous round of DNA sequencing.

product rule states that the probability of two or more independent events occurring together is the product of the probabilities that each event will occur by itself.

programmed cell death (PCD) see *apoptosis*.

prokaryotes one of the three major evolutionary lineages of living organisms known as domains; characterized by the lack of a nuclear membrane. Contrast with *eukaryotes*.

prokaryotic gene regulation control of gene expression in a bacterial cell via mechanisms to increase or decrease the *transcription* and *translation* of specific genes or groups of genes.

prometaphase the stage of *mitosis* or *meiosis* just after the breakdown of the nuclear envelope, when the chromosomes connect to the *spindle apparatus* and begin to move toward the *metaphase plate*.

promoters DNA sequences near the beginning of genes that signal RNA polymerase where to begin transcription. Compare with *terminators*.

prophage the integrated phage genome.

prophase the phase of the cell cycle marked by the emergence of the individual chromosomes from the undifferentiated mass of chromatin, indicating the beginning of mitosis.

protein domains discrete functional units of a genome.

proteins large polymers composed of hundreds to thousands of amino-acid subunits strung together in a specific order into long chains. Proteins are required for the structure, function, and regulation of the body's cells, tissues, and organs.

proteome the complete set of proteins encoded by a genome.

protooncogene a *gene* that can mutate into an *oncogene*—an *allele* that causes a cell to become cancerous.

prototroph a microorganism (usually wild type) that can grow on minimal medium in the absence of one or more growth factors. Compare with *auxotroph.*

pseudodominance expression of a phenotype caused by a recessive allele in a deletion heterozygote because the other homolog has no copy of that gene.

pseudogenes a nonfunctioning gene. The result of duplication and divergence events in which one copy of an originally functioning gene has undergone mutations such that it no longer has an intact polypeptide coding sequence.

pseudolinkage characteristic of a heterozygote for a *reciprocal translocation*, in which genes located near the translocation breakpoint behave as if they are linked even though they originated on nonhomologous chromosomes.

pulse-field gel electrophoresis a special type of electrophoretic protocol that allows an extreme extension of the normal separating capacity of the gels. In this protocol, the DNA sample is subjected to pulses of electrical current that alternate between two directions. The range of sizes separated in this manner is a function of pulse length. Compare *electrophoresis.*

punctuated equilibrium the tendency of evolution to proceed through long periods of stasis (lack of change) followed by short periods of explosive change.

pure-breeding lines families of organisms that produce offspring with specific parental traits that remain constant from generation to generation.

***p* value** numerical probability that a particular set of observed experimental results represents a chance deviation from the values predicted by a particular hypothesis.

Q

quantitative trait See *continuous traits.*

quantitative trait loci (QTLs) loci that control the expression of *continuous traits.*

quaternary structure structure made up of the three-dimensional configuration of subunits in a multimeric protein.

R

radial loop-scaffold model the model of looping and gathering of DNA by nonhistone proteins that results in high compaction of chromosomes at mitosis.

random amplification of polymorphic DNA (RAPD) a protocol designed to detect single-base changes at polymorphic loci throughout a genome.

random walk description of movement of bacteria to achieve chemotaxis reflecting nonpredictable changes in the direction of movement. Contrast *biased random walk.*

reading frame the partitioning of groups of three nucleotides from a fixed starting point such that the sequential interpretation of each succeeding triplet generates the correct order of amino acids in the resulting polypeptide chain.

rearrangements events in which the genome is reshaped by the reorganization of DNA sequences within one or more chromosomes.

receptors proteins embedded in the membrane of a cell that bind to *growth factors* and other *ligands;* initiating a *signal transduction pathway* in the cell.

recessive allele an allele whose phenotype is not expressed in a heterozygote.

recessive epistasis special case of epistasis, in which the allele causing the epistasis is recessive. Compare with *dominant epistasis.*

recessive lethal allele an allele that prevents the birth or survival of homozygotes, though heterozygotes carrying the allele survive.

recipient during gene transfer in bacteria, the cell that receives the genetic material. See *donor, transformation, conjugation,* and *transduction.*

reciprocal crosses crosses in which the traits in the males and females are reversed, thereby controlling whether a particular trait is transmitted by the egg or the pollen.

reciprocal translocation results when two breaks, one in each of two chromosomes, yield DNA fragments that do not relegate to their chromosome of origin; rather they switch places and become attached to the other chromosome. Compare with *translocation.*

recombinant DNA molecules a combination of DNA molecules of different origin that are joined using recombinant DNA technologies.

recombinant DNA technology modern tools of genetic analysis based on the isolation, analysis, and use of DNA fragments purified from complex genomes. Recombinant DNA technology is used in basic research to characterize genes and other genomic elements, and also as a tool of biotechnology to produce important medical products.

recombinants chromosomes that carry a mix of alleles derived from different homologs.

recombinant type a phenotype reflecting a new combination of genes occurring during gamete formation.

recombination the process by which offspring derive a combination of genes different from that of either parent; the generation of new allelic combinations. In higher organisms, this can occur by *crossing-over.*

recombination frequency (RF) the percentage of recombinant progeny; can be used as an indication of the physical distance separating any two genes on a chromosome. See *centimorgan.*

regulative development developmental process in which blastomeres can regulate and adjust their development to make up for missing cells. Contrast with *mosaic development.*

release factors proteins that recognize *stop codons* and help end translation.

replicon (replication unit) the DNA running both ways from one origin of replication to the endpoints where it merges with DNA from adjoining replication forks.

repressor a type of *transcription factor* that can bind to specific *cis*-acting elements such as *operator sites* and thereby diminish or prevent *transcription.*

reproductive development period in plant development that produces flower and seed. Contrast *vegetative development.*

response to selection (R) the amount of evolution, or change in mean trait value, resulting from selection. Compare with *selection differential.*

restriction the bacterial capacity for limiting viral growth.

restriction enzymes proteins made by bacteria that recognize specific, short nucleotide sequences and cut DNA at those sites.

restriction fragment length polymorphism (RFLP) variation between individuals in DNA fragment size cut by specific *restriction enzymes; polymorphic* sequences that result in RFLPs are used as markers on both *physical maps* and *genetic linkage* maps.

restriction map a linear diagram that illustrates the positions of *restriction enzyme* recognition sites along a DNA molecule.

restrictive condition an environmental condition that prevents the survival of an individual with a *conditional lethal* allele. Contrast with *permissive condition.*

retroposons genetic elements that transpose via reverse transcription of an RNA intermediate. One class of *transposable elements.* Contrast with *transposons.*

retroviruses viruses that hold their genetic information in a single strand of RNA and carry the enzyme *reverse transcriptase* to convert that RNA into DNA within a host cell.

reverse mutation a mutation that causes a novel mutant to revert back to wild type.

reverse transcriptase an RNA-dependent DNA *polymerase* that synthesizes DNA strands complementary to an RNA *template*. The product of reverse transcriptase is a *cDNA* molecule.

reverse transcription the process by which *reverse transcriptase* synthesizes DNA strands complementary to an RNA *template*. The product of reverse transcription is a cDNA molecule.

reversion See *reverse mutation*.

ribonucleic acid (RNA) a *polymer* of ribonucleotides found in the nucleus and cytoplasm of cells; it plays an important role in protein synthesis. There are several classes of RNA molecules, including *messenger RNA (mRNA)*, *transfer RNA (tRNA)*, ribosomal RNA (rRNA), and other small RNAs, each serving a different purpose.

ribosome binding sites regions on *procaryotic mRNAs* containing both an *initiation codon* and a *Shine-Dalgarno box*; ribosomes bind to these sites to start *translation*.

ribosomes cytoplasmic structures composed of ribosomal RNA (rRNA) and protein; the sites of protein synthesis.

ribozymes RNA molecules that can act as enzymes to catalyze specific chemical reactions.

RNA See *ribonucleic acid*.

RNA editing specific alteration of the genetic sequence carried within an RNA molecule after transcription is completed.

RNA-like strand strand of a double-helical DNA molecule that has the same nucleotide sequence as an *mRNA* (except for the substitution of T for U) and that is complementary to the *template strand*.

RNA-mediated interference *phenocopies* that result from the activity of double-stranded RNA that interferes with the expression of the corresponding gene.

RNA polymerases enzymes that transcribe a DNA sequence into an RNA *transcript. Eukaryotes* have three types of RNA polymerases called pol I, pol II, and pol III that are responsible for transcribing different classes of genes.

RNA splicing a process that deletes *introns* and joins together adjacent *exons* to form a mature mRNA consisting only of exons.

RNA world hypothetical primordial world in which RNA became the first replicator.

Robertsonian translocation translocation arising from breaks at or near the centromeres of two acrocentric chromosomes. The reciprocal exchange of broken parts generates one large metacentric chromosome and one very small chromosome.

S

satellite DNAs blocks of repetitive, simple noncoding sequences, usually around centromeres; these blocks have a different chromatin structure and different higher-order packaging than other chromosomal regions.

saturation mutagenesis an attempt to isolate mutations in all of the genes that direct a particular biological process.

scaffold-associated regions (SARs) special, irregularly spaced repetitive base sequences of DNA that associate with nonhistone proteins to define chromatin loops. SARs are most likely the sites at which DNA is anchored to the condensation scaffold.

secondary oocytes/spermatocytes germ-line cells in which meiosis II occurs.

secondary structure localized region of a *polypeptide* chain with a characteristic geometry, such as an α-helix or β-pleated sheet.

second-division segregation pattern a *tetrad* in which the arrangement of *ascospores* indicates that the two alleles of a gene segregated from each other in the second meiotic division.

sectors portions of a growing *colony* of microorganisms that have a different *genotype* than the remainder of the colony.

segmentation gene in *Drosophila*, a large group of genes responsible for subdividing the body into an array of body segments.

segregation equal separation of alleles for each trait during gamete formation, in which one allele of each gene goes to each gamete.

selectable markers vector genes that make it possible to pick out cells harboring a *recombinant DNA molecule* constructed with that vector.

selection a process that establishes conditions in which only the desired mutant will grow.

selection differential (S) measure of the strength of selection on a trait. Compare *response to selection*.

self-fertilization (selfing) fertilization in which both egg and pollen come from the same plant.

semiconservative replication a pattern of double helix duplication in which complementary base pairing followed by the linkage of successive nucleotides yields two daughter double helices that each contain one of the original DNA strands intact (conserved) and one completely new strand.

semisterility a condition in which the capacity of generating viable offspring is diminished by at least 50%.

sepals green leaflike structure composing the first *whorl* of a flower.

sequence tagged sites (STSs) one-of-a-kind markers that tag positions along the DNA molecule.

sequencing determining the order of *nucleotides* (base sequences) in a DNA or RNA molecule or the order of amino acids in a *protein*.

sex chromosomes the X and Y chromosomes in human beings, which determine the sex of an individual. Compare with *autosome*.

sex-influenced traits traits that can show up in both sexes but are expressed differently in each sex due to hormonal differences.

sex-limited traits traits that affect a structure or process that is found in one sex but not the other.

Shine-Dalgarno box a sequence of six nucleotides in *mRNA* that is one of two elements comprising a *ribosome binding site* (the other element is the *initiation codon*).

signal transducers cytoplasmic proteins that relay signals inside the cell.

signal transduction the activation and inhibition of intracellular targets after binding of *growth factors* or other *ligands* to *receptors*.

silent mutations mutations without effects on phenotype; usually denotes *point mutations* that change one of the three bases in a *codon*, but that do not change the identity of the specified amino acid because of the *degeneracy* of the *genetic code*.

SINE short *interspersed elements*; one of the two major classes of transposable elements in mammals. Contrast with *LINE*.

sister chromtids the two identical copies of a chromosome, connected through a *centromere*, that exist immediately after chromosome replication.

somatic cells any cell in an organism except gametes and their precursors. Compare with *germ cells*.

somatic gene therapy remedial measures in which a replacement gene is inserted into affected tissue to compensate for a faulty gene.

Southern blot protocol for transferring DNA sequences separated by *gel electrophoresis* onto a nitrocellulose filter paper for analysis by *hybridization* with a *DNA probe*. Compare *Northern blot analysis*.

specialized transducing phages *bacteriophage* carrying mainly phage DNA but also one or a few of the bacterial genes that lie near the site of *prophage* insertion. They can transfer these genes to another bacterium in the process known as specialized transduction.

spermatids haploid cells produced at the end of meiosis that will mature into sperm.

spermatogenesis the production of sperm.

spermatogonia diploid germ cells in the testes.

S phase stage of the cell cycle during which chromosome replication occurs.

spindle apparatus microtubule-based structure responsible for chromosome movements and *segregation* during cell division.

splice acceptors   nucleotide sequences in a *primary transcript* at the border between an *intron* and the downstream *exon* that follows it; required for proper *RNA splicing.*

splice donors   nucleotide sequences in a *primary transcript* at the border between an *intron* and the upstream *exon* that precedes it; required for proper *RNA splicing.*

spliceosome   a complicated intranuclear machine that ensures that all of the splicing reactions take place in concert.

stamen   structure comprising the third *whorl* of a flower that bears the male gametes in the form of pollen.

stem cells   are relatively undifferentiated cells that undergo asymmetric mitotic divisions. One of the daughter cells produced by such a division is another stem cell, while the other daughter cell can differentiate. In this way, stem cells are self-renewing but can also give rise to differentiated cells.

sticky end   the result achieved after digestion by many *restriction enzymes* that break the *phosphodiester bonds* on the two strands of a double helix DNA molecule at slightly different locations. The resulting double-stranded DNA molecule has a single protruding strand at each end that is usually one to four bases in length.

stigma   pollen-receptive structure at the top of the pistil of a flower.

stop codons   see *nonsense codons.*

subcloning   cloning smaller pieces of DNA that are enriched for sequences of interest.

substitution   occurs when a base at a certain position in one strand of the DNA molecule is replaced by one of the other three bases.

substrate   the target of an enzyme, usually cellular proteins.

subunit   a single *polypeptide* that is a constituent of a *multimeric protein.*

sum rule   the probability that any of two or more mutually exclusive events will occur is the sum of their individual probabilities.

supercoiling   additional twisting of the DNA molecule caused by movement of the replication fork during unwinding. Refer to *DNA topoisomerases.*

suppressor mutations   mutations that alleviate the phenotypic abnormality caused by another mutation. The two mutations can be in the same gene (see *intragenic suppression*), or the suppressor mutation can be in a second, *modifier gene.*

suspensor   in plants, a structure analogous to the umbilical cord in mammals.

synaptonemal complex   structure that helps align homologous chromosomes during prophase of meiosis I.

syncytial blastoderm   in *Drosophila* embryos, formed when most of the nuclei migrate out to the cortex just under the surface of the embryo.

syncytium   an animal cell with two or more nuclei.

syntenic   relationship of two or more loci found to be located on the same chromosome. Compare with *conserved synteny.*

syntenic blocks   blocks of linked loci within a *genome.*

synthetic lethality   situation when an individual carrying mutations in two different genes dies, even though individuals carrying either mutation alone survive.

systems biology   the global study of multiple components of biological systems and their simultaneous interactions.

T

TAFs   *T*BP-*associated f*actors; one type of basal factors.

tandem duplications   repeats of a chromosomal region that lie adjacent to each other, either in the same order or in reverse order. Contrast with *nontandem duplications.*

TBP   *TATA box-binding protein;* key basal factor that assists the binding of RNA polymerase II to the promoter and the initiation of basal levels of transcription.

telomerase   an enzyme critical to the successful replication of *telomeres* at chromosome ends.

telomeres   specialized terminal structures on eukaryotic chromosomes that ensure the maintenance and accurate replication of the two ends of each linear chromosome.

telophase   the final stage of mitosis in which the daughter chromosomes reach the opposite poles of the cell and re-form nuclei.

temperate bacteriophages   after infecting the host, these phages can enter either the lytic cycle or the alternative lysogenic cycle, during which their DNA integrates into the host genome. Compare with *virulent bacteriophages.*

template   a strand of DNA or RNA that is used as a model by DNA or RNA *polymerase* or by *reverse transcriptase* for the creation of a new complementary strand of DNA or RNA.

template strand   the strand of the double helix that is complementary to both the RNA-like DNA strand and the mRNA.

teratype (T)   a tetrad carrying four kinds of haploid cells; two different parental class spores and two different recombinants.

terminalization   shifting of the chiasmata from their original position at the centromere toward the chromosome end or *telomere.* Refer to *chiasmata.*

terminators   sequences in the RNA products that tell RNA polymerase where to stop transcription. Compare with *promoters.*

tertiary structure   ultimate three-dimensional shape of a polypeptide.

testcross   a cross used to determine the *genotype* of an individual showing a *dominant phenotype* by mating with an individual showing the *recessive phenotype.*

tetrad   the assemblage of four ascospores (resulting from meiosis) in a single *ascus.*

tetrad analysis   the use of *tetrads* to study gene *linkage* and *recombination* during *meiosis* in fungi.

tetrasomic   otherwise *diploid* organism with four copies of a particular chromosome.

thymine   a nitrogenous base, one member of the base pair A–T (adenine–thymine).

torpedo stage embryo   stage of embryonic development in plants where two protuberances expand and differentiate into two well-defined, discrete cotyledons.

total phenotype variance (VP)   population deviation calculated as the sum of the *genetic variance* and the *environmental variance.*

totipotent   description of cell state during early embryonic development in which the cells have not yet differentiated and retain the ability to produce every type of cell found in the developing embryo and adult animal.

***trans*-acting element**   a gene that codes for a *transcription factor.*

transcript   the product of *transcription.*

transcription   the conversion of DNA-encoded information to its RNA-encoded equivalent.

transcriptional silencing   hypercondensation of chromatin domains makes it impossible to activate genes within those domains, no matter what transcription factors are active in the cell.

transcription bubble   the region of DNA unwound by RNA *polymerase.*

transcription factor   a type of protein that can associate directly or indirectly with a cis-*acting element* and, in doing so, control or modulate the expression of transcriptional initiation. Categories of transcription factors include *activators, repressors, coactivators,* and *corepressors.*

transcriptome   the population of mRNAs expressed in a single cell or cell type.

transduction   one of the mechanisms by which bacteria transfer genes from one strain to another; donor DNA is packaged within the protein coat of a bacteriophage and transferred to the recipient when the phage particle infects it. Recipient cells are known as transductants.

transfection   transformation of mammalian cells via the uptake of DNA from the medium.

transfer RNA (tRNA)   small RNA adaptor molecules that place specific amino acids at the correct position in a growing polypeptide chain.

transformation one of the mechanisms by which bacteria transfer genes from one strain to another; occurs when DNA from a donor is added to the bacterial growth medium and is then taken up from the medium by the recipient. The recipient cell is known as a transformant. Contrast with *conjugation.*

transgene any piece of foreign DNA that researchers have inserted into the genome of an organism.

transgenic any individual carrying a *transgene.*

transient- (or triangular-) stage embryo stage following *globular stage* in plant development; embryo becomes self-sufficient for growth, having used up most of its maternally deposited reserves.

transitions a type of *substitution* mutation that occurs when one purine (A or G) replaces the other purine, or one pyrimidine (C or T) replaces the other pyrimidine. Contrast with *transversions.*

translation the process in which the codons carried by mRNA direct the synthesis of proteins from amino acids according to the *genetic code.* Compare with *transcription.*

translocation a rearrangement that occurs when parts of two nonhomologous chromosomes change places. Compare with *reciprocal translocation.*

*trans***-splicing** a rare type of *RNA splicing* that joins together *exons* of the *primary transcripts* of two different genes.

transposable elements all DNA segments that move about in the genome, regardless of mechanism.

transposition the movement of *transposable elements* from one position on the genome to another.

transposons units of DNA that move from place to place within the genome without an RNA intermediate, sometimes causing a change in gene function when they insert themselves in a new chromosomal location.

transversions a type of *substitution* mutation that occurs when a purine (A or G) replaces a pyrimidine (C or T) or when a pyrimidine replaces a purine.

trisomic individual having one extra chromosome in addition to the normal diploid set of the species.

truncated proteins polypeptides with fewer amino acids than normal encoded by genes containing *nonsense mutations.*

twin spots adjacent patches of tissue that are phenotypically distinct from each other and from the surrounding tissue; can be produced as a result of mitotic recombination.

U

unequal crossing-over change in DNA caused by erroneous recombination in which one homologous chromosome ends up with a duplication while the other homolog sustains a deletion.

uniparental inheritance transmission of genes via one parent. Most species transmit mtDNA and cpDNA through the mother.

unordered tetrads *tetrads* in yeast in which the four *ascospores* are randomly arranged in the ascus.

5′ untranslated region (5′ UTR) portion of an *mRNA* between its 5′ end and the *initiation codon;* the 5′ UTR is thus by definition contained within one or more *exons* but does not encode any amino acids in the mRNA's protein product.

upstream movement opposite the direction RNA follows when moving along a gene. Compare with *downstream.*

uracil a nitrogenous base normally found in RNA but not in DNA; uracil is capable of forming a *base pair* with adenine.

V

variance statistical measurement of deviation from the mean (middle); typically expressed in plus or minus terms referring to the relationship to the mean.

vector a specialized DNA sequence that can enter a living cell, signal its presence to an investigator by conferring a detectable property on the host cell, and provide a means of replication for itself and the foreign DNA inserted into it. A vector must also possess distinguishing physical traits by which it can be purified away from the host cell's genome. See *cloning vector.*

vegetative development in plants, the period of growth before flowering. Contrast with *reproductive development.*

vernalization the process of exposing plants to cold early in the vegetative development to promote earlier flowering.

virulent bacteriophages after infecting the host, these phages always enter the lytic cycle, multiply rapidly, and kill the host. Compare with *temperate.*

virus a noncellular biological entity that can reproduce only within a host cell. Viruses consist of nucleic acid covered by protein. Inside an infected cell, the virus uses the synthetic capability of the host to produce progeny virus.

W

whorls in the structure of a flower, the concentric regions of modified leaves including *sepals, petals, stamens,* and *carpels.*

wild-type allele an allele whose frequency is greater than 1%, designated by a superscript plus sign ($^+$). See *allele frequency.*

wobble ability of the 5′-most nucleotide of an anticodon to interact with more than one nucleotide at the 3′-end of codons; helps explain the *degeneracy* of the *genetic code.*

Y

YAC See *yeast artificial chromosome.*

yeast artificial chromosome (YAC) a vector used to clone DNA fragments up to 400 kb; it is constructed from telomeric, centromeric, and replication origin sequences needed for replication in yeast cells. Compare with *cloning vector, cosmid.*

Z

Z-form DNA DNA in which the nucleotide sequences cause the structure to assume a zigzag shape due to the helices spiraling to the left. Compare with *B-form DNA.* The significance of this variation on DNA structure is unknown at this time.

zygote the diploid cell formed by the fertilization of the egg by the sperm during sexual reproduction.

Text Permissions

Chapter 1

Fig. 1.13 / Stephen H. Friedn, The Magic of Microarrays, Feb 2002 Scientific American, courtesy Jared Schneidman Designs.

Chapter 3

Genetics & Society / Provided by the Institut National d'Etudes Demographiques. Reprinted with permission from SCIENCE: "Fragment of a Family Tree Showing the Transmission of Juvenile Glaucoma" April 19, 1991, p. 369 © American Association for the Advancement of Science.

Chapter 4

Problem 4-25 / From E.J. Mange & A.P. Mange, Basic Human Genetics, 2e 1999, Sunderland, MA: Sinauer Assc.

Chapter 7

Figure 7.10 / From Molecular Cell Biology by Lodish, et al. © 1986, 1990, 1995 by Scientific American Books. Used with permission by W.H. Freeman and Company. **Fig. 7.12c** / From Principles of Genetics, 2e by Snustad, Simmons © 2000 with permission. John Wiley & Sons, Inc. New York.

Chapter 8

Fig. 8.11c From An Introduction to Genetic Analysis, Griffiths et al © 2000 with permission WH Freeman, New York. **Fig. 8.12** / From Molecular Cell Biology, 4e © 1999 by Lodish, Darnell et al © 1999 with permission WH Freeman, New York.

Chapter 9

Fig. 9.1b / Weatherall DJ, Clegg JB: *The Thalassemia Syndrome, 3e* © 1981 with permission of Blackwell Scientific Publications.

Chapter 10

Figures 10.13, 10.15, 10.16, 10.17, 10.18, 10.19 / "Initial Sequencing and Analysis of the Human Genome", *Nature:* Vol 409, February 15th, 2001, with permission of the International Human Genome Sequencing Consortium. **Fig. 10.20** / Glusman, Lancet: "The Complete Human Olfactory Subgenome", *Genome Research,* May 2001, Vol 11 (3): 685-702 Courtesy Cold Spring Harbor Laboratory Press. **Fig. 10.24** / Hood, Hunkapiller, Smith, "Automated DNA Sequencing and Analysis of the Human Genome" *Genomics,* 1:201-212, © 1987 with permission Academic Press. **Fig. 10.26** / Reprinted with permission Gusella, JF et al: "DNA markers for nervous system diseases" from Science 225: 1320 © 1984 by the AAAS. **Fig. 10.33** / Courtesy Stan Fields, University of Washington Yeast Resource Center.

Chapter 11

Fig. 11.18a / Gusella JF et al: DNA Markers for Nervous System Disease, 1984, Science 225: 1320; with permission AAAS. **Fig. 11.18b** / From American Journal of Human Genetics, 39: 382-391 © 1980 by Magenis et al (the James Gusella Lab). Published by the University of Chicago Press.

Chapter 12

Fig. 12.11b / Courtesy of Alberts et al, *Molecular Biology of the Cell, 3rd edition.* New York: Garland Publishing, 1994. Courtesy of Alberts et al, *Essential Cell Biology.* New York: Garland Publishing, 1998.

Chapter 13

Fig. 13.1 / Derived in part from M.F. Seldin, K.W. Hunter, and M.L. Watson, *Mammalian Genome* 4, S10 (1993): M.F. Seldin, J.B. Prins, N. Rodrigues, J.A. Todd and M.H. Meisler, Mammalian Genome 4, S47 (1993) Courtesy of Springer-Verlag Publishing. **Fig. 13.35** / Snijders et al; Assembly of microarrays for genome-wide measurement of DNA copy number; with permission Nature Genetics Vol. 29 No. 3: 263-269 (2001).

Chapter 14

Fig. 14.24 / H.C. Berg; "The helical filaments of the thin flagella that propel bacteria do not wave or beat but instead rotate rigidly like propellers" With permission Sci Am. 1975 Aug; 233(2):36-44.

Chapter 15

Fig. 15-15b / Shoffner, Wallace et al; Myoclonic epilepsy and ragged-red fiber disease (MERRF) is associated with a mitochondrial DNA tRNA (Lys) mutation. *Cell* 1990 Jun 15;61(6):931-7. with permission.

Chapter 16

Fig. 16.7c / Reprinted with permission of Science, "View of the lac repressor-DNA complex monomer" Vol 271, p. 250, March 1996; American Association for the Advancement of Science. **Fig. 16.13** / Courtesy of Alberts, et al., Molecular Biology of the Cell, 3rd ed, New York: Garland Publishing, 1994.

Chapter 18

Table 18.2 / Adapted from V.T. DeVita et al., Cancer: Principles and Practices of Oncology, 4e © 1993 With permission Lippencott Williams & Wilkins. **Table 18.3** / Adapted from Lewin, Genetics 1e, Oxford University Press, Inc. by permission. **Fig. 18.17** / With permission from Elsevier Science; TIG, April 1993, vol. 9, no. 4. **Fig. 18.19** / Reprinted by the permission of the American Cancer Society, Inc. **Fig. 18.21** / Scientific American, March 1995, p. 77; Jared Schneidman Design, with permission of Scientific American. **Fig. 18.24b** / *Introduction to Genetic Analysis* by Anthony J.F. Griffiths, Jeffrey H. Miller, David Suzuki, Richard Lewontin, William M. Gelbart © 1996, 1993, 1989, 1986, 1981, 1976 by W.H. Freeman and Company. Used with permission. **Fig. 18.28** / *The Development of Drosophila Melanogaster* / "Mitosis and Morphogenesis in Drosophila Embryo: Point and Counterpoint" by V.E. Foe, G.M. Odell, B.A. Edgar, © 1993 Cold Springs Harbor Press. With permission.

Chapter 19

Fig. 19.7b / Gilbert, *Developmental Biology,* 5e courtesy Sinauer Associates, Inc. **Fig. 19.26a** / Gilbert, *Developmental Biology,* 5e courtesy Sinauer Associates, Inc. **Fig. 19.28** / Gilbert, *Developmental Biology,* 5e courtesy Sinauer Associates, Inc. **Fig. 19.A, 19B** / Wolpert, *Principles of Development,* 2e, © 2002 Oxford University Press, with permission.

Reference B

Table B.1 / With permission from Elsevier Science; "To Be or Not to Be a Flower—Control of Meristem Identity" TIG January 1998 Vol. 14 No. 1 by H. Ma. **Table B.2** / With permission from Elsevier Science; Coupland "Genetic and environmental control of flowering time in Arabidopsis" TIG, October 1995 Vol. 11, No. 10 p. 395. **Fig. B.4** / with permission The Arabidopsis Genome Initiative "Analysis of the genome sequence of the flowering plant *Arabidopsis thaliana*"; Nature 408, 796-815 (14 Dec 2000). **Fig. B.6** / "Double fertilization in *Arabidopsis*" TIG, April 1995, p. 148-9, vol. 11, no. 4 with permission Elsevier Science. **Fig. B.12** / Waterhouse, Wang, Lough "Gene Silencing as an adaptive defense against viruses" *Nature* 411, 834-842 (14 June 2001). **Fig. B.16** / Bleeker, Kende: Ethylene: A gaseous signal molecular in plants Annual Review / 2000 Annual Review of Cell Developmental Biology Vol 16: 1-18. **Fig. B.19** / Courtesy of George Resteck, with permission of Scientific American.

Reference C

Fig. C.5b / With permission; Springer-Verlag "Development: The Molecular Genetic Approach, Embryogenesis *in Caenorhabditis elegans*", fig. 1 p. 273 by Lois G. Edgar. **Fig. C.6** / Cold Springs Harbor Symposium on Quantitative Biology, vol. 58 (1983) Cold Spring Harbor, NY: Cold Springs Harbor Laboratory Press. **Fig. C.7a** / Gilbert, *Developmental* Biology, 5e with permission Sinauer Associates, Inc. **Fig. C.7b** / Adapted from William B. Wood, 1988 Ch 8, Embryology in the Nematode *Caenorhabditis Elegans,* p. 220 Originally from Proceedings of the National Academy of Science US 79; 1558-1562 (1982). **Fig. C.12** / Kemphues et al, 'Identification of Genes Required for Cytoplasmic Localization in Early *C. elegans* embryos', Cell, Vol. 52, 311-320, February 12, 1988, fig. 7, p. 318. **Fig. C.17** / With permission of Current Opinion in Genetics & Development; Kayne, Sternberg 'Ras pathways in *Caenorhabditis elegans*' 1995.

Reference D

Fig. D.16 / Dickson, "Targeted knockouts by homologous recombination" Nature, Vol. 405, June 22, 2002, Fig. 1, p. 897.

Reference E

Fig. E.3 / This figure was adapted with permission (J.T. Eppig, Mouse Genome Database [MGD], The Jackson Laboratory, Bar Harbor, ME). Data depicted are from MGD as of 4/98. Current information may be accessed at http://www.infomratics.jax.org. **Fig. E.10** / Leder, Pattengal, Kuo; Consequences of Widespread Deregulation of the c-myc Gene in Transgenic Mice: Multiple Neoplasms and Normal Development / Cell Vol. 45, fig. 1 and 2, 485-495, May 23, 1986. **Fig. E.12a** / Silver, Ewulonu, Snyder, Schimenti; "Promoter Mapping of the Mouse *Tcp-10b^t* Gene in Transgenic Mice Identifies Essential Male Germ Cell Regulatory Sequences' From Molecular Reproduction and Development Vol. 43, pp. 290-297; fig. 4, p. 294, © 1996 courtesy John Wiley & Sons, Inc. **Fig. E.15** / McGinnis & Kurmlauf 'Summary of Mouse Development' Cell, Vol. 68, January 1992, p. 283-302. **Fig. E.16** / McGinnis & Krumlauf 'Summary of Mouse Development' Cell, Vol. 68, January 1992, p. 283-302.

Photo Credits

Table of Contents

1: © Doug Sokell/Visuals Unlimited; **2:** © Keith V. Wood/Visuals Unlimited; **3:** © Joseph Hacia, National Human Genome Research Institute/National Institute of Health; **4:** © Mark Harmel/Stone/Getty; **5:** © Chad Ehlers/Stone/Getty; **6:** © Dr. Jeremy Burgess/SPL/Photo Researchers, Inc.

Chapter 1

Opener: © James Strachan/Stone/Getty; **1.1a:** © David M. Phillips/Visuals Unlimited; **1.1b:** © T.E. Adams, Visuals Unlimited; **1.1c:** © Vol. 44/Photo Disc; **1.1d:** © Doug Sokell/Visuals Unlimited; **1.1e:** GenomeSystems Inc./Photo provided by Kearns Communication Group; **1.1f:** © Vol. 24/PhotoDisc; **1.3:** Applied Biosystems/Peter Arnold, Inc.; **1.4:** © Biophoto Associates/Photo Researchers, Inc.; **1.9a:** © Carolina Biological/Photo Researchers, Inc.; **1.9b:** © Vol. OS02/PhotoDisc; **T1.1:** © J. William Schopf, UCLA. Reprinted with permission from *Science* 260: 640-646, 1993. "Microfossils of the Early Archean Apex Chert: New Evidence of the Antiquity of Life." © AAAS; **1.11 a,b:** Edward B. Lewis, CIT; **1.12(1):** © David M. Phillips/Visuals Unlimited; **1.12(2):** Lee Hartwell; **1.12(3):** © Sinclair Stammers/ SPL/Photo Researchers, Inc.; **1.12(4):** Courtesy Debra Nero/Cornell University; **1.12(5):** © Myung Shin/Bergman Collection; **1.14:** Vol. 103/PhotoDisc.

Chapter 2

Opener: © Brian Stablyk/Stone/Getty; **2.1:** © Bruce Aryes/Stone/Getty; **2.2:** © SPL/Photo Researchers, Inc.; **2.3:** © Saudjie Cross Siino/ Weathertop Labradors; **2.4:** © The Metropolitan Museum of Art, Gift of John D. Rockefeller, Jr., 1932 (32.143.2), Photograph ©1996 The Metropolitan Museum of Art; **2.5 a,b:** © Mendelianum Institute, Moravian Museum; **2.6:** © Klaus Gulbrandsen/SPL/Photo Researchers, Inc.; **2.7a:** © Dwight Kuhn Photography; **2.19a:** © SPL/Photo Researchers, Inc.; **2.19 b-d:** © Mendelianum Institute, Moravian Museum.

Chapter 3

Opener: © Tom & Pat Leeson/Photo Researchers, Inc.; **3.1:** © The Bergman Collection; **3.3a:** © John D. Cunningham/Visuals Unlimited; **3.7a (left):** © McGraw-Hill Higher Education Group, Inc./Jill Birschbach, Photographer; Arranged by Alexandra Dove, McArdle Laboratory, University of Wisconsin-Madison; **3.7a (right):** © Charles River Laboratories; **3.7a (bottom):** © Charles River Laboratories; **3.10a:** © Stanley Flegler/Visuals Unlimited; **3.12a:** © William H. Allen, Jr./Allen Stock Photography; **3.18a:** © Lynn McLaren/Photo Researchers, Inc.; **3.19a:** © Renee Lynn/Photo Researchers, Inc.; **3.20a:** © Rudi Von Briel/Photo Edit.

Chapter 4

Opener: © Adrian T. Sumner/Stone/Getty; **4.1:** © Richard Hutchings/Photo Researchers, Inc.; **4.2:** © L. West/Photo Researchers, Inc.; **4.3c:** © Biophoto Association/Photo Researchers, Inc.; **4.6:** © Scott Camazine/Photo Researchers, Inc.; **4.8 a-f:** Photographs by Dr. Conly L. Reider, Division of Molecular Medicine, Wadsworth Center, NYS Dept. of Health, Albany, NY; **4.9a:** © David M. Phillip/Visuals Unlimited; **4.9b:** © R. Calentine/Visuals Unlimited; **4.10:** © Dr. Byron Williams/Cornell University; **A:** © Dr. Michael Goldberg/Cornell University; **B:** © Dr. Michael Goldberg/Cornell University; **4.16:** © Dr. Leona Chemnick, Dr. Oliver Ryder/San Diego Zoo, Center for Reproduction of Endangered Species; **4.22:** Color deficit simulation courtesy of Vischeck (*www.vischeck.com*). Source image courtesy of NASA; **4.24:** © Gene Trindl/MPTV Photo Archive.

Chapter 5

Opener: © Rudy Von Briel/Photo Edit; **5.16 a-1:** © J. Forsdyke, Gene Cox/ SPL/Photo Researchers, Inc.; **5.16 b-1:** © James W. Richardson/Visuals Unlimited; **5.21:** © Dr. Eric Alani/Cornell University; **5.27:** Image courtesy of B.A. Montelone, Ph.D. and T.R. Manney, Ph.D.

Chapter 6

Opener: © Tony Stone Imaging/Getty; **6.1a:** © George Bernard/Animals Animals; **6.1b:** © William Hauswirth; **6.1c:** © Archivo Iconografico, S.A./CORBIS; **6.3a (both):** © Evanston Northwestern Healthcare, Evanston, IL; **6.4b:** © The Bergman Collection; **6.6:** © Science Source/Photo Researchers, Inc.; **6.9a:** © A. Barrington Brown/Science Source/Photo Reserachers, Inc.; **6.11a:** © Biophoto Assoicates/Photo Researchers, Inc.; **6.11b:** © Microworks/Dan/Phototake; **6.11c:** © Ross Inman & Maria Schnös, University of Wisconsin, Madison, WI; **6.11d:** © Jack D. Griffith/University of North Carolina Lineberger Comprehensive Cancer Center.

Chapter 7

Opener: © Milkie Studio, Inc./Stone/Getty; **7.1:** © Dr. Don Fawcett/J.R. Paulson & U.K. Laemmli/Photo Researchers, Inc.; **7.3 (both):** © Charles River Laboratories; **7.7b:** © The Bergman Collection; **A:** © Science VU/Visuals Unlimited; **7.14 (all):** © Carolina Biological/Photo Researchers, Inc.; **7.17 a-1:** © The Bergman Collection; **7.17 b-1:** © Seymour Benzer; **7.27a:** Courtesy of Dr. Patricia Olds-Clarke and Dr. Stephen Pilder, Temple University School of Medicine, Philadelphia, PA; **7.27c:** © Tom Vasicek; **7.27d:** © Science VU/P.R. Turner/Visuals Unlimited; **7.30:** Color deficit simulation courtesy of Vischeck (*www.vischeck.com*). Source image courtesy of NASA.

Chapter 8

8.1: © Sinclair Stammers, SPL, Photo Researchers, Inc.; **8.11 b-2:** © Professor Oscar Miller/SPL/Photo Researchers, Inc.; **8.17b:** © Dr. Thomas Maniatis, Thomas H. Lee Professor of Molecular and Cellular Biology, Harvard University.

Chapter 9

Opener: © Cytographics/Visuals Unlimited; **9.1a:** © Stanley Flegler/Visuals Unlimited; **9.5 b1-2:** © Lee Silver, Princeton University; **9.14a:** © Overseas/Phototake; **9.15 (all):** © Lee Silver, Princeton University; **9.18b:** © Jean Claude Revy/Phototake; **9.21 a-3:** © Omikron/Science Source/Photo Researchers, Inc; **9.21 b-3:** © Professor Sir David Weatherall, University of Oxford.

Chapter 10

10.1 (left): Reprinted by permission from Nature 409:15th February 2001 © 2001 Macmillan Publishers Ltd.; **10.1 (right):** Reprinted Full cover of *Science* 2/16/01. © 2001 American Association for the Advancement of Science; **10.5:** © Lee Silver, Princeton University; **10.6b:** © Dr. Gopal Murti/SPL/Photo Researchers, Inc.; **10.25b:** Image courtesy Patrick O. Brown and Joe Derisi; **10.29:** © Phototake; **10.35:** © Scimat/ Photo Researchers, Inc.

Chapter 11

Opener: © Lee Silver, Princeton University; **11.1a:** Courtesy of Ronald Carson, The Reproductive Science Center of Boston/IntegraMed America, Inc.; **11.6b:** © Lee Silver, Princeton University; **11.10:** Reprinted by permission of Oxford University Press. *Nucleic Acids Research*, Vol 26(16), 1998, fig 1 "Two Color Hybridization Analysis using High Density Oligonucleotide Arrays and Energy Transfer Dyes," Hacia et al. Image courtesy Joseph G. Hacia, University of Southern California; **11.14b:** © Lee Silver, Princeton University; **11.15:** Reprinted by permission from *Nature*, 394, figure 1, (1998), © 1998 Macmillan Publishers Ltd. Image courtesy Esther N. Signer; **11.20b:** Reprinted with permission from *Nature* 1990 July 19;346(6281):216-7, Sinclair et al. © 1990 Macmillian Magazines Limited; **11.21 (top):** © Brigid Hogan, Vanderbilt University; **11.21 (bottom):** Reproduced with permission from Dr. Robin Lovell-Badge/MRC National Institute for Medical Research; **11.22b:** © Johanna Rommens/Hospital for Sick Children, Toronto Reprinted from *Science* 245:1066, 1989. © 2001 American Association for the Advancement of Science; **11.24a:** © Calgene, Inc.; **Problem 11.11:** Lee Silver from *http://www.geninfo.no/undervisning/DNA-fingerprinting.asp*; **Problem 11.12:** © Lee Silver, Princeton University.

Chapter 12

Opener: © David M. Phillips/Visuals Unlimited; **12.1:** © Biophoto Associates/Photo Researchers, Inc.; **12.2a:** © Dr. Don Fawcett, U.K. Laemmli/Photo Researchers, Inc.; **12.2b:** © Daniel A. Starr/University of Colorado; **12.3a:** © Ada L. Olins, University of Tennessee/Biological Photo

Service; **12.3c:** © Dr. Gerard J. Bunick, Oak Ridge National Laboratory; **12.4 a-1:** © Dr. Barbara Hamkalo/University of California-Irvine, Department of Biochemistry; **12.4 a-2:** © Dr. Don Fawcett/H. Ris and A. Olins/Photo Researchers, Inc.; **12.5:** © Dr. Don Fawcett/J.R. Paulson, U.K. Laemmli/Photo Researchers, Inc.; **12.6a:** © Jacques Giltay, Center for Medical Genetics, Utrecht; **12.7a:** © H. Kreigstein and D.S. Hogness "Mechanism of DNA Replication in Drosophila Chromosomes: Structure of Replication Forks and Evidence of Bidirectionality", *Proceedings of the National Academy of Sciences* USA, 71(1974): 135-139; **12.8:** © Dr. Robert Moyzis/University of California-Irvine, Department of Biochemistry; **12.11a-2:** © Dr. Jeremy David Pickett Heaps, School of Botany, University of Melbourne; **12.13a:** © Stanislav Fakan, University of Lausanne; **12.13b:** © Doug Chapman, University of Washington Medical Center Cytogenetics Laboratory; **12.14a (both):** © Dr. Clinton Bishop, Department of Biology, West Virginia University; **12.15a-1:** © Jack M. Bostrack/Visuals Unlimited; **12.15 a-2:** © David M. Phillips/Visuals Unlimited; **12.15b:** © David M. Phillips/Visuals Unlimited; **12.15c:** © Dr. Michael Goldberg, Cornell University; **12.16b:** © O.L. Miller, B.R. Beatty, D.W. Fawcett/Visuals Unlimited.

Chapter 13

Opener: © Michael Goldberg/Cornell University; **C:** © Courtesy of The Centers for Disease Control; **13.4b:** © Dr. Ross MacIntyre, Cornell University; **13.10 (both):** © Cabisco/Visuals Unlimited; **13.12:** © Provided by M.J. Moses, Duke University, from Poorman, Moses, Davisson and Roderick: *Chromosoma* (Berl.) 83:419-429 (1981); **13.15b:** © Lisa G. Shaffer, PhD/Baylor College of Medicine; **13.17a-1:** © Dr. E. Walker/SPL/Photo Researchers, Inc.; **13.17 a-2:** © Photo Researchers, Inc.; **13.18d:** © M.G. Neuffer, University of Missouri; **13.20:** © CORBIS; **13.21 (both):** © Dr. Michael Goldberg, Cornell University; **13.22b:** © Dr. Nina Fedoroff; **13.31:** © Leonard Lessin/Peter Arnold, Inc.; **13.34b:** © Davis Barber/PhotoEdit.

Chapter 14

Opener: © Pat O'Hara/Stone/Getty; **14.2a:** © Dr. Jeremy Burgess/SPL/Photo Researchers, Inc.; **14.2b:** © Stephen Frish Photography; **14.3:** © David M. Phillips/Visuals Unlimited; **14.4a:** © Dr. Gopal Murti/SPL/Photo Researchers, Inc.; **14.8a:** © SPL/Science Source/Photo Researchers, Inc.; **14.22a:** © Jack D. Griffith/University of North Carolina Lineberger Comprehensive Cancer Center; **14.26:** © Julius Adler, University of Wisconsin; **A:** © Veronika Burmeister/Visuals Unlimited: **B:** © Oliver Meckes/Photo Researchers, Inc.

Chapter 15

Opener: © Newcomb & Wergin/Stone/Getty; **15.1:** © Eric L. Heyer/Grant Heilman Photography; **15.4:** Electron micrograph by Dr. Stephen Hajduk/University of Alabama at Birmingham; **15.11a** © Carolina Biological Supply/Phototake; **15.12a (both):** © Jim Strawser/Grant Heilman Photography; **15.14:** © Dr. John Sanford/Sanford Scientific, Inc.; **15.15a:** © Reprinted from *Trends in Genetics,* January 1989, Vol. 5, No.1, Dr. Douglas C. Wallace, p.11. © 1989, with permission from Elsevier Science.

Chapter 16

Opener: © Illustration courtesy of Mitchell Lewis, University of Pennsylvania. Reprinted with permission from *Science* 271:1247, from M. Lewis, G. Chang, N.C. Horton, M.A. Kercher, H.C. Pace, M.A. Schumacher, R.G. Brennan and P. Lu, © 1996 American Association for the Advancement of Science; **16.1:** © London School of Hygieyne & Topical Medicine/SPL/Photo Researchers, Inc.; **16.4:** © Bettmann/UPI/CORBIS; **A:** © Dr. Ann Hirsch, UCLA.

Chapter 17

Opener: Reprinted by permission from *Nature,* Vol. 391:103-106 © 1998 Macmillan Publishers Ltd. Image courtesy Carlos R. Escalante; **17.4:** Reprinted with permission from Nature, Vol. 377, 14 Sept. 1995, fig. 2, p. 124 © 1995 Macmillan Magazines Limited; **17.5:** © S.L. McKnight; **17.10d:** © Andreas Bauxevanis, National Human Genome Research Institute, National Institute of Health.

Chapter 18

Opener: © Lennart Nilsson/Albert Bonniers Forlag/Boehringer Ingelheim International GmbH; **18.4 (both):** © Lee Hartwell, Fred Hutchinson Cancer Research Center; **18.11b (both):** © Thea Tlsty, University of California-San Francisco Medical Center-Pathology; **18.13b:** © Michael R. Speicher and David C. Ward, "The Coloring of Cytogenetics." *Nature Genetics,* 2:1046-1048, 1996, figs 2 and 3. Photos courtesy of David C. Ward; **18.16 b-2:** © Dr. Joe Gray, University of California-San Francisco; **18.24:** © Custom Medical Stock Photo; **18.25a:** © Nancy Kedersha/SPL/Photo Researchers, Inc.; **18.25b:** © Dr. John Zajicek/SPL/Photo Researchers, Inc.; **18.25c:** © Martin M. Rotker/Photo Researchers, Inc.: **18.25d:** © Frederick C. Skvara, M.D./The Bergman Collection; **18.25e:** © Frederick C. Skvara, M.D./The Bergman Collection; **18.27 (all):** © David N. Louis, M.D./Massachusetts General Hospital.

Chapter 19

Opener: Courtesy of The Indiana Hand Center and Gary Schnitz; **19.1a:** © Dennis Kunkel Microscopy, Inc.; **19.1b:** © Lennart Nilsson/Albert Bonniers Forlag, From *A Child is Born,* Dell Publishing Company; **19.2:** © Photomicrographs by F. Rudolph Turner/The FlyBase Consortium, 1999; **19.3:** © Sinclair Stammers/SPL/Photo Researchers, Inc; **19.5a:** Courtesy Dr. Walter Gehring; **19.5b(both):** © Helen Pearson, Western General Hospital/MRC Human Genetics Unit; **19.7c:** © EMBO J 1998 Mar 17:17 (6):1642-55, Fig 3 parts B&C By permission of Oxford University Press. Image courtesy of Glen Merlino, National Institute of Health; **19.8b(both):** Courtesy of Daryl Hurd and Ken Kemphues, Cornell University; **19.9 a,b:** Reprinted from *Developmental Biology,* Vol. 96, No 8, Wang et al: "A mouse model for Achondroplasia produced by targeting fibroblast growth factor receptor 3", pp 4455-4460. © 1999, with permission from Elsevier Science; **19.10:** Courtesy Dr. Walter Gehring; **19.13a-d:** © Glenn Doughty and S. Artavanis-Tsakonas; **19.16:** Courtesy Thomas R. Bürglin, based on the data of Otting et al. EMBO J 1990 Oct. 9 (10):3085-92; **19.17:** Photo courtesy of Robert Hill. Reprinted from Development 1995 121:(5) 1433-1442, © 1995 with permission from Company of Biologists Ltd.; **19.18 b&d:** Courtesy Andras Nagy PhD; **19.19a:** Reprinted from Development 1998 125: (23) 4303-4312, © 1998 reprinted with permission from Company of Biologists Ltd.; **19.22 b,c:** Reprinted from CELL, Vol. 87 No 5, Basler , Zecca, Struhl, "Direct and Long-range action of a Wingless Morphogen Gradient", pp. 833-844, © 1996, with permission from Elsevier Science; **19.23:** © Dennis Kunkel Microscopy, Inc.; **19.24:** © Scott Weatherbee; **19.25:** © John Reinitz and David Kosman; **19.26b:** Courtesy of Katia Litiere and Daniel St Johnston, University of Cambridge; **19.27b (both):** Reprinted from Development 128: 1137-1145, © 2001 with permission from Company of Biologists Ltd; Image courtesy Silvia Bonaccorsi; **19.29:** © AP/Wide World Photo.

Chapter 20

Opener: © Jim Pikerell/Stone/Getty; **20.1a:** © A.M. Siegelman/Visuals Unlimited; **20.4:** © Schafer & Hill/Peter Arnold, Inc.; **20.10a:** © PhotoDisc; **20.11a:** © Robert Noonan/Photo Researchers, Inc.; **20.13a:** © Dr. Eckart Pott/OKAPIA/Photo Researchers, Inc.; **20.14a:** © Frans Lanting/Photo Researchers, Inc.

Chapter 21

21.3a: © J. William Schopf, UCLA. Reprinted with permission from *Science* 260: 640-646, 1993. "Microfossils of the Early Archean Apex Chert: New Evidence of the Antiquity of Life." © 1993 AAAS; **21.5a:** © Daniel J. Cox/Tony Stone; **21.5b:** © Joe McDonald/Animals Animals; **21.5c:** © Renee Lynn/Stone/Getty; **21.5d:** © Roger De La Harpe/Animals Animals.

Reference A

Opener: © E. Koneman/Visuals Unlimited; **A.1:** Lee Hartwell; **A.2b:** © Dan Gottschling, Fred Hutchinson Cancer Research Center; **A.4:** © Reprinted fig 1 from *Genome Research,* Basari et al Vol 7: 768-771 © 1997. Image courtesy of Munira Basrai, Phil Hieter and Jef Boeke; **A.5:** Reprinted by permission from *Nature,* "The Yeast Genome Directory," Vol. 387, Chart of Duplications supplement 33-34 (29 May 1997), H.W. Mewes © 1997 Macmillan Publishers Ltd.; **A.16 a,b:** © Reprinted fig. 2 with permission from *Science,* Roberts et al. Vol. 287, 875 © 2000 AAAS. Image courtesy Chris Roberts, Ph.D., Rosetta Inpharmatics, Inc.; **A.16c:** © Reprinted fig 1 with permission from *Science* Ren et al Vol 290, 2306 © AAAS. Image courtesy of Richard Young.

Reference B

Opener: © Illustration-Katherine Sutliff/Reprinted with permission from *Science,* Oct. 23, 1998, Genetic and Physical Maps-Genome Maps 9. © 1998 AAAS; **B.1:** © Dr. Elliot Meyerowitz, California Institute of Technology; **B.2 (both):** © Dr. Elliot Meyerowitz, California Institute of Technology; **B.3:** © Reprinted with permission from *Science,* October 23, 1998, Genetic and Physical Maps-Genome Maps 9. © 1998 AAAS; **B.5 b-d:** © John Bowman; **B.6a (left):** Reprinted from Trends in Genetics, Vol. 5, No 1, Hauser et al, pp 11, © 1989, with permission from Elsevier Science. Image courtesy of Bernard A. Hauser & Charles S. Gasser, University of California-Davis; **B.6a (right):** © Sarah Patterson, Tony Bleecker, University Wisconsin-Madison; **B.7 b-f:** John Harada, University of California-Davis. Copyrighted by the American Society of Plant Physiologists and is reprinted with permission. *Plant Cell* 6:1731-1745; **B.8a:** © Christian Fankhauser/Salk Institute; **B.8b:** © Amasino; **B.9:** © Sarah Patterson, Tony Bleecker, University Wisconsin-Madison; **B.10b:** © John Celenza, Boston University; **B.13a (both):** © John Harada, University of California-Davis. Copyrighted by the American Society of Plant Physiologists and is reprinted with permission. *Plant Cell* 6:1731-1745; **B.13b (both):** © Photographs courtesy of Daniel M. Vernon, Whitman College, Department of Biology; **B.14:** © John Bowman; **B.15:** © Dr. Joe Kieber, University of Illinois, Reprinted with permission from *Cell* 427-431, Kieber et al. 1993; **B.17a:** © Bridey Maxwell/Salk Institute; **B.17b:** © Michael Neff/Salk Institute; **B.18 (all):** © Dr. Elliot Meyerowitz,